T0093580

METEOROLOGICAL MONOGRAPHS

VOLUME 57 2017

THE ATMOSPHERIC RADIATION MEASUREMENT (ARM) PROGRAM: THE FIRST 20 YEARS

Edited by

D. D. Turner
R. G. Ellingson

American Meteorological Society
45 Beacon Street, Boston, Massachusetts 02108

ISBN 978-1-944970-05-5
ISSN 0065-9401

Published by the American Meteorological Society
45 Beacon St., Boston, MA 02108

As of 2015, AMS Meteorological Monographs are available as open access online, with a print-on-demand option. Volumes issued before 2015, numbers 1–55, are available electronically through Springer Nature, of which a selection of casebound volumes are available for purchase at the AMS online bookstore (www.ametsoc.org/bookstore).

Printed in the United States of America
by Allen Press, Inc., Lawrence, KS

TABLE OF CONTENTS

Introduction

D. D. TURNER

National Severe Storms Laboratory, Norman, Oklahoma

R. G. ELLINGSON

The Department of Earth, Ocean and Atmospheric Science, The Florida State University, Tallahassee, Florida

The Atmospheric Radiation Measurement (ARM) Program is by all measures a unique enterprise. The U.S. Department of Energy (DOE) conceived the Program in 1990 as its primary contribution to the then newly created U.S. Global Change Research Program. Ari Patrinos, who was a DOE program manager in 1990, said in his forward to the ARM Program Plan that

> [the] recent heightened concern about global warming from an enhanced greenhouse effect has prompted the department (DOE) to accelerate the research to improve predictions of climate change. The emphasis is on the timing and magnitude of climate change as well as on the regional characteristics of this change. The Atmospheric Radiation Measurement (ARM) Program was developed to supply an improved predictive capability, particularly as it relates to the cloud–climate feedback.

ARM was a natural evolution of two separate research efforts within the DOE that were evaluating radiation parameterizations used in general circulation models (GCMs) and assessing the differences in cloud feedback in GCMs. Both of these activities required data, and the ARM Program was designed to provide that data. The original science vision of the ARM Program was for it to be a 10-yr endeavor, but the success of ARM has resulted in the Program continuing for more than 25 years and counting.

In many ways, ARM should be considered the model basic research program. It was formulated to address the largest uncertainties that were restricting progress in improving the fidelity of GCMs, namely the representation of radiation and clouds in these models. It had an ambitious goal to markedly improve our understanding within a decade. To do this, the founders of the Program realized that a new paradigm was needed. Continuous multiyear datasets that spanned the entire range of atmospheric conditions were required. Sites had to be established in climatically important regions of the world; organizational structures and procedures had to be developed; scientific and infrastructure staff had to be selected; instrumentation needed to be chosen and built or purchased; and the process of designing, building, and running the Program had to be conducted.

And so the Program was born. But like virtually all endeavors, the original plans did not anticipate every contingency. One of ARM's great accomplishments was (and is) its ability to grow and evolve, adjusting to accommodate new scientific questions and priorities and to match changes in the funding structure provided by DOE. ARM is now a very complex \$65M per year activity,[1] supporting the activities of approximately 100 scientists, associates, and students together with over 150 scientific and engineering staff within the Program's infrastructure. The Program's story is a remarkable one, and one that needed to be told.

The ARM history is also a story about people. The vision of the Program was developed by a relatively small group of people, but then augmented and implemented by many. The Program has benefitted from excellent leadership from DOE management (such as Ari Patrinos, Peter Lunn, Pat Crowley, and Wanda Ferrell) and staff from most of the DOE national laboratories (such as Marv Wesely, Ric Cederwall, John Vitko, Joyce Tichler, Bill Clements, and many, many others) who helped lead different components of the Program. The

Corresponding author address: D. D. Turner, NOAA/National Severe Storms Laboratory, Forecast Research and Development Division, National Weather Center, 120 David L. Boren Blvd., Norman, OK 73072.
E-mail: dave.turner@noaa.gov, rellingson@fsu.edu

DOI: 10.1175/AMSMONOGRAPHS-D-16-0001.1

[1] A careful reader of this monograph will note that the current annual budget, which is at its highest ever, still does not reach that anticipated by the earliest program visionaries in 1990.

scientific direction of the Program was guided over the first 20 years by three chief scientists and members of the Science Team Executive Committee (STEC) and later the Science and Infrastructure Steering Committee (SISC). However, many of the real impacts to the ARM Program rose from the rank-and-file members of the Program, as scientific guidance often came from funded Principal Investigators (PIs) and their colleagues through the different scientific working groups used to organize the Science Team. Many of the components and improvements of the infrastructure originated from ideas of people within the infrastructure at all levels. There have also been many dozens of people working behind the scenes on contracts, budgeting, technical editing of ARM documents, organizing ARM's yearly meetings, public relations, outreach, and countless other tasks that are necessary for any enterprise as large as ARM. While it is virtually impossible to mention everyone who played a significant role in ARM, Figs. 1–3 show some of the more prominent people from the Science Team, infrastructure, and management of the ARM Program over the last two decades.

The infrastructure of the Program has provided the bedrock upon which the scientific achievements of the Program are based. Compared to other large research programs such as NASA's Earth Observing System (EOS), one could argue that the ARM infrastructure is woefully understaffed given the breadth of the activities that the Program is supporting. However, the infrastructure team's hard work and pride in the Program have overcome many obstacles.

There is a strong sense of community around ARM. Some of this sense of community can be attributed to the members of the infrastructure, who have remained as relatively constant contributors at ARM meetings for the last two decades. The general organization of the Science Team via the use of working groups, which are led by a selected member from that working group, also has contributed to this feeling of community. As such, the Program has also been a great incubator for young scientists, and many of the scientific leaders in the Program today began as graduate students supported by PIs who were funded by ARM. This has created a feeling of "ownership" by many scientists and infrastructure members in the Program, and that vested interest has benefitted the ARM tremendously.

The impact of ARM has been large and far-reaching. For the last two decades, ARM-funded scientists have been very visible reporting their results at American Meteorological Society (AMS) meetings (e.g., at the Cloud Physics and Atmospheric Radiation conferences), the fall American Geophysical Union (AGU) meeting, International Union of Geodesy and Geophysics (IUGG) meetings, and other national and international meetings. ARM scientists are on the International Radiation Commission, actively participate on the Intergovernmental Panel on Climate Change (IPCC), and organize and engage in the Global Energy and Water Cycle Experiment (GEWEX) activities, among many other endeavors. Additionally, ARM has actively participated in outreach activities to engage and educate elementary and high school students, communities, and other "non-scientists." These outreach activities were very important in the early days when the Program was establishing the long-term sites and wanted to develop a positive relationship with the surrounding communities.

This monograph provides the history of the birth and growth of the ARM Program. While ARM-funded scientists and others using ARM data have published many hundreds of papers in scientific journals, only three previously published papers provide any broad overview of the Program. Gerry Stokes and Steve Schwartz discussed the original conception and birth of the Program in an early paper that appeared shortly after the Southern Great Plains (SGP) site started collecting data (Stokes and Schwartz 1994), Tom Ackerman and Gerry Stokes provided a high-level overview of some of the main accomplishments of Program in its first decade (Ackerman and Stokes 2003), and Jim Mather and Jimmy Voyles described more recent activities that have greatly broadened the scope of the Program near the end of its second decade (Mather and Voyles 2013). However, these three papers only provide snapshots and are unable to delve into myriad of accomplishments, challenges, and decision points that the Program has faced and overcome along the way.

This monograph aims to provide those details. To tell the story of ARM, a collection of ARM scientists and infrastructure members were solicited to provide their thoughts on various components of the Program. The first four chapters provide a high-level view of the origin, birth, and maturation of the ARM Program. Chapters 5 through 12 cover different components of the ARM infrastructure, including how the sites were selected, an overview of each of the sites including how the ARM Mobile Facility came to be, the history of airborne observations in the Program, a synopsis of the ARM data system and its evolution, and the Program's data quality

FIG. 1. Some people contributing to the ARM Program between 1990 to 2009. (a) Tom Ackerman; (b) Bob Ellingson; (c) Sally McFarlane, Rob Wood, Allison McComiskey, Matthew Shupe, Steve Ghan, Jian Wang, and Ashley Williamson; (d) Minghua Zhang; (e) Steve Ghan; (f) Matt Macduff, Dick Eagan, and Rick Wagener; (g) John Goldsmith; (h) Jimmy Voyles and Wanda Ferrell; (i) Maike Ahlgrimm; (j) Christian Jakob and Ted Cress; (k) Mark Miller; (l) Dave Turner and Eli Mlawer; (m) Kuo-Nan Liou and Andy Lacis; (n) Randy Peppler; (o) Steve Klein; (p) Raymond McCord; (q) Ken Moran and Kevin Widener; (r) Marv Wesely; (s) Mark Ivey, Doug Sisterson, Bob Ellingson, and Warren Wiscombe; (t) Roger Marchand; (u) Jim Mather; (v) Warren Wiscombe, Chuck Long, Jay Mace, Matthew Shupe, and Dave Turner; (w) Sally McFarlane.

FIG. 2. More people contributing to the ARM Program between 1990 and 2009. (a) John Vitko; (b) Dave Randall; (c) Graham Feingold; (d) Steve Krueger; (e) Larry Berg; (f) Sharon Zuhoski, Nancy Burleigh, and Jane McKinney; (g) Hank Revercomb and Bob Ellingson; (h) Doug Sisterson, Warren Wiscombe, Raymond McCord, Wanda Ferrell, Jimmy Voyles, and Sylvia Edgerton; (i) Tony Clough; (j) Warren Wiscombe; (k) Eva Baroni; (l) Rich Ferrare; (m) Andy Vogelmann; (n) Joyce Penner; (o) Tony Del Genio; (p) Jennifer Comstock; (q) Pavlos Kollias; (r) Catherine Gautier; (s) Eugene Clothiaux; (t) Wanda Ferrell, Pete Lamb, and Jimmy Voyles; (u) Joe Michalsky; (v) Hans Verlinde; (w) Ric Cederwall; (x) Giri Palanisamy; (y) Anthony Davis, Eugene Clothiaux, and Howard Barker; (z) Matthew Shupe and Jay Mace; (aa) Dave Turner, Kevin Widener, Fairly Barnes, Joyce Tichler, Barry Lesht, Warren Wiscombe, Ric Cederwall, Tim Tooman, Raymond McCord, and Randy Peppler; (bb) Tom Ackerman; (cc) Kelle Smith and Pat Nelson.

INTRODUCTION

FIG. 3. Still more people contributing to the ARM Program between 1990 to 2009. (a) Ted Cress and Steve Schwartz; (b) Ric Cederwall; (c) Gerry Stokes; (d) Tom Stoffel; (e) Zhanqing Li; (f) Beat Schmid; (g) Marv Dickenson; (h) Richard Somerville; (i) Jay Mace; (j) Peter Lunn; (k) Bernie Zak; (l) Jim Spinhirne; (m) Maria Cadeddu; (n) Joe Michalsky; (o) Warren Wiscombe, Graeme Stephens, and Bob Ellingson; (p) Harvey Melfi; (q) Ed Westwater and Chuck Long; (r) Christine Chiu; (s) Ari Patrinos; (t) Pat Crowley and Wanda Ferrell; (u) Ells Dutton; (v) Bill Clement, Ted Cress, and Fairly Barnes; (w) Gerry Stokes, Ted Cress, and Tom Ackerman; (x) Marv Wesely, Doug Sisterson, Weigang Gao, and Rich Coulter; (y) Sally McFarlane and Jim Barnard; (z) Bob Cess; (aa) Hank Revercomb.

program. Chapters 13 through 30 capture the progress ARM has made on various scientific topics. The range of topics covered is impressive!

We hope you enjoy reading this monograph on the Atmospheric Radiation Measurement (ARM) Program's first twenty years.

This monograph is dedicated to Pete Lamb and Marv Wesely, who passed before their time, and all of the scientists, engineers, managers, and staff that contributed their energy and ideas to the formulation and evolution of the ARM Program.

REFERENCES

Ackerman, T. P., and G. M. Stokes, 2003: The Atmospheric Radiation Measurement Program. *Phys. Today*, **56**, 38–44, doi:10.1063/1.1554135.

Mather, J. H., and J. W. Voyles, 2013: The ARM climate research facility: A review of structure and capabilities. *Bull. Amer. Meteor. Soc.*, **94**, 377–392, doi:10.1175/BAMS-D-11-00218.1.

Stokes, G. M., and S. E. Schwartz, 1994: The Atmospheric Radiation Measurement (ARM) Program: Programmatic background and design of the cloud and radiation test bed. *Bull. Amer. Meteor. Soc.*, **75**, 1201–1221, doi:10.1175/1520-0477(1994)075<1201:TARMPP>2.0.CO;2.

Chapter 1

The Atmospheric Radiation Measurement Program: Prelude

ROBERT G. ELLINGSON

Department of Earth, Ocean and Atmospheric Science, Florida State University, Tallahassee, Florida

ROBERT D. CESS

School of Marine and Atmospheric Sciences, State University of New York at Stony Brook, Stony Brook, New York

GERALD L. POTTER

NASA Goddard Space Flight Center, Greenbelt, Maryland

1. Introduction

The U.S. Department of Energy (DOE) was already concerned about the potential impact of the increasing content of CO_2 in the atmosphere on future climate in the early 1970s (see Riches 1983) and commissioned a set of six state-of-the-art reports (e.g., MacCracken and Luther 1985) that attempted to highlight the uncertainties in general circulation models (GCMs) and their underlying parameterizations well ahead of the Intergovernmental Panel on Climate Change (IPCC) program. Of considerable concern was that, although greenhouse warming is forced entirely by a radiative perturbation of a few watts per square meter or less, neither field measurements nor radiation model intercomparisons had ever achieved anywhere near this level of accuracy. Furthermore, there was a considerable range of GCM predictions due to greenhouse warming, and it was not entirely clear how much of the differences could be traced to differences in the initial radiative forcing used in the models and how much was due the range of parameterizations and their resulting feedbacks.

To obtain a better understanding of the GCM conundrum, the DOE instituted several intercomparison studies, two of which have particular importance. One, an intercomparison of longwave radiation codes (wavelengths > 4 microns) was initiated by Fred Luther in 1982. The second, a GCM intercomparison project, was begun by Robert Cess and Gerald Potter in 1984. These projects eventually grew to major international intercomparison studies that led the DOE to the conclusion that cloud–radiative feedback is the single most important effect determining the magnitude of possible climatic responses to human activity. This conclusion, in turn, led to the establishment of the Atmospheric Radiation Measurements (ARM) Program, the subject of this monograph. The DOE conclusion that the role of clouds was a critical knowledge gap was subsequently echoed in the first IPCC report (IPCC 1990).

In light of the importance of these intercomparison studies to the establishment of ARM, a summary of each of the studies and their findings is provided below.

2. ICRCCM

The Intercomparison of Radiation Codes used in Climate Models (ICRCCM) resulted from the unification of independent U.S. and European projects begun almost simultaneously in 1982. Frederick Luther initiated a comparison study of longwave radiative transfer models for the DOE's Carbon Dioxide Research Division. In Europe, Yves Fouquart and Jean-Francois Gelyn proposed a model comparison study that focused on both longwave and solar radiative transfer parameterizations in climate models. The World Climate

Corresponding author address: Robert G. Ellingson, Department of Earth, Ocean and Atmospheric Science, Florida State University, Tallahassee, FL 32306.
E-mail: rellingson@fsu.edu

DOI: 10.1175/AMSMONOGRAPHS-D-15-0029.1

Research Programme (WCRP) and the International Radiation Commission (IRC) established a joint working group on ICRCCM in 1984, and ICRCCM was officially started under the leadership of Frederick Luther and Yves Fouquart. Many of the details concerning ICRCCM are well documented in the literature (see Luther 1984; Luther et al. 1988; Ellingson and Fouquart 1991), and this document repeats or summarizes material contained in them.

The objectives of the ICRCCM studies during 1982–88 were the following:

1) to develop a better understanding of the differences in radiation model approaches,
2) to understand how these differences affect model sensitivity,
3) to evaluate the effects of simplifying assumptions,
4) to evaluate the ability of the radiation models to simulate the real atmosphere, and
5) to evaluate the effect of using different sources of spectral line data in the radiation codes.

The initial focus of ICRCCM was on comparisons of clear-sky longwave radiation calculations using identical atmospheric profiles of radiatively important variables. The models included in the study ranged from very simplified ones used in GCMs to the most sophisticated line-by-line (LBL) models. The study was subsequently expanded to include clear-sky shortwave (solar) and cloudy-sky shortwave and longwave cases. Workshops to discuss preliminary results were held in Frascatti, Italy, in 1984 and in College Park, Maryland, in 1986. A third workshop to finalize conclusions of the various calculations was held in Paris, France, in August 1988. Details concerning the final results may be found in Ellingson and Fouquart (1991), Ellingson et al. (1991), and Fouquart et al. (1991), part of an issue of the *Journal of Geophysical Research–Atmospheres* dedicated to ICRCCM.

Unfortunately, Dr. Fredrick Luther, one of the initial ICRCCM cochairmen, became ill and died in September 1986. His death was a major setback, but his vision and plans for ICRCCM allowed the project to continue under the leadership of Robert Ellingson and Yves Fouquart.

Going into ICRCCM, the assumption was that the physics and absorption line data were well known and that therefore the various models could not possibly disagree much, regardless of the type of parameterization (i.e., the details of the implementation). This complacency was shattered by the actual results of ICRCCM. For the longwave clear cases, with about 40 participants representing almost all the world's major modeling groups, ICRCCM revealed intermodel disagreements ranging from 30 to 70 W m^{-2} (Luther et al.

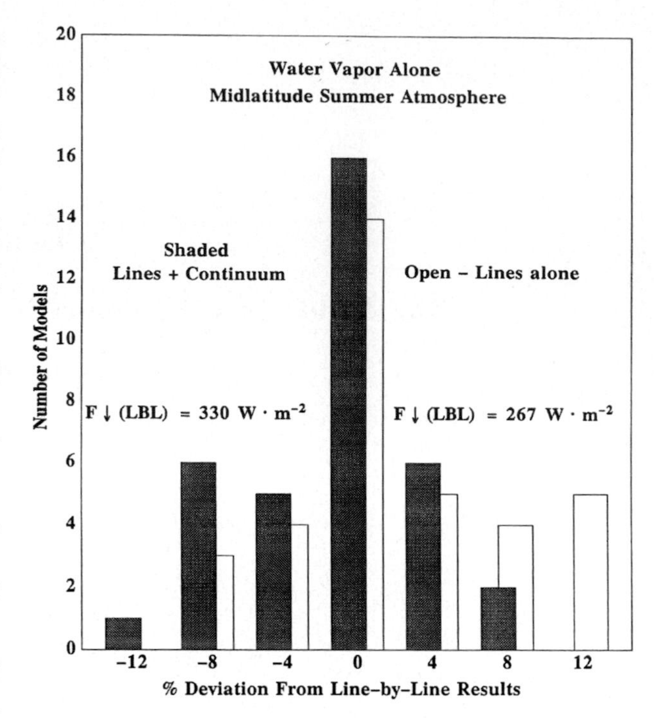

FIG. 1-1. Number distributions of downward fluxes at the surface relative to the GFDL line-by-line results when water vapor lines only and the lines plus the continuum are included in the calculations with the Air Force Geophysical Laboratory (AFGL) midlatitude summer atmosphere. Note that the histogram bars cover a range of ±2% centered on the given value, and they are displaced slightly for better viewing. [From Ellingson et al. (1991).]

1988). The disagreements were worse when pure H_2O (Fig. 1-1) and pure CO_2 atmospheres were considered, indicating that the better agreement found in the all-gas cases (Fig. 1-2) was partly accidental. Furthermore, the disagreements proved remarkably robust, surviving for years during which modelers combed their models for errors, omissions, and overly crude numerical procedures. Subsequent ICRCCM calculations, involving cloudy longwave cases, and clear and cloudy shortwave cases, revealed equally large or larger disagreements, ranging up to 20%–30% in fluxes and up to 70% in flux sensitivity to constituent changes (Ellingson et al. 1990). Table 1-1 provides, as an example, the range of agreement between band and LBL model calculations of total shortwave absorption of solar radiation from different participants.

After four years during which participants were allowed to revise their results, the participating LBL models did manage to agree to approximately 1% in clear-sky cases, and the number of less detailed models that agreed with the LBL models to within 2% increased. This good agreement among LBL models was achieved, however, only after the LBL modelers agreed

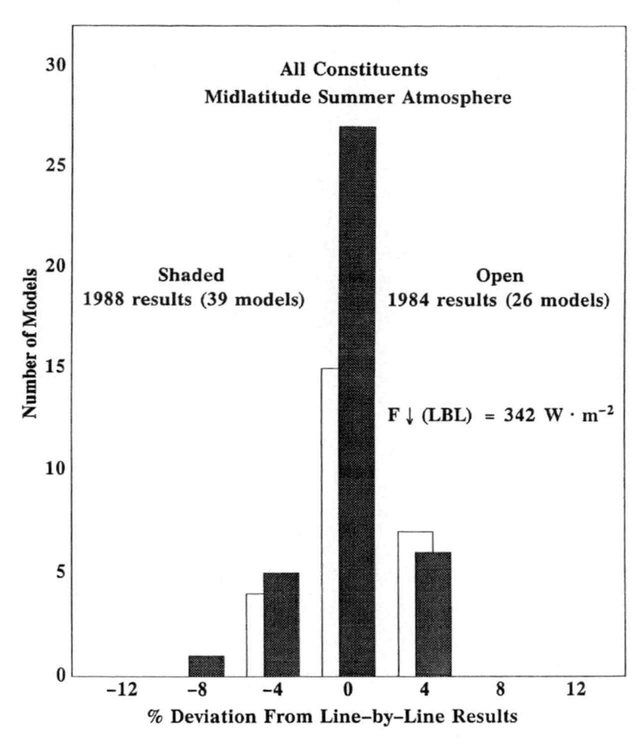

FIG. 1-2. As in Fig. 1-1, but with all constituents included in the calculations. The solid and open bars represent the results from the 1988 and 1994 workshops, respectively [from Ellingson et al. (1991)]. The mean and RMS differences relative to the line-by-line downward flux at the surface were 1.0 and 8.7 W m^{-2}, respectively.

to cut off their lines the same distance from line center and to use the same continuum absorption model. Thus, the ICRCCM community reluctantly concluded that

> [u]ncertainties in the physics of line wings and in the proper treatment of the continuum make it impossible for line-by-line models to provide an absolute reference (Luther et al. 1988, p. 46).

The ICRCCM community considered the gamut of the then available laboratory and atmospheric observations that might be used to validate the LBL models (e.g., broadband hemispheric flux data, aircraft or surface-based spectral data, satellite spectrometers or laboratory spectra). Each was found lacking for a variety of reasons, such as poor calibration, the lack of detailed measurements of the radiatively important variables, or incomplete range of variables found in the atmosphere (laboratory data).

The ICRCCM participants concluded that existing field observations, while they shed light on various issues facing ICRCCM, could not decisively resolve the large intermodel disagreements. As such they recommended a more sophisticated observational strategy, as follows (Luther 1984, p. 31):

> A dedicated field measurement program is recommended for the purpose of obtaining accurate spectral radiances rather than integrated fluxes as a basis for evaluating model performance.

Following the 1988 ICRCCM workshop, the IRC and the WCRP endorsed a second phase of ICRCCM with the primary purpose of validating radiation models through comparison with observations. A primary term of reference of this second phase was stated as follows (Ellingson and Fouquart 1990, p. 37; see also, Bolle 2008, p. 89):

> [The goal is to] determine the requirements for real in situ data for validation of high spectral resolution models and other radiative transfer computations and explore ways of obtaining these data by either a specific dedicated measurement programme or by appropriate enhancement of other experimental activities, such as may be part of ISLSCP [International Satellite Land Surface Climatology Project] and ISCCP [International Satellite Cloud Climatology Project] regional experiments.

3. SPECTRE

Robert Ellingson and Warren Wiscombe took the ICRCCM recommendations and shaped them into a very specific program called SPECTRE (Spectral

TABLE 1-1. Total atmospheric shortwave absorption for the six cases concerning the absorption by water vapor only. The abbreviations MLS, TRO, and SAW denote the midlatitude summer, tropical, and subarctic winter Air Force Geophysical Laboratory atmospheres, respectively. The band models are grouped into two classes according to their spectral resolution (HR and LR for high resolution and low resolution, respectively). Here θ_0 is the solar zenith angle in degrees. [From Fouquart et al. (1991).]

		LBL (W m^{-2})		Median (W m^{-2})		Range (%)		RMS differences (%)		No. of models	
Case	θ_0	1	2	HR	LR	HR	LR	HR	LR	HR	LR
MLS	30	172.4	178.2	171.0	165.7	12	11	6	7	10	11
MLS	75	67.1	69.6	62.4	64.3	21	14	11	10	10	11
TRO	30	187.8	195.4	181.9	181.2	16	12	7	8	10	11
TRO	75	72.3	75.7	66.9	69.2	29	14	14	10	10	11
SAW	30	94.5	99.7	101.7	97.1	49	14	14	4	10	11
SAW	75	39.3	41.1	41.1	39.5	17	10	6	4	10	11

Radiance Experiment; Ellingson et al. 1990) with emphasis on accurate measurements of emission spectra and proper quantitative characterization of the atmosphere. The idea was to do the following:

- integrate the use of the radiation and profiling technology,
- obtain detailed radiation and atmospheric data for a variety of important conditions,
- distribute the data to the international ICRCCM community, and
- intercompare model results with the data for the purpose of calibrating the various radiation codes used in climate models.

The key features of the proposed experiment were the following:

- spectral radiation measurements in the form of continuous spectra, not broadband measurements nor disjointed spectral bands;
- radiance (intensity) rather than flux measurements;
- redundant measurements of radiance;
- frequent and careful radiometric calibration in the field against known standards; and
- simultaneous, instantaneous profiles of temperature, humidity, aerosol and cloud.

SPECTRE was funded under the DOE "Quantitative Links Program" and by the National Aeronautics and Space Administration (NASA) and was carried out in conjunction with the NASA-sponsored First ISCCP Regional Experiment (FIRE) Cirrus II intensive observation period from 13 November to 7 December 1991, in Coffeyville, Kansas. Details concerning the experiment are given by Ellingson and Wiscombe (1996).

During SPECTRE, highly experienced scientists from different universities and government agencies carried out three main functions: spectrometer, remote, and in situ measurements. The spectrometers were Fourier transform interferometers characterized by a large wavelength range (3–18 μm), high spectral resolution (1 cm^{-1} or better), cryogenic cooling of the detectors, and routine blackbody calibration in the field (Ellingson and Wiscombe 1996). The spectrometers and resolutions included the NASA Stratospheric Infrared Interferometer Spectrometer (SIRIS; 0.06 cm^{-1}), the University of Wisconsin Atmospheric Emitted Radiance Interferometer (AERI; (0.5 cm^{-1}), the University of Denver interferometer (1.0 cm^{-1}), and the NOAA Fourier Infrared Spectrometer (FIRS; 0.5 cm^{-1}). The Wisconsin AERI eventually became a prototype ARM instrument designed for autonomous surface operation that calibrates against two internal temperature-controlled black bodies every 10 min. The mature AERI is now operating continuously at the ARM fixed and mobile sites (Turner et al. 2016, chapter 13).

The in situ measurements included frequent radiosonde launches, flask samples of the concentration of trace gases (CO_2, CH_4, N_2O, CO, and Freons F11 and F12) near the surface, and vertical profiles of O_3 using ozonesondes (Ellingson and Wiscombe 1996). The remote sensing measurements included water vapor profiles with a calibrated Raman lidar operated by Harvey Melfi, total column O_3 with a Dobson spectrometer, and vertical profiles of virtual temperature with a Radio Acoustic Sounding System (RASS). The experiment benefited greatly from the deployment of a Millimeter Cloud Radar (MMCR) by Thomas Ackerman as part of the FIRE Cirrus II project that helped identify clear-sky periods. As configured with FIRE Cirrus II, the SPECTRE deployment constituted a supersatellite on the ground for the observing period, and this was the first time this instrument complement was assembled.

Following a scrutiny of the SPECTRE data by the various principal investigators (PIs), atmospheric data from four meteorologically different days were released to the ICRCCM working group to perform calculations for comparison with AERI observations, after which the working group met from May 24 to 26 May 1995, at the University of Maryland at College Park to discuss the comparisons.

Overall, 22 participants submitted 29 sets of calculations, including 10 LBL, 12 narrowband (NB), and 7 broadband (climate model) models. For intercomparison purposes, Ellingson (1995) integrated the various NB and LBL model calculations and the observations to a common grid. A comparison of the average of the four comparison spectra with the average NB spectra is shown in Fig. 1-3, whereas Fig. 1-4 shows a comparison of the distribution of the mean observed minus calculated radiance at the common resolution. To estimate the effects of these differences on the downwelling flux at the surface, the differences were integrated over the 520 to 2600 cm^{-1} interval and converted to flux using model-calculated angular corrections. The distribution of the flux differences is shown in Fig. 1-5.

In general, the ICRCCM participants were very pleased with the level of average agreement between the observed and model-calculated radiances and spectrally integrated downward flux at the surface (4.5 $W m^{-2}$ root-mean-square error), a marked improvement over the 1988 results (cf. Figs. 1-2 and 1-5). However, the quality of the comparisons is somewhat misleading because of the small ranges of water vapor and temperature observed during the one-month experiment in late

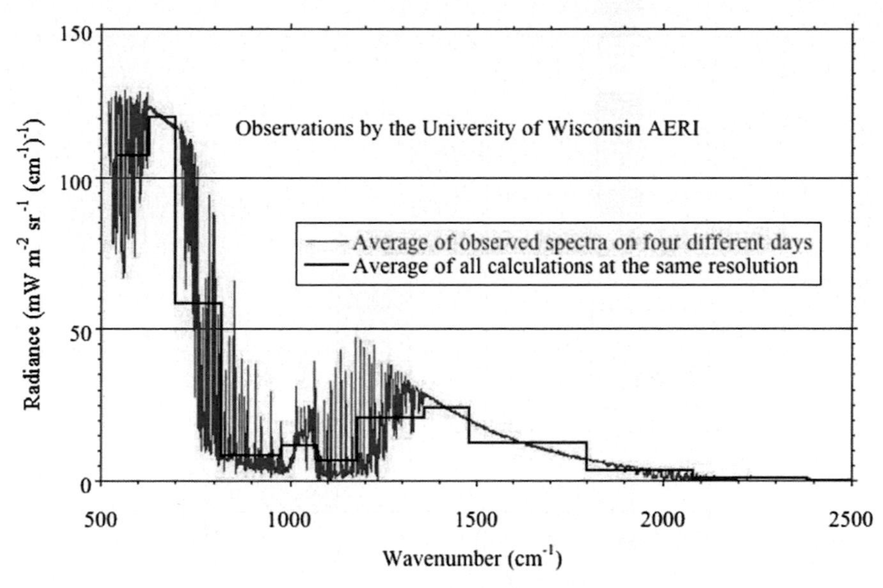

FIG. 1-3. Average AERI spectrum for the four SPECTRE–ICRCCM test cases (red line) and the average spectrum from the model calculations for the intercomparison resolution.

autumn, the lack of observations in the 0 to 500 cm^{-1} portion of the spectrum, the absence of many GCM-type models in the spectral comparisons, and the similarity among several LBL models. The large range of disagreement between the observations and calculations in spectral intervals with large radiance variability (e.g., 800–1000 cm^{-1}) highlights regions that are studied extensively during ARM (Mlawer and Turner 2016, chapter 14). Unfortunately, the SPECTRE observations would not allow estimates on the accuracy of clear-sky fluxes and cooling rate calculations at other levels of the atmosphere nor at other climatologically important locations across the globe.

The 1995 ICRCCM workshop also reviewed the state of shortwave modeling and concluded that the problems associated with shortwave radiation remained as they were following the 1988 workshop. As a consequence, the working group recognized the need for and recommended a SPECTRE-like experiment to ferret out the cause for disagreement between the shortwave models.

Plans for SPECTRE were well underway when the DOE decided to launch the ARM Program, and the SPECTRE proposal was used in large part for the ARM Program Plan (DOE 1990). SPECTRE was viewed as a prototype for what became the ARM Central Facility a

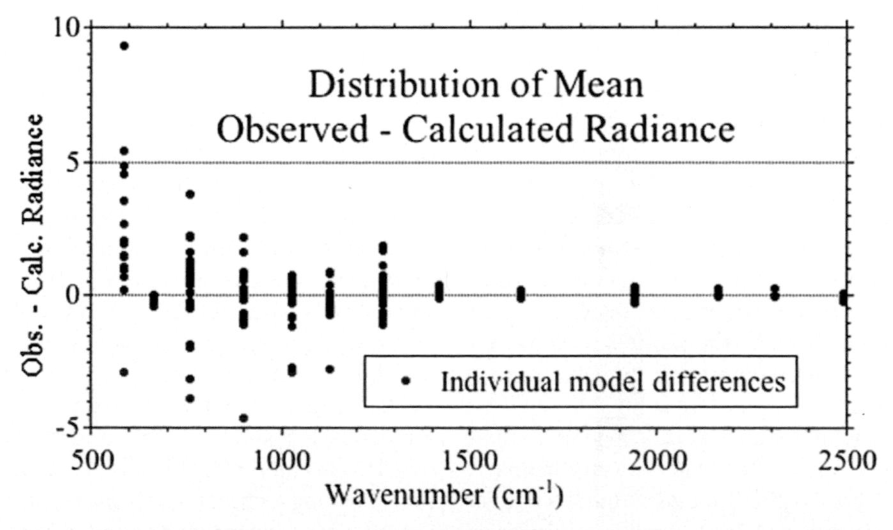

FIG. 1-4. Spectral distribution of the average AERI-observed − calculated radiance spectrum calculated from the four SPECTRE–ICRCCM test cases.

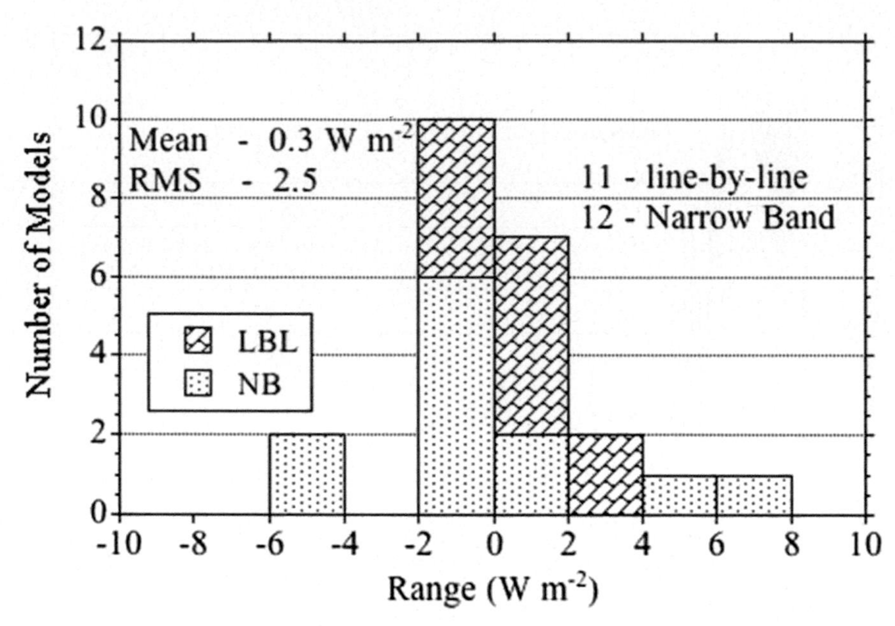

FIG. 1-5. Distribution of the average observed − calculated clear-sky downward fluxes at the surface for the four SPECTRE–ICRCCM test cases.

few years later, and several ARM investigators used data from it for the first case of the ARM Instantaneous Radiative Flux (IRF) experiment (Mlawer and Turner 2016, chapter 14) before the regular ARM data flow commenced.

4. FANGIO

In 1984, the DOE asked Robert Cess and Gerald Potter to organize a GCM intercomparison project. The motivation for this project was to understand the large differences that existed for climate-change simulations among the various GCMs at that time. This evolved into the FANGIO (Feedback Analysis of GCMs and in Observations) project. Cess and Potter devised a procedure for diagnosing cloud feedback in GCMs (see Cess and Potter 1988) that was adopted at the first FANGIO workshop in 1987. There were eight FANGIO workshops held in Europe and the United States from 1987 to 1996. The FANGIO project constituted the first structured intercomparison of GCMs, and the principal conclusion of this study was that model-to-model differences in cloud feedback were largely the cause of the significant differences in climate-change simulations by the models used at that time. In the following, we summarize the salient findings from FANGIO.

To understand cloud feedback, it is useful to first demonstrate how clouds affect the present climate. Figure 1-6 shows Earth's global-mean radiation budget at the top of the atmosphere (TOA) and also a fictitious situation for which there are no clouds, but all else remains unchanged. Numbers have been rounded for the purpose of illustration. Because clouds are bright, their presence increases reflection of shortwave radiation by 50 W m^{-2} (cooling), while the greenhouse effect of clouds results in a longwave warming of 30 W m^{-2}. Thus the net effect is a 20 W m^{-2} cooling, conventionally expressed as a cloud-radiative forcing (CRF) of −20 W m^{-2} (Ramanathan et al. 1989). With the subscript c referring to clear-sky fluxes, then

$$\mathrm{CRF} = (F_c - F) - (Q_c - Q), \quad (1\text{-}1)$$

where F and Q respectively denote the global-mean emitted infrared and net downward solar fluxes at the TOA. It is the change in CRF (ΔCRF), associated with a change in climate, that constitutes cloud feedback. For example, a doubling of atmospheric CO_2 produces roughly a 4 W m^{-2} direct radiative forcing G of the climate system. If the ensuing climate change altered CRF from −20 to −16 W m^{-2}, so that $\Delta\mathrm{CRF} = G$, then ΔCRF would amplify the direct radiative forcing by a factor of 2 (a twofold positive feedback). Zero cloud feedback corresponds to $\Delta\mathrm{CRF}/G = 0$, while $\Delta\mathrm{CRF}/G < 0$ denotes negative feedback. Thus $\Delta\mathrm{CRF}/G$ quantifies the net cloud feedback (Cess et al. 1989, 1990).

The direct radiative forcing G of the surface–atmosphere system is evaluated by holding all other climate parameters fixed. It is this quantity that induces the ensuing climate change, and physically it represents a change in the net (shortwave plus longwave) radiative flux at the TOA. For an increase in the

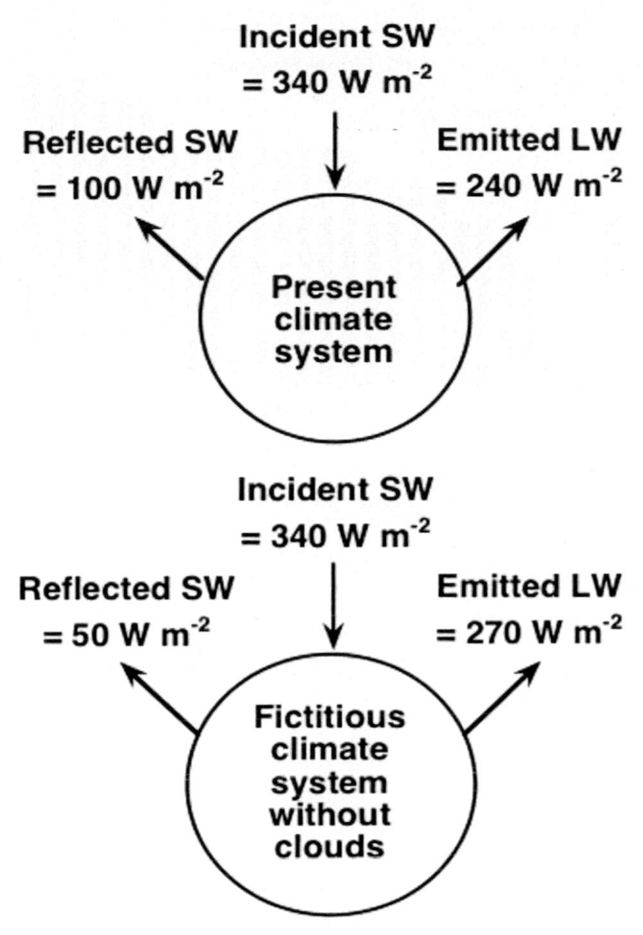

FIG. 1-6. Earth's TOA radiation budget together with a fictitious planet in which there are no clouds, but all else remains the same. LW denotes longwave (infrared) radiation and SW denotes shortwave (solar) radiation. Numbers have been rounded for the purpose of illustration. The fictional planet is used solely as an aid in understanding the radiative impact of clouds and as such is not in equilibrium; i.e., the sum of the emitted LW and the reflected SW do not add to 340 W m^{-2} [from Cess et al. (1996)].

CO_2 concentration of the atmosphere, G is the reduction in the emitted TOA longwave flux resulting solely from the CO_2 increase, and this reduction results in a heating of the surface–atmosphere system. The response process is the change in climate that is then necessary to restore the TOA radiation balance, such that

$$G = \Delta F - \Delta Q. \tag{1-2}$$

Thus ΔF and ΔQ represent the climate-change TOA responses to the direct radiative forcing G, and these are the quantities that are impacted by climate feedback mechanisms. Furthermore, the change in surface climate, expressed as the change in global-mean surface temperature ΔT_s, is related to the direct radiative forcing G by

$$\Delta T_s = \lambda G, \tag{1-3}$$

where λ is the climate sensitivity parameter:

$$\lambda = (\Delta F/\Delta T_s - \Delta Q/\Delta T_s)^{-1}. \tag{1-4}$$

An increase in λ thus represents an increased climate change due to a given climate forcing G.

The methodology proposed by Cess and Potter (1988) was employed by Cess et al. (1990) to quantify cloud feedback in a suite of 19 GCMs. This methodology consisted of imposing ±2-K sea surface temperature (SST) perturbations, in conjunction with a perpetual July simulation, as a surrogate climate change for the sole purpose of intercomparing climate sensitivity. This procedure is in essence an inverse climate change simulation. Rather than introducing a forcing G into the models and then letting the climate respond to this forcing, they instead prescribed the climate change and let the models in turn produce their respective forcing in accordance with Eq. (1-1). This procedure eliminated the substantial computer time required for equilibration of the models. The second advantage was that, since the same SSTs were prescribed, all of the models had essentially the same control climate because land temperatures are tightly coupled, through atmospheric transport, to the SSTs. The models then all produced a global-mean ΔT_s between the −2-K and +2-K SST perturbation simulations that was close to 4 K, and different model sensitivities in turn resulted in different values for G.

The perpetual July simulation eliminated another problem. This study focused solely on atmospheric feedback mechanisms, and inspection of output from all the models showed that climate feedback caused by changes in snow and ice coverage was suppressed through use of a fixed sea ice constraint and because the perpetual July simulations produced little snow cover in the Northern Hemisphere.

As discussed by Cess et al. (1990), there exist differing definitions of cloud feedback. For example, Wetherald and Manabe (1988) have addressed cloud feedback by performing two simulations, one with computed clouds and the other holding clouds fixed at their control climate values. Thus in this definition cloud feedback is referenced to the simulation in which clouds are invariant to the change in climate, while all other feedback processes are operative. For their CO_2 doubling simulations, Wetherald and Manabe (1988) found that cloud feedback amplified global warming by a factor of 1.3. Hansen et al. (1984), again for a CO_2 doubling, employed a radiative–convective model to diagnose three categories of feedback mechanisms within the

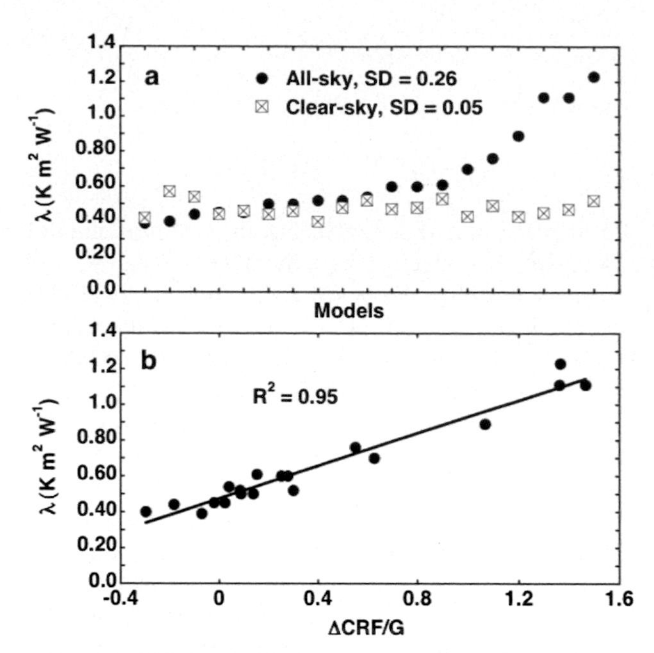

FIG. 1-7. (a) The sensitivity parameter λ, both for Earth with clouds and for the fictitious planet without clouds, for the 19 GCMs. SD denotes the standard deviation. (b) The sensitivity parameter λ, as a function of the cloud feedback parameter ΔCRF, for the 19 GCMs. The solid line is a linear fit, and R^2 is the square of the correlation coefficient.

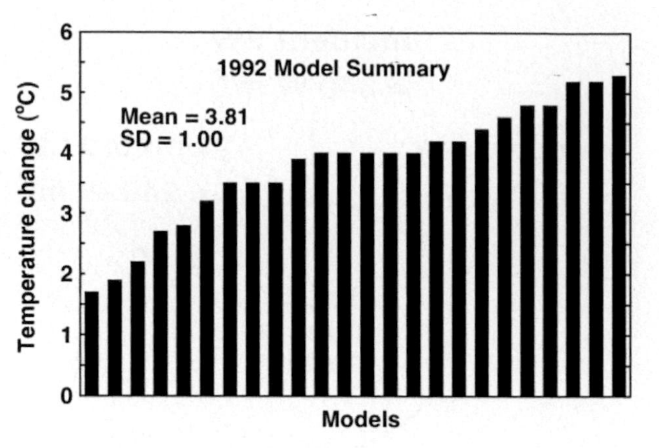

FIG. 1-8. Summary of the increase in global mean equilibrium surface temperature caused by a doubling of atmospheric CO_2 concentrations. These results are from simulations with atmospheric GCMs with a seasonal cycle, a mixed-layer ocean, and interactive clouds. This figure is constructed from Table 3.2(a) of IPCC (1990) and Table B2 of IPCC (1992).

Goddard Institute for Space Studies (GISS) GCM: water vapor, snow/ice–albedo, and cloud feedbacks. As did Wetherald and Manabe (1988), Hansen et al. found that cloud feedback produced a factor of 1.3 amplification. However, their feedback definition differs from that of Wetherald and Manabe; it is referenced not only to fixed clouds but also to the absence of both water vapor feedback and snow/ice–albedo feedback. When their results are reformulated in terms of Wetherald and Manabe's definition, their cloud feedback amplification factor is 1.8. The study by Cess et al. (1990) adopted yet a third definition of cloud feedback as discussed above using $\Delta\mathrm{CRF}/G$ as a feedback parameter.

Cess et al. (1990) analyzed output from 19 GCMs, and their results for λ are summarized in Fig. 1-7a based on both global-mean all-sky (clear plus clouds) and clear-sky TOA fluxes. The procedure for evaluating the clear-sky fluxes is described by Cess et al. (1990). The clear-sky λ values from the models agree well with one another, but for the all-sky λ there is roughly a threefold variation in climate sensitivity. Since the only difference between the clear-sky and all-sky sensitivities is the inclusion of clouds in the latter, this demonstrates that cloud feedback is a major cause of the differences in model sensitivity. This conclusion is strengthened by the results shown in Fig. 1-7b. Clearly the model-to-model differences in climate sensitivity are mostly caused by model-to-model differences in $\Delta\mathrm{CRF}/G$ (i.e., cloud feedback).

This study by Cess et al. (1990) thus helped to explain that model-to-model differences in cloud feedback were largely the cause of the significant differences in climate-change simulations by the models used at that time. Shown in Fig. 1-8 is the increase in global-mean surface temperature, caused by a doubling of atmospheric carbon dioxide, as computed by 24 different GCMs up to 1992. Assuredly, much of these differences can be attributed to differences in cloud feedback among the models, and we emphasize that the FANGIO project was the first to elucidate this important point.

5. Summary

By the late 1980s it was quite clear that there were major questions concerning the accuracy of the radiation codes used in climate models and in the predictions by GCMs due to differences in cloud feedback between the models. It was under this backdrop that Dr. Robert Hunter, former director of the DOE Office of Energy Research, commissioned a one-day seminar in November 1988 to summarize the DOE's climate activities as an aid for setting the stage for future initiatives. Hunter and his review team were adamant on devising climate programs that allowed testing hypotheses with observations. Working with Hunter, Dr. Aristides Patrinos, former Associate Director for Biological and Environmental Research for the DOE Office of Science, began the Quantitative Links Program and the subsequent ARM Program that expanded the approach of SPECTRE to include testing of models of the properties and occurrence of clouds from which to glean additional information about cloud feedback in GCMs.

REFERENCES

Bolle, H.-J., 2008: International Radiation Commissions 1896 to 2008: Research into atmospheric radiation from IMO to IAMAS. IAMAS Publ. Series 1, International Association of Meteorology and Atmospheric Sciences, 150 pp. [Available online at http://www.irc-iamas.org/files/IRC_History_Bolle_2.pdf.]

Cess, R. D., and G. L. Potter, 1988: A methodology for understanding and intercomparing atmospheric climate feedback processes in general circulation models. *J. Geophys. Res.*, **93**, 8305–8314, doi:10.1029/JD093iD07p08305.

——, and Coauthors, 1989: Interpretation of cloud–climate feedback as produced by 14 atmospheric general circulation models. *Science*, **245**, 513–516, doi:10.1126/science.245.4917.513.

——, and Coauthors, 1990: Intercomparison and interpretation of climate feedback processes in 19 atmospheric general circulation models. *J. Geophys. Res.*, **95**, 16 601–16 615, doi:10.1029/JD095iD10p16601.

——, and Coauthors, 1996: Cloud feedback in atmospheric general circulation models: An update. *J. Geophys. Res.*, **101**, 12 791–12 794, doi:10.1029/96JD00822.

DOE, 1990: Atmospheric Radiation Measurement Program Plan. DOE/ER-0441, U.S. Department of Energy, Office of Energy Research, 116 pp.

Ellingson, R. G., 1995: The Intercomparison of Radiation Codes Used in Climate Models (ICRCCM): Summary of a workshop. Report to the International Radiation Commission, 10 pp. [Available online at http://www.atmos.umd.edu/~bobe/icrccm95.pdf.]

——, and Y. Fouquart, 1990: The Intercomparison of Radiation Codes in Climate Models (ICRCCM). World Climate Research Program Rep. WCRP-39, WMO/TD 371, World Meteorological Organization, 47 pp.

——, and ——, 1991: The intercomparison of radiation codes in climate models (ICRCCM): An overview. *J. Geophys. Res.*, **96**, 8925–8927, doi:10.1029/90JD01618.

——, and W. J. Wiscombe, 1996: The Spectral Radiance Experiment (SPECTRE): Project description and sample results. *Bull. Amer. Meteor. Soc.*, **77**, 1967–1985, doi:10.1175/1520-0477(1996)077<1967:TSREPD>2.0.CO;2.

——, ——, D. Murcray, W. Smith, and R. Strauch, 1990: ICRCCM phase II: Verification and calibration of radiation codes in climate models. DOE/ER/60971, 12 pp.

——, J. Ellis, and S. Fels, 1991: The intercomparison of radiation codes in climate models: Long wave results. *J. Geophys. Res.*, **96**, 8929–8953, doi:10.1029/90JD01450.

Fouquart, Y., B. Bonnel, and V. Ramaswamy, 1991: Intercomparing shortwave radiation codes for climate studies. *J. Geophys. Res.*, **96**, 8955–8968, doi:10.1029/90JD00290.

Hansen, J., A. Lacis, D. Rind, G. Russel, P. Stone, I. Fung, R. Ruedy, and J. Lerner, 1984: Climate sensitivity: Analysis of feedback mechanisms. *Climate Processes and Climate Sensitivity, Geophys. Monogr.*, Vol. 29, Amer. Geophys. Union, 130–163.

IPCC, 1990: *Climate Change: The IPCC Scientific Assessment.* J. T. Houghton, G. J. Jenkins, and J. J. Ephraums, Eds., Cambridge University Press, 365 pp.

——, 1992: *Climate Change 1992: The Supplementary Report to the IPCC Scientific Assessment.* J. T. Houghton, B. A. Callander, and S. K. Varney, Eds., Cambridge University Press, 218 pp.

Luther, F. M., Ed., 1984: The Intercomparison of Radiation Codes in Climate Models (ICRCCM). World Climate Program Rep. WCP-93, World Meteorological Organization, 37 pp.

——, R. G. Ellingson, Y. Fouquart, S. Fels, N. A. Scott, and W. J. Wiscombe, 1988: Intercomparison of radiation codes in climate models (ICRCCM): Longwave clear-sky results—A workshop summary. *Bull. Amer. Meteor. Soc.*, **69**, 40–48.

MacCracken, M. C., and F. M. Luther, 1985: Projecting the climatic effects of increasing carbon dioxide. U.S. Department of Energy, DOE/ER-0237, OL14836795M, 381 pp.

Mlawer, E. J., and D. D. Turner, 2016: Spectral radiation measurements and analysis in the ARM Program. *The Atmospheric Radiation Measurement (ARM) Program: The First 20 Years, Meteor. Monogr.*, No. 57, Amer. Meteor. Soc., doi:10.1175/AMSMONOGRAPHS-D-15-0027.1.

Ramanathan, V., R. D. Cess, E. F. Harrison, P. Minnis, B. R. Barkstrom, E. Ahmad, and D. Hartmann, 1989: Cloud-radiative forcing and climate: Results from the Earth Radiation Budget Experiment. *Science*, **243**, 57–63, doi:10.1126/science.243.4887.57.

Riches, M., 1983: CO_2 Climate Research Plan. U.S. Department of Energy Rep. DOE/ER-0186, 22 pp.

Turner, D. D., E. J. Mlawer, and H. E. Revercomb, 2016: Water vapor observations in the ARM Program. *The Atmospheric Radiation Measurement (ARM) Program: The First 20 Years, Meteor. Monogr.*, No. 57, Amer. Meteor. Soc., doi:10.1175/AMSMONOGRAPHS-D-15-0025.1.

Wetherald, R. T., and S. Manabe, 1988: Cloud feedback processes in a general circulation model. *J. Atmos. Sci.*, **45**, 1397–1415, doi:10.1175/1520-0469(1988)045<1397:CFPIAG>2.0.CO;2.

Chapter 2

Original ARM Concept and Launch

GERALD M. STOKES

Stony Brook University, State University of New York, Stony Brook, New York

1. Introduction

The foregoing chapter (Ellingson et al. 2016, chapter 1) lays out the scientific foundation for the initiation of the Atmospheric Radiation Measurement (ARM) Program. In parallel with the development of this scientific foundation, public and political attention was being focused on the climate issue. Even broader scientific findings in climate came together with this heightened public interest and created a large expansion of the funding for climate research. These funds allowed ARM, a ground-based program, to be put together on a scale more comparable to the large satellite programs of NASA and larger in budget than previous ground-based climate research efforts. ARM was developed in parallel with major efforts from other agencies, such as NASA's Earth Observing System (EOS), during a time of considerable expansion of global climate research.

This chapter is not written as a chronology. It is rather a description of the key intellectual threads of ARM that were developed concurrently and that are described here as interrelated stories. In a sense, it is a guide to the balance of the monograph, describing primarily how the program came together. Most of the references are to internal ARM documents and the other chapters in this monograph, where the topics introduced here are discussed in more detail. Ackerman et al. (2016, chapter 3) provides a description of how these initial steps were transformed into the mature program that exists today, and Mather et al. (2016, chapter 4) provides an overview of some of the more significant scientific accomplishments of ARM.

Corresponding author address: Gerald M. Stokes, Department of Technology and Society, Stony Brook University, Stony Brook, NY 11794.
E-mail: gerry.stokes@gmail.com

DOI: 10.1175/AMSMONOGRAPHS-D-15-0021.1

2. The challenge from the Committee on Earth Sciences

The period of 1988–92 saw a huge increase in the research effort of the United States on understanding Earth's climate. An important precipitating event was the public and congressional perception of and response to the heat wave of the summer of 1988, which has been estimated to have caused 5000–10 000 deaths and more than $70 billion in weather-related economic losses in the central and eastern United States.[1] In the same year, the Intergovernmental Panel on Climate Change (IPCC)[2] was formed by the World Meteorological Organization and the United Nations Environment Program (UNEP), and its first assessment report was published in 1990. This period culminated with Earth Summit in Rio in 1992 and the negotiation of the UN Framework Convention on Climate Change in the same year.

At this time, the U.S. Global Change Research Program (USGCRP) began as a presidential initiative and was defined further by the Global Change Research Act of 1990.[3] This ongoing program coordinates the efforts of federal agencies on global environmental change,[4] which are described annually in the report

[1] The public perception of the climate change issue, which is reflected in an online article (http://www.csmonitor.com/USA/2011/0711/As-much-of-US-swelters-here-are-5-worst-heat-waves-of-past-30-years/Summer-1988), drove much of the discussion of climate both in the presidential election and throughout the presidential term of George H. W. Bush.

[2] https://www.ipcc.ch/organization/organization_history.shtml

[3] Global Change Research Act of 1990, Pub. L. No. 101-606, 104 Stat. 309 (1990).

[4] http://www.globalchange.gov/about

"Our Changing Planet."[5] This program was in its formative stages in 1989, and each agency sought to gain approval for its specific contributions to the national research effort. In the spring and summer of 1989, the Department of Energy (DOE), one of the participating agencies, was designing its contribution to the USGCRP. The DOE had sponsored climate research programs starting in 1978 as part of its Carbon Dioxide Research and Assessment Program. The DOE carbon dioxide research program had produced some very important products, including six state-of-the-art reports[6] that were a comprehensive review of the state of climate research in the mid-1980s, an interesting predecessor to the IPCC assessment reports half a decade later. In preparing its USGCRP efforts, the DOE focused on results that were emerging from the Intercomparison of Radiation Codes in Climate Models (ICRCCM; Ellingson and Fouquart 1991).[7] As noted in Ellingson et al. (2016, chapter 1), ICRCCM was part of a DOE effort to reconcile differences in climate modeling results. The approach was to compare radiation parameterizations used in these models and to understand why and how the models might differ.

The DOE's early proposals to pursue what was later called the ARM Program were not agreed to by the governing body of the USGCRP—the Committee on Earth Sciences (CES). The disagreements among the CES principals revolved largely around the size of the program. The program was perceived as too large at $200 million per year, and the view was that the atmospheric science community was not large enough to take on the work even if funded at that level. After some intensive negotiation, the DOE agreed in mid-August of 1989 to produce a revised Program Plan by 1 November of that year for peer review and approval.

Ari Patrinos, Director of the Atmospheric and Climate Research Division within the Office of Energy Research of the DOE, and who had represented the DOE at the CES, was given the charge by the DOE to develop this plan. He selected a team of three people—Gerry Stokes of the Pacific Northwest National Laboratory (PNNL); Bob Ellingson of the University of Maryland, College Park; and Dave Sowle of Mission Research Corporation—to write the plan. Patrinos also convened representatives from most of the DOE laboratories to support the preparation of the plan. This represented the first time that the DOE laboratories had the opportunity for systematic input into the program planning process for ARM. Prior to this time, federal rules prohibited the involvement of DOE laboratory staff, who were not federal employees, from participating in a process that was the part of the formulation of the federal budget. The selection of Stokes, Ellingson, and Sowle was intended to give the laboratories, the academic community, and the private sector a voice in the formulation of ARM.

Stokes, who became ARM's first Chief Scientist, was an astrophysicist and head of the Global Studies Program at PNNL and had experience in atmospheric spectroscopy and radiometry. Sowle was a representative of the private sector whose experience was largely in defense-related remote sensing applications.

Bob Ellingson, then a professor at the University of Maryland, College Park, played the key role in the early phase of the proposal. As one of the architects and key investigators of ICRCCM, he had already prepared a plan together with Warren Wiscombe of NASA (who would later become the third ARM Chief Scientist) for the Spectral Radiance Experiment (SPECTRE; Ellingson and Wiscombe 1996; Ellingson et al. 2016, chapter 1) to add real observations to the comparison of the radiative modeling algorithms. Their approach called for the characterization of the physical properties of the atmospheric column concurrent with the measurement of the downward spectral radiance and flux. The flux observations would be compared with calculations of the spectral radiance and flux based on concurrent measurements of the atmospheric characteristics that governed the radiation: temperature and composition

[5] The web page http://www.globalchange.gov/browse/reports has connections to each volume of this series of documents. They are prepared as part of the annual budget submission to the U.S. Congress by the President of the United States. These reports chronicle the growth and programmatic evolution of the USGCRP as told by the federal agencies.

[6] There were six reports: four primary reports ("Atmospheric carbon dioxide and the global carbon cycle," DOE/ER-0239; "Direct effects of increasing carbon dioxide on vegetation," DOE/ER-0238; "Detecting the climatic effects of increasing carbon dioxide," DOE/ER-0235; and "Projecting the climatic effects of increasing carbon dioxide," DOE/ER-0237) and two specialized reports ("Characterization of information requirements for studies of CO_2 effects: Water resources, agriculture, fisheries, forests and human health," DOE/ER-0236, and "Glaciers, ice sheets, and sea level: Effect of a CO_2 induced climatic change," DOE/ER/60235-1). They were peer reviewed by a large team convened by Roger Revelle, serving as the chairman of the Committee on Climate of the American Association for the Advancement of Science (AAAS).

[7] This decision was made by Robert Hunter, the assistant secretary of energy, responsible for the Office of Energy Research as part of the focus on the Quantitative Links Program as described in Ellingson et al. (2016, chapter 1).

(most critically water)[8] as a function of altitude and pressure.

This approach was seen as being critical to ensuring that the radiative transfer in climate models was represented correctly. ICRCCM had shown that a convergence of results could be gained between the high-spectral-resolution radiative transfer models and the lower-resolution parameterizations used in general circulation models (GCMs) of Earth's climate that were required for good computational performance. At this time in the development of climate models, a significant fraction of computational time was dedicated to radiative transfer. The parameterization of detailed radiative transfer results to lower resolution was an essential part of the modeling process, which allowed the climate models to be practical tools for research. ICRCCM then pointed out that the next step was to ensure that the higher-resolution models were correct by comparing them to actual observations. This paradigm of ensuring that there was a convergence between parameterizations of physical processes in climate models and higher-fidelity models of the physics became a signature of the ARM design and approach.

The path to accomplishing this for the clear sky case, as outlined by Ellingson and Wiscombe (1996), was seen as straightforward. The cloudy sky cases were seen as somewhat more complicated, but the team writing the proposal felt that the general paradigm of first assuring that the parameterization of the cloudy sky conditions matched higher-resolution models and then observationally confirming the fidelity of the higher resolution models would work. This path to an effective treatment of the radiative transfer in these situations also seemed appropriate, but it was clear that there were key issues that would have to be addressed in both three-dimensional radiative transfer and the measurement of the corresponding radiative properties of the clouds.

The earliest discussions among the laboratory team working on the Program Plan focused not only on the measurement of the radiative properties of clouds, but also on the larger question of whether those properties could be predicted in climate models. This discussion led to a critical decision, proposed by Stokes and the team to DOE, in which they supported the idea that, in order to solve the radiation problem and advance climate modeling, the issue of both cloud formation and characterization must be addressed. This decision shaped much of ARM and has probably led to its enduring mission, well beyond the 10 years originally envisaged. The importance of the cloud–radiation interaction problem to climate research was highlighted in the IPCC First Assessment Report released in 1990 and in all subsequent assessments.[9]

An important consequence of the decision to work on the "greater cloud problem" was the operational decision that ARM was to be a continuously operating facility. Much of the cloud research prior to ARM was conducted on a campaign basis. In this mode, a large number instruments and investigators were assembled in one place for a few weeks to perhaps 2 months. As a result, the data collected were quite literally at the mercy of the weather. This method led to some excellent results, but it did not result in a climatically representative dataset. The concept evolved that ARM would provide a continuous data record on climatologically appropriate time scales—years rather than weeks. The consequences of this were many and were a central, and highly controversial, feature of the Program Plan that was submitted for review on 1 November 1989; these consequences will be discussed later in this chapter.

The Program Plan was circulated on 1 November 1989 to more than a dozen scientific reviewers in the private, government, and university communities. A public meeting on the program was held in Washington, D.C., later that month to elicit additional comments. A response to the various comments was prepared and presented in January, which included a proposed path forward for the ARM Program. Probably the most important aspect of the review was the very thoughtful commentary provided on solving the "whole cloud problem." While the original Program Plan was extremely well informed by the intellectual work of Ellingson and others on what parameters needed to be measured to test radiation codes, the plan was quite weak on what was needed to be measured in order to address the larger cloud problem. The response to the review provided assurance that work would begin to strengthen the approach to cloud formation and characterization. In particular, Tom Ackerman, who became the second ARM Chief Scientist, and Bruce Albrecht at

[8] In the view of the author, one of the key contributions of ARM was to the measurement of water vapor, which is covered in Turner et al. (2016b, chapter 13).

[9] The series of assessment reports are available at https://www.ipcc.ch/publications_and_data/publications_and_data_reports.shtml. In the IPCC First Assessment Report, McBean and McCarthy (1990, p. 319) make the statement that "further modelling and observational research will nevertheless be necessary to achieve accurate representation in climate models of the role of clouds and radiation." This call for better representation of clouds continues in the Third Assessment Report, where it was stated that "the role of clouds in the climate system continues to challenge the modelling of climate" (Moore et al. 2001, p. 775).

Penn State University, who had been developing a draft plan on how to advance the study of cloud properties using millimeter-wavelength cloud radar, provided extensive input during this peer review that greatly augmented the nascent ARM cloud observing strategy.

On a snowy December day, when the government was closed because of weather, during a discussion involving Robert Corell of the National Science Foundation (NSF); Michael Hall and Lester Machta of NOAA; Robert Schiffer of NASA; Richard Anthes, then president of the University Corporation for Atmospheric Research (UCAR), representing the CES; and the DOE team led by Ari Patrinos, the project was approved to proceed. The consequence of this was dramatic for the program, because funds had already been appropriated for a DOE effort as part of the USGCRP and $9 million were immediately available for use a few weeks later on 1 January 1990.

3. A systems approach and the first Program Plan

One stipulation imposed by the CES was that a revised Program Plan be prepared that incorporated the reviewers' comments and the DOE responses to those comments. To this end, work began in early January 1990 with a systems design session lasting 3 days that set the top-level organization for the project. The process that Gerry Stokes selected was a highly disciplined enterprise modeling approach developed by the WISDM Corporation of Issaquah, Washington, based on previous experience with the company.[10] The robust system engineering approach at the core of the WISDM process allowed the project team to use the high-level design, developed in the first 3-day meeting, as the basis for more detailed designs of individual components of the project. Over the course of the first 2 years of the project, about a dozen more detailed, but similarly structured, facilitated-planning sessions were held. These sessions resulted in what could be termed the ARM Enterprise Model—a self-consistent and detailed operational description of the entire ARM project.[11] Key concepts that persist today in the project were defined at those meetings.

The ARM Enterprise Model included the basic elements of site selection, operations, instrumentation selection, and the data system. These elements were then described in the revised Program Plan DOE/ER-0441 [the executive summary is in ARM (2016a, appendix A)], which was released in February 1990.[12] Key features of the final Program Plan were the description of the basic experimental approach, the rationale for multiple sites, the motivation for 24/7 operation, the inclusion of an instrument development program, and the role of the Science Team. These elements are described in great detail in other chapters of this monograph. In the balance of this chapter, the top-level ideas that shaped the launch of the ARM Program and their motivation will be described.

The success of the intensive planning sessions that created the ARM Enterprise Model and the initial launch of the ARM Program was a product of the ability of the DOE National Laboratories to act as a team. The DOE atmospheric sciences program had promoted the laboratories working together on large field programs for many years, and it was therefore natural that individual laboratories took responsibility for major components of the ARM system. Under the overall leadership of Gerry Stokes at PNNL (as the ARM Program's first Chief Scientist) and Mike Riches of the DOE Office of Health and Environmental Research, laboratory leads were named for the various elements of the program. Argonne National Laboratory and the late Marv Wesley were in charge of instrumentation, Lawrence Livermore National Laboratory and Marv Dickerson assumed responsibility for atmospheric modeling, Los Alamos National Laboratory and Sumner Barr assumed responsibility for Operations, and Brookhaven National Laboratory (BNL) and Steve Schwartz led the site selection process. Wesley, Dickerson, and Barr were senior managers in the atmospheric science programs at their respective laboratories, each with more than 20 years of experience in the atmospheric sciences, and Schwartz was and is a leading atmospheric chemist at BNL. The data management and particularly the data archive were the responsibility of Oak Ridge National Laboratory and Paul Kanciruk of the Carbon Dioxide Information and Analysis Center (CDIAC). As time evolved, operations and data management were subdivided and again individual laboratories were given responsibility for key components of the associated systems. The history of collaboration among the DOE laboratories in these

[10] The current approach of the WISDM Corporation is described on their website (http://www.wisdm.com). The process used by ARM to develop the requirements for the ARM Enterprise Model is described in this material.

[11] Within the project, the term "ARM Enterprise Model" was rarely used—the team more frequently used the terms "WISDM process" and "WISDM model." The design sessions were very demanding and created a deep and common understanding of what the goals were of the program and the methods that would be used to achieve them.

[12] The entire Program Plan is available at https://www.arm.gov/publications/doe-er-0441.pdf.

areas and the flexibility of funding granted by the DOE also allowed the program leadership to draw easily on highly qualified individuals from all of the DOE laboratories in executing the detailed plans for the broad areas of the program. This was particularly true in instrumentation and operations.

An important property of the ARM Enterprise Model was that it was strictly hierarchical. The design that was generated in 3 days in January 1990 was a very careful and, in retrospect, surprisingly complete description of the relationship among the major elements of the program and their relationship to the outside community with which the ARM Program interacted. As a result, when groups met to provide further design for site operations, the relationship to the data system, the instrument selection process, and the relationship with the Science Team were already defined. The definition of these relationships not only included specific requirements but also general principles that guided the path forward, such as 24/7 operations and the overriding principle that ARM was being built to serve the Science Team, not to generate data that the operations staff thought might be useful.

From the perspective of the leadership of the program, this design allowed the program to maintain discipline as its implementation went forward on many parallel paths, the goal of which was to get ARM in the field as quickly and as productively as possible. In the first year of the program, two important additions to the management team were made that were key to this aggressive schedule for deployment: Ted Cress became the technical director of the program and Peter Lunn was named the ARM program manager at DOE. Cress, a meteorologist, had just retired from the Air Force, where he had been a member of CES representing the Department of Defense (DOD), and Lunn was a senior program manager who came from the Defense Nuclear Agency. These two brought critical experience to the management of the program and were very much responsible for its early success.

4. The surrogate science team and development of "experiments"

As noted above, the ARM Program was designed to serve a set of investigators from the broader scientific community, competitively selected, to advance the representation of clouds and radiation in climate models—this set of investigators was the ARM Science Team. During the early development of ARM, this team did not yet exist. The scientific breadth needed did not exist within the DOE laboratories, and thus another approach was needed.

The formal solicitation for Science Team proposals began almost immediately after the program was funded and continued through the summer of 1990. In the interim, the nature of the design of the program required that there be input from the scientific community. This input was required ahead of the selection process that would create the more permanent Science Team structure. To this end, a series of meetings were convened to further define the broad scientific approach to the project. The individuals invited were drawn from the broader climate research community, many of whom were members of the peer review panel for the original Program Plan, and others who were long-standing investigators in the DOE climate research community. These individuals were invited to be part of what was termed the surrogate Science Team. One of the more pressing matters for these early Science Team meetings was gaining further clarity on what kind of measurements the scientific community would need, what instruments they saw as providing the data, and how they would use the data. Among the most important issues continued to be what experimental approach would be taken to address the cloud formation problem. The revised Program Plan of February 1990 (ARM 2016a, appendix A) added material on this topic but still had an incomplete view of how ARM would approach this problem.

The meetings were structured around an idea that grew out of the ARM Enterprise Model design sessions—that of an "experiment." For ARM, an experiment was framed with a hypothesis, which was a particular calculation performed in a climate or radiative transfer model. In the Program Plan, this approach was described as the Cloud and Radiation Testbed (CART). The groups were asked what information was required as input to do the calculation—observed properties of the atmosphere as opposed to model-derived quantities—and what observations were necessary to confirm the validity of the calculation. The prototype for this approach was termed the instantaneous radiative flux (IRF) experiment, for which the results and conclusions of ICRCCM and the design of SPECTRE laid the intellectual foundation. Additional information on the IRF and its radiative closure experiment is provided in Turner et al. (2016b, chapter 13) and Mlawer and Turner (2016, chapter 14).

The surrogate Science Team meetings led to the definition of a second experiment, the single column model (SCM). This approach, advocated by Dave Randall of Colorado State University and Richard Somerville of the Scripps Institution of Oceanography, became the basis for addressing the cloud formation problem. In the SCM, one would not only specify the state of the atmosphere as a function of altitude averaged over a climate grid model scale volume but also define the fluxes at the boundaries of the cells of quantities such as water vapor and other atmospheric quantities. This changed the design of the basic ARM site as originally envisaged in the Program Plan. Most specifically, it led to the need for boundary sites that would facilitate the measurement and estimation of the advective tendencies at the edges of what was shaping up to be a fully instrumented GCM grid cell. Additional information on the SCM experiment and the development of the concept is provided in Zhang et al. (2016, chapter 24).

These two experiments drove much of the early work in ARM. They set the scale of the first site, the southern Great Plains (SGP) site in north-central Oklahoma that covered roughly 150 000 km^2, and led to both the choices of instrumentation and their siting—leading to the design of the central facility, boundary facilities, and extended facilities described in Cress and Sisterson (2015, chapter 5) and Sisterson et al. (2016, chapter 6) of this monograph. The organization of these early meetings also helped shape the approach of convening working groups within the Science Team after it was selected. The first four working groups of the program would eventually solidify into the IRF, SCM, aerosol properties (McComiskey and Ferrare 2016, chapter 21), and cloud properties working groups, the last of which focused on both radar observations (Kollias et al. 2016, chapter 17) and retrievals (Shupe et al. 2016, chapter 19).

An important historical note is that during this period of time, the early 1990s, an unfortunate event took place that initially was seen as a setback, but eventually led to some very important and innovative work. In February 1990, the final scanner of the Earth Radiation Budget Experiment (ERBE) failed. This satellite had been important to experiments like the Feedback Analysis of GCMs and Intercomparison with Observations (FANGIO) described in Ellingson et al. (2016, chapter 1). ERBE had been seen as the tool to be used to measure the top of the atmosphere radiative fluxes, both downward and upwelling, for the ARM scientists. Its loss sparked the thinking that eventually led to the ARM Unmanned Aerial Vehicle (ARM-UAV) Program (Schmid et al. 2016, chapter 10).

5. Site selection process and outcomes: Where would ARM be sited?

The ARM Program Plan envisaged several fixed sites in well-chosen, climatologically significant areas. It spoke broadly about the challenges of different kinds of sites, noting, for example, the significant problems that would face an ocean site. The Program Plan was silent on what particular sites would be occupied, but it laid out some broad ideas on selection criteria. It also suggested that there should be a "mobile" site that could be deployed in areas of particular interest for shorter periods of time.

In the spring of 1990, the WISDM system engineering process was used to design a site selection procedure that became far more specific about what characteristics the sites should have. Principles for identifying broad regions of interest were first delineated, invoking important ideas such as identifying locations that would provide a range of physical conditions that would "stress" models, where the logistics were not prohibitive, and finally in areas where cooperation was possible with other complementary programs. This process and approach is documented in DOE (1991). An important feature of the process outlined in this report was that once a site was selected, it affected the evaluation criteria for the subsequent sites. The philosophy was that ARM would be a collection of sites that provided the diversity of physical conditions and therefore cloud properties that would be used to stress the models in different climatic regimes.

The site selection process, which is detailed in Cress and Sisterson (2016, chapter 5), identified five locales of interest. In order of priority, the locales were the U.S. SGP, the tropical western Pacific (TWP), the eastern North Atlantic or eastern North Pacific, the North Slope of Alaska (NSA), and the Gulf Stream. Several other sites were recommended for short-term or campaign occupancy: central Australia or the Sonoran Desert; the northwest United States or southwest Canadian coast; the Amazon or Congo basin; and the Beaufort, Bering, or Greenland Seas.

Once the locales were identified, the process of site selection within each of the locales began; these details are provided by Sisterson et al. (2016, chapter 6), Long et al. (2016, chapter 7), and Verlinde et al. (2016, chapter 8). Concurrently, the design of the instrumentation and data systems (McCord and Voyles 2016, chapter 11) was proceeding at a rapid rate.

6. The Instrument Development Program

With the help of the results of the surrogate Science Team meetings, a vision for the nature of the instrumentation needs and performance requirements evolved beyond the list of instruments found in the ARM Program Plan. It became clear that, if ARM were going to meet its observational goals, a significant evolution was needed in instrumentation. In this context, the importance of an Instrument Development Program

(IDP) became increasingly obvious. The IDP was described briefly in the Program Plan as a long-term investment program. However, the need for a 24/7 operational paradigm highlighted the fact that many existing instruments could meet the measurements needs, but very few of them could operate on the largely unattended basis required. The history of using the campaign mode of research for atmospheric science had allowed many fine instruments to be developed; however, they were not ready for ARM-style operational deployment. As a result the IDP, under the leadership of Jeff Griffin (a leader in instrumentation development from PNNL), instrument developers were given the means and motivation to focus on making the instruments ready for facility deployment. Perhaps the most successful IDP projects, which were critical to the advancement of ARM's programmatic science, were the atmospheric emitted radiance interferometer (AERI; Turner et al. 2016b, chapter 13), the millimeter wavelength cloud radar (MMCR; Kollias et al. 2016, chapter 17), and the Raman lidar (Turner et al. 2016a, chapter 18).

Given the community's experience in campaign modes of operation, the transformation was not always an easy one. In fact, it was uncharted ground for the DOE team that led the process as well. They too were used to the campaign mode of operation. The key differences were several. First, in campaign mode, the individuals responsible for an instrument were often the developer, who could diagnose problems and issues as well as fix them quickly and sometimes easily. Someone tended the instruments practically all the time. Data were collected and calibrated by the team and many times thoroughly massaged after the completion of the campaign. Generally, the data would be delivered to the balance of the campaign participants weeks or months later after the campaign ended.

At an ARM facility, operations staff would be tending, repairing, and calibrating the instruments. The data would be delivered as much as possible in near–real time to the data system. There would be no respite, such as the end of the campaign, to go through the data.

Interestingly, early on it became clear that many of the more mature instruments, capable of independent operation, were not routinely used in that fashion. More specifically, the exigencies of 24/7 operation for periods beyond a few weeks seemed to be an issue. The instrument team developed a concept of and approach for the management of instruments that not only brought more mature instruments online quickly but also facilitated the movement of instruments from the IDP to routine operations. By assigning mentors for all of the instruments, the instrument team had an individual responsible for both shepherding each instrument through field deployment and to ensure the data coming from them met the quality standards of ARM. Ackerman et al. (2016, chapter 3) describes how these roles evolved as the program matured. More details on the role of the instrument mentors are provided in Cress and Sisterson (2016, chapter 5) and their impact on data quality in Peppler et al. (2016, chapter 12).

While the core data collection strategy for the ARM Program called for continuous measurements of relevant data, it was clear from the outset that some critical measurements could not be made continuously. The most important of these were direct sampling measurements such as those made from aircraft, or periods where a significantly larger number of radiosondes needed to be launched. The recognition of this need evolved into what were called intensive observational periods (IOPs). Initially, IOPs were seen as regularly scheduled events where aircraft, for example, could be brought in to provide critical measurements. However, the concept evolved rapidly to include periods of time where, for example, instruments that were part of the IDP could be brought in for testing. The success of the IOPs became so critical to the ARM programmatic goals that they became part of the standard ARM budget and the debate was not about whether IOPs would happen, but rather what could be done to optimize the science that came from them. Probably the most important early IOPs were early cloud remote sensing IOPs where millimeter-wavelength cloud radars were evaluated (Kollias et al. 2016, chapter 17), the water vapor IOPs (Turner et al. 2016b, chapter 13), and the first use of unmanned aerospace vehicles (UAVs) for scientific observations (Schmid et al. 2016, chapter 10). Ackerman et al. (2016, chapter 3) describes how the IOP proposal, planning, and execution evolved in time and how these IOPs became part of ARM's aggressive interagency collaboration mechanism.

7. Data, data, data

One of the mantras of the early ARM Program was that it is easier to recover from a bad analysis than from bad data. The focus therefore became one of collecting data of known and reasonable quality and the preservation of data in readily usable and retrievable standard formats. This mandated that all raw instrument data, and the associated metadata for the observations, were saved so that reprocessing (if needed) could be performed. Next was the requirement to ensure the productivity of the Science Team. Rather than designing a generic system for collection and retrieval of data from which the principal investigator (PI) would do further data reduction and analysis prior to its use in experiments that combined data from other sources, ARM would deliver all the data needed by a PI directly to the PI. Data transfer was particularly challenging in the early days of the program, when the Internet was still in its infancy. Finally, as described in Ackerman et al. (2016, chapter 3), there appeared over time a focus on the creation of "value-added data products" by ARM. An example was the creation of a merged water vapor profile that included many sources of data. These consensus data products were and are viewed as critical to meeting the goals of ARM. Allowing the PIs to develop their own water vapor or temperature profile products from the varied in situ and remote sensors, for example, defeated the concepts and purpose of the ARM approach. Its goal was rather to give the best estimation of physical conditions and test the model predictions against real observations. It is the comparison of the results of multiple models driven by the same data compared with the same observations that is critical. Differences deriving from differences in the handling or processing of observational data were not relevant.

All of the above considerations put a tremendous premium on two things—getting the data needed to the Science Team and making sure that it was of the highest quality. Ackerman et al. (2016, chapter 3) has a description of how this evolved, but in the early stages of the program, the relationship between the ARM infrastructure and the Science Team was managed by what was termed the Experiment Center. This team, based at PNNL, supported the coordination of data for the PIs from the ARM sites and carried out some data reductions to ensure there were common products for derived quantities such as temperature and water vapor profiles. This coordination applied not only to the data collected by the ARM Program, but other datasets as well, such as satellite and National Weather Service data. The acquisition of the related data from other sources was managed as ancillary data but was provided as part of the service to the Science Team by a team based at BNL. The eventual value of this approach was that testing different models was done using common input data and common comparison data, putting the focus on the physics in the models where it belonged.

The involvement of the Science Team in ARM infrastructure activities led to what could be considered the first real demonstration of the power of this approach in the development of the consensus water vapor profile (Turner et al. 2016b, chapter 13), the importance of which is noted above. At the recommendation of the instrument mentors and the working group focused on the instantaneous radiative flux experiment (Mlawer and Turner 2016, chapter 14), several IOPs were dedicated to the development and refinement of the methods for both collecting and processing the several methods used for measuring both the integrated water vapor column and the water vapor profile.

Finally, the focus on both value-added products and data quality meant that the continual review of data, the consequent improvement of the products, and the direct connection with the PIs on the Science Team meant that the ARM Data Archive not only was regularly reprocessing data but had to keep track of who had received previous data versions. This placed stringent requirements on the ARM Data Archive, which impacted how it was developed over time (McCord and Voyles 2016, chapter 11).

8. Science and reviews

The ARM Science Team came together in the fall of 1990 with the selection of the first investigators in the program. These investigators became entrained in the new approach that was being developed in the program. A key part of the membership of the initial Science Team included several reviewers from the original Program Plan and participants in the surrogate Science Team. These individuals were critical to the early formation of the Science Team. The Science Team met on an annual basis, and the proceedings of those meetings, beginning with the Second ARM Science Team Meeting, are available online (http://www.arm.gov/publications/proceedings). The initial meeting, held in Las Vegas, was largely an informational meeting for the newly selected PIs and no proceedings were produced. These proceedings provide an excellent way to view the year-by-year progress in the program, covering not only the science but the instrument development program and the development of the ARM sites.

The Science Team was organized in a variety of ways. Initially, there was a focus on the IRF and SCM experiments, with working groups of laboratory staff and Science Team members meeting to review instrumentation and data processing necessary to support the work of the Science Team (Mather et al. 2016, chapter 4). Other working groups formed to support the instrument development program, aerosol measurements, cloud properties, and mesoscale modeling, the latter seen as the analog for cloud formation to the high-resolution radiative transfer models for radiative transfer.

Another focus for the Science Team was around the individual sites and the site scientists: Pete Lamb of the University of Oklahoma for the SGP, Tom Ackerman of Pennsylvania State University in the TWP, and Knut Stamnes of the University of Alaska for the NSA. The site scientists were selected through a limited competitive process managed by the DOE. The site scientists were expected to provide scientific input into the operations of the site, oversee the data quality program (Peppler et al. 2016, chapter 12) for the site, provide an educational outreach effort, and sustain their own research program at the sites. A site manager from the DOE laboratories supported each of these scientists: Doug Sisterson of Argonne National Laboratory (SGP), Bill Clements of Los Alamos National Laboratory (TWP), and Bernie Zak of Sandia National Laboratories (NSA). While the groups of interested scientists for these sites were not as large as the other working groups, these small groups of scientists were important as instrumentation was selected, unique data quality issues were identified and worked on, and the sites became operational.

A few years into the program a subset of the Science Team, which consisted largely of the leaders of the working groups, the site scientists, and the ARM chief scientist, was created and identified as the Science Team Executive Committee (STEC). As the program evolved beyond the initial focus of occupying sites and operational details, the STEC took an active role in selecting targets for IOPs and setting overall scientific priorities. The first ARM Science Plan[13] was prepared under their guidance and its publication in 1996 marked the end of the initial build out of ARM and a transition to its more scientific operation. The STEC met regularly with ARM program managers and DOE laboratory staff to continue to provide scientific guidance of ARM as the program matured.

A final, but key, element of the formative years was a review process organized by DOE using JASON, an independent scientific advisory group, which was frequently used by DOD and DOE to review large multidisciplinary programs and ideas. Initially organized by Gordon MacDonald of JASON and the MITRE Corporation, these reviews proved incredibly valuable for the senior leadership of the program as ARM moved toward and through its initial deployment. An example of the value of the JASON reviews came as the ARM Program struggled to deal with the loss of the ERBE scanner noted earlier. Working groups within ARM proposed a variety of ideas, including having ARM launch its own satellite to the use of UAVs to make the measurement. A JASON team reviewed the various proposals, and eventually ARM began a new program to use UAVs for the first time in atmospheric research.

9. Challenges of a rapid deployment, and 43 April 1992

As Field Marshall Helmuth Karl Bernhard Graf von Moltke once noted, "no plan survives first contact with the enemy." ARM was and is a large complex operation. The von Moltke quote was used by Stokes as the program was pushed to rapid deployment. The view was that plans would be shaped by experience, and planning had its limits.

As the instrument selection and data system concepts came together, the site selection and operational system development proceeded in parallel. Relatively early in the process, laboratories and individuals were identified to manage site operations. While the surrogate Science Team helped ARM organize its approach to measurements, instrument selection, and data management, there are other sets of activities developed in the early days of ARM that have continued to the present. ARM as conceived had laid out an aggressively novel approach with many new methods and systems. The fear of being paralyzed by the need for the perfect plan led to an early decision to set the start of deployment for April 1992—the thought being that we really would not know what we were doing until we actually had something in the field. Therefore, in advance of the first deployment of ARM, the program benefitted from other programs where experience could be gained by the ARM team in operating in different climates and countries. These included a NASA program in Coffeyville, Kansas (Ellingson et al. 2016, chapter 1); a deployment to Kavieng in the TWP during the Tropical Ocean and Global Atmosphere Coupled Ocean–Atmosphere Response Experiment (TOGA COARE; Long et al. 2016, chapter 7); and deployments on the ice during the Surface Heat Budget of the Arctic Ocean (SHEBA) experiment (Verlinde et al. 2016, chapter 8). In general, ARM provided value in the form of basic radiation measurements for these campaigns, but through them it gained valuable experience.

The activities and concepts above set the stage for the first deployment, with instrumentation borrowed from NCAR, at the first ARM site near Lamont, Oklahoma. It was a modest step, which took place, somewhat facetiously, on 43 April 1992—not quite meeting the goal of an April start. But that target of a rapid deployment less than 30 months after the approval of the program drove the team to extraordinary levels of innovation and commitment. While the details of individual plans and ideas fell by the wayside—in keeping with von Moltke's assertion—the initial Program Plan, forged largely in the fall of 1989, has had surprising durability and consequence, but the experience in the field shaped and molded the program as it went forward. The following chapters outline the details of those efforts and results.

Shortly after the opening of the SGP site, it was dedicated to the memory of, and named after, Fred Luther,

[13] Available from https://www.arm.gov/publications/programdocs/doe-er-0670t.pdf. The executive summary of this Science Plan is included in ARM (2016b, appendix B).

the original head of ICRCCM, whose leadership of the laboratory community and partnership with university and international collaborators led to ARM.

10. Epilogue

The research in ARM continued throughout the writing of this manuscript, and one recent result is of particular note. Part of the original ARM proposal by DOE was to actually measure the change in radiative forcing by CO_2. This has been done recently by Feldman et al. (2015). Their result, a 0.2 W m^{-2} $decade^{-1}$ increase in forcing at the surface measured both at the SGP site and the NSA site, involved the use of one of the most successful instruments to come out of the instrument development program, the AERI and the detailed line-by-line radiative transfer models validated through ARM. This result is a testimony not only to the work of the authors of that paper, but also to the tremendous dedication of more than a generation of scientists who created the instrumentation and the high quality that made the work possible.

REFERENCES

ARM, 2016a: Appendix B: Executive summary: Science Plan for the Atmospheric Radiation Measurement Program (ARM). *The Atmospheric Radiation Measurement (ARM) Program: The First 20 Years, Meteor. Monogr.*, No. 57, Amer. Meteor. Soc., doi:10.1175/AMSMONOGRAPHS-D-15-0035.1.

——, 2016b: Appendix C: Executive summary: Atmospheric Radiation Measurement Program Science Plan: Current status and future directions of the ARM Science Program. *The Atmospheric Radiation Measurement (ARM) Program: The First 20 Years, Meteor. Monogr.*, No. 57, Amer. Meteor. Soc., doi:10.1175/AMSMONOGRAPHS-D-15-0034.1.

Ackerman, T. P., T. S. Cress, W. Ferrell, J. H. Mather, and D. D. Turner, 2016: The programmatic maturation of the ARM Program. *The Atmospheric Radiation Measurement (ARM) Program: The First 20 Years, Meteor. Monogr.*, No. 57, Amer. Meteor. Soc., doi:10.1175/AMSMONOGRAPHS-D-15-0054.1.

Cress, T. S., and D. Sisterson, 2016: Deploying the ARM sites and supporting infrastructure. *The Atmospheric Radiation Measurement (ARM) Program: The First 20 Years, Meteor. Monogr.*, No. 57, Amer. Meteor. Soc., doi:10.1175/AMSMONOGRAPHS-D-15-0049.1.

DOE, 1991: Identification, recommendation and justification of potential locales for ARM sites. Rep. DOE-ER-0494T, U.S. Department of Energy, 14 pp. [Available online at https://www.arm.gov/publications/programdocs/doe-er-0494t.pdf.]

Ellingson, R. G., and Y. Fouquart, 1991: The intercomparison of radiation codes in climate models: An overview. *J. Geophys. Res.*, **96**, 8925–8927, doi:10.1029/90JD01618.

——, and W. J. Wiscombe, 1996: The Spectral Radiance Experiment (SPECTRE): Project description and sample results. *Bull. Amer. Meteor. Soc.*, **77**, 1967–1985, doi:10.1175/1520-0477(1996)077<1967:TSREPD>2.0.CO;2.

——, R. D. Cess, and G. L. Potter, 2016: The Atmospheric Radiation Measurement Program: Prelude. *The Atmospheric Radiation Measurement (ARM) Program: The First 20 Years, Meteor. Monogr.*, No. 57, Amer. Meteor. Soc., doi:10.1175/AMSMONOGRAPHS-D-15-0029.1.

Feldman, D. R., W. D. Collins, P. J. Gero, M. S. Torn, E. J. Mlawer, and T. R. Shippert, 2015: Observational determination of surface radiative forcing by CO_2 from 2000 to 2010. *Nature*, **519**, 339–343, doi:10.1038/nature14240.

Kollias, P., and Coauthors, 2016: Development and applications of ARM millimeter-wavelength cloud radars. *The Atmospheric Radiation Measurement (ARM) Program: The First 20 Years, Meteor. Monogr.*, No. 57, Amer. Meteor. Soc., doi:10.1175/AMSMONOGRAPHS-D-15-0037.1.

Long, C. N., J. H. Mather, and T. P. Ackerman, 2016: The ARM Tropical Western Pacific (TWP) sites. *The Atmospheric Radiation Measurement (ARM) Program: The First 20 Years, Meteor. Monogr.*, No. 57, Amer. Meteor. Soc., doi:10.1175/AMSMONOGRAPHS-D-15-0024.1.

Mather, J. H., D. D. Turner, and T. P. Ackerman, 2016: Scientific maturation of the ARM Program. *The Atmospheric Radiation Measurement (ARM) Program: The First 20 Years, Meteor. Monogr.*, No. 57, Amer. Meteor. Soc., doi:10.1175/AMSMONOGRAPHS-D-15-0053.1.

McBean, G., and J. McCarthy, 1990: Narrowing the uncertainties: A scientific action plan for improved prediction of global climate change. *Climate Change: The IPCC Scientific Assessment*, J. T. Houghton, G. J. Jenkins, and J. J. Ephraums, Eds., Cambridge University Press, 311–328.

McComiskey, A., and R. A. Ferrare, 2016: Aerosol physical and optical properties and processes in the ARM Program. *The Atmospheric Radiation Measurement (ARM) Program: The First 20 Years, Meteor. Monogr.*, No. 57, Amer. Meteor. Soc., doi:10.1175/AMSMONOGRAPHS-D-15-0028.1.

McCord, R., and J. W. Voyles, 2016: The ARM data system and archive. *The Atmospheric Radiation Measurement (ARM) Program: The First 20 Years, Meteor. Monogr.*, No. 57, Amer. Meteor. Soc., doi: 10.1175/AMSMONOGRAPHS-D-15-0043.1.

Mlawer, E. J., and D. D. Turner, 2016: Spectral radiation measurements and analysis in the ARM Program. *The Atmospheric Radiation Measurement (ARM) Program: The First 20 Years, Meteor. Monogr.*, No. 57, Amer. Meteor. Soc., doi:10.1175/AMSMONOGRAPHS-D-15-0027.1.

Moore, B., III, W. L. Gates, L. J. Mata, A. Underdal, R. J. Stouffer, B. Bolin, and A. Ramirez Rojas, 2001: Advancing our understanding. *Climate Change: The Scientific Basis*, J. T. Houghton et al., Eds., Cambridge University Press, 769–785.

Peppler, R., K. Kehoe, J. Monroe, A. Theisen, and S. Moore, 2016: The ARM data quality program. *The Atmospheric Radiation Measurement (ARM) Program: The First 20 Years, Meteor. Monogr.*, No. 57, Amer. Meteor. Soc., doi:10.1175/AMSMONOGRAPHS-D-15-0039.1.

Schmid, B., R. G. Ellingson, and G. McFarquhar, 2016: ARM aircraft measurements. *The Atmospheric Radiation Measurement (ARM) Program: The First 20 Years, Meteor. Monogr.*, No. 57, Amer. Meteor. Soc., doi: 10.1175/AMSMONOGRAPHS-D-15-0042.1.

Shupe, M. D., J. M. Comstock, D. D. Turner, and G. G. Mace, 2016: Cloud property retrievals in the ARM Program. *The Atmospheric Radiation Measurement (ARM) Program: The First 20 Years, Meteor. Monogr.*, No. 57, Amer. Meteor. Soc., doi:10.1175/AMSMONOGRAPHS-D-15-0030.1.

Sisterson, D., R. Peppler, T. S. Cress, P. Lamb, and D. D. Turner, 2016: The ARM Southern Great Plains

(SGP) site. *The Atmospheric Radiation Measurement (ARM) Program: The First 20 Years, Meteor. Monogr.*, No. 57, Amer. Meteor. Soc., doi:10.1175/AMSMONOGRAPHS-D-16-0004.1.

Turner, D. D., J. E. M. Goldsmith, and R. A. Ferrare, 2016a: Development and applications of the ARM Raman lidar. *The Atmospheric Radiation Measurement (ARM) Program: The First 20 Years, Meteor. Monogr.*, No. 57, Amer. Meteor. Soc., doi:10.1175/AMSMONOGRAPHS-D-15-0026.1.

——, E. J. Mlawer, and H. E. Revercomb, 2016b: Water vapor observations in the ARM Program. *The Atmospheric Radiation Measurement (ARM) Program: The First 20 Years, Meteor. Monogr.*, No. 57, Amer. Meteor. Soc., doi:10.1175/AMSMONOGRAPHS-D-15-0025.1.

Verlinde, H., B. Zak, M. D. Shupe, M. Ivey, and K. Stamnes, 2016: The ARM North Slope of Alaska (NSA) sites. *The Atmospheric Radiation Measurement (ARM) Program: The First 20 Years, Meteor. Monogr.*, No. 57, Amer. Meteor. Soc., doi:10.1175/AMSMONOGRAPHS-D-15-0023.1.

Zhang, M., R. C. J. Somerville, and S. Xie, 2016: The SCM concept and creation of ARM forcing datasets. *The Atmospheric Radiation Measurement (ARM) Program: The First 20 Years, Meteor. Monogr.*, No. 57, Amer. Meteor. Soc., doi:10.1175/AMSMONOGRAPHS-D-15-0040.1.

Chapter 3

The Programmatic Maturation of the ARM Program

THOMAS P. ACKERMAN

Joint Institute for the Study of the Atmosphere and Ocean, University of Washington, Seattle, Washington

TED S. CRESS*

Pacific Northwest National Laboratory, Richland, Washington

WANDA R. FERRELL

U.S. Department of Energy, Germantown, Maryland

JAMES H. MATHER

Pacific Northwest National Laboratory, Richland, Washington

DAVID D. TURNER

NOAA/National Severe Storms Laboratory, Norman, Oklahoma

1. Introduction

The early years of the Atmospheric Radiation Measurement (ARM) Program (see Stokes 2016, chapter 2) were devoted to the establishment of ground-based remote sensing sites at the Southern Great Plains (SGP), Tropical Western Pacific (TWP), and North Slope of Alaska (NSA; see Sisterson et al. 2016, chapter 6; Long et al. 2016, chapter 7; Verlinde et al. 2016, chapter 8). The ARM Program focused a great deal of its activity on site selection, instrument choices, site development, data management, and basic software development and implementation, which is described in Cress and Sisterson (2016, chapter 5). The period around 1998 represented a transition period for ARM. The SGP site was largely completed (as initially conceived), the TWP Manus site was up and running and the Nauru site was about to be installed, and the NSA Barrow site was up and running. Data were being collected and archived routinely and data rates were increasing exponentially by year.[1]

During this period, the Science Team was actively engaged in learning how to use effectively the available and expected data. Initially, thc ARM Science Team was organized around two broad research themes. One was the instantaneous radiative flux (IRF) concept and the other was the single-column model (SCM) approach. The IRF approach was derived from the SPECTRE experience (Ellingson et al. 2016, chapter 1) and focused on measuring all the radiatively active components in an atmospheric vertical column, computing downwelling radiative fluxes from these components, and then asking if the computations matched observations. The SCM approach attempted to represent experimentally a single-grid column in a climate

* Retired.

Corresponding author address: Thomas P. Ackerman, JISAO, University of Washington, 3737 Brooklyn Ave. NE, Seattle, WA 98105.
E-mail: tpa2@uw.edu

DOI: 10.1175/AMSMONOGRAPHS-D-15-0054.1

[1] An excellent summary of the state of the ARM Program in 1998 was produced by W. Ferrell, P. Crowley, T. S. Cress, and G. M. Stokes, "History and Status of the Atmospheric Radiation Measurement Program March 1998" (available at http://www.arm.gov/publications/proceedings/conf08/extended_abs/history.pdf?id=59; accessed July 2013). Additional background information was published by Ackerman and Stokes (2003).

model, including the advective forcing, and the clouds that were generated in such a column. It became apparent that both these concepts required expansion and redefinition. The IRF concept worked well for noncloudy skies but had significant problems in cloudy skies because the cloud radiative properties could not be adequately summarized from the measurements alone. Similarly, ground-based measurements of aerosol optical depth were insufficient to resolve the impact of aerosol on solar fluxes. Boundary conditions for the SCM were a significant challenge, as was determining the area-averaged cloud properties for comparison with model output. In addition, there was a perceived need for higher-resolution (2D and 3D) cloud modeling.

Technological progress also had a significant impact on the ARM Program and science during this early period. Electronic data transmission was in its infancy with low transmission rates and high costs, especially for large data files from relatively remote sites. Thus, data were transported on physical media, often with relatively long delay times. Computers and storage media were expensive and limited by today's standards. During the 1990s, computers and the Internet were in a rapid expansion mode, which required plans to be continually revised and updated by program management. This created both the opportunity to reduce costs, as computer prices fell, but also the demand for purchase of new systems with greater capability.

By the late 1990s, ARM had grown from an initial vision into a vigorous and dynamic program. It had expanded, however, to the limits of its financial resources, in part because the proposed budget was never actually allocated. The completion of the initial site development, the increased availability of ARM data, and the expansion of the Science Team in new directions began to shift management needs and interests. In the early stages, ARM project management was carried out largely by the original Chief Scientist, Gerry Stokes, and the technical management at the Pacific Northwest National Laboratory (PNNL) along with advice and guidance from the Science Team Executive Committee (STEC), which comprised senior members of the ARM Science Team (Mather et al. 2016, chapter 4).

In 1999, Gerry Stokes left the program for a different position, and Thomas Ackerman was hired as the ARM Chief Scientist.[2] At the same time, the Department of Energy (DOE) commissioned the Chief Scientist and the STEC to produce an assessment of ARM accomplishments, to carry out an evaluation of the science direction of ARM and to identify areas that required improvement with the preparation of a revised science plan for ARM. The goal of these activities was to create a stable model for the ARM Program as it transitioned from a developing to a mature program.

The planning for a mature ARM Program occurred under the constraints of a fixed budget and a time horizon of roughly another decade of data collection. The ARM Program received its first funding in 1990, but its funding profile plateaued after a few years and remained relatively flat for the remainder of the decade. The budget was sufficient to establish only three fixed locales (SGP, TWP, and NSA), and expansion plans for both the TWP and NSA were curtailed. The initial program plan had called for a decade of data collection; substantive data collection, however, started in 1996 with the deployment of the millimeter cloud radar (MMCR; Kollias et al. 2016, chapter 17) at the SGP and the deployment of the TWP Manus site and with the deployment of the NSA Barrow site by 1998. Science and budget realities forced the ARM management to find operational efficiencies, increase synergy across the program, and shift research emphases to focus on critical science issues.

2. The ARM questions

The fundamental goal of the ARM Program was to improve cloud parameterizations in global climate models (GCMs) through improved understanding of cloud and radiation processes obtained from a combination of modeling and data analysis (Mather et al. 2016, chapter 4). In 1999, the DOE program management asked the Chief Scientist to assess how well the ARM Program was doing in answering the ARM questions. To a large degree, this question was concerned with process. The question was not whether the program had obtained the answers but whether the program scientists had the tools and structure required to seek the answers. The obvious follow-on was then to decide what could be done to improve the tools, structure, and process.

The ARM Chief Scientist summarized the initial results of this analysis at the 1999 ARM Science Team Meeting. Assessments included a qualitative scoring as to whether a dataset or process understanding was 1) sufficient for ARM purposes, 2) good but requiring some further work, 3) reasonably well understood but requiring substantial work, or 4) important but lacking a well-conceived idea of how to proceed. It is worth a brief look at this assessment to understand what drove ARM development over roughly the next decade and where progress has been made.

Clear-sky[3] radiative flux understanding was deemed to be largely sufficient in the infrared but requiring

[2] The ARM Program had three Chief Scientists: Gerry Stokes from 1990 to 1998, Tom Ackerman from 1999 to 2004, and Warren Wiscombe from 2005 to 2009.

[3] "Clear-sky" here refers actually to non-cloudy sky (i.e., aerosol effects are included in the clear sky). We use the term as a simple differentiation from "cloudy" sky.

some additional work on the solar side. Cloudy-sky radiative flux understanding was likewise deemed to be largely sufficient in the infrared but requiring research on aspects of broken cloudiness. Interestingly, solar radiative transfer on the grid box or SCM domain was placed in the last category: important but lacking good ideas on how to proceed. ARM continued to carry out research on the infrared problem (see Mlawer and Turner 2016, chapter 14) and one of its huge successes has been the resolution of many of the questions associated with this issue.

Water vapor column and liquid water path measurements were deemed to be sufficient as well. This was largely the result of considerable improvement in microwave radiometry (see Turner et al. 2016, chapter 13) brought about by ARM research and partnership with manufacturers of microwave radiometers. Water vapor profiling in the lower atmosphere was deemed sufficient, but much less so in the upper troposphere. Ice water path and its spatial distribution were both placed in the fourth category. The recently deployed scanning millimeter radars (Mather and Voyles 2013) may finally bring some much needed clarity to this latter question.

Cloud retrievals from ground-based sensors were in their infancy at this point and were generally considered to be in the bottom two categories (categories 3 and 4). The only exception was stratus clouds, which were placed in category 2, largely because of an increased ability to measure liquid water path and retrieve optical depth values. ARM research on retrievals is another area of huge success (Shupe et al. 2016, chapter 19), including extensions to satellite instruments (Marchand 2016, chapter 30).

SCM investigations required a set of both surface fluxes and lateral boundary fluxes (Zhang et al. 2016, chapter 24). Radiative fluxes were deemed largely sufficient. Temperature and water fluxes were seen as good, but requiring further work at the SGP site. For the TWP and NSA sites, however, these fluxes were placed in category 4, a reflection of the fact that these sites were located in areas where there were limited options for carrying out extended site measurements. The final category of parameterization testing and development was deemed to be largely in category 4. This was and is an ongoing issue for ARM.

This assessment and the discussions that followed led to the identification of several issues/questions that shaped the direction of the ARM Program over the next decade: 1) data quality and continuity, 2) data fusion and value-added products, 3) intensive observing period (IOP) process and management, 4) ARM Mobile Facilities, 5) Science Team refocusing, and 6)improved parameterization development process.

Intertwined throughout these issues was the need for instrument development. The long-term focus of the ARM Program allowed scientists time to analyze and critique data streams, which in turn led to identification of issues and proposed improvements to existing instruments or new instruments. Specific examples of this process are provided elsewhere in this monograph.[4] The constrained resources of the program during this period, however, required management to weigh instrument priorities and find innovative ways of partnering with other agencies and private companies.

The bulk of this article is devoted to a discussion of how these issues molded ARM Program development over the subsequent decade (1999–2009). For reasons of clarity, we have chosen to discuss them largely as separate themes but they were, and are, clearly interconnected. This connectivity required ARM management to engage in a continuous trade-off between recognized and defined program needs and programmatic and financial feasibility. Setting priorities and making decisions was difficult and sometimes contentious (e.g., deciding the relatively priority of repairing and/or purchasing new instruments that were requested by the different working groups). One of the real hallmarks of the ARM Program during this period, however, was a genuine and thoughtful partnership among program management, infrastructure personnel, and Science Team members. The success of the ARM Program was built largely upon this partnership and acceptance of shared sacrifices to accomplish a desired end, namely the advancement of ARM science.

a. Program organization and evaluations

Before turning to the issues identified above, we think it useful to review briefly the organizational structure of the ARM Program and program evaluations that were carried out during its formative years. In February 1996, the ARM Program released a Science Plan, written primarily by the STEC (U.S. Department of Energy 1996; see appendix B therein). The 1996 program plan provides a view of the ARM Program as it prepared to launch into a fully operational phase. The science strategy is focused largely on two problems: the IRF problem (question 1 above) and the SCM approach (question 2), and resulted in two working groups within the ARM Science Team to focus on these issues (Mather et al. 2016, chapter 4). This Science Plan, along with several program evaluations, laid the foundation for the next phase of the program.

[4] Chapters 13 to 18 in this monograph all are connected with ARM-related instrument development. In addition, the topic shows up in a number of other chapters such as those dealing with the sites, the mobile facility, and the aircraft program. Stokes (2016, chapter 2) describes the Instrument Development Program.

Early on, as mentioned by Stokes (2016, chapter 2), the ARM Program management turned to a group of prominent scientists, known as the JASONS, for a periodic review of the Program. One of their earliest reports on the ARM Program appeared in 1995 and is fairly brief. In 1997, the Washington Advisory Group (WAG, chaired by Robert M. White) provided a program review that was quite detailed in its recommendations. Both of these reports are generally laudatory reviews of ARM Program progress and stature within the research community, and both note the pressures inherent in maintaining sustained funding and suggest the need to address this problem. Both reports also comment on the important role of intensive field campaigns and the need to manage these activities, a subject we address below. It is interesting that this tension regarding support for continuous ground-based measurements, routine in situ observations, and/or intensive campaigns has existed from the very beginning of the ARM Program and continues through to the present. The WAG report also identified the need for maintaining data integrity including calibration and data management, another issue of continuing importance to ARM.

The reports, the ongoing concerns of the Science Team about science management, and the appointment of a new ARM Chief Scientist in 1999 led to a reorganization of the Science Team and the Science Team Executive Committee around 2000–01. By now, the STEC representation had been built around four large working group themes, two large groups focused on IRF and cloud parameterization and modeling, and two small groups focused on cloud properties and aerosols (see Fig. 4-1 in Mather et al. 2016, chapter 4). The growth of ARM science and the associated scientific community made these groups too large for easy communication and sometimes unresponsive to new needs. The STEC appointed formal working group steering committees for each of these four groups with the intent of distributing work and increasing communication. In addition, it provided a mechanism for the creation of smaller, temporary working groups for specific projects, particularly those that cut across the four major groups. This structure worked well and remained in place for much of the next decade.

From a data and data flow point of view, the computational component of the ARM Program was also undergoing significant change. As discussed by Stokes (2016, chapter 2), the Experiment Center functioned as the data acquisition focal point for the program, acquiring data from the ARM sites as well as external sources. Early on, the Experiment Center was tasked with virtually all things data (with the exception of the archival of the data), including ascertaining data continuity and quality. As data streams from sophisticated instruments began to be received and technically involved data processing and quality control became required, this had to change and evolve. This became a key element of the evolution of the program and is discussed in detail below and in McCord and Voyles (2016, chapter 11).

DOE program management also experienced some significant changes during this period. The growth of the program required the addition of management staff. In the late 1990s, management responsibilities were divided principally between the Science Team management under the direction of Pat Crowley and the ARM infrastructure under the direction of Wanda Ferrell. With the sudden retirement of Crowley in 2000, Ferrell inherited responsibility for the management of the entire ARM Program, a very challenging endeavor that she then managed for much of the next decade.

The Chief Scientist and the reorganized STEC began the process of developing a revised Science Plan in about 2001 and produced a final version of that new plan in 2004 (U.S. Department of Energy 2004; see appendix C therein). It provided the blueprint for ARM development from the early 2000s to the formation of the Atmospheric Science Research (ASR) program (described later in this chapter). The 2004 plan identified a set of key science goals:

Maintain the data record at the remote sites at least through the next 5-yr period.

Improve significantly our understanding of and ability to parameterize the 3D cloud-radiation problem at scales from the local atmospheric column to the GCM grid square.

Develop new techniques to retrieve the properties of ice clouds and mixed-phase clouds and thereby improve our understanding of the life cycle processes in these clouds and their interaction with atmospheric radiation (see Shupe et al. 2016, chapter 19).

Develop a focused research effort on the indirect aerosol problem that spans observations, physical models, and climate model parameterizations (see Feingold and McComiskey 2016, chapter 22; Ghan and Penner 2016, chapter 27).

Implement and evaluate an operational methodology to calculate broadband heating rates in atmospheric columns at the ARM sites (see McFarlane et al. 2016, chapter 20).

Develop and implement methodologies to use ARM data more effectively to confront atmospheric models, both at the CRM and the GCM scale (see Zhang et al. 2016, chapter 24; Krueger et al. 2016, chapter 25; Randall et al. 2016, chapter 26).

These goals have guided the program for a decade. This chapter focuses on the program developments that were intended to enable the research community to address these important questions.

b. Data quality and continuity

From an observational perspective, the conceptual vision of ARM was to define a suite of necessary ground-based measurements to address the ARM science questions, deploy instruments that could make these measurements, and then continuously collect data of known quality. Not surprisingly, translating this vision into reality was far more complicated than most of the ARM community realized at the beginning of the program. In some cases, definition of needed measurements led to a call for new instruments; however, for both existing and new instruments, the difficulties centered on the twin issues of data continuity and quality.

Data continuity was fairly easy to obtain for measurements of standard meteorological quantities such as surface meteorology and broadband radiation. Achieving data continuity for more complex instruments was hampered by the fact that many of these instruments were not commercially available and had never been deployed for continuous measurements or, in some cases, instruments for some required observations did not exist (or existed only in some relatively primitive state). The ARM-related histories of specific instruments (e.g., microwave radiometers, infrared interferometers, Raman lidar, and cloud radars) are discussed elsewhere in this monograph; here we comment only on the broad ARM approach.

Early on, the ARM management recognized the need to have instrument experts available to the program and devised the idea of instrument mentors (Stokes 2016, chapter 2). Each instrument (or instrument class) was intended to have a mentor who would be supported for some fraction of his/her time out of the ARM infrastructure budget to monitor instrument performance and be available to consult with on-site staff when instrument problems arose. This approach seemed particularly well suited to the ARM concept of deploying similar instruments at multiple sites. The instrument mentor approach worked well, except for cost. As the ARM Program moved toward maturity, both the number of different instruments and the number of each type of instrument grew, but the pot of money available for mentors did not. The obvious consequence was that instrument mentoring suffered, producing corresponding problems with data continuity and quality.

The data continuity problem gradually sorted itself out in various ways. Some instruments were simply dropped from the ARM suite because they were deemed too unreliable and/or expensive to maintain. This instrument triage was often painful because it required an explicit statement of ARM research priorities and an implicit rejection of certain types of research through instrument decommissioning (e.g., the decommissioning of the whole sky imager because it never lived up to its goal to provide automated cloud cover routinely without substantial human interaction). Other instruments moved from prototypes to more reliable production models supported in large part by DOE investments directly through the ARM Program or by Small Business Innovative Research (SBIR) initiatives (e.g., the GVR and GVRP , which are microwave radiometers that observe downwelling radiance around the 183.3-GHz water vapor line). Other instruments (both hardware and software components) went through an evolutionary development based on science and engineering efforts supported within the ARM Program and through external contracts.

The instrument evolutionary development efforts, formalized early in the program as the Instrument Development Program, are among the great successes of the ARM Program. Scientists funded by ARM used portions of their research efforts to diagnose problems and investigate possible solutions. Engineering staff, funded by the ARM infrastructure, consulted freely with the scientists and invested precious time and effort in carrying out specific instrument tests and operational procedures as part of the development. Program management supported both groups in these activities because of the need to solve the instrument problems. We think it fair to say that failure to solve some of these really difficult instrument problems would have produced a far smaller and less useful ARM Program.

ARM management and scientists recognized from the very beginning that data quality and assessment was a critical element of the program but it took some time to find the right approach. In the early days, ARM tried a two-pronged approach that combined examination of data streams by instrument mentors and scientific users (primarily the site scientist teams). This approach had the virtue of relatively low cost but ultimately failed because the process was too uneven. The analogy used at the time (probably based on looking at agricultural areas around the SGP site) was that ARM had a giant field of data that required plowing and some areas were plowed deeply while in other areas the surface was barely scratched. The solution was the establishment of a data quality office at the University of Oklahoma in 2000. The office was charged with implementing data assessment procedures that could identify data quality issues in a timely fashion and then communicating that information to the appropriate operations and engineering personnel. In addition, the office served and serves as a point of contact for data quality issues raised by data users in the scientific community. [See Peppler et al. (2016, chapter 12) for more details about the Data Quality Office.]

The establishment of the Data Quality Office was extremely important for the long-term health of the ARM Program. The Data Quality Office now serves as a crucial link between site operations, the ARM Data Archive, and the user community. Office personnel devise quantitative tests of data quality and incorporate information on data quality in the ARM Data Archive through data flags and metadata. They respond to inquiries from users about data quality and track these questions to resolution whenever possible. By formalizing this process, ARM has created a "uniformly plowed" field. This is not to say that there are no data problems; it is almost impossible to envisage a program the size of ARM that has no data issues. (Today ARM operates over 350 instruments that produce over 1500 unique data streams across six ground-based facilities.) The Data Quality Office, however, has provided ARM with a mechanism to reduce data problems through regular, consistent, and careful examination of the data.

c. *Data fusion and value-added products*

The initial focus of the ARM Program was to obtain simultaneous datasets at a single site from multiple instruments. Within a very short time, the focus broadened to include the idea of data fusion in two distinct ways. The first is that multiple instruments may be measuring the same or very closely related atmospheric quantities, so these measurements should be merged into a single estimate of that atmospheric quantity. For example, water vapor is a critical component of the atmosphere and many investigations require water vapor profiles (e.g., water vapor concentration as a function of height). The ARM data suite includes water vapor measurements made by conventional radiosondes, microwave radiometers, infrared radiometers, Raman lidar, and instruments at the surface and on meteorological towers. Each of these data streams has strengths and weaknesses and a highly sophisticated user may be aware of them. For most users, however, this requires a level of understanding that requires too much time to acquire and is not a productive use of precious research time. The question then is how to provide a merged (or fused) dataset for users that represents an optimal data product.

The second type of data fusion combines measurements that provide complementary information about some quantity of interest. For example, the ARM community embraced the use of millimeter wavelength cloud radar and lidars to study cloud properties. Lidar is a very sensitive probe that responds to very small concentrations of hydrometeors but saturates quickly at large concentrations. Radar lacks the sensitivity to small concentrations but is able to provide information through deep layers and multiple layers of clouds. Thus, one can combine the information from these two quite different sensors to produce continuous profiles of cloud occurrence. A related feature is that multiple measurements from a single or multiple instruments can be combined into a mathematical retrieval that provides information about some atmospheric quantity that is not readily observed in a direct way. The retrieval process was well understood in atmospheric sciences and had been applied to satellite data beginning in the late 1970s to retrieve profiles of temperature and humidity. There had been very little application, however, to ground-based systems prior to ARM and little of the satellite experience translated directly to the ARM instruments.[5]

The need for creating value-added products or "VAPs" (the generic name that ARM applied to these data fusion activities) was well appreciated from the onset of ARM, and is called out specifically in the 1996 ARM Science Plan (U.S. Department of Energy 1996). The only method in place to create them, however, was the activity of members of the Science Team. One of the earliest examples of this effort was the creation of the Active Remote Sensing of Clouds (ARSCL) VAP to provide, initially, profiles of cloud occurrence as a function of time. The science behind this VAP grew out of research at the Pennsylvania State University, funded in large part by DOE, to understand cloud radar data using the Penn State radar (Clothiaux et al. 1995). When ARM put a micropulse lidar and MMCR in place at the SGP site, the Penn State group and colleagues took the logical path of extending their research and creating a product that anyone could use (Clothiaux et al. 1998, 1999, 2000; Kollias et al. 2016, chapter 17).

The initial VAPs were built around science ideas and codes developed by science investigators. The codes were then adapted at one of the participating DOE laboratories for operational conditions and run by the ARM infrastructure. A large number of VAPs are in operation today; an example of a subset of these is given in Table 3-1. While this process was highly efficient in

---

[5] One of the successful outcomes of the ARM Program has been much closer contact between satellite and ground-based retrieval groups leading to shared retrieval algorithms. This has been particularly apparent for active sensors (radar and lidar). For example, the NASA *CloudSat* team and ARM millimeter-wavelength radar scientists have worked closely together on the development of radar products and similar retrieval algorithms have been applied to the NASA A-Train constellation of instruments and the ARM ground-based instruments (Marchand 2016, chapter 30).

TABLE 3-1. Examples of ARM value-added products.

Name	Description
AERIPROF	AERI-retrieved thermodynamic profiles in cloud-free conditions (Feltz et al. 2003)
ARMBE	ARM best estimate—an aggregate of core parameters (e.g., radiative fluxes, cloud fraction, water vapor path) from multiple ARM data sources, averaged to 1-h time resolution (Xie et al. 2010).
ARSCL	Active remote sensing of cloud layers—a retrieval of cloud layers in the vertical column based on multiple instruments including radar and lidar (Clothiaux et al. 2000).
MERGESONDE	Merged radiosondes—interpolates radiosonde temperature, humidity and wind profiles in time using the microwave radiometer, surface observations, and numerical weather prediction model output as constraints.
MFRSRAOD	Aerosol optical depth derived from the Multifilter Rotating Shadowband Radiometer (Michalsky et al. 2010).
MWRRET	MWR-retrieved precipitable water vapor and liquid water path using a physical-iterative method (Turner et al. 2007b)
QCRad	Quality controlled radiative fluxes—an analyzed radiative flux product that includes quality control flags, a best estimate from several possible flux sources (Long and Shi 2008).
VarAnal	Constrained variational analysis—Dynamic and thermodynamic tendencies in a vertical column derived from a radiosonde horizontal array or numerical weather model analysis. (Zhang and Lin 1997).

development cost, it was inefficient in operational cost and organization. Scientists tend to write programs that serve their own particular scientific purposes and are relatively unconcerned with issues of generic reliability and code versioning and traceability.[6] As the ARM Program matured, it became apparent that these latter problems were proving to be a significant barrier to VAP production and overall data usage and therefore needed to be addressed.

The STEC and ARM management decided to institutionalize a VAP process within the program structure that relied on a combination of science input and testing and engineering development and implementation. The key concept was a "translator," usually a scientist, who served as the liaison between scientists and software engineers. The basic features of the process were and are the following:

Identification of VAPs. The scientific working groups and subgroups are the focal point of the science and therefore are the originators of VAPs. They are in the best position to know which VAPs are possible and most useful to the community, where "possible" implies the requirement of writing down an algorithm that can be implemented.

Prioritization. VAP development requires money and, clearly, the number of VAPs desired by the science community can outstrip the available resources. In the initial implementation of the process, the STEC [later the Science and Infrastructure Steering Committee (SISC)] and program management were and are charged with establishing priorities.

Development and implementation. Once a VAP is approved, implementation is assigned to a translator and a software engineer. They, along with the originating scientists, are responsible for establishing VAP criteria, estimating resources, defining a schedule, and implementing the algorithm in a framework that allows it to be run in an autonomous manner.

Evaluation. After the VAP is implemented, a test dataset spanning a relatively long time period is generated and made available to sponsoring scientists and the working group that promoted the VAP. This allows a subgroup of the community to analyze the dataset to ensure that the algorithm is working as desired. Frequently, the application of a new algorithm to a longer-term dataset reveals problems that the originating scientist did not anticipate, which requires that the algorithm be moved back into a development/implementation state until the problems are resolved.

Operations. Once the VAP is implemented using the computing resources available at the ARM Data Archive and the sponsoring working group gives its approval of the evaluation dataset, the VAP moves to an operational status. It is then the responsibility of the originating scientists and the translator to ensure that it is performing correctly and evaluate the product.

[6] One of the recurring themes in the development of ARM has been the need to merge science and engineering. Building continuously operating facilities required ARM to develop engineering standards for instrument and site operations. Similarly, VAPs required engineering standards for code and documentation. The science demands, however, were continuously evolving so the engineering standards had to be flexible enough to permit change and evolutionary growth. These paired requirements forged a unique partnership between engineers and scientists on a program-wide basis that was simultaneously challenging and rewarding. One of ARM's outstanding achievements was the creation of a relatively seamless team of scientists, software engineers, and hardware engineers. Many of the articles in this volume are coauthored by combinations of these groups, along with the management that endorsed and encouraged that collaboration.

This VAP process began in 2001 with the appointment of four translators, one assigned to each of the four main working groups. As with any new initiative, a bit of time was required to put in place the right combination of translators and software engineers and make the process run smoothly. Within a year or two, the VAP process became integral to the ARM Program and a distinct success. The process created robust codes that provided reliable value-added products, especially for the merged measurement products or "best estimates" as they become known. Relying on expert scientists to provide the scientific backbone for these products gave them credibility. Implementing engineered codes gave them reliability. Of course, the codes are always evolving as the science and instrumentation evolves, leading to changes, but ones that are managed and recorded.

Ideas for VAPs rapidly outpaced available program resources, a problem that was exacerbated by the fact that development often took longer than expected due to the complexity of the process, and that most VAPs were evolutionary in principle. In addition, the translators found that some VAPs were simply not ready for implementation because the science was insufficiently developed or the algorithms were too complex to be implemented as robust operational codes. This was particularly a problem for retrieval algorithms for cloud properties. These algorithms were mathematically sensitive, causing program crashes and breaks in the automated processing. However, many of these retrievals were of interest to many members of the Science Team. After much discussion, the ARM Program took the unusual step of creating space in the ARM Data Archive for what became known as Principal Investigator (PI) datasets. These datasets are generated by PIs using documented algorithms but are produced by the PI's research group rather than by ARM infrastructure members. While the ARM Program dictates the data format and metadata for these PI datasets, it makes no statement about quality or robustness. This process has provided a helpful middle ground for the program by providing community access to interesting datasets while reducing the time and cost for the program itself to produce them. The popularity and number of PI-produced datasets started slowly, but has continually grown with time; as a consequence, the program has improved methods to collect metadata about these datasets and make them more visible in the ARM Data Archive.

d. *IOP process and management*

A central element of the ARM vision is to provide continuous ground-based remote sensing at multiple sites. In the early days of the program, ARM did not yet have sites in full operation so the program supported participation by ARM scientists in planned field programs such as the NASA FIRE [First International Satellite Cloud Climatology Project (ISCCP) Regional Experiment] campaigns and TOGA COARE (Tropical Ocean and Global Atmosphere Coupled Ocean–Atmosphere Response Experiment). When the ARM Program began to develop its observing sites, not all instruments could be put in place simultaneously, so ARM sponsored intensive observing periods during which investigators brought their own instruments to the ARM sites to supplement existing ARM instrumentation, or ARM supported other agencies by taking ARM instruments to their campaigns [e.g., the Surface Heat Budget of the Arctic Ocean (SHEBA) experiment; see Verlinde et al. (2016, chapter 8)]. Furthermore, as instruments from the Instrument Development Program (IDP; Stokes 2016, chapter 2) were placed at the sites, IOPs were typically used to confirm that the IDP instruments were indeed operating as desired. The conceptual idea during the early ARM days was that IOPs for various purposes would be required, but as the sites became populated with the permanent ARM instrument set, the need for these IOPs would be reduced (Stokes 2016, chapter 2). This proved not to be true. For example, the 1996 Science Plan identifies IOPs as a way to provide special measurements such as aircraft sampling.

IOPs (or campaigns, as they are sometimes called) presented an additional challenge for the ARM Program in terms of management and allocation of resources. On the one hand, the ARM paradigm was based on continuous observations; on the other hand, IOPs were needed to obtain science-defined measurements that were too costly to be obtained routinely or could not yet be done in a quasi-unattended mode. Because cloud and radiation measurements had previously been done almost exclusively in field programs, some part of the science community felt that this was the only useful way in which they could be done. Thus the ARM science and program management was forced to chart a difficult course between conflicting demands that allowed the continuous measurement program to grow and mature while still enabling enhanced measurements during IOP periods.

By the end of the 1990s, IOPs had become an integral part of ARM science and operations and their annual number, complexity, and cost was increasing.[7] Some

---

[7] A timeline summary of ARM IOPs can be found on the ARM website (http://www.arm.gov/campaigns/; scroll to bottom; accessed 4 Sep 2013), as well as a table of IOPs by year (http://www.arm.gov/about/stats/campaigns; accessed 4 Sep 2013).

IOPs were simple instrument comparisons lasting a few weeks to a month;[8] some IOPs lasted for a year or more; some involved the use of expensive assets such as airplanes and ships; some were ARM-centric and some were joint national and/or international operations. As with other ARM activities, it was clear that some process needed to be put in place to manage IOPs so that allocation of resources fit within program science goals and available funding.

The IOP process established in the mid-1990s and early-2000s was fairly simple. A group of scientists (typically from one of the original scientific working groups) proposed an IOP to ARM through the creation of a short science plan that stated objectives, an activity plan, and resource requirements. The plan was vetted by the ARM Technical Director regarding resources and cost and then passed to the STEC for evaluation. The STEC then evaluated the various proposals in terms of program science objectives and budget impact and approved those most highly ranked. This process worked well in large part because it provided for an iterative discussion among the proposers, the ARM Program management, and the scientific leadership. In many cases, these discussions sharpened objectives and honed resources, including bringing in funding from other participants.

One of the interesting events that occurred during this period was the creation of an ongoing IOP to measure in situ aerosol profiles (McComiskey and Ferrare 2016, chapter 21). In some sense this was an effort to merge the conflicting demands of routine measurements and an IOP. A rugged set of instruments to measure aerosol properties was designed to fit into the passenger seat of a small Cessna aircraft. The Oklahoma company that owned the Cessna was instructed to fly a couple of times a week over the SGP site when permitted by weather and airplane availability. Data were downloaded after each flight and sent to the science group that proposed the IOP [e.g., see Andrews et al. (2004) for early results]. This approach has now been applied to other measurements [e.g., liquid water cloud properties in Vogelmann et al. (2012)], blurring the distinction between continuous measurements and IOPs.

In 2004, the ARM sites were designated as a National Scientific User Facility called the ARM Climate Research Facility, which posed new considerations for the ARM science community. As a User Facility, the ARM sites were now seen as serving the larger international science community, not just the ARM science community. While this created the opportunity for additional funding and resources, it also changed many past processes. The collective ARM site managers (the Infrastructure Management Board) and DOE program managers now assumed responsibility for reviewing and approving the smaller campaigns. For large campaigns, an ARM Science Board, under the direction of DOE program managers, was created and charged with determining the scientific merit of the proposed campaigns. The ARM Science Board is composed of both ARM and non-ARM scientists to reflect the user base for the facility, and would meet once yearly to evaluate the proposals for the larger campaigns. This Science Board reviewed and discussed the scientific merits of the proposals and their likely impacts. The final decision on which of the large campaigns would be supported by the program were (and are) made by DOE program managers.

e. ARM Mobile Facility

As described by Cress and Sisterson (2016, chapter 5), the original ARM vision proposed the establishment of five permanent sites and one movable site. By the late 1990s, ARM had developed three fixed locales but lacked resources to develop any more fixed locales. It had become increasingly apparent, however, that there was a scientific need and programmatic desire to sample cloud and radiation properties at other locations. The primary science driver was to study cloud properties and radiative effects in climate regimes not sampled at the three fixed sites (e.g., marine stratus and orographic clouds). In addition, there was a realization that ARM could benefit by partnering with field campaigns organized by national and international groups. In many cases, these field campaigns were built around aircraft sampling but had limited ground-based capability. By providing the latter, ARM could benefit from the former.

The ARM Chief Scientist resurrected the idea of a deployable facility in 2000, which in turn led to a feasibility study and a white paper and, eventually, a commitment to build an ARM Mobile Facility [AMF; see Miller et al. (2016, chapter 9) for a complete discussion]. The AMF concept and design followed closely from experiences and lessons learned from deploying instruments at the fixed sites of the TWP and NSA. The AMF performed well from its very first deployment in 2005, which can be attributed to the maturity of operations at the fixed sites, including instrument robustness and data integrity, and the ability and skill of the AMF team itself. The first AMF proved to be so popular that a second was built

[8] The ARM sites have become perhaps the most heavily instrumented atmospheric research sites in the world, and many IOPs are where investigators bring their new instrument to run side by side with operational ARM instruments to help evaluate their new technology.

and both are now in essentially continuous operation at sites around the world.

One of the thorniest issues associated with the AMFs (and to some extent, IOPs) was the need for end-to-end support that included funding for the PIs and subsequent scientific analysis. In the early days of the ARM Program, IOPs were proposed by ARM Science Team members who had science research grants. Some of these IOPs included deployments of PI instruments (such as millimeter radars and different types of lidars) that were early prototypes of an AMF. In these cases, the IOP data augmented the continuous ARM data and were analyzed and used by the proposers using their existing funding. AMF deployments (and IOPs) fell under the auspices of the ARM Climate Research User Facility (because the AMF was established shortly after ARM became a User Facility) and were approved largely independent of Science Team funding. The result was that there is no explicit linkage between IOP and science funding. DOE management realized that this situation was undesirable and addressed it by providing some direct funding for the lead investigators of the IOP, after the IOP was reviewed by the ARM Science Board and selected for funding by DOE.

f. Science team refocusing

The rapid growth of the ARM Program throughout the decade of the 1990s also impacted the activities of the Science Team. During the latter half of the 1990s, ARM science efforts really blossomed, attracting more attention leading to an increase in the number and scientific diversity of proposals. Given that Science Team funding had largely plateaued by the end of this period, the greater diversity of topics implied less depth in some areas. This diversity also made it more difficult to link the broad ARM research community with the ongoing infrastructure activities because the links between science investigations and infrastructure activity were more tenuous in some cases. Priority was given to improve this linkage.

Around 1999, ARM management addressed the Science Team issues in two ways, one focusing on its general culture and a second that involved pruning of certain activities. In the first case, the ARM Program management, including the Chief Scientist, mandated a tighter coupling between Science Team research projects and ARM Program goals. While this connection had been implicitly considered, it now became an explicit consideration in proposal reviews. Proposals were expected to have defined scientific goals that were congruent with ARM goals. In addition, principal investigators and coinvestigators were strongly encouraged to see themselves as a resource for the ARM Program in activities such as defining ARM science, consulting with ARM infrastructure personnel on specific science needs and data usage, and participating in data quality and VAP activities. These were not new ideas—they were part of the original vision of ARM—but they had become diffused as the program grew. Restating and emphasizing them was really an attempt to refocus the Science Team.

The pruning effort was considerably more painful. The program evaluation discussed above identified several scientific areas in which the program had been very successful. Water vapor profiling and clear-sky radiative transfer (particularly in the thermal infrared) were two areas that were readily identified as huge ARM successes. ARM had expended significant resources in these two areas (see Turner et al. 2016, chapter 13; Mlawer and Turner 2016, chapter 14) and science papers had demonstrated that it was possible to compute infrared radiative transfer in the clear sky to an accuracy of a percent or better [Turner et al. (2004) report an accuracy of better than 2 W m^{-2} in downward, spectrally resolved radiance compared to Atmospheric Emitted Radiance Interferometer (AERI) measurements]. Doing so required knowledge of the water vapor profile to a similar level of accuracy, which was also possible. Thus, as management looked at the broad program needs, a decision was made to deemphasize research in these two areas that had been key efforts from the beginning of the program. This decision was not popular with all, but was necessary in order to find resources to attack other problems.

The results of the Science Team refocusing were largely positive. ARM research scientists responded well to the call for greater involvement in ARM programmatic issues. The restructuring of the working group structure to allow for smaller, ad hoc projects created a new avenue for participation and brought together small teams of scientists, engineers, and site personnel to work on specific issues. The Clouds of Low Water Optical Depth (CLOWD) project, which started around 2003, provides an excellent example of such a project (Turner et al. 2007a). Decisions to shift scientific resources resulted in losing some science team members but opened the door for participation by other scientists working in areas such as cloud retrievals, 3D radiative transfer, aerosol physics, and high-resolution cloud modeling. The flourishing of ARM science that occurred in the 2000s is due in no small part to these efforts.

g. Improved parameterization development process

Any scientist who has been engaged in cloud parameterization can attest to the extreme difficulty of the parameterization problem. ARM was put in place specifically to attack this problem through measurements, data analysis, physical process modeling, and parameterization development and testing. As ARM progressed in the early 1990s, science efforts were predominantly in the areas of measurement and data analysis and process modeling, particularly related to cloud processes. This is not to say that there were no efforts in climate model parameterization development, but those efforts were limited in scope. Toward the end of the 1990s, research efforts using global models increased. These included the development of a forecasting system for a global climate model and the use of a multiscale climate model. Details of these efforts are included in chapters by Krueger et al. (2016, chapter 25) and Randall et al. (2016, chapter 26).

However, there was still a strong feeling that more needed to be done to promote the use of ARM data in the parameterization problem and that ARM science funding could not be stretched much further. One largely unexpected development that occurred in the late 1990s was the use of ARM data for evaluation of weather forecasting models, led particularly by the European Center for Medium-Range Weather Forecasts (ECMWF). After consultation with leadership at the National Centers for Environmental Prediction (NCEP) and ECMWF, ARM decided to fund a postdoctoral research position called the ARM Fellow at each of these institutions. The ARM Fellow was recruited by the forecasting center, and approved and directly funded by DOE management. The only constraint placed on research was that it should involve the application of ARM data to forecast model improvement and validation. This program was particularly successful at ECMWF where several outstanding young scientists were recruited over time and worked on a variety of parameterization development and model evaluation problems (see Ahlgrimm et al. 2016, chapter 28).

h. Links with other programs

In its very early days, the ARM Program was focused largely on its own internal development, but that perspective changed rapidly. As the SGP site developed into a robust remote sensing site, other programs requested use of the data and the site. Joint field campaigns were held with NCAR, satellite validation campaigns with NASA, and international collaborations under the aegis of the Global Energy and Water Cycle Experiment (GEWEX). The logistical constraints and costs of working at the more remote NSA and TWP sites encouraged collaborative research projects such as SHEBA at the NSA (Uttal et al. 2002) and Nauru99 at the TWP (Westwater et al. 2003). These projects involved national partners such as the NSF and NOAA as well as international partners (e.g., Canada in SHEBA and Japan in Nauru99).

We could generate a long list of these interactions, but it is more useful here to consider the overall impact of these collaborations on the program. ARM profited in many ways from these linkages. It gained credibility in the broader scientific arena, enhanced science productivity by leveraging resources, and helped drive the agenda of the U.S. Climate Research Program. These interactions in the decade of the 2000s helped lay the foundations for the use of the AMF in international programs like the African Monsoon Multidisciplinary Analysis (AMMA) and COPS (see Miller et al. 2016, chapter 9). Early discussions between ARM scientists and their counterparts in Europe have now led to a formal collaboration between the ARM Climate User Facility and European scientists to develop shared data portals and algorithms[9] (Haeffelin et al. 2016, chapter 29).

3. ARM as a DOE Science User Facility

From the beginning of the ARM Program, data collected from the ARM sites were considered to be a community resource and the research done with these data was highly collaborative across government agencies, universities, private companies, and international institutions. This open character of ARM was due in part to the distributed nature of the observation facility and the global nature of climate research. At the core of this open architecture was the ARM Data Archive that has always made the ARM data freely available to anyone (McCord and Voyles 2016, chapter 11). Formally recognizing this open character of ARM, the infrastructure component of ARM was designated as a National Scientific User Facility in 2004 called the ARM Climate Research Facility (Mather and Voyles 2013). As a User Facility, ARM was to serve as a resource for the broad climate research community. While ARM had already been acting in this regard to a significant extent, the designation as a User Facility formalized this role. Prior to designation as a User Facility, the ARM infrastructure was coupled to, but distinct from, the ARM Science Team. Following the User Facility designation, the management split between the infrastructure and science became more distinct; however, the structure of the ARM Science Team remained unchanged. The STEC did evolve to

[9] U.S./European Workshop on Climate Change Challenges and Observations, 6–8 Nov 2012; workshop report available at http://science.energy.gov/~/media/ber/pdf/CESD_EUworkshop_report.pdf; accessed 23 Oct 2013.

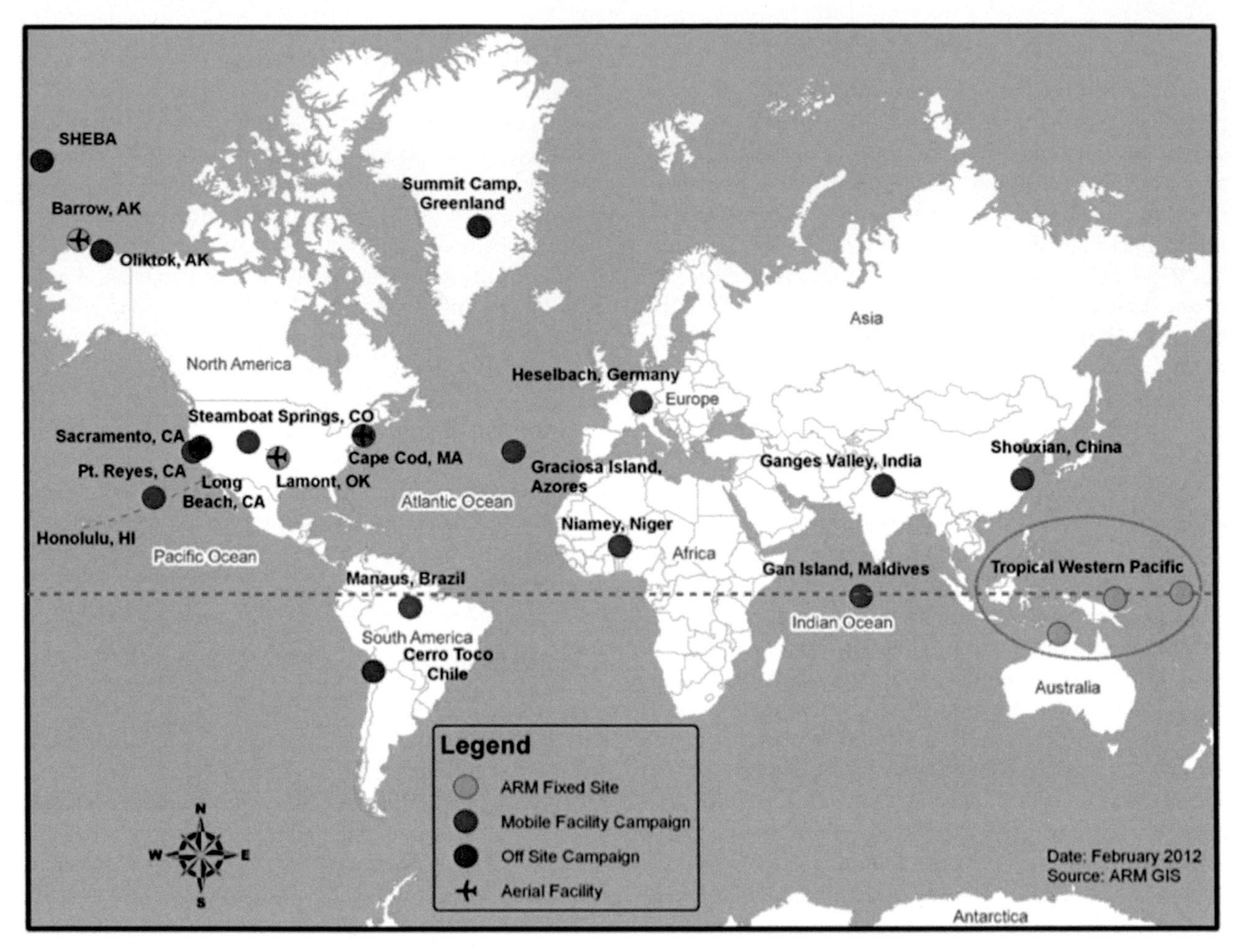

FIG. 3-1. Location of the permanent ARM sites, ARM Mobile Facility deployments, and field campaigns as of 2012.

the SISC and served as the link between the Science Team and User Facility.

The designation as a User Facility required a change in how DOE selected and approved field campaigns at the ARM sites. Previously, that process had been internal to ARM and DOE, and all of the larger field campaigns had their roots in the ARM Science Team. After becoming a User Facility, DOE appointed a Science Board whose primary responsibility was to review proposals for field campaigns and make recommendations to DOE management. The proposal process begins each year with a call for proposals in the first half of the year followed by review by the Science Board over the summer, with the final selections by DOE management in the fall (Voyles and Chapman 2012). This process is followed for so-called facility proposals involving large field campaigns at any of the permanent sites, aircraft operations, or deployment of an ARM Mobile Facility. As a result, scientists outside of the ARM Program have led many of the more recent field campaigns.

Mobile facilities—facilities that are relocatable for short-term deployments—were part of the original ARM plan (U.S. Department of Energy 1990, appendix A therein) but took many years to become a reality (Miller et al. 2016, chapter 9). The first ARM Mobile Facility was completed in 2005. It was designed to be portable and to include the same basic suite of instruments found at any of the fixed remote-location ARM sites (e.g., Nauru). The first AMF deployment was to the Point Reyes National Seashore near San Francisco in 2005 and was immediately followed by a deployment to Niamey, Niger (Miller and Slingo 2007), the first in a series of international deployments (Fig. 3-1). Since then, AMF proposals have represented the largest portion of facility requests that are reviewed each year by the Science Board, although the augmentation of measurements at the fixed sites continues to be important. Miller et al. (2016, chapter 9) provide a further discussion on the history, development, and accomplishments of the AMFs, including an overview of each deployment.

Aircraft operations have always been important for supplementing the routine ground-based observations at ARM sites. For many years, aircraft support was obtained through a combination of contracted and collaborative activities. The latter included both interagency collaborations (e.g., with NASA for the ARM-FIRE Water Vapor Experiment in 2000; Ferrare et al. 2004) and support from the DOE Unmanned Aerial Vehicles (UAV) program [see further discussion in Schmid et al. (2016, chapter 10)]. In 2006, DOE ended the UAV program and initiated a new formalized aerial component of ARM. This was initially referred to as the ARM Aerial Vehicle Program, but soon was renamed and is now known as the ARM Aerial Facility (AAF). For several years, the AAF primarily served a coordinating and contracting role; however, beginning in 2010, the AAF began to make use of the Battelle Gulfstream G1 research aircraft.

With the deployment of the first AMF in 2005, the field campaign landscape began to change. In particular, there was a shift in scientific emphasis from field campaigns at fixed-location sites to AMF deployments, since the AMF enabled the program to collect data in other climatically important regions of the world. Given the time scale of several years to plan and execute a major field campaign, this transition occurred over a period of several years. This does not mean that field campaigns at the fixed ARM sites were no longer considered important nor supported; however, there was a shift in emphasis and resources to the new ARM Mobile Facilities as well as the Aerial Facility.

This change had both positive and negative effects on the science community. For the ARM Science Team, the change meant less direct involvement in the planning of field campaigns. This change was inevitable due to ARM's new status as a User Facility and the associated requirement to subject facility proposals to a peer-review process via the ARM Science Board. The other side effect of this change is that field campaign proposals were suddenly equally accessible to anyone regardless of their affiliation. Through this open process, the first AMF was deployed for five campaigns in a row outside of the United States to address key science issues. The PI for two of those five campaigns had no direct ties to DOE; those campaigns were the Radiative Atmospheric Divergence Using the AMF, Geostationary Earth Radiation Budget (GERB) Data, and AMMA Stations (RADAGAST) campaign in Niamey, Niger, in 2006 (Miller and Slingo 2007) and the COPS deployment to the Black Forest in Germany in 2007 (Wulfmeyer et al. 2011). All of these AMF deployments included a significant contribution from ARM's international colleagues. Of course, scientists connected with the ARM Program continued to be very involved in these deployments, often as PIs, but now the doors were open to the international climate science community. A result of this was to significantly broaden the interest in ARM both in the United States and internationally.

4. Going beyond the soda straw, and other new measurement needs

Part of the challenge in evaluating climate models directly with ARM measurements (vs. through process study analysis and associated improvement of representation of those processes in models) is that ARM measurements are inherently local while most global models have horizontal resolutions of about 100 km (although this resolution continues to improve). This issue is mitigated to a certain extent at the SGP because that facility includes a network of extended facilities (Sisterson et al. 2016, chapter 6). The measurements at those extended sites, which covered 142 000 km^2 circa 2008, were somewhat site dependent but included at least broadband radiative fluxes and surface meteorology. Therefore, information about the spatial variability of some geophysical fields is available at the SGP site, although this information is not as detailed or complete as is available at the central facility.

During this period in the mid-2000s following the designation of ARM as a User Facility, certain topics related to bridging this spatial scale gap were being discussed with increasing frequency. Perhaps foremost among these was the idea of using scanning radars (at frequencies ranging from the microwave at 35–94 GHz to frequencies in the 1- to 10-GHz range appropriate for studying precipitation and cloud dynamics) to study the three-dimensional distribution of clouds and precipitation around the ARM sites. In 2010, one of the ARM 94-GHz cloud radars was modified to permit scanning (Kollias et al. 2016, chapter 17); the radar was first deployed to the ARM Mobile Facility in the Azores.

There were many discussions during this period of using small Unmanned Aerial Systems (UASs) to obtain better spatial representation on a routine basis. UASs were not new to ARM (Stephens et al. 2000; Schmid et al. 2016, chapter 10) but the idea of using small systems on a routine basis had not yet been implemented because of federal aviation restrictions. A big step toward the goal of flying UASs routinely came in 2004 when the NSA site management team obtained permission for a small (4 nautical mile diameter) warning area centered on Oliktok, Alaska.

While the goal of all these measurement efforts was to move from a one-dimensional (1D) soda-straw view of the atmosphere to a 3D spatial view, providing a better spatial or volumetric match with models was not the only intended application of this information. Many atmospheric processes, such as convective initiation, are hypothesized to have dependencies on small-scale inhomogeneities in the atmospheric state field, and evaluating and improving 3D radiative transfer models requires 3D cloud observations. In addition, this spatial information will provide improved better information about cloud life cycle processes, which are essential to capturing these processes correctly within models.

Meanwhile, there had also been ongoing discussions that additional information about a variety of geophysical parameters and processes that could not be adequately probed with the instrumentation then available. These included better measurements of ice crystal habit, the partitioning of liquid and ice in mixed-phase clouds, precipitation and cloud properties in the presence of precipitation, boundary layer dynamics, and aerosol properties and processes such as absorption, formation from precursor gases, and the ability to serve as ice nuclei. Together these discussions precipitated a series of meetings designed to extract from the science community a clear sense of unfulfilled measurement needs.

5. Facility workshops and the 2009 Recovery Act

ARM Program management tries to be responsive to the needs of the scientific community, especially when these needs are well articulated in reports authored by both ARM scientists and those outside of the ARM family. In 2007 and 2008, DOE convened two workshops that were designed to assess the state of the ARM User Facility and needs to address critical gaps in climate science. The first workshop, held in the fall of 2007, discussed the priorities for potential future ARM sites and facility needs. Participants at the first workshop were split between representatives of the ARM science community and the broader climate research community. One of the main results from the report (U.S. Department of Energy 2007) was the identification of key areas for future measurements; this list included the Azores, Greenland, South Asia, the Amazon, and the Southern Ocean. Additionally, if DOE were to develop a second Mobile Facility, this report recommended that it should be sufficiently compact and flexible that it could be deployed on a ship or difficult-to-reach locations.

A second workshop was held a year later that focused on identifying key science issues and missing measurements (U.S. Department of Energy 2008). Recommendations from the second workshop included expanding radar capabilities to include features such as multiple frequencies, dual polarization, scanning to improve measurements of microphysics, and longer wavelengths to better characterize precipitation processes. Other recommendations included better measurements of surface and boundary layer properties, aerosol properties, and upper tropospheric water vapor. DOE management led by Wanda Ferrell had a remarkable amount of foresight in holding these workshops because both proved to be very timely.

In 2008, the baseline funding for the ARM User Facility was increased specifically to implement a second Mobile Facility to partially address some of the recommendations from the 2007 workshop and to specifically support marine deployments. While the first several deployments of this second AMF were land-based to help the developing facility mature, in 2012/13 the AMF2 was deployed on a ship that repeatedly transited between Los Angeles and Honolulu to sample boundary layer clouds in this region.

While the AMF2 was under development, the ARM User Facility was awarded funds through the 2009 American Reinvestment and Recovery Act to significantly enhance the program's measurement capabilities. The Recovery Act was an economic stimulus package implemented widely across the Federal Government to projects that had tasks that were ready to be worked on right away. The recommendations from the workshop held less than six months earlier, combined with ongoing interactions between the ARM User Facility and the ARM Science Team, put ARM management in the position to be able to react very quickly (as was required) when the Recovery Act opportunity came along. Through the Recovery Act, a broad variety of instruments were added to the facility for improving the measurement of cloud, aerosol, and precipitation properties and for measuring surface radiative and heat fluxes. Significant additions included scanning radars at millimeter and centimeter wavelengths, several types of advanced lidars for profiling aerosol extinction, water vapor concentration, and clear air motion, and aerosol instruments to provide improved measurements of physical and chemical properties (Mather and Voyles 2013). A listing of core ARM instruments including those added through the Recovery Act is provided in Table 3-2.

The significant improvement in the instrumentation at the ARM sites via the Recovery Act coincided with a desire by DOE management to more closely link different climate research programs within DOE's Climate and Environmental Sciences Division. This led to the merging of the ARM science program with another program that focused on aerosol processes and properties to develop a new scientific program called

TABLE 3-2. Recovery Act instruments. The sites where each instrument is deployed are identified with the following key: S = SGP, N = NSA/Barrow, T1 = TWP/Manus, T3 = TWP/Darwin, A1 = AMF1, A2 = AMF2, AF = ARM Aerial Facility, MA = Mobile Aerosol Observing System.

Category	Instruments
Radars	Scanning 35-GHz/94-GHz cloud radar (S, N, A1), scanning 35-GHz/10-GHz cloud radar (M, D, A2), scanning 5-GHz Radar (S, T1), scanning 10-GHz radar (S, N), zenith-pointing 35-GHz cloud radar (S, N, T1, T3, A2)
Lidars	Raman lidar (D), Raman lidar upgrade (S), high spectral resolution lidar (B, A2), Doppler lidar (S, D, A1), micropulse lidar upgrades (S, B, M, D, A1, A2)
Radiometers	Atmospheric emitted radiance interferometers (S, N, T1, T3, A1), 3-channel (23 , 31, and 90 GHz) microwave radiometers (S, N, T1, T3, A1, A2), solar array spectrometer – hemispheric viewing (S, A1), solar array spectrometer – zenith viewing (S, A1)
Aerosol–gas sampling instruments	Aerosol observing system (including optical properties and number concentrations; T3,A2), hygroscopic tandem differential mobility analyzer (T3, A2, MA), single particle soot photometer (MA, AF), particle into liquid sampler (MA, AF), proton transfer mass spectrometer (MA), aerosol chemical speciation monitor (S, T3, MA), trace gas system (MA, AF), scanning mobility particle sizer (AF), ultrahigh scanning aerosol sizer (MA, AF)
Cloud sampling instruments (in situ)	Fast forward scattering spectrometer probe (AF), two-dimensional imaging probe (AF), high-volume precipitation spectrometer (AF), cloud spectrometer and impactor (AF), fast cloud droplet probe (AF), liquid/ice water sensor (AF)

Atmospheric System Research (ASR; U.S. Department of Energy 2010; Mather et al. 2016, chapter 4). This new program focuses on the better characterization of the myriad of processes associated with clouds and aerosols, their interaction with radiation, and their impact on climate.

To support ARM and ASR research efforts, it is very important to develop the expertise, datasets, and software tools to make the best use of the new facility measurement capabilities. These measurements have the potential of providing remarkable insights into processes associated with cloud and aerosol life cycles but will require persistent effort to fully realize these benefits. In particular, the scanning cloud radars have not been used previously on a continuous basis and there are few examples where they have been used at all. Optimizing the use of these instruments including their scanning strategy and derived data products will require close collaboration between ARM and the science user community.

Another resource for the ASR scientific community will be the data from the recent deployment of two new sites. In 2013, ARM began operating a new fixed-location site in the Azores where the AMF was deployed in 2009–10. The purpose for this site is to continue to explore the life cycle of marine stratus clouds, which is very uncertain yet is critical in regulating Earth's energy balance (Bony and Dufresne 2005). To aid in this work, the site will be equipped with additional instrumentation relative to the earlier AMF deployment in the Azores, including an X-band centimeter wavelength radar for studying drizzle, a Doppler lidar for characterizing below-cloud vertical air motion, and an Aerosol Observing System (AOS). ARM has also developed a third AMF that will first be deployed for an extended term at Oliktok, Alaska, one of the ancillary sites used during the Mixed-Phase Arctic Cloud Experiment (M-PACE; Verlinde et al. 2007). As noted earlier, a key attribute of Oliktok is the restricted airspace managed by DOE centered on that site. That restricted airspace opens the possibility of operating Tethered Balloon Systems and Unmanned Aerial Systems in conjunction with the AMF. Such a combination would allow links to be made between the ground-based observations along the Arctic Ocean coast and the adjacent ocean–sea ice, which was one of the original goals for the NSA site (Stamnes et al. 1999).

As ARM undertakes measurements in new locales, it has ended operations in the TWP in 2014 after 18 yr. The cessation of operations in the tropics is enabling a redistribution of instruments and resources, which is intended to accelerate the application of ARM observations and data processing for the understanding of key atmospheric processes and the representation of these processes in global climate models. This reconfiguration of the ARM User Facility is focused on enhancements of the SGP and NSA sites and has three main facets (U.S. Department of Energy 2014):

1) enhancing ARM observations and measurement strategies to enable the routine operation of high-resolution models and to optimize the use of ARM data for the evaluation of these models.
2) undertaking the routine operation of high-resolution models at ARM sites; and

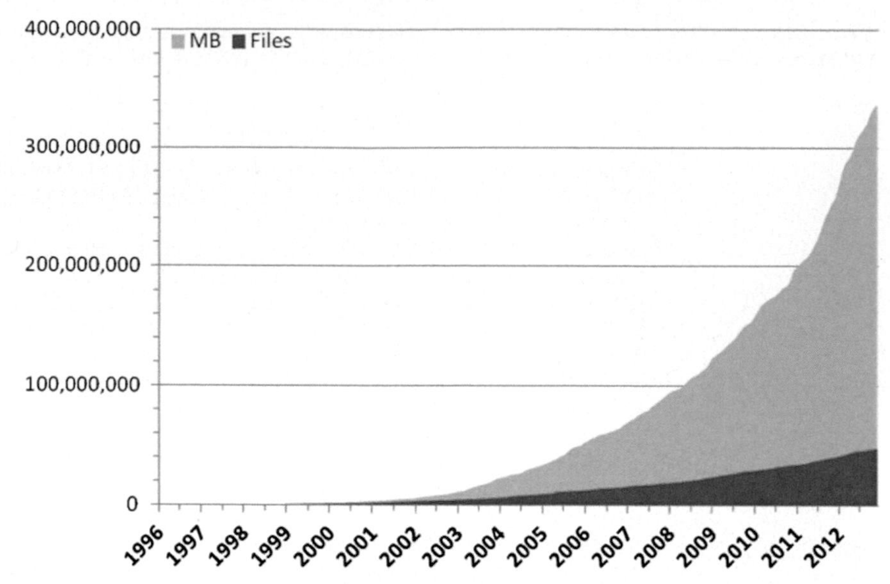

FIG. 3-2. Cumulative growth in files and megabytes requested from the ARM Data Archive.

3) developing data products and analysis tools that enable the evaluation of models using ARM data.

ARM is undergoing some significant changes; however, its mission to provide comprehensive observation datasets for climate research and the improvement of climate models remains unchanged. As the ASR community makes progress in understanding aerosol and cloud life cycles and the interaction between clouds, aerosols, and precipitation, the measurement needs will evolve and thus areas of emphasis within ARM will also evolve. But the core emphasis on clouds and aerosols, and their effect on radiation and now precipitation, will continue for some time.

6. Continued evolution and growth of ARM

The ARM Program began with a vision largely unconstrained by prior experience. It grew prolifically in the 1990s in many different directions simultaneously and experimentally. When questions arose about what to do, the oft-repeated mantra was to do what made sense. Those of us who were part of the program in that decade remember the enthusiasm, the brainstorming, the successes, and the failures of those early days. By the later years of the decade of the 1990s, it was clear that ARM was moving into a new phase. As a maturing program, ARM needed to transition to a longer time horizon for program planning, develop a better sense of process in order to make sure that all the science voices were heard, set priorities in the face of constrained resources, and increase the interaction between Science Team and infrastructure personnel. The mantra now became how to ensure that we focused on what made sense and weed out what did not.

The ARM Program made a series of decisions around 2000–01 that put processes in place to further the goals of ARM. These included the data quality office, the VAP process, and an IOP planning process. These changes were implemented to create a situation in which priorities could be set and resources allocated fairly. One of the important and recognized consequences was the pruning of the program to eliminate certain instruments, reduce the resources applied to some science questions, and reduce the footprint of some operational components. Given the fixed resources available to the program in the period of the 2000s, these changes were inevitable in order to make room for an expanded instrument set and new science questions. This process would be repeated again as the ARM Program's observational capability grew substantially with the Recovery Act, and when the scientific priorities changed as ARM science transitioned to the ASR.

The mature ARM Program of the 2000s was in some sense less exciting than the growing program of the 1990s, but was considerably more productive in terms of the breadth of the research and publications. Changes in the program infrastructure were put in place to enhance the ability of the Science Team to conduct research and they largely met that goal. In the case of VAPs, the program cost was greater than anticipated, mainly because it proved to be more difficult to create "bullet-proof" code than had been expected. Overall, the cost of infrastructure activities increased, but much of that increase was the result of increased scope rather than a failure to contain costs. In fact, ARM Program management fought doggedly and successfully throughout this period to find ways to economize.

During this period, the ARM Program also gained international visibility because of its success. It served as an exemplar that promoted the growth of several similar sites in Europe, such as the Chilbolton Observatory in the United Kingdom, Cabauw in the Netherlands, Lindenberg Observatory in Germany, and Paliseau in France (see Haeffelin et al. 2016, chapter 29). It forged links with other science programs in the United States, particular those sponsored by NASA, by serving as an instrument test bed, a key satellite validation site (or sites), and target point for aircraft observations. These connections can most easily be identified in the list of IOPs mentioned previously but are also noted elsewhere in this volume (Marchand 2016, chapter 30). Modeling groups began to make increased use of ARM data, especially as ARM undertook the creation of datasets particularly targeted toward their needs. (See Fig. 3-2 for overall growth in data usage.) Attendance at ARM Science Team Meetings grew into the hundreds, attracting not only those funded directly by ARM but also those using the ARM data or sites for these purposes. The ARM Program has continued to grow, pushing the observational boundaries with new state-of-the-art instruments, conducting IOPs at both its fixed and AMF sites, and garnering new users.

When the ARM Program was first proposed toward the latter part of 1989, there was a considerable amount of skepticism raised about whether the program could be carried out as envisaged and whether the results would be worth the cost. The first decade of ARM was devoted to proving that the program could indeed be carried out. The vision had to be reduced somewhat, largely because the proposed budget never materialized. The second decade of ARM was devoted to demonstrating that the results were worth the cost. The mature ARM Program produced a record of solid scientific achievement, including significant breakthroughs in our understanding of cloud and aerosol processes and radiative forcing. Perhaps the ARM Program's most important legacy is the recognition that continuous, ground-based remote sensing is a critically important tool for understanding the complex interactions between clouds, aerosols, atmospheric radiation, weather, and climate.

REFERENCES

Ackerman, T. P., and G. M. Stokes, 2003: The Atmospheric Radiation Measurement Program. *Phys. Today*, **56**, 38–44, doi:10.1063/1.1554135.

Ahlgrimm, M., R. Forbes, J.-L. Morcrette, and R. Neggers, 2016: ARM's impact on numerical weather prediction at ECMWF. *The Atmospheric Radiation Measurement (ARM) Program: The First 20 Years, Meteor. Monogr.*, No. 57, Amer. Meteor. Soc., doi:10.1175/AMSMONOGRAPHS-D-15-0032.1.

Andrews, E., P. J. Sheridan, J. A. Ogren, and R. Ferrare, 2004: In situ aerosol profiles over the Southern Great Plains cloud and radiation test bed site: 1. Aerosol optical properties. *J. Geophys. Res.*, **109**, D06208, doi:10.1029/2003JD004025.

Bony, S., and J.-L. Dufresne, 2005: Marine boundary layer clouds at the heart of tropical cloud feedback uncertainties in climate models. *Geophys. Res. Lett.*, **32**, L20806, doi:10.1029/2005GL023851.

Clothiaux, E. E., M. A. Miller, B. A. Albrecht, T. P. Ackerman, J. Verlinde, D. M. Babb, R. M. Peters, and W. J. Syrett, 1995: An evaluation of a 94-GHz radar for remote sensing of cloud properties. *J. Atmos. Oceanic Technol.*, **12**, 201–229, doi:10.1175/1520-0426(1995)012<0201:AEOAGR>2.0.CO;2.

——, G. G. Mace, T. P. Ackerman, T. J. Kane, J. D. Spinhirne, and V. S. Scott, 1998: An automated algorithm for detection of hydrometeor returns in micro pulse lidar data. *J. Atmos. Oceanic Technol.*, **15**, 1035–1042, doi:10.1175/1520-0426(1998)015<1035:AAAFDO>2.0.CO;2.

——, and Coauthors, 1999: The Atmospheric Radiation Measurement program cloud radars: Operational modes. *J. Atmos. Oceanic Technol.*, **16**, 819–827, doi:10.1175/1520-0426(1999)016<0819:TARMPC>2.0.CO;2.

——, T. P. Ackerman, G. G. Mace, K. P. Moran, R. T. Marchand, M. A. Miller, and B. E. Martner, 2000: Objective determination of cloud heights and radar reflectivities using a combination of active remote sensors at the ARM CART sites. *J. Appl. Meteor.*, **39**, 645–665, doi:10.1175/1520-0450(2000)039<0645:ODOCHA>2.0.CO;2.

Cress, T. S., and D. L. Sisterson, 2016: Deploying the ARM sites and supporting infrastructure. *The Atmospheric Radiation Measurement (ARM) Program: The First 20 Years, Meteor. Monogr.*, No. 57, Amer. Meteor. Soc., doi:10.1175/AMSMONOGRAPHS-D-15-0049.1.

Ellingson, R. G., R. D. Cess, and G. L. Potter, 2016: The Atmospheric Radiation Measurement Program: Prelude. *The Atmospheric Radiation Measurement (ARM) Program: The First 20 Years, Meteor. Monogr.*, No. 57, Amer. Meteor. Soc., doi:10.1175/AMSMONOGRAPHS-D-15-0029.1.

Feingold, G., and A. McComiskey, 2016: ARM's aerosol–cloud–precipitation research (aerosol indirect effects). *The Atmospheric Radiation Measurement (ARM) Program: The First 20 Years, Meteor. Monogr.*, No. 57, Amer. Meteor. Soc., doi:10.1175/AMSMONOGRAPHS-D-15-0022.1.

Feltz, W. F., W. L. Smith, H. B. Howell, R. O. Knuteson, H. Woolf, and H. E. Revercomb, 2003: Near-continuous profiling of temperature, moisture, and atmospheric stability using the Atmospheric Emitted Radiance Interferometer (AERI). *J. Appl. Meteor.*, **42**, 584–595, doi:10.1175/1520-0450(2003)042<0584:NPOTMA>2.0.CO;2.

Ferrare, R. A., and Coauthors, 2004: Characterization of upper tropospheric water vapor measurements during AFWEX using LASE. *J. Atmos. Oceanic Technol.*, **21**, 1790–1808, doi:10.1175/JTECH-1652.1.

Ghan, S., and J. Penner, 2016: ARM-led improvements in aerosols in climate and climate models. *The Atmospheric Radiation Measurement (ARM) Program: The First 20 Years, Meteor. Monogr.*, No. 57, Amer. Meteor. Soc., doi:10.1175/AMSMONOGRAPHS-D-15-0033.1.

Haeffelin, M., and Coauthors, 2016: Parallel developments and formal collaboration between European atmospheric profiling observatories and the U.S. ARM research program. *The*

Atmospheric Radiation Measurement (ARM) Program: The First 20 Years, *Meteor. Monogr.*, No. 57, Amer. Meteor. Soc., doi:10.1175/AMSMONOGRAPHS-D-15-0045.1.

Kollias, P., and Coauthors, 2016: Development and applications of ARM millimeter-wavelength cloud radars. *The Atmospheric Radiation Measurement (ARM) Program: The First 20 Years*, *Meteor. Monogr.*, No. 57, Amer. Meteor. Soc., doi:10.1175/AMSMONOGRAPHS-D-15-0037.1.

Krueger, S. K., H. Morrison, and A. M. Fridlind, 2016: Cloud-resolving modeling: ARM and the GCSS story. *The Atmospheric Radiation Measurement (ARM) Program: The First 20 Years*, *Meteor. Monogr.*, No. 57, Amer. Meteor. Soc., doi:10.1175/AMSMONOGRAPHS-D-15-0047.1.

Long, C. N., and Y. Shi, 2008: An automated quality assessment and control algorithm for surface radiation measurements. *Open Atmos. Sci. J.*, **2**, 23–37, doi:10.2174/1874282300802010023.

——, J. H. Mather, and T. P. Ackerman, 2016: The ARM Tropical Western Pacific (TWP) sites. *The Atmospheric Radiation Measurement (ARM) Program: The First 20 Years*, *Meteor. Monogr.*, No. 57, Amer. Meteor. Soc., doi:10.1175/AMSMONOGRAPHS-D-15-0024.1.

Marchand, R., 2016: ARM and satellite cloud validation. *The Atmospheric Radiation Measurement (ARM) Program: The First 20 Years*, *Meteor. Monogr.*, No. 57, Amer. Meteor. Soc., doi:10.1175/AMSMONOGRAPHS-D-15-0038.1.

Mather, J. H., and J. W. Voyles, 2013: The ARM climate research facility: A review of structure and capabilities. *Bull. Amer. Meteor. Soc.*, **94**, 377–392, doi:10.1175/BAMS-D-11-00218.1.

——, D. D. Turner, and T. P. Ackerman, 2016: Scientific maturation of the ARM Program. *The Atmospheric Radiation Measurement (ARM) Program: The First 20 Years*, *Meteor. Monogr.*, No. 57, Amer. Meteor. Soc., doi:10.1175/AMSMONOGRAPHS-D-15-0053.1.

McComiskey, A., and R. A. Ferrare, 2016: Aerosol physical and optical properties and processes in the ARM Program. *The Atmospheric Radiation Measurement (ARM) Program: The First 20 Years*, *Meteor. Monogr.*, No. 57, Amer. Meteor. Soc., doi:10.1175/AMSMONOGRAPHS-D-15-0028.1.

McCord, R., and J. W. Voyles, 2016: The ARM data system and archive. *The Atmospheric Radiation Measurement (ARM) Program: The First 20 Years*, *Meteor. Monogr.*, No. 57, Amer. Meteor. Soc., doi:10.1175/AMSMONOGRAPHS-D-15-0043.1.

McFarlane, S. A., J. H. Mather, and E. J. Mlawer, 2016: ARM's progress on improving atmospheric broadband radiative fluxes and heating rates. *The Atmospheric Radiation Measurement (ARM) Program: The First 20 Years*, *Meteor. Monogr.*, No. 57, Amer. Meteor. Soc., doi:10.1175/AMSMONOGRAPHS-D-15-0046.1.

Michalsky, J., F. Denn, C. Flynn, G. Hodges, P. Kiedron, A. Koontz, J. Schemmer, and S. E. Schwartz, 2010: Climatology of aerosol optical depth in north-central Oklahoma: 1992–2008. *J. Geophys. Res.*, **115**, D07203, doi:10.1029/2009JD012197.

Miller, M. A., and A. Slingo, 2007: The ARM Mobile Facility and its first international deployment. *Bull. Amer. Meteor. Soc.*, **88**, 1229–1244, doi:10.1175/BAMS-88-8-1229.

——, K. Nitschke, T. P. Ackerman, W. R. Ferrell, N. Hickmon, and M. Ivey, 2016: The ARM mobile facilities. *The Atmospheric Radiation Measurement (ARM) Program: The First 20 Years*, *Meteor. Monogr.*, No. 57, Amer. Meteor. Soc., doi:10.1175/AMSMONOGRAPHS-D-15-0051.1.

Mlawer, E. J., and D. D. Turner, 2016: Spectral radiation measurements and analysis in the ARM Program. *The Atmospheric Radiation Measurement (ARM) Program: The First 20 Years*, *Meteor. Monogr.*, No. 57, Amer. Meteor. Soc., doi:10.1175/AMSMONOGRAPHS-D-15-0027.1.

Peppler, R., K. Kehoe, J. Monroe, A. Theisen, and S. Moore, 2016: The ARM data quality program. *The Atmospheric Radiation Measurement (ARM) Program: The First 20 Years*, *Meteor. Monogr.*, No. 57, Amer. Meteor. Soc., doi:10.1175/AMSMONOGRAPHS-D-15-0039.1.

Randall, D. A., A. D. Del Genio, L. J. Donner, W. D. Collins, and S. A. Klein, 2016: The impact of ARM on climate modeling. *The Atmospheric Radiation Measurement (ARM) Program: The First 20 Years*, *Meteor. Monogr.*, No. 57, Amer. Meteor. Soc., doi:10.1175/AMSMONOGRAPHS-D-15-0050.1.

Schmid, B., R. G. Ellingson, and G. M. McFarquhar, 2016: ARM aircraft measurements. *The Atmospheric Radiation Measurement (ARM) Program: The First 20 Years*, *Meteor. Monogr.*, No. 57, Amer. Meteor. Soc., doi:10.1175/AMSMONOGRAPHS-D-15-0042.1.

Shupe, M. D., J. M. Comstock, D. D. Turner, and G. G. Mace, 2016: Cloud property retrievals in the ARM Program. *The Atmospheric Radiation Measurement (ARM) Program: The First 20 Years*, *Meteor. Monogr.*, No. 57, Amer. Meteor. Soc., doi:10.1175/AMSMONOGRAPHS-D-15-0030.1.

Sisterson, D., R. Peppler, T. S. Cress, P. Lamb, and D. D. Turner, 2016: The ARM Southern Great Plains (SGP) site. *The Atmospheric Radiation Measurement (ARM) Program: The First 20 Years*, *Meteor. Monogr.*, No. 57, Amer. Meteor. Soc., doi:10.1175/AMSMONOGRAPHS-D-16-0004.1.

Stamnes, K., R. G. Ellingson, J. A. Curry, J. E. Walsh, and B. D. Zak, 1999: Review of science issues, deployment strategy, and status for the ARM North Slope of Alaska–Adjacent Arctic Ocean climate research site. *J. Climate*, **12**, 46–63, doi:10.1175/1520-0442-12.1.46.

Stephens, G., and Coauthors, 2000: The Department of Energy's Unmanned Aerospace Vehicle (UAV) Program. *Bull. Amer. Meteor. Soc.*, **81**, 2915–2938, doi:10.1175/1520-0477(2000)081<2915:TDOESA>2.3.CO;2.

Stokes, G. M., 2016: Original ARM concept and launch. *The Atmospheric Radiation Measurement (ARM) Program: The First 20 Years*, *Meteor. Monogr.*, No. 57, Amer. Meteor. Soc., doi:10.1175/AMSMONOGRAPHS-D-15-0021.1.

Turner, D. D., and Coauthors, 2004: The QME AERI LBLRTM: A closure experiment for downwelling high spectral resolution infrared radiance. *J. Atmos. Sci.*, **61**, 2657–2675, doi:10.1175/JAS3300.1.

——, and Coauthors, 2007a: Thin liquid water clouds: Their importance and our challenge. *Bull. Amer. Meteor. Soc.*, **88**, 177–190, doi:10.1175/BAMS-88-2-177.

——, S. A. Clough, J. C. Liljegren, E. E. Clothiaux, K. E. Cady-Pereira, and K. L. Gaustad, 2007b: Retrieving liquid water path and precipitable water vapor from the Atmospheric Radiation Measurement (ARM) microwave radiometers. *IEEE Trans. Geosci. Remote Sens.*, **45**, 3680–3689, doi:10.1109/TGRS.2007.903703.

——, E. J. Mlawer, and H. E. Revercomb, 2016: Water vapor observations in the ARM Program. *The Atmospheric Radiation Measurement (ARM) Program: The First 20 Years*, *Meteor. Monogr.*, No. 57, Amer. Meteor. Soc., doi:10.1175/AMSMONOGRAPHS-D-15-0025.1.

U.S. Department of Energy, 1990: Atmospheric Radiation Measurement Program Plan. DOE/ER-0441, 121 pp.

——, 1996: Science Plan for the Atmospheric Radiation Measurement (ARM) Program. DOE/ER-0670T, 74 pp.

——, 2004: Atmospheric Radiation Measurement Program Science Plan. DOE/ER-0402, 62 pp.

——, 2007: Report on the ARM Climate Research Facility Expansion Workshop. DOE/SC-ARM-0707, 50 pp.

——, 2008: ARM Climate Research Facility Workshop Report. DOE/SC-ARM-0804, 23 pp.

——, 2010: Atmospheric System Research (ASR) Science and Program Plan. DOE/SC-ASR-10-001, 77 pp. [Available online at http://science.energy.gov/~/media/ber/pdf/Atmospheric_system_research_science_plan.pdf.]

——, 2014: Atmospheric Radiation Measurement Climate Research Facility Decadal Vision. DOE/SC-ARM-14-029, 21 pp.

Uttal, T., and Coauthors, 2002: Surface heat budget of the Arctic Ocean. *Bull. Amer. Meteor. Soc.*, **83**, 255–275, doi:10.1175/1520-0477(2002)083<0255:SHBOTA>2.3.CO;2.

Verlinde, J., and Coauthors, 2007: The Mixed-Phase Arctic Cloud Experiment (M-PACE). *Bull. Amer. Meteor. Soc.*, **88**, 205–221, doi:10.1175/BAMS-88-2-205.

——, B. Zak, M. D. Shupe, M. Ivey, and K. Stamnes, 2016: The ARM North Slope of Alaska (NSA) sites. *The Atmospheric Radiation Measurement (ARM) Program: The First 20 Years, Meteor. Monogr.*, No. 57, Amer. Meteor. Soc., doi:10.1175/AMSMONOGRAPHS-D-15-0023.1.

Vogelmann, A. M., and Coauthors, 2012: RACORO extended-term aircraft observations of boundary layer clouds. *Bull. Amer. Meteor. Soc.*, **93**, 861–878, doi:10.1175/BAMS-D-11-00189.1.

Voyles, J. W., and L. A. Chapman, 2012: Field campaign guidelines. DOE/SC-ARM-11-003, 20 pp.

Westwater, E. R., B. B. Stankov, D. Cimini, Y. Han, J. A. Shaw, B. M. Lesht, and C. N. Long, 2003: Radiosonde humidity soundings and microwave radiometers during Nauru99. *J. Atmos. Oceanic Technol.*, **20**, 953–971, doi:10.1175/1520-0426(2003)20<953:RHSAMR>2.0.CO;2.

Wulfmeyer, V., and Coauthors, 2011: The Convective and Orographically Induced Precipitation Study (COPS): The scientific strategy, the field phase, and first highlights. *Quart. J. Roy. Meteor. Soc.*, **137**, 3–30, doi:10.1002/qj.752.

Xie, S., and Coauthors, 2010: Clouds and more: ARM climate modeling best estimate data. *Bull. Amer. Meteor. Soc.*, **91**, 13–20, doi:10.1175/2009BAMS2891.1.

Zhang, M., and J. L. Lin, 1997: Constrained variational analysis of sounding data bases on column-integrated budgets of mass, heat, moisture, and momentum: Approach and application to ARM measurements. *J. Atmos. Sci.*, **54**, 1503–1524, doi:10.1175/1520-0469(1997)054<1503:CVAOSD>2.0.CO;2.

——, R. C. J. Somerville, and S. Xie, 2016: The SCM concept and creation of ARM forcing datasets. *The Atmospheric Radiation Measurement (ARM) Program: The First 20 Years, Meteor. Monogr.*, No. 57, Amer. Meteor. Soc., doi:10.1175/AMSMONOGRAPHS-D-15-0040.1.

CHAPTER 4

Scientific Maturation of the ARM Program

J. H. MATHER

Pacific Northwest National Laboratory, Richland, Washington

D. D. TURNER

NOAA/National Severe Storms Laboratory, Norman, Oklahoma

T. P. ACKERMAN

University of Washington, Seattle, Washington

1. Introduction

There has been a tremendous evolution in both the technical and scientific capabilities of the Atmospheric Radiation Measurement (ARM) Program during its first 20 years. The Program was "born" in early 1990 and by mid-1992 the first observations were being collected at its Southern Great Plains (SGP) site (Stokes 2016). Over the next several years, additional instrumentation was deployed at the SGP site; by mid-1996 the site was considered fully instrumented, as per the scope that had been originally planned (U.S. Department of Energy 1990; appendix A, Stokes and Schwartz 1994), with the installation of the millimeter-wave cloud radar (MMCR; see Kollias et al. 2016). The following years saw the deployment of the first and second Tropical Western Pacific (TWP) sites at Manus and Nauru (Long et al. 2016) and the deployment of ARM instrumentation during the year-long Surface Heat Budget of the Arctic (SHEBA) experiment and at the North Slope of Alaska (NSA) sites (Verlinde et al. 2016). The programmatic aspects of ARM were evolving also in order to deal with the challenges of managing such a diverse program (Ackerman et al. 2016), especially as the ARM Program consolidated its airborne components into a facility (Schmid et al. 2006); the ARM Mobile Facility was conceived, developed, and deployed (Miller et al. 2016); and the Program was designated a Department of Energy (DOE) Science User Facility (Mather and Voyles 2013).

The scientific goals of the ARM Program from the beginning were to determine the effect of atmospheric composition and structure on the earth's radiative energy balance and to understand the processes that affected those atmospheric properties—with a particular emphasis on clouds (Ellingson et al. 2016; Stokes 2016; U.S. Department of Energy 1990). The fundamental goal of the ARM Program was to improve cloud parameterizations in general circulation models (GCMs) through improved understanding of cloud and radiation processes obtained from a combination of modeling and data analysis. The essential ARM questions were defined as follows:

1) If we can specify a cloud field, can we compute the radiative fluxes associated with that field?
2) If we can specify the large-scale atmospheric state variables, can we predict the cloud field properties associated with that state?

Answering the first question requires detailed knowledge of cloud properties, such as 3D structure, ice and liquid water path, hydrometeor concentration and size (and shape in the case of ice), and accurate, spectrally resolved, radiance and flux measurements. Answering the second requires knowledge of the 3D state properties and area-wide cloud field properties. These central themes defined the ARM measurement priorities but the scientific emphasis has continually shifted and evolved over the years,

Corresponding author address: James Mather, Pacific Northwest National Laboratory, 3200 Innovation Boulevard, Richland, WA 99354.
E-mail: jim.mather@pnnl.gov

DOI: 10.1175/AMSMONOGRAPHS-D-15-0053.1

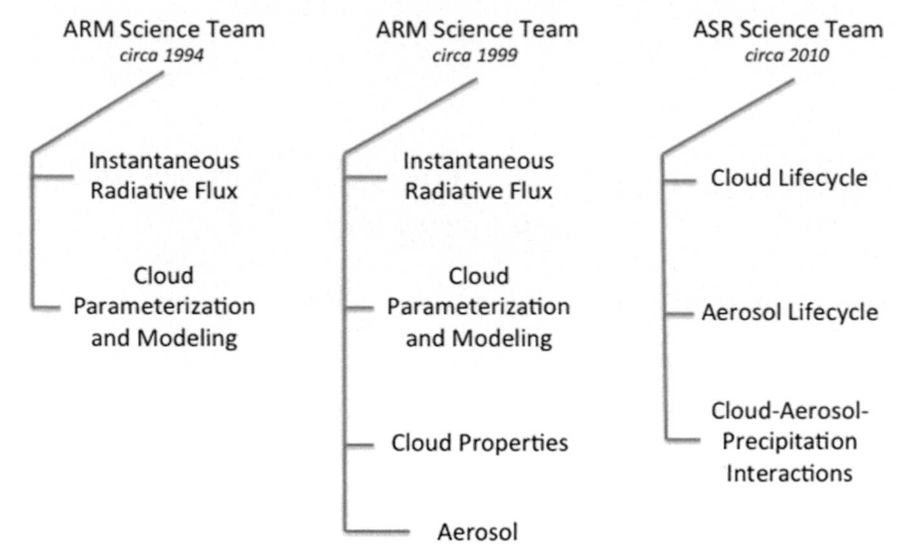

FIG. 4-1. The evolution of the working groups within the ARM (and later ASR) Science Team.

as have some of the details about how instruments are deployed and how data are processed and used.

From the beginning, DOE management looked to the ARM Science Team for input into the scientific priorities of the Program. In the ARM Program's formative stages, there was even a "surrogate" Science Team that was hand selected from the broader community to provide this input until the peer-review process had the time for proposals to be solicited, evaluated, and selected (Stokes 2016). To better facilitate communication among the different members of the ARM Science Team and to coordinate efforts of many individually funded research teams, working groups with specific foci were established. The Program benefitted as many grass-roots initiatives (ranging from ideas for field experiments to new instruments to deploy and scientific focus areas for the Program) were developed by ARM-funded scientists and flowed upward to ARM management after being vetted through the working groups. As the Program matured scientifically and programmatically, new challenges arose that required that the Science Team adapt; one way to illustrate this is by noting how the working groups evolved over time (Fig. 4-1). The working groups often had both formal and informal subgroups within them to focus on more specific topics. The working group structure worked extremely well for both coordinating the research within the group, but also for aggregating recommendations from individual principal investigators (PIs) to ARM management on how to improve the infrastructure to better serve the needs of the ARM Science Team.

After the creation of the formal Science Team, a subset of this community was selected to serve as the Science Team Executive Committee (STEC). Typically, the chairs of the working groups were on the STEC, which was augmented with other scientists from the program to provide additional perspective. The STEC met regularly with infrastructure managers and DOE management to discuss the Program's scientific direction and to prioritize the recommendations that were made by the working groups. This was especially important in the early years. The STEC would evolve into the Science and Infrastructure Steering Committee (SISC) after the ARM infrastructure was designated as a DOE User Facility (Ackerman et al. 2016).

The ARM Program also had three Chief Scientists to help guide it during its formative phase through its maturity: Gerry Stokes from 1990 to 1998, Tom Ackerman from 1999 to 2004, and Warren Wiscombe from 2005 to 2009 (Fig. 4-2). The Chief Scientists worked closely with DOE and infrastructure management to advance ARM science, helped to organize and conduct the annual Science Team Meeting in the spring, provided input on infrastructure priorities from the scientific perspective, and served as the ambassadors of the Program. The ARM Chief Scientists worked with the STEC to develop the ARM Science Plan in 1996 (U.S. Department of Energy 1996, appendix B therein) and its update in 2004 (U.S. Department of Energy 2004, appendix C therein), and with the SISC to develop the Science Plan for the Atmospheric System Research (ASR) program that was created in 2010 (more on ASR at the end of this chapter).

The ARM Program's scientific growth was tremendous. One simple metric of this is the number of peer-reviewed journal articles supported by ARM over the last 20 years (Fig. 4-3). Other measures of the scientific impact of the ARM Program are the increase in the number of high-quality science proposals submitted to

FIG. 4-2. The Chief Scientists of the ARM Program: (left to right) Gerry Stokes, Tom Ackerman, and Warren Wiscombe.

DOE each year, and the diversity of the topics spanned by the published peer-reviewed papers in the later years (relative to the earlier time periods) with ARM support. It is important to note that the plateau seen in Fig. 4-3 is largely due to the plateau in the scientific funding that had occurred by the late 1990s.

This chapter highlights some of the broad scientific accomplishments of the ARM Program during its first 20 yr. This story could have been told multiple ways. We have elected to organize the story into thematic sections; while the presentation of each section is largely chronological, much of the scientific work was occurring in parallel. Chapters 13 through 30 of this monograph provide much more detail on the scientific accomplishments in the various areas than could be covered here.

2. Using ARM observations to improve our understanding of radiation, aerosol, and clouds

a. Making a good measurement

Deploying automated instruments into these unique locations to collect long-term datasets challenged the Program in many ways. Prior to ARM, most complex observational datasets collected in the field were limited to relatively short time frames of several weeks to perhaps several months. The PI and cohorts would deploy the instruments, collect the data, and then spend potentially many months calibrating and analyzing the data before they were made available to the community. Often, it would also be difficult to collect the different datasets from the various PIs, and it would be challenging to sort out the various differences in the way the different PIs quality-controlled and organized their data. The ARM Program's decision to collect long-term datasets precluded this model, and thus the Program decided to make the data available almost immediately[1] after it was collected and enlist the help of the ARM Science Team and site scientists to help with the quality control (Peppler et al. 2016).

However, first the instruments had to be able to run continuously; the ARM Program's Instrument Development Program (IDP; Stokes 2016) invested heavily to quickly advance several remote sensing systems so that they could run in an unattended way. Three of the most obvious examples are the Atmospheric Emitted Radiance Interferometer (AERI; Turner et al. 2016b), the MMCR (Kollias et al. 2016), and the Raman lidar (Turner et al. 2016a); however, instruments such as the micropulse lidar (MPL; Campbell et al. 2002), microwave radiometer (MWR; Liljegren and Lesht 1996), and many others benefitted from ARM investments to make them more robust for long-term operations (U.S. Department of Energy 2004). However, the investments in these instruments did not immediately result in usable datasets; it often took multiple years of interactions between observationalists and the instrument mentors before the data from an instrument were useful for the modeling community.

Just because an instrument is running does not ensure that its data are useful; there must be calibration methods in place to ensure that the data are accurate and the uncertainties quantified. The long-term operational paradigm—together with the deployment of instruments that measured similar or complementary geophysical

[1] The Internet was in its infancy when the ARM Program started, and thus the phrase "almost immediately" has different connotations depending on the reference point. In the mid-1990s, data collected at the SGP site were delivered to ARM Science Team members within a week of collection, and data from the TWP sites were typically delivered every few months due to the lack of efficient and affordable communication with the remote sites. By the early 2000s, almost all the data collected from ARM instruments were made available to the ARM Science Team within 2 days.

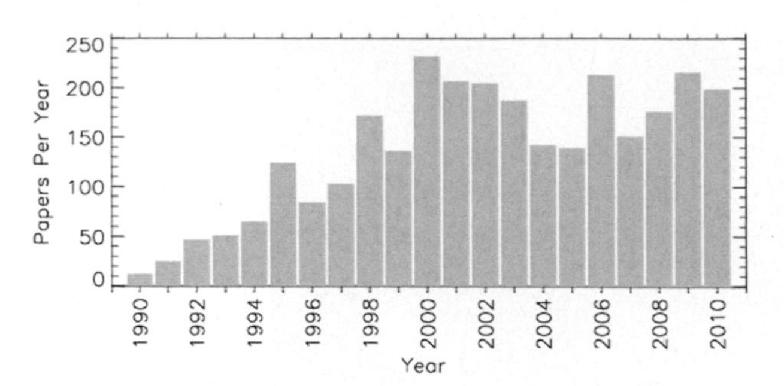

FIG. 4-3. Number of papers published in peer-reviewed journals supported by the ARM Program from its inception until 2010. Data are from the ARM publication database.

variables that could be intercompared, perhaps using a model as a transfer standard, via quality measurements experiments (QMEs; Peppler et al. 2016; Turner et al. 2004)—highlighted many issues associated with calibration that were not well understood before. Many of the calibration issues were first identified from longer-term analysis, and then intensive observation periods (IOPs) were used to further elucidate the source of the error in the calibration method and develop improved methods.

One example of this was the development of an automated method to maintain the calibration of the MWRs (Liljegren 2000), which drew heavily from results garnered from the water vapor IOPs (Turner et al. 2016b) that, among other things, investigated the various error sources that might impact the calibration of the MWRs (Han and Westwater 2000). A second example was the discovery that there were systematic errors in radiosonde relative humidity measurements that were associated with how the manufacturer was calibrating them in batches (Liljegren and Lesht 1996; Turner et al. 2016b); prior to this, radiosondes were typically assumed to be the gold standard for profiles of water vapor and temperature and this "batch" behavior was unnoticed because of the short duration of previous field campaigns. A third example is the need for the Radiometric Calibration Facility at the SGP site, which was built in the late 1990s to calibrate dozens of broadband solar and infrared radiometers simultaneously against radiometric standards (Michalsky and Long 2016), thereby ensuring consistent calibration across large numbers of instruments that are deployed at the many ARM sites.

The investment in hardening the instruments and running them autonomously for long time periods, and especially the development of improved calibration approaches that were subsequently applied to the long-term measurements, made the ARM datasets uniquely valuable. ARM also deployed complimentary instruments, such as the laser ceilometer and the MMCR to measure different aspects of cloud structure, which led to the development of more sophisticated analyses and cleaner interpretation of atmospheric structure and phenomena. The standardization of the format of the data files using netCDF tools and an easy-to-use data archives for storage also contributed to the long-term success of ARM (McCord and Voyles 2016). These are considered some of the primary technical accomplishments of the ARM Program, which subsequently led to many of the scientific accomplishments below.

b. Instantaneous radiative fluxes in the clear sky

One of the primary ARM goals was to understand the interaction of radiation with aerosols and clouds. However, early investigations showed that more fundamental clear-sky radiative transfer issues needed to be resolved first before significant progress with clouds could be made; these issues are discussed in detail in Ellingson et al. (2016).

Early on, the Instantaneous Radiative Flux (IRF) working group, building on previous radiation-focus programs (such as the Spectral Radiance Experiment) to address the findings of the Intercomparison of Radiation Codes used in Climate Models (ICRCCM) effort (Ellingson et al. 2016), undertook a number of radiative closure studies (Mlawer and Turner 2016) in which investigators used ARM observations of parameters such as temperature and humidity profiles as inputs to a radiative transfer model, and then compared the radiation calculations to radiation measurements at the surface. An important tool in these early clear-sky infrared closure studies was the AERI (Knuteson et al. 2004a,b; Turner et al. 2016b). The AERI provided high-spectral-resolution radiance measurements at infrared wavelengths. Comparisons with simulations from the AER line-by-line radiative transfer model (Clough et al. 1992) revealed that errors and uncertainties in water vapor measurements were a limiting factor in constraining surface infrared fluxes (Revercomb et al. 2003).

The problems with water vapor profiles led to an intensive study of water vapor measurements including a series of field campaigns beginning with the Water Vapor Intensive Operation Period in 1996 (Revercomb et al. 2003; Turner et al. 2016b). These studies led to innovative techniques for measuring water vapor including use of a two-channel microwave radiometer to mitigate differences across batches of radiosondes (Turner et al. 2003), use of the AERI to retrieve profiles of water vapor and temperature in the boundary layer (Feltz et al. 2003), and the use of the Raman lidar to obtain more continuous water vapor profiles (Turner and Goldsmith 1999).

In addition to these observation-oriented activities, there was also a great deal of activity related to the improvement of infrared radiative transfer models going on within the IRF working group. Particular areas of focus were improvements in the water vapor continuum, which led to the ability to compute downwelling radiation to better than 1.5 W m^{-2} (Tobin et al. 1999; Turner et al. 2004). These improved high-spectral-resolution line-by-line radiative transfer models were used to develop an infrared radiative model suitable for climate models (Mlawer et al. 1997; Mlawer and Turner 2016). The Rapid Radiative Transfer Model (RRTM) represented a significant improvement in accuracy over radiative transfer models used within GCMs at the time and was a direct outcome of the strong emphasis that ARM placed on clear-sky infrared radiative transfer through the 1990s (Iacono et al. 2000).

As in the infrared, there were challenges in closing the surface radiation budget in the solar shortwave part of the spectrum. Well-calibrated cavity radiometers gave confidence in measurements of the direct solar beam; however, constraining the diffuse solar flux proved to be more challenging (Dutton et al. 2001; Michalsky and Long 2016). And as with efforts to close the longwave radiation budget, there were also significant uncertainties in environmental parameters required to calculate shortwave fluxes. In the clear (or noncloudy) sky, key environmental uncertainties were the surface spectral albedo, the distribution and radiative properties of aerosols, and the absorptivity of gaseous constituents (Kato et al. 1997; Michalsky and Long 2016).

Characterization of the surface spectral albedo is complicated because the downwelling solar flux is sensitive to the albedo over a region spanning on the order of a few kilometers and the albedo is typically heterogeneous spatially (on land) and variable in time. Early calculations tended to gloss over these effects but they do introduce errors in radiation calculations. The realization of these effects, and subsequent aerial measurements to investigate the spatial distribution of the albedo over different surfaces, led to the quantification of these uncertainties and provided the means to reduce their magnitude (Michalsky et al. 2003).

Discrepancies between modeled and measured shortwave fluxes at the surface in the mid-1990s pointed to a large amount of anomalous solar absorption in the column (e.g., Kato et al. 1997; Valero et al. 1997; Zender et al. 1997). This effect was particularly pronounced in cloudy skies but there were also challenges in achieving closure in clear skies. There was a great deal of speculation regarding the source of this apparent absorption: aerosols, gaseous species, or measurement errors. The challenges in achieving shortwave closure in both clear and cloudy skies led to the ARM Enhanced Shortwave Experiment (ARESE) aircraft experiments in 1995 and 2000 (Valero et al. 2003; Schmid et al. 2006; Michalsky and Long 2016), and to improvements in the measurements of solar diffuse radiation (Michalsky et al. 2005). The results from the ARESE-2 experiment demonstrated that, while there is still a small bias between the observed radiative flux and the computed flux, the bias could be explained largely by the uncertainties associated with the measured flux, mismatch between the ground and airborne radiometers, and uncertainties in both the radiative transfer models and the inputs used to drive it and in particular surface albedo (Ackerman et al. 2003; Li 2004).

c. *Aerosol optical properties*

Reducing uncertainties in shortwave measurements required developing a good understanding of the optical properties of aerosols. A great deal of work on aerosols initially in ARM was focused on the direct effect of aerosols on, primarily, shortwave radiation. Measurements of aerosol properties included both in situ observations at the surface and remote sensing. Obtaining the vertical profile of aerosol optical properties was, and remains, a challenging measurement problem.

ARM sites include an Aerosol Observing System (AOS) that provides in situ measurements of aerosol optical and microphysical properties (Sheridan et al. 2001). The suite of AOS instruments provides a continuous view of the aerosol population and optical properties at the surface, but says nothing directly about the profile of aerosol properties. In a daytime convective environment, it is often assumed that aerosols are well mixed through the boundary layer and that the surface-based observations are representative of the aerosol aloft; however, even in that case, one must account for the dependence of aerosol optical properties on relative humidity as many species of aerosols grow hygroscopically as relative humidity increases.

Several tools have been used to extend aerosol measurements away from the surface beginning with passive and active remote sensing. The primary measurement required for evaluating the direct effect of aerosols on shortwave radiation is the total column aerosol optical depth (AOD), and the primary tool to measure AOD at the ARM sites is the Multifilter Rotating Shadow Band Radiometer (MFRSR; Harrison et al. 1994). The autonomous derivation of AOD from continuous MFRSR irradiance measurements requires careful cloud screening. This process is now largely automated but can still be difficult in some environments (Michalsky et al. 2010).

The MFRSR provides the total AOD but it does not provide any information about the vertical distribution of aerosols. The vertical distribution is not necessary to evaluate the impact of aerosol on the surface radiation budget but it is important for determining the profile of absorption, and therefore radiative heating, within the column, and it is critical for determining the effect of aerosols on clouds.

Lidars are sensitive to aerosols and can be used to obtain information about vertical profiles. However, a simple elastic scatter lidar cannot provide optical extinction (scattering + absorption) without making significant assumptions, the main assumption being that the extinction-to-backscatter ratio is constant through the troposphere (e.g., Welton 2000; Schmid et al. 2006). The ARM Raman lidar includes measurements of total backscatter as well as profiles of Raman scattering by nitrogen, which provides the means of deriving aerosol extinction (Turner et al. 2002, 2016a). Using the SGP Raman lidar observations, Ferrare et al. (2001) found that the extinction-to-backscatter ratio cannot be considered constant more than 30% of the time.

From 2000 to 2007, a small aircraft outfit with basic aerosol in situ instruments was flown over the ARM SGP site several times per week. These flights provided an important record of the variability in the aerosol vertical structure up to about 5 km above sea level (Andrews et al. 2004). The size of the aircraft and the goal of flying on a routine basis significantly limited the number of instruments the aircraft could carry as part of the In Situ Aerosol Profile extended campaign. To more fully test the relationship of surface observations to the column, in 2003 the ARM Program conducted an expansive Aerosol IOP (Ferrare et al. 2006; McComiskey and Ferrare 2016). This IOP, together with some smaller IOPs and long time series analysis at the SGP site, concentrated heavily on improving measurements of aerosol absorption, understanding processes that affect the absorbing properties of aerosol, and quantifying the radiative impact of these aerosols (e.g., Andrews et al. 2004, 2011; Sheridan et al. 2005; Arnott et al. 2006). Comparisons of aerosol extinction profiles observed by the Raman lidar, airborne sun photometers, and in situ scattering and absorption measurements demonstrated initially large differences (Schmid et al. 2006), but the upgrade in the Raman lidar (Turner et al. 2016a) resulted in much better agreement with airborne sun-photometer extinction measurements (Schmid et al. 2009).

d. The distribution and radiative impact of clouds

From the beginning of the ARM Program, there were several tools available for obtaining cloud macrophysical (location) and microphysical (particle size, shape, phase, number concentration, etc.) properties. The MPL (Campbell et al. 2002) and laser ceilometer both provided cloud-base information, sometimes for multiple cloud layers if the layers were optically thin, while instruments like the Whole Sky Imager (WSI), which was used by ARM from the mid-1990s until it was retired in 2004 (Sisterson et al. 2016), and the Total Sky Imager (TSI), which was used by ARM from 2000 until present, provided a hemispheric view of cloud cover (Long et al. 2006). Passive longwave or shortwave broadband measurements could also be used to provide information about the optical properties of both liquid and ice clouds in combination with microwave (e.g., Dong et al. 1997) or lidar (Comstock and Sassen 2001) measurements. However, work on clouds began to accelerate in late 1996 when the MMCR was deployed at the ARM SGP site (Moran et al. 1998; Kollias et al. 2016).

The MMCR was a 35-GHz pulse-Doppler radar that provided vertical profiles of reflectivity and Doppler velocity from cloud and precipitation particles. Because of its short (8.6 mm) wavelength, the radar was sensitive to most cloud particles, even in the presence of light to moderate precipitation (which attenuates the MMCR's signal), although it was not always sufficiently sensitive to detect very small (smaller than the order of 10 μm) cloud droplets because of the strong sensitivity (D^6) of the radar backscatter to particle diameter D. The MMCR is very complementary to lidars, which are particularly sensitive to small particles with high number density. Together, the MMCR and MPL provided unprecedented information about vertical cloud structure on a continuous basis (e.g., Mace and Benson 2008). This information was made readily accessible to the broad community through the Active Remote Sensing of Cloud Layers (ARSCL) cloud mask, which quickly became one of the most widely used ARM data products (Clothiaux et al. 2000; Kollias et al. 2016).

The MMCR also gave rise to a variety of cloud property retrievals based on the radar observations (reflectivity and Doppler velocity) alone, or in combination with passive radiometers or in combination with other instruments. Much of the work over the first decade of the MMCR operation focused on single-phase liquid (e.g., Dong and Mace 2003; Turner et al. 2007a) or ice clouds (e.g., Mace et al. 1998; Comstock et al. 2007). Mixed-phase conditions are commonly found in deep convection and in Arctic stratus; however, there was a strong sense in the community that progress had to be made with the single-phase clouds before the more challenging mixed-phase clouds could be tackled in earnest. The availability of a large range of different complementary observations that provided information on cloud macrophysical and microphysical properties resulted in a huge number of different retrieval algorithms being developed; this history is captured by Shupe et al. (2016).

The improvements in cloud retrievals allowed the Program to also investigate how changes in the aerosol concentration impacts cloud properties (i.e., the aerosol indirect effect). ARM scientists were the first to measure the aerosol indirect effect using ground-based sensors (Feingold et al. 2003); they also developed methods to derive information on cloud condensation nuclei from ARM observations (Feingold et al. 1998; Ghan et al. 2006), quantified aerosol hygroscopicity using in situ and remote sensing data (Pahlow et al. 2006), and performed experiments to quantify the aerosol impact on cloud properties in stratiform (e.g., McComiskey et al. 2009) and cumulus (e.g., Berg et al. 2011) clouds.

e. Three-dimensional radiative transfer

From the very beginning of ARM, the Program supported efforts to better characterize and model 3D radiative transport in various media (e.g., Davis and Marshak 2001; O'Hirok and Gautier 1998), to develop improved 3D radiative transfer models and intercompare these models with each other and observational datasets (e.g., Barker and Marshak 2001; Han and Ellingson 2000; Kablick et al. 2011), and to determine how well 1D solar radiative transfer models handle unresolved clouds in cloud resolving model output (Barker et al. 2003). Much of the ARM Program's success in 3D radiative transfer is neatly summarized in the textbook edited by Marshak and Davis (2006).

f. Cloud modeling

One of the central goals of the ARM Program is to improve the representation of clouds and their radiative effects in climate models. This goal could have been handled by simply making ARM data available in anticipation that the modeling community would use the data, but the DOE program managers decided to not leave this to chance. Instead, modeling activities were built into ARM from the beginning. The general strategy has been that of the Global Energy and Water Cycle Experiment (GEWEX) Cloud System Study (Randall et al. 2003)—namely, to use observations, and sometimes numerical weather prediction model reanalyses, in the region around an ARM site on the scale of a GCM grid box to provide the initial dynamical forcing conditions for a single column model (a single column from a GCM; Zhang et al. 2016) or a cloud ensemble or cloud-resolving model (Randall et al. 1996, 2003; Krueger et al. 2016). The model is then run and the output cloud field and other fields are compared with observations—either direct comparison of time series or statistical distributions.

The development of model forcing data using the constrained variational analysis technique is another of the primary accomplishments of the ARM Program (Zhang and Lin 1997; Zhang et al. 2016). This model forcing dataset is dependent on having high temporal resolution profiles of temperature, humidity, and wind that are typically provided by radiosondes, but uses surface and top-of-the-atmosphere measurements to constrain the energy and water sources and sinks (with precipitation being perhaps the most important sink).

Unfortunately, the 2–4 radiosonde launches per day, typical for ARM sites, are often not adequate for properly capturing and representing the advective tendencies of temperature and water vapor to get the large-scale forcing dataset. Therefore, ARM has a long history of conducting IOPs in which the frequency of radiosonde launches is increased to 8 per day at 4 to 6 sites around the SGP domain to support modeling activities (Zhang et al. 2016). The first of these so-called SCM IOPs was held early in 1995, just a few years after the beginning of data collection, and there have been many campaigns since focused on supporting modeling activities. Variational forcing datasets have also been created for a number of major field campaigns away from the SGP site, such as in the cases of the Mixed-Phase Arctic Cloud Experiment (M-PACE; Verlinde et al. 2007) and the Tropical Warm Pool International Cloud Experiment (TWP-ICE; May et al. 2008). Additionally, a continuous forcing dataset has been developed (Xie et al. 2004) that allows long-term SCM or cloud-resolving model simulations to be conducted and evaluated against ARM observations (e.g., Henderson and Pincus 2009) or can evaluate reanalysis data over multiple years (Kennedy et al. 2011).

These forcing datasets are critical inputs for the modeling community, allowing single-column models, cloud-resolving models, and large-eddy simulation models to be run over the ARM site in a manner that allows output from these models to be evaluated with other ARM observations (e.g., cloud fraction profiles). The M-PACE modeling studies (e.g., Klein et al. 2009; Morrison et al. 2009) illustrated the challenges in simulating mixed-phase clouds, the critical need to have accurate ice nuclei (IN) concentration observations, and the importance of having accurate cloud macro- and microphysical properties in the Arctic environment. The uncertainties in the observed IN concentrations and processes associated with ice nucleation have proved to be a central issue for analyses of M-PACE and the Indirect and Semi-Direct Aerosol Campaign (ISDAC; McFarquhar et al. 2011) data, and continue to be a central issue in the study of arctic clouds. TWP-ICE led to a variety of studies related to precipitation (Fridlind et al. 2011), diabatic heating (Xie et al. 2010a), and the effects of model resolution (Boyle and Klein 2010), among others. TWP-ICE also demonstrated that cloud-resolving models overpredict updraft speeds and reflectivities in convection (Varble et al. 2014), and that uncertainty in the measurement of precipitation compromises the ability of SCMs to respond correctly in weakly forced environments (Davies et al. 2013).

Other modeling activities with roots in the ARM Program are the Cloud-Associated Parameterizations

Testbed (CAPT; Phillips et al. 2004; Williamson et al. 2005) and the "super-parameterization" framework (Khairoutdinov et al. 2005; Ovtchinnikov et al. 2006). In the CAPT framework, a climate model is initialized and run like a numerical weather prediction model and then evaluated against observations, typically over an ARM site. The object of the CAPT framework is to evaluate the tendency of the climate model to develop errors over short time scales (a few days). Additionally, the ARM observations were used to help evaluate parameterizations in the European Center for Medium-Range Weather Forecasts (ECMWF) model (Ahlgrimm et al. 2016); both this activity and the CAPT activity helped make progress on the second major ARM question: "If the large-scale variables are specified properly in the model, will it predict the proper cloud properties?" In the superparameterization (which is also known as the multiscale modeling) framework, a two-dimensional cloud-resolving model is embedded in each grid cell of a GCM. This technique is computationally expensive but much less expensive than a full global cloud-resolving model while providing information about subgrid-scale cloud structure and a better representation of a large number of moist processes.

g. Site-specific science

The SGP site (Sisterson et al. 2016) was the first CART site. Its designation as a test bed was appropriate both because of the role the site served to evaluate models and because the SGP also served as a testing ground for other sites. Most of the activities described so far in this chapter were carried out—or first carried out—at the SGP site. But ARM expanded to other sites to sample other regions of the world, some of which were permanent and others of more limited duration.

After the SGP, the next site to be installed was the TWP site on Manus Island, Papua New Guinea, in 1996 (Mather et al. 1998; Long et al. 2016). This was followed by the NSA site in Barrow, Alaska, in 1998 (Stamnes et al. 1999; Verlinde et al. 2016). The Manus and Barrow sites did not have all the instruments found at the SGP, but they had a critical core set including the MMCR, MPL, AERI, MWR, and radiosondes. With the establishment of the Manus and Barrow sites, ARM was collecting measurements in three major climate regimes. These sites greatly expanded the range of science topics that could be addressed with ARM data.

At Manus, and later at the ARM TWP sites deployed on Nauru Island and at Darwin, Australia, the science focus was on tropical maritime convection and associated cirrus outflow (Long et al. 2016). Although the SGP is convectively active during part of the year, the tropical atmosphere and the associated convection are inherently different than found at the SGP. In the tropics, the thermal structure varies very little in comparison with the midlatitudes and convection tends to be much more stochastic in nature, although it does organize on large scales by tropical waves such as the Madden–Julian oscillation and on longer time scales such as El Niño–Southern Oscillation (ENSO). Tropical phenomena such as ENSO have a strong influence on the global circulation so understanding the physical processes that affect these phenomena is critical for improving simulations of global climate.

At the NSA site, stratiform clouds are prevalent, and these clouds very often are mixed-phase clouds (Verlinde et al. 2016). The dynamics and processes at work in these mixed-phase Arctic clouds are not well understood and only recently have there been good measurements of these cloud systems. These new observations suggest that there are many processes operating in a fine balance that enable mixed-phase stratiform clouds to exist for hours to days at a time (Morrison et al. 2012). The rate at which the ice in the Arctic Ocean has been decreasing over the last decade has exceeded model predictions (Stroeve et al. 2012), and thus it is important to understand the role that clouds, aerosols, and the atmospheric thermal structure play in this change. Additionally, the small amount of precipitable water vapor (PWV) in the winter at the NSA site results in the thermal infrared spectrum being much more transparent than at other ARM sites. ARM observations in the Arctic have been used to evaluate and improve spectral radiative transfer models in both the traditional 8–12-μm infrared window and the less transparent 18–25-μm window (Tobin et al. 1999; Delamere et al. 2010), which was one of the original goals of the research at the NSA site (Stamnes et al. 1999).

The observations made at the ARM sites are very complementary to satellite observations; the ARM observations are much more detailed with higher vertical and temporal resolution and capture the diurnal cycle well, whereas the satellite observations provide a much larger spatial view. Hence, satellite observations have always been very important for ARM, and ARM's observations have been immensely valuable for the various satellite programs (Marchand 2016).

3. Shifting strategy

With the establishment of the full set of fixed-location sites and the development of an integrated measurement and modeling strategy, the ARM Program evolved to become a fully mature scientific program. The TWP Darwin site was deployed in 2002, and the ARM measurement infrastructure had reached a plateau. All of the fixed sites were up and running, there were no plans for additional permanent observation facilities and, with a few exceptions, the core set of measurement capabilities had been implemented at all the measurement sites. It took a decade after the first measurements were made at the SGP site to reach this point, which was coincidentally the originally anticipated length of the ARM Program. The extended deployment schedule was primarily driven by budget considerations, but there were many lessons to be learned and difficult problems to solve in getting the Program to this point that also took time.

Throughout the first decade of the ARM Program, the Science Team and the infrastructure were tightly coupled. This coupling manifested itself in the planning and implementation of field campaigns, optimization of instrument configurations, and the development of advanced data products. However, this tight coupling also tended to limit the breadth of users of ARM data. ARM data were publicly available from the ARM Data Archive very early in the Program (McCord and Voyles 2016); however, there was a sense at the time by some scientists outside of ARM that ARM was a closed program. This perception was likely due in part to the close coupling between the Science Team and the infrastructure. This situation changed when ARM was designated a User Facility (Ackerman et al. 2016). This designation reinforced the notion that ARM data are useful for a much wider community than just the ARM Science Team, and led to a wider range of science being conducted with ARM data.

a. ARM science program and its relationship with the new ARM facility

In many respects, there were no significant changes in the organization of the ARM science program during the first years after ARM was designated as a User Facility (i.e., from 2004 to 2008). There was a shift away from clear-sky radiative studies toward evaluating radiative and microphysical properties of clouds and aerosols, and improving the representation of clouds and aerosols within climate models. Scientific understanding continued to evolve, and during this period the Science Team began to consider the need to expand the core measurement capabilities to obtain more detailed information on cloud and aerosol properties and to obtain some spatial information of these properties, with the focus of improving the representation of clouds in GCMs.

Research investigating cloud properties and processes accelerated greatly during the 2004–10 period. The cloud radar and its primary data product ARSCL were mature (Kollias et al. 2016) and a large number of PIs within the ARM Program were working on cloud property retrieval algorithms (Shupe et al. 2016). Major accomplishments included a better understanding of the vertical distribution of clouds and their radiative effects (Mace and Benson 2008; Mather and McFarlane 2009), and microphysical properties of low liquid water stratus clouds (e.g., Dong and Mace 2003; McComiskey et al. 2009) and cirrus clouds (e.g., McFarquhar et al. 2007; Deng and Mace 2008; Protat et al. 2010). Mixed-phase clouds present a particular set of measurement challenges, but a great deal of progress was also made using radar Doppler spectra, depolarized lidar backscatter profiles, and spectral infrared radiances (e.g., Turner 2005; Shupe et al. 2008; de Boer et al. 2009; Luke et al. 2010). The ARM Program began saving Doppler spectra from the cloud radars around 2000; this was a seminal moment for the Program because it provided a tremendous increase in information content and enabled vastly more complex cloud microphysical retrievals (Shupe et al. 2016).

Scientists within the ARM Program also realized that the main workhorse for quantifying the liquid water path (LWP) in the overhead clouds, the MWR, did not have the required sensitivity to accurately measure LWP when the LWP was small. This uncertainty was extremely important because (a) the median LWP for clouds is less than $100\,\mathrm{g\,m^{-2}}$ at most global locations (Turner et al. 2007b); (b) the uncertainty in the MWR-retrieved LWP was approximately $25\,\mathrm{g\,m^{-2}}$, which is fractionally quite large for a large fraction of liquid-bearing clouds (Marchand et al. 2003); (c) the longwave and shortwave radiative impact of biases in the observed LWP was largest when the LWP was small (i.e., less than $100\,\mathrm{g\,m^{-2}}$); and (d) aerosol effects on clouds are typically the largest when the LWP is small (Turner et al. 2007a). To illustrate the importance of getting an accurate measure of LWP, the error in the computed downwelling shortwave flux was between 5 and $15\,\mathrm{W\,m^{-2}}$ (depending on the effective radius of the liquid droplets) for a $1\,\mathrm{g\,m^{-2}}$ error in the LWP when the cloud LWP was less than $50\,\mathrm{g\,m^{-2}}$. This led to the formation of a focus group called Clouds of Low Water Depth (CLOWD; Vogelmann et al. 2012) that evaluated different techniques and organize and conduct a couple of field experiments, and ultimately led to the recommendation that the program needed to acquire microwave radiometers with an additional channel near 90 GHz, which would provide 3 times more sensitivity to LWP than the original two-channel MWRs that were being used by the Program at all of its sites.

Many different cloud property retrieval algorithms were developed by ARM PIs during the early 2000s, and it was clear that each algorithm had different strengths and weaknesses, in part because they used different input datasets and made different assumptions. One of the activities performed by the CLOWD group was a large, organized intercomparison of cloud properties in single-layer warm stratiform clouds from different algorithms over some specified cases. There was also a large intercomparison effort that focused on retrieved cloud properties from single-layer cirrus clouds, and another that focused on single-layer mixed-phase clouds. These intercomparisons demonstrated huge differences in the retrieved properties among the different algorithms (Turner et al. 2007a; Comstock et al. 2007; Shupe et al. 2008), even when exactly the same input was used in the different algorithms. These findings forced the Science Team to reevaluate the assumptions made in the retrievals to and investigate new ways to determine the uncertainties in these microphysical retrievals and communicate them to the data users (Zhao et al. 2012).

Another research area that gained momentum during this time period was the routine measurement of small-scale vertical motions above the ARM sites, since cloud and precipitation properties are intricately coupled with the dynamical motions in the cloud. ARM cloud radars are configured to record the first three moments of the Doppler velocity spectra (Kollias et al. 2007). These data had been used for some time to derive information about vertical motion in clouds, often for the purpose of deriving cloud properties (Deng and Mace 2008). However, in precipitating conditions the cloud radar signal is attenuated, making the accurate determination of vertical motions more difficult. Furthermore, cloud and precipitation particles are falling and have a terminal velocity, and thus algorithms had to be developed that accounted for the fall speed of the hydrometeors in order to properly determine the atmospheric motion in the cloud. Faced with these challenges, a new focus group was formed within the ARM Science Team in 2008 to evaluate methods for measuring vertical motion in clouds over a range of cloud types, from shallow warm boundary layer clouds to deep convective precipitating cloud systems.

ARM scientists worked hard to improve retrievals of aerosol optical properties in broken cloud fields and to compensate for bias errors that often emerge in these retrievals (Kassianov and Ovtchinnikov 2008). There were also studies of dust as part of the Niger AMF deployment [discussed below and in Miller et al. (2016)], where the AMF site experienced heavy dust loading at times (Slingo et al. 2006; Turner 2008; McFarlane et al. 2009). In addition to optical properties, effects of aerosols on cloud properties were also a topic of considerable study at both the fixed ARM sites (e.g., Lubin and Vogelmann 2006; Feingold et al. 2006; Guo et al. 2007; Klein et al. 2009; McFarquhar et al. 2011) and AMF deployments (e.g., McComiskey et al. 2009; J. Liu et al. 2011).

During this period, many ARM scientists began to demonstrate the value of the long-term operational nature of ARM's observations to develop robust multiyear analyses. For example, Mace and Benson (2008) developed a multiyear distribution of cloud properties and radiative heating rate profiles at the SGP, Michalsky et al. (2010) constructed a climatology of aerosol optical thickness and Angstrom exponent, Gero and Turner (2011) investigated the trends in downwelling infrared radiance at the SGP site over a 14-yr period, and Andrews et al. (2004, 2011) derived climatological profiles of aerosol scattering and absorption from in situ observations made by a small aircraft over a multiyear period. These climatologies are useful for both evaluating the variability of GCM output to ensure that it captures the entire dynamic range of the observations, as well as putting the observations from any particular field experiment into context relative to the long-term dataset (i.e., are the data from the IOP representative of what is normally observed at that site?).

The airborne measurements of aerosol optical properties were ARM's first experiment with routine airborne measurements (Andrews et al. 2004, 2011). During this period, there was an increasing sentiment in the ARM community that while intensive aircraft campaigns could be quite valuable, it was important to find ways to obtain larger sample sizes from airborne platforms. The ARM Program has now conducted multiple long-term aircraft campaigns, including the Airborne Carbon Measurement Experiment (ACME; Biraud et al. 2013), the Routine ARM Aerial Facility (AAF) CLOWD Optical Radiative Observations (RACORO; Vogelmann et al. 2012) campaign, and the Small Particles in Cirrus (SPARTICUS; Deng et al. 2013) campaign.

b. Field campaigns and the transition to the mobile facility

While some investigators were beginning to make use of the long-term record available from ARM sites, there also continued to be a great deal of effort focused on observations during intensive field campaigns. Many large field campaigns occurred at the ARM fixed sites between 2004 and 2010. The AMF, which was developed both to help sample climatic regions away from the permanent ARM sites and allow ARM to participate in large field experiments organized by other agencies and nations (Miller et al. 2016), came of age during this period and has now been deployed many times both nationally and internationally.

Major field campaigns that were held at the fixed ARM sites during this period included the 2004 M-PACE (Verlinde et al. 2007) over NSA, the 2006 TWP-ICE (May et al. 2008) above the Darwin site, the 2007 Cloud and Land Surface Interaction Campaign (CLASIC) at SGP (Miller et al. 2007), and the 2008 ISDAC (McFarquhar et al. 2011) at NSA. These field experiments were perhaps the most comprehensive IOPs that had been conducted at the ARM sites yet, with many additional ground-based instruments deployed to augment the standard ARM instruments, aircraft in situ observations, and a large number of ARM scientists and national and international participants. All of these IOPs had significant modeling efforts associated with them, with perhaps a dozen or more groups using models of all scales (from large-eddy simulation to cloud-resolving models to single-column models) focused on cases during each IOP. The infrastructure worked hard to produce forcing datasets to drive these models (Zhang et al. 2016) and to produce the observational datasets needed to evaluate the accuracy of the model simulations. Each of these IOPs had multiple breakout sessions at the spring Science Team meeting and fall Working Group meetings to coordinate research efforts and discuss results.

Although work on clear-sky radiation was diminishing during this period, there was an effort to characterize radiative absorption at far-infrared (far-IR) wavelengths that ultimately required an "off-site" field campaign. At wavelengths longer than about 20 μm, the lower atmosphere is highly absorbing under most conditions so it is difficult to measure detailed absorption properties of water vapor from most surface-based locations. However, the accurate treatment of water vapor absorption in climate models is essential to get the outgoing longwave radiation budget and atmospheric radiative heating correct. In regions with very low amounts of water vapor and particularly at high altitudes, the atmospheric opacity due to water vapor absorption starts to diminish. A pair of experiments called the Radiative Heating in Underexplored Bands Campaigns (RHUBC) were carried out first on the North Slope of Alaska in 2007 and then in the Atacama desert of Chile at an altitude of 5320 m in 2009 to study this far-IR region (Turner and Mlawer 2010). RHUBC-II in Chile was the first atmospheric science experiment to measure spectrally resolved radiances at the surface across the entire thermal infrared portion of the electromagnetic spectrum (3–1000 μm; Turner et al. 2012a). The marked improvements in the modeling of the absorption of water vapor in the far IR that resulted from the RHUBC experiments was shown to have an impact on the simulation of atmospheric dynamic properties in a global climate model simulation (Turner et al. 2012b).

With the deployment of the first AMF in 2005, the field campaign landscape began to change. In particular, there was a shift in emphasis from field campaigns at fixed-location sites to AMF deployments, since the AMF enabled the Program to collect data in other climatically important regions of the world. Given the time scale of several years to plan and execute a major field campaign, this transition occurred over a period of several years. This does not mean that field campaigns at the primary SGP, TWP, and NSA sites were no longer considered important or were not supported; however, there was a shift in emphasis and resources to the new AMFs as well as the Aerial Facility (Schmid et al. 2006).

This change had multiple and varied effects on the science community. For the ARM Science Team, the change meant less direct involvement in the planning of field campaigns. This change was inevitable due to ARM's new status as a user facility and the associated requirement to subject facility proposals to a peer-review process via the ARM Science Board (Ackerman et al. 2016). The other side effect of this change is that field campaign proposals were suddenly equally accessible to anyone regardless of their affiliation. Through this open process, the first AMF was deployed for five campaigns in a row outside of the United States to address key science issues (Miller et al. 2016). The PI for two of those five campaigns had no direct ties to DOE; those campaigns were the RADAGAST [Radiative Atmospheric Divergence using ARM Mobile Facility, Geostationary Earth Radiation Budget (GERB) data, and African Monsoon Multidisciplinary Analysis (AMMA) stations] campaign in Niamey, Niger, in 2006 (Miller and Slingo 2007) and the Convective- and Orographically-induced-Precipitation Study (COPS) deployment to the Black Forest in Germany in 2007 (Wulfmeyer et al. 2011). All of these AMF deployments included a significant contribution from ARM's international colleagues. Of course, scientists connected with the ARM Program continued to be very involved in these deployments, often as PIs, but now the doors were open to the international climate science community. A result of this was to significantly broaden the interest in ARM both in the United States and internationally.

Another fundamental change associated with the shift toward operating the AMFs was the duration of field campaigns. Consistent with ARM's goal of collecting climatologically significant datasets, the AMFs are typically deployed for about a year at a time. Traditional intensive field campaigns provide enhanced measurements focused on a specific problem for a short time but AMF deployments offer a standard set of ARM measurements for (typically) an annual cycle. Analyses of AMF data have shown that these extended observation sets can reveal characteristic of a locale that are not apparent from the short time series typical of intensive field campaigns (e.g., Slingo et al. 2008; Rémillard et al. 2012).

c. *Linking observations with models*

As noted earlier, improving climate models is a central goal of the ARM Program. The process of improving parameterizations in climate models is a complicated one that often involves a combination of observations, analysis, high-resolution modeling, and global-scale modeling. Through the application of special datasets and through ongoing efforts to link model developers with observation data, significant improvements were made in model parameterizations during the 2000s, which is summarized nicely by Randall et al. (2016). As a single example, in the most recent version of the NCAR–DOE Community Atmosphere Model version 5 (CAM5; Neale et al. 2010), several improvements have made important use of ARM data and associated research, including using an improved radiation scheme (Mlawer et al. 2016) and better representation of cloud and aerosol processes (e.g., X. Liu et al. 2011b).

In climate models, radiative transfer involves the interaction of solar radiation and terrestrial infrared radiation with the earth's atmosphere and land surface. Radiative transfer models tend to take a lot of computer time. A version of the RRTM radiative transfer model, grounded in many years of ARM science, was developed to run efficiently in GCMs (RRTMG; Iacono et al. 2008) and included in the CAM5 along with several other global models. RRTMG has also been integrated with a novel technique for capturing the interaction of radiation with clouds that improved both efficiency and accuracy (Pincus et al. 2003; Mlawer et al. 2016).

Microphysical parameterization changes in CAM5 included the addition of a three-mode aerosol scheme and a two-mode microphysics scheme. The three-mode aerosol scheme represents the diversity of aerosols in three size ranges. This parameterization has proved to be efficient and accurate when compared to more complex aerosol schemes that would not be computationally feasible to run in a climate model (Ghan et al. 2012). A two-moment cloud scheme describes cloud droplets in terms of water mass per unit volume as well as the number of cloud droplets in that volume. This represents a significant advance over single-moment treatments of clouds that only determine the water mass (Gettelman et al. 2008). Additionally, Bretherton and Park (2009) developed a new moist turbulence scheme tested with a variety of field observation data including data from M-PACE. These parameterization developments represent a significant advance in the realism with which these processes are represented in climate models, which is expected to improve the model's ability to accurately simulate climate change.

To accelerate the use of ARM data for climate model development, the ARM Best Estimate (ARMBE) product was developed during the mid-2000s (Xie et al. 2010b). The ARMBE product provides simplified access to ARM data by packaging critical geophysical quantities derived from ARM observations into a convenient file structure with a consistent temporal resolution. The ARMBE product made ARM data drastically more assessable to the modeling community, and thus facilitated the evaluation of model simulations with ARM data.

The development of the ARMBE product was in response to an increasing emphasis on connecting the climate modeling community to ARM data. Using ARM data to improve climate models had always been a central goal of ARM but during this period (c. 2004–08) this goal took on a greater sense of urgency. The ARMBE product has proved to be very popular within the climate modeling community where it is used both by individual investigators and as a model evaluation dataset (e.g., Ahlgrimm and Forbes 2012; Song et al. 2013).

4. The Atmospheric System Research Program

In 2010, as the first Recovery Act instruments were being deployed in the field, the ARM Science Team also underwent a significant change. The DOE Climate and Environmental Science Division (CESD), which is the division of DOE within which ARM resides, was working to better integrate its climate research programs. At that time, CESD included two atmospheric observation programs with related objectives: the ARM Program and the Atmospheric Science Program (ASP). The ASP focused on the life cycle of aerosol particles including aerosol nucleation, mixing states, aggregation, and growth. ARM also included an aerosol component but it was largely restricted to radiative properties of aerosols and the role of aerosols in cloud formation. However, with this dual program structure in CESD, there was some risk of activities falling between the two programs or of having duplicated effort. So, in 2010, DOE merged the ARM Science Program and the ASP to form the Atmospheric System Research (ASR) program. In many ways, the structure of ASR was modeled after the former ARM Science Program but encompassed the full scope of ARM and ASP. DOE

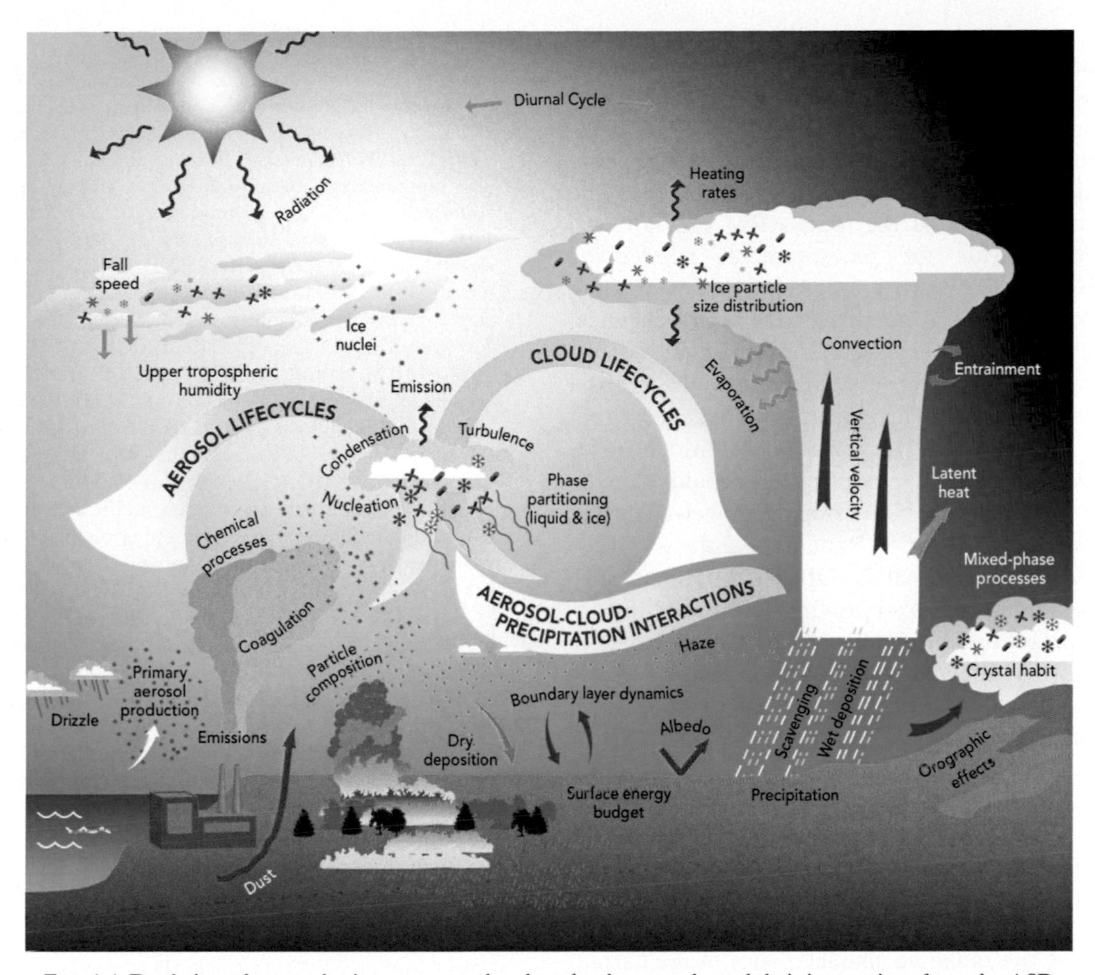

FIG. 4-4. Depiction of atmospheric processes related to clouds, aerosols, and their interactions from the ASR Science Plan (U.S. Department of Energy 2010).

management solicited the input from the ARM and ASP chief scientists, the SISC, and other scientists both within and external to the two programs in the development of the new ASR Program.

The scope of the new ASR Program was articulated in a Science Plan published in 2010 (U.S. Department of Energy 2010). In addition to an increased emphasis on aerosols, there was also a distinct shift toward better understanding of the range of atmospheric processes that need to be captured by numerical models (many of these processes are illustrated in Fig. 4-4, which is the cover of the ASR Science Plan). Previously, the emphasis had been on characterizing the atmospheric state, cloud properties, and aerosol properties, and in particular their radiative characteristics. With the shift to processes, questions such as how the atmospheric state and the associated cloud, aerosol, and precipitation fields evolved came to the fore. To help focus on process-related issues, the ASR Science Team was organized into three working groups: cloud life cycle, aerosol life cycle, and cloud–aerosol–precipitation interactions (Fig. 4-1). These working groups illustrate the centrality of clouds and aerosols to the ASR Program and to ARM. Unlike the previous organization of the ARM Science Team, there were no longer working groups specifically focused on modeling or radiation. Instead, these activities were integrated into each of the three working groups. Each working group has two chairpersons, one representing the observation community and one representing the modeling community. These working group chairs serve as the ASR representatives on the SISC.

These changes were not without their challenges, however. While the ARM science community had several years to adjust to the user facility model, the ASP Science Team, which had been working with a field campaign measurement model up to that time, had to make more rapid changes. Furthermore, the ASP measurement approach tended to rely more on research-grade instruments deployed for short periods of time to obtain the needed detailed information on composition and structure of aerosols; these instruments were not always well suited to the ARM model of continuous operation.

The Recovery Act (Mather and Voyles 2013) provided the necessary means to greatly enhance the aerosol observations at the fixed ARM sites and the AMFs to provide the measurements needed by the aerosol life cycle working group. However, there was also a recognized need that additional chemical measurements may be required for some experiments. Thus, through the Recovery Act, a Mobile Aerosol Observing System was developed to provide a core set of aerosol chemical measurements. Additional measurements for particular applications could then be requested for intensive field campaigns, such as for the Carbonaceous Aerosols Radiative Effects Study (CARES; Zaveri et al. 2012). Other instruments, such as Doppler lidars, are also proving very useful in connecting surface measurements of aerosol properties to those aloft by allowing the evolution of the mixing within the boundary layer over the diurnal cycle to be better understood.

5. ARM–ASR future directions

The scope of the ARM facility and its science partner, the ASR, has expanded considerably since the early days of the program when the focus was on understanding the radiative characteristics of the current state of the atmosphere. However, even at the outset, the stated goals of the ARM Program included a need to improve the understanding of the distribution of clouds and their impact on the earth's radiation balance (U.S. Department of Energy 1990). Understanding this distribution requires a study of a wide range of atmospheric processes, and this has become the focus for ASR and is being supported by measurements from ARM.

The emphasis on characterizing and improving our ability to observe and model the life cycles of clouds and aerosol is beginning to gain traction within the ASR science community. A great deal of work is still needed in important areas such as the properties and life cycle of mixed-phase clouds; the organization of tropical convection; the role of dynamics (especially small scale) in cloud, aerosol, and precipitation processes; and the evolution of aerosols from precursor gases and nucleation through their growth phase and ultimately their impact on clouds. Making progress in these areas is critical to improving the representation of clouds and aerosol in climate models and so will likely continue to garner a great deal of attention over the coming years.

REFERENCES

Ackerman, T. P., D. M. Flynn, and R. T. Marchand, 2003: Quantifying the magnitude of anomalous solar absorption. *J. Geophys. Res.*, **108**, 4273, doi:10.1029/2002JD002674.

——, T. S. Cress, W. Ferrell, J. H. Mather, and D. D. Turner, 2016: The programmatic maturation of the ARM Program. *The Atmospheric Radiation Measurement (ARM) Program: The First 20 Years, Meteor. Monogr.*, No. 57, Amer. Meteor. Soc., doi:10.1175/AMSMONOGRAPHS-D-15-0054.1.

Ahlgrimm, M., and R. Forbes, 2012: The impact of low clouds on surface shortwave radiation in the ECMWF model. *Mon. Wea. Rev.*, **140**, 3783–3794, doi:10.1175/MWR-D-11-00316.1.

——, ——, J.-J. Morcrette, and R. Neggers, 2016: ARM's impact on numerical weather prediction at ECMWF. *The Atmospheric Radiation Measurement (ARM) Program: The First 20 Years, Meteor. Monogr.*, No. 57, Amer. Meteor. Soc., doi:10.1175/AMSMONOGRAPHS-D-15-0045.1.

Andrews, E., P. J. Sheridan, J. A. Ogren, and R. Ferrare, 2004: In situ aerosol profiles over the Southern Great Plains cloud and radiation test bed site: 1. Aerosol optical properties. *J. Geophys. Res.*, **109**, D06208, doi:10.1029/2003JD004025.

——, ——, and ——, 2011: Seasonal differences in the vertical profiles of aerosol optical properties over rural Oklahoma. *Atmos. Chem. Phys.*, **11**, 10 661–10 676, doi:10.5194/acp-11-10661-2011.

Arnott, W. P., and Coauthors, 2006: Photoacoustic insight for aerosol light absorption aloft from meteorological aircraft and comparison with particle soot absorption photometer measurements: DOE Southern Great Plains climate research facility and the coastal stratocumulus imposed perturbation experiments. *J. Geophys. Res.*, **111**, D05S02, doi:10.1029/2005JD005964.

Barker, H. W., and A. Marshak, 2001: Inferring optical depth of broken clouds above green vegetation using solar radiometric measurements. *J. Atmos. Sci.*, **58**, 2989–3006, doi:10.1175/1520-0469(2001)058<2989:IODOBC>2.0.CO;2.

——, and Coauthors, 2003: Assessing 1D atmosphere solar radiative transfer models: Interpretation and handling of unresolved clouds. *J. Climate*, **16**, 2676–2699, doi:10.1175/1520-0442(2003)016<2676:ADASRT>2.0.CO;2.

Berg, L. K., C. M. Berkowitz, J. C. Barnard, G. Senum, and S. R. Springston, 2011: Observations of the first aerosol indirect effect in shallow cumuli. *Geophys. Res. Lett.*, **38**, L03809, doi:10.1029/2010GL046047.

Biraud, S. C., M. S. Torn, J. R. Smith, C. Sweeney, W. J. Riley, and P. P. Tans, 2013: A multi-year record of airborne CO_2 observations in the US Southern Great Plains. *Atmos. Meas. Tech.*, **6**, 751–763, doi:10.5194/amt-6-751-2013.

Boyle, J. S., and S. A. Klein, 2010: Impact of horizontal resolution on climate model forecasts of tropical precipitation and diabatic heating for the TWP-ICE period. *J. Geophys. Res.*, **115**, D23113, doi:10.1029/2010JD014262.

Bretherton, C. S., and S. Park, 2009: A new moist turbulence parameterization in the Community Atmosphere Model. *J. Climate*, **22**, 3422–3448, doi:10.1175/2008JCLI2556.1.

Campbell, J. R., D. L. Hlavka, E. J. Welton, C. J. Flynn, D. D. Turner, J. D. Spinhirne, V. S. Scott, and I. H. Hwang, 2002: Full-time, eye-safe cloud and aerosol lidar observations at Atmospheric Radiation Measurement program sites: Instruments and data processing. *J. Atmos. Oceanic Technol.*, **19**, 431–442, doi:10.1175/1520-0426(2002)019<0431:FTESCA>2.0.CO;2.

Clothiaux, E. E., T. P. Ackerman, G. G. Mace, K. P. Moran, R. T. Marchand, M. Miller, and B. E. Martner, 2000: Objective determination of cloud heights and radar reflectivities using a combination of active remote sensors at the ARM CART sites. *J. Appl. Meteor.*, **39**, 645–665, doi:10.1175/1520-0450(2000)039<0645:ODOCHA>2.0.CO;2.

Clough, S. A., M. J. Iacono, and J.-L. Moncet, 1992: Line-by-line calculation of atmospheric fluxes and cooling rates: Application to water vapor. *J. Geophys. Res.*, **97**, 15 761–15 785, doi:10.1029/92JD01419.

Comstock, J. M., and K. Sassen, 2001: Retrieval of cirrus cloud radiative and backscattering properties using combined lidar and infrared radiometer (LIRAD) measurements. *J. Atmos. Oceanic Technol.*, **18**, 1658–1673, doi:10.1175/1520-0426(2001)018<1658:ROCCRA>2.0.CO;2.

——, and Coauthors, 2007: An intercomparison of microphysical retrievals for upper tropospheric ice clouds. *Bull. Amer. Meteor. Soc.*, **88**, 191–204, doi:10.1175/BAMS-88-2-191.

Davies, L., and Coauthors, 2013: A single-column model ensemble approach applied to the TWP-ICE experiment. *J. Geophys. Res. Atmos.*, **118**, 6544–6563, doi:10.1002/jgrd.50450.

Davis, A. B., and A. Marshak, 2001: Multiple scattering in clouds: Insights from three-dimensional diffusion/P1 theory. *Nucl. Sci. Eng.*, **137**, 251–280.

de Boer, G., E. W. Eloranta, and M. D. Shupe, 2009: Arctic mixed-phase stratiform cloud properties from multiple years of surface-based measurements at two high-latitude locations. *J. Atmos. Sci.*, **66**, 2874–2887, doi:10.1175/2009JAS3029.1.

Delamere, J. S., S. A. Clough, V. Payne, E. J. Mlawer, D. D. Turner, and R. Gamache, 2010: A far-infrared radiative closure study in the Arctic: Application to water vapor. *J. Geophys. Res.*, **115**, D17106, doi:10.1029/2009JD012968.

Deng, M., and G. G. Mace, 2008: Cirrus cloud microphysical properties and air motion statistics using radar Doppler moments: Water content, particle size, and sedimentation relationships. *Geophys. Res. Lett.*, **35**, L17808, doi:10.1029/2008GL035054.

——, ——, Z. Wang, and R. P. Lawson, 2013: Evaluation of several A-Train ice cloud retrieval products with in situ measurements collected during the SPARTICUS campaign. *J. Appl. Meteor. Climatol.*, **52**, 1014–1030, doi:10.1175/JAMC-D-12-054.1.

Dong, X., and G. G. Mace, 2003: Profiles of low-level stratus cloud microphysics deduced from ground-based measurements. *J. Atmos. Oceanic Technol.*, **20**, 42–53, doi:10.1175/1520-0426(2003)020<0042:POLLSC>2.0.CO;2.

——, T. P. Ackerman, E. E. Clothiaux, P. Pilewskie, and Y. Han, 1997: Microphysical and radiative properties of boundary stratiform clouds deduced from ground-based measurements. *J. Geophys. Res.*, **102**, 23 829–23 843, doi:10.1029/97JD02119.

Dutton, E. G., J. J. Michalsky, T. Stoffel, B. W. Forgan, J. Hickey, D. W. Nelson, T. L. Alberta, and I. Reda, 2001: Measurement of broadband diffuse solar irradiance using current instrumentation with a correction for thermal offset errors. *J. Atmos. Oceanic Technol.*, **18**, 297–314, doi:10.1175/1520-0426(2001)018<0297:MOBDSI>2.0.CO;2.

Ellingson, R. G., R. D. Cess, and G. L. Potter, 2016: The Atmospheric Radiation Measurement Program: Prelude. *The Atmospheric Radiation Measurement (ARM) Program: The First 20 Years, Meteor. Monogr.*, No. 57, Amer. Meteor. Soc., doi:10.1175/AMSMONOGRAPHS-D-15-0029.1.

Feingold, G., S. Yang, R. M. Hardesty, and W. R. Cotton, 1998: Feasibility of retrieving cloud condensation nucleus properties from Doppler cloud radar, microwave radiometer, and lidar. *J. Atmos. Oceanic Technol.*, **15**, 1188–1195, doi:10.1175/1520-0426(1998)015<1188:FORCCN>2.0.CO;2.

——, W. L. Eberhard, D. E. Veron, and M. Previdi, 2003: First measurements of the Twomey effect using ground-based remote sensors. *Geophys. Res. Lett.*, **30**, 1287, doi:10.1029/2002GL016633.

——, R. Furrer, P. Pilewskie, L. Remer, Q. Min, and H. Jonsson, 2006: Aerosol indirect effect studies at Southern Great Plains during the May 2003 Intensive Operations Period. *J. Geophys. Res.*, **111**, D05S14, doi:10.1029/2004JD005648.

Feltz, W. F., W. L. Smith, H. B. Howell, R. O. Knuteson, H. Woolf, and H. E. Revercomb, 2003: Near-continuous profiling of temperature, moisture, and atmospheric stability using the Atmospheric Emitted Radiance Interferometer (AERI). *J. Appl. Meteor.*, **42**, 584–597, doi:10.1175/1520-0450(2003)042<0584:NPOTMA>2.0.CO;2.

Ferrare, R. A., D. D. Turner, L. A. Heilman Brasseur, W. F. Feltz, O. Dubovik, and T. P. Tooman, 2001: Raman lidar observations of the aerosol extinction-to-backscatter ratio over the Southern Great Plains. *J. Geophys. Res.*, **106**, 20 333–20 348, doi:10.1029/2000JD000144.

——, G. Feingold, S. Ghan, J. Ogren, B. Schmid, S. E. Schwartz, and P. Sheridan, 2006: Preface to special section: Atmospheric Radiation Measurement Program May 2003 Intensive Operations Period examining aerosol properties and radiative influences. *J. Geophys. Res.*, **111**, D05S01, doi:10.1029/2005JD006908.

Fridlind, A. M., and Coauthors, 2011: A comparison of TWP-ICE observational data with cloud-resolving model results. *J. Geophys. Res.*, **117**, D05204, doi:10.1029/2011JD016595.

Gero, P. J., and D. D. Turner, 2011: Long-term trends in downwelling spectral infrared radiance over the U.S. Southern Great Plains. *J. Climate*, **24**, 4831–4843, doi:10.1175/2011JCLI4210.1.

Gettelman, A., H. Morrison, and S. Ghan, 2008: A new two-moment bulk stratiform cloud microphysics scheme in the Community Atmosphere Model, version 3 (CAM3). Part II: Single-column and global results. *J. Climate*, **21**, 3660–3679, doi:10.1175/2008JCLI2116.1.

Ghan, S. J., and Coauthors, 2006: Use of in situ cloud condensation nuclei, extinction, and aerosol size distribution measurements to test a method for retrieving cloud condensation nuclei profiles from surface measurements. *J. Geophys. Res.*, **111**, D05S10, doi:10.1029/2004JD005752.

——, X. Liu, C. Easter, R. Zaveri, P. J. Rasch, J.-H. Yoon, and B. Eaton, 2012: Toward a minimal representation of aerosols in climate models: Comparative decomposition of aerosol direct, semidirect, and indirect radiative forcing. *J. Climate*, **25**, 6461–6476, doi:10.1175/JCLI-D-11-00650.1.

Guo, H., J. E. Penner, M. Herzog, and S. Xie, 2007: Investigation of the first and second aerosol indirect effects using data from the May 2003 intensive operational period at the Southern Great Plains. *J. Geophys. Res.*, **112**, D15206, doi:10.1029/2006JD007173.

Han, D., and R. G. Ellingson, 2000: An experimental technique for testing the validity of cumulus cloud parameterizations for longwave radiation calculations. *J. Appl. Meteor.*, **39**, 1147–1159, doi:10.1175/1520-0450(2000)039<1147:AETFTT>2.0.CO;2.

Han, Y., and E. R. Westwater, 2000: Analysis and improvement of tipping calibration for ground-based microwave radiometers. *IEEE Trans. Geosci. Remote Sens.*, **38**, 1260–1276, 10.1109/36.843018.

Harrison, L., J. Michalsky, and J. Berndt, 1994: Automated multifilter rotating shadow-band radiometer: An instrument for optical depth and radiation measurements. *Appl. Opt.*, **33**, 5118–5125, doi:10.1364/AO.33.005118.

Henderson, P. W., and R. Pincus, 2009: Multiyear evaluations of a cloud model using ARM data. *J. Atmos. Sci.*, **66**, 2925–2936, doi:10.1175/2009JAS2957.1.

Iacono, M. J., E. J. Mlawer, S. A. Clough, and J.-J. Morcrette, 2000: Impact of an improved longwave radiation model, RRTM, on the energy budget and thermodynamic properties of the NCAR Community Climate Model, CCM3. *J. Geophys. Res.*, **105**, 14 873–14 890, doi:10.1029/2000JD900091.

——, J. Delamere, E. Mlawer, M. Shephard, S. Clough, and W. Collins, 2008: Radiative forcing by long-lived greenhouse gases: Calculations with the AER radiative transfer models. *J. Geophys. Res.*, **113**, D13103, doi:10.1029/2008JD009944.

Kablick, G. P., III, R. G. Ellingson, E. E. Takara, and J. Gu, 2011: Longwave 3D benchmarks for inhomogeneous clouds and comparisons with approximate methods. *J. Climate*, **24**, 2192–2205, doi:10.1175/2010JCLI3752.1.

Kassianov, E. I., and M. Ovtchinnikov, 2008: On reflectance ratios and aerosol optical depth retrieval in the presence of cumulus clouds. *Geophys. Res. Lett.*, **35**, L06807, doi:10.1029/2008GL033231.

Kato, S., T. P. Ackerman, E. E. Clothiaux, J. H. Mather, G. G. Mace, M. L. Wesely, F. Murcray, and J. Michalsky,1997: Uncertainties in modeled and measured clear-sky surface shortwave irradiances. *J. Geophys. Res.*, **102**, 25 881–25 898, doi:10.1029/97JD01841.

Kennedy, A. D., X. Dong, B. Xi, S. Xie, Y. Zhang, and J. Chen, 2011: A comparison of MERRA and NARR reanalyses with the DOE ARM SGP data. *J. Climate*, **24**, 4541–4557, doi:10.1175/2011JCLI3978.1.

Khairoutdinov, M., D. Randall, and C. DeMott, 2005: Simulations of atmospheric general circulation using a cloud-resolving model as a superparameterization of physical processes. *J. Atmos. Sci.*, **62**, 2136–2154, doi:10.1175/JAS3453.1.

Klein, S. A., and Coauthors, 2009: Intercomparison of model simulations of mixed-phase clouds observed during the ARM Mixed-Phase Arctic Cloud Experiment. Part I: Single-layer cloud. *Quart. J. Roy. Meteor. Soc.*, **135**, 979–1002, doi:10.1002/qj.416.

Knuteson, R. O., and Coauthors, 2004a: The Atmospheric Emitted Radiance Interferometer (AERI) Part I: Instrument design. *J. Atmos. Oceanic Technol.*, **21**, 1763–1776, doi:10.1175/JTECH-1662.1.

——, and Coauthors, 2004b: The Atmospheric Emitted Radiance Interferometer (AERI) Part II: Instrument performance. *J. Atmos. Oceanic Technol.*, **21**, 1777–1789, doi:10.1175/JTECH-1663.1.

Kollias, P., E. E. Clothiaux, M. A. Miller, E. P. Luke, K. L. Johnson, K. P. Moran, K. B. Widener, and B. A. Albrecht, 2007: The Atmospheric Radiation Measurement Program cloud profiling radars: Second-generation sampling strategies, processing, and cloud data products. *J. Atmos. Oceanic Technol.*, **24**, 1199–1214, doi:10.1175/JTECH2033.1.

——, and Coauthors, 2016: Development and applications of ARM millimeter-wavelength cloud radars. *The Atmospheric Radiation Measurement (ARM) Program: The First 20 Years, Meteor. Monogr.*, No. 57, Amer. Meteor. Soc., doi:10.1175/AMSMONOGRAPHS-D-15-0037.1.

Krueger, S. K., H. Morrison, and A. M. Fridlind, 2016: Cloud-resolving modeling: ARM and the GCSS story. *The Atmospheric Radiation Measurement (ARM) Program: The First 20 Years, Meteor. Monogr.*, No. 57, Amer. Meteor. Soc., doi:10.1175/AMSMONOGRAPHS-D-15-0047.1.

Li, Z., 2004: On the solar radiation budget and the cloud absorption anomaly debate. *Observation, Theory, and Modeling of the Atmospheric Variability*, World Scientific, 437–456.

Liljegren, J. C., 2000: Automatic self-calibration of ARM microwave radiometers. *Microwave Radiometry and Remote Sensing of the Earth's Surface and Atmosphere*, P. Pampaloni and S. Paloscia, Eds., VSP Press, 433–443.

——, and B. M. Lesht, 1996: Measurements of integrated water vapor and cloud liquid water from microwave radiometers in the DOE ARM Cloud and Radiation Testbed in the U.S. Southern Great Plains. *Proc. IGARSS '96*, Lincoln, NE, 1675–1677, doi:10.1109/IGARSS.1996.516767.

Liu, J., J. Zheng, Z. Li, and M. Cribb, 2011: Analysis of cloud condensation nuclei properties at a polluted site in southeastern China during the AMF-China Campaign. *J. Geophys. Res.*, **116**, D00K35, doi:10.1029/2011JD016395.

Liu, X., and Coauthors, 2011: Testing cloud microphysics parameterizations in NCAR CAM5 with ISDAC and M-PACE observations. *J. Geophys. Res.*, **116**, D00T11, doi:10.1029/2011JD015889.

Long, C. N., J. M. Sabburg, J. Calbó, and P. Pagès, 2006: Retrieving cloud statistics from ground-based daytime color all-sky images. *J. Atmos. Oceanic Technol.*, **23**, 633–652, doi:10.1175/JTECH1875.1.

——, J. H. Mather, and T. P. Ackerman, 2016: The ARM Tropical Western Pacific (TWP) sites. *The Atmospheric Radiation Measurement (ARM) Program: The First 20 Years, Meteor. Monogr.*, No. 57, Amer. Meteor. Soc., doi:10.1175/AMSMONOGRAPHS-D-15-0024.1.

Lubin, D., and A. Vogelmann, 2006: A climatologically significant aerosol longwave indirect effect in the Arctic. *Nature*, **439**, 453–456, doi:10.1038/nature04449.

Luke, E. P., P. Kollias, and M. D. Shupe, 2010: Detection of supercooled liquid in mixed-phase clouds using radar Doppler spectra. *J. Geophys. Res.*, **115**, D19201, doi:10.1029/2009JD012884.

Mace, G. G., and S. Benson, 2008: The vertical structure of cloud occurrence and radiative forcing at the SGP ARM site as revealed by 8 years of continuous data. *J. Climate*, **21**, 2591–2610, doi:10.1175/2007JCLI1987.1.

——, T. P. Ackerman, P. Minnis, and D. F. Young, 1998: Cirrus layer microphysical properties derived from surface-based millimeter radar and infrared interferometer data. *J. Geophys. Res.*, **103**, 23 207–23 216, doi:10.1029/98JD02117.

Marchand, R., 2016: ARM and satellite cloud validation. *The Atmospheric Radiation Measurement (ARM) Program: The First 20 Years, Meteor. Monogr.*, No. 57, Amer. Meteor. Soc., doi:10.1175/AMSMONOGRAPHS-D-15-0038.1.

——, T. P. Ackerman, E. R. Westwater, S. A. Clough, K. Cady-Pereira, and J. C. Liljegren, 2003: An assessment of microwave absorption models and retrievals of cloud liquid water using clear sky data. *J. Geophys. Res.*, **108**, 4773, doi:10.1029/2003JD003843.

Marshak, A., and A. B. Davis, Eds., 2006: *3D Radiative Transfer in Cloud Atmospheres*. Springer, 686 pp.

Mather, J. H., and S. A. McFarlane, 2009: Cloud classes and radiative heating profiles at the Manus and Nauru Atmospheric Radiation Measurement (ARM) sites. *J. Geophys. Res.*, **114**, D19204, doi:10.1029/2009JD011703.

——, and J. W. Voyles, 2013: The ARM Climate Research Facility: A review of structure and capabilities. *Bull. Amer. Meteor. Soc.*, **94**, 377–392, doi:10.1175/BAMS-D-11-00218.1.

——, T. P. Ackerman, W. E. Clements, F. J. Barnes, M. D. Ivey, L. D. Hatfield, and R. M. Reynolds, 1998: An atmospheric radiation and cloud station in the tropical western Pacific. *Bull. Amer. Meteor. Soc.*, **79**, 627–642, doi:10.1175/1520-0477(1998)079<0627:AARACS>2.0.CO;2.

May, P. T., J. H. Mather, G. Vaughan, C. Jakob, G. M. McFarquhar, K. N. Bower, and G. G. Mace, 2008: The Tropical Warm Pool International Cloud Experiment. *Bull. Amer. Meteor. Soc.*, **89**, 629–645, doi:10.1175/BAMS-89-5-629.

McComiskey, A., and R. A. Ferrare, 2016: Aerosol physical and optical properties and processes in the ARM Program. *The Atmospheric Radiation Measurement (ARM) Program: The First 20 Years, Meteor. Monogr.*, No. 57, Amer. Meteor. Soc., doi:10.1175/AMSMONOGRAPHS-D-15-0028.1.

——, G. Feingold, A. S. Frisch, D. D. Turner, M. A. Miller, J. C. Chiu, Q. Min, and J. A. Ogren, 2009: An assessment of aerosol–cloud interactions in marine stratus clouds based on surface remote sensing. *J. Geophys. Res.*, **114**, D09203, doi:10.1029/2008JD011006.

McCord, R., and J. W. Voyles, 2016: The ARM data system and archive. *The Atmospheric Radiation Measurement (ARM) Program: The First 20 Years, Meteor. Monogr.*, No. 57, Amer. Meteor. Soc., doi:10.1175/AMSMONOGRAPHS-D-15-0043.1.

McFarlane, S. A., E. I. Kassianov, J. Barnard, C. Flynn, and T. P. Ackerman, 2009: Surface shortwave forcing during the Atmospheric Radiation Measurement Mobile Facility deployment in Niamey, Niger. *J. Geophys. Res.*, **114**, D00E06, doi:10.1029/2008JD010491.

McFarquhar, G. M., J. Um, M. Freer, D. Baumgardner, G. L. Kok, and G. G. Mace, 2007: Importance of small ice crystals to cirrus properties: Observations from the Tropical Warm Pool-International Cloud Experiment (TWP-ICE). *Geophys. Res. Lett.*, **34**, L13803, doi:10.1029/2007GL029865.

——, and Coauthors, 2011: Indirect and Semi-Direct Aerosol Campaign (ISDAC): The impact of Arctic aerosols on clouds. *Bull. Amer. Meteor. Soc.*, **92**, 183–201, doi:10.1175/2010BAMS2935.1.

Michalsky, J. J., and C. N. Long, 2016: ARM solar and infrared broadband and filter radiometry. *The Atmospheric Radiation Measurement (ARM) Program: The First 20 Years, Meteor. Monogr.*, No. 57, Amer. Meteor. Soc., doi:10.1175/AMSMONOGRAPHS-D-15-0031.1.

——, Q. Min, J. Barnard, R. Marchand, and P. Pilewskie, 2003: Simultaneous spectral albedo measurements near the Atmospheric Radiation Measurement Southern Great Plains (ARM SGP) central facility. *J. Geophys. Res.*, **108**, 4254, doi:10.1029/2002JD002906.

——, and Coauthors, 2005: Toward the development of a diffuse horizontal shortwave irradiance working standard. *J. Geophys. Res.*, **110**, D06107, doi:10.1029/2004JD005265.

——, F. Denn, C. Flynn, G. Hodges, P. Kiedron, A. Koontz, J. Schemmer, and S. E. Schwartz, 2010: Climatology of aerosol optical depth in north-central Oklahoma: 1992–2008. *J. Geophys. Res.*, **115**, D07203, doi:10.1029/2009JD012197.

Miller, M. A., and A. Slingo, 2007: The Atmospheric Radiation Measurement (ARM) Mobile Facility (AMF) and its first international deployment: Measuring radiative flux divergence in West Africa. *Bull. Amer. Meteor. Soc.*, **88**, 1229–1244, doi:10.1175/BAMS-88-8-1229.

——, and Coauthors, 2007: SGP Cloud and Land Surface Interaction Campaign (CLASIC): Science and implementation plan. DOE/SC-ARM-0703, 14 pp.

——, K. Nitschke, T.P. Ackerman, W. R. Ferrell, N. Hickmon, and M. Ivey, 2016: The ARM Mobile Facilities. *The Atmospheric Radiation Measurement (ARM) Program: The First 20 Years, Meteor. Monogr.*, No. 57, Amer. Meteor. Soc., doi:10.1175/AMSMONOGRAPHS-D-15-0051.1.

Mlawer, E. J., and D. D. Turner, 2016: Spectral radiation measurements and analysis in the ARM Program. *The Atmospheric Radiation Measurement (ARM) Program: The First 20 Years, Meteor. Monogr.*, No. 57, Amer. Meteor. Soc., doi:10.1175/AMSMONOGRAPHS-D-15-0027.1.

——, S. J. Taubman, P. D. Brown, M. J. Iacono, and S. A. Clough, 1997: Radiative transfer for inhomogeneous atmospheres: RRTM, a validated correlated-k model for the longwave. *J. Geophys. Res.*, **102**, 16 663–16 682, doi:10.1029/97JD00237.

——, M. Iacono, R. Pincus, H. Barker, L. Oreopoulos, and D. Mitchell, 2016: Contributions of the ARM Program to radiative transfer modeling for climate and weather applications. *The Atmospheric Radiation Measurement (ARM) Program: The First 20 Years, Meteor. Monogr.*, No. 57, Amer. Meteor. Soc., doi:10.1175/AMSMONOGRAPHS-D-15-0041.1.

Moran, K. P., B. E. Martner, M. J. Post, R. A. Kropfli, D. C. Welsh, and K. B. Widener, 1998: An unattended cloud-profiling radar for use in climate research. *Bull. Amer. Meteor. Soc.*, **79**, 443–455, doi:10.1175/1520-0477(1998)079<0443:AUCPRF>2.0.CO;2.

Morrison, H., and Coauthors, 2009: Intercomparison of model simulations of mixed-phase clouds observed during the ARM Mixed-Phase Arctic Cloud Experiment, Part II: Multilayered cloud. *Quart. J. Roy. Meteor. Soc.*, **135**, 1003–1019, doi:10.1002/qj.415.

——, G. de Boer, G. Feingold, J. Harrington, M. D. Shupe, and K. Sulia, 2012: Resilience of persistent Arctic mixed-phase clouds. *Nat. Geosci.*, **5**, 11–17, doi:10.1038/ngeo1332.

Neale, R. B., and Coauthors, 2010: Description of the NCAR Community Atmosphere Model (CAM 5.0). NCAR/TN-486+STR, 268 pp.

O'Hirok, W., and C. Gautier, 1998: A three-dimensional radiative transfer model to investigate the solar radiation within a cloudy atmosphere: Part II. Spectral effects. *J. Atmos. Sci.*, **55**, 3065–3076, doi:10.1175/1520-0469(1998)055<3065:ATDRTM>2.0.CO;2.

Ovtchinnikov, M., T. P. Ackerman, R. Marchand, and M. Khairoutdinov, 2006: Evaluation of the multiscale modeling framework using data from the Atmospheric Radiation Measurement program. *J. Climate*, **19**, 1716–1729, doi:10.1175/JCLI3699.1.

Pahlow, M., and Coauthors, 2006: Comparison between lidar and nephelometer measurements of aerosol hygroscopicity at the Southern Great Plains Atmospheric Radiation Measurement site. *J. Geophys. Res.*, **111**, D05S15, doi:10.1029/2004JD005646.

Peppler, R., K. Kehoe, J. Monroe, A. Theisen, and S. Moore, 2016: The ARM data quality program. *The Atmospheric Radiation Measurement (ARM) Program: The First 20 Years, Meteor. Monogr.*, No. 57, Amer. Meteor. Soc., doi:10.1175/AMSMONOGRAPHS-D-15-0039.1.

Phillips, T. J., and Coauthors, 2004: Evaluating parameterizations in general circulation models: Climate simulation meets weather prediction. *Bull. Amer. Meteor. Soc.*, **85**, 1903–1915, doi:10.1175/BAMS-85-12-1903.

Pincus, R., H. W. Barker, and J.-J. Morcrette, 2003: A fast, flexible, approximation technique for computing radiative transfer in inhomogeneous cloud fields. *J. Geophys. Res.*, **108**, 4376, doi:10.1029/2002JD003322.

Protat, A., J. Delanoe, A. Plana-Fattori, P. T. May, and E. J. O'Connor, 2010: The statistical properties of tropical ice clouds generated by the West African and Australian monsoons, from ground-based radar–lidar observations. *Quart. J. Roy. Meteor. Soc.*, **136**, 345–363, doi:10.1002/qj.490.

Randall, D. A., K.-M. Xu, R. J. Somerville, and S. Iacobellis, 1996: Single-column models and cloud ensemble models as links between observations and climate models. *J. Climate*, **9**, 1683–1697, doi:10.1175/1520-0442(1996)009<1683:SCMACE>2.0.CO;2.

——, and Coauthors, 2003: Confronting models with data: The GEWEX Cloud Systems Study. *Bull. Amer. Meteor. Soc.*, **84**, 455–469, doi:10.1175/BAMS-84-4-455.

——, A. D. Del Genio, L. J. Donner, W. D. Collins, and S. A. Klein, 2016: The impact of ARM on climate modeling. *The Atmospheric Radiation Measurement (ARM) Program: The First 20 Years, Meteor. Monogr.*, No. 57, Amer. Meteor. Soc., doi:10.1175/AMSMONOGRAPHS-D-15-0050.1.

Rémillard, J., P. Kollias, E. Luke, and R. Wood, 2012: Marine boundary layer cloud observations in the Azores. *J. Climate*, **25**, 7381–7398, doi:10.1175/JCLI-D-11-00610.1.

Revercomb, H. E., and Coauthors, 2003: The ARM Program's water vapor intensive observation periods: Overview, initial accomplishments, and future challenges. *Bull. Amer. Meteor. Soc.*, **84**, 217–236, doi:10.1175/BAMS-84-2-217.

Schmid, B., and Coauthors, 2006: How well do state-of-the-art techniques measuring the vertical profile of tropospheric aerosol extinction compare? *J. Geophys. Res.*, **111**, D05S07, doi:10.1029/2005JD005837.

——, and Coauthors, 2009: Validation of aerosol extinction profiles from routine Atmospheric Radiation Measurement Program Climate Research Facility measurements. *J. Geophys. Res.*, **114**, D22207, doi:10.1029/2009JD012682.

——, R. G. Ellingson, and G. M. McFarquhar, 2016: ARM aircraft measurements. *The Atmospheric Radiation Measurement (ARM) Program: The First 20 Years, Meteor. Monogr.*, No. 57, Amer. Meteor. Soc., doi:10.1175/AMSMONOGRAPHS-D-15-0042.1.

Sheridan, P., D. Delene, and J. Ogren, 2001: Four years of continuous surface aerosol measurements from the Department of Energy's Atmospheric Radiation Measurement Program Southern Great Plains Cloud and Radiation Testbed site. *J. Geophys. Res.*, **106**, 20 735–20 748, doi:10.1029/2001JD000785.

——, and Coauthors, 2005: The Reno aerosol optics study: An evaluation of aerosol absorption measurement methods. *Aerosol Sci. Technol.*, **39**, 1–16, doi:10.1080/027868290901891.

Shupe, M. D., and Coauthors, 2008: A focus on mixed-phase clouds: The status of ground-based observational methods. *Bull. Amer. Meteor. Soc.*, **89**, 1549–1562, doi:10.1175/2008BAMS2378.1.

——, J. M. Comstock, D. D. Turner, and G. G. Mace, 2016: Cloud property retrievals in the ARM Program. *The Atmospheric Radiation Measurement (ARM) Program: The First 20 Years, Meteor. Monogr.*, No. 57, Amer. Meteor. Soc., doi:10.1175/AMSMONOGRAPHS-D-15-0030.1.

Sisterson, D., R. Peppler, T. S. Cress, P. Lamb, and D. D. Turner, 2016: The ARM Southern Great Plains (SGP) site. *The Atmospheric Radiation Measurement (ARM) Program: The First 20 Years, Meteor. Monogr.*, No. 57, Amer. Meteor. Soc., doi:10.1175/AMSMONOGRAPHS-D-16-0004.1.

Slingo, A., and Coauthors, 2006: Observations of the impact of a major Saharan dust storm on the atmospheric radiation balance. *Geophys. Res. Lett.*, **33**, L24817, doi:10.1029/2006GL027869.

——, and Coauthors, 2008: Overview of observations from the RADAGAST experiment in Niamey, Niger. Part 1: Meteorology and thermodynamic variables. *J. Geophys. Res.*, **113**, D00E01, doi:10.1029/2008JD009909.

Song, H., W. Lin, Y. Lin, A. B. Wolf, R. Neggers, L. J. Donner, A. D. Del Genio, and Y. Liu, 2013: Evaluation of precipitation simulated by seven SCMs against the ARM observations at the SGP site. *J. Climate*, **26**, 5467–5492, doi:10.1175/JCLI-D-12-00263.1.

Stamnes, K., R. G. Ellingson, J. A. Curry, J. E. Walsh, and B. D. Zak, 1999: Review of science issues, deployment strategy, and status for the ARM North Slope of Alaska–Adjacent Arctic Ocean climate research site. *J. Climate*, **12**, 46–63, doi:10.1175/1520-0442-12.1.46.

Stokes, G. M., 2016: Original ARM concept and launch. *The Atmospheric Radiation Measurement (ARM) Program: The First 20 Years, Meteor. Monogr.*, No. 57, Amer. Meteor. Soc., doi:10.1175/AMSMONOGRAPHS-D-15-0021.1.

——, and S. E. Schwartz, 1994: The Atmospheric Radiation Measurement (ARM) Program: Programmatic background and design of the cloud and radiation test bed. *Bull. Amer. Meteor. Soc.*, **75**, 1201–1221, doi:10.1175/1520-0477(1994)075<1201:TARMPP>2.0.CO;2.

Stroeve, J. C., V. Kattsov, A. Barrett, M. C. Serreze, T. Pavlova, M. Holland, and W. N. Meier, 2012: Trends in Arctic sea ice extent from CMIP5, CMIP3, and observations. *Geophys. Res. Lett.*, **39**, L16502, doi:10.1029/2012GL052676.

Tobin, D. C., and Coauthors, 1999: Downwelling spectral radiance observations at the SHEBA ice station: Water vapor continuum measurements from 17 to 26 μm. *J. Geophys. Res.*, **104**, 2081–2092, doi:10.1029/1998JD200057.

Turner, D. D., 2005: Arctic mixed-phase cloud properties from AERI-lidar observations: Algorithm and results from SHEBA. *J. Appl. Meteor.*, **44**, 427–444, doi:10.1175/JAM2208.1.

——, 2008: Ground-based retrievals of optical depth, effective radius, and composition of airborne mineral dust above the Sahel. *J. Geophys. Res.*, **113**, D00E03, doi:10.1029/2008JD010054.

——, and J. E. M. Goldsmith, 1999: Twenty-four-hour Raman lidar water vapor measurements during the Atmospheric Radiation Measurement Program's 1996 and 1997 water vapor intensive observation periods. *J. Atmos. Oceanic Technol.*, **16**, 1062–1076, doi:10.1175/1520-0426(1999)016<1062:TFHRLW>2.0.CO;2.

——, and E. J. Mlawer, 2010: The Radiative Heating in Underexplored Bands Campaigns. *Bull. Amer. Meteor. Soc.*, **91**, 911–923, doi:10.1175/2010BAMS2904.1.

——, R. A. Ferrare, L. A. Heilman Brasseur, W. F. Feltz, and T. P. Tooman, 2002: Automated retrieval of water vapor and aerosol profiles from an operational Raman lidar. *J. Atmos. Oceanic Technol.*, **19**, 37–50, doi:10.1175/1520-0426(2002)019<0037:AROWVA>2.0.CO;2.

——, B. M. Lesht, S. A. Clough, J. C. Liljegren, H. E. Revercomb, and D. C. Tobin, 2003: Dry bias and variability in Vaisala radiosondes: The ARM experience. *J. Atmos. Oceanic Technol.*, **20**, 117–132, doi:10.1175/1520-0426(2003)020<0117:DBAVIV>2.0.CO;2.

——, and Coauthors, 2004: The QME AERI LBLRTM: A closure experiment for downwelling high spectral resolution infrared radiance. *J. Atmos. Sci.*, **61**, 2657–2675, doi:10.1175/JAS3300.1.

——, and Coauthors, 2007a: Thin liquid water clouds: Their importance and our challenge. *Bull. Amer. Meteor. Soc.*, **88**, 177–190, doi:10.1175/BAMS-88-2-177.

——, S. A. Clough, J. C. Liljegren, E. E. Clothiaux, K. Cady-Pereira, and K. L. Gaustad, 2007b: Retrieving liquid water path and precipitable water vapor from Atmospheric Radiation Measurement (ARM) microwave radiometers. *IEEE Trans. Geosci. Remote Sens.*, **45**, 3680–3690, doi:10.1109/TGRS.2007.903703.

——, and Coauthors, 2012a: Ground-based high spectral resolution observations of the entire terrestrial spectrum under extremely dry conditions. *Geophys. Res. Lett.*, **39**, L10801, doi:10.1029/2012GL051542.

——, A. Merrelli, D. Vimont, and E. J. Mlawer, 2012b: Impact of modifying the longwave water vapor continuum absorption model on community Earth system model simulations. *J. Geophys. Res.*, **117**, D04106, doi:10.1029/2011JD016440.

——, J. E. M. Goldsmith, and R. A. Ferrare, 2016a: Development and applications of the ARM Raman lidar. *The Atmospheric*

Radiation Measurement (ARM) Program: The First 20 Years, Meteor. Monogr., No. 57, Amer. Meteor. Soc., doi:10.1175/AMSMONOGRAPHS-D-15-0026.1.

——, E. J. Mlawer, and H. E. Revercomb, 2016b: Water vapor observations in the ARM program. *The Atmospheric Radiation Measurement (ARM) Program: The First 20 Years, Meteor. Monogr.*, No. 57, Amer. Meteor. Soc., doi:10.1175/AMSMONOGRAPHS-D-15-0025.1.

U.S. Department of Energy, 1990: Atmospheric Radiation Measurement Program Plan. DOE/ER-0441, 121 pp.

——, 1996: Science Plan for the Atmospheric Radiation Measurement Program (ARM). DOE/ER-0670T, 74 pp.

——, 2004: Atmospheric Radiation Measurement Program Science Plan: Current status and future directions of the ARM Science Program. DOE/ER-ARM-0402, 62 pp.

——, 2010: Atmospheric System Research (ASR) Science and Program Plan. DOE/SC-ASR-10-001, 77 pp.

Valero, F. P. J., R. D. Cess, M. Zhang, S. K. Pope, A. Bucholtz, B. Bush, and J. Vitko Jr., 1997: Absorption of solar radiation by the cloudy atmosphere: Interpretations of collocated aircraft measurements. *J. Geophys. Res.*, **102**, 29 917–29 927, doi:10.1029/97JD01782.

——, and Coauthors, 2003: Absorption of solar radiation by the clear and cloudy atmosphere during the Atmospheric Radiation Measurement Enhanced Shortwave Experiments (ARESE) I and II: Observations and models. *J. Geophys. Res.*, **108**, 4016, doi:10.1029/2001JD001384.

Varble, A., and Coauthors, 2014: Evaluation of cloud-resolving and limited area model intercomparison simulations using TWP-ICE observations. Part 1: Deep convective updraft properties. *J. Geophys. Res. Atmos.*, **119**, 13 891–13 918, doi:10.1002/2013JD021371.

Verlinde, J., and Coauthors, 2007: The Mixed-Phase Arctic Cloud Experiment. *Bull. Amer. Meteor. Soc.*, **88**, 205–221, doi:10.1175/BAMS-88-2-205.

——, B. Zak, M. D. Shupe, M. Ivey, and K. Stamnes, 2016: The ARM North Slope of Alaska (NSA) sites. *The Atmospheric Radiation Measurement (ARM) Program: The First 20 Years, Meteor. Monogr.*, No. 57, Amer. Meteor. Soc., doi:10.1175/AMSMONOGRAPHS-D-15-0023.1.

Vogelmann, A., and Coauthors, 2012: RACORO extended-term aircraft observations of boundary layer clouds. *Bull. Amer. Meteor. Soc.*, **93**, 861–878, doi:10.1175/BAMS-D-11-00189.1.

Welton, E. J., 2000: Ground-based lidar measurements of aerosols during ACE-2: Instrument description, results, and comparisons with other ground-based and airborne measurements. *Tellus.*, **52B**, 636–651, doi:10.1034/j.1600-0889.2000.00025.x.

Williamson, D., J. S. Boyle, R. Cederwall, M. Fiorino, J. Hnilo, J. Olson, G. Potter, and S. Xie, 2005: Moisture and temperature balances at the Atmospheric Radiation Measurement Southern Great Plains site in forecasts with the Community Atmospheric Model (CAM2). *J. Geophys. Res.*, **110**, D15S16, doi:10.1029/2004JD005109.

Wulfmeyer, V., and Coauthors, 2011: The Convective and Orographically induced Precipitation Study (COPS): The scientific strategy, the field phase, and first highlights. *Quart. J. Roy. Meteor. Soc.*, **137**, 3–30, doi:10.1002/qj.752.

Xie, S., R. T. Cederwall, and M. Zhang, 2004: Developing long-term single-column model/cloud system-resolving model forcing data using numerical weather prediction products constrained by surface and top of the atmosphere observations. *J. Geophys. Res.*, **109**, D01104, doi:10.1029/2003JD004045.

——, T. Hume, C. Jakob, S. Klein, R. McCoy, and M. Zhang, 2010a: Observed large-scale structures and diabatic heating and drying profiles during TWP-ICE. *J. Climate*, **23**, 57–79, doi:10.1175/2009JCLI3071.1.

——, and Coauthors, 2010b: ARM climate modeling best estimate data: A new data product for climate studies. *Bull. Amer. Meteor. Soc.*, **91**, 13–20, doi:10.1175/2009BAMS2891.1.

Zaveri, R., and Coauthors, 2012: Overview of the 2010 Carbonaceous Aerosols and Radiative Effects Study (CARES). *Atmos. Chem. Phys.*, **12**, 7647–7687, doi:10.5194/acp-12-7647-2012.

Zender, C. S., B. Bush, S. K. Pope, A. Bucholtz, W. D. Collins, J. T. Kiehl, F. P. J. Valero, and J. Vitko, 1997: Atmospheric absorption during the Atmospheric Radiation Measurement (ARM) Enhanced Shortwave Experiment (ARESE). *J. Geophys. Res.*, **102**, 29 901–29 915, doi:10.1029/97JD01781.

Zhang, M. H., and J.-L. Lin, 1997: Constrained variational analysis of sounding data bases on column-integrated budgets of mass, heat, moisture, and momentum: Approach and application to ARM measurements. *J. Atmos. Sci.*, **54**, 1503–1524, doi:10.1175/1520-0469(1997)054<1503:CVAOSD>2.0.CO;2.

——, R. C. J. Somerville, and S. Xie, 2016: The SCM concept and creation of ARM forcing datasets. *The Atmospheric Radiation Measurement (ARM) Program: The First 20 Years, Meteor. Monogr.*, No. 57, Amer. Meteor. Soc., doi:10.1175/AMSMONOGRAPHS-D-15-0040.1.

Zhao, C., and Coauthors, 2012: Towards understanding of differences in current cloud retrievals of ARM ground-based measurements. *J. Geophys. Res.*, **117**, D10206, doi:10.1029/2011JD016792.

Chapter 5

Deploying the ARM Sites and Supporting Infrastructure

TED S. CRESS*
Pacific Northwest National Laboratory, Richland, Washington

DOUGLAS L. SISTERSON
Argonne National Laboratory, Lemont, Illinois

1. Background

Ellingson et al. (2016, chapter 1) discusses the scientific stimuli that led to the ARM Program. Stokes (2016, chapter 2) discusses the DOE efforts to develop the program's concept for a proposal to the Committee on Earth Sciences (CES) of the U.S. Global Change Research Program (USGCRP) in 1990. Two of the most influential scientists driving the proposed program forward were Fred Luther of Lawrence Livermore National Laboratory and Bob Ellingson of Florida State University. As a result of the Intercomparison of Radiation Codes in Climate Models (ICRCCM) activity (Ellingson and Fouquart 1991), Luther (1984) was among those who concluded that "a dedicated field measurement program is recommended for the purpose of obtaining accurate spectral radiances rather than integrated fluxes as a basis for evaluating model performance."

Ellingson and Wiscombe took it a step further, proposing a field program (Ellingson and Wiscombe 1996; Ellingson et al. 2016, chapter 1) of "real observations" to evaluate radiative modeling results. At the same time, it was realized that observations were needed to evaluate the accuracy of cloud properties that were being predicted by models and that there was much work needed to improve GCMs in this regard. As discussed by Stokes (2016, chapter 2), these influences led the DOE to direct the preparation of a proposal for the incipient USGCRP.

Stokes (2016, chapter 2) discusses the preparation of the initial DOE proposal submitted to the CES and its rejection; it was considered too big and too costly—nearly equaling the then total investment in atmospheric basic research by all departments of the Federal government.[1] The CES requested a revised, less expensive proposal (Stokes 2016, chapter 2) that was accepted, which became the ARM Program Plan (DOE 1990; ARM 2016a, appendix A) and the primary guidance for the implementation of the ARM Program. For a field program, that approval was not without its doubters. The proposed program of long-term observations called for the use, in many cases, of instruments that were still experimental or laboratory grade. The proposal also called for data to be acquired from instruments in real time and to be shared in near–real time. The proposed approach was viewed as extremely aggressive, requiring a substantially new approach and potentially new technical capabilities as compared to the historical approach of time-limited campaigns. For the proposed DOE program, the data communications challenges alone appeared nearly insurmountable. As Stokes (2016, chapter 2) observed, the resultant features of the program were "highly controversial." To the doubters, it appeared to promise a "step too far" with an attendant high potential to fall short.

* Retired.

Corresponding author address: Ted Cress, 279 Hillwood St., Richland, WA 99352.
E-mail: cressts@earthlink.net

DOI: 10.1175/AMSMONOGRAPHS-D-15-0049.1

[1] The initial DOE proposal was about $200 million per year. The total federal budget for "supporting research" in meteorology for fiscal year 1990 was $445 million, of which $210 million was for NASA, mostly for satellite research and development. The remaining $235 million included engineering development efforts as well as basic research (Ramirez 1990).

Indeed, the approved DOE program was aggressive. The Program Plan (ARM 2016a, appendix A) clearly presented objectives and goals for the program but, for deployment purposes, provided no blueprint or roadmap to successfully get there. The proposed duration and complexity was unprecedented. It was clear at the time that the challenges would be many and that a successful deployment would be dependent on effectively recognizing and dealing with issues as they arose. It became a hallmark of the deployment effort that, as new issues or requirements were identified (and there were many), they were, for the most part, effectively addressed, producing robust solutions that typically stood the test of time. It was also clear that the proposed program would not succeed on DOE efforts alone. DOE would need to foster an extraordinary level of interagency collaboration and cooperation, extending from simple acquisition of routine data [e.g., National Weather Service (NWS) weather and radar data] to the acquisition of less accessible data (e.g., raw radar data from NWS or research Doppler radars or satellite and aircraft data from other field programs). These requirements had to be factored into deployment planning for the implementation of a robust data system.

2. Approved Program Plan guidance

In chapter 2, Stokes (2016) described the "intellectual threads" leading to ARM and the evolution of the Program Plan. This section summarizes the key requirements that were identified and thus drove and guided the deployment effort.

Stokes and Schwartz (1994) present the primary ARM Program Plan objectives as the following:

1) To relate observed radiative fluxes in the atmosphere, spectrally resolved and as a function of position and time, to the atmospheric temperature, composition (specifically including water vapor and clouds), and surface radiative properties.
2) To develop and test parameterizations that describe atmospheric water vapor, cloud and the surface properties governing atmospheric radiation in terms of relevant prognostic variables, with the objective of incorporating these parameterizations into general circulation and related models.

These objectives were to be pursued with a particular focus on atmospheric radiative properties as a consequence of the presence of clouds and aerosols. Since acquired measurements would be used as input to drive GCM radiation parameterizations and as the validation data for GCM calculations, Ackerman et al. (2016, chapter 3) viewed these objectives in a more direct way:

1) If we can specify a cloud field, can we compute the radiative fluxes associated with that field?
2) If we can specify the large-scale atmospheric state variables, can we predict the cloud field properties associated with that state?

This restatement of ARM's aim to provide the data critical to improve radiation and cloud parameterizations gives an immediate insight into what was required to translate the approved ARM Program Plan into a roadmap leading to a siting strategy, the design and deployment of the measurement sites, and the processing of the data. For GCM applicability, these questions had to be asked for every climate regime.

a. Concept development (fleshing out the skeleton of the Program Plan)

After the approval of the DOE proposal by the CES, DOE requested Gerry Stokes, Pacific Northwest National Laboratory (PNNL), to assume leadership of the effort as ARM's first chief scientist to translate the ARM Program Plan into an implementable strategy. In chapter 2, Stokes (2016) discusses the facilitated planning methodology (commonly referred to as WISDM) used to develop the necessary plans and guidance. Facilitated planning workshops (i.e., WISDM sessions) were convened repeatedly over the next few years to prescribe virtually all aspects of the field experiment structure, including site operations, instrument operations, and data processing and management. The workshops identified the programmatic elements and their functions, fleshed out the details of what was needed, and determined how information and data would be captured, distributed, and archived within the program and made available to the ARM Science Team and the general research community. The workshops were not constrained by budget considerations and current technology. The intent was to allow participants to look far enough into the future to deploy a capability that would not become obsolete within a decade. The sentiment was, "If you can imagine it, you can build it." The results of these workshops are discussed in the following sections.

At the same time, Gerry Stokes was addressing the need to marshal the technical and management manpower with the skills necessary to map out the ARM deployment and to put ARM into the field with the full operational capability envisioned in the ARM Program Plan. To this end, eight DOE laboratories (Argonne National Laboratory, Brookhaven National Laboratory, Oak Ridge National Laboratory, Los Alamos National Laboratory, Sandia National Laboratories, Lawrence Livermore National Laboratory, Pacific Northwest National Laboratory, and the National Renewable Energy Laboratory) were requested to participate and assume responsibilities for various aspects of the program. Management and technical personnel from this family of DOE laboratories constituted the nucleus of the facilitated planning workshops and then effectively collaborated to put workshop results into motion.

It is important to appreciate that this collaboration among the DOE laboratories was a culture shift for the laboratories and for DOE. Although the laboratories certainly had cooperated in field programs before ARM [e.g., the Atmospheric Studies in Complex Terrain (ASCOT) project], they typically had to compete against one another for DOE funding. For the ARM Program, DOE decided to fund the Program through a single management office, which managed and funded the functional efforts of the laboratories. The culture shift was challenging, but the laboratories readily recognized that together they had the required expertise across the range of tasks involved in implementation of the Program, whereas individually they did not. In this respect, the ARM Program was directly a DOE "corporate program," with the DOE family of laboratories working as a distributed, yet integrated, corporate resource. Programmatic funding did not include funding for the research. Proposed research efforts at those same laboratories were funded directly by DOE via the Science Team solicitation and proposal review process used for all Science Team efforts. But projects like ASCOT did create some spirit of cooperation among many of the scientists at the different laboratories.

b. *Requirements (what had to happen?)*

The ARM Program Plan provided details, circa 1990, about the types of measurements that could be foreseen as required to support the program's stated objectives. To be applicable to GCM parameterizations, two measurement approaches were recognized as required:[2]

1) A concentration of instruments and support facilities to provide measurements of the vertical column of the atmosphere (commonly referred to as a "soda straw" view of the atmosphere over a point)
2) An extended network to measure the three-dimensional structure of the atmospheric column on the scale of a GCM grid cell

Combined with the requirement to gather data to address these two questions neatly summarized by Ackerman et al. (2016, chapter 3) above, plus the requirement for real-time data acquisition and quality control, a set of measurement requirements important to deployment planning could be specified:

- Continuous (24/7) measurements
- Measurements of solar and infrared radiation, both spectrally resolved and broadband, for a range of climatically different meteorological conditions to constrain detailed, line-by-line radiative calculations under clear, cloudy, and overcast conditions for global application
- Measurements of surface and overlying meteorological variables, including cloud type and distribution, wind, and temperature
- Measurements of clouds, radiative properties, and atmospheric properties over a wide range of scales
- Measurements of the microphysical properties of clouds
- Measurements of atmospheric water vapor, aerosols, and atmospheric trace gases
- State-of-the-art calibration capabilities

Considering the measurements required, the facility requirements identified were just as important, and highly significant for planning and deployment, including:

- Sites to be installed in climatologically representative regimes across the globe
- Sites typically operating for years to acquire statistically significant data for seasonal and annual cycles in the climate system with shorter campaigns in additional areas as deemed necessary
- Field-hardened instruments
- Extensibility—the ability to extend and/or adjust measurement facilities to accommodate permanent or temporal adjustments of the operational measurement scheme or the fielding of new instruments, some with potentially high data volume rates
- A robust data environment

c. *The Cloud and Radiation Test Bed*

To meet the data and facility requirements summarized above, and consonant with the proposed budget, the Program Plan introduced the concept of the Cloud and Radiation Test Bed (CART) and specified it to have the following features:

- Four to six permanent field sites, with each site comprising a central facility and an extended network
- An extended network of 16–25 surface observing sites distributed over a representative GCM grid cell area (additional sites for characterizing the three-dimensional structure of the atmosphere over the central facility important to radiative fluxes being measured at the central facility were added later)
- An in situ sampling capability
- A mobile site
- Capability to host specialized observational campaigns

[2] The surrogate science team meetings described by Stokes (2016, chapter 2) termed these efforts the instantaneous radiative flux measurement strategy and the single-column model experiment measurement strategy, respectively.

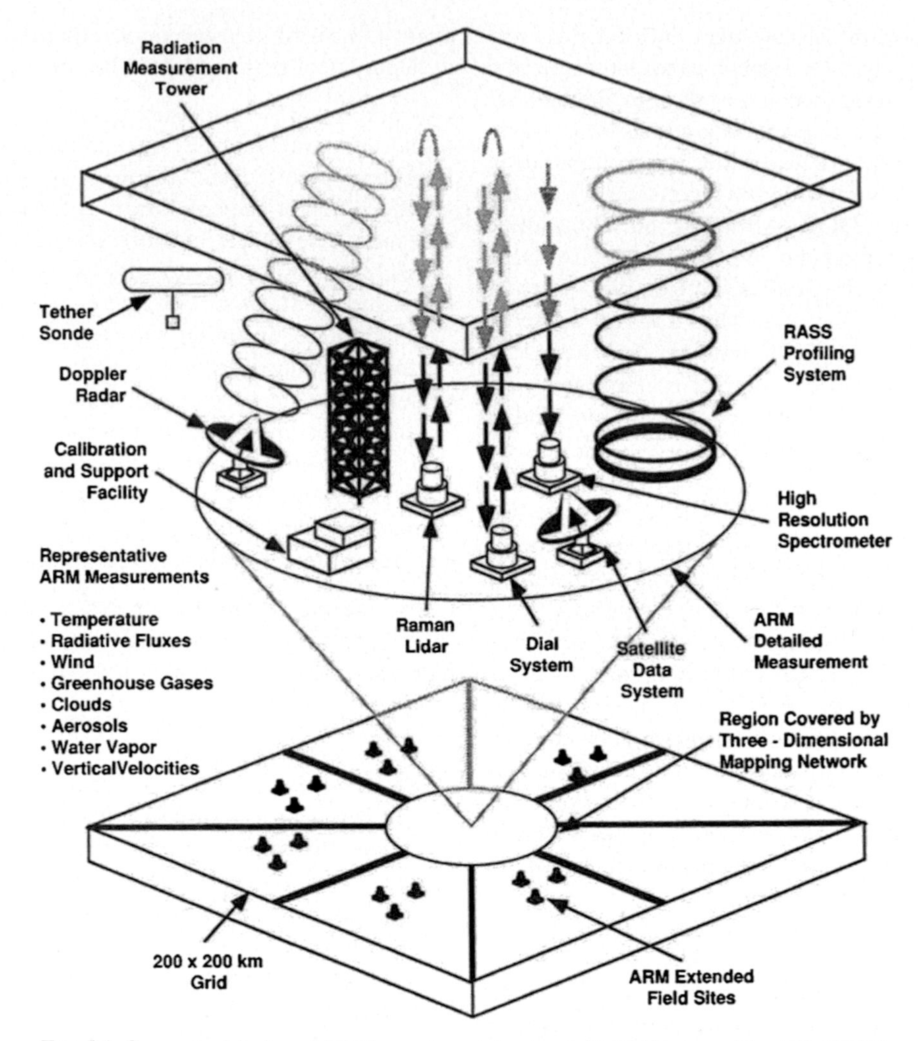

FIG. 5-1. Conceptual design of ARM experimental network [ARM Program Plan (DOE 1990, Fig. 12)].

- A data environment meeting the requirements of the Program Plan
- State-of-the-art calibration capability at each site

The term "test bed" was selected deliberately, because the overall goal of the ARM Program was to "develop and test parameterizations of important atmospheric processes, particularly cloud and radiative processes" (Stokes and Schwartz 1994). The goals would be addressed by comparing radiation measurements against calculations and vice versa, in accordance with questions posed by Ackerman et al. (2016, chapter 3) quoted above.

d. Conceptual CART measurement site

The basic conceptual design for a CART measurement site is shown in Fig. 5-1. The idealized CART site, reflective of the listed requirements, included a central facility to acquire data to support the instantaneous radiative flux (IRF) measurement strategy (i.e., the soda straw view overhead) and a surrounding network of sites to meet the requirements for the single-column model (SCM) measurement strategy. The extended surface observing facilities would be distributed over an area comparable to a GCM grid cell. Finally (not shown in the illustration), four to six boundary facilities established at the edges of the CART site boundaries would obtain sufficient vertical measurements of the overlying atmosphere to characterize the advective tendencies across the boundaries of the grid cell (Zhang et al. 2016, chapter 24). Also not shown, three to four auxiliary sites would be established around the central facility (approximately 20 km away) equipped to map the three-dimensional structure of the atmosphere over the central facility. The boundary and auxiliary sites were improvements to the original Program Plan scheme and were suggested by the ARM Science Team. The entire site was designed to enable acquisition of sufficient data to represent the full range of processes acting within the grid cell and to permit the desired calculations to evaluate radiative and cloud parameterizations being used in GCMs.

e. *CART site instrumentation*

With the configuration conceptualized, what instrumentation was required to provide the observations needed to address the two questions summarized by Ackerman et al. (2016, chapter 3)? The Program Plan presented a rather comprehensive look at the types of instruments that would likely be required. The selection of instruments actually deployed was influenced heavily by recommendations from the Science Team (Stokes 2016, chapter 2) and through the efforts of the instrument team (see below). The following is a brief list of instrument types envisioned for each site facility.

1) CENTRAL FACILITY

- Spectral radiation: For longwave (infrared) radiation: spectrometers and interferometers for measurements at wavelengths between 3 and 25 μm at high spectral resolution (1 cm^{-1} or better) using a field-proven, rugged design; For visible radiation (sunlight): spectrophotometers for spectrally resolved measurements.
- Broadband radiometers: A duplication, to the extent possible, of the pyranometers, pyrgeometers, and pyrheliometers selected by the World Climate Research Program (WCRP) for the Global Baseline Surface Radiation Network (GBSRN) to provide measurements of direct normal shortwave (solar beam), diffuse horizontal shortwave (sky), global horizontal shortwave (total hemispheric), upwelling shortwave (reflected), downwelling longwave (atmospheric infrared), and upwelling longwave (surface infrared) radiation.
- Meteorological instrumentation: Surface-based or tower (i.e., near surface) measurements of meteorological variables [temperature, humidity (i.e. water vapor), pressure, wind speed and direction, precipitation amount, surface fluxes, and cloud cover]. Measurements of aerosols, trace gases, aerosol optical depth, and soil moisture. Vertical profile measurements of meteorological variables associated with radiative transfer. These include in situ measurements from balloonborne sensors or aircraft. The program also considered remote sensors, such as Raman lidar, differential absorption lidar, and microwave radiometry for water vapor and radio acoustic soundings for winds and temperature critical, with the idea that these remote sensors would ultimately replace the need to launch radiosondes (Turner et al. 2016a, chapter 18).
- Calibration instrumentation: A state-of-the-art calibration facility to ensure radiometric calibration with standards referenced to World Radiation Center instruments. A field calibration capability for balloonborne sensors.

2) EXTENDED FIELD SITES

- Broadband radiometric instrumentation identical to the central facility reflective of the WCRP GBSRN network (pyranometers, pyrgeometers, and pyrheliometers).
- Surface meteorological, surface flux, and soil moisture instrumentation as at the central facility.
- Surface reflectivity measurements.
- Cloud cover, perhaps with whole-sky imagery.

3) BOUNDARY SITES

- Perhaps collocated with an extended field site.
- Additional instruments include a balloonborne sounding system for measurements of advective fluxes germane to the SCM experiment.

4) AUXILIARY SITES

- Perhaps collocated with an extended field site.
- Instrumentation suited to map the three-dimensional structure of the atmospheric properties in the vicinity of the soda straw experiment at the central facility.
- Sites would be 10–20 km from the central facility and instrumented with, perhaps, radio acoustic sounders, scanning radars, and whole-sky cloud imagers.

f. *Planning for laboratory-grade or developmental instruments in CART—The IDP*

As Stokes (2016, chapter 2) discussed, many effective instruments had been developed and used in short- and midterm observational campaigns, with operation and data quality being dependent on principal investigator interaction and data reduction. Some of these kinds of instruments were identified specifically in the ARM Program Plan as being desirable for CART measurements (DOE 1990). The ARM Instrument Development Program (IDP) (Stokes and Schwartz 1994; DOE 1996; Stokes 2016, chapter 2; Ackerman et al. 2016, chapter 3) was implemented to support continuing engineering development and technical evaluation of selected instruments to develop a capability for unattended, routine 24/7 operations within the CART environment. Several IDP-supported efforts were highly successful, providing instruments for CART that would not have been available. Some of the successes were the atmospheric emitted radiance interferometer (AERI; Turner et al. 2016b, chapter 13), the multifilter rotating shadowband radiometer (MFRSR; Michalsky and Long 2016, chapter 16), the micropulse lidar, the millimeter cloud radar (MMCR; Kollias et al. 2016, chapter 17), and the Raman lidar (Turner et al. 2016a, chapter 18). Not all instrument development efforts under the IDP were successful in producing instruments for routine, unattended 24/7 operation. Only limited success was achieved for a solar radiance transmission interferometer, an absolute solar transmittance interferometer, and a rotating shadowband spectrometer, for example.

g. Planning for the CART data system

The ARM Program Plan provided a view of what was required for data acquisition, processing, and archival, but the real meat of the CART Data Environment (as the data system was termed in those years) was put on the skeleton during a series of facilitated planning workshops. The skeleton provided by the ARM Program Plan guidance called for the CART Data Environment to have the following characteristics and features (DOE 1990; Stokes and Schwartz 1994; McCord and Voyles 2016, chapter 11):

- Development to be kept to a minimum, with the system depending on existing data centers for data distribution and archival functions
- Real-time processing of instrument data streams while executing primary quality control with feedback to site operations for instrument maintenance and operations
- Data transmission to a central location for conversion to a standard format for distribution to science members and the archive, where it would be available to the general scientific community
- Data fully documented and archived in raw and processed form to facilitate reprocessing, if required, at later dates
- Routine acquisition of data from external sources, such as weather and sounding data from NWS, satellite data from NASA and from ongoing field programs, such as the WCRP GBSRN
- An extensible data system with the capability to grow to accommodate additional instruments, increases in data volume, and to support short- or midterm intensive field programs conducted by ARM

The effective implementation and reliable operation of the data system was recognized to constitute a critical element of the program, as clearly stated in the Program Plan: "The design of the ARM project presupposes a well-designed, smoothly functioning, research data management system."

3. Deployment and implementation of CART

As discussed previously, the ARM Program Plan provided a rather detailed high-level view of what the deployed CART would look like, with requirements for operations, instruments, and data processing and archival. However, it did not provide the roadmap to get it done. Facilitated planning workshops were utilized, essentially, to convert the Program Plan into a hierarchical structure of functional requirements for the deployment and operation of CART. Personnel with the required technical backgrounds from the DOE laboratories were tasked to participate as part of the teams addressing each of the functional elements identified earlier. These teams began their efforts in parallel in 1990.

Stokes (2016, chapter 2) points out that the relatively complete description of the relationships between elements of the program provided in the Program Plan enabled the teams' efforts. The teams were able to take those understood relationships and develop detailed approaches (and therefore plans) for what had to be done to put CART in the field, make it operational, and provide the path to meeting the data needs of the Science Team.

The following sections discuss the deployment activity of those teams focused on site selection, operations, instrumentation, and the data. With respect to the modeling team, the program made the decision not to run its own model and subsequently relied, for the most part, upon the efforts of the Science Team's SCM working group for modeling activity. The SCM working group led the development of the scheme for estimating advective tendencies using the data from a CART site's measurement network (Zhang et al. 2016, chapter 24).

a. Site/locale selection

At the outset, one of the highest priority efforts was to identify where CART field sites should be established. The site selection process devolved into a two-step process, distinguishing between climatologically significant locales and a specific site selection within the locales. The task of identifying and evaluating climatically significant locales was led by Steve Schwartz from Brookhaven National Laboratory. DOE Technical Report ER 0495T (DOE 1991) presents the findings and recommendations and

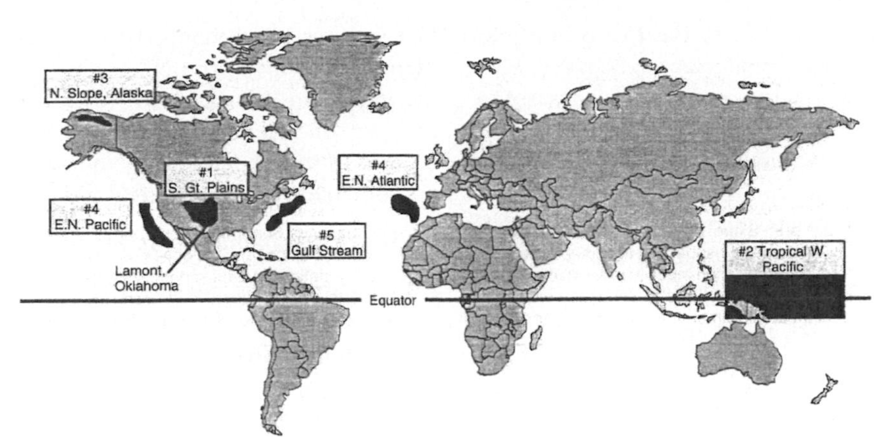

FIG. 5-2. Geographical distribution of recommended locales circa 1991 (Stokes and Schwartz 1994).

rationale for the locales identified. Specific siting considerations and decisions for the three sites that were ultimately established in the initial deployment are detailed in Sisterson et al. (2016, chapter 6), Long et al. (2016, chapter 7), and Verlinde et al. (2016, chapter 8).

To start the process of locale selection, the team first established a set of principles derived from the purposes of the ARM Program (DOE 1991; Stokes and Schwartz 1994). Briefly stated, these were:

- The set of locales should stress the radiation models, spanning the domain of radiation-influencing attributes (latitude, altitude, clouds, humidity, aerosols, etc.).
- Climatological and surface-property attributes should be as homogeneous as possible across the locale (with deliberate exceptions).
- Establishing a site within the locale must be logistically feasible.
- The opportunity for collaboration with other programs in a given locale gives additional weight to the significance of that locale.

Based on these principles, nominally 20 locales were identified for further evaluation. Five locales (Fig. 5-2) were recommended and ranked based on operational feasibility and the anticipated budget for establishing the sites as well as the locale's scientific value (for GCM modeling). The locales and the rationale for each were as follows:

1) U.S. Southern Great Plains (SGP): logistics; synergism; a wide variety of cloud types; a wide range of temperature and specific humidity; large annual, synoptic, and diurnal variations
2) Tropical Western Pacific Ocean (TWP): deep tropical convection; cirrus clouds; interannual variability in sea surface temperature; high sea surface temperature; high specific humidity
3) Eastern North Pacific Ocean (ENP) or eastern North Atlantic Ocean: marine stratus; transition between marine stratus and broken cloud fields; high specific humidity
4) North Slope of Alaska (NSA): large seasonal variations in surface properties; distinct surface properties from other locales
5) Gulf Stream off eastern North America, extending eastward: extreme values and variation in surface heat fluxes; marine stratus clouds; altostratus clouds; mature cyclonic storms; genesis region for cumulonimbus and widespread layered clouds associated with large synoptic storms

In addition to the five primary locales for permanent installations, four supplementary locales intended for episodic occupation by the ARM Mobile Facility were identified with the caveat that other sites might be added or substituted based on specific Science Team needs. The four potential ARM mobile facility deployment sites were:

1) Central Australia or Sonoran Desert: high temperature; low specific humidity; frequent totally clear skies
2) U.S. Northwest–southwest Canada coast: coastal and orographic inhomogeneity; marine stratus and nimbostratus clouds
3) Amazon basin or Congo basin: deep convection; large latent heat fluxes; high specific humidity; large seasonal variation in rainfall
4) Beaufort Sea, Bering Sea, or Greenland Sea: sea ice; sea ice edge; fog and marine stratus clouds

On the basis of the locale recommendations, the ARM management team identified authors to provide locale-specific reports for each of the five primary locales, evaluating the operational issues to be faced. Each of the reports examined the research purposes that each site could address, as well as the specific instrumentation, modeling, implementation, and operational issues that emerged. The authors had the freedom to explore the full set of issues encountered in addressing the following operational questions, establishing the content of these reports:

- Why conduct operations within this locale?
- What measurements must be taken?
- What are the logistic and operational problems for CART operations at this locale?
- How can these problems be resolved?
- What are the logical linkages to other candidate locales and the most appropriate extensions of the primary mission?
- How would the measurement strategies outlined in the ARM Program Plan (DOE 1990) be implemented at this CART locale?

Significantly, the narrowing of a locale to a site depended on ease of site access, transportation, supplies, and services. Additionally, since the SGP and NSA locales were in the United States, National Environmental Policy Act (NEPA) requirements dictated screening to ensure that the proposed CART site and related activities would not adversely affect any environmentally sensitive areas, such as historic or cultural resources, protected areas such as parks, or other ecologically significant areas. Furthermore, the CART site and activities must not threaten animal species that were considered endangered or threatened or have adverse effects on the designated critical habitats for these species.

Other significant considerations for site placement included restrictions on the use of airspace for aircraft and balloons, operation of lidars, and access to the appropriate radio transmission frequencies and telephone/Internet for operations and communications. Another important discriminant was the opportunity for synergism with other field programs where instruments and data could be shared.

Relatedly, the ARM Program scope was evaluated against budget expectations on an annual basis. Budget limitations, recognized circa 1994–95, resulted in the reduction of planned site deployments from five to three, deferred any development or planning for a mobile facility (possibly permanently), and slowed deployment activity at the primary sites and in the capability growth of the data system (Lunn et al. 1995). The sites to be established were the SGP, TWP, and NSA. The eastern Pacific/Atlantic Ocean and Gulf Stream sites were changed from primary sites to supplementary status. The NSA was moved ahead of the ocean margins (i.e., eastern North Pacific/Atlantic) site because of recent substantial field programs in the eastern North Atlantic and the scarcity of radiation- and/or cloud-related research programs in the Arctic.

b. Site deployment/operations planning

Planning for the physical deployment and operation of the sites was the function of the site operations team led by Sumner Barr of Los Alamos National Laboratory. Like the other teams, the site operations team participated in facilitated planning workshops to flesh out the site deployment and operations management functionality. Included in the defined functionalities for each site were roles for a site manager and a site scientist. Planning the functionality of the sites created the roadmap or blueprint for them, which was something that was missing from the ARM Program Plan.

Based on recommendations from the ARM management team and to kick start deployment of the first site, DOE named Doug Sisterson of Argonne National Laboratory as the site manager for the SGP. April 1992 was targeted as the start-up date for data to be generated at that site. Setting a target date accomplished two things: first, it removed the fear of delaying site deployment for the development of a perfect plan (Stokes 2016, chapter 2); second, it provided a firm timeline for the infrastructure to map actual deployment milestones. Proposals were requested from the laboratories for nominees to manage the other sites and to propose how these other sites might be configured considering the conceptual CART site design, the functionality being developed in the planning workshops, and the exigencies of the specific locale. Proposals were reviewed by ARM management and recommendations provided to DOE. Consequently, Bill Clements from Los Alamos National Laboratory (LANL) was named site manager for the TWP and Bernie Zak from the Sandia National Laboratory (SNL) for the NSA. In addition, Paul Michael from Brookhaven National Laboratory (BNL) was given responsibility for the Gulf Stream site, and Mike Reynolds from BNL for the eastern oceans margins site. Shortly thereafter these two locales were removed from the deployment schedule.

While the first site was to be operational in 1992, the conceptual plan was to activate one site per year, with the five sites operational by 1997. As discussed, budget realities forced the scaling back of the intended project to three sites, deferred/cancelled the Mobile Facility, and dictated a phased deployment of the sites in accordance with what the annual funding allowed. A mantra remained to get all of the sites established and operational as soon as it could be accomplished. It was the site managers who bore the brunt of the effort to meet the timeline, and it was they who saw the task to a successful end despite the plethora of hurdles to overcome. The result was that first measurements at the SGP were started in 1992 (Stokes 2016, chapter 2), but extended facilities were still being deployed two years later. The TWP site achieved initial operational status in 1996 (Long et al. 2016, chapter 7) and the NSA in 1997 (Verlinde et al. 2016, chapter 8).

Site deployment began when the site managers converged on a general blueprint for their sites. Identifying which IRF and SCM requirements could be addressed at each site was the long pole in the planning tent. It was clear that, except for the SGP, neither of the other sites could implement the ideal CART site depicted in Fig. 5-1 with a central facility surrounded by a network of extended sites because of remoteness and geography. Each site had to conceive of a facility and instrument configuration that made the most sense. This challenge alone was major and involved working with ARM management and the Science Team to achieve an optimum and achievable site plan. As usual, however, the devil was in the details, and these details were, at times, major hurdles. A few examples of the types of hurdles that were recognized included NEPA and similar requirements at state and local levels; contracting with foreign governments; and developing local support in the country (i.e., around the U.S. sites) and in foreign lands (not as easy as it sounds). As discussed by Long et al. (2016, chapter 7), the limited availability of scientific and technical talent at the TWP island sites posed a significant challenge and required substantial efforts to reach out to local governments and communities.

Timetables were constructed, and deliverables were determined. Detailed planning went into all aspects of site design, instrumentation, operations, and data quality. The efforts have proven robust, with remarkably few changes being required over years of operation.

c. Concrete poured

> No plan survives first contact with the enemy.
> —Field Marshall Helmuth Carl Bernard Graf von Moltke [courtesy of Stokes (2016, chapter 2)]

Moltke's comment was never so apt as when applied to actually putting sites on the ground in each of the locales. Each of the site locale analysis reports started with the idealized CART site (Fig. 5-1) but then married the ideal with reality. On the basis of synergism with existing measurements from other programs and logistic constraints (e.g., terrain, power, communications, and politics), actual locations for CART sites were identified and the instrumentation and facilities rescoped to fit. The SGP site was the least complicated and closest to the idealized site—about the size of a GCM grid cell ($\sim$300 km $\times$ 300 km), more-or-less homogeneous across its surface, relatively easily accessible, and there was a road to it. It had the additional benefit of being in the midst of a high density of measurements being made by other programs that would be beneficial to the ARM Science Team.

The TWP and NSA sites, however, were unique. The warm pool in the TWP locale would be best characterized by establishing a central facility–type site (i.e., primarily supporting the IRF, or soda straw, measurement strategy) in the middle of the warm pool with similar facilities at additional sites located closer to the east and west peripheries of the warm pool. These would necessarily be island sites, limiting SCM characterization capability. As an example of the type of out of the box thinking that was required for these remote sites, it was decided that the best deployment model for the TWP would be to build the central facilities using sea containers configured prior to deployment and delivered as units to the selected sites (Long et al. 2016, chapter 7). The NSA site was somewhere in between the SGP and TWP in complexity, being on land but with extremely limited siting options. A central facility focused on the IRF measurement strategy was feasible, but the gridcell characterization would be limited to a few sites at best supplemented by temporal field programs, probably in collaboration with other programs (Verlinde et al. 2016, chapter 8).

d. Local site operations

The facilitated planning workshops mapped site operations down to an implementable level of detail. This mapping determined the functions of on-site personnel, their qualifications, what information and reports were needed (corrective maintenance, preventative maintenance, calibrations, an operations log book, etc.), and who needed to receive them. The actual staffing of personnel and establishment of operational protocols (e.g., facility and instrument maintenance) was left to the individual site managers. Some debate did arise about operating a site with a permanent scientific staff as opposed to hiring local people and training them. A qualified scientific staff was perceived as desirable because of the complexity and diversity of instrumentation, and ARM was a research program. Hiring locally would require training and routine visits from technical experts but would be more cost effective and, perhaps, make the site more valued by the surrounding community. The latter scenario was implemented and proved to be very effective.

One of the most critical functionalities identified in the planning process was data acquisition, processing, and communications. While data acquisition and processing will be discussed in a later section, data communication capability was a critical element for site operations, requiring either the provision of sufficient bandwidth for the anticipated data flows [nominally 7 GB day^{-1} per site (DOE 1990)] or a plan to physically transport storage media (sneakernet)[3] if required. The advent of widespread Internet access during this period greatly facilitated data transfer capabilities (McCord and Voyles 2016, chapter 11), with site-specific attributes as discussed in the chapters that follow. For example, in the beginning of the TWP's history, the sneakernet was critical to getting most data back from these tropical sites to the data system computers on the U.S. mainland.

e. *Site scientists*

One of the functional roles identified in the facilitated planning workshops was the need for a site scientist. It was envisioned that a site scientist would be named for each fixed site, ideally a researcher from a university near the site or who was deeply involved in research in the geographic area of the ARM site. The overall responsibility of the site scientists would be to ensure that local site operations did not jeopardize data quality (e.g., by regularly driving diesel trucks past the aerosol intake stack), assist with field campaigns at the site, review onsite changes to physical structures, and oversee the data quality practices for the site. The site scientists would also ensure that ARM efforts did not become insular from the interests of the general atmospheric research community. As the program evolved, the role of the SGP site scientist's office, in particular, took on a greater role in data quality, as discussed by Ackerman et al. (2016, chapter 3) and Peppler et al. (2016, chapter 12).

A natural tension existed between the site manager and the site scientist, because the ideal support for research objectives was often in conflict with logistical or budgetary reality. The tension ensured that all feasible options were always taken into account in site operations planning.

The selection of the site scientists was through a request for proposals issued by DOE. DOE and ARM management reviewed all proposals, selecting Dr. Peter Lamb from University of Oklahoma as the site scientist for the SGP, Dr. Thomas Ackerman from Pennsylvania State University for the TWP, and Dr. Knut Stamnes from the University of Alaska for the NSA as the original site scientists for the three primary sites.

[3] Sneakernet—a term coined to denote the physical recording of data and the transport of that recorded data to a processing center by personnel returning from the remote site.

4. Instrumentation

Instrumentation planning for the CART sites followed a path similar to site selection and site operations functional areas. Facilitated planning workshops identified the key functional activities that would be required to acquire, deploy, and operate instruments in the CART site environment. Commercially available and routinely used instruments posed one set of questions for acquisition and implementation, but other desired instruments were not as mature and not as readily incorporated into the 24/7 operation of the CART facility. Additional instruments were anticipated to evolve from the IDP, as discussed earlier, and required a more fundamental approach. The instrument team, under the leadership of Marv Wesley at Argonne National Laboratory, developed a methodology to deploy each class of instrument with primary attention being paid to reliable operation and data quality. Interfacing the instruments to the data system for continuous operation, data quality control, and data distribution will be discussed later. To deal with the spectrum of issues for putting an instrument into the CART environment, instrument mentors were assigned to each instrument. Site-specific instrument operational issues that required instrument mentor support are discussed in the site-specific chapters.

a. *Instrument acquisition*

Instrument acquisition within the DOE environment both is and is not a straightforward task. The straightforward task is the contracting with a vendor for an instrument for delivery meeting a specified set of requirements (e.g., environmental hardening, performance limits, calibration requirements, and delivery schedule). Critical deliverables for each procurement needed to address finer points of data processing, such as the formats of the data output and available quality control checks, just to name two. The not-so-straightforward issue concerned instrument acquisition funding. The funding to support the ARM Program infrastructure in deploying and maintaining the site is received from the DOE operations and maintenance (O&M) budget line, where DOE managers have the discretion to determine the funding for individual programs under their oversight. Instruments, however, are considered capital property (if over a modest cost threshold) and must be acquired with capital funds. Capital funding is a different process that requires identification of capital needs, review against other DOE capital needs, and finally a designation of the capital funds that will be allocated to the individual programs. Requestors (programs) will likely not get all the funds requested and, most likely, not in the time frame that they would like to receive them. In this context, capital funding for ARM instruments was not received in time to permit acquisition and delivery in time to meet the April 1992 start-up date. To meet that date, the first instruments to arrive at the SGP site were borrowed from NCAR and replaced with ARM property later in the year. It was largely the foreseen capital funding in future years that proved to be a key factor in the decision to reduce the number of sites and phase the instrument deployment schedule, as has been discussed previously.

b. *Instrument mentors*

As mentioned, ARM recognized that technical experts were needed for each instrument. These experts not only had to know the instrument hardware, but they also needed to know instrument software and, most importantly, had to have used the data from that instrument previously in their own research. ARM instrument mentors were critically important and impactful. Their specific tasks evolved as deployment phased into operation, but new instruments were always on the horizon (or closer), and as such the list of responsibilities remained relatively static during the deployment years:

- Develop the technical specifications for instruments and spare components to be procured.
- Develop procedures for instrument operations (e.g., daily rounds, maintenance, and calibration).
- Assess instrument (and measurement) uncertainty, status, and quality.
- Manage instrument repairs.
- Work with site ops to upgrade instrument performance or upgrade an instrument as appropriate.
- Work with ARM data system personnel on data product requirements, the specification of appropriate operating ranges, the determination of appropriate data flags when data fall outside the range, and the development of any additional data quality control procedures that are feasible in the course of real-time data acquisition and processing.
- Alert the Data Quality Office, site operations, and science data users to data quality problems.
- For the balloonborne sounding system (BBSS) and other in situ sampling system, provide for a continuing in situ sampling program.
- Participate in intensive operational periods (IOPs)[4] as appropriate.

[4] "IOP" has been an ambiguous reference in ARM since the beginning. Some use it to refer to an "intensive operational period," and some prefer to use it to mean an "intensive observational period." Both uses are found regularly on the ARM website.

5. The CART Data Environment

The ARM data system (McCord and Voyles 2016, chapter 11), as initially established, was known as the CART Data Environment (CDE). The elemental structure of the CDE and how the data flowed in the system are shown in Fig. 5-3. Recognizing that data was to be made available to the Science Team as soon as instruments were in the field, planning for the data system and development of some elements of the CDE were begun immediately upon approval and funding of the program. One aim of the CDE developmental effort was, to the degree feasible, to use existing facilities and data centers for processing, dissemination, and archiving of the data. Underscoring what proved to be an inadequate ability to foresee the pace of technological advance, the plan also specified that "existing technology in software and hardware" would be used (to control cost). Of course, one could argue over the definition of the word "existing" (i.e., at the time or at a time in the future).

The mantra for the program was that the data would have "known and reasonable quality" (Stokes 2016, chapter 2). Accordingly, the Program Plan discussed various operational requirements for the data system. The following requirements distilled from the Program Plan were central to the design and implementation of the CDE:

- Real-time quality control (for every instrument data stream)
- Data availability on site for instrument monitoring
- "Ready availability" of data to data centers
- Data to be "well documented"
- Data (raw and processed) and data documentation (metadata) to be archived for future reference and the possible necessity of reprocessing in the future
- Data converted to a "standard format" for ease of use by the research community

The CDE schematic implies data moving smoothly through the system, but as has been discussed, adequate bandwidth needed to be established at the SGP site, and data from the remote sites had to be physically transported on storage media. The data flow was not necessarily smooth or without delay, at least at first.

The deployment era contained great challenges, a lot of time and travel, and experts who came together to learn the software and hardware to actually design and implement the CDE. They used, for the most part, strict software system design protocols to design the various aspects of the system and develop the coding required. To that group, the data management gurus, the data system probably appeared to be simply a very large but

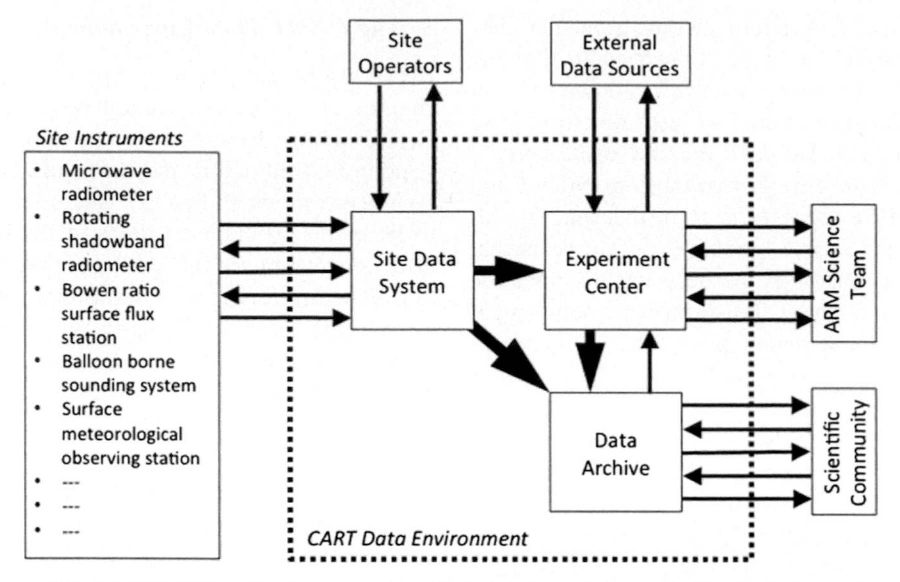

FIG. 5-3. CART Data Environment circa 1992 showing data flow. The Site Data System is a component of the CDE established at each site's central facility for the purpose of acquiring data in real time and completing initial processing, which included conversion to standard units, application of calibrations, passing data through quality control checks, and running measurement-related algorithms for evaluating instrument performance. Processed data were forwarded to the Experiment Center, and both raw and processed data were sent to the archive. The Experiment Center was a two-component data center, with a program data center at PNNL and an external data center at BNL. The PNNL facility was responsible for receiving data from the sites, creating higher-order data products, creating data packages tailored to the needs of Science Team members, and "pushing" those packages to them. The BNL facility was responsible for acquiring data from external sources, creating data packages for the Science Team, and pushing the packages to them. The ARM data archive was established at Oak Ridge National Laboratory and was responsible for receiving and archiving all raw and processed data from all field sites and the Experiment Center. It also had reprocessing responsibility and acted as the user interface for the general scientific community (Stokes and Schwartz 1994).

tractable data system design challenge. To the nongurus, it had the appearance of an overwhelming task with innumerable opportunities to fail. The truth, fortunately, was in between, and a functional and reliable system was developed with the first elements of the system in place to transmit data successfully from the SGP site in April 1992.

a. Data and Science Integration Team

The Data and Science Integration Team (DSIT)[5] was the evolutionary group of data system developers and other members of the ARM Program with responsibility to develop, field, and manage the various elements that made up the CDE. For the purposes of this discussion, references to DSIT and its responsibilities embody the spectrum of an instrument's activity from development to deployment to operation regardless of whether or not DSIT was the formal name at the time or not. Like the name, leadership of this activity changed with time, beginning with Ron Melton at PNNL and then Jimmy Voyles at PNNL. Paul Kanciruk, Paul Singly, and Raymond McCord at Oak Ridge National Laboratory (ORNL), with primary responsibility for the ARM data archive, were, in essence, partners with Melton and Voyles in managing this very large and challenging aspect of the ARM Program. Gerry Stokes, the first ARM chief scientist, felt that the DSIT (and its predecessor the Experiment Support Team) was important because an observatory had to support ongoing experiments, which required coordination.

The DSIT had many faces. On one hand, it developed the software and computer systems for site operations and the data centers. On the other hand, it had a

[5] Before the DSIT, there was an Experiment Support Team and a Data Management Team. Marv Dickerson and Ric Cederwall were leaders of the former, and Ron Melton and Jimmy Voyles led the latter. These two teams were merged to create the DSIT.

responsibility to interact with members of the Science Team, leading the way to translating science needs into data needs, with data quality as a fundamental objective. For data acquisition, DSIT involved the instrument mentors to ensure that the instruments were producing data of expected precision and accuracy and to ensure that a complete record of instrument calibrations and operational history was maintained. In another context, the DSIT was that element of ARM that had responsibility for responding to what might be called special data needs. One example was the development of showcase datasets. Early in the program, there was concern about how to make ARM data more accessible and useable by both the Science Team and by the general scientific community. One option was to create more-or-less complete datasets for a given question. The DSIT worked with Science Team members to create and organize the dataset, making it available as a singular entity.

In essence, in the data world of ARM, DSIT "carried the water." They were at the heart of ARM and served a critical function for the Science Team. As deployment was completed, various aspects of DSIT's role split off to be part of other functions, detailed, in part, by Ackerman et al. (2016, chapter 3), although they do not call out DSIT by name.

b. *Value-added products*

Value-added products (VAPs) were not, conceptually, an identified element of the initial deployment in the early 1990s. VAPs are discussed in substantial detail by Ackerman et al. (2016, chapter 3). Antecedents to VAPs, however, were developed almost immediately upon instrument deployment. To ensure the highest data quality, measurements from different sources were merged to create an integrated data product specifically intended to evaluate and improve measurement capabilities of deployed instruments. As the project matured, these merged products became more complex and involved but still served their purpose to "make it easier for Science Team members to use ARM data, or to reprocess the original dataset to improve the quality of the data" (DOE 1996; ARM 2016b, appendix B). If the results of the VAPs could point to an instrument or operational improvement, the results were fed back to the DSIT and instrument mentors for evaluation and the implementation of corrective action, if feasible.

c. *External and IOP data*

It was recognized, even during the preparation of the Program Plan, that ARM could not unilaterally obtain all of the data that was going to be required by the Science Team. Data from programs like the NOAA Wind Profiler Demonstration Network (e.g., Weber et al. 1990) and the WCRP BSRN sites (e.g., Ohmura et al. 1998), as well as routine surface observations and vertical soundings from the NWS would all be necessary and would involve acquisition and distribution to meet Science Team data requirements. In addition, data from field programs conducted in collaboration with ARM [e.g., FIRE and the Spectral Radiance Experiment (SPECTRE)] or conducted in the vicinity of ARM, such as the TWP island sites (e.g., TOGA COARE), would be needed. The planning and development of the functionality of the ARM data centers were undertaken in full recognition of these realities, which then became part of the routine operation of the program. While ARM was basically designed to provide routine measurements continuously, those measurements were not always going to be sufficient. The capability was planned into ARM and its data system to ramp up for high-intensity efforts for limited periods of time to acquire data during IOPs that might be "too expensive or personnel intensive to be conducted continuously" (Stokes and Schwartz 1994).

6. Conclusions

The ARM Program was conceived in the wake of research efforts as part of ICRCCM that concluded that one of the largest sources of error in the GCMs being used for climate modeling was in the radiation parameterization components of those models. The recommended path to improving the parameterizations was a long-term measurement program involving permanent (more than a decade) ground sites measuring a full spectrum of radiation-influencing parameters of the atmosphere. DOE opted to adopt this issue as the primary focus of the department's contribution to the USGCRP. To organize and implement the necessary field observational program, DOE tapped the technical strength of its family of laboratories, using technical expertise in those laboratories as a corporate resource. During planning discussions, it was recognized that, on the surface, while the ICRCCM recommendation suggested a soda straw experiment (measurements in a column over a point), in reality, clouds and their representation in the models were part of the larger picture involving radiation processes in the atmosphere and needed to be addressed as well. The laboratory planning sessions produced a concept involving a highly instrumented permanent facility surrounded by smaller groups of instruments to document the three-dimensional structure of the atmosphere over a soda straw measurement facility at the center. A network of five measurement sites was proposed with additional sites for short-term data acquisition efforts. Budget considerations limited the project to three permanent sites: the SGP, TWP, and NSA. The first and most comprehensive site was the SGP, which produced its first data in 1992, with deployment continuing into 1996. The TWP initial operational capability was 1996, and the NSA began operating continuously in 1997. The suite of sites comprised what was termed the Cloud and Radiation Test Bed. The data system for the project was designed to acquire data in real time, ensure data quality control by several methods, and transfer the data to the Science Team and to the ARM data archive. The ARM data archive then functioned as the entry point (user facility) for the general scientific community who desired to access and use ARM data. These capabilities evolved with time away from the initial deployment, but remain very much in operation to the current time.

REFERENCES

ARM, 2016a: Appendix B: Executive summary: Science Plan for the Atmospheric Radiation Measurement Program (ARM). *The Atmospheric Radiation Measurement (ARM) Program: The First 20 Years, Meteor. Monogr.*, No. 57, Amer. Meteor. Soc., doi:10.1175/AMSMONOGRAPHS-D-15-0035.1.

——, 2016b: Appendix C: Executive summary: Atmospheric Radiation Measurement Program Science Plan: Current status and future directions of the ARM Science Program. *The Atmospheric Radiation Measurement (ARM) Program: The First 20 Years, Meteor. Monogr.*, No. 57, Amer. Meteor. Soc., doi:10.1175/AMSMONOGRAPHS-D-15-0034.1.

Ackerman, T. P., T. S. Cress, W. R. Ferrell, J. H. Mather, and D. D. Turner, 2016: The programmatic maturation of the ARM Program. *The Atmospheric Radiation Measurement (ARM) Program: The First 20 Years, Meteor. Monogr.*, No. 57, Amer. Meteor. Soc., doi:10.1175/AMSMONOGRAPHS-D-15-0054.1.

DOE, 1990: Atmospheric Radiation Measurement Program Plan. DOE Tech. Rep. DOE/ER-04411990, 121 pp. [Available at http://www.arm.gov/publications/doe-er-0441.pdf.]

——, 1991: Identification, recommendation, and justification of potential locales for ARM sites. U.S. DOE Tech. Rep. DOE/ER 0495T, 160 pp.

——, 1996: Science Plan for the Atmospheric Radiation Measurement Program. U.S. DOE Tech. Rep. DOE/ER-0670T, 174 pp.

Ellingson, R. G., and Y. Fouquart, 1991: The intercomparison of radiation codes in climate models: An overview. *J. Geophys. Res.*, **96**, 8925–8927, doi:10.1029/90JD01618.

——, and W. J. Wiscombe, 1996: The Spectral Radiance Experiment (SPECTRE): Project description and sample results. *Bull. Amer. Meteor. Soc.*, **77**, 1967–1985, doi:10.1175/1520-0477(1996)077<1967:TSREPD>2.0.CO;2.

——, R. D. Cess, and G. L. Potter, 2016: The Atmospheric Radiation Measurement Program: Prelude. *The Atmospheric Radiation Measurement (ARM) Program: The First 20 Years, Meteor. Monogr.*, No. 57, Amer. Meteor. Soc., doi:10.1175/AMSMONOGRAPHS-D-15-0029.1.

Kollias, P., and Coauthors, 2016: Development and applications of ARM millimeter-wavelength cloud radars. *The Atmospheric Radiation Measurement (ARM) Program: The First 20 Years, Meteor. Monogr.*, No. 57, Amer. Meteor. Soc., doi:10.1175/AMSMONOGRAPHS-D-15-0037.1.

Long, C. N., J. H. Mather, and T. P. Ackerman, 2016: The ARM Tropical Western Pacific (TWP) sites. *The Atmospheric Radiation Measurement (ARM) Program: The First 20 Years, Meteor. Monogr.*, No. 57, Amer. Meteor. Soc., doi:10.1175/AMSMONOGRAPHS-D-15-0024.1.

Lunn, P., T. S. Cress, and G. M. Stokes, 1995: History and Status of the Atmospheric Radiation Measurement Program—March 1995. *Proc. Fifth Atmospheric Radiation Measurement (ARM) Science Team Meeting*, San Diego, CA, U.S. DOE, iii–vii. [Available online at https://www.arm.gov/publications/proceedings/conf05/extended_abs/history.pdf.]

Luther, F., Ed., 1984: The Intercomparison of Radiation Codes in Climate Models. World Climate Program Rep. WCP-93, 37 pp.

McCord, R., and J. W. Voyles, 2016: The ARM data system and archive. *The Atmospheric Radiation Measurement (ARM) Program: The First 20 Years, Meteor. Monogr.*, No. 57, Amer. Meteor. Soc., doi:10.1175/AMSMONOGRAPHS-D-15-0043.1.

Michalsky, J. J., and C. N. Long, 2016: ARM solar and infrared broadband and filter radiometry. *The Atmospheric Radiation Measurement (ARM) Program: The First 20 Years, Meteor. Monogr.*, No. 57, Amer. Meteor. Soc., doi:10.1175/AMSMONOGRAPHS-D-15-0031.1.

Ohmura, A., and Coauthors, 1998: Baseline Surface Radiation Network (BSRN/WCRP): New precision radiometry for climate research. *Bull. Amer. Meteor. Soc.*, **79**, 2115–2136, doi:10.1175/1520-0477(1998)079<2115:BSRNBW>2.0.CO;2.

Peppler, R., K. Kehoe, J. Monroe, A. Theisen, and S. Moore, 2016: The ARM data quality program. *The Atmospheric Radiation Measurement (ARM) Program: The First 20 Years, Meteor. Monogr.*, No. 57, Amer. Meteor. Soc., doi:10.1175/AMSMONOGRAPHS-D-15-0039.1.

Ramirez, A., Ed., 1990: The federal plan for meteorological services and supporting research: Fiscal year 1990. U.S. Office of the Federal Coordinator for Meteorological Services and Supporting Research Rep. FCM-P1-2015, 290 pp. [Available online at http://www.ofcm.gov/fedplan/FY2016/pdf/FCM-P1-2015.pdf.]

Sisterson, D. L., R. Peppler, T. S. Cress, P. Lamb, and D. D. Turner, 2016: The ARM Southern Great Plains (SGP) site. *The Atmospheric Radiation Measurement (ARM) Program: The First 20 Years, Meteor. Monogr.*, No. 57, Amer. Meteor. Soc., doi:10.1175/AMSMONOGRAPHS-D-16-0004.1.

Stokes, G. M., 2016: Original ARM concept and launch. *The Atmospheric Radiation Measurement (ARM) Program: The First 20 Years, Meteor. Monogr.*, No. 57, Amer. Meteor. Soc., doi:10.1175/AMSMONOGRAPHS-D-15-0021.1.

——, and S. E. Schwartz, 1994: The Atmospheric Radiation Measurement (ARM) Program: Programmatic background and design of the cloud and radiation test bed. *Bull. Amer. Meteor. Soc.*, **75**, 1201–1221, doi:10.1175/1520-0477(1994)075<1201:TARMPP>2.0.CO;2.

Turner, D. D., J. E. M. Goldsmith, and R. A. Ferrare, 2016a: Development and applications of the ARM Raman lidar. *The*

Atmospheric Radiation Measurement (ARM) Program: The First 20 Years, Meteor. Monogr., No. 57, Amer. Meteor. Soc., doi:10.1175/AMSMONOGRAPHS-D-15-0026.1.

——, E. J. Mlawer, and H. E. Revercomb, 2016b: Water vapor observations in the ARM Program. *The Atmospheric Radiation Measurement (ARM) Program: The First 20 Years, Meteor. Monogr.*, No. 57, Amer. Meteor. Soc., doi:10.1175/AMSMONOGRAPHS-D-15-0025.1.

Verlinde, H., B. Zak, M. D. Shupe, M. Ivey, and K. Stamnes, 2016: The ARM North Slope of Alaska (NSA) sites. *The Atmospheric Radiation Measurement (ARM) Program: The First 20 Years, Meteor. Monogr.*, No. 57, Amer. Meteor. Soc., doi:10.1175/AMSMONOGRAPHS-D-15-0023.1.

Weber, B. L., and Coauthors, 1990: Preliminary evaluation of the first NOAA demonstration network wind profiler. *J. Atmos. Oceanic Technol.*, **7**, 909–918, doi:10.1175/1520-0426(1990)007<0909:PEOTFN>2.0.CO;2.

Zhang, M., R. C. J. Sommerville, and S. Xie, 2016: The SCM concept and creation of ARM forcing datasets. *The Atmospheric Radiation Measurement (ARM) Program: The First 20 Years, Meteor. Monogr.*, No. 57, Amer. Meteor. Soc., doi:10.1175/AMSMONOGRAPHS-D-15-0040.1.

Chapter 6

The ARM Southern Great Plains (SGP) Site

D. L. SISTERSON

Argonne National Laboratory, Argonne, Illinois

R. A. PEPPLER

Cooperative Institute for Mesoscale Meteorological Studies, University of Oklahoma, Norman, Oklahoma

T. S. CRESS*

Pacific Northwest National Laboratory, Richland, Washington

P. J. LAMB[+]

Cooperative Institute for Mesoscale Meteorological Studies, University of Oklahoma, Norman, Oklahoma

D. D. TURNER

NOAA/National Severe Storms Laboratory, Norman, Oklahoma

1. Introduction

At its very core, the Atmospheric Radiation Measurement (ARM) Program's objective was to make a wide range of atmospheric measurements that would support the science needed to improve the treatment of clouds and atmospheric radiation in global climate models (e.g., Stokes and Schwartz 1994; Ackerman and Stokes 2003; Ellingson et al. 2016, chapter 1; Stokes 2016, chapter 2). Early ARM planning indicated that this ambitious objective would require the establishment of several comprehensive measurement facilities in key climate regions across the globe (Cress and Sisterson 2016, chapter 5). Further, it was clear that one of those facilities should be placed in a midcontinental, midlatitude location in the Northern Hemisphere, because of the importance of those extensive areas for the functioning of the global climate system and ultimately for society through crucial agricultural production and water resources (U.S. Department of Energy 1991).

The necessity for ARM to have its first and most comprehensive measurement site in a midlatitude and midcontinent location had a strong meteorological basis. Such locations were considered to experience the broadest range of cloud and atmospheric state conditions because of their rich variety of migratory disturbances and air masses along with strong diurnal and annual cycles of surface and atmospheric conditions. The obvious choice of the central United States in the above context stemmed from its quasi-uniform surface conditions and avoidance of terrain complications, along with its logistical simplicity that involved ease of access, proximity to sources and routes of supply, and availability of troubleshooting expertise (Barr and Sisterson 2000).

The Southern Great Plains (SGP) site went from an idealized concept to reality in a very short period of time in the 1990s. This chapter describes the site, some of the people who made it happen, the logistics of building up the physical site, and some of the more unique scientific studies that have been made with the SGP observations.

* Retired.
[+] Deceased.

Corresponding author address: Douglas L. Sisterson, Environmental Science Division, Argonne National Laboratory, 9700 South Cass Avenue, EVS/Building 240, Argonne, IL 60439.
E-mail: dlsisterson@anl.gov

DOI: 10.1175/AMSMONOGRAPHS-D-16-0004.1

Dr. Fred Luther was one of the motivating forces that led ultimately to the establishment of the ARM Program (Ellingson et al. 2016, chapter 1; Cress and Sisterson 2016, chapter 5), and as such the SGP site was dedicated in Luther's name in 1992. The dedication plaque reads "During his all-too-short career, Dr. Luther made many outstanding contributions to the field of atmospheric research, particularly to furthering the understanding of atmospheric radiation and its interactions with clouds, aerosols, and gases. Talented individuals, researchers, and scientists join here to promote man's understanding of the physical processes that make his environment." This is a succinct statement of the purpose of the SGP site (and all of the ARM sites).

2. Site selection

a. Southern Great Plains locale

Initially, three "similar but different" candidate locales were identified in the central United States (appendix A; U.S. Department of Energy 1990): 1) a Midwest locale that included Illinois, Indiana, western Ohio, northeastern Missouri, and eastern Iowa; 2) a southern Great Plains locale that initially spanned northern Texas, Oklahoma, Kansas, and eastern Colorado; and 3) a northern Great Plains locale that extended across Kansas, Nebraska, North and South Dakota, and the eastern halves of Wyoming and Montana. A preliminary assessment favored the Midwest locale because of its greater range of surface energy fluxes and pollutant aerosol variation that can influence cloud optical properties. Crucial logistical issues in the Midwest locale arose that diminished its attractiveness. These issues included the use of airspace for aircraft and balloons, operation of lidar (some non-eye-safe) systems, and access to appropriate radio transmission frequencies for operations and communications. Furthermore, the scientific advantage for the Midwest locale over the other two locales was small. There was a clear recognition of advantages in shifting the selection to either the northern or southern Great Plains related to superior logistics and potential synergisms with other programs or facilities that were or ultimately would be in place. The SGP was chosen because of its considerably greater potential for synergism.

There were several synergistic opportunities that led the ARM Program to favor the SGP site. Briefly summarized, these included the following:

- The National Oceanic and Atmospheric Administration (NOAA) Wind Profiler Demonstration Network (WPDN; Smith and Benjamin 1993; Barth et al. 1994; Ralph et al. 1995) would provide profiles of wind direction and speed to altitudes through the depth of the troposphere with a temporal resolution of 15 min at sites located in the central part of the country, with its densest cluster in Oklahoma and Kansas.
- The Oklahoma Mesonet (Brock et al. 1995) was beginning to install its then 109-station network of instruments to provide continuous temperature, wind speed and direction, pressure, humidity, and other meteorological data for long-term climate studies across the state of Oklahoma, and could potentially be used to augment ARM data for a higher density of surface measurements.
- The developing National Weather Service (NWS) Weather Surveillance Radar (WSR-88D) Doppler radar network and its proposed facilities (Crum and Alberty 1993), which had radars located in a manner that would provide almost complete areal coverage over the SGP locale.
- The Tropical Rainfall Measuring Mission (TRMM; Simpson et al. 1996) satellite would be a valuable source of remote sensing data on clouds and precipitation. Because the satellite's orbit was designed for tropical systems, it was in a 35° inclination orbit, which means it was able to collect data between 35° south and 35° north. Thus, its orbit would permit data acquisition over a significant amount of the SGP site (Kummerow et al. 1998).
- The First ISCCP Regional Experiment (FIRE) program had conducted a field campaign in southeastern Kansas (Coffeyville) in the late 1980s to study cloud radiation feedback in cirrus clouds (FIRE-Cirrus) that provided invaluable insights and background for ARM instrumentation and facilities (Ackerman et al. 1990).
- Project Storm-Scale Operational and Research Meteorology (STORM)-Fronts Experimental Systems Test (STORM-FEST) was planning a field campaign in the spring of 1992 to study mesoscale convective complexes in the Southern Great Plains during the spring thunderstorm season, providing an early opportunity for collaboration (Szoke et al. 1994).
- A GEWEX Global Continental-Scale International Project (GCIP) was planned for 1994 and was expected to take advantage of the STORM deployment in Oklahoma and Kansas as well as watershed studies in Oklahoma (Lawford 1999).

The SGP was given the highest priority as the first ARM site and received the earliest development support from the ARM Program. This decision was driven primarily for the ease of access to the site (relative to the other sites being considered for the ARM Program in the tropics and Arctic). Furthermore, the high density of the vertical atmospheric structure data from the WPDN and the high density of surface characterization sites from the Oklahoma Mesonet were extremely attractive.

In addition to the synergistic value of collaborating with other research programs and the ease of access, there were two additional significant considerations related to the SGP locale, namely airspace and National Environmental Policy Act (NEPA) constraints. A primary concern was the potential for airspace restrictions over potential sites; although routine aircraft measurements initially would be cost prohibitive, short-term intensive operations periods or field campaigns were likely to use aircraft. Limitations of the airspace for scientific missions at all altitudes would be a critical factor for selecting specific site locations. Over much of the SGP locale, Vance Air Force Base had a restricted airspace for pilot training. As would be learned, Vance managed their airspace by dividing their controlled airspace into specified blocks. After tentative siting of the SGP Central Facility (CF), discussions with Vance resulted in an agreement to allow the airspace block over the site to be used by ARM aircraft (and to "flight-follow" them) during ARM field campaigns as long as it did not interfere with the Vance training mission. There would be occasions when Vance would need to preclude ARM use of the airspace, but that impact could be almost entirely mitigated by timely planning. This cooperation would turn out to be a major advantage for the SGP site.

The other major consideration for locale selection was the potential for limitations imposed by NEPA requirements. For example, there were specific guidelines not only regarding wetlands and historical sites, but also the impact of sound, light, and instrument frequency on local residents and communities that had to be evaluated. The physical layout and remoteness of the SGP location made addressing the NEPA requirements easier than would have been possible at the other candidate U.S. midlatitude sites.

b. The Cloud and Radiation Testbed

As discussed in Cress and Sisterson (2016, chapter 5), the SGP was selected as the first Cloud and Radiation Testbed (CART) locale. Once the SGP locale was chosen, the actual physical location of the SGP site facilities needed to be identified. In late 1991, Sumner Barr and Doug Sisterson completed an exhaustive study that was officially published much later (Barr and Sisterson 2000), which identified specific measurement locations, potential ARM Program collaborations, and airspace limitations within the SGP locale.

The primary mission of the SGP site was to meet the data needs of the instantaneous radiative flux (IRF; Mlawer and Turner 2016, chapter 14) and single-column model (SCM; Zhang et al. 2016, chapter 24) measurement strategies described in Stokes (2016, chapter 2). The IRF approach required the collection of data at and above the CF on the vertical distribution of radiation and radiatively active constituents of the atmosphere, and on the radiative properties of the lower boundary (Table 6-1). To that end, vertical profiles and integrated measures of temperature and water vapor were to be observed at regular intervals with traditional balloon-borne sounding systems, and continuously with state-of-the-science remote sensing systems (microwave radiometer, radar acoustic sounding system, Raman lidar). Cloud cover was to be quantified continuously by contemporary remote sensing systems (day-night whole sky imager, laser ceilometer, and a micropulse lidar system). Profiles of cloud microphysical properties would be derived from the millimeter-wave cloud radar (MMCR) and lidar systems. Components of the surface radiation budget were monitored continuously, in both a broadband manner (with traditional instrumentation) and with spectrally resolved state-of-the-science instruments. Satellite and aircraft (during field campaigns) platforms provided additional information on the vertical distribution of radiation. The near-surface aerosol content of the atmosphere was sampled by an optical particle counter, integrating nephelometers, an optical absorption system, and cloud condensation nuclei counters. Ozone was monitored continuously at the ground level, with vertical profiles obtained during field campaigns. The CF was designed to serve as the figurative center of the SGP site domain and was, in short, the location for the vertically pointing or "soda-straw"-type measurements through the depth of the atmosphere.

The SCM data requirements were another matter altogether. The general approach to SGP data collection for the ARM SCM research effort was through the network of facilities, providing routine data, arrayed over the grid cell area (Fig. 6-1). However, significant amounts of data were needed from field campaigns designed to support the estimation of large-scale vertical motion and temperature and moisture tendencies due to horizontal advection (appendix B; U.S. Department of Energy 1996; Zhang et al. 2016, chapter 24). In particular, the SGP SCM data were intended to permit investigation of a wide range of site-specific questions:

- What processes control the formation, evolution, and dissipation of cloud systems?
- What relative roles do the advection of air mass properties and variation in surface characteristics play in cloud development?
- How do these roles vary with season and short-term climatic regime?
- What aspects of cloud development are controlled by the low-level jet, the moisture return flow from the Gulf of Mexico during winter and early spring, the development of mesoscale convective complexes, and frontal passages?
- What are the effects on radiative fluxes of regional northwest-to-southeast gradients of elevation, soil type, vegetation, temperature, and precipitation?
- How important for atmospheric energy transport processes are seasonally varying distributions of aerosols

TABLE 6-1. An abridged timeline of major events at the SGP site.

Time	Event
May 1992	First (borrowed) instrument (a surface meteorological station) deployed at the CF
By Jul 1993	Surface meteorology and radiation and fluxes, 915-MHz radar wind profiler and RASS, 60-m tower meteorology, AERI-00, microwave radiometer, and 60-m tower all operational. Radiosondes are being launched 3 per day Monday through Friday at the CF. Surface meteorology and fluxes instruments operational at 7 EFs
By Dec 1993	Aerosol observing system, 50-MHz radar wind-profiler and RASS, whole-sky imager, and ceilometer installed at CF. Surface radiation installed at the above EFs, and additional EFs installed. Microwave radiometers and radiosonde systems installed at the 3 of the 4 BFs, with 1 sonde per day at the BFs Monday through Friday
Apr 1994	Remote cloud sensing IOP
By Dec 1994	SORTI and MPL installed at CF. All 22 EFs were operational by now. The BFs and CF were launching 4 sondes per day at 0000, 0600, 1200, and1800 UTC Monday through Friday
Jul–Oct 1995	First SCM IOP conducted, Surface Energy Exchange intensive observing period (IOP), UAV IOP, ARESE-I IOP
By Dec 1995	Radiometer calibration facility operational. New BF at Purcell (BF6) installed. Radiosonde launches at the BFs reduced to 1 per day Monday through Friday. AERI-01 replaced the AERI-00
Aug-Sep 1996	Raman lidar installed, Boundary Layer Experiment IOP, Water Vapor IOP, UAV IOP
By Dec 1996	CIMEL sunphotometer installed, MMCR installed, surface water and temperature sensors (SWATS) installed at most EFs, and radar wind profilers with RASS installed at three IFs
Sep 1997	Mega-IOP (combination of a cloud, water vapor, aerosol, SCM, and UAV IOPs)
Apr 1998	AERIs installed at all 4 BFs
Mar 2000	Cloud IOP, ARESE-II IOP
Sep 2000	ARM-FIRE Water Vapor Experiment
By Dec 2002	Routine launching of radiosondes at BFs stopped
Mar 2003	Aerosol IOP; Cloud IOP; AERIs removed from all four BFs
By May 2006	Whole-sky imager and 50-MHz radar wind profiler decommissioned
By Dec 2009	All observations stopped at the BFs; EFs relocated to make SGP footprint smaller
By Mar 2011	Most ARRA procured instruments (e.g., scanning cloud radar, scanning precipitation radar, Doppler lidar) operational at the SGP
Jun 2011	Mid-Continental Convective Cloud Experiment (MC3E) IOP
Mar 2016	New BFs deployed

and particulates that could emanate regionally from oil refineries and wheat field burning?

Against these requirements, six Intermediate Facilities (IFs, sometimes called "Auxiliary Facilities" in older ARM documentation) were intended to provide three-dimensional mapping of the atmosphere above the CF from about 10 km away. The network of approximately 25 Extended Facilities (EFs) was designed to obtain a distribution of surface meteorological, broadband radiometric, and surface flux variables across the extent of the site domain The EFs were to be situated in a weighted distribution according to land use rather than a simple geometric pattern. To the greatest extent possible, EF sites were to be augmented by Oklahoma Mesonet measurement sites, which were to be nearly identical in instrumentation. Finally, the four Boundary Facilities (BFs) were to be placed approximately 200 km away from the CF on the sides of the SGP domain to establish the general large-scale motion of the atmosphere passing into and out of the domain—in short, the advection terms. Instruments at the BFs were to provide profiles of the atmospheric state. One EF was to be collocated at each BF, unless an Oklahoma Mesonet site was located within 5 km of the BF.

3. Evolution of the SGP site

The SGP site experienced three distinct phases in its evolution:

1) The establishment of the site, which took several years as new instruments from the Instrument Development Program (Stokes 2016, chapter 2) took time to mature and be deployed.
2) A mature phase that was marked by a reorganization of the infrastructure in the late 1990s and becoming a National Scientific User Facility in 2004.
3) A large reconfiguration of the SGP that resulted from input from scientists through a series of DOE-sponsored workshops.

a. Establishment of the SGP site

The SGP site was deployed slowly over time due to a wide range of programmatic issues associated with identifying instruments, arranging land leases, establishing the infrastructure at each of the facilities, and (primarily) budget. The first SGP instrument was installed at the CF on 13 May 1992 using a borrowed portable automated meteorological (PAM) station from the National Center for Atmospheric Research

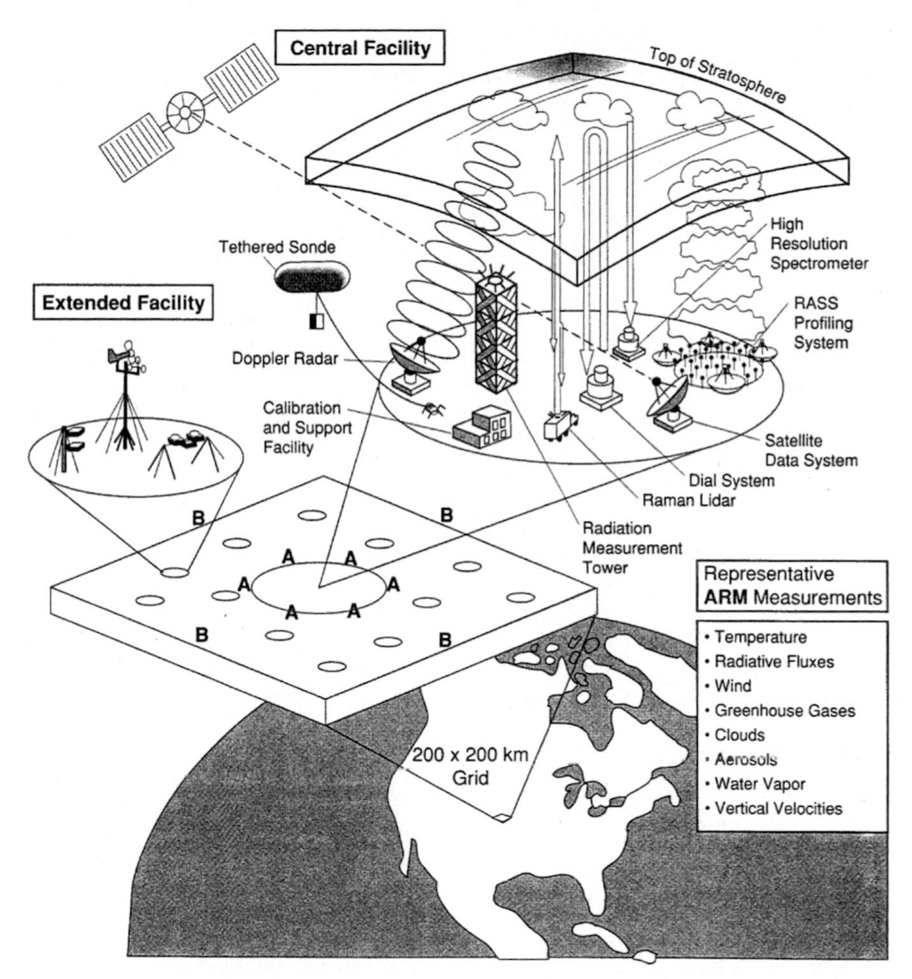

FIG. 6-1. Conceptual implementation design circa 1992 for the 1 Central, 6 Auxiliary, 25 Extended, and 4 Boundary Facilities for the SGP CART Site (after Barr and Sisterson 2000).

(Stokes 2016, chapter 2), and the CF was not deemed to be "complete" until the MMCR (Kollias et al. 2016, chapter 17) was installed in 1996. The four BFs, which were critical for providing observations of the advective tendencies over the domain for the SCM experiment (Zhang et al. 2016, chapter 24), were installed between 1993 and 1995, although the Atmospheric Emitted Radiance Interferometers (AERIs; Turner et al. 2016b, chapter 13) were not deployed at these facilities until 1998. Ultimately, 23 EFs (including one at the CF) were established, with the last being installed in 1995. Three IFs were established by 1996. At that time, the SGP site was considered complete and spanned nearly 90 000 km^2 across south-central Kansas and central Oklahoma.

However, before any instruments could be deployed, there were many tasks that needed to be completed first, and these tasks consumed an immense amount of time and energy. The first and most critical phase of establishing the SGP site was the completion and approval of an Environmental Assessment (EA) that was required by NEPA. The assessment had to identify all instrument locations a priori so that their physical locations could be screened for approval. Figure 6-2 shows the first detailed map of where the SGP facilities were to be located. This assessment was required not only by DOE, but also by the states of Oklahoma and Kansas. No work could begin on the implementation of the SGP site until a "finding of no significant impact" was issued. Furthermore, landowner permissions needed to be obtained before the EA would be approved. As part of this process, the impacts of instrument noise, frequency, and location, as well as impacts from possible aircraft overflights, all had to be documented. Items that needed to be presented in the EA included impacts to wildlife due to instrument towers and guy cables, the amount of soil disrupted by buildings, shelters, pathways, and pads, and more. During the early 1990s, large EAs typically took 12–24 months for approval, but the EA for the SGP established an unofficial record of

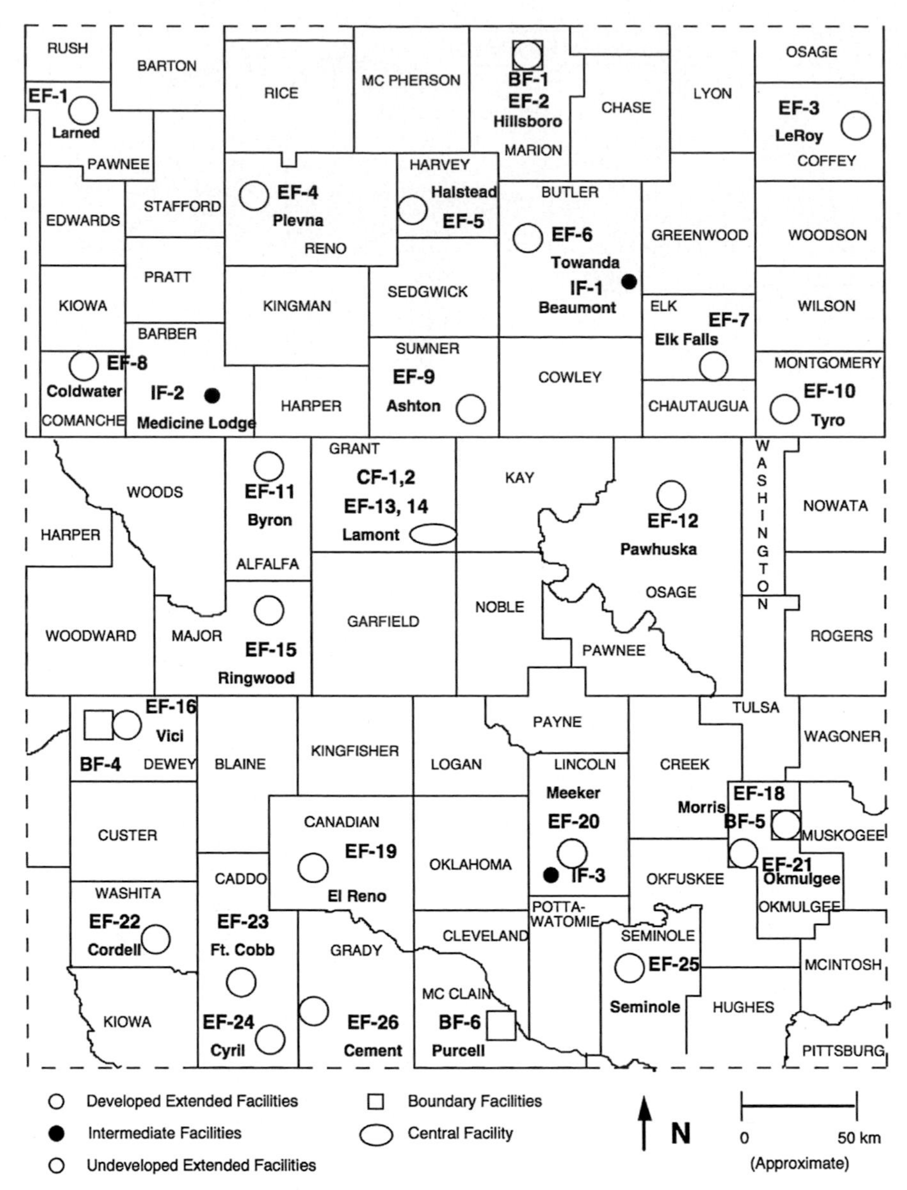

FIG. 6-2. Proposed locations of the SGP Central, Extended, and Auxiliary (name later changed to Intermediate) Facilities circa 1992 (after Barr and Sisterson 2000).

just 108 days between submission and when the permission to proceed was given.

As discussed above, the physical locations for each of the sites needed to be identified before the EA could proceed. So during the summer of 1990, Doug Sisterson and Peter Lunn, who was the DOE ARM Program Manager at the time, spent a week driving in Oklahoma to attend community town hall meetings in the area determined to be the best location for the CF. Meetings were set up with local residents to inform them about the ARM Program and its potential for local residents. Lunn made it clear that ARM was looking for a community that would be receptive to scientific instruments and the scientists, as well as jobs for local residents. He and Sisterson met local residents in churches, taverns, and city halls. They met people who discussed how the government had come in to "help" them before, in one instance regarding establishing potential nuclear waste disposal sites in south central Kansas in the early 1970s. Therefore, trying to sell the DOE Atmospheric "Radiation" Measurement Program ended up being a challenge, but this challenge was overcome by describing the establishment of ARM as a "sunshine" study. Lunn was gifted at addressing the widely diverse and outspoken local residents, and his honesty and patience won over nearly every person that attended the meetings.

Meanwhile, Jack Shannon from Argonne National Laboratory was asked to help identify locations for all of the Central, Intermediate, Extended, and Boundary Facilities and to contact the landowners to start the conversation about arranging leases for these facilities. Jack was an atmospheric scientist and modeler, but most importantly, was born and raised in Oklahoma, attended the University of Oklahoma, and "spoke Oklahoman." Jack spent considerable time driving around Oklahoma and Kansas, knocking on doors to find agreeable landowners. Most of his leads were identified during his frequent stops at local Farm Bureau offices; they were instrumental in identifying farmers who were 1) in need of financial assistance, 2) well-known and respected in their local communities, and 3) willing to host scientific instruments and scientists for weather-related research.

Doug Sisterson negotiated the lease for the 160-acre CF (Fig. 6-3) with Vicki and Stan Schulein over a home-cooked, family dinner in their home and officially signed it in May 1992, only days before the first instrument was scheduled to be deployed.

One unique physical aspect of the location of the CF was that it had a hollowed-out shale pit, which was surrounded by a berm on a couple of sides. This was very important as the Program was going to deploy radar wind profilers (RWPs) with radio acoustic sounding system (RASS) units. The RWPs are able to provide profiles of wind, but when the RASS units were activated, the radars tracked the propagating speed of the sound wave, thereby providing a direct measurement of the virtual temperature profile. Two frequencies were envisioned: one at 915 MHz, which was suitable for low-altitude profiling, and one at 50 MHz for profiling in the middle-to-upper troposphere. These sound waves emitted by the two units sounded like a high-pitched whistle and a deeper warbling, respectively. As part of the EA, the Program needed to ensure that these acoustical instruments did not irritate or harm humans or animals. By placing the RWPs and their RASS units in this depression, the Program was able to meet the EA requirements.[1]

The observational goal of the CF was to provide the most comprehensive view of the overlying state of the atmosphere from surface to tropopause in the world. It needed to be staffed daily and have offices, storage space, and facilities for instrument calibration and repair and to house the site data computer system. The goal was to staff the facility not with on-site Ph.D. researchers, but with local personnel who had a basic skill set that, after training, documentation, and oversight, would enable them to become skilled technicians. It was envisioned that the on-site technicians would learn both by experience and with guidance from Instrument Mentors (Cress and Sisterson 2016, chapter 5), who were to be on site as needed. The technical staff would service CF instruments daily and would visit all other facilities once every two weeks. Part-time staff were hired at the BFs to conduct routine radiosonde launches, and more frequently when those facilities were launching balloons in support of field campaigns. This model proved to be an effective approach both scientifically and economically. Hiring and training people from the local community exceeded expectations and the SGP site became seen as a welcomed neighbor. The on-site SGP staff grew as site activity grew. By 2010 the site employed about 30 staff, making it the third largest employer in Grant County, Oklahoma.

The distribution of the EFs in the original idealized ARM planning illustrations indicated a uniform geometric pattern. In reality, it was determined that the distribution of these sites would be better sited scientifically if they were distributed by land use category. The primary land use classification throughout the SGP area was agriculture and pasture with a small portion characterized as forest. In addition, a significant northwest-to-southeast gradient in both temperature and precipitation (lower to higher) further impacted siting decisions (Fig. 6-4).

The instrumentation at the IFs was envisioned to include scanning instruments (e.g., atmospheric radars), instruments with a wide field of view (e.g., whole-sky imagers), or instruments that would provide a wider view of the atmospheric state in the vicinity of the CF. In 1996, 915-MHz radar wind profilers with RASS units were installed at the IFs to provide wind and virtual temperature profiles as well as boundary layer height measurements around the CF. The continuous measurements coupled with the detailed CF radiosonde information increased the view of the atmospheric state from "soda straw" to a 4D view on the order of 15–30 km.

Instrumentation at the BFs was initially limited to a balloon-borne sounding system, a ceilometer, and a microwave radiometer; however, these facilities were

[1] A public demonstration was required to satisfy the EA. On the day of the event, one of the local farmers came up to the crew that was setting up for the test and wanted to know the exact time the RASS test would start. Then he went down to the pasture where his cattle were feeding. The crew started the test, and the farmer came up about 20 minutes later and asked when the test was starting again. The crew pointed to the RASS, which was warbling away, and the farmer asked if that was all it was? The crew said yes, it would do that for about 10 min every hour or so. He remarked that farm tractors were louder than that. And the cattle never looked up once.

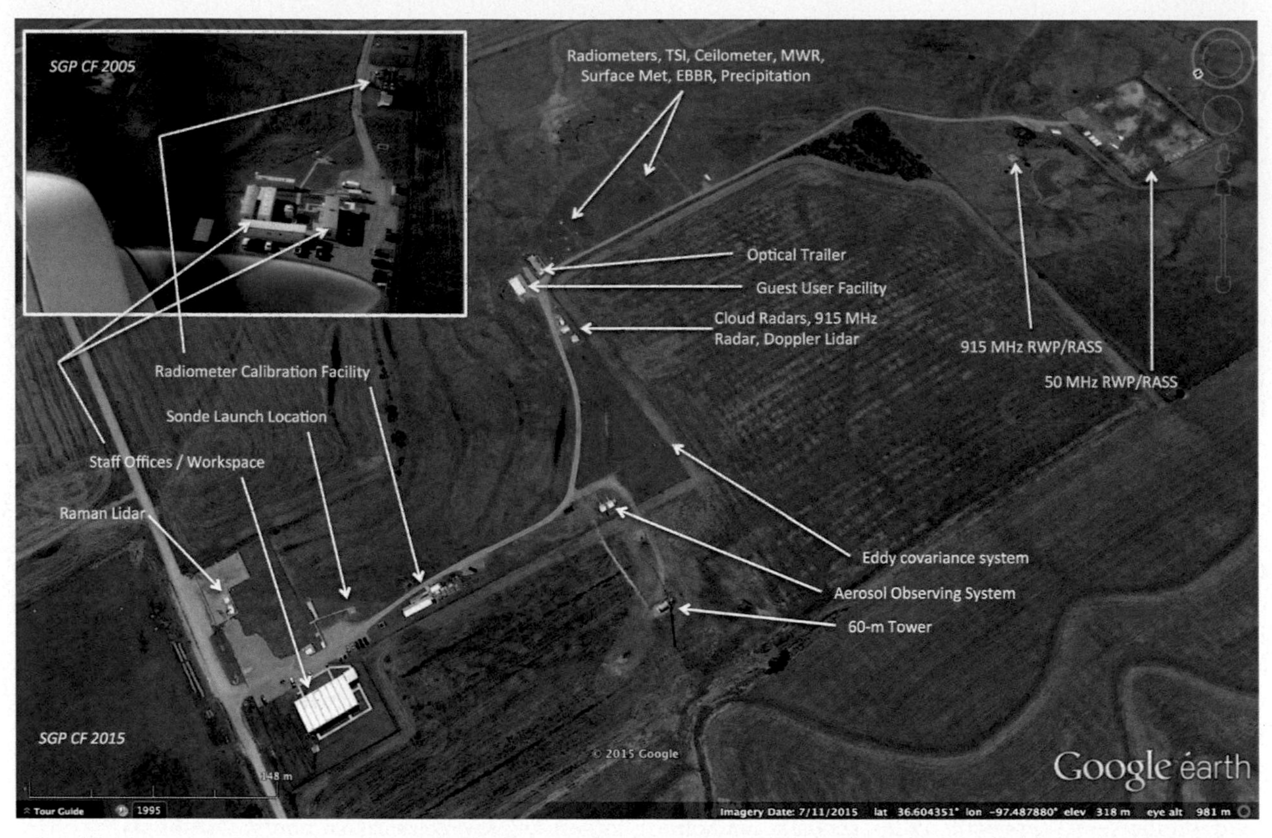

FIG. 6-3. Photo of the SGP CF in 2015 with the locations of some of the observing systems indicated. North is to the lower left. The 915- and 50-MHz radar wind profiler (RWP) and RASS locations on the upper right of the image are no longer used; the 50-MHz system was retired and the 915-MHz system was moved next to the cloud radars. The inset photo, taken in 2002, shows the mobile homes that were used for the primary working spaces for the SGP staff. (Photo from Google Earth, downloaded December 2015. Inset photo courtesy of the ARM Program).

later augmented with AERI systems. These instruments provided for wind, temperature, and water vapor profiles on which to base estimations of the lateral fluxes of moisture and energy into and out of the SGP domain, along with the divergence and tendencies of atmospheric properties for that volume; these observations were required by the SCM measurement strategy.

Site personnel and scientific oversight were identified quickly. Doug Sisterson was selected as the SGP Site Manager in late 1990. Jim Teske, a retired Operations Manager for the NOAA WPDN, and David Breedlove, who provided technical support for the WPDN, became the first on-site staff at the SGP site and were based at the CF. Teske's communication and configuration management skills, commitment to safety, knowledge of instrumentation, and experience as a U.S. Navy officer contributed significantly to the early success of the SGP site in his role as On-Site Operations Manager. Together, Teske, Breedlove, and Sisterson identified other local talent to join the SGP operations team to maintain and operate the site.

One of the really impactful activities at the SGP site was the early implementation of a safety program. During site implementation, the smallest of incidents could have caused a shutdown or, worse, crippled the deployment with ripple effects to other parts of ARM. Monte Brandner of Argonne National Laboratory not only provided oversight for the overall construction of the SGP, but also implemented a Continuous Quality Improvement Program that mandated inspections of instruments and facilities to evaluate the adequacy of the communications equipment, site structures, and procedures being used by the SGP operations team. These activities helped reveal potential issues before they became problems and really illustrated the benefits of an integrated safety management effort for a field program, especially for a long-term program like ARM.

The ARM Program was established to do science, and there was a genuine desire to keep scientists integrated with the operations that were occurring at each of the sites. Scientific oversight of the SGP site was established in 1992 when Peter Lamb at the University

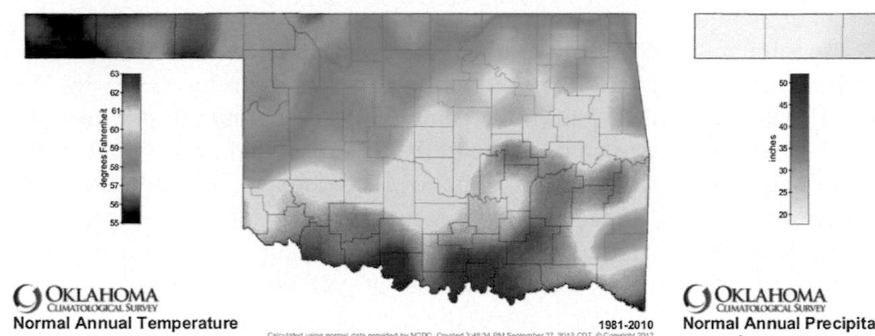

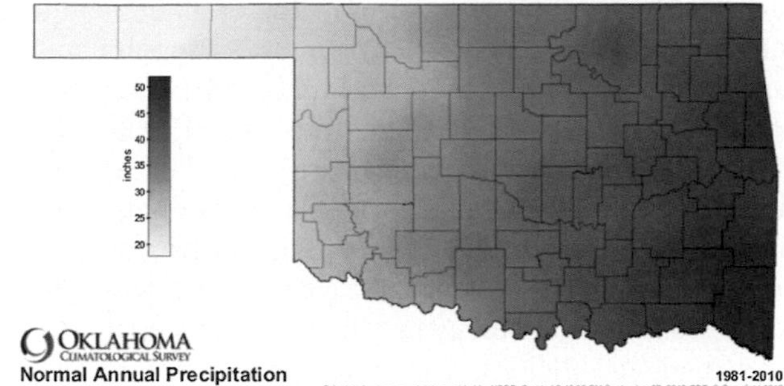

FIG. 6-4. Annual mean values of the (left) air temperature (°F) at 2-m height and (right) precipitation (in.) from 1981 to 2010. Images courtesy of the Oklahoma Climatological Survey.

of Oklahoma, which is located about 100 miles south of the CF, was selected to serve as the SGP Site Scientist. Pete Lamb's Site Scientist Team and the Site Manager wrote a Site Scientific Mission Plan every six months from 1993 through 1999 (Schneider et al. 1993a,b; 1994a,b; Splitt et al. 1995; Peppler et al. 1996a,b; 1997a,b; 1998a,b; 1999a,b; 2000) that detailed priorities for scientific activities, summarized scientific goals, described new instrumentation, outlined field campaigns, and listed upcoming expectations for the next six months. These reports[2] are rich with the details of the development of the SGP site and are available through the ARM Program website.[3]

The deployment of instruments at the SGP site was not done at once, but was phased over many years with the site being considered fully instrumented by late 1996 by Tom Ackerman (who was the second ARM Chief Scientist). Table 6-1 provides an abridged timeline of the development of the infrastructure and deployment of instruments at SGP, as well as some of the more significant IOPs. The slow deployment of instruments to the SGP site was due in a large degree to the time it took for the Instrument Development Program (IDP; Stokes 2016, chapter 2; Cress and Sisterson 2016, chapter 5) to provide the robust, hardened instruments needed for ARM's operational paradigm. Almost all of ARM's IDP instruments were initially deployed and evaluated at the SGP. These IDP instruments provided truly unique (especially in the mid-1990s) measurements for atmospheric science, and are now an important part of the core instruments seen at almost every ARM site. These instruments include the AERI (Turner et al. 2016b, chapter 13), the world's first automated Raman lidar (Turner et al. 2016a, chapter 18), the multifilter rotating shadowband radiometer (Michalsky and Long 2016, chapter 16), the micropulse lidar (Campbell et al. 2002), and the millimeter cloud radar (Kollias et al. 2016, chapter 17). Each time that a major IDP instrument was deployed initially at the SGP site, an IOP was conducted to bring other similar research-grade instruments to the SGP to help evaluate the abilities of the new IDP-developed instrument. The instrument intercomparison effort was a large component of many of the IOPs that occurred at the SGP before 2000. However, today the focus has largely changed, wherein instrument developers bring newly developed instruments to the SGP in order to characterize them relative to the operational (and well understood) ARM instruments.

b. The mature SGP site

By the late 1990s, it was clear that ARM was entering a new phase. The various sites were largely established (with the exception of the TWP site at Darwin, which was to be installed in 2003; Long et al. 2016, chapter 7) and ARM no longer was faced with a hurried need to field instrumentation and race to meet short-fuse deadlines. DOE instigated a formal review of the ARM Infrastructure in mid-1999, and this review would result in an internal reorganization of the Program's infrastructure (Cress and Sisterson 2016, chapter 5).

This reorganization shifted the ARM Program to a longer-term operational paradigm involving a more formalized infrastructure to ensure overall Program efficiencies. Steps were taken to make the various ARM sites and facilities more uniform in both instrument operations and procedures used to process, document, and handle the data. An ARM Operations Manager for all sites was established, and Doug Sisterson assumed

[2] These reports every 6 months were no longer required by ARM Program management after mid-2000, and hence the reports were no longer produced.

[3] Many of the early reports published by the ARM Program were produced only in hardcopy. Most of these have since been rescanned into PDF format and placed upon the ARM web page, and can be found using the search engine on the "publications" tab of the ARM web page http://www.arm.gov. However, any ARM report can be retrieved by sending an e-mail message to info@arm.gov.

this role. James Liljegren, who was the original mentor for the microwave radiometers in the Program and heavily involved in many of the initial SGP activities, replaced Sisterson to become the second SGP Site Program Manager; he held this role until 2005. During this time, Liljegren's main focus was to formalize the processes that dealt with configuration management (such as storing spare instruments and components, arranging for localized instrument repairs, improving procedures for instrument calibrations, and developing better defined roles and procedures for Instrument Mentors and on-site technical support and their interactions), improved operational efficiencies particularly regarding support for field campaigns, and tighter budgeting for leaner economic times. One significant improvement made at the SGP during Liljegren's tenure was the construction of the building with a stable observation deck to support guest scientists and their instruments during IOPs; this facility was used heavily during the 2003 Aerosol IOP (McComiskey and Ferrare 2016, chapter 21).

In 2004, Jim Teske retired as the SGP on-site Operations Manager and was succeeded by Dan Rusk, who assumed the role through 2011. John Schatz, who had been the full-time on-site SGP Safety Officer since 1993, became the third on-site Manager to date. However, the core operations and engineering staff at the SGP site have remained relatively constant over the decades. This stability has benefitted the Program tremendously, because the SGP staff have developed incredible expertise with the various ARM instruments and have developed a different set of skills that complements those of the instrument mentors nicely.

In 2004, the infrastructure component of the ARM Program was designated a DOE Biological and Environmental Research (BER) Scientific User Facility and became known as the ARM Climate Research Facility (ACRF; Ackerman et al. 2016, chapter 3). From the mission statement, the ACRF "provides the climate research community with strategically located in situ and remote sensing observatories designed to improve the understanding and representation, in climate and earth system models, of clouds and aerosols as well as their interactions and coupling with the Earth's surface." The SGP was the flagship of ACRF and continues to be one of the key sites for long-term measurements of radiative fluxes, cloud and aerosol properties, and related atmospheric properties. This change, along with the relatively easy access by the scientific community to the SGP due to its location in the center of the United States, has resulted in a large number of small IOPs conducted at the SGP site that have focused on a huge range of topics (see www.arm.gov/campaigns/table for a full listing of IOPs conducted at each ARM site).

After about a decade of operations at the SGP, the cost of operating aging instruments was starting to increase. In addition, there was a strong desire by the ARM scientific community to continue to add new instruments to the SGP site that would keep ARM at the forefront of climate observations. Therefore, ARM formed the instrument "Sunset Committee" that was charged with evaluating the scientific utility against the operational demands of various instruments to determine if any of the current ARM instruments could be retired, thereby providing funding that could be used for other instruments. This determination proved to be a rather difficult and emotional task since researchers were used to the data from instruments they knew well (including their quirks) and retiring (or "sunsetting") of instruments was a new concept that was hard to embrace. Ultimately, a few instruments were retired (listed in Table 6-1), but the impact on the operational budget was relatively minimal.

In late 2005, Brad Orr became the third SGP Site Manager and was focused primarily on how the site could be nimbly reconfigured as climate models and research needs evolved. For instance, one question was "Could the SGP footprint be changed to reduce the coarse EF site spacing of the original grid with a finer spacing over an area of 150×150 km?" The horizontal resolution of climate models is continuously improving, with typical model grid sizes now smaller than the original size of the SGP domain. The finer resolution of the models required finer resolution of the data inside to domain to identify inhomogeneities across the domain. Furthermore, the original horizontal extent of the SGP required staff to drive many hundreds of miles in order to maintain the full complement of EFs across the domain, which resulted in increased costs to maintain all of these facilities. Thus there was a desire to reduce the size of the domain from a programmatic point of view to save money and effort, but only if this would not have a negative impact on the science.

In an iterative relationship typical to many of ARM's scientific endeavors, a conversation was begun between the SCM community, and in particular Shaocheng Xie, who was the translator supporting that community [see Ackerman et al. (2016, chapter 3) for a description of a "translator"], to see if an optimal reduced-sized SGP footprint could be found. Xie performed a set of variational analyses (Zhang et al. 2016, chapter 24) using data from different subsets of EF locations to determine an optimal array of site locations. Based on Xie's runs, a new SGP footprint was conceived in 2006 and approved in 2007, although the reconfigured site array would take 24 months to complete. As part of this restructuring, the BFs were decommissioned in 2009, and several of the original EFs in Oklahoma that were farthest from the CF were retired although the EFs in Kansas were maintained. Much of the instrumentation from those sites was used to populate the new denser array of new EFs that were closer to the CF.

c. *Renewed growth at the SGP site*

In 2007 and 2008, the DOE Program Managers held workshops to obtain community input on scientific problems that ARM should be pursuing and recommendations for new ARM instrumentation and new locales that were needed to address these problems (Ackerman et al. 2016, chapter 3). The timing of these workshops was fortuitous as the American Recovery and Reinvestment Act (ARRA) came into being in 2009 to help improve the U.S. economy after the recession of 2008. However, ARRA desired "shovel-ready" projects in order to jump-start the economy, and the workshop reports (along with input from the ARM scientists that was accumulated through the yearly ARM Science Team Meetings) provided DOE Management the ability to present a cohesive plan on how to greatly upgrade the ACRF to address some pressing scientific problems. As a result, the U.S. Department of Energy's Office of Science allocated $60 million from ARRA to the ARM Climate Research Facility. These funds were distributed across the ARM infrastructure, allowing all of the sites to acquire new instruments, improvements to the computational and physical infrastructure, and more.

At the CF, two of the changes were very noticeable (Fig. 6-3). The first was the construction of a modular building at the CF to house the site's staff. The original buildings used by site staff were single-wide mobile homes that were deployed in 1992; mobile homes were used because at the time the ARM Program was considered a 10-yr program and the site would be decommissioned afterward. However, by 2009 these mobile homes were in very poor shape and more space was required to support the staff needed for the increased instrument complement that would soon be at the site. The new building provided much needed space and comfort, as the mobile trailers were hot in the summer and cold in the winter!

The second obvious change at the CF was the deployment of the scanning dual-frequency cloud radar. Before installation of this radar, virtually all of the instruments at the ARM sites appeared to the untrained visitor to just sit there with the exception of the radiometers on solar trackers or with shadowbands or the anemometers that measure wind speed. The scanning cloud radar gave visitors something to look at while they were at the site.

The ARRA funds also allowed ARM to transition from a constrained soda-straw perspective to a more three-dimensional perspective (Mather and Voyles 2013). The scanning cloud radars allowed the Program to get statistics on the spatial distribution of the clouds, instead of having to rely on the frozen turbulence assumption to convert the high-temporal-resolution zenith observations to spatial statistics. Additionally, four new scanning precipitation radars that operate at longer wavelengths (10 cm and 3 cm) were deployed at the IFs around the CF, providing for the first time spatial measurements of precipitation and allowing dual and triple Doppler retrievals of the horizontal wind field to be derived. Last but not least, ARRA also enabled ARM to greatly improve its ability to measure the size distribution and composition of aerosols at the site, thereby enabling a wide range of research to investigate new aerosol particle formation and growth processes and better understand how aerosols can evolve into cloud condensation nuclei. These new instruments have greatly expanded the science from ARM observations, and were an important component of the process-level research that is at the core of the Atmospheric System Research (ASR) Program that began in 2010 (Mather et al. 2016, chapter 4).

There have also been some changes in the leadership of the SGP site. In 2012, Doug Sisterson resumed overall oversight of the SGP site again, working closely with John Schatz to continue the implementation of the new instruments that started with ARRA. He continued in this role until 2014, when Nicki Hickmon assumed the SGP Site Manager role.

In a sense, the end of the first 20 years of the ARM Program at the SGP is similar to when the ARM Program was just starting the SGP by implementing state-of-the-art instruments and new facilities intended to attack an ever-widening scope of scientific questions. Many of the current challenges are to develop improved methods to operate and calibrate these new instruments, process and distribute the voluminous data that comes from them, and improve the scientific utility of the Program by merging together data from both original and new ARM instruments.

4. Scientific contributions

Between 1992 and 2010, many different scientific and engineering studies had been conducted using data from the SGP site. The topics include characterizing IDP and other instruments; evaluating different water vapor measurements and developing a more accurate water vapor product; improving clear-sky radiative transfer models; investigating different ways to construct an objective analysis from ARM observations that could be used to drive SCM models; deriving statistics on cloud overlap; characterizing the absorption of shortwave radiation by aerosols and clouds; developing new aerosol and cloud property retrieval algorithms; evaluating cloud and precipitation microphysical parameterizations in SCMs; improving understanding of land surface properties and the impact on the atmosphere above; and much more. These topics are covered in detail in chapters 13 through 27 of this monograph. Instead of trying to summarize all of the scientific advancements made using data from the SGP site, we focus here on one unique aspect of the SGP: its multidecadal record of continuous observations.

One of the motivations of the ARM Program was to provide a dataset that captured all of the major modes of variability in the atmosphere, including diurnal, synoptic, seasonal, and yearly variations. The ARM Program instruments have done exactly this, and several studies have looked at how various geophysical variables have evolved over multiyear periods. These studies include the development of a decade-long dataset that can be used to drive SCM and cloud-resolving models (Xie et al. 2004), a climatology of cirrus and the relationship with atmospheric state (Mace et al. 2006), an analysis of the vertical structure of cloud occurrence and overlap (Mace and Benson 2008), determination of a significant increase in the all-sky downwelling shortwave radiation over the central United States (Long et al. 2009), the variation of aerosol optical depth and Angstrom exponent (Michalsky et al. 2010), the variability in the vertical profile of aerosol scattering and absorption properties (Andrews et al. 2011), a climatology and trend analysis of the downwelling spectral infrared radiation (Turner and Gero 2011; Gero and Turner 2011), and an observational analysis of the surface radiative forcing by carbon dioxide (Feldman et al. 2015). These studies are interesting in their own right, but also provide the foundation for detailed case studies because the conditions of the case study can be placed into perspective with the longer-term climatology.

5. Summary

The SGP site has been the observational centerpiece and anchor of the ARM Program since 1992. It represents the scientifically required midlatitude, midcontinental observing facility. It was selected over two other U.S. locales that were similarly attractive from a weather/climate regime standpoint, but the ultimate choice of the SGP site was made due to its substantially greater potential for synergy with other developing state and federal observational networks and research programs.

Entering its third decade, the mission of the SGP site remains the same—to provide a continuous multivariable observationally based dataset that can be used to understand atmospheric processes and to evaluate and improve how these processes are represented in climate models. The transition to a National Scientific User Facility in the mid-2000s enlarged the users to include scientists that were outside the ARM Program and placed new demands on the operational aspects of the site. The merging of the DOE's Atmospheric Sciences Program with the ARM Scientific Program to create the new Atmospheric System Research (ASR) Program together with the Recovery Act funding in the late 2000s resulted in an expanded site, with new instruments to focus on quantifying aerosol processes and the spatial variability of clouds and precipitation (Mather and Voyles 2013). The SGP provides perhaps the most complete and comprehensive set of observations of any of the large-scale atmospheric observatories located around the world.

In 2016, the SGP will transform again by installing new BFs, albeit much closer to the CF than they were before, to support routine high-resolution modeling based on data from the SGP site (U.S. Department of Energy 2014; Table 6-1). This expansion keeps the SGP site and its increasing capabilities at the forefront of the effort to provide the long-term multivariable datasets needed for climate system model improvements. The SGP will continue to evolve to provide the data necessary to address emerging issues of the scientific community. That was its origin, and that is its role.

REFERENCES

Ackerman, S. A., W. L. Smith, J. D. Spinhirne, and H. E. Revercomb, 1990: The 27–28 October 1986 FIRE IFO cirrus case study: Spectral properties of cirrus clouds in the 8–12-μm window. *Mon. Wea. Rev.*, **118**, 2377–2388, doi:10.1175/1520-0493(1990)118<2377:TOFICC>2.0.CO;2.

Ackerman, T. P., and G. Stokes, 2003: The Atmospheric Radiation Measurement Program. *Phys. Today*, **56**, 38–45, doi:10.1063/1.1554135.

——, T. S. Cress, W. R. Ferrell, J. H. Mather, and D. D. Turner, 2016: The programmatic maturation of the ARM Program. *The Atmospheric Radiation Measurement (ARM) Program: The First 20 Years, Meteor. Monogr.*, No. 57, Amer. Meteor. Soc., doi:10.1175/AMSMONOGRAPHS-D-15-0054.1.

Andrews, E., P. J. Sheridan, and J. A. Ogren, 2011: Seasonal differences in the vertical profiles of aerosol optical properties over rural Oklahoma. *Atmos. Chem. Phys.*, **11**, 10 661–10 676, doi:10.5194/acp-11-10661-2011.

Barr, S., and D. L. Sisterson, 2000: Locale analysis report for the Southern Great Plains. ARM Rep. 00-001, 55 pp. [Available online at http://www.arm.gov/publications/site_reports/sgp/arm-00-001.pdf.]

Barth, M. F., R. B. Chadwick, and D. W. van de Kamp, 1994: Data processing algorithms used by NOAA's Wind Profiler

Demonstration Network. *Ann. Geophys.*, **12**, 518–528, doi:10.1007/s00585-994-0518-1.

Brock, F. V., K. C. Crawford, R. L. Elliott, G. W. Cuperus, S. J. Stadler, H. L. Johnson, and M. D. Eilts, 1995: The Oklahoma Mesonet: A technical overview. *J. Atmos. Oceanic Technol.*, **12**, 5–19, doi:10.1175/1520-0426(1995)012<0005:TOMATO>2.0.CO;2.

Campbell, J. R., D. L. Hlavka, E. J. Welton, C. J. Flynn, D. D. Turner, J. D. Spinhirne, V. S. Scott, and I. H. Hwang, 2002: Full-time, eye-safe cloud and aerosol lidar observations at Atmospheric Radiation Measurement program sites: Instruments and data processing. *J. Atmos. Oceanic Technol.*, **19**, 431–442, doi:10.1175/1520-0426(2002)019<0431:FTESCA>2.0.CO;2.

Cress, T. S., and D. L. Sisterson, 2016: Deploying the ARM sites and supporting infrastructure. *The Atmospheric Radiation Measurement (ARM) Program: The First 20 Years, Meteor. Monogr.*, No. 57, Amer. Meteor. Soc., doi:10.1175/AMSMONOGRAPHS-D-15-0049.1.

Crum, T. D., and R. L. Alberty, 1993: The WSR-88D and the WSR-88D operational support facility. *Bull. Amer. Meteor. Soc.*, **74**, 1669–1687, doi:10.1175/1520-0477(1993)074<1669:TWATWO>2.0.CO;2.

Ellingson, R. G., R. D. Cess, and G. L. Potter, 2016: The Atmospheric Radiation Measurement Program: Prelude. *The Atmospheric Radiation Measurement (ARM) Program: The First 20 Years, Meteor. Monogr.*, No. 57, Amer. Meteor. Soc., doi:10.1175/AMSMONOGRAPHS-D-15-0029.1.

Feldman, D. R., W. D. Collins, P. J. Gero, M. S. Torn, E. J. Mlawer, and T. R. Shippert, 2015: Observational determination of surface radiative forcing by CO_2 from 2000 to 2010. *Nature*, **519**, 339–343, doi:10.1038/nature14240.

Gero, P. J., and D. D. Turner, 2011: Long-term trends in downwelling spectral infrared radiance over the U.S. Southern Great Plains. *J. Climate*, **24**, 4831–4843, doi:10.1175/2011JCLI4210.1.

Kollias, P., and Coauthors, 2016: Development and applications of ARM millimeter-wavelength cloud radars. *The Atmospheric Radiation Measurement (ARM) Program: The First 20 Years, Meteor. Monogr.*, No. 57, Amer. Meteor. Soc., doi:10.1175/AMSMONOGRAPHS-D-15-0037.1.

Kummerow, C., W. Barnes, T. Kozu, J. Shiue, and J. Simpson, 1998: The Tropical Rainfall Measuring Mission (TRMM) sensor package. *J. Atmos. Oceanic Technol.*, **15**, 809–817, doi:10.1175/1520-0426(1998)015<0809:TTRMMT>2.0.CO;2.

Lawford, R. G., 1999: A midterm report on the GEWEX Continental-Scale International Project (GCIP). *J. Geophys. Res.*, **104**, 19 279–19 292, doi:10.1029/1999JD900266.

Long, C. N., E. G. Dutton, J. A. Augustine, W. Wiscombe, M. Wild, S. A. McFarlane, and C. J. Flynn, 2009: Significant decadal brightening of downwelling shortwave in the continental United States. *J. Geophys. Res.*, **114**, D00D06, doi:10.1029/2008JD011263.

——, J. H. Mather, and T. P. Ackerman, 2016: The ARM Tropical Western Pacific (TWP) sites. *The Atmospheric Radiation Measurement (ARM) Program: The First 20 Years, Meteor. Monogr.*, No. 57, Amer. Meteor. Soc., doi:10.1175/AMSMONOGRAPHS-D-15-0024.1.

Mace, G. G., and S. Benson, 2008: The vertical structure of cloud occurrence and radiative forcing at the SGP ARM site as revealed by 8 years of continuous data. *J. Climate*, **21**, 2591–2610, doi:10.1175/2007JCLI1987.1.

——, ——, and E. Vernon, 2006: Cirrus clouds and the large-scale atmospheric state: Relationships revealed by six years of ground-based data. *J. Climate*, **19**, 3257–3278, doi:10.1175/JCLI3786.1.

Mather, J. H., and J. W. Voyles, 2013: The ARM Climate Research Facility: A review of structure and capabilities. *Bull. Amer. Meteor. Soc.*, **94**, 377–392, doi:10.1175/BAMS-D-11-00218.1.

——, D. D. Turner, and T. P. Ackerman, 2016: Scientific maturation of the ARM Program. *The Atmospheric Radiation Measurement (ARM) Program: The First 20 Years, Meteor. Monogr.*, No. 57, Amer. Meteor. Soc., doi:10.1175/AMSMONOGRAPHS-D-15-0053.1.

McComiskey, A., and R. A. Ferrare, 2016: Aerosol physical and optical properties and processes in the ARM Program. *The Atmospheric Radiation Measurement (ARM) Program: The First 20 Years, Meteor. Monogr.*, No. 57, Amer. Meteor. Soc., doi:10.1175/AMSMONOGRAPHS-D-15-0028.1.

Michalsky, J. J., and C. N. Long, 2016: ARM solar and infrared broadband and filter radiometry. *The Atmospheric Radiation Measurement (ARM) Program: The First 20 Years, Meteor. Monogr.*, No. 57, Amer. Meteor. Soc., doi:10.1175/AMSMONOGRAPHS-D-15-0031.1.

——, F. Denn, C. Flynn, G. Hodges, P. Kiedron, A. Koontz, J. Schlemmer, and S. E. Schwartz, 2010: Climatology of aerosol optical depth in north-central Oklahoma: 1992–2008. *J. Geophys. Res.*, **115**, D07203, doi:10.1029/2009JD012197.

Mlawer, E. J., and D. D. Turner, 2016: Spectral radiation measurements and analysis in the ARM Program. *The Atmospheric Radiation Measurement (ARM) Program: The First 20 Years, Meteor. Monogr.*, No. 57, Amer. Meteor. Soc., doi:10.1175/AMSMONOGRAPHS-D-15-0027.1.

Peppler, R. A., P. J. Lamb, and D. L. Sisterson, 1996a: Site scientific mission plan for the Southern Great Plains CART site January–June 1996. ARM Rep. ARM-96-001, 86 pp.

——, ——, and ——, 1996b: Site scientific mission plan for the Southern Great Plains CART site July–December 1996. ARM Rep. ARM-96-002, 100 pp.

——, ——, and ——, 1997a: Site scientific mission plan for the Southern Great Plains CART site January–June 1997. ARM Rep. ARM-97-001, 95 pp.

——, D. L. Sisterson, and P. J. Lamb, 1997b: Site scientific mission plan for the Southern Great Plains CART site July–December 1997. ARM Rep. ARM-97-002, 74 pp.

——, ——, and ——, 1998a: Site scientific mission plan for the Southern Great Plains CART site January–June 1998. ARM Rep. ARM-98-002, 85 pp.

——, ——, and ——, 1998b: Site scientific mission plan for the Southern Great Plains CART site July–December 1998. ARM Rep. ARM-98-003, 79 pp.

——, ——, and ——, 1999a: Site scientific mission plan for the Southern Great Plains CART site January–June 1999. ARM Rep. ARM-99-001.

——, ——, and ——, 1999b: Site scientific mission plan for the Southern Great Plains CART site July–December 1999. ARM Rep. ARM-99-002, 89 pp.

——, ——, and ——, 2000: Site scientific mission plan for the Southern Great Plains CART site January–June 2000. ARM Rep. ARM-00-006, 44 pp.

Ralph, F. M., P. J. Neiman, D. W. van de Kamp, and D. C. Law, 1995: Using spectral moment data from NOAA's 404-MHz radar wind profilers to observe precipitation. *Bull. Amer. Meteor. Soc.*, **76**, 1717–1739, doi:10.1175/1520-0477(1995)076<1717:USMDFN>2.0.CO;2.

Schneider, J. M., P. J. Lamb, and D. L. Sisterson, 1993a: Site scientific mission plan for the Southern Great Plains CART site July–December 1993. ARM Rep. ARM-93-001, 34 pp.

——, ——, and ——, 1993b: Site scientific mission plan for the Southern Great Plains CART site January–June 1994. ARM Rep. ARM-94-001, 80 pp.

——, ——, and ——, 1994a: Site scientific mission plan for the Southern Great Plains CART site July–December 1994. ARM Rep. ARM-94-002, 58 pp.

——, ——, and ——, 1994b: Site scientific mission plan for the Southern Great Plains CART site January–June 1995. ARM Rep. ARM-95-001, 62 pp.

Simpson, J., C. Kummerow, W.-K. Tao, and R. F. Adler, 1996: On the Tropical Rainfall Measurement Mission (TRMM). *Meteor. Atmos. Phys.*, **60**, 19–36, doi:10.1007/BF01029783.

Smith, T. L., and S. G. Benjamin, 1993: Impact of network wind profiler data on a 3-h data assimilation system. *Bull. Amer. Meteor. Soc.*, **74**, 801–807, doi:10.1175/1520-0477(1993)074<0801:IONWPD>2.0.CO;2.

Splitt, M. E., P. J. Lamb, and D. L. Sisterson, 1995: Site scientific mission plan for the Southern Great Plains CART site July–December 1995, 74 pp.

Stokes, G. M., 2016: Original ARM concept and launch. *The Atmospheric Radiation Measurement (ARM) Program: The First 20 Years, Meteor. Monogr.*, No. 57, Amer. Meteor. Soc., doi:10.1175/AMSMONOGRAPHS-D-15-0021.1.

——, and S. E. Schwartz, 1994: The Atmospheric Radiation Measurement (ARM) Program: Programmatic background and design of the Cloud and Radiation Test bed. *Bull. Amer. Meteor. Soc.*, **75**, 1201–1221, doi:10.1175/1520-0477(1994)075<1201:TARMPP>2.0.CO;2.

Szoke, E. J., J. M. Brown, J. A. McGinley, and D. Rodgers, 1994: Forecasting for a large field program: STORM-FEST. *Wea. Forecasting*, **9**, 593–605, doi:10.1175/1520-0434(1994)009<0593:FFALFP>2.0.CO;2.

Turner, D. D., and P. J. Gero, 2011: Downwelling 10 μm infrared radiance temperature climatology for the Atmospheric Radiation Measurement Southern Great Plains site. *J. Geophys. Res.*, **116**, D08212, doi:10.1029/2010JD015135.

——, J. E. M. Goldsmith, and R. A. Ferrare, 2016a: Development and applications of the ARM Raman lidar. *The Atmospheric Radiation Measurement (ARM) Program: The First 20 Years, Meteor. Monogr.*, No. 57, Amer. Meteor. Soc., doi:10.1175/AMSMONOGRAPHS-D-15-0026.1.

——, E. J. Mlawer, and H. E. Revercomb, 2016: Water vapor observations in the ARM Program. *The Atmospheric Radiation Measurement (ARM) Program: The First 20 Years, Meteor. Monogr.*, No. 57, Amer. Meteor. Soc., doi:10.1175/AMSMONOGRAPHS-D-15-0025.1.

U.S. Department of Energy, 1990: Atmospheric Radiation Measurement Program plan. DOE/ER-0441, Environmental Sciences Division, Office of Health and Environmental Research, Office of Energy Research, U.S. Department of Energy, 116 pp.

——, 1991: Identification, recommendation, and justification of potential locales for ARM sites. DOE/ER-0495T, Atmospheric and Climate Research Division, Office of Health and Environmental Research, Office of Energy Research, U.S. Department of Energy, 160 pp.

——, 1996: Science plan for the Atmospheric Radiation Measurement (ARM) Program. DOE/ER-0670T, Environmental Sciences Division, Office of Health and Environmental Research, Office of Energy Research, U.S. Department of Energy, 86 pp.

——, 2014: Climate and Environmental Sciences Division ARM/ASR high-resolution modeling workshop. DOE/SC-0169, U.S. Department of Energy, 46 pp.

Xie, S., R. T. Cederwall, and M. Zhang, 2004: Developing long-term single-column model/cloud system-resolving model forcing data using numerical weather prediction products constrained by surface and top of the atmosphere observations. *J. Geophys. Res.*, **109**, D01104, doi:10.1029/2003JD004045.

Zhang, M., R. C. J. Somerville, and S. Xie, 2016: The SCM concept and creation of ARM forcing datasets. *The Atmospheric Radiation Measurement (ARM) Program: The First 20 Years, Meteor. Monogr.*, No. 57, Amer. Meteor. Soc., doi:10.1175/AMSMONOGRAPHS-D-15-0040.1.

Chapter 7

The ARM Tropical Western Pacific (TWP) Sites

C. N. LONG* AND J. H. MATHER

Pacific Northwest National Laboratory, Richland, Washington

T. P. ACKERMAN

University of Washington, Seattle, Washington

1. Introduction

One of the earliest and liveliest discussions of the ARM Program was about the number and possible locations of the proposed ground-based observing sites. Early versions of the ARM Program Plan (U.S. DOE 1990) envisaged placing ground-based remote sensing facilities at five or more locations, but the locations were not specified. As soon as ARM was approved in the Department of Energy budget, determining these locales became one of the highest priorities. The workshops and discussion sessions that were held to address this issue focused on two related questions:

1) What are the most climatically important regimes to sample?
2) For which regimes do we have the least available information?

Scientific and logistical considerations led to establishing the first locale in the continental interior of the United States, which led to the Southern Great Plains (SGP) site [see Sisterson et al. (2016, chapter 6)]. There was strong consensus that the next two locales should be established in the tropical warm pool and in the Arctic to span, as it were, the extremes of global climate. The selection of the warm pool locale, which led to the establishment of the Tropical Western Pacific (TWP) sites, occurred because of the recognized importance of the TWP in tropical and extratropical climate variability, about which relatively little was known at the time. The TWP area is typified by a strong east-to-west gradient in various climate characteristics, including sea surface temperature, column water vapor amounts, and frequency of convection (Ackerman et al. 1999) and is also characterized by strong solar heating.

a. *Choosing the TWP sites*

ARM Program management appointed site managers for each of these three locales (as well as for two locales that were never subsequently built because of financial considerations) and then selected site scientists through a competitive proposal process. The site scientist and manager were charged jointly with implementation of the locale choice, including choosing an actual site within the designated locale, identifying required instruments, and building the site. The TWP team of Site Scientist Thomas Ackerman (The Pennsylvania State University) and Site Manager William Clements [Los Alamos National laboratory (LANL)] convened several meetings to discuss the site location, most notably a workshop held in Santa Fe, New Mexico, in May of 1992. The consensus strategy that emerged for the TWP was that measurements should occur at three to five locations along the equator selected to sample the shifts in convection and the Walker circulation associated with El Niño–Southern Oscillation (ENSO) events; this strategy was later revised to three sites along the equator. The first site was to be located in the heart of the tropical warm pool, the second in the area of high variability in atmospheric and oceanic properties

* Current affiliation: University of Colorado, Boulder, Colorado.

Corresponding author address: C. N. Long, University of Colorado, NOAA/ESRL GMD/CIRES, 325 Broadway St., Boulder, CO 80305.
E-mail: chuck.long@noaa.gov

DOI: 10.1175/AMSMONOGRAPHS-D-15-0024.1

associated with ENSO cycles, and the third in the subsidence region of the eastern Pacific. The proposal included two additional sites, one north and one south of the equator in the region of high variability, in order to sample the movement of the intertropical convergence zone (ITCZ) and its consequences. These recommendations were summarized in two reports. The first report, entitled "Science and Siting Strategy for the Tropical Western Pacific ARM CART Locale," was authored by Ackerman et al. (1993). The second report is a final updating of that document issued in 1999, coauthored by the same individuals, and its title was expanded to "Tropical Western Pacific Cloud and Radiation Testbed: Science, Siting, and Implementation Strategies," because it includes the actual site design information (Ackerman et al. 1999).

At the same time that the TWP plan was being developed, extensive activities were underway to carry out the Tropical Ocean and Global Atmosphere Coupled Ocean–Atmosphere Response Experiment (TOGA COARE; Webster and Lukas 1992) in the warm pool region of the equatorial Pacific between 140°E and the date line. Much as the joint Spectral Radiance Experiment (SPECTRE)/FIRE experiments in Kansas provided a prototype for the SGP site [see Ellingson et al. (2016, chapter 1)], TOGA COARE provided a unique opportunity to test scientific ideas and deployment concepts for the proposed TWP site. The ARM TWP team participated in COARE by operating a ground-based remote sensing facility at Kavieng, Papua New Guinea (PNG), from November 1992 to February 1993. This Pilot Radiation Observation Experiment (PROBE) campaign (Renne et al. 1994) was the first ARM field campaign and TWP science contribution, providing useful radiation data for TOGA COARE (Waliser et al. 1996; Long 1996) and producing a landmark paper by Westwater et al. (1999), which won the 2000 Professor Dr. Vilho Vaisala Award in Atmospheric Sciences from the World Meteorological Organization. PROBE included high-spectral-resolution IR observations from a Fourier transform infrared radiometer that were used to update the then-current Clough–Kniezys–Davis (CKD) water vapor continuum formulation (Clough et al. 1989, 1992) and a dual-channel microwave water radiometer (MWR) for continuous column water vapor amount observations (Westwater et al. 1999).

TOGA COARE also provided an opportunity for the TWP team to assess possible locations for a permanent site in the TWP warm pool region. Based on a climatological assessment, largely done using satellite observations, the team settled on Kavieng or Manus Island, PNG. Both places are within a few degrees of the equator in the heart of the warm pool. After careful consideration, the team chose Manus Island primarily for two reasons. The ARM site on Manus could be located at Momote Airport, which is on the eastern shore of the island in a geographically flat area. This location was deemed superior to the possible site near Kavieng because it was next to the open ocean and largely undeveloped, other than the airport runway itself. Second, and equally important, the provincial government of Manus is very supportive of educational activities on the island and was willing to work closely with the ARM team in developing the necessary logistics. In retrospect, Manus has proved to be an excellent choice in terms of science, political stability, and long-term relationships.

The focus of the team then shifted to the selection of a site farther to the east. Using long-time series of outgoing longwave radiation (OLR) data, the team identified the region around 170°E longitude as being the most variable with regard to OLR (and presumably cloud occurrence). For logistical reasons, the team narrowed its options to either Nauru or Tarawa Atoll. Nauru is an independent republic, while Tarawa is part of the far-flung Republic of Kirabati. Both sites presented challenges. Nauru is a small rocky island once covered by guano deposits but now almost completely devastated in the interior by strip mining. The mining generated considerable revenue at the time, but the island was suffering economically by the mid-1990s. Finding a good site was constrained by the relatively high population density along the island shore. This latter problem was even greater at Tarawa. Tarawa is a small atoll with a very high population density and little available land. It also has severe environmental problems with freshwater availability and sanitation. The consensus was to locate in Nauru in large part because the only available site in Tarawa was too small and too impaired by surrounding development.

As a result of financial constraints within the ARM Program, the additional sites in the equatorial Pacific were never developed. A third site, however, was developed in Darwin, Australia, in conjunction with the Australian Bureau of Meteorology. This site arose in part from the need to provide a more local staging area for maintenance of Manus and Nauru and in part from the excellent existing scientific infrastructure and data collection in the area. The Darwin climate is driven heavily by the Australian monsoon and thus provides an interesting contrast to the more oceanic convection typical at Manus and the ENSO-driven variability at Nauru. Figure 7-1 shows the three ARM TWP site locations.

b. Establishing the sites

The first challenge in establishing the TWP was how to construct a continuously operating facility in a remote

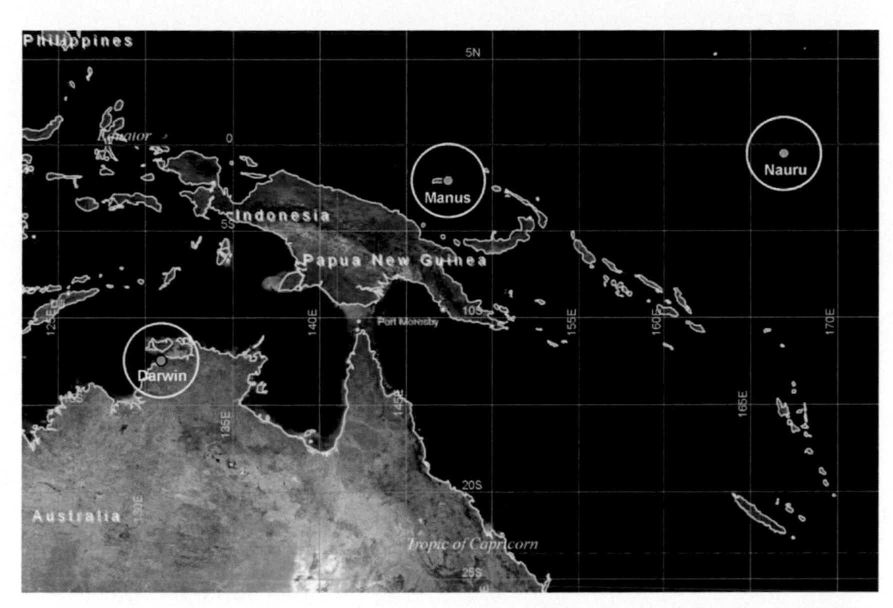

FIG. 7-1. ARM TWP sites. Image courtesy of the U.S. Department of Energy ARM Climate Research Facility.

location that had no long-term scientific or technical personnel posted to the area. Besides the landmark observations, the lessons learned from PROBE had a significant role in design of the Atmospheric Radiation and Cloud Station (ARCS). As noted in Ackerman et al. (1993, 9–18): "The logistical issues sharpened in our perception as a result of our experiences in PROBE [. . .]. A good deal of thought has gone into the proposed ARCS. The current conceptual design has been heavily influenced by the design of the Integrated Sounding Systems developed and deployed by the National Center for Atmospheric Research (NCAR) and the National Oceanographic and Atmospheric Administration (NOAA), and by our experiences with PROBE." Important elements of the ARCS design included the use of International Standards Organization (ISO)-certified shipping containers as instrument shelters, integrated backup power, a centralized redundant site data system (SDS), connectivity for the site management team, and near-real-time health and status monitoring.

The TWP team decided to build the site in the United States using sea containers and then ship the rebuilt containers to the TWP. This added some complexity to the choice of sites, since ports and cranes were now required to offload the containers, as well as heavy equipment to position them at their destination; however, it allowed the containers to be handled relatively easily at commercial ports. These containers did double duty as shipping containers and, in the field, as shelters for instruments, computer systems, and work space.

Another key element of the site infrastructure was integrated backup power. The ARCS was designed with a combination of a backup diesel generator that could support the entire site and uninterruptible power supplies that would keep instruments and computers running during a power outage while the generator kicked on. This provided much better operational continuity than would have been available otherwise.

The SDS was designed to provide a single virtual point of entry to the site. The SDS managed data collection from all the instruments, as well as environmental monitoring systems, control of off-site communications, bundling of data, and the creation of compact health and status files. Originally, there was very limited bandwidth to the tropical sites. Routine transmission of data was managed over a geostationary satellite link and was limited to about a kilobyte per hour. Careful thought was given to packing information from instruments and infrastructure systems into that tiny data packet, but it was sufficient to gain a remarkable amount of information about the site conditions and was used to identify system failures.

Instrument selection was based on a combination of scientific needs and instrument robustness. The instrument suite closely matched those previously deployed at the SGP site. Some of the more experimental instruments, like the Raman lidar, were not deployed. The ARCS deployment was also much more compact than the SGP. Whereas the SGP site was distributed over a large portion of north-central Oklahoma and south-central Kansas, the ARCS concept mirrored the SGP central facility. Among other things, this compact design enabled continuous remote contact with all the instruments.

Mark Ivey (Sandia National Laboratories), working closely with Site Managers Bill Clements and Fairley Barnes (of Los Alamos National Laboratory), was largely responsible for the organization and supervision of the ARCS development for both the Manus and Nauru sites. The TWP development team included the site management team, data system developers, instrument technical leads, communications specialists, and representatives from the TWP site science office and met regularly in Albuquerque, New Mexico, to develop and implement the ARCS concept.

After several years of dedicated effort, the TWP team established the first TWP site at Momote Airport on Manus Island in August 1996. Members of the team spent more than 6 weeks on Manus establishing the site and making sure that all systems were working properly. The installation period was a time of intense effort for everyone, with long days and many hours spent out in hot, humid weather. It was also a time of great satisfaction watching the system come to operational life after so many years of preparation. The site at Nauru followed in November of 1998 with a similar period of installation and testing.

The TWP operations were, for many years, the only part of the ARM Program that dealt with foreign governments. The Manus and Nauru sites were established in collaboration with the governments of PNG and Nauru, respectively. This process was long and occasionally frustrating but ultimately necessary and extremely helpful. The initial step was to negotiate permission for an ongoing U.S. scientific presence in these countries. Both countries were wary of a western presence, having previous experiences with broken promises and a perceived lack of reciprocity. The ARM Program and TWP team came to the negotiations with two specific offers: namely, training and employment of local individuals to operate and maintain the sites, including launching of meteorological balloons and open sharing of data. Both were very important to the national governments and led to an acceptance of the ARM Program and permission to install the sites. The radiosonde data were critical because they fulfilled obligations that PNG and Nauru had to the World Meteorological Organization to provide soundings for forecast model initialization. At the level of the local government in Manus, engagement with their secondary education program was the key factor in accepting the ARM Program. It is important to recognize the positive role that both governments played in the site selection and the approval process. Without their participation, the ARM Program could not have established these sites and would not have had the ongoing success in operations and data collection.

c. Outreach

The principal outreach activity of the TWP site has been its interactions with the secondary education systems on Manus Island and Nauru. As part of the early siting negotiations in Manus, a commitment was made to the Provincial Governor to provide educational resources to the secondary schools on the island. As the TWP effort progressed, responsibility for this education program was given to Fairley Barnes (LANL), who had a tremendous heart for the students and the education program. She worked with school administrators and teachers to determine how the ARM Program could help them, organized presentations (by herself and others with relevant expertise) at the schools, arranged tours of the ARM facility for teachers and students, and facilitated the donation of personal computers to the schools.

The outreach program organized more formally in 1998 and adopted three goals: 1) inform and enrich primary, secondary, and college programs in the tropical western Pacific region; 2) focus on basic science concepts, through activities and study of climate, climate change, and the effects of climate change relevant to the region; and 3) foster career goals in science for students in the region. It organized workshops for teachers in both countries, working with several regional partners in the South Pacific island nations and Australia. It also worked to develop curriculum units and a variety of teaching tools, including a traveling kiosk that could be used for interactive learning and viewing the data.

In the mid-2000s, in response to a general lack of funding for ARM activities, the ARM Program moved to consolidate its education activities into a single education office rather than funding them through the site offices. While this achieved a desired fiscal efficiency, it had a negative impact on TWP outreach and education. The TWP is historically the only permanent ARM site located outside the United States, and the teachers and students in its locale have substantially different interests and needs. With the consolidation, these were given lower priority to other considerations; as a result, the current educational outreach program has a reduced flexibility and capability to address the unique issues in the TWP and became less engaged with the TWP nations.

2. The TWP sites

a. Manus

Developing an ideal measurement strategy for the TWP is challenging because multiple factors play roles in determining the distribution of convection: most

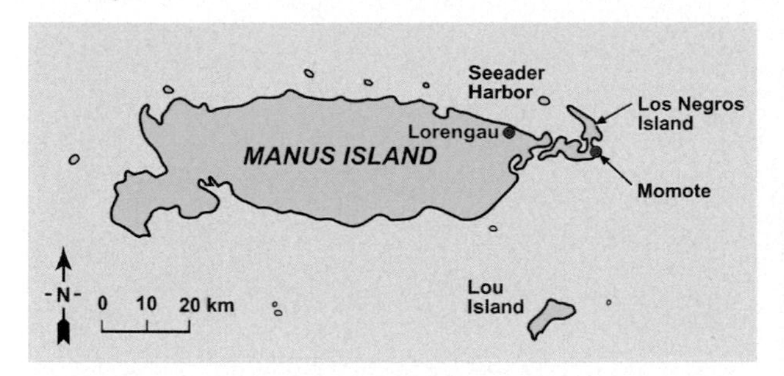

FIG. 7-2. Map of Manus and Los Negros Islands showing the Momote site, where the ARM Manus facility is located.

FIG. 7-3. October 1996 picture of the Manus site, as originally configured.

notably, sea surface temperature gradients and surface heating associated with the Maritime Continent (e.g., Ramage 1968; Neale and Slingo 2003; Shibagaki et al. 2006). Furthermore, the Maritime Continent is characterized by a broad spectrum of island sizes ranging up to the large islands of New Guinea, Borneo, and Sumatra and the adjacent continents of Australia and Asia. The first ARM TWP site was established on Manus Island (2.1°S, 147.4°E; Fig. 7-1) in Papua New Guinea and represents a moderate-size island at 100 km long on the eastern edge of the Maritime Continent (see Fig. 7-2).

The Manus site was the first TWP site deployed; thus, it became the first test of the remote ARCS operations paradigm. The operation of the Manus site is a collaborative effort between ARM and the Papua New Guinea National Weather Service, who supply the onsite observers, who are trained to handle day-to-day and weekly maintenance, such as cleaning radiometer domes, changing desiccant in radiometers, and swapping data system tapes. For maintenance and support that required more specialized knowledge and experience, a regional service team (RESET) visited the site nominally every 6 months or more frequently, as needed (Mather et al. 1998a).

The Manus deployment was planned to occur in early 1995 (Ackerman et al. 1993) in plenty of time to participate in the primarily NOAA-sponsored Combined Sensor Program (CSP) campaign (Post et al. 1997) but was delayed when an intensive predeployment internal review led the design team to postpone the installation. Because of the expected challenges of deploying a permanent station at a remote location (e.g., it would be very difficult to get any forgotten items), a beta test of the installation was carried out. In the test, the entire ARCS was packed up at Sandia National Laboratories, where it had originally been developed, and then unpacked and deployed for several weeks in San Diego, near the ultimate port of departure. This beta test provided an important dry run for the set up process and a review to ensure that the team would have everything it needed for the real deployment. During this exercise, it was decided that some technical issues with integration of this first ARCS system warranted a delay in the deployment scheduling. For the CSP campaign, a subset of ARCS instruments was deployed in collaboration with the NOAA–NCAR long-term Integrated Sounding System (ISS) operated at the Momote Airfield site, as the land-based ARM contribution to the CSP campaign. The technical issues were addressed, and the ARCS was deployed; observations began in October 1996 (Mather et al. 1998a), with GPS rawinsondes added in 1997, a Whole Sky Imager in 1998, and the Millimeter Cloud Radar (MMCR) in 1999. Figure 7-3 is a photo from October 1996 showing the original site configuration.

Deep within the warm pool, Manus experiences persistent cloudiness and convective activity and is influenced by the Madden–Julian oscillation (MJO; Madden and Julian 1994; Zhang 2005), as shown by Wang et al. (2011). Because the Manus area is so deeply embedded in the warm pool, it shows little intraseasonal or interannual variability in sky cover, downwelling radiative fluxes, and surface cloud radiative effect due to ENSO (McFarlane et al. 2013). The Manus observations were of immediate interest to the ARM research community and led to early studies in subjects such as cloud regimes (Jakob and Tselioudis 2003) and atmospheric state and surface radiation budget (Mather et al. 1998b).

The TWP area is primarily oceanic in nature. From the start, there was concern that observations made on a tropical island would be influenced so significantly by the land presence that they would not be representative of the larger surrounding oceanic area. One aspect of the CSP campaign was to compare the ship and land site measurements as the ship got closer to the site. Unfortunately, the ship time near the island combined with the limited land instrumentation due to the ARCS

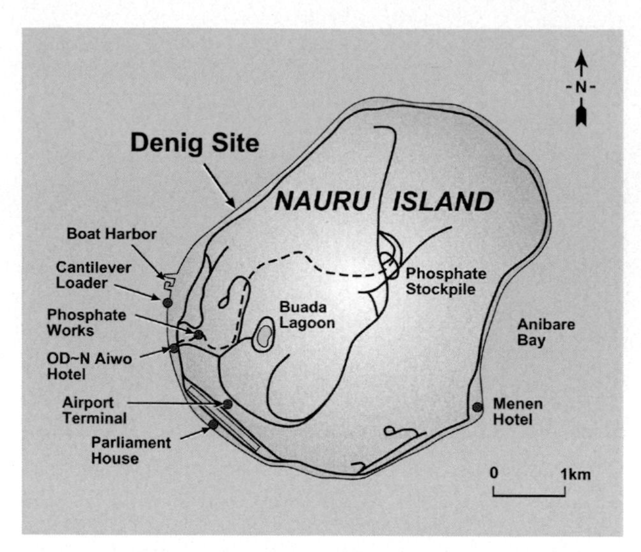

FIG. 7-4. Map of Nauru Island showing the location of the ARM site at Denig.

FIG. 7-5. Picture of the Nauru site less than a year after start of operations during the Nauru99 campaign.

deployment delay was inadequate to answer the question but did suggest that there was no major land influence on the limited site observations that were available. To better understand the local representativeness of the Manus ARM site, the ARM-funded Manus Variability Study campaign collected surface radiation, meteorological, and cloud-base height observations at the ARM C-band scanning radar site just over 7 km away at the naval base at Lombrum. The C-band site was installed as part of the Recovery Act upgrades (see next paragraph) and started operations in November of 2010. The 15 months of data collected there (August 2011–November 2012) in conjunction with the again ARM-funded ARM MJO Investigation Experiment on Manus (AMIE-Manus; Long et al. 2010) are currently being analyzed to identify any statistical differences between the two sites in variables such as low-cloud-base heights, fractional sky cover, downwelling shortwave (SW) and longwave (LW) radiation, and cloud radiative effects.

As noted in Long et al. (2013), the new challenge at the TWP sites is to expand from the vertically pointing, or "soda straw," view that the ARM remote sensors have used since operations began to examine the local spatial variability in clouds and precipitation. Meeting this challenge requires enhanced measurement capabilities, which were facilitated by funding from the American Recovery and Reinvestment Act,[1] along with development of new data processing and retrieval algorithms. The C-band (6 GHz) scanning precipitation radar, along with scanning X-band (9.5 GHz) and Ka-band (35 GHz) radars (Bharadwaj et al. 2011) deployed on Manus, will enhance the upgraded vertically pointing cloud radars and lidars. The temporal integration of this expanded spatial view will allow new genres of process-oriented data analysis and modeling studies of cloud field evolution.

b. Nauru

The second TWP site was established on Nauru Island (0.5°S, 166.9°E; Fig. 7-1), which is a small island located on the eastern edge of the warm pool, with virtually all ARCS instruments operating by the start of 1999. As noted above, land is at a premium on the small island, which, as shown in the island map (Fig. 7-4), is only 6 km long by 4 km wide. Consequently, the physical location literally had to be built up in order to give sufficient area for the site (Fig. 7-5; photo taken from the Flinders University's Cessna aircraft during the Nauru99 campaign). Although our preference was for a site on the southern coast of the island near the airport, negotiations for that location broke down when a local family objected to the need for dump trucks to unload crushed coral and phosphate material from the island center (referred to locally as "top side") next door to their homes. We then obtained a second location on the western shore, where the site was built, becoming operational in late November of 1998. Figure 7-5 shows the white surface of crushed coral material that was used to build up the site area. The interior center of the island is now almost completely devastated by strip mining of the nearly depleted guano deposits for phosphate, leaving behind mostly unvegetated and impassible karst fields (Fig. 7-6).

Like those from Manus, the Nauru observations were of immediate interest for the ARM research community

[1] For more information on all Recovery Act instruments and upgrades, see Mather and Voyles (2013) and http://www.arm.gov/about/recovery-act/instruments.

FIG. 7-6. Picture of a Nauru karst field remaining after phosphate mining in the central part of the island taken in April of 2004. The vertical structures remaining are extremely hard petrified coral, with the deep intervening spaces virtually impossible to walk through.

and led to early studies in subjects such as albedo (Matthews et al. 2002), atmospheric humidity (Westwater et al. 2003), tropical cirrus (Comstock et al. 2002), and Kelvin waves (Holton et al. 2002). Unlike Manus, Nauru exhibits strong variability associated with ENSO; thus, it is an important site for documenting this variability and its effect on cloudiness and the surface radiation budget. Time series of OLR data show the region around the Nauru area as being the most variable with regard to OLR and, as found in studies using Nauru data (e.g., Jensen and Del Genio 2006; Porch et al. 2006; McFarlane et al. 2013), in cloud occurrence as well. Nauru experiences convectively active periods (including El Niño) when cloudiness characteristics are similar to those of Manus and times of suppressed convection when the area is embedded in the descending branch of the Walker circulation (including La Niña). Nauru experiences about 40% overcast skies during daylight hours for convectively active El Niño conditions, but, during convectively suppressed La Niña conditions, overcast occurs only 10% of the time, and the sky cover is 50% or less during 67% of the observations (McFarlane et al. 2013).

The Nauru99 campaign occurred in June and July of 1999, several months after the Nauru site became operational. Nauru99 was an international research collaboration conducted on and around the island of Nauru (Reynolds 1998).[2] Participants, along with ARM, included the NOAA R/V *Ronald H. Brown* and the Japan Agency for Marine-Earth Science and Technology (JAMSTEC) R/V *Mirai*, which measured surface and radiation fluxes at sea for comparison with the land-based ARCS systems and the Tropical Atmosphere Ocean (TAO) buoy array. During the campaign, which occurred during an episode of convectively suppressed conditions, a stream of small clouds was observed to start forming over the island in late afternoon and advect downstream. This "cloud street" was studied using satellite data by Nordeen et al. (2001) and was shown to stretch up to more than 200 km downstream of Nauru. The Nauru cloud street is formed by low-level flow over and around the small island when a heat island occurs because of solar heating of the center of the island, where the phosphate has been removed, leaving mostly bare ground (Savijarvi and Matthews 2004). The downstream cloud street is maintained through a pair of island-generated vortices maintained by the convection that occurs between them (Matthews et al. 2007) and does have some impact on the ARM site measurements.

The evidence of the Nauru cloud street prompted the ARM-funded Nauru Island Effect Study (NIES; Long 2001), which deployed a set of instruments on the eastern side of the island for comparison to the ARM site data on the western side in order to attempt to quantify what measurements might be affected and to what extent. McFarlane et al. (2005) showed that only low-level cloud amounts and downwelling SW radiation values were affected when an island effect was occurring. Additionally, a method of detecting the occurrence of the island effect was developed from the NIES data, prompting the deployment of a small set of radiometers on the southern end of the island near the airport. These data have been used by the TWP site scientist team to produce a Nauru Island effect dataset,[3] which denotes the times of occurrence. Five years of data from this dataset were then used to quantify the Nauru Island effect by Long and McFarlane (2012), showing that the island effect occurs about 11% of the time during daylight hours. Over the long term, the effect increases the low-level (500–1000 m) cloud occurrence by 1% overall and decreases the overall average downwelling SW by 1%. However, for shorter-term studies or those that separate data by conditions such as convectively active or suppressed regimes, the effect can have significant impacts, and use of the Nauru Island effect dataset is advised to determine occurrence and impact.

While the Nauru site observations have been shown to be scientifically useful, continuing economic decline and deteriorating infrastructure has made working there too

[2] Additional details are available at http://www.arm.gov/campaigns/twp1999nauru.

[3] See http://www.arm.gov/data/pi/45.

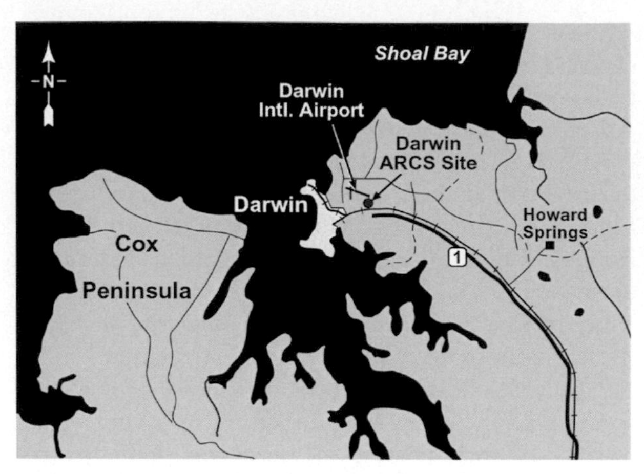

FIG. 7-7. Map of the Darwin, Australia, area showing the location of the ARM (ARCS) site near the Darwin airport and the nearby seas.

FIG. 7-8. November 2004 picture of the Darwin site viewed from a landing flight at the Darwin airport. ARM facilities are located in the right half of the photo; on the left side are the Darwin Airport Australian Bureau of Meteorology Office facilities.

difficult. The active sensors (cloud radar, lidar, etc.) at the ARM site at Nauru were removed in February of 2009, and, sadly, all ARM observations on Nauru have now ended. The remaining site infrastructure was handed off to the Nauruan government as the foundation for establishing a meteorological station in August of 2013.

c. *Darwin*

While the original sampling strategy called for distributing additional sites to more fully explore the prevalent north–south and east–west convection gradients, logistics also plays an important role. Identifying suitable sites in the TWP region that meet the science requirements and have the infrastructure necessary to support an ARM site was challenging. Further adding to the logistical challenges was a strong interest in having the capability of periodically making airborne measurements in conjunction with the site, as was often done at the SGP. From this perspective, Darwin, Australia (12.4°S, 130.9°E; Fig. 7-1), is an ideal location in the region. Darwin has been a cornerstone of tropical meteorological measurements since the beginning of the last century (e.g., Allan et al. 1991) and boasts an excellent support structure for long-term measurements and intensive field campaigns, including a major airport, a regional Bureau of Meteorology forecasting office, and a research-grade C-band radar (Keenan et al. 1998).

Darwin is firmly within the tropical latitude band; however, while the variability of convection is dominated by ENSO and the MJO at Nauru and Manus, respectively, the dominant source of variability at Darwin is the annual Australia monsoon (Wang et al. 2011; May et al. 2012; McFarlane et al. 2013). The Australia monsoon is characterized by a dry season extending from approximately May through October, a transition or buildup season from October through November–December, and a wet season from December through April (e.g., Holland 1986; Drosdowsky 1996; Pope et al. 2008; Evans et al. 2012). The wet season is modulated by active and suppressed convection periods that are related to other modes of tropical variability (e.g., the MJO; Evans et al. 2014), but the dominant signal is the monsoon.

While Darwin is a coastal site and not an island site, as shown in the map of the Darwin area (Fig. 7-7), an argument can be made that, in addition to the logistic attractions, it also serves as a good location for studying west Paçific tropical convection from a scientific perspective. During the convectively active period of the monsoon, widespread cloud cover reduces surface radiative heating (May et al. 2012). Consequently, instead of land–sea circulations and associated coastal convection that is common in the coastal tropics, convection is driven by large-scale dynamical forcing associated with the monsoon trough. Convection in these conditions has similar characteristics to periods of widespread convection (e.g., during the active phase of the MJO) over the open ocean (May et al. 2008). Meanwhile, during periods when local heating dominates forcing, often referred to as break periods, conditions are not unlike those found along the coasts of large islands in the Maritime Continent (Pope et al. 2008). Therefore, during the wet season, Darwin can be, and often has been, used as a base for studying tropical convection. Prior to the deployment of the ARM site at Darwin, campaigns focusing on the Darwin area have included the Island Thunderstorm Experiment (ITEX; Keenan et al. 1989), the Stratosphere–Troposphere Exchange Experiment (STEP; Russell et al. 1993), the Maritime Continent Thunderstorm Experiment (MCTEX; Keenan et al. 2000), and the Darwin Area Wave Experiment (DAWEX; Hamilton et al. 2004). MCTEX, in particular, was a collaboration with the DOE that deployed millimeter-wavelength radars, a microwave radiometer, and a suite of radiometers in the Tiwi Islands.

Data collection at the ARM Darwin site began in March of 2002. An aerial photo from November 2004 (Fig. 7-8) shows the site layout on the right side of the photo, while the left side is the Darwin Airport Australian Bureau of Meteorology facilities. Operations at the site began somewhat slowly because it is a significant challenge to operate instruments continuously at the tropical sites. For the first several years, the Manus and Nauru sites were assigned a higher priority than the Darwin site, so some of the key instruments at Darwin, notably the cloud radar and micropulse lidar, saw reduced operation time during the first few years because parts were used to keep equipment running at the other sites. However, beginning in approximately 2003, planning began for a major field campaign largely driven by ARM to be based in Darwin. The Tropical Warm Pool–International Cloud Experiment (TWP-ICE)[4] was held in January and February 2006 and brought together a number of measurement components and also coincided with two European-supported campaigns: Aerosol and Chemical Transport in Tropical Convection (ACTIVE) and Stratosphere–Climate Links with Emphasis on the Upper Troposphere and Lower Stratosphere (SCOUT-O3; Vaughan et al. 2008). TWP-ICE was designed to study the life cycle and properties of tropical convective systems and included the ARM Darwin site, the Bureau of Meteorology measurement assets in the Darwin region, a five-site radiosonde array centered on Darwin (in addition to the operational soundings at Darwin), three aircraft, a ship, and a variety of ground-based guest instruments (May et al. 2008).

The ACTIVE and SCOUT campaigns had different science goals than ARM's TWP-ICE but provided a very useful set of complementary measurements. ACTIVE was focused on aerosol processes and added two aircraft and a series of ozonesondes launched from Darwin (Vaughan et al. 2008). ACTIVE was actually held in two phases: the first phase was in November–December 2005 during the so-called buildup monsoon phase and the early part of the wet season; and the second phase overlapped with TWP-ICE. These two periods have distinctly different aerosol loading with much more aerosol typically present during the buildup period than during the later stages of the wet season (Bouya et al. 2010), as well as somewhat different convective characteristics, as the buildup phase tends to be more characterized by coastal-heating-type convection (Pope et al. 2008).

SCOUT, meanwhile, was focused on upper-tropospheric humidity and the exchange of water vapor, aerosols, and chemicals between the upper troposphere and lower stratosphere (Vaughan et al. 2008). Because of the focus on stratosphere–troposphere exchange, there was interest by the SCOUT planning team in particularly deep and, if possible, predictable convection. These considerations led the SCOUT group to focus on the buildup period, during which they could study deep, island-based convection just north of Australia in a relatively predictable natural laboratory.

While it would have been desirable to completely align all three campaigns, the scientific focus of TWP-ICE was on the maritime convection associated with the active monsoon periods [see Evans et al. (2012) for a monthly summary of synoptic states and convective activity]. Consequently, SCOUT and TWP-ICE occurred at different times during the 2005/06 wet season. By spanning both periods, the ACTIVE campaign provides context for the full period.

The diversity of measurements has provided input for a broad range of research topics, including the boundary layer (May et al. 2012), convective clouds (e.g., Frederick and Schumacher 2008; Cetrone and Houze 2009; Collis et al. 2013), gravity waves (Hecht et al. 2009), precipitation (Giangrande et al. 2014), and satellite evaluation studies (Liu et al. 2010). A particular goal of the TWP-ICE campaign was to provide a dataset that could be used for climate model evaluation. One key component to support this type of research is the development of a model forcing dataset [Zhang and Lin (1997); Zhang et al. (2016, chapter 24)] that provides dynamical boundary conditions for running cloud-resolving models. The model output can then be compared with observations of cloud life cycle and properties. The radiosonde array provided a key component of this forcing dataset, as did surface heat fluxes and area precipitation from the C-band radar (Xie et al. 2010).

The combination of a well-constrained forcing dataset with extensive observations has led to a number of modeling studies that include analyses of cloud-resolving models (e.g., Fridlind et al. 2012; Mrowiec et al. 2012; Wang et al. 2009) and climate models (e.g., Wu et al. 2009; Song and Zhang 2011) to study the effects of model resolution (Lin et al. 2012) and processes such as entrainment and diabatic heating (Xie et al. 2010; Del Genio and Wu 2010). These data and model simulations point to the problems inherent in simulating subgrid-scale phenomena, such as vertical velocity (Collis et al. 2013) and precipitation (Varble et al. 2011).

Similar to our experiences with Manus and Nauru, interest in the areal representativeness of the ARM Darwin site motivated the ARM-funded Darwin ARM Climate Research Facility (ACRF) Representativeness Experiment (DARE). As in the Manus Variability

[4] For more information about TWP-ICE, see http://www.arm.gov/campaigns/twp2006twp-ice.

Study, a small radiometer system was deployed at the Gunn Point C-band polarimetric radar system (C-POL) site located about 25 km northeast of the ARM site. Data were collected from October 2009 through June 2011 and are currently being analyzed to identify any statistical differences between the two sites in variables such as fractional sky cover, downwelling SW and LW radiation, and cloud transmissivity and radiative effects.

3. Scientific contributions

The measurements at the ARM TWP sites have resulted in significant scientific contributions to our understanding of tropical processes and their variability. In a recent paper, Long et al. (2013) summarized these contributions, focusing on the use of the Manus and Nauru observations to study the tropical radiation budget, cloud properties, and cloud radiative effects, which are all components of the original science questions that drove ARM to deploy sites in the TWP. In addition, the paper identifies contributions to model evaluation, parameterization development, and satellite comparison and validation. While many of these research areas were included in the original ARM science vision for the TWP, the scope of the research has notably surpassed expectations.

For decades, Darwin has been an important center of respected scientific observations and tropical research, certainly long before the ARM site was deployed at the Darwin airport. The ARM Darwin site is now contributing to this legacy in unique ways. In particular, the addition of cloud radar data and support for aircraft measurements has provided new and important avenues for the study of cloud properties and processes. The examples cited here serve to illustrate the types of additional studies enabled by ARM but are certainly not an exhaustive accounting.

As with all of ARM, research studies in the TWP are based on both the continuous measurements and shorter-term, more intensive campaign efforts. For the TWP, most field campaign efforts have had significant TWP site scientist team participation and leadership. Larger campaigns, such as Nauru99 (Tom Ackerman, TWP site scientist at the time as co-primary investigator), TWP-ICE (Jim Mather, associate site scientist at the time as co-primary investigator), and AMIE-Manus and the corresponding AMIE-Gan (Chuck Long, current TWP site scientist as primary investigator for both) have focused on the larger science questions of the tropical regime. But there have also been more targeted campaigns, such as the NIES, DARE, and the Manus Variability Study all with the TWP site scientist as PI that are geared toward making the TWP data more useful and understandable to the science community. From its inception, the TWP site scientist team has worked both to advance our scientific understanding and to enable the TWP observations to be more useful to the entire community.

4. Summary

The TWP features the warmest sea surface temperatures on Earth and is typified by strong solar heating, plentiful evaporation, and abundant convection and precipitation. Significant inter- and intraseasonal variability driven by phenomena such as ENSO and the MJO, along with migration of the intertropical convergence zone, drives teleconnections that affect many other parts of the globe. Thus it is no surprise that the ARM Program targeted the TWP for long-term cloud and radiation measurement sites.

As discussed here and in summary papers (e.g., Long et al. 2013), significant use has been made of the TWP site data to improve our understanding of the TWP regime. This substantial body of work has targeted many of the original science questions of the initial ARM plan (Ackerman et al. 1999), making considerable progress. The research has addressed many of the original questions but has also shaped new and refined older science questions.

With time and maturity, factors that influence the practical and technical aspects of long-term observational activities have evolved. For instance, for Manus and Nauru, communications available to monitor instrument health and status in near-real time were restricted to a small transmit package time slot of the GOES satellite system once per hour. Thus, the site data system produced a packet of hourly averages of a subset of the measurements and system information for transmission back to TWP Operations. This was certainly a significant limitation for monitoring data quality, since reasonable-looking hourly averages could mask a multitude of issues, especially for the active sensors. For example, only an hourly average of the cloud radar transmission power could be included as the monitoring variable for radar performance, which gave no information on the receivers or calibration. The full data were shipped back to Operations initially via 4-mm tapes and later via hard drive every several months. Verbal communications with on-site observer staff also were limited, initially through a landline telephone that was often inoperable, particularly at Manus, where there was initially an emergency satellite phone with limited minutes but that had fax and 1200-bps modem capability. At Nauru, there was an Internet link through Nauru Telco via a link to Australia for a while before Nauru dropped the link. But this Internet had a very low data rate because the phone lines between the Nauru site and the Telco were poor. One solution that was installed in late 1999 to early 2000 was an Inmarsat B satellite phone on site to be used for site communications when the landline phones were down. This amounted to a significant cost, however, and was soon abandoned. Application for Very Small Aperture Terminal (VSAT) licenses was started in July 2002, and satellite Internet installations were done in October (Manus) and November 2002 (Nauru). Needless to say, having satellite Internet connectivity available amounted to a quantum improvement in the ability to monitor and remotely operate the on-site instruments and systems. With satellite internet communications came a significant improvement in overall data quality and continuity. We note that the Darwin site came online in 2002 with good quality Internet availability from the start. Nonetheless, the large volumes of data generated by some instruments, such as the cloud radar, continue to be shipped on hard disk periodically back to the United States.

The deterioration of the economic and political situation on Nauru, along with the changing funding and operational demands of the ARM Program, prompted the withdrawal of ARM activities and instrumentation from Nauru. The Nauru data have been beneficial to the scientific community in studies of the tropical regime (e.g., see Long et al. 2013) and will undoubtedly see continued application. One of the most critical losses for the atmospheric science and weather forecasting communities, however, is likely to be the absence of sonde launches in this data-sparse, yet critically important, region. The nearest upper-air station to Nauru that is currently active is located at Majuro Atoll, more than 950 km to the northeast, with the next nearest at Pohnpei, Micronesia, more than 1250 km to the northwest. One hope is that programs such as the WMO Global Climate Observing System (GCOS) Reference Upper-Air Network (GRUAN) effort might sponsor and fund sonde launches by the Nauruans.

At the same time as the loss of the Nauru site, a new phase of tropical cloud studies commenced at the Manus and Darwin sites. New scanning cloud and precipitation radars will allow expansion from the vertically pointing view that has typified ARM sites and data from the beginning, adding a critical spatial view for characterization of the cloud and precipitation fields. One weakness of the TWP observational efforts with respect to model evaluation has been the lack of adequate dynamic and thermodynamic context needed as input to drive cloud-resolving and single-column modeling studies. Results from TWP-ICE show that the combination of large-scale precipitation estimation from scanning precipitation radar, centrally located soundings, and reanalysis products may provide an adequate constraint to construct useful model forcing datasets (Xie et al. 2010) using the variational analysis methodology [Zhang and Lin (1997); Zhang et al. (2001, 2016, chapter 24)]. Forcing data can be further augmented by enhanced sounding periods. Thus, the Darwin site and now, with the progress being made regarding single-sonde-site forcing datasets, the Manus site with its scanning C-band radar, when combined with extended periods of frequent sonde launches and likely NWP data, offers the opportunity of producing model-forcing datasets for single-column and cloud-resolving modeling efforts. While the existing long-term measurements of clouds, radiation, and atmospheric state provide a distinctive observational record that will continue to be utilized by the research community, the coming decades offer significant new opportunities for atmospheric research in this challenging region thanks to ongoing observations and the new improved instrumentation.

Acknowledgments. We thank all of those responsible for the operation and maintenance of the sites and instruments down through the years, including personnel of the TWP Operations Office, all of the on-site local observers, and the RESET technicians, both former U.S. personnel and, beginning with the development of the site in Darwin, personnel from the Australian Bureau of Meteorology. Their diligent and dedicated efforts are much of the reason for the high quality and continuity of the data collected in such challenging circumstances, and these efforts are often underappreciated and unrecognized.

REFERENCES

Ackerman, T. P., W. E. Clements, and D. S. Renne, 1993: Science and siting strategy for the Tropical Western Pacific ARM CART locale. ARM Tech. Rep. TWP93.0100104, 49 pp. [Available online at https://www.arm.gov/publications/site_reports/twp/TWP93.0100104.pdf.]

——, ——, F. J. Barnes, and D. S. Renne, 1999: Tropical Western Pacific Cloud and Radiation Testbed: Science, siting, and implementation strategies. ARM Tech. Rep. ARM-99-004, 71 pp. [Available online at www.arm.gov/publications/site_reports/twp/arm-99-004.pdf.]

Allan, R. J., N. Nicholls, P. D. Jones, and I. J. Butterworth, 1991: A further extension of the Tahiti-Darwin SOI, early ENSO events and Darwin pressure. *J. Climate*, **4**, 743–749, doi:10.1175/1520-0442(1991)004<0743:AFEOTT>2.0.CO;2.

Bharadwaj, N., K. B. Widener, K. L. Johnson, S. Collis, and A. Koontz, 2011: ARM radar infrastructure for global and regional climate study. *35th Conf. on Radar Meteorology*, Pittsburgh, PA, Amer. Meteor. Soc., 16B.4. [Available online at https://ams.confex.com/ams/35Radar/webprogram/Paper191707.html.]

Bouya, Z., G. P. Box, and M. A. Box, 2010: Seasonal variability of aerosol optical properties in Darwin, Australia. *J. Atmos. Sol.-Terr. Phys.*, **72**, 726–739, doi:10.1016/j.jastp.2010.03.015.

Cetrone, J., and R. A. Houze Jr., 2009: Anvil clouds of tropical mesoscale convective systems in monsoon regions. *Quart. J. Roy. Meteor. Soc.*, **135**, 305–317, doi:10.1002/qj.389.

Clough, S. A., F. X. Kneizys, and R. W. Davies, 1989: Line shape and the water vapor continuum. *J. Atmos. Res.*, **23**, 229–241, doi:10.1016/0169-8095(89)90020-3.

——, M. J. Iacono, and J.-L. Moncet, 1992: Line-by-line calculation of atmospheric fluxes and cooling rates: Application to water vapor. *J. Geophys. Res.*, **97**, 15 761–15 785, doi:10.1029/92JD01419.

Collis, S., A. Protat, P. T. May, and C. Williams, 2013: Statistics of storm updraft velocities from TWP-ICE including verification with profiling measurements. *J. Appl. Meteor. Climatol.*, **52**, 1909–1922, doi:10.1175/JAMC-D-12-0230.1.

Comstock, J. M., T. P. Ackerman, and G. G. Mace, 2002: Ground-based lidar and radar remote sensing of tropical cirrus clouds at Nauru Island: Cloud statistics and radiative impacts. *J. Geophys. Res.*, **107**, 4714, doi:10.1029/2002JD002203.

Del Genio, A. D., and J. Wu, 2010: The role of entrainment in the diurnal cycle of continental convection. *J. Climate*, **23**, 2722–2738, doi:10.1175/2009JCLI3340.1.

Drosdowsky, W., 1996: Variability of the Australian summer monsoon at Darwin: 1957–1992. *J. Climate*, **9**, 85–96, doi:10.1175/1520-0442(1996)009<0085:VOTASM>2.0.CO;2.

Ellingson, R. G., R. D. Cess, and G. L. Potter, 2016: The Atmospheric Radiation Measurement Program: Prelude. *The Atmospheric Radiation Measurement (ARM) Program: The First 20 Years, Meteor. Monogr.*, No. 57, Amer. Meteor. Soc., doi:10.1175/AMSMONOGRAPHS-D-15-0029.1.

Evans, S. M., R. T. Marchand, T. P. Ackerman, and N. Beagley, 2012: Identification and analysis of atmospheric states and associated cloud properties for Darwin, Australia. *J. Geophys. Res.*, **117**, D06204, doi:10.1029/2011JD017010.

——, ——, and ——, 2014: Variability of the Australian monsoon and precipitation trends at Darwin. *J. Climate*, **27**, 8487–8500, doi:10.1175/JCLI-D-13-00422.1.

Frederick, K., and C. Schumacher, 2008: Anvil characteristics as seen by C-POL during the Tropical Warm Pool International Cloud Experiment (TWP-ICE). *Mon. Wea. Rev.*, **136**, 206–222, doi:10.1175/2007MWR2068.1.

Fridlind, A. M., and Coauthors, 2012: A comparison of TWP-ICE observational data with cloud-resolving model results. *J. Geophys. Res.*, **117**, D05204, doi:10.1029/2011JD016595.

Giangrande, S. E., M. J. Bartholomew, M. Pope, S. Collis, and M. P. Jensen, 2014: A summary of precipitation characteristics from the 2006–11 northern Australian wet seasons as revealed by ARM disdrometer research facilities (Darwin, Australia). *J. Appl. Meteor. Climatol.*, **53**, 1213–1231, doi:10.1175/JAMC-D-13-0222.1.

Hamilton, K., R. A. Vincent, and P. T. May, 2004: Darwin Area Wave Experiment (DAWEX) field campaign to study gravity wave generation and propagation. *J. Geophys. Res.*, **109**, D20S01, doi:1029/2003JD004393.

Hecht, J. H., and Coauthors, 2009: Imaging of atmospheric gravity waves in the stratosphere and upper mesosphere using satellite and ground-based observations over Australia during the TWP-ICE campaign. *J. Geophys. Res.*, **114**, D18123, doi:10.1029/2008JD011259.

Holland, G. J., 1986: Interannual variability of the Australian summer monsoon at Darwin: 1952–82. *Mon. Wea. Rev.*, **114**, 594–604, doi:10.1175/1520-0493(1986)114<0594:IVOTAS>2.0.CO;2.

Holton, J. R., M. J. Alexander, and M. T. Boehm, 2002: Evidence for short vertical wavelength Kelvin waves in the Department of Energy–Atmospheric Radiation Measurement Nauru99 radiosonde data. *J. Geophys. Res.*, **107**, 20 125–20 129, doi:10.1029/2001JD900108.

Jakob, C., and G. Tselioudis, 2003: Objective identification of cloud regimes in the Tropical Western Pacific. *Geophys. Res. Lett.*, **30**, 2082, doi:10.1029/2003GL018367.

Jensen, M. P., and A. D. Del Genio, 2006: Factors limiting convective cloud-top height at the ARM Nauru Island Climate Research Facility. *J. Climate*, **19**, 2105–2117, doi:10.1175/JCLI3722.1.

Keenan, T. D., M. J. Manton, and G. J. Holland, 1989: The Island Thunderstorm Experiment (ITEX)—A study of tropical thunderstorms in the Maritime Continent. *Bull. Amer. Meteor. Soc.*, **70**, 152–159, doi:10.1175/1520-0477(1989)070<0152:TITESO>2.0.CO;2.

——, K. Glasson, F. Cummings, T. S. Bird, J. Keeler, and J. Lutz, 1998: The BMRC/NCAR C-Band polarimetric (C-POL) radar system. *J. Atmos. Oceanic Technol.*, **15**, 871–886, doi:10.1175/1520-0426(1998)015<0871:TBNCBP>2.0.CO;2.

——, and Coauthors, 2000: The Maritime Continent Thunderstorm Experiment (MCTEX): Overview and some results. *Bull. Amer. Meteor. Soc.*, **81**, 2433–2455, doi:10.1175/1520-0477(2000)081<2433:TMCTEM>2.3.CO;2.

Lin, Y., and Coauthors, 2012: TWP-ICE global atmospheric model intercomparison: Convection responsiveness and resolution impact. *J. Geophys. Res.*, **117**, D09111, doi:10.1029/2011JD017018.

Liu, Z., R. T. Marchand, and T. P. Ackerman, 2010: A comparison of observations in the Tropical Western Pacific from ground-based and satellite millimeter wavelength cloud radars. *J. Geophys. Res.*, **115**, D24206, doi:10.1029/2009JD013575.

Long, C. N., 1996: Surface radiative energy budget and cloud forcing: Results from TOGA COARE and techniques for identifying and calculating clear sky irradiance. Ph.D. dissertation, The Pennsylvania State University, 193 pp.

——, 2001: The Nauru Island Effect Study (NIES) IOP science plan. ARM Tech. Doc. DOE-SC-ARM-0505, 13 pp. [Available online at https://www.arm.gov/publications/programdocs/doe-sc-arm-0505.pdf.]

——, and S. A. McFarlane, 2012: Quantification of the Nauru Island influence on ARM measurements. *J. Appl. Meteor. Climatol.*, **51**, 628–636, doi:10.1175/JAMC-D-11-0174.1.

——, and Coauthors, 2010: AMIE (ARM MJO Investigation Experiment): Observations of the Madden–Julian oscillation for modeling studies science plan. DOE/ARM Tech. Rep. DOE/SC-ARM-10-007, 20 pp. [Available online at https://www.arm.gov/publications/programdocs/doe-sc-arm-10-007.pdf.]

——, and Coauthors, 2013: ARM research in the equatorial western Pacific: A decade and counting. *Bull. Amer. Meteor. Soc.*, **94**, 695–708, doi:10.1175/BAMS-D-11-00137.1.

Madden, R. A., and P. R. Julian, 1994: Observations of the 40–50-day tropical oscillation—A review. *Mon. Wea. Rev.*, **122**, 814–837, doi:10.1175/1520-0493(1994)122<0814:OOTDTO>2.0.CO;2.

Mather, J. H., and J. W. Voyles, 2013: The ARM Climate Research Facility: A review of structure and capabilities. *Bull. Amer. Meteor. Soc.*, **94**, 377–392, doi:10.1175/BAMS-D-11-00218.1.

——, T. P. Ackerman, W. E. Clements, F. J. Barnes, M. D. Ivey, L. D. Hatfield, and R. M. Reynolds, 1998a: An atmospheric radiation and cloud station in the tropical western Pacific. *Bull. Amer. Meteor. Soc.*, **79**, 627–642, doi:10.1175/1520-0477(1998)079<0627:AARACS>2.0.CO;2.

——, ——, M. P. Jensen, and W. E. Clements, 1998b: Characteristics of the atmospheric state and the surface radiation budget at the tropical western Pacific ARM site. *Geophys. Res. Lett.*, **25**, 4513, doi:10.1029/1998GL900196.

Matthews, S., P. Schwerdtfeger, and J. M. Hacker, 2002: Use of albedo modeling and aircraft measurements to examine the albedo of Nauru. *Aust. Meteor. Mag.*, **51**, 229–236.

——, J. M. Hacker, J. Cole, J. Hare, C. N. Long, and R. M. Reynolds, 2007: Modification of the atmospheric boundary layer by a small island: Observations from Nauru. *Mon. Wea. Rev.*, **135**, 891–905, doi:10.1175/MWR3319.1.

May, P. T., J. H. Mather, G. Vaughan, and C. Jakob, 2008: Field research: Characterizing oceanic convective cloud systems: The Tropical Warm Pool International Cloud Experiment. *Bull. Amer. Meteor. Soc.*, **89**, 153–155, doi:10.1175/BAMS-89-2-153.

——, C. N. Long, and A. Protat, 2012: The diurnal cycle of the boundary layer, convection, clouds, and surface radiation in a coastal monsoon environment (Darwin, Australia). *J. Climate*, **25**, 5309–5326, doi:10.1175/JCLI-D-11-00538.1.

McFarlane, S. A., C. N. Long, and D. M. Flynn, 2005: Impact of island-induced clouds on surface measurements: Analysis of the ARM Nauru Island Effect Study data. *J. Appl. Meteor.*, **44**, 1045–1065, doi:10.1175/JAM2241.1.

——, ——, and J. Flaherty, 2013: A climatology of surface cloud radiative effects at the ARM Tropical Western Pacific sites. *J. Appl. Meteor. Climatol.*, **52**, 996–1013, doi:10.1175/JAMC-D-12-0189.1.

Mrowiec, A. A., C. Rio, A. M. Fridlind, A. S. Ackerman, A. D. Del Genio, O. M. Pauluis, A. C. Varble, and J. Fan, 2012: Analysis of cloud-resolving simulations of a tropical mesoscale convective system observed during TWP-ICE: Vertical fluxes and draft properties in convective and stratiform regions. *J. Geophys. Res.*, **117**, D19201, doi:10.1029/2012JD017759.

Neale, R., and J. Slingo, 2003: The Maritime Continent and its role in the global climate: A GCM study. *J. Climate*, **16**, 834–848, doi:10.1175/1520-0442(2003)016<0834:TMCAIR>2.0.CO;2.

Nordeen, M. L., P. Minnis, D. R. Doelling, D. Pethick, and L. Nguyen, 2001: Satellite observations of cloud plumes generated by Nauru. *Geophys. Res. Lett.*, **28**, 631–634, doi:10.1029/2000GL012409.

Pope, M., C. Jakob, and M. J. Reeder, 2008: Convective systems of the north Australian monsoon. *J. Climate*, **21**, 5091–5112, doi:10.1175/2008JCLI2304.1.

Porch, W. M., S. Olsen, P. Chylek, M. Dubey, B. G. Henderson, and W. Clodius, 2006: Satellite and surface observations of Nauru Island clouds: Differences between El Niño and La Niña periods. *Geophys. Res. Lett.*, **33**, L13804, doi:10.1029/2006GL026339.

Post, M. J., and Coauthors, 1997: The Combined Sensor Program: An air–sea science mission in the central and western Pacific Ocean. *Bull. Amer. Meteor. Soc.*, **78**, 2797–2815, doi:10.1175/1520-0477(1997)078<2797:TCSPAA>2.0.CO;2.

Ramage, C. S., 1968: Role of a tropical "Maritime Continent" in the atmospheric circulation. *Mon. Wea. Rev.*, **96**, 365–370, doi:10.1175/1520-0493(1968)096<0365:ROATMC>2.0.CO;2.

Renne, D. S., T. A. Ackerman, and W. E. Clements, 1994: PROBE: The Pilot Radiation Observation Experiment. Preprints, *Eighth Conf. on Atmospheric Radiation*, Nashville, TN, Amer. Meteor. Soc., 270–271.

Reynolds, R. M., 1998: Science and implementation plan (draft)—NAURU-99: An international study of tropical climate in the vicinity of Nauru Island in the tropical western Pacific Ocean, ARM Tech. Doc., 11 pp. [Available online at http://www.arm.gov/science/nauru99/scienceplan.pdf.]

Russell, P. B., L. Pfister, and H. B. Selkirk, 1993: The tropical experiment of the Stratosphere–Troposphere Exchange Project (STEP): Science objectives, operations, and summary findings. *J. Geophys. Res.*, **98**, 8563–8589, doi:10.1029/92JD02521.

Savijarvi, H., and S. Matthews, 2004: Flow over small heat islands: A numerical sensitivity study. *J. Atmos. Sci.*, **61**, 859–868, doi:10.1175/1520-0469(2004)061<0859:FOSHIA>2.0.CO;2.

Shibagaki, Y., and Coauthors, 2006: Multiscale aspects of convective systems associated with an intraseasonal oscillation over the Indonesian Maritime Continent. *Mon. Wea. Rev.*, **134**, 1682–1696, doi:10.1175/MWR3152.1.

Sisterson, D. L., R. A. Peppler, T. S. Cress, P. Lamb, and D. D. Turner, 2016: The ARM Southern Great Plains (SGP) site. *The Atmospheric Radiation Measurement (ARM) Program: The First 20 Years, Meteor. Monogr.*, No. 57, Amer. Meteor. Soc., doi:10.1175/AMSMONOGRAPHS-D-16-0004.1.

Song, X. L., and G. J. Zhang, 2011: Microphysics parameterization for convective clouds in a global climate model: Description and single-column model tests. *J. Geophys. Res.*, **116**, D02201, doi:10.1029/2010JD014833.

U.S. DOE, 1990: Atmospheric Radiation Measurement Program Plan. DOE Tech. Doc. DOE/ER-04411990, 121 pp. [Available online at http://www.arm.gov/publications/doe-er-0441.pdf.]

Varble, A. C., and Coauthors, 2011: Evaluation of cloud-resolving model intercomparison simulations using TWP-ICE observations: Precipitation and cloud structure. *J. Geophys. Res.*, **116**, D12206, doi:10.1029/2010JD015180.

Vaughan, G., K. Bower, C. Schiller, A. R. MacKenzie, T. Peter, H. Schlager, N. R. P. Harris, and P. T. May, 2008: SCOUT-O3/ACTIVE High-altitude aircraft measurements around deep tropical convection. *Bull. Amer. Meteor. Soc.*, **89**, 647–662, doi:10.1175/BAMS-89-5-647.

Waliser, D. E., W. D. Collins, and S. P. Anderson, 1996: An estimate of the surface shortwave cloud forcing over the western Pacific during TOGA COARE. *Geophys. Res. Lett.*, **23**, 519–522, doi:10.1029/96GL00245.

Wang, Y., C. N. Long, L. R. Leung, J. Dudhia, S. A. McFarlane, J. H. Mather, S. J. Ghan, and X. Liu, 2009: Evaluating regional cloud-permitting simulations of the WRF model for the Tropical Warm Pool International Cloud Experiment (TWP-ICE), Darwin, 2006. *J. Geophys. Res.*, **114**, D21203, doi:10.1029/2009JD012729.

——, ——, J. H. Mather, and X. D. Liu, 2011: Convective signals from surface measurements at ARM Tropical Western Pacific site: Manus. *Climate Dyn.*, **36**, 431–449, doi:10.1007/s00382-009-0736-z.

Webster, P. J., and R. Lukas, 1992: TOGA COARE: The Coupled Ocean–Atmosphere Response Experiment. *Bull. Amer. Meteor. Soc.*, **73**, 1377–1416, doi:10.1175/1520-0477(1992)073<1377:TCTCOR>2.0.CO;2.

Westwater, E. R., and Coauthors, 1999: Ground-based remote sensor observations during PROBE in the tropical western Pacific. *Bull. Amer. Meteor. Soc.*, **80**, 257–270, doi:10.1175/1520-0477(1999)080<0257:GBRSOD>2.0.CO;2.

——, B. B. Stankov, D. Cimini, Y. Han, J. A. Shaw, B. M. Lesht, and C. N. Long, 2003: Radiosonde humidity soundings and microwave radiometers during Nauru99. *J. Atmos. Oceanic Technol.*, **20**, 953–971, doi:10.1175/1520-0426(2003)20<953:RHSAMR>2.0.CO;2.

Wu, J., A. D. Del Genio, M. S. Yao, and A. B. Wolf, 2009: WRF and GISS SCM simulations of convective updraft properties during TWP-ICE. *J. Geophys. Res.*, **114**, D04206, doi:10.1029/2008JD010851.

Xie, S., T. Hume, C. Jakob, S. Klein, R. McCoy, and M. Zhang, 2010: Observed large-scale structures and diabatic heating and drying profiles during TWP-ICE. *J. Climate*, **23**, 57–79, doi:10.1175/2009JCLI3071.1.

Zhang, C., 2005: Madden–Julian oscillation. *Rev. Geophys.*, **43**, RG2003, doi:10.1029/2004RG000158.

Zhang, M. H., and J. L. Lin, 1997: Constrained variational analysis of sounding data based on column-integrated budgets of mass, heat, moisture, and momentum: Approach and application to ARM measurements. *J. Atmos. Sci.*, **54**, 1503–1524, doi:10.1175/1520-0469(1997)054<1503:CVAOSD>2.0.CO;2.

——, ——, R. T. Cederwall, J. J. Yio, and S. C. Xie, 2001: Objective analysis of ARM IOP data: Method and sensitivity. *Mon. Wea. Rev.*, **129**, 295–311, doi:10.1175/1520-0493(2001)129<0295:OAOAID>2.0.CO;2.

——, R. C. J. Somerville, and S. Xie, 2016: The SCM concept and creation of ARM forcing datasets. *The Atmospheric Radiation Measurement (ARM) Program: The First 20 Years, Meteor. Monogr.*, No. 57, Amer. Meteor. Soc., doi:10.1175/AMSMONOGRAPHS-D-15-0040.1.

Chapter 8

The ARM North Slope of Alaska (NSA) Sites

J. VERLINDE

The Pennsylvania State University, University Park, Pennsylvania

B. D. ZAK*

Sandia National Laboratories, Albuquerque, New Mexico

M. D. SHUPE

Cooperative Institute for Research in Environmental Sciences, University of Colorado Boulder, and NOAA/Earth System Research Laboratory, Boulder, Colorado

M. D. IVEY

Sandia National Laboratories, Albuquerque, New Mexico

K. STAMNES

Stevens Institute of Technology, Hoboken, New Jersey

1. Introduction

When the ARM Program embarked on its quest to select sites for the proposed ground-based atmospheric observatories, there was a broad consensus that more than one site was needed. One site had to be in the equatorial regions, where a disproportionate share of the solar energy fueling Earth's atmospheric general circulation is received, and the other in a polar location, where radiant energy lost to space greatly exceeds the energy received by the sun. While the latitudinal energy imbalance suggested a polar site, an Arctic location was preferred, because the Arctic plays a stronger role in the general circulation than does the Antarctic (Crowley and North 1991). As far back as 1896, the Swedish scientist Svante Arrhenius suggested that changes in Earth's atmospheric composition would lead to faster changes in the Arctic compared to the rest of the globe (Arrhenius 1896), a process confirmed by general circulation models and now called Arctic amplification [e.g., see summary in Serreze and Barry (2011)]. The high sensitivity of the Arctic to climate change is a result of the susceptibility of the cryosphere (sea ice, land ice, and permafrost) to changes in energy fluxes and influential feedback processes. For example, over the Arctic sea ice, the strong sea ice–albedo feedback amplifies the observed decline in ice extent (e.g., Wendler et al. 2010). Faster changes in the Arctic relative to lower latitudes disturb the latitudinal energy balance and, hence, the general circulation patterns in both the oceans and atmosphere, impacting the entire globe. It may be said that what happens in the Arctic does not stay in the Arctic!

Low temperatures in the Arctic result in low water vapor mixing ratios in the atmosphere, low enough that additional spectral regions, such as the water vapor rotational absorption bands in Earth's emission spectrum, become semitransparent in the cloudless atmosphere, increasing radiative energy losses from the surface (Stamnes et al. 1999). The presence of clouds in the atmosphere alters the atmospheric radiative absorption spectrum and thus has a strong influence on the surface energy budget and radiative losses to space. Most Arctic

* Retired.

Corresponding author address: J. Verlinde, The Pennsylvania State University, Dept. of Meteorology, Walker Bldg. 502, University Park, PA 16802.
E-mail: jxv7@psu.edu

DOI: 10.1175/AMSMONOGRAPHS-D-15-0023.1

clouds are found in the lower troposphere, where they interact with strong and persistent near-surface temperature inversions, a characteristic of the Arctic environment (Serreze et al. 1992). Most of the uncertainties in the Arctic radiation budget are associated with an incomplete understanding of the process interactions between Earth's surface and the atmosphere through this interfacial layer. Complex interactions involving surface–atmosphere energy and water vapor exchange, multiphase cloud processes, the stratified lower troposphere, and radiative energy profiles combine to make the high latitudes challenging to represent accurately in numerical weather prediction and climate models. These challenges together motivated the ARM Program to establish an Arctic observatory to study the many processes that regulate the flow of radiative energy through the atmospheric system.

2. Site selection

a. The North Slope of Alaska

The initial site scientist for the North Slope of Alaska, Knut Stamnes (University of Alaska, Fairbanks), and Site Manager Bernie Zak (Sandia National Laboratories) were tasked to identify specific candidate ARM site locations within the Arctic region. The early decisions for the Arctic site were driven by analyses that revealed surface warming trends over Arctic land areas (Chapman and Walsh 1993) but no, or perhaps even weak cooling, trends over the central Arctic ice pack (Kahl et al. 1993). The ice pack results conflicted with general circulation model simulations (Walsh 1993), which predicted warming trends in the central Arctic Ocean. These differences between observed and simulated trends suggested that high-latitude ocean–atmosphere–ice interactions were represented poorly in general circulation models (Walsh and Crane 1992) and that processes over both land and sea warranted study. Thus, it became apparent that the single Arctic site had to be in an area that straddles an Arctic coast away from significant topography. Among the areas in the Arctic that meet these criteria, the North Slope of Alaska and the adjacent Arctic Ocean (NSA/AAO) was the area that permitted the most cost-effective scientific operations. As a result, the ARM Arctic facility, consisting of a comprehensive suite of ground-based atmospheric instruments, was established in Barrow, Alaska (71°19′N, 156°37′W).

Barrow is situated well away from the industrial activities associated with the North Slope oil fields around Prudhoe Bay. For three decades (roughly 1950–80), Barrow had been the site of the Naval Arctic Research Laboratory (NARL; Shelesnyak 1948). After the closure of NARL, the National Science Foundation (NSF), National Oceanic and Atmospheric Administration (NOAA), and other agencies continued Arctic research in and around Barrow, eventually on and adjacent to 11 square miles of land set aside for that purpose as the Barrow Environmental Observatory (BEO). A site north of the town of Barrow and next to the BEO, located on federal land controlled by the NOAA Climate Monitoring and Diagnostics Laboratory (CMDL) and within 100 m of their clean air laboratory, was selected for the ARM facility. This site is a few kilometers from the northernmost point of U.S. territory and close to water (Chukchi Sea and Elson Lagoon) through ~270°, ranging from the southwest to southeast (Fig. 8-1). The winds at the site come predominantly from the ocean (Zak et al. 2002). A supporting office and light laboratory space is located ~2 km west of the primary instrument facility, within the former NARL complex, and leased from the Ukpeagvik Iñupiat Corporation (UIC). UIC is a corporation owned by the native people of Barrow, and scientific facilities located at NARL were administered for UIC by the Barrow Arctic Science Consortium (BASC).

In addition to Barrow, an auxiliary North Slope site was established at the inland village of Atqasuk, approximately 100 km south of Barrow. The objective with this site was to study the ocean-to-coast-to-inland transition. This inland site was anticipated to have a more continental character than Barrow, being colder and drier in winter and warmer in summer.

The Barrow site was formally dedicated on 1 July 1997 in a ceremony that included Martha Krebbs, director of the DOE Office of Energy Research; Peter Lunn, DOE ARM program manager; Ben Nageak, mayor of the North Slope Borough; Max Ahgeak, president of UIC; and other luminaries. The dedication ceremony was coordinated with the celebration of the 25th anniversary of the founding of the North Slope Borough. The joint festivities included traditional refreshments, native dances, and dedication speeches. Routine data acquisition began early in 1998 at Barrow and in June 2000 at Atqasuk.

b. The adjacent Arctic Ocean

Getting measurements in the central Arctic requires expensive logistics, because the Arctic sea ice can be unstable during the summer months and impenetrable by ship in the winter months. Long-term measurements using sophisticated instruments, such as those utilized by the ARM Program, place high demands on infrastructure and power resources. The Surface Heat Budget of the Arctic Ocean (SHEBA) project (Moritz et al. 1993), led by the National Science Foundation and the Office of Naval Research, provided an opportunity to take advantage of logistical support funded by other programs. The ARM Program decided to collaborate on this interagency project that was being organized on the same time frame as the new facility

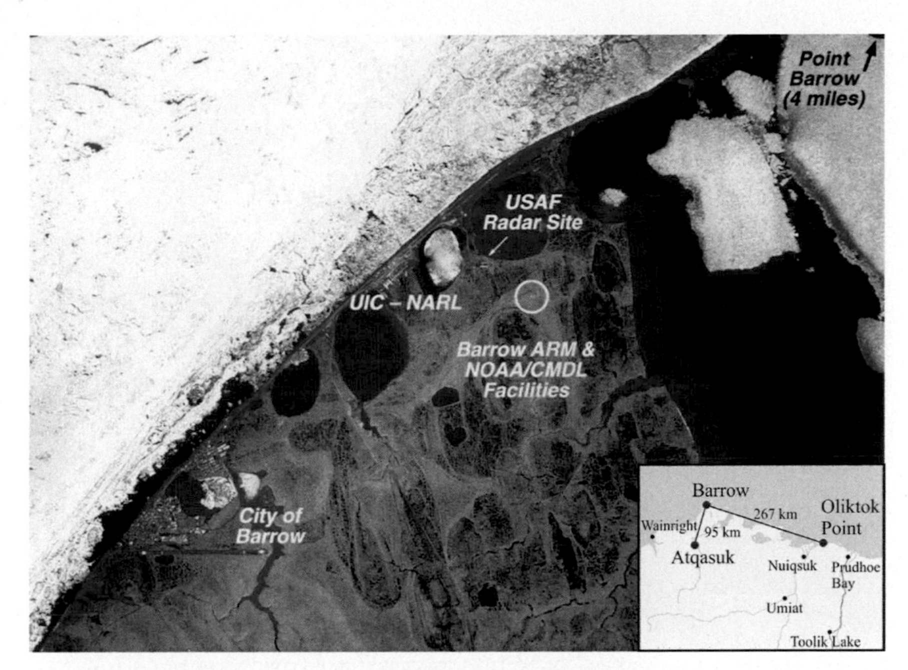

FIG. 8-1. Satellite view of the area around the Barrow NSA facility. The city of Barrow, the location of the old Naval Arctic Research Laboratory (where the ARM duplex is situated), and Point Barrow are indicated. The distance from the facility to the city of Barrow is approximately 6 mi. Elson Lagoon can be seen to the east of the facility. The insert shows the three ARM North Slope of Alaska locations (red dots) along with other population centers.

in Barrow. SHEBA focused on climate-relevant processes in the perennial Arctic ice pack and was centered on the Canadian Coast Guard ice breaker *Des Groseilliers*, which was intentionally frozen into the Arctic ice pack for a full annual cycle starting in fall 1997 (Fig. 8-2). The SHEBA science objectives comprised studying the relationships among radiative fluxes (especially as affected by surface– and cloud–radiative interactions), the mass balance of sea ice, and the storage and release of energy and salt in the ocean mixed layer. These objectives complemented the Arctic-specific ARM research objectives, making ARM participation logical from both scientific and logistical considerations.

The ARM Program provided a comprehensive set of up- and downwelling radiation measurements to complement and expand the NSF- and ONR-funded instruments.

Specific ARM objectives for participating in SHEBA were to make Barrow atmospheric and radiative data an adequate surrogate for similar data over the Arctic Ocean by doing the following:

- addressing the disparities between model and observational trends along the coast and over the central pack ice;
- investigating if relevant radiative processes and phenomena in the vicinity of Barrow were sufficiently similar to the same processes and phenomena within the central Arctic Ocean (the Arctic ice pack).

Participation in SHEBA also brought with it the benefit of collaboration with the NASA-led First International Satellite Cloud Climatology Project (ISCCP) Regional Experiment (FIRE) Phase III (Curry et al. 2000), which focused on Arctic clouds using satellite and airborne data.

c. *Outreach*

The local Iñupiat community in Barrow was familiar with scientific research because virtually every Barrow family had members who had worked for NARL. In addition, the local elders were very interested in climate, having seen their environment change over the decades. As a result, ARM outreach took place in an atmosphere of understanding and could draw on local people and knowledge for help.

A person who was pivotal in this regard was Tom Albert, at the time the chief scientist for the North Slope Borough Department of Wildlife Management. Through a creative, multiyear research effort involving several U.S. universities, Tom was able to show that the bowhead whale population was much more numerous than other researchers had thought. This allowed the International Whaling Commission to regulate, but not

FIG. 8-2. The *Des Groseilliers* and the SHEBA ice camp (photo courtesy of Kevin Widener).

ban, the hunting of bowhead whales. Since subsistence on the bowhead whale was (and remains) a cornerstone of Iñupiat culture on the North Slope, this established Tom as the most trusted scientist by the regional community.

Tom Albert introduced Bernie Zak and Knut Stamnes to the influential organizations in the community, as well as to the local, state, and federal agencies that function on the North Slope, and suggested who they should contact and brief on NSA/AAO plans. A series of public meetings were held in Barrow, Atqasuk, and the other potentially affected North Slope villages in order to ensure that the local people would be aware of what the ARM Program planned to do, why and how, and so that they could influence the plans. With the help of Tom and the Borough, the ARM Program even had Iñupiaq translators participate in the meetings to assure that the older Iñupiat people, some of whom struggled with English, were well informed. Through these efforts, the ARM NSA/AAO became a local North Slope project in which the community was heavily invested, not something simply imposed on the community by outside interests. The ARM Program also developed and funded an interactive educational kiosk featuring interviews by ARM scientists and community elders about climate change, which was placed in the Barrow Iñupiat Heritage Center for many years.

d. Science objectives

The growing understanding that the Arctic region was particularly vulnerable to a changing climate dictated the science objectives for the North Slope site. There was a realization that Arctic system physical processes, in many ways unique compared to other regions around the globe, were not represented well in models used to study climate (Tao et al. 1996). The presence of the ice-covered ocean through most of the year greatly impacts air–sea exchanges and the atmospheric processes in the interfacial layers [see summary in Curry et al. (1996)]. Evaluation of model results showed large differences in surface temperatures, pressures, and atmospheric cloud fractions produced by different models (Tao et al. 1996). Tao et al. (1996) attributed these intermodel differences to different specifications of sea ice, a lack of physically based links between cloudiness and air temperature, differences in model resolution and in formulation of various physical processes, and uncertainties in the observational database. They concluded that the highest priorities for improved simulation of the Arctic environment were the proper treatment of cloud–radiative interactions and local surface–atmosphere interactions. These and other studies (e.g., Curry et al. 1996; Randall et al. 1998) motivated the primary broad objective of the NSA site to collect data on radiation–climate feedbacks, important drivers in Arctic amplification (Stamnes et al. 1999). However, it was also recognized that the unique characteristics of the Arctic environment offered the opportunity to study other fundamental processes important to understanding Earth's climate.

The importance of the snow/ice–albedo feedback in the Arctic had long been recognized (e.g., Budyko 1969). This feedback is envisioned as a warming (cooling) climate leading to increased (decreased) melting of surface snow/ice, reducing (increasing) the surface albedo and leading to increased (decreased) solar radiation absorption by the surface, which in turn leads back to enhanced warming (cooling). However, from the early design of the NSA site, the importance of atmospheric processes on this radiative feedback was understood (e.g., Curry et al. 1996). Stamnes et al. (1999, 54–55) summarized the complexity of the radiation–climate feedbacks as follows:

> A perturbation to the Arctic Ocean radiation balance may arise from increased greenhouse gas concentrations and/or increasing amounts of aerosol. A perturbation in the surface radiation balance of the sea ice results in a change in sea ice characteristics (i.e., ice thickness and areal distribution, surface temperature and surface albedo). These changes in sea ice characteristics, particularly the surface temperature and fraction of open water, will modify fluxes of radiation and surface sensible and latent heat, which will modify the atmospheric temperature, humidity, and dynamics. Modifications to the atmospheric thermodynamic and dynamic structure will modify cloud properties (e.g., cloud fraction, cloud optical depth), which will in turn modify the radiative fluxes.

The presence/absence of clouds and aerosol particles in the atmosphere constitutes the greatest source of variability in the radiative energy flow through the atmosphere. Widespread and persistent low-level clouds impart a net warming effect on the Arctic surface energy budget throughout most of the year (e.g., Curry and Ebert 1992). This warming effect derives from downwelling longwave radiation by these clouds and plays a critical role in modulating the snow/ice–albedo feedback by impacting the timing and rate of surface albedo changes induced by melting snow. During sunlit months, the clouds also serve to shade the surface from solar radiation. The seasonal interplay of the warming and cooling effects is related to the sun angle, surface albedo, cloud properties, and temperature (e.g., Shupe and Intrieri 2004). These considerations highlight the need for a proper understanding of the processes that determine cloud, atmosphere, and surface characteristics, their interactions, and seasonal evolution.

Cloud processes have a big impact on general circulation model simulations because of the significant role clouds play in regulating the radiative energy fluxes through the atmosphere. Yet much uncertainty existed in our understanding of the processes responsible for the formation, evolution, and dissipation of clouds in the stable Arctic lower troposphere. The net effect of changes in cloudiness (warming or cooling) on climate depends on cloud microphysical properties, such as water phase partitioning, hydrometeor size and shape, and macrophysical properties, such as cloud morphology, altitude, thickness, and spatial inhomogeneity (Stamnes et al. 1999). All these cloud properties are strong functions of atmospheric aerosol particles, the numbers and composition of which undergo large seasonal changes. Curry et al. (1996) concluded that the general lack of understanding of fundamental cloud processes precluded a determination of the role of cloud feedbacks on modulating Arctic amplification.

Cloud–radiation feedback processes are coupled closely to atmospheric temperature and water vapor feedbacks. Changes in the ocean–atmosphere mean temperature and water vapor fluxes resulting from changes in the surface state (liquid/solid) imply large changes in the absolute water vapor content of the atmosphere. Such changes alter the radiative fluxes through the atmosphere, with a warmer atmosphere with more water vapor leading to enhanced warming at the surface, which is a positive feedback. Curry et al. (1995, 1996) argued that the complicated structure of the cold lower troposphere may lead to an enhanced water vapor feedback. They suggested that this enhanced feedback is the result of reduced vertical mixing through the stable lower troposphere and the role of cold-cloud precipitation processes in keeping the relative humidity close to ice saturation. Curry et al. (1996) also suggested that the magnitude of this important feedback is uncertain, because it depends on accurate modeling of the role of low-level clouds on the vertical temperature and humidity profiles.

Because the snow/ice–albedo is inextricably linked to atmospheric feedback processes, studying one in isolation from the others may be quite misleading. Stamnes et al. (1999) argued that it is essential to improve our physical understanding of the component processes of the radiation–climate feedback in order to characterize the interdependence among these feedback processes to reduce the uncertainty in their combined effects. This understanding motivated the primary scientific objectives (Stamnes et al. 1999) for the ARM NSA site to focus on improving:

- the treatment of radiative transfer in the coupled atmosphere–snow–ice–surface system;
- the treatment of radiative effects of mixed-phase and ice-phase clouds, aerosols, and cloud–aerosol mixtures;
- the description of basic cloud microphysical properties and how these are influenced by atmospheric and aerosol characteristics;
- a better understanding of the relative importance of surface and advective fluxes of moisture for the formation of clouds;
- a better understanding of the interactions among turbulence, radiation, and cloud microphysical processes in the evolution of the cloudy atmosphere.

In addition to these Arctic-centric objectives, it was also realized that environmental conditions at Barrow presented the opportunity for easier access to address several climate-important phenomena. Clouds forming in the cold and stable Arctic environment tend to be predominantly stratiform mixed phase (liquid water and ice in the same general volume) or fully glaciated. These frequently occurring low-altitude clouds were viewed as an opportunity to study marine stratus, in general, and cold cloud microphysics processes with application well beyond just Arctic clouds. Arctic summer marine stratus processes may be taken as a surrogate for eastern ocean margin marine stratus. Similarly, the glaciated wintertime cloud processes may provide insight into ice clouds that are present around the globe at much higher altitudes that are less accessible to researchers. The low water vapor amounts in the winter Arctic atmosphere also offered the opportunity to study processes contributing to global longwave radiative losses through the cold and dry upper troposphere.

3. Establishment of the sites

The NSA facilities implementation took place in several phases to take maximum advantage of lessons learned from the Southern Great Plains and Tropical Western Pacific sites and the opportunity offered by the interagency SHEBA project. The first phase (1997–98) sought to acquire experimental data to study radiative transfer processes through the atmosphere (in which both radiative energy flows and the surface and atmospheric characteristics that influence them are measured). These data were to be collected in the coastal environment of Barrow and simultaneously within the perennial Arctic ice pack as part of the SHEBA project to study radiative transfer. In the second phase (1999), the ARM instruments that were part of SHEBA were moved to Atqasuk to complete the ocean–coastal transition–inland transect of radiometric and atmospheric/surface experimental data. An anticipated third phase of multiple distributed sites to broaden the focus on cloud formation/dissipation studies was never implemented because of fiscal constraints. However, the science objectives envisioned for phase three were pursued by a series of short-term intensive observing periods.

In addition to more traditional instruments (e.g., radiosondes, surface meteorology, and surface broadband radiometers; see Stamnes et al. 1999 for a full list), the ARM Program deployed some very unique remote sensors to the NSA site. These instruments, many of which were developed as part of the ARM's Instrument Development Program (Stokes 2016, chapter 2), include the extended-range atmospheric emitted radiance interferometer (Tobin et al. 1999) to explore the strong water vapor absorption and emission lines in the 16–25-μm (400–600 cm^{-1}) portion of the spectrum and a polarization-sensitive micropulse lidar (Flynn et al. 2007) for phase discrimination in optically thin clouds.

Operating a suite of advanced instruments in the Arctic presented a series of challenges. While the SHEBA instruments were deployed for a limited period on the flight deck of the ice breaker *Des Groseilliers*, which provided a stable platform and where technical support personnel was readily available, the two North Slope facilities presented greater challenges. These facilities had to be located close to, but with limited impact from, two local communities. The North Slope land areas consist mostly of tundra, the top of the Arctic permafrost. In the short summer months, the top layer of the permafrost melts, leaving the ground soggy and wet. Any structure must be erected on pilings driven into the permafrost layer or on a gravel pad, which prevents the underlying tundra from melting. The Barrow facility rests on several piling platforms, whereas the Atqasuk facility was built on a gravel pad next to the road between the village and the local airport (virtually every village in rural Alaska has an airport). Originally, most of the inside equipment in Barrow resided in the "Great White" shelter (Fig. 8-3), while at Atqasuk, equipment was sheltered in the "Pumpkin" (Fig. 8-4), both named for the primary colors of the shelters. The Barrow facility had two additional instrument platforms added after the initial deployment. In the early 2000s, the need for additional space to host guest instruments and the facility data system led to the construction of a two-room additional shelter at Barrow, which was completed in 2005. The final expansion at Barrow came through the American Recovery and Reinvestment Act funding in 2010, when two additional shelters were required to host the new scanning cloud radar and lidar systems. Additional laboratory space and the new 3-cm precipitation radar were located in the Barrow Arctic Research Center (BARC), which is 2.2 km west from the Great White (see Fig. 8-1). The NSA site administrative office is located in a duplex in the old NARL complex, close to the BARC, rented from UIC to serve the dual purpose of office space and housing for visiting scientists. The availability of six bedrooms to accommodate visitors proved to be highly advantageous in the late 2000s with the increasing pressure on housing in Barrow coming from an expanding oil industry presence.

The second major challenge faced during the implementation was maintaining and servicing the cutting-edge technology instrumentation throughout the year. The need was met with the appointment of Walter Brower at Barrow as facility manager and initially Jimmy Ivanov and later Doug Whiteman as site operators at Atqasuk. Being native to the North Slope, these

FIG. 8-3. The Great White in Barrow in the early 2000s. The large container housed most of the instruments and all computers (photo courtesy of Mark Ivey).

men brought a wealth of practical experience to the program along with a natural ability to find a solution to almost any problem. Hence, they quickly accomplished several tasks that otherwise would have required an instrument engineer to travel to Barrow. At the same time, they provided a source of local information to visitors unfamiliar with the dangers inherent to life in the Arctic.

The final challenge faced in establishing the sites was related to operating instruments not specifically designed for the harsh Arctic environment. Prior to SHEBA, the NSA/AAO site scientist team (Knut Stamnes working with Rune Storvold) deployed selected ARM instruments at Fairbanks to observe problems and develop solutions, after which the ARM/SHEBA hardware was deployed for a cold test at Barrow during February–April 1997. The effects of low temperatures and the formation of hoarfrost on optical instrumentation were recognized early. To minimize measurement errors, instruments were sheltered to protect them from the cold, while ventilation and modest heating were added to several instruments to prevent frost accumulation. The site scientist team developed routines for near-real-time data quality inspection of all the data by dedicated personnel: first, by graduate students within the field during SHEBA and, later, in Fairbanks, until the ARM Data Quality Office was established.

4. Scientific contributions

The SHEBA project provided an early focus for much of the scientific effort. The SHEBA suite of instruments (Uttal et al. 2002) included cloud radar and lidar, a microwave radiometer, a spectral infrared radiometer, radiosonde soundings, and longwave and shortwave broadband radiometers. These instruments complemented similarly sophisticated measurements in the sea ice and ocean. While there have been shorter-term experiments operating similar suites of atmospheric instruments within the sea ice since that time (i.e., Tjernström et al. 2014), SHEBA is the only example to date of year-round, comprehensive cloud–atmosphere measurements in the central Arctic. As a result, many groundbreaking findings from SHEBA provided a first look into cloud and atmospheric processes over the central Arctic sea ice and thus account for a major component of the community knowledge in that area.

For the first time, the annual cycle of cloud occurrence fraction and type was derived from objective, comprehensive ground-based instruments (e.g., Intrieri et al. 2002b; Shupe et al. 2005). Mixed-phase clouds were characterized in detail and found to be particularly frequent and persistent (Turner 2005; Shupe et al. 2006), with these liquid-containing clouds playing a previously unknown dominant role in the surface radiation budget over the sea ice (Shupe and Intrieri 2004). A variety of remote sensor techniques was used to develop the first characterization of Arctic cloud microphysical properties over an annual cycle (Westwater et al. 2001; Lin et al. 2003; Turner 2005; Shupe et al. 2001, 2005, 2006), which could be linked to the annual evolution of surface cloud radiative forcing (Intrieri et al. 2002a) and net surface energy budget (Persson et al. 2002). With such unique, first and one-of-a-kind findings, SHEBA has been the basis for many Arctic regional model parameterization evaluation and development activities (Khvorostyanov et al. 2001; Zhang and Lohmann 2003; Morrison et al. 2003, 2005; Fu and Hollars 2004; Wyser and Jones 2005; Rinke et al. 2006; Tjernström et al. 2008; Wyser et al. 2008; Du et al. 2011; de Boer et al. 2012; Fridlind et al. 2012).

The extreme environmental conditions at the North Slope provided an early focus for scientific activities at Barrow. Early investigations into the performance of instruments to accurately measure downwelling atmospheric radiative fluxes continued with the International

FIG. 8-4. The Pumpkin at Atqasuk with the local mode of transportation (photo courtesy of Will Shaw and Jim Barnard).

Pyrgeometer and Absolute Sky-Scanning Radiometer Comparison (IPASRC II) in 2001, which focused on both measurement and modeling of downward longwave irradiance (Marty et al. 2003). The next challenge was accurate characterization of the typically low precipitable water vapor and integrated cloud liquid.

The Millimeter-Wave Radiometric Arctic Winter Measurements Experiment Intensive Observing Period in 2004 (Racette et al. 2005) led to the addition of a high-frequency radiometer (183 GHz) to the NSA complement of instruments in 2005 (Cadeddu et al. 2007). The addition of this new instrument improved the accuracy of retrieved precipitable water and liquid water path in winter, when these quantities can drop to very low values.

The Radiative Heating in Underexplored Bands Campaign (RHUBC) in 2007 at Barrow took advantage of the low wintertime water vapor contents as a proxy to explore radiative cooling and heating in the mid-to-upper troposphere globally (Turner and Mlawer 2010) while also offering another opportunity to confirm the precision of the precipitable water retrievals (Cimini et al. 2009).

The transition of the NSA site scientist responsibilities from Knut Stamnes to Hans Verlinde in 2002 resulted in a greater focus on cloud processes. The Mixed-Phase Arctic Cloud Experiment (M-PACE) in fall of 2004 (Verlinde et al. 2007) sought to characterize mixed-phase cloud properties and processes. The design of the experiment built on the original NSA third-phase plans by establishing an extensive observing facility at Oliktok Point on the coast to the east of Barrow and a radiosonde site at Toolik Field Station just north of the Brooks Range along the haul road between Fairbanks and Prudhoe Bay. This experiment was the first extensive ARM aircraft campaign on the North Slope and brought the University of North Dakota Citation and the ARM-UAV *Proteus* aircraft. In situ observations from this campaign (McFarquhar et al. 2007) served as the basis for new surface remote sensing methods (i.e., Shupe et al. 2008a,b; Luke et al. 2010; de Boer et al. 2011; Rambukkange et al. 2011) and many cloud modeling studies (i.e., Fridlind et al. 2007; Solomon et al. 2009; Fan et al. 2009) of these fall transition season mixed-phase clouds. These combined efforts from many research groups resulted in a better description of the eddy structure and microphysical processes associated with single-layer mixed-phase clouds, as summarized in Morrison et al. (2012), and rejuvenated an interest in ice nucleation and ice growth processes in these clouds.

The detailed and distributed M-PACE measurements were used for evaluation of cloud process parameterizations. Xie et al. (2006) evaluated the European Centre for Medium-Range Weather Forecasts (ECMWF) model for the M-PACE period and found that, while it successfully represented the large-scale dynamic and thermodynamic structure and near-surface conditions, the model underpredicted liquid water path and thus downwelling longwave radiation. A two-part series of papers (Klein et al. 2009; Morrison et al. 2009) evaluated microphysical parameterizations of varying complexity in single- and multilayer mixed-phase clouds and found that more detailed parameterizations produced better comparisons with observations but that agreement depended on the type (single-/multilayer) of cloud modeled. Xie et al. (2008) evaluated a community climate model and found that improving the microphysical parameterizations produced better agreement with observed cloud properties and longwave radiative fluxes.

The impacts of aerosol particles on mixed-phase clouds were explored in another aircraft campaign: the Indirect and Semi-Direct Aerosol Campaign (ISDAC) conducted around Barrow in the spring transition season in 2008 (McFarquhar et al. 2011). The spring transition period, with sea ice covering the ocean adjacent to Barrow, stands in contrast to the fall, open ocean conditions observed during M-PACE, in addition to higher aerosol concentrations in the spring versus the fall. The extensive suite of aerosol and cloud instruments on the aircraft for ISDAC allowed in-depth studies of the aerosol particle characteristics over Barrow (e.g., Shantz et al. 2014), their relationship to cloud microphysical properties (e.g., Jackson et al. 2012; Ervens et al. 2011), and impact on cloud evolution (e.g., Avramov et al. 2011; Solomon et al. 2011). The combined SHEBA, M-PACE, and ISDAC datasets are still being explored to gain a better understanding of the fundamental processes that determine cloud formation, evolution, and dissipation.

Not many studies have addressed the coast-to-inland transition captured in the Barrow and Atqasuk data. Summertime clouds over Atqasuk exhibited generally larger liquid water paths (Doran et al. 2002), greater optical depths, and smaller ratios of measured to clear-sky irradiances (Doran et al. 2006) than clouds over Barrow. These differences were attributed to greater upward heat and water vapor fluxes over the wet tundra and lakes.

Although much was gained from the shorter-duration campaigns at the North Slope, the ongoing, long-term measurements are also of great value. Dong and Mace (2003) analyzed stratus cloud properties in liquid (or liquid-dominated) clouds observed in a five-month study from May through September 2000 to show a transition in microphysical characteristics from early May going into the summer season as the lower troposphere destabilizes and more effective precipitation processes reduce the aerosol concentrations. This analysis was followed by a 10-yr study providing a climatology of the cloud radiative forcing and cloud fraction (Dong et al. 2010). That study found that Barrow cloud occurrence fractions are comparable to those derived from ground-based radar–lidar observations during SHEBA and from satellite observations over the western Arctic regions. Furthermore, they found that, as a result of differences in latitude and surface conditions, clouds have a more pronounced and extended period of net surface radiative cooling at Barrow relative to SHEBA, demonstrating one strength of having observations in both regions. Shupe et al. (2011) and Shupe (2011) compiled data on cloud occurrence fraction, persistence, phase, and phase–temperature dependence at Barrow and Atqasuk and compared them with similar measurements at other Arctic observing sites. All sites exhibited a clear annual cycle of cloud occurrence with clouds least frequent in winter and most frequent in the late-summer-to-fall transition season at most sites. By comparing observations from Barrow with similar ones from Eureka, Canada, Cox et al. (2012) showed the influence of larger-scale climatological flow patterns on the radiative fluxes. Advection from Greenland and the central Arctic causes a drier and colder atmosphere at Eureka in the Canadian Archipelago compared to Barrow, where advection is predominantly from warmer locations leading to increased longwave radiative fluxes.

The longer-term measurements also proved useful to study details of Arctic aerosol–cloud interactions and their net effect on the surface radiative properties of low clouds. Using coordinated long-term ARM-funded aerosol measurements at the NOAA facility directly adjacent to the Barrow NSA site, Garrett et al. (2004) found that the microphysical properties of low clouds at Barrow were strongly sensitive to the long-range transport of pollution. The effect of these aerosols can impact the cloud emissivity and can lead to changes in downwelling longwave radiation of 3–5 $W m^{-2}$ (Lubin and Vogelmann 2006; Garrett and Zhao 2006). When combined with aerosol indirect shortwave effects (Lubin and Vogelmann 2007), the total indirect effect of aerosols at Barrow was shown to vary annually from a maximum warming of $+3 W m^{-2}$ in March to a cooling of $-11 W m^{-2}$ in May (Lubin and Vogelmann 2010).

Comparisons of numerical weather prediction and climate model cloud and radiation simulations against Barrow observations have revealed deficiencies in model parameterizations. Zhao and Wang (2010), using nine years of Barrow observations for the period 1999–2007, found that although the ECMWF model captured the general seasonal variation of surface fluxes and low-level cloud fraction, it experienced difficulty representing the boundary layer temperature inversion height and strength, overestimated the cloud fraction by 20% or more, and underestimated the liquid water path by over 50% in the cold season. In similar studies, but looking at different models, de Boer et al. (2012) found similar problems with the critical lower-troposphere processes and phenomena in the Community Climate System Model, version 4, and Walsh et al. (2009) found biases in radiative fluxes and cloud radiative forcing in reanalyses products.

5. Summary

The decision to place the ARM polar observing site in the Arctic proved to have been timely, in light of recent dramatic regional changes since its establishment. The rate of summer Arctic sea ice decline has exceeded that predicted by most of the models that contributed to the Intergovernmental Panel on Climate Change (IPCC) process (Stroeve et al. 2007), reaching new sea ice extent minima in 2007 and 2012. At the same time, ice volume declines even more precipitously (Kwok et al. 2009) with the loss of perennial ice (Maslanik et al. 2011). The urgency for understanding the physical causes of these larger-than-predicted ice losses is great (Stroeve et al. 2012). There is a growing community-wide understanding of the important role of lower-tropospheric cloud processes (Francis and Hunter 2006; Kay et al. 2008; Stroeve et al. 2012). This realization motivated the recent decision to deploy an ARM Mobile Facility at Oliktok Point starting in 2013, effectively completing the original phase-three plans. Taking advantage of a restricted air zone established for M-PACE, Oliktok Point will be a base for tethered balloon and unmanned aerial system operations exploring the lower-troposphere structure across the land-to-Arctic Ocean transition.

The recent addition of scanning polarimetric precipitation and dual-frequency cloud radars will expose new avenues for cloud process research to new dimensions. Not only will researchers be able to have a first look at the three-dimensional mesoscale structure of cloud and precipitating systems, but exploitation of the three frequency and polarimetric measurements will allow for detailed characterization of cloud and hydrometeor properties in the context of those structures. Such studies will help to develop the physical understanding needed to adapt the lower-troposphere parameterizations currently employed in large-scale models, mostly developed from lower- and midlatitude observations, to represent better processes in the stable polar air. The recent establishment of the Department of Energy Next-Generation Ecosystem Experiments (NGEE) site at Barrow will also provide a comprehensive set of measurements to study terrestrial–lower atmosphere interactions in even greater detail.

If imitation is the highest form of flattery, the establishment of several additional circum-Arctic sites is a testimony to the farsightedness of the original DOE-ARM planning. Since the establishment of the NSA site, the Canadians established Eureka on Ellesmere Island in Nunavut, the northernmost of Canada's three territories; the National Science Foundation is supporting similar measurements at Summit Station on the Greenland ice sheet; and at the Tiksi Hydrometeorological Observatory located in the Russian Far East, a partnership of U.S./Russian/Finnish agencies supports a subset of the typical DOE-ARM suite.

Acknowledgments. The authors would like to acknowledge the dedication and ingenuity of the local site operators, Walter Brower, Jimmy Ivanov, and Doug Whiteman, in maintaining the NSA sites in a very harsh environment. Early contributions by Rune Storvold and Hans Eide, and later by Dana Truffer-Moudra, Jessica Cherry, and Martin Stuefer from the University of Alaska Fairbanks and Scot Richardson from the Pennsylvania State University, established high data quality control standards.

REFERENCES

Arrhenius, S., 1896: On the influence of carbonic acid in the air upon the temperature on the ground. *Philos. Mag. J. Sci.*, **41**, 237–276, doi:10.1080/14786449608620846.

Avramov, A., and Coauthors, 2011: Toward ice formation closure in Arctic mixed-phase boundary layer clouds during ISDAC. *J. Geophys. Res.*, **116**, D00T08, doi:10.1029/2011JD015910.

Budyko, M. I., 1969: The effect of solar radiation variations on the climate of the Earth. *Tellus*, **21A**, 611–619, doi:10.1111/j.2153-3490.1969.tb00466.x.

Cadeddu, M. P., V. H. Payne, S. A. Clough, K. Cady-Pereira, and J. Liljegren, 2007: Effect of the oxygen line-parameter modeling on temperature and humidity retrievals from ground-based radiometers. *IEEE Trans. Geosci. Remote Sens.*, **45**, 2216–2223, doi:10.1109/TGRS.2007.894063.

Chapman, W. L., and J. E. Walsh, 1993: Recent variations of sea ice and air temperatures in high latitudes. *Bull. Amer. Meteor. Soc.*, **74**, 33–47, doi:10.1175/1520-0477(1993)074<0033:RVOSIA>2.0.CO;2.

Cimini, D., F. Nasir, E. R. Westwater, V. H. Payne, D. D. Turner, E. J. Mlawer, M. L. Exner, and M. P. Cadeddu, 2009: Comparison of ground-based millimeter-wave observations in the Arctic winter. *IEEE Trans. Geosci. Remote Sens.*, **47**, 3098–3106, doi:10.1109/TGRS.2009.2020743.

Cox, C. J., V. P. Walden, and P. Rowe, 2012: A comparison of the atmospheric conditions at Eureka, Canada and Barrow, Alaska (2006–2008). *J. Geophys. Res.*, **117**, D12204, doi:10.1029/2011JD017164.

Crowley, T. J., and G. R. North, 1991: *Paleoclimatology. Oxford Monogr. Geol. Geophys.*, Vol. 18, Oxford University Press, 360 pp.

Curry, J. A., and E. E. Ebert, 1992: Annual cycle of radiative fluxes over the Arctic Ocean: Sensitivity to cloud optical properties. *J. Climate*, **5**, 1267–1280, doi:10.1175/1520-0442(1992)005<1267:ACORFO>2.0.CO;2.

——, J. L. Schramm, M. C. Serreze, and E. E. Ebert, 1995: Water vapor feedback over the Arctic Ocean. *J. Geophys. Res.*, **100**, 14 223–14 229, doi:10.1029/95JD00824.

——, ——, W. B. Rossow, and D. Randall, 1996: Overview of Arctic cloud and radiation characteristics. *J. Climate*, **9**, 1731–1764, doi:10.1175/1520-0442(1996)009<1731:OOACAR>2.0.CO;2.

——, and Coauthors, 2000: FIRE Arctic Clouds Experiment. *Bull. Amer. Meteor. Soc.*, **81**, 5–29, doi:10.1175/1520-0477(2000)081<0005:FACE>2.3.CO;2.

de Boer, G., W. D. Collins, S. Menon, and C. N. Long, 2011: Using surface remote sensors to derive radiative characteristics of mixed-phase clouds: An example from M-PACE. *Atmos. Chem. Phys.*, **11**, 11 937–11 949, doi:10.5194/acp-11-11937-2011.

——, W. Chapman, J. Kay, B. Medeiros, M. D. Shupe, S. Vavrus, and J. Walsh, 2012: A characterization of the present-day Arctic atmosphere in CCSM4. *J. Climate*, **25**, 2676–2695, doi:10.1175/JCLI-D-11-00228.1.

Dong, X., and G. G. Mace, 2003: Arctic stratus cloud properties and radiative forcing derived from ground-based data collected at Barrow, Alaska. *J. Climate*, **16**, 445–461, doi:10.1175/1520-0442(2003)016<0445:ASCPAR>2.0.CO;2.

——, B. Xi, K. Crosby, C. N. Long, R. S. Stone, and M. D. Shupe, 2010: A 10 year climatology of Arctic cloud fraction and radiative forcing at Barrow, Alaska. *J. Geophys. Res.*, **115**, D17212, doi:10.1029/2009JD013489.

Doran, J. C., S. Zhong, J. C. Liljegren, and C. Jakob, 2002: A comparison of cloud properties at a coastal and inland site at the North Slope of Alaska. *J. Geophys. Res.*, **107**, 4120, doi:10.1029/2001JD000819.

——, J. C. Barnard, and W. J. Shaw, 2006: Modification of summertime Arctic cloud characteristics between a coastal and inland site. *J. Climate*, **19**, 3207–3219, doi:10.1175/JCLI3782.1.

Du, P., E. Girard, A. K. Bertram, and M. D. Shupe, 2011: Modeling of the cloud and radiation processes observed during SHEBA. *Atmos. Res.*, **101**, 911–927, doi:10.1016/j.atmosres.2011.05.018.

Ervens, B., G. Feingold, K. Sulia, and J. Harrington, 2011: The impact of microphysical parameters, ice nucleation mode, and habit growth on the ice/liquid partitioning in mixed-phase Arctic clouds. *J. Geophys. Res.*, **116**, D17205, doi:10.1029/2011JD015729.

Fan, J., M. Ovtchinnikov, J. M. Comstock, S. A. McFarlane, and A. Khain, 2009: Ice formation in Arctic mixed-phase clouds: Insights from a 3-D cloud-resolving model with size-resolved aerosol and cloud microphysics. *J. Geophys. Res.*, **114**, D04205, doi:10.1029/2008JD010782.

Flynn, C. J., A. Mendoza, Y. Zheng, and S. Mathur, 2007: Novel polarization-sensitive micropulse lidar measurement technique. *Opt. Express*, **15**, 2785–2790, doi:10.1364/OE.15.002785.

Francis, J. A., and E. Hunter, 2006: New insight into the disappearing Arctic sea ice. *Eos, Trans. Amer. Geophys. Union*, **87**, 509–511, doi:10.1029/2006EO460001.

Fridlind, A. M., A. S. Ackerman, G. McFarquhar, G. Zhang, M. R. Poellot, P. J. DeMott, A. J. Prenni, and A. J. Heymsfield, 2007: Ice properties of single-layer stratocumulus during the Mixed-Phase Arctic Cloud Experiment: 2. Model results. *J. Geophys. Res.*, **112**, D24202, doi:10.1029/2007JD008646.

——, B. van Diedenhoven, A. S. Ackerman, A. Avramov, A. Mrowiec, H. Morrison, P. Zuidema, and M. D. Shupe, 2012: A FIRE-ACE/SHEBA case study of mixed-phase Arctic boundary-layer clouds: Entrainment rate limitations on rapid primary ice nucleation processes. *J. Atmos. Sci.*, **69**, 365–389, doi:10.1175/JAS-D-11-052.1.

Fu, Q., and S. Hollars, 2004: Testing mixed-phase cloud water vapor parameterizations with SHEBA/FIRE-ACE observations. *J. Atmos. Sci.*, **61**, 2083–2091, doi:10.1175/1520-0469(2004)061<2083:TMCWVP>2.0.CO;2.

Garrett, T. J., and C. Zhao, 2006: Increased Arctic cloud longwave emissivity associated with pollution from mid-latitudes. *Nature*, **440**, 787–789, doi:10.1038/nature04636.

——, ——, X. Dong, G. G. Mace, and P. V. Hobbs, 2004: Effects of varying aerosol regimes on low-level Arctic stratus. *Geophys. Res. Lett.*, **31**, L17105, doi:10.1029/2004GL019928.

Intrieri, J. M., C. W. Fairall, M. D. Shupe, P. O. G. Persson, E. L Andreas, P. S. Guest, and R. E. Moritz, 2002a: An annual cycle of Arctic surface cloud forcing at SHEBA. *J. Geophys. Res.*, **107**, 8039, doi:10.1029/2000JC000439.

——, M. D. Shupe, T. Uttal, and B. J. McCarty, 2002b: An annual cycle of Arctic cloud characteristics observed by radar and lidar at SHEBA. *J. Geophys. Res.*, **107**, 8030, doi:10.1029/2000JC000423.

Jackson, R. C., and Coauthors, 2012: The dependence of ice microphysics on aerosol concentration in arctic mixed-phase stratus clouds during ISDAC and M-PACE. *J. Geophys. Res.*, **117**, D15207, doi:10.1029/2012JD017668.

Kahl, J. D., D. J. Charlevoix, N. A. Zaftseva, R. C. Schnell, and M. C. Serreze, 1993: Absence of evidence for greenhouse warming over the Arctic Ocean in the past 40 years. *Nature*, **361**, 335–337, doi:10.1038/361335a0.

Kay, J. E., T. L'Ecuyer, A. Gettelman, G. Stephens, and C. O'Dell, 2008: The contribution of cloud and radiation anomalies to the 2007 Arctic sea ice extent minimum. *Geophys. Res. Lett.*, **35**, L08503, doi:10.1029/2008GL033451.

Khvorostyanov, V. I., J. A. Curry, J. O. Pinto, M. Shupe, B. A. Baker, and K. Sassen, 2001: Modeling with explicit spectral water and ice microphysics of a two-layer cloud system of altostratus and cirrus observed during the FIRE Arctic Clouds Experiment. *J. Geophys. Res.*, **106**, 15 099–15 112, doi:10.1029/2000JD900521.

Klein, S. A., and Coauthors, 2009: Intercomparison of model simulations of mixed-phase clouds observed during the ARM Mixed-Phase Arctic Cloud Experiment. I: Single-layer cloud. *Quart. J. Roy. Meteor. Soc.*, **135**, 979–1002, doi:10.1002/qj.416.

Kwok, R., G. F. Cunningham, M. Wensnahan, I. Rigor, H. J. Zwally, and D. Yi, 2009: Thinning and volume loss of the Arctic Ocean sea ice cover: 2003–2008. *J. Geophys. Res.*, **114**, C07005, doi:10.1029/2009JC005312.

Lin, B., P. Minnis, and A. Fan, 2003: Cloud liquid water path variations with temperature observed during the Surface Heat Budget of the Arctic Ocean (SHEBA) experiment. *J. Geophys. Res.*, **108**, 4427, doi:10.1029/2002JD002851.

Lubin, D., and A. M. Vogelmann, 2006: A climatologically significant aerosol longwave indirect effect in the Arctic. *Nature*, **439**, 453–456, doi:10.1038/nature04449.

——, and ——, 2007: Expected magnitude of the aerosol shortwave indirect effect in springtime Arctic liquid water clouds. *Geophys. Res. Lett.*, **34**, L11801, doi:10.1029/2006GL028750.

——, and ——, 2010: Observational quantification of a total aerosol indirect effect in the Arctic. *Tellus*, **62B**, 181–189, doi:10.1111/j.1600-0889.2010.00460.x.

Luke, E. P., P. Kollias, and M. D. Shupe, 2010: Detection of supercooled liquid in mixed-phase clouds using radar Doppler spectra. *J. Geophys. Res.*, **115**, D19201, doi:10.1029/2009JD012884.

Marty, C., and Coauthors, 2003: Downward longwave irradiance uncertainty under Arctic atmospheres: Measurements and modeling. *J. Geophys. Res.*, **108**, 4358, doi:10.1029/2002JD002937.

Maslanik, J., J. Stroeve, C. Fowler, and W. Emery, 2011: Distribution and trends in Arctic sea ice age through spring 2011. *Geophys. Res. Lett.*, **38**, L13502, doi:10.1029/2011GL047735.

McFarquhar, G. M., G. Zhang, M. R. Poellot, G. L. Kok, R. McCoy, T. P. Tooman, A. Fridlind, and A. J. Heymsfield, 2007: Ice properties of single-layer stratocumulus during the Mixed-Phase Arctic Cloud Experiment: 1. Observations. *J. Geophys. Res.*, **112**, D24201, doi:10.1029/2007JD008633.

——, and Coauthors, 2011: Indirect and semi-direct aerosol campaign: The impact of Arctic aerosols on clouds. *Bull. Amer. Meteor. Soc.*, **92**, 183–201, doi:10.1175/2010BAMS2935.1.

Moritz, R. E., J. A. Curry, A. S. Thorndike, and N. Untersteiner, 1993: Surface heat budget of the Arctic Ocean. ARCSS/OAII S Tech. Rep. 3, 33 pp.

Morrison, H., M. D. Shupe, and J. A. Curry, 2003: Modeling clouds observed at SHEBA using a bulk microphysics parameterization implemented into a single-column model. *J. Geophys. Res.*, **108**, 4255, doi:10.1029/2002JD002229.

——, J. A. Curry, M. D. Shupe, and P. Zuidema, 2005: A new double-moment microphysics parameterization for application in cloud and climate models. Part II: Single-column modeling of Arctic clouds. *J. Atmos. Sci.*, **62**, 1678–1693, doi:10.1175/JAS3447.1.

——, and Coauthors, 2009: Intercomparison of model simulations of mixed-phase clouds observed during the ARM Mixed-Phase Arctic Cloud Experiment. II: Multilayer cloud. *Quart. J. Roy. Meteor. Soc.*, **135**, 1003–1019, doi:10.1002/qj.415.

——, G. de Boer, G. Feingold, J. Harrington, M. Shupe, and K. Sulia, 2012: Resilience of persistent Arctic mixed-phase clouds. *Nat. Geosci.*, **5**, 11–17, doi:10.1038/ngeo1332.

Persson, P. O. G., C. W. Fairall, E. L Andreas, P. S. Guest, and D. K. Perovich, 2002: Measurements near the Atmospheric Surface Flux Group tower at SHEBA: Near-surface conditions and surface energy budget. *J. Geophys. Res.*, **107**, 8045, doi:10.1029/2000JC000705.

Racette, P. E., and Coauthors, 2005: Measurement of low amounts of precipitable water vapor using ground-based millimeterwave radiometry. *J. Atmos. Oceanic Technol.*, **22**, 317–337, doi:10.1175/JTECH1711.1.

Rambukkange, M. P., J. Verlinde, E. W. Eloranta, C. J. Flynn, and E. E. Clothiaux, 2011: Using Doppler spectra to separate hydrometeor populations and analyze ice precipitation in multilayered mixed-phase clouds. *IEEE Trans. Geosci. Remote Sens.*, **8**, 108–112, doi:10.1109/LGRS.2010.2052781.

Randall, D., and Coauthors, 1998: Status of and outlook for large-scale modeling of atmosphere–ice–ocean interactions in the Arctic. *Bull. Amer. Meteor. Soc.*, **79**, 197–219, doi:10.1175/1520-0477(1998)079<0197:SOAOFL>2.0.CO;2.

Rinke, A., and Coauthors, 2006: Evaluation of an ensemble of Arctic regional climate models: Spatiotemporal fields during the SHEBA year. *Climate Dyn.*, **26**, 459–472, doi:10.1007/s00382-005-0095-3.

Serreze, M. C., and R. G. Barry, 2011: Processes and impacts of Arctic amplification: A research synthesis. *Global Planet. Change*, **77**, 85–96, doi:10.1016/j.gloplacha.2011.03.004.

——, R. C. Schnell, and J. D. Kahl, 1992: Low-level temperature inversions of the Eurasian Arctic and comparisons with Soviet drifting station data. *J. Climate*, **5**, 615–629, doi:10.1175/1520-0442(1992)005<0615:LLTIOT>2.0.CO;2.

Shantz, N. C., I. Gultepe, E. Andrews, A. Zelenyuk, M. E. Earle, A. M. Macdonald, P. S. K. Liu, and W. R. Leaitch, 2014: Optical, physical, and chemical properties of springtime aerosol over Barrow Alaska in 2008. *Int. J. Climatol.*, **34**, 3125–3138, doi:10.1002/joc.3898.

Shelesnyak, M. C., 1948: The history of the Arctic Research Laboratory, Point Barrow, Alaska. *Arctic*, **1**, 97–106.

Shupe, M. D., 2011: Clouds at Arctic atmospheric observatories. Part II: Thermodynamic phase characteristics. *J. Appl. Meteor. Climatol.*, **50**, 645–661, doi:10.1175/2010JAMC2468.1.

——, and J. M. Intrieri, 2004: Cloud radiative forcing of the Arctic surface: The influence of cloud properties, surface albedo, and solar zenith angle. *J. Climate*, **17**, 616–628, doi:10.1175/1520-0442(2004)017<0616:CRFOTA>2.0.CO;2.

——, T. Uttal, S. Y. Matrosov, and A. S. Frisch, 2001: Cloud water contents and hydrometeor sizes during the FIRE Arctic Clouds Experiment. *J. Geophys. Res.*, **106**, 15 015–15 028, doi:10.1029/2000JD900476.

——, ——, and ——, 2005: Arctic cloud microphysics retrievals from surface-based remote sensors at SHEBA. *J. Appl. Meteor.*, **44**, 1544–1562, doi:10.1175/JAM2297.1.

——, S. Y. Matrosov, and T. Uttal, 2006: Arctic mixed-phase cloud properties derived from surface-based sensors at SHEBA. *J. Atmos. Sci.*, **63**, 697–711, doi:10.1175/JAS3659.1.

——, J. S. Daniel, G. de Boer, E. W. Eloranta, P. Kollias, E. Luke, C. N. Long, D. D. Turner, and J. Verlinde, 2008a: A focus on mixed-phase clouds: The status of ground-based observational methods. *Bull. Amer. Meteor. Soc.*, **89**, 1549–1562, doi:10.1175/2008BAMS2378.1.

——, P. Kollias, P. O. G. Persson, and G. M. McFarquhar, 2008b: Vertical motions in Arctic mixed-phase stratiform clouds. *J. Atmos. Sci.*, **65**, 1304–1322, doi:10.1175/2007JAS2479.1.

——, V. P. Walden, E. Eloranta, T. Uttal, J. R. Campbell, S. M. Starkweather, and M. Shiobara, 2011: Clouds at Arctic atmospheric observatories. Part I: Occurrence and macrophysical properties. *J. Appl. Meteor. Climatol.*, **50**, 626–644, doi:10.1175/2010JAMC2467.1.

Solomon, A., H. Morrison, P. O. G. Persson, M. D. Shupe, and J.-W. Bao, 2009: Investigation of microphysical parameterizations of snow and ice in Arctic clouds during MPACE through model–observation comparisons. *Mon. Wea. Rev.*, **137**, 3110–3128, doi:10.1175/2009MWR2688.1.

——, M. D. Shupe, P. O. G. Persson, and H. Morrison, 2011: Moisture and dynamical interactions maintaining decoupled Arctic mixed-phase stratocumulus in the presence of a humidity inversion. *Atmos. Chem. Phys.*, **11**, 10 127–10 148, doi:10.5194/acp-11-10127-2011.

Stamnes, K., R. G. Ellingson, J. A. Curry, J. E. Walsh, and B. D. Zak, 1999: Review of science issues, deployment strategy, and status for the ARM North Slope of Alaska–Adjacent Arctic Ocean climate research site. *J. Climate*, **12**, 46–63, doi:10.1175/1520-0442-12.1.46.

Stokes, G. M., 2016: Original ARM concept and launch. *The Atmospheric Radiation Measurement (ARM) Program: The First 20 Years, Meteor. Monogr.*, No. 57, Amer. Meteor. Soc., doi:10.1175/AMSMONOGRAPHS-D-15-0021.1.

Stroeve, J. C., M. M. Holland, W. Meier, T. Scambos, and M. Serreze, 2007: Arctic sea ice decline: Faster than forecast. *Geophys. Res. Lett.*, **34**, L09501, doi:10.1029/2007GL029703.

——, M. C. Serreze, M. M. Holland, J. E. Kay, J. Malanik, and A. P. Barrett, 2012: The Arctic's rapidly shrinking sea ice cover: A research synthesis. *Climatic Change*, **110**, 1005–1027, doi:10.1007/s10584-011-0101-1.

Tao, X., J. E. Walsh, and W. L. Chapman, 1996: An assessment of global climate model simulations of Arctic air temperatures. *J. Climate*, **9**, 1060–1076, doi:10.1175/1520-0442(1996)009<1060:AAOGCM>2.0.CO;2.

Tjernström, M., J. Sedlar, and M. D. Shupe, 2008: How well do regional climate models reproduce radiation and clouds in the Arctic? An evaluation of ARCMIP simulations. *J. Appl. Meteor. Climatol.*, **47**, 2405–2422, doi:10.1175/2008JAMC1845.1.

——, and Coauthors, 2014: The Arctic Summer Cloud Ocean Study (ASCOS): Overview and experimental design. *Atmos. Chem. Phys. Discuss.*, **14**, 2823–2869, doi:10.5194/acp-14-2823-2014.

Tobin, D. C., and Coauthors, 1999: Downwelling spectral radiance observations at the SHEBA ice station: Water vapor continuum measurements from 17 to 26μm. *J. Geophys. Res.*, **104**, 2081–2092, doi:10.1029/1998JD200057.

Turner, D. D., 2005: Mixed-phase cloud properties from AERI lidar observations: Algorithm and results from SHEBA. *J. Appl. Meteor.*, **44**, 427–444, doi:10.1175/JAM2208.1.

——, and E. J. Mlawer, 2010: The Radiative Heating in Underexplored Bands Campaigns (RHUBC). *Bull. Amer. Meteor. Soc.*, **91**, 911–923, doi:10.1175/2010BAMS2904.1.

Uttal, T., and Coauthors, 2002: Surface heat budget of the Arctic Ocean. *Bull. Amer. Meteor. Soc.*, **83**, 255–275, doi:10.1175/1520-0477(2002)083<0255:SHBOTA>2.3.CO;2.

Verlinde, J., and Coauthors, 2007: The Mixed-Phase Arctic Cloud Experiment (M-PACE). *Bull. Amer. Meteor. Soc.*, **88**, 205–221, doi:10.1175/BAMS-88-2-205.

Walsh, J. E., 1993: Climate change: The elusive Arctic warming. *Nature*, **361**, 300–301, doi:10.1038/361300a0.

——, and R. G. Crane, 1992: A comparison of GCM simulations of Arctic climate. *Geophys. Res. Lett.*, **19**, 29–32, doi:10.1029/91GL03004.

——, W. L. Chapman, and D. H. Portis, 2009: Arctic cloud fraction and radiative fluxes in atmospheric reanalyses. *J. Climate*, **22**, 2316–2334, doi:10.1175/2008JCLI2213.1.

Wendler, G., M. Shulski, and B. Moore, 2010: Changes in the climate of the Alaskan North Slope and the ice concentration of the adjacent Beaufort Sea. *Theor. Appl. Climatol.*, **99**, 67–74, doi:10.1007/s00704-009-0127-8.

Westwater, E. R., Y. Han, M. D. Shupe, and S. Y. Matrosov, 2001: Analysis of integrated cloud liquid and precipitable water vapor retrievals from microwave radiometers during the Surface Heat Budget of the Arctic Ocean project. *J. Geophys. Res.*, **106**, 32 019–32 030, doi:10.1029/2000JD000055.

Wyser, K., and C. G. Jones, 2005: Modeled and observed clouds during Surface Heat Budget of the Arctic Ocean (SHEBA). *J. Geophys. Res.*, **110**, D09207, doi:10.1029/2004JD004751.

——, and Coauthors, 2008: An evaluation of Arctic cloud and radiation processes during the SHEBA year: Simulation results from eight Arctic regional climate models. *Climate Dyn.*, **30**, 203–223, doi:10.1007/s00382-007-0286-1.

Xie, S., S. A. Klein, J. J. Yio, A. C. M. Beljaars, C. N. Long, and M. Zhang, 2006: An assessment of ECMWF analyses and model forecasts over the North Slope of Alaska using observations from the ARM Mixed-Phase Arctic Cloud Experiment. *J. Geophys. Res.*, **111**, D05107, doi:10.1029/2005JD006509.

——, J. Boyle, S. A. Klein, X. Liu, and S. Ghan, 2008: Simulations of Arctic mixed-phase clouds in forecasts with CAM3 and AM2 for M-PACE. *J. Geophys. Res.*, **113**, D04211, doi:10.1029/2007JD009225.

Zak, B. D., H. W. Church, K. Stamnes, and K. Widener, 2002: North Slope of Alaska and adjacent Arctic Ocean Cloud and Radiation Testbed: Science, siting, and implementation strategies. U.S. Department of Energy Tech. Rep. ARM-02-001, 60 pp. [Available online at http://www.arm.gov/publications/site_reports/nsa/arm-02-001.pdf.]

Zhang, J., and U. Lohmann, 2003: Sensitivity of single column model simulations of Arctic springtime clouds to different cloud cover and mixed phase cloud parameterizations. *J. Geophys. Res.*, **108**, 4439, doi:10.1029/2002JD003136.

Zhao, M., and Z. Wang, 2010: Comparison of Arctic clouds between European Center for Medium-Range Weather Forecasts simulations and Atmospheric Radiation Measurement Climate Research Facility long-term observations at the North Slope of Alaska Barrow site. *J. Geophys. Res.*, **115**, D23202, doi:10.1029/2010JD014285.

Chapter 9

The ARM Mobile Facilities

M. A. MILLER,* K. NITSCHKE,+ T. P. ACKERMAN,# W. R. FERRELL,@
N. HICKMON,& AND M. IVEY**

** Rutgers, The State University of New Jersey, New Brunswick, New Jersey*
+ Los Alamos National Laboratory, Los Alamos, New Mexico
 # University of Washington, Seattle, Washington
@ U.S. Department of Energy, Germantown, Maryland
& Argonne National Laboratory, Lemont, Illinois
*** Sandia National Laboratories, Albuquerque, New Mexico*

1. Beginnings

One of the earliest and liveliest discussions in the Atmospheric Radiation Measurement (ARM) Program was about site locations. Prior to the ARM Program, cloud research was carried out largely via field campaigns, initially by aircraft and then coupled with short-term deployments of ground-based instruments. Field campaign locations were, of course, selected to meet the requirements of the particular research program. Thus, the idea of having permanently located sites created a great deal of debate about criteria for determining locations and how to apply those criteria.

The initial discussions on site locations were captured in an ARM report entitled "Identification, recommendation, and justification of potential locales for ARM SITES" (Schwartz 1991; Cress and Sisterson 2016, chapter 5). The report divided its recommended locales into two categories: primary and supplementary. The five primary locales were meant to span the range of climatically important cloud types and their influence on atmospheric radiation. The report also stated:

> *...there was agreement that sufficient differences exist between the recommended primary locale categories and other radiatively or climatically important locale categories* ***to require a short-duration occupancy of certain supplementary locales*** *to assess the ability of the primary sites to adequately capture and describe the key phenomena controlling the transfer of radiation in the atmosphere* [emphasis added].

Maps of the primary locales and suggested supplementary locales are shown in Fig. 9-1, which is a modified form of the figure in the original report.

During the early days of the program, site managers were chosen for the five permanent sites and for a relatively unspecified movable facility that could be used to address the need for measurements in the supplementary locales. However, the funding profile for the program failed to meet the original requests and it became clear that the ARM Program could not support five primary sites, let alone the occupation of supplementary sites, so the movable facility became an early casualty.

2. Revisiting the concept of a movable facility

By 2000, roughly a decade after the ARM Program began, the three permanent sites functioned well and were regularly delivering data (Ackerman et al. 2016, chapter 3). However, two issues with the ARM plan had surfaced. The first followed from the earlier discussions on supplementary locales, which had become even more important with the loss of two of the proposed primary sites. The three permanent sites infrequently (or never) sampled tropical continental clouds, midlatitude coastal clouds, and marine low-level clouds, particularly the large stratus/stratocumulus sheets. Second, there was an obvious synergy between ground-based and aircraft measurements that had first

Corresponding author address: Mark A. Miller, Institute for Earth, Ocean, and Atmospheric Sciences, Rutgers, The State University of New Jersey, 14 College Farm Road, New Brunswick, NJ 08901.
E-mail: m.miller@envsci.rutgers.edu

DOI: 10.1175/AMSMONOGRAPHS-D-15-0051.1

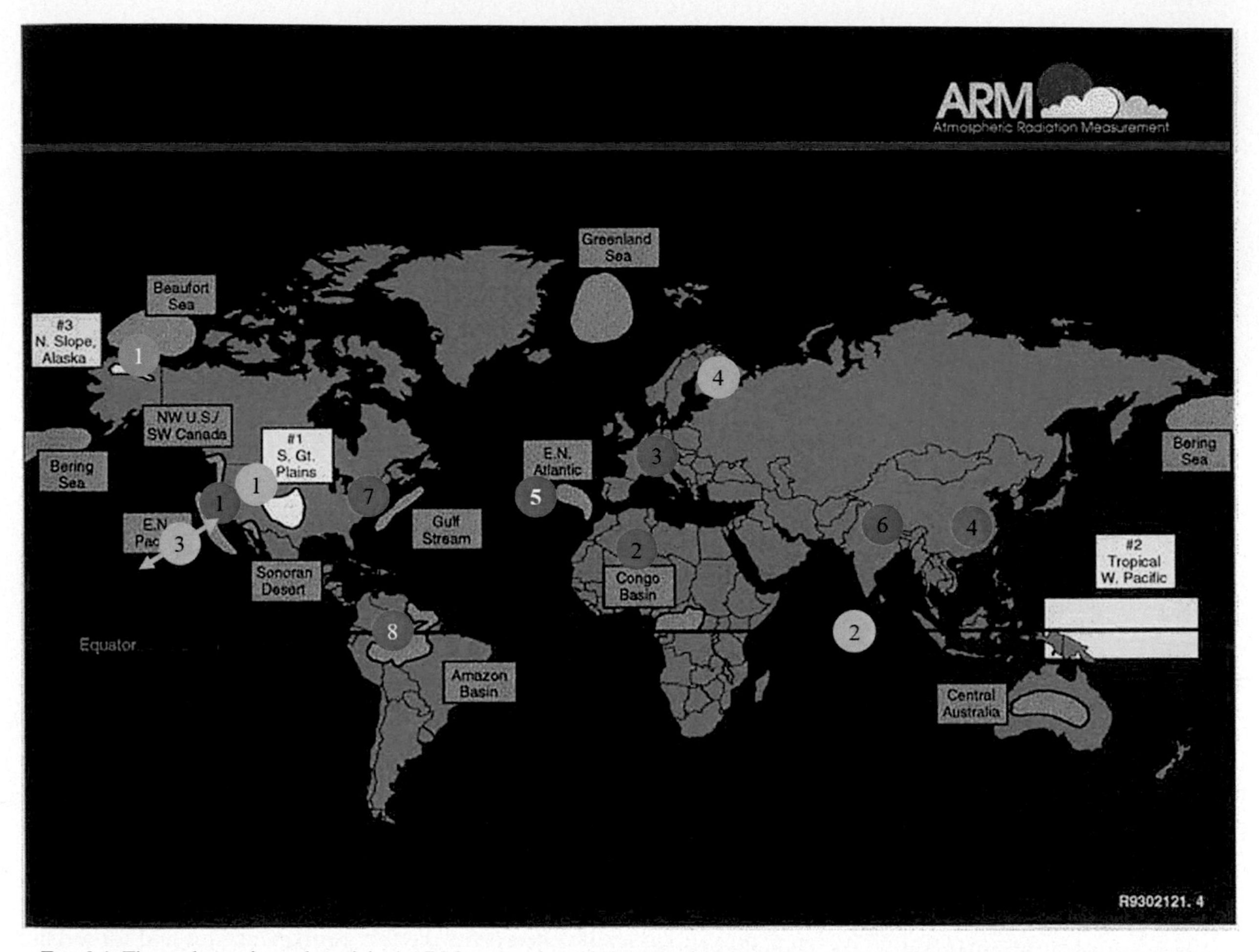

FIG. 9-1. Figure drawn from the original ARM report circa 1992 suggesting possible locales for AMF deployments. The yellow boxes indicate the locations of the ARM fixed sites and the light blue boxes indicate suggested locations for additional fixed sites and mobile deployments. Orange circles containing numbers indicate the locations and chronology of AMF1 deployments and the circles with white numbers indicate the deployment was 2 years in duration. Note that the AMF1 deployment location in Niger is south of the indicated location, beneath the blue box. Also shown are locations for the initial AMF2 (tan circles) and AMF3 (gray circle) with numbers indicating the deployment order. The arrows attached to the third AMF2 deployment indicate the approximate trajectory of the container ship *Spirit* and the 5-yr deployment period for AMF3 is indicated with a white number.

become apparent in the FIRE[1] cirrus experiments, held in Madison, Wisconsin, and in Coffeyville, Kansas, in 1991, and in the Atlantic Stratocumulus Transition Experiment (ASTEX) conducted in the Azores in 1992. While this synergy could have been exploited occasionally at the permanent sites, there were many greater opportunities to do so if an ARM facility were to be collocated with planned field campaigns. In addition, early analysis of data from the Southern Great Plains (SGP) and Tropical Western Pacific (TWP) sites suggested that, although the sites experienced significant interannual variability, a significant range of the climatological cloud properties was sampled in a single year, so single-year deployments in other locales would have provided valuable information.

The early discussions of a movable facility were relatively unstructured because the program had not yet developed a working permanent site. By 2000, the perception of this problem had been altered radically by ARM's participation in the Surface Heat Budget of the Arctic Ocean (SHEBA; Uttal et al. 2002; Verlinde et al. 2016, chapter 8) experiment during 1997–98, and by the work of the Pennsylvania State University TWP group, which had been responsible for conducting ARM science using the TWP site (Long et al. 2016, chapter 7; Mather

[1] First ISCCP Regional Experiment (FIRE) Langley DAAC Project/Campaign Document (https://eosweb.larc.nasa.gov/project/fire/guide/fire_project.pdf).

et al. 1998). The site development approach at both the NSA and TWP sites used modified 20-foot-long sea containers as housing and constructed the site instrumentation and data systems within them in the United States. These containers were then shipped to the experiment locations and assembled as an observing site there (Mather et al. 1998). The Chief Scientist at the time, Tom Ackerman, had developed a scientific rationale and a deployment strategy for the ARM Mobile Facility (AMF) using the TWP and SHEBA models as the basis for a movable facility.

New funding was requested to address this scientific gap, and at the ARM Science Team Meeting in 2001 a new initiative to create an ARM deployable facility was announced. This initiative had been conceived during an initial discussion at the previous year's Science Team Meeting and subsequent discussions at a workshop hosted at the Pacific Northwest National Laboratory (PNNL). The plans were fairly rudimentary and refinements were required. Further discussions were held during the next year, but funding issues delayed the effort. At the ARM Science Team Meeting in 2003, the Chief Scientist announced that the ARM Program expected to have the first phase of an ARM Mobile Facility One (AMF1) completed in fiscal year 2004 (FY04) and that the first deployment was expected in 2005. The expected benefits of the AMF1 were to be additional datasets from other climatic regimes, improved understanding of cloud and radiation processes by testing relationships developed at the permanent sites in new locations, new regionally specific science questions, and improved interactions with other science programs both nationally and internationally. A prototype design and estimated integration schedule was presented at the meeting (Widener 2003). Mark Miller, who had participated in the NASA FIRE and ASTEX projects, was designated as the Site Scientist for the new AMF1 in 2004, Larry Jones [Los Alamos National Laboratory (LANL)] as its Operations Manager, and Kim Nitschke (LANL) as its Chief Operations Engineer. A formal design review meeting for AMF1 was held at the headquarters of the American Geophysical Union (AGU) in 2004 and attendees included Department of Energy (DOE) ARM operations staff (Doug Sisterson and Larry Jones), members of the ARM management team (Tom Ackerman and Jimmy Voyles), the AMF1 Site Scientist (Mark Miller), and external reviewers (Steve Rutledge and Mark Ivey). The AMF1 was constructed at PNNL and upon completion LANL assumed responsibility for its operation and logistical support.

The initial deployment of the AMF1 took place at Point Reyes, California, (#1 in Fig. 9-1) between March and September 2005 in support of the Marine Stratus, Radiation, and Drizzle (MASRAD) experiment proposed by Mark Miller, which was designed to sample coastal marine stratocumulus clouds and serve as a burn-in for the new system. The Pt. Reyes National Seashore was chosen as the initial deployment location because it was a long, linear coastline nearly orthogonal to the mean northwesterly wind direction that carried marine stratocumulus inland. The National Park Service had helped identify potential deployment locations for the new AMF1 that met the noise restrictions and visual standards for a national park. A standard suite of ARM instruments were deployed along with a 95-GHz cloud radar borrowed from the University of Miami.

The experiment lasted six months, and the AMF1 ran relatively smoothly. One important measurement lesson was learned at Pt. Reyes: clouds with bases near the ground have the advantage of strong connections with the aerosol measurements, but often fall below the first measurement height of the cloud radar. Scientifically, the coastal stratocumulus observed at Pt. Reyes possessed turbulence levels that suggested considerable land influence (Ching et al. 2010) and exhibited observable changes in radiation throughput as a consequence of differing levels of pollution (McComiskey et al. 2009). They were also found to have predictable amounts of liquid water because they were highly insulated from the much warmer and drier air above cloud top. This insulation enabled them to respond more predictably to the presence of pollution (Kim et al. 2012).

In 2004, the DOE designated ARM as a national scientific user facility (Ackerman et al. 2016, chapter 3), and a proposal competition was implemented to decide the location of subsequent deployments of AMF1. Proposals were (and still are) reviewed for scientific merit by the ARM Science Board and for logistical and cost feasibility by the AMF1 operations staff. The final decision was based on scientific merit, cost, logistical considerations, and relevance to the ARM mission.

3. Africa and beyond: The AMF1 program goes global

The initial winning AMF1 proposal was submitted by Anthony Slingo, who proposed an experiment using the AMF1 to be conducted in West Africa as part of a larger, multiyear experiment known as the African Monsoon Multidisciplinary Analysis (AMMA). The Radiative Divergence using AMF1, GERB and AMMA Stations (RADAGAST) served as an embedded component of AMMA. West Africa was the second deployment location for the AMF1 and was the first real test of the concept (Miller and Slingo 2007).

Just as MASRAD was beginning in Pt. Reyes, California, an AMF1 advance team consisting of Mark Miller, Larry Jones, and Kim Nitschke traveled to Niamey,

Niger, accompanied by Peter Lamb and Doug Sisterson to find a location where the AMF1 could be deployed in support of RADAGAST (#2 in Fig. 9-1). Niger exposed many of the challenges that would be faced when deploying in underdeveloped regions. The team visited the U.S. Embassy and three different governing agencies during its initial trip to Niger and struggled to understand which of these agencies controlled the airport where it seemed obvious that the AMF1 had to be deployed. The airport had generator power and security and its landscape was similar to the surrounding Sahelian landscape except for its single runway. Eventually, the airport was made available for the deployment and the AMF1 arrived via a Boeing 747 cargo jet. Some months after the advance team's trip to Niamey and the AMF1 deployment planning was in full swing, Mark Miller, Tony Slingo, and Peter Lamb attended a planning meeting for AMMA in Dakar, Senegal.

It had been suggested by Mark Miller during the initial site survey that an ancillary site that collected basic radiation and surface meteorological measurements be established away from the airport to help assess the regional representativeness of the AMF1 deployment location. As a result, a small solar-powered site was established near Banizoumbou, Niger (about 60 km away from the main AMF1 site) in a radiatively natural environment.

The RADAGAST deployment was extremely successful, and there is no question that this deployment is one of the most significant achievements of the ARM and AMF1 program (Slingo et al. 2006; Miller and Slingo 2007). The only setback in the deployment was that the AMF1 cloud radar was still under construction during the first three months, but this period was during the dry season when clouds were less prevalent over the Sahel. The RADAGAST deployment produced the first cross-atmosphere radiation budget on a time scale compatible with cloud development (15 min) and the first comparison of this budget with simulations from GCMs (Miller et al. 2012).

Successfully completing the RADAGAST deployment catapulted the AMF1 program into the scientific spotlight. Confidence that was gained during this first international deployment in a challenging location facilitated a string of consecutive international and national deployments using AMF1. Each was guided by a different scientific mission and presented a new set of logistical challenges. High demand for AMF1 deployments led to the development of two additional AMFs, as described forthwith. Below is a survey of the key scientific foci and results for each of the AMF missions and a synopsis of some of the key issues encountered in the deployments.

a. *The Convective Orographic Precipitation Study experiment*

Germany has a serious problem with flash flooding in the Black Forest region. Models had failed to forecast major flash flooding events that had resulted in significant losses of life and property, and there was a basic lack of understanding of the organizing principles of the convection that was responsible. A proposal submitted by Volker Wulfmeyer to simultaneously collect data from aircraft and the AMF1 in the Black Forest region during the formative stages of convection was selected by the ARM Science Board (#3 in Fig. 9-1).

The AMF1 advance team arrived in the Black Forest Region in March 2006 and the deployment in support of the Convective Orographic Precipitation Study (COPS) began in March 2007. The AMF1 was again embedded in a large international experiment and its mission was to address the nature and structure of orographic flows and the microphysical morphology of associated convection. The plan included contributions from scientific groups in Germany who deployed instrumentation such as a scanning Doppler lidar and several specialized microwave radiometers alongside the AMF1. The expanded capabilities that were provided by these guest instruments served as a blueprint for the development of AMF1 over the next few years. It was operated in the Black Forest for one year in concert with an array of sensors within a dense COPS network. The data collected captured the position and characteristics of convergence zones and other convective initiation processes and how they related to the resulting convection (Wulfmeyer et al. 2008). It was found that flows induced by orographic forcing and channeling were the principal mechanisms that initiated convection. It was also discovered that these flows strongly modulated the precipitation distributions within and downstream of the COPS domain. Observations also showed that the latent and sensible heat fluxes in the COPS domain were driven primarily by vegetation rather than by soil moisture, which was erroneously serving as the dominant driver of these fluxes in mesoscale models.

b. *The study of aerosol indirect effects in China*

The scientific rationale for the AMF1 deployment in China evolved from curious satellite observations. Evidently, liquid clouds in southeastern China contained much more liquid water than liquid clouds with the same thickness in other parts of the world. Zhanqing Li and Graham Stevens hypothesized that these clouds were potentially subject to the second aerosol indirect effect (Albrecht 1989). This effect was based upon a hypothesis suggesting that clouds formed in polluted regions produced precipitation less efficiently than in pristine regions. This encumbrance was thought to have increased cloud liquid water over southeastern China. But validation of these satellite observations was required to confirm that the second aerosol indirect effect was the culprit. Such a validation required a state-of-the-art remote sensing system, so Zhanqing Li submitted a successful proposal to deploy the AMF1 in the vicinity of Lake Taihu, which is a large lake in southern China. Proximity to this lake was important because it could be used to assist in the interpretation of accompanying satellite measurements.

While modern China was beginning to embrace western science, it was unclear if the geopolitical boundaries were relaxed enough to allow for scientific measurements of this type sponsored by the U.S. DOE. Furthermore, the experiment plan was complicated and called for three measurement sites: the AMF1 itself, an ancillary facility borrowed from NASA that was deployed at the edge of the Taklimakan Desert in northwest China, and a small collection of instruments situated near Beijing, which hosted the Olympic Games during the same year as the AMF1 deployment.

The AMF1 advance team traveled to China for an 11-day exploration trip in March 2007. The trip began in Beijing. Following meetings at the Chinese Academy of Science and other agencies, the team traveled to Zhangye in western China, mostly by car and literally following the Great Wall to the edge of the Taklimakan Desert northwest of Zhangye, where a site for the northern deployment was chosen. Next the team traveled to Nanjing for meetings and subsequently to Wuxi, near Lake Taihu. Everything seemed set for the deployment of the AMF1 in China during 2008.

A key issue at Taihu was the ability to launch radiosondes in proximity to a Chinese military base, which was nearby. Upon return to the United States, the advance team learned a few weeks later that the request to launch radiosondes at Taihu had been declined. This event required switching to a contingency plan, but this setback was one of many. A few nearby locations were suggested by the Chinese government, but none was optimal for an AMF1 deployment. Finally, a site near Shouxian, which is 500 km west of Shanghai, was chosen (#4 in Fig. 9-1). It was small and the AMF1 had to be deployed in a compressed configuration. An ancillary site at Taihu that consisted of a microwave radiometer, surface radiometers, and surface meteorology was established, since the radiometer was a passive sensor and provided some limited information about the atmospheric column.

Before the AMF1 instrumentation was to be deployed, a data transmission and sharing agreement was developed with the Chinese government. Normally, AMF1 data were transmitted daily to the Data Management Facility (DMF) in the United States and, shortly thereafter, to the ARM Data Archive (McCord and Voyles 2016, chapter 11). This was not consistent with Chinese data sharing policy, so concessions were necessary. It was mutually agreed that data would be supplied directly to the Chinese government, who would inspect it and, ultimately, release it to the DMF. Unfortunately, this agreement led to a sequestration of AMF1 data in China for nearly the entire deployment, but a more restrictive limitation was the prohibition of any Internet connectivity to the AMF1 instrumentation. This precluded the usual ARM quality assurance process. Shortly after the AMF1 began operating in China, its cloud radar failed and was shipped back to the United States for repair. As a result, only two months of cloud radar data were collected during the autumn of 2008. Sadly, observations of the low stratus liquid clouds, which were desired to satisfy the scientific objectives of the deployment, were virtually absent from the cloud radar data, though an analysis of these data in conjunction with radiosonde data was used to produce a cloud climatology (Zhang et al. 2010, 2013). The wealth of general and specialized measurements that were collected using AMF1 in China combined with data collected by coincident programs has led to successful investigations of the optical, physical, chemical, and cloud nucleating properties (Liu et al. 2011) of anthropogenic, natural, and mixed aerosols and interactions with the East Asian monsoon system (Li et al. 2010).

Inasmuch as the deployment in Africa was a crowning achievement of the AMF1 program, the AMF1 deployment in China was limited by geopolitical and logistical considerations. On the bright side, the development of new scientific partnerships may have opened the door for additional scientific collaborations with China.

c. *Clouds, Aerosol, and Precipitation in the Marine Boundary Layer*

There was a general consensus that varying depictions of the cloud feedbacks associated with boundary layer stratocumulus clouds in global climate models were partly responsible for the spread in predictions of global warming. As a consequence of their large areal coverage on Earth's surface, their radiative feedbacks must be portrayed accurately in GCMs, and the original plan for ARM fixed sites had included a marine boundary layer site in the Azores.

Recognizing the need for long-term and comprehensive measurements of marine stratocumulus clouds, Robert Wood had proposed the Clouds, Aerosol, and Precipitation in the Marine Boundary Layer (CAP-MBL) experiment and his proposal had been approved. Earlier surface-based remote sensor measurements collected during ASTEX from the islands of Santa Maria and Terceira laid the groundwork for the CAP-MBL experiment and the ARM Program in general. Island effects had been documented during ASTEX (Miller and Albrecht 1995) and had been linked to the significant terrain on the island of Santa Maria, which implied that Graciosa Island in the Azores was a preferable location for an AMF1 deployment (#5 in Fig. 9-1). It was smaller in size, had lower terrain, was slightly farther north than the islands used during ASTEX, and it had a small airport.

The AMF1 advance team visited San Miguel, Terciera, and Graciosa Islands during March 2008. Landing at the airport on Graciosa had suggested that the airport itself would serve as an excellent AMF1 deployment location. It was located on the northern shoreline so the prevailing northerly winds during the summer stratocumulus season would have carried clouds over the site before encountering any significant terrain. Breaking waves were heard from the runway by the advance team. Only 90 km south-southeast of Graciosa was the enormous volcanic peak on the island of Pico, which unbeknownst to the advance team during the initial visit played a role in the CAP-MBL project. Data collection began on Graciosa in May 2009. It had been a long time coming given that the Azores had been designated as one of the original fixed sites.

A few months after AMF1 data collection began at Graciosa, it became clear that marine stratocumulus clouds were being observed in a new light. The AMF1 provided a new level of detail about the diversity of marine clouds over the north central Atlantic Ocean. So encouraging were these initial data that Robert Wood submitted a second proposal to the ARM Science Board to extend the deployment for an additional year. Perhaps having sensed the pent-up demand for these data, ARM accepted Dr. Wood's follow-up proposal and the CAP-MBL deployment was extended for an unprecedented second year (Wood et al. 2015). During this extended deployment period, the ARM Program tested a new scanning cloud radar at Graciosa, which was a harbinger of things to come for the AMF1.

When opportunity knocked, the door had been opened; this was how the deployment of several solar and infrared radiometers near the top of Pico during CAP-MBL began. Many years prior to the AMF1 deployment an International Chemical Observatory (ICO) station had been established at 2225 m above sea level on the summit caldera of Pico Mountain. The ICO had collected trace gas and related data during the summer for many years, and Mark Miller submitted a proposal to ARM to deploy radiometers in coincidence with the ICO site. The scientific objective of this deployment was to measure the radiative fluxes and aerosol optical thickness above marine boundary layer clouds and thereby provide a constraint to be used in conjunction with surface radiation measurements from the AMF1 at Graciosa to directly measure the cloud optical thickness in broken cloud fields. There was no road to the top of Pico, only a narrow, steep trail. So the AMF1 technician, Carlos Sousa, carried a small radiation platform on his back to the peak and installed it during the summer of 2010. This system provided extremely unique data and had demonstrated the cleanliness of the air mass above Pico, which rivaled the low aerosol loads observed at Mauna Loa.

As always, there were a few problems. The most significant was the loss of the Atmospheric Emitted Radiance Interferometer (AERI; Turner et al. 2016, chapter 13) for six months of the CAP-MBL deployment. The AERI provided invaluable, nearly continuous information about the details of infrared emissions and the thermodynamic profile. This information was a crucial supplement to radiosonde data and was used to measure the liquid water content of extremely thin clouds (Turner 2007). Another problem discovered after the deployment ended was that the rainfall measurements at Graciosa were biased. This bias was due to improper mounting of the optical rain gauge and the present weather detector. An investigation by Mark Miller's research team revealed that the instrument mounting was altered to accommodate the deployment in Shouxian, China. When the system was shipped to Graciosa, an operations team comprised of some new members reconstructed the configuration that they had disassembled in China rather than reverting to the original, correct configuration. The lesson learned in light of the rain gauge mounting issue was configuration control: when changes were made to the deployment configuration to accommodate a peculiarity they had to be carefully recorded and reversed during the following deployment.

Results from CAP-MBL altered some traditional views of stratocumulus clouds and of the marine boundary layer clouds over the eastern North Atlantic (ENA) in general. The first long-term study of the climatology of cloud structure and its links to drizzle demonstrated that stratocumulus clouds with a depth exceeding 250 m and a liquid water path exceeding 60 $\mathrm{g\,m^{-2}}$ produced drizzle over the ENA (Rémillard et al. 2012). Known mesoscale circulations in marine stratocumulus, termed mesoscale cellular convection, were implicated as being partly related to air mass type. Cold air outbreaks over the ENA were linked to high concentrations of a certain type of mesoscale cellular convection (open cell), which in turn was shown to be associated with reduced pollutant loads. Comparisons of the morphology of marine stratocumulus observed at ENA with those observed over other parts of the world had exposed differences in the thermodynamic environments in which they formed (Ghate et al. 2015). And the first measurements of the updraft mass flux in a large population of small marine cumulus clouds provided new information about cumulus dynamics (Ghate et al. 2011).

d. The Ganges Valley Aerosol Experiment

The Indian monsoon was the lifeblood of India because it supplied rain for agriculture and snowfall to the Himalayas that eventually melted into freshwater that supplied many of India's rivers. There were known and hypothesized links between the Indian monsoon and other important atmospheric circulations. And economic growth in India produced a juxtaposition of this monsoon circulation and the byproducts of manufacturing in northern India, which included large quantities of highly absorbing black carbon aerosols. In addition to the obvious health issues caused by this black carbon, it had been thought to deposit on snow surfaces in the Himalayas thereby reducing the snow's albedo and potentially altering the dynamics of the Indian monsoon. Measurements were needed to evaluate the impacts of black carbon in the region.

Rao Kotamarthi submitted a successful proposal to the ARM Science Board that outlined a plan to obtain measurements of clouds, precipitation, and complex aerosols and study their impact on cloud formation and monsoon activity in the vicinity of the Ganges Valley. The experiment plan combined aircraft measurements collected over the valley itself with AMF1 measurements from the nearby foothills of the Himalayas. The Aries Astronomical Observatory located high on a mountain near the mountain resort city of Nainital, India, was selected as the AMF1 deployment site for the Ganges Valley Aerosol Experiment (GVAX; #6 in Fig. 9-1). The AMF1 was positioned at a location and height that enabled measurements of aerosols spilling through the foothills into the snow-covered peaks of the Himalayas.

In February 2010, the AMF1 advance team traveled to India to organize the three components of the GVAX deployment: the AMF1 itself, a satellite measurement site near the city of Pantnagar, and a Mobile Aerosol Observing System (MAOS) aerosol monitoring facility and aircraft deployment located in Lucknow. The team visited the U.S. Embassy in New Delhi and the headquarters of the Indian Science Agency before it embarked on a long train ride to Pantnagar. With the help of Indian hosts, the team toured a small regional university with enough infrastructure to support a satellite deployment. Next the team traveled by car to the Aries Observatory at Nainital, which was located at 1951 m elevation. The observatory was situated atop a steep mountain and, while it possessed excellent communications infrastructure, space was extremely limited. A platform was constructed atop one of the buildings and the microwave radiometers and surface meteorological and downwelling radiation measurement systems were mounted on this platform. The situation was not optimal and there were concerns about measuring radiation at this site because adjacent to the platform location was a large dome associated with an Aries telescope. But there was not a better location available because the observatory was surrounded by steep mountain slopes.

The GVAX experiment began in June 2011. Its scientific objective was to measure clouds, precipitation, and complex aerosols to study the relationship between the aerosols, cloud formation, and monsoon activity in the region. Siting and logistical difficulties prevented the vertically pointing cloud radar and a new dual-wavelength scanning cloud radar from being deployed, but two important new instruments were added: a scanning Doppler lidar and a solar spectrometer. The AMF1 was deployed in support of a planned aircraft campaign in the Ganges Valley with ARM Aerial Facility (AAF; see Schmid et al. 2016, chapter 10) aircraft to have served a pivotal role in the experiment, but ARM was denied flight clearance for a U.S. research aircraft. Indian research aircraft surveyed pollution in the Ganges Valley, but lack of AAF support was a serious detriment to GVAX. As in the deployment in China, data were sequestered by the Indian scientific agencies. This caused a significant delay in data access, but all data collected were released at the conclusion of the experiment.

e. The Two-Column Aerosol Project

The principal theme of the Two-Column Aerosol Project (TCAP) campaign proposed by Carl Berkowitz and Larry Berg was to quantify the impacts of aerosol mixing state and optical properties upon aerosol radiative effects. To meet these scientific objectives, AMF1 and the Mobile Aerosol Observing System were deployed on Cape Cod, Massachusetts, for one year beginning in the summer of 2012 (#7 in Fig. 9-1). These observations were supplemented by two aircraft intensive observation periods (IOPs), one during the summer of 2012 and a second in the winter. Each IOP required two aircraft.

The AMF1 was deployed at a site along the Cape Cod National Seashore along the bluffs near North Truro,

FIG. 9-2. The AMF1 in its initial deployment at Pt. Reyes, California, in 2005. It was housed in four shipping containers and included 30 instruments.

Massachusetts, and data collection began on 22 July 2012. The experiment, while still in the analysis phase, was extremely successful, and initial results suggested that the experiment objectives were met (Titos et al. 2014; Kassianov et al. 2013). In addition, the TCAP deployment marked another critical turning point in the history of the AMF program because several new cutting-edge instruments purchased through the American Recovery and Reinvestment Act (ARRA) were deployed. Prominent among these instruments was a scanning microwave radiometer and a scanning Ka-W band dual wavelength cloud radar that complemented the existing vertically pointing W-Band cloud radar. The addition of these scanning systems marked a quantum leap in the observation footprint of AMF1 from its original soda straw–like column view.

f. GOAmazon2014/15

The first deployment in the Southern Hemisphere is in progress at the time of this writing. Scot Martin proposed the Green Ocean Amazon 2014–15 (GOAmazon2014/15) experiment, and the AMF1 is currently deployed in a pasture surrounded by Amazonian jungle near the river village of Manicapuru, Brazil (#8 in Fig. 9-1). The experiment is designed to study how aerosol and cloud life cycles are influenced by pollutant outflow from a tropical megacity. A main objective of the experiment is to examine the interplay between biogenic and anthropogenic aerosols. Early results are encouraging and the deployment of AMF1 in support of GOAmazon2014/15 exemplifies its metamorphosis over the course of the programs. The increasing size and scope of AMF1 deployments has significantly altered its appearance and footprint. Figure 9-2 shows AMF1 in its initial deployment in Pt. Reyes, California, in 2005 and Fig. 9-3 its latest deployment nine years later in support of GOAmazon2014/15.

4. AMF2 and AMF3

The AMF program was conceived around the fundamental notion that data from remote and undersampled regions would improve our understanding and simulation of Earth's climate system. When Hans Verlinde of the Pennsylvania State University proposed that AMF1 be deployed on an icebreaker, the need for additional AMFs came into focus. The original design of AMF1 had consisted of extended lightweight military-specification shelters that were seaworthy and came with connector systems used on offshore oil rigs. Also, AMF1 was originally designed with a small footprint so that it could be located on ships of differing sizes based on experience at SHEBA and discussions with engineers at the Woods Hole Oceanographic Institute. But through time AMF1 had been modified and expanded so that it was no longer ship deployable. Thus, there was a clear need for systems designed to be operated in marine, shipboard environments and in harsh land-based environments, and increased demand for AMF deployments of this type led to the construction of the second and third ARM Mobile Facilities (AMF2 and AMF3). These AMFs differed slightly

FIG. 9-3. The most recent deployment of AMF1 near Manicapuru, Brazil, in 2014 in support of GOAmazon (includes the Mobile Aerosol Observing Site, which is currently deployed alongside AMF1). The current version of AMF1 consists of 10 shipping containers and contains over 150 instruments.

in their missions in that AMF2 was designed for traditional one- or two-year deployments and AMF3 for deployments of up to five years. They were constructed with stainless steel modules and flexible power options so that they could be reconfigured in many ways and AMF2 included two stabilized platforms to remediate wave motion: one for the W-band radar and another for the radiometric instruments. All instruments included in these AMFs could be collocated or deployed away from the main site depending upon the science requirements or siting challenges. With ARRA additions, the portable design of these AMFs was expanded to include additional instruments that were required to further satisfy the science goals of their campaigns. Argonne National Laboratory successfully competed for the building and operating of AMF2, and Richard Coulter and Brad Orr became the operators. The AMF3 was constructed after AMF2 was completed, but at Sandia National Laboratories with Mark Ivey as its operator.

a. STORMVEX

The AMF2 was first deployed in support of the Storm Peak Laboratory Cloud Property Validation Experiment (STORMVEX) in Steamboat Springs, Colorado, which was successfully proposed by Dr. Gerald Mace of the University of Utah and colleagues (tan #1 in Fig. 9-1). The objective was to collect data to enhance our understanding of liquid and mixed-phase clouds by using AMF2 instruments in conjunction with Storm Peak Laboratory (SPL), which is a cloud and aerosol research facility operated by the Desert Research Institute. The designed flexibility of the AMF2 allowed deployment of instruments in three locations in addition to instruments deployed at the SPL, which was located at 3203 m in elevation. The SPL and Steamboat Resort Mountain collaborated with the AMF2 team and installed instruments near Thunderhead Lodge, at an elevation of 2759 m, the Aerosol Observing System at Christie Peak at an elevation of 2440 m, and the majority of the instruments in the valley at an elevation of 2078 m (Fig. 9-4). The AMF2 was tested during this maiden deployment by environmental conditions including high winds, temperatures below 0°F, and heavy snowfall.

This initial deployment of AMF2 was successful and the data collected continue to be analyzed. Some of the initial results from STORMVEX showed that nonspherical atmospheric ice particles can enhance radar backscattering and beam attenuation above that expected from spheres of the same mass (Marchand et al. 2013). Initial results also demonstrated that polarization measurements between zenith and slant viewing angles can be used to infer information about

FIG. 9-4. AMF2 instruments in the valley floor at Steamboat Springs, Colorado, during STORMVEX.

the geometry of planar ice crystals (Matrosov et al. 2012).

b. AMIE-GAN

A proposal by Charles Long of PNNL and colleagues was approved for the AMF2 to take its first international voyage to Addu Atoll, Maldives (tan #2 in Fig. 9-1). The ARM Madden–Julian oscillation (MJO) Investigation Experiment on Gan Island (AMIE-Gan) was a part of a larger campaign, DYNAMO/CINGY2011, which included AMIE-Manus. The experiment was designed to test several current hypotheses regarding the mechanisms responsible for MJO initiation and propagation in the Indian Ocean area (Yoneyama et al. 2013; Gottschalck et al. 2013). The AMF2 deployment was organized with almost all instrumentation in a main site, but with a supplementary site where the Ka-X scanning ARM cloud radar (SACR) was deployed to enable radar scanning over the main site (Fig. 9-5). Analysis of AMIE-Gan data is ongoing.

c. MAGIC

Ernest Lewis of Brookhaven National Laboratory and colleagues successfully proposed the Marine ARM GPCI Investigation of Clouds[2] (MAGIC) campaign, which required the AMF2 to operate aboard the Horizon Lines container ship, *Spirit*. Beginning in October 2012, the *Spirit* hosted the AMF2 on round trips between Los Angeles, California, and Honolulu, Hawaii, and provided a moving platform for nearly 200 days of continuous onboard measurements (tan #3 in Fig. 9-1). The primary objective of MAGIC was to improve the representation of the stratocumulus-to-cumulus transition in climate models. The MAGIC science team is currently using AMF2 data to document the small-scale physical processes associated with turbulence, convection, and radiation in a variety of marine cloud types. These variables are currently unrealistic in climate models due to a lack of adequate observational data.

Ernie Lewis, Brookhaven National Laboratory (BNL), and Brad Orr, Argonne National Laboratory (ANL), led a siting cruise prior to the AMF2 deployment to understand the engineering and environmental challenges. Nicki Hickmon, ANL, assumed the role of Site Manager for AMF2 in May 2012 and directed the installation. Modifications to the ship's bridge deck were made to support the weight and mounting of the AMF2. Working in the Port of Los Angeles during regular docking times required a high degree of organization to accomplish modifications, approvals, and lifts while other operations involving other ships were underway. At one point, issues with the ship caused the entire AMF2 to be offloaded with two days' notice prior to docking and a single day for the actual offload procedure. The AMF2 technicians worked exhaustively and the entire AMF2 was offloaded within six hours of the ship docking with no damage to instruments or equipment. After reinstalling the AMF2 on the ship, the observation period was completed. Launching radiosondes was particularly challenging with the wind dynamics around the ship and cargo, but the AMF2 technicians perfected the procedure, leading to an acceptable success rate (Fig. 9-6).

Observations from MAGIC are being analyzed currently and initial results have reinforced the scientific approach that was taken. Data collected during MAGIC have led to estimates of the entrainment rate and water and energy budgets that have helped unravel the changing mix of processes that accompanies the transition from stratocumulus to cumulus conditions (Kalmus et al. 2014).

d. BAECC

Tuukka Petäjä of the University of Helsinki and colleagues submitted a successful AMF2 deployment proposal to the ARM Science Board to measure biogenic aerosols emitted from forests and determine their effects on clouds, precipitation, and climate. The AMF2 operated from February to September 2014 in Hyytiälä, Finland, at the University of Helsinki's Station for Measuring Ecosystem–Atmosphere Relations (SMEAR-II; Fig. 9-7; location tan #4 in Fig. 9-1). Once again the

[2] GPCI = GCSS Pacific Cross-Section Intercomparison, a working group of GCSS.

FIG. 9-5. AMF2 instruments on Addu Atoll, Maldives, during AMIE-GAN.

flexibility of the AMF2 to be reconfigured for each deployment given the site and environmental conditions proved important. The Ka-X SACR was elevated to minimize the forest blocking and the Aerosol Observing System was operated at a site away from the main AMF2 deployment location to enable coincident data collection with the SMEAR-II aerosol system. Analysis of data collected during the Biogenic Aerosols: Effects on Clouds and Climate project (BAECC) is ongoing.

The AMF3 deployment at Oliktok Point, Alaska, was part of an ongoing need to understand the representativeness of the North Slope of Alaska (NSA) fixed site and was planned to last five years (gray #1 in Fig. 9-1). The newly constructed AMF3 was installed in two phases starting in summer of 2013. The first phase of the installation included shelters that were specifically designed for the Arctic, with a shared entry space that links individual shelters and additional insulation for each shelter. The first phase of installation included the same "baseline" instruments for cloud, solar, thermal, and meteorological measurement that have been used at the NSA site since 1997. The second phase, completed in 2014, includes scanning cloud radars and a Raman lidar. Oliktok Point is part of the U.S. Air Force Oliktok Point Long Range Radar station and it was the location of previous ARM field campaign activities, including the Mixed-Phase Arctic Cloud Experiment (MPACE) in 2004 and the Arctic Lower Troposphere Observed Structure (ALTOS) in 2010. The current agreement with the Air Force includes provisions for limited billeting of ARM staff at the station. The Oliktok Point site is linked by the Prudhoe Bay road system and the TransAlaska pipeline road to the lower 48 states.

A unique advantage of the AMF3 deployment at Oliktok Point is access to airspace for research purposes. Starting in 2004, a restricted area was established over Oliktok Point by the Federal Aviation Administration (FAA) for DOE. This restricted area gives DOE and the ARM Program control of a 4-mile-diameter circle of airspace for operations that include tethered balloons and unmanned aerial systems. A warning area that would enable research operations in a corridor of airspace extending roughly 700 miles north of Oliktok is being reviewed with approval expected early next year. Plans include operating tethered balloons, unmanned (Fig. 9-8) and manned aircraft, and related aerial systems at Oliktok to support ARM science.

FIG. 9-6. Sonde launched from the Horizon Lines *Spirit* cargo vessel in 2012.

FIG. 9-7. The AMF2 deployed in Hyytiälä, Finland, at the University of Helsinki's Station for Measuring Ecosystem–Atmosphere Relations.

5. A perspective on AMF deployment challenges through the life of the program

Through the course of the AMF program many challenges have been overcome, some of which are documented here, and during its 10-yr history the AMF program has steadily grown in size and complexity. Accompanying this growth has been a proportional increase in the size and complexity of AMF operations, especially transportation. Originally, five or six technical staff prepared AMF1 for transporting; now it takes longer, requires more staff, and must include some instrument specialists to handle the more complicated remote sensing systems like the scanning cloud radar.

Logistical challenges faced during the myriad of AMF1 deployments may be roughly grouped by deployment location. International deployments have involved many more considerations and logistical challenges than domestic deployments. They have required regulatory approvals from the host country, which have taken time to acquire, and the waiving of standard import duties. One of the most difficult issues has been data export from the host country and establishing the rules of data availability. Common practice in the ARM Program has been to make data freely available as soon as it has been collected (McCord and Voyles 2016, chapter 11), but this policy has been inconsistent with the international data export requirements of some countries. Sometimes the data have been embargoed (the China and India deployments, for example) for a period to allow the host country to insure that no sensitive information was contained in the data files.

Safety and security has been a major issue in underdeveloped areas of the world for both personnel and equipment. In these cases, special accommodation has been necessary for the on-site technician, any technical staff that visits the site, and the facility itself. Airports have been favored deployment locations in underdeveloped areas because they already possess a modicum of safety and security infrastructure, have associated regional transportation hubs, and have been able to accommodate research aircraft. In the case of the deployment in Niamey, Niger, AMF1 was transported to the site by cargo aircraft and deployed at the international airport in Niamey because other travel routes to the region were insecure (and known to be dangerous) and the airport was the only secure

FIG. 9-8. Prototype of an Unmanned Aerial Vehicle that was deployed with AMF3.

TABLE 9-1. ARM Mobile Facility Deployments up to 2014.

Deployment location	Facility	Dates	Scientific focus
Point Reyes, California	AMF1	Mar 2005–Sep 2005	Marine stratocumulus
Niamey, Niger	AMF1	Jan 2006–Dec 2006	Monsoon convection and dust
Black Forest, Germany	AMF1	Apr 2007–Jan 2008	Orographic and convective precipitation
Shouxian, China	AMF1	May 2008–Dec 2008	Aerosol–cloud interactions
Graciosa Island, Portugal	AMF1	May 2009–Dec 2010	Marine stratocumulus
Steamboat Springs, Colorado	AMF2	Nov 2010–Apr 2011	Orographic clouds/remote sensing
Nainital, India	AMF1	Jun 2011–Apr 2012	Aerosol–cloud interactions
Gan Island, Maldives	AMF2	Oct 2011–Apr 2012	Tropical convection
Cape Cod, Massachusetts	AMF1	Jun 2012–May 2013	Aerosol–cloud interactions
California–Hawaii Transect	AMF2	Oct 2012–Sep 2013	Marine cumulus/stratocumulus
Manaus, Brazil	AMF1	Jan 2014–Dec 2014	Cloud–aerosol–precipitation interactions
Hyytiala, Finland	AMF2	Feb 2014–Sep 2014	Aerosol–cloud interactions

option. These regions have also presented unique challenges in locating appropriate, affordable long-term housing for the resident technical support staff that have performed on-site maintenance and offered 24/7 responsiveness.

An issue that has been intertwined in all AMF activities is the language barrier in non-English-speaking countries. To navigate this barrier and many other administrative issues that have arisen, the AMF operations staff has relied heavily on the deployment's principal investigator, who may have local contacts, and on the in-country scientific host. A key component of each AMF deployment has been the engagement of the regional scientific and local host communities. Engaging the latter has proven to be particularly crucial because it has been necessary to identify and work with local and in-county subcontractors to assist in preparing and maintaining the site, and because round-the-clock radiosonde launches have been done by local resident operators trained by the AMF technical staff. Knowledge transfer and scientific development also has been an important goal of the AMF deployments in underdeveloped areas.

The addition of the AMFs added many challenges to the ARM infrastructure, including the DMF, the data quality (DQ) office, the archive, and the instrument mentors. The workloads of these components of the ARM infrastructure were stretched before the addition of the AMFs and remain stretched at this time. While the stretching of these resources was recognized when the AMFs were added, resources not currently available are required to ameliorate this problem. In the meantime, AMF data quality and access remain viable, but could be improved in the future.

Over the years the role of the AMF1 site scientist has varied depending on the deployment. In some cases, the scientist who proposed the experiment was well versed in the remote sensing techniques employed in the AMFs, while in other cases not. And the increasing breadth and sophistication of the AMF's measurement suite has challenged the knowledge base of even the savviest scientist. So the position of site scientist has changed with the AMF1 and in the latter years the focus has been upon learning new instrumentation, performing ASR cloud and radiation science when the experiment focus did not address the basic programmatic needs of each deployment, and performing high-level quality control.

6. Conclusions and recommendations

The AMF concept has been part of the ARM Program from its beginning. Implementation required a combination of programmatic opportunity and technical readiness that was reached in the early 2000s.

The AMF has proven to be of high value to the program, in particular because it has allowed synergistic collaborations with other international programs and countries. Also, ARM has been able to acquire datasets in climate regimes that are not otherwise sampled by the permanent sites (Table 9-1). Excellent examples include the deep dust layers observed during the Niamey deployment (Slingo et al. 2006) and the marine stratus during the Azores deployment. In both cases, data collected by the AMF have been used in new and exciting ways to challenge models of all varieties.

A serpentine path across the Northern Hemisphere during the past decade has netted some of the most unique data streams in the ARM Data Archive and has produced some of the most interesting and unique science in the climate research community. The ARM Data Archive typically records an increase of approximately 100–120 new users as a result of an AMF deployment. This number translates to an increase of about 10%–15% in the number of active data users. In total, users of AMF data constitute 15%–20% of all requests to the archive and when location is considered, the AMF program produces ~50% of the unique measurements stored there. Growth in the AMF program has challenged the capacity of the archive, which now receives multiple requests for terabytes of data during a given month. Many of these requests for large datasets are the result of AMF deployments.

Building on the past success of the AMF program has led to a future in which three AMFs with different capabilities are available for worldwide deployment by any international investigator. But there is always room for improvement. One recommendation based on past experience is that the AMF deployment period never be less than two years without appropriate justification. Data from a single year are extremely useful, but some scientific questions require a two-year dataset to provide a snapshot of regional variability. Another recommendation is that the position of AMF Site Scientist be preserved to insure adequate scientific input to the program and to data users.

The AMF program has become one of the most successful and visible components of the ARM Program. Its scientific reach has transcended international borders and it has become a high-value global asset. Most importantly, the expansion of the AMF program reflects the increasing influence and demand for the science that it enables.

Acknowledgments. We dedicate this chapter to the memories of Dr. Anthony Slingo and Dr. Peter J. Lamb whose seminal efforts during the first AMF1 international deployment in West Africa helped pave the way for the future success of the AMF program. The authors also wish to express their deep appreciation to the technicians, instrument mentors, scientists, and all other members of the ARM community who have made the AMF program what it is today.

REFERENCES

Ackerman, T. P., T. S. Cress, W. Ferrell, J. H. Mather, and D. D. Turner, 2016: The programmatic maturation of the ARM Program. *The Atmospheric Radiation Measurement (ARM) Program: The First 20 Years, Meteor. Monogr.*, No. 57, Amer. Meteor. Soc., doi:10.1175/AMSMONOGRAPHS-D-15-0054.1.

Albrecht, B. A., 1989: Aerosols, cloud microphysics, and fractional cloudiness. *Science*, **245**, 1227–1230, doi:10.1126/science.245.4923.1227.

Ching, J., N. Riemer, M. Dunn, and M. A. Miller, 2010: In-cloud turbulence structure of marine stratocumulus. *Geophys. Res. Lett.*, **37**, L21808, doi:10.1029/2010GL045033.

Cress, T. S., and D. L. Sisterson, 2016: Deploying the ARM sites and supporting infrastructure. *The Atmospheric Radiation Measurement (ARM) Program: The First 20 Years, Meteor. Monogr.*, No. 57, Amer. Meteor. Soc., doi:10.1175/AMSMONOGRAPHS-D-15-0049.1.

Ghate, V. P., M. A. Miller, and L. DiPretore, 2011: Vertical velocity structure of marine boundary layer trade wind cumulus clouds. *J. Geophys. Res.*, **116**, D16206, doi:10.1029/2010JD015344.

——, ——, B. A. Albrecht, and C. W. Fairall, 2015: Thermodynamic and radiative structure of stratocumulus-topped boundary layers. *J. Atmos. Sci.*, **72**, 430–451, doi:10.1175/JAS-D-13-0313.1.

Gottschalck, J., P. E. Roundy, C. J. Schreck III, A. Vintzileos, and C. Zhang, 2013: Large-scale atmospheric and oceanic conditions during the 2011-12 DYNAMO Field Campaign. *Mon. Wea. Rev.*, **141**, 4173–4196, doi:10.1175/MWR-D-13-00022.1.

Kalmus, P., M. Lebsock, and J. Teixeira, 2014: Observational boundary layer energy and water budgets of the stratocumulus-to-cumulus transition. *J. Climate*, **27**, 9155–9170, doi:10.1175/JCLI-D-14-00242.1.

Kassianov, E., J. Bernard, M. Pekour, L. K. Berg, J. Fast, J. Michalsky, K. Lantz, and G. Hodges, 2013: Temporal variability of aerosol properties during TCAP: Impact on radiative forcing. *Remote Sensing of Clouds and the Atmosphere XVIII and Optics in Atmospheric Propagation and Adaptive Systems XVI*, A. Comeron et al., Eds., International Society for Optical Engineering (SPIE Proceedings, Vol. 8890), doi:10.1117/12.2029355.

Kim, Y.-J., B.-G. Kim, M. A. Miller, Q. Min, and C.-K. Song, 2012: Enhanced aerosol–cloud relationships in more stable adiabatic clouds. *Asia-Pac. J. Atmos. Sci.*, **48**, 283–293, doi:10.1007/s13143-012-0028-0.

Li, Z., and Coauthors, 2010: East Asian Studies of Tropospheric Aerosols and their Impact on Regional Climate (EAST-AIRC): An overview. *J. Geophys. Res.*, **116**, D00K34, doi:10.1029/2010JD015257.

Liu, J., Y. Zheng, Z. Li, and M. Cribb, 2011: Analysis of cloud condensation nuclei properties at a polluted site in southeastern China during the AMF-China Campaign. *J. Geophys. Res.*, **116**, D00K35, doi:10.1029/2011JD016395.

Long, C. N., J. H. Mather, and T. P. Ackerman, 2016: The ARM Tropical Western Pacific (TWP) sites. *The Atmospheric Radiation Measurement (ARM) Program: The First 20 Years, Meteor. Monogr.*, No. 57, Amer. Meteor. Soc., doi:10.1175/AMSMONOGRAPHS-D-15-0024.1.

Marchand, R., G. G. Mace, A. G. Hallar, I. B. McCubbin, S. Y. Matrosov, and M. D. Shupe, 2013: Enhanced radar backscattering due to oriented ice particles at 95 GHz during STORMVEX. *J. Atmos. Oceanic Technol.*, **30**, 2336–2351, doi:10.1175/JTECH-D-13-00005.1.

Mather, J. H., T. P. Ackerman, W. E. Clements, F. J. Barnes, M. D. Ivey, L. D. Hatfield, and R. M. Reynolds, 1998: An atmospheric radiation and cloud station in the tropical western Pacific. *Bull. Amer. Meteor. Soc.*, **79**, 627–642, doi:10.1175/1520-0477(1998)079<0627:AARACS>2.0.CO;2.

Matrosov, S. Y., G. G. Mace, R. Marchand, M. D. Shupe, A. G. Hallar, and I. B. McCubbin, 2012: Observations of ice crystal habits with a scanning polarimetric W-band radar at slant linear depolarization ratio mode. *J. Atmos. Oceanic Technol.*, **29**, 989–1008, doi:10.1175/JTECH-D-11-00131.1.

McComiskey, A., G. Feingold, S. Frisch, D. Turner, M. A. Miller, J. C. Chiu, Q. Min, and J. Ogren, 2009: An assessment of aerosol–cloud interactions in marine stratus clouds based on surface remote sensing. *J. Geophys. Res.*, **114**, D09203, doi:10.1029/2008JD011006.

McCord, R., and J. W. Voyles, 2016: The ARM data system and archive. *The Atmospheric Radiation Measurement (ARM)*

Program: The First 20 Years, *Meteor. Monogr*., No. 57, Amer. Meteor. Soc., doi:10.1175/AMSMONOGRAPHS-D-15-0043.1.

Miller, M. A., and B. A. Albrecht, 1995: Surface-based observations of mesoscale cumulus–stratocumulus interaction during ASTEX. *J. Atmos. Sci.*, **52**, 2809–2826, doi:10.1175/1520-0469(1995)052<2809:SBOOMC>2.0.CO;2.

——, and A. Slingo, 2007: The Atmospheric Radiation Measurement (ARM) Mobile Facility (AMF) and its first international deployment: Measuring radiative flux divergence in West Africa. *Bull. Amer. Meteor. Soc.*, **88**, 1229–1244, doi:10.1175/BAMS-88-8-1229.

——, V. P. Ghate, and R. Zahn, 2012: The radiation budget of the West African Sahel and its controls: A perspective from observations and global climate models. *J. Climate*, **25**, 5976–5996, doi:10.1175/JCLI-D-11-00072.1.

Rémillard, J., P. Kollias, E. Luke, and R. Wood, 2012: Marine boundary layer cloud observations at the Azores. *J. Climate*, **25**, 7381–7398, doi:10.1175/JCLI-D-11-00610.1.

Schmid, B., R.G. Ellingson, and G. M. McFarquhar, 2016: ARM aircraft measurements. *The Atmospheric Radiation Measurement (ARM) Program: The First 20 Years*, *Meteor. Monogr*., No. 57, Amer. Meteor. Soc., doi:10.1175/AMSMONOGRAPHS-D-15-0042.1.

Schwartz, S. E., 1991: Identification, recommendation and justification of potential locales for ARM sites. DOE/ER-0495T, 19 pp. [Available online at http://www.arm.gov/publications/programdocs/doe-er-0495t.pdf.]

Slingo, A., and Coauthors, 2006: Observations of the impact of a major Saharan dust storm on the atmospheric radiation balance. *Geophys. Res. Lett.*, **33**, L24817, doi:10.1029/2006GL027869.

Titos, G., A. Jefferson, P. Sheridan, E. Andrews, H. Lyamani, L. Alados-Arboledas, and J. A. Ogren, 2014: Aerosol light-scattering enhancement due to water uptake during the TCAP campaign. *Atmos. Chem. Phys.*, **14**, 7031–7043, doi:10.5194/acp-14-7031-2014.

Turner, D. D., 2007: Improved ground-based liquid water path retrievals using a combined infrared and microwave approach. *J. Geophys. Res.*, **112**, D15204, doi:10.1029/2007JD008530.

——, E. J. Mlawer, and H. E. Revercomb, 2016: Water vapor observations in the ARM Program. *The Atmospheric Radiation Measurement (ARM) Program: The First 20 Years*, *Meteor. Monogr*., No. 57, Amer. Meteor. Soc., doi:10.1175/AMSMONOGRAPHS-D-15-0025.1.

Uttal, T., and Coauthors, 2002: Surface Heat Budget of the Arctic Ocean. *Bull. Amer. Meteor. Soc.*, **83**, 255–275, doi:10.1175/1520-0477(2002)083<0255:SHBOTA>2.3.CO;2.

Verlinde, J., B. Zak, M. D. Shupe, M. Ivey, and K. Stamnes, 2016: The ARM North Slope of Alaska (NSA) sites. *The Atmospheric Radiation Measurement (ARM) Program: The First 20 Years*, *Meteor. Monogr*., No. 57, Amer. Meteor. Soc., doi:10.1175/AMSMONOGRAPHS-D-15-0023.1.

Widener, K. B., 2003: ARM Mobile Facility—Design and schedule for integration. *Proc. 13th ARM Program Science Team Meeting*, Broomfield, CO, U.S. Department of Energy. [Available online at http://www.arm.gov/publications/proceedings/conf13/extended_abs/widener-kb.pdf.]

Wood, R., and Coauthors, 2015: Clouds, Aerosols, and Precipitation in the Marine Boundary Layer: An ARM Mobile Facilities deployment. *Bull. Amer. Meteor. Soc.*, **96**, 419–440, doi:10.1175/BAMS-D-13-00180.1.

Wulfmeyer, V., and Coauthors, 2008: The Convective and Orographically Induced Precipitation Study. *Bull. Amer. Meteor. Soc.*, **89**, 1477–1486, doi:10.1175/2008BAMS2367.1.

Yoneyama, K., C. Zhang, and C. N. Long, 2013: Tracking pulses of the Madden–Julian Oscillation. *Bull. Amer. Meteor. Soc.*, **94**, 1871–1891, doi:10.1175/BAMS-D-12-00157.1.

Zhang, J., H. Chen, Z. Li, X. Fan, L. Peng, Y. Yu, and M. Cribb, 2010: Analysis of cloud layer structure in Shouxian, China using RS92 radiosonde aided by 95 GHz cloud radar. *J. Geophys. Res.*, **115**, D00K30, doi:10.1029/2010JD014030.

——, Z. Li, H. Chen, and M. Cribb, 2013: Validation of a radiosonde-based cloud layer detection method against a ground-based remote sensing method at multiple ARM sites. *J. Geophys. Res. Atmos.*, **118**, 846–858, doi:10.1029/2012JD018515.

Chapter 10

ARM Aircraft Measurements

BEAT SCHMID
Pacific Northwest National Laboratory, Richland, Washington

ROBERT G. ELLINGSON
Florida State University, Tallahassee, Florida

GREG M. MCFARQUHAR
University of Illinois at Urbana–Champaign, Urbana, Illinois

1. Introduction

Airborne observations enhance the surface-based ARM measurements by providing vertical and horizontal context for surface-based measurements, evaluation of remote sensing measurements made from space or the surface, data for development of model parameterizations, and information necessary for process studies that is not available from surface- or space-based remote sensing methods.

Over the years, ARM has carried out manned and unmanned aircraft campaigns under different organizational and operational paradigms. The separately funded ARM Unmanned Aerospace Vehicle (ARM-UAV) program carried out 12 missions between 1993 and 2006 relying on UAVs and piloted aircraft. The ARM-UAV program was established originally to develop measurement techniques and instruments suitable for use with a new class of high-altitude, long-endurance UAVs and to demonstrate these instruments and measurement techniques in a series of field campaigns designed to support the climate change community with valuable datasets. ARM-UAV also supported field campaigns using piloted aircraft when platform or instrument constraints made the use of UAV platforms incompatible with the objectives of dedicated intensive operations periods (IOPs) that were needed to achieve specific programmatic or scientific goals. Thus, ARM-UAV sponsored a number of campaigns using piloted aircraft.

In parallel, the main ARM Program also sponsored several piloted aircraft campaigns led by ARM principal investigators outside the ARM-UAV program. These campaigns focused on in situ observations of aerosols over the ARM Southern Great Plains site.

In October 2006, the ARM Aerial Facility was established formally as an integral part of ARM with the mandate of executing all future ARM aircraft campaigns under one organizational umbrella.

In what follows, we describe the major outcomes of airborne campaigns carried out in these different phases of ARM airborne research.

2. ARM-UAV program description

ARM-UAV arose from the DOE Atmospheric Remote Sensing and Assessment Program (ARSAP), a joint program with the Department of Defense (DOD) begun as part of the Strategic Environmental Research and Development Program (SERDP) that was aimed at identifying science problems and associated observations that could be studied with UAVs. During the ongoing planning for ARM, it was recognized that UAVs offered ARM the unprecedented potential of making in-atmosphere measurements at high altitudes (up to and above 20 km) for long periods of time (in excess of 24 h) and at reasonably low air speeds (less than $100 \mathrm{m\,s^{-1}}$) that were suited ideally for problems that could not be

Corresponding author address: Beat Schmid, Atmospheric Sciences and Global Change, Pacific Northwest National Laboratory, 902 Battelle Blvd., Richland, WA 99352.
E-mail: beat.schmid@pnnl.gov

DOI: 10.1175/AMSMONOGRAPHS-D-15-0042.1

studied completely from ground-based measurements (e.g., radiative heating rates, grid-square radiation budgets, cloud profiling, etc.) Thus, the ARM-UAV program was born, albeit with funding and management separate from ARM but with overlapping science teams (see below). John Vitko of Sandia National Laboratories (SNL) led the technical management of the UAV program under the direction of the DOE UAV program manager, Patrick Crowley.

The initial planning for the ARM-UAV program was influenced by a December 1992 meeting of an Interim Science Team (IST) drawn primarily from the existing ARM community (see Crowley and Vitko 1994). The meeting identified two broad classes of missions for ARM-UAV: "quasi-continuous missions," which emphasize consistent long-term observations of key radiation-cloud parameters as part of a continuous ARM data stream, and "investigative missions," which change with time and focus on testing specific hypotheses. The basic quasi-continuous missions were identified as the continuous/near-continuous measurement of radiative fluxes along with both in situ and remote sensing measurements of water vapor and cloud properties. Some proposed representative investigative missions included testing of specific hypotheses on the role of deep convection in the tropical Pacific, the drying/moistening of the upper troposphere, the source of the asymmetry in water vapor concentrations in the Northern and Southern Hemispheres, and ozone chemistry near the tropopause (Crowley and Vitko 1994).

In 1992, an ARM-UAV Science Team comprised of leading atmospheric scientists from 10 universities and government research centers was formed with the assistance of the SERDP funding. The Science Team was formed to assist with the development of new measurement capabilities and the measurement of atmospheric heating in a well-defined layer and then to relate it to cloud properties and water vapor content [see Tooman (1997) for a complete list of the Science Team and their projects]. As the program evolved, identification of scientific missions was obtained from the ARM Science Team. The measurement capabilities of the ARM-UAV program were developed from scratch, because UAVs had not been used for science observations or in controlled civilian airspace. In particular, the weight, power, and volume requirements of many instruments necessitated the redesign of the instruments to meet the payload capabilities of available UAVs. ARM-UAV, under SERDP, initiated an instrument development program to produce prototypes of instruments that could be flown on UAVs in support of ARM science. The new instruments that were developed included: the cloud detection lidar by Lawrence Livermore Laboratory, the UAV atmospheric emitted radiance interferometer by the University of Wisconsin–Madison, the hemispheric optimized net radiometer by Los Alamos National Laboratory, the multispectral pushbroom imaging radiometer by Sandia National Laboratories, the frost point and laser diode hygrometers by Brookhaven National Laboratory, the scanning spectral polarimeter by Colorado State University, and a UAV-mounted 95-GHz radar system by the University of Massachusetts [see Tooman (1997) for instrument details]. SNL, Livermore, California, carried out oversight of the development of the instrumentation under the leadership of John Vitko and Tim Tooman.

Since UAV-mounted instruments do not offer the in-flight capability of direct onboard human intervention of the instruments, techniques had to be developed to monitor and control instruments in flight and to record the data on the ground. SNL developed a system that is capable of operation on an unpressurized manned or unmanned aircraft at an altitude of 20 km with flight duration of 72 h. The system allows data to be transmitted from the instrumentation on multiple aircraft to a ground receiving station where the data are recorded, and operational commands are sent to the aircraft science payload via an uplink. No data are stored on the UAV.

The SNL telemetry system also feeds quick-look data to an on-site data management system that allows the data to be viewed in real time by the mission controller, mission scientist, instrument mentors, and on-site technicians. The system also allows simultaneous gathering and display of data from a local ARM site. These capabilities allow the instrument mentors to determine the state of the instruments onboard and to restart instruments that appear to have been compromised since takeoff. Further, the real-time review of the telemetered data allowed in-flight monitoring and modification of flight patterns associated with individual science missions. Note that the telemetry and data collection system was made deliberately flexible so that it could accommodate use with unmanned and/or manned aircraft.

Every deployment of the ARM-UAV facility included a science-team-led effort to identify possible experiments that might be conducted during the deployment. The aircraft to be used for the deployment were then specified based upon the scientific needs and the availability of necessary aircraft (UAV and/or a manned aircraft). This was followed by the publication of a detailed experiment and science plan that served as a mission selection document while in the field (e.g., Ellingson and Tooman 1999). Thus, ARM-UAV is somewhat of a misnomer because the program used a combination of manned or unmanned aircraft, depending upon the science requirements and platform suitability.

The flight-planning document was also used to request approval from the DOE and the Federal Aviation Administration (FAA). The FAA and the DOE required much more detailed flight and operation plans to meet the safety concerns associated with unmanned aircraft, as compared to manned aircraft. In particular, when UAVs were used in experiments within the continental United States, the program was required to use a manned aircraft to escort the UAV between the surface and 5.5 km and return. Will Bolton of SNL led this lengthy approval process for each of the ARM-UAV IOPs.

Following deployment of personnel to the aircraft operations base, mission planning and selection meetings took place daily following the mission scientist's analysis of National Weather Service (NWS) data and forecasts. Preflight meetings were generally held a few hours before each flight to make final selection of a day's mission based on the most recent analysis of the NWS information.

The mission controller directed operations of the instruments during the flight and coordinated aircraft operations with the aircraft operator using input from the mission scientist and instrument mentors. Each flight was followed by a postflight debriefing led by the mission controller that reviewed the actual mission flown, instrument deficiencies, if any, and mission accomplishments. The mission controller and scientist wrote mission summaries for the data record following the meeting. If flights were planned on successive days, a mission planning meeting followed the postflight meeting. A flight day work period typically lasted 12 h or more for most of the ARM-UAV team. Tim Tooman and Roger Busby of SNL served as mission controller and director of ground operations, respectively. Robert Ellingson of the University of Maryland and Florida State University and Greg McFarquhar of the University of Illinois served as mission scientists for the 1993–2002 and 2002–06 periods, respectively.

Postflight data management processed the telemetered data along with the mission logs, summaries, and instrument lists into a complete, well-documented dataset of known quality. When the dataset was completed, it was transferred to the ARM archive for access by the community, as has been done for all ARM data (McCord and Voyles 2016, chapter 11). The reader is referred to Tooman (1997) for specific details concerning the ARM-UAV instrumentation, operations, and data processing.

SERDP funding of ARM-UAV continued through 1997, after which the program funding transitioned to ARM, effective fiscal year 1997, at a lower level. The ARM-UAV science activities transitioned simultaneously. The management of ARM-UAV remained relatively unchanged through 2006, when all ARM aircraft campaigns operations were merged into one program, the ARM Aerial Facility (AAF).

3. Noteworthy ARM-UAV achievements

ARM-UAV carried out 12 different campaigns (Fig. 10-1) generally centered on three classes of scientific experiments:

- Radiative fluxes, in which aircraft were used to make high-accuracy measurements (~1%) of the solar and infrared radiative transport throughout the troposphere under a variety of clear-sky, cloud aerosol, and water vapor conditions
- Cloud properties, in which remote sensing techniques were used to develop and validate techniques for obtaining cloud reflectivity, phase (ice or water), effective droplet size, etc.
- Satellite calibration and validation, where the instruments were used to indirectly calibrate sensors on operational satellites as well as to validate retrieval algorithms for such derived quantities as flux divergence, cloud properties, and water vapor profiles.

The following material summarizes the major achievements of the ARM-UAV program from several campaigns before the program was integrated with other aircraft activities to form the AAF. The reader may find details concerning all of the science campaigns and instrumentation in a series of UAV campaign summaries and science plans available at the ARM website (http://www.arm.gov/sites/aaf/uavcampaigns) and in Stephens et al. (2000).

a. UAV development flights

The first two missions in 1993 and 1994, called the UAV development flights (UDF), focused on measuring vertical profiles of long- and shortwave radiation fluxes under clear-sky conditions with a General Atomics Gnat 750, a midsize UAV. The operable payload contained four broadband radiometers plus a downwelling total direct diffuse radiometer developed by Francisco Valero (see Valero et al. 1982, 1989; Valero and Pilewskie 1992). Valero et al. (1996) showed that radiation model calculations based on the UAV-measured in situ atmospheric thermodynamic data during this series yielded upward and downward flux profiles that agreed closely with those observed by the onboard radiometers. The importance of this set of flights is that they marked the first scientific use of a UAV and the first aircraft profiling over the highly instrumented ARM Southern Great Plains (SGP) Central Facility in Lamont, Oklahoma.

b. 24-h mission

Demonstration of the capability of the UAV for extended scientific data gathering operations near the tropopause was the major emphasis of a 3-week deployment during fall 1996. The General Atomics Altus,

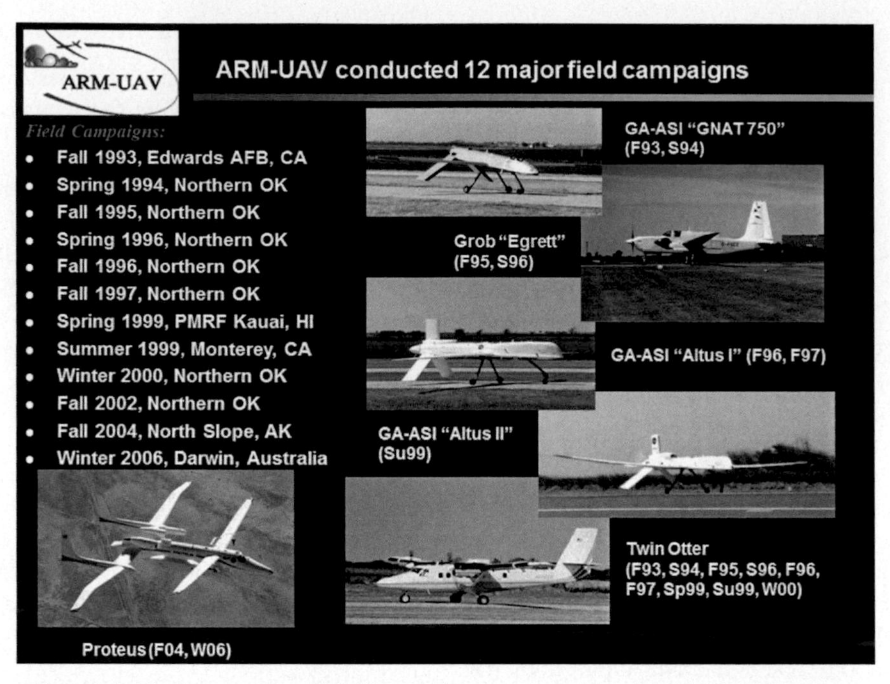

FIG. 10-1. Platforms used and field campaigns conducted by the ARM-UAV program. Seasons are denoted as follows: summer (S), spring (Sp), fall (F), winter (W), and summer (Su). Years are given as follows: 9x and 0x denote 199x and 200x, respectively.

capable of carrying a 150-kg payload to a 10-km altitude, was utilized in this campaign. Its payload included the Gnat 750 instruments plus a scanning spectral solarimeter (SSP), cloud detection lidar (CDL), and a multispectral pushbroom imaging radiometer (MPIR). The campaign operations were organized to ramp up to extended operations from short-duration flights. The shorter flights gathered data to support the ARM water vapor IOP (Turner et al. 2016a, chapter 18) and to aid in the development of new models for calculating the transfer of solar and terrestrial radiation in clear and cloudy conditions. The successful completion of a >26-h UAV flight on 5 October 1996 highlighted the completion of this deployment and marked the first >24-h mission of a UAV for gathering data for science applications.

c. Extended operations near the tropopause

The spring 1999 campaign operated from the Pacific Missile Range Facility, Kauai, Hawaii. This experiment was designed as a two-aircraft cirrus cloud experiment and deployed the Altus II UAV for the first time. This UAV has a dual-stage turbo charger that can lift the aircraft above 16 km. During this campaign, the Altus II operated for 16.5 h at or above 15.24 km (50 000 ft) and 1.75 h above 16.75 km (55 000 ft), generally above all cirrus, and flew in tight formation with a manned aircraft, a de Havilland Canada DHC-6 Twin Otter, that flew below the cloud base. The Altus provided measurements of spectral and broadband radiative fluxes, spectral radiances, and lidar backscattering properties of tropical and subtropical cirrus. The Twin Otter provided similar radiometric measurements below the cloud as well as radar reflectivity measurements obtained with UAV-developed 94-GHz airborne cloud radar (from Stephens et al. 2000). This campaign was the first to demonstrate the capabilities of a UAV to remotely sense cirrus tropical properties. Further, the analyses shown by Stephens et al. (2000) showed a remarkable degree of similarity between the radar reflectivity profile predicted by both a cloud-resolving model and an ECMWF 24-h forecast and the profiles measured by the cloud radar and the onboard cloud detection lidar, thus implying a manner of realism of the modeled ice water contents.

d. ARM Enhanced Shortwave Experiments (ARESE I and II)

In the mid-90s a scientific controversy in the ARM Program concerning the magnitude of absorption of solar radiation by clouds (Mather et al. 2016, chapter 4; Michalsky and Long 2016, chapter 16) provided the first opportunity for an ARM-UAV mission to address a major science question. Analyses by Cess et al. (1995), Ramanathan et al. (1995), and Pilewskie and Valero (1995) had concluded that the absorption by the entire atmospheric column in the presence of clouds exceeds model predictions of absorption by about $35\,\mathrm{W\,m^{-2}}$ (daytime average) over the Pacific warm pool. Those authors recognized that what appeared to be small errors in calculating absorption by the atmosphere might have huge consequences in tropical atmospheric dynamics. Furthermore, if the modeling of solar absorption by clouds was in error, remote sensing data used to infer cloud microphysical properties were also likely in error.

1) ARESE I

The ARM-UAV campaign in fall 1995 [commonly known as the ARM Enhanced Shortwave Experiment (ARESE I)] was flown with manned aircraft. ARESE I focused on two scientific objectives: (i) the direct measurement of the absorption of solar radiation by clear and cloudy atmospheres and the placement of uncertainty bounds on these measurements; and (ii) the investigation of the possible causes of absorption in excess of the model predictions.

To accomplish these objectives, the experiment used a combination of satellite, manned aircraft, and ground observations to make solar flux measurements at different altitudes throughout the atmospheric column. At the heart of this were a carefully "stacked" DHC-6 Twin Otter and Grob Egrett "cloud sandwich" with the Otter at 0.4–1.5 km and the Egrett at 13 km. This was overflown on occasions by the NASA ER-2 flying at about 20 km, which because of its much higher speed, did not stay in constant alignment with the Twin Otter/Egrett stack but did provide periodic coincidences with these other aircraft. All three aircraft carried identical up- and down-looking Valero broadband radiometers (Valero et al. 1982) and flew over identical up-looking radiometers at the ARM SGP Central Facility and extended facilities. Radiance measurements from the GOES satellites were used to retrieve top-of-the-atmosphere fluxes. These flux measurements were supplemented by a variety of cloud property measurements from the ground, the Egrett, and the ER-2, including radar, lidar, and multispectral measurements. Additionally, information such as water vapor profiles, aerosol optical depths, cloud structure, and ozone profiles, needed as input in radiative transfer calculations, was acquired from a variety of ARM ground-based observing systems and sonde launches.

The ARESE flights were conducted at the SGP site from 25 September through 1 November 1995. Approximately 60 h of in-flight data were accumulated by 12 scientific data flights under a variety of atmospheric conditions ranging from clear to solid overcast. The analyses of these data by Zender et al. (1997) and Valero et al. (1997) again suggested that cloudy skies absorb more shortwave radiation than predicted by current models. Unfortunately, these studies were not able to define a region in the solar spectrum where the anomalous absorption was taking place. In the end, uncertainties concerning the instrumentation, the small number of overcast samples, and the meteorological conditions led to a variety of challenges to the conclusions (see Smith et al. 1997; Li et al. 1999; O'Hirok et al. 2000). Therefore, questions concerning cloud shortwave absorption continued to dominate the science issues. Nonetheless, the experiences gained from ARESE I and new theoretical studies led to the belief that the various shortcomings of ARESE I could be overcome by a new ARESE experiment—ARESE II—conducted during a 6-week period in spring 2000.

2) ARESE II

The ARESE I objectives were adapted for ARESE II in 2000, but the sampling strategy was changed to use observations from one aircraft and the ground following the analyses of Marshak et al. (1997, 1999). Marshak's analyses focus on methods to remove the effects of horizontal fluxes in two-aircraft (or aircraft-ground) measurements of cloud absorption obtained by taking the difference between time-averaged net fluxes at two levels. The Marshak et al. (1997) simulated observations found that the optimal flight patterns depend on cloud structure and horizontal distance between the aircraft, with the ideal case being when both sets of radiometers see the same piece of cloud most of the time for extensive overcast stratus conditions. This condition is satisfied when the aircraft is within about 1-km distance centered on the ground radiometers.

Marshak et al. (1999) demonstrated that the apparent absorption obtained from the net flux differences correspond to true absorption if the data are conditionally sampled to periods for which the horizontal fluxes are zero. These conditions occur when there is little or no flux divergence in nonabsorbing portions of the spectrum. This led in part to the selection of the radiometers used in the experiment (discussed below).

Limiting the observations to the vicinity of the ARM SGP Central Facility allowed the use of cloud radar data and signals from various ceilometers to characterize the cloud structure, including microphysics, as a function of time. In short, the measurement strategy used a single aircraft repeatedly overflying the ARM Central Facility to provide the top-of-the-cloud fluxes and combined those with surface-based measurements to determine both broadband and spectrally resolved vertical flux divergences. Details concerning the ARESE II planning may be found in Ellingson and Tooman (1999).

The aircraft was equipped with several different sets of zenith- and nadir-looking radiation instruments.

These included the Valero radiometers suite [total solar, fractional solar, and total direct diffuse radiometer (TDDR), covering 400–700 nm in 6 contiguous bands of 50-nm plus a 10-nm band at 500 nm in a nonabsorbing region], a complementary set of radiometers provided by the Meteorological Research Institute of Japan (MRI), the solar spectral flux radiometer (SSFR; 300–2500 nm in ~300 channels), and the spectrally scanning polarimeters (SSP2; 400–2500 nm in 120 channels). Identical instruments were located on the ground in a zenith-looking mode at the Central Facility and, in the case of the MRI radiometers, nadir looking as well. The continuous spectral coverage of the shortwave region, and from multiple radiometers, was one of the major improvements of ARESE II over ARESE I. Pre- and postmission radiometer instrument intercomparison and calibration were performed to identify inconsistencies between the various measurement systems, and spectral albedo measurements were obtained near the Central Facility by Michalsky et al. (2003).

Detailed descriptions of ARESE II and findings are given by Ackerman et al. (2003), O'Hirok and Gautier (2003), Valero et al. (2003) and Asano et al. (2004). The various cloud absorption studies found the observed absorption was greater than that calculated but smaller than the cloud absorption and discrepancies noted during ARESE I. By including aerosols with modest absorbing properties in their 3D model computations, O'Hirok and Gautier (2003) isolated the discrepancy to the near-infrared spectral region (0.68–3.9 mm). Overall, the measurements and radiative transfer model comparisons did not unambiguously support the occurrence of anomalous absorption, as the differences between model calculations and observations were within the range of uncertainties in the observations and model calculations.

e. Mixed-Phase Arctic Cloud Experiment (M-PACE)

Prior to the M-PACE project, Greg McFarquhar replaced Bob Ellingson as chief scientist of the ARM-UAV program and served as mission scientist during M-PACE, with Tim Tooman still serving as mission controller. During M-PACE (fall 2004, North Slope, Alaska), mixed-phase clouds were measured in situ by the University of North Dakota Citation and remotely by the Scaled Composites *Proteus* aircraft (Verlinde et al. 2007, 2016, chapter 8). Although both the Citation and the *Proteus* were piloted aircraft, the data systems on the *Proteus* were run in a virtual mode, as if they were on a UAV, as they were controlled from the ground and real-time data could be downloaded and inspected. An extensive database on the vertical profiles of the mixed-phase clouds occurring in both single and multiple layers was constructed (McFarquhar et al. 2007b) using data acquired during spiral ascents and descents over remote sensing instruments at Barrow and Oliktok Point and during ramped ascents and descents between these two locations. The single-layer mixed-phase clouds were dominated by liquid, but McFarquhar et al. (2013) showed that some small particles in these mixed-phase clouds were ice, contrary to prior assumptions that all small particles in such clouds were water.

The biggest impact of M-PACE data was its use for evaluating large-eddy simulation (LES), cloud-resolving (CRM), and general circulation model (GCM) simulations that contribute to the fundamental understanding of microphysical processes in mixed-phase clouds and for evaluating and improving remote sensing retrievals. For example, Fridlind et al. (2007) showed the formation of ice nuclei from drop evaporation residuals, and drop freezing during evaporation could explain why ice crystal concentrations are greater than ice nuclei concentrations. Model intercomparison studies (e.g., Klein et al. 2009; Morrison et al. 2009) showed that more detailed representations of microphysics generally gave model simulations more consistent with observations and compared the performance of double- and single-moment parameterization schemes (Solomon et al. 2009), where double-moment schemes prognosed both mass and number, whereas single-moment schemes only prognosed mass. Several other modeling studies also used M-PACE data in conjunction with modeling simulations to gain better understanding of Arctic mixed-phase clouds. Parameterizations for large-scale models (Liou et al. 2008) and cloud-resolving models (Morrison et al. 2008) have also been developed with M-PACE data. Remote sensing studies also showed the relationship between mixed-phase cloud characteristics and vertical motions (Shupe et al. 2008).

f. Tropical Warm Pool–International Cloud Experiment (TWP-ICE)

The Scaled Composites *Proteus* was used as the major airborne platform during TWP-ICE, where again the instruments were operated in a mode amenable to remotely piloted aircraft, with the instruments controlled and monitored from a ground station; there were no instrument operators on board. This was made possible because of the considerable investment SNL had placed into the development of an infrastructure to do this as part of the ARM-UAV program. Data collected during TWP-ICE [winter 2006, Darwin, Australia; May et al. (2008); Long et al. (2016, chapter 7)] have led to breakthroughs in understanding how processes in tropical ice clouds affect cloud feedbacks and climate. The

TWP-ICE data also represented the first time that cloud data were collected on an aircraft where the in situ probes were controlled remotely. McFarquhar et al. (2007a) used these data to quantify the contributions of small ice crystals with diameter $< 50\,\mu$m to the mass and radiative properties of cirrus. Quantifying the importance of small ice crystals has been a controversial and unsolved problem for the last 20 years, yet this knowledge is critically needed to quantify cirrus effects on longwave and shortwave radiation and, hence, to better represent cloud feedbacks in GCMs. TWP-ICE data suggested that the shattering of large ice crystals on protruding components of forward scattering probes used to crystal size distributions produced hundreds of smaller crystals. To illustrate the impact of this finding for climate studies, TWP-ICE data motivated a study whereby the NCAR Community Atmosphere Model (CAM3.0) was used to examine the impact of different assumptions about small ice crystal concentrations on cloud radiative forcing (Mitchell et al. 2008): it was found the inclusion of small ice crystals could produce 12% differences in cloud ice amount and 5.5% in cirrus cloud coverage, corresponding to a net cloud forcing of $-5\,\mathrm{W\,m^{-2}}$ and upper-troposphere temperatures of over 3°C.

The TWP-ICE data have also had important applications for motivating and evaluating modeling studies, whereby TWP-ICE observational data were compared with results of CRM simulations, with the CRM simulations subsequently compared against the results of limited-area models, single-column models, and larger-scale models. Thus, like M-PACE, the TWP-ICE project served as a model for showing how in situ and remote sensing data collected during a field campaign could motivate a myriad of modeling and remote sensing studies.

4. Other ARM aircraft missions prior to 2007

Observation of aerosols at the ARM sites was limited initially to in situ measurements on the ground. Vertical profile information, however, is necessary for radiative forcing calculations. Limited information was available from lidars, but there was dire need of validation (see McComiskey and Ferrare 2016, chapter 21). To this end, ARM conducted several aircraft campaigns focused on airborne in situ observation of aerosol properties. These studies were conducted outside the ARM-UAV umbrella by ARM principal investigators and prior to ARM merging all aerial efforts into the AAF in 2007.

a. In Situ Aerosol Profiles (IAP)

From 2000 to 2007, ARM carried out twice-per-week routine flights over the SGP site with a small aircraft (initially a Cessna 172 but using a Cessna 206 after 2005) measuring nearly 600 vertical profiles of aerosol optical properties. The resulting 8-yr record reveals significant seasonal differences in aerosol scattering and absorption as a function of altitude. As one example, the single-scattering albedo was found to be lowest in winter, particularly aloft. A detailed analysis of this unique long-term record has been presented by Andrews et al. (2011). Validity of the measurements has been tested with interplatform comparisons and closure studies (Andrews et al. 2004, 2006; Hallar et al. 2006; Ferrare et al. 2006b; Kassianov et al. 2007; Schmid et al. 2009). The data record also has been used for the evaluation of modeled black carbon profiles (Skeie et al. 2011). A complete list of publications resulting from this effort can be found in Schmid et al. (2014).

b. Aerosol Intensive Operation Period

The May 2003 Aerosol Intensive Operations Period (AIOP), examined the properties and radiative influences of aerosols over SGP using ground-based (in situ, radiometer, and lidar), and airborne measurements [Center for Interdisciplinary Remotely-Piloted Aircraft Studies (CIRPAS) Twin Otter and Greenwood Cessna 172]. As stated in the preface to the special issue resulting from AIOP (Ferrare et al. 2006a, p. 1), "The scientific hypotheses that were investigated during this IOP were posed as 'closure experiments' in which an observable quantity [was] measured in two or more different ways, or [was] measured as well as calculated (modeled) using other measured quantities. Closure [was] achieved if the several measures agree within their mutual uncertainties." The observable quantities examined in that fashion included aerosol absorption (Arnott et al. 2006), scattering (Hallar et al. 2006), extinction (Schmid et al. 2006; Ferrare et al. 2006b; Strawa et al. 2006), hygroscopicity (Pahlow et al. 2006), asymmetry parameter (Andrews et al. 2006), shortwave radiation (Michalsky et al. 2006), and CCN concentrations (Gasparini et al. 2006; Ghan et al. 2006). A complete list of papers contained in the AIOP special issue can be found online (http://www.agu.org/journals/ss/DOEARM1/). Data from the AIOP also have contributed to numerous other studies since (e.g., Kassianov et al. 2007; Wang et al. 2007; de Boer et al. 2013). The AIOP is discussed further in this monograph by McComiskey and Ferrare (2016, chapter 21).

c. Aerosol Lidar Validation Experiment

A major goal of the 2003 AIOP was to assess accuracy of the aerosol (and water vapor) retrievals of the ARM Raman lidar at SGP [at the time, the only Raman lidar in the world making around-the-clock autonomous measurements in an operational setting (see Turner et al. 2016b, chapter 13)]. During AIOP, the Raman lidar was found to overestimate aerosol extinction by 54% (at $\lambda = 355$ nm) when compared to the NASA Ames Airborne Tracking Sunphotometer (AATS-14), an instrument frequently used as a standard for aerosol extinction measurements (Schmid et al. 2006; Ferrare et al. 2006b). The large bias in the Raman lidar measurements during AIOP stemmed from a gradual loss of lidar sensitivity starting about the end of 2001 and going unnoticed until after AIOP. As a result the Raman lidar underwent major refurbishments and upgrades (Ferrare et al. 2006b; Newsom et al. 2009; Turner et al. 2016b, chapter 13) after AIOP, necessitating a new validation campaign—the Aerosol Lidar Validation Experiment (ALIVE) conducted in September 2005. Again, AATS-14 was used as aerosol extinction standard. The refurbished and upgraded Raman lidar, along with improvements to its data processing algorithm (Newsom et al. 2009), led to a very small extinction bias in ALIVE of 6% (Schmid et al. 2009)—a major improvement over the situation in AIOP.

For this focused 11-day campaign, AATS-14 was flown on a Jetstream 31 (Sky Research, Ashland, Oregon) along with the Research Scanning Polarimeter (RSP; Cairns et al. 1999), providing multispectral measurements of the upwelling polarized radiance. Synergistic use of RSP and AATS-14 data from ALIVE has been made to retrieve aerosol and surface properties (Knobelspiesse et al. 2008; Waquet et al. 2009).

5. The ARM Aerial Facility (2007–)

A meeting between ARM Science Team members interested in airborne observations was held at the 2006 ARM Science Team Meeting to discuss future directions for the ARM-UAV program. At that time, it was recognized that ARM-UAV had reached a mature state where the original long-term goal of making routine observations could be pursued in coordination with supporting IOPs, both measurement strategies designed to answer questions addressing the largest source of uncertainty in global warming: namely, the interaction of clouds with solar and thermal radiation. McFarquhar (2006) developed a whitepaper describing new directions for ARM-UAV: namely, the development of a routine observational program with piloted and unpiloted aerospace vehicles. The goal of the new program, which was first called the ARM Aerial Vehicle Program (AAVP) before being renamed the ARM Aerial Facility, was to support the following three types of activities:

- Routine observations of cloud, aerosol, and radiative properties;
- Participation in IOPs designed to contribute to the fundamental understanding of cloud properties and processes;
- Foster an instrument development program whereby miniaturized in situ and remote sensing instruments would be purchased or developed, with the small size of the instruments ultimately allowing them to be used on UAV platforms.

The whitepaper also outlined several research questions that would be pertinent for the AAF to investigate.

The AAF was formally established in October 2006 with the mandate of executing all future ARM aircraft campaigns under one organizational umbrella as an integral part of the ARM Climate Research Facility (Ackerman et al. 2016, chapter 3). Operation of AAF was awarded to the Pacific Northwest National Laboratory (PNNL) following a competitive process. James Mather served as AAF technical director briefly before transferring this role to Beat Schmid in 2007. Greg McFarquhar continued as the chief scientist of AAF until 2009, when that role was abolished.

As part of the instrument development effort, AAF cosponsored a workshop in 2008 called "Advances in Airborne Instrumentation for Measuring Aerosol, Cloud, Radiation and Atmospheric State Parameters" (McFarquhar et al. 2011b). The workshop identified state-of-the-art measurement techniques, emerging instruments, and technologies that would be flight ready within approximately one year as well as deficiencies in airborne instrumentation slowing research, and it promoted dialogue between scientists and instrument makers on measurement needs. One of the outcomes of the workshop was the issue of a call for maturation and hardening of aircraft instruments, which resulted in the awarding of five different proposals to mitigate some of the measurement gaps identified in the workshop.

A detailed description of the achievements of AAF is presented by Schmid et al. (2014), so only a very brief summary is given here. Schmid et al. (2014) also provide a complete list of publications resulting from the campaigns carried out by AAF.

a. ARM Airborne Carbon Measurement Experiment (ACME)

The still ongoing airborne carbon measurements at SGP started in 2002 when a flask sampler was added to the Cessna 172 used for the IAP campaign discussed earlier. The carbon measurements were enhanced gradually, and, in 2008, the Cessna 206 became the dedicated aircraft for ACME after the aerosol instruments had been removed from its payload. A review of the 10-yr record of carbon cycle gases above SGP is presented by Biraud et al. (2013). The measurements also have been used to validate ground- and satellite-based column measurements of CO_2 (e.g., Wunch et al. 2011; Kulawik et al. 2013; Kuai et al. 2013) as well as airborne lidar measurements of CO_2 profiles (Abshire et al. 2010) and an earth system model (Keppel-Aleks et al. 2013).

b. The AAF virtual hangar

With the exception of the small aircraft used for IAP and ACME, AAF did not initially have a dedicated aircraft. In addition, only a small number of aircraft instruments (inherited from the ARM-UAV program) were owned by AAF. In this virtual hangar mode, AAF successfully carried out several campaigns working with organizations and investigators who provided their research aircraft and most of the instrumentation. These included the Cloud and Land Surface Interaction Campaign (CLASIC; Oklahoma, 2007), the Indirect and Semi-Direct Aerosol Campaign (ISDAC; Alaska, 2008), the Routine AAF Clouds with Low Optical Water Depths (CLOWD) Optical Radiative Observations (RACORO; Oklahoma, 2009) and the Small Particles in Cirrus (SPARTICUS; Oklahoma, 2010) campaign.

CLASIC and ISDAC were both large campaigns involving several aircraft from different agencies. While CLASIC research was impacted by the unusually wet weather during the campaign (Lamb et al. 2012) and resource challenges, a large number of publications resulted from ISDAC [McFarquhar et al. (2011a); see references in Schmid et al. (2014)]. The relatively polluted conditions expected and found in ISDAC in April 2008 nicely contrasted the relatively pristine conditions observed during M-PACE in October 2004. This allowed numerous authors to investigate the roles of aerosols in modifying cloud properties (Fan et al. 2011; Larson et al. 2011; Liu et al. 2011; Jackson et al. 2012; McFarquhar et al. 2013).

RACORO and SPARTICUS represent, to the best of our knowledge, the first and only 6-month routine cloud sampling campaigns ever carried out, creating an extensive database of properties of water and ice clouds, respectively (see, e.g., Vogelmann et al. 2012; Deng et al. 2013; Lawson 2011). These projects represent a new paradigm for collection of airborne data, and required a different operating model from typical, short-term, intensive aircraft field programs. Thus, RACORO adopted the IAP program's practice of simplifying both operations and instrument selection, successfully applying these practices to a collection of in situ cloud as well as aerosol data. SPARTICUS then applied the same paradigms to the collection of ice crystal data in cirrus.

c. The current AAF

In 2009, AAF started managing operations of the Battelle-owned Gulfstream I (G-1) large twin-turboprop research aircraft (Fig. 10-2). Furthermore, the American Recovery and Reinvestment Act of 2009 provided funding for the procurement of over 20 new instruments to be used aboard the G-1 and AAF contracted aircraft. AAF is also engaged in the maturation and testing of newly developed airborne sensors to help foster the next generation of airborne instruments (e.g., Lu et al. 2012; Dunagan et al. 2013; Leen et al. 2013).

The AAF has matured into a facility with extensive capabilities available for research. At this time, AAF has over 50 state-of-the-art instruments at its disposal, which typically is further augmented by leading-edge guest instrumentation based on science requirements (see Schmid et al. 2014). As an example, each of the AAF cloud probes is now equipped with knife-edge tips designed to reduce shattering of ice crystals and droplets into the sampling volume, which was a problem that plagued the TWP-ICE (see discussion above) and many other missions carried out during the last 20 years (e.g., Korolev et al. 2013; Jackson and McFarquhar 2014). The G-1 has become the dedicated aircraft for AAF and has undergone extensive upgrades to carry out a broad array of campaigns. AAF campaigns carried out to date with the G-1 include the Carbonaceous Aerosols and Radiative Effects Study (CARES; Zaveri et al. 2012), the Two-Column Aerosol Project (TCAP; Shinozuka et al. 2014), the Biomass Burning Observation Project (BBOP), and Green Ocean Amazon (GOAmazon).

In the future, AAF will continue to carry out missions proposed to and approved by the ARM science board (Ackerman et al. 2016, chapter 3). The aircraft platform and instrumentation will be driven by science needs outlined in the proposal. The AAF will continue to use contracted aircraft where warranted, particularly for cirrus research, where a higher operating ceiling is required. Approved future missions currently include the ARM Cloud Aerosol Precipitation Experiment (ACAPEX) in California, continued routine flights over the ARM site in Oklahoma, routine flights between the ARM Barrow and Oliktok Point sites at the North Slope of Alaska, and flights with small UAVs at Oliktok Point.

6. Summary

Focusing on cloud, aerosol, radiation, and gas phase measurements and their interactions, ARM has used aircraft to enhance its surface-based measurements almost since its inception in 1992.

FIG. 10-2. The G-1 aircraft in flight equipped for a cloud mission.

The separately funded ARM Unmanned Aerospace Vehicle (UAV) program carried out 12 missions between 1993 and 2006 relying on both UAVs and piloted aircraft. ARM-UAV has really pioneered the use of UAVs for scientific research. Particularly noteworthy is the successful completion of a >26-h UAV flight on 5 October 1996, which marked the first >24-h mission of a UAV gathering data for science applications. Equally impressive are high-altitude flights in 1999 with the Altus II above Hawaii. This marked the first time a UAV performed nadir-looking remote sensing of cirrus clouds by flying at altitudes above all clouds.

ARM also has funded aircraft campaigns outside the UAV program, such as the 2003 Aerosol Intensive Operation Period (AIOP) and the 2005 Aerosol Lidar Validation Experiment (ALIVE). In 2000, ARM also started twice-per-week routine flights with a small aircraft measuring vertical profiles of aerosol optical properties (through 2007). Measurements of carbon cycle gases were added to the routine flights in 2002, were enhanced in 2007, and continue to this date as the ARM Airborne Carbon Measurement Experiment (ACME).

The ARM Aerial Facility (initially named the ARM Aerial Vehicle Program) was formally established in October 2006 with the mandate of executing all future ARM aircraft campaigns under one organizational umbrella. Initially starting out as a virtual hangar only, AAF has matured into a facility for the ARM science community with extensive capabilities, including two dedicated aircraft and over 50 state-of-the-art instruments for atmospheric science research. Since 2006, AAF has carried out nine aircraft campaigns enabling research on aerosols, clouds, aerosol–cloud interactions, and trace gases. During this period, AAF has produced numerous scientific and logistical firsts, including the 8-yr in situ aerosol optical properties record (IAP), the 12-yr (and growing) carbon cycle gas profile record (ACME), and two 6-month airborne cloud sampling campaigns (RACORO and SPARTICUS).

REFERENCES

Abshire, J. B., and Coauthors, 2010: Pulsed airborne lidar measurements of atmospheric CO_2 column absorption. *Tellus*, **62B**, 770–783, doi:10.1111/j.1600-0889.2010.00502.x.

Ackerman, T. P., D. M. Flynn, and R. T. Marchand, 2003: Quantifying the magnitude of anomalous solar absorption. *J. Geophys. Res.*, **108**, 4273, doi:10.1029/2002JD002674.

——, T. S. Cress, W. R. Ferrell, J. H. Mather, and D. D. Turner, 2016: The programmatic maturation of the ARM Program. *The Atmospheric Radiation Measurement (ARM) Program: The First 20 Years, Meteor. Monogr.*, No. 57, Amer. Meteor. Soc., doi:10.1175/AMSMONOGRAPHS-D-15-0054.1.

Andrews, E., P. J. Sheridan, J. A. Ogren, and R. Ferrare, 2004: In situ aerosol profiles over the Southern Great Plains cloud and radiation test bed site: 1. Aerosol optical properties. *J. Geophys. Res.*, **109**, D06208, doi:10.1029/2003JD004025.

——, and Coauthors, 2006: Comparison of methods for deriving aerosol asymmetry parameter. *J. Geophys. Res.*, **111**, D05S04, doi:10.1029/2004JD005734.

——, P. J. Sheridan, and J. A. Ogren, 2011: Seasonal differences in the vertical profiles of aerosol optical properties over rural Oklahoma. *Atmos. Chem. Phys.*, **11**, 10 661–10 676, doi:10.5194/acp-11-10661-2011.

Arnott, W. P., and Coauthors, 2006: Photoacoustic insight for aerosol light absorption aloft from meteorological aircraft and comparison with particle soot absorption photometer measurements: DOE Southern Great Plains Climate Research Facility and the coastal stratocumulus imposed perturbation experiments. *J. Geophys. Res.*, **111**, D05S02, doi:10.1029/2005JD005964.

Asano, S., A. Uchiyama, A. Yamazaki, and K. Kuchiki, 2004: Solar radiation budget from the MRI radiometers for clear and cloudy air columns within ARESE II. *J. Atmos. Sci.*, **61**, 3082–3096, doi:10.1175/JAS-3288.1.

Biraud, S. C., M. S. Torn, J. R. Smith, C. Sweeney, W. J. Riley, and P. P. Tans, 2013: A multi-year record of airborne CO_2 observations in the US Southern Great Plains. *Atmos. Meas. Tech.*, **6**, 751–763, doi:10.5194/amt-6-751-2013.

Cairns, B., E. E. Russell, and L. D. Travis, 1999: The Research Scanning Polarimeter: Calibration and ground-based measurements. *Polarization: Measurement, Analysis, and Remote Sensing II*, D. H. Goldstein and D. B. Chenault, Eds., International Society for Optical Engineering (SPIE Proceedings, Vol. 3754), 186–197, doi:10.1117/12.366329.

Cess, R. D., and Coauthors, 1995: Absorption of solar radiation by clouds: Observations versus models. *Science*, **267**, 496–499, doi:10.1126/science.267.5197.496.

Crowley, P. A., and J. Vitko Jr., 1994: The Atmospheric Radiation Measurement Unmanned Aerospace Vehicle Program: An overview. *Proc. Third Atmospheric Radiation Measurement (ARM) Science Team Meeting*, Norman, OK, U.S. Department of Energy, 345–346. [Available online at https://www.arm.gov/publications/proceedings/conf03/extended_abs/crowley_pa.pdf.]

de Boer, G., S. E. Bauer, T. Toto, S. Menon, and A. M. Vogelmann, 2013: Evaluation of aerosol–cloud interaction in the GISS ModelE using ARM observations. *J. Geophys. Res.*, **118**, 6383–6395, doi:10.1002/jgrd.50460.

Deng, M., G. G. Mace, Z. Wang, and R. P. Lawson, 2013: Evaluation of several A-Train ice cloud retrieval products with in-situ measurements collected during the SPARTICUS campaign. *J. Appl. Meteor. Climatol.*, **52**, 1014–1030, doi:10.1175/JAMC-D-12-054.1.

Dunagan, S., and Coauthors, 2013: Spectrometer for Sky-Scanning Sun-Tracking Atmospheric Research (4STAR): Instrument technology. *Remote Sens. Environ.*, **5**, 3872–3895, doi:10.3390/rs5083872.

Ellingson, R. G., and T. P. Tooman, 1999: Science and experiment plan for the second Atmospheric Radiation Measurement enhanced shortwave experiment. Sandia National Laboratories, Livermore, CA.

Fan, J., S. Ghan, M. Ovchinnikov, X. Liu, P. J. Rasch, and A. Korolev, 2011: Representation of Arctic mixed-phase clouds and the Wegener–Bergeron–Findeisen process in climate models: Perspectives from a cloud-resolving study. *J. Geophys. Res.*, **116**, D00T07, doi:10.1029/2010JD015375.

Ferrare, R., G. Feingold, S. Ghan, J. Ogren, B. Schmid, S. E. Schwartz, and P. Sheridan, 2006a: Preface to special section: Atmospheric Radiation Measurement Program May 2003 Intensive Operations Period examining aerosol properties and radiative influences. *J. Geophys. Res.*, **111**, D05S01, doi:10.1029/2005JD006908.

——, and Coauthors, 2006b: Evaluation of daytime measurements of aerosols and water vapor made by an operational Raman lidar over the Southern Great Plains. *J. Geophys. Res.*, **111**, D05S08, doi:10.1029/2005JD005836.

Fridlind, A. M., A. S. Ackerman, G. M. McFarquhar, G. Zhang, M. R. Poellot, P. J. DeMott, A. J. Prenni, and A. J. Heymsfield, 2007: Ice properties of single-layer stratocumulus during the Mixed-Phase Arctic Cloud Experiment (MPACE): 2. Model results. *J. Geophys. Res.*, **112**, D24202, doi:10.1029/2007JD008646.

Gasparini, R., D. R. Collins, E. Andrews, P. J. Sheridan, J. A. Ogren, and J. G. Hudson, 2006: Coupling aerosol size distributions and size-resolved hygroscopicity to predict humidity-dependent optical properties and cloud condensation nuclei spectra. *J. Geophys. Res.*, **111**, D05S13, doi:10.1029/2005JD006092.

Ghan, S. J., and Coauthors, 2006: Use of in situ cloud condensation nuclei, extinction, and aerosol size distribution measurements to test a method for retrieving cloud condensation nuclei profiles from surface measurements. *J. Geophys. Res.*, **111**, D05S10, doi:10.1029/2004JD005752.

Hallar, A. G., and Coauthors, 2006: Atmospheric Radiation Measurements Aerosol Intensive Operating Period: Comparison of aerosol scattering during coordinated flights. *J. Geophys. Res.*, **111**, D05S09, doi:10.1029/2005JD006250.

Jackson, R. C., and G. M. McFarquhar, 2014: An assessment of the impact of antishattering tips and artifact removal techniques on bulk cloud ice microphysical and optical properties measured by the 2D Cloud Probe. *J. Atmos. Oceanic Technol.*, **31**, 2131–2144, doi:10.1175/JTECH-D-14-00018.1.

——, and Coauthors, 2012: The dependence of ice microphysics on aerosol concentration in Arctic mixed-phase stratus clouds during ISDAC and M-PACE. *J. Geophys. Res.*, **117**, D15, doi:10.1029/2012JD017668.

Kassianov, E. I., C. J. Flynn, T. P. Ackerman, and J. C. Barnard, 2007: Aerosol single-scattering albedo and asymmetry parameter from MFRSR observations during the ARM Aerosol IOP 2003. *Atmos. Chem. Phys.*, **7**, 3341–3351, doi:10.5194/acp-7-3341-2007.

Keppel-Aleks, G., and Coauthors, 2013: Atmospheric carbon dioxide variability in the Community Earth System Model: Evaluation and transient dynamic during the twentieth and twenty-first centuries. *J. Climate*, **26**, 4447–4475, doi:10.1175/JCLI-D-12-00589.1.

Klein, S. A., and Coauthors, 2009: Intercomparison of model simulations of mixed-phase clouds observed during the ARM Mixed-Phase Arctic Cloud Experiment. I: Single-layer cloud. *Quart. J. Roy. Meteor. Soc.*, **135**, 979–1002, doi:10.1002/qj.416.

Knobelspiesse, K. D., B. Cairns, B. Schmid, M. O. Román, and C. B. Schaaf, 2008: Surface BRDF estimation from an aircraft compared to MODIS and ground estimates at the Southern Great Plains site. *J. Geophys. Res.*, **113**, D20105, doi:10.1029/2008JD010062.

Korolev, A., E. Emery, and K. Creelman, 2013: Modification and tests of particle probe tips to mitigate effects of ice shattering. *J. Atmos. Oceanic Technol.*, **30**, 690–708, doi:10.1175/JTECH-D-12-00142.1.

Kuai, L., and Coauthors, 2013: Profiling tropospheric CO_2 using Aura TES and TCCON instruments. *Atmos. Meas. Tech.*, **6**, 63–79, doi:10.5194/amt-6-63-2013.

Kulawik, S. S., and Coauthors, 2013: Comparison of improved Aura Tropospheric Emission Spectrometer CO_2 with HIPPO and SGP aircraft profile measurements. *Atmos. Chem. Phys.*, **13**, 3205–3225, doi:10.5194/acp-13-3205-2013.

Lamb, P. J., D. H. Portis, and A. Zangvil, 2012: Investigation of large-scale atmospheric moisture budget and land surface interactions over U.S. Southern Great Plains including for CLASIC (June 2007). *J. Hydrometeor.*, **13**, 1719–1738, doi:10.1175/JHM-D-12-01.1.

Larson, V. E., B. J. Nielsen, J. Fan, and M. Ovchinnikov, 2011: Parameterizing correlations between hydrometeor species in mixed-phase Arctic clouds. *J. Geophys. Res.*, **116**, D00T02, doi:10.1029/2010JD015570.

Lawson, R. P., 2011: Effects of ice particles shattering on the 2D-S probe. *Atmos. Meas. Tech.*, **4**, 1361–1381, doi:10.5194/amt-4-1361-2011.

Leen, B. L., X.-Y. Yu, M. Gupta, D. Baer, J. M. Hubbe, C. D. Kluzek, J. M. Tomlinson, and M. R. Hubbell, 2013: Fast in situ airborne measurement of ammonia using a mid-infrared off-axis ICOS spectrometer. *Environ. Sci. Technol.*, **47**, 10 446–10 453, doi:10.1021/es401134u.

Li, Z., A. P. Trishchenko, H. W. Barker, G. L. Stephens, and P. Partain, 1999: Analyses of Atmospheric Radiation Measurement (ARM) Program's Enhanced Shortwave Experiment (ARESE) multiple data sets for studying cloud absorption. *J. Geophys. Res.*, **104**, 19 127–19 134, doi:10.1029/1999JD900308.

Liou, K. N., Y. Gu, Q. Yue, and G. McFarquhar, 2008: On the correlation between ice water content and ice crystal size and its application to radiative transfer and general circulation models. *Geophys. Res. Lett.*, **35**, L13805, doi:10.1029/2008GL033918.

Liu, X., and Coauthors, 2011: Testing cloud microphysics parameterizations in NCAR CAM5 with ISDAC and M-PACE observations. *J. Geophys. Res.*, **116**, D00T11, doi:10.1029/2011JD015889.

Long, C. N., J. H. Mather, and T. P. Ackerman, 2016: The ARM Tropical Western Pacific (TWP) sites. *The Atmospheric Radiation Measurement (ARM) Program: The First 20 Years, Meteor. Monogr.*, No. 57, Amer. Meteor. Soc., doi:10.1175/AMSMONOGRAPHS-D-15-0024.1.

Lu, J., R. Shaw, and W. Yang, 2012: Improved particle size estimation in digital holography via sign matched filtering. *Opt. Express*, **20**, 12 666–12 674, doi:10.1364/OE.20.012666.

Marshak, A., A. Davis, W. Wiscombe, and R. Cahalan, 1997: Inhomogeneity effects on cloud shortwave absorption measurements: Two-aircraft simulations. *J. Geophys. Res.*, **102**, 16 619–16 637, doi:10.1029/97JD01153.

——, W. Wiscombe, A. Davis, L. Oreopoulous, and R. Cahalan, 1999: On the removal of the effect of horizontal fluxes in two-aircraft measurements of cloud absorption. *Quart. J. Roy. Meteor. Soc.*, **125**, 2153–2170, doi:10.1002/qj.49712555811.

Mather, J. H., D. D. Turner, and T. P. Ackerman, 2016: Scientific maturation of the ARM Program. *The Atmospheric Radiation Measurement (ARM) Program: The First 20 Years, Meteor. Monogr.*, No. 57, Amer. Meteor. Soc., doi:10.1175/AMSMONOGRAPHS-D-15-0053.1.

May, P. T., J. H. Mather, G. Vaughan, C. Jacob, G. M. McFarquhar, and K. N. Bower, 2008: The Tropical Warm Pool International Cloud Experiment (TWPICE). *Bull. Amer. Meteor. Soc.*, **89**, 629–645, doi:10.1175/BAMS-89-5-629.

McComiskey, A., and R. A. Ferrare, 2016: Aerosol physical and optical properties and processes in the ARM Program. *The Atmospheric Radiation Measurement (ARM) Program: The First 20 Years, Meteor. Monogr.*, No. 57, Amer. Meteor. Soc., doi:10.1175/AMSMONOGRAPHS-D-15-0028.1.

McCord, R., and J. W. Voyles, 2016: The ARM data system and archive. *The Atmospheric Radiation Measurement (ARM) Program: The First 20 Years, Meteor. Monogr.*, No. 57, Amer. Meteor. Soc., doi:10.1175/AMSMONOGRAPHS-D-15-0043.1.

McFarquhar, G. M., 2006: The development of a routine observational program with piloted and unpiloted aerospace vehicles: New directions for ARM UAV. ARM Tech. Rep., 11 pp. [Available online at http://www.atmos.illinois.edu/~mcfarq/avpp.whitepaperoverview.pdf.]

——, J. Um, M. Freer, D. Baumgardner, G. L. Kok, and G. Mace, 2007a: The importance of small ice crystals to cirrus properties: Observations from the Tropical Western Pacific International Cloud Experiment (TWP-ICE). *Geophys. Res. Lett.*, **34**, L13803, doi:10.1029/2007GL029865.

——, G. Zhang, M. R. Poellot, G. L. Kok, R. McCoy, T. Tooman, and A. J. Heymsfield, 2007b: Ice properties of single-layer stratocumulus during the Mixed-Phase Arctic Cloud Experiment: 1. Observations. *J. Geophys. Res.*, **112**, D24201, doi:10.1029/2007JD008633.

——, and Coauthors, 2011a: Indirect and Semi-Direct Aerosol Campaign. *Bull. Amer. Meteor. Soc.*, **92**, 183–201, doi:10.1175/2010BAMS2935.1.

——, B. Schmid, A. Korolev, J. A. Ogren, P. B. Russell, J. Tomlinson, D. D. Turner, and W. Wiscombe, 2011b: Airborne instrumentation needs for climate and atmospheric research. *Bull. Amer. Meteor. Soc.*, **92**, 1193–1196, doi:10.1175/2011BAMS3180.1.

——, J. Um, and R. C. Jackson, 2013: Small cloud particle shapes in mixed-phase clouds. *J. Appl. Meteor. Climatol.*, **52**, 1277–1293, doi:10.1175/JAMC-D-12-0114.1.

Michalsky, J. J., and C. N. Long, 2016: ARM solar and infrared broadband and filter radiometery. *The Atmospheric Radiation Measurement (ARM) Program: The First 20 Years, Meteor. Monogr.*, No. 57, Amer. Meteor. Soc., doi:10.1175/AMSMONOGRAPHS-D-15-0031.1.

——, Q. Min, J. Barnard, R. Marchand, and P. Pilewskie, 2003: Simultaneous spectral albedo measurements near the Atmospheric Radiation Measurement Southern Great Plains (ARM SGP) central facility. *J. Geophys. Res.*, **108**, 4254, doi:10.1029/2002JD002906.

——, and Coauthors, 2006: Shortwave radiative closure studies for clear skies during the Atmospheric Radiation Measurement 2003 Aerosol Intensive Observation Period. *J. Geophys. Res.*, **111**, D14S90, doi:10.1029/2005JD006341.

Mitchell, D. L., P. Rasch, D. Ivanova, G. M. McFarquhar, and T. Nousianen, 2008: The impact of controversial small ice crystals on GCM simulations. *Geophys. Res. Lett.*, **35**, L09806, doi:10.1029/2008GL033552.

Morrison, H., J. O. Pinto, J. A. Curry, and G. M. McFarquhar, 2008: Sensitivity of modeled arctic mixed-phase stratocumulus to cloud condensation and ice nuclei over regionally varying surface conditions. *J. Geophys. Res.*, **113**, D05203, doi:10.1029/2007JD008729.

——, and Coauthors, 2009: Intercomparison of model simulations of mixed-phase clouds observed during the ARM Mixed-Phase Arctic Cloud Experiment. II: Multilayer cloud. *Quart. J. Roy. Meteor. Soc.*, **135**, 1003–1019, doi:10.1002/qj.415.

Newsom, R. K., D. D. Turner, B. Mielke, M. Clayton, R. Ferrare, and C. Sivamaran, 2009: Simultaneous analog and photon counting detection for Raman lidar. *Appl. Opt.*, **48**, 3903–3914, doi:10.1364/AO.48.003903.

O'Hirok, W., and C. Gautier, 2003: Absorption of shortwave radiation in a cloudy atmosphere: Observed and theoretical estimates during the second Atmospheric Radiation Measurement Enhanced Shortwave Experiment (ARESE). *J. Geophys. Res.*, **108**, 4412, doi:10.1029/2002JD002818.

——, ——, and P. Ricchiazzi, 2000: Spectral signature of column solar radiation absorption during the Atmospheric Radiation Measurement Enhanced Shortwave Experiment (ARESE). *J. Geophys. Res.*, **105**, 17 471–17 480, doi:10.1029/2000JD900190.

Pahlow, M., and Coauthors, 2006: Comparison between lidar and nephelometer measurements of aerosol hygroscopicity at the Southern Great Plains Atmospheric Radiation

Measurement site. *J. Geophys. Res.*, **111**, D05S15, doi:10.1029/2004JD005646.

Pilewskie, P., and F. P. J. Valero, 1995: Direct observations of excess solar absorption by clouds. *Science*, **267**, 1626–1629, doi:10.1126/science.267.5204.1626.

Ramanathan, V., B. Subasilar, G. J. Zhang, W. Conant, R. D. Cess, J. T. Kiehl, H. Grassl, and L. Shi, 1995: Warm pool heat budget and shortwave cloud forcing: A missing physics? *Science*, **267**, 499–503, doi:10.1126/science.267.5197.499.

Schmid, B., and Coauthors, 2006: How well do state-of-the-art techniques measuring the vertical profile of tropospheric aerosol extinction compare? *J. Geophys. Res.*, **111**, D05S07, doi:10.1029/2005JD005837.

——, and Coauthors, 2009: Validation of aerosol extinction and water vapor profiles from routine Atmospheric Radiation Measurement Program Climate Research Facility measurements. *J. Geophys. Res.*, **114**, D22207, doi:10.1029/2009JD012682.

——, and Coauthors, 2014: The DOE ARM Aerial Facility. *Bull. Amer. Meteor. Soc.*, **95**, 723–742, doi:10.1175/BAMS-D-13-00040.1.

Shinozuka, Y., and Coauthors, 2014: Correction to "Hyperspectral aerosol optical depths from TCAP flights." *J. Geophys. Res.*, **119**, 1692–1693, doi:10.1002/jgrd.51089.

Shupe, M. D., P. Kollias, O. G. Persson, and G. M. McFarquhar, 2008: Vertical motions in Arctic mixed-phase stratus. *J. Atmos. Sci.*, **65**, 1304–1322, doi:10.1175/2007JAS2479.1.

Skeie, R. B., T. Berntsen, G. Myhre, C. A. Pedersen, J. Ström, S. Gerland, and J. A. Ogren, 2011: Black carbon in the atmosphere and snow, from pre-industrial times until present. *Atmos. Chem. Phys.*, **11**, 6809–6836, doi:10.5194/acp-11-6809-2011.

Smith, W. L., Jr., L. Nguyen, and P. Minnis, 1997: Cloud radiative forcing derived from ARM surface and satellite measurements during ARESE and the Spring ARM/UAV IOP. *Proc. Ninth Conf. on Atmospheric Radiation*, Long Beach, CA, Amer. Meteor. Soc., 1–4.

Solomon, A., H. Morrison, O. Persson, M. D. Shupe, and J.-W. Bao, 2009: Investigation of microphysical parameterizations of snow and ice in arctic clouds during M-PACE through model–observation comparisons. *Mon. Wea. Rev.*, **137**, 3110–3128, doi:10.1175/2009MWR2688.1.

Stephens, G. L., and Coauthors, 2000: The Department of Energy's Atmospheric Radiation Measurement (ARM) Unmanned Aerospace Vehicle (UAV) Program. *Bull. Amer. Meteor. Soc.*, **81**, 2915–2937, doi:10.1175/1520-0477(2000)081<2915:TDOESA>2.3.CO;2.

Strawa, A. W., and Coauthors, 2006: Comparison of in situ aerosol extinction and scattering coefficient measurements made during the Aerosol Intensive Operating Period. *J. Geophys. Res.*, **111**, D05S03, doi:10.1029/2005JD006056.

Tooman, T., 1997: SERDP: Atmospheric remote sensing and assessment program—Final report, Part 1: The lower atmosphere. Sandia National Laboratories Tech. Rep. SAND97-8221, 92 pp.

Turner, D. D., J. E. M. Goldsmith, and R. A. Ferrare, 2016a: Development and applications of the ARM Raman lidar. *The Atmospheric Radiation Measurement (ARM) Program: The First 20 Years, Meteor. Monogr.*, No. 57, Amer. Meteor. Soc., doi:10.1175/AMSMONOGRAPHS-D-15-0026.1.

——, E. J. Mlawer, and H. E. Revercomb, 2016b: Water vapor observations in the ARM Program. *The Atmospheric Radiation Measurement (ARM) Program: The First 20 Years, Meteor. Monogr.*, No. 57, Amer. Meteor. Soc., doi:10.1175/AMSMONOGRAPHS-D-15-0025.1.

Valero, F. P. J., and P. Pilewskie, 1992: Latitudinal survey of spectral optical depths of the Pinatubo volcanic cloud-derived particle sizes, columnar mass loadings, and effects on planetary albedo. *Geophys. Res. Lett.*, **19**, 163–166, doi:10.1029/92GL00074.

——, W. J. Gore, and L. P. Giver, 1982: Radiative flux measurements in the troposphere. *Appl. Opt.*, **21**, 831–838, doi:10.1364/AO.21.000831.

——, T. P. Ackerman, and W. J. Y. Gore, 1989: The effects of the Arctic haze as determined from airborne radiometric measurements during AGASP II. *J. Atmos. Chem.*, **9**, 225–244, doi:10.1007/BF00052834.

——, S. K. Pope, R. G. Ellingson, A. W. Strawa, and J. Vitko Jr., 1996: Determination of clear-sky radiative flux profiles, heating rates, and optical depths using unmanned aerospace vehicles as a platform. *J. Atmos. Oceanic Technol.*, **13**, 1024–1030, doi:10.1175/1520-0426(1996)013<1024:DOCSRF>2.0.CO;2.

——, A. Bucholtz, B. C. Bush, S. Pope, W. Collins, P. Flatau, A. Strawa, and W. Gore, 1997: The Atmospheric Radiation Measurements Enhanced Shortwave Experiment (ARESE): Experimental and data details. *J. Geophys. Res.*, **102**, 29 929–29 937, doi:10.1029/97JD02434.

——, and Coauthors, 2003: Absorption of solar radiation by the clear and cloudy atmosphere during the Atmospheric Radiation Measurement Enhanced Shortwave Experiments (ARESE) I and II: Observations and models. *J. Geophys. Res.*, **108**, 4016, doi:10.1029/2001JD001384.

Verlinde, J., and Coauthors, 2007: The Mixed-Phase Arctic Cloud Experiment (M-PACE). *Bull. Amer. Meteor. Soc.*, **88**, 205–221, doi:10.1175/BAMS-88-2-205.

——, B. Zak, M. D. Shupe, M. Ivey, and K. Stamnes, 2016: The ARM North Slope of Alaska (NSA) sites. *The Atmospheric Radiation Measurement (ARM) Program: The First 20 Years, Meteor. Monogr.*, No. 57, Amer. Meteor. Soc., doi:10.1175/AMSMONOGRAPHS-D-15-0023.1.

Vogelmann, A., and Coauthors, 2012: RACORO extended-term aircraft observations of boundary layer clouds. *Bull. Amer. Meteor. Soc.*, **93**, 861–878, doi:10.1175/BAMS-D-11-00189.1.

Wang, J., and Coauthors, 2007: Observation of ambient aerosol particle growth due to in-cloud processes within boundary layers. *J. Geophys. Res.*, **112**, D14207, doi:10.1029/2006JD007989.

Waquet, F., B. Cairns, K. Knobelspiesse, J. Chowdhary, L. D. Travis, B. Schmid, and M. I. Mishchenko, 2009: Polarimetric remote sensing of aerosols over land. *J. Geophys. Res.*, **114**, D01206, doi:10.1029/2008JD010619.

Wunch, D., and Coauthors, 2011: A method for evaluating bias in global measurements of CO_2 total columns from space. *Atmos. Chem. Phys.*, **11**, 12 317–12 337, doi:10.5194/acp-11-12317-2011.

Zaveri, R. A., and Coauthors, 2012: Overview of the 2010 Carbonaceous Aerosols and Radiative Effects Study (CARES). *Atmos. Chem. Phys.*, **12**, 7647–7687, doi:10.5194/acp-12-7647-2012.

Zender, C., B. Bush, S. Pope, A. Bucholtz, W. Collins, J. Kiehl, F. Valero, and J. Vitko, 1997: Atmospheric absorption during the Atmospheric Radiation Measurement (ARM) Enhanced Shortwave Experiment (ARESE). *J. Geophys. Res.*, **102**, 29 901–29 915, doi:10.1029/97JD01781.

Chapter 11

The ARM Data System and Archive

RAYMOND MCCORD
Oak Ridge National Laboratory, Oak Ridge, Tennessee

JIMMY VOYLES
Pacific Northwest National Laboratory, Richland, Washington

1. Introduction

Every observationally based research program needs a way to collect data from instruments, convert the data from its raw format into a more usable format, apply quality control, process it into higher-order data products, store the data, and make the data available to its scientific community. This data flow is illustrated pictorially in Fig. 11-1. These are the basic requirements of any scientific data system, and ARM's data system would have to address these requirements and more. This chapter provides one view of the development of the ARM data system, which includes the ARM Data Archive, and some of the notable decisions that were made along the way.

It is impossible to talk about the development of the ARM data system without first placing it in context of the evolution of computers and associated infrastructure. At the start of ARM in 1990, the Intel 486 computer with 25-MHz processing speed had just been released, internal hard drives were about 40–100 MB in size, national networks were very loosely connected but had significantly less reliability and capability than today's Internet, and the World Wide Web technology was an experimental concept. However, it was already clear that computers and associated technology were developing at a rapid pace and that any design of the ARM data system would have to be flexible enough to accommodate new technology as it came along. Balancing the need for flexibility was the assumption that ARM was initially envisioned to be a 10-yr research program (Stokes 2016, chapter 2; Cress and Sisterson 2016, chapter 5), and thus at the beginning it was anticipated that the data system might only go through one generation of updates. The growth of the program into a multidecadal program with significantly increasing computational and storage demands placed tremendous stress upon the ARM data system, causing it to be reorganized and reconfigured several times. Today, the data system continues to be a living system, evolving as needed to support the ARM mission and user community.

Figure 11-1 shows a phased and linear view of the ARM data flow. However, the data flow is related to the data life cycle, which occurs when researchers make discoveries with data and identify new science questions and data needs. These questions often require new measurements, new sampling schemes, new advanced data products, and new evaluation procedures (see Fig. 11-2). Because of the close coupling of the science and infrastructure in the ARM Program, the data system was continually adapted to support the new data requirements. This cyclic pattern of change is common in scientific data systems but is often insignificant when the project duration is only 3–5 yr. The ARM data system is unusual since it has undergone a 20-yr history of continual evolution to support the significant increases in data volume, types of data, and configurations of field sites. In addition, the interactions between the researchers and the data system changed in ways that could not be anticipated at the start of ARM.

2. Initial requirements

Data system–related activities for ARM were a major component of the programmatic scope from the very

Corresponding author address: Raymond McCord, Oak Ridge National Laboratory, 1 Bethel Valley Rd., Oak Ridge, TN 37831.
E-mail: mccordra@ornl.gov

DOI: 10.1175/AMSMONOGRAPHS-D-15-0043.1

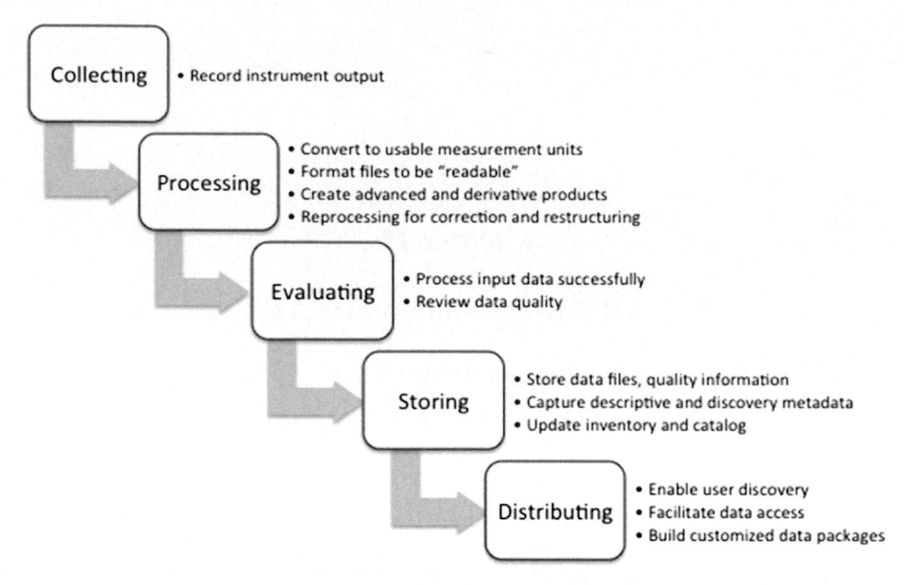

FIG. 11-1. Functional scope of the ARM data system.

beginning of the planning and implementation (Cress and Sisterson 2016, chapter 5). The major components of the original ARM Program plan (U.S. Department of Energy 1990; ARM 2016, appendix A) included

- the ARM science team (researchers);
- instrument team (acquisition, implementation, and development);
- site operations (development and operations); and
- data systems (design, development, and operations).

In this discussion, data system refers to all portions of the hardware, software, data processing, and data flow. The ARM Program's longstanding research objective of comparing atmospheric process models with measurements for improving models was critically dependent on the data system. Early inclusion of the data system in the programmatic planning process enabled the ARM to define durable requirements and goals (Cress and Sisterson 2016, chapter 5). This very early planning strategy assured readiness of the systems when the first site was operational.

The early period of design and implementation of the data system spanned 1990–93. The early requirements for the ARM data system were divided into the following:

- data life cycle: the flow and processing of data (Table 11-1);
- data heterogeneity: structure, size, and source (Table 11-2); and
- design strategy: arrangement of hardware and software, and development approach (Table 11-3).

The overall structure of the data system was originally divided into three logical groups of systems: site data systems, a data processing facility, and the ARM Data Archive. The data systems remain in these logical groups after the 20 yr of ARM history (Fig. 11-3). The details of the data systems evolved with the maturity of the experimental design and the reality of the first budgets, early instruments, and first site operations.

3. Early advantages and unknowns

The maturity of scientific data management was an important influence on the initial and early designs of the ARM data system. Many of the original ARM data system staff had 5–15 yr of experience from smaller and shorter research programs. They had a good vision for the overall scope of the data system and a good sense of feasibility for the options. The early data managers also knew that an incremental design approach was needed for successful implementation.

At its inception, the size of the ARM data system was near the upper limits of large systems for environmental research. The primary unknowns in the initial data system design were as follows:

- How do we organize, store, and access this very large data volume?
  - Online data storage was very efficient (high speed, easy access) but very expensive with short technology life cycles.
  - Offline data storage was cost effective, but data access was slow and labor intensive. Automated tape libraries were new and had limited reliability.
- How do we transfer the very large data volumes between the field sites, processing centers, quality reviewers, and researchers in a timely manner?

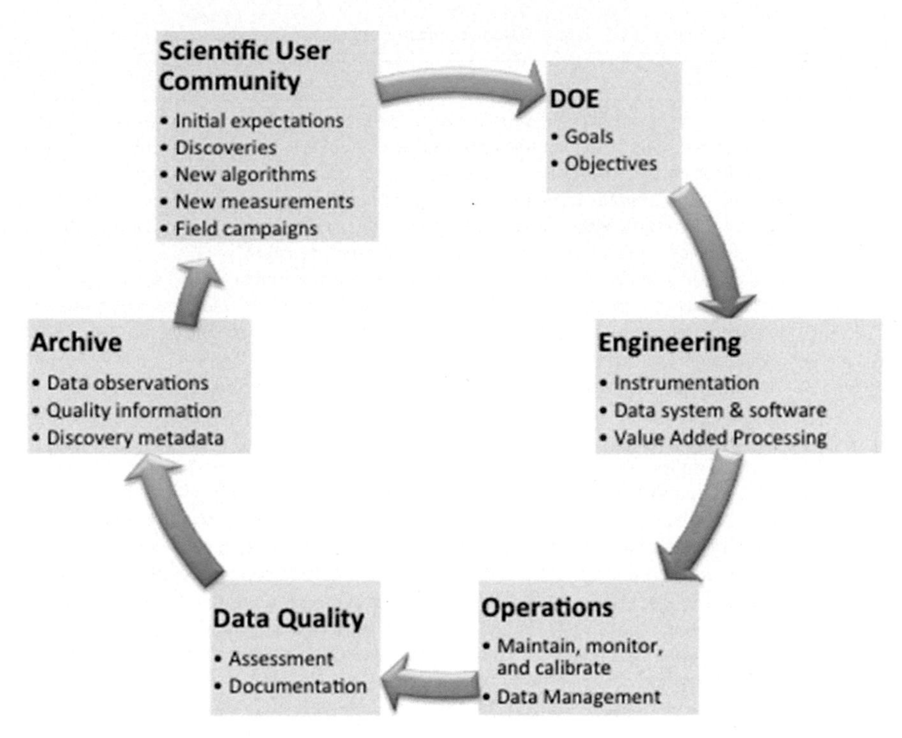

FIG. 11-2. ARM data life cycle includes programmatic, infrastructure, and external elements.

- ARM was expected to be a geographically distributed program of field sites, data processing systems, infrastructure staff, and research users.
- High-speed and long-distance networks were relatively new and had limited stability. The software for transferring data was primitive and the user community had limited experience.
- The data flow needed to be timely to keep the systems stable and to provide a reasonable timeline for initial data quality review and problem resolution.

4. The early period: 1992–97

During this early operational period from 1992 to 1997, the scope for ARM expanded from a few instruments in a single location to several facilities at a single site [Southern Great Plains (SGP)] to a second facility in operation [tropical western Pacific (TWP)] with a third facility [North Slope of Alaska (NSA)] under implementation. The data system was pushed through several changes in this period. The transition from planning to reality (real instruments in operation with output data and a user community) caused many changes. Other changes occurred as ARM implemented different system designs and data products for each of the sites. Advances in information technology also enabled significant changes during the period. The primary changes away from the original requirements for the data system during this period are summarized in the following sections.

a. The data stream

When ARM started, there was hope that a single unified information system or database could be used to handle all of the data collected by the ARM Program. However, it was quickly realized that the variety of the instruments and sites resulted in a very diverse set of datasets. This diversity made it extremely difficult to store all of the data in a single database or to use a single unified information system. Thus, the program needed another approach.

Ultimately, the basic structure of ARM data became data streams of files, where a data stream was a set of files that came from a single source (e.g., an instrument or algorithm) with a fixed file structure. The data stream became a fundamental aspect of the ARM data structure. Nearly all data stream files contained measurements spanning a single day from a single instrument and location (e.g., from a surface meteorological station at the NSA site).

b. Adoption of NetCDF

ARM, like most scientific programs, understood the need to have metadata and documentation reliably associated with its data. For example, it is important to document the SI unit and minimum and maximum

TABLE 11-1. Early ARM requirements: data life cycle.

Requirement	Explanation
Continuous data	This research expected to compare measurements with the results of models or use measurements for input to models. The plan for a continuous span of ARM measurements was a significant contrast to the dominant history of field campaigns for other observational programs.
Seamless data	The collection of measurements should approach the same completeness as model results (every value for every time step). Minimized data gaps also ensure the cooccurrence of multiple measurements needed for systems research (e.g., "closure" in radiative transfer).
Cumulative data	All data collected by ARM were expected to have long-term value. A cumulative view of data accommodates research on interannual variations and long-term assessment of model/data comparisons. The ARM data system should provide cumulative data retention and access for the users. Cumulative data schemes were unusual when ARM was starting.
10 yr of measurements	The minimum time range for ARM observations was proposed to be 10 yr. This duration was needed to obtain a range of climatologically significant conditions for each location. Yet, this duration was known to span multiple generations of technology.
Decades of data use	Research on models and their revisions was expected to be a long-term process. The ARM collection was also anticipated to be valuable in numerous unforeseen ways.
Data quality	ARM data were expected to have "known and reasonable" quality. All data should be screened with quality control limits. Intercomparisons should be made when possible. The proposed instrument design included redundancy of measurements. This provided robustness of operations and a better understanding of data quality. These quality processes should be "near term" to minimize duration of quality issues and insure that all data users were also provided quality information (see Peppler et al. 2016, chapter 12). The data system should provide resources for these data reviews.
VAPs	VAP processing was proposed for three primary purposes: • combining data from multiple data streams into a single integrated product (intended to facilitate easy use by researchers); • data quality intercomparisons between similar or redundant measurements (also called QMEs); and • derivation of physical quantities that were not directly measured by the instruments (e.g., cloud base height from the profiles of micropulse lidar backscatter). VAPs were expected to have a wide and ongoing timeline for development. Some VAPs were clearly anticipated. Others were expected to result from future research. The data system should provide the software development and processing capabilities for VAP creation.

values for each geophysical variable in the data stream. As a consequence, ARM adopted the NetCDF file structure because it was a self-describing, machine-independent, and efficient binary data format with an extensive set of open-source software libraries for data access and file manipulation (Rew and Davis 1990; http://www.unidata.ucar.edu/software/netcdf/). A self-describing data format was very important because ARM data files would be moved and restructured in many ways during research use. This format allowed for global metadata about the entire data file (the date of creation, source of the information, programmatic citations, etc.) and metadata for specific data fields (descriptive strings about each field, valid minimums and maximums, precision, etc.). Machine independence was important because the research community used a variety of hardware (UNIX machines, PCs running windows, etc.) and software. Furthermore, binary data formats are more efficient from a storage perspective (i.e., the binary files are smaller than ASCII files).

ARM data were intended to be widely available, and open-source software, such as the NetCDF package, insured the ability to maintain access to the data without proprietary restrictions and prevented the software from becoming abandoned because of shifting corporate interests. The NetCDF library included routines that enabled users to directly access data fields by name with their own software without knowing detailed information about file layout and field position. Tools that came with the NetCDF software distribution readily enabled users to concatenate files across time for a single data stream and retain only a portion of the data fields (to reduce the number of data files and the data volume).

The decision to select NetCDF was easy because the data format was being developed specifically to support atmospheric research. NetCDF readily supported the time, spectral, and height dimensions needed for ARM data, thereby supporting the wide variety of data structures that were stored in NetCDF format. Data streams range from simple constructs such as simple time series of radiometric data to complex multidimensional arrays from cloud radars.

The selection of NetCDF had many of the positive features listed above. However, this decision also had negative impacts on the ARM user community. NetCDF libraries required the user to write software in a programming language (e.g., C, C++, and

TABLE 11-2. Early ARM data system requirements: data heterogeneity.

Requirement	Explanation
Homogeneous structures	ARM data would be used in numerous combinations. The data format was required to be common across all of the measurements. The data structure was initially proposed to be a database and/or files with a common format.
External data	The data system should import data from other programs (NOAA/NASA satellites and weather forecast operations). These data would be used as input for VAPS and would be essential to ARM research. Processes to acquire and incorporate external data would be diverse. External data should be restructured into an ARM data format.
Open access	Accessibility should follow U.S. Global Change Research Program policies for open data sharing and maximize the use of ARM data (U.S. Global Change Research Program 1991). The user community should extend beyond ARM researchers.
Very large volumes	Remote sensing (both uplooking radars and lidars and downlooking satellites) and spectral data would have very large data volumes. In addition to the large data volumes, ARM data would also include very large numbers of small data files (e.g., surface meteorology). The design should include a mixture of online and offline storage. The cumulative volume would be much larger than most environmental data collections.
Routine and field campaign data	The data system should be able to process both files collected routinely (daily for many years) and for special studies (field campaigns and IOPs).

FORTRAN) to access and process the data, which was a significant challenge in the early days of ARM because NetCDF was a newly developed package. Not many in the atmospheric scientific community were familiar with it. Higher-level languages like Perl, Java, R, Python, MATLAB, and IDL with routines to access/read/write NetCDF files became more available in the mid-to-late 1990s, making it easier for scientists to use ARM data.

Each of these data access and processing methods required an investment in training, software development or software purchase to use the data. DOE management was initially concerned that the selection of NetCDF might restrict the use of ARM data by

TABLE 11-3. Early ARM data system requirements: design strategy.

Requirement	Explanation
Data delivery	Initial plans for ARM included coupled operation of field sites and models with quasi-real-time intercomparisons for ARM research. Field campaigns were expected to need near-term data processing for forecasting. Data delivery should be timely and uninterrupted. Data transfer to users would use high-speed networks (even before the Internet as we know it today). When network capacity was insufficient, the data transfer would include express-shipped storage media. Delays in data flow could potentially delay discovery of data quality and operational problems.
Incremental implementation	The initial design of the data system focused on rapid implementation. The data system should be functional by the start of field operations and should provide immediate turnaround for data quality assessment. The system design should use existing technology when possible.
Independent software components	Because of the wide disparity in data volumes between the data products, the software modules for data processing should be designed for independent development and operations (between sites and instruments).
Flexibility and scalability	The needs of ARM would evolve within the scope of the planned 10 yr of operations. Also, the span of ARM would exceed the hardware life cycle for data systems. Independent systems and processes were needed to assure that the data system could easily grow with development of new instruments (a major initial component of ARM).
Geographically distributed	The comparison of observations and models would have a global scope. Initial locations for field sites would be isolated by thousands of kilometers. The data system was distributed to enable parallel development of the components. Each site system should have independent operations to prevent large impacts from problems at a single location. The development of the data system was distributed across five national laboratories. The data system design accommodated the geographic spread of field sites, data systems, and data system developers.

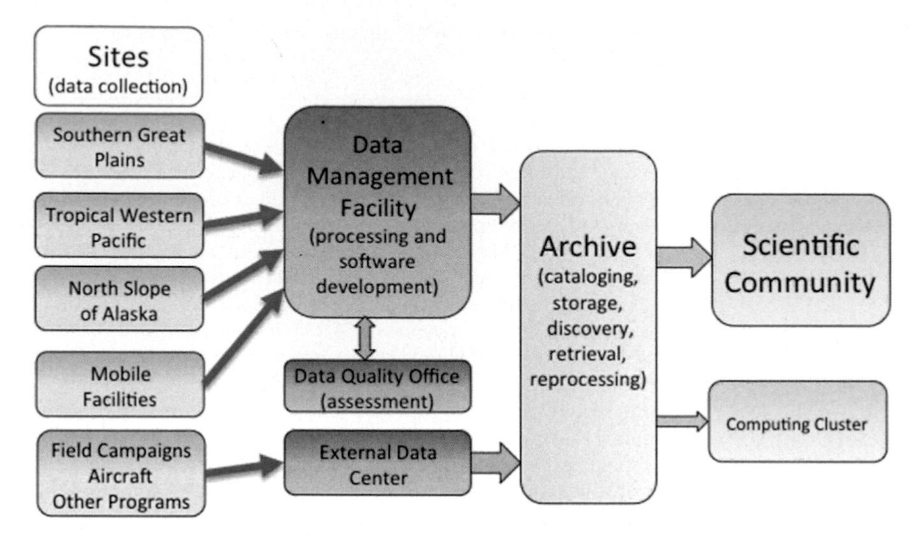

FIG. 11-3. ARM data system overview.

requiring the scientists to have programming skills to access the data, thereby limiting data use by scientists who could only use interactive tools like spreadsheets or text editors to "look at" data or create simple data plots. Thus, in the early 1990s, an experiment support team, which was a small group of data system staff, would work with ARM scientists to help them learn to manipulate ARM data in NetCDF files. This group was later called the ARM Experiment Center and also gathered requirements for commonly needed advanced data products. These efforts were the origin of ARM's value-added products (VAPs; Ackerman et al. 2016, chapter 3). As newer tools became available and more scientists learned how to use NetCDF, this group was no longer needed to support ARM scientists, but efforts to design and create VAPs continued.

In retrospect, the decision to organize the data into daily NetCDF files and data streams as the basic elements for the ARM data structure has been a very robust and durable decision.

c. *Raw versus processed data*

There was an axiom that was popular in the early days of the ARM Program that is typically attributed to Gerry Stokes, the first chief scientist of the ARM Program: "one can survive bad analysis but cannot survive bad data" (G. Stokes 1996, personal communication). This idea resulted in the decision to save the raw output from each instrument in the ARM Data Archive as well as the more highly processed (i.e., calibrated, value added) data. While this approach consumed much more space in the ARM Data Archive, it also allowed problems in both simple ingest codes (i.e., processing codes that converted raw data to NetCDF) and VAP codes to be fixed. While this might seem obvious now, it was less so when the price of storage was markedly higher and the rate of data storage much slower than today.

d. *Real-time to near-real-time availability*

The goal of autonomous 24/7/365 operations required new thinking for making the data available to the scientific community; as Stokes (2016, chapter 2) indicated, there would be no respite at the end of the campaign for which to process/calibrate/quality control (QC) the data. Thus, the program's original intent was to make the data available in real time. However, because of the relative infancy of the Internet and processing systems, data collection and processing took a finite amount of time. For a site that was well connected to the Internet such as the SGP, data could be made available to the scientific community within 2 days; however, for remote sites such as the TWP, the data would not be available for potentially several months because it was shipped back from the remote sites via transportable media (like tapes or portable disks).

The goal of making the data available in real time or near–real time was considered a positive paradigm shift in the atmospheric community, and relaxing this goal by making the data available a few days after receiving it at the primary data system was initially seen by some in the ARM community as not meeting the goal. But over the first 5 yr of the program, the scientific usage pattern of ARM data shifted from near-term comparison with models to an evaluation of a continuous series of case studies. These case studies were usually a series of intermittent days that were specifically suited for the evaluation of a common scientific question. The scientific users typically needed to access a period of data spanning a few weeks to several months. The immediate need to access recent data was generally irrelevant.

So while the scientific need for near-real-time data availability became lower priority, the program maintained the "available within 2 days" approach. This near-real-time approach allowed the infrastructure to identify instrument failures, processing errors, measurement bias, and other data quality issues more quickly (Peppler et al. 2016, chapter 12). This short delay assured that researchers were provided a stable data product with known quality.

e. Data quality

One of the program's mantras is that "ARM data have known and reasonable quality" (Gracio et al. 1996). Prior to ARM, atmospheric data were typically collected in relatively short duration field experiments, and instrument investigators would have several months after the field campaign to QC their data. ARM's operational paradigm prohibited this because there was no respite in the data collection for QC. Thus, the program had to develop automated approaches to apply the QC. Three distinctly different approaches were (are) used.

The simplest form of QC applies a set of logical rules based upon the specific data stream and its origin (e.g., the surface radiation measurements at SGP) during the data processing to set a QC flag in the NetCDF data stream file. These QC flags identified values that were too low, too high, too erratic, or failed other expectations.

A higher level of QC could be applied by comparing observations from different sources or by comparing observations to a model simulation where the model was driven by a different observed dataset. The quality measurement experiments (QMEs) were very popular in the ARM Program in the mid-1990s, with many of them developed to help evaluate radiometric instrument data quality (Mlawer and Turner 2016, chapter 14; Peppler et al. 2016, chapter 12). QMEs were automated in the data system, producing another data stream that was analyzed by scientific users and could find more subtle problems with the measurements than could be found using more simple minimum/maximum/delta checks.

Often, the best description of data quality came in the form of manually entered data quality reports (DQRs). These DQRs are more dynamic than QC flags and contain a description of an "event" that altered the normal quality of the data. Events include a wide range of conditions such as instrument degradation and contamination and temporary operating conditions (power failures, frozen or snow covered sensors, etc.). The quality impacts from these events are defined for a specific time range, list of data products, and specific measurements. DQRs are typically submitted by either the instrument mentor [see Cress and Sisterson (2016, chapter 5) for a description] or the Data Quality Office (Peppler et al. 2016, chapter 12). DQRs are now provided as "companion" information with the data files when researchers request data from the archive.

f. Pushing versus pulling data

The initial design of ARM's data flow expected researchers to define relatively static combinations of data products to be delivered to them. When data files were ready for distribution, the ARM Experiment Center would "push" the data files to the researcher's system, either over the Internet (via a program like FTP) or by copying data onto computer media and shipping these to the researcher. The "pushing of data" was implemented to meet the initial requirement for near-term comparison between measurements and modeling results and to ensure that the scientific users were getting ARM data. However, the pattern of data use quickly evolved to retrospective analyses of special case study days that occurred during the history of data collection and were relevant to the researcher's needs. Pushing a steady stream of data to researchers was not necessary for this pattern of use, and this role of the Experiment Center was discontinued. Furthermore, other advancements in Internet accessibility and web applications (see below) made it easier for scientists to select the data they wanted and to "pull" it to their machines. The flow of data to the researchers evolved from a more labor-intensive push to a more efficient and selective pull of data. With the pull method, users only worked with the required data instead of all of it.

g. The growing web and its applications

The rapidly advancing field of web applications, search engines, and the continuing development of the Internet provided new opportunities for the ARM data system to evolve. User interfaces at the ARM Data Archive allowed users to specify dynamic selections for their data needs. These selections created customized lists of data products and data ranges that resulted in data requests from the archive. The archive retrieved the requested data files from the mass storage and put them online for the researchers to download during the next week. Dynamic selections of data from the archive were more adaptable to the evolving needs of the researchers and less labor intensive for the data system staff.

The initial user interface for accessing the data inventory information of the ARM Data Archive required server class hardware and the use of X windows display capabilities on a researcher's computer. This generally required computers, network capacity, and expertise more substantial than a typical PC and scientist in the early 1990s. The software to interactively display data availability and forms to capture user data requests was tedious to develop and had limited scalability.

The invention of Hypertext Markup Language (HTML) in the early 1990s allowed an unlimited structure of text, pictures, and drawings. The addition of HTML forms enabled users to enter selection criteria into web pages. Developing software to display dynamic content on subsequent web pages based on the user's criteria was a major boon to ARM. This enabled ARM to communicate to the user manageable portions of its instrument documentation, operations records, data inventory, and data quality information. Web applications and related data retrieval processes enabled more automated processes and allowed for the initial scalability of the ARM Data Archive.

Soon after web technology became available, displaying simple lists or tables summarizing ARM's scope became more content than most users would absorb. Dynamic web displays of information were very important as the types of instruments moved into tens, the types of data products moved into hundreds, the different measurements across several locations became greater than 1000, and ARM accumulated hundreds of thousands of user accessible files.

The unlimited linkage of web-based information was important for documenting nonroutine operations from the intensive operational period (IOP) and the resulting nonroutine data products. Web links could easily accommodate the layers of detail needed to describe complex combinations of instruments and data products associated with field campaigns.

The web pages were an immediate advantage to ARM and greatly facilitated the creation of large quantities of information. However, they quickly became opportunities for inconsistent organization of information and confusion for ARM users. The ease of creating web information in common word processor software resulted in many people creating content in many different styles with inconsistent keywords and organization structure. In response to this complexity, a staff member was assigned to coordinate web-based information.

The inconsistencies shown in the web information revealed inconsistencies during many steps across the data life cycle. Many other factors led ARM into a period of improved organizational structure (or reorganization). The management, structure, and logic of the data system were also soon reorganized.

h. Dropping operational models from ARM's scope

In the earliest days, the program had planned on running spectral radiation and single-column models (SCMs) as part of the data system's infrastructure. However, it quickly became apparent that the challenges of collecting and managing the field measurements required all of the available resources, and thus running SCMs was dropped because the level of effort was too high. However, the spectral radiative transfer models were still run as part of the ongoing QME effort.

5. Reorganization and growth: 1998–2007

The experiences gained in the first epoch of the ARM data system highlighted many things that worked well and some that did not. In particular, many aspects of the data system were separated in ways that were not very efficient or made little sense. During this period from 1998 to 2007, the ARM Program continued to expand, and ARM managers addressed issues that arose from this growth and design decisions made in the early period.

a. Consolidation

The early data system management structure of ARM was site-centric, and each field site started independent and parallel implementations. This structure was efficient for getting started, but also resulted in independent designs by each site for data systems, data quality review, and data products. The implementation of multiple solutions to similar problems limited the data system's ability to keep up with the required growth needed for new sites and instruments. The data systems were a venue where many unnecessary complexities and inconsistencies became apparent from divergent evolution of site data systems and processes. As ARM accumulated more data history than the normal 3–5-yr project, continuing to support the resulting complexity was difficult. Presenting the data collection to the researchers with the archive user interface was problematic.

An infrastructure review committee recommended in 2000 that the overall structure of ARM be changed from site-centric to a structure emphasizing science-centric themes across the array of ARM measurements (ARM 2001). The revised ARM infrastructure included groups with thematic responsibilities (technical coordination, publications and communications, operations, engineering, data quality, data processing, etc.). The new programmatic structure focused common processes for developing site data systems, ARM software, data quality review procedures, and data products across all of the locations. Common solutions enabled the data system to respond more readily to continued growth from new science themes. These changes were applied not only to new instruments and locations, but also to significant redesign of existing processes and products.

Variations across the ARM data systems should now only occur to accommodate the environmental and logistical differences between the sites.

Several changes in the organizational structure of ARM impacted the data system. The production-scale processing of ARM data from all sites and VAPs, which was spread over many different locations, was concentrated into a single data management facility (DMF). The DMF succeeded the Experiment Center as a centralized location for ARM software development. The DMF component of the data system also enabled the development of extensive tools for tracking data processing and data flow. The ARM DSView tool provided numerous interrelated reports on processing status and was implemented in 2001 and updated in 2007 and 2011 (Macduff et al. 2007; ARM 2011). The same tool was also used to monitor data system operations at the sites. Data processing at the field sites was limited to the needs for operational oversight and short-term needs for field campaigns. The site data systems and instrument computers were also standardized as they were replaced during this period of ARM history.

The Data Quality Office was established in 2000 to review all ARM data and to address concerns about inconsistent data quality (Peppler et al. 2016, chapter 12), which led to several changes in the management of data quality information. A major update in most of data products in spring 2001 included a standardized structure for QC flags for most measurements. These flags included the results of tests for minimum, maximum, offset limits, and other tests specified by the instrument mentors. The DQRs were expanded to identify specific measurements and data products that were associated with a data quality event. The DQR distribution was restructured from plain text into portable web pages that included a table of contents showing the DQR titles and a tabular format for the DQR details. During this same time period, a new process at the archive began distributing DQRs that retroactively described data quality issues. A customized distribution of DQRs was prepared for each researcher based on their history of requests for data files. Attempts also were made to define a "quality color" (red, yellow, and green) from both the subjective judgment in the DQRs and an automated algorithm based on the prevalence of QC flags. The quality color from DQRs was attainable. Enabling auto quality color with a stable, automated process and comprehendible presentation to the user was elusive.

Other changes from ARM's reorganization impacted the structure of the overall data system. Gerry Stokes, the first ARM chief scientist, predicted that "all data in the archive would need to be reprocessed seven times" (G. Stokes 1997, personal communication). In the early era, reprocessing was done on the individual data system that originally processed the data (e.g., the specific site data system). However, this was very inefficient and cumbersome. Thus, a reprocessing center was set up to standardize the procedures and records for the reprocessing tasks. Formalized reprocessing enabled ARM to improve data by removing inconsistencies between sites and across time for a single measurement type (e.g., surface meteorology). Reprocessing also applied updated and more robust algorithms to the data and improved the completeness of the data collection.

VAPs are an important source of ARM data. As indicated above, VAPs combine observations from different instruments, sometimes with model output, to create new data streams. Just like the instrumentation in the program, VAP algorithms need to be robust and able to run in all conditions and thus are typically programmed by staff trained in computer science to create robust, well-documented codes. In the early years, these software developers worked directly with ARM principal investigators (PIs) to develop VAPs; however, this was not efficient because often a communication gap existed between the PIs and the developers. Thus, ARM started pairing VAP software developers with "translators": infrastructure scientists who could help the software developers interpret the scientific algorithms for VAPs and review interim results (Ackerman et al. 2016, chapter 3). The translators regularly coordinated the scope of the VAPs and their priority to improve the consistency of the data collection. Data distribution to all researchers (both ARM scientists and the broader research community) was assigned to the ARM Data Archive. All of these changes improved the efficiency of each of these activities.

b. New approaches to attain consistency

The continually expanding collection of data streams and more powerful and web-based tools to display the ARM data collection revealed yet other challenges for the ARM data system. As the number of data products increased with new locations, instruments, and VAPs, it was increasingly more difficult to maintain consistency for the design of the data products. Detailed aspects of the data product design such as data field names and quality control limits were inconsistent between similar data (e.g., across all of the meteorological or radiometric measurements). This inconsistency was confusing and annoying to the data users. In 2004, efforts began to put all details for the data fields from all data products into a database. This change enabled differences between similar measurements to be readily identified and consistency easier to attain. This database became a production feature of the data system in 2006.

The continually increasing scope of ARM data made it more difficult to determine the details of the inputs required for VAPs and their availability. A database was developed to define the dependencies between the VAPs and their input data streams. This dependency information included the details about data fields needed for input into each resulting VAP data field. This dependency database was added to the data stream configuration information previously described.

Inconsistent ARM language became apparent as the ARM data products began to converge toward consistency. The size and scope of the data collection expanded rapidly in 2000–04 and also contributed to inconsistent language. For example, the thematic categories of measurements used for data discovery at the archive and for the web-based ARM documentation were similar but different enough to be confusing to the researchers. A collaborative effort in early 2006 defined and adopted a metadata structure to be used by both data discovery and web organization. After this metadata consolidation, many of the archive user interfaces and the ARM web pages could be composed dynamically from the common metadata content stored in a single database. These changes assured that ARM users encountered the same language and logic as they interacted with ARM information.

The incompleteness of the metadata was a critical problem when the display of ARM information and access to data became dependent on the common metadata. Throughout this entire period, efforts were made to encourage the creation and recording of metadata about the data products, measurements, instruments, and sites. The initial solution included educating many members of the infrastructure about the scope of metadata and its role in communicating ARM information and data to the researchers. The infrastructure staff contributing metadata ranged from instrument mentors to software developers and translators to data quality staff. The next step included formally defining the metadata elements and developing templates to record and transmit the metadata. Establishing informal processes to referee the consistency of the metadata was the final step. Timely creation of metadata was an elusive objective. ARM shared this challenge with most other research programs during this era.

During this midportion of ARM's history, the number of data products from field campaigns grew beyond a short list that could be shown on a single web page. The complexity of the data collection from field campaigns was challenging. The structure and scope of field campaign data varied with each campaign. Standardization across the field campaign data was limited because the collaborating researchers used specialized methods for processing and organizing the data. Standardizing the data structures required too many resources. Evolving standards for data structures also limited standardization. Most of the field campaign data followed a consistent structure for each campaign and instrument. More detailed data structures were unconstrained. To accommodate this complexity and retain a maximum amount of these data, the field campaign part of the data system was designed with the following principles:

- When requested, researchers were provided guidance on file names and directory structures.
- A simple FTP procedure or shipped media was used for data transfer from the researchers to ARM.
- The data were organized in a master directory structured by year, site, campaign, instrument, and PI name.
- Documentation for each subdirectory was requested to explain the scope of additional data directories and filenames.
- A web-based user interface was developed that could navigate and provide access to any amount of documentation, complexity, or organizational depth for the data collection.

These design principles were found to be robust as the field campaign data collection grew. Both data providers and data users readily accepted this design.

c. *Communicating the scope of ARM's data collection*

The increased size and complexity of the data collection during this period also challenged the researcher's ability to understand what data were available from the ARM Program over its duration. Simple query logic that searched for the availability of data based on criteria of location, measurement type, and date range failed to find data because the incremental implementation of sites, instruments, and VAPs. The failed queries provided no information about why data were not found (the time range was invalid for a specific location, instrument was offline for repair, etc.). Additional interfaces were developed for the archive that presented hierarchical summaries or catalogs of data availability (McCord et al. 1999).

The expanding data collection also challenged the researcher's ability to understand the ARM results. Comprehension was limited because scanning the contents of the data files was not possible. Web-based displays showing plots of ARM measurements were developed. In 2004, a user interface (thumbnail browser) went into production at the archive that displayed tens of thumbnails (standardized, postage stamp size, and data plots) per web page. The data plots (both small thumbnails and labeled page-size graphs) were precomputed to assure a responsive and predictable performance of this user interface as the size of the ARM dataset expanded. This user interface allowed researchers to customize displays of specific measurements from multiple data products and visually scan the results in monthly increments (ARM 2004).

A web-based tool for interactively plotting small amounts of NetCDF data (NCVweb) was also developed during the same time period (Bottone and Moore 2003). This NCVweb tool allowed users to select relatively small time ranges of data and "zoom in" on the details of individual measurements. The NCVweb tool was expanded to display the details of the data fields and export small portions of data into text files or smaller NetCDF files. Both the NCVweb and thumbnail browser user interfaces for the archive are still used by researchers. Routinely generated data plots and interactive visualization of data with NCVweb also are used during data quality review [see Peppler et al. (2016, chapter 12) for more details].

The primary themes contained in the midperiod of ARM's data system history included

- attaining efficiency through the consolidation of data system functions;
- expanding the creation and communication of data quality information;
- converging on consistency in the detailed structure and description for ARM data products and documentation; and
- improving the communications with the researchers through new interfaces to present the scope of the data collection and visualization of ARM's results.

All of these changes continued to help ARM's success in the final era of the first 20 yr.

6. Continual growth and improving durability: 2008–present

During the period between 2008 and the present, ARM's data system had to adapt to increases to the number of new instruments and sites. Also, the designation of the ARM facility as a national scientific user facility resulted in ARM no longer being purely a research program but now becoming a truly operational facility. To sustain operations and durability during these changes, it was necessary for the data system to be more robust and nimble. The changes and continued growth of ARM's data system are summarized in the following sections.

a. Challenging growth opportunities

Persistent themes for ARM during this most recent period of ARM's history were growth and accelerating growth. These themes originated from implementing several new sites (e.g., the new permanent site in the Azores) and instruments as well as more tools for efficient data product development. The primary events that triggered growth during this period include the addition of

- a second mobile facility (2008–10);
- the purchase of many new types of instruments with Recovery Act (stimulus) funding (2009–11; Mather and Voyles 2013; Voyles and Mather 2010);
- acquisition of scanning instruments with new types of data products (2010–12);
- implementation of new fixed location in the Azores (2012–14); and
- implementation of a third mobile facility (2012–14).

Many of the Recovery Act instruments included the ability to measure profiles, three-dimensional (3D) scans, spectral views of solar radiation or Doppler shifts in radar, and lidar signals. For example, the daily data volume for the cloud radar dataset increased from 15 to 20 to 80 to 100 GB day^{-1}. Improvements in instrument, sensor, and computing technologies also contributed to rapid increases in growth rate in the overall data volume collected by ARM. All of these changes contributed to dramatic increases in the volumes of data moving daily from the sites through the DMF to the ARM Data Archive (Fig. 11-4).

The designation of ARM as a national scientific user facility in 2004 (Ackerman et al. 2016, chapter 3) converted ARM from a 10-yr project to a permanent operation for atmospheric measurements, data collection, and data distribution. Performance metrics for the facility were defined for data completeness and increasing the number of users. This designation encouraged ARM to aggressively identify scientific collaborators for mobile facility implementations and other smaller-scale field campaigns. A significant number of new users are added to the ARM user community with each new mobile facility location. All of these factors increased the diversity of data users. The longevity of ARM encouraged researchers to evaluate larger quantities of data. The rate of data usage by the researchers also accelerated (Fig. 11-5).

As the increase in the outgoing and incoming data volume ranged from 10 to 120 times respectively during this period, the data system required continuing

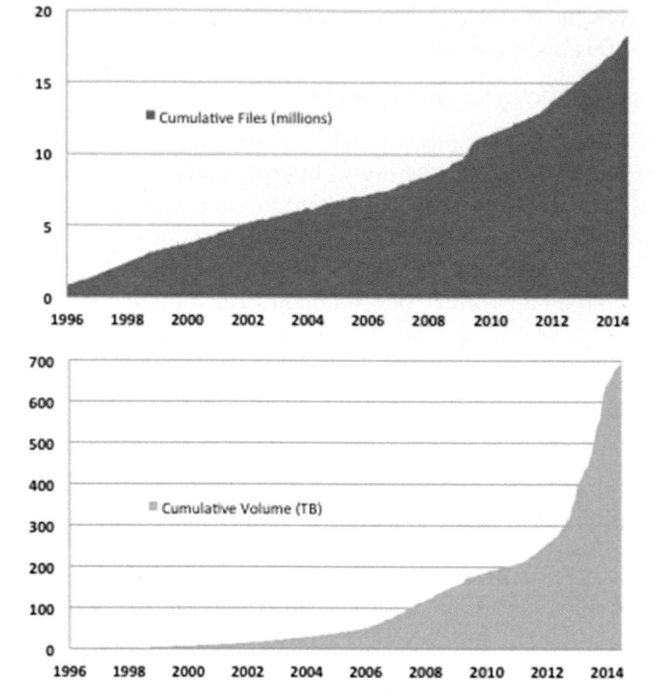

FIG. 11-4. Cumulative data volume stored in the ARM Data Archive in terms of (top) number of files (millions) and (bottom) volume (terabytes).

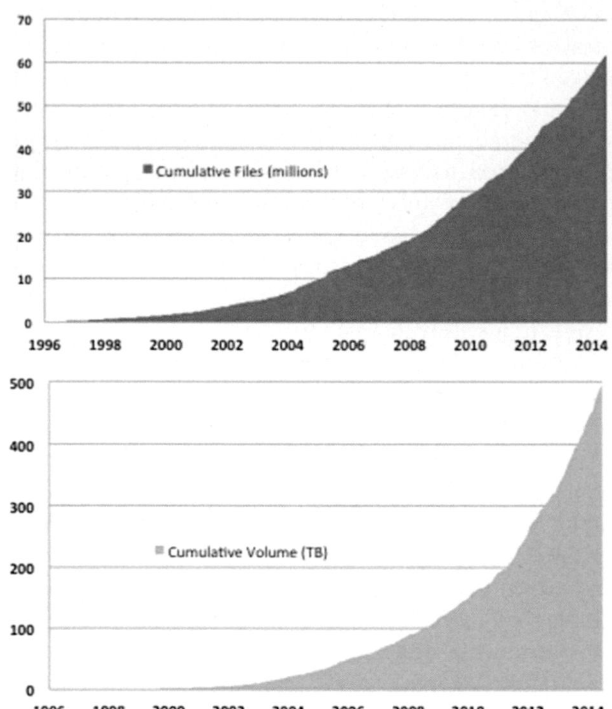

FIG. 11-5. Cumulative data volume requested from the ARM Data Archive in terms of (top) number of files (millions) and (bottom) volume (terabytes).

revisions to its design in order to remain nimble. Several changes occurred in the data system to help it reduce the complexity in its design and the chaos in its performance. The changes summarized in the remainder of this section enabled the system to more quickly adapt to the changing requirements.

b. Expansion and coupling of hardware

Expansion for data volume was accommodated primarily with increased data storage at each point in the data life cycle. Data access was increased with high-performance networks and high-performance file servers. The file servers concurrently used several hardware interfaces and numerous storage devices. In 2012, the DMF hardware was moved from Pacific Northwest National Laboratory, which had operated the DMF and its predecessor the Experiment Center from the beginning of the ARM Program, to Oak Ridge National Laboratory. This move enabled the DMF network and storage devices to be combined with the archive. This change enabled quicker adaption and reallocation of the combined hardware capacity to different parts of the data life cycle as needed. In an analogous manner, the site data system has been modified by consolidating individual instrument computers into virtual machines on a single redundant pair of computers with replicated storage. This design reduces maintenance efforts and provides the capability to transition processing from one set of computers to another if the first fails at any of the sites. All of these changes allow the various ebbs and surges of data flow to share common storage buffering resources that are more robust and durable than can be implemented for each process.

c. Consolidation and software efficiency tools

The very large number of new instruments purchased with the Recovery Act funding resulted in the need to build an integrated software development environment (ISDE) as part of the data system. The vision for ISDE included providing standardized tools for the common software tasks. The availability of ISDE would reduce the effort for new software and allow development to focus on the specialized portions of the software. ISDE was also intended to be portable so that software developed by researchers would more readily follow ARM standards for software and data products. After 3 yr of experience with ISDE, its scope was reduced to handling the data input, combining data from multiple inputs into a common interval for sampling rate, data product design specifications (from the existing database), and handling the output. Most of the software that uses ARM data requires some form of these functions. ISDE was renamed the ARM Data Integrator (ADI) to better represent the scope of its functions. ADI was valuable because it reduced the effort for developing new data products (i.e., VAPs). It also focused most of the software development effort on the scientific logic because the reading and writing, as well as the time syncing of different data streams, was now done internally by ADI. ADI also enabled researchers to develop more standardized software and data products when collaborating with ARM.

d. Reducing the complexity

The expansion and requirements for the durability of ARM data processes resulted in the formalization of the VAP development process. In the previous periods of ARM, VAP development could be a very lengthy series of iterations between the translators, the software developers, and the ARM PIs. Earlier styles of VAP development included incomplete processes for defining the data product, selecting advanced algorithms, evaluating the results, and declaring completion of the development. These earlier styles sometimes resulted in endless development processes for VAPs. The formalized process defined a specific series of evaluations for the following:

- confirmation of algorithms to be used and knowledge of their logic;
- decisions to release interim versions of the VAPs data for review by the researchers;
- assessment of the interim progress on the VAP development; and
- creation of documentation of the finalized data product.

The process also recognized that some VAP developments might be abandoned or left as is because further progress depended on new research beyond the scope of ARM. In some instances, these intermediate VAPs were contributed by researchers as PI data products. These PI data products enabled the research community to make additional discoveries until the methods for a final product were developed. These revisions to the VAP development process enabled ARM to meet continuing challenges from a very large number of possible VAPs associated with the expanding number of sites and instruments.

The program realized that PI data products offered an excellent way to get additional higher-order processed data into the ARM Data Archive, thus both capturing it and making it available to other scientists. The program created an area in the archive specifically for value-added datasets provided by ARM PIs. With this procedure, a translator worked with the PI to move the data into the archive. Then, a metadata editor was developed allowing the PI to include metadata about the contributed dataset that both described the algorithm used to create the dataset and provided key words so that other PIs could more easily find the dataset within the archive holdings.

The reprocessing of different datasets over time led to multiple versions of data products being used in research projects, especially for those PIs performing long-term analyses. Standards for data products were needed to reduce the complexity of the data processing by the researchers. ARM developed a standards document (and revision process) for data product names and data product design as an extension of the standardization theme in 2012/13. An infrastructure committee representing all portions of the data system, the data life cycle, and a few data users developed these standards. Sustainable standards have enabled the ARM data system to readily expand. They also facilitated the long-term durability of the data products and related software.

Sustaining standards for the names and descriptions of ARM data required improvements for creating and reviewing the metadata. Development of a web application to formally track the proposal, review, and approval of the metadata began in 2012. This application enabled ARM to improve the completeness of the metadata and expand the scope of metadata creation to include the field campaign data. Formalized metadata for these data were needed to consolidate the discovery of routine and field campaign data into a single user interface for the archive. This consolidation has improved the durability of the user interface and helped to expand the researcher's data discovery.

e. Improving the use of very large ARM datasets

Some of the new instruments acquired in the late 2000s as part of the Recovery Act, such as the scanning cloud radars, Doppler lidars, and aerosol mass spectrometers, create huge amounts of data every day. These files are sufficiently large (many hundreds of gigabytes per day) that transferring these files to PI computers was challenging and very inefficient. To address this, special computers were added to the ARM data system to provide the researchers a means of exploring these new and very large data. These systems (the ARM computing cluster) were implemented in late 2010. They included a cluster of processors, very large amounts of shared memory, and tens of terabytes of storage and workspace. Popular scientific software applications were installed on these systems to facilitate the analysis of these datasets, and dozens of researchers explored the large data products. This addition was another revision to the data system that supported expansion and durability.

Another strategy toward expansion and durability of the data system was making model-ready, condensed, and very specialized (showcase) data products. The development of these showcase data products started in 2007. The processing for showcase data products included new steps such as best-estimate logic that replaced missing values from a primary instrument with the next best choice of an equivalent measurement from a different instrument. Additional quality evaluation and summarization provided the best possible long-term representation of ARM's core measurements. A showcase, best-estimate product was produced for the full duration of ARM's history for all sites (Xie et al. 2010). Several data products with extensive summarization, integration, and preparation were expected to further engage the researcher community with ARM data.

The expansion and durability of ARM's data systems were assured by the consolidation of hardware, software, and development resources into configurations that supported increased flexibility. The consolidation enabled the sharing of common and standardized components. During consolidation, changes occurred to data storage, user interfaces, software development tools, showcase data products, and review processes for interim data products and metadata. All of these revisions contributed to the efficiencies needed for the data system design to adapt to the dramatic growth of ARM's measurement resources from new instruments at new locations in this most recent period of ARM's history.

7. Summary

Plans for the ARM data system began very early in the planning for the ARM Program. This planning strategy was motivated partially by extensive and complex expectations for the data system needed to accomplish ARM's research objectives. The early participants in ARM also knew that the capacity requirements for the data system were very large and complex when compared to contemporary systems supporting environmental measurements in the early 1990s. Many of the tasks in the early history of ARM were organized into parallel and incremental steps to expedite the implementation of sites, instruments, and data systems. This organizational strategy benefited ARM, but also accidentally resulted in the implementation of different designs for data systems, data review practices, and data products for each site.

The divergence of many aspects of ARM's infrastructure became problematic when the data accumulation extended beyond the duration of a normal research project life cycle (3–5 yr) and the implementation of multiple sites occurred. The divergence was especially critical because the inconsistency and complexity of the ARM data collection made research analyses more difficult. Even communicating the scope of ARM datasets became challenging.

The midperiod of ARM's data system history included significant reorganization of the ARM infrastructure and the data systems. The independent development of the data system and data products by each site was consolidated into centralized groups for many of the tasks. Centralization allowed ARM to address many of the implementation challenges with a smaller number of solutions. Many of the components of ARM's information about configuration, quality, and descriptions were migrated into shared databases for use in many ARM functions. This consolidation of information provided a consistent view of ARM documentation and data to the researchers. These databases also enabled the development of dynamic (customized) displays of information so that researchers could interact with smaller portions of the ARM documentation and data collection. Visualization of ARM's measurements were created for large portions of the data collection and made accessible on the Internet. The researchers could more quickly scan and understand ARM's results.

In the final period of ARM's data system history, the data system adapted to the very large jumps in the number of new instruments and sites. Furthermore, the designation of the ARM facility as a national user facility resulted in a change of thinking; ARM was no longer purely a research program but had to be concerned with running as a truly operational facility. To sustain operations and durability during these changes, it was necessary for the data system to be more robust and nimble. These characteristics of the data system were attained by revisions that enabled more sharing of computing resources between the DMF and ARM Data Archive. Streamlined management and configuration of site data systems resulted from the implementation of virtual machine technology. Establishing standards for the naming and design of the ARM data products reduced the growth of the data complexity. Formalized processes for VAP software development and review of metadata improved the efficiency of these efforts. Implementation of shared software tools improved the development of ARM software. Expanded computing resources dedicated to the analysis of data without download enabled the researchers to effectively use ARM data and keep up with its rapid growth. The development of highly integrated and best-estimate data products facilitated the use of ARM data by a broader user community.

Acknowledgments. The authors thank the DOE ARM Program for supporting their effort to create this publication and for their many interesting opportunities and experiences while working with ARM. Dave Turner is also thanked for his encouragement and suggestions during the revision and finalization of this publication. Thanks also to the numerous and talented ARM staff members who have created and operated the ARM data system throughout its history.

REFERENCES

Ackerman, T. P., T. S. Cress, W. R. Ferrell, J. H. Mather, and D. D. Turner, 2016: The programmatic maturation of the ARM Program. *The Atmospheric Radiation Measurement (ARM) Program: The First 20 Years, Meteor. Monogr.*, No. 57, Amer. Meteor. Soc., doi:10.1175/AMSMONOGRAPHS-D-15-0054.1.

ARM, 2001: The Atmospheric Radiation Measurement Program Infrastructure Review Report (AIR): Summary of recommendations. U.S. DOE Rep. DOE/SC-ARM-0001, 3 pp. [Available online at http://www.arm.gov/publications/programdocs/doe-sc-arm-0001.pdf.]

——, 2004: New thumbnail browser for data archive gives users more options. ARM Climate Research Facility News, accessed 13 August 2014. [Available online at http://www.arm.gov/news/facility/post/1109.]

——, 2011: Welcome to the matrix: Overhaul of data system status viewer hits bull's-eye. ARM Climate Research Facility News, accessed 13 August 2014. [Available online at http://www.arm.gov/news/facility/post/14503.]

——, 2016: Appendix A: Executive summary: Atmospheric Radiation Measurement Program Plan. *The Atmospheric Radiation Measurement (ARM) Program: The First 20 Years, Meteor. Monogr.*, No. 57, Amer. Meteor. Soc., doi:10.1175/AMSMONOGRAPHS-D-15-0036.1.

Bottone, S., and S. Moore, 2003: Tools for viewing and quality checking ARM data. *Proc. 13th Atmospheric Radiation Measurement (ARM) Science Team Meeting*, Broomfield, CO, ARM, 1–12. [Available online at http://www.arm.gov/publications/proceedings/conf13/extended_abs/bottone-s.pdf.]

Cress, T. S., and D. L. Sisterson, 2016: Deploying the ARM sites and supporting infrastructure. *The Atmospheric Radiation Measurement (ARM) Program: The First 20 Years, Meteor. Monogr.*, No. 57, Amer. Meteor. Soc., doi:10.1175/AMSMONOGRAPHS-D-15-0049.1.

Gracio, D.K., and Coauthors, 1996: Data systems for science integration within the Atmospheric Radiation Measurement Program. *Proc. 12th Int. Conf. on Interactive Information Processing Systems (IIPS)*, Atlanta, GA, Amer. Meteor. Soc., 327–336.

Macduff, M., S. Choudhury, and J. Daily, 2007: The new DSView based on portal technology. *17th Atmospheric Radiation Measurement (ARM) Science Team Meeting*, Monterey, CA, ARM. [Available at http://www.arm.gov/publications/proceedings/conf17/poster/P00144.pdf.]

Mather, J. H., and J. W. Voyles, 2013: The ARM climate research facility: A review of structure and capabilities. *Bull. Amer. Meteor. Soc.*, **94**, 377–392, doi:10.1175/BAMS-D-11-00218.1.

McCord, R. A., D. J. Strickler, B. M. Horwedel, R. C. Ward, and S. W. Christensen, 1999: A catalog-based user interface enhances data selection from the ARM archive. *Proc. 15th Int. Conf. on Interactive Information and Processing Systems (IIPS)*, Amer. Meteor. Soc., Dallas, TX, 105–108.

Mlawer, E. J., and D. D. Turner, 2016: Spectral radiation measurements and analysis in the ARM Program. *The Atmospheric Radiation Measurement (ARM) Program: The First 20 Years, Meteor. Monogr.*, No. 57, Amer. Meteor. Soc., doi:10.1175/AMSMONOGRAPHS-D-15-0027.1.

Peppler, R., K. Kehoe, J. Monroe, A. Theisen, and S. Moore, 2016: The ARM data quality program. *The Atmospheric Radiation Measurement (ARM) Program: The First 20 Years, Meteor. Monogr.*, No. 57, Amer. Meteor. Soc., doi:10.1175/AMSMONOGRAPHS-D-15-0039.1.

Rew, R. K., and G. P. Davis, 1990: NetCDF: An interface for scientific data access. *IEEE Comput. Graphics Appl.*, **10**, 76–82, doi:10.1109/38.56302.

Stokes, G. M., 2016: Original ARM concept and launch. *The Atmospheric Radiation Measurement (ARM) Program: The First 20 Years, Meteor. Monogr.*, No. 57, Amer. Meteor. Soc., doi:10.1175/AMSMONOGRAPHS-D-15-0021.1.

U.S. Department of Energy, 1990: Atmospheric radiation measurement program plan. U.S. DOE Rep. DOE/ER-04411990, 121 pp. [Available online at http://www.arm.gov/publications/doe-er-0441.pdf.]

U.S. Global Change Research Program, 1991: Our changing planet: The FY 1992 U.S. Global Change Research Program. Committee on Earth and Environmental Sciences, 90 pp. [Available online at http://data.globalchange.gov/report/usgcrp-ocpfy1992.]

Voyles, J. W., and J. H. Mather, 2010: Recovery Act instruments: Deployment and data processing plans. *Extended Abstracts, First Science Team Meeting of the Atmospheric System Research (ASR)*, Bethesda, MD, ARM. [Available online at http://asr.science.energy.gov/meetings/stm/posters/poster_pdf/2010/P000257.pdf.]

Xie, S., and Coauthors, 2010: Clouds and more: ARM climate modeling best estimate data. *Bull. Amer. Meteor. Soc.*, **91**, 13–20, doi:10.1175/2009BAMS2891.1.

Chapter 12

The ARM Data Quality Program

RANDY A. PEPPLER, KENNETH E. KEHOE, JUSTIN W. MONROE, AND ADAM K. THEISEN

Cooperative Institute for Mesoscale Meteorological Studies, University of Oklahoma, Norman, Oklahoma

SEAN T. MOORE

Orbital ATK Inc., San Francisco, California

1. Introduction

As of this writing, nearly 7000 ARM Climate Research Facility data fields from 400 instruments are monitored for data quality control on a daily basis. This chapter reviews the history and evolution of ARM Program data quality assurance since the beginning of the program and describes the processes in place today. It also provides advice to those who collect field data, especially in an operational context. ARM's infrastructure was charged to produce data of "known and reasonable quality" for use by climate researchers. This is challenged by the fact that there are hundreds of different instruments of varying types in different climatic locations, translating into thousands of individual data (variable) streams. Some of these variables are geophysical variables (or will be processed by some algorithm to be such), but many are intended purely to help characterize the state of the instrument that made them (e.g., instrument temperature). The goal of the data quality program is to assess the quality of all of these variables.

To better complete this data quality mission, an ARM Data Quality Office (DQO) was formed in July 2000 to provide overall guidance and management of a program to assure that the data collected at ARM sites meet the data quality objectives and tolerances as defined by the science user community and to make estimates of that assurance publicly available. The DQO is accountable to the ARM Technical Director and works daily with the ARM Infrastructure, Atmospheric Systems Research (ASR) science team members, and the broader ARM user community to develop an end-to-end data quality assurance system that results in continuous, consistent quantitative assessment and continual improvement of ARM data streams through improved instrument performance based on what has been learned. The DQO leads the development and implementation of data quality algorithms and visualizations, analysis of results, and the reporting of the results both to the program and to the scientific community. It is responsible for achieving efficiencies within instrument suites and across collection sites with respect to data-checking algorithms, metadata collection, and data quality reporting. It works closely with instrument mentors, site scientists, site operators, and data and engineering staff to develop the data quality tools and analyses needed.

2. History and evolution of ARM data quality inspection and assessment

a. Early programmatic efforts

The reader is referred to Peppler et al. (2008a) for more detail on ARM's data quality assessment history. Early programmatic efforts in data quality inspection and assessment focused on the first field site, the Southern Great Plains (SGP). These efforts included the development of self-consistency checks for individual data streams (Blough 1992) and quality measurement experiments (QME; Miller et al. 1994) for comparing multiple data streams. Self-consistency checking involved not only simple range and rate-of-change tests, but also automated statistical assessment of individual

Corresponding author address: Randy A. Peppler, CIMMS, University of Oklahoma, 120 David L. Boren Blvd., Norman, OK 73072. E-mail: rpeppler@ou.edu

DOI: 10.1175/AMSMONOGRAPHS-D-15-0039.1

data streams for internal anomalies—this was done both to detect outliers and to identify instrument failure. In each case, flags were created to notify instrument operators and data users of the issues. Some statistical assessment was accomplished using a Bayesian dynamic linear model. Early applications of these checks were made for the detection of moisture on radiometer domes, and for the detection of signal attenuation, side-lobe leakage, presence of birds, and other interference on wind profilers.

A QME concept was developed at the beginning of the program to compare multiple data streams against a set of expected outcomes of the comparison, including experimental hypothesis. The multiple data streams that served as QME input included direct observations from instruments, measurements derived from multiple instrument observations and the subsequent application of algorithms to them, and model output. The idea behind this concept was that comparisons involving multiple data streams should reveal more information about quality than single data stream self-consistency checks could allow. As such, a major function of the automated QMEs was to identify data anomalies in near–real time and to help data quality analysts identify the root cause of unusual behavior. The measurements produced by the QME were treated as official data products and were archived. An early QME example compared vertically integrated water vapor from microwave radiometers with the output of a microwave radiometer instrument performance model that used thermodynamic profiles from radiosondes to drive the model. Another early QME made hourly comparisons between infrared spectral radiances observed by a Fourier transform interferometer and the output of a line-by-line radiative transfer model (Turner et al. 2004).

Substantial effort also was made early in the program toward day-to-day data quality assurance by instrument mentors (Stokes 2016, chapter 2; Cress and Sisterson 2016, chapter 5). Instrument mentors played and continue to play a vital role by 1) independently monitoring the data produced by their assigned instruments using various analytical and interpretive techniques and 2) reporting their findings on potential problems, suggesting solutions to site operators, and actively participating in the problem-resolution process. Instrument mentors were and continue to be a first line of defense in data quality assessment and problem diagnosis and solution. During the 1990s, instrument mentor and site data quality efforts often were independent and sometimes duplicative.

b. Southern Great Plains site efforts

Site scientists for the SGP site (Sisterson et al. 2016, chapter 6) at the University of Oklahoma assisted instrument mentors by developing methods to facilitate the graphical, automated display of data and within-file limit checks (Splitt 1996; Peppler and Splitt 1997). The idea behind these diagnostics was to make them available for viewing by instrument mentors and site operators on the web within 2 days of data ingest, regardless of the physical location of the sites or the data viewer someone used. Among the earliest diagnostic tools developed by the SGP site scientists were comparisons between hemispheric broadband solar irradiances and modeled clear-sky estimates, and respective comparisons of shortwave albedo estimates and broadband longwave observations from multiple SGP collection sites. Interpretive guidance was developed by instrument mentors to aid site scientists in evaluating plots, and the site scientists developed an e-mail reporting system for alerting site operators and instrument mentors in near–real time about possible problems. In time, the volume of data collected at SGP sites caught up with the capabilities of the SGP site scientists to review them in a timely manner, which spurred more attempts at automation that came to fruition once the DQO was formed. Scientists under contract to ARM at Mission Research Corporation also led early efforts to create automated diagnostic algorithms to evaluate some data streams.

c. Tropical Western Pacific site efforts

Efforts to display and assess data collected at Tropical Western Pacific (TWP) sites (Long et al. 2016, chapter 7) were undertaken initially by site scientists at Pennsylvania State University (PSU) with less instrument mentor involvement after initial siting and operation. This was partly because of the remote TWP locale presenting unique communication complications. Its first site installed at Manus Island in October 1996 included the core instrumentation found at the SGP Central Facility site, but unlike SGP collection sites in Oklahoma and Kansas, the extremely limited bandwidth of the network connection between Manus and ARM's Data Management Facility (DMF) led to delayed data delivery in the early years.

During this period, data examination by site scientists occurred in two stages. The first stage identified potential instrument and site maintenance issues and was directed toward on-site operations staff. The second stage involved a more detailed review of the data and was directed toward the science user community. To address the operations requirement, a compact data status message was constructed that included hourly statistics from most of the instruments along with environmental parameters such as the temperatures of instrument enclosures. These messages were sent via the Geostationary Operational Environmental Satellites (GOES) link each hour. Each day, plots of these hourly data were generated and posted on a website at PSU. Initially, this process was carried out by the site scientists but eventually transitioned to site operators. The plots proved useful for identifying gross errors in the data, which were fed back to site operations' technicians, allowing them to plan on-site repair visits.

Once a full TWP dataset was delivered to the DMF, the site scientists produced daily data graphics and performed diagnostic tests. Such tests included the closure of the solar direct and diffuse components, net radiative flux, and comparison of integrated radiosonde water vapor with the water vapor derived from a microwave radiometer. Data gaps also were cataloged, which led to the uncovering of problems with dataloggers. After the full dataset arrived and data were examined, the site scientist produced a report describing issues with the data, and at that point the data were released to the public. This report was submitted to the DMF, though there was not a mechanism at the time to include all of its information to data users. Most of the information was subsequently converted to forms suitable for wide distribution (explanatory text, tables, or figures). The procedure for reviewing the data prior to their release ended about the time the second TWP site was installed at Nauru in November 1998. With two sites running, it became impractical to review all of the data prior to release.

As seen during this early phase of operation at the tropical sites, the TWP site scientist took the lead role in the examination of the data collected. When a question regarding a specific instrument arose, the site scientist typically contacted the appropriate instrument mentor and worked with the mentor to solve the problem. If, on the other hand, the source of a problem was already known or suspected by the site scientists, they contacted site operators directly to work on resolution. This model of putting the site scientist at the front line of data review had distinct advantages as well as disadvantages. An advantage was that site scientists have a vested interest in the instruments at the site, and looking at multiple instrument data streams provided a holistic view of site performance that was useful for problem solving. However, this system was inefficient and time consuming, limiting site scientist time for activities such as promotion of the data to the scientific community and planning and implementing TWP field campaigns. With the establishment of DQO, the role of TWP site scientist in routine data review gradually but dramatically changed.

d. North Slope of Alaska site efforts

At the North Slope of Alaska (NSA) site (Verlinde et al. 2016, chapter 8), still another model for data quality assurance was used. Site scientists and site operators jointly subjected data to a systematic program of quality checks (e.g., Delamere et al. 1999). Data streams were visually inspected on a daily basis; from these visual inspections, metadata documenting the overall quality of the data streams were generated. Such inspections facilitated detection of instrument malfunction at the Barrow site as it was spinning up in 1998. A web-based archive of graphical images was developed and maintained to facilitate visual inspection by NSA site scientists at the Geophysical Institute of the University of Alaska, Fairbanks. These graphics were updated and made available daily. In addition to visual inspections, site scientists interpreted limits that were applied to the data. Instrument mentors had relatively little involvement in data quality assurance activities at this site after instrument installation and official data release. NSA site scientists and operators played a crucial role in discussions during 1999–2000 on how to better automate data quality checking at the site and across ARM, and how to place this information and other metadata both within data files and on the web.

e. Efforts after the World Wide Web

It should be noted that the evolution of the web, which took place during the first few years of the ARM Program, was a transformation point, especially for data quality efforts. Use of the web began as a grassroots effort and grew quickly, especially as browser technology and the Internet evolved. However, it was unevenly adopted by the three sites, which was another reason for uneven data quality treatment, especially with respect to data quality reporting and data quality coordination between different parts of the ARM infrastructure. Site scientists at the University of Oklahoma, ARM scientists at Lawrence Livermore National Laboratory, and contracted scientists at Mission Research Corporation were among the first in the program to embrace the web and develop automated algorithms that generated quick-look images that were published on the web for instrument mentors, site scientists, and others to review. There was some contention at the time about the eventual role of the web in ARM infrastructure efforts, but ultimately the use of the web for the data quality program in particular became a cornerstone of the effort.

f. Establishment of the Data Quality Office and beyond

As described above, while site-based instrument mentor, site scientist, and site operator efforts were crucial for detecting instrument malfunction and minimizing the amount of poor data collected, these efforts were unevenly developed and applied across the sites, and oftentimes were independent of each other and duplicative. This often led to varying treatments of like measurements taken at different locations, leading to uneven data quality reporting and resulting in uneven levels of data user confidence. A key finding of the ARM Program Infrastructure Review, conducted in summer 1999, stated that "a primary mission of the ARM Infrastructure is to produce a 'legacy data set' that is invaluable for research on global change. We are particularly concerned about the coordination and completeness of the quality assurance information describing ARM data" (ARM Program Infrastructure Review Committee 2001, p. 2). The review recommended that the program's data quality activities should be consolidated and coordinated, and it recommended the creation of a new position of "Data Quality Manager." This recommendation evolved in July 2000 into the establishment of the DQO at the University of Oklahoma, as described at the beginning of this chapter.

The DQO has since coordinated the program's data quality assurance activities, in continued close consultation and participation with instrument mentors, site scientists, site operators, and others. The DQO incorporated the best practices from past site efforts to create what is seen today. From the SGP, it incorporated ideas regarding self-consistency checks for individual data streams and QMEs, data intercomparisons beginning with radiation measurements, and initial ideas for web display of diagnostic plotting and display of within-file quality control limits. These actions led to the framework for a web-based display system that began as a way to compare radiation measurements and analyze soil water and temperature sensors (Bahrmann and Schneider 1999). From the TWP, the DQO took ideas for more sophisticated data plotting, including measurement intercomparisons, and the relationships it established between the site scientists and instrument mentors. The TWP site also was the first site to interact directly with data users, something that the DQO has tried to emulate, and was the first site to coordinate data quality assurance as a whole site as opposed to individual instrument mentors distributing their own assessments. From the NSA, the DQO incorporated its systematic regimen of integrating quality control checks within a file and metadata organization structure, which was novel and elements of which are used throughout different parts of the ARM infrastructure. The NSA also was an early participant in creating a web repository for graphical products. And, as described above, individuals at the University of Oklahoma and Lawrence Livermore National Laboratory were the first to bring ARM into the web age (mid-1990s) by creating web repositories of "quicklooks" in thumbnail tabular form that allowed full graphical viewing once selected. This model, novel for its time as the powers of the web were being discovered, has been emulated countless times by those both inside and outside of the ARM Program.

Instrument mentors, site scientists, and site operators still retain strong, complementary roles in the quality assurance process. These roles are embodied though problem discovery and resolution efforts, and various weekly coordination teleconferences scheduled to discuss and resolve pressing data quality issues. Instrument mentors, as the technical authorities for their instruments, continue to provide an in-depth instrument-specific perspective on data quality, responsibility for helping resolve problems, and expert help in identifying problematic long-term data trends. They also are the final arbiters of data quality to the public as embodied in their writing of data quality reports. Site scientists, as authorities on their locale and its scientific mission, provide a broad perspective on data quality spanning the full range of site instrumentation. They also help oversee their site's problem-resolution process and perform targeted research on topics related to site data quality issues. Site scientists interact directly with the scientific community to plan and conduct field campaigns at their sites, which have at times identified previously unknown data quality issues (see below). Site operators implement the problem-resolution process by orchestrating and conducting the corrective maintenance actions requested. They are key in ensuring the smooth, routine operation of the sites and their instruments through regular preventative maintenance and the application of periodic instrument mentor-specified calibration checks.

The next section describes the workflow of the data quality review process as it exists today, with an emphasis on what the DQO does. This process involves creation of data plots and displays of within-file data quality information as data are collected and ingested by ARM, routine data inspection, assessment, and status reporting by data quality analysts, problem reporting, problem resolution, and finally communication of data quality results to data users. Consultation of long-term data trends (to put current measurements in context), maintenance and calibration reports, and the development of data quality documentation (as interesting issues are discovered) to serve as pattern recognition are part of this process.

3. Data stream inspection and assessment, problem reporting and resolution, and reprocessing

We refer readers to Peppler et al. (2008b) for more details on the processes described here. Given the data volume described at the beginning of this chapter, data inspection and assessment activities must be automated and efficient, although human inspection of the results still remains a high priority.

The quality assurance model has three components. The first component is a "rapid evaluation and response" piece involving data inspection and assessment that is designed to identify gross and some more subtle issues within the data streams as fast as possible and relay that information to site operators and the instrument mentors so that the (potential) problem-resolution process can begin. The goal of this component is to minimize the amount of data that is affected by the problem. The second component involves documenting and reporting data quality issues for the scientific user; this is primarily done via text-based but machine-readable data quality reports (see below). The third component involves reprocessing of data after known problems have been identified and solved to provide end users with the best products available.

a. Inspection and assessment

The main objective of near-real-time data inspection and assessment is to quickly identify data issues and report them to instrument mentors, site scientists, and site operators so that corrective maintenance actions can be scheduled and performed, limiting the amount of unacceptable data collected. Data quality analysts at the DQO perform much of the routine data inspection, assessment, and initial problem reporting on a daily to weekly basis. This analysis is conducted not only by DQO full-time staff but also by University of Oklahoma School of Meteorology undergraduate student employees who have an interest in meteorological observations and instrumentation. These student analysts have been paramount to the DQO's success over the years, and many have gone on to graduate school and, in some cases, became faculty at other institutions of higher education. This tasking allows full-time DQO staff to spend more time on the development of data quality checking algorithms, an activity that is done in coordination with the technical guidance of instrument mentors and site scientists. This activity has resulted in the development of a broad suite of automated tools and procedures packaged into a web-based system (http://dq.arm.gov/dq-explorer/cgi-bin/main), an evolution of a forerunner system described in Peppler et al. (2005). The ARM network configuration provides the DQO with the computing power and file services (both at the DMF) needed to facilitate data quality algorithm processing. As mentioned earlier, a system prototype was created in the late 1990s by SGP site scientists as a way to monitor solar trackers for radiometers, and was later expanded to monitor the then-new soil water and temperature system. After formation of the DQO, that system was formalized into a program-wide, web-based data quality tool that has been modernized over time.

Inspection and assessment are accomplished in a three-tier process. The first tier is the application of simple consistency checks like minimum, maximum, and delta (a comparison of consecutive values collected to detect abrupt changes) checks, and whether data exists. The second tier relates to the question "do the data make sense" for the current meteorological setting (e.g., does the temperature follow the normal diurnal cycle, do cloud base height estimates from different instruments agree to within some nominal bounds, and are there artifacts in the data that are nonphysical from a meteorological perspective?). The third tier is a more advanced statistical analysis using techniques and longer time series data that can help find more subtle problems (see section 6). This last tier can be quite powerful, but care has to be taken not to flag real signals that might be outside of a 3-sigma limit, for example.

As a process, every hour the latest available data collected by fielded instruments are ingested by the DMF, processed by the DQO algorithms, and displayed in a web-based system. The DQO processing creates a graphical summary of quality control fields (flags) within each file, a graphical summary of additional DQO-generated quality control tests, and a suite of graphical depictions of data (Fig. 12-1). Student data quality analysts primarily access the system by selecting the site, data stream, and date range of interest for each data product in their purview. Prior to final submission, a color-coded request queue alerts the analyst if the dates for the data product(s) in their submission have not been processed by the DQO. A color table then is displayed showing hourly flagging summaries for each measurement (both in-file testing and additional tests performed by the DQO). This helps an analyst to quickly screen potential problem areas. All tables and graphics are updated hourly as data arrive to the DMF.

Data quality analysts visually inspect all flagging results and data graphics. Diagnostic plots, including cross-instrument comparisons that in many ways mimic the early QMEs, display daily but are updated hourly. Inspection of these plots of primary and diagnostic variables helps identify data abnormalities not always detected through automated flagging or analysis of a data stream in isolation. A succession of daily diagnostic plots in thumbnail form, which can illustrate trends in data, is available as well (http://plot.dmf.arm.gov/plotbrowser/; Fig. 12-2). Analysts may select a site, one or more data streams, and a date to view thumbnails of data plots for up to 30 days at a time. The thumbnail format facilitates comparison of different instruments that measure like quantities and can provide a view of near-term trends. The example in Fig. 12-2 highlights a

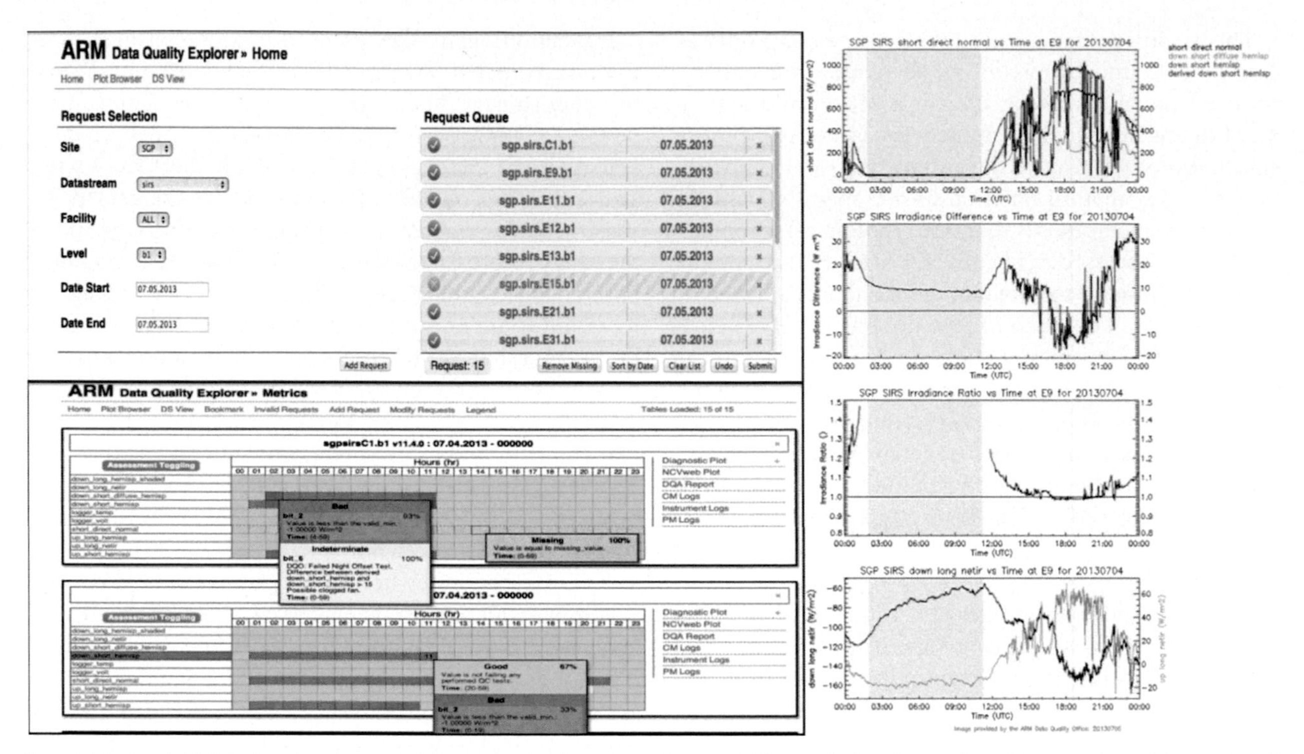

FIG. 12-1. (top left) Display showing initial data selection interface, (bottom left) sample hourly metrics table overview with associated error messages, and (right) related diagnostics plots.

case in the top panels where an unanticipated data transition takes place with 915-MHz Radar Wind Profiler precipitation consensus data, while the bottom panels as a comparison show data from another site for the same type of profiler that do not show the transition.

Another plotting capability provides the analyst with an interactive, web-based tool (Fig. 12-3). Key features include focusing on data periods of less than or more than one day, and particular data fields of interest can be specified from pulldown menus. Plots may compare data from multiple facilities, show comparisons to reference measurements, show slices of multispectral data such as atmospheric emitted radiance interferometer (AERI) radiances, or even display color-filled contours of radar reflectivity and lidar backscatter. For closer inspection, data values can be displayed in tabular form or downloaded in an ASCII comma-delimited format for easy importation into spreadsheet applications. Analysts can view file headers to obtain direct access to metadata or a summary of data field descriptions and basic data statistics. In the example shown in Fig. 12-3, temperatures are plotted for all SGP surface meteorological (MET) instrumentation sites on the same graph. In this case, a problem with the temperature sensor at the E33 facility can easily be detected.

Finally, another utility can be used to quickly create large batches of diagnostic plots that provide a detailed view of the within-file quality control describing all relevant fields in a data stream, as well as basic statistics about the data. As shown in Fig. 12-4, the top panel of the default plot for one-dimensional time series includes the data values color coded by the assessment level, while the bottom panel provides a color-coded view of the descriptions of the individual quality control (tests) applied to the data in the top panel. This example shows a plot of aerosol optical depth from the aerosol optical depth value-added product (VAP; Koontz et al. 2013). Green indicates that no tests were failed and the data are "good," yellow indicates that the data failed a test with an assessment level of "indeterminate," and red indicates that the data failed a test with an assessment level of "bad." This utility also provides a number of command-line options for customizing plot generation, such as the ability to plot all fields in a file automatically, plot along a specific coordinate in a two- or three-dimensional field, plot a two-dimensional field as multiple stacked line plots, and plot short time periods covering a few hours or long time periods covering multiple years. It has proven very useful during the evaluation of ARM data products under development, and it often allows the DQO to quickly detect errors in the automated quality control algorithms.

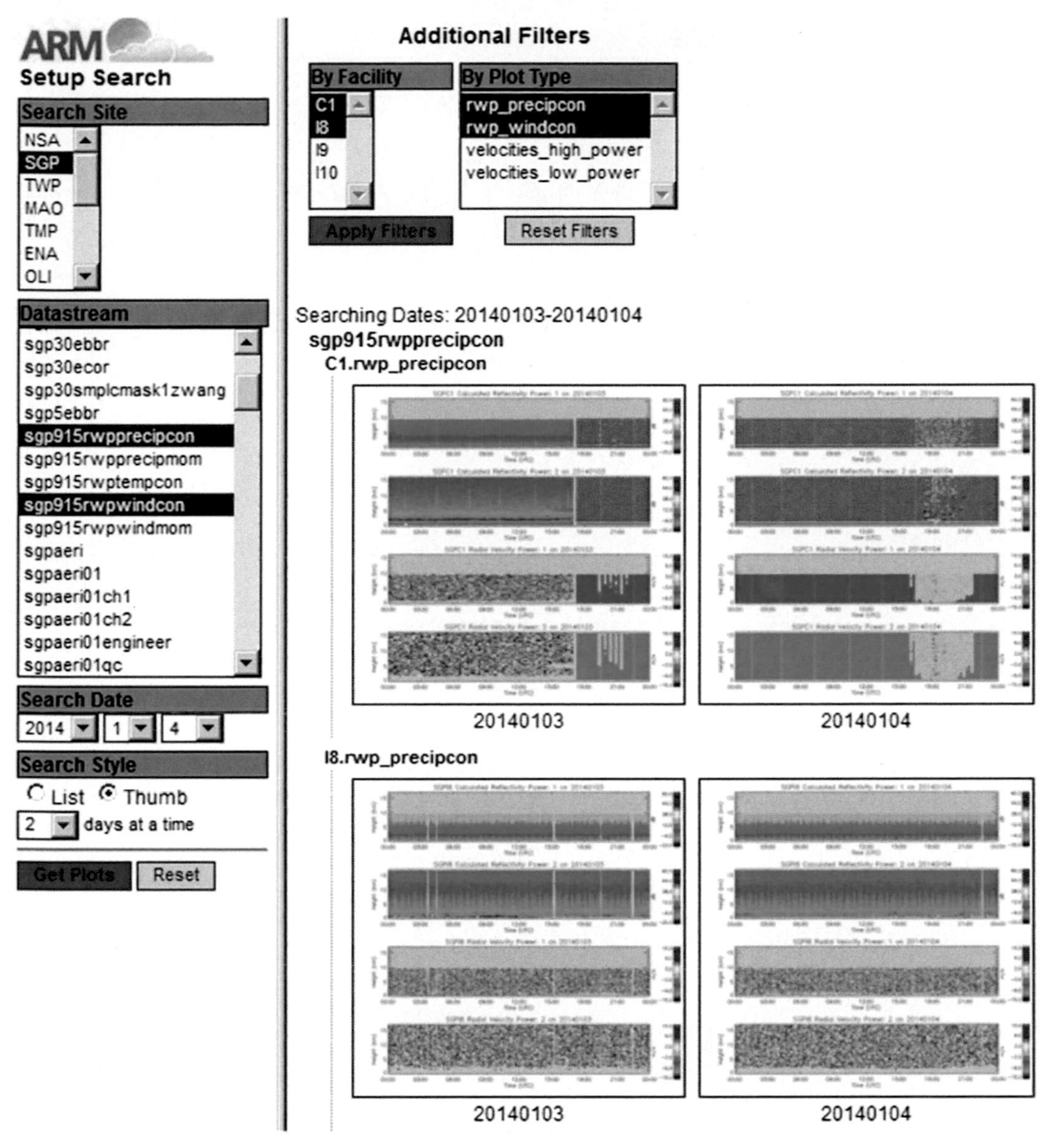

FIG. 12-2. Plot browser display showing multiday thumbnails from the SGP site. Shown are (top) two plots of 915-MHz Radar Wind Profiler (RWP) precipitation consensus data for a site showing an unanticipated data transition later in the first day that continues into the next day and (bottom) two plots from another site for the same type of profiler on the same day that does not show the transition. Clicking on a thumbnail brings up a full-resolution image.

b. Problem reporting and resolution

Once data have been inspected and assessed, a variety of reporting mechanisms allow the data quality analysts to inform instrument mentors, site operators, and site scientists of their findings. Data quality reporting mechanisms are based on web searchable and accessible databases that allow the various pieces of information produced during the quality assurance process to be neatly conveyed to problem solvers in a timely manner. The history of ARM data problem reporting is complex and is beyond the scope of this chapter (see Peppler et al. 2008a,b). However, early on in the program, a problem identification form system was implemented to allow ARM science team members to help document data quality issues encountered beyond those identified by site scientists and instrument mentors. This system provided scientists a mechanism to notify infrastructure members when a problem in the data was encountered. However, the system was used inconsistently and produced an ''uneven plowing'' of ARM data fields. The successor system described

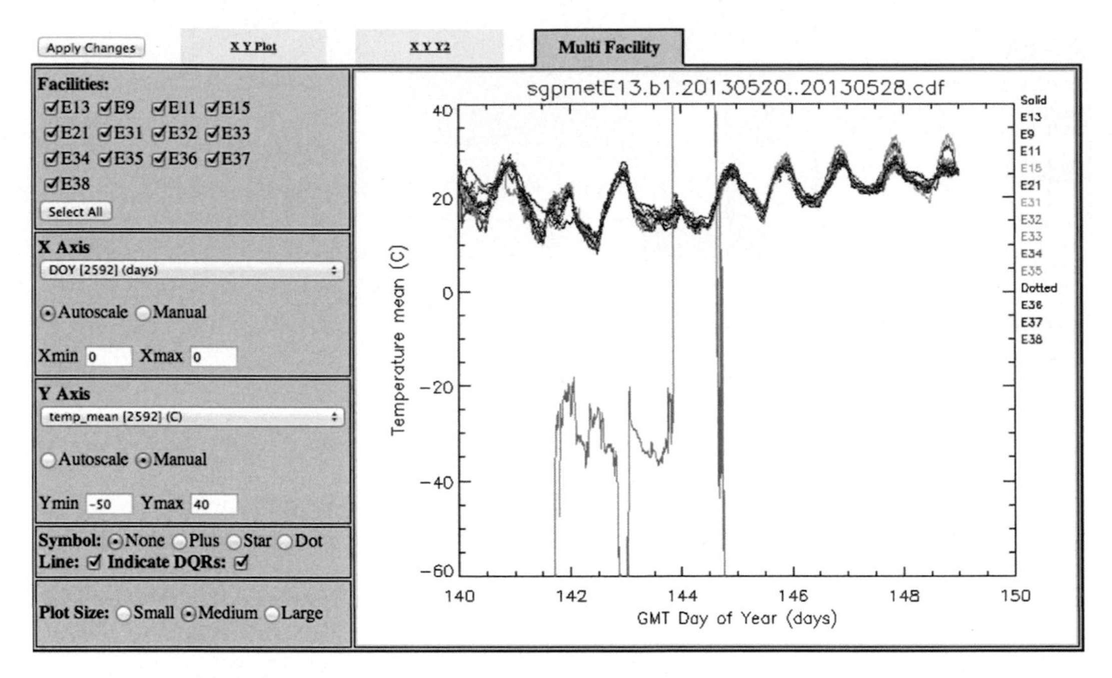

FIG. 12-3. Interactive plotting example showing 2-m temperature plotted for all SGP MET sites on the same graph. In this case, a problem temperature sensor at site E33 can be easily detected.

below has resulted in a more even and consistent treatment of ARM problem reporting and resolution.

The problem reporting system is divided into three linked processes:

1) Weekly reports are issued on data inspection and assessment results by DQO data quality analysts and distributed internally to instrument mentors, site operators, and site scientists.
2) Reports are issued describing problems discovered by data quality analysts or instrument mentors and distributed internally to instrument mentors, site operators, and site scientists so they can initiate a problem-resolution process—these online reports document the progress and status of the actions proposed and implemented.
3) Data quality reports documenting a known problem and its resolution as written by instrument mentors are distributed publicly to the data user community.

These data quality reports are provided to the data user along with the data they describe when an official data request is made to the ARM Data Archive. The history of a problem, including its discovery, the corrective actions taken to resolve it, and a report on its effects on data quality are typically included in these public reports and are database searchable on many criteria. The linked databases allow for the tracking of problem trends and help identify problematic instrument systems or facilities. An ARM Data Archive service has been implemented to filter ordered data based on data quality reports. A data user may decide upfront which types of problems warrant data removal and can receive a custom product based on their particular needs. If a data user prefers more fine-grained control of data filtering, they can have their own processing codes query a data quality report web service in order to receive the details regarding any data problems in a machine-readable form.

c. Reprocessing

Last, the ARM infrastructure conducts an extensive data reprocessing program that is informed by the data quality assessment process. Reprocessing is performed to fix known data issues and has been used extensively throughout the lifetime of ARM. Reprocessing requires the modification or elimination of previous data quality reports and the subsequent reissuing of data to all who may have downloaded the data from the Data Archive (the Data Archive tracks all users of all data streams; McCord and Voyles 2016, chapter 11). Reprocessing is not able to fix all problems, but it has been helpful in providing data users with the best products available. As an example of reprocessing, the MET instrumentation at the SGP Central Facility site has been known by different names since the beginning of ARM [Surface Meteorological Observation Station (SMOS) and then MET]. Each has had changing variable names over time, which has been confusing to data users. These data

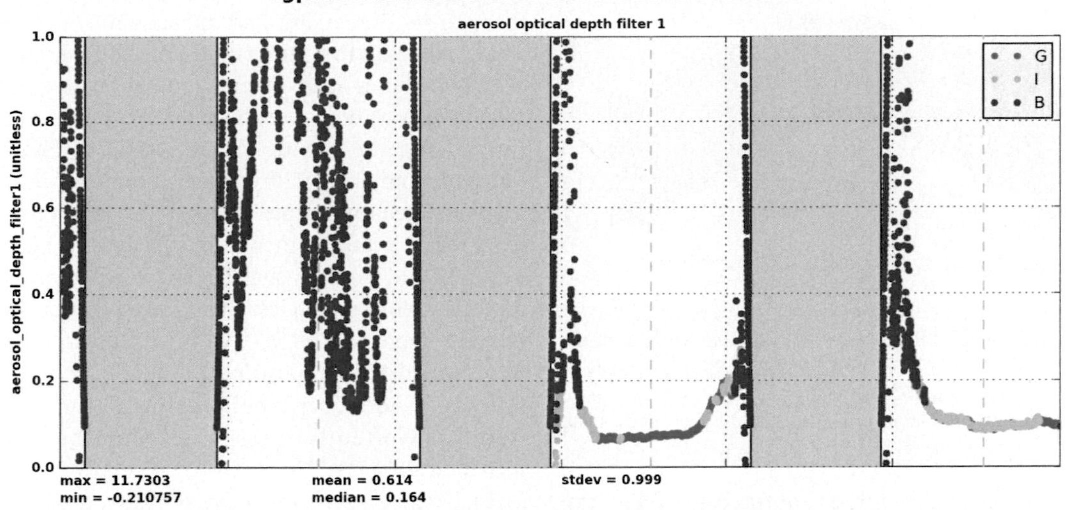

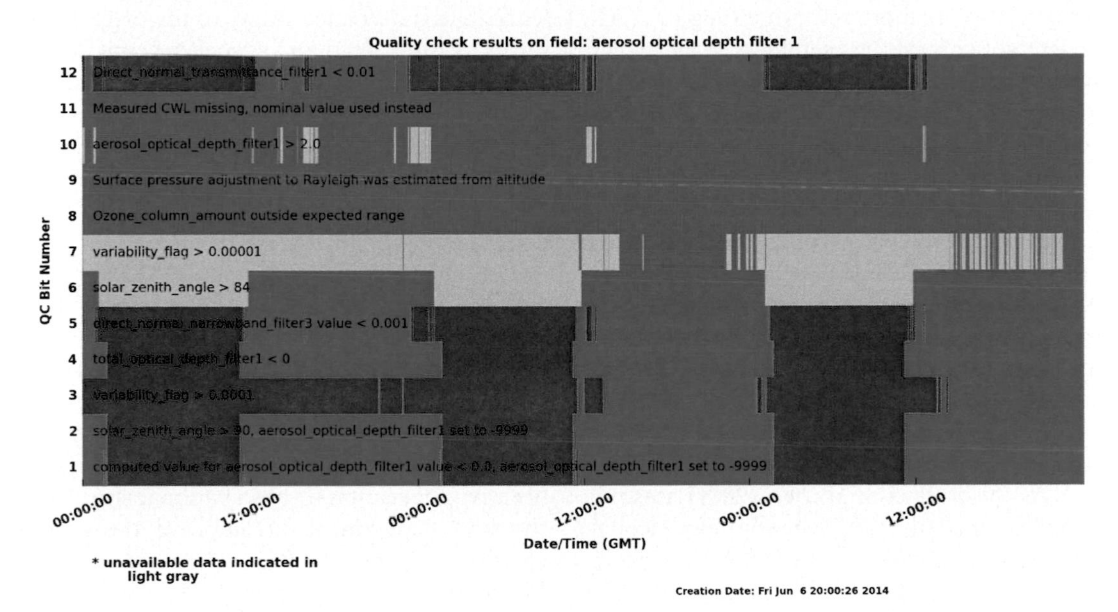

FIG. 12-4. Plot of aerosol optical depth from the SGP aerosol optical depth VAP (Koontz et al. 2013). The gray and yellow background in the top panel represents nighttime and daytime, respectively. Green indicates that no tests were failed and the data are "good," yellow indicates that the data failed a test with an assessment level of "indeterminate," and red indicates that the data failed a test with an assessment level of "bad." Missing values (−9999) are automatically masked from the top-panel plot, and the y axis is scaled from 0 to 1 to show additional detail.

streams were reprocessed to provide one consistent dataset running from 1993 to the present time, greatly improving the ease of working with a long time series of the data (see Fig. 12-5). In another case, the European Centre for Medium-Range Weather Forecasts (ECMWF) model output retrieved from the ARM External Data Center was improperly ingested into the Merged Sounding (MERGESONDE) VAP (Troyan 2012). This issue adversely impacted moisture fields such as relative humidity and vapor pressure, causing the values within the boundary layer at the TWP Darwin site to essentially drop to zero during time periods when sounding data was unavailable. A software update and subsequent reprocessing corrected this problem.

4. Role of VAPs in data quality characterization

Some of the scientific data needs of ARM data users are met through the creation of VAPs (http://www.arm.gov/data/vaps; Ackerman et al. 2016, chapter 3). Despite the extensive instrumentation deployed at the field sites, some measurements of interest are either impractical or

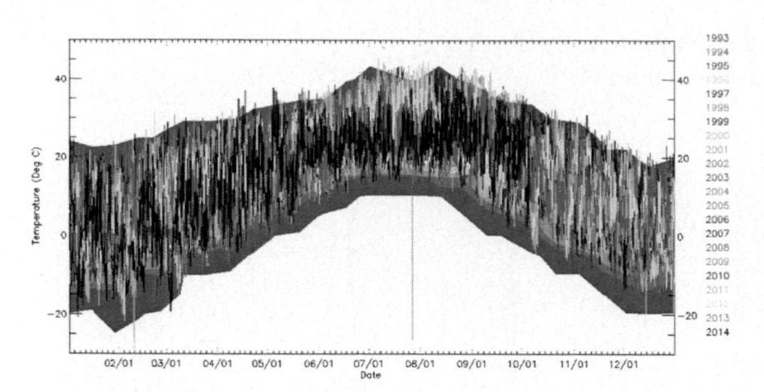

FIG. 12-5. Twenty years of temperature data from one SGP MET plotted over a PDF to indicate where varying percentages of the data normally lie (50%, 75%, 90%, 95%, 99.9%). A couple of outliers that have not been documented by data quality reports or removed by mentor supplied limits can be seen as dropouts in the data.

impossible to make directly or routinely—VAPs have filled this void. The creation and processing of VAPs have shed light not only on the usefulness of the higher-level products produced but also have provided information about the quality of the input data streams and the operation of the instruments that produced them. VAP processing and the analysis that is needed going into creating the VAPs has allowed detection of more subtle measurement inaccuracies that often defy detection through standard near-real-time data quality approaches such as limits testing or cross-measurement plotting comparisons. Two examples of VAP analysis aiding ARM data quality assurance efforts are described next.

a. Cloud radar

The Active Remotely-Sensed Cloud Locations (ARSCL) VAP (Clothiaux et al. 2000, 2001) uses millimeter cloud radar data as its primary input (Kollias et al. 2016, chapter 17). While the amount of power broadcast by the radar and returned by targets can be monitored, there are many factors involved in the operation of this complex radar that can affect data quality. ARSCL processing has revealed both radar measurement issues and radar operating characteristics. ARSCL output serves as input to the Baseline Cloud Microphysical Retrievals (MicroBase) VAP (Dunn et al. 2011), where retrievals are scrutinized both in terms of their consistency with other measurements and their relevance to the physical circumstances within which they are embedded. Consideration of consistency and situational context has been powerful for determining data quality to a degree not possible when analyzing measurements or retrievals in isolation.

b. Radiative transfer

The Broadband Heating Rate Profiles (BBHRP) VAP (Mlawer et al. 2002; McFarlane et al. 2016, chapter 20) allowed the discovery of an unforeseen problem through its processing and subsequent user feedback. This VAP takes the output of the ARSCL and MicroBase VAPs and uses it in detailed radiative transfer model calculations to compute the radiative fluxes at both the surface and the top of the atmosphere. The BBHRP output is compared with surface and top-of-atmosphere irradiance measurements in a closure experiment framework (Mlawer and Turner 2016, chapter 14). This comparison revealed a subtle shift in model-minus-measurement flux difference statistics with direct normal shortwave measurements at the SGP Central Facility, a shift that was caused by human error when two digits of the normal incidence pyrheliometer calibration factor were inadvertently transposed while being entered into a datalogger. This error resulted in a roughly 2% error in the direct shortwave measurements, which is within the stated uncertainty of the calibrations themselves (Stoffel 2005) and as such was not detectable by standard limits and cross-comparison testing. This finding resulted in a reprocessing task to correct for this mistake, and led to a much improved direct normal solar flux dataset.

5. Role of field campaigns in data quality, measurement, and site characterization

ARM's data collection sites, including mobile facility deployments, host field campaigns to address specific scientific questions, augment routine data collections, and test and validate new instruments (http://www.arm.gov/campaigns). An emphasis of many campaigns has been on application of observational strategies and instrument deployments to improve the accuracy and quality of key measurements. A few are described here, which in some cases have produced climate community-wide ramifications on field measurement characterization.

a. Water vapor

Given the importance of water vapor as a greenhouse gas and its role in the life cycle of clouds and precipitation, the transfer of latent and sensible heat, and atmospheric chemistry, ARM has expended considerable observational effort, particularly at the SGP site, on the measurement of water vapor (Turner et al. 2016, chapter 13). Water vapor experiments held in 1996, 1997, and 2000, and a lidar experiment in 1999 provided key information on the quality and accuracy of onsite water vapor instrumentation (Revercomb et al. 2003). Dual-radiosonde launches revealed significant variability across and within calibration batches and showed that differences between any two radiosondes act as an altitude-independent scale factor in the lower troposphere, such that a well-characterized reference can be used to reduce the variability. An approach subsequently was adopted by ARM to scale the radiosonde's moisture profile to agree with the precipitable water vapor observed by the microwave radiometer. This scaling significantly reduced the sonde-to-sonde variability by a factor of 2 (Turner et al. 2003). These water vapor experiments were able to verify that 1) 60-m tower-mounted in situ sensors can serve as an absolute measurement reference, 2) the SGP site's unique Raman lidar can serve as a stable transfer standard, and 3) the sensitivity of the site's microwave radiometers was excellent over a wide range of integrated water vapor. Data from the 1997 experiment figured strongly in an effort to evaluate retrievals of column water vapor and liquid water amounts from microwave radiometers (Ivanova et al. 2002).

During the first water vapor experiment in 1996, on-site humidity measurements were verified through laboratory intercomparison of in situ moisture sensors (including both capacitive chip and chilled mirror sensors) using Oklahoma Mesonet calibration facilities. Tests were made both before and after the experiment, making it possible to detect instrument problems prior to the experiment and instrument failure or drift during the experiment (Richardson et al. 2000). As a result of this work, modifications were made to humidity sensor calibration procedures and redundant humidity and temperature sensors were fielded to better detect sensor drift and calibration error.

While the aforementioned water vapor experiments were concerned with characterization of water vapor in the lower troposphere, the ARM First International Satellite Cloud Climatology Project (ISCCP) Regional Experiment (FIRE) Water Vapor Experiment (AFWEX), conducted with the National Aeronautics and Space Administration (NASA) in November–December 2000, attempted to better characterize the measurement of upper tropospheric water vapor (Ferrare et al. 2004). Results showed excellent agreement between satellite and Raman lidar observations of upper tropospheric humidity with systematic differences of about 10%; radiosondes, conversely, were found to be systematically drier by 40% relative to both satellite and lidar measurements (Soden et al. 2004). Existing strategies for correcting radiosonde dry biases were found to be inadequate in the upper troposphere, and an alternative method was suggested that considerably improved radiosonde measurement agreement with lidar observations. The alternative method was recommended as a strategy to improve the quality of the global historical record of radiosonde water vapor observations during the satellite era.

b. Atmospheric radiation

Some field campaigns have helped characterize the measurement of atmospheric radiation, especially broadband radiation (Michalsky and Long 2016, chapter 16). The second ARM Enhanced Shortwave Experiment (ARESE-II), conducted in February–April 2000 at the SGP site (Michalsky et al. 2002), focused on broadband shortwave calibration using both ground-based and aircraft-mounted radiometers and a special radiometer that could be considered a reference standard (Michalsky and Long 2016, chapter 16). A diffuse horizontal shortwave irradiance experiment held during September–October 2001 at the SGP site (Michalsky et al. 2003) characterized a nighttime offset by comparing diffuse irradiance measurements among most commercial pyranometers and a few prototypes, with the goal of reducing the uncertainty of shortwave diffuse irradiance measurements in lieu of a standard or reference for the measurement. An international pyrgeometer and absolute sky-scanning radiometer comparison held during September–October 1999 at the SGP site (Philipona et al. 2001) shed light on the reliability and consistency of atmospheric longwave radiation measurements and calculations and determined their uncertainties, also in lieu of an existing absolute standard.

c. Site characterization

Field campaigns also have contributed understanding on the representativeness of the ARM sites with respect to how well the data that are collected accomplish their scientific intent relative to desired measurement needs. The Manus Island and Nauru TWP sites had been established to make measurements representative of the surrounding oceanic area (Long et al. 2016, chapter 7). A goal of the Nauru99 field campaign was to investigate whether the small island, producing a cloud street phenomenon, was influencing the measurements made there (Post and Fairall 2000). The affirmative result led to a yearlong Nauru Island Effects Study in which a quantification of the island effect on measurements was made (Long 1998) and a method to detect the effect's ongoing occurrence and influence on collected data was developed (McFarlane et al. 2005). This study led to an explanation of the cloud street phenomenon (Matthews et al. 2007). These activities were able to quantify how well the measurements characterized the surrounding oceanic area and more generally illustrated the importance of considering spatial scales as part of the quality assurance process for siting instrumentation to measure the intended target environment.

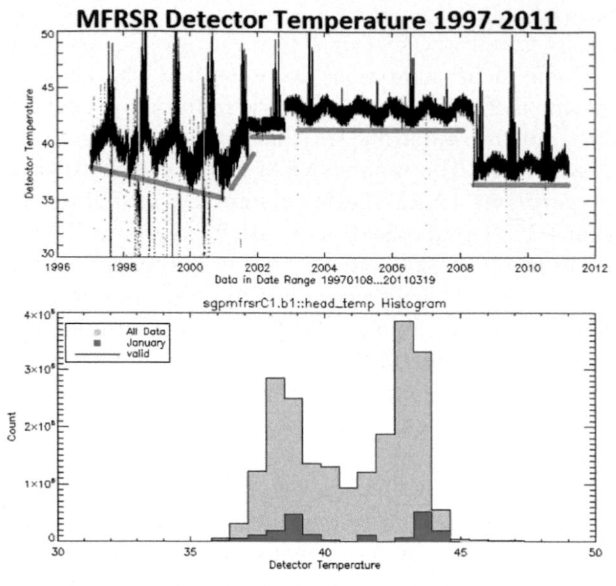

FIG. 12-6. (top) Long time series plots and (bottom) associated frequency distributions are used to detect trends and significant shifts in data that may indicate data quality problems. This example displays SGP MFRSR values over a 15-yr period at one site and indicates five distinct shifts in the data.

6. Use of ARM's historical data archive to improve current and past assessments

With over 20 years of continuous data amassed for some measurements at the ARM Data Archive (McCord and Voyles 2016, chapter 11), it is becoming possible to conduct statistical analysis on specific time scales; this work should provide valuable context for real-time measurements being made. Historical data are mined (Moore et al. 2007) to identify site-specific and time-varying (monthly or seasonal) quality control flagging limits and to facilitate better detection of subtle trends and abrupt changes in data (Fig. 12-6) that are difficult to understand when not considered in a broader context. This allows the incorporation of departures from the ARM climatology to inform the quality assurance process. Frequency distributions categorized by month and season help establish better data range limits specific to those time periods. Time series that alert analysts to outliers allow them to better distinguish bad data from unusual but valid data. Shown in Fig. 12-5 is 20 years of temperature data from one SGP MET plotted over a probability density function (PDF) to indicate where varying percentages of the data normally lie (50%, 75%, 90%, 95%, 99.9%). Figure 12-6 displays a long time series plot and associated frequency distribution of 15 years of multifilter rotating shadowband radiometer (MFRSR) values at one site—these indicate five distinct shifts in the data over that time.

In this analysis system, statistics and intelligent data range limits are used in a feedback mechanism to help data quality analysts and instrument mentors better manage validation checks. Web-based applications allow an analyst to request a particular data analysis and to view the results, allowing the dynamic creation of statistics and parametric analyses over any custom time range (e.g., day, month, year, or years).

7. Summary

The ARM data quality program necessarily has evolved from solid site-specific efforts as sites were established and matured, to a comprehensive, coordinated system that inspects, assesses, and reports on data quality incidents across all instrument systems and data collection sites. This coordinates the efforts not only of the DQO but also those of the instrument mentors, site scientists, site operators, and data system engineers. The role of scientific analysis, embodied primarily through the efforts to create VAPs and to conduct field campaigns, has allowed ARM to improve data quality by better characterizing measurements needed to improve the treatment of clouds and radiation in climate models, and better characterizing the sites where the measurements are taken so that they better fulfill the needs of the research community. The cumulative effect of these is a story of much individual and group effort that has taken place since the early 1990s.

What lessons has ARM learned over 20 years that are valuable to other field observation programs? A key lesson learned is that field measurement realities in sometimes harsh operational settings often deviate very quickly from intended instrument functioning in terms of what had been prescribed through laboratory calibration and created through inherent instrument measurement preciseness. Continuous calibration checking and routine scheduled maintenance therefore are essential to allow measurement systems to produce data as close as possible to their intended form. Also, intercomparison of like measurements at a particular location, to the extent possible and even for brief time periods, is key for helping establish the fidelity of the particular measurements being taken; the water vapor field campaigns, for example, were invaluable in helping ARM establish how to best measure water vapor, which had community-wide ramifications.

A comprehensive data quality assessment program is essential for documenting the quality assurance process and ultimately in producing a dataset of some prescribed known quality and usability. The program must collect and track data about the system (metadata) at every point along the assessment path, from instrument selection and procurement to initial fielding and beta testing; to field operation, data collection, and quality inspection and assessment; to problem reporting and resolution; and to data distribution and communication of information about those data. If the process and details of the data created cannot be described, the data will have limited scientific value.

Going forward, the ARM data quality assessment process will take even greater advantage of automation opportunities as new tools and ways of thinking are developed. This automation process is necessary because the ARM data collection volume is anticipated to increase over time. This process needs to take increasing advantage of the sometimes 20-years-long ARM climatology that has been developed to better place current measurements into longer-term perspective and to create dynamic quality control flagging limits on various time scales to make them more meaningful. Another future goal is to better characterize current data in terms of inherent measurement accuracy and defined instrument precision, in concert with establishment of data spread over the long-term collection horizon.

REFERENCES

Ackerman, T. P., T. S. Cress, W. R. Ferrell, J. H. Mather, and D. D. Turner, 2016: The programmatic maturation of the ARM Program. *The Atmospheric Radiation Measurement (ARM) Program: The First 20 Years, Meteor. Monogr.*, No. 57, Amer. Meteor. Soc., doi:10.1175/AMSMONOGRAPHS-D-15-0054.1.

ARM Program Infrastructure Review Committee, 2001: The Atmospheric Radiation Measurement Program Infrastructure Review Report (AIR): Summary of Recommendations. U.S. Dept. of Energy ARM Program Doc. DOE/SC-ARM-0001, 3 pp. [Available online at http://www.arm.gov/publications/programdocs/doe-sc-arm-0001.pdf?id=55.]

Bahrmann, C. P., and J. M. Schneider, 1999: Near real-time assessment of SWATS data quality, resulting in an overall improvement in pre- sent-day SWATS data quality. *Proc. Ninth Atmospheric Radiation Measurement (ARM) Science Team Meeting*, San Antonio, TX, U.S. Dept. of Energy, 6 pp. [Available online at http://www.arm.gov/publications/proceedings/conf09/extended_abs/bahrmann_cp.pdf.]

Blough, D. K., 1992: Real-time statistical quality control and ARM. *Proc. 46th Annual ASQC Quality Congress*, Nashville, TN, American Society for Quality Control, 484–490.

Clothiaux, E. E., T. P. Ackerman, G. G. Mace, K. P. Moran, R. T. Marchand, M. A. Miller, and B. E. Martner, 2000: Objective determination of cloud heights and radar reflectivities using a combination of active remote sensors at the ARM CART sites. *J. Appl. Meteor.*, **39**, 645–665, doi:10.1175/1520-0450(2000)039<0645:ODOCHA>2.0.CO;2.

——, and Coauthors, 2001: The ARM Millimeter Wave Cloud Radars (MMCRs) and the Active Remote Sensing of Clouds (ARSCL) Value Added Product (VAP). DOE Tech. Memo. ARM VAP-002.1, 56 pp. [Available online at http://www.arm.gov/publications/tech_reports/arm-vap-002-1.pdf.]

Cress, T. S., and D. L. Sisterson, 2016: Deploying the ARM sites and supporting infrastructure. *The Atmospheric Radiation Measurement (ARM) Program: The First 20 Years, Meteor. Monogr.*, No. 57, Amer. Meteor. Soc., doi:10.1175/AMSMONOGRAPHS-D-15-0049.1.

Delamere, J. S., and Coauthors, 1999: The first year of operation of the North Slope of Alaska/Adjacent Arctic Ocean ARM site: An overview of instrumentation, data streams, and data quality assurance procedures. *Proc. Ninth Atmospheric Radiation Measurement (ARM) Science Team Meeting*, San Antonio, TX, U.S. Dept. of Energy, 4 pp. [Available online at http://www.arm.gov/publications/proceedings/conf09/extended_abs/delamere1_js.pdf.]

Dunn, M., K. Johnson, and M. Jensen, 2011: The Microbase Value-Added Product: A baseline retrieval of cloud microphysical properties. Tech. Rep. DOE/SC-ARM/TR-095, 34 pp. [Available online at http://www.arm.gov/publications/tech_reports/doe-sc-arm-tr-095.pdf.]

Ferrare, R. A., and Coauthors, 2004: Characterization of upper-troposphere water vapor measurements during AFWEX using LASE. *J. Atmos. Oceanic Technol.*, **21**, 1790–1808, doi:10.1175/JTECH-1652.1.

Ivanova, K., E. E. Clothiaux, H. N. Shirer, T. P. Ackerman, J. C. Liljegren, and M. Ausloos, 2002: Evaluating the quality of ground-based microwave radiometer measurements and retrievals using detrended fluctuations and spectral analysis methods. *J. Appl. Meteor.*, **41**, 56–68, doi:10.1175/1520-0450(2002)041<0056:ETQOGB>2.0.CO;2.

Kollias, P., and Coauthors, 2016: Development and applications of ARM millimeter-wavelength cloud radars. *The Atmospheric Radiation Measurement (ARM) Program: The First 20 Years, Meteor. Monogr.*, No. 57, Amer. Meteor. Soc., doi:10.1175/AMSMONOGRAPHS-D-15-0037.1.

Koontz, A., C. Flynn, G. Hodges, J. Michalsky, and J. Barnard, 2013: Aerosol optical depth value-added product. Tech. Rep. DOE/SC-ARM/TR-129, 32 pp. [Available online at http://www.arm.gov/publications/tech_reports/doe-sc-arm-tr-129.pdf.]

Long, C. N., 1998: Nauru Island Effect Study (NIES) IOP Science Plan. Tech. Doc. DOE/SC-ARM-0505, 17 pp. [Available online at http://www.arm.gov/publications/programdocs/doe-sc-arm-0505.pdf.]

——, J. H. Mather, and T. P. Ackerman, 2016: The ARM Tropical Western Pacific (TWP) sites. *The Atmospheric Radiation Measurement (ARM) Program: The First 20 Years, Meteor. Monogr.*, No. 57, Amer. Meteor. Soc., doi:10.1175/AMSMONOGRAPHS-D-15-0024.1.

Matthews, S., J. M. Hacker, J. Cole, J. Hare, C. N. Long, and R. M. Reynolds, 2007: Modification of the atmospheric boundary layer by a small island: Observations from Nauru. *Mon. Wea. Rev.*, **135**, 891–905, doi:10.1175/MWR3319.1.

McCord, R., and J. W. Voyles, 2016: The ARM data system and archive. *The Atmospheric Radiation Measurement (ARM) Program: The First 20 Years, Meteor. Monogr.*, No. 57, Amer. Meteor. Soc., doi:10.1175/AMSMONOGRAPHS-D-15-0043.1.

McFarlane, S. A., C. N. Long, and D. M. Flynn, 2005: Impact of island-induced clouds on surface measurements: Analysis of the ARM Nauru Island Effect Study data. *J. Appl. Meteor.*, **44**, 1045–1065, doi:10.1175/JAM2241.1.

——, J. H. Mather, and E. J. Mlawer, 2016: ARM's progress on improving atmospheric broadband radiative fluxes and heating rates. *The Atmospheric Radiation Measurement (ARM) Program: The First 20 Years, Meteor.*

Monogr., No. 57, Amer. Meteor. Soc., doi:10.1175/AMSMONOGRAPHS-D-15-0046.1.

Michalsky, J. J., and C. N. Long, 2016: ARM solar and infrared broadband and filter radiometry. *The Atmospheric Radiation Measurement (ARM) Program: The First 20 Years, Meteor. Monogr.*, No. 57, Amer. Meteor. Soc., doi:10.1175/AMSMONOGRAPHS-D-15-0031.1.

——, and Coauthors, 2002: Broadband shortwave calibration results from the Atmospheric Radiation Measurement Enhanced Shortwave Experiment II. *J. Geophys. Res.*, **107**, 4287, doi:10.1029/2001JD001231.

——, and Coauthors, 2003: Results from the first ARM diffuse horizontal shortwave irradiance comparison. *J. Geophys. Res.*, **108**, 4108, doi:10.1029/2002JD002825.

Miller, N. E., J. C. Liljegren, T. R. Shippert, S. A. Clough, and P. D. Brown, 1994: Quality measurement experiments within the Atmospheric Radiation Measurement Program. Preprints, *Fifth Symp. on Global Change Studies*, Nashville, TN, Amer. Meteor. Soc., 35–39.

Mlawer, E. J., and D. D. Turner, 2016: Spectral radiation measurements and analysis in the ARM Program. *The Atmospheric Radiation Measurement (ARM) Program: The First 20 Years, Meteor. Monogr.*, No. 57, Amer. Meteor. Soc., doi:10.1175/AMSMONOGRAPHS-D-15-0027.1.

——, and Coauthors, 2002: The Broadband Heating Rate Profile (BBHRP) VAP. *Proc. 12th Atmospheric Radiation Measurement (ARM) Science Team Meeting*, St. Petersburg, FL, U.S. Dept. of Energy, 12 pp. [Available online at http://www.arm.gov/publications/proceedings/conf12/extended_abs/mlawer-ej.pdf.]

Moore, S. T., K. Kehoe, R. Peppler, and K. Sonntag, 2007: Analysis of historical ARM measurements to detect trends and assess typical behavior. *16th Conf. on Applied Climatology*, San Antonio, TX, Amer. Meteor. Soc., P2.6. [Available online at http://ams.confex.com/ams/pdfpapers/119946.pdf.]

Peppler, R. A., and M. E. Splitt, 1997: SGP Site Scientist Team data quality assessment activities. *Proc. Seventh Atmospheric Radiation Measurement (ARM) Science Team Meeting*, San Antonio, TX, U.S. Dept. of Energy, 403–406. [Available online at http://www.arm.gov/publications/proceedings/conf07/extended_abs/peppler_ra.pdf.]

——, K. E. Kehoe, K. L. Sonntag, S. T. Moore, and K. J. Doty, 2005: Improvements to and status of ARM's Data Quality Health and Status System. *15th Conf. on Applied Climatology*, Savannah, GA, Amer. Meteor. Soc., J3.13. [Available online at http://ams.confex.com/ams/pdfpapers/91618.pdf.]

——, and Coauthors, 2008a: Quality Assurance of ARM Program Climate Research Facility data. Tech. Rep. DOE/SC-ARM/TR-082, 65 pp. [Available online at http://www.arm.gov/publications/tech_reports/doe-sc-arm-tr-082.pdf.]

——, and Coauthors, 2008b: An overview of ARM Program Climate Research Facility data quality assurance. *Open Atmos. Sci. J.*, **2**, 192–216, doi:10.2174/1874282300802010192.

Philipona, R., and Coauthors, 2001: Atmospheric longwave irradiance uncertainty: Pyrgeometers compared to an absolute sky-scanning radiometer, atmospheric emitted radiance interferometer, and radiative transfer model calculations. *J. Geophys. Res.*, **106**, 28 129–28 141, doi:10.1029/2000JD000196.

Post, M. J., and C. F. Fairall, 2000: Early results from the Nauru99 campaign on NOAA ship Ronald H. Brown. *Proc. Int. Geoscience and Remote Sensing Symp.*, Honolulu, HI, IEEE, 1151–1153, doi:10.1109/IGARSS.2000.858052.

Revercomb, H. E., and Coauthors, 2003: The ARM Program's water vapor intensive observation periods. *Bull. Amer. Meteor. Soc.*, **84**, 217–236, doi:10.1175/BAMS-84-2-217.

Richardson, S. J., M. E. Splitt, and B. M. Lesht, 2000: Enhancement of ARM surface meteorological observations during the fall 1996 water vapor intensive observation period. *J. Atmos. Oceanic Technol.*, **17**, 312–322, doi:10.1175/1520-0426(2000)017<0312:EOASMO>2.0.CO;2.

Sisterson, D., R. Peppler, T. S. Cress, P. Lamb, and D. D. Turner, 2016: The ARM Southern Great Plains (SGP) site. *The Atmospheric Radiation Measurement (ARM) Program: The First 20 Years, Meteor. Monogr.*, No. 57, Amer. Meteor. Soc., doi:10.1175/AMSMONOGRAPHS-D-16-0004.1.

Soden, B. J., D. D. Turner, B. M. Lesht, and L. M. Miloshevich, 2004: An analysis of satellite, radiosonde, and lidar observations of upper tropospheric water vapor from the Atmospheric Radiation Measurement Program. *J. Geophys. Res.*, **109**, D04105, doi:10.1029/2003JD003828.

Splitt, M. E., 1996: Data quality display modules—Assessment of instrument performance at the Southern Great Plains Cloud and Radiation Testbed site. *Proc. Sixth Atmospheric Radiation Measurement (ARM) Science Team Meeting*, San Antonio, TX, U.S. Dept. of Energy, 3 pp. [Available online at http://www.arm.gov/publications/proceedings/conf06/extended_abs/splitt_me.pdf.]

Stoffel, T., 2005: Solar Infrared Radiation Station (SIRS) Handbook. Tech. Rep. ARM TR-025, 27 pp. [Available online at http://www.arm.gov/publications/tech_reports/handbooks/sirs_handbook.doc.]

Stokes, G. M., 2016: Original ARM concept and launch. *The Atmospheric Radiation Measurement (ARM) Program: The First 20 Years, Meteor. Monogr.*, No. 57, Amer. Meteor. Soc., doi:10.1175/AMSMONOGRAPHS-D-15-0021.1.

Troyan, D., 2012: Merged sounding value-added product. Tech. Rep. DOE/SC-ARM/TR-087, 19 pp. [Available online at http://www.arm.gov/publications/tech_reports/doe-sc-arm-tr-087.pdf.]

Turner, D. D., B. M. Lesht, S. A. Clough, J. C. Liljegren, H. E. Revercomb, and D. C. Tobin, 2003: Dry bias and variability in Vaisala RS80-H radiosondes: The ARM experience. *J. Atmos. Oceanic Technol.*, **20**, 117–132, doi:10.1175/1520-0426(2003)020<0117:DBAVIV>2.0.CO;2.

——, and Coauthors, 2004: The QME AERI LBLRTM: A closure experiment for downwelling high spectral resolution infrared radiance. *J. Atmos. Sci.*, **61**, 2657–2675, doi:10.1175/JAS3300.1.

——, E. J. Mlawer, and H. E. Revercomb, 2016: Water vapor observations in the ARM Program. *The Atmospheric Radiation Measurement (ARM) Program: The First 20 Years, Meteor. Monogr.*, No. 57, Amer. Meteor. Soc., doi:10.1175/AMSMONOGRAPHS-D-15-0025.1.

Verlinde, J., B. Zak, M. D. Shupe, M. Ivey, and K. Stamnes, 2016: The ARM North Slope of Alaska (NSA) sites. *The Atmospheric Radiation Measurement (ARM) Program: The First 20 Years, Meteor. Monogr.*, No. 57, Amer. Meteor. Soc., doi:10.1175/AMSMONOGRAPHS-D-15-0023.1.

Chapter 13

Water Vapor Observations in the ARM Program

D. D. TURNER

NOAA/National Severe Storms Laboratory, Norman, Oklahoma

E. J. MLAWER

Atmospheric and Environmental Research, Inc., Lexington, Massachusetts

H. E. REVERCOMB

Space Science and Engineering Center, University of Wisconsin–Madison, Madison, Wisconsin

1. Introduction

From the earliest days of the ARM Program, water vapor and temperature measurements were considered among the highest priority measurements made at the ARM sites. The program founders recommended that these observations "be performed on a continuing, real-time basis throughout the experiment" so that the radiation models and cloud parameterizations could be evaluated over a wide range of atmospheric conditions (DOE 1990; ARM 2016, appendix A). Furthermore, the program founders believed that these measurements should be made on a time scale appropriate for the study of radiative properties, which necessitated a temporal resolution on the order of minutes (DOE 1990).

In the early 1990s, many different instruments were available to provide information on the atmospheric water vapor concentration and temperature. Some of these instruments had been used for years in the operational community (e.g., radiosonde and in situ sensors), whereas some were very much research instruments (e.g., Raman lidars, global positioning systems, and microwave radiometers). However, because of the critical need to measure water vapor with the precision necessary to improve the accuracy of radiative transfer models, the program decided to deploy multiple instruments sensitive to water vapor at each site. This strategy provided opportunities to compare the different technologies and develop new methods to combine observations to create more accurate water vapor products.

This chapter details the history and evolution of water vapor measurements within the ARM Program. Most of the early work focused around the needs of the Instantaneous Radiative Flux (IRF) working group within the ARM science team. The IRF was striving to advance the work started with the Intercomparison of Radiation Codes in Climate Models (ICRCCM; see Ellingson et al. 2016, chapter 1): namely, the evaluation and improvement of the accuracy of detailed radiative transfer models used to build radiation parameterizations used in climate models. Although temperature observations are just as critical to advance ARM science, the uncertainties in temperature observations were considered smaller and less important and, consequently, received less attention in the earlier days of the program.

2. The earliest days

In the early years of the program, the IRF was arguably the most organized working group because 1) the Spectral Radiance Experiment (SPECTRE) science objectives were a large part of the ARM science plan (Ellingson and Wiscombe 1996; Ellingson et al. 2016, chapter 1); 2) the instruments needed to accomplish some of the IRF goals, in particular the Atmospheric Emitted Radiance Interferometer (AERI), were already available and were among the first instruments deployed at the Southern Great Plains (SGP) site; and 3) a large fraction of the IRF membership participated

Corresponding author address: D. D. Turner, NOAA/National Severe Storms Lab, Forecast Research and Development Division, National Weather Center, 120 David L. Boren Blvd., Norman, OK 73072.
E-mail: dave.turner@noaa.gov

DOI: 10.1175/AMSMONOGRAPHS-D-15-0025.1

in SPECTRE and thus were already familiar with each other and the datasets they would analyze.

The original goal established by the IRF, because of the instrumentation available in the early 1990s, was to improve the accuracy of clear-sky infrared radiative transfer models, and in particular, the water vapor continuum model component [see Turner and Mlawer (2010) for a simple overview of the water vapor continuum, or Mlawer et al. (2012) for a more detailed discussion]. Toward this end, the IRF developed the idea of the Qualitative Measurement Experiment (QME; Turner et al. 2004). The QME was a radiance closure experiment designed to evaluate three critical components simultaneously: 1) the spectral radiance observations, 2) the radiative transfer model, and 3) the temperature and humidity observations, as well as any other observations required by the radiative transfer model. While there were several QMEs defined by the IRF, the primary one compared AERI spectral downwelling radiance observations (Knuteson et al. 2004a,b; see the appendix) against those computed by the Line-by-Line Radiative Transfer Model (LBLRTM; Clough et al. 1992, 2005). A large number of IRF and ARM infrastructure members would work on this closure exercise for over a decade (Turner et al. 2004). The next chapter in this monograph (Mlawer and Turner 2016, chapter 14) provides additional details about the AERI/LBLRTM QME effort and some of the significant improvements made as a result of those activities.

The initial challenge of this QME was which, if any, of the three components of this radiative closure study could be trusted enough to serve as a foundation for evaluating the other components. In many ways, the QME is like building a three-legged stool; if any of the legs are weak or insufficient, then the stool will not stand properly until that leg is improved. The original QME idea (Miller et al. 1994) proposed to take advantage of portions of the electromagnetic spectrum with better-known (i.e., less uncertain) spectroscopy, which, together with the perspective that the measurement–calculation results should be spectrally consistent, would provide a suitable basis to evaluate the observations used in the QME. Then, after there was confidence that the observations were good, improvements could be made to the radiative transfer model itself. Key to this approach was a robust approach to specifying the thermodynamic profile in the radiating column above the AERI.

The original programmatic desire was that the thermodynamic profile observations that would be used as input to the LBLRTM would come from the Raman lidar and radio acoustic sounding system (RASS; May et al. 1988). These instruments are active profiling systems, meaning that they transmit electromagnetic energy and observe the backscattered signal, which is then calibrated to provide profiles of water vapor and virtual temperature, respectively. Even though these systems take advantage of radiative transfer principles, these observations were considered independent of uncertainties that affect radiative transfer models in general circulation models (GCMs). However, it would take many years before the Raman lidar was developed fully into an operational system (Turner et al. 2016, chapter 18), and the ability of the RASS to provide temperature profiles over much of the troposphere was never truly realized. Thus, the ARM Program, and the IRF in particular, depended heavily on the radiosondes launched by the program. However, radiosondes turned out to have their own issues with respect to reliability and accuracy.

3. Characterizing the radiosonde and MWR

One of the earliest comparisons done to assess water vapor measurements was to compare the precipitable water vapor (PWV) measured by the ascending radiosonde with that retrieved from the two-channel ARM microwave radiometers (MWRs). This was, in fact, the first QME; the radiosonde profile was input to a microwave radiative transfer model, and the calculated brightness temperatures were compared to those measured by the MWR. The multiyear coincident observations from the MWR and radiosondes (and the AERI) would turn out to be very important for truly characterizing these sensors.

The comparisons of PWV from the MWR and radiosonde demonstrated significant bias between the two instruments, again bringing to the forefront the question of which components of this QME could be trusted. Initially, this PWV bias, which could be 10% or more, was attributed to errors in the radiative transfer model used to invert the MWR's brightness temperatures to PWV, and thus tuning functions were developed and applied to the MWR observations (Liljegren 1994). These tuning functions were linear functions used to modify the observed brightness temperatures to account for supposed deficiencies in the microwave water vapor absorption model, and were determined typically from a large number of clear-sky radiosonde/MWR observations. However, it seemed that the tuning functions needed to change with some periodic interval, and thus there were questions about whether this was really an issue with the absorption model[1] or if it was

[1] The program was using the Liebe and Layton (1987) model for gaseous absorption in the microwave portion of the spectrum at the time. The other commonly used model at the time was Rosenkranz (1998). Tony Clough and his colleagues would subsequently develop the MonoRTM for the microwave region of the spectrum with ARM support (Clough et al. 2005; Payne et al. 2008), and this would eventually become the primary microwave absorption model used by the program (e.g., Turner et al. 2007).

due to changes in the calibration in either the MWR or perhaps the radiosonde.

Tony Clough was a strong voice in the IRF who advocated that the absorption model was accurate and was not responsible for the differences in the PWV between the sonde and MWR. He had performed laboratory measurements to measure the Stark effect and was able to determine that the strength of the 22.2-GHz water vapor absorption line (which is the water vapor spectral feature that was being observed by the ARM MWRs) was known to better than 1% (Clough et al. 1973). Furthermore, PWV is more sensitive to MWR-measured downwelling radiation at 23.8 GHz, which is on the side of the 22.2-GHz water vapor line, than its measurement at 31.4 GHz, which is in a window between absorption lines. The 23.8-GHz observation is close to the "hinge" point of the 22.2-GHz water vapor absorption line, and thus the modeled downwelling radiance at 23.8 GHz is relatively insensitive to errors in the half-width of this absorption line. For these reasons, Tony Clough argued that the differences between the PWV observed by the MWR and radiosondes were due to instrument calibration.

It had become clear that uncertainty in the PWV was the limiting factor in improving the accuracy of the infrared radiative transfer models (Revercomb et al. 2003). The need to greatly improve the accuracy of the water vapor observations, and in particular the PWV measurements, led to the development and execution of a series of water vapor intensive observation periods (WVIOPs) by the ARM Program. These WVIOPs, which were led by Hank Revercomb, occurred at the SGP site in 1996, 1997, 1999, and 2000 and brought a wide range of different instruments to the site to help characterize the accuracy and precision of operational and experimental instruments (Revercomb et al. 2003).

The WVIOPs, together with the operational nature of the ARM Program that collected a massive amount of coincident observations over many years, helped to improve the accuracy of these water vapor measurements. The improvements spanned many aspects of the measurement problem: better understanding of instrument biases, calibration improvements, retrieval algorithm and forward model improvements, improvements in sampling, and more. Indeed, the discussions among the WVIOP participants during the ARM science team meetings and fall working meetings in the mid-to-late 1990s were far ranging and exciting, and it took several years to fully appreciate how the many changes worked together to improve the overall accuracy of all of the water vapor observations.

The simultaneous side-by-side comparison of many MWRs was one of the activities that occurred during the WVIOPs. Jim Liljegren, the MWR instrument mentor at the time, deployed up to six of the ARM two-channel MWRs side by side to evaluate the consistency of the measurements from these systems (Revercomb et al. 2003). Ed Westwater, an ARM science team member, deployed his large trailer-based MWR at the site also. These activities demonstrated that small details, such as ensuring that the radiometer was level and accounting for the finite beamwidth of the radiometer, were important to obtaining accurate calibrations from the tip-curve technique (Han and Westwater 2000), which was used to determine the calibration of all of the ARM MWRs in the field. Jim Liljegren incorporated these ideas, together with additional findings on the temperature-dependent aspects of the calibration, into a new algorithm to automatically (without human intervention) calibrate the MWRs (Liljegren 2000). These advances in the MWR calibration removed small biases but, more importantly, created a dataset that was consistently calibrated over time by removing the subjective elements from the process.

The long-term collocated MWR and radiosonde dataset also highlighted several problems with the radiosonde measurements. In particular, an early analysis by Jim Liljegren and Barry Lesht (the radiosonde instrument mentor) demonstrated that factory calibration errors could impact the accuracy of radiosonde humidity measurements. ARM has always utilized Vaisala radiosondes, starting with the RS80-H, transitioning to the RS90 in the early 2000s, and then to the RS92 around 2002. Vaisala calibrates dozens of radiosondes simultaneously in a batch at its factory, and because of the large number of sondes that ARM uses, ARM typically would receive many, or even all, of the radiosondes from a calibration batch. Liljegren and Lesht (1996), in an attempt to understand the differences between the sonde- and MWR-derived PWV values, separated PWV biases by the radiosonde serial number and discovered that one particular batch of sondes produced by Vaisala in the mid-1990s had a very poor humidity calibration. This finding spurred a deeper investigation, and significant mean bias differences (up to 20%) in the radiosonde calibration as a function of batch were found (Turner et al. 2003b; Fig. 13-1).

To better understand the bias between sonde and MWR PWV, many dual launches, where two radiosondes were flown on the same balloon in order to sample the same volume, were performed during the WVIOPs. These dual launches demonstrated that the humidity calibration of sondes from the same batch were generally in good agreement, with water vapor mixing ratio differences within 1%–2%, but that sondes from different calibration batches could have differences as

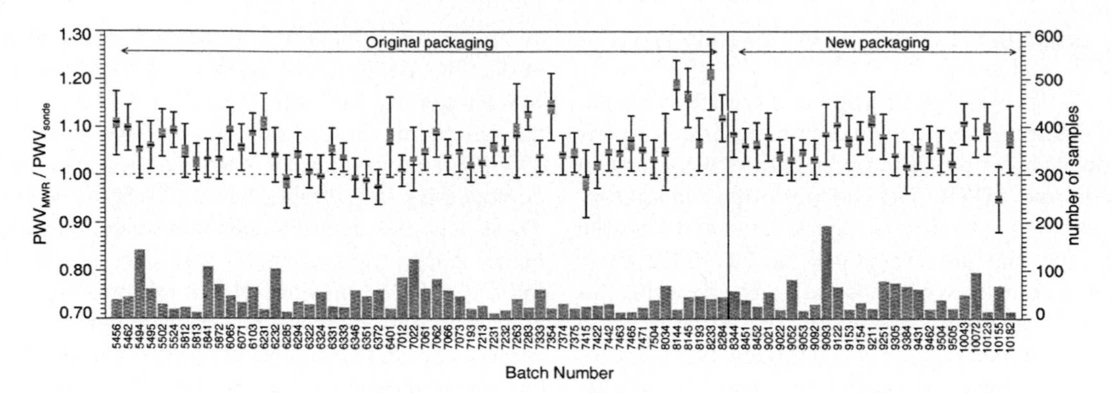

FIG. 13-1. Comparisons of the ratio of PWV from the MWR to that from the radiosonde from April 1994 to July 2000, separated by calibration batch, which shows the batch-to-batch variability of the results. Data are included only if the radiosonde achieved a height of at least 10 km, the cloud liquid water path was less than 50 g m^{-2}, and the number of samples in the batch was at least 10. The histograms at the bottom (gray bars) indicate the number of comparisons in each batch. The error bars indicate ± 1 standard deviation about the mean, whereas the gray boxes are ± 1 standard error of the mean. The thick black horizontal line is the median for each batch. The packaging of the radiosondes was changed midway through the period in an attempt to reduce/eliminate the dry bias, but it did not change the character of the differences. From Turner et al. (2003b).

great as 20% (Revercomb et al. 2003; Turner et al. 2003b). Figure 13-2 shows two dual-sonde examples that illustrate the differences in the moisture profiles from two sondes from different calibration batches. These dual launches suggested that the bias was to first order height independent, suggesting that the sonde data could be corrected by the use of a single-height independent scale factor. [The height-independent nature of the radiosonde humidity calibration bias was noted earlier by Ferrare et al. (1995), although the dataset at the time was not large enough to posit a hypothesis on the reason for the bias.] These findings agreed well with the findings by Wang et al. (2002), who determined that the source of the radiosonde humidity bias was due to outgassing of the RS80-H sonde packaging, which then effectively reduced the surface area of the capacitive humidity sensor (which would translate into a height-independent bias).

The dry bias induced by the radiosonde packaging was one source of error that was identified by the ARM Program during its WVIOPs. Another systematic error was discovered with long-term MWR versus sonde PWV comparisons. This activity demonstrated a diurnal difference in the bias between the two instruments, with the daytime results showing a larger bias (sonde more dry) than at night. The question at the time was simply the following: is there some temperature- or solar-energy-dependent calibration artifact impacting the accuracy of the MWR, or are the sondes the source of the issue? To investigate this issue, IRF members used comparisons between the AERI and the LBLRTM. The LBLRTM was used as a transfer standard, wherein radiosonde profiles were used as input to the model to compute downwelling radiance that could be compared to the AERI. Two sets of inputs were used to drive the LBLRTM: the original radiosonde profiles and radiosonde profiles where the humidity field was multiplied by a height-independent factor to yield the same PWV as was observed by the MWR. This resulted in two sets of infrared spectral residuals. Since the LBLRTM could potentially have spectral biases, the analysis would hinge more on the consistency of the spectral bias and RMS difference between daytime and nighttime results. [The focus on using the RMS difference to evaluate different inputs into a radiative transfer model would be used again when evaluating different cloud retrieval algorithms in radiative closure exercises (McFarlane et al. 2016, chapter 20).] The comparisons with the AERI demonstrated conclusively that the MWR was a stable instrument and that the diurnal difference between the sonde and MWR PWV was due to the radiosonde sensor (Turner et al. 2003b). Ultimately, this diurnal bias would be shown to be due to solar heating of the relative humidity sensor on the radiosonde (e.g., Vömel et al. 2007; Miloshevich et al. 2006; Miloshevich et al. 2009; Cady-Pereira et al. 2008).

4. The Rosetta Stone

While the program had come a long way in better understanding the radiosonde and the calibration of the MWR, there was still concern that the MWR-retrieved PWV may still be biased because a forward radiative transfer model was used to invert the observed brightness temperature observations to PWV. Thus, one of the primary objectives was to compare first-principle

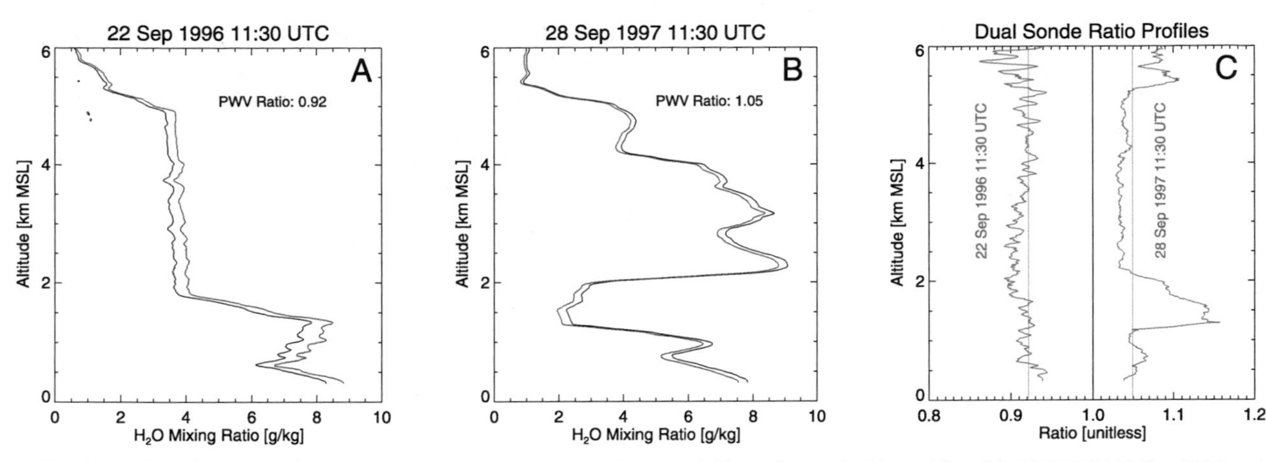

FIG. 13-2. The RH profiles from two radiosonde packages on the same balloon (i.e., a dual launch) at (a) 1130 UTC 22 Sep 1996 and (b) 1130 UTC 28 Sep 1997, where the radiosondes were from different calibration batches. (c) The ratio of these profiles [green and brown for the cases in (a) and (b), respectively] as a function of height.

measurement techniques in an attempt to define a "rock" that could serve as the reference for other observations.

Chilled-mirror hygrometers were considered to be one of these rocks, as the principle of the measurement is straightforward and related directly to the amount of water vapor in the atmosphere. This in situ sensor was not really designed for automated, long-term observations. However, the three-week comparisons during the 1996 and 1997 WVIOPs with another instrument, an in situ capacitive humidity sensor, at the surface demonstrated that well-maintained capacitive sensors could measure water vapor to better than 2% accuracy (Richardson et al. 2000). This result had two ramifications. First, since these capacitive in situ sensors are relatively cheap, the program installed two of these sensors for redundancy at both the 25- and 60-m levels on the tower at the SGP site. Second, because the radiosonde uses these same capacitive humidity sensors, the radiosonde variability was concluded not to be due to the sensor approach but rather to calibration, implementation of the sensor into the radiosonde package (e.g., should it have a cap to prevent wetting by cloud liquid or not), storage time, or other factors (Revercomb et al. 2003).

It was important to compare the two water vapor measurements that were thought to be accurate: traceable in situ sensors (chilled mirrors and capacitive sensors) on the tower and the PWV retrieved from the MWR based on underlying spectroscopy of the 22.2-GHz water vapor absorption line. To extend the in situ measurements throughout the column, the scanning Raman lidar from the National Aeronautics and Space Administration (NASA) Goddard Space Flight Center (GSFC) was used (Whiteman et al. 1992). This system, which played an important role in the development of the ARM Raman lidar (Turner et al. 2016, chapter 18), was able to scan in elevation and thus could make measurements directly next to the in situ sensors on the 60-m tower. Since the scanning Raman lidar, like the ARM Raman lidar, needs only a height-independent calibration factor, this factor was derived from the chilled-mirror hygrometer on the tower. The calibrated Raman lidar then collected zenith observations of water vapor mixing ratio, which were vertically integrated and compared to the PWV retrieved from the MWR. These observations were only done at night in cloud-free conditions during the WVIOPs, conditions under which the scanning Raman lidar was able to profile water vapor throughout the entire troposphere. The comparison, which spanned PWV values from less than 1 cm to over 4 cm, showed virtually identical agreement in water vapor sensitivity between the chilled-mirror hygrometer and MWR observations of the 22.2-GHz line with a slope of almost exactly 1.0 (Fig. 13-3; Revercomb et al. 2003). This finding was robust using both the statistical retrieval used in the mid-to-late 1990s by the ARM (red points in Fig. 13-3) and the newer physical retrieval developed in the early 2000s by Turner et al. (2007; blue points in Fig. 13-3). This finding was considered the "Rosetta Stone" that allowed the program to have faith in the accuracy of the MWR's PWV observations and to use these observations to begin the work of evaluating and improving the LBLRTM.

The demonstrated stability and accuracy of the MWR-retrieved PWV was a great boon for ARM science. The program began to scale all of its operational products to agree with the MWR in PWV. For example, the program created a value-added product called the Liebe-scaled sonde (LSSONDE; a new sonde data

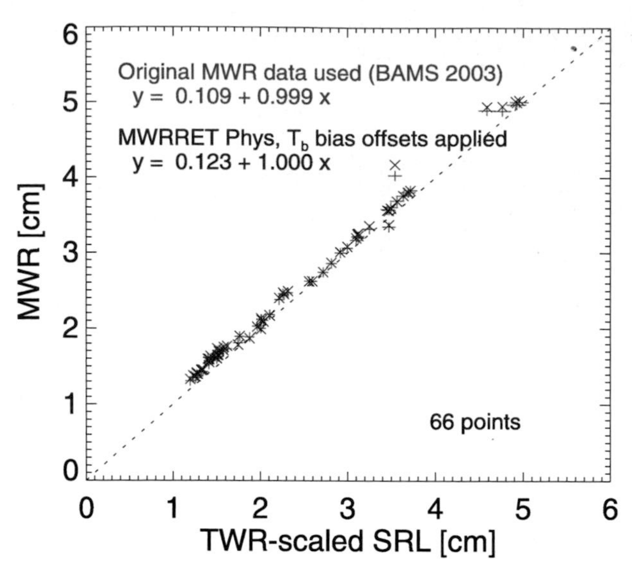

FIG. 13-3. The comparison of the MWR-retrieved PWV (y axis) with PWV computed from the chilled-mirror calibrated NASA scanning Raman lidar (x axis) during the 1997 WVIOP. The excellent agreement in sensitivity (i.e., the slope of exactly 1.0) demonstrated that these two first-principle measurements agreed and were termed the Rosetta Stone. From Turner et al. (2007).

product where the humidity profile was scaled to agree with the MWR), to use the MWR as the calibration source for the Raman lidar (Turner and Goldsmith 1999) and to scale all humidity profiles in its atmospheric state best-estimate value-added product (MERGESONDE) so that its PWV agreed with that value measured by the MWR.

5. Improving Arctic water vapor observations

The range of PWV at the SGP site is typically between 0.5 and 5.0 cm. This large dynamic range is due to the dry and cold Arctic airmass events that occasionally propagate that far south in the winter, as well as the occasional near-tropical-like events that come from the south in the summer. However, the typical radiometric uncertainty (i.e., random noise) in the MWR brightness temperature observations is 0.35 K, which translates into a PWV uncertainty of 0.025 cm (Turner et al. 2007); this is a relatively large percentage error for dry cases. At the North Slope of Alaska (NSA) site, the PWV can be as low as 0.1 cm, so the radiometric uncertainty in the MWR translates into a 25% random error. Since one of the primary objectives of the NSA site (Stamnes et al. 1999; Verlinde et al. 2016, chapter 8) was to evaluate the accuracy of infrared radiative transfer models in spectral regions such as the far-infrared that are normally opaque at typical midlatitude sites, more accurate values of PWV were needed than could be obtained from the MWR. For example, Tobin et al. (1999) noted that the strength of the foreign-broadened water vapor continuum model in the 17–26-μm spectral region was about three times too strong and made some modifications based upon data collected during the Surface Heat Budget of the Arctic Ocean (SHEBA) field campaign, but the 25% uncertainty in the water vapor observations resulted in approximately the same amount of uncertainty in the modeled strength of this absorption.

The program needed to have a more accurate water vapor measurement in the Arctic. Thus, there was a desire to conduct WVIOPs at the NSA site. Ed Westwater led the charge, with the support of many within the IRF. Tony Clough again was a forceful player, advocating for microwave radiometer observation on the 183.3-GHz water vapor absorption line. This line is significantly stronger than the 22.2-GHz water vapor line, and its strength is also known to better than 1% from Stark effect observations (Clough et al. 1973). The first Arctic WVIOP occurred in 1999 (Racette et al. 2005), a second in 2004 (Mattioli et al. 2007), and a third [under the auspices of the first Radiative Heating in Underexplored Bands Campaign (RHUBC)] in 2007 (Turner and Mlawer 2010). All three of these campaigns took place in the late winter (i.e., February–March), when the PWV is climatologically the lowest and the skies are more likely to be cloud free (thereby simplifying the analysis).

As was done during the WVIOPs at the SGP site, several guest instruments were operated at the NSA site to complement the operational ARM instruments. One guest instrument that was common in all three of these experiments was the National Oceanic and Atmospheric Administration (NOAA) ground-based scanning radiometer (GSR; Cimini et al. 2007). The first WVIOP also employed a NASA radiometer with multiple channels in the microwave and submillimeter spectral regions (Racette et al. 2005), and multichannel microwave radiometer profilers (MWRP) from Radiometrics (the same company that built the operational two-channel MWRs) were deployed during the 2004 and 2007 IOPs.

The 1999 and 2004 IOPs demonstrated that the observations at 183 GHz were indeed much more sensitive to the PWV in the dry Arctic atmosphere than the 22.2-GHz observations (e.g., Racette et al. 2005; Mattioli et al. 2007), and, when combined with the 22.2-GHz observation, accurate PWV retrievals could be performed for the entire range of PWVs that exists at the NSA site. Furthermore, there was good agreement in the PWVs retrieved individually from the 22-GHz and 183-GHz radiometers in the PWV range, where both had sensitivity (i.e., from roughly 0.4 to 1.0 cm). When taken together with the fact that the strengths of both of these water vapor absorption lines were determined the same way via measurements of the Stark effect, the program could conclude that the 183-GHz measurements were included in the Rosetta Stone conclusions (i.e., that they could serve as a PWV reference). These studies were so conclusive that DOE issued a small business innovation research (SBIR) funding opportunity to give vendors the opportunity to develop operational 183-GHz radiometers; ultimately, two different vendors would seize this opportunity and develop radiometers that would be evaluated at the NSA site during RHUBC-I in 2007. These radiometers were then evaluated side by side, together with the GSR, during RHUBC-I and were found to agree very well; Cimini et al. (2009) concluded that the ARM Program was now able to measure PWV in the Arctic to better than 3%. This significant improvement in water vapor measurements allowed Delamere et al. (2010) to further refine the modeled strength of the infrared water vapor continuum absorption model. These new 183.3-GHz radiometers are operational at the NSA site, and automated PWV and liquid water path retrievals have been developed for them (Cadeddu et al. 2009).

An additional finding of the first two WVIOPs concerned the accuracy of radiosonde humidity measurements in the Arctic. The first two WVIOPs analyzed the ARM-launched RS80-H (during the 1999 experiment) and the RS90 (during the 2004 campaign) radiosondes, as well as the VIZ/Sippican resistive hygristor radiosondes (Wade 1994) that the National Weather Service was launching twice daily approximately 5 km away from the NSA site. These comparisons demonstrated that the VIZ/Sippican moisture sensors were very poor, with wet biases in the lower troposphere and dry biases in the upper troposphere (Mattioli et al. 2007, 2008). However, the comparisons with both types of Vaisala sondes demonstrated that, as at SGP, the Vaisala sondes were able to measure the profile of the water vapor mixing ratio relatively accurately if a height-independent scale factor, such as the ratio of the PWVs from the 183-GHz radiometer and the sonde itself, was applied to the entire humidity profile (Racette et al. 2005; Mattioli et al. 2007, 2008).

6. Improvements to the microwave radiative transfer model

The microwave radiative transfer models used by the ARM Program have sufficient accuracy for PWV retrievals because the strengths of the relevant water vapor absorption lines are known with high accuracy, and the microwave observations were performed at the spectral location at which errors in the half-width of the absorption line have minimal impact. However, the continued development of microwave radiometer instrumentation would provide new observations that would be used by the program to continue to evaluate and improve the accuracy of the half-widths of the 22.2- and 183.3-GHz water vapor absorption lines, as well as the underlying water vapor continuum absorption model.

The first microwave radiometer with more than two channels that was operated routinely by the program was the 12-channel MWRP (Solheim et al. 1998a), which was developed in the late 1990s; the MWRP had five channels along the side of the 22.2-GHz water vapor absorption line. MWRP observations began at the SGP site in February 2000. The SBIR process also resulted in two new radiometers that made multiple observations along the 183.3-GHz line, both of which were deployed at the NSA site prior to RHUBC-I: the G-band microwave radiometer (GVR; Cadeddu et al. 2007b) and the G-band microwave radiometer profiler (GVRP; Cimini et al. 2009). Finally, the desire to have increased sensitivity to low amounts of liquid water path led to the program acquiring microwave radiometers with channels at 90 and 150 GHz (Shupe et al. 2016, chapter 19).

Line-by-line radiative transfer models like the LBLRTM require a spectroscopic database with the position, strength, width, and other properties of each absorption line for all gases that absorb radiation in Earth's atmosphere. The High-Resolution Transmission (HITRAN) database, for example, is a compilation of these parameters and has been updated many times over the decades as new laboratory and field observations, theoretical calculations, and other insights have improved our understanding of these parameters (e.g., Rothman et al. 1992, 2005). Since there are only a relatively small number of absorption lines in the microwave portion of the spectrum, microwave absorption models typically have these line parameters hardcoded, and thus the model reflects the state of the knowledge at the time the model was created.

Liljegren et al. (2005) were the first to use the new MWRP observations to evaluate and characterize the line strength and half-width of the 22.2-GHz line. They confirmed that the line strength suggested by the Stokes effect matched the MWRP observations the best but that the newly determined line width in the 2004 version of HITRAN was more accurate than the widths used in either the Rosenkranz (1998) or the Liebe and Layton (1987) models. Payne et al. (2008) used the MWRP at the SGP and the GVR at the NSA to evaluate the line widths of the 22.2- and 183.3-GHz lines and demonstrated that the line widths predicted by a new theoretical calculation matched the widths derived from the observations to within 3%. The excellent agreement in the widths between observations and theory gave confidence to the theory, and the monochromatic radiative transfer model (MonoRTM) was modified to include both these widths and the temperature dependence of these widths that were suggested by the theory (Payne et al. 2008). These changes to the modeled widths of these spectral lines greatly improved the accuracy of humidity profiles that were retrieved from these multichannel microwave radiometers (Liljegren et al. 2005; Payne et al. 2008). The multichannel MWRP was also useful in evaluating the oxygen spectroscopy at 60 GHz (Cadeddu et al. 2007a); this spectral region is used for profiling atmospheric temperature.

These multichannel observations, together with a long-term analysis of the two-channel MWR data at the SGP site and the 150-GHz data that were collected during an ARM Mobile Facility (AMF) deployment to the Black Forest in Germany (Wulfmeyer et al. 2011), were used to make important refinements to the water vapor continuum absorption model used in the microwave. The water vapor continuum model has two components: the self-broadened component, where water vapor molecules interact with other water vapor molecules and the foreign-broadened component, where water vapor molecules interact with nitrogen and oxygen molecules (Mlawer et al. 2012). These observations, which spanned a large range of PWV as a result of the observations coming from different climatic locations, helped to more accurately separate the contributions from the two components and determine the strength of the absorption of each more accurately (Turner et al. 2009; Payne et al. 2011; Fig. 13-4), leading to improvement in the Mlawer–Tobin–Clough–Kneizys–Davies (MT_CKD) water vapor continuum model from v2.1 to v2.4. This improvement in the modeled strength of the water vapor continuum in the microwave portion of the spectrum would have a large impact on the strength of the modeled absorption by the continuum in the far-infrared (Delamere et al. 2010; Mlawer et al. 2012), which would, in turn, have a significant impact on both the radiative and dynamical evolution of global climate model simulations (Turner et al. 2012).

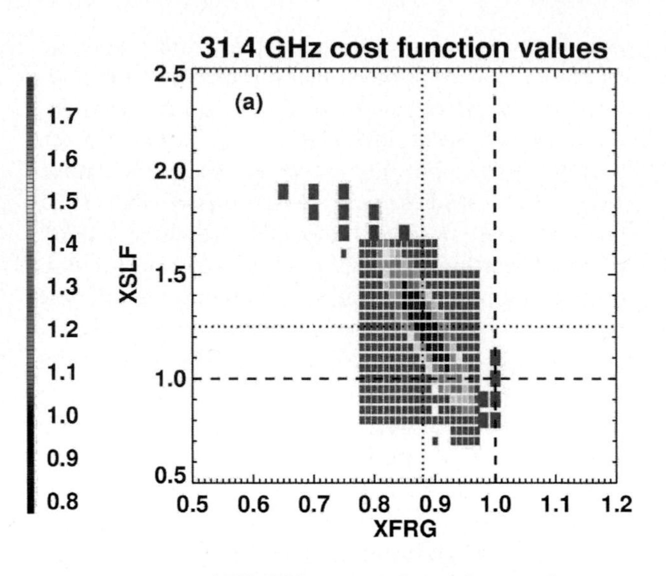

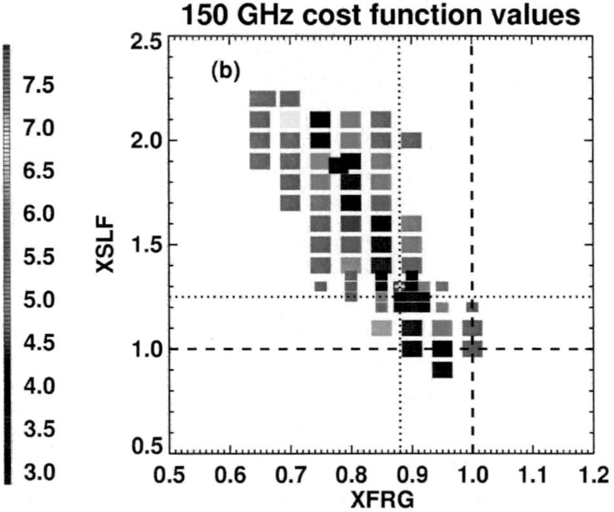

FIG. 13-4. Cost function surfaces for different scaling factors applied to the MT_CKD v2.1 self- (XSLF) and foreign- (XFGN) broadened water vapor continuum coefficients at (a) 31.4 and (b) 150 GHz. The purple regions indicate better fits of the coefficients to the two datasets (31.4 and 150 GHz, respectively). Crosshairs with dashed lines show the positions of the v2.1 values, while crosshairs with dotted lines show the new values of the water continuum that were incorporated into MT_CKD v2.4, which were values that gave the best results at the two frequencies simultaneously. Modified from Payne et al. (2011).

7. Other technologies

A wide range of other water vapor technologies played an important role during the early ARM years and were part of the WVIOPs. These technologies included the Raman lidar, global positioning system (GPS) PWV estimates, and solar attenuation measurements of PWV. The development of the ARM Raman lidar is a colorful story and one of the many successes of the ARM Program; a full discussion is provided by Turner et al. (2016, chapter 18). The GPS and solar attenuation methods provided a relatively affordable way to measure PWV routinely at many sites, such as the SGP extended facilities, and thereby provide a measure of the spatial variability of the water vapor field over the large SGP domain. Furthermore, several different retrieval algorithms were developed to provide thermodynamic profiles from remote sensors at the ARM sites; these profiles would have coarser vertical resolution but much higher temporal resolution than radiosondes. Finally, different analysis approaches were suggested on how to merge these different datasets together to provide humidity profiles throughout the troposphere over the diurnal cycle by blending together different instruments.

a. GPS PWV observations

The GPS sensor measures the propagation delay that occurs as the signal emitted by an orbiting satellite reaches the receiver on the ground. This delay, after accounting for the frequency-dependent delay caused by the ionosphere, is the result of the refractivity of the (primarily) lower troposphere, and the total delay is often separated into hydrostatic (sometimes called the dry) and wet components. The GPS measurement is usually accompanied by a measure of atmospheric pressure at the surface; this observation allows the hydrostatic component of the delay to be accurately estimated. The wet delay, which is the sum of the delay from both water vapor and liquid water, is computed as the residual between the total observed delay and the computed dry delay. However, the wet delay is dominated by water vapor in most conditions because of the difference in the mass of water vapor relative to the mass of liquid water in a column of air. Thus, the computed hydrostatic delay (from the surface pressure measurement), together with the GPS-observed delays, can be translated into a measure of PWV along the slant path to the satellite (i.e., SPWV) using a wet delay "mapping function," which is essentially a characteristic water vapor profile that can be scaled (Niell 1996; Niell et al. 2001). Since the GPS sensor is able to track multiple satellites simultaneously, many different SPWV values can be derived, and these are typically processed to provide one PWV estimate over the GPS sensor every 30 min.

The technique to derive PWVs from GPS observations was first described by Bevis et al. (1992), demonstrated in the GPS/STORM experiment in the spring of 1993, and integrated at the ARM SGP site shortly thereafter (Gutman et al. 1994). The technology was very exciting, as GPS sensors were much more affordable than MWRs and thus could be deployed en masse to form a PWV observation network. Thus, many different groups performed comparisons between sondes, MWRs, solar radiometers, and GPS PWV observations (e.g., Liljegren and Lesht 1996; Wolfe and Gutman 2000; Braun et al. 2001; Liou et al. 2001; Braun et al. 2003; Mattioli et al. 2005; Champollion et al. 2009; Alexandrov et al. 2009). The ARM Program played a significant role advancing the science behind the GPS PWV measurements, both in improving the accuracy of the algorithms and in using these retrievals to advance our understanding of atmospheric water vapor.

As with all retrieval algorithms, there were multiple ways to derive the PWV from the GPS observations, and hence different software analysis tools were developed, such as GPS Analysis at Massachusetts Institute of Technology (GAMIT; King and Bock 1996) and Bernese (Rothacher 1992). These software packages are essentially forward models, and each makes somewhat different assumptions and approximations. In particular, there were two important decisions that needed to be made and were debated by the GPS analysis community. The first decision concerned the lowest cutoff angle that should be used in the analysis: i.e., what is the lowest elevation of the GPS satellite above the horizon that should be allowed to contribute to the solution? The GPS signal-to-noise ratio is very small (i.e., less than 1), and thus collecting data at lower-elevation angles when the satellite is near the horizon increases the contribution to the wet delay (and hence the signal-to-noise) because the atmospheric pathlength is longer, and the water vapor concentration is the highest in the boundary layer. However, the propagating GPS signal could multipath (i.e., reflect off the ground) when the satellite was close to the horizon, which would result in overestimates of the wet delay and hence SPWV because of inappropriately long pathlengths. Ultimately, through long-term comparisons at the SGP site and WVIOP results, the compromise between good sensitivity to boundary layer moisture and minimal multipath errors suggested that a 7° elevation angle was the best cutoff angle to use (Fang et al. 1998). The second decision facing the GPS community was the choice of mapping function, as different mapping functions became more important for lower-elevation angles especially below 5°. Different mapping functions could result in changes of up to several percent in PWV, and the impact of the mapping function depended heavily on the cutoff angle used. (Thus, the cutoff angle decision helped largely to mitigate this source of uncertainty.) Furthermore, WVIOPs and long-term analysis at the ARM site with the MWRs demonstrated that using a single mapping function for both day and night processing often resulted in small seasonal and diurnal biases in the derived PWV (Rocken et al. 2001). The focus offered by the WVIOPs (and other experiments) provided opportunities to directly compare GPS processing algorithms and choices, which ultimately led to a convergence between the different algorithms (Revercomb et al. 2003).

One of the possible applications of a dense network of GPS sensors was to derive profiles of water vapor from the SPWV observations using tomographic retrieval techniques. Many GPS sensors were installed around the SGP domain, thus providing the opportunity to perform the tomographic retrieval and to compare the results with independent water vapor profiles observed by instruments at the SGP Central Facility. Champollion et al. (2009) did exactly this and demonstrated that the GPS-derived water vapor field, while smoother in height than that observed by other instruments, had sufficient accuracy to provide useful spatial information during a rapidly changing water vapor field. This helped improve our understanding of the evolution of the boundary layer during a convective initiation event. That said, it has become the practice of the operational numerical modeling community to directly assimilate the observed GPS tropospheric delays instead of the SPWV values or the tomographic reconstructions (Smith et al. 2007).

b. *Solar PWV observations*

Solar attenuation methods offer a different way to derive PWV. These techniques usually make use of direct-beam measurements of solar irradiance in a spectrally narrow (order 5–10 nm wide) water vapor absorption band; the most typically used band is centered at 940 nm. The accuracy of the PWV retrievals from these methods depends strongly on the instrument calibration, the characterization of the instrument parameters (e.g., the bandpass of the instrument), and the accuracy of the radiative transfer model used in the retrieval. The WVIOPs provided excellent opportunities to bring additional solar radiometers to the SGP site and evaluate the different retrieval methods to get PWVs from these observations. However, since the solar technique measures the absorption of radiation in water vapor spectral bands, the presence of clouds and optically thick aerosol layers that extinguish the solar radiation in these bands hampers these retrievals.

Another instrument that matured with ARM funding was the multifilter rotating shadowband radiometer (MFRSR; Harrison et al. 1994). The MFRSR was deployed at many of the ARM sites and, in particular, was deployed at both the Central and extended facilities at the SGP site. Michalsky et al. (1995) derived a method to derive total column PWV along the slant path (i.e., SPWV toward the sun) from the MFRSR observations, which could be converted to PWV by assuming that the water vapor field was horizontally homogeneous. Halthore et al. (1997) and Kiedron et al. (2003) also developed algorithms to retrieve PWV from 940-nm solar observations.

During the 1997 WVIOP, other narrowband solar radiometers were brought to the SGP site, including the six-channel Ames airborne tracking sunphotometer (AATS-6; Schmid et al. 2001), the Cimel sunphotometer (Holben et al. 1998), and the rotating shadowband spectrometer (RSS; Harrison et al. 1999). PWV retrievals from these three radiometers and the MFRSR were compared to the MWR. Initially, good results were obtained from all four solar radiometers and the MWR; however, a deeper analysis suggested that there were compensating errors that were affecting the comparison (Schmid et al. 2001). For example, the various retrieval techniques used different radiative transfer models to derive PWV from the solar observations. When the same radiative transfer model was used for all retrievals, the spread in the PWV results among the solar radiometers increased (Schmid et al. 2001). Furthermore, Giver et al. (2000) found an error in the absorption line information for the 940-nm water vapor band in the HITRAN database used by the radiative transfer models, and using the updated line information resulted in the PWV derived from the solar radiometers being 6%–14% drier than the MWR-retrieved value (Schmid et al. 2001). These 6%–14% differences were thought to be related to uncertainties associated with instrument calibration and incomplete knowledge of the spectral bandpass (MFRSR and AATS-6) or slit-function (RSS) profiles, although uncertainties in either the water vapor spectroscopy and/or the water vapor continuum absorption model in the near-infrared were also candidates.

An updated analysis was performed by Alexandrov et al. (2009) using long-term MFRSR and MWR data at the SGP site, together with Cimel and AATS-6 radiometer data collected at the SGP during the 2000 WVIOP. This new comparison used the updated spectroscopy that was available in the HITRAN 2004 database. This study demonstrated that the agreement among the PWVs retrieved from the solar radiometers was within 4% but that again the solar radiometer retrievals were about 10% drier than the MWR-retrieved PWV, with the difference being larger when the PWV was more than 2 cm (Alexandrov et al. 2009). Like Schmid et al. (2001), the Alexandrov et al. (2009) study concluded that the differences are likely not caused by the radiometric calibration of the solar radiometers, but are more likely due to spectroscopy or errors in the spectral bandpass of the solar radiometers.

c. *Ground-based thermodynamic profile retrievals*

The original plan of the ARM Program was for limited use of radiosondes, with remote sensors providing the routine water vapor profiles required by the program. Furthermore, the remote sensors would provide higher-time-resolution profiles of water vapor than are possible with radiosondes, which would be useful in characterizing the temporal evolution of the moisture in the boundary layer as atmospheric fronts and other disturbances passed overhead. It was hoped that the

Raman lidar would provide these observations at the SGP site, but given the expense of this system, only a single Raman lidar would be deployed by the program at the SGP site.[2] Thus, other remote sensors would have to provide the higher-temporal-resolution observations at the other ARM sites.

Spectrally resolved radiance observations from passive microwave and infrared radiometers have been used to profile the troposphere from space since the 1970s. The original microwave profiling (MWRprof) algorithm was developed by Ed Westwater and colleagues, using the two-channel MWR data together with the temperature profiles from the radar wind profiler/RASS to get profiles of temperature and humidity in the boundary layer (Han and Westwater 1995; Stankov et al. 1996; Westwater 1997). It provided very low-vertical-resolution profiles of humidity because of the limited spectral information available to the retrieval. Nonetheless, this algorithm was one of the earliest value-added products implemented within the ARM Data Management Facility (McCord and Voyles 2016, chapter 11). The development of the MWRP provided additional spectral information for the microwave retrieval community, and newer and more advanced algorithms were developed (e.g., Solheim et al. 1998b; Ware et al. 2003; Liljegren et al. 2005). The MWRP and radiometers like it are able to provide low-vertical-resolution but high-temporal-resolution temperature and humidity profiles in most weather situations aside from medium-to-heavy precipitation and thus are well suited for operational networks like ARM (e.g., Hardesty and Hoff 2012). Similar retrieval algorithms also were developed for the multifrequency observations from the GVR and GVRP (e.g., Racette et al. 2005; Cimini et al. 2010). In addition, since clouds are usually optically thin at microwave frequencies, profiles of temperature and humidity can be retrieved from MWRP observations at and above the cloud height, wherein other methods, such as the Raman lidar, are largely limited to altitudes below cloud base.

In the mid-1990s, Bill Smith and Wayne Feltz developed a retrieval algorithm that used infrared radiance data observed by the AERI to get profiles of temperature and humidity (AERIprof; Feltz et al. 1998; Smith et al. 1999). These AERIprof retrievals were used to characterize frontal passages and drylines over the SGP (e.g., Feltz et al. 1998; Turner et al. 2000). The ARM Program deployed five AERI systems over the SGP domain from 1998 to 2003, and the thermodynamic retrieval data from this network were used to evaluate the water vapor and temperature evolution and convective stability indices during tornadic and nontornadic storms (Feltz and Mecikalski 2002; Wagner et al. 2008). The AERI thermodynamic retrievals are also used as part of the Raman lidar processing algorithms (Turner et al. 2002) and for providing input to other retrieval algorithms, such as the cumulus entrainment algorithm of Wagner et al. (2013). Löhnert et al. (2009) demonstrated that the information content in the AERI observations was 2–4 times higher than the MWRP for both temperature and water vapor, but the Smith–Feltz AERI algorithm was limited to cases where there are no clouds in the AERI field of view. However, recent developments have demonstrated the ability to retrieve thermodynamic properties from the AERI under clouds (Turner and Löhnert 2014).

d. Merging water vapor datasets together

The ultimate goal of the program with respect to water vapor (and temperature) fields is to specify these properties at all times and heights above its facilities. This type of data product is needed for a wide range of research topics that are important to the ARM Program. With no single instrument able to provide these profiles in all weather conditions at high temporal [$O(10)$ min] and vertical resolution, this data product must result from a combination of observations from several sensors and perhaps output from numerical models.

Research on how to best marry together observations from different remote sensors began in earnest in the program long before the remote sensors themselves were mature. For example, Stankov et al. (1996) presented different ways to integrate data from remote sensors, such as MWRs, wind profilers, and more. Han et al. (1997) used a Kalman filter to merge water vapor data from the MWR and (NASA) Raman lidar in order to provide profiles where the Raman lidar observations were either limited by clouds or by solar noise (the latter during the daytime).

ARM has developed a simple "merged sonde" data product called MERGESONDE to provide the baseline thermodynamic profiles at all times and heights above the ARM sites that are needed to compute radiative heating rates in clear and cloudy atmospheres (McFarlane et al. 2016, chapter 20). This algorithm combines radiosonde, MWR, surface meteorological data, and numerical weather prediction model output to derive these thermodynamic fields. However, this product is relatively simple and currently does not include any of the measurements or retrievals

[2] The success of the SGP Raman lidar, together with the funding received in 2009 from the American Recovery and Reinvestment Act (ARRA), allowed the program to construct a nearly identical Raman lidar and deploy it to the Darwin TWP site in late 2010. The program recently constructed a third Raman lidar that was deployed at the new ARM site in Oliktok Point, Alaska, in 2014.

from the Raman lidar, AERI, or profiling MWRs, although upgrades are planned for future versions.

Tobin et al. (2006) have developed a merged thermodynamic product that combines radiosondes, AERIprof retrievals, and MWR PWV observations to derive a best-estimate temperature and water vapor profile for validating similar products from NASA and EUMETSAT satellites [e.g., products derived from the Atmospheric Infrared Sounder (AIRS), Infrared Atmospheric Sounding Interferometer (IASI), and Cross-Track Infrared Sounder (CrIS) sensors]. The ARM Program, in collaboration with NASA and in the spirit of being a user facility (Ackerman et al. 2016, chapter 3), has conducted several dedicated multimonth IOPs at the SGP, NSA, and Tropical Western Pacific (TWP) sites to support these satellite-validation IOPs; these IOPs have been extremely useful in continually improving the operational data products derived from these satellites (e.g., Tobin et al. 2006; Bedka et al. 2010).

One of the most important value-added products in the ARM Program is the variational analysis product that was developed by Minghua Zhang and colleagues. This algorithm takes observational data from a large number of sensors (e.g., surface meteorological stations, MWRs, radiosondes, and more) and merges them together to create a dataset that can be used to drive single-column and cloud-resolving models (Zhang and Lin 1997; Zhang et al. 2001, 2016, chapter 24).

8. Current status and future outlook

Improving water vapor measurements was an extremely high priority in the first decade of the ARM Program. A large number of investigators worked on developing, characterizing, and improving different technologies as part of this effort. Many different instruments matured into operational instruments as a result of this funding: the MWR, AERI, Raman lidar, GPS, and GVR/GVRP.

Several WVIOPs were organized at the SGP and NSA sites to provide opportunities to bring together these investigators, along with additional research-grade instruments, to evaluate the ARM operational and research-grade instruments. The results from these IOPs, along with the longer-term comparisons of the operational ARM instruments, greatly improved our understanding of these instruments, including details in calibration, the underlying radiative transfer models, and sensitivity and precision of the observations themselves. Ultimately, these efforts reduced the uncertainty in the profile of the water vapor mixing ratio in the lower-to-middle troposphere from 15%–20% in the early days of the ARM Program to approximately 3% in both the SGP and NSA regions—a remarkable improvement. This new level of accuracy in the water vapor measurement opened the door to making important improvements in the absorption spectroscopy in microwave, infrared, and near-infrared radiative transfer models (Mlawer and Turner 2016, chapter 14).

This large improvement in the program's ability to measure the water vapor profile accurately, which was achieved largely in the late 1990s for midlatitude sites, allowed the program managers to reprioritize how the program was utilizing its research funds. Thus, starting around 2002, there was a decrease in the number of investigators working on profiling water vapor so that additional investigators could be funded to support other ARM programmatic objectives (Mather et al. 2016, chapter 4).

There is more work that can be done. The accuracy of the water vapor spectroscopy in the visible and near-infrared is still uncertain; this is quite possibly the reason why the solar PWV retrievals that use the 940-nm water vapor absorption band do not agree with the MWR's PWV retrievals. Another area that needs work concerns water vapor measurements in the upper troposphere. There has been some work to evaluate the accuracy of water vapor observations in this region (e.g., Ferrare et al. 2004; Soden et al. 2004; Miloshevich et al. 2006), but additional work (especially in the Arctic and tropical atmospheres) is needed. Accurate upper-tropospheric water vapor observations are very important to understand processes at work in the upper troposphere, such as ice nucleation mechanisms (e.g., Comstock et al. 2004) and radiative processes that control outgoing infrared emission to space (e.g., Ferrare et al. 2004).

Finally, the program needs to develop improved data synthesis products to create water vapor datasets to better cover all heights and times above the ARM sites and to provide some spatial context (e.g., for gradients and smaller-scale inhomogeneities) around the sites. These improvements will require utilizing the remote sensors at the sites, any sensors that are distributed around the sites (e.g., surface meteorological observations, MFRSRs, and GPS), and satellite observations. Tomographic techniques that utilize slant path observations from a large number of sensors also may contribute to this challenge. The deployment of Raman lidars at several of the ARM sites will help address some of this need, as well as the efforts currently underway to get the AERIprof retrieval algorithm (or its successor, which is able to work in cloudy scenes) ported to the tropics and Arctic. However, it will be very important that these merged products are well characterized and have quantified uncertainty estimates so that the scientific community can use these data products properly.

While there is much that still remains, the program has made tremendous strides in its ability to measure water vapor. The improvements in this area are certainly among the highlights of the first 20 years of the ARM Program.

APPENDIX

The AERI

Accurate spectrally resolved downwelling infrared radiance measurements were required to address the problems raised by ICRCCM (Ellingson et al. 2016, chapter 1; Ellingson and Fouquart 1991). The AERI was developed under the auspices of the ARM Instrument Development Program (Stokes 2016, chapter 2) to provide these observations. The AERI traces its origins back to both the ICRCCM requirement and to the desire to retrieve thermodynamic profiles from space using passive spectral infrared measurements. The AERI is one of the observational success stories of ARM and would play a key role in characterizing the accuracy of water vapor measurements made by the ARM Program (Turner et al. 2003b), improving the accuracy of the LBLRTM (Tobin et al. 1999; Turner et al. 2004; Delamere et al. 2010; Mlawer et al. 2012), and serving as input to cloud and aerosol retrieval algorithms (e.g., Mace et al. 1998; Comstock et al. 2007; Turner 2008).

The AERI was designed and fabricated by the Space Science and Engineering Center (SSEC) at the University of Wisconsin–Madison (Knuteson et al. 2004a,b). SSEC has a long history in developing satellite observation systems and using these data to characterize clouds and thermodynamic profiles in the atmosphere (e.g., Suomi and Vonder Haar 1969; Vonder Haar and Suomi 1971; Smith et al. 1981). SSEC developed an airborne infrared interferometer called the High-Resolution Infrared Sounder (HIS) in the mid-1980s to demonstrate the improved vertical resolution of retrieved temperature and humidity profiles that could be achieved if a satellite sensor had the high spectral resolution of an interferometer (Revercomb et al. 1988). In a proof-of-concept experiment, SSEC operated the HIS in an uplooking mode during the Ground-Based Atmospheric Profiling Experiment (GAPEX; Smith et al. 1990). The success of GAPEX led to the development of the first dedicated ground-based interferometer by SSEC, which was called the ground-based HIS (GB-HIS). The GB-HIS demonstrated the feasibility of the instrument design and calibration approach (Knuteson et al. 2004a). SSEC learned many lessons from the development and operation of the GB-HIS; these lessons and ARM funding led to the development of the prototype AERI system (called AERI-00).

The AERI-00 was first deployed in the SPECTRE field campaign (Ellingson and Wiscombe 1996; Ellingson et al. 2016, chapter 1). Based upon the results from this experiment, a set of instrument specifications was developed for the ARM Program. Principal among these requirements was the need for the instrument to work operationally without manual intervention. The detectors of these interferometers are cooled cryogenically to get the required signal-to-noise ratio. Early versions of these interferometers used liquid nitrogen to cool the detectors, which required manual refilling every 8 or 24 h. SSEC developed the first truly operational AERI (called the AERI-01) with ARM support; this system used a mechanical Sterling cooler to keep the detectors cold (Knuteson et al. 2004a). The AERI-01 was first deployed to the SGP site in the summer of 1995, replacing the AERI-00 that had been operating there since 1992. Continued experience with the AERI-01 led to improvements in the AERI technology that were incorporated into the AERI-v2 systems that were subsequently deployed at all of the primary ARM sites (e.g., SGP, NSA, TWP, and AMF). In the meantime, the AERI-01 has continued to operate at the SGP site for nearly 20 years, allowing the first long-term evaluation of spectral downwelling IR radiation to be performed (Gero and Turner 2011).

The strength of the AERI is its calibration. The use of two well-characterized (in both temperature and emissivity) blackbodies, together with the quantification of the nonlinear behavior of the detector, results in a radiometric calibration accuracy of better than 1% of the ambient radiance (3σ) (Knuteson et al. 2004b). The instrument also performs a spectral calibration and corrects for self-apodization artifacts that result from the finite field of view (Knuteson et al. 2004b), which makes it significantly easier to compare the AERI-observed radiances to those computed with a line-by-line model (e.g., Turner et al. 2004; Delamere et al. 2010) or to use them in a thermodynamic profiling retrieval algorithm (e.g., Feltz et al. 1998, 2003; Turner and Löhnert 2014).

The original temporal resolution of the AERI was set to 10 min; this was a compromise between achieving a good signal-to-noise ratio in the observed radiance for spectroscopic validation and profiling while maintaining the temporal resolution needed to resolve evolving atmospheric conditions. However, the AERI radiance observations soon were being used in retrieval algorithms to characterize cloud properties (e.g., Mace et al. 1998; Deslover et al. 1999; Turner et al. 2003a), which can change very rapidly as the cloud advects over the instrument. Thus, the program changed the temporal sampling strategy of the AERI so that a radiance spectrum was collected every 20–30 s, and a principal component-based noise filter was used to reduce the increased random error back to approximately the same noise level that was inherent in the 10-min data (Turner et al. 2006).

REFERENCES

Ackerman, T. P., T. S. Cress, W. R. Ferrell, J. H. Mather, and D. D. Turner, 2016: The programmatic maturation of the ARM Program. *The Atmospheric Radiation Measurement (ARM) Program: The First 20 Years, Meteor. Monogr.*, No. 57, Amer. Meteor. Soc., doi:10.1175/AMSMONOGRAPHS-D-15-0054.1.

Alexandrov, M. D., and Coauthors, 2009: Columnar water vapor retrievals from multifilter rotating shadowband radiometer data. *J. Geophys. Res.*, **114**, D02306, doi:10.1029/2008JD010543.

ARM, 2016: Appendix A: Executive summary: Atmospheric Radiation Measurement Program Plan. *The Atmospheric Radiation Measurement (ARM) Program: The First 20 Years, Meteor. Monogr.*, No. 57, Amer. Meteor. Soc., doi:10.1175/AMSMONOGRAPHS-D-15-0036.1.

Bedka, S., R. O. Knuteson, H. E. Revercomb, D. C. Tobin, and D. D. Turner, 2010: An assessment of the absolute accuracy of the Atmospheric Infrared Sounder v5 precipitable water vapor product at tropical, midlatitude, and Arctic ground-truth sites: September 2002 through August 2008. *J. Geophys. Res.*, **115**, D17310, doi:10.1029/2009JD013139.

Bevis, M., S. Businger, T. A. Herring, R. A. Anthes, C. Rocken, and R. H. Ware, 1992: GPS meteorology: Remote sensing of atmospheric water vapor using the global positioning system. *J. Geophys. Res.*, **97**, 15 787–15 801, doi:10.1029/92JD01517.

Braun, J., C. Rocken, and R. Ware, 2001: Validation of line-of-sight water vapor measurements with GPS. *Radio Sci.*, **36**, 459–472, doi:10.1029/2000RS002353.

——, ——, and J. Liljegren, 2003: Comparisons of line-of-sight water vapor observations using the global positioning system and a pointing microwave radiometer. *J. Atmos. Oceanic Technol.*, **20**, 606–612, doi:10.1175/1520-0426(2003)20<606:COLOSW>2.0.CO;2.

Cadeddu, M. P., S. A. Clough, V. H. Payne, K. Cady-Pereira, and J. C. Liljegren, 2007a: Effect of the oxygen line-parameter modeling on temperature and humidity retrievals from ground-based microwave radiometers. *IEEE Trans. Geosci. Remote Sens.*, **45**, 2216–2223, doi:10.1109/TGRS.2007.894063.

——, J. C. Liljegren, and A. Pazmany, 2007b: Measurements and retrievals from a new 183-GHz water-vapor radiometer in the Arctic. *IEEE Trans. Geosci. Remote Sens.*, **45**, 2217–2223, doi:10.1109/TGRS.2006.888970.

——, D. D. Turner, and J. C. Liljegren, 2009: A neural network for real-time retrievals of PWV and LWP from Arctic millimeter-wave ground-based observations. *IEEE Trans. Geosci. Remote Sens.*, **47**, 1887–1900, doi:10.1109/TGRS.2009.2013205.

Cady-Pereira, K., M. W. Shephard, D. D. Turner, E. J. Mlawer, S. A. Clough, and T. J. Wagner, 2008: Improved daytime column-integrated precipitable water vapor from Vaisala radiosonde humidity sensors. *J. Atmos. Oceanic Technol.*, **25**, 873–883, doi:10.1175/2007JTECHA1027.1.

Champollion, C., C. Flamant, O. Bock, F. Masson, D. D. Turner, and T. Weckwerth, 2009: Mesoscale GPS tomography applied to the 12 June 2002 convective initiation event of IHOP_2002. *Quart. J. Roy. Meteor. Soc.*, **135**, 645–662, doi:10.1002/qj.386.

Cimini, D., E. R. Westwater, A. J. Gasiewski, M. Klein, V. Leuski, and S. Dowlatshahi, 2007: The Ground-Based Scanning Radiometer (GSR): A powerful tool for the study of the Arctic atmosphere. *IEEE Trans. Geosci. Remote Sens.*, **45**, 2759–2777, doi:10.1109/TGRS.2007.897423.

——, F. Nasir, E. R. Westwater, V. H. Payne, D. D. Turner, E. J. Mlawer, M. L. Exner, and M. Cadeddu, 2009: Comparison of ground-based millimeter-wave observations in the Arctic winter. *IEEE Trans. Geosci. Remote Sens.*, **47**, 3098–3106, doi:10.1109/TGRS.2009.2020743.

——, E. R. Westwater, and A. J. Gasiewski, 2010: Temperature and humidity profiling in the Arctic using ground-based millimeter-wave radiometry and 1DVAR. *IEEE Trans. Geosci. Remote Sens.*, **48**, 1381–1388, doi:10.1109/TGRS.2009.2030500.

Clough, S. A., Y. Beers, G. P. Klein, and L. S. Rothman, 1973: Dipole moment of water vapor from Stark measurements of H_2O, HDO, and D_2O. *J. Chem. Phys.*, **59**, 2254–2259, doi:10.1063/1.1680328.

——, M. J. Iacono, and J.-L. Moncet, 1992: Line-by-line calculation of atmospheric fluxes and cooling rates: Application to water vapor. *J. Geophys. Res.*, **97**, 15 761–15 781, doi:10.1029/92JD01419.

——, M. W. Shephard, E. J. Mlawer, J. S. Delamere, M. J. Iacono, K. Cady-Pereira, S. Boukabara, and P. D. Brown, 2005: Atmospheric radiative transfer modeling: A summary of the AER codes. *J. Quant. Spectrosc. Radiat. Transfer*, **91**, 233–244, doi:10.1016/j.jqsrt.2004.05.058.

Comstock, J. M., T. P. Ackerman, and D. D. Turner, 2004: Evidence of high ice supersaturation in cirrus clouds using ARM Raman lidar measurements. *Geophys. Res. Lett.*, **31**, L11106, doi:10.1029/2004GL019705.

——, and Coauthors, 2007: An intercomparison of microphysical retrieval for upper-tropospheric ice clouds. *Bull. Amer. Meteor. Soc.*, **88**, 191–204, doi:10.1175/BAMS-88-2-191.

Delamere, J. S., S. A. Clough, V. Payne, E. J. Mlawer, D. D. Turner, and R. Gamache, 2010: A far-infrared radiative closure study in the Arctic: Application to water vapor. *J. Geophys. Res.*, **115**, D17106, doi:10.1029/2009JD012968.

Deslover, D. H., W. L. Smith, P. K. Piironen, and E. W. Eloranta, 1999: A methodology for measuring cirrus cloud visible-to-infrared spectral optical depth ratios. *J. Atmos. Oceanic Technol.*, **16**, 251–262, doi:10.1175/1520-0426(1999)016<0251:AMFMCC>2.0.CO;2.

DOE, 1990: Atmospheric Radiation Measurement Program Plan. DOE Tech. Doc. DOE/ER-0441, 121 pp. [Available online at https://www.arm.gov/publications/doe-er-0441.pdf.]

Ellingson, R. G., and Y. Fouquart, 1991: The intercomparison of radiation codes in climate models: An overview. *J. Geophys. Res.*, **96**, 8925–8927, doi:10.1029/90JD01618.

——, and W. J. Wiscombe, 1996: The Spectral Radiance Experiment (SPECTRE): Project description and sample results. *Bull. Amer. Meteor. Soc.*, **77**, 1967–1985, doi:10.1175/1520-0477(1996)077<1967:TSREPD>2.0.CO;2.

——, R. D. Cess, and G. L. Potter, 2016: The Atmospheric Radiation Measurement Program: Prelude. *The Atmospheric Radiation Measurement (ARM) Program: The First 20 Years, Meteor. Monogr.*, No. 57, Amer. Meteor. Soc., doi:10.1175/AMSMONOGRAPHS-D-15-0029.1.

Fang, P., M. Bevis, Y. Bock, S. Gutman, and D. Wolfe, 1998: GPS meteorology: Reducing systematic errors in geodetic estimates for zenith delay. *Geophys. Res. Lett.*, **25**, 3583–3586, doi:10.1029/98GL02755.

Feltz, W. F., and J. R. Mecikalski, 2002: Monitoring high-temporal-resolution convective stability indices using the ground-based Atmospheric Emitted Radiance Interferometer (AERI) during the 3 May 1999 Oklahoma–Kansas tornado outbreak. *Wea. Forecasting*, **17**, 445–455, doi:10.1175/1520-0434(2002)017<0445:MHTRCS>2.0.CO;2.

——, W. L. Smith, R. O. Knuteson, H. E. Revercomb, H. M. Woolf, and H. B. Howell, 1998: Meteorological applications of temperature and water vapor retrievals from the ground-based Atmospheric Emitted Radiance Interferometer (AERI). *J. Appl. Meteor.*, **37**, 857–875, doi:10.1175/1520-0450(1998)037<0857:MAOTAW>2.0.CO;2.

——, ——, H. B. Howell, R. O. Knuteson, H. Woolf, and H. E. Revercomb, 2003: Near-continuous profiling of temperature, moisture, and atmospheric stability using the Atmospheric Emitted Radiance Interferometer (AERI). *J. Appl. Meteor.*, **42**, 584–597, doi:10.1175/1520-0450(2003)042<0584:NPOTMA>2.0.CO;2.

Ferrare, R. A., S. H. Melfi, D. N. Whiteman, K. D. Evans, F. J. Schmidlin, and D. Starr, 1995: A comparison of water vapor measurements made by Raman lidar and radiosondes. *J. Atmos. Oceanic Technol.*, **12**, 1177–1195, doi:10.1175/1520-0426(1995)012<1177:ACOWVM>2.0.CO;2.

——, and Coauthors, 2004: Characterization of upper-tropospheric water vapor measurements during AFWEX using LASE. *J. Atmos. Oceanic Technol.*, **21**, 1790–1808, doi:10.1175/JTECH-1652.1.

Gero, P. J., and D. D. Turner, 2011: Long-term trends in downwelling spectral infrared radiance over the U.S. Southern Great Plains. *J. Climate*, **24**, 4831–4843, doi:10.1175/2011JCLI4210.1.

Giver, L. P., C. Chackerian Jr., and P. Varanasi, 2000: Visible and near-infrared $H_2^{16}O$ line intensity corrections for HITRAN-96. *J. Quant. Spectrosc. Radiat. Transfer*, **66**, 101–105, doi:10.1016/S0022-4073(99)00223-X.

Gutman, S. I., R. B. Chadwick, D. W. Wolf, A. Simon, T. Van Hove, and C. Rocken, 1994: Toward an operational water vapor remote sensing system using the global positioning system. *Proc. Fourth Atmospheric Radiation Measurement (ARM) Science Team Meeting*, Charleston, SC, U.S. DOE, 173–180. [Available online at http://www.arm.gov/publications/proceedings/conf04/extended_abs/gutman_si.pdf?id=29.]

Halthore, R. N., T. F. Eck, B. N. Holben, and B. L. Markham, 1997: Sun photometric measurements of atmospheric water vapor column abundance in the 940-nm band. *J. Geophys. Res.*, **102**, 4343–4352, doi:10.1029/96JD03247.

Han, Y., and E. R. Westwater, 1995: Remote sensing of tropospheric water vapor and cloud liquid water by integrated ground-based sensors. *J. Atmos. Oceanic Technol.*, **12**, 1050–1059, doi:10.1175/1520-0426(1995)012<1050:RSOTWV>2.0.CO;2.

——, and ——, 2000: Analysis and improvement of tipping calibration for ground-based microwave radiometers. *IEEE Trans. Geosci. Remote Sens.*, **38**, 43–52, doi:10.1109/36.843018.

——, ——, and R. A. Ferrare, 1997: Applications of Kalman filtering to derive water vapor profiles from Raman lidar and microwave radiometers. *J. Atmos. Oceanic Technol.*, **14**, 480–487, doi:10.1175/1520-0426(1997)014<0480:AOKFTD>2.0.CO;2.

Hardesty, R. M., and R. Hoff, 2012: Thermodynamic Profiling Technologies Workshop report to the National Science Foundation and the National Weather Service. NCAR Tech. Note NCAR/TN-448+STR, 80 pp, doi:10.5065/D6SQ8XCF.

Harrison, L., J. Michalsky, and J. Berndt, 1994: Automated multifilter rotating shadow-band radiometer: An instrument for optical depth and radiation measurements. *Appl. Opt.*, **33**, 5118–5125, doi:10.1364/AO.33.005118.

——, M. Beauharnois, J. Berndt, P. Kiedron, J. Michalsky, and Q. Min, 1999: The rotating shadowband spectroradiometer (RSS) at SGP. *Geophys. Res. Lett.*, **26**, 1715–1718, doi:10.1029/1999GL900328.

Holben, B. N., and Coauthors, 1998: AERONET—A federated instrument network and data archive for aerosol characterization. *Remote Sens. Environ.*, **66**, 1–16, doi:10.1016/S0034-4257(98)00031-5.

Kiedron, P., J. Berndt, J. Michalsky, and L. Harrison, 2003: Column water vapor from diffuse irradiance. *Geophys. Res. Lett.*, **30**, 1565, doi:10.1029/2003GL016874.

King, R. W., and Y. Bock, 1996: Documentation for the GAMIT GPS analysis software, version 9.4. Massachusetts Institute of Technology and Scripps Institution of Oceanography Tech. Note, 192 pp.

Knuteson, R. O., and Coauthors, 2004a: Atmospheric Emitted Radiance Interferometer. Part I: Instrument design. *J. Atmos. Oceanic Technol.*, **21**, 1763–1776, doi:10.1175/JTECH-1662.1.

——, and Coauthors, 2004b: Atmospheric Emitted Radiance Interferometer. Part II: Instrument performance. *J. Atmos. Oceanic Technol.*, **21**, 1777–1789, doi:10.1175/JTECH-1663.1.

Liebe, H. J., and D. H. Layton, 1987: Millimeter-wave properties of the atmosphere: Laboratory studies and propagation modeling. National Telecommunications and Information Administration Tech. Rep. 87-224, 74 pp. [Available online at http://www.its.bldrdoc.gov/publications/tr-87-224.aspx.]

Liljegren, J. C., 1994: Two-channel microwave radiometer for observations of total column precipitable water vapor and cloud liquid water path. *Proc. Fifth Symposium on Global Change Studies*, Nashville, TN, Amer. Meteor. Soc., 262–269.

——, 2000: Automatic self-calibration of ARM microwave radiometers. *Microwave Radiometry and Remote Sensing of the Earth's Surface and Atmosphere*, P. Pampaloni and S. Paloscia, Eds., VSP Press, 433–443.

——, and B. M. Lesht, 1996: Measurements of integrated water vapor and cloud liquid water from microwave radiometers at the DOE ARM Cloud and Radiation Testbed in the U.S. Southern Great Plains. *Proc. Int. Geoscience Remote Sensing Symp.* 1996, Lincoln, NE, IEEE, 1675–1677, doi:10.1109/IGARSS.1996.516767.

——, S. A. Boukabara, K. Cady-Pereira, and S. A. Clough, 2005: The effect of the half-width of the 22-GHz water vapor line on retrievals of temperature and water vapor profiles with a 12-channel microwave radiometer. *IEEE Trans. Geosci. Remote Sens.*, **43**, 1102–1108, doi:10.1109/TGRS.2004.839593.

Liou, Y., Y. Teng, T. Van Hove, and J. C. Liljegren, 2001: Comparison of precipitable water vapor observations in the near tropics by GPS, microwave radiometer, and radiosondes. *J. Appl. Meteor.*, **40**, 5–15, doi:10.1175/1520-0450(2001)040<0005:COPWOI>2.0.CO;2.

Löhnert, U., D. D. Turner, and S. Crewell, 2009: Ground-based temperature and humidity profiling using spectral infrared and microwave observations. Part I: Simulated retrieval performance in clear-sky conditions. *J. Appl. Meteor. Climatol.*, **48**, 1017–1032, doi:10.1175/2008JAMC2060.1.

Mace, G. G., T. P. Ackerman, P. Minnis, and D. F. Young, 1998: Cirrus layer microphysical properties derived from surface-based millimeter radar and infrared interferometer data. *J. Geophys. Res.*, **103**, 23 207–23 216, doi:10.1029/98JD02117.

Mather, J. H., D. D. Turner, and T. P. Ackerman, 2016: Scientific maturation of the ARM Program. *The Atmospheric Radiation Measurement (ARM) Program: The First 20 Years, Meteor. Monogr.*, No. 57, Amer. Meteor. Soc., doi:10.1175/AMSMONOGRAPHS-D-15-0053.1.

Mattioli, V., E. R. Westwater, S. I. Gutman, and V. R. Morris, 2005: Forward model studies of water vapor using scanning microwave radiometers, global positioning system, and radiosondes during the Cloudiness Intercomparison Experiment. *IEEE Trans. Geosci. Remote Sens.*, **43**, 1012–1021, doi:10.1109/TGRS.2004.839926.

——, ——, D. Cimini, J. Liljegren, B. M. Lesht, S. I. Gutman, and F. J. Schmidlin, 2007: Analysis of radiosonde and ground-based remotely sensed PWV data from the 2004 North Slope of Alaska Arctic Winter Radiometric Experiment. *J. Atmos. Oceanic Technol.*, **24**, 415–431, doi:10.1175/JTECH1982.1.

——, ——, ——, A. J. Gasiewski, M. Klein, and V. Leuski, 2008: Microwave and millimeter-wave radiometric and radiosonde observations in an arctic environment. *J. Atmos. Oceanic Technol.*, **25**, 1768–1777, doi:10.1175/2008JTECHA1078.1.

May, P., R. Strauch, and K. Moran, 1988: The altitude coverage of temperature measurements using RASS with wind profiling radars. *Geophys. Res. Lett.*, **15**, 1381–1384, doi:10.1029/GL015i012p01381.

McCord, R., and J. W. Voyles, 2016: The ARM data system and archive. *The Atmospheric Radiation Measurement (ARM) Program: The First 20 Years, Meteor. Monogr.*, No. 57, Amer. Meteor. Soc., doi:10.1175/AMSMONOGRAPHS-D-15-0043.1.

McFarlane, S. A., J. H. Mather, and E. J. Mlawer, 2016: ARM's progress on improving atmospheric broadband radiative fluxes and heating rates. *The Atmospheric Radiation Measurement (ARM) Program: The First 20 Years, Meteor. Monogr., No. 57, Amer. Meteor. Soc.*, doi:10.1175/AMSMONOGRAPHS-D-15-0046.1.

Michalsky, J. J., J. C. Liljegren, and L. C. Harrison, 1995: A comparison of Sun photometer derivations of total column water vapor and ozone to standard measures of same at the Southern Great Plains Atmospheric Radiation Measurement site. *J. Geophys. Res.*, **100**, 25 995–26 003, doi:10.1029/95JD02706.

Miller, N. E., J. C. Liljegren, T. R. Shippert, S. A. Clough, and P. D. Brown, 1994: Quality measurement experiments within the Atmospheric Radiation Measurement Program. *Proc. Fourth Atmospheric Radiation Measurement (ARM) Science Team Meeting*, Charleston, SC, U.S. DOE, 5–9. [Available online at http://www.arm.gov/publications/proceedings/conf04/extended_abs/miller_ne.pdf?id=49.]

Miloshevich, L. M., H. Vömel, D. N. Whiteman, B. M. Lesht, F. J. Schmidlin, and F. Russo, 2006: Absolute accuracy of water vapor measurements from six operational radiosonde types launched during AWEX-G and implications for AIRS validation. *J. Geophys. Res.*, **111**, D09S10, doi:10.1029/2005JD006083.

——, ——, ——, and T. Leblanc, 2009: Accuracy assessment and correction of Vaisala RS92 radiosonde water vapor measurements. *J. Geophys. Res.*, **114**, D11305, doi:10.1029/2008JD011565.

Mlawer, E. J., and D. D. Turner, 2016: Spectral radiation measurements and analysis in the ARM Program. *The Atmospheric Radiation Measurement (ARM) Program: The First 20 Years, Meteor. Monogr.*, No. 57, Amer. Meteor. Soc., doi:10.1175/AMSMONOGRAPHS-D-15-0027.1.

——, V. H. Payne, J.-L. Moncet, J. S. Delamere, M. J. Alvarado, and D. C. Tobin, 2012: Development and recent evaluation of the MT_CKD model of continuum absorption. *Philos. Trans. Roy. Soc. London*, **A370**, 2520–2556, doi:10.1098/rsta.2011.0295.

Niell, A. E., 1996: Global mapping functions for the atmosphere delay at radio wavelengths. *J. Geophys. Res.*, **101**, 3227–3246, doi:10.1029/95JB03048.

——, A. J. Coster, F. S. Solheim, V. B. Mendes, P. C. Toor, R. B. Langely, and C. A. Upham, 2001: Comparison of measurements of atmospheric wet delay by radiosonde, water vapor radiometer, GPS, and VLBI. *J. Atmos. Oceanic Technol.*, **18**, 830–850, doi:10.1175/1520-0426(2001)018<0830:COMOAW>2.0.CO;2.

Payne, V. H., J. S. Delamere, K. E. Cady-Pereira, R. R. Gamache, J.-L. Moncet, E. J. Mlawer, and S. A. Clough, 2008: Air-broadened half-widths of the 22 GHz and 183 GHz water vapor lines. *IEEE Trans. Geosci. Remote Sens.*, **46**, 3601–3617, doi:10.1109/TGRS.2008.2002435.

——, E. J. Mlawer, K. E. Cady-Pereira, and J.-L. Moncet, 2011: Water vapor continuum absorption in the microwave. *IEEE Trans. Geosci. Remote Sens.*, **49**, 2194–2208, doi:10:1109/TGRS.2010.2091416.

Racette, P. E., and Coauthors, 2005: Measurement of low amounts of precipitable water vapor using ground-based millimeterwave radiometry. *J. Atmos. Oceanic Technol.*, **22**, 317–337, doi:10.1175/JTECH1711.1.

Revercomb, H. E., H. Buijs, H. B. Howell, D. D. LaPorte, W. L. Smith, and L. A. Sromovsky, 1988: Radiometric calibration of IR Fourier transform spectrometers: Solution to a problem with the High-Resolution Interferometer Sounder. *Appl. Opt.*, **27**, 3210–3218, doi:10.1364/AO.27.003210.

——, and Coauthors, 2003: The Atmospheric Radiation Measurement Program's water vapor intensive observation periods: Overview, accomplishments, and future challenges. *Bull. Amer. Meteor. Soc.*, **84**, 217–236, doi:10.1175/BAMS-84-2-217.

Richardson, S. J., M. E. Splitt, and B. M. Lesht, 2000: Enhancement of ARM surface meteorological observations during the fall 1996 water vapor intensive observation period. *J. Atmos. Oceanic Technol.*, **17**, 312–322, doi:10.1175/1520-0426(2000)017<0312:EOASMO>2.0.CO;2.

Rocken, C., S. Sokolovskiy, J. M. Johnson, and D. Hunt, 2001: Improved mapping of tropospheric delays. *J. Atmos. Oceanic Technol.*, **18**, 1205–1213, doi:10.1175/1520-0426(2001)018<1205:IMOTD>2.0.CO;2.

Rosenkranz, P. W., 1998: Water vapor continuum absorption: A comparison of measurements and models. *Radio Sci.*, **33**, 919–928, doi:10.1029/98RS01182.

Rothacher, M., 1992: Orbits of satellite systems in space geodesy. Ph.D. thesis, University of Bern, 243 pp.

Rothman, L. S., and Coauthors, 1992: The HITRAN molecular database: Editions of 1991 and 1992. *J. Quant. Spectrosc. Radiat. Transfer*, **48**, 469–507, doi:10.1016/0022-4073(92)90115-K.

——, and Coauthors, 2005: The HITRAN 2004 molecular spectroscopic database. *J. Quant. Spectrosc. Radiat. Transfer*, **96**, 139–204, doi:10.1016/j.jqsrt.2004.10.008.

Schmid, B., and Coauthors, 2001: Comparison of columnar water-vapor measurements from solar transmittance methods. *Appl. Opt.*, **40**, 1886–1896, doi:10.1364/AO.40.001886.

Shupe, M. D., J. M. Comstock, D. D. Turner, and G. G. Mace, 2016: Cloud property retrievals in the ARM Program. *The Atmospheric Radiation Measurement (ARM) Program: The First 20 Years, Meteor. Monogr.*, No. 57, Amer. Meteor. Soc., doi:10.1175/AMSMONOGRAPHS-D-15-0030.1.

Smith, T. L., S. G. Benjamin, S. I. Gutman, and S. R. Sahm, 2007: Forecast impact from assimilation of the GPS-IPW observations into the Rapid Update Cycle. *Mon. Wea. Rev.*, **135**, 2914–2930, doi:10.1175/MWR3436.1.

Smith, W. L., and Coauthors, 1981: First sounding results from VAS-D. *Bull. Amer. Meteor. Soc.*, **62**, 232–236.

——, and Coauthors, 1990: GAPEX: A Ground-Based Atmospheric Profiling Experiment. *Bull. Amer. Meteor. Soc.*, **71**, 310–318, doi:10.1175/1520-0477(1990)071<0310:GAGBAP>2.0.CO;2.

——, W. F. Feltz, R. O. Knuteson, H. E. Revercomb, H. M. Woolf, and H. B. Howell, 1999: The retrieval of planetary boundary layer structure using ground-based infrared spectral radiance observations. *J. Atmos. Oceanic Technol.*, **16**, 323–333, doi:10.1175/1520-0426(1999)016<0323:TROPBL>2.0.CO;2.

Soden, B. J., D. D. Turner, B. M. Lesht, and L. M. Miloshevich, 2004: An analysis of satellite, radiosonde, and lidar observations of upper tropospheric water vapor from the Atmospheric Radiation Measurement Program. *J. Geophys. Res.*, **109**, D04105, doi:10.1029/2003JD003828.

Solheim, F. S., J. R. Godwin, and R. Ware, 1998a: Passive, ground-based remote sensing of temperature, water vapor, and cloud liquid profiles by a frequency-synthesized microwave radiometer. *Meteor. Z.*, **7**, 370–376.

——, ——, ——, E. R. Westwater, and Y. Han, 1998b: Radiometric temperature, water vapor, and cloud liquid water profiling with various inversion methods. *Radio Sci.*, **33**, 393–404, doi:10.1029/97RS03656.

Stamnes, K., R. G. Ellingson, J. A. Curry, J. E. Walsh, and B. D. Zak, 1999: Review of science issues, deployment strategy, and status for the ARM North Slope of Alaska – Adjacent Arctic Ocean climate research site. *J. Climate*, **12**, 46–63, doi:10.1175/1520-0442-12.1.46.

Stankov, B. B., E. Gossard, and E. R. Westwater, 1996: High vertical resolution humidity profiling from combined remote sensors. *J. Atmos. Oceanic Technol.*, **13**, 1285–1290, doi:10.1175/1520-0426(1996)013<1285:UOWPEO>2.0.CO;2.

Stokes, G. M., 2016: Original ARM concept and launch. *The Atmospheric Radiation Measurement (ARM) Program: The First 20 Years, Meteor. Monogr.*, No. 57, Amer. Meteor. Soc., doi:10.1175/AMSMONOGRAPHS-D-15-0021.1.

Suomi, V. E., and T. H. Vonder Haar, 1969: Geosynchronous meteorological satellite. *J. Spacecr.*, **6**, 342–344, doi:10.2514/3.29604.

Tobin, D. C., and Coauthors, 1999: Downwelling spectral radiance observations at the SHEBA ice station: Water vapor continuum measurements from 17 to 26 μm. *J. Geophys. Res.*, **104**, 2081–2092, doi:10.1029/1998JD200057.

——, and Coauthors, 2006: Atmospheric Radiation Measurement site atmospheric state best estimates for Atmospheric Infrared Sounder temperature and water vapor retrieval validation. *J. Geophys. Res.*, **111**, D09S14, doi:10.1029/2005JD006103.

Turner, D. D., 2008: Ground-based retrievals of optical depth, effective radius, and composition of airborne mineral dust above the Sahel. *J. Geophys. Res.*, **113**, D00E03, doi:10.1029/2008JD010054.

——, and J. E. M. Goldsmith, 1999: Twenty-four-hour Raman lidar water vapor measurements during the Atmospheric Radiation Measurement Program's 1996 and 1997 water vapor intensive observation periods. *J. Atmos. Oceanic Technol.*, **16**, 1062–1076, doi:10.1175/1520-0426(1999)016<1062:TFHRLW>2.0.CO;2.

——, and E. J. Mlawer, 2010: The Radiative Heating in Underexplored Bands Campaigns (RHUBC). *Bull. Amer. Meteor. Soc.*, **91**, doi:10.1175/2010BAMS2904.1.

——, and U. Löhnert, 2014: Retrieving temperature and humidity profiles from the ground-based Atmospheric Emitted Radiance Interferometer (AERI). *J. Appl. Meteor. Climatol.*, **53**, 752–771, doi:10.1175/JAMC-D-13-0126.1.

——, W. F. Feltz, and R. A. Ferrare, 2000: Continuous water vapor profiles from operational ground-based active and passive remote sensors. *Bull. Amer. Meteor. Soc.*, **81**, 1301–1317, doi:10.1175/1520-0477(2000)081<1301:CWBPFO>2.3.CO;2.

——, R. A. Ferrare, L. A. Heilman Brasseur, W. F. Feltz, and T. P. Tooman, 2002: Automated retrievals of water vapor and aerosol profiles from an operational Raman lidar. *J. Atmos. Oceanic Technol.*, **19**, 37–50, doi:10.1175/1520-0426(2002)019<0037:AROWVA>2.0.CO;2.

——, S. A. Ackerman, B. A. Baum, H. E. Revercomb, and P. Yang, 2003a: Cloud phase determination using ground-based AERI observations at SHEBA. *J. Appl. Meteor.*, **42**, 701–715, doi:10.1175/1520-0450(2003)042<0701:CPDUGA>2.0.CO;2.

——, B. M. Lesht, S. A. Clough, J. C. Liljegren, H. E. Revercomb, and D. C. Tobin, 2003b: Dry bias and variability in Vaisala radiosondes: The ARM experience. *J. Atmos. Oceanic Technol.*, **20**, 117–132, doi:10.1175/1520-0426(2003)020<0117:DBAVIV>2.0.CO;2.

——, and Coauthors, 2004: The QME AERI LBLRTM: A closure experiment for downwelling high spectral resolution infrared radiance. *J. Atmos. Sci.*, **61**, 2657–2675, doi:10.1175/JAS3300.1.

——, R. O. Knuteson, H. E. Revercomb, C. Lo, and R. G. Dedecker, 2006: Noise reduction of Atmospheric Emitted Radiance Interferometer (AERI) observations using principal component analysis. *J. Atmos. Oceanic Technol.*, **23**, 1223–1238, doi:10.1175/JTECH1906.1.

——, S. A. Clough, J. C. Liljegren, E. E. Clothiaux, K. Cady-Pereira, and K. L. Gaustad, 2007: Retrieving liquid water path and precipitable water vapor from the Atmospheric Radiation Measurement (ARM) microwave radiometers. *IEEE Trans. Geosci. Remote Sens.*, **45**, 3680–3690, doi:10.1109/TGRS.2007.903703.

——, U. Löhnert, M. Cadeddu, S. Crewell, and A. Vogelmann, 2009: Modifications to the water vapor continuum in the microwave suggested by ground-based 150-GHz observations. *IEEE Trans. Geosci. Remote Sens.*, **47**, 3326–3337, doi:10.1109/TGRS.2009.2022262.

——, A. Merrelli, D. Vimont, and E. J. Mlawer, 2012: Impact of modifying the longwave water vapor continuum absorption model on community Earth system model simulations. *J. Geophys. Res.*, **117**, D04106, doi:10.1029/2011JD016440.

——, J. E. M. Goldsmith, and R. A. Ferrare, 2016: Development and applications of the ARM Raman lidar. *The Atmospheric Radiation Measurement (ARM) Program: The First 20 Years, Meteor. Monogr.*, No. 57, Amer. Meteor. Soc., doi:10.1175/AMSMONOGRAPHS-D-15-0026.1.

Verlinde, J., B. Zak, M. D. Shupe, M. Ivey, and K. Stamnes, 2016: The ARM North Slope of Alaska (NSA) sites. *The Atmospheric Radiation Measurement (ARM) Program: The First 20 Years, Meteor. Monogr.*, No. 57, Amer. Meteor. Soc., doi:10.1175/AMSMONOGRAPHS-D-15-0023.1.

Vömel, H., and Coauthors, 2007: Radiation dry bias of the Vaisala RS92 humidity sensor. *J. Atmos. Oceanic Technol.*, **24**, 953–963, doi:10.1175/JTECH2019.1.

Vonder Haar, T. H., and V. E. Suomi, 1971: Measurements of the Earth's radiation budget from satellites during a five-year period. Part I: Extended time and space means. *J. Atmos. Sci.*, **28**, 305–314, doi:10.1175/1520-0469(1971)028<0305:MOTERB>2.0.CO;2.

Wade, C. G., 1994: An evaluation of problems affecting the measurement of low relative humidity on the U.S. radiosonde. *J. Atmos. Oceanic Technol.*, **11**, 687–700, doi:10.1175/1520-0426(1994)011<0687:AEOPAT>2.0.CO;2.

Wagner, T. J., W. F. Feltz, and S. A. Ackerman, 2008: The temporal evolution of convective indices in storm-producing environments. *Wea. Forecasting*, **23**, 786–794, doi:10.1175/2008WAF2007046.1.

——, D. D. Turner, L. K. Berg, and S. K. Krueger, 2013: Ground-based remote retrievals of cumulus entrainment rates. *J. Atmos. Oceanic Technol.*, **30**, 1460–1471, doi:10.1175/JTECH-D-12-00187.1.

Wang, J., H. L. Cole, D. J. Carlson, E. R. Miller, K. Beierle, A. Paukkunen, and T. K. Laine, 2002: Corrections of humidity measurement errors from the Vaisala RS80 radiosonde—Application to TOGA COARE data. *J. Atmos. Oceanic Technol.*, **19**, 981–1002, doi:10.1175/1520-0426(2002)019<0981:COHMEF>2.0.CO;2.

Ware, R., R. Carpenter, J. Güldner, J. C. Liljegren, T. Nehrkorn, F. Solheim, and F. Vandenberghe, 2003: A multichannel radiometric profiler of temperature, humidity, and cloud liquid. *Radio Sci.*, **38**, 8079, doi:10.1029/2002RS002856.

Westwater, E. R., 1997: Remote sensing of tropospheric temperature and water vapor by integrated observing systems. *Bull. Amer. Meteor. Soc.*, **78**, 1991–2006.

Whiteman, D. N., S. H. Melfi, and R. A. Ferrare, 1992: Raman lidar system for measurement of water vapor and aerosols in the Earth's atmosphere. *Appl. Opt.*, **31**, 3068–3082, doi:10.1364/AO.31.003068.

Wolfe, D. E., and S. I. Gutman, 2000: Developing an operational, surface-based, GPS, water vapor observing system for NOAA: Network design and results. *J. Atmos. Oceanic Technol.*, **17**, 426–439, doi:10.1175/1520-0426(2000)017<0426:DAOSBG>2.0.CO;2.

Wulfmeyer, V., and Coauthors, 2011: The Convective and Orographically-induced Precipitation Study (COPS): The scientific strategy, the field phase, and research highlights. *Quart. J. Roy. Meteor. Soc.*, **137**, 3–30, doi:10.1002/qj.752.

Zhang, M. H., and J. L. Lin, 1997: Constrained variational analysis of sounding data based on column-integrated budgets of mass, heat, moisture, and momentum: Approach and application to ARM measurements. *J. Atmos. Sci.*, **54**, 1503–1524, doi:10.1175/1520-0469(1997)054<1503:CVAOSD>2.0.CO;2.

——, ——, R. T. Cederwall, J. J. Yio, and S. C. Xie, 2001: Objective analysis of ARM IOP data: Method and sensitivity. *Mon. Wea. Rev.*, **129**, 295–311, doi:10.1175/1520-0493(2001)129<0295:OAOAID>2.0.CO;2.

——, R. C. J. Somerville, and S. Xie, 2016: The SCM concept and creation of ARM forcing datasets. *The Atmospheric Radiation Measurement (ARM) Program: The First 20 Years, Meteor. Monogr.*, No. 57, Amer. Meteor. Soc., doi:10.1175/AMSMONOGRAPHS-D-15-0040.1.

Chapter 14

Spectral Radiation Measurements and Analysis in the ARM Program

E. J. MLAWER

Atmospheric and Environmental Research, Lexington, Massachusetts

D. D. TURNER

National Severe Storms Laboratory, NOAA, Norman, Oklahoma

1. Introduction

The spectral signatures of radiation produced by various atmospheric constituents are key to our understanding of many issues related to climate and weather. Ground-based measurements of spectrally resolved radiation are particularly rich sources of information on atmospheric gases, clouds, and aerosol properties. For there to be confidence in simulations by atmospheric models, including general circulation models (GCMs), it is essential that calculations by the most accurate radiative transfer codes be able to reproduce these spectral measurements for a broad range of conditions. This perspective was central to the founding objectives of the ARM Program, provided an essential focus of the program during its early years, and was at the core of many of the program's important accomplishments during its history.

A critical motivation for establishing the Atmospheric Radiation Measurement (ARM) Program was to develop the capability to evaluate and improve line-by-line radiation codes, which are the most physically based radiative transfer algorithms, through extensive comparisons with high-quality spectral radiation measurements. In particular, results from the Intercomparison of Radiation Codes in Climate Models (ICRCCM; Ellingson and Fouquart 1991; Ellingson et al. 2016, chapter 1), although directed at the evaluation of the performance of fast radiation parameterizations, were key to establishing the impetus for a program such as ARM with a spectral radiation focus. A key conclusion from the analysis of longwave ICRCCM results (Ellingson et al. 1991) was that, although many fast radiation codes used within climate models had spectral errors that partially canceled out when fluxes over a wide spectral range were computed, line-by-line modelers did not have sufficient confidence in their own models to advocate using them as references. The participants in this study therefore recommended that "a program be organized to simultaneously measure the spectral radiance at high spectral resolution along with the atmospheric variables necessary to calculate the radiance, particularly for clear-sky conditions" (p. 8952). The ARM Program was developed as the answer to this challenge, and this chapter (along with other related chapters in this monograph) details the research program that was followed toward its successful resolution.

The initial response to the ICRCCM recommendation to improve radiative transfer parameterizations through the analysis of field observations was the organization of the Spectral Radiation Experiment (SPECTRE; Ellingson and Wiscombe 1996; Ellingson et al. 2016, chapter 1). This one-month field experiment deployed several infrared interferometers to Coffeyville, Kansas, to measure the downwelling infrared spectral radiance along with a range of sensors, both in situ (e.g., radiosonde, flask measurements of trace gases like carbon dioxide and methane, etc.) and remote (e.g., Raman lidar, Radio Acoustic Sounding System, cloud radar), to characterize the atmospheric state needed as input to drive the radiation models. SPECTRE, although limited, had a number of successes. The Atmospheric Emitted Radiance Interferometer (AERI; Knuteson et al. 2004a,b), which was developed by the University of Wisconsin–Madison, was demonstrated to have a robust

Corresponding author address: E. J. Mlawer, Atmospheric and Environmental Research Inc., 131 Hartwell Ave., Lexington, MA 02421.
E-mail: emlawer@aer.com

DOI: 10.1175/AMSMONOGRAPHS-D-15-0027.1

calibration, and the initial comparisons showed a much better level of agreement between the observed and line-by-line calculated radiances than the range between calculations that was demonstrated during ICRCCM. The SPECTRE dataset, however, was from a single location and from a short duration campaign, a limitation that the ARM Program, which was developed based on the SPECTRE experience and science plan, was designed to overcome. A central objective of the ARM Program was to "relate observed radiative fluxes in the atmosphere, spectrally resolved and as a function of position and time, to the atmospheric temperature, composition (specifically including water vapor and clouds), and surface radiative properties" (Stokes and Schwartz 1994, p. 1203). ARM data would provide the observations at the different climatic locations over longer time periods to fulfill the critical recommendation of ICRCCM.

This chapter provides a history of some of the ARM Program's accomplishments in the analysis of spectral radiation measurements. The program has had greater successes in the thermal infrared spectral region than in the solar, primarily due to the development of the AERI in the early years of the program. [See Turner et al. (2016, chapter 13) for a short history of the development of this instrument.] Progress in the shortwave portion of the spectrum was more challenging due to the relatively few groups developing instruments to measure spectrally resolved radiance observations in that spectral region; instrumental challenges (especially for accurate radiative and spectral calibration) were more daunting than for the longwave. Much of the analysis of spectral radiation focused on clear skies in order to improve the accuracy of the emission and absorption of atmospheric gases and aerosols, with valuable results obtained from ARM spectral radiometers deployed in a number of diverse locations. As will be seen below, a recurring theme of these investigations is the nature of continuum absorption due to water vapor, an important factor in the flow of radiant energy in our atmosphere. As the program's cloud observations matured and new algorithms were developed to retrieve cloud macro- and microphysical properties (Kollias et al. 2016, chapter 17; Shupe et al. 2016, chapter 19), characterizing and improving the spectral radiative transfer algorithms in cloudy atmospheres has taken on a more prominent role. This topic is addressed in section 5 of this chapter.

2. Clear-sky longwave studies

The paradigm used in the ARM Program for analysis of spectral radiation observations is the radiative closure study, which involves the simultaneous critical evaluation of 1) the spectral radiance observations, 2) the physics and implementation of the radiative transfer model, and 3) the specification of the temperature, humidity, and other atmospheric and surface properties relevant to radiation. The evaluation includes any observations on which these properties are based and are required by the radiative transfer model. The efforts to evaluate and improve radiative transfer models in the ARM Program through this approach were undertaken and led by members of the Instantaneous Radiative Flux (IRF) working group, most notably Bob Ellingson, Tony Clough, and Hank Revercomb, within the ARM Science Team. Since the ARM data were collected routinely, the IRF developed the idea of the Quality Measurement Experiment (QME; Turner et al. 2004) to routinely perform the radiative transfer calculations and some higher-order processing needed for the radiative closure experiment.

With respect to the evaluation and improvement of line-by-line radiative transfer codes, the ARM closure studies focused mostly on the spectroscopic properties of water vapor. Although the strengths, widths, and other properties of H_2O absorption lines were scrutinized in these studies, a great deal of attention was focused on analysis related to H_2O self and foreign continuum absorption. This was anticipated in the longwave ICRCCM study, which concluded that uncertainty in the H_2O continuum led to significant limitations in climate studies (Ellingson et al. 1991). As will be seen, efforts within the ARM Program directed at attaining a comprehensive understanding of the properties of the longwave H_2O continuum ended up requiring spectrally resolved observations from several different locations. Although the program did provide some support for directed spectroscopic studies (e.g., Rothman 1992; Varanasi 1998), the primary avenue within ARM for advancement in this area relied on the analysis of field observations of spectrally resolved radiances.

a. *Improvements to the modeling of water vapor absorption*

The primary QME organized by the IRF compared the spectral infrared radiances observed by the AERI and computed by the line-by-line radiative transfer model (LBLRTM; Clough et al. 1992; 2005). Atmospheric state measurements were a critical component of this QME. In particular, the uncertainty in water vapor profile measurements was determined to be the limiting factor in improving the clear-sky longwave radiative transfer model (Revercomb et al. 2003). To reduce these uncertainties to a level where the water vapor profiles could be used with sufficient confidence, a concentrated effort and a series of field experiments were conducted at the ARM Southern Great Plains (SGP) and North Slope of Alaska (NSA) sites. Turner et al. (2016, chapter 13) provides a history of this work, without which improvements in the infrared radiation modeling would not have been possible.

The prototype AERI-00, which used liquid nitrogen to cryogenically cool its detectors and thus required manual attention to refill the liquid nitrogen dewar, was installed at the SGP site in late 1993. Even with sizable uncertainties in the H_2O profiles, the IRF anticipated similar results from AERI-LBLRTM comparisons from this deployment as were experienced during SPECTRE. However, the comparison between the AERI and LBLRTM in clear skies showed a much larger bias than was seen in SPECTRE; importantly, this large bias was seen in all seasons. The change in the size of the bias between the SPECTRE and early ARM results led to many heated discussions on the source of the bias: was it a problem with the AERI observations, aerosols that needed to be included in the calculation (perhaps the SPECTRE results were just "lucky" to have very low aerosol loading), or some problem with the atmospheric state (temperature and humidity) observations? This problem was not resolved until the AERI-01—the first AERI that could be run operationally by using a mechanical Sterling cooler to keep the detectors at cryogenic temperatures—was deployed to the SGP site. The side-by-side comparison of the AERI-01 and the AERI-00 showed that the bias was in the AERI-00 observations, which was traced to a small obscuration in the field of view of the AERI-00 (Knuteson et al. 1999). This served as an important lesson in deploying AERI instruments operationally, and all future AERIs (from the AERI-01 onward) were deployed differently than the AERI-00 in order to eliminate the possible obscuration bias that affected the original AERI.

While the IRF was trying to understand the bias between the AERI-00 and LBLRTM, the Pilot Radiation Observation Experiment (PROBE) was conducted in 1993 in Kavieng, Papua New Guinea. A Fourier transform infrared (FTIR) spectrometer from the National Oceanic and Atmospheric Administration (NOAA) was among the instruments deployed as part of PROBE, and initial comparisons between the FTIR and LBLRTM showed very large biases between 800 and 1000 cm^{-1} in the very moist tropical atmosphere above Kavieng. This ARM-funded analysis suggested that the absorption in version 1 of the Clough–Kneizys–Davies (CKD) H_2O continuum model[1] used within the LBLRTM was too weak, and a modified version of the continuum model (version 2.1) was created. A subsequent cruise to the western Pacific by the research ship *Discoverer* in 1996, which again included the NOAA FTIR and a newer generation AERI, confirmed the results from PROBE (Han et al. 1997). Thus, the ARM Program had made its first major advancement in our understanding of H_2O continuum absorption.

Meanwhile, the IRF continued to develop the QME at the SGP site, with comparisons between AERI-01 observations and LBLRTM calculations that were presented in a series of ARM Science Team Meeting presentations from 1994 to 1999 by Pat Brown, Tony Clough, and colleagues. A great deal of effort was focused upon the improvement of the H_2O observations; ultimately the uncertainty in the accuracy of the measurement of precipitable water vapor (PWV) at SGP would drop from order 15% in the early 1990s to 3% by the early 2000s (Turner et al. 2016, chapter 13). The extended dataset at the SGP, together with these improved PWV observations, allowed the CKD model to be further refined in the 800–1300 cm^{-1} spectral region (Turner et al. 2004), the "atmospheric window" so important to planetary energy balance. The SGP QME data also were extremely useful in evaluating new versions of the high-resolution transmission molecular absorption (HITRAN) absorption line database, and especially the H_2O line parameters in this database (Turner et al. 2004). The improvements that were made in the main infrared window (800–1300 cm^{-1}) from the PROBE, *Discoverer*, and SGP datasets removed a significant amount of spectral cancelation of error, and contributed significantly to the improvement of the downwelling longwave flux calculation by the LBLRTM by approximately 5 $W\,m^{-2}$ (Turner et al. 2004).

The SGP QME and tropical PROBE and *Discoverer* results were instrumental in improving the modeling of H_2O absorption, most notably H_2O continuum absorption in the atmospheric window. However, these results did not address spectroscopic issues within strong H_2O absorption bands, such as the pure rotation band from 0 to 625 cm^{-1} ($>16\,\mu m$). With its participation in the Surface Heat Budget of the Arctic Ocean (SHEBA) field campaign (Uttal et al. 2002), ARM radiative closure studies were extended to these spectral regions. An extended range AERI,[2] which was modified to have sensitivity to downwelling radiation at wavenumbers as low as 400 cm^{-1} ($25\,\mu m$), was deployed on the icebreaker that SHEBA used as its

[1] The water vapor continuum model used within LBLRTM during the first decade of the ARM Program was the CKD model (Clough et al. 1989), which was widely used throughout the community. In the early 2000s this model was revised and renamed the MT_CKD model (Mlawer et al. 2012). Details on these continuum formulations and their evolution over time can be found in these references; a simple explanation of the water vapor continuum is given in Turner and Mlawer (2010).

[2] The typical AERI is sensitive to radiation at wavenumbers as low as 530 cm^{-1} ($19\,\mu m$).

floating base. In the very dry Arctic atmosphere during the winter, a portion of the rotational water vapor band in the far-infrared between 100 and 625 cm^{-1} (100 to 16 μm) is partially transparent. Thus, the accuracy of the radiative transfer model can be evaluated using surface-based radiation measurements. (At larger PWV amounts, such as those seen in midlatitudes or in the tropics, this portion of the spectrum is opaque and the accuracy of the water vapor continuum or line parameters cannot be evaluated.) Tobin et al. (1999) used the AERI observations at SHEBA to demonstrate that the absorption of the CKD foreign continuum model in the far-infrared was nearly a factor of 3 too strong, leading to a significant modification to the CKD model in this spectral region.

The far-infrared accounts for nearly 40% of the total outgoing longwave radiation emitted by the planet and is very important for the radiative atmospheric heating/cooling rate profiles in the middle to upper troposphere (Clough et al. 1992; Harries et al. 2008). Thus, the improvements made by Tobin et al. (1999) had a big effect on the atmospheric heating rate profiles. However, the uncertainty in the water vapor profiles used by Tobin et al. was on the order of 25%, and this uncertainty translated directly into the same level of uncertainty in the continuum model adjustment. Furthermore, the extended-range AERI did not provide measurements in a significant portion of the far-infrared spectrum, so the accuracy of radiative transfer models in this unobserved spectral region was unable to be evaluated in SHEBA. This led ARM to organize the Radiative Heating in Underexplored Bands Campaigns (RHUBC; Turner and Mlawer 2010) to collect and analyze additional spectral radiation measurements in dry environments.

After SHEBA, the ARM Program made solid progress improving the accuracy of PWV observations in very dry climates, and by the time of the first RHUBC campaign new operational microwave radiometers at 183 GHz had been developed for deployment to the ARM NSA site (Turner et al. 2016, chapter 13). RHUBC-I deployed three of these microwave radiometers and three infrared interferometers to the NSA site in late winter 2007 to evaluate and refine the Tobin et al. modification with the improved water vapor observations. The experiment was a success. RHUBC-I demonstrated good agreement between all three of the 183-GHz radiometers, each of which used different technologies and calibration approaches (Cimini et al. 2009). Analysis of data from RHUBC-I, along with additional data collected before and after the experiment using one 183-GHz radiometer that was running operationally at NSA, demonstrated that adjustments were needed to the strengths of both the H_2O foreign and self-continuum models in the far-IR, and, importantly, refined the widths of some of the more prominent H_2O absorption lines in the 400–625 cm^{-1} region (Delamere et al. 2010).

The Delamere et al. (2010) study provided information on the strength of continuum absorption at the high wavenumber end of the H_2O pure rotation band (400–650 cm^{-1}). At the low wavenumber edge of the band (i.e., around 5 cm^{-1}), radiative closure studies at a number of microwave frequencies by Payne et al. (2011) provided analogous and consistent information on the continuum. The microwave analysis included observations at 5 cm^{-1} (150 GHz) from the deployment of the ARM Mobile Facility (AMF) to Germany in 2007 that were also analyzed by Turner et al. (2009). Based on these studies, an updated version of the continuum model (MT_CKD_2.4) was developed, with a nearly 50% change in the strength of the water vapor foreign continuum at 200 cm^{-1} (50 μm). GCM simulations demonstrated that this large change to the water vapor continuum model had a significant radiative and dynamic impact on a global climate model simulation (Turner et al. 2012a).

The success of RHUBC-I led to the second RHUBC experiment, which was conducted in fall 2009 at a high-altitude site (5.3 km AGL) in the Chilean Andes. The PWV amounts in RHUBC-II were nearly 5 times drier than the driest conditions experienced at the NSA site, and thus a larger portion of the far-infrared region was partially transparent. In particular, radiative transfer model calculations at wavenumbers as small as 220 cm^{-1} (45 μm) could be evaluated with the RHUBC-II dataset. Five interferometers were deployed during RHUBC-II, along with one of the 183-GHz radiometers from RHUBC-I. This experiment provided the first complete spectral measurement of the entire downwelling terrestrial infrared spectrum from the ground (Turner et al. 2012b) and demonstrated that to first order the current water vapor continuum model, which includes the big change at 200 cm^{-1} indicated above, is more accurate than the continuum model that existed before the RHUBC experiments.

ARM-related studies also contributed to advancements in our knowledge of the water vapor self-continuum at the short wavelength end of the AERI spectral domain. Using AERI measurements from SGP, two independent studies (Strow et al. 2006; Mlawer et al. 2012) determined that the MT_CKD self-continuum was significantly too weak from 2400 to 2600 cm^{-1}. This later study also showed that the result in this spectral region is consistent with certain laboratory and field observations of the continuum in the near-infrared.

b. Other modeling advances in the infrared

Gaseous species other than water vapor also had their infrared spectroscopy evaluated and improved using ARM AERI observations. In 2002, AERI measurements from the NSA site were used, along with measurements from the University of Wisconsin's airborne High-resolution Interferometer Sounder (HIS) and Scanning HIS (S-HIS), to modify the carbon dioxide line shape and carbon dioxide continuum used in the 500–900 cm^{-1} region in LBLRTM. After the introduction in LBLRTM of P- and R-branch line coupling for CO_2, AERI measurements from the ARM Tropical Western Pacific (TWP) site were used to determine that these previous adjustments were no longer necessary (Payne et al. 2007). More recently, SGP AERI measurements were used along with satellite measurements to modify the CO_2 continuum and line coupling implementation near 2400 cm^{-1} in LBLRTM (Mlawer et al. 2012).

c. Spectral trends

Among the many achievements of the AERI/LBLRTM QME was the establishment of confidence in the accuracy of the AERI radiance observations. Furthermore, since the AERI design allows it to monitor both its calibration and sensitivity every 10 min, it is an ideal sensor for long-term trend analysis, which was one of the original design goals of the multidecadal observations by the ARM Program. Using 14 years of data (1997–2010) collected at the SGP site by the AERI-01 instrument, an analysis was performed to characterize the distribution of the downwelling radiance as a function of clear sky, opaque cloud, and "thin cloud" (Turner and Gero 2011), and then to determine if there were any trends in these classifications over the entire record or when analyzed as a function of season or diurnal cycle. Gero and Turner (2011) identified numerous statistically significant trends in the downwelling radiance over the 14-yr period (e.g., Fig. 14-1, bottom). The trends in the spectrally resolved AERI observations, such as that shown for the autumn in the top panel of Fig. 14-1, allowed these trends to be attributed to trends in PWV (in clear-sky scenes) or in cloud properties (in cloudy scenes).

3. Radiative transfer model development

It is clear from the previous section that the ARM Program has been instrumental in establishing LBLRTM as a state-of-the-science radiative transfer model. This benefits both the ARM scientific community and the larger community as well. LBLRTM is used to train forward models utilized in operational satellite retrievals (Clough et al. 2006; Clerbaux et al. 2007) and data assimilation schemes, to develop radiation codes for climate applications (Mlawer et al. 1997), and to provide reference calculations for model intercomparison studies (Barker et al. 2003; Oreopoulos et al. 2012). The breadth and importance of these applications attests to the value of ARM spectral observations and related research in advancing atmospheric and climate science.

Figure 14-2 shows the spectral improvements made in the longwave radiative transfer model over a period corresponding to ARM's first 20 years for moist (middle panel) and dry (bottom panel) conditions. ARM-related advances are responsible for a large fraction of this improvement. As can be seen in Fig. 14-3, which illustrates the impact on vertical flux of improvements to LBLRTM implemented between 1999 and 2009, significant enhancement in model quality continued into ARM's second decade (Delamere et al. 2010). These model improvements have reduced the overall residuals to a level such that elevated levels of trace gases can be observed (Fig. 14-4; Shephard et al. 2003) and more accurate satellite-based retrievals of temperature and species abundances can be attained (Alvarado et al. 2013). Further discussion of ARM accomplishments related to radiative fluxes can be found in McFarlane et al. (2016, chapter 20).

Although not every application in the infrared requires that scattering be included, the lack of scattering in the radiative transfer solution in LBLRTM posed a problem for conditions in which scattering is an important consideration. For example, the scattering contribution from a single-layer liquid water cloud in the downwelling radiance observed by the AERI can be as large as 12% of the total signal, depending on the liquid water path and the wavelength of the radiation (Turner and Löhnert 2014). This limitation prompted ARM-funded development of two codes that utilize gaseous optical depths from LBLRTM and scattering properties of the medium (e.g., liquid or ice clouds, aerosol layers) to perform scattering calculations at high spectral resolution. The Code for High-resolution Accelerated Radiative Transfer and Scattering (CHARTS; Moncet and Clough 1997) multiple-scattering model uses the adding/doubling technique to perform computationally efficient calculations in plane-parallel atmospheres. For accuracies appropriate for remote sensing applications (e.g., ~0.1% in radiances throughout the thermal region), the computational gain achieved in the radiance calculations may be as high as 3000 compared to other existing multiple scattering algorithms. Among its applications, CHARTS was applied to the modeling of observations from the ground-based AERI interferometer in

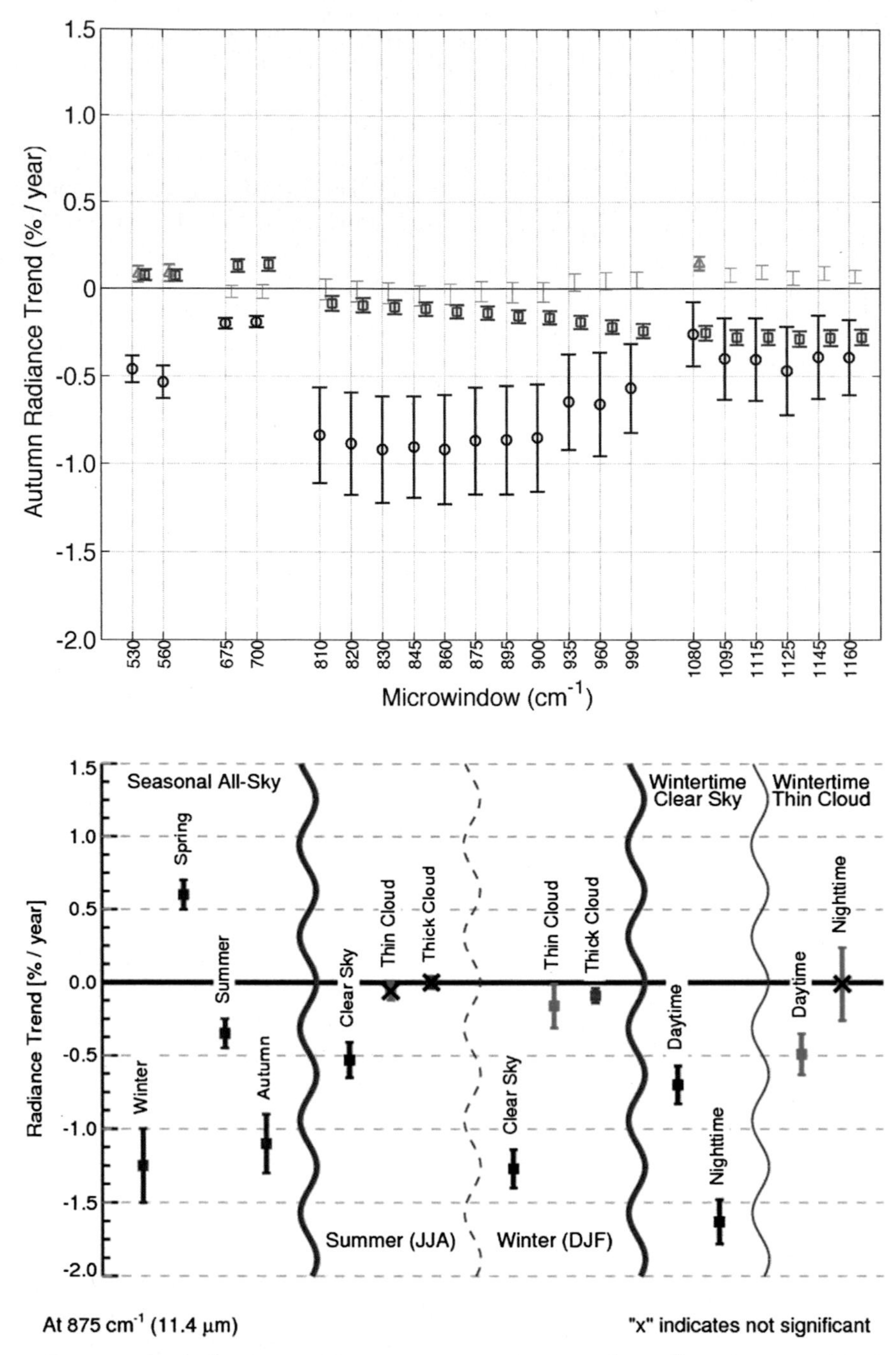

FIG. 14-1. (top) The downwelling spectral radiance trends (% yr^{-1}) for clear sky (blue), opaque clouds (red), and "thin clouds" (green) for the autumn (SON) at the SGP [Fig. 7d from Gero and Turner (2011)]. (bottom) A subset of the trends at 875 cm^{-1} from Gero and Turner (2011) that highlights some of the more interesting findings.

cloudy-sky conditions, used for reference calculations in shortwave radiation code intercomparison studies (e.g., Barker et al. 2003; Oreopoulos et al. 2012), and used for shortwave radiative closure studies (see below).

Another model developed was LBLDIS (Turner et al. 2003), which is a combination of LBLRTM and the Discrete Ordinate Radiative Transfer (DISORT) algorithm (Stamnes et al. 1988). LBLDIS is a flexible model, and is able to compute radiance and flux across a spectrum or in specified spectral intervals (such as a selection of microwindows between absorption lines). This model is the backbone for the

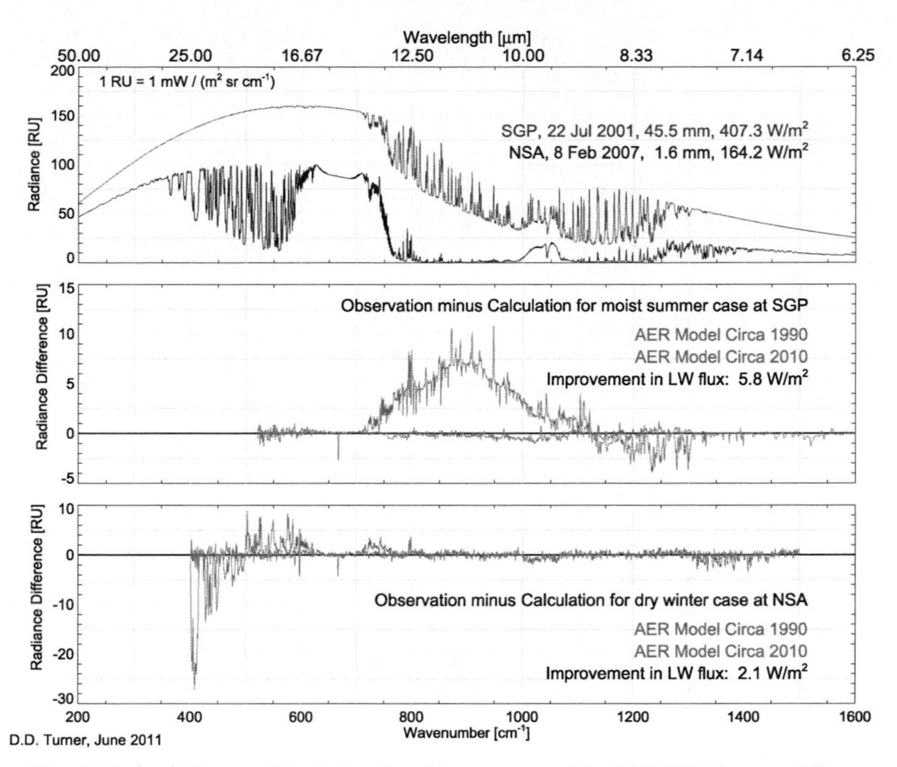

FIG. 14-2. (top) Downwelling infrared radiance computed by LBLRTM for two different cases: a warm, wet case observed at the SGP site (red) and a cold, dry case observed at the NSA site (blue). (middle) The AERI-observed minus LBLRTM-calculated radiance residuals for the warm, wet SGP case, where the LBLRTM calculation was performed using the version of the model that was available at the start of the program in 1990 (brown) and using the version of the model available in 2010 (green). (bottom) As in (middle), but for the cold, dry case using the extended range AERI observations at the NSA site. Note that the improvement in the downwelling longwave (LW) flux includes cancelation of error.

Mixed-Phase Cloud Retrieval Algorithm (MIXCRA; Turner 2005), which is able to retrieve liquid water and ice water optical depths and effective radii of both liquid water and ice particles simultaneously from AERI-observed radiance spectra. Figure 14-5 presents the application of LBLDIS to an AERI observation under extremely dusty conditions observed when the AMF was deployed to Niamey, Niger (Miller and Slingo 2007; Turner 2008). Using an external mixture of kaolinite and hematite spheres, the left panel indicates that good agreement was found between the LBLDIS calculation and the AERI observed radiance in the 600–1400 cm^{-1} region, with the inclusion of scattering in the calculation of radiation making a small, but noticeable, difference. The poor agreement in the 2000–3000 cm^{-1} region most likely reflects inaccurate specification of the aerosol scattering phase function, although the importance of including scattering in the line-by-line calculation is clearly seen.

ARM also contributed to the development of the Santa Barbara DISORT Atmospheric Radiative Transfer (SBDART) program (Ricchiazzi et al. 1998), an easy-to-use moderate resolution code designed for general (including scattering) radiative transfer problems in remote sensing and radiation studies. Although SBDART did not have high enough spectral resolution to compare with individual spectral elements of instruments like the AERI, its convenience for broadband studies or ones involving some amount of spectral resolution made it very popular among researchers. For example, Dufresne et al. (2002) used SBDART to compute the spectral dependence of the radiative forcing, including the effects of scattering, due to mineral aerosols.

4. Clear-sky shortwave studies

The focus of the IRF during the early years of the program was on spectral radiative closure studies in the longwave, a spectral region in which the main absorption sources had clear spectral signatures and the program possessed a well-calibrated radiometric

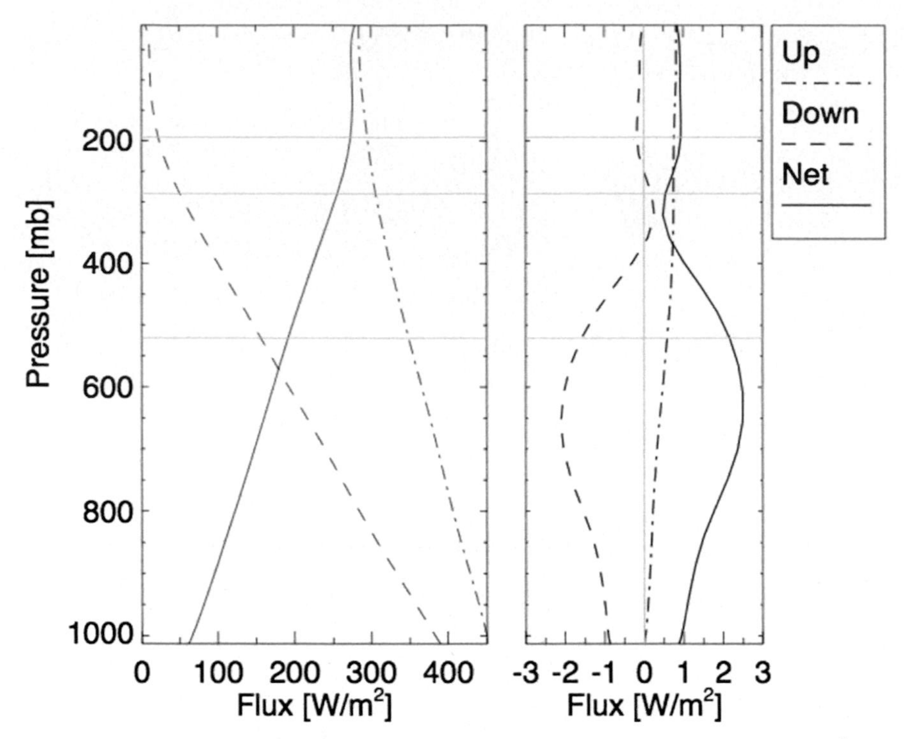

FIG. 14-3. Impact of revisions to LBLRTM on clear-sky longwave fluxes (10–2000 cm^{-1}) for a standard tropical atmosphere. (left) Downwelling, upwelling, and net (up minus down) fluxes computed using LBLRTM from 2009. (right) Differences in computed fluxes (LBLRTM circa 1999 minus LBLRTM from 2009) due to model upgrades.

instrument. Interest and activity in the shortwave region was stimulated after the publication of the broadband radiative closure study of Kato et al. (1997). This study found that, for several clear-sky cases, the modeled shortwave surface downwelling radiation was ~5% greater than the surface irradiance measured by broadband radiometers. This discrepancy also was shown to be mostly in the diffuse component of the shortwave irradiance. This study's analysis of measurement and modeling uncertainties eliminated many possible causes of this lack of agreement, leading the authors to speculate that absorption was due to an unknown "gas X" that may be missing in the model. Subsequent broadband closure studies (e.g., Halthore et al. 1997; Halthore and Schwartz 2000) also demonstrated similar lack of agreement. Given that gaseous absorption occurs primarily in bands consisting of evident absorption lines, this issue was well suited to be addressed through the comparison of spectral shortwave measurements with line-by-line radiative transfer calculations. Although ARM did not possess a single well-calibrated instrument that measured spectrally resolved radiation throughout the entire spectral region in which the majority of solar irradiance occurs, the IRF, under the leadership of Warren Wiscombe and Tony Clough, helped resolve this issue by analyzing measurements from multiple spectral instruments.

Early in the program, support was provided to the University of Denver for development of two instruments to measure spectrally resolved solar radiation, the Absolute Solar Transmittance Interferometer (ASTI; Hawat et al. 2002) and the Solar Radiance Transmission Interferometer (SORTI), the latter of which was configured to provide spectra from 4000 to 13 000 cm^{-1} (750 nm to 2400 nm) with 0.035 cm^{-1} resolution. Unfortunately, the SORTI suffered from a variety of technical issues and was not used heavily by ARM scientists. The ASTI, which measures the direct-solar beam (central 16% of the solar disk) over the spectral range 2000 to 10 000 cm^{-1} with a resolution of 0.6 cm^{-1} (HWHM), was deployed at SGP for an extended period in the 1990s. The instrument was calibrated with a reference tungsten lamp that had a maximum temperature of 2800 K through the same optical path traveled by the solar radiation; this provided the ASTI with an absolute calibration uncertainty of better than 5%. The analysis of the near-infrared spectra from this instrument established that LBLRTM was missing a number of collision-induced oxygen absorption bands, and a parameterization of these bands was developed for the

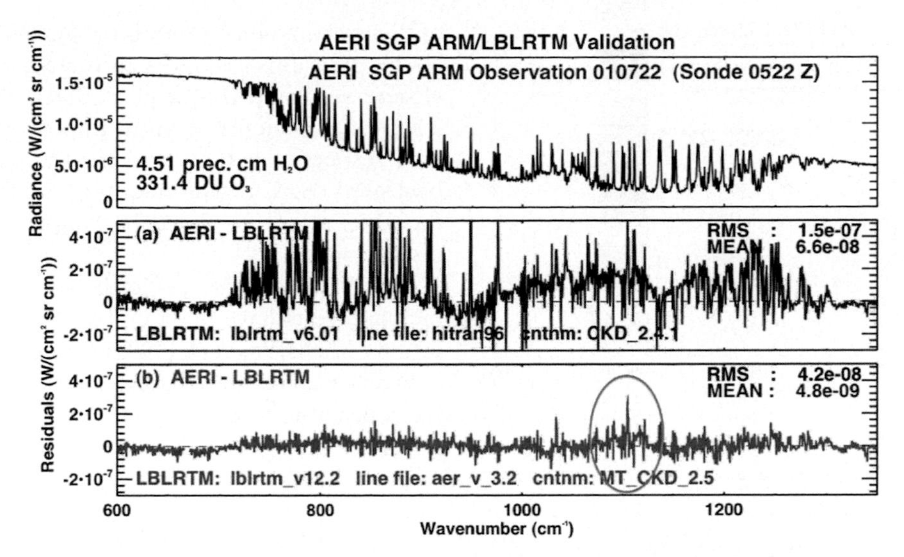

FIG. 14-4. (top) AERI radiance observation from SGP on 22 Jul 2001 for a case with 4.5 cm of PWV; (middle) radiance differences between AERI and LBLRTM calculation using older spectroscopy (HITRAN 96, CKD_2.4.1). (bottom) Radiance differences between AERI and LBLRTM with recent spectroscopic updates (line file - aer_v3.2, MT_CKD_2.5). A spectral feature (the green circle in the bottom panel) associated with formic acid (HCOOH) is identified in the residuals, and is consistent with an atmospheric abundance of this gas equal to ~9 times its assumed background concentration. [This is an updated version of a figure appearing in Shephard et al. (2003).]

CKD continuum model (Mlawer et al. 1998). These absorption bands, which account for less than $1\,\mathrm{W\,m^{-2}}$ of direct-beam absorption for a solar zenith angle of 60°, were the only large residual features observed in the analysis of ASTI spectra (Brown et al. 1998). Figure 14-6 shows a $\sim 2000\,\mathrm{cm^{-1}}$ portion of an ASTI observation and the corresponding ASTI-LBLRTM residuals, which indicates overall good agreement despite certain identifiable issues. [See Mlawer et al. (1998) for residuals in this region before and after the inclusion of the collision-induced oxygen bands.] It must be noted that comparisons such as these are not able to uncover spectroscopic errors within saturated bands, nor are they able to detect issues related to very broad absorbers due to the piecewise linear scaling that has been performed on the ASTI spectra to remove the impact of aerosols and instrument calibration error. For all the ASTI cases analyzed, however, there was very good agreement between measurement and calculations for significantly different solar zenith angles and PWV amounts.

Analysis of clear-sky absorption in the remainder of the near-infrared region and the entire visible spectral region was made possible by the deployment at SGP of the Rotating Shadowband Spectroradiometer (RSS; Harrison et al. 1999), developed at the State University of New York at Albany. The RSS detector is either a silicon-diode array or a charge-coupled device (CCD) that is tilted to correct for chromatic aberration and bring all of the dispersed wavelengths at the detector surface to a focus. The out-of-band rejection of stray light is about 10^5, which is considerably better than that of a single grating spectrometer. The RSS uses the shadowband technique to simultaneously measure diffuse and total (and thereby direct) irradiance from 9500 to $28\,000\,\mathrm{cm^{-1}}$ (360–1050 nm) and can be calibrated with high accuracy using Langley regression in cloud-free and horizontally homogeneous scenes. Even though this instrument's resolution was too coarse to resolve individual absorption lines, analysis of RSS spectra was able to eliminate the possibility of significant unmodeled absorption in the spectral regions observed.

Radiative closure analysis using an early version of the RSS instrument, measuring in 512 channels with spectral resolution ranging from $91\,\mathrm{cm^{-1}}$ in the near-IR to $65\,\mathrm{cm^{-1}}$ in the ultraviolet, demonstrated that good agreement with LBLRTM/CHARTS calculations could be attained for both direct and diffuse irradiances across the entire spectrum (Mlawer et al. 2001). A sensitivity analysis indicated that the technique used would have detected a source of molecular absorption of the magnitude of the Chappuis band of ozone had it been unknown, which is responsible for less absorbed solar irradiance than the missing absorption that had been speculated in earlier broadband studies. Figure 14-7 presents a case from the extension of this analysis to the 1024-channel RSS. Good overall agreement

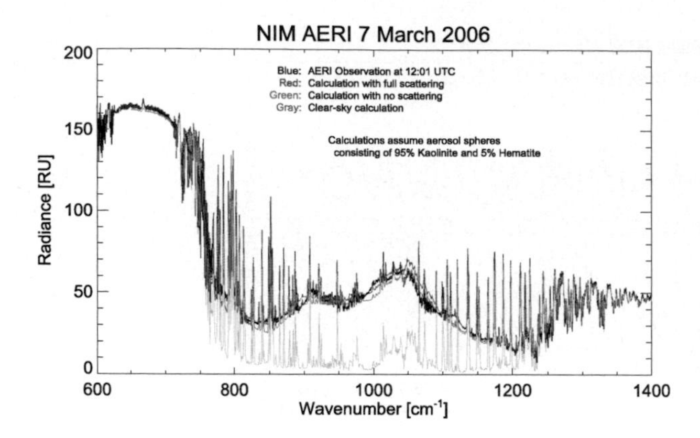

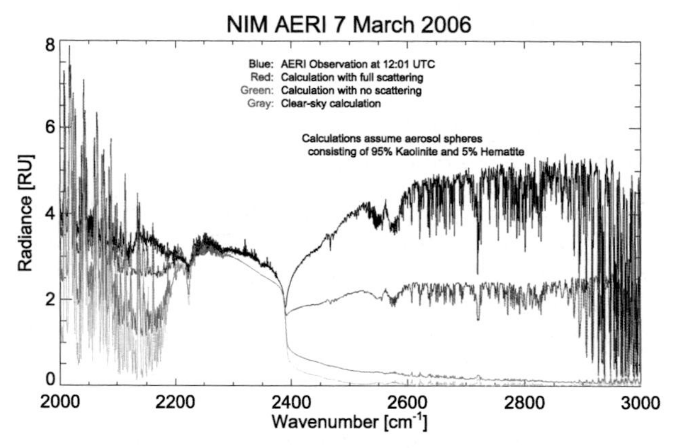

FIG. 14-5. Downwelling radiance observed by the AERI (blue) and computed with the LBLDIS with scattering (red) and without scattering (green) during a dust storm at Niamey for two spectral regions (top: 600–1400 cm^{-1}; bottom: 2000–3000 cm^{-1}) observed by the AERI, where the downwelling radiance in the larger wavenumbers (shorter wavelengths) includes a substantial amount of scattered solar radiation. The calculation assumed that the dust composition was an external mixture of kaolinite and hematite spheres. The gray lines indicate the downwelling radiance that would be observed if there were no dust aerosols in the sky (i.e., a pristine-sky LBLDIS calculation). 1 radiance unit (RU) is 1 mW $(m^2\ sr\ cm^{-1})^{-1}$.

between the measurement and calculation is seen, although direct irradiance residuals associated with the water vapor band at 10 600 cm^{-1} and the oxygen A-band (13 000 cm^{-1}) can be observed. The diffuse residuals in the near-IR are thought to be due to deviation of the aerosol optical depth from the assumed Angstrom relation and/or possible spectral dependence of the aerosol single-scattering albedo.

Through these closure studies with the ASTI and RSS, members of the IRF demonstrated that spectrally resolved measurements and calculations were in basic agreement and there were no unknown absorption bands in the shortwave. Other studies determined that the primary cause of the lack of closure in the broadband studies was flawed radiometric calibration of the instruments (Haeffelin et al. 2001; Dutton et al. 2001). After this issue was corrected, later broadband closure studies (Mlawer et al. 2003; Michalsky et al. 2006), also benefitting from improved model inputs with respect to aerosol and surface properties, demonstrated solid measurement-calculation agreement. [See McFarlane et al. (2016, chapter 20) for further discussion.]

In addition to radiative closure studies, clear-sky RSS measurements have been utilized for numerous applications. Michalsky et al. (1999) used Langley analysis to obtain optical depths from RSS observations, identified six collision-induced oxygen absorption bands in the optical depth spectra, and found no unidentified absorption bands. Kiedron et al. (2001) used RSS measurements in the 10 500 cm^{-1} (940 nm) water vapor band to perform PWV retrievals in dry and cold conditions, and found good correlation with PWV values retrieved from a collocated microwave radiometer. Gianelli et al. (2005) determined from RSS measurements that aerosol size distributions at SGP were bimodal and developed a retrieval to derive information about both the fine and coarse modes, as well as column amounts of NO_2. [Further discussion about the accomplishments in the program with respect to the determination of aerosol properties can be found in McComiskey and Ferrare (2016), chapter 21.] A long-term dataset of extraterrestrial solar irradiance obtained from Langley analysis of RSS spectra was used by Harrison et al. (2003) to analyze and identify issues with commonly used specifications of solar spectral irradiance.

Analyses of measurements of diffuse solar radiation motivated the development of an approach to specify the spectral surface albedo in the vicinity of the SGP site (McFarlane et al. 2011). First, measurements from each channel of two downlooking Multifilter Radiometers (MFRs) at the SGP site (mounted on the 10- and 25-m towers) are used with the corresponding measurement of total irradiance from the uplooking Multifilter Rotating Shadowband Radiometer (MFRSR) to obtain surface albedo values in the five MFR/MFRSR channels. Based upon observed spectral surface reflectances in the Bowker et al. (1985) atlas, the surface albedo values are used to identify a surface type under each tower as snow-covered, green vegetation, nonvegetated (e.g., soil), or partial vegetation, which is a linear combination of green and nonvegetated surfaces given by a Normalized Difference Vegetation Index (NDVI) obtained from the surface albedo values in two of the channels. Based on the same atlas of spectral surface reflectances, the surface albedo values are then used with the identified surface type to obtain a piecewise linear surface albedo function covering the

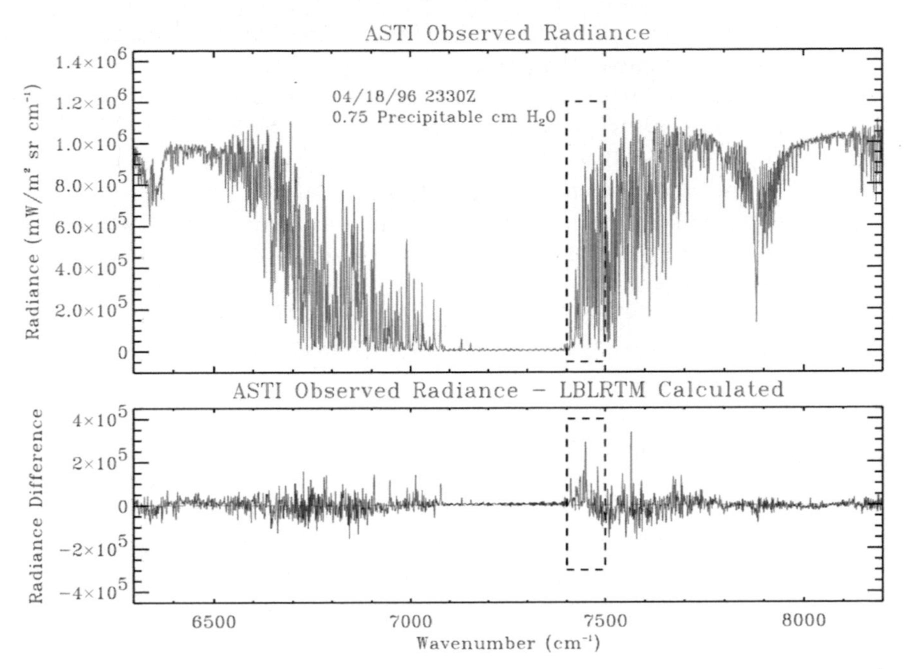

FIG. 14-6. From Brown et al. (1998), (top) ASTI radiance spectra for 6300–8200 cm^{-1} measured on 18 Apr 1996 at SGP for a low PWV case at a solar zenith angle of 71.5°. Notable absorption features seen in this panel are a saturated water vapor band centered at 7200 cm^{-1} and a collision-induced oxygen band at 7800 cm^{-1}. (bottom) Spectral differences between this ASTI observation and a corresponding LBLRTM calculation. Dotted rectangle indicates a region in which the spectral residuals indicated issues with line intensities and widths.

entire shortwave region. Validations of these spectral albedo functions with respect to a spectroradiometer (Trishchenko et al. 2003) showed good agreement for all surface types for wavelengths less than 1280 nm and extending farther into the near-IR for green and nonvegetated surfaces, although a fair amount of variability was seen throughout the near-IR for all surface types. Radiative transfer calculations at SGP typically employ an equal weighting of the spectral surface albedo functions associated with the two MFRs [available as a Value Added Product (VAP) at the ARM Data Archive].

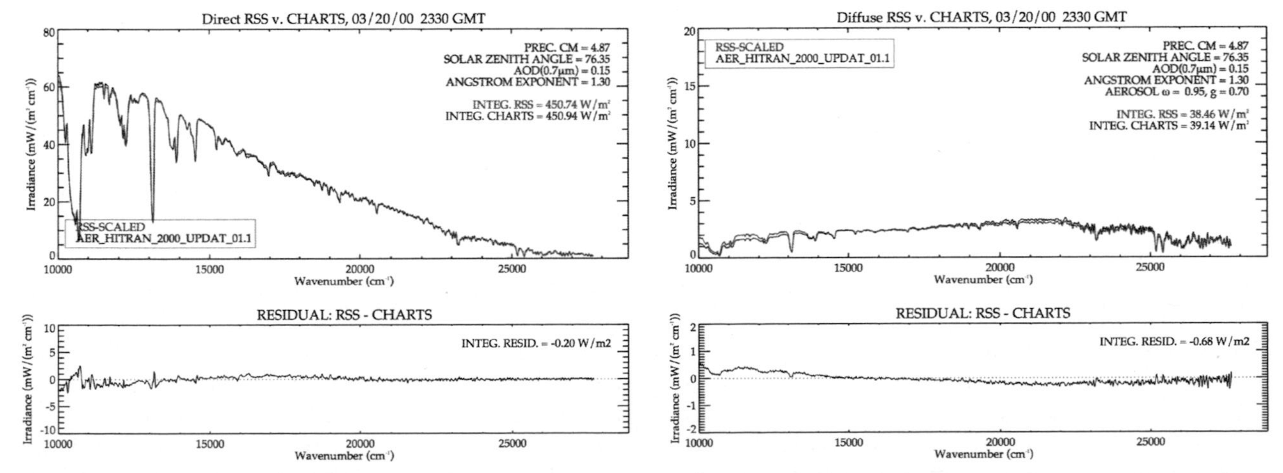

FIG. 14-7. (left top) RSS measurement (red) and LBLRTM/CHARTS calculation (black) of direct irradiance $[mW (m^2 cm^{-1})^{-1}]$ for a clear-sky case on 20 Mar 2000 at SGP. (left bottom) Direct irradiance differences between measurement and calculation. (right top) RSS measurement and LBLRTM/CHARTS calculation of diffuse irradiance for this case. (right bottom) Measurement–calculation differences for diffuse irradiance. Various specifics of the calculations are listed on the plots.

Even with the productive applications of ASTI and RSS measurements, the ARM Radiative Processes Working Group (RPWG; formerly the IRF) identified as a high priority the deployment of a spectral shortwave instrument with greater spectral coverage than both the ASTI and RSS. This led to the development of the Shortwave Spectrometer (SWS), which provides zenith radiance measurements from 4600 to 28 500 cm^{-1} (350–2170 nm). The SWS was deployed alongside the RSS at SGP in the mid-2000s, and a shortwave (SW) QME was established to evaluate these instruments, along with other inputs needed for the SW closure exercise (e.g., the spectral surface albedo product from above and aerosol properties). While radiative closure analysis of a large number of RSS cases with LBLRTM/CHARTS indicated agreement similar to that shown in Fig. 14-7, there were sizeable differences between the SWS measurements and calculations that could not be explained by any reasonable deficiency in either the model or the data used as input to the model (Delamere et al. 2009). Subsequently, field tests determined that the SWS measurement of zenith radiance is susceptible to contamination from direct solar irradiance incident on the instrument fore-optics (i.e., a light leak existed in the instrument). The SWS fore-optics were subsequently redesigned by the instrument mentor team and the instrument was redeployed (Flynn et al. 2010). Evaluation of measurements from the modified SWS is ongoing.

5. Cloudy-sky spectral analyses

Although ARM spectral measurements have not been exploited as extensively for cloudy-sky applications as they have for clear-sky studies, there have been several notable uses of AERI and RSS spectra to further our understanding of cloud optical and microphysical properties.

One advantage of having spectral infrared radiance observations, such as from the AERI, is that they are very sensitive to changes in cloud properties as long as the cloud is not optically thick. The AERI is extremely sensitive to liquid water path (LWP) and effective radius (Reff) when the LWP is less than $\sim$60 g m^{-2} and provides a markedly better retrieval of LWP than the microwave radiometer for these smaller LWP values (Turner 2007). This advantage is extremely important since a large fraction of liquid-bearing clouds at all climatic locations have LWP less than 100 g m^{-2} (Turner et al. 2007a), and the accurate determination of LWP is critical in order to accurately compute the radiative impact of these clouds (Turner et al. 2007b; Sengupta et al. 2003). Furthermore, since liquid water and ice absorb in different spectral regions observed by the AERI, these observations can be used to retrieve properties of mixed-phase clouds. The MIXCRA retrieval described above has been used extensively to study Arctic clouds (Turner 2005), has been validated with direct measurements of liquid and ice cloud optical depths observed by the polarization-sensitive high-spectral-resolution lidar (Turner and Eloranta 2008), and has been used to retrieve the cloud fraction in the AERI's field of view in broken cumulus cases (Turner and Holz 2005).

Figure 14-8 shows an example of the sensitivity of the infrared radiance to cloud properties. The AERI observed downwelling radiance during the deployment of the AMF at Pt. Reyes, California, in 2005 (Miller et al. 2016, chapter 9). MIXCRA was used to retrieve the cloud properties, and for this case determined that the LWP was 42.2 g m^{-2} and Reff was 8.2 μm. The sensitivity of the AERI radiance to changes in LWP and Reff is shown in the bottom panel of Fig. 14-8, as well as the impact on the downwelling radiance if scattering was not included in the calculation. When the AERI-retrieved cloud properties were used to compute the downwelling shortwave radiative flux, the results agreed much better with the flux observation from a collocated pyranometer than flux calculations that used the cloud properties derived from the collocated microwave radiometer (Turner 2007).

An extension of radiative closure studies using RSS spectra to liquid clouds concluded that good spectral measurement–model agreement for diffuse irradiance could be attained with a reasonable choice for effective radius for a homogeneous single-layer cloud (Mlawer et al. 2001). This agreement, plus the impossibility of attaining spectral agreement from adding a small amount of cloud absorption in the near-IR, provided evidence that liquid clouds did not have unknown absorption in the near-IR, as had been speculated (e.g., Valero et al. 1997). The difficulty in finding sufficiently homogeneous clouds limited the applicability of this type of analysis, in which a single set of cloud properties accurately represents the entire cloud in the hemispheric field of view of the RSS. A more sophisticated application of RSS spectra for the study of clouds was established in a series of papers by Min and collaborators that explored the information contained in RSS measurements in the oxygen A-band concerning photon path length and scattering (Harrison and Min 1997; Min and Harrison 1999; Min et al. 2001). This work was later extended to develop a detection method for multilayer clouds (Li and Min 2010), an approach to retrieve vertical profiles of liquid water content, optical depth, and effective size (Li and Min 2013), and motivated the development of the High-Resolution Oxygen A-Band

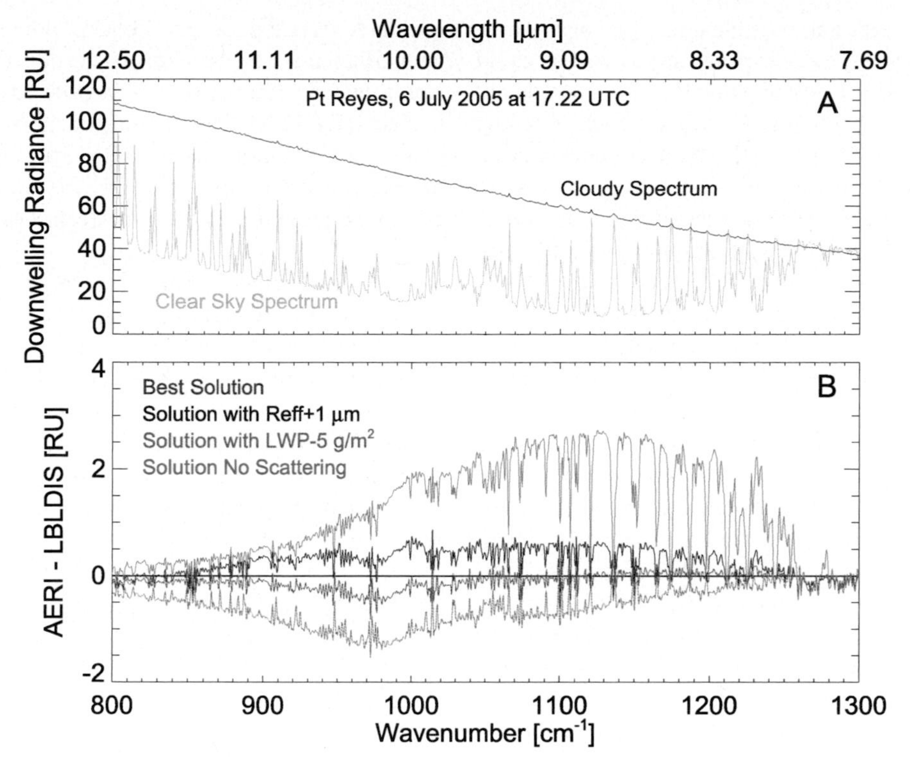

FIG. 14-8. (a) Downwelling radiance (purple) observed by the AERI at Pt. Reyes, California, on 6 Jul 2005, with a simulated clear-sky spectrum (gray). (b) The observed minus computed radiance where the "best solution" (red) used the retrieved cloud properties from the MIXCRA algorithm (LWP of 42.2 g m^{-2} and Reff of 8.2 μm). The blue and brown traces show the residuals that would result if the Reff was increased by 1 μm or the LWP was decreased by 5 g m^{-2}, respectively. The green trace shows the residual that would result if the calculation that used the best solution did not include scattering.

Spectrometer (HABS) at the University of Albany (Min et al. 2011).

Other ARM-related spectral instruments also have led to notable results. For example, spectra from a spectroradiometer measuring from 4500 to 28 600 cm^{-1} (350–2200 nm, from Analytical Spectral Devices) that was deployed at NSA in spring 2008 as part of the Indirect and Semi-Direct Aerosol Campaign (ISDAC) were used to study single-layer liquid and mixed-phased clouds. Irradiance measurements in different spectral regions allowed determination of cloud phase and optical depth, and the impact on surface shortwave irradiance of the presence of ice along with liquid in a cloud of fixed optical depth was determined to be typically 5 W m^{-2}, although it could be as great as 8–10 W m^{-2} (Lubin and Vogelmann 2011). Also, in a series of papers and conference presentations (e.g., Marshak et al. 2009; Chiu et al. 2009), measurements from the SWS at SGP were used to explore an intriguing linear relationship exhibited by zenith radiances measured in the transition zone between cloudy and clear regions. This relationship allows a straightforward determination of the optical depth and effective size of liquid clouds (McBride et al. 2013).

6. Summary and looking ahead

The scientific advances that have resulted from analysis of ARM spectral radiometric measurements have been impressive and wide-ranging. Still, several unresolved topics remain for which a spectral perspective will be beneficial, and the program is well positioned to contribute to further progress. As discussed above, the far-IR spectral region had been relatively underexplored, and ongoing analysis of spectral radiance measurements from the ARM RHUBC-I and -II field campaigns should provide a conclusive resolution. There has been preliminary discussion of a third RHUBC campaign, with a focus on cloud-radiative processes in the radiatively potent far-IR region.

The near-IR spectral region also has not received a great deal of attention in the program despite its significant amount of solar radiation, mix of gaseous absorption bands and uncertain amount of water vapor continuum absorption, and possible utility for constraining aerosol properties and size distributions, as well as cloud property retrievals. The design issues that were uncovered with respect to the SWS (see above) set back potential progress in this area, but occurred at a fortuitous time since funding from the American Recovery and Reinvestment Act (ARRA) of 2009 enabled a significant upgrade of ARM spectral radiometric instrumentation measuring in the near-IR shortly thereafter. The reconfigured SWS, which measures zenith radiance from 4650 to 28 500 cm^{-1} (350–2150 nm) at moderate spectral resolution, was redeployed to SGP in 2011. ARRA also allowed the addition of four Shortwave Array Spectrometers (SAS) to ARM's suite of instruments. The SAS measures from 5900 to 29 400 cm^{-1} at moderate resolution. One version of the SAS measures zenith (SAS-Ze) radiance, while a second version measures hemispheric irradiance (SAS-He); both types have been deployed for a period of time at SGP and during various AMF campaigns. Although the effort to calibrate the SAS has not yet been completed, Lubin et al. (2013) has exploited the spectral content of the SAS-Ze to retrieve cloud phase, optical depth, and effective size for single-phase clouds. Measurements from the SWS and SAS are expected to be valuable in validating ice optical property parameterizations, a topic of notable complexity that has not been the subject of comprehensive radiative closure analysis.

There have been significant recent advances in specifying the spectral optical properties of ice clouds, such as the modified anomalous diffraction approximation (MADA) of Mitchell et al. (1996) and subsequent publications and an approach that combines geometric optics and the finite difference time domain technique (Yang et al. 2000, 2005). The suite of ARM collocated zenith-pointing instruments that span much of the solar and thermal spectrum (SAS, SWS, AERI) will provide information that will be key to their evaluation.

The accomplishments of the ARM Program with respect to the analysis of spectral radiation measurements have fulfilled the aspirations of the ICRCCM participants and the objectives of the program's founders. ARM radiative closure studies have greatly improved our knowledge of the absorption properties of atmospheric constituents and significantly lowered the uncertainties associated with clear-sky radiative transfer, thereby establishing a confident foundation on which to build fast radiative transfer codes for use in atmospheric prediction models. The ARM Program contributed greatly in this area, as detailed in Mlawer et al. (2016, chapter 15). Spectrally resolved radiation measurements at the ARM sites, when coupled with observations from the program's suite of collocated instruments that measure thermodynamic, aerosol, and cloud properties, have also precipitated substantial advances in our understanding of atmospheric processes. These accomplishments, which are due to the efforts of many ARM scientists and infrastructure members, have perhaps even exceeded the ICRCCM recommendations and the original goals of ARM. In the future, ARM spectral measurements are likely to lead to further advances in our knowledge, and add to the program's catalog of spectral successes.

Acknowledgments. The authors would like to acknowledge the leadership of the ARM Instantaneous Radiative Flux (IRF) Working Group and the Water Vapor Intensive Observation Period (WVIOP)—Tony Clough, Bob Ellingson, Hank Revercomb, and Warren Wiscombe—for all their contributions to the work described in this section. Their efforts were essential to the success of spectral radiation research in the ARM Program.

REFERENCES

Alvarado, M. J., V. H. Payne, E. J. Mlawer, G. Uymin, M. W. Shephard, K. E. Cady-Pereira, J. Delamere, and J. Moncet, 2013: Performance of the line-by-line radiative transfer model (LBLRTM) for temperature, water vapor, and trace gas retrievals: Recent updates evaluated with IASI case studies. *Atmos. Chem. Phys.*, **13**, 6687–6711, doi:10.5194/acp-13-6687-2013.

Barker, H. W., and Coauthors, 2003: Assessing 1D atmospheric solar radiative transfer models: Interpretation and handling of unresolved clouds. *J. Climate*, **16**, 2676–2699, doi:10.1175/1520-0442(2003)016<2676:ADASRT>2.0.CO;2.

Bowker, D. E., R. E. Davis, D. L. Myric, K. Stacy, and W. T. Jones, 1985: Spectral reflectances of natural targets for use in remote sensing. NASA Ref. Publ. 1139, 181 pp. [Available online at ntrs.nasa.gov/archive/nasa/casi.ntrs.nasa.gov/19850022138.pdf.]

Brown, P. D., S. A. Clough, E. J. Mlawer, T. R. Shippert, and F. J. Murcray, 1998: High resolution validation in the shortwave: ASTI/LBLRTM QME. *Proc. Eighth ARM Science Team Meeting*, Richland, WA, U.S. Department of Energy, 101–108.

Chiu, C. J., A. Marshak, Y. Knyazikhin, P. Pilewskie, and W. J. Wiscombe, 2009: Physical interpretation of the spectral radiative signature in the transition zone between cloud-free and cloudy regions. *Atmos. Chem. Phys.*, **9**, 1419–1430, doi:10.5194/acp-9-1419-2009.

Cimini, D., F. Nasir, E. R. Westwater, V. H. Payne, D. D. Turner, E. J. Mlawer, M. L. Exner, and M. Cadeddu, 2009: Comparison of ground-based millimeter-wave observations in the Arctic winter. *IEEE Trans. Geosci. Remote Sens.*, **47**, 3098–3106, doi:10.1109/TGRS.2009.2020743.

Clerbaux, C., and Coauthors, 2007: The IASI/MetOp Mission: First observations and highlights of its potential contribution to GMES. *Space Res. Today*, **168**, 19–24, doi:10.1016/S0045-8732(07)80046-5.

Clough, S. A., F. X. Kneizys, and R. W. Davies, 1989: Line shape and the water vapor continuum. *Atmos. Res.*, **23**, 229–241, doi:10.1016/0169-8095(89)90020-3.

——, M. J. Iacono, and J.-L. Moncet, 1992: Line-by-line calculation of atmospheric fluxes and cooling rates: Application to water vapor. *J. Geophys. Res.*, **97**, 15 761–15 785, doi:10.1029/92JD01419.

——, M. W. Shephard, E. J. Mlawer, J. S. Delamere, M. J. Iacono, K. Cady-Pereira, S. Boukabara, and P. D. Brown, 2005: Atmospheric radiative transfer modeling: A summary of the AER codes. *J. Quant. Spectrosc. Radiat. Transfer*, **91**, 233–244, doi:10.1016/j.jqsrt.2004.05.058.

——, and Coauthors, 2006: Forward model and Jacobians for tropospheric emission spectrometer retrievals. *IEEE Trans. Geosci. Remote Sens.*, **44**, 1308–1323, doi:10.1109/TGRS.2005.860986.

Delamere, J., E. Mlawer, J. Michalsky, P. Kiedron, C. Flynn, and C. Long, 2009: Update on shortwave spectral radiative closure studies at the ACRF SGP site. *Proc. 19th ARM Science Team Meeting*, Louisville, KY, U.S. Department of Energy. [Abstract available online at https://www.arm.gov/publications/proceedings/conf19/display?id=NjIw.]

——, S. A. Clough, V. H. Payne, E. J. Mlawer, D. D. Turner, and R. R. Gamache, 2010: A far-infrared radiative closure study in the Arctic: Application to water vapor. *J. Geophys. Res.*, **115**, D17106, doi:10.1029/2009JD012968.

Dufresne, J.-L., C. Gautier, P. Ricchiazzi, and Y. Fouquart, 2002: Longwave scattering effects of mineral aerosols. *J. Atmos. Sci.*, **59**, 1959–1966, doi:10.1175/1520-0469(2002)059<1959:LSEOMA>2.0.CO;2.

Dutton, E. G., J. J. Michalsky, T. Stoffel, B. W. Forgan, J. Hickey, D. W. Nelson, T. L. Alberta, and I. Reda, 2001: Measurement of broadband diffuse solar irradiance using current commercial instrumentation with a correction for thermal offset errors. *J. Atmos. Oceanic Technol.*, **18**, 297–314, doi:10.1175/1520-0426(2001)018<0297:MOBDSI>2.0.CO;2.

Ellingson, R. G., and Y. Fouquart, 1991: The intercomparison of radiation codes in climate models: An overview. *J. Geophys. Res.*, **96**, 8925–8927, doi:10.1029/90JD01618.

——, and W. J. Wiscombe, 1996: The Spectral Radiance Experiment (SPECTRE): Project description and sample results. *Bull. Amer. Meteor. Soc.*, **77**, 1967–1985, doi:10.1175/1520-0477(1996)077<1967:TSREPD>2.0.CO;2.

——, J. Ellis, and S. Fels, 1991: The intercomparison of radiation codes used in climate models: Long wave results. *J. Geophys. Res.*, **96**, 8929–8953, doi:10.1029/90JD01450.

——, R. D. Cess, and G. L. Potter, 2016: The Atmospheric Radiation Measurement Program: Prelude. *The Atmospheric Radiation Measurement (ARM) Program: The First 20 Years, Meteor. Monogr.*, No. 57, Amer. Meteor. Soc., doi:10.1175/AMSMONOGRAPHS-D-15-0029.1.

Flynn, C., and Coauthors, 2010: New shortwave array spectrometers for the ARM Climate Research Facility. *First Atmospheric System Research (ASR) Science Team Meeting*, Bethesda, MD, U.S. Department of Energy. [Abstract available online at http://asr.science.energy.gov/meetings/stm/posters/view?id=238.]

Gero, P. J., and D. D. Turner, 2011: Long-term trends in downwelling spectral infrared radiance over the U.S. Southern Great Plains. *J. Climate*, **24**, 4831–4843, doi:10.1175/2011JCLI4210.1.

Gianelli, S. M., B. E. Carlson, and A. A. Lacis, 2005: Aerosol retrievals using rotating shadowband spectroradiometer data. *J. Geophys. Res.*, **110**, D05203, doi:10.1029/2004JD005329.

Haeffelin, M., S. Kato, A. M. Smith, C. K. Rutledge, T. P. Charlock, and J. R. Mahan, 2001: Determination of the thermal offset of the Eppley precision spectral pyranometer. *Appl. Optics*, **40**, 472–484, doi:10.1364/AO.40.000472.

Halthore, R. N., and S. E. Schwartz, 2000: Comparison of model-estimated and measured diffuse downward irradiance at surface in cloud-free skies. *J. Geophys. Res.*, **105**, 20 165–20 177, doi:10.1029/2000JD900224.

——, ——, J. J. Michalsky, G. P. Anderson, R. A. Ferrare, B. N. Holben, and H. M. Ten Brink, 1997: Comparison of model estimated and measured direct-normal solar irradiance. *J. Geophys. Res.*, **102** (D25), 29 991–30 002, doi:10.1029/97JD02628.

Han, Y., J. A. Shaw, J. H. Churnside, P. D. Brown, and S. A. Clough, 1997: Infrared spectral radiance measurements in the tropical Pacific atmosphere. *J. Geophys. Res.*, **102** (D4), 4353–4356, doi:10.1029/96JD03717.

Harries, J., and Coauthors, 2008: The far-infrared Earth. *Rev. Geophys.*, **46**, RG4004, doi:10.1029/2007RG000233.

Harrison, L., and Q. Min, 1997: Photon pathlength distributions from O_2 A-band absorption. *IRS'96: Current Problems in Atmospheric Radiation*, W. L. Smith and K. Stamnes, Eds., Deepak Publishing, 594–598.

——, M. Beauharnois, J. Berndt, P. Kiedron, J. Michalsky, and Q. Min, 1999: The rotating shadowband spectroradiometer (RSS) at SGP. *Geophys. Res. Lett.*, **26**, 1715–1718, doi:10.1029/1999GL900328.

——, P. Kiedron, J. Berndt, and J. Schlemmer, 2003: Extraterrestrial solar spectrum 360– 1050 nm from Rotating Shadowband Spectroradiometer measurements at the Southern Great Plains (ARM) site. *J. Geophys. Res.*, **108**, 4424, doi:10.1029/2001JD001311.

Hawat, T., T. Stephen, and F. Murcray, 2002: Absolute solar transmittance interferometer for ground-based measurements. *Appl. Opt.*, **41**, 3582–3589, doi:10.1364/AO.41.003582.

Kato, S., T. P. Ackerman, and E. E. Clothiaux, 1997: Uncertainties in modeled and measured clear-sky surface shortwave irradiances. *J. Geophys. Res.*, **102**, 25 881–25 898, doi:10.1029/97JD01841.

Kiedron, P., and Coauthors, 2001: A robust retrieval of water vapor column in dry Arctic conditions using the rotating shadowband spectroradiometer. *J. Geophys. Res.*, **106**, 24 007–24 016, doi:10.1029/2000JD000130.

Knuteson, R. O., B. Whitney, H. E. Revercomb, and F. A. Best, 1999: The history of the University of Wisconsin Atmospheric Emitted Radiance Interferometer (AERI) prototype during the period April 1994 through July 1995. ARM Tech. Rep. TR-001.1, 43 pp. [Available online at http://www.arm.gov/publications/tech_reports/arm-tr-001.1.pdf.]

——, and Coauthors, 2004a: The Atmospheric Emitted Radiance Interferometer (AERI). Part I: Instrument design. *J. Atmos. Oceanic Technol.*, **21**, 1763–1776, doi:10.1175/JTECH-1662.1.

——, and Coauthors, 2004b: The Atmospheric Emitted Radiance Interferometer (AERI). Part II: Instrument performance. *J. Atmos. Oceanic Technol.*, **21**, 1777–1789, doi:10.1175/JTECH-1663.1.

Kollias, P., and Coauthors, 2016: Development and applications of ARM millimeter-wavelength cloud radars. *The Atmospheric Radiation Measurement (ARM) Program: The First 20 Years, Meteor. Monogr.*, No. 57, Amer. Meteor. Soc., doi:10.1175/AMSMONOGRAPHS-D-15-0037.1.

Li, S., and Q. Min, 2010: Diagnosis of multilayer clouds using photon path length distributions. *J. Geophys. Res.*, **115**, D20202, doi:10.1029/2009JD013774.

——, and ——, 2013: Retrievals of vertical profiles of stratus cloud properties from combined oxygen A-band and radar observations. *J. Geophys. Res. Atmos.*, **118**, 769–778, doi:10.1029/2012JD018282.

Lubin, D., and A. Vogelmann, 2011: The influence of mixed-phase clouds on surface shortwave irradiance during the Arctic spring. *J. Geophys. Res.*, **116**, D00T05, doi:10.1029/2011JD015761.

——, ——, and C. Flynn, 2013: Retrieval of cloud microphysical properties from the new shortwave array spectroradiometer. *Proc. Fourth Atmospheric System Research (ASR) Science Team Meeting*. Potomac, MD, U.S. Department of Energy. [Abstract available online at http://asr.science.energy.gov/meetings/stm/posters/view?id=808.]

Marshak, A., Y. Knyazikhin, J. C. Chiu, and W. J. Wiscombe, 2009: Spectral invariant behavior of zenith radiance around cloud edges observed by ARM SWS. *Geophys. Res. Lett.*, **36**, L16802, doi:10.1029/2009GL039366.

McBride, P., A. Marshak, Y. Knyazkhin, J. Chiu, and W. Wiscombe, 2013: What can be learned from ARM shortwave hyperspectral observations? *Fourth Atmospheric System Research (ASR) Science Team Meeting*. Potomac, MD, U.S. Department of Energy. [Abstract available online at http://asr.science.energy.gov/meetings/stm/posters/view?id=918.]

McComiskey, A., and R. A. Ferrare, 2016: Aerosol physical and optical properties and processes in the ARM Program. *The Atmospheric Radiation Measurement (ARM) Program: The First 20 Years, Meteor. Monogr.*, No. 57, Amer. Meteor. Soc., doi:10.1175/AMSMONOGRAPHS-D-15-0028.1.

McFarlane, S. A., K. L. Gaustad, E. J. Mlawer, C. N. Long, and J. Delamere, 2011: Development of a high spectral resolution surface albedo product for the ARM Southern Great Plains central facility. *Atmos. Meas. Tech.*, **4**, 1713–1733, doi:10.5194/amt-4-1713-2011.

——, J. H. Mather, and E. J. Mlawer, 2016: ARM's progress on improving atmospheric broadband radiative fluxes and heating rates. *The Atmospheric Radiation Measurement (ARM) Program: The First 20 Years, Meteor. Monogr.*, No. 57, Amer. Meteor. Soc., doi:10.1175/AMSMONOGRAPHS-D-15-0046.1.

Michalsky, J. J., M. Beauharnois, J. Berndt, L. Harrison, P. Kiedron, and Q. Min, 1999: O_2-O_2 absorption band identification based on optical depth spectra of the visible and near-infrared. *Geophys. Res. Lett.*, **26**, 1581–1584, doi:10.1029/1999GL900267.

——, and Coauthors, 2006: Shortwave radiative closure studies for clear skies during the Atmospheric Radiation Measurement 2003 Aerosol Intensive Observation Period. *J. Geophys. Res.*, **111**, D14S90, doi:10.1029/2005JD006341.

Miller, M. A., and A. Slingo, 2007: The ARM Mobile Facility and its first international deployment: Measuring radiative flux divergence in West Africa. *Bull. Amer. Meteor. Soc.*, **88**, 1229–1244, doi:10.1175/BAMS-88-8-1229.

——, K. Nitschke, T. P. Ackerman, W. R. Ferrell, N. Hickmon, and M. Ivey, 2016: The ARM Mobile Facilities. *The Atmospheric Radiation Measurement (ARM) Program: The First 20 Years, Meteor. Monogr.*, No. 57, Amer. Meteor. Soc., doi:10.1175/AMSMONOGRAPHS-D-15-0051.1.

Min, Q., and L. C. Harrison, 1999: Joint statistics of photon pathlength and cloud optical depth. *Geophys. Res. Lett.*, **26**, 1425–1428, doi:10.1029/1999GL900246.

——, ——, and E. E. Clothiaux, 2001: Joint statistics of photon pathlength and cloud optical depth: Case studies. *J. Geophys. Res.*, **106**, 7375–7385, doi:10.1029/2000JD900490.

——, B. Yin, J. Berndt, L. Harrison, and P. Kiedron, 2011: A high-resolution oxygen A-band spectrometer (HABS) and photon path length distribution. *Fall AGU Meeting*, San Francisco, CA, Amer. Geophys. Union, Abstract A11H-0200.

Mitchell, D. L., A. Macke, and Y. Liu, 1996: Modeling cirrus clouds. Part II: Treatment of radiative properties. *J. Atmos. Sci.*, **53**, 2967–2988, doi:10.1175/1520-0469(1996)053<2967:MCCPIT>2.0.CO;2.

Mlawer, E. J., S. J. Taubman, P. D. Brown, M. J. Iacono, and S. A. Clough, 1997: Radiative transfer for inhomogeneous atmospheres: RRTM, a validated correlated-k model for the longwave. *J. Geophys. Res.*, **102**, 16 663–16 682, doi:10.1029/97JD00237.

——, S. A. Clough, P. D. Brown, T. M. Stephen, J. C. Landry, A. Goldman, and F. J. Murcray, 1998: Observed atmospheric collision-induced absorption in near-infrared oxygen bands. *J. Geophys. Res.*, **103**, 3859–3863, doi:10.1029/97JD03141.

——, and Coauthors, 2001: Comparisons between RSS measurements and LBLRTM/CHARTS calculations for clear and cloudy conditions. *Proc. 11th ARM Science Team Meeting*, Atlanta, GA, U.S. Department of Energy. [Available online at http://www.arm.gov/publications/proceedings/conf11/extended_abs/mlawer_ej.pdf.]

——, and Coauthors, 2003: Recent developments on the broadband heating rate profile value-added product. *Proc. 13th ARM Science Team Meeting*, Broomfield, CO, U.S. Department of Energy. [Available online at http://www.arm.gov/publications/proceedings/conf13/extended_abs/mlawer-ej.pdf.]

——, V. H. Payne, J.-L. Moncet, J. S. Delamere, M. J. Alvarado, and D. C. Tobin, 2012: Development and recent evaluation of the MT_CKD model of continuum absorption. *Philos. Trans. Roy. Meteor. Soc.*, **370A**, 2520–2556, doi:10.1098/rsta.2011.0295.

——, M. J. Iacono, R. Pincus, H. Barker, L. Oreopoulos, and D. Mitchell, 2016: Contributions of the ARM Program to radiative transfer modeling for climate and weather applications. *The Atmospheric Radiation Measurement (ARM) Program: The First 20 Years, Meteor. Monogr.*, No. 57, Amer. Meteor. Soc., doi:10.1175/AMSMONOGRAPHS-D-15-0041.1.

Moncet, J.-L., and S. A. Clough, 1997: Accelerated monochromatic radiative transfer for scattering atmospheres: Application of a new model to spectral radiance observations. *J. Geophys. Res.*, **102**, 21 853–21 866, doi:10.1029/97JD01551.

Oreopoulos, L., and Coauthors, 2012: The continual intercomparison of radiation codes: Results from Phase I. *J. Geophys. Res.*, **117**, D06118, doi:10.1029/2011JD016821.

Payne, V. H., J.-L. Moncet, and S. A. Clough, 2007: Improved spectroscopy for microwave and infrared satellite data assimilation. *JCSDA Fifth Workshop on Satellite Data Assimilation,* College Park, MD, Joint Center for Satellite Data Assimilation. [Available online at http://www.jcsda.noaa.gov/documents/meetings/wkshp7/Session1_RT_Clouds/payne_jcsda_meeting_may2007.pdf.]

——, E. J. Mlawer, K. E. Cady-Pereira, and J.-L. Moncet, 2011: Water vapor continuum absorption in the microwave. *IEEE Trans. Geosci. Remote Sens.*, **49**, 2194–2208, doi:10.1109/TGRS.2010.2091416.

Revercomb, H. E., and Coauthors, 2003: The ARM Program's water vapor intensive observation periods: Overview, accomplishments, and future challenges. *Bull. Amer. Meteor. Soc.*, **84**, 217–236, doi:10.1175/BAMS-84-2-217.

Ricchiazzi, P., S. Yang, C. Gautier, and D. Sowle, 1998: SBDART: A research and teaching software tool for plane-parallel radiative transfer in the Earth's atmosphere. *Bull. Amer. Meteor. Soc.*, **79**, 2101–2114, doi:10.1175/1520-0477(1998)079<2101:SARATS>2.0.CO;2.

Rothman, L. S., 1992: The HITRAN atmospheric molecular spectroscopic database. *Proc. Second ARM Science Team Meeting*, Washington, D.C., U.S. Department of Energy, 15–20.

Sengupta, M., E. E. Clothiaux, T. P. Ackerman, S. Kato, and Q. Min, 2003: Importance of accurate liquid water path for estimation of solar radiation in warm boundary layer clouds: An observational study. *J. Climate*, **16**, 2997–3009, doi:10.1175/1520-0442(2003)016<2997:IOALWP>2.0.CO;2.

Shephard, M. W., A. Goldman, S. A. Clough, and E. J. Mlawer, 2003: Spectroscopic improvements providing evidence of formic acid in AERI-LBLRTM validation spectra. *J. Quant. Spectrosc. Radiat. Transfer*, **82**, 383–390, doi:10.1016/S0022-4073(03)00164-X.

Shupe, M. D., J. M. Comstock, D. D. Turner, and G. G. Mace, 2016: Cloud property retrievals in the ARM Program. *The Atmospheric Radiation Measurement (ARM) Program: The First 20 Years, Meteor. Monogr.*, No. 57, Amer. Meteor. Soc., doi:10.1175/AMSMONOGRAPHS-D-15-0030.1.

Stamnes, K., S. C. Tsay, W. J. Wiscombe, and K. Jayaweera, 1988: A numerically stable algorithm for discrete-ordinate-method radiative transfer in multiple scattering and emitting layered media. *Appl. Opt.*, **27**, 2502–2509, doi:10.1364/AO.27.002502.

Stokes, G. E., and S. E. Schwartz, 1994: The Atmospheric Radiation Measurement (ARM) Program: Programmatic background and design of the cloud and radiation test bed. *Bull. Amer. Meteor. Soc.*, **75**, 1201–122, doi:10.1175/1520-0477(1994)075<1201:TARMPP>2.0.CO;2.

Strow, L. L., S. E. Hannon, S. De-Souza Machado, H. E. Mottler, and D. C. Tobin, 2006: Validation of the atmospheric infrared sounder radiative transfer algorithm. *J. Geophys. Res.*, **111**, D09S06, doi:10.1029/2005JD006146.

Tobin, D. C., and Coauthors, 1999: Downwelling spectral radiance observation at the SHEBA ice station: Water vapor continuum measurements from 17 to 26 μm. *J. Geophys. Res.*, **104**, 2081–2092, doi:10.1029/1998JD200057.

Trishchenko, A. P., Y. Luo, M. Cribb, Z. Li, and K. Hamm, 2003: Surface spectral albedo intensive operational period at the ARM SGP site in August 2002: Results, analysis, and future plans. *Proc. 13th ARM Program Science Team Meeting*, Richland, WA, U.S. Department of Energy. [Available online at http://www.arm.gov/publications/proceedings/conf13/extended_abs/trishchenko-ap.pdf.]

Turner, D. D., 2005: Arctic mixed-phase cloud properties from AERI lidar observations: Algorithm and results from SHEBA. *J. Appl. Meteor.*, **44**, 427–444, doi:10.1175/JAM2208.1.

——, 2007: Improved ground-based liquid water path retrievals using a combined infrared and microwave approach. *J. Geophys. Res.*, **112**, D15204, doi:10.1029/2007JD008530.

——, 2008: Ground-based infrared retrievals of optical depth, effective radius, and composition of airborne mineral dust above the Sahel. *J. Geophys. Res.*, **113**, D00E03, doi:10.1029/2008JD010054.

——, and R. E. Holz, 2005: Retrieving cloud fraction in the field-of-view of a high-spectral-resolution infrared radiometer. *IEEE Geosci. Remote Sens. Lett.*, **2**, 287–291, doi:10.1109/LGRS.2005.850533.

——, and E. W. Eloranta, 2008: Validating mixed-phase cloud optical depth retrieved from infrared observations with high spectral resolution lidar. *IEEE Geosci. Remote Sens. Lett.*, **5**, 285–288, doi:10.1109/LGRS.2008.915940.

——, and E. J. Mlawer, 2010: The Radiative Heating in Underexplored Bands Campaigns (RHUBC). *Bull. Amer. Meteor. Soc.*, **91**, 911–923, doi:10.1175/2010BAMS2904.1.

——, and P. J. Gero, 2011: Downwelling 10 μm radiance temperature climatology for the Atmospheric Radiation Measurement Southern Great Plains site. *J. Geophys. Res.*, **116**, D08212, doi:10.1029/2010JD015135.

——, and U. Löhnert, 2014: Information content and uncertainties in thermodynamic profiles and liquid cloud properties retrieved from the ground-based Atmospheric Emitted Radiance Interferometer (AERI). *J. Appl. Meteor. Climatol.*, **53**, 752–771, doi:10.1175/JAMC-D-13-0126.1.

——, S. A. Ackerman, B. A. Baum, H. E. Revercomb, and P. Yang, 2003: Cloud phase determination using ground-based AERI observations at SHEBA. *J. Appl. Meteor.*, **42**, 701–715, doi:10.1175/1520-0450(2003)042<0701:CPDUGA>2.0.CO;2.

——, and Coauthors, 2004: The QME AERI LBLRTM: A closure experiment for downwelling high spectral resolution infrared radiance. *J. Atmos. Sci.*, **61**, 2657–2675, doi:10.1175/JAS3300.1.

——, S. A. Clough, J. C. Liljegren, E. E. Clothiaux, K. Cady-Pereira, and K. L. Gaustad, 2007a: Retrieving liquid water path and precipitable water vapor from Atmospheric Radiation Measurement (ARM) microwave radiometers. *IEEE Trans. Geosci. Remote Sens.*, **45**, 3680–3690, doi:10.1109/TGRS.2007.903703.

——, and Coauthors, 2007b: Thin liquid water clouds: Their importance and our challenge. *Bull. Amer. Meteor. Soc.*, **88**, 177–190, doi:10.1175/BAMS-88-2-177.

——, U. Loehnert, M. Cadeddu, S. Crewell, and A. Vogelmann, 2009: Modifications to the water vapor continuum in the microwave suggested by ground-based 150 GHz observations. *IEEE Trans. Geosci. Remote Sens.*, **47**, 3326–3337, doi:10.1109/TGRS.2009.2022262.

——, A. Merrelli, D. Vimont, and E. J. Mlawer, 2012a: Impact of modifying the longwave water vapor continuum absorption model on community Earth system model simulations. *J. Geophys. Res.*, **117**, D04106, doi:10.1029/2011JD016440.

——, and Coauthors, 2012b: Ground-based high spectral resolution observations of the entire terrestrial spectrum under extremely dry conditions. *Geophys. Res. Lett.*, **39**, L10801, doi:10.1029/2012GL051542.

——, E. J. Mlawer, and H. E. Revercomb, 2016: Water vapor observations in the ARM Program. *The Atmospheric Radiation Measurement (ARM) Program: The First 20 Years, Meteor. Monogr.*, No. 57, Amer. Meteor. Soc., doi:10.1175/AMSMONOGRAPHS-D-15-0025.1.

Uttal, T., and Coauthors, 2002: Surface heat budget of the Arctic Ocean. *Bull. Amer. Meteor. Soc.*, **83**, 255–275, doi:10.1175/1520-0477(2002)083<0255:SHBOTA>2.3.CO;2.

Valero, F. P. J., R. D. Cess, M. Zhang, S. K. Pope, A. Bucholtz, B. Bush, and J. Vitko Jr., 1997: Absorption of solar radiation by the cloudy atmosphere: Interpretations of collocated aircraft measurements. *J. Geophys. Res.*, **102** (D25), 29 917–29 927, doi:10.1029/97JD01782.

Varanasi, P., 1998: Laboratory spectroscopy in support of the Atmospheric Radiation Measurement program. *Proc. Seventh ARM Science Team Meeting,* Washington, D.C., U.S. Department of Energy, 277.

Yang, P., K. N. Liou, K. Wyser, and D. Mitchell, 2000: Parameterization of the scattering and absorption properties of individual ice crystals. *J. Geophys. Res.*, **105**, 4699–4718, doi:10.1029/1999JD900755.

——, H. Wei, H.-L. Huang, B. A. Baum, Y. X. Hu, G. W. Kattawar, M. I. Mishchenko, and Q. Fu, 2005: Scattering and absorption property database for nonspherical ice particles in the near through far-infrared spectral region. *Appl. Opt.*, **44**, 5512–5523, doi:10.1364/AO.44.005512.

Chapter 15

Contributions of the ARM Program to Radiative Transfer Modeling for Climate and Weather Applications

ELI J. MLAWER AND MICHAEL J. IACONO

Atmospheric and Environmental Research, Lexington, Massachusetts

ROBERT PINCUS

University of Colorado, Boulder, and NOAA/Earth System Research Laboratory/Physical Sciences Division, Boulder, Colorado

HOWARD W. BARKER

Environment and Climate Change Canada, Toronto, Ontario, Canada

LAZAROS OREOPOULOS

NASA Goddard Space Flight Center, Greenbelt, Maryland

DAVID L. MITCHELL

Desert Research Institute, Reno, Nevada

1. Introduction

Accurate climate and weather simulations must account for all relevant physical processes and their complex interactions. Each of these atmospheric, ocean, and land processes must be considered on an appropriate spatial and temporal scale, which leads these simulations to require a substantial computational burden. One especially critical physical process is the flow of solar and thermal radiant energy through the atmosphere, which controls planetary heating and cooling and drives the large-scale dynamics that moves energy from the tropics toward the poles. Radiation calculations are therefore essential for climate and weather simulations, but are themselves quite complex even without considering the effects of variable and inhomogeneous clouds. Clear-sky radiative transfer calculations have to account for thousands of absorption lines due to water vapor, carbon dioxide, and other gases, which are irregularly distributed across the spectrum and have shapes dependent on pressure and temperature. The line-by-line (LBL) codes that treat these details have a far greater computational cost than can be afforded by global models. Therefore, the crucial requirement for accurate radiation calculations in climate and weather prediction models must be satisfied by fast solar and thermal radiation parameterizations with a high level of accuracy that has been demonstrated through extensive comparisons with LBL codes.

The calculation of a vertical profile of radiative fluxes and heating rates for each spatial grid cell and time step in a global model involves computations of

- absorption and scattering properties of the gases, clouds, and aerosols present in the cell at that time, but in a parameterized form that circumvents the consideration of the full complexity of the physics;
- values related to sources of solar and thermal radiation;
- solution of the radiative transfer equation for each subelement of the parameterization;

Corresponding author address: Eli Mlawer, Atmospheric and Environmental Research, 131 Hartwell Ave., Lexington, MA 02421.
E-mail: mlawer@aer.com

DOI: 10.1175/AMSMONOGRAPHS-D-15-0041.1

• and finally integrating the solutions for these subelements to obtain spectrally broadband fluxes and heating rates.

Numerous simplifications and approximations are typically made in these parameterizations, such as assuming that each grid cell is homogeneous and plane parallel (i.e., not dependent on clouds in neighboring cells and ignoring planetary curvature). This complexity makes it difficult to build a parameterization that is both fast and accurate enough to be useful.

The challenge of improving radiative transfer calculations in climate simulations was a central motivation for the establishment of the Atmospheric Radiation Measurement (ARM) Program and, consequently, one of the most critical objectives of the program's first decade. In Stokes and Schwartz (1994, p. 1203), the primary objectives of the ARM Program are 1) "to relate observed radiative fluxes in the atmosphere, spectrally resolved . . . to the atmospheric temperature, composition . . . and surface radiative properties" and 2) "to develop and test parameterizations that describe atmospheric water vapor, clouds, and the surface properties governing atmospheric radiation . . . with the objective of incorporating these parameterizations into general circulation and related models." The second of these objectives depends on the first. Confidence that a computationally fast radiative transfer parameterization used for climate simulations accounts accurately for the radiative effects associated with all relevant atmospheric conditions must spring from confidence that we have a detailed understanding of these radiative processes. This, in turn, necessitates a rigorous evaluation of our ability to compute spectrally resolved radiative fluxes for this range of conditions. Accomplishments in the ARM Program led to substantial advances with respect to both objectives. Mlawer and Turner (2016, chapter 14) address the program's accomplishments with regard to the observation and modeling of spectrally resolved radiation, while this chapter details achievements related to the modeling of radiative processes for climate and weather applications.

The results of the Intercomparison of Radiation Codes in Climate Models (ICRCCM) effort (Ellingson and Fouquart 1991) provided important motivation for the objectives of the ARM Program (Ellingson et al. 2016, chapter 1). ICRCCM was directed at understanding and evaluating the differences between radiative transfer models, both spectrally resolved (i.e., LBL) and fast parameterizations for climate applications. The longwave (LW) intercomparisons (Ellingson et al. 1991) determined that the uncertainties in spectroscopic parameters prevented any LBL model from being considered a reference, but progress was noted

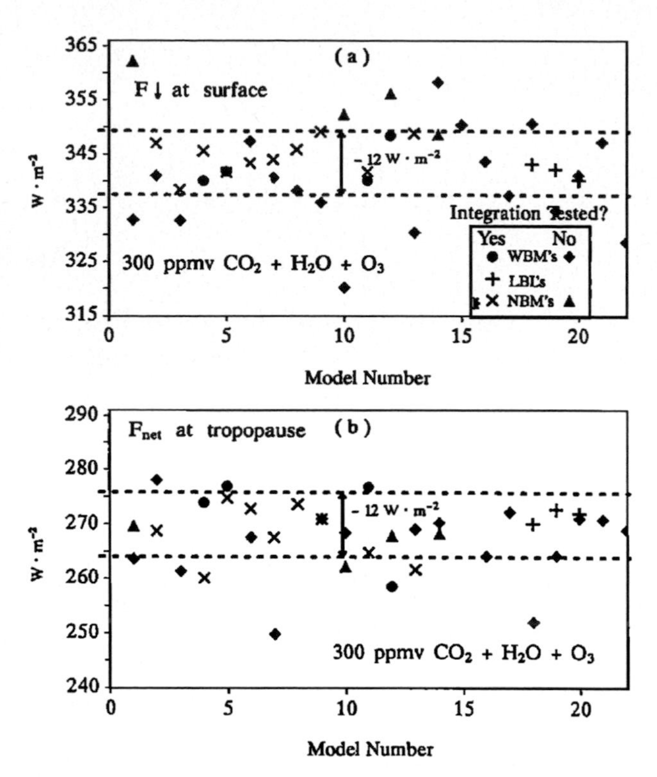

FIG. 15-1. For longwave radiation codes participating in ICRCCM, calculations of (top) downwelling surface flux and (bottom) net flux at the tropopause for the midlatitude summer atmosphere (H_2O and O_3 abundances only) and 300 ppmv of CO_2. For further explanation, see caption of Fig. 15 and related discussion in Ellingson et al. (1991).

relative to previous comparisons. The conclusions from ICRCCM with respect to longwave radiation codes suitable for climate applications were sobering, with a 5%–10% rms differences from LBL results and "poorer agreement" for the sensitivity to changes in abundances of absorbing gases. Figure 15-1 (Fig. 15 from Ellingson et al. 1991) provides evidence that supports this conclusion. For an atmosphere with reasonable profiles of H_2O, CO_2, and O_3, the best-performing "band models" have their respective computed downward surface fluxes (Fig. 15-1a) and net fluxes at the tropopause (Fig. 15-1b) fall within $12\,W\,m^{-2}$ (3.5%) of each other, and within approximately $\pm 6\,W\,m^{-2}$ (1.7%) of reference LBL calculations. Band models that perform more poorly also can be seen in this figure. For comparison, doubling the abundance of CO_2 in the atmosphere causes a radiative forcing of $\sim 4\,W\,m^{-2}$, while the radiative forcing by individual long-lived greenhouse gases over the last two centuries is less than $1\,W\,m^{-2}$. Shortwave (SW) results from ICRCCM were not any better, with "a considerable spread in the responses of different codes to a set of well-defined atmospheric profiles" (Fouquart et al. 1991, p. 8955). The SW ICRCCM study concluded that errors in the solar radiation calculations of these band models were greater than those required for accurate climate simulations.

The dual issues of LBL model uncertainties and broadband model errors led to major ARM initiatives in the early years of the program. Shepard (Tony) Clough of Atmospheric and Environmental Research (AER) was a key figure in many of the program's accomplishments that helped resolve these issues. His 1990 research proposal to ARM resolved to attack on both fronts, proposing "to provide a highly accurate RT model for scattering and nonscattering atmospheres, to use calculations from this model for the parameterization required for GCM RT codes . . . and to validate this code against measurements and calculations from the high accuracy model for a wide range of atmospheric regimes." This research program led to the establishment of AER's Line-By-Line Radiative Transfer Model (LBLRTM; Clough et al. 1992) as "a highly accurate RT model" through extensive validation with high-quality measurements of spectrally resolved radiation, which is detailed in Mlawer and Turner (2016, chapter 14). With this result, Clough was able to confidently undertake the crucial goal of developing a new, fast radiation code that effectively reproduced the flux and cooling rate calculations of LBLRTM, thereby ensuring that the calculations of this parameterized code would be directly traceable to ARM spectral radiation measurements. Developed from ARM funding and using ARM measurements, this fast radiation code, RRTM for GCM applications (RRTMG; Mlawer et al. 1997; Iacono et al. 2000), which was subsequently incorporated in numerous climate and weather prediction models throughout the world, represents a major triumph for the program.

Section 2 of this chapter discusses the development of RRTMG and its implementation in various general circulation models (GCMs), most notably the Integrated Forecast System of the European Centre for Medium-Range Weather Forecasts and the Community Earth System Model (CESM1) of the National Center for Atmospheric Research. Although the core of RRTMG is its stored tables of gaseous absorption coefficients and algorithms that operate on these coefficients, the application of this code (and other fast RT codes) to climate or weather problems also must consider cloudy conditions. As a result, the ARM Program also has given rise to major accomplishments in cloudy-sky radiative transfer within GCMs. This includes a development of the ice optical property parameterization (Mitchell 2002) integrated in RRTMG for use in CESM1. ARM support also led to the Monte Carlo Independent Column Approximation (McICA; Pincus et al. 2003; Barker et al. 2008), a method to treat subgrid-scale variability in cloud properties, including the variability introduced by cloud vertical correlations (overlap). Both these accomplishments relevant to cloudy-sky RT in GCMs are discussed in section 3 of this chapter. In addition, sections 3 and 4 detail initiatives supported by ARM that extended and updated previous RT code intercomparisons, the ICRCCM-III effort (Barker et al. 2003), and the Continual Intercomparison of Radiation Codes (Oreopoulos and Mlawer 2010; Oreopoulos et al. 2012).

2. Development of RRTMG

The development of RRTMG progressed in two steps. First, a fast radiation code was built with the general goal of achieving accuracy effectively equivalent to state-of-the-art LBL models with respect to the impact on climate simulations. Ad hoc but ambitious accuracy targets of 1 W m^{-2} for net flux at all altitudes and 0.1 K day^{-1} for tropospheric hearting rate were adopted; there were no known studies that established that this level of accuracy would have an undetectable effect on simulations. The name of the code that was developed was motivated by the phrase Rapid Radiative Transfer Model (its formal name is simply RRTM). This code serves as a reference fast radiation code for scientific applications and further development, but it is not fast enough for use in GCMs. The second step was to increase the computational efficiency of RRTM without significantly degrading its accuracy, thereby enabling the use of this accelerated version, RRTMG, in GCMs (both codes are available at http://rtweb.aer.com).

a. Development of RRTM

In the years immediately preceding the development of RRTM, a number of compelling papers (e.g., Goody et al. 1989; Lacis and Oinas 1991; Fu and Liou 1992) established the capability of the correlated-k method (Ambartzumian 1936) for fast and accurate calculations of radiative fluxes and heating rates in the atmosphere, including the effects of multiple scattering. In the correlated-k method, absorption coefficients as a function of wavenumber are reordered monotonically (creating a k distribution), thereby allowing spectral elements with similar opacities to be grouped together and treated as a single monochromatic element. This technique reduces the number of needed individual radiation calculations by $\sim 10^5$ relative to LBL calculations, thereby allowing a dramatic increase in computational speed while maintaining a level of accuracy believed to be acceptable for climate and weather simulations. This combination of speed and accuracy motivated the choice of the correlated-k method for RRTM instead of other band model approaches available at the time.

One aspect of the correlated-k method that limits its accuracy is that the mapping from spectral space to a space where absorption coefficients are monotonically ordered (g space, where the ordering variable g ranges from 0 to 1) is not fixed for a given spectral region, but depends on pressure, temperature, and, importantly, atmospheric composition. Therefore, a range of g values in a k distribution at one vertical level generally does not correspond to the same spectral elements as the same g values at a different level. The k distributions stored in RRTM were computed for a range of values of pressure, temperature, and ratio of key absorbing gases and then averaged for chosen subintervals in g space. Radiative transfer calculations, with each subinterval being treated in the same manner as a single monochromatic element in an LBL calculation, are performed using these stored values despite the possible absence of correlation. This potential lack of spectral correlation is an important contributor to RRTM errors, as is the fact that the approach used to combine spectral elements with similar opacities does not necessarily preserve the average transmittance of those elements. Extensive validation (see below) has shown that this approach provides impressive accuracy. Details about RRTM's bands and their respective absorbing gases, grid of stored absorption coefficients, subintervals in g space, etc. can be found in Mlawer et al. (1997) and Mlawer and Clough (1998).

RRTM relies on a number of innovative algorithmic features to obtain this high level of accuracy. Spectral regions in which two gases have spectrally overlapping absorption bands pose an issue for correlated-k models since varying abundances of gases involved in such an overlap (e.g., H_2O) allow a wide variety of mappings from spectral space to g space. This overlapping makes it difficult for the code to store a reasonable number of absorption coefficients from which to accurately compute optical depths for all abundance combinations that might be encountered. Also, overlapping absorption bands can cause a lack of spectral correlation in g space between different levels in a vertical profile, leading to the accuracy issue discussed above. RRTM handles overlapping absorbing bands through the use of a "binary species parameter" in affected spectral regions, which varies from 0 (second species is dominant) to 1 (first species is dominant) in such a way that most abundances encountered have values of this parameter between 1/8 and 7/8, allowing accurate linear interpolation of stored coefficients. A more detailed interpolation method is performed near the extreme values of this parameter. Another feature of RRTM is that spectrally dependent values, such as the optical depths due to minor absorbing gases in a band, the Planck function in the thermal infrared spectral region, and the extraterrestrial solar irradiance in the shortwave, are handled in a manner that respects their respective correlations with the major absorbing gases. This approach, which is implemented by applying the mapping from spectral space to g space that defines the k distribution to other spectrally dependent quantities (for a subset of the parameters for which the main absorption coefficients are stored), allows greater consistency with monochromatic radiative transfer. Figure 15-2 provides an example of this approach.

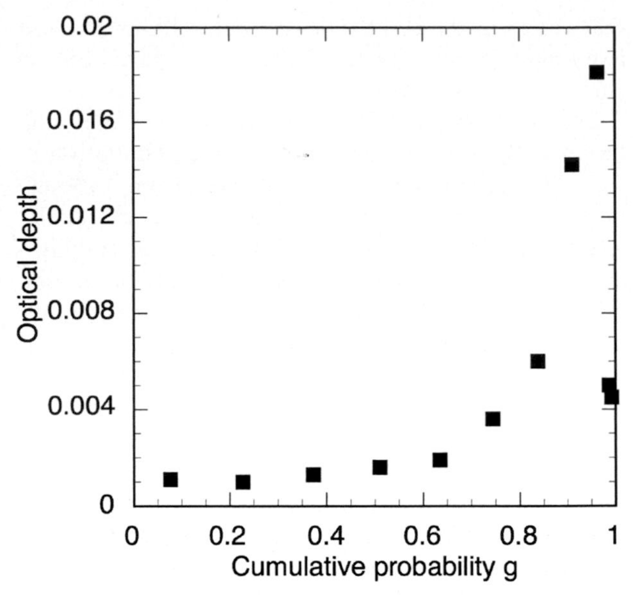

FIG. 15-2. For a layer in the midlatitude summer atmosphere profile and the spectral range 980–1080 cm^{-1}, optical depths were obtained by applying the mapping from spectral space to g space that defines the layer's k distribution (for all absorbing gases) to the optical depths for this layer due to the minor absorbing gas CO_2. The results shown were then averaged over the subintervals in g space used in RRTM. The optical depths are not monotonic with respect to g. The use of this approach in the code ensures a reasonable spectral correspondence between the optical depths from major and minor absorbing species in a band, while allowing an accurate calculation when the abundance of each species is independently varied (from Mlawer et al. 1997).

Liquid cloud optical properties derived from the Hu and Stamnes (1993) parameterization were implemented in RRTM, as were several parameterizations of ice cloud optical properties, most notably Fu (1996) and Fu et al. (1998). A user can also specify band-by-band cloud optical properties directly. For RRTM_LW a fast maximum-random overlap algorithm (called two-layer memory) is available while the shortwave code assumes that each cloudy layer is completely overcast. Spectral variability of aerosol optical depths is handled in RRTM_SW by a generalized Ångström relation (Molineaux et al. 1998), with aerosol scattering properties allowed to vary with spectral band. Great care was taken to validate the radiative effect of aerosols computed by the code with respect to LBL calculations for a range of aerosol properties, water vapor loadings, and solar zenith angles.

Not only does RRTM obtain its stored absorption coefficients from LBLRTM, but it also has been extensively validated with respect to flux and heating rate calculations performed with LBLRTM. Therefore, the performance of RRTM is traceable directly to the numerous validations of LBLRTM that have been performed with high-quality spectrally resolved measurements, including those detailed in Mlawer and Turner (2016, chapter 14). Mlawer et al. (1997) presented validations with respect to LBLRTM for six standard atmospheres, but this set proved to not be sufficiently broad to establish the code's accuracy for the full range of conditions encountered in global simulations. In subsequent evaluations, a suite of 42 atmospheres (Garand et al. 2001) was utilized successfully. Detailed validation statistics for RRTMG are provided in section 2b(2) below.

b. RRTMG and its application to GCMs

1) RRTM TO RRTMG

To provide a radiative transfer model that can be applied directly to GCMs with an accuracy that remains traceable to measurements, RRTM was modified to produce RRTMG (Iacono et al. 2003; Morcrette et al. 2008). The former model retains the highest accuracy relative to LBL results for single-column calculations, while the latter provides improved computational efficiency with minimal loss of accuracy for GCM applications. RRTMG shares the same basic physics and absorption coefficients as RRTM, but it incorporates several modifications that improve computational efficiency and represent subgrid-scale cloud variability. In particular, the total number of quadrature points (*g* points) used to calculate radiances in the longwave was reduced from the standard 256 in RRTM_LW to 140 in RRTMG_LW. For each spectral band, the particular reduction implemented was based on minimizing the impact on flux and heating rate accuracy, resulting in 2–16 *g* points per band. In the shortwave, the number of *g* points was reduced from the 224 in RRTM_SW to 112 in RRTMG_SW. In addition, the multiple-scattering code Discrete Ordinates Radiative Transfer Program for a Multi-Layered Plane-Parallel Medium (DISORT; Stamnes et al. 1988) employed by RRTM_SW was replaced with a much faster two-stream radiative transfer solver (Oreopoulos and Barker 1999) in RRTMG_SW. (RRTMG does not include the effects of scattering in the LW.) The complexity of representing fractional cloudiness in the presence of multiple scattering was eventually addressed in RRTMG_LW and SW with the addition of McICA (Barker et al. 2002; Pincus et al. 2003), which is a statistical technique for representing subgrid-scale cloud variability including cloud overlap. This method is described in detail in section 3b(2).

2) RRTMG VALIDATION

A critical component of applying RRTMG to atmospheric models is the validation of its accuracy. Figure 15-3 presents an analysis of RRTMG_LW accuracy relative to LBLRTM (as of 2007) for a set of 42 clear atmospheric profiles spanning a wide range of temperature and moisture values. For most cases, the accuracy of RRTMG_LW for clear-sky net flux is better than 1.5 $W m^{-2}$ at all levels, and heating rates agree to within 0.2 $K day^{-1}$ in the troposphere and 0.4 $K day^{-1}$ in the stratosphere. RRTMG_SW accuracy in clear sky relative to RRTM_SW is within 3 $W m^{-2}$ for flux at all levels, and heating rates agree to within 0.1 $K day^{-1}$ in the troposphere and 0.35 $K day^{-1}$ in the stratosphere. Motivated by interactions with the GCM community, Iacono et al. (2008) evaluated RRTMG using the methodology of the Radiative Transfer Model Intercomparison Project (RTMIP; Collins et al. 2006), which involved model calculations "forced" by increased abundances of greenhouse gases for a set of scenarios relevant to climate change. This study reasserted the overall excellent performance of RRTMG, most notably at the surface, where Collins et al. (2006) found the largest discrepancies between GCM and LBL radiative transfer codes. In all RTMIP cases except one, RRTMG longwave forcings were within a range of −0.20 to 0.23 $W m^{-2}$ of those calculated by LBLRTM, with more than half of the results within 0.10 $W m^{-2}$. In the shortwave, for all RTMIP cases except one, RRTMG shortwave forcings were within a range of −0.16 to 0.38 $W m^{-2}$ of the spectral multiple-scattering model Code for High-Resolution Atmospheric Radiative Transfer with Scattering (CHARTS). Since radiative forcing by individual long-lived greenhouse gases over the last two centuries is on the order of 1 $W m^{-2}$ or less, the results in Iacono et al. (2008) were key to establishing that RRTMG has sufficient accuracy to be used by GCMs to properly model the radiative contribution of these gases to global climate change.

It is important to view these validation results, later confirmed by the Continual Intercomparison of Radiation Codes (CIRC) intercomparison (see section 4), in the context of the overall unsatisfying results of ICRCCM that led to the birth of the ARM Program. In less than a decade, efforts performed under the program's auspices had led to the development of a fast radiation code with impressive clear-sky accuracy that was traceable to ARM high-quality spectral radiation measurements. The next

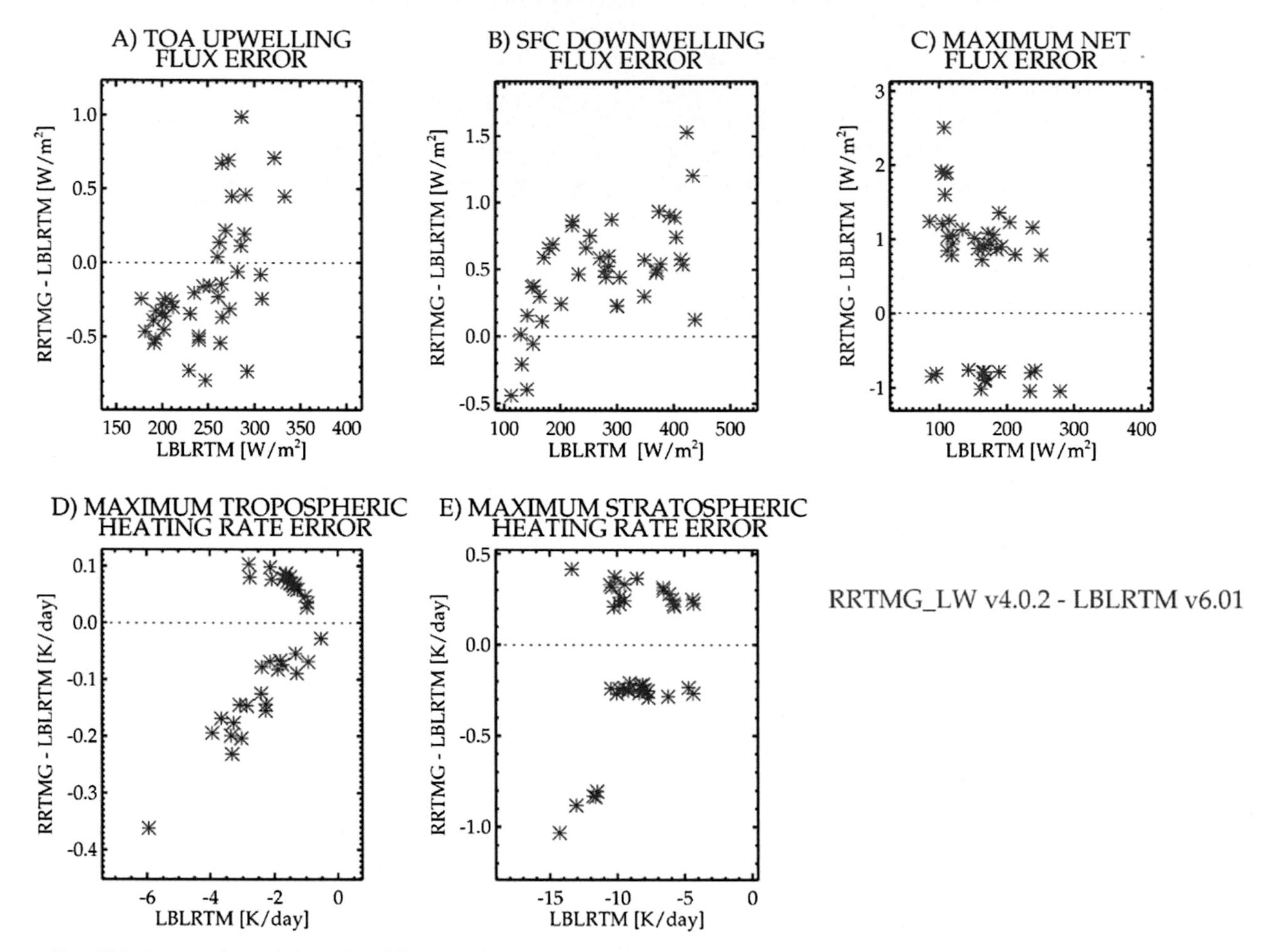

FIG. 15-3. Scatterplots of clear-sky differences between RRTMG_LW and the LBL model LBLRTM plotted as a function of the LBLRTM calculation over the 10–3250 cm^{-1} spectral range for (a) TOA upwelling flux, (b) surface downwelling flux, (c) maximum net flux difference, (d) maximum tropospheric heating rate difference, and (e) maximum stratospheric heating rate difference. Calculations are for the 42 diverse profiles of Garand et al. (2001). Since RRTMG_LW calculations are intended to reproduce those of LBLRTM as closely as possible, the differences between the two models are referred to as "errors" in this figure.

chapter in this success story was the use of this code to advance climate and weather simulations.

3) Implementation in atmospheric models

Because of its high accuracy and computational efficiency, RRTMG has been implemented in numerous national and international atmospheric models to provide validated radiative transfer for improved weather forecasts and climate change predictions. The European Centre for Medium-Range Weather Forecasts (ECMWF) became the first modeling center to make operational use of RRTMG_LW in 2000 to improve radiative processes within the Integrated Forecast System (IFS) weather forecast model (Morcrette et al. 2001). This forecast system was used to generate the ERA-40 reanalysis (Uppala et al. 2005) as well as the ERA-Interim reanalysis. Further reduction in shortwave and cloudy-sky radiation biases was realized with the application of both RRTMG_SW and McICA in the IFS in 2007 (Morcrette et al. 2008). Simulations with this configuration showed remarkable improvement in a number of radiative and dynamical fields due to the application of RRTMG and McICA (Morcrette et al. 2008; Ahlgrimm et al. 2016, chapter 28). Particular improvement was seen in the simulation of longwave (see Fig. 15-4) and shortwave cloud radiative forcing (now usually referred to as "cloud radiative effect").

The National Centers for Environmental Prediction (NCEP) first began using RRTMG_LW in the Global Forecast System (GFS) for operational forecasts in 2003, and RRTMG_SW in 2010, though the GFS does not

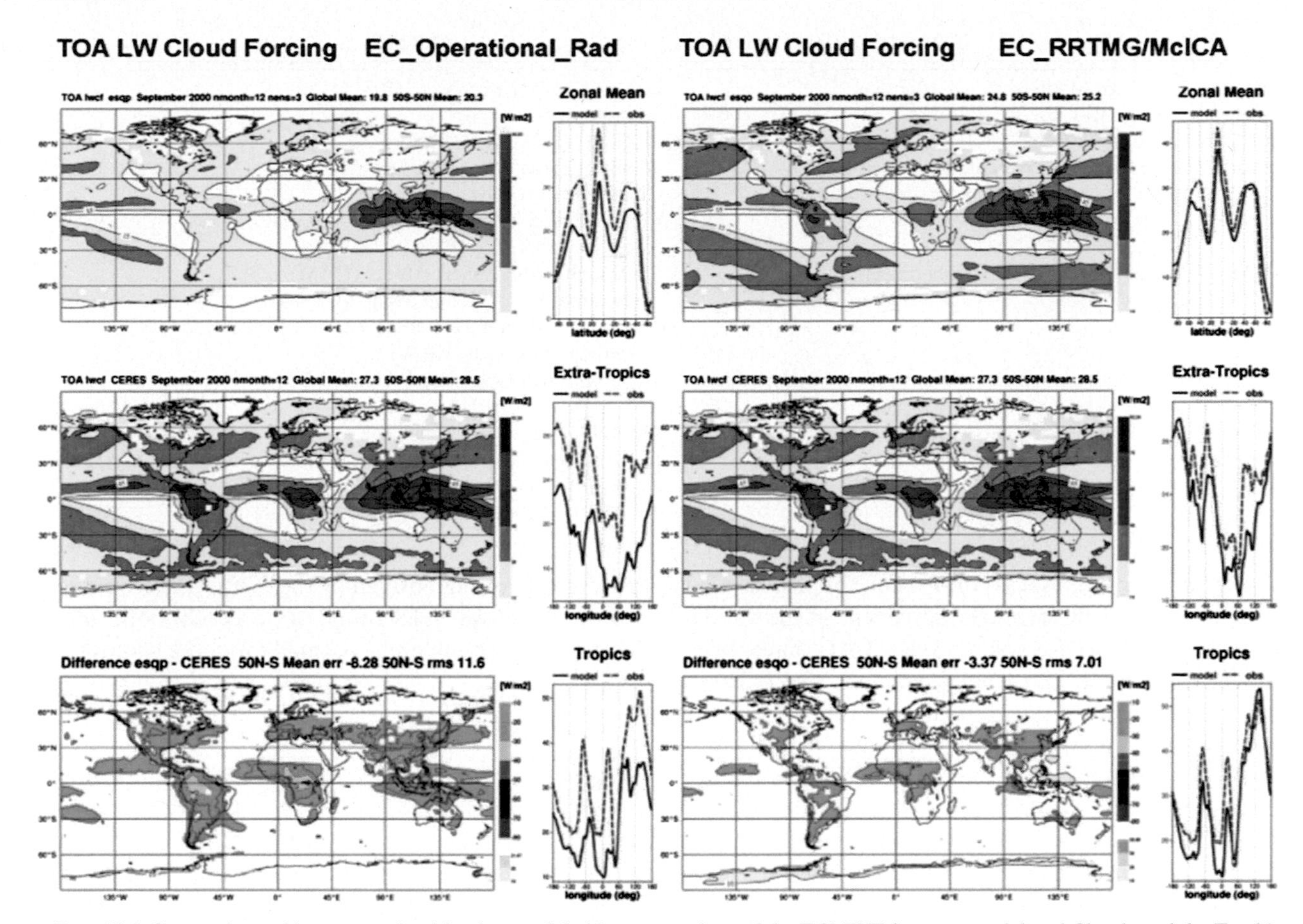

FIG. 15-4. Comparison of longwave cloud forcing modeled by two versions of the ECMWF forecast model and Clouds and the Earth's Radiant Energy System (CERES) measurements averaged for 1 year ending in September 2000. Longwave cloud forcing is shown for the (top left) operational ECMWF model, (top right) the ECMWF model with RRTMG/McICA, (center) the CERES measurement, and (bottom) the model minus observed differences. Also shown are zonal mean and longitudinal line plots averaged over the tropics and extratropics for each model (black) and the CERES observed values (red) (Morcrette et al. 2008). All units are $\mathrm{W\,m^{-2}}$.

currently utilize McICA. The Climate Forecast System (CFS), which is based on GFS but adapted for longer simulations (Saha et al. 2006), first began using the longwave code in 2004 and the shortwave code in 2010. The recently updated CFS version 2 has implemented McICA with the latest versions of RRTMG (Saha et al. 2014). Application of the new radiation code showed particular improvement in the significant upper stratospheric cold bias in the operational GFS and a notable reduction in sea surface temperature anomalies in the CFS. NCEP's coupled global reanalysis covering the last three decades, the Climate Forecast System Reanalysis (CFSR; Saha et al. 2010) also uses RRTMG within its atmospheric component.

Experiments with the original National Center for Atmospheric Research (NCAR) Community Climate Model (CCM) suggested that this global model would benefit from improvements in radiative transfer (Iacono et al. 2000, 2003). Application of RRTMG and McICA into the NCAR Community Atmosphere Model (CAM) was realized with the public release of CAM5 in 2010 (Neale et al. 2010), now the atmospheric component of the coupled Community Earth System Model (CESM1). The radiation enhancement was accompanied by several additional major changes to the physics parameterizations in the NCAR climate model, including the treatment of cloud microphysics (see next section) and aerosols. Both the longwave and shortwave RRTMG codes are also radiation options in the NCAR-supported Advanced Research version of the Weather Research and Forecasting (WRF) Model (Skamarock et al. 2008), one of the most widely used regional weather forecast models.

Global and regional models currently utilizing RRTMG are listed in Table 15-1.

3. Advances in cloudy-sky radiative transfer

The development of RRTMG, built on the foundation of extensive validations of LBLRTM with spectral radiation measurements, resolved the major clear-sky radiative transfer issues in GCMs that were a key

TABLE 15-1. Global and regional model applications of RRTMG.

Institution	Model
European Centre for Medium-Range Weather Forecasts (ECMWF)	Integrated Forecast System (IFS)
	ERA-40 reanalysis
Max Planck Institute (MPI)	ECHAM (ECMWF-Hamburg)
National Centers for Environmental Research (NCEP)	Global Forecast System (GFS)
	Climate Forecast System (CFS)
	Climate Forecast System Reanalysis (CFSR)
	Rapid Update Cycle (RUC)
NCAR (National Center for Atmospheric Research)	Community Atmosphere Model (CAM)
	Community Earth System Model (CESM)
	Advanced Research version of Weather Research and Forecasting Model (ARW)
National Aeronautics and Space Administration/Goddard Space Flight Center (NASA/GSFC)	Goddard Earth Observing System (GEOS)
Laboratory for Dynamical Meteorology (LMD)	LMDZ
China Meteorological Administration (CMA)	Global/Regional Assimilation and Prediction System (GRAPES)
French National Meteorological Service (Météo-France)	Nonhydrostatic Mesoscale Model (Meso-NH)

motivation for the ARM Program. This assessment rests, in part, on the representativeness of the validations to which LBLRTM and RRTMG have been subjected. In large part, the assessment also is based on a solid theoretical understanding of the propagation of radiation in a clear atmosphere, including the underlying molecular spectroscopy and observations of the homogeneity of absorbing gas abundances at GCM scales. This combination of systematic validation and theoretical understanding is conclusive in establishing the accuracy of the radiation model for all clear-sky conditions that occur in the atmosphere.

Establishing a similar level of confidence in radiative calculations in cloudy atmospheres is not nearly as straightforward. ARM-supported research has had impacts at both the scale of individual cloud particles and on the scales of cloud systems. Both scales exhibit variability that hinders radiation parameterization development and its systematic validation with radiometric observations. One focus in the program has been on methods to represent the optical properties of ice clouds, which are difficult because ice crystal shapes and sizes can vary widely. The material properties of ice (i.e., refractive indices) are well-known, but shape and habit have profound impacts on the spectral absorption and scattering of radiation that have historically been hard to generalize.

Though the single-scattering properties of spherical drops are much more certain, both liquid and ice clouds exhibit substantial horizontal and vertical variability across a wide range of scales, which undermines the use of plane-parallel, homogenous RT methods. One consequence is that it is extremely challenging to construct cloudy-sky comparisons between LBL models and spectral observations since the models cannot account for the impacts of inhomogeneity. This inability also is present in the radiation calculations in GCMs and introduced first-order biases quantified by the ICRCCM-III project described below. ARM-funded research both highlighted this problem and eventually found a solution.

a. Development of a parameterization of ice optical properties for CAM

The optical properties of spherical liquid cloud droplets are predicted accurately by Mie theory. Prior to 1990 there was hope that Mie theory could treat the optical properties of ice clouds adequately using an equivalent area sphere approach. However, aircraft observations of the microphysical and radiative properties of cirrus clouds showed that, for a given observed downward thermal emittance from a cirrus cloud, a Mie calculation predicting this emittance also produced an albedo value that was at least a factor of 2 smaller than that was observed (Stackhouse and Stephens 1991). This discrepancy could not be explained microphysically [e.g., by adding unmeasured small ice crystals to the ice particle size distribution (PSD)], and it became apparent that new approaches were needed to treat the scattering and absorption properties of nonspherical ice particles. The ARM-funded research described in this section advanced the understanding of ice optical properties, eventually leading to a new parameterization incorporated in CAM5's implementation of RRTMG.

Ice optical properties depend strongly on ice particle shape (e.g., Mitchell and Arnott 1994), but an important question is how they can be formulated for any particle shape. Using ARM funding, Mitchell et al. (1996) proposed a solution by representing ice particle mass and projected area (on which optical properties depend) as area- and mass-dimensional power laws (henceforth A–D and m–D relationships). By combining a form of van de Hulst's (1981) anomalous diffraction approximation (ADA) as described by Bryant and

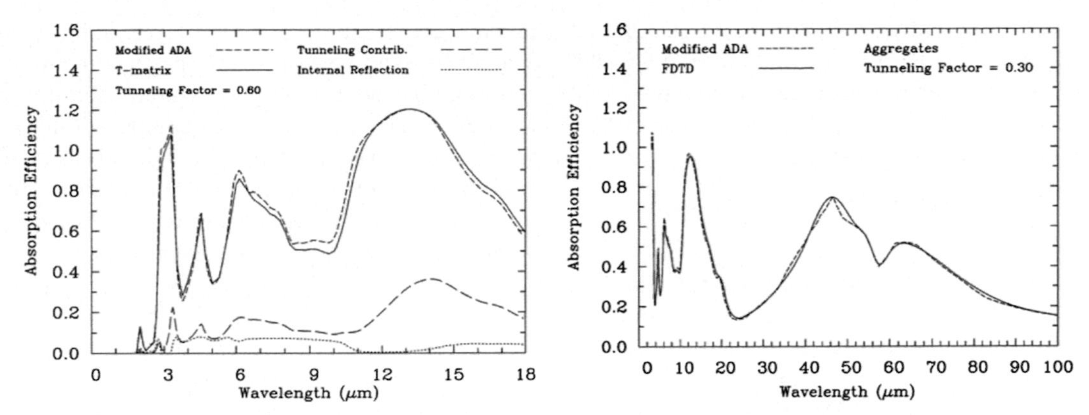

FIG. 15-5. (left) Comparison of MADA with T matrix for Q_{abs} based on a measured PSD of hexagonal columns. Absorption contributions predicted by MADA for photon tunneling and internal reflection/refraction are indicated. (right) Comparison of MADA with the FDTD method at terrestrial wavelengths for a cirrus PSD at −55°C. Ice particles are aggregates of hexagonal columns and plates.

Latimer (1969) with *m–D* and *A–D* relationships, analytical solutions for the extinction and absorption coefficients, β_{ext} and β_{abs}, were obtained as a function of the ice PSD parameters, ice particle shape, and wavelength. The asymmetry parameter for the PSD was estimated from ray-tracing calculations and parameterized in terms of ice particle shape, size, and wavelength (Mitchell et al. 1996). While the effects of internal reflection and refraction were accounted for in the solutions for β_{ext} and β_{abs}, wave resonance effects (also referred to as photon tunneling) were not. Subsequent ARM-funded research (Mitchell 2000) then parameterized wave resonance effects into a modified ADA (MADA) by relating resonance to the index of refraction and size parameter x (where $x = \pi d_e/\lambda$, d_e = effective photon path diameter of particle and λ = wavelength). Two types of wave resonance were parameterized: one increasing the ray path within a particle through internal "resonating" reflections (affecting both β_{ext} and β_{abs}) and another responsible for surface waves (depending solely on x and affecting only β_{ext}). The latter are sometimes referred to as edge effects. Although first tested against Mie theory and applied to liquid water clouds, MADA was applied subsequently to ice clouds as described in Mitchell (2002).

The left panel of Fig. 15-5 shows the PSD absorption efficiency Q_{abs} calculated by T matrix and MADA for a laboratory-grown ice cloud having a narrow PSD, with contributions from internal reflection/refraction and tunneling shown as predicted by MADA. MADA is compared against the finite difference time domain (FDTD) method over a greater wavelength range in the right panel of Fig. 15-5 for ice crystal aggregates and a PSD typical of cirrus clouds. The difference between MADA and these other methods is never more than 15% for any PSD size parameter ≥1 (based on PSD effective diameter) for any ice particle shape assumed. MADA also has been tested successfully against laboratory extinction efficiency (Q_{ext}) measurements (Mitchell et al. 2001, 2006), and an earlier version of MADA (Mitchell et al. 1996) used observed cirrus microphysical measurements to successfully predict the radiometric measurements mentioned above (Stackhouse and Stephens 1991).

MADA formed the basis of the cloud (both liquid water and ice) optics scheme of Harrington and Olsson (2001), and it is used in the Regional Atmospheric Modeling System (RAMS; Cotton et al. 2003) and in cloud-process models (e.g., Liu et al. 2003). More recently, this scheme was implemented in RRTMG to provide the ice optics in CAM5 (Gettelman et al. 2010). This allows the microphysical and radiative processes in CAM5 to be based on common assumptions about ice particle mass and projected area (i.e., *m–D* and *A–D* expressions) and a common ice PSD. The latter is important since, for a given effective diameter D_e and ice water content, β_{ext} and β_{abs} may vary up to 42% and 33%, respectively, for different PSD shapes (e.g., degree of bimodality) at thermal infrared wavelengths (Mitchell et al. 2011).

Prior to CAM5, the ice optical properties in CAM were treated using the projected area equivalent spheres approach of Ebert and Curry (1992, hereinafter EC). In Fig. 13 of Mitchell et al. (2006), the flux weighted mass absorption coefficient predicted in the IR window region (8.0–12.5 μm) by EC for cirrus clouds ($D_e <$ 150 μm; hexagonal columns assumed) is ~50% greater than that predicted by MADA and the ice optics schemes of Fu et al. (1998) and Yang et al. (2001). These three other schemes produce almost identical results. In the SW for a given D_e, both the mass normalized β_{ext} and g are greater in the EC scheme than in MADA.

Evaluation of the radiative effect of the MADA-based parameterization in RRTMG used in CAM5 determined that the globally averaged shortwave cloud radiative effect changed by $\sim$4.5 W m^{-2} compared to RRTMG with the EC ice optics parameterization.

b. The role of cloud variability and the development of McICA

By the late 1990s, it was widely recognized that fluxes computed by 1D multilayer, plane-parallel, homogeneous (PPH) radiative transfer models used in large-scale atmospheric models (LSAMs) contained significant (local values > 10 W m^{-2}) biases that depend strongly on cloud regime. Such models typically relied on basic two-stream approximations to compute fluxes in the clear and cloudy portions of a layer, each assumed to be homogeneous, and weighted the fluxes transported between layers using some analytic representation of an idealized vertical overlap assumption. Both sources of error, horizontal variability and the handling of vertical overlap, were originally considered as separate problems, although they are both fundamentally issues of small-scale heterogeneity.

The neglect of horizontal variability of cloud extinction attracted interest because it was known to introduce biases in both shortwave (Stephens 1988; Cahalan et al. 1994) and longwave (Fu et al. 2000) calculations. A variety of methods were used to address the impact of variability, all of which attempted to fold descriptions of cloud structure, or cloud-radiation interactions, directly into the 1D radiative transfer model. The simplest approach, applicable to shortwave radiation, was to continue to use the two-stream approximation but with a reduction in cloud optical depth that depended on cloud fraction (Tiedtke 1996). Others sought to analytically rescale the cloud optical properties based on assumptions about isotropic variability (e.g., Stephens 1988; Cairns et al. 2000). Barker (1996) introduced an elegant solution to a very specific problem by weighting single-layer two-stream solutions with lognormal density functions of optical depth (this choice of distribution being inspired by satellite retrievals and cloud-scale models). Oreopoulos and Barker (1999) extended this approach to multiple layers, though the handling of cloud overlap was simplistic. Each of these methods had inherent drawbacks or limiting assumptions, ranging from physical implausibility (in the case of tuned values of parameters) to inflexibility with respect to assumptions made in other parts of the model.

The small-scale heterogeneity caused by cloud overlap was treated almost uniformly by analytically mixing clear- and cloudy-sky fluxes within radiation codes. Collins (2001) introduced the idea of enumerating all possible combinations of cloudy layers and their relative frequencies and weighting the fluxes computed for the columns that most strongly affected radiation by that frequency. This approach was computationally expensive because it required many individual cloudy-column calculations and could not be extended to internally inhomogeneous clouds because the number of possible combinations quickly became enormous.

1) Identifying the problems: ICRCCM-III

With no approach giving satisfactory results, the feeling among modelers was that the existing paradigm of searching for closed-form solutions had hit an impasse. It seemed that the most productive way forward was to corral those models that were, or could be, used in atmospheric models and perform an intercomparison resembling that of Fouquart et al. (1991), but focusing on their handling of cloudy rather than clear-sky spectral fluxes. This intercomparison, designated ICRCCM-III (Barker et al. 2003) and supported using ARM funding, had as its primary motivations to 1) assess 1D solar radiative transfer models suitable for use in atmospheric models for complicated cloudy atmospheres and 2) demonstrate the relative importance to those models of addressing unresolved vertical and horizontal cloud fluctuations.

Participating codes were assessed first for clear-sky and single-layer, plane-parallel, homogeneous cloudy atmospheres in order to establish baseline differences in broadband fluxes before moving to complicated cloud cases. Benchmark fluxes were produced by a range of 3D Monte Carlo algorithms operating on cloud fields produced by cloud-resolving models (CRMs), which established the results that all 1D codes strived for. A single 3D model was used to establish "conditional" benchmarks for simplified versions of the CRM fields, such as precise overlap of horizontally homogenized plane-parallel clouds. The conditional benchmarks allowed a modeler to verify that their code, which they knew was incomplete, was at least addressing properly what it intended to address. An example of the methodology used in this intercomparison is shown in Fig. 15-6.

The main result of the study involved the range of conditional benchmarks, for they demonstrated clearly that overlap and horizontal fluctuations of cloud have to be dealt with together. The secondary result was that classes of 1D models had wide ranges of performance relative to their appropriate conditional benchmark—often they were not doing what was expected of them. Finally, no single 1D model, or 1D modeling strategy, stood out as the clear choice for use in atmospheric models, confirming what many had suspected at the outset of ICRCCM-III.

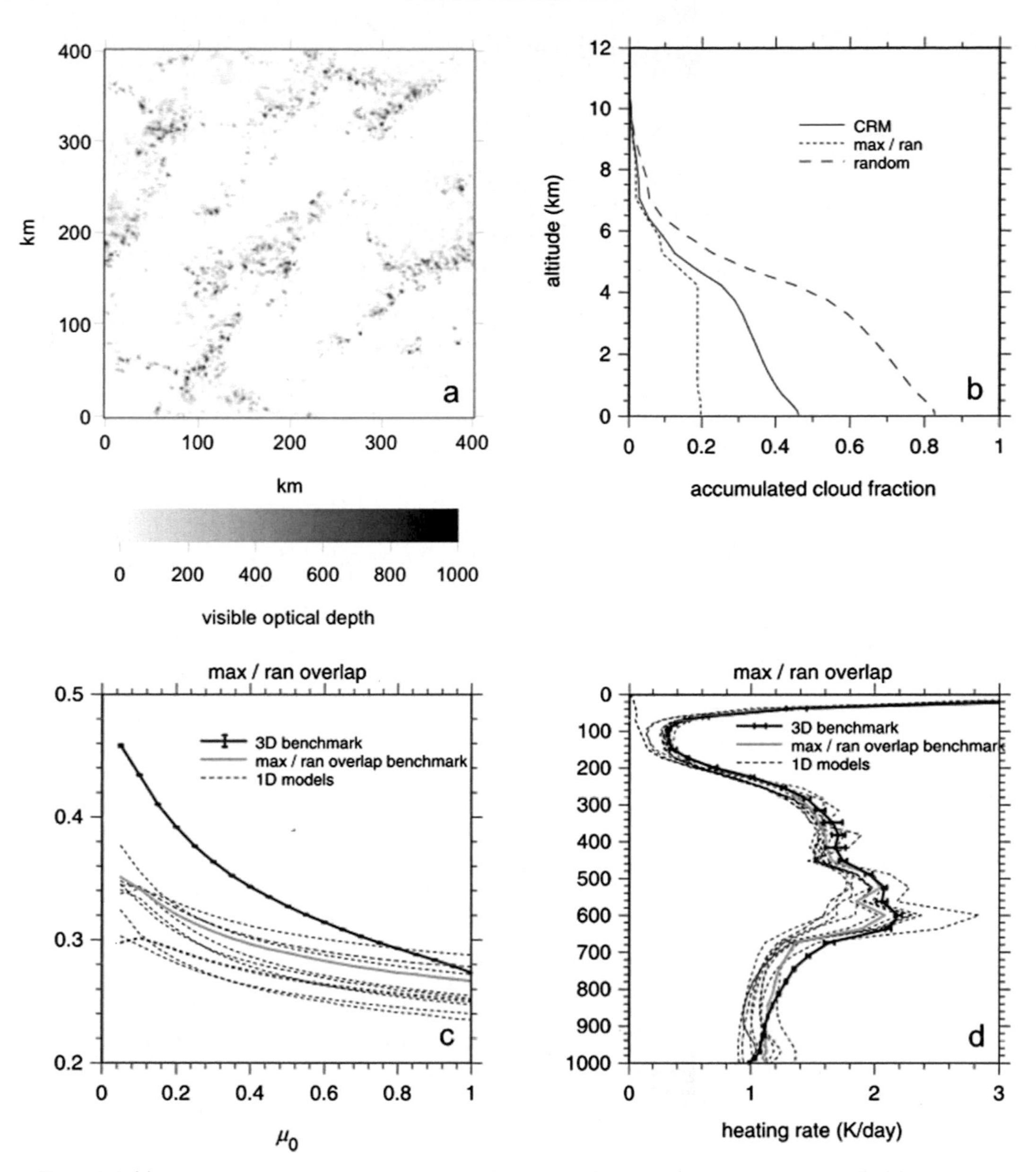

FIG. 15-6. (a) Model-generated cloud field that could be inside a GCM cell (Grabowski et al. 1998). (b) Downward accumulated cloud fractions for the cloud field shown in (a). CRM represents the actual function, while max/ran and random are corresponding functions assuming that layer cloud fractions follow the maximum-random and random overlap assumptions, respectively. (c) TOA albedo as a function of cos(SZA) for the field shown in (a). The 3D benchmark is the mean of four Monte Carlo models with error bars representing standard deviations. Gray line is from a Monte Carlo model acting on this field's correct maximum-random overlap rendition with horizontally homogeneous clouds. Dashed lines are from several 1D codes that all claimed to be doing max/ran overlap with horizontally homogeneous clouds. (d) Domain-averaged heating rates for the calculations described in (c) (from Barker et al. 2003).

ICRCCM-III highlighted a significant challenge for climate model radiation codes, namely, the treatment of variability in cloudy atmospheres. The study demonstrated that both horizontal variability and vertical structure are important in determining domain-mean fluxes, and also that methods existing at that time to describe overlap did not reproduce benchmark Independent Column Approximation (ICA) calculations (whose utility ICRCCM-III demonstrated), particularly for solar radiation calculations in which multiple scattering is important. Climate modelers understood that representing this variability was important for radiation [among other processes (Pincus and Klein 2000)], so these results were timely, as they arrived soon

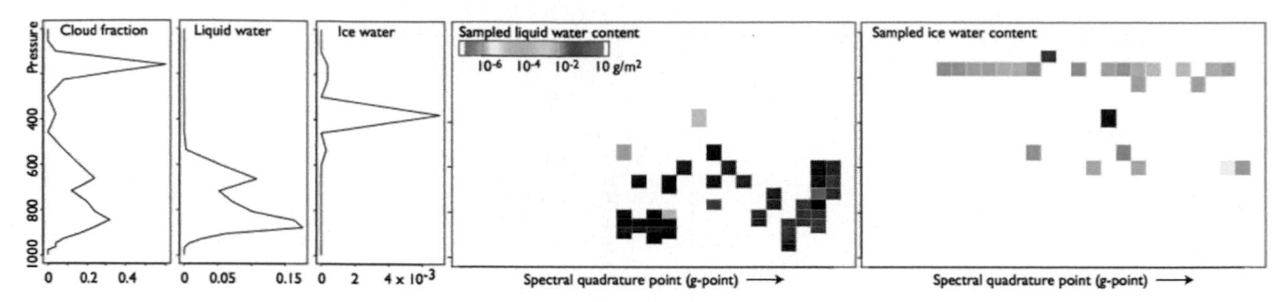

FIG. 15-7. An example of generating cloud samples (subcolumns) from atmospheric states for use with McICA. In this example, clouds are represented within the model with values for cloud fraction and liquid and ice water contents; profiles for an example column are shown. These values are used to infer distributions of total water at each level (see Pincus et al. 2005). An overlap assumption describes the probability of cloudiness and cloud condensate amount at each level depending on the value in the level above. Samples, shown in the two panels on the right, are constructed randomly using probabilities consistent with the statistics within each layer and the overlap assumption, so that large ensembles of columns reproduce the input properties.

after initial efforts to predict small-scale cloud variability in climate models (Tompkins 2002; Golaz et al. 2002) and new formulations of overlap from ground-based ARM-like instruments (Hogan and Illingworth 2000).

A subsequent ARM-funded analysis in the longwave spectral region supported the conclusion of the shortwave-based ICRCCM-III study that 1D approaches faced great challenges in modeling inherently inhomogeneous clouds. For several complex cloud fields, Kablick et al. (2011) evaluated the ability of various approximate methods to compute radiative fluxes and heating rates with respect to benchmark calculations by a 3D Monte Carlo algorithm. Results of the study indicated that overlap schemes used in GCMs such as maximum/random resulted in large errors in domain-averaged fluxes and heating rates, while ICA calculations were consistently more accurate. The authors concluded that "there is an inherent deficiency in the ability of 1D models to accurately calculate radiative quantities" in inhomogeneous cloud fields.

2) A FLEXIBLE SOLUTION: McICA

At the 2002 ARM Science Team Meeting, just two days after the ICRCCM-III manuscript was submitted for publication, an entirely new method for treating variability of all kinds was conceived. McICA weaves together two threads of prior experience with a new insight. The first thread was the idea that arbitrarily complicated cloud structures, no matter how they arose, could be represented as a set of discrete, internally homogeneous samples (or subcolumns). This idea, which came from the synthetic pixels used by the "ISCCP simulator" (Yu et al. 1996; Klein and Jakob 1999), greatly simplifies the radiative transfer. Simply expanding the number of radiation calculations by the number of samples would incur far too much computational cost. Experiences in numerical weather prediction, however, had already demonstrated that model forecasts are remarkably resilient to grid-scale noise (Buizza et al. 1999). It was therefore suggested that the new method could exploit this resilience by drawing as many samples of cloud states as there are spectral integration points in the host radiation scheme. As schematically represented in Fig. 15-7, each cloud sample is associated with a different, randomly chosen spectral point (further randomized at each time step and grid column) and the spectral fluxes are summed as in a normal ICA calculation. This means that any variability in cloud properties is sampled incompletely for any given radiative flux calculation, but that this limited sampling introduces random noise, not bias, with respect to the reference calculation (i.e., the ICA on an infinitely large set of samples). Initial experiments with the European Centre for Medium-Range Weather Forecasts model (Pincus et al. 2003) suggested that the amount of noise introduced by McICA would not affect model forecasts.

McICA has conceptual appeal because it offers a method for treating cloud variability in an unbiased way regardless of the cause or form of the variability. It thus fits naturally with interest in predicting in-cloud variability through a time-evolving probability distribution (Shupe et al. 2016, chapter 19), and with formulations for cloud overlap that account for in-cloud variability (Räisänen et al. 2004; Pincus et al. 2005) and aim for insensitivity to vertical resolution. It has structural appeal because the radiative transfer routines are separated perfectly from the description of cloud variability and may be simplified by removing overlap and other complex treatments; the routines can then track the governing equation sets closely. Had McICA been available at the time of ICRCCM-III, it would have replicated the ICA benchmarks (subject to statistical noise), which for domain averages agree very well with full 3D solutions, and been an obvious path forward.

As a practical matter, McICA is also relatively easy to implement and effective. Experiments were first tried with the climate models developed by NCAR (Räisänen et al. 2005) and GFDL (Pincus et al. 2006); this number soon expanded to half a dozen (Barker et al. 2008), none of which seemed to be affected by the noisy sampling of small-scale cloud variability introduced by McICA. Medium-range weather forecasts were improved (Morcrette et al. 2008), and McICA became operational at ECMWF on 5 June 2007. McICA was introduced along with RRTMG for shortwave calculations; longwave calculations had used RRTMG since June 2000. The ECWMF implementation was built in-house along with vectorized versions of RRTMG.

McICA also was adopted as the only means of treating overlap in the versions of RRTMG supported by AER starting in August 2007 (with version 4.1 of the longwave and 3.1 of the shortwave). The AER implementation generates samples following Räisänen et al. (2004), although facilities also are available for using externally generated samples. Compared to other fast codes with fewer monochromatic points, RRTMG is especially well-suited to utilize McICA since its relatively high number of *g* points allows for good sampling of the distribution of cloud properties within each column.

At the same time, it became clear that it was too easy to introduce errors into radiation parameterizations, especially for the treatment of absorption by gases (Collins et al. 2006; Oreopoulos et al. 2012), so modeling centers were strongly motivated to adopt parameterizations like RRTMG that are routinely tested against spectrally detailed observations and reference LBL calculations. Over time RRTMG has replaced custom radiative transfer packages in many models (the NCAR CESM1, for example) and McICA has been adopted even more widely. (It is used by the Met Office HadGEM series with a completely different radiation package; see Hill et al. 2011.) The increasing uniformity of radiative transfer treatments across models does not reduce the value of multimodel ensembles, however. Model diversity in an ensemble is used to represent uncertainty, and the relatively small uncertainty in radiative transfer permits a single parameterization to be used for this physical process for all ensemble members.

4. Evaluation of GCM RT codes

While the findings and lessons of ICRCCM were major motivations for developing the concept of the ARM Program, no organized effort to establish a new radiative transfer code intercomparison that would take advantage of ARM radiative and other observations emerged for many years. This may at first seem surprising given that the Spectral Radiance Experiment (SPECTRE), an experimental field program considered in some ways a predecessor of ARM (Ellingson et al. 2016, chapter 1), was designed specifically to establish reference standards against which to compare radiative transfer models (Ellingson and Wiscombe 1996). Undoubtedly, part of the reluctance to embark on an intercomparison using ARM data was the desire for a certain level of maturity to be reached with regard to understanding instrument capabilities and the limitations of retrieval algorithms. By the time of the GEWEX Radiation Panel meeting in 2003, participants with close involvement in ARM deemed that such maturity had been reached. An additional reason that using ARM data for an RT intercomparison gained momentum was the existence and success of the Broadband Heating Rate Profile (BBHRP) effort (Mlawer et al. 2002; McFarlane et al. 2016, chapter 20) which required assembling and synthesizing multiple ARM data streams to enable production of an RRTM-based radiative flux and heating rate product. The birth of the Continual Intercomparison of Radiation Codes (CIRC; Oreopoulos and Mlawer 2010) can be traced back to those discussions. As the name implies, one of the central ideas was that the project would become the source of an evolving and regularly updated permanent reference database for evaluation of radiative transfer codes used in a variety of Earth system models.

During the initial stages of CIRC planning, it became apparent that choosing only cases with homogeneous atmospheric conditions best supported the need for CIRC cases to be well characterized and easily understood by participants. This condition also allowed the intercomparison to be inclusive of all approaches to handling the radiative effects of clouds. In addition, for ideal CIC cases, radiative closure had to be achieved for the measured radiative flux at two physical and two spectral domain boundaries, that is, the SW and LW upward irradiances at the top of the atmosphere and the downward fluxes at the surface. Another major criterion was that the set of cases would have to span a wide variety of conditions, not only with regard to the presence or absence of clouds, but also with respect to atmospheric moisture and aerosol content, surface properties, and illumination conditions. Because the single-scattering properties of ice crystals are not defined uniquely for a given effective size, cloudy cases containing ice (including those of mixed thermodynamic phase) were not considered for the first phase of CIRC.

The BBHRP data stream contained sufficient information to identify the candidate cases and further test their suitability by matching (per SPECTRE rationale) LW spectral measurements at the surface from the

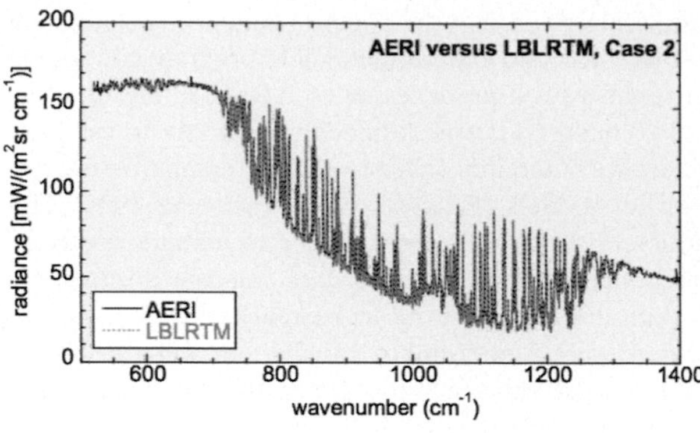

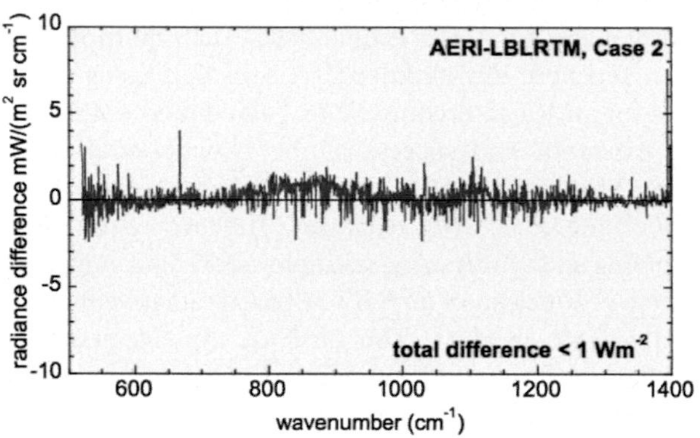

FIG. 15-8. (top) Spectral radiances for an extensive range of the radiatively important thermal spectrum as measured by AERI and calculated with LBLRTM and (bottom) their differences for CIRC baseline case 2. When converted to fluxes the differences correspond to less than 1 W m^{-2}. Such comparisons were essential for assuring the quality of atmospheric input and of the reference LBL thermal calculations of CIRC cases (from Oreopoulos and Mlawer 2010).

atmospheric emitted radiance interferometer (AERI) instrument with LBLRTM calculations (Fig. 15-8). These LBL calculations and their SW counterparts were then used as the reference calculations to evaluate the approximate models (see CIRC website http://circ.gsfc.nasa.gov), and thus the cases for the initial phase of CIRC were born almost 5 years after the idea for such a project was put forth. Six of the seven "baseline" cases were based purely on observations and supported by flux closure, while the remaining case was a 2xCO_2 extension of the driest and coldest observed case. Additional cases were generated by simplifying these baseline cases with successive removals of aerosol, cloud, and the spectral dependence of surface albedo (details can be found in Oreopoulos et al. 2012). These extra cases not only enabled testing model performance for simpler atmospheres, but also allowed assessments of cloud and aerosol forcing, and the impact of spectral surface albedo variations.

Figures 15-9 and 15-10 summarize the performance of the radiative transfer codes that participated in CIRC phase I in terms of percentage errors. Blue shades (negative values) indicate underestimates by the approximate codes. RRTM and RRTMG—models 1 and 2 in these figures, respectively—demonstrated notable accuracy in this intercomparison. An overall finding of CIRC was that approximate codes match LW reference calculations (Fig. 15-9) better than those in the SW (Fig. 15-10), echoing one of the main ICRCCM findings. Diffuse and absorbed SW fluxes are particular areas of concern. Obtaining the correct breakdown of total downward flux into direct and diffuse may be important for the simulation of surface processes such as vegetation growth and the carbon cycle in advanced Earth system models with interactive land components and should receive more attention in the future. The underestimate of SW absorption by less spectrally detailed models was recognized previously (e.g., Ackerman et al. 2003) and seems to have persisted for most approximate codes. This is consistent with a general tendency to overestimate the total downward SW flux reaching the surface, which makes obtaining a global balance between net radiative warming at the surface and cooling by turbulent fluxes problematic when modeling the surface radiation imbalance. The details of how surface albedo was averaged in very wide bands were found to be relevant for the SW TOA fluxes of the CIRC experiments, suggesting that more care should be given to proper representations of spectral albedo variations. While performance with regard to LW CO_2 forcing calculations was good overall with a couple of exceptions, estimation of SW CO_2 forcing was found to be a capability that some codes either did not have at all or was restricted to downwelling surface fluxes. For codes with full SW CO_2 forcing capabilities, performance was often quite poor, consistent with Collins et al. (2006). These and additional findings are described in greater detail in Oreopoulos et al. (2012).

So, how much progress have approximate radiation codes made since the ICRCCM era? ICRCCM had many more participating codes and its cases were conceived differently; for example, some experiments were based on synthetic atmospheres of a single absorber. ICRCCM's results therefore likely exposed different model deficiencies than CIRC did. Despite these difficulties in quantifying improvement, Oreopoulos et al. (2012) attempted a simple comparison with Ellingson et al. (1991) and Fouquart et al. (1991) and provided measures that indicated the CIRC generation of models is indeed better than those of the ICRCCM era.

Obviously, many aspects of approximate radiative transfer model performance were not addressed by CIRC. The cloudy cases, for instance, were as simple as possible since it is still challenging to produce reference LBL fluxes for complex cloud microphysics and structures. CIRC also did not explore whether the accuracy of the participating models with respect to the reference LBL calculations

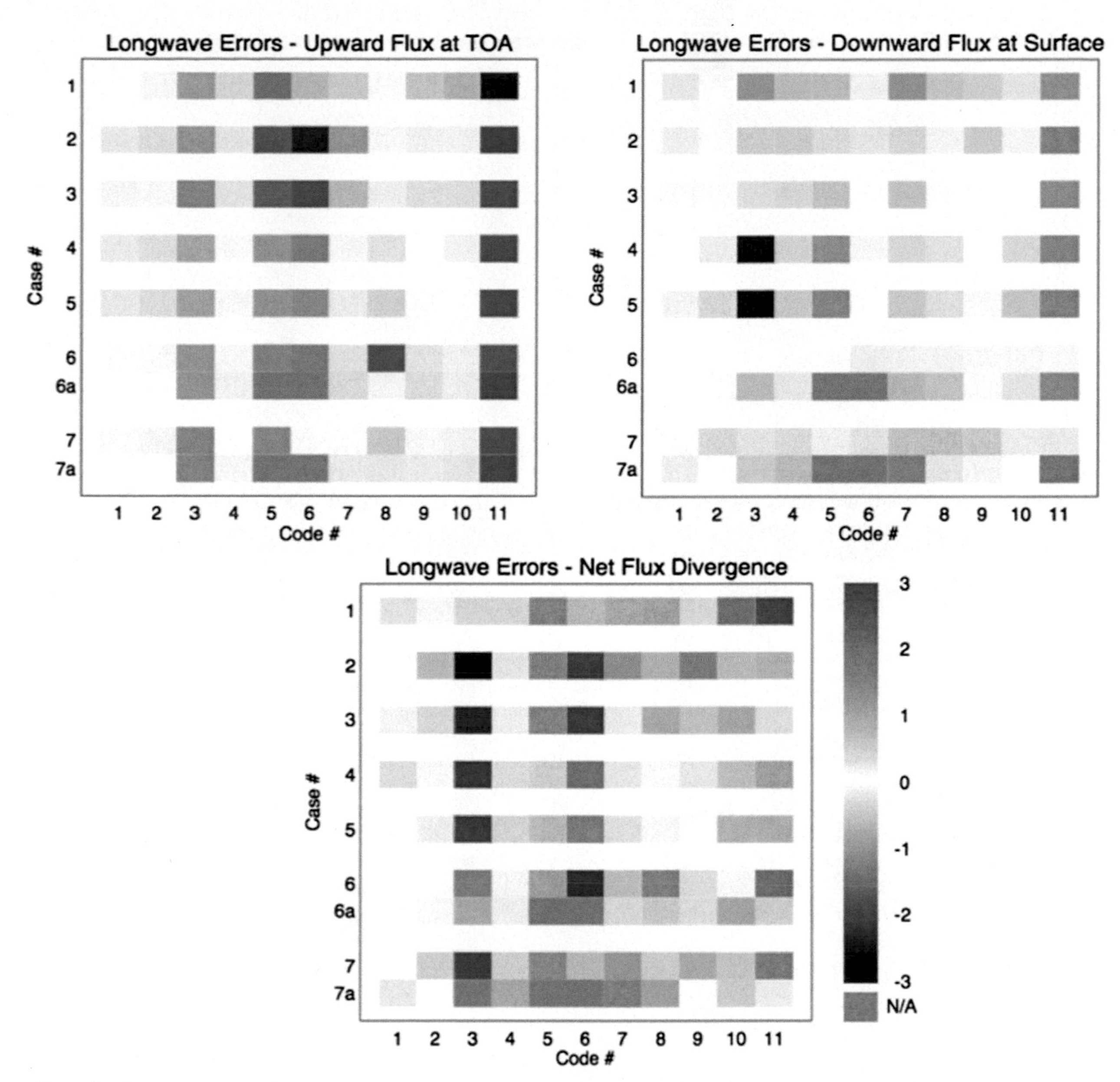

FIG. 15-9. Percentage errors for each participating model and CIRC phase I case for thermal infrared fluxes (upward flux at TOA, downward flux at the surface and atmospheric flux divergence). Gray indicates unavailable submissions (from Oreopoulos et al. 2012).

correlates (inversely) with their computational efficiency. A proper assessment of computational efficiency is feasible as a future endeavor as long as a common computational platform can be made available for performing all calculations. Undertakings such as the CIRC-tested Cloud–Aerosol–Radiation (CAR) Ensemble Modeling System (http://car.umd.edu) that assemble different radiation codes under a unified modeling infrastructure potentially can facilitate such computational speed assessments.

Regardless of what exact direction future radiative transfer code intercomparisons will take, CIRC has paved the way on how to use a variety of observations and the concepts of broadband and spectral closure to build cases, accentuate the reliability of LBL calculations, and create reference datasets that can be maintained, updated, and expanded periodically. These accomplishments would not be possible without the infrastructure and seed datasets available because of ARM. As ARM radiation measurements continue to be analyzed and their accuracy becomes better characterized, a consensus may be reached on the fundamental problem of determining acceptable levels of performance for the approximate radiation codes used in either earth model or operational flux product generation environments.

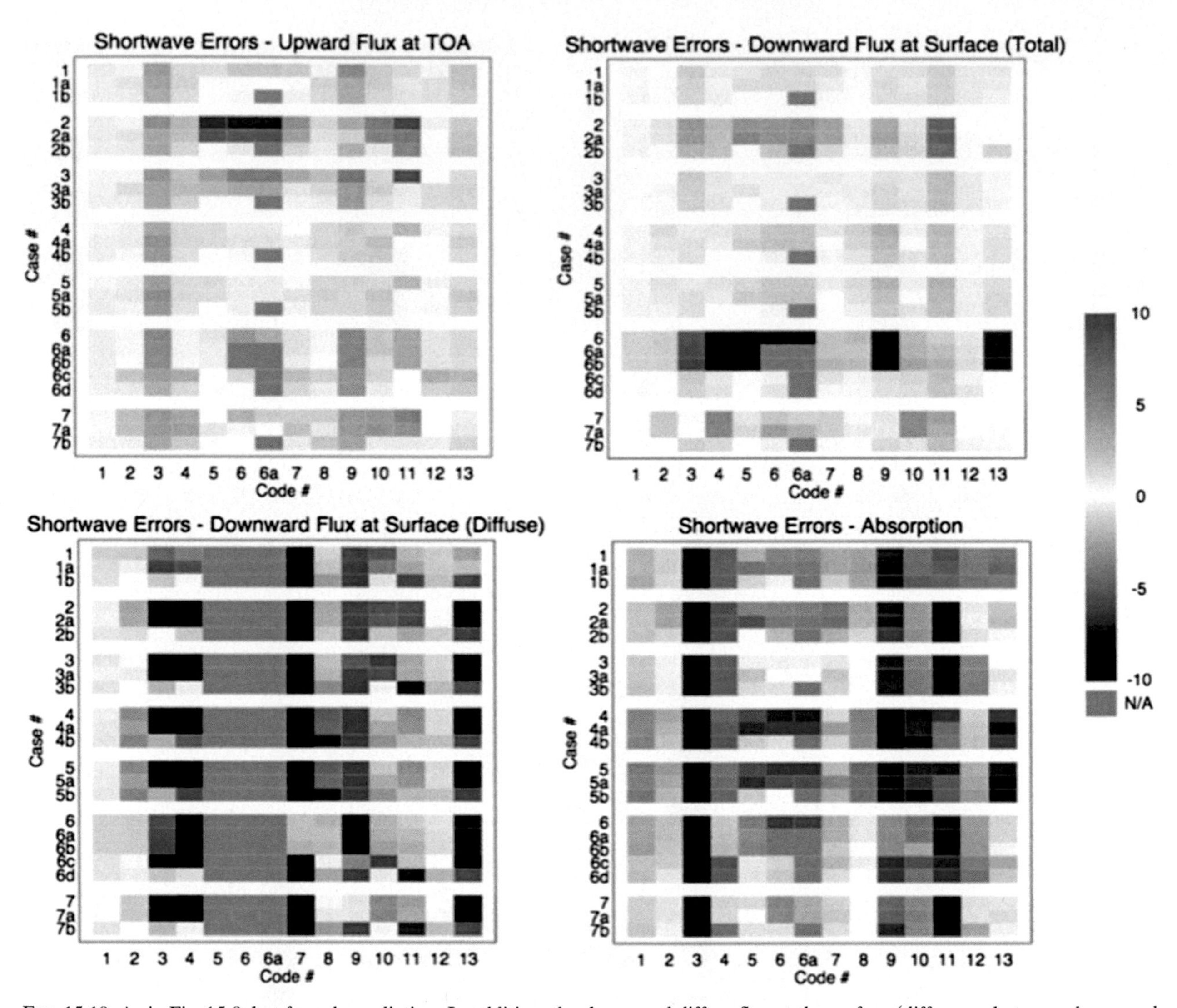

FIG. 15-10. As in Fig. 15-9, but for solar radiation. In addition, the downward diffuse flux at the surface (difference between downward total and direct solar flux) is shown (from Oreopoulos et al. 2012).

5. Conclusions

The accomplishments precipitated by the ARM Program with respect to improved radiative transfer in atmospheric models have been profound, leading the program to significantly lower its priority on radiation-related research. Ultimately, the mission of ARM was redefined in tandem with the creation of a new science-focused program (Atmospheric Systems Research; Ackerman et al. 2016, chapter 3) without "radiation" in its title. Despite this reordering of priorities, ARM's history of radiative transfer research is viewed rightfully as one of the program's greatest successes.

This reordering of the program's priorities should be seen, however, as a temporary, and perhaps premature, declaration of victory. Prior to ICRCCM, the atmospheric modeling community thought that there was a solid understanding of atmospheric radiative fluxes and heating rates, only to be proven wrong. Progress during the past decades, including the substantial accomplishments detailed in this chapter and Mlawer and Turner (2016, chapter 14), has placed our understanding on a more solid foundation, but future global models are likely to have a much higher level of sophistication. In particular, consideration of 3D radiative effects in climate and weather simulations is likely to gain prominence, especially as the spatial scale of models continues to decrease. More ambitious objectives for radiation calculations in climate and weather prediction models perhaps will reinvigorate the need for improvements to radiation parameterizations, leading to a new cycle of investment, investigation, and accomplishment.

REFERENCES

Ackerman, T. P., D. M. Flynn, and R. T. Marchand, 2003: Quantifying the magnitude of anomalous solar absorption. *J. Geophys. Res.*, **108**, 4273, doi:10.1029/2002JD002674.

——, T. S. Cress, W. R. Ferrell, J. H. Mather, and D. D. Turner, 2016: The programmatic maturation of the ARM Program. *The Atmospheric Radiation Measurement (ARM) Program: The First 20 Years, Meteor. Monogr.*, No. 57, Amer. Meteor. Soc., doi:10.1175/AMSMONOGRAPHS-D-15-0054.1.

Ahlgrimm, M., R. Forbes, J.-J. Morcrette, and R. Neggers, 2016: ARM's impact on numerical weather prediction at ECMWF. *The Atmospheric Radiation Measurement (ARM) Program: The First 20 Years, Meteor. Monogr.*, No. 57, Amer. Meteor. Soc., doi:10.1175/AMSMONOGRAPHS-D-15-0032.1.

Ambartzumian, V., 1936: The effect of the absorption lines on the radiative equilibrium of the outer layers of the stars. *Publ. Obs. Astron. Univ. Leningrad*, **6**, 7–18.

Barker, H. W., 1996: A parameterization for computing grid-averaged solar fluxes for inhomogeneous marine boundary layer clouds. 1. Methodology and homogeneous biases. *J. Atmos. Sci.*, **53**, 2289–2303, doi:10.1175/1520-0469(1996)053<2289:APFCGA>2.0.CO;2.

——, R. Pincus, and J.-J. Morcrette, 2002: The Monte Carlo Independent Column Approximation: Application within large-scale models. *Proc. GCSS-ARM Workshop on the Representation of Cloud Systems in Large-Scale Models*, Kananaskis, AB, Canada, ARM Program, 10 pp.

——, and Coauthors, 2003: Assessing 1D atmospheric solar radiative transfer models: Interpretation and handling of unresolved clouds. *J. Climate*, **16**, 2676–2699, doi:10.1175/1520-0442(2003)016<2676:ADASRT>2.0.CO;2.

——, J. N. S. Cole, J.-J. Morcrette, R. Pincus, P. Räisänen, K. von Salzen, and P. A. Vaillancourt, 2008: The Monte Carlo Independent Column Approximation: An assessment using several global atmospheric models. *Quart. J. Roy. Meteor. Soc.*, **134**, 1463–1478, doi:10.1002/qj.303.

Bryant, F. D., and P. Latimer, 1969: Optical efficiencies of large particles of arbitrary shape and orientation. *J. Colloid Interface Sci.*, **30**, 291–304, doi:10.1016/0021-9797(69)90396-8.

Buizza, R., M. Miller, and T. N. Palmer, 1999: Stochastic representation of model uncertainties in the ECMWF ensemble prediction system. *Quart. J. Roy. Meteor. Soc.*, **125**, 2887–2908, doi:10.1002/qj.49712556006.

Cahalan, R. F., W. Ridgway, W. J. Wiscombe, T. L. Bell, and J. B. Snider, 1994: The albedo of fractal stratocumulus clouds. *J. Atmos. Sci.*, **51**, 2434–2455, doi:10.1175/1520-0469(1994)051<2434:TAOFSC>2.0.CO;2.

Cairns, B., A. A. Lacis, and B. E. Carlson, 2000: Absorption within inhomogeneous clouds and its parameterization in general circulation models. *J. Atmos. Sci.*, **57**, 700–714, doi:10.1175/1520-0469(2000)057<0700:AWICAI>2.0.CO;2.

Clough, S. A., M. J. Iacono, and J.-L. Moncet, 1992: Line-by-line calculation of atmospheric fluxes and cooling rates: Application to water vapor. *J. Geophys. Res.*, **97**, 15 761–15 785, doi:10.1029/92JD01419.

Collins, W. D., 2001: Parameterization of generalized cloud overlap for radiative calculations in general circulation models. *J. Atmos. Sci.*, **58**, 3224–3242, doi:10.1175/1520-0469(2001)058<3224:POGCOF>2.0.CO;2.

——, and Coauthors, 2006: Radiative forcing by well-mixed greenhouse gases: Estimates from climate models in the Intergovernmental Panel on Climate Change (IPCC) Fourth Assessment Report (AR4). *J. Geophys. Res.*, **111**, D14317, doi:10.1029/2005JD006713.

Cotton, W. R., and Coauthors, 2003: RAMS 2001: Current status and future directions. *Meteor. Atmos. Phys.*, **82**, 5–29, doi:10.1007/s00703-001-0584-9.

Ebert, E. E., and J. A. Curry, 1992: A parameterization of ice-cloud optical-properties for climate models. *J. Geophys. Res.*, **97**, 3831–3836, doi:10.1029/91JD02472.

Ellingson, R. G., and Y. Fouquart, 1991: The intercomparison of radiation codes in climate models: An overview. *J. Geophys. Res.*, **96**, 8925–8927, doi:10.1029/90JD01618.

——, and W. J. Wiscombe, 1996: The Spectral Radiance Experiment (SPECTRE): Project description and sample results. *Bull. Amer. Meteor. Soc.*, **77**, 1967–1985, doi:10.1175/1520-0477(1996)077<1967:TSREPD>2.0.CO;2.

——, J. Ellis, and S. Fels, 1991: The intercomparison of radiation codes used in climate models: Longwave results. *J. Geophys. Res.*, **96**, 8929–8953, doi:10.1029/90JD01450.

——, R. D. Cess, and G. L. Potter, 2016: The Atmospheric Radiation Measurement Program: Prelude. *The Atmospheric Radiation Measurement (ARM) Program: The First 20 Years, Meteor. Monogr.*, No. 57, Amer. Meteor. Soc., doi:10.1175/AMSMONOGRAPHS-D-15-0029.1.

Fouquart, Y., B. Bonnel, and V. Ramaswamy, 1991: Intercomparing shortwave radiation codes for climate studies. *J. Geophys. Res.*, **96**, 8955–8968, doi:10.1029/90JD00290.

Fu, Q., 1996: An accurate parameterization of the solar radiative properties of cirrus clouds for climate models. *J. Climate*, **9**, 2058–2082, doi:10.1175/1520-0442(1996)009<2058:AAPOTS>2.0.CO;2.

——, and K. N. Liou, 1992: On the correlated k-distribution method for radiative transfer in nonhomogeneous atmospheres. *J. Atmos. Sci.*, **49**, 2139–2156, doi:10.1175/1520-0469(1992)049<2139:OTCDMF>2.0.CO;2.

——, P. Yang, and W. B. Sun, 1998: An accurate parameterization of the infrared radiative properties of cirrus clouds for climate models. *J. Climate*, **11**, 2223–2237, doi:10.1175/1520-0442(1998)011<2223:AAPOTI>2.0.CO;2.

——, B. Carlin, and G. Mace, 2000: Cirrus horizontal inhomogeneity and OLR bias. *Geophys. Res. Lett.*, **27**, 3341–3344, doi:10.1029/2000GL011944.

Garand, L., and Coauthors, 2001: Radiance and Jacobian intercomparison of radiative transfer models applied to HIRS and AMSU channels. *J. Geophys. Res.*, **106**, 24 017–24 031, doi:10.1029/2000JD000184.

Gettelman, A., and Coauthors, 2010: Global simulations of ice nucleation and ice supersaturation with an improved cloud scheme in the Community Atmosphere Model. *J. Geophys. Res.*, **115**, D18216, doi:10.1029/2009JD013797.

Golaz, J.-C., V. E. Larson, and W. R. Cotton, 2002: A PDF-based model for boundary layer clouds. Part I: Method and model description. *J. Atmos. Sci.*, **59**, 3540–3551, doi:10.1175/1520-0469(2002)059<3540:APBMFB>2.0.CO;2.

Goody, R. M., R. West, L. Chen, and D. Crisp, 1989: The correlated k-distribution method for radiation calculation in nonhomogeneous atmospheres. *J. Quant. Spectrosc. Radiat. Transfer*, **42**, 539–550, doi:10.1016/0022-4073(89)90044-7.

Grabowski, W. W., X. Wu, M. W. Moncrieff, and W. D. Hall, 1998: Cloud-resolving modeling of tropical cloud systems during Phase III of GATE. Part II: Effects of resolution and the third spatial dimension. *J. Atmos. Sci.*, **55**, 3264–3282, doi:10.1175/1520-0469(1998)055<3264:CRMOCS>2.0.CO;2.

Harrington, J. Y., and P. Q. Olsson, 2001: A method for the parameterization of cloud optical properties in bulk and bin microphysical models. Implications for Arctic cloudy boundary layers. *Atmos. Res.*, **57**, 51–80, doi:10.1016/S0169-8095(00)00068-5.

Hill, P. G., J. Manners, and J. C. Petch, 2011: Reducing noise associated with the Monte Carlo Independent Column Approximation for weather forecasting models. *Quart. J. Roy. Meteor. Soc.*, **137**, 219–228, doi:10.1002/qj.732.

Hogan, R. J., and A. J. Illingworth, 2000: Deriving cloud overlap statistics from radar. *Quart. J. Roy. Meteor. Soc.*, **126**, 2903–2909, doi:10.1002/qj.49712656914.

Hu, Y. X., and K. Stamnes, 1993: An accurate parameterization of the radiative properties of water clouds suitable for use in climate models. *J. Climate*, **6**, 728–742, doi:10.1175/1520-0442(1993)006<0728:AAPOTR>2.0.CO;2.

Iacono, M. J., E. J. Mlawer, S. A. Clough, and J.-J. Morcrette, 2000: Impact of an improved longwave radiation model, RRTM, on the energy budget and thermodynamic properties of the NCAR community climate model, CCM3. *J. Geophys. Res.*, **105**, 14 873–14 890, doi:10.1029/2000JD900091.

——, J. S. Delamere, E. J. Mlawer, and S. A. Clough, 2003: Evaluation of upper tropospheric water vapor in the NCAR Community Climate Model, CCM3, using modeled and observed HIRS radiances. *J. Geophys. Res.*, **108**, 4037, doi:10.1029/2002JD002539.

——, ——, ——, M. W. Shephard, S. A. Clough, and W. D. Collins, 2008: Radiative forcing by long-lived greenhouse gases: Calculations with the AER radiative transfer models. *J. Geophys. Res.*, **113**, D13103, doi:10.1029/2008JD009944.

Kablick, G. P., R. G. Ellingson, E. E. Takara, and J. Gu, 2011: Longwave 3D benchmarks for inhomogeneous clouds and comparisons with approximate methods. *J. Climate*, **24**, 2192–2205, doi:10.1175/2010JCLI3752.1.

Klein, S. A., and C. Jakob, 1999: Validation and sensitivities of frontal clouds simulated by the ECMWF model. *Mon. Wea. Rev.*, **127**, 2514–2531, doi:10.1175/1520-0493(1999)127<2514:VASOFC>2.0.CO;2.

Lacis, A. A., and V. Oinas, 1991: A description of the correlated k-distribution method for modeling nongray gaseous absorption, thermal emission, and multiple scattering in vertically inhomogeneous atmospheres. *J. Geophys. Res.*, **96**, 9027–9074, doi:10.1029/90JD01945.

Liu, H.-C., P. K. Wang, and R. Schlesinger, 2003: A numerical study of cirrus clouds. Part I: Model description. *J. Atmos. Sci.*, **60**, 1075–1084, doi:10.1175/1520-0469(2003)60<1075:ANSOCC>2.0.CO;2.

McFarlane, S. A., J. H. Mather, and E. J. Mlawer, 2016: ARM's progress on improving atmospheric broadband radiative fluxes and heating rates. *The Atmospheric Radiation Measurement (ARM) Program: The First 20 Years, Meteor. Monogr.*, No. 57, Amer. Meteor. Soc., doi:10.1175/AMSMONOGRAPHS-D-15-0046.1.

Mitchell, D. L., 2000: Parameterization of the Mie extinction and absorption coefficients for water clouds. *J. Atmos. Sci.*, **57**, 1311–1326, doi:10.1175/1520-0469(2000)057<1311:POTMEA>2.0.CO;2.

——, 2002: Effective diameter in radiation transfer: General definition, applications and limitations. *J. Atmos. Sci.*, **59**, 2330–2346, doi:10.1175/1520-0469(2002)059<2330:EDIRTG>2.0.CO;2.

——, and W. P. Arnott, 1994: A model predicting the evolution of ice particle size spectra and the radiative properties of cirrus clouds. Part II: Dependence of absorption and extinction on ice crystal morphology. *J. Atmos. Sci.*, **51**, 817–832, doi:10.1175/1520-0469(1994)051<0817:AMPTEO>2.0.CO;2.

——, A. Macke, and Y. Liu, 1996: Modeling cirrus clouds. Part II: Treatment of radiative properties. *J. Atmos. Sci.*, **53**, 2967–2988, doi:10.1175/1520-0469(1996)053<2967:MCCPIT>2.0.CO;2.

——, W. P. Arnott, C. Schmitt, A. J. Baran, S. Havemann, and Q. Fu, 2001: Photon tunneling contributions to extinction for laboratory grown hexagonal columns. *J. Quant. Spectrosc. Radiat. Transfer*, **70**, 761–776, doi:10.1016/S0022-4073(01)00044-9.

——, A. J. Baran, W. P. Arnott, and C. Schmitt, 2006: Testing and comparing the modified anomalous diffraction approximation. *J. Atmos. Sci.*, **63**, 2948–2962, doi:10.1175/JAS3775.1.

——, R. P. Lawson, and B. Baker, 2011: Understanding effective diameter and its application to terrestrial radiation in ice clouds. *Atmos. Chem. Phys.*, **11**, 3417–3429, doi:10.5194/acp-11-3417-2011.

Mlawer, E. J., and S. A. Clough, 1998: Shortwave and longwave enhancements in the rapid radiative transfer model. *Proc. Seventh Atmospheric Radiation Measurement (ARM) Science Team Meeting*, San Antonio, TX, U.S. Dept. of Energy, 409–413. [Available online at https://www.arm.gov/publications/proceedings/conf07/extended_abs/mlawer_ej.pdf.]

——, and D. D. Turner, 2016: Spectral radiation measurements and analysis in the ARM Program. *The Atmospheric Radiation Measurement (ARM) Program: The First 20 Years, Meteor. Monogr.*, No. 57, Amer. Meteor. Soc., doi:10.1175/AMSMONOGRAPHS-D-15-0027.1.

——, S. J. Taubman, P. D. Brown, M. J. Iacono, and S. A. Clough, 1997: Radiative transfer for inhomogeneous atmospheres: RRTM, a validated correlated-*k* model for the longwave. *J. Geophys. Res.*, **102**, 16 663–16 682, doi:10.1029/97JD00237.

——, and Coauthors, 2002: The broadband heating rate profile (BBHRP) VAP. *Proc. 12th Atmospheric Radiation Measurement (ARM) Science Team Meeting*, St. Petersburg, FL, U.S. Dept. of Energy, 12 pp. [Available online at https://www.arm.gov/publications/proceedings/conf12/extended_abs/mlawer-ej.pdf.]

Molineaux, B., P. Ineichen, and N. O'Neill, 1998: Equivalence of pyrheliometric and monochromatic aerosol optical depths at a single key wavelength. *Appl. Opt.*, **37**, 7008–7018, doi:10.1364/AO.37.007008.

Morcrette, J.-J., E. J. Mlawer, M. J. Iacono, and S. A. Clough, 2001: Impact of the radiation-transfer scheme RRTM in the ECMWF forecast system. Technical report in the *ECMWF Newsletter*, No. 91, ECMWF, Reading, United Kingdom, 2–9. [Available online at http://www.ecmwf.int/sites/default/files/elibrary/2001/14633-newsletter-no91-summer-2001.pdf.]

——, H. W. Barker, J. N. S. Cole, M. J. Iacono, and R. Pincus, 2008: Impact of a new radiation package, McRad, in the ECMWF Integrated Forecasting System. *Mon. Wea. Rev.*, **136**, 4773–4798, doi:10.1175/2008MWR2363.1.

Neale, R. B., and Coauthors, 2010: Description of the NCAR Community Atmosphere Model (CAM5.0). NCAR Tech. Note NCAR/TN-486+STR, 268 pp. [Available online at www.cesm.ucar.edu/models/cesm1.1/cam/docs/description/cam5_desc.pdf.]

Oreopoulos, L., and H. W. Barker, 1999: Accounting for subgrid-scale cloud variability in a multi-layer 1D solar radiative transfer algorithm. *Quart. J. Roy. Meteor. Soc.*, **125**, 301–330, doi:10.1002/qj.49712555316.

——, and E. Mlawer, 2010: The continual intercomparison of radiation codes (CIRC). *Bull. Amer. Meteor. Soc.*, **91**, 305–310, doi:10.1175/2009BAMS2732.1.

——, and Coauthors, 2012: The continual intercomparison of radiation codes: Results from Phase I. *J. Geophys. Res.*, **117**, D06118, doi:10.1029/2011JD016821.

Pincus, R., and S. A. Klein, 2000: Unresolved spatial variability and microphysical process rates in large-scale models. *J. Geophys. Res.*, **105**, 27 059–27 065, doi:10.1029/2000JD900504.

——, H. W. Barker, and J.-J. Morcrette, 2003: A fast, flexible, approximate technique for computing radiative transfer in inhomogeneous cloud fields. *J. Geophys. Res.*, **108**, 4376, doi:10.1029/2002JD003322.

——, C. Hannay, S. A. Klein, K. M. Xu, and R. S. Hemler, 2005: Overlap assumptions for assumed probability distribution function cloud schemes in large-scale models. *J. Geophys. Res.*, **110**, D15S09, doi:10.1029/2004JD005100.

——, R. S. Hemler, and S. A. Klein, 2006: Using stochastically generated subcolumns to represent cloud structure in a large-scale model. *Mon. Wea. Rev.*, **134**, 3644–3656, doi:10.1175/MWR3257.1.

Räisänen, P., H. W. Barker, M. F. Khairoutdinov, J. Li, and D. A. Randall, 2004: Stochastic generation of subgrid-scale cloudy columns for large-scale models. *Quart. J. Roy. Meteor. Soc.*, **130**, 2047–2067, doi:10.1256/qj.03.99.

——, ——, and J. N. S. Cole, 2005: The Monte Carlo Independent Column Approximation's conditional random noise: Impact on simulated climate. *J. Climate*, **18**, 4715–4730, doi:10.1175/JCLI3556.1.

Saha, S., and Coauthors, 2006: The NCEP Climate Forecast System. *J. Climate*, **19**, 3483–3517, doi:10.1175/JCLI3812.1.

——, and Coauthors, 2010: The NCEP Climate Forecast System Reanalysis. *Bull. Amer. Meteor. Soc.*, **91**, 1015–1057, doi:10.1175/2010BAMS3001.1.

——, and Coauthors, 2014: The NCEP Climate Forecast System Version 2. *J. Climate*, **27**, 2185–2208, doi:10.1175/JCLI-D-12-00823.1.

Shupe, M. D., J. M. Comstock, D. D. Turner, and G. G. Mace, 2016: Cloud property retrievals in the ARM Program. *The Atmospheric Radiation Measurement (ARM) Program: The First 20 Years, Meteor. Monogr.*, No. 57, Amer. Meteor. Soc., doi:10.1175/AMSMONOGRAPHS-D-15-0030.1.

Skamarock, W. C., and Coauthors, 2008: A description of the Advanced Research WRF version 3. NCAR Tech. Note NCAR/TN-475+STR, 113 pp., doi:10.5065/D68S4MVH.

Stackhouse, P. W., Jr., and G. L. Stephens, 1991: A theoretical and observational study of the radiative properties of cirrus: Results from FIRE 1986. *J. Atmos. Sci.*, **48**, 2044–2059, doi:10.1175/1520-0469(1991)048<2044:ATAOSO>2.0.CO;2.

Stamnes, K., S. C. Tsay, W. Wiscombe, and K. Jayaweera, 1988: Numerically stable algorithm for discrete-ordinate-method radiative transfer in multiple scattering and emitting layered media. *Appl. Opt.*, **27**, 2502–2509, doi:10.1364/AO.27.002502.

Stephens, G. L., 1988: Radiative transfer through arbitrarily shaped optical media. Part I: A general method of solution. *J. Atmos. Sci.*, **45**, 1818–1836, doi:10.1175/1520-0469(1988)045<1818:RTTASO>2.0.CO;2.

Stokes, G. E., and S. E. Schwartz, 1994: The Atmospheric Radiation Measurement (ARM) Program: Programmatic background and design of the cloud and radiation test bed. *Bull. Amer. Meteor. Soc.*, **75**, 1201–1221, doi:10.1175/1520-0477(1994)075<1201:TARMPP>2.0.CO;2.

Tiedtke, M., 1996: An extension of cloud-radiation parameterization in the ECMWF model: The representation of subgrid-scale variations of optical depth. *Mon. Wea. Rev.*, **124**, 745–750, doi:10.1175/1520-0493(1996)124<0745:AEOCRP>2.0.CO;2.

Tompkins, A. M., 2002: A prognostic parameterization for the subgrid-scale variability of water vapor and clouds in large-scale models and its use to diagnose cloud cover. *J. Atmos. Sci.*, **59**, 1917–1942, doi:10.1175/1520-0469(2002)059<1917:APPFTS>2.0.CO;2.

Uppala, S. M., and Coauthors, 2005: The ERA-40 Re-Analysis. *Quart. J. Roy. Meteor. Soc.*, **131**, 2961–3012, doi:10.1256/qj.04.176.

van de Hulst, H. C., 1981: *Light Scattering by Small Particles*. Dover, 470 pp.

Yang, P., B.-C. Gao, B. A. Baum, Y. X. Hu, W. J. Wiscombe, S.-C. Tsay, D. M. Winker, and S. L. Nasiri, 2001: Radiative properties of cirrus clouds in the infrared (8–13 μm) spectral region. *J. Quant. Spectrosc. Radiat. Transfer*, **70**, 473–504, doi:10.1016/S0022-4073(01)00024-3.

Yu, W., M. Doutriaux, G. Seze, H. Treut, and M. Desbois, 1996: A methodology study of the validation of clouds in GCMs using ISCCP satellite observations. *Climate Dyn.*, **12**, 389–401, doi:10.1007/BF00211685.

CHAPTER 16

ARM Solar and Infrared Broadband and Filter Radiometry

JOSEPH J. MICHALSKY

NOAA Earth System Research Laboratory, Boulder, Colorado

CHARLES N. LONG

Pacific Northwest National Laboratory, Richland, Washington

1. Introduction

Two papers published in the early 1990s comparing radiation transfer codes for the infrared (Ellingson et al. 1991) and for the solar (Fouquart et al. 1991) irradiance concluded that many of the radiation transfer codes (parameterized to reduce run time) used in climate models did not agree with state-of-the-art line-by-line radiative transfer codes; for the most part line-by-line codes agreed with one another. However, the measurements to confirm that the radiative fluxes produced by these line-by-line codes represented truth were unavailable. The Spectral Radiation Experiment (SPECTRE), a 25-day experiment in the fall of 1991, (Ellingson and Wiscombe 1996) was conducted near Coffeyville, Kansas, to simultaneously obtain surface radiation measurements and the most important of the inputs needed for these radiative transfer models, including temperature, humidity, aerosol, and cloud profiles. The ARM Program greatly expanded this initial effort to include a range of climates and to acquire at least 10 years of measurements with a focus on improving the number and the quality of the measured inputs needed for the models and improving the quality of the radiation measurements and the radiative transfer models.

The ARM Program and Baseline Surface Radiation Network (BSRN; Ohmura et al. 1998) matured together. Ellsworth Dutton, who played a significant role in the ARM Program, also served as the BSRN project manager for BSRN's first 20 years. ARM was focused on all aspects of trying to close the problem between radiative transfer models and radiation measurements using measured model inputs. BSRN was focused primarily on providing state-of-the-art broadband solar and infrared measurements at sites throughout the world that represented every type of climate. The BSRN data were intended for satellite and climate model validation and for the detection of long-term trends. Consequently, both programs benefited from each other's research efforts to improve solar and infrared radiometry. In 1990, estimates of the standard uncertainty for global horizontal, direct normal, diffuse horizontal solar irradiance, and downwelling global infrared irradiance were 15, 3, 10, and 30 $\mathrm{W\,m^{-2}}$, respectively. The goal of BSRN was to reduce these broadband measurement uncertainties to 5, 2, 5, and 20 $\mathrm{W\,m^{-2}}$ (Ohmura et al. 1998). Most of these goals have been met, and in some cases exceeded, through efforts in both ARM and BSRN. The radiation measurements of ARM adhere to the strict specifications of the BSRN, and five ARM sites report their data to the BSRN archive making up 10% of the total number of BSRN sites.

This chapter describes the history of ARM's ground-based measurements of broadband solar and broadband infrared radiation first, followed by a description of ARM's spectral solar measurements that are made with interference filter radiometers. Spectral measurements made using spectrometers are discussed in Mlawer and Turner (2016, chapter 14).

Measurements of the broadband (i.e., the spectrally integrated solar spectrum or shortwave; 280–4000 nm) that come directly from the sun without being absorbed or scattered are measurements of direct normal irradiance

Corresponding author address: Joseph J. Michalsky, NOAA/ University of Colorado Cooperative Institute for Research in Environmental Sciences, 325 Broadway, Boulder, CO 80305.
E-mail: joseph.michalsky@noaa.gov

DOI: 10.1175/AMSMONOGRAPHS-D-15-0031.1

(DNI) and are made with a pyrheliometer mounted on a solar tracker to follow the sun. Measurements of skylight [diffuse horizontal irradiance (DHI)] arising from sunlight scattered by molecules, aerosols, and clouds are made with a pyranometer that has the DNI blocked by a shadow ball or disk mounted on a solar tracker. Unshaded pyranometers are mounted on a separate stand to measure all downwelling [global horizontal irradiance (GHI)] or all upwelling (reflected) solar radiation.

Broadband infrared (spectrally integrated from about 4000 to 50000 nm, or longwave) radiation from the sky is measured with a pyrgeometer that has a hemispheric field of view and an (approximate) cosine response. Broadband infrared emitted by the surface is measured with a horizontally mounted down-facing pyrgeometer. Only broadband infrared measurements are covered in this chapter; high-resolution spectral infrared measurements are discussed in Mlawer and Turner (2016, chapter 14).

Shortwave spectral measurements are made with interference-filter radiometers that cover selected portions of the solar spectrum where silicon-based detectors are responsive (300–1100 nm). One exception to this is the ARM Aeronet Robotic Network (AERONET) Cimel sunphotometers that have one channel at 1640 nm that uses an indium gallium arsenide (InGaAs) detector. The ARM Program primarily uses these narrowband spectral measurements to determine the spectral dependence of aerosol optical depth, cloud optical depth, and surface spectral albedo.

These broadband solar and infrared upwelling and downwelling irradiance measurements are made at the ARM's central facilities and at extended sites. Downwelling narrowband spectral measurements are made at all of the central and extended facilities. At the Southern Great Plains (SGP) central facility only, upwelling measurements are made with multifilter rotating shadowband radiometer (MFRSR) heads, referred to as the multifilter radiometer (MFR), in order to measure spectral albedo. At the SGP there were 23 extended facilities covering 142 000 km^2 until 2009 when the areal coverage shrank to 22 500 km^2 and 16 sites. At the North Slope of Alaska (NSA) site, there was the central facility near Barrow and one extended facility near Atqasuk. The Tropical Western Pacific (TWP) site had only central facilities on Nauru Island; in Manus, Papua New Guinea; and in Darwin, Australia. The ARM Mobile Facilities (AMFs) spend months to a year at selected sites. AMFs have the same radiation measurement suite as found at the central facilities of the fixed ARM sites.

NSA and AMF deployments in cold regions present special problems for radiometry. Ice buildup on instruments is the primary issue. A 2-yr intensive operational period (IOP), led by Scott Richardson, was conducted beginning in 2007 and ending in 2009 to study different techniques to heat and ventilate radiometers to keep the ice from forming yet minimizing the effects on the integrity of the measurements. Although improvements have been made, Arctic radiometry continues to be a challenge.

This chapter includes a discussion of specific broadband instruments used for measurements, their calibrations, and changes to the instrumentation including the data logging. Uncertainties in the field measurements are covered. A few of the major results using these measurements will be highlighted. Spectral measurements are discussed following the broadband sections including the instruments used, uncertainties, and a few significant results based on these measurements. Albedo measurements are covered only briefly since these calculated quantities are only recently beginning to be used. Note, some of the topics discussed in this chapter are also covered, in more detail or with a different emphasis, by Mlawer and Turner (2016, chapter 14), McFarlane et al. (2016, chapter 20), and McComiskey and Ferrare (2016, chapter 21).

2. Broadband solar measurements

Up-looking pyranometers are mounted about 2 m above the surface and their domes are ventilated with ambient air to reduce dust and dew buildup. The broadband solar that is reflected by the surface is measured with a down-facing, horizontally mounted pyranometer that is not ventilated, and albedo is determined by taking the ratio of this measurement to the downwelling GHI. The most accurate GHI is obtained, not from the unshaded pyranometer, but from summing DNI and DHI components:

$$\mathrm{GHI} = \mathrm{DNI} \times \cos(\mathrm{SZA}) + \mathrm{DHI}, \qquad (16\text{-}1)$$

where solar zenith angle (SZA) is the angle between the zenith and the sun's direction. The unshaded pyranometer measures global irradiance; however, this measurement is less accurate and used mainly to compare to the component sum in Eq. (16-1) for the purpose of detecting tracker failure or sensor blockage. The down-facing pyranometer is mounted near the top of a 10-m tower. The surface under the tower is chosen to be uniform in the near vicinity (out to about 4 times the height of the sensor). The solar spectrum includes all radiation between 280 and 4000 nm. Pyrheliometers use a glass window that includes this range, but pyranometer measurements often respond over a somewhat smaller range (typically, 300–3000 nm). There are small solar contributions from wavelengths that are shorter than 300 nm and longer than 3000 nm. However, since pyranometers are calibrated by comparing to a full-spectrum instrument, their response is very nearly proportional to the full solar spectrum. Pyranometers measure hemispheric radiation with a sensor that has an (approximate) cosine response, and pyrheliometers in ARM measure direct beam radiation with a field of view of 5.7° that necessarily includes some diffuse radiation in the sun's aureole.

Measurements of global (i.e., total) horizontal solar radiation are made with the Eppley model precision spectral pyranometer (PSP). PSPs are designated first-class instruments according to World Meteorological Organization (WMO) specifications (Coulson and Howell 1980). ARM measurements of diffuse radiation used Eppley PSPs until it was (re)discovered (Gulbrandsen 1978) that there is a significant offset with this instrument (Bush et al. 2000; Dutton et al. 2001). In 2001 all of the Eppley model PSPs were replaced with the Eppley model 8–48s for DHI measurements at all ARM sites worldwide. The 8–48 (i.e., black and white) pyranometer has the desirable property that its offset is nearly zero (Michalsky et al. 2003). The Eppley PSP is now only used for the GHI backup measurement to the BSRN-preferred component summation. The preference for summation of observations from two radiometers [Eq. (16-1)] versus a single pyranometer measurement is explained in chapters 5 and 6 of Vignola et al. (2012). In essence, the uncertainties are reduced by at least a factor of 50% when using the component sum for global irradiance even compared to GHI measurements obtained with top-of-the-line pyranometers [see chapter 6 in Vignola et al. (2012)]. The current estimates of 95% uncertainties of GHI are about ±3% under very good conditions; therefore, a reasonable estimate of uncertainty for routine field operations is higher at ±4%. The DHI is measured with a current 95% uncertainty of about ±2% for very good conditions and ±3% under field conditions.

The Eppley model normal incidence pyrheliometer (NIP) is used to measure broadband direct normal solar irradiance in ARM. This instrument also received a first-class designation from the WMO (Coulson and Howell 1980). This instrument is mounted on a solar tracker (Kipp and Zonen 2AP; http://www.kippzonen.com/?product/2141/2AP.aspx) to follow the sun to better than 1°. This same tracker is used to shade the pyranometer used for DHI. In a recent paper Michalsky et al. (2011) quantified the uncertainty of measurements made with several pyrheliometers including the Eppley NIP used in the ARM Program. The comparison was noteworthy in that the measurements were made in typical midlatitude field conditions in all seasons. The results indicated that the Eppley NIP has a 95% uncertainty of ±1.3%, which is actually better than the 2% uncertainty suggested by the manufacturer. However, three commercial pyrheliometers in this study had 95% uncertainties of around 0.7%.

The ARM Data Archive (www.archive.arm.gov) contains broadband solar radiation data that were taken using a datalogger that sampled once every 20 s at the SGP sites until 1997. New Campbell loggers were exchanged for these early ones to allow 2-s sampling with 60-s averaging beginning in 1997. In 2012 a newer version of the loggers permitted a step up to 1-s sampling with 60-s averaging at the SGP sites. All TWP, NSA, and ARM Mobile Facility sites used loggers with 1-s sampling and 60-s averaging from the outset since they sampled fewer signals than the radiation facilities at the SGP. As stated above, the diffuse measurements were made using the Eppley 8–48 after 2001; the PSP measurements of diffuse and global irradiance that appear in the archive have been offset corrected to the beginning of the datasets. For the best set of measurements with all corrections implemented, the archive data streams that include "qcrad" in the data stream title should be retrieved; the title "sgpqcradbrs1longC1.s1," for example, contains the last few years of quality-controlled data up to the current measurements.

3. Broadband solar calibrations

All solar measurements in ARM are ultimately tied to the World Radiometric Reference (WRR) that is maintained at the Physikalisch-Meteorologisches Observatorium Davos and World Radiation Center (http://www.pmodwrc.ch). The WRR for DNI consists of several absolute cavity radiometers that have been carefully characterized (Fröhlich et al. 1995); currently, it includes six radiometers from five different manufacturers. Operationally, an absorbing cavity within the radiometer is heated when the sun impinges on the cavity through an aperture with a precisely defined area. A shutter is closed and then the cavity is heated by an electrical current to the same temperature it had with the sun shining into the cavity. The solar irradiance is then calculated as the electrical power to heat the cavity in watts divided by the area of the aperture with small corrections applied, based on the characterization of the cavity radiometer.

Every five years ARM cavity radiometers, which serve as secondary standards for ARM shortwave measurements, travel to Davos, Switzerland, and are calibrated against the WRR at the International Pyrheliometer Comparison (IPC). On the off years, the National Renewable Energy Laboratory (NREL) in Golden, Colorado, hosts a comparison [NREL Pyrheliometer Comparison (NPC)] using several cavity radiometers that are calibrated at the quinquennial IPCs in Davos. Cavity radiometers are not used in ARM because they are more than 10 times the cost of thermopile pyrheliometers and

FIG. 16-1. The Radiometric Calibration Facility at the ARM Southern Great Plains site near Billings, Oklahoma; close-up of pyranometers being calibrated at the facility.

operate with an open cavity, which makes it impractical to use for continuous unattended measurements because dust and other debris may get into the cavity and contaminate the measurements.

The diffuse reference consists of three Eppley model 8–48 pyranometers that are calibrated against the WRR-traceable cavity radiometers using a procedure known as the shade/unshade technique. In this method, a horizontally mounted pyranometer under clear and stable irradiance conditions is shaded and then unshaded from the direct sun using a blocker of the same angular diameter as the angular diameter of a cavity radiometer (5°). The difference in voltages from the pyranometer for the two configurations is equated to the cavity-measured direct irradiance after multiplication by the cosine (SZA). These measurements are repeated at several SZAs, but the average value near the SZA of 45° is typically used for the single-number calibration factor for the pyranometer. The average of three 8–48s is used for the diffuse horizontal standard irradiance.

The ARM Radiometric Calibration Facility (RCF; see Fig. 16-1) was constructed to calibrate the many pyranometers and pyrheliometers used at the permanent and mobile ARM facilities. These radiometers are calibrated en masse in Broadband Outdoor Radiation Calibrations (BORCALs; Myers et al. 2002) performed at the ARM SGP site by comparing pyrheliometer voltage outputs with the ARM cavity radiometer irradiances and by comparing pyranometer voltage outputs with the component sum irradiance, that is,

$$\text{global irradiance reference} = \text{cavity} \times \cos(\text{SZA}) + \text{reference diffuse}. \qquad (16\text{-}2)$$

Measurements are made under clear, stable conditions with the DNI over 700 W m^{-2}, but the single calibration factor is based on measurements near 45°SZA. All broadband calibrations for ARM are performed at the SGP RCF and instruments are shipped to the other ARM sites for deployment. In fact, the RCF is a true user facility serving as the site for the first International Pyrgeometer and Absolute Sky-scanning Radiometer Comparison (Philipona et al. 2001) and the site of three diffuse irradiance IOPs (Michalsky et al. 2003, 2005, 2007).

4. Broadband infrared measurements

Downwelling infrared radiation from the sky and upwelling infrared from the surface is measured using Eppley model precision infrared radiometer (PIR) pyrgeometers oriented horizontally. The down-looking pyrgeometer is mounted at about the 10-m level. Up-looking pyrgeometers are mounted on solar trackers at 2 m above the ground and are shaded from direct solar radiation to minimize solar contributions to the signals that may arise from pinholes in the silicon dome covering these instruments or from shortwave leakage of the interference filter covering the PIR dome. The dome of the up-looking PIR is ventilated with ambient air.

5. Broadband infrared calibrations

The current calibrations used in ARM for the pyrgeometers are those provided by the manufacturer Eppley Laboratory, Inc. A pyrgeometer produces three electrical signals: a thermopile signal, a dome temperature signal, and a case temperature signal. The data are converted to infrared irradiance I_{ir} using

$$I_{\text{ir}} = V_{\text{thermo}}/C + \sigma T_B^4 - k\sigma(T_D^4 - T_B^4), \qquad (16\text{-}3)$$

where σ is the Stefan–Boltzmann constant, V_{thermo} is the thermopile voltage, T_B and T_D are temperatures (K) of the case (or body) and dome, respectively, and C and k are calibration constants. Eppley's calibration procedure

PIRs – AERI irradiance vs. surface temperature (2000)

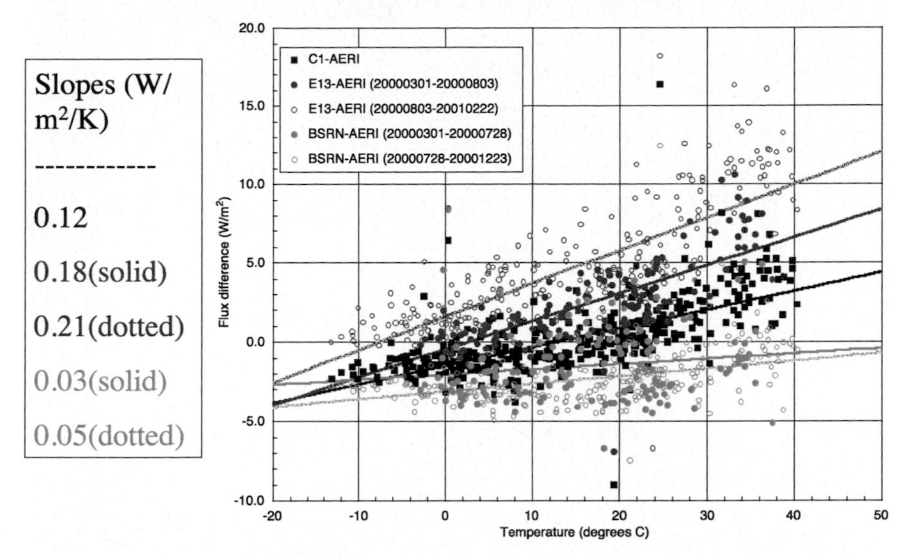

FIG. 16-2. Collocated AERI-based estimated LW irradiance subtracted from PIR-based measured irradiances for three PIRs designated C1, E13 (two time periods), and BSRN (two time periods) as a function of ambient temperature. Slopes are to the bottom left of the figure.

is outlined in chapter 10 of Vignola et al. (2012). The pyrgeometer peers into a blackbody whose temperature is controlled by a circulating water bath first at 5°C and then at 15°C. In about five minutes after being positioned to look into the blackbody, the instrument and blackbody come to equilibrium and T_B and T_D are at the same temperature. Therefore,

$$C = \frac{V_{\text{thermo}}}{\sigma(T_{\text{BB}}^4 - T_B^4)}, \tag{16-4}$$

where T_{BB} is the blackbody temperature at 5°C and subsequently 15°C. The calibration constant is based on the average of its value at the two temperature settings of the blackbody. The constant k in Eq. (16-2) is assigned a value of 4 by Eppley.

Using a constant value of 4 for k may not be appropriate since that value was assigned for KRS-5 domes that are no longer used for the Eppley PIR; all PIR domes in ARM are silicon, and the manufacturer (via private communication) now suggests that 3.5 may be a more appropriate constant for k. Other blackbody calibration methods solve for the value of k [e.g., see chapter 10 of Vignola et al. (2012)]. In one study, for example, Philipona et al. (2001) found a range for k of 2.75 to 3.64 for a set of six PIRs that were calibrated at the World Radiation Center in Davos, Switzerland (www.pmodwrc.ch).

Figure 16-2 above is a plot of an Atmospheric Emitted Radiance Interferometer (AERI)-based broadband irradiance estimate subtracted from PIR measurements of downwelling irradiance. The AERI (Knuteson et al. 2004; Turner et al. 2016, chapter 13) measures spectral radiance that is converted to infrared irradiances by filling in missing portions of the spectrum (minor corrections) and estimating the angular dependence of the radiance (Marty et al. 2003). While there are uncertainties in the AERI-based irradiance estimates, and this method cannot to be considered a standard, the fact that there are differences in slopes in Fig. 16-2 for the three independent PIR measurements collocated with the AERI points to an issue with pyrgeometer calibration since they are not producing the same irradiances on average. We suspect that the issue is the assignment of a fixed value for k in Eq. (16-3). This suspicion is based on our ability to reproduce the value of the other calibration constant c with great reliability, and based on our and others' experience that PIRs calibrated in other black bodies often have k values different from the fixed value of 4 used for the ARM PIRs, as indicated in the previous paragraph.

ARM's attempt to perform its own blackbody calibrations in early 2001 (Reda and Stoffel 2001), using a newly constructed blackbody calibrator, produced an unrealistically large offset that was traced to nonuniform temperatures in the new blackbody cavity (Stoffel et al. 2006). The ARM Radiation Focus Group decided to change to an outdoor calibration technique that compares to pyrgeometers that have been calibrated against the World Infrared Standard Group (WISG) in Davos, Switzerland. However, as of this writing the PIR

FIG. 16-3. The radiometer cluster at the ARM Southern Great Plains Central Facility.

measurements in the archive continue to be based on the original Eppley calibrations.

An interesting side note to the broadband infrared work is that the RCF at SGP was used in the First International Pyrgeometer and Absolute Sky-scanning Radiometer Comparison (IPASRC I; Philipona et al. 2001) and the ARM NSA site was used in IPASRC II (Marty et al. 2003) to establish the WISG that ARM will use for calibrations and that many radiation installations are already using for pyrgeometer calibrations.

6. Significant achievements in broadband radiometry

Figure 16-3, from the SGP site, shows most of the instruments that make the broadband and narrowband radiation measurements discussed in this chapter. The Cimel sunphotometer is shown in the foreground. A UV-MFRSR, a UVB broadband radiometer, and a visible MFRSR from Colorado State University are to the left. Diffuse and infrared downwelling measurements are made on the tracker identified by the two arms with black spheres at their ends (center right) that block the sun; this tracker also carries the solar direct normal instrument. To the right of the tracker is an unshaded pyranometer that makes the GHI measurement. The tower behind the tracker makes upwelling solar and infrared measured at 10 m. The archive of BSRN-quality radiation measurements beginning in 1993 at the SGP site and beginning in 1998 at the NSA and TWP sites is a significant achievement.

In the early years of the ARM Program, the Instantaneous Radiative Flux (IRF) working group was dedicated to improving measurements and radiative transfer models. Often this was done through quality measurement experiments where, for example, measured inputs to radiative transfer models were used to predict the measured irradiance. Finding explanations for any differences led to improved measured inputs, improved models, and improved radiation measurements. An example is provided in chapter 14 for spectral infrared radiance (Mlawer and Turner 2016, chapter 14; Turner et al. 2004), but comparisons of longwave and shortwave flux calculations with observations were also a prominent focus (e.g., Clough et al. 2000; Ricchiazzi et al. 1998; Michalsky et al. 2006). A few achievements are highlighted below.

a. Diffuse irradiance

Using data from the first ARM Enhanced Shortwave Experiment (ARESE), Kato et al. (1997) performed a careful study where they compared clear-sky radiation models with ARM measurements of direct normal and diffuse horizontal shortwave irradiances. Direct irradiances agreed well with the models with correct inputs of aerosol optical depth (AOD), water vapor column, and ozone column. Modeled diffuse irradiances using reasonable aerosol optical properties of absorption and scattering were consistently higher than DHI measurements by a significant margin. Halthore et al. (1998)

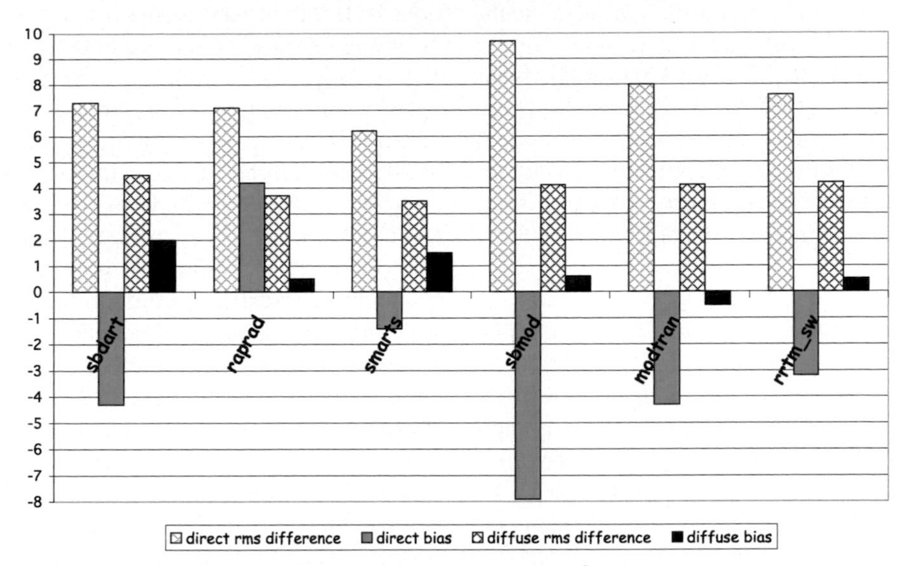

FIG. 16-4. Model minus measurement irradiance ($W m^{-2}$) for six models showing direct (green) and diffuse (blue) rms differences (hatched) and biases (solid). Note the mean direct and diffuse irradiances for the 30 cases are given in the figure title. From Michalsky et al. (2006).

reached a similar conclusion, finding consistently higher modeled diffuse irradiance than measured using sensible model inputs for the aerosol load and its optical properties. These results created a significant debate within the ARM (and larger) community: Was this bias between the observations and the model due to some missing absorption in the radiative transfer model or was this a problem with the radiation observations or the observations used to drive the radiative transfer model?

Papers by Bush et al. (2000) and Dutton et al. (2001) looked into offset problems with thermopile sensors and found that black disk detector pyranometers, such as the Eppley model PSP, had significant offsets. Dutton et al. (2001) found that these offsets were largest during the midafternoons of clear-sky days. In both Bush et al. (2000) and Dutton et al. (2001), a much earlier effort by Gulbrandsen (1978) was cited that clearly pointed to an early, but forgotten, recognition of this problem; this offset was largely ignored until measurements and models reached the level of accuracy occurring in the BSRN/ARM era. The Dutton et al. (2001) study led to the replacement of all Eppley model PSPs used for ARM DHI measurements with Eppley model 8–48s, which do not have significant offsets. PSPs were retained for the GHI backup measurements, and all measurements of GHI made with PSPs were corrected in the ARM archive. A detailed explanation of the correction used for the ARM PSPs is given in Younkin and Long (2004).

Halthore et al. (2004) followed these improvements to the diffuse measurements with a study that concluded that model and measurement agreement was much improved, but that there were differences at times that could only be explained by invoking highly absorbing but perhaps unrealistic aerosol absorption properties. In a comprehensive study, Michalsky et al. (2006) compared six shortwave radiation models with direct and diffuse measurements for 30 cases over a wide range of aerosol loads and SZAs during an aerosol intensive campaign at the SGP Central Facility. The 30-case average bias (model − measurement; see Fig. 16-4) for the direct normal ranged from a low of −1.0% to a high of +0.7% for an averaged DNI of $762 W m^{-2}$. For the diffuse irradiance, the range was from −0.6% to +1.9% for an averaged DHI of $109 W m^{-2}$. The improved agreement compared to Halthore et al. (2004) was attributed to having accurate AODs, reasonable single scattering albedos (SSA), and asymmetry parameters (AP) from in situ measurements, and good water vapor inputs for the model. However, the main cause of the good agreement in Michalsky et al. (2006) between the measurements and the model calculations was the change from using a broadband albedo, which was the same for all wavelengths, in the calculation to a six-band spectral albedo measurement with reasonable assumptions for the wavelength dependent interpolation and extrapolation; this change led to a 7% reduction of the diffuse irradiance in the model calculations.

A major improvement in radiation measurements was ARM's development of a standard procedure for calibrating the DHI; this standard did not exist prior to the

ARM Program. After three IOPs to find pyranometers with consistent behavior for diffuse measurements, Michalsky et al. (2007) proposed and demonstrated a procedure for developing a diffuse working standard that reduced the uncertainty of diffuse measurements. Those results apply to clear-sky measurements, which are arguably the most difficult diffuse measurements to make reliably. Clear skies with low aerosol burden have the lowest irradiance since aerosols increase the diffuse irradiance.

b. *Direct irradiance*

Under ideal circumstances, direct irradiance would be measured by windowless absolute cavity radiometers (Vignola et al. 2012). Since operating with an open cavity is not practical for long-term operations and cavity radiometers are expensive, pyrheliometers based on thermopile detectors are used for routine measurements. To determine the uncertainty of these instruments a comparison was conducted for nearly a year to compare pyrheliometers for all seasons under realistic field conditions. The results are reported in Michalsky et al. (2011). In general, DNIs are measured with better accuracy than is claimed by the manufacturer. The good news for ARM is that Eppley pyrheliometers that are used in ARM were found to have a 95% measurement uncertainty of $\pm 1.3\%$, much better than the $\pm 2.0\%$ claimed by the manufacturer. However, three competitively priced pyrheliometer manufacturers had instruments with 95% uncertainties of about $\pm 0.7\%$. This study, for the first time, provided a clear understanding of the uncertainty of this primary measurement made by ARM and the BSRN community as a whole.

c. *ARESE and ARESE II*

In 1995, three journal articles by Cess et al. (1995), Ramanathan et al. (1995), and Pilewskie and Valero (1995) appeared that suggested absorption in clouds was far greater than theoretical models predicted. Shortly after these publications, ARM mounted a campaign called the ARM Enhanced Shortwave Experiment (ARESE; Valero et al. 1997a) to measure radiation above and below clouds from aircraft flying in tandem and then to compare those measurements with radiation measurements when there were no clouds. The experiment, which was conducted in the fall of 1995, was marked by exceptionally clear weather; therefore, the analysis relied heavily on a single overcast day to estimate absorption in the clouds. Analyzing those data, Valero et al. (1997b) and Zender et al. (1997) found excess absorption consistent with the 1995 papers. Other experiments conducted elsewhere at about the same time by British researchers (Francis et al. 1997) and Japanese researchers (Asano et al. 2000) found that the measured absorption differed from theoretical calculations by less than 10% compared to the more than 50% deviation reported by Valero et al. (1997b) and Zender et al. (1997).

To resolve the issue, a second experiment called ARESE II (Valero et al. 2003), conducted in the winter of 2000, used three sets of radiometers on the aircraft that flew above the clouds of the ARM SGP central facility. Multiple sets of radiation measurements also were made at the surface of the central facility to calculate the net radiation at the surface that was compared with the net radiation measured at the aircraft. All ground and aircraft radiometers were calibrated to the same standard simultaneously (Michalsky et al. 2002) after the offset issues were found and dealt with by using zero offset radiometers or offset corrections. Ackerman et al. (2003), using two state-of-the-art radiative transfer models, analyzed the results on three cloudy days of ARESE II and found that measurements and models for the atmospheric absorption in the clouds agreed to within 10%, which was within the estimated model and measurement accuracy. Cloud absorption on the order of 50% more than model predictions was clearly not observed in this experiment. A good discussion of this appears in Kerr (2003) that emphasizes that both measurement and model improvement led to this result.

d. *Radiative flux analysis*

Long and Ackerman (2000) used ARM data to develop a technique to understand the effect of clouds on downwelling solar irradiance (GHI) at the surface. The technique uses the GHI and DHI time series to detect periods of clear (i.e., cloudless) skies. If a day's detected clear-sky period spans a sufficient range of SZA, the data are used to fit functions with the cosine of the SZA as the independent variable. The coefficients of the fit are then linearly interpolated between clear periods to produce continuous estimates of clear-sky GHI and DHI over that period. Since the predicted clear-sky irradiances are affected by whatever water vapor, aerosols, and other constituents are present, the irradiance differences (radiative effects) when clouds are present are only due to clouds; the assumption is that aerosol, water vapor, and the other constituents remain relatively constant over the day.

Figure 16-5 demonstrates the technique developed by Long and Ackerman (2000). The clear-sky global solar (GHI) model closely matches the measured global solar in the morning and the extrapolation assumes that a clear sky prevails the rest of the day. The deviations of the measured GHI from this model represent the clouds' effects on the radiation budget. The same argument holds for the diffuse solar (DHI) plotted in the blue colors.

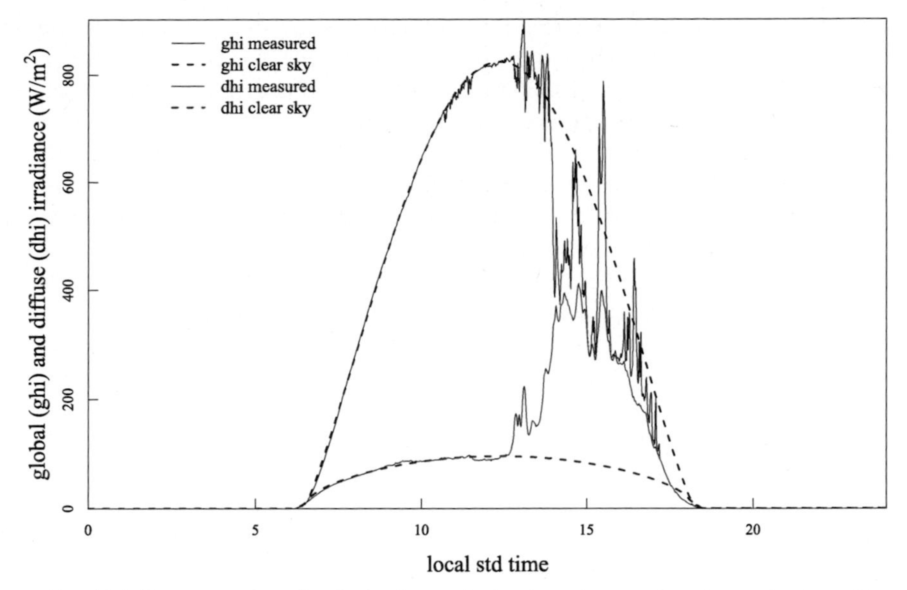

FIG. 16-5. Demonstration of radiative flux analysis where clear portion of morning is used to estimate what the rest of a clear-sky day's pattern in global and diffuse would be expected to follow if it remained clear. Deviations from the clear-sky global and diffuse are attributed to cloud radiative forcing.

Later, techniques were developed to produce continuous estimates of clear-sky downwelling infrared I_{ir} (Long and Turner 2008) irradiance, and upwelling solar and infrared (Long 2005) irradiance. These clear-sky estimates are used along with the corresponding measurements in various combinations to calculate the complete net radiative cloud forcing, and infer cloud macrophysical properties such as daylight fractional sky cover (Long et al. 2006), I_{ir} effective sky cover (Dürr and Philipona 2004), cloud optical depth (Barnard et al. 2008), cloud transmissivity, effective clear-sky emissivity, and sky brightness temperature.

Using the long-term surface radiation datasets available, several papers have used radiative flux analysis (RFA) estimates in model-measurement comparisons of surface radiation and cloud amounts. Wild et al. (2006) used RFA clear-sky results to test the general circulation models (GCMs) that participated in the atmospheric model intercomparison projects (AMIP I and AMIP II) and the model intercomparisons for the Fourth Assessment Report (AR4) of the Intergovernmental Panel on Climate Change (IPCC 2007), while Qian et al. (2012) tested the AR4 models' accuracy for cloud fraction and radiative effects. Zhang et al. (2010) show that the RFA-related suite of retrieved parameters allows for comparison of satellite and surface data sorted by when both are experiencing similar cloudiness, thus getting around many of the previous problems inherent in comparing values that represent differing spatial coverage, an age-old satellite–surface comparison issue.

The RFA data have been used in analyses of surface cloud effects and trends. Long et al. (2009) showed significant decadal brightening over the continental United States that is attributed to changes in cloudiness rather than changes in aerosol loading. For clear sky, the increase was manifested in the diffuse shortwave rather than the direct shortwave that would be expected for the documented decreased aerosol loading across the decade of the study. Mace et al. (2006) studied cloud radiative forcing at the ARM SGP site and comparisons to satellites, while Berg et al. (2011) studied surface radiative forcing specifically by single-layer shallow cumulous clouds at SGP. At the ARM NSA site Dong et al. (2010) studied the cloud radiative forcing climatology. In the ARM TWP sites Wang et al. (2011) used RFA cloud effects and sky-cover retrievals to show that the Madden–Julian oscillation is detected and quantified in the Manus, Papua New Guinea, site data; May et al. (2012) investigated the diurnal cycle of the Australian monsoon including cloudiness and radiative effects; and McFarlane et al. (2013) analyzed the climatology of surface cloud radiative effects, by cloud type, at all three TWP sites.

7. Spectral solar measurements for aerosol retrievals

a. *Cimel sunphotometers*

ARM solar spectral measurements have focused mainly on the retrievals of AOD and aerosol optical properties. In the late 1990s ARM became part of

AERONET (Holben et al. 1998) with the introduction of Cimel sunphotometers at the central facility sites in Oklahoma, Alaska, and on Nauru. The Cimel sunphotometers (see Fig. 16-3) are now at all ARM permanent central facilities and ARM mobile facilities; in addition, spares are kept for replacements. The spares are critical because it takes a month or more to calibrate the sunphotometers, which is done by sending the instrument to AERONET headquarters at the Goddard Space Flight Center in Greenbelt, Maryland. These spares fill significant gaps in the data stream that would otherwise be created. Calibration is via comparisons to other Cimel sunphotometers that have been calibrated at Mauna Loa Observatory in Hawaii using Langley plot analyses.

The wavelengths that are standard in all of the ARM Cimels are 340, 380, 440, 500, 675, 870, 940, 1020, and 1640 nm. All are used for AOD retrievals except the 940-nm channel, which is used to estimate the amount of water vapor in the column. Measurements to calculate AOD are made at every half air mass if the air mass is greater than 2, or every 15 min if the air mass is less than 2. The uncertainties for AOD are estimated at 0.01 to 0.02 (Gregory 2011). The four channels at 440, 675, 870, and 1020 nm are used to scan in the solar almucanter and in the solar-zenith plane measuring radiance at defined angles. These scans, occurring four times daily for each scan plane, are used, along with the direct spectral irradiance measurements, to retrieve AP (a measure of the forward versus backscattered radiation), SSA (a measure of the absorption by the aerosol), and aerosol size distributions (Dubovik et al. 2000).

b. Multifilter rotating shadowband radiometer

Several variants of the MFRSR are used in the ARM Program. The MFRSR was developed with ARM funding in the early 1990s (Harrison et al. 1994) and was licensed to Yankee Environmental Systems, Inc., in the mid-1990s. While the Cimel is a traditional pointing sunphotometer with a 1.2° field of view and a filter wheel that cycles to sequentially sample all nine channels, the MFRSR makes simultaneous measurements in seven temperature-controlled channels. The GHI is measured with the rotating band out of the field of view. The band rotates about an axis parallel to Earth's rotation allowing the required steps in hour angle to be calculated to block the diffuser from direct solar radiation in order to measure the DHI (http://yesinc.com/products/radvis.html). Finally, the DNI is calculated from these two measurements [see Harrison et al. (1994) for complete details]. Six of these channels contain 10-nm filters; five are used for AOD measurements and one for water vapor. The seventh channel is an unfiltered silicon detector that is used to estimate broadband solar after corrections for spectral response are applied. This unfiltered channel's measurements can be used as backup to the broadband thermopile measurements of shortwave irradiance.

The MFRSR samples wavelengths at 415, 500, 615, 673, and 870 nm for AOD determinations. Instruments are calibrated for AOD retrievals after first screening all of the data for acceptable Langley plots as outlined in Harrison and Michalsky (1994). Robust estimates of the calibration constants using all of the acceptable Langley plots are obtained following the procedures described in Michalsky et al. (2001), which provide uncertainties in the derived AOD, due to uncertainties in calibration, of 0.01 to 0.02. These methods use field measurements to continuously update the calibration; therefore, the instruments are not removed from the field for calibration unless there is an instrument failure. Several studies (e.g., Michalsky et al. 2010) have compared the Cimel sunphotometer estimates of AOD with the MFRSR AODs and found results clearly within the uncertainties cited above. In that paper, a 15-yr AOD climatology (through the early part of 2008) was produced for the SGP aerosols based on the continuous monitoring possible using the field calibration procedure just outlined; the importance of this result is discussed in McComiskey and Ferrare (2016, chapter 21). Although the 940-nm channel measurements are made, the data have not been used to routinely retrieve water vapor in the ARM Program, although comparisons with other methods to retrieve precipitable water vapor have been performed (e.g., Alexandrov et al. 2008, and references therein; Turner et al. 2016, chapter 13). The MFRSR sampling is every 20 s although some early data were sampled every 15 s. While Cimel almucanter and solar-zenith plane scans are used to retrieve AP and SSA, as explained above, the MFRSR diffuse and direct irradiances are used in a different procedure to retrieve the same parameters. Kassianov et al. (2007) demonstrated these retrievals for AP and SSA and compared their retrievals with in situ techniques and AERONET retrievals with good agreement. The most extensive effort to use MFRSRs for the characterization of aerosols was by Alexandrov and his colleagues at the Goddard Institute for Space Studies in New York City (Alexandrov et al. 2008, and many references therein). Their focus was on retrieving size distributions including the separation of fine and coarse modes. They found good agreement when comparing derivations using two different MFRSRs and when comparing MFRSR retrievals to the Cimel sun–sky-based retrievals.

A key input for the retrieval of the SSA and AP is the spectral albedo of the land surface. Later in this chapter we note that surface albedo is key to deriving cloud optical depths. As discussed earlier in this chapter, surface spectral albedo was a key factor in achieving radiative closure between clear-sky broadband models and measurements (Michalsky et al. 2006). The albedo measurements used for that study were made with the MFRSR for downwelling irradiance; the upwelling measurements were made with the detector head of the MFRSR, the MFR. The filters for the MFR and MFRSR are at the same wavelengths. At the SGP central facility one MFR is placed near the top of a 10-m tower not far from the MFRSRs. This site is pastureland and is considered representative of fallow land around the ARM site. Another MFR is sited at the 25-m level of the 60-m meteorological tower that is within a cultivated field some 350 m WNW of the 10-m tower. This field was planted mostly with winter wheat in the early years of ARM, but after the fall of 2004 corn, soybeans, and wheat have been rotated (McFarlane et al. 2011). The field is tilled before planting and after harvest revealing bare soil until weeds or planted crops emerge. McFarlane et al. (2011) developed a value added product (VAP), useful in radiative transfer calculations, to derive a continuous surface spectral albedo dataset using the MFRs and MFRSRs plus a calculation of Normalized Difference Vegetative Index (NDVI) from these albedos to provide an understanding of the surface type.

c. *Normal incidence multifilter radiometer*

Two straightforward modifications of the MFRSR were made for specific purposes in ARM. The normal incidence multifilter radiometer (NIMFR) has the same filter-detector package as the MFRSR, but is equipped with a baffled tube to restrict the field of view to 5.7° to match the Eppley model NIP pyrheliometer. Samples in all channels are obtained simultaneously every 20 s. It is calibrated in the same way as the MFRSR using multiple Langley regressions and deriving a robust estimate of the extraterrestrial response for each aerosol channel from these. The AODs derived from the NIMFR should have less uncertainty than those obtained with the MFRSR because there are no cosine response corrections to apply, which add uncertainty. Michalsky et al. (2010) show agreement within 0.014 with the MFRSR, which is slightly better than the agreement of the MFRSR with AERONET, since the sampling and wavelengths are better matched in the former case.

8. Spectral solar measurements for cloud retrievals

a. *Multifilter rotating shadowband radiometer*

The MFRSR can be used to measure cloud optical depths for totally overcast skies filled with water clouds (no ice). The instrument is calibrated when the sky is cloud free using Langley plots as described earlier, which calibrates the direct beam for top-of-the-atmosphere response. Moreover, the calibration applies to the diffuse and global irradiance as well, since the same sensor is used in all three measurements. Measurements of irradiance under clouds then can be used to calculate the transmission by taking the ratio of this measurement to the top-of-the-atmosphere global irradiance (i.e., the direct irradiance multiplied by the cosine of the solar zenith angle). A radiative transfer model that has estimated or measured surface albedo and assumed or obtained independently cloud droplet radius can be run iteratively changing the optical depth until the transmission matches the measurement. Min and Harrison (1996) have shown that there is only a small added uncertainty in the cloud optical depth if a fixed reasonable estimate of cloud droplet radius is assumed. Their procedure uses measurements at the shortest MFRSR wavelength, 415 nm, because the surface albedo is low there, and small differences from the true value have little effect on the estimated cloud optical depth. If the liquid water path is measured by a collocated microwave radiometer (see, e.g., Shupe et al. 2016, chapter 19), then this algorithm can retrieve the cloud droplet radius and produce an improved estimate of the cloud optical depth. In later studies, Min et al. (2008) demonstrated the retrieval of cloud fraction from spectral measurements made with the MFRSR. Wang and Min (2008) further developed methods to retrieve thin and mixed-phase cloud optical depths using the MFRSR.

b. *Two-channel narrow field-of-view and Cimel radiometers in cloud mode*

The narrow field-of-view (NFOV) radiometer was developed to measure zenith radiances at 1-s temporal resolution with two channels, one in the red (673 nm) and one in the near-infrared (870 nm), for the purpose of retrieving cloud properties. Cloud optical properties at these two wavelengths are nearly identical, but green vegetative surfaces reflect these two wavelengths very differently. This instrument uses some MFRSR technology, but is different in that it does not use a diffuser over the two separate receivers since that would cut the signal to unusably low levels. This instrument also uses higher gain for the low radiance signals. The current field of view is 1.2°, down from the original 5.7°, with the smaller field of view matching the Cimel, allowing comparisons of the two instruments' cloud retrievals. This smaller field of view allows successful sampling of geometrically smaller clouds; for the technique to work the cloud must fill the field of view. Chiu et al. (2006) demonstrated that 1-s data captures the frequency at which clouds evolve naturally. They estimate that cloud optical depths using this technique are measured with an uncertainty of about 15%. Chiu et al. (2012) recently used the ARM Cimel in its zenith radiance mode measuring with visible and near-infrared (1640 nm) wavelengths to retrieve cloud optical depth, effective cloud droplet radius, and liquid water path.

9. The future of broadband and narrowband radiometry in ARM

Remarkable improvements have been made in the accuracy of radiation measurements over the past 20 years based on research in the ARM Program and BSRN. The DNI was measured with good accuracy before, but the accuracy of commercial pyrheliometers is now better understood (Michalsky et al. 2001). These DNI measurements can now be made with accuracies of 0.7% using an all-weather instrument. The measurement of DHI is much better now since a procedure to produce a DHI standard has been developed (Michalsky et al. 2007); DHI accuracies of around 2% are possible. Infrared measurements have shown the most improvement. IPASRC I/II studies conducted at SGP and NSA were pivotal in establishing the WISG. These measurements can be made at the 2%–3% level, where they were at about the 10% uncertainty level at the outset of the ARM Program.

a. Multifilter rotating shadowband radiometer

MFRSRs were refurbished with new filters and better temperature control in the late 2000s. The new filter sets have better throughput resulting in better signal-to-noise ratios. Close attention was paid to reducing the out-of-band light getting through the filters, which was adding nonnegligible uncertainty to AOD retrievals. The temperature is controlled very closely near the set point except on a few hot summer days when the internal temperature exceeds the set point; the improved temperature control improves the calibration stability of the instrument for long time periods between successful Langley calibrations. A Campbell datalogger controls the MFRSR, allowing them to be operated in more flexible modes. ARM is considering a slight change to the filter set where the 615-nm channel would be swapped for a 1625-nm channel to enable several new scientific studies, and the feasibility of this change has been demonstrated. This change would permit a better estimate of the shortwave surface albedo in the near-infrared; it would allow a two-channel retrieval of cloud optical depth and cloud particle radius simultaneously, and it would allow a better retrieval of large aerosol sizes.

b. Broadband

In the comparison of pyrheliometers reported in section 6b (Michalsky et al. 2011) it was found that the current ARM pyrheliometers measured direct irradiance with an estimated accuracy of 1.3% within 95% confidence limits. It would seem prudent to upgrade the ARM pyrheliometers to one of the instruments in that comparison that measured with an accuracy near 0.7%. This would result in a substantial uncertainty improvement of the measurements since direct beam is the largest contributor to downwelling shortwave irradiance.

Infrared measurements should improve significantly with the impending improvement in the calibration procedure. Outdoor calibrations to the WISG should reduce the absolute uncertainty of these measurements to around $\pm 5\,\mathrm{W\,m^{-2}}$. Given the long-term stability of the Eppley model PIR, the calibrations should be valid from the earliest measurements in ARM. This will allow for an interesting analysis of surface net infrared changes over the past 20 years in the case of the SGP and NSA central facilities.

Acknowledgments. Anthony Bucholtz, Dave Turner, and an anonymous referee offered insightful comments that led to considerable improvement of the final version of this chapter.

REFERENCES

Ackerman, T. P., D. M. Flynn, and R. T. Marchand, 2003: Quantifying the magnitude of anomalous solar absorption. *J. Geophys. Res.*, **108**, 4273, doi:10.1029/2002JD002674.

Alexandrov, M. D., A. A. Lacis, B. E. Carlson, and B. Cairns, 2008: Characterizations of atmospheric aerosols using MFRSR measurements. *J. Geophys. Res.*, **113**, D08204, doi:10.1029/2007JD009388.

Asano, S., A. Uchiyama, Y. Mano, M. Murakami, and Y. Takayama, 2000: No evidence for solar absorption anomaly by marine water clouds through collocated aircraft radiation measurements. *J. Geophys. Res.*, **105**, 14 761–14 775, doi:10.1029/2000JD900062.

Barnard, J. C., C. N. Long, E. I. Kassianov, S. A. McFarlane, J. M. Comstock, M. Freer, and G. M. McFarquhar, 2008: Development and evaluation of a simple algorithm to find cloud optical depth with emphasis on thin ice clouds. *Open Atmos. Sci. J.*, **2**, 46–55, doi:10.2174/1874282300802010046.

Berg, L. K., E. I. Kassianov, C. N. Long, and D. L. Mills Jr., 2011: Surface summertime radiative forcing by shallow cumuli at the Atmospheric Radiation Measurement Southern Great Plains site. *J. Geophys. Res.*, **116**, D01202, doi:10.1029/2010JD014593.

Bush, B. C., F. P. J. Valero, A. S. Simpson, and L. Bignone, 2000: Characterization of thermal effects in pyranometers: A data correction algorithm for improved measurement of surface insolation. *J. Atmos. Oceanic Technol.*, **17**, 165–175, doi:10.1175/1520-0426(2000)017<0165:COTEIP>2.0.CO;2.

Cess, R. D., and Coauthors, 1995: Absorption of solar radiation by clouds: Observations versus models. *Science*, **267**, 496–499, doi:10.1126/science.267.5197.496.

Chiu, J. C., A. Marshak, Y. Knyazikhin, W. J. Wiscombe, H. W. Barker, J. C. Barnard, and Y. Luo, 2006: Remote sensing of cloud properties using ground-based measurements of zenith radiance. *J. Geophys. Res.*, **111**, D16201, doi:10.1029/2005JD006843.

——, and Coauthors, 2012: Cloud droplet size and liquid water path retrievals from zenith radiance measurements: Examples from the Atmospheric Radiation Measurement Program and the Aerosol Robotic Network. *Atmos. Chem. Phys.*, **12**, 10 313–10 329, doi:10.5194/acp-12-10313-2012.

Clough, S. A., and Coauthors, 2000: A longwave broadband QME based on ARM pyrometer and AERI measurements. *Extended Abstracts, 10th ARM Science Team Meeting*, San Antonio, TX, ARM Program. [Available online at http://www.arm.gov/publications/proceedings/conf10/extended_abs/clough_sa.pdf.]

Coulson, K. L., and Y. Howell, 1980: Solar radiation instruments. *Sunworld*, **4**, 87–94.

Dong, X., B. Xi, K. Crosby, C. N. Long, R. S. Stone, and M. D. Shupe, 2010: A 10 year climatology of Arctic cloud fraction and radiative forcing at Barrow, Alaska. *J. Geophys. Res.*, **115**, D17212, doi:10.1029/2009JD013489.

Dubovik, O., A. Smirnov, B. N. Holben, M. D. King, Y. J. Kaufman, T. F. Eck, and I. Slutsker, 2000: Accuracy assessments of aerosol optical properties retrieved from Aerosol Robotic Network (AERONET) sun and sky radiance measurements. *J. Geophys. Res.*, **105**, 9791–9806, doi:10.1029/2000JD900040.

Dürr, B., and R. Philipona, 2004: Automatic cloud amount detection by surface longwave downward radiation measurements. *J. Geophys. Res.*, **109**, D05201, doi:10.1029/2003JD004182.

Dutton, E. G., J. J. Michalsky, T. Stoffel, B. W. Forgan, J. Hickey, D. W. Nelson, T. L. Alberta, and I. Reda, 2001: Measurement of broadband diffuse solar irradiance using current commercial instrumentation with a correction for thermal offset errors. *J. Atmos. Oceanic Technol.*, **18**, 297–314, doi:10.1175/1520-0426(2001)018<0297:MOBDSI>2.0.CO;2.

Ellingson, R. G., and W. J. Wiscombe, 1996: The Spectral Radiance Experiment (SPECTRE): Project description and sample results. *Bull. Amer. Meteor. Soc.*, **77**, 1967–1985, doi:10.1175/1520-0477(1996)077<1967:TSREPD>2.0.CO;2.

——, J. Ellis, and S. Fels, 1991: The intercomparison of radiation codes used in climate models: Longwave results. *J. Geophys. Res.*, **96**, 8929–8953, doi:10.1029/90JD01450.

Fouquart, Y., B. Bonnel, and V. Ramaswamy, 1991: Intercomparing shortwave radiation codes for climate studies. *J. Geophys. Res.*, **96**, 8955–8968, doi:10.1029/90JD00290.

Francis, P. N., J. P. Taylor, P. Hignett, and A. Slingo, 1997: On the question of enhanced absorption of solar radiation by clouds. *Quart. J. Roy. Meteor. Soc.*, **123**, 419–434, doi:10.1002/qj.49712353809.

Fröhlich, C., R. Philipona, J. Romero, and C. Wehrli, 1995: Radiometry at the Physikalisch-Meteorologisches Observatorium Davos and World Radiation Centre. *Opt. Eng.*, **34**, 2757–2766, doi:10.1117/12.205682.

Gregory, L., 2011: Cimel Sunphotometer (CSPHOT) Handbook. ARM Tech. Rep. DOE/SC-ARM TR-056, 17 pp. [Available online at https://www.arm.gov/instruments/csphot.]

Gulbrandsen, A., 1978: On the use of pyranometers in the study of spectral solar radiation and atmospheric aerosols. *J. Appl. Meteor.*, **17**, 899–904, doi:10.1175/1520-0450(1978)017<0899:OTUOPI>2.0.CO;2.

Halthore, R. N., S. Nemesure, S. E. Schwartz, D. G. Imre, A. Berk, E. G. Dutton, and M. H. Bergin, 1998: Models overestimate diffuse clear-sky surface irradiance: A case for excess atmospheric absorption. *Geophys. Res. Lett.*, **25**, 3591–3594, doi:10.1029/98GL52809.

——, M. A. Miller, J. A. Ogren, P. J. Sheridan, D. W. Slater, and T. Stoffel, 2004: Further developments in closure experiments for the surface diffuse irradiance under cloud-free skies at a continental site. *Geophys. Res. Lett.*, **31**, L07111, doi:10.1029/2003GL019102.

Harrison, L., and J. Michalsky, 1994: Objective algorithms for the retrieval of optical depths from ground-based measurements. *Appl. Opt.*, **33**, 5126–5132, doi:10.1364/AO.33.005126.

——, ——, and J. Berndt, 1994: Automated multifilter rotating shadow-band radiometer: An instrument for optical depth and radiation measurements. *Appl. Opt.*, **33**, 5118–5125, doi:10.1364/AO.33.005118.

Holben, B. N., and Coauthors, 1998: AERONET—A federated instrument network and data archive for aerosol characterization. *Remote Sens. Environ.*, **66**, 1–16, doi:10.1016/S0034-4257(98)00031-5.

IPCC, 2007: *Climate Change 2007: The Physical Science Basis*. Cambridge University Press, 996 pp.

Kassianov, E., C. J. Flynn, T. P. Ackerman, and J. C. Barnard, 2007: Aerosol single-scattering albedo and asymmetry parameter form MFRSR observations during the ARM aerosol IOP 2003. *Atmos. Chem. Phys.*, **7**, 3341–3351, doi:10.5194/acp-7-3341-2007.

Kato, S., T. P. Ackerman, E. E. Clothiaux, J. H. Mather, G. G. Mace, M. L. Wesely, F. Murcray, and J. Michalsky, 1997: Uncertainties in modeled and measured clear-sky surface shortwave irradiances. *J. Geophys. Res.*, **102**, 25 881–25 898, doi:10.1029/97JD01841.

Kerr, R. A., 2003: Making clouds darker sharpens cloudy climate models. *Science*, **300**, 1859–1860, doi:10.1126/science.300.5627.1859a.

Knuteson, R. O., and Coauthors, 2004: The Atmospheric Emitted Radiance Interferometer (AERI) Part I: Instrument design. *J. Atmos. Oceanic Technol.*, **21**, 1763–1776, doi:10.1175/JTECH-1662.1.

Long, C. N., 2005: On the estimation of clear-sky upwelling shortwave and longwave. *Extended Abstracts, 15th ARM Science Team Meeting*, Daytona Beach, FL, ARM Program. [Available online at http://www.arm.gov/publications/proceedings/conf15/extended_abs/long_cn.pdf.]

——, and T. P. Ackerman, 2000: Identification of clear skies from broadband pyranometer measurements and calculation of downwelling shortwave cloud effects. *J. Geophys. Res.*, **105**, 15 609–15 626, doi:10.1029/2000JD900077.

——, and D. D. Turner, 2008: A method for continuous estimation of clear-sky downwelling longwave radiative flux developed using ARM surface measurements. *J. Geophys. Res.*, **113**, D18206, doi:10.1029/2008JD009936.

——, T. P. Ackerman, K. L. Gaustad, and J. N. S. Cole, 2006: Estimation of fractional sky cover from broadband shortwave radiometer measurements. *J. Geophys. Res.*, **111**, D11204, doi:10.1029/2005JD006475.

——, E. G. Dutton, J. A. Augustine, W. Wiscombe, M. Wild, S. A. McFarlane, and C. J. Flynn, 2009: Significant decadal brightening of downwelling shortwave in the continental United States. *J. Geophys. Res.*, **114**, D00D06, doi:10.1029/2008JD011263.

Mace, G. G., and Coauthors, 2006: Cloud radiative forcing at the Atmospheric Radiation Measurement Program Climate Research Facility: 1. Technique, validation, and comparison to satellite-derived diagnostic quantities. *J. Geophys. Res.*, **111**, D11S90, doi:10.1029/2005JD005921.

Marty, C., and Coauthors, 2003: Downward longwave irradiance uncertainty under Arctic atmospheres: Measurements and modeling. *J. Geophys. Res.*, **108**, 4358, doi:10.1029/2002JD002937.

May, P. T., C. N. Long, and A. Protat, 2012: The diurnal cycle of the boundary layer, convection, clouds, and surface radiation in a coastal monsoon environment (Darwin, Australia). *J. Climate*, **25**, 5309–5326, doi:10.1175/JCLI-D-11-00538.1.

McComiskey, A., and R. A. Ferrare, 2016: Aerosol physical and optical properties in the ARM Program. *The Atmospheric Radiation Measurement (ARM) Program: The First 20 Years, Meteor. Monogr.*, No. 57, Amer. Meteor. Soc., doi:10.1175/AMSMONOGRAPHS-D-15-0028.1.

McFarlane, S. A., K. L. Gaustad, E. J. Mlawer, C. N. Long, and J. Delamere, 2011: Development of a high spectral resolution surface albedo product for the ARM Southern Great Plains central facility. *Atmos. Meas. Tech.*, **4**, 1713–1733, doi:10.5194/amt-4-1713-2011.

——, C. N. Long, and J. Flaherty, 2013: A climatology of surface cloud radiative effects at the ARM Tropical Western Pacific sites. *J. Appl. Meteor. Climatol.*, **52**, 996–1013, doi:10.1175/JAMC-D-12-0189.1.

——, J. H. Mather, and E. J. Mlawer, 2016: ARM's progress on improving atmospheric broadband radiative fluxes and heating rates. *The Atmospheric Radiation Measurement (ARM) Program: The First 20 Years, Meteor. Monogr.*, No. 57, Amer. Meteor. Soc., doi:10.1175/AMSMONOGRAPHS-D-15-0046.1.

Michalsky, J. J., J. Schlemmer, W. Berkheiser III, J. Berndt, L. Harrison, N. Laulainen, N. Larson, and J. Barnard, 2001: Multiyear measurements of aerosol optical depth in the Atmospheric Radiation Measurement and Quantitative Links programs. *J. Geophys. Res.*, **106**, 12 099–12 107, doi:10.1029/2001JD900096.

——, and Coauthors, 2002: Broadband shortwave calibration results from the Atmospheric Radiation Measurement Enhanced Shortwave Experiment II. *J. Geophys. Res.*, **107** (D16), doi:10.1029/2001JD001231.

——, and Coauthors, 2003: Results from the first ARM diffuse horizontal shortwave irradiance comparison. *J. Geophys. Res.*, **108**, 4108, doi:10.1029/2002JD002825.

——, and Coauthors, 2005: Toward the development of a diffuse horizontal shortwave irradiance working standard. *J. Geophys. Res.*, **110**, D06107, doi:10.1029/2004JD005265.

——, and Coauthors, 2006: Shortwave radiative closure studies for clear skies during the Atmospheric Radiation Measurement 2003 Aerosol Intensive Observation Period. *J. Geophys. Res.*, **111**, D14S90, doi:10.1029/2005JD006341.

——, C. Gueymard, P. Kiedron, L. J. B. McArthur, R. Philipona, and T. Stoffel, 2007: A proposed working standard for the measurement of diffuse horizontal shortwave irradiance. *J. Geophys. Res.*, **112**, D16112, doi:10.1029/2007JD008651.

——, F. Denn, C. Flynn, G. Hodges, P. Kiedron, A. Koontz, J. Schlemmer, and S. E. Schwartz, 2010: Climatology of aerosol optical depth in north-central Oklahoma: 1992–2008. *J. Geophys. Res.*, **115**, D07203, doi:10.1029/2009JD012197.

——, and Coauthors, 2011: An extensive comparison of commercial pyrheliometers under a wide range of routine observing conditions. *J. Atmos. Oceanic Technol.*, **28**, 752–766, doi:10.1175/2010JTECHA1518.1.

Min, Q., and L. C. Harrison, 1996: Cloud properties derived from surface MFRSR measurements and comparison with GOES results at the ARM SGP site. *Geophys. Res. Lett.*, **23**, 1641–1644, doi:10.1029/96GL01488.

——, T. Wang, C. N. Long, and M. Duan, 2008: Estimating fractional sky cover from spectral measurements. *J. Geophys. Res.*, **113**, D20208, doi:10.1029/2008JD010278.

Mlawer, E. J., and D. D. Turner, 2016: Spectral radiation measurements and analysis in the ARM Program. *The Atmospheric Radiation Measurement (ARM) Program: The First 20 Years, Meteor. Monogr.*, No. 57, Amer. Meteor. Soc., doi:10.1175/AMSMONOGRAPHS-D-15-0027.1.

Myers, D. R., T. L. Stoffel, I. Reda, S. M. Wilcox, and A. Andreas, 2002: Recent progress in reducing the uncertainty in and improving pyranometer calibrations. *J. Solar Energy. Eng.*, **124**, 44–49, doi:10.1115/1.1434262.

Ohmura, A., and Coauthors, 1998: Baseline Surface Radiation Network (BSRN/WCRP): New precision radiometry for climate research. *Bull. Amer. Meteor. Soc.*, **79**, 2115–2136, doi:10.1175/1520-0477(1998)079<2115:BSRNBW>2.0.CO;2.

Philipona, R., and Coauthors, 2001: Atmospheric longwave irradiance uncertainty: Pyrgeometers compared to an absolute sky-scanning radiometer, atmospheric emitted radiance interferometer, and radiative transfer model calculations. *J. Geophys. Res.*, **106**, 28 129–28 141, doi:10.1029/2000JD000196.

Pilewskie, P., and F. P. J. Valero, 1995: Direct observations of excess absorption by clouds. *Science*, **267**, 1626–1629, doi:10.1126/science.267.5204.1626.

Qian, Y., C. N. Long, H. Wang, J. Comstock, S. A. McFarlane, and S. Xie, 2012: Evaluation of cloud fraction and its radiative effect simulated by IPCC AR4 global models against ARM surface observations. *Atmos. Chem. Phys.*, **12**, 1785–1810, doi:10.5194/acp-12-1785-2012.

Ramanathan, V., B. Subasilar, B, G. J. Zhang, W. Conant, R. D. Cess, J. T. Kiehl, H. Grassl, and L. Shi, 1995: Warm pool heat budget and shortwave cloud forcing: A missing physics? *Science*, **267**, 499–503, doi:10.1126/science.267.5197.499.

Reda, I., and T. Stoffel, 2001: Pyrgeometer calibrations at NREL. *Proc. 11th ARM Science Team Meeting*, Atlanta, GA, ARM Program. [Available online at http://www.arm.gov/publications/proceedings/conf11/poster_abs/P00005.]

Ricchiazzi, P., S. Yang, C. Gautier, and D. Sowle, 1998: SBDART: A research and teaching software tool for plane-parallel radiative transfer in the Earth's atmosphere. *Bull. Amer. Meteor. Soc.*, **79**, 2101–2114, doi:10.1175/1520-0477(1998)079<2101:SARATS>2.0.CO;2.

Shupe, M. D., J. M. Comstock, D. D. Turner, and G. G. Mace, 2016: Cloud property retrievals in the ARM Program. *The Atmospheric Radiation Measurement (ARM) Program: The First 20 Years, Meteor. Monogr.*, No. 57, Amer. Meteor. Soc., doi:10.1175/AMSMONOGRAPHS-D-15-0030.1.

Stoffel, T., I. Reda, J. Hickey, E. Dutton, and J. Michalsky, 2006: Pyrgeometer calibrations for the Atmospheric Radiation Measurement Program: Updated approach. *Extended Abstracts, 16th ARM Science Team Meeting*, Albuquerque, NM, ARM Program. [Available online at http://www.arm.gov/publications/proceedings/conf16/extended_abs/stoffel_t.pdf.]

Turner, D. D., and Coauthors, 2004: The QME AERI LBLRTM: A closure experiment for downwelling high spectral resolution infrared radiance. *J. Atmos. Sci.*, **61**, 2657–2675, doi:10.1175/JAS3300.1.

——, E. J. Mlawer, and H. E. Revercomb, 2016: Water vapor observations in the ARM Program. *The Atmospheric Radiation Measurement (ARM) Program: The First 20 Years,*

Meteor. Monogr., No. 57, Amer. Meteor. Soc., doi:10.1175/AMSMONOGRAPHS-D-15-0025.1.

Valero, F. P. J., A. Bucholtz, B. C. Bush, S. K. Pope, W. D. Collins, P. Flatau, A. Strawa, and W. J. Y. Gore, 1997a: Atmospheric Radiation Measurements Enhanced Shortwave Experiment (ARESE): Experimental and data details. *J. Geophys. Res.*, **102**, 29 929–29 937, doi:10.1029/97JD02434.

——, R. D. Cess, M. Zhang, S. K. Pope, A. Bucholtz, B. Bush, and J. Vitko, 1997b: Absorption of solar radiation by the cloudy atmosphere: Interpretations of collocated aircraft measurements. *J. Geophys. Res.*, **102**, 29 917–29 927, doi:10.1029/97JD01782.

——, and Coauthors, 2003: Absorption of solar radiation by the clear and cloudy atmosphere during the Atmospheric Radiation Measurement Enhanced Shortwave Experiments (ARESE) I and II: Observations and models. *J. Geophys. Res.*, **108**, 4016, doi:10.1029/2001JD001384.

Vignola, F., J. Michalsky, and T. Stoffel, 2012: *Solar and Infrared Radiation Measurements.* CRC Press, 410 pp.

Wang, T., and Q. Min, 2008: Retrieving optical depths of optically thin and mixed-phase clouds from MFRSR measurements. *J. Geophys. Res.*, **113**, D19203, doi:10.1029/2008JD009958.

Wang, Y., C. N. Long, J. H. Mather, and X. D. Liu, 2011: Convective signals from surface measurements at ARM Tropical Western Pacific site: Manus. *Climate Dyn.*, **36**, 431–449, doi:10.1007/s00382-009-0736-z.

Wild, M., C. N. Long, and A. Ohmura, 2006: Evaluation of clear-sky solar fluxes in GCMs participating in AMIP and IPCC-AR4 from a surface perspective. *J. Geophys. Res.*, **111**, D01104, doi:10.1029/2005JD006118.

Younkin, K., and C. N. Long, 2004: Improved correction of IR loss in diffuse shortwave measurements: An ARM value added product. ARM Tech. Rep. ARM TR-009, 47 pp.

Zender, C. S., B. Bush, S. K. Pope, A. Bucholtz, W. D. Collins, J. T. Kiehl, F. P. J. Valero, and J. Vitko, 1997: Atmospheric absorption during the Atmospheric Radiation Measurement (ARM) Enhanced Shortwave Experiment (ARESE). *J. Geophys. Res.*, **102**, 29 901–29 915, doi:10.1029/97JD01781.

Zhang, Y., C. N. Long, W. B. Rossow, and E. G. Dutton, 2010: Exploiting diurnal variations to evaluate the ISCCP-FD flux calculations and radiative-flux-analysis-processed surface observations from BSRN, ARM and SURFRAD. *J. Geophys. Res.*, **115**, D15105, doi:10.1029/2009JD012743.

Chapter 17

Development and Applications of ARM Millimeter-Wavelength Cloud Radars

PAVLOS KOLLIAS,[a,k] EUGENE E. CLOTHIAUX,[b] THOMAS P. ACKERMAN,[c] BRUCE A. ALBRECHT,[d]
KEVIN B. WIDENER,[e] KEN P. MORAN,[f] EDWARD P. LUKE,[g] KAREN L. JOHNSON,[g] NITIN BHARADWAJ,[e]
JAMES B. MEAD,[h] MARK A. MILLER,[i] JOHANNES VERLINDE,[b] ROGER T. MARCHAND,[c]
AND GERALD G. MACE[j]

[a] *McGill University, Montreal, Quebec, Canada*
[b] *The Pennsylvania State University, University Park, Pennsylvania*
[c] *University of Washington, Seattle, Washington*
[d] *University of Miami, Miami, Florida*
[e] *Pacific Northwest National Laboratory, Richland, Washington*
[f] *National Oceanic and Atmospheric Administration, Boulder, Colorado*
[g] *Brookhaven National Laboratory, Upton, New York*
[h] *ProSensing, Inc., Amherst, Massachusetts*
[i] *Rutgers, The State University of New Jersey, New Brunswick, New Jersey*
[j] *University of Utah, Salt Lake City, Utah*

1. Introduction

As the ARM Program was getting underway in the early 1990s, studies by Ramanathan et al. (1989) and Cess et al. (1990) highlighted the importance of cloud and radiation interactions to climate. Ramanathan et al. (1989) demonstrated that, on average, clouds cool the climate system but that different cloud types can have different influences upon it. Cess et al. (1990) showed that general circulation models have an array of different responses to the same sea surface temperature change that result from differences in model clouds and their interactions with radiation. In their papers discussing the ARM Program, Stokes and Schwartz (1994) and later Ackerman and Stokes (2003) emphasized the importance of characterizing clouds throughout a vertical column in order to fully understand the radiation field associated with them. They made clear that coupling of high-fidelity observations of clouds and radiation were necessary to improving model parameterizations of them, which were in turn necessary for improving prognostic models of future climate.

Lidar and radar are two key technologies for remote sensing of cloud properties through vertical columns of the atmosphere. Lidar remote sensing of cirrus cloud properties was already a well-developed discipline in the early 1990s (e.g., Sassen et al. 1990) with a focus on understanding cirrus microphysical and radiative properties and their importance to climate (e.g., Platt et al. 1987). Stokes and Schwartz (1994) recognized the importance of lidar to the ARM Program; in fact, many of the coauthors of the Sassen et al. (1990) and Platt et al. (1987) papers, as well as their institutional colleagues, were listed by Stokes and Schwartz (1994) as early investigators in the ARM Program.

While lidar is the optimal instrument for studying optically (in the visible and infrared regions of the electromagnetic spectrum) thin clouds, this technology is limited for study of optically thick clouds, which attenuate the lidar beams. Optically thick clouds are often transparent at the longer microwave wavelengths, so the importance of microwave radar remote sensing of clouds was also clear at this time. As the ARM Program started, radar remote sensing of cloud properties in support of climate studies was not nearly as mature as that for lidar. The development of millimeter-wavelength radar technology and its conversion

[k] Current affiliation: School of Marine and Atmospheric Sciences, Stony Brook University, State University of New York, Stony Brook, New York.

Corresponding author address: Pavlos Kollias, Institute of Terrestrial and Planetary Atmospheres, School of Marine and Atmospheric Sciences, Stony Brook University, State University of New York, Stony Brook, NY 11794-5000.
E-mail: pavlos.kollias@stonybrook.edua

DOI: 10.1175/AMSMONOGRAPHS-D-15-0037.1

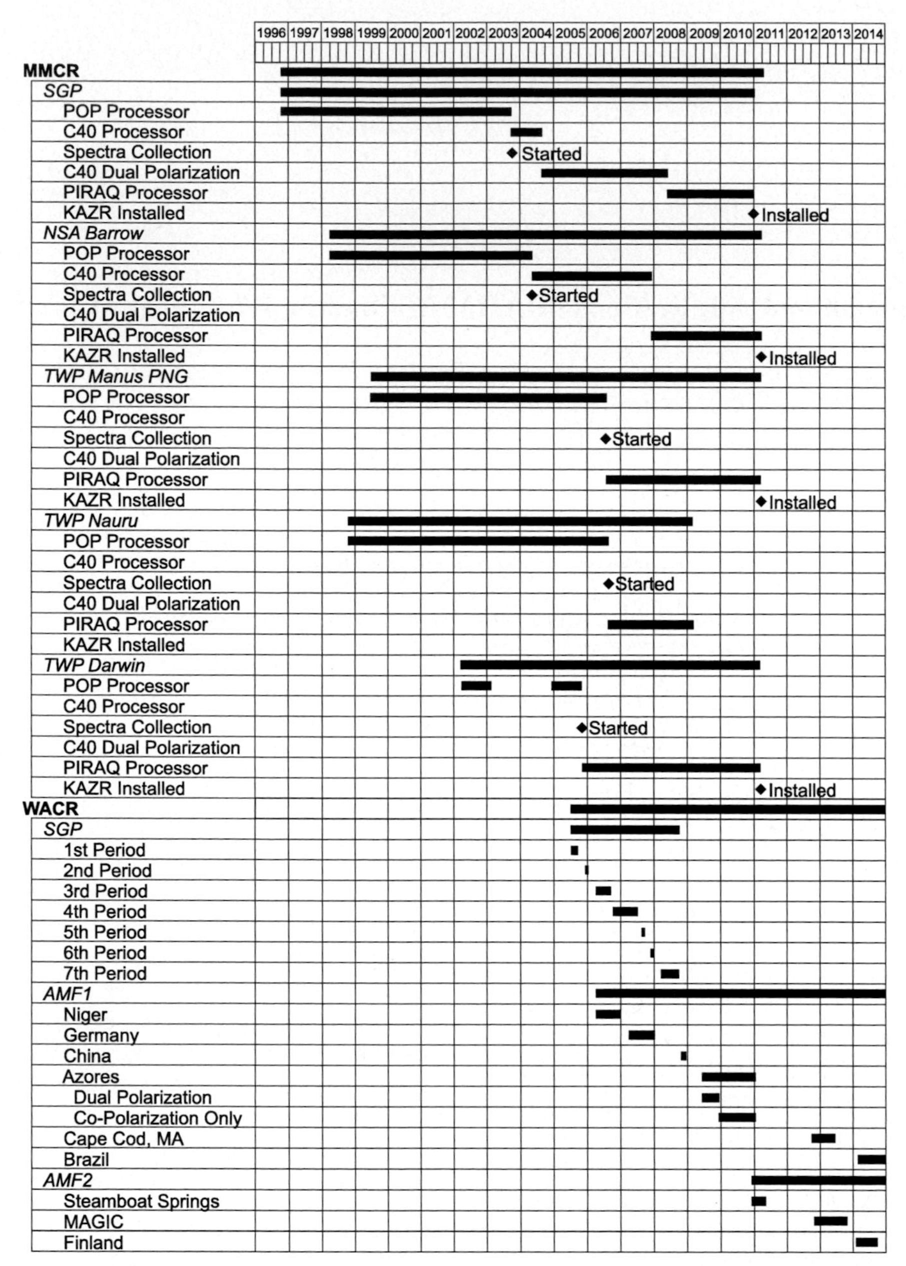

FIG. 17-1. Timeline of the key developments in the 20-yr history of the ARM cloud radar.

to an automated, operational system is a notable achievement of the ARM Program.

When the ARM Program began, the only millimeter-wavelength radars available for meteorological research deployment were a 35-GHz system managed by the National Oceanic and Atmospheric Administration (NOAA) Wave Propagation Laboratory and a 94-GHz system developed by Roger Lhermitte at the University of Miami. In addition, the Pennsylvania State University research group headed by Bruce Albrecht was in the process of building a deployable system funded by the Office of Naval Research. Now, 20 years later, the ARM Program supports a suite of millimeter-wavelength radars at each of its permanent sites and mobile facilities (Fig. 17-1). The data from these radars have, in many ways, revolutionized our knowledge of cloud structure and cloud processes, and we expect that new data will continue to expand our understanding in these areas (e.g., Kollias et al. 2007a). The success of the ARM Program has fostered the development and deployment of additional millimeter-wavelength radars at several sites in Europe and Asia. The proposal that ultimately led to the launch of the National Aeronautics and Space Administration (NASA)

TABLE 17-1. Various types of radars used in the atmospheric sciences today.

Designation	Band	Frequency (GHz)	Wavelength (mm)
Cloud mm-wave	W	75.0–110.0 (WACR: 95.04; W-SACR: 93.93)	2.7–4.0 (WACR: 3.15; W-SACR: 3.19)
Cloud mm-wave	Ka	26.5–40.0 (MMCR/KAZR: 34.86; Ka-SACR: 35.29)	7.5–11.3 (MMCR/KAZR: 8.6; Ka-SACR: 8.5)
Cloud/precipitation mm-wave	Ku	12.0–18.0	16.7–25.0
Cloud/precipitation cm-wave	X	8.0–12.0 (X-SACR: 9.73)	25.0–37.5 (X-SACR: 30.8)
Cloud/precipitation cm-wave	C	4.0–8.0	37.5–75.0
Weather cm-wave	S	2.0–4.0 (NOAA 3-GHz profiler: 2.835)	75.0–150.0 (NOAA 3-GHz profiler: 106)

CloudSat radar relied heavily on results obtained by the ARM Program. The goal of this chapter is to describe the evolution and current status of the ARM Program millimeter-wavelength radar effort.

2. A brief primer on cloud radar terminology and parameters

As the operating wavelength of radar becomes longer, its capability to detect small particles (e.g., cloud drops or ice crystals) decreases. This decrease is because the scattering efficiency is inversely proportional to wavelength to the fourth power for small particles. This loss in scattering efficiency with increasing wavelength can be offset to an extent by increased transmitter power and antenna size at the longer wavelengths. Furthermore, attenuation due to liquid water increases roughly as frequency squared. Taking into consideration all of these effects, practical ground-based cloud radar systems operating at Ka band have the highest sensitivity, followed by W band. In contrast, weather radars, which are used to track precipitation, utilize long wavelengths because their beams must penetrate to far distances and their targets of interest are large, precipitating particles. Table 17-1 illustrates the range of centimeter- and millimeter-wavelength radars commonly used in the atmospheric sciences today, together with ways of identifying them.

Millimeter-wavelength cloud radars (MMCRs) specific to the ARM Program are listed in Table 17-2. The ARM radars use shorter wavelengths in order to enhance their sensitivity to small cloud drops and ice crystals. However, as Table 17-2 illustrates, there are many other features of the ARM radars that are relevant to their sensitivity. For example, as the antenna diameter of a radar increases, so does its sensitivity; as the length of the pulses that it transmits increases, so does the power that it is sending up into the atmosphere, leading to enhanced sensitivity. Similarly, if the pulse repetition frequency (PRF) increases, more pulses are transmitted into the atmosphere per unit time, leading to an enhancement of radar sensitivity through signal integration. While transmitting longer pulses into the atmosphere increases sensitivity, doing so leads to a degradation of spatial resolution along the beam. To circumvent this trade-off, pulse compression, or pulse coding, techniques are implemented on some ARM radars. These techniques are ones for which the radar transmits long pulses, but with modulations within them to encode extra information, to enhance sensitivity but in which additional signal processing techniques that utilize the encoded information in the pulses are applied to the returned powers to maintain high spatial resolution.

Each pulse transmitted by radar leads to continuous (in time) power scattered back to the radar by the atmospheric constituents encountered by the radar pulse. In this case, time is equivalent to range (or distance) from the radar, and each sample of this return power by the radar receiver leads to an output signal corresponding to a particular range, or range gate, from the radar. A time series of output signals corresponding to different distances from the radar results from a single pulse, and multiple pulses lead to multiple time series of these range-dependent output signals. Radars are designed to analyze the output signals from multiple pulses that correspond to the same range gate (i.e., the same distance from the radar). These signals originating from the same range gate, called in-phase/quadrature-phase (I/Q) data, can be analyzed via fast Fourier transforms (FFTs) to produce spectra from which power (the zero-Doppler moment), mean Doppler velocity (the first Doppler moment), and Doppler spectral width (the second Doppler moment) estimates of the cloud particles at this range can be inferred.

A technique called coherent integration was used by ARM in the earlier days, in which up to 10 successive I/Q samples at a particular range gate were summed prior to FFT processing to improve signal-to-noise ratio. Although coherent integration is computationally efficient, the signal-to-noise improvement rolls off with increasing Doppler shift. Later signal processors in the W-Band ARM Cloud Radar (WACR) and Ka-Band ARM Zenith-Pointing Radar (KAZR) achieved the same overall processing gain without Doppler velocity–dependent errors by increasing the length of the FFT.

TABLE 17-2. Specifications of ARM Program profiling cloud radars. NLFM—nonlinear frequency modulation.

Parameter	MMCR (1995–2004)	MMCR (2004–10)	WACR (2005–present)	KAZR (2011–present)
Transmitter type	TWTA	TWTA	EIKA	TWTA
Frequency (GHz)	34.86	34.86	95.00	34.86
Pulse compression	Barker	Barker	No	NLFM
Pulse length (ns)	300–19 200	300–19 200	300	300–12 000
Antenna size (m)	2.0 (3.0)[a]	2.0 (3.0)[a]	1.8	2.0 (3.0)[a]
3-dB beamwidth	0.31 (0.19)[a]	0.31 (0.19)[a]	0.19	0.31 (0.19)[a]
PRF range (kHz)	6.0–14.0	6.0–14.0	7.5–10.0	3.0–10.0
Tx polarization	Linear	Linear	Linear	Linear
Rx polarization	Copolar	Co/cross-polar	Co/cross-polar	Co/cross-polar
Saturation (d*BZe*) at 1 km	+20	+20	+10	+35
Coherent integration	Yes	Yes	No	No
Signal dwell (sec)	9	2	2	2
Number of operational modes	4 (BL, CI, GE, PR)	5 (BL, CI, GE, PR, PO)	2 (GE, PO)	2 (GE, PO)
Mode sequence period (sec)	36	12–14	4	2
Receiver efficiency (%)	3–25	60–70	100	100
FFT length	64	128–256	256	256
Spectra recording	No	Yes	Yes	Yes

[a] At the SGP site.

Although several combinations of radar parameters can be selected, as discussed above, each comes with strengths and weaknesses. Specific combinations of radar parameters implemented for the ARM radars are called modes, and the more modes that the radar cycles through, the longer it takes before a mode is revisited. Table 17-2 illustrates that, as the ARM Program has progressed, the capability of the ARM radars has improved so that fewer modes need to be run to accomplish specific scientific objectives. Moreover, the efficiency (i.e., the percentage of pulses that the radar transmits that are processed by the data system—a measure of the data system's ability to keep up with the radar transmitter) of the ARM radars has improved, finally reaching 100%. This increase in efficiency has led to smaller time periods, called dwell times, necessary for the radar to collect and process the same number of pulses, leading to significant improvements in temporal resolution without a loss of sensitivity.

One final notable feature evident in Table 17-2 is the gradual implementation of polarization diversity on the ARM radars. The first ARM radars transmitted (Tx) linearly polarized electromagnetic radiation and received (Rx) only the copolarized signal (i.e., the signal that has the same polarization as the transmitted wave). In 2004, the ARM radars were upgraded to receive both the copolarized and cross-polarized signals.

3. The beginnings of the DOE ARM Program cloud radar

The first cloud radar, the MMCR, developed by the ARM Program was deployed at the ARM Southern Great Plains (SGP) site in 1996. A search through the ARM data archive yields the date of 8 November 1996 for the first file generated by this MMCR. Many technical achievements over the decades after World War II enabled the generation of this first file. So what are the watershed events that led to the development of the ARM Program's cloud radar activities? From our perspective, these events were the First ISCCP Regional Experiments (FIRE) Second Cirrus Intensive Field Observation (IFO) campaign in southern Kansas from 13 November 1991 through 7 December 1991 (e.g., Uttal et al. 1995) and the FIRE Second Marine Stratocumulus IFO campaign in the Azores Islands in the eastern North Atlantic from 1 June 1992 through 28 June 1992 (Albrecht et al. 1995).

The FIRE Second Cirrus IFO campaign in Kansas was the prototype for the ARM SGP site (Sisterson et al. 2016, chapter 6). Although it was only a month-long field campaign, the FIRE experiment brought together for the first time at a single site many instruments, each with a separate heritage and representing state-of-the-art remote sensing of cloud properties, to make measurements in the same or adjacent volumes of atmospheric air in support of cloud and radiation studies. These instruments included the dual-polarimetric, scanning Ka-band (35 GHz) Doppler radar developed by NOAA (e.g., Pasqualucci et al. 1983), the W-band (94 GHz) Doppler radar developed by the Pennsylvania State University (Clothiaux et al. 1995) using the design of Lhermitte (1987), and a suite of four lidars (e.g., Sassen et al. 1995). The FIRE Second Cirrus IFO in Kansas was collocated and coincident with the DOE- and NASA-funded Spectral Radiance Experiment (SPECTRE), which collected state-of-the-art surface radiation measurements (Ellingson and Wiscombe 1996; Ellingson et al. 2016, chapter 1). One of the key outcomes of the FIRE Second Cirrus IFO campaign was that the Pennsylvania State University W-band radar was run quasi operationally for the entire IFO period. The FIRE Second Marine Stratocumulus IFO campaign that followed six months later provided additional support for the use of millimeter-wavelength radar by demonstrating that the radar could be transported easily to a somewhat remote location and used again in a quasi-operational mode in conjunction with other instruments, including the NOAA Ka-band radar, to investigate cloud properties. Equally importantly, the Pennsylvania State University and NOAA groups, among others, demonstrated the scientific utility of the radar data and began the development of analysis software for the radar, including an early version of a cloud mask algorithm (e.g., Clothiaux et al. 1995; Miller and Albrecht 1995; Uttal et al. 1995; Danne et al. 1996).

These two field campaigns, particularly the FIRE Second Cirrus IFO campaign in conjunction with SPECTRE, represented the early beginnings of the ARM Program's radar activities that focused on clouds and radiation measurements in support of understanding cloud processes and improving climate models (Stokes 2016, chapter 2). With ARM sites envisioned in remote areas of the world, ARM's cloud radar faced numerous challenges beyond those early field campaigns, including going beyond quasi-continuous operation of cloud radars to full 24/7 operations with no onsite technical support.

4. The need for enhanced radar sensitivity

Measurements from the FIRE Second Cirrus IFO campaign made one thing abundantly clear. Although MMCRs could detect many types of clouds, they were not able to match the sensitivity of the lidars in detecting thin cirrus. Inspection of the displays of the polarization diversity lidar during cirrus case study periods showed that, at times, the upper troposphere contained an abundance of ice crystals that were below the detection limit of the Ka- and W-band radars. In the case of thicker cirrus, the radar and lidar observations of cirrus extent often coincided. These early observations during the FIRE Second Cirrus IFO campaign established the necessity of building cloud radars for the ARM Program with as much sensitivity as technically possible.

5. Mitigating the impacts of attenuation at cloud radar operating frequencies

During the FIRE Second Marine Stratocumulus IFO campaign in the Azores Islands, high-powered lidars were not available. The cloud radars, low-powered laser ceilometers, and passive microwave radiometers were the tools of the day. The reasons for the change from the FIRE Second Cirrus IFO campaign were obvious: the marine clouds that shrouded the Azores Islands during the campaign were thick liquid water clouds, the numerous droplets of which rapidly attenuated lidar signals, making these clouds of less interest to the lidar community. These clouds were detected readily by the cloud radars, which consistently profiled the location of cloud liquid water from cloud base to cloud top (Frisch et al. 1995; Miller and Albrecht 1995). In the marine boundary layer during this campaign, as well as for the FIRE Second Cirrus IFO campaign, attenuation of the cloud radar signals was not a dominant issue. In the case of water vapor and liquid water attenuation, Ka-band cloud radars have a clear advantage over W-band radars because absorption is significantly less at the longer wavelength; W-band attenuation is often severe in rain or clouds of high liquid water content or when scanning horizontally through regions of high humidity.

Nowhere was the attenuation difference better demonstrated than during the Maritime Continent Thunderstorm Experiment (MCTEX; Keenan et al. 2000) in the Tiwi Islands, Northern Territory, Australia, November and December 1995. These islands are famous for Hector, which are massive thunderstorms that develop over them every early afternoon at this time of the year. For the MCTEX experiment, the University of Massachusetts deployed a dual-frequency Ka/W-Band cloud profiling radar system (CPRS; Mead et al. 1994), while NOAA deployed a vertically pointing S-band radar (Ecklund et al. 1999). On one occasion, a Hector formed right over the radar installation; at this time, the W-band radar signal completely attenuated by 200 m and the Ka-band by 2 km, but the S-band radar detected hydrometeors up to 18 km of altitude. These observations indicate that, while shorter-wavelength radars are better able to detect small hydrometeors, longer-wavelength radars have value in detecting larger hydrometeors in cases where attenuation of the shorter-wavelength radar beams is severe.

6. The ARM Program MMCRs

One of the early discussions in the ARM Program was whether to deploy Ka- or W-band radars to its sites. W-band radar signals have greater sensitivity to small hydrometeors, while Ka-band radar signals suffer less attenuation. The decision was made to go with Ka-band radar for several reasons of roughly equal weight. First, Ka-band wavelengths suffer less attenuation. Second, Ka-band technology was considered more robust than W-band technology because the radar community had

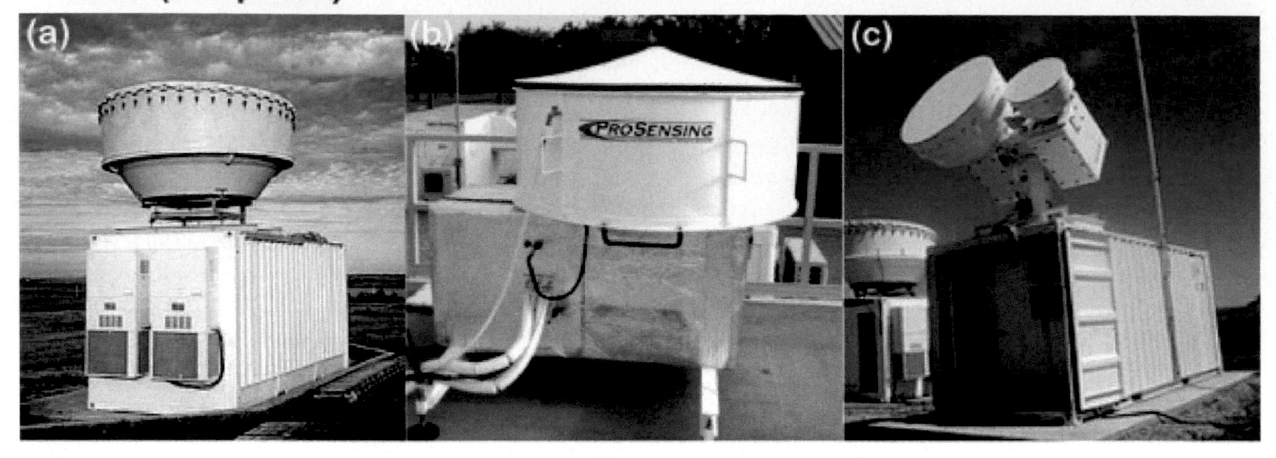

FIG. 17-2. (a) The MMCR with its 10-ft antenna at the ARM SGP site; the KAZR has a similar external design to the MMCR; (b) the WACR; and (c) the Ka/W-SACR.

more experience with it. (As it turned out, W-band components were equally robust, but that was unclear at the time.) Third, W-band components were more expensive than Ka-band components.[1]

The NOAA radar group was quite familiar with Ka-band radar technology and was selected to develop a Ka-band cloud radar system to be deployed at each of the ARM fixed sites. The first system was deployed at the SGP site in 1996 (Fig. 17-2a). The design focus was on enhancing radar sensitivity. Moran et al.'s (1998) paper on the design of the ARM MMCR emphasized the importance of sensitivity and design features that were intended to enhance it, primarily the pulse-coded radar signals that delivered both sensitivity and high spatial resolution (Figs. 17-3a,b). Because the pulse-coded waveforms had detrimental effects close to the surface, especially evident in Fig. 17-3a, Moran et al. (1998) designed several different operational modes for the radar (Figs. 17-3a–d), each one intended for detection of specific cloud types. The coded long-pulse "cirrus" mode (Fig. 17-3a) was designed for maximum sensitivity but with contamination issues close to the surface. The coded short-pulse "boundary layer" mode (Fig. 17-3b) was designed for as much sensitivity as possible to thin boundary layer clouds while minimizing clutter near the surface. The "general" mode (Fig. 17-3c) was designed to be an artifact-free mode but with saturation issues during precipitation (see Table 17-2 for the reflectivities at 1 km that lead to saturation, or too much power for the receiver to handle linearly). The "precipitation" mode was designed as a robust mode that minimized all known artifacts (e.g., velocity folding) but unfortunately did not address receiver saturation. Sensitivity of all modes was also enhanced by the use of a large antenna with a 10-ft diameter at the SGP site. Smaller, 6-ft-diameter antennas were used at the remote sites because the large 10-ft antenna could not fit into the sea containers used for transportation.

Another innovation sought from the beginning of the ARM Program was collection of spectral data in addition to moments. Unfortunately, the data volumes associated with spectral data were too high to transmit routinely over the Internet when the ARM Program got underway. This was true of the ARM SGP site where communications were best, as well as for the remote ARM Tropical Western Pacific (TWP) and North Slope of Alaska (NSA) sites, where communications were much more severely limited. The ARM Program has worked continuously to improve data handling and with much success, as will become clear later.

7. Identifying weak returns and dealing with atmospheric plankton

Associated with specialized hardware for maximizing radar sensitivity came the need for processing software that built upon it. No longer was it sufficient to apply a simple threshold to radar return signals in identification of hydrometeors, as is often done with lower-frequency precipitation radars, where signal-to-noise ratio is typically high. Cloud radar signals, especially for thin clouds,

[1] At an early ARM Program science team meeting, the ARM Chief Scientist Gerald Stokes sent Thomas Ackerman, Robert Kropfli, and Robert McIntosh off to a corner and told them to come back with a final recommendation for the millimeter-wavelength system. Their consensus was a Ka-band radar.

FIG. 17-3. MMCR mode data on 9 Apr 1997 for the (a) coded long-pulse cirrus (CI) mode, (b) coded short-pulse boundary layer (BL) mode, (c) general mode, and (d) precipitation (PR) mode. (e) The ARSCL VAP for 9 Apr 1997. (f) Particularly severe atmospheric plankton in the general mode on 19 Jun 1997. Mean Doppler velocities on 11 Apr 1997 for the (g) general and (h) precipitation modes.

are often buried in the noise, and effort is required to isolate signal from noise. Clothiaux et al. (1995) investigated two different spatial filters for extracting signals from noise, and Clothiaux et al. (2000) implemented spatial filtering in the operational processing of MMCR Doppler moments. The output of the signal extraction, mode merging, and plankton clearing processing came to be called the Active Remote Sensing of Clouds (ARSCL) value-added product (VAP; Fig. 17-3e). As they pointed out, the two most important elements of operational processing of MMCR data are extracting weak signals from the noise and distinguishing signals from insects, spider webs, floating agricultural debris, and any other nonhydrometeor return, or what came to be called "atmospheric plankton" after Lhermitte (1966).

The complicating effects of atmospheric plankton should not be underestimated, especially during the warm months at the ARM sites. Many of the examples of MMCR signal returns in Moran et al. (1998) were from relatively warm months—12 May 1997 (their Fig. 3), 10 April 1997 (their Fig. 4), and 19 June 1997 (their Fig. 5)—and these examples clearly demonstrate strong radar signals from atmospheric plankton below approximately 2 km of altitude, where it often dominates the signals. A worst-case example occurred on 19 June 1997 (their Fig. 5 and reproduced here as Fig. 17-3f) when the atmospheric plankton dominated the signal up to 2.5 km of altitude and could be seen sporadically as high as 3 km of altitude. These returns from nonmeteorological targets severely complicate the operational processing of MMCR signals to isolate the cloud and precipitation returns. A procedure was developed for operational processing of MMCR data using lidar signal returns, which were unaffected by atmospheric plankton, to aid in identifying returns from hydrometeors and atmospheric plankton. However, this approach was not optimal, only allowing for the identification of plankton below cloud base.

The following techniques have been used with some success to identify plankton:

- Examination of linear depolarization ratio (LDR; i.e., the ratio of the power received in the orthogonal, or cross-polarized, channel to that received in the transmission, or copolarized, channel of a dual-receiver channel radar when a linearly polarized signal is transmitted);
- Processing of high-frequency (pulse-by-pulse) radar returns (i.e., the I/Q data);
- Examination of the shape of the radar Doppler spectra (Luke et al. 2008).

Nonetheless, the effect of plankton is much smaller at W band, which has been one of the factors prompting the ARM Program to deploy W-band radars in recent years.

The significant contributions of atmospheric plankton to observed signal returns at Ka-band frequencies probably should not have come as a surprise. In a series of papers from the late 1980s and early 1990s, Joseph R. Riley of the Radar Entomology Unit of the Natural Resources Institute, United Kingdom, and others made clear the value of 8-mm-wavelength radars in tracking insects. Unfortunately, it sometimes takes time for information to cross disciplinary boundaries.

8. The Active Remote Sensing of Clouds value-added product

In the early years of the ARM Program, the transitioning of software developed by scientists to operational applications within the ARM Program infrastructure was not a structured activity. Recognizing the need for operational radar processing, Eugene Clothiaux, working with ARM scientists Jimmy Voyles and David Turner in the early 2000s, created an initial operational version of the processing software of Clothiaux et al. (2000). Eventually, toward the middle of the decade, responsibility for the code was transferred to the ARM Program, where it continues to be maintained and developed. Under ARM Program guidance, the active remote sensing of clouds value-added product (ARSCL VAP) became the successful product that it has been and still is.

At its core, the ARSCL VAP contains the results of processing efforts to separate weak signals from noise, merge the different operational modes into holistic views of the vertical columns above the ARM sites, and mitigate the effects of atmospheric plankton (Fig. 17-3e). The primary goal of the ARSCL VAP, along with the MMCR, is to provide a high-temporal- (10 s) and spatial- (~45 m) resolution, long-term view of the vertical distributions of hydrometeors above the ARM sites. As noted in the article by Moran et al. (1998), the MMCR is intended for "climate research," with the intent of mapping in time the macroscopic properties of clouds and the radar reflectivities associated with them.

The focus on macroscopic cloud properties is reflected in the ARSCL VAP. It contains a series of three height-versus-time cloud radar reflectivity (zero-Doppler moment) fields in units of decibels of reflectivity (i.e., Reflectivity, ReflectivityNoClutter, ReflectivityBestEstimate) that are a blend of the reflectivities from the different operational modes running on the MMCR and tuned to the detection of different cloud types. Each field is the result of additional processing applied to the reflectivities in order to identify clutter, with the parameter

ReflectivityBestEstimate representing the final outcome of the processing. The product also contains cloud-base heights versus time from a proprietary retrieval algorithm applied to laser ceilometer backscattering data (e.g., CloudBaseCeilometerStd), cloud-base heights (e.g., CloudBaseMplCloth to be replaced with CloudBaseMplZwang), and height-versus-time masks of all significant detections above molecular scattering (e.g., CloudMaskMplCloth to be replaced with CloudMaskMplZwang) from micropulse lidar backscattering data. Cloud-base heights retrieved from laser ceilometers and micropulse lidars are merged to form the best-estimate cloud-base height field CloudBaseBestEstimate. A series of fields called CloudLayerBottomHeights and CloudLayerTopHeights are derived from the ReflectivityBestEstimate field together with CloudBaseBestEstimate and one of the micropulse-lidar-derived cloud mask fields and represent the most succinct information on the time evolution of the vertical structure of clouds contained in the ARSCL VAP. In fact, the cloud-base height and cloud layer information contained in the main ARSCL VAP product are extracted from it and placed into two separate data products that contain files with much smaller sizes.

The ARSCL VAP also contains blended height-versus-time fields of the first (MeanDopplerVelocity) and second (SpectralWidth) radar Doppler moments, as well as a measure (SignaltoNoiseRatio) of the strengths of the returns leading to all three Doppler moments. Because many analyses benefit from knowing whether or not precipitation is reaching the surface, time series of a surface precipitation flag (CloudBasePrecipitation) were created from optical rain gauge data or a sensor on the microwave radiometers that detects liquid water.

Because interpolation of the operational mode datasets onto a single time grid is not optimal for all studies, the ARSCL VAP also contains a series of files that contain the original operational mode Doppler moments together with the CloudBaseBestEstimate field interpolated from the 10-s grid onto the operational mode time grid. These mode-based files also contain height-versus-time parameters called qc_RadarArtifacts and qc_ReflectivityClutterFlag. These two parameters contain flags constructed from the 10-s ARSCL VAP product and interpolated to the operational mode time grid that inform the user of issues in the individual mode data gleaned from an analysis of all of the mode data.

As we consider briefly at the end of the chapter, the MMCRs and the ARSCL VAP have gone a long way toward satisfying the intended purpose of providing useful observations on the macroscopic properties of clouds and have expanded the reach of cloud radar data into modeling studies. However, every operational product that has value encourages use and increased scrutiny, which in turn leads to the discovery of additional problems and the desire for additional features. The result is an accumulation of an ongoing set of tasks to improve the process. This has certainly been our experience with the ARSCL VAP.

9. Fixing problems in the ARSCL VAP

By the mid-2000s plenty of ARSCL data files were available for scientists to analyze. As they did so, several problems with the ARSCL VAP were identified (e.g., Kollias et al. 2005). Among the more important problems were the following:

1) ARSCL VAP radar reflectivities were not calibrated.
2) Collection of water on the MMCR radome affected calibration in unknowable ways.
3) All MMCR modes (including the precipitation mode) were often saturated during precipitation events (Fig. 17-3d).
4) Scientific usability of the Doppler moments was compromised significantly by the long 9-s averages used to create them.
5) Atmospheric plankton contamination of the returns for the SGP site MMCR was so severe during the warm months (Fig. 17-3f) that it rendered long-term studies of boundary layer clouds at this site problematic.
6) Doppler moment accuracies were compromised by frequent velocity aliasing in the general mode (cf. Fig. 17-3g to Fig. 17-3h, in which positive, correct velocities in the "likely saturated" area of Fig. 17-3h have wrapped, or aliased, into incorrect negative velocities in the "velocity artifacts" area of Fig. 17-3g), boundary layer mode, and cirrus mode.
7) Doppler spectra during periods of strong returns had mirror images at oppositely signed Doppler velocities that, while 30 dB weaker than the desired returns, nonetheless corrupted the spectra sufficiently to preclude their use in cloud studies.
8) Doppler spectra had processing artifacts near zero Doppler velocities that made their use in cloud studies problematic.
9) Doppler moment accuracies were compromised by use of operational mode data in the final VAP that were from low signal-to-noise ratio measurements at times when other modes had higher signal-to-noise ratio measurements.

Inspection of this long list of issues with the MMCR output and associated ARSCL VAP might be viewed

FIG. 17-4. The waveguide from the NSA MMCR has worked its way free from its connection to the antenna port. This was a slow process that spanned many months of the harsh Arctic environment at the NSA site.

as a failure of the concept and/or its implementation. Not surprisingly, we have a very different perspective. For the first time, cloud radar signals ranging over seven orders of magnitude of returned power, from −50 to 20 dBZe (where dBZe is a popular unit for characterizing the power backscattered to a radar represented by the equivalent radar reflectivity factor Ze), were being scrutinized closely by scientists and engineers driven to use the results for understanding atmospheric processes at high resolution and in great detail. As a result of this scrutiny, sets of interlocking, often nonlinear, problems were found, resulting from a combination of technical (hardware) issues, software issues, computer limitations, and boundary layer meteorology. These issues were challenges that had to be overcome, one by one, to improve the ARM Program's capability to characterize cloud properties at each of its sites.

Each of the issues in the list above has served to launch efforts within the ARM Program to solve them and improve ARM radar products. As this process has evolved, the ARM Program's cloud radar activities have grown so that today the program is at the threshold of not only mapping the macroscopic properties of clouds but obtaining highly detailed information on cloud microphysical processes inside them as well. This evolution within the program is best illustrated by consideration of how these problems have been addressed and solved within the program.

10. Calibrating vertically pointing cloud radars

From the beginning of the ARM Program, calibration of the cloud radars has been a key consideration. The first version of the MMCR contained hardware for closely monitoring the transmit power and injecting a noise signal into the receiver as part of normal operations. However, this left several passive radio-frequency components and the antenna out of the calibration loop. The ARM Program's radar engineers periodically measured the loss in these radio-frequency components during annual site visits. Typically, the MMCR antennas were thoroughly characterized once in their life cycle just after their manufacture. However, it was not possible to measure the performances (e.g., gain, beamwidth, cross-polarization isolation, etc.) of these 2–3-m-diameter Cassegrain antennas in the field. Attempts were made to place calibration targets (e.g., BBs, ball bearings shot out of a paintball gun, and dangling a metallic sphere under a helicopter) in the radar beam. However, with only 0.3° beam widths, getting these targets into exact regions within the radar field of view is extremely difficult and certainly not feasible on a routine basis.

In 2004, the ARM Program began the procurement of the WACR (Fig. 17-2b) to be collocated with the MMCR. Because the WACR had a much smaller (24-in diameter) antenna, a splash plate was incorporated into its design so that its beam could be deflected to point at a trihedral corner reflector target. For the first time, the ARM Program gained the capability to calibrate one of its radars with an external calibration target that had a known reflectivity value.

An example of why it is important to calibrate with an external target was evidenced by a component failure at the ARM NSA site in Barrow, Alaska, during 2006 through 2007. In addition to being in a harsh Arctic environment with severe cold, the radar is sited 600 m from a saltwater lagoon. Corrosion slowly acted on the waveguide that was attached to the antenna feed, eventually breaking the machine screws that attached the waveguide to the feed (Fig. 17-4). This was such a slow process that diminishing radar returns went unobserved until an ARM radar engineer arrived at the site for an inspection. He was amazed that the radar was detecting any return power at all!

This inability to calibrate a vertically pointing MMCR is perhaps its greatest weakness. This problem was recognized during the early design discussions in the 1990s, but no one realized that it would be such a significant, insidious problem for operational cloud radars, because there were no such systems at the time. The possibility of scanning the cloud radar (for both calibration and science) was considered but was dropped, because it was a significant challenge to build an operational vertically pointing system in itself, scanning hardware was expensive and operationally problematic at the remote sites,[2] and slow operational degradation was not seen as a critical issue. (Degradation of the power source was considered, but there were ways to monitor this process without the need for a total system calibration via calibration target measurements.) The view now shared widely across the program is that calibration is about proving to oneself that every piece of the radar is working and, to the extent that a piece is not working optimally, quantifying the impact of that degradation on the overall estimates of the power backscattered by hydrometeors within each radar sample volume. In retrospect, this view should have been implemented much earlier, but it is unclear how the program could have addressed this problem during its first decade of operations, given the limitations and cost of scanning technology at the time.

These painful lessons about calibration have not been lost on the program. In January 2011, the SGP MMCR was replaced with the new KAZR (which is similar in outward appearance to the MMCR, as Fig. 17-2a illustrates). By itself, the KAZR, like the MMCR, cannot be calibrated from the antenna to the receiver. But every deployment of a KAZR will come with the deployment of a dual-frequency Scanning ARM Cloud Radar (SACR; Fig. 17-2c), one frequency of which is in the Ka band. At the SGP site, the first Ka-band SACR (Ka-SACR) was deployed in May 2011. The program has adopted the view that the deployment of every SACR must include placement of a corner reflector nearby. The SACRs will be calibrated both internally via transmitter and receiver measurements and externally via corner reflector measurements. The calibration of the SACRs will then be transferred to the calibration of the KAZRs, with comparisons of the SACR calibration to the KAZR's internal transmitter and receiver measurements used to identify problems with the KAZR that require referencing in some way to measurements from a known target.

[2] The discussions at the time included the trade-off among antenna size, radar sensitivity, and pointing (or scanning). The SGP Ka band with its 10-ft antenna dish was designed to achieve maximum sensitivity to low hydrometeor concentrations, and it was enormously successful from that perspective. But the large size of the antenna precluded pointing because of weight and cost, not to mention the question of alteration in the antenna shape if it were tipped from horizontal to vertical orientation.

In the fall of 2012, the ARM Program radar operations and engineering group reported its first findings using corner reflector measurements to calibrate the SACRs. The findings were promising. By scanning across the corner reflector using a set of stacked raster scans, they were not only able to estimate overall SACR calibration but were also able to map out rough outlines of the antenna beam pattern. In this way, they were able to identify problems with the SACR antenna first deployed to the Two-Column Aerosol Project (TCAP) on Cape Cod, Massachusetts. They were also able to use corner reflector measurements to track in time the impacts of icy precipitation on the SGP SACR's performance. As ice accumulated on the SACR's radome, attenuation increased, and the beam pattern of the SACR changed considerably. As the ice melted and the radome dried out, they were able to watch the SACR return to nominal operating characteristics. As a result of these early successes, corner reflector raster scans are being implemented as part of normal SACR scanning operations on a time scale sufficiently fine to track changes in calibration during precipitation events.

11. Avoiding cloud radar reflectivity saturation from precipitation

The amount of radar signal power backscattered from a single spherical liquid drop that is small compared to the radar wavelength is approximately proportional to the drop diameter to the sixth power. The total power backscattered from a collection of spherical drops all the same size is the number of drops in the illuminated volume times the backscattered power per drop. Consider a liter of air with one precipitating drop with a diameter of 1 mm and 10^5 cloud droplets each with a diameter of 10 μm. The ratio of the power backscattered by the single precipitating drop to that from all of the cloud droplets is 10^7. (Is that not amazing? One precipitating drop backscatters 10^7 more power than 10^5 cloud droplets.) Another challenge to consider is that the MMCR is sensitive to echoes as weak as -50 dBZe, and their intensity can reach 20 dBZe. But accurately measuring the power over such a dynamic range is no easy engineering task to accomplish (consider holding a single penny versus 10^7 of them), and the MMCR saturates on the high end of this power range. Each time a return power saturates the receiver, the bias in the estimate (in this case underestimation) of the returned power increases.

To mitigate this problem, the return power to the radar needs to be reduced by a known amount before it gets to the hardware in the receiver that saturates. In 2004, at a meeting at the NOAA Environmental Technology

Laboratory (now known as the NOAA Earth System Research Laboratory in Boulder, Colorado, ARM Program and NOAA radar engineers and scientists agreed on a possible mitigation strategy. The MMCR receiver has four switches that protect it during transmission of the high-power pulse. During transmission, the switches are closed, and any power that leaks from the transmitter into the receiver is blocked. When a transmit pulse has cleared the radar, the switches are opened, allowing backscattered power from the atmosphere to enter the radar receiver. Each switch attenuates signals by an amount of 20–25 dB. It was decided to create a new operational mode whereby one of the switches was kept closed (protection mode) even when the radar received signals from the atmosphere. By keeping the switch closed, the received atmosphere signal was attenuated by an amount equal to 20–25 dB, thereby extending the MMCR receiver dynamic range by 20–25 dB. Knowing how much the signal was attenuated by the switch, an equal amount was added during data postprocessing, thus producing a calibrated reflectivity. This approach was implemented only during the operation of the MMCR precipitation mode and was put into operation within the MMCRs at the SGP and TWP sites over the period from mid-2004 through mid-2006. After a period of experimentation, during which time the precipitation mode data of the MMCR's with the innovation were corrupted by timing issues, this approach was implemented successfully, and the quality of the precipitation mode data was vastly improved.

Leaving a switch closed in the precipitation mode when the radar was receiving atmospheric returns is a clever, but not elegant, solution to saturation. Despite the additional protection, the MMCR receiver still saturated in heavy precipitation (above 5 mm hr^{-1}). As importantly, the software (ARSCL) that was designed to select the mode with the highest signal-to-noise ratio (SNR) was never adjusted to routinely select the lower SNR but nonsaturated, radar reflectivities of the precipitation mode.

In the design of the new KAZRs and SACRs, saturation of the receiver during periods of heavy precipitation is dealt with more elegantly. The KAZR features an improved radar receiver and a larger dynamic range for the signal powers within it; thus, its nominal saturation level is 15 dB higher than that of the MMCR (see Table 17-2). This is, of course, not enough. The current plan is to implement amplitude tapering on the transmitted pulse to reduce the return signal power. In addition, the ARM Program has strengthened its ability to profile precipitation by collocating 915-MHz radar wind profilers with the profiling cloud radars (Tridon et al. 2013).

Over the years, interactions between scientists using ARM radar data (and identifying problems within it) and ARM radar engineers have led to substantially improved radar systems. These interactions are no less important going forward, given the growth of the number of radars in the ARM Program and the complexity of the radar data products produced by them.

12. The rise of the digital receiver: Temporal resolution and Doppler spectra

In the early design and implementation of the ARM cloud radars, the need for continuous and robust operations of the MMCR led to a decision to use two digital signal processors (DSPs) based on those in use by the NOAA wind-profiler network along with the OS/2 operating system-based profiler online program (POP) software that came with them. The POP was the standard wind-profiler processor that provided reliable moments data (i.e., reflectivities, Doppler velocities, and spectral widths) but at a low collection efficiency (5%–30%). The initial temporal resolution of the MMCR data was 9 s; however, most of the time in the 9-s window was dedicated to signal processing rather than signal integration, which is the definition of low efficiency. This initial choice of the DSP and POP software, made largely for reasons of cost and a shorter development cycle, eventually led to issues that created strong pressure for innovation.

The pressure for innovation was driven by several significant science issues. The 9-s temporal resolution of the original MMCR Doppler moments was dictated by the need for averaging in order to increase the SNR of the measurements and to keep the data volume down. The emphasis on SNR was again in response to the need for sensitivity in support of cloud macrophysical studies. The time interval and the data volume restrictions, however, were driven by the available computer technology in the 1990s. The cost of computer power, especially data storage and networking, was much higher early in the 1990s. Despite the data limitations, several ARM scientists were eager for higher-resolution data both in time and in spectral space because they recognized the research potential of these data. For example, Kollias et al. (2001) demonstrated that 1-s temporal resolution was required for dynamical studies of updrafts, downdrafts, and turbulence in fair weather cumuli. Additional research by Kollias et al. (2005) demonstrated the importance of time scales shorter than 9 s that could only be captured by a more efficient processor. These science considerations, together with the availability of more capable radar and computer hardware, led to a processor upgrade effort that was undertaken in the period from 2003 to 2005.

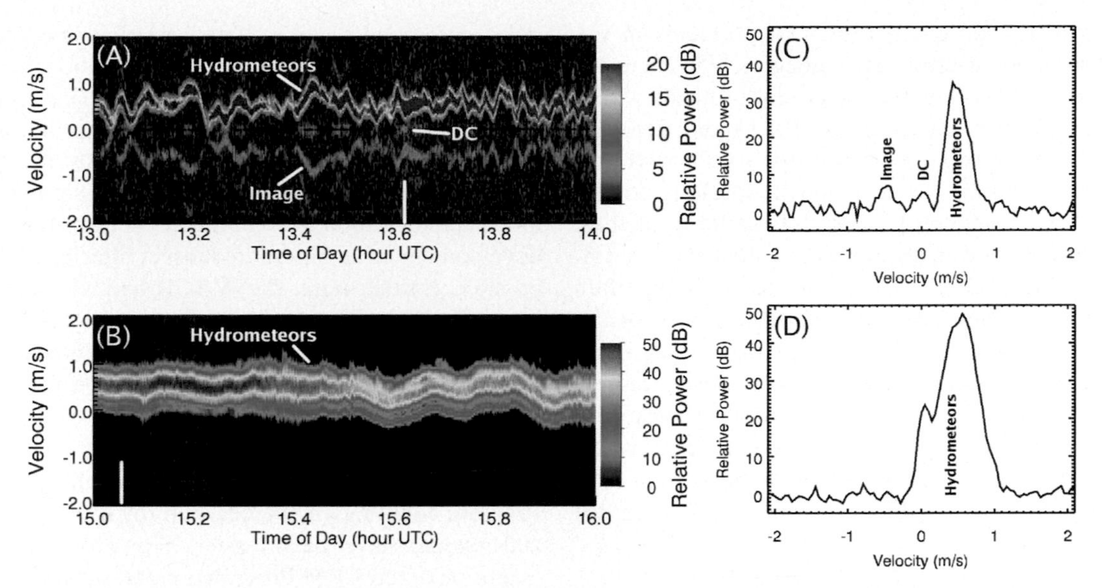

FIG. 17-5. (a) Example of an MMCR Doppler spectrogram collected over 1 h at constant height in a cloud at the NSA in 2009. (b) Example of a KAZR Doppler spectrogram collected over 1 h at constant height in a cloud at the NSA in 2012. (c) One of the Doppler spectra shown in (a), but as relative power vs velocity. The spectral image is labeled as "image" in this figure and is not the result of any physical phenomena in the atmosphere; rather, the spectral image results from hydrometeor contributions to powers in the spectrum at positive velocities artificially bleeding into powers at negative velocities because of imperfections in the radar receiver. (d) One of the KAZR Doppler spectra shown in (b), but as relative power vs velocity.

With support from the ARM Program and NOAA, Radian, Inc., (subsequently purchased by Vaisala) began the design of a new DSP board for the MMCR. The new board, based on the C-40 DSP from Texas Instruments, was a five-DSP-on-the-board processor. The advantages of the new design were the use of higher clock frequencies and multiple DSPs working in parallel to accelerate radar processing power. The hardware ran under a new radar control software program called Lower Atmosphere Profiler—Extensible Markup (LAP-XM), which in turn ran in the Windows 2000 environment. This new processor was eventually installed at the SGP and NSA sites as well as on the NOAA MMCR that was deployed to Eureka, Canada. The processor efficiency of the new board was 50%. Unfortunately, Texas Instruments manufactured only three such processor boards before this line was discontinued.

Around the same time that the C-40 was discontinued, the National Center for Atmospheric Research licensed its PC-Integrated Radar Acquisition System (PIRAQ-III) technology to Vaisala. The new PIRAQ-III was selected as the upgrade for the TWP sites. The benefit of the PIRAQ-III was that the boards and software were supported products of Vaisala (Widener et al. 2004). Moreover, the processor efficiency of the PIRAQ-III was close to 70%, an improvement over the C-40 processors. The first PIRAQ-III upgrade was completed at Darwin in November 2005. The upgrades for Manus and Nauru were completed the following year. In an attempt to enhance processor efficiency for the NSA Indirect and Semi-Direct Aerosol Campaign (ISDAC; McFarquhar et al. 2011) in April 2008, the C-40 processor at NSA, which was installed just before the NSA Mixed-Phase Arctic Cloud Experiment (M-PACE; Verlinde et al. 2007) in October 2004, was replaced by a PIRAQ-III in December 2007.

During the same period from 2004 to 2005, the MMCR sampling modes were revised. The new modes introduced wider velocity ranges for the Doppler spectra and shorter (1.5 s) integration times. Most importantly, the ARM Program started to record radar Doppler spectra on more than a case-study basis. Preliminary analysis of Doppler spectra recorded by the C-40 and PIRAQ-III boards showed the presence of spectral images when the peak spectral power was higher than 25–30 dB above the noise (Fig. 17-5). This artificial feature was a real nuisance to radar meteorologists and scientists who used the Doppler spectra for cloud microphysical retrievals. As it turned out, features like this in the NSA ISDAC PIRAQ-III Doppler spectra rendered them no better, and often far worse, than the NSA MPACE C-40 Doppler spectra.

As MMCR Doppler spectra became available throughout the mid-2000s, the scientists who analyzed them found that they contained a wealth of information, especially regarding the study of precipitating liquid drops and ice crystals in the presence of cloud liquid water droplets. But there were two problems in the MMCR Doppler spectra that immediately became apparent: removal of low-level constant voltages within the radar receiver led to artifacts near zero-Doppler velocities, and strong power returns from downward-moving precipitating particles led to spurious powers showing up at oppositely signed Doppler velocities. These spurious power returns were called spectral images (see Fig. 17-5c). While these images were often 30 dB down (or one one-thousandth) in power of the actual returns from precipitating particles, they often fell at Doppler velocities where cloud power returns occurred. As a consequence, the Doppler spectra were compromised for study of cloud power returns, which require either no images in the Doppler spectra or ones that are at least 40 dB down (or one ten-thousandth) in power.

13. The WACR as a test bed for future MMCR improvements

While the MMCR digital receiver upgrade saga was unfolding, a parallel effort from 2003 to 2005 was underway to acquire a 94-GHz (W band) radar at the ARM SGP site. The WACR acquisition was motivated by the need to improve detectability of clouds in the boundary layer at the ARM SGP site during the warm season, a period when insect echoes were making the qualitative and quantitative estimation of hydrometeor returns very challenging.

The improved capability of W-band radars to separate hydrometeor returns from those of insects is based on the suppression of the insect radar returns by 20 dB compared to the MMCR because of the drop in backscattered powers (in this case, from the insects) as their sizes approach and exceed the radar wavelength (Luke et al. 2008). Several of the authors of this chapter were present in an intense meeting that took place at one of the DOE laboratories in 2004 to decide on the utility of a WACR as an insect-free radar. The determination of the ARM chief scientist at the time, Thomas Ackerman, played a critical role in the final decision to build and deploy the WACR. Preliminary comparison of WACR and MMCR reflectivities at the SGP site in 2005 demonstrated that a 94-GHz radar has excellent sensitivity, suppresses the clear-air clutter from insects, and is highly sensitive to weak returns from boundary layer clouds and cirrus layers.

However, this is just half of the story. Unlike the MMCR, the WACR does not use pulse coding and operates in only copolarization [transmits H polarization and receives H-polarization radiation in a general (GE) mode] and cross-polarization [transmits H polarization and receives V polarization in a polarization (PO) mode], each with a 2-s integration, or dwell, time. Having fewer (two) modes eased the development of the WACR ARSCL VAP compared to the MMCR ARSCL VAP, which merges six modes over a span of 16 s. The frequent repetition of the cross-polarization mode and the suppression of the insect returns made the development of an automated insect filtering algorithm possible. Furthermore, the WACR had a sophisticated internal calibration mode, an intermediate-frequency (IF) digital receiver with 100% data efficiency, and produced high-quality radar Doppler spectra with no artifacts.

If we knew in advance the many different ways the acquisition of the WACR would benefit the ARM Program cloud radar, the decision for its acquisition in 2004 would have been easier. However, the future evolution of the ARM Program's cloud radars was clear: it was to be based on an evolution of WACR technology.

14. Doppler moments from spectra with complications

For MMCR processing, Doppler moments computed from Doppler spectra with velocity aliasing are problematic because the radar's processing does not account for the velocity aliasing. Doppler moments computed from Doppler spectra with both precipitation and cloud drops are problematic in that the Doppler moments characterize simultaneously the return powers from both of them and not the properties of each of the two separate types of drops individually. If return powers from atmospheric plankton make significant contributions to the Doppler spectra, the Doppler moments that result from them contain information from both hydrometeors and the plankton. To address all of these problems there is one way forward: collect the spectra so that detailed investigations of the spectra can lead to methods for processing them that remove all of these problems (or at least reduce them to a manageable level). In support of this goal, the ARM Program started collecting MMCR spectra continuously as the MMCR C-40 and MMCR PIRAQ-III upgrades were made and collecting WACR spectra in late 2005, extending this procedure to all KAZR and SACR Doppler spectra obtained while the radars are pointing vertically.

These Doppler spectra have been put to their intended purposes. Luke et al. (2008) demonstrated that a neural network trained on insect contributions to Doppler spectra is successful in identifying insect contributions to novel Doppler spectra in about 92% of cases. They further demonstrated that, when polarization information is included in the analysis of Doppler spectra, detection of atmospheric plankton gets even better. At the same time, Kollias et al. (2007b) showed that availability of Doppler spectra makes analysis of velocity aliasing, multiple hydrometeor phase or size spectrum contributors to the return power, and other features possible, thereby providing a path to much improved Doppler moments. Ongoing efforts within the ARM Program have expanded upon the study of Kollias et al. (2007b), leading to novel processing strategies that will use polarization-diverse Doppler spectra to mitigate past processing issues in creation of the Doppler moments.

Implementation of many of the ARM Program's new strategies requires access to Doppler spectra and then processing each and every one of them. Doppler spectra from the MMCRs, KAZRs, and SACRs now fill the ARM data archive with upward of 150 TB of data. This staggering amount of data places constant stress on collection, transmission, and storage technologies and will continue to increase in size with each passing day of ARM cloud radar operations. That said, modern technology brings many novel techniques to deal with these extremely large datasets. Ed Luke adapted one of them—graphics processing units (GPUs)—to the processing of radar Doppler spectra. With the implementation of his new processing strategies on the GPUs they can work their way through the entire ARM data archive holdings of Doppler spectra within a month of wall clock time. In addition to processing power, Ed Luke developed a sophisticated radar Doppler spectra visualization and analysis toolkit that enhances the ability of scientists to access, analyze and synthesize radar Doppler spectra from multiple radars and other ARM data (http://www.gim.bnl.gov/armclouds/specvis_java_toolkit/).

15. The fourth-generation MMCR, or KAZR, and the new KA/W/X-SACRs

From the first day of MMCR operations at the ARM SGP site, our community has been on a steep learning curve related to the challenges of continuous operations and data processing. The sluggish pace in MMCR digital receiver hardware and software upgrades between 1996 and 2011, which could not run ahead of technological advances, provided time to focus on these challenges not directly associated with the radar itself. However, propelled by the experience acquired over the last 18 years, advances in technology, and the boost to the radar infrastructure through the American Recovery and Reinvestment Act of 2009 (Mather and Voyles 2013), ARM has enhanced its cloud and precipitation observational capabilities and is again facing a daunting challenge as it attempts to operate dozens of radars and process the complex data from them. Kevin Widener, the leader of the ARM radar operations and engineering group at the time, was instrumental in managing the complex process of developing technical specifications, going through the public tender, and acquiring, testing and deploying these new ARM cloud radars.

The MMCR was replaced with the ProSensing, Inc., KAZR. The new KAZR uses several of the core technologies and capabilities of the MMCR: 1) the travel wave tube amplifier (TWTA) technology, 2) the frequency (35-GHz), 3) some of the radio-frequency (RF) electronics, 4) the antenna, and 5) the environmental enclosure. The biggest improvement in the KAZR relative to the MMCR can be found in the new digital receiver that replaces the C-40 and PIRAQ-III receivers, enabling the KAZR to transmit sophisticated long-pulse waveforms with pulse compression along with short pulses for probing the lower atmosphere. Along with superior-quality radar Doppler spectra free of image artifacts (Fig. 17-5) and higher 1-s temporal resolution, the KAZR provides researchers with the capability to study cloud dynamics and microphysics using the recorded radar Doppler spectra (e.g., Kollias et al. 2011).

In addition to the upgrades to the vertically profiling cloud radar, the ARM Program acquired eight SACRs that were developed and fabricated by ProSensing, Inc. The SACRs are sophisticated, dual-frequency (five SACRs have Ka/W-band frequency pairs and three have Ka/X-band frequency pairs) radar systems mounted on a single pedestal (Fig. 17-2c). The Ka/W-SACRs are intended mostly for midlatitude and Arctic regions where the impact of attenuation by atmospheric water vapor is less severe than in the tropics. The beams of the Ka- and W-band radars within an SACR are almost matched to each other and very narrow (less than 0.33°). Contrary to the KAZR, the Ka/W-SACR uses a relatively high-power extended interaction klystron amplifier (EIKA) transmitter with peak power over 1.6 kW. The Ka- and W-band radar systems have dual polarization but only transmit horizontal polarization states and receive signals in both horizontal and vertical polarization states. The last two of the eight Ka/W-SACR systems came online in 2015 and are fully polarization diverse; that is, they transmit both horizontal and vertical polarization states (though not simultaneously) and measure both horizontal and vertical polarization returns, which are invaluable for the study of ice clouds.

The Ka/W-SACRs use a digital waveform generator capable of producing arbitrary waveforms, which enables the use of frequency diversity and pulse compression waveforms. They use spectral processing for filtering and parameter estimation. The systems routinely store the full Doppler spectrum when operating in zenith-pointing mode and are capable of storing the base-band in-phase and quadrature-phase signals. They use corner reflector targets located on towers at a range of 400–500 m for calibration of themselves and the nearby KAZRs and/or WACRs. They can be remotely configured and have been designed to implement adaptive scanning strategies (Kollias et al. 2014).

The X/Ka-SACRs are mostly geared toward tropical regions where atmospheric water vapor will not have as severe an impact on X- and Ka-band beams as compared to a W-band beam. The beams of the X- and Ka-band radars are not matched to each other. The X-SACRs use a much (about 3.5 times) wider beam compared with the Ka-SACR, as mounting an X-band antenna with a 0.33° beamwidth (comparable to that of a Ka-SACR) on the same pedestal is not practical. Their receiver and transmitter units are antenna mounted, which results in low power losses in the radio-frequency sections of the transmitter and receiver. The X-SACRs are fully polarimetric, simultaneously transmitting and receiving both horizontal and vertical polarization states.

The SACRs are the primary instruments for the detection of cloud properties (boundaries, water contents, dynamics, etc.) beyond the vertical columns directly above the ARM sites. Having scanning capabilities with two frequencies and polarization allows for more accurate probing of a variety of cloud systems (e.g., drizzle and shallow, warm rain), better correction for attenuation, use of attenuation for liquid water content retrievals, polarimetric characterization of nonspherical particles, and habit identification.

16. Reprocessing of the ARSCL VAP and wind profilers as part of cloud observations

The last file produced by the SGP MMCR occurred in January 2011, that for the NSA MMCR in March 2011, and those for the TWP MMCRs in February and March 2011. The lifetime of the MMCRs within the ARM Program spanned November 1996 through March 2011 for a total approaching 15 years. Its record within the ARM Program is now complete.

In May 2013, the ARM Program organized a meeting to discuss all the problems and issues in MMCR data and its related ARSCL VAP. The suboptimal ways in which the operational modes of the MMCRs were merged into the ARSCL VAP were discussed and ways to improve the merging considered. Newer, better methods for assessing the quality of MMCR mode data emerged as a result.

One set of ideas put forth in the meeting showed how specific consistency checks between different MMCR modes could be used to identify saturation within the MMCR receiver for that mode. Knowing whether or not a mode is saturated provides important information on the quality of the mode data and its suitability for inclusion within the ARSCL VAP (Galletti et al. 2014). The last question remaining in this investigation is whether or not saturation can be detected in the precipitation mode itself, both before and after the switch approach to mitigating saturation was implemented.

Another set of ideas put forth by several of the participants of the meeting demonstrated that there is simply no way to assess attenuation in real time through the MMCR radome and atmosphere above without measurements at lower nonattenuating (i.e., wind profiler) wavelengths (Tridon et al. 2013). These arguments further strengthen the arguments for placing a 915-MHz wind profiler and a disdrometer beside each KAZR and SACR pair in order to complete the instrument package with today's technology that can best be used to probe atmospheric hydrometeors. It is conceivable that, in the future, the wind profiler will become one of the key input data streams to the ARSCL VAP.

Once the ARM cloud radar program resolves the issues related to the quality of data from each mode and how best to handle attenuation in real time, it will be time to reprocess the ARSCL VAP in its entirety. The complete MMCR dataset, its overall quality, and the quality of the related ARSCL VAPs for all sites and times that emanate from the reprocessing will be discussed in a review article. This article will represent the definitive statement on the quality of the MMCR data record within the ARM Program.

17. Closing remarks

The ARM Program's decision to invest in the relatively new and immature technology of millimeter-wavelength cloud radars was both bold and brave. This decision was based on the determined advocacy of a small group of ARM scientists and engineers[3] backed up by a handful of scientific studies using data primarily from research radars built and operated by groups at the Pennsylvania State University (Thomas Ackerman, Bruce Albrecht, and colleagues), the University of Massachusetts (Robert McIntosh, James Mead, and colleagues), and the NOAA Environmental Research Laboratory (Robert Kropfli, Kenneth Moran, and colleagues). In retrospect, it may

[3] A number of the authors of this chapter were members of that group. We would like to acknowledge the contributions of the late Dr. Robert McIntosh of the University of Massachusetts and Dr. Robert Kropfli of NOAA Environmental Technology Laboratory (ETL) to the initiation of and support for the ARM radar program.

seem that it was an easy decision, but at the time there were serious questions about the reliability of millimeter-wavelength technology, the high cost of building instruments for which there were no commercial vendors, and the extent to which continuous cloud radar data could be collected and used. Development, particularly of processing and storage software, was challenging because of unexpected problems and sometimes slower than hoped. ARM Program management is to be congratulated for persevering in its vision to deploy cloud radars to its sites in spite of these difficulties.

The operation of the ARM millimeter-wavelength cloud radars beginning in 1996 and continuing to the present has provided the scientific community with an unprecedented quantitative view of cloud properties and processes over the ARM fixed sites and the new mobile facilities. To those of us who remember the excitement of acquiring a few hours of continuous millimeter-wavelength radar data during a field campaign, this wealth of radar data is truly stunning. The nearly 15 years of MMCR observations, now being extended by the new generation of radars, have been critically important in advancing our capability to evaluate model simulations, develop and evaluate cloud parameterizations, and provide key insights into cloud processes that are central to the climate modeling puzzle.

While discussion of cloud property retrievals and their use in model process and parameterization development is beyond the scope of this chapter, these topics are the focus of Shupe et al. (2016, chapter 19). Moreover, accessible discussions on liquid cloud, ice cloud, and mixed-phase cloud property retrievals can be found in the articles by Turner et al. (2007), Comstock et al. (2007), and Shupe et al. (2008), with many other published articles highlighting the importance of cloud radar data in this ongoing process (e.g., Botta et al. 2011; Deng and Mace 2006; Dong and Mace 2003; Frisch et al. 1995; Kollias et al. 2001, 2002, 2007a, 2011; Luke et al. 2010; Luke and Kollias 2013; Matrosov et al. 2012; O'Connor et al. 2005; Shupe et al. 2004).

The development and advancement of millimeter-wavelength radar hardware and software components and support for scientific use of radar data are a lasting legacy of the ARM Program. There is no doubt that the ARM Program played the leading role in this field for the past 20 years. Its new radars that are now being integrated into the fixed sites and mobile facilities are evidence that the ARM Program expects to sustain its leadership role. Although there have been many obstacles to reaching the full potential of millimeter-wavelength cloud radars, our vision for the role that these radars could and should play in atmospheric cloud research and improving climate simulations is being realized.

REFERENCES

Ackerman, T. P., and G. M. Stokes, 2003: The Atmospheric Radiation Measurement Program. *Phys. Today*, **56**, 38–44, doi:10.1063/1.1554135.

Albrecht, B. A., C. S. Bretherton, D. Johnson, W. H. Schubert, and A. S. Frisch, 1995: The Atlantic stratocumulus transition experiment—ASTEX. *Bull. Amer. Meteor. Soc.*, **76**, 889–904, doi:10.1175/1520-0477(1995)076<0889:TASTE>2.0.CO;2.

Botta, G., K. Aydin, J. Verlinde, A. E. Avramov, A. S. Ackerman, A. M. Fridlind, G. M. McFarquhar, and M. Wolde, 2011: Millimeter wave scattering from ice crystals and their aggregates: Comparing cloud model simulations with X- and Ka-band radar measurements. *J. Geophys. Res.*, **116**, D00T04, doi:10.1029/2011JD015909.

Cess, R. D., and Coauthors, 1990: Intercomparison and interpretation of climate feedback processes in 19 atmospheric general circulation models. *J. Geophys. Res.*, **95**, 16 601–16 615, doi:10.1029/JD095iD10p16601.

Clothiaux, E. E., M. A. Miller, B. A. Albrecht, T. P. Ackerman, J. Verlinde, D. M. Babb, R. M. Peters, and W. J. Syrett, 1995: An evaluation of a 94-GHz radar for remote sensing of cloud properties. *J. Atmos. Oceanic Technol.*, **12**, 201–229, doi:10.1175/1520-0426(1995)012<0201:AEOAGR>2.0.CO;2.

——, T. P. Ackerman, G. G. Mace, K. P. Moran, R. T. Marchand, M. A. Miller, and B. E. Martner, 2000: Objective determination of cloud heights and radar reflectivities using a combination of active remote sensors at the ARM CART sites. *J. Appl. Meteor.*, **39**, 645–665, doi:10.1175/1520-0450(2000)039<0645:ODOCHA>2.0.CO;2.

Comstock, J. M., and Coauthors, 2007: An intercomparison of microphysical retrieval algorithms for upper-tropospheric ice clouds. *Bull. Amer. Meteor. Soc.*, **88**, 191–204, doi:10.1175/BAMS-88-2-191.

Danne, O., G. G. Mace, E. E. Clothiaux, X. Dong, T. P. Ackerman, and M. Quante, 1996: Observing structures and vertical motions within stratiform clouds using a vertical pointing 94-GHz cloud radar. *Contrib. Atmos. Phys.*, **69**, 229–237.

Deng, M., and G. G. Mace, 2006: Cirrus microphysical properties and air motion statistics using cloud radar Doppler moments. Part I: Algorithm description. *J. Appl. Meteor. Climatol.*, **45**, 1690–1709, doi:10.1175/JAM2433.1.

Dong, X., and G. G. Mace, 2003: Profiles of low-level stratus cloud microphysics deduced from ground-based measurements. *J. Atmos. Oceanic Technol.*, **20**, 42–53, doi:10.1175/1520-0426(2003)020<0042:POLLSC>2.0.CO;2.

Ecklund, W. L., C. R. Williams, and P. E. Johnston, 1999: A 3-GHz profiler for precipitating clouds. *J. Atmos. Oceanic Technol.*, **16**, 309–322, doi:10.1175/1520-0426(1999)016<0309:AGPFPC>2.0.CO;2.

Ellingson, R. G., and W. J. Wiscombe, 1996: The Spectral Radiance Experiment (SPECTRE): Project description and sample results. *Bull. Amer. Meteor. Soc.*, **77**, 1967–1985, doi:10.1175/1520-0477(1996)077<1967:TSREPD>2.0.CO;2.

——, R. D. Cess, and G. L. Potter, 2016: The Atmospheric Radiation Measurement Program: Prelude. *The Atmospheric Radiation Measurement (ARM) Program: The First 20 Years*, *Meteor. Monogr.*, No. 57, Amer. Meteor. Soc., doi:10.1175/AMSMONOGRAPHS-D-15-0029.1.

Frisch, A. S., C. W. Fairall, and J. B. Snider, 1995: Measurement of stratus cloud and drizzle parameters in ASTEX with a Ka-band Doppler radar and a microwave radiometer. *J. Atmos. Sci.*, **52**, 2788–2799, doi:10.1175/1520-0469(1995)052<2788:MOSCAD>2.0.CO;2.

Galletti, M., D. Huang, and P. Kollias, 2014: Zenith/nadir pointing mm-wave radars: Linear or circular polarization? *IEEE Trans. Geosci. Remote Sens.*, **52**, 628–639, doi:10.1109/TGRS.2013.2243155.

Keenan, T., and Coauthors, 2000: The Maritime Continent Thunderstorm Experiment (MCTEX): Overview and some results. *Bull. Amer. Meteor. Soc.*, **81**, 2433–2455, doi:10.1175/1520-0477(2000)081<2433:TMCTEM>2.3.CO;2.

Kollias, P., B. A. Albrecht, R. Lhermitte, and A. Savtchenko, 2001: Radar observations of updrafts, downdrafts, and turbulence in fair-weather cumuli. *J. Atmos. Sci.*, **58**, 1750–1766, doi:10.1175/1520-0469(2001)058<1750:ROOUDA>2.0.CO;2.

——, ——, and F. Marks Jr., 2002: Why Mie? Accurate observations of vertical air velocities and raindrops using a cloud radar. *Bull. Amer. Meteor. Soc.*, **83**, 1471–1483, doi:10.1175/BAMS-83-10-1471.

——, E. E. Clothiaux, B. A. Albrecht, M. A. Miller, K. P. Moran, and K. L. Johnson, 2005: The Atmospheric Radiation Measurement program cloud profiling radars: An evaluation of signal processing and sampling strategies. *J. Atmos. Oceanic Technol.*, **22**, 930–948, doi:10.1175/JTECH1749.1.

——, ——, M. A. Miller, B. A. Albrecht, G. L. Stephens, and T. P. Ackerman, 2007a: Millimeter-wavelength radars: New frontier in atmospheric cloud and precipitation research. *Bull. Amer. Meteor. Soc.*, **88**, 1608–1624, doi:10.1175/BAMS-88-10-1608.

——, ——, ——, E. P. Luke, K. L. Johnson, K. P. Moran, K. B. Widener, and B. A. Albrecht, 2007b: The Atmospheric Radiation Measurement Program cloud profiling radars: Second-generation sampling strategies, processing, and cloud data products. *J. Atmos. Oceanic Technol.*, **24**, 1199–1214, doi:10.1175/JTECH2033.1.

——, J. Rémillard, E. Luke, and W. Szyrmer, 2011: Cloud radar Doppler spectra in drizzling stratiform clouds: 1. Forward modeling and remote sensing applications. *J. Geophys. Res.*, **116**, D13201, doi:10.1029/2010JD015237.

——, N. Bharadwaj, K. Widener, I. Jo, and K. Johnson, 2014: Scanning ARM cloud radars. Part I: Operational sampling strategies. *J. Atmos. Oceanic Technol.*, **31**, 569–582, doi:10.1175/JTECH-D-13-00044.1.

Lhermitte, R. M., 1966: Probing air motion by Doppler analysis of radar clear air returns. *J. Atmos. Sci.*, **23**, 575–591, doi:10.1175/1520-0469(1966)023<0575:PAMBDA>2.0.CO;2.

——, 1987: Small cumuli observed with a 3 mm wavelength Doppler radar. *Geophys. Res. Lett.*, **14**, 707–710, doi:10.1029/GL014i007p00707.

Luke, E. P., and P. Kollias, 2013: Separating cloud and drizzle radar moments during precipitation onset using Doppler spectra. *J. Atmos. Oceanic Technol.*, **30**, 1656–1671, doi:10.1175/JTECH-D-11-00195.1.

——, ——, K. L. Johnson, and E. E. Clothiaux, 2008: A technique for the automatic detection of insect clutter in cloud radar returns. *J. Atmos. Oceanic Technol.*, **25**, 1498–1513, doi:10.1175/2007JTECHA953.1.

——, ——, and M. D. Shupe, 2010: Detection of supercooled liquid in mixed-phase clouds using radar Doppler spectra. *J. Geophys. Res.*, **115**, D19201, doi:10.1029/2009JD012884.

Mather, J. H., and J. W. Voyles, 2013: The ARM Climate Research Facility: A review of structure and capabilities. *Bull. Amer. Meteor. Soc.*, **94**, 377–392, doi:10.1175/BAMS-D-11-00218.1.

Matrosov, S. Y., G. G. Mace, R. T. Marchand, M. D. Shupe, A. G. Hallar, and I. B. McCubbin, 2012: Observations of ice crystal habits with a scanning polarimetric W-band radar at slant linear depolarization ratio mode. *J. Atmos. Oceanic Technol.*, **29**, 989–1008, doi:10.1175/JTECH-D-11-00131.1.

McFarquhar, G. M., and Coauthors, 2011: Indirect and Semi-Direct Aerosol Campaign: The impact of Arctic aerosols on clouds. *Bull. Amer. Meteor. Soc.*, **92**, 183–201, doi:10.1175/2010BAMS2935.1.

Mead, J. B., A. L. Pazmany, S. M. Sekelsky, and R. E. McIntosh, 1994: Millimeter-wave radars for remotely sensing clouds and precipitation. *Proc. IEEE*, **82**, 1891–1905, doi:10.1109/5.338077.

Miller, M. A., and B. A. Albrecht, 1995: Surface-based observations of mesoscale cumulus–stratocumulus interaction during ASTEX. *J. Atmos. Sci.*, **52**, 2809–2826, doi:10.1175/1520-0469(1995)052<2809:SBOOMC>2.0.CO;2.

Moran, K. P., B. E. Martner, M. J. Post, R. A. Kropfli, D. C. Welsh, and K. B. Widener, 1998: An unattended cloud-profiling radar for use in climate research. *Bull. Amer. Meteor. Soc.*, **79**, 443–455, doi:10.1175/1520-0477(1998)079<0443:AUCPRF>2.0.CO;2.

O'Connor, E. J., R. J. Hogan, and A. J. Illingworth, 2005: Retrieving stratocumulus drizzle parameters using Doppler radar and lidar. *J. Appl. Meteor.*, **44**, 14–27, doi:10.1175/JAM-2181.1.

Pasqualucci, F., B. W. Bartram, R. A. Kropfli, and W. R. Moninger, 1983: A millimeter-wavelength dual-polarization Doppler radar for cloud and precipitation studies. *J. Climate Appl. Meteor.*, **22**, 758–765, doi:10.1175/1520-0450(1983)022<0758:AMWDPD>2.0.CO;2.

Platt, C. M. R., J. C. Scott, and A. C. Dilley, 1987: Remote sounding of high clouds. Part IV: Optical properties of midlatitude and tropical cirrus. *J. Atmos. Sci.*, **44**, 729–747, doi:10.1175/1520-0469(1987)044<0729:RSOHCP>2.0.CO;2.

Ramanathan, V., R. D. Cess, E. F. Harrison, P. Minnis, B. R. Barkstrom, E. Ahmad, and D. Hartmann, 1989: Cloud-radiative forcing and climate: Results from the Earth Radiation Budget Experiment. *Science*, **243**, 57–63, doi:10.1126/science.243.4887.57.

Sassen, K., C. J. Grund, J. D. Spinhirne, M. H. Hardesty, and J. M. Alvarez, 1990: The 27–28 October FIRE IFO cirrus case study: A five lidar overview of cloud structure and evolution. *Mon. Wea. Rev.*, **118**, 2288–2311, doi:10.1175/1520-0493(1990)118<2288:TOFICC>2.0.CO;2.

——, and Coauthors, 1995: The 5–6 December 1991 FIRE IFO II jet stream cirrus case study: Possible influences of volcanic aerosols. *J. Atmos. Sci.*, **52**, 97–123, doi:10.1175/1520-0469(1995)052<0097:TDFIIJ>2.0.CO;2.

Shupe, M. D., P. Kollias, S. Y. Matrosov, and T. L. Schneider, 2004: Deriving mixed-phase cloud properties from Doppler radar spectra. *J. Atmos. Oceanic Technol.*, **21**, 660–670, doi:10.1175/1520-0426(2004)021<0660:DMCPFD>2.0.CO;2.

——, and Coauthors, 2008: A focus on mixed-phase clouds: The status of ground-based observational methods. *Bull. Amer. Meteor. Soc.*, **89**, 1549–1562, doi:10.1175/2008BAMS2378.1.

——, J. M. Comstock, D. D. Turner, and G. G. Mace, 2016: Cloud property retrievals in the ARM Program. *The Atmospheric Radiation Measurement (ARM) Program: The First 20 Years, Meteor. Monogr.*, No. 57, Amer. Meteor. Soc., doi:10.1175/AMSMONOGRAPHS-D-15-0030.1.

Sisterson, D., R. Peppler, T. S. Cress, P. Lamb, and D. D. Turner, 2016: The ARM Southern Great Plains (SGP) site. *The Atmospheric Radiation Measurement (ARM) Program: The First*

20 Years, Meteor. Monogr., No. 57, Amer. Meteor. Soc., doi:10.1175/AMSMONOGRAPHS-D-16-0004.1.

Stokes, G. M., 2016: Original ARM concept and launch. *The Atmospheric Radiation Measurement (ARM) Program: The First 20 Years, Meteor. Monogr.*, No. 57, Amer. Meteor. Soc., doi:10.1175/AMSMONOGRAPHS-D-15-0021.1.

——, and S. E. Schwartz, 1994: The Atmospheric Radiation Measurement (ARM) Program: Programmatic background and design of the cloud and radiation test bed. *Bull. Amer. Meteor. Soc.*, **75**, 1201–1221, doi:10.1175/1520-0477(1994)075<1201:TARMPP>2.0.CO;2.

Tridon, F., A. Battaglia, P. Kollias, E. Luke, and C. R. Williams, 2013: Signal postprocessing and reflectivity calibration of the Atmospheric Radiation Measurement Program 915-MHz wind profilers. *J. Atmos. Oceanic Technol.*, **30**, 1038–1054, doi:10.1175/JTECH-D-12-00146.1.

Turner, D. D., and Coauthors, 2007: Thin liquid water clouds: Their importance and our challenge. *Bull. Amer. Meteor. Soc.*, **88**, 177–190, doi:10.1175/BAMS-88-2-177.

Uttal, T., E. E. Clothiaux, T. P. Ackerman, J. M. Intrieri, and W. L. Eberhard, 1995: Cloud boundary statistics during FIRE II. *J. Atmos. Sci.*, **52**, 4276–4284, doi:10.1175/1520-0469(1995)052<4276:CBSDFI>2.0.CO;2.

Verlinde, J., and Coauthors, 2007: The Mixed-Phase Arctic Cloud Experiment. *Bull. Amer. Meteor. Soc.*, **88**, 205–221, doi:10.1175/BAMS-88-2-205.

Widener, K. B., A. S. Koontz, K. P. Moran, K. A. Clark, C. Chander, M. A. Miller, and K. L. Johnson, 2004: MMCR upgrades: Present status and future plans. *Proc. 14th Atmospheric Radiation Measurement (ARM) Science Team Meeting*, Albuquerque, NM, U.S. DOE. [Available online at http://www.arm.gov/publications/proceedings/conf14/extended_abs/widener1-kb.pdf.]

Chapter 18

Development and Applications of the ARM Raman Lidar

D. D. TURNER

NOAA/National Severe Storms Laboratory, Norman, Oklahoma

J. E. M. GOLDSMITH

Sandia National Laboratories, Livermore, California

R. A. FERRARE

NASA Langley Research Center, Hampton, Virginia

1. Introduction

From the earliest days of the Atmospheric Radiation Measurement (ARM) Program, measurements of water vapor profiles at high temporal and vertical resolution were deemed to be critical for both the radiative transfer and cloud processes studies that the ARM Program planned to undertake (DOE 1990). The dream of the ARM Program founders was that ground-based remote sensors would measure these profiles routinely, and that the program would be able to move away from the routine launching of radiosondes to characterize the thermodynamic profile above the ARM sites. This is reflected in the original program plan, in which the first two profiling systems listed were Raman lidar or differential absorption lidar (DIAL) to profile water vapor, and radio acoustic sounding systems with radar wind profiles to measure temperature and wind profiles (DOE 1990).

Here, we discuss the history of using lidars to profile water vapor within the ARM Program, including the early studies to determine what type of lidar would be best suited for the program, the development of the ARM Raman lidar, the initial challenges that were experienced with the construction of the first fully automated system, and examples that demonstrate the scientific utility of the ARM Raman lidar.

Corresponding author address: D. D. Turner, NOAA/National Severe Storms Laboratory, Forecast Research and Development Division, National Weather Center, 120 David L. Boren Blvd., Norman, OK 73072.
E-mail: dave.turner@noaa.gov

DOI: 10.1175/AMSMONOGRAPHS-D-15-0026.1

2. Using lidars to profile water vapor before ARM

The development of the laser in the early 1960s led to much excitement in the scientific community, and it was not long before the laser was used to study the atmosphere. This was especially true after the development of reliable lasers that were able to emit high-energy, short (~ns) pulses of narrowband radiation. The first lidar remote sensing experiment that used these new lasers was used to determine the distance between Earth and the moon (Smullin and Fiocco 1962).

Raman lidars take advantage of the Raman energy shift that occurs when a photon scatters inelastically off a molecule in the atmosphere, resulting in a lower energy (longer wavelength) photon. The energy shifts associated with this process are molecule specific and are relatively large (e.g., the vibrational-rotational shifts are 3652 and 2331 cm^{-1} for water vapor and nitrogen, respectively); this makes the selection of the return from various molecules relatively easy using dichroic beam splitters and interference filters. The first Raman lidar measurements occurred in 1966, as measurements of nitrogen (Cooney 1968) and oxygen (Leonard 1967)—the gases with the highest concentrations in the atmosphere—were made. The first Raman lidar water vapor measurements of the atmosphere were made in 1969 by Harvey Melfi, who showed good agreement with coincident radiosonde profiles of water vapor (Melfi 1972). Turner and Whiteman (2002) provide a history of using Raman lidars to profile trace gases and aerosols in the atmosphere.

DIAL systems profile water vapor in a different manner. DIALs transmit laser energy at two wavelengths: one tuned to the absorption line of the gas of interest, and one at a nearby wavelength where the absorption by that gas is markedly smaller. In these systems, the laser beam is typically scattered back to the system by aerosol particles, since the wavelengths used are typically in the near-infrared where molecular scattering is very weak. The analysis then proceeds to measure the amount of relative attenuation that occurred between the online and offline returns, and since the relative strengths of the absorption at the two wavelengths is known, the concentration of the gas at each range cell can be determined. Ed Browell and his colleagues (Browell et al. 1979) were among the pioneers who profiled water vapor and ozone using ground-based and airborne DIAL systems.

The invention of the laser led to an intense period of activity as the laser was utilized to probe the atmosphere. However, lasers were still immature and detectors were inefficient. Raman scattering is a very weak process (approximately 3–4 orders of magnitude smaller than Rayleigh scattering), and thus the signal-to-noise ratio hampered many atmospheric studies using this technique. The stringent requirements of frequency control on the online wavelength made the laser transmitters in DIAL systems extremely complicated, and hindered their development. It was not until the late 1980s, when higher-power and better-quality lasers were developed as well as greatly improved detection technologies, that the atmospheric Raman lidar and DIAL observations regained the "luster" they had in the late 1960s and early 1970s.

The advances in both laser and detector technology in the mid-to-late 1980s led to a rebirth of the lidar atmospheric sensing of water vapor. For example, Harvey Melfi and his group at the National Aeronautics and Space Administration (NASA) Goddard Space Flight Center (GSFC) used Raman lidar measurements to provide detailed information on the structure of water vapor during frontal passages (Melfi et al. 1989) and some airborne DIAL systems were being developed at this time (e.g., Ehret et al. 1993; Ismail and Browell 1989). However, because of the weak Raman scattering process and high solar noise during the daytime, Raman lidars were only used at night. Furthermore, both Raman lidars and DIAL systems were still manually intensive research instruments, and only small, sporadic datasets were being collected. Thus, the ARM Program had two viable options and needed to determine which of the two avenues would more likely lead to a successful automated system.

3. The decision process

Measuring the evolution of water vapor in the boundary layer over the entire diurnal cycle was deemed to be a critical measurement for the ARM Program. The two technologies, Raman lidar and DIAL, both had attractive features yet both also had some drawbacks. The primary strengths and weaknesses, in a relative sense, are listed in Table 18-1 and were also provided in a review article by Grant (1991); a survey of the accuracy and resolution of different Raman lidars and DIAL systems was provided in Weckwerth et al. (1999). A proposal to pursue both technologies was submitted to the ARM Instrument Development Program (Stokes 2016, chapter 2), but the complexity of the DIAL laser transmitter and concerns about its stability, which must be frequency-stabilized to a selected water vapor absorption line, led to a decision to only fund the Raman lidar work as a collaboration between Sandia National Laboratories (SNL) and the GSFC group.

TABLE 18-1. Relative advantages of the Raman lidar vs DIAL, as perceived in the early 1990s.

System	Advantages
Raman lidar	Simpler laser transmitter Able to profile aerosol extinction and backscatter directly (without assumptions) Higher nighttime maximum altitude limit
DIAL	Better signal-to-noise in boundary layer No need for external calibration standard

However, this decision then led to additional questions. Raman lidars measure the energy shift associated with Raman scattering by water vapor and nitrogen molecules, and the strength of this scattering is proportional to λ^{-4}, where λ is the wavelength of the laser. Thus, the shorter the wavelength of the laser, the more intense the Raman scattering signal is (all other things equal), which is an advantage since Raman scattering is a weak process. Photomultiplier detectors are also more efficient and have lower background levels at shorter wavelengths.

Therefore, lasers that operate in the UV are desirable. There were two choices:

1) Should the program use a solid-state or excimer (gas) laser?
2) What was the best system configuration to maximize performance in the daytime where there would be significant background signal from the sun?

Excimer lasers were relatively popular in the early 1990s because they had higher output power levels than solid-state lasers. Initial Raman lidar development within the ARM Program evaluated, among other things, different types of excimer lasers (Goldsmith et al. 1994). However, the gases used within these lasers were corrosive and difficult to handle. Additionally, the beam quality was poorer for excimer lasers than for solid-state lasers;

this is important if the detector uses a narrow field of view (FOV). For these reasons, the ARM Program ultimately elected to use a solid-state neodymium-doped yttrium–aluminum–garnet (Nd:YAG) laser in its operational system.

Two studies proceeded simultaneously by groups at SNL and GSFC to evaluate the best system configuration to make daytime Raman lidar water vapor measurements. GSFC investigated making water vapor observations in the so-called solar-blind region of the spectrum (Whiteman et al. 1993), where the laser transmitted in a spectral region of the UV (248 nm) and there was significant absorption of the solar energy by ozone, which reduced the solar signal measured by the detectors; this approach had been used in the late 1970s (Renaut et al. 1980). SNL investigated the use of a very narrow FOV with narrow interference filter bandpasses for the receiver to reduce the amount of solar background (Bisson and Goldsmith 1993), using a laser that operated at 308 nm where ozone absorption is minimal in the troposphere. Note that the narrow bandpass interference filters utilized very new technology in order to maintain a reasonable transmission, and there were concerns about leakage of out-of-band radiation that would contaminate the measurements. These studies helped to confirm a modeling study that indicated both solar-blind and narrow bandpass/narrow FOV systems had similar daytime capabilities (Goldsmith and Ferrare 1992).

The initial studies demonstrated fairly rapidly that the narrow bandpass, narrow FOV system offered similar range capability as the solar blind method during the day, but was more accurate because the differential absorption of ozone in the boundary layer did not need to be determined. Furthermore, the narrow bandpass, narrow FOV systems were much superior during the nighttime, allowing water vapor to be profiled throughout the troposphere. However, there were still questions to be answered regarding the optimization of the configuration of the systems, and thus personnel at both GSFC (Whiteman et al. 1992) and SNL (Bisson et al. 1999) built systems to investigate these issues. These two systems were then deployed side by side and numerous radiosondes, which served as truth, were launched in order to evaluate the different technologies (Goldsmith et al. 1994). The SNL "big lidar" served as the prototype for the Cloud and Radiation Testbed (CART) Raman lidar (CARL), while the GSFC scanning Raman lidar (SRL) provided extremely useful observations in future water vapor campaigns (Turner et al. 2016, chapter 13) and evolved with time over the next decade, serving as an experimental platform for subsequent upgrades to CARL.

4. Development and evolution of the ARM Raman lidar

a. Building the Raman lidar

The results from all of these studies were used in the development of the operational CART Raman lidar. The lidar would use a solid-state Nd:YAG laser (output at 355 nm) and a narrow bandpass, narrow FOV architecture. To improve the observations in the near field (i.e., in the lowest several hundred meters just above the Raman lidar), the lidar would have a second FOV that had a larger aperture and would make measurements simultaneously with the narrow FOV. The narrow and wide FOV of the lidar would be set to 0.3 and 2 mrad, respectively; these are often referred to as the high-altitude (or high) and low-altitude (low) channels. All channels would use interference filters that had a nominal bandpass of 0.3 nm. The full details of the CARL system are given in Goldsmith et al. (1998).

The Raman lidar community had demonstrated that aerosol extinction could be measured directly by a Raman lidar system without the need for any assumptions relating aerosol backscatter to extinction (Ansmann et al. 1990), and a number of experimental Raman lidar systems were built with the capability to measure both water vapor and aerosol extinction (e.g., Whiteman et al. 1992; Ansmann et al. 1992; Reichardt et al. 1996). Thus, both FOVs in the aft optics of the operational Raman lidar would be configured with channels sensitive to the vibrational-rotational Raman scattering by water vapor (408 nm) and nitrogen (387 nm), as well as channels sensitive to the elastic return at the laser wavelength (Fig. 18-1, top). Thus, two lidar systems would share the same laser transmitter and telescope. This redundancy in the aft optics would prove to be useful in later analyses.

To make this an autonomous system, it would be necessary to monitor and adjust several parameters that were formerly performed manually. In particular, the laser output power and the alignment of the outgoing laser beam within the narrow FOV fluctuated with the temperature inside the enclosure due to thermal expansion and vibrations; it would be necessary to develop automated ways to maintain the optimal alignment of each of these components.

Routines were developed in LabView (the software that ran the CARL system) to optimally adjust the alignment of the frequency doubler and trippler crystals within the laser, thereby maintaining maximum laser output power, and the orientation of the final steering mirror that directed the laser beam into the sky. These routines were run at regular intervals (e.g., every 3 h) to "tweak" the laser power or alignment. During each alignment tweak, the system would stop collecting data

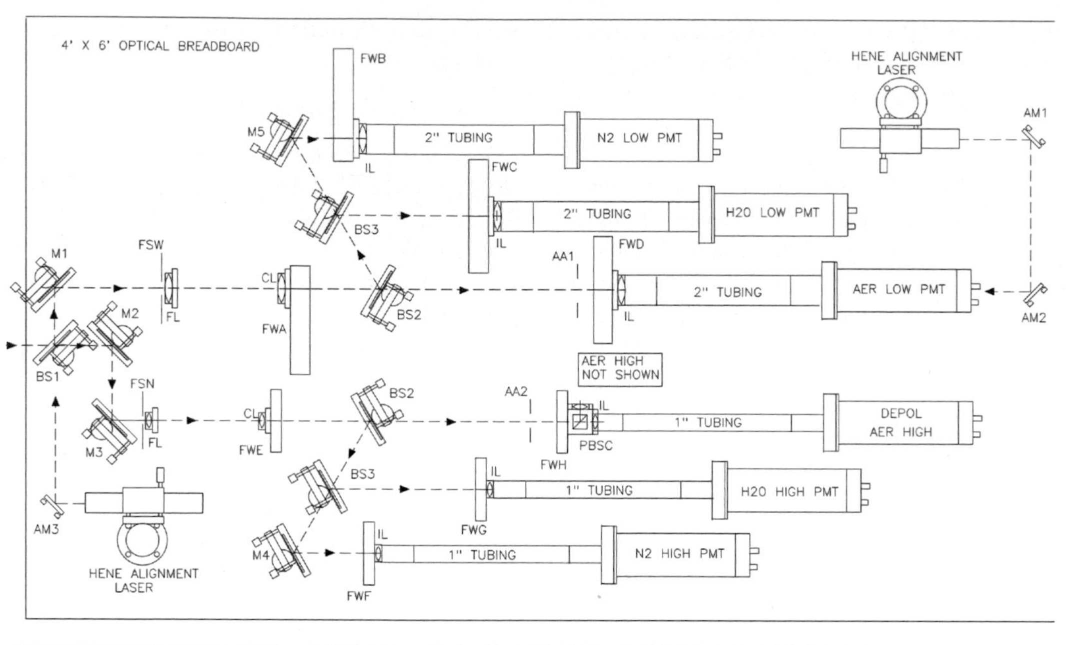

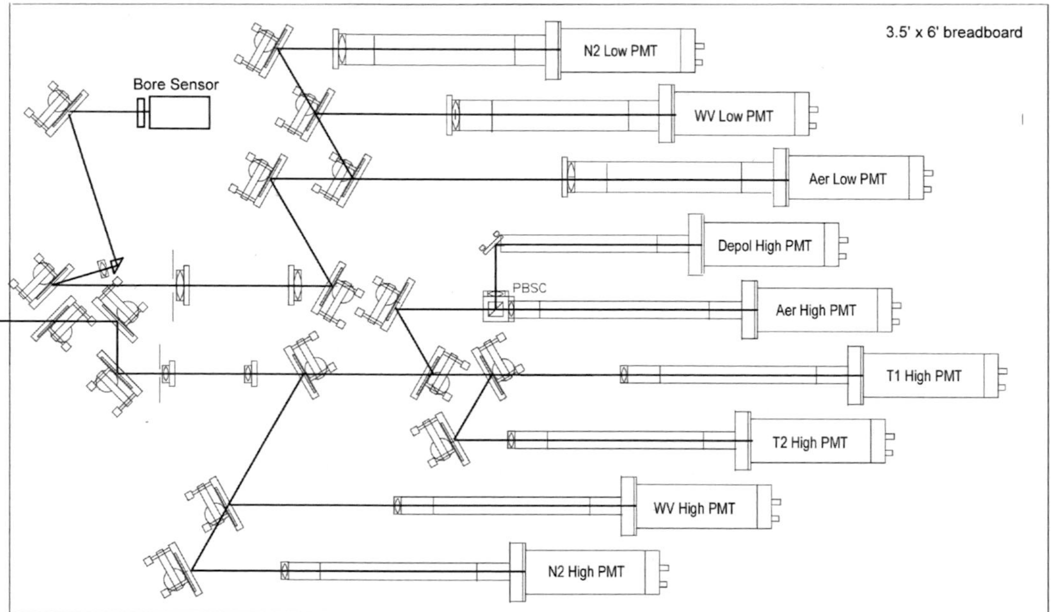

FIG. 18-1. The layout of the aft optics of the ARM Raman lidar (top) as originally deployed in 1996 and (bottom) after the 2004 upgrade. The receiver telescope is located to the left of the optical breadboard, with the light coming into the aft optics through beam splitter "BS1" (only identified in the top layout).

momentarily, and then scan the laser beam through the FOV along the north–south axis, fit a polynomial to the data, and then return to the location that maximized the return signal. The process was then repeated along the east–west axis, and then the system would return to operations. Ideally, this would align the laser within the center of the FOV; however, in practice hysteresis in the micromotors that moved the mirror would result in some minor amount of misalignment. This would turn out to have a negligible effect on the derived water vapor mixing ratio, but would be a serious challenge to the aerosol processing scripts for reasons described below. Note that a similar tweaking strategy was used to optimize the orientation of the doubling and tripling crystals within the laser.

While the alignment tweak logic typically worked well in clear skies, the presence of clouds, especially if a cloud advected over the lidar during the tweak process, could result in very poor results. Thus, the LabView software needed to have logic built into it to identify clouds before the tweak process occurred and abort it. The software also needed to identify periods when the tweak was unsuccessful for any reason (typically due to clouds); if this occurred then the software needed to be able to return to the original alignment position and continue making measurements. This procedure usually worked well, but occasionally the system would end up misaligned. If the misalignment was not bad, then the next alignment tweak would correct it; however, on the rare occasions it would be necessary for the operators at the Southern Great Plains (SGP) site to manually restore the alignment. (These situations where the operators needed to perform a manual adjustment were easily determined by eye in cloud-free conditions because the raw signal in the narrow FOV nitrogen channel would decay too quickly with height.)

It was essential for an operational system to be eyesafe. CARL operates in the UV region of the spectrum where the maximum permissible exposure (MPE) is higher than in the visible or near-infrared portion of the spectrum. However, since Raman scattering is such a weak signal, CARL uses a high-energy laser that transmits 300–400 mJ pulses at 30 Hz. This could only be achieved by expanding the laser beam significantly to reduce the energy density of the outgoing beam. This was done with a 15× beam expander prior to the final steering mirror. Calculations indicated that any aircraft overhead would have to be flying slower than 6 miles per hour in order to exceed the MPE at any point. This eliminated the need to have a manual spotter or automated radar to disable the system if an aircraft passed above the lidar (Goldsmith et al. 1998).

While CARL was an automated system, there were three things that greatly reduced its operational uptime after it was initially deployed in the mid-1990s. The first issue was the threat of hail; a large hailstone could easily shatter the protecting window above the telescope and thus the telescope itself. Thus, site operations staff shut down CARL and closed its hardened hatch whenever severe weather was imminent; the threat of weather like this occurs relatively often in north-central Oklahoma. In the spring of 1998, a stainless steel wire mesh was placed above the output window to protect the system from hail, which allowed continuous lidar operations in these weather conditions. The installation of the hail shield decreased the strength of the backscattered signal by about 18%, but this was deemed to be acceptable in order to maintain a higher uptime without the risk to the system.

The second issue that affected operational uptime was power continuity. Small disruptions in electrical service, even for a fraction of a second, would cause the laser to shut down. The laser could not be restarted automatically, and required manual intercession to physically turn a key to restart the system. Murphy's Law would dictate that these power bumps would happen most in the evenings or on weekends when the SGP operators where not on site, and resulted in significant periods of downtime. This issue was solved with the installation of a large uninterruptible power supply that conditions the power for the lidar and is able to run the system for several minutes in the event of a power fluctuation. These changes, as well as a variety of other minor system improvements (e.g., improvements in the air conditioning system in the lidar's enclosure, continued improvement of the laser and its components by the vendor, etc.), led to a continued increase in the operational uptime as the lidar matured (Fig. 18-2).

The ARM Program realized early on that it would need scientific experts to serve as "mentors" for the various instruments in the program. These mentors would provide guidance for the site operations staff and help develop analysis techniques to improve the use of the data from their system throughout the program. The Raman lidar was the first ARM instrument to have a dual-mentor model, wherein one mentor was responsible for the hardware/system aspects of the instrument and the second mentor was responsible for the development of analysis routines and calibration. This model has been highly successful and is still used within ARM today both for the Raman lidar and other instruments (e.g., the millimeter cloud radar).

b. Measuring water vapor

The Raman lidar was delivered to the SGP site during the fall of 1995, but it took many months before the system was truly ready to collect data. During this period, the Instantaneous Radiative Flux (IRF) working group organized a series of water vapor intensive operational periods (WVIOPs) that would, in part, help to evaluate the accuracy of the Raman lidar's water vapor measurements. The IRF working group was extremely interested in getting remotely sensed profiles of water vapor in order to improve their radiative transfer models, because the uncertainty in the radiosonde water vapor profiles was the limiting factor in improving the accuracy of clear-sky infrared radiative transfer models at the time (Revercomb et al. 2003). The first two WVIOPs, which were conducted in the fall of 1996 and 1997, brought a wide range of instruments to the SGP site, including the SRL from GSFC, tethersondes and aircraft measuring water vapor in situ, and other instruments. These datasets were used to develop an automated calibration strategy for the CARL and to

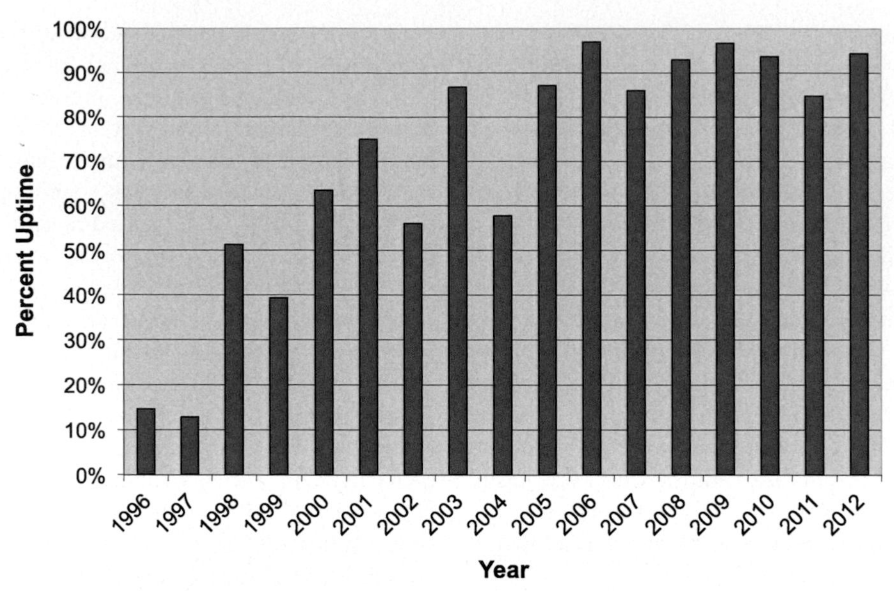

FIG. 18-2. Operational "uptime" of the SGP Raman lidar (CARL) per year.

evaluate the accuracy of this calibration over the diurnal cycle.

Raman lidar water vapor profiles normally are calibrated to an external measurement of water vapor, due to uncertainty in the Raman backscattering cross section of water vapor and difficulties associated with characterizing the wavelength dependence of the system throughput. Prior to these WVIOPs, the most common calibration standard was the radiosonde; the Raman lidar profile would be calibrated using a height-independent scale factor to match the radiosonde measurements over some altitude region. However, one of the main findings of the WVIOPs (Revercomb et al. 2003; Turner et al. 2003) was that the calibration of the radiosonde's water vapor sensor can vary significantly between calibration batches, and also changed with the age of the radiosonde (i.e., the time period between when the sonde was calibrated in the factor to when it was actually launched into the atmosphere). Furthermore, Ferrare et al. (1995) had demonstrated that some types of radiosondes have very poor performance in low relative humidity conditions (RH $< 30\%$), and using these data would result in very poor calibration for the Raman lidar. These calibration uncertainties were the source of the IRF's frustrations with radiosondes. Thus, a different calibration standard for the Raman lidar was needed.

The WVIOPs demonstrated that the microwave radiometer (MWR) observed precipitable water vapor (PWV) value was much more robust and accurate than the corresponding value-derived radiosonde water vapor data (Turner et al. 2016, chapter 13), and a decision was made to calibrate CARL against this measurement. During the night in clear-sky scenes, this was perfectly acceptable because CARL was able to profile water vapor from near the surface (the in situ observations from the surface and 60-m tower were used to fill in the lowest levels where the lidar was blind) to above 10 km with only a 10-min average; thus the lidar was seeing over 99% of the water vapor in the column and comparisons with the MWR could be performed. However, in cloudy scenes where the laser beam was attenuated, or in the daytime when the solar noise limited the maximum altitude of the lidar water vapor observations to about 3 km, a different approach was needed. Furthermore, in order to prevent saturation of the detectors in the 408-nm water vapor channels, the lidar used a "bright" mode during the daytime, wherein the lidar automatically inserted neutral density filters into these channels to attenuate the signal. Thus, it was not obvious originally how to transfer the nighttime calibration into the daytime.

The approach that was initially used was to determine separate calibration factors for the nighttime and daytime periods, where the latter used an estimate of the fraction of the PWV that the lidar did see (based upon a radiosonde climatology relating fraction of PWV to height above the surface) to scale the MWR-derived PWV to the lidar value (Turner and Goldsmith 1999). This initial approach yielded acceptable intercomparison results between the lidar and radiosondes, in situ observations from tethersondes, and aircraft observations. However, the approach also demonstrated some large errors (5%–10%) at sunrise and sunset in the 1996 WVIOP results (Turner and Goldsmith 1999). It was believed that these errors were due to nonlinear effects in the lidar detection system, and a variety of different correction factors were unable to satisfactorily remove the error. A simple change of switching to (from) the bright mode earlier (later) in the day at sunrise (sunset) was tested in the 1997 WVIOP and showed marked improvements.

However, these comparisons still did not yield great faith in the consistency of the nighttime versus daytime calibration of CARL's water vapor measurements. The WVIOPs did demonstrate conclusively that the CARL and SRL systems agreed to better than 5% in water vapor mixing ratio if using the same data source for their height-independent calibrations (Revercomb et al. 2003), which was an excellent result as there are a range of lidar-dependent correction factors that need to be applied to the data (e.g., corrections for photon pulse-pileup, overlap, etc.). To evaluate the consistency of the night-vs-day calibrations, the water vapor DIAL from the Max Planck Institute for Meteorology (Wulfmeyer and Bösenberg 1998) was deployed to the SGP site in 1999 for the so-called lidar WVIOP. The comparison of the DIAL with CARL demonstrated that the method of calibrating the daytime and nighttime data independently resulted in an approximate 10% diurnal bias in the Raman lidar calibration (Linne et al. 2000). We also realized that we could simply translate the nighttime calibration factor into the daytime by simply accounting for the attenuation of the bright mode neutral density filter; this made the daytime and nighttime data fully consistent.

Since CARL is an automated system, we needed to be able to calibrate data when an evening was cloudy and there had to be consistency in calibration for consecutive days. Thus, an approach was adopted that used nighttime data from three consecutive days to determine the calibration factor for the central day (Turner et al. 2002). This rolling calibration approach would also be used for calibrating other CARL data products.

The water vapor profiles in the boundary layer and midtroposphere were used in several different studies. In the first example, the CARL observations were used to evaluate the accuracy of the water vapor profiles retrieved from the Atmospheric Emitted Radiance Interferometer (AERI) when the structure of the boundary layer was evolving rapidly, such as during the passage of atmospheric fronts or drylines (Turner et al. 2000). CARL data were also used in the evaluation of infrared radiative transfer models during clear-sky periods (Turner et al. 2004); however, lack of coincident temperature profiles from a measurement system other than the radiosonde proved to be a limiting factor in this application. Another interesting application used CARL water vapor observations to investigate horizontal convective rolls in the convective boundary layer (Mecikalski et al. 2006); this would be the first paper of several that would start to take advantage of the high temporal and spatial resolution of the Raman lidar observations to look at boundary layer structure and phenomena.

The advances from the 1996, 1997, and 2000 WVIOPs demonstrated that CARL is able to measure water vapor well in the boundary layer and midtroposphere, but how well could it measure water vapor in the upper troposphere (UT) at night? To answer this question, and to evaluate that accuracy of other water vapor measurements in the UT, ARM and NASA conducted the ARM–FIRE Water Vapor Experiment (AFWEX) over the SGP site in late 2000. Like the previous WVIOPs, the GSFC SRL was deployed near CARL, but AFWEX also benefited from the deployment of the airborne NASA Langley Research Center Laser Atmospheric Sensing Experiment (LASE) water vapor DIAL (Browell et al. 1997) on the NASA DC-8 aircraft and the launching of chilled mirror frostpoint hygrometers during the campaign. AFWEX demonstrated very good agreement between CARL, SRL, and LASE throughout the entire troposphere (Ferrare et al. 2004). AFWEX analyses also demonstrated that Vaisala RS80-H radiosondes (the model being launched by the ARM Program at the time) had a significant dry bias (10%–30%) in the UT, but that a known radiosonde calibration model correction and accounting for the time lag of the sonde sensor removed this bias (Ferrare et al. 2004). These nighttime comparisons provided high confidence in the CARL observations of humidity throughout the troposphere, and set the stage for a couple of interesting UT projects.

The large dry bias in the radiosonde UT data has significant implications for outgoing longwave radiation calculations and modeling (e.g., Ferrare et al. 2004), satellite radiance and product validation, and modeling studies investigating cirrus processes and properties. However, CARL was only at a single location, and thus there was a desire to transfer the knowledge acquired during AFWEX to a larger domain. Soden et al. (2004) compared nighttime measurements from CARL, the radiance observations in the 6.7-μm water vapor channel on the Geostationary Operational Environmental Satellite (*GOES-8*), and radiosonde profiles over a multimonth period that encompassed all of the WVIOP periods. This was the extension of an earlier work that used the NASA SRL observations collected during the FIRE campaign in Coffeyville (Soden et al. 1994). Using a simple model to convert between radiance space and profile space, they showed that the GOES and

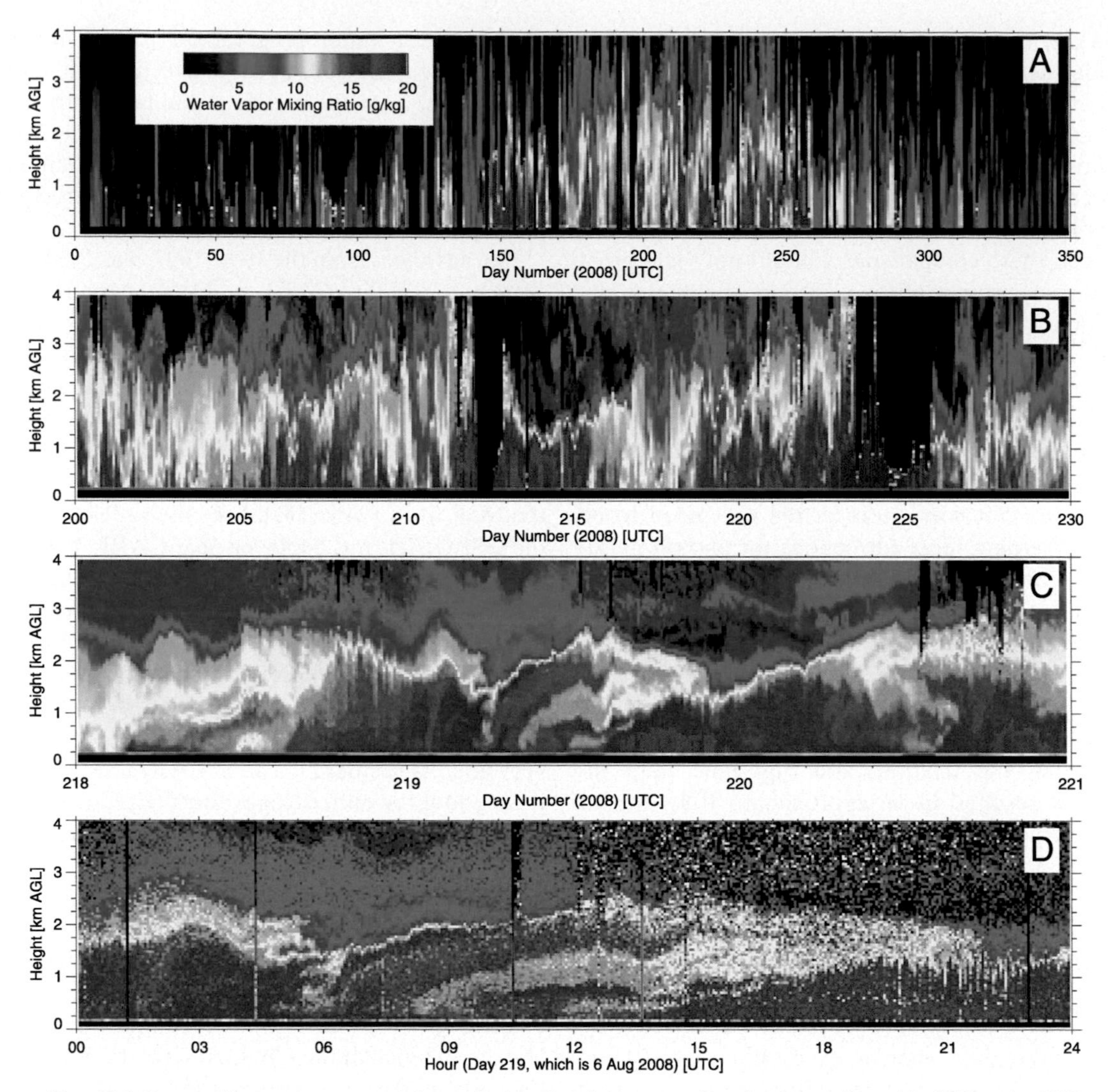

FIG. 18-3. Raman lidar water vapor mixing ratio time–height cross sections for data collected in 2008 at the SGP, with a range of different "zooms": (a) 350 days, (b) 30 days, (c) 3 days, and (d) 1 day. The data in the first three panels are at 10-min, 75-m resolution, whereas the resolution of the data in (d) is 10 s, 37.5 m.

Raman lidar observations agreed to within 10%, but that the radiosonde was systematically drier by ~40%. The authors then developed a variational assimilation method that used GOES radiances to correct the radiosonde dry bias, demonstrated its accuracy, and discussed the utility of this method to correct radiosondes launched at other locations beyond the SGP.

Cirrus clouds play an important role in both the energy and moisture budgets of the atmosphere. Upper tropospheric humidity plays an important role in the nucleation (i.e., homogeneous vs heterogeneous nucleation) and maintenance of cirrus clouds, yet is a difficult variable to measure for long periods of time across a range of conditions. Comstock et al. (2004) used a year of CARL UT water vapor observations to demonstrate that ice supersaturation occurs more than 40% of the time in the uppermost portions of midlatitude cirrus clouds, which supports the theory that homogeneous nucleation occurs frequently.

c. Measuring aerosol properties

The automated nature of the Raman lidar provided wonderful multiday views of water vapor mixing ratio (e.g., Fig. 18-3) and aerosol scattering ratio and extinction (Fig. 18-4). The initial analysis of the aerosol data showed tremendous promise in the observations, but there were some artifacts in the data. The ARM Raman lidar was designed to measure water vapor first and foremost; aerosol and cloud observations were considered of secondary importance. This ultimately led to

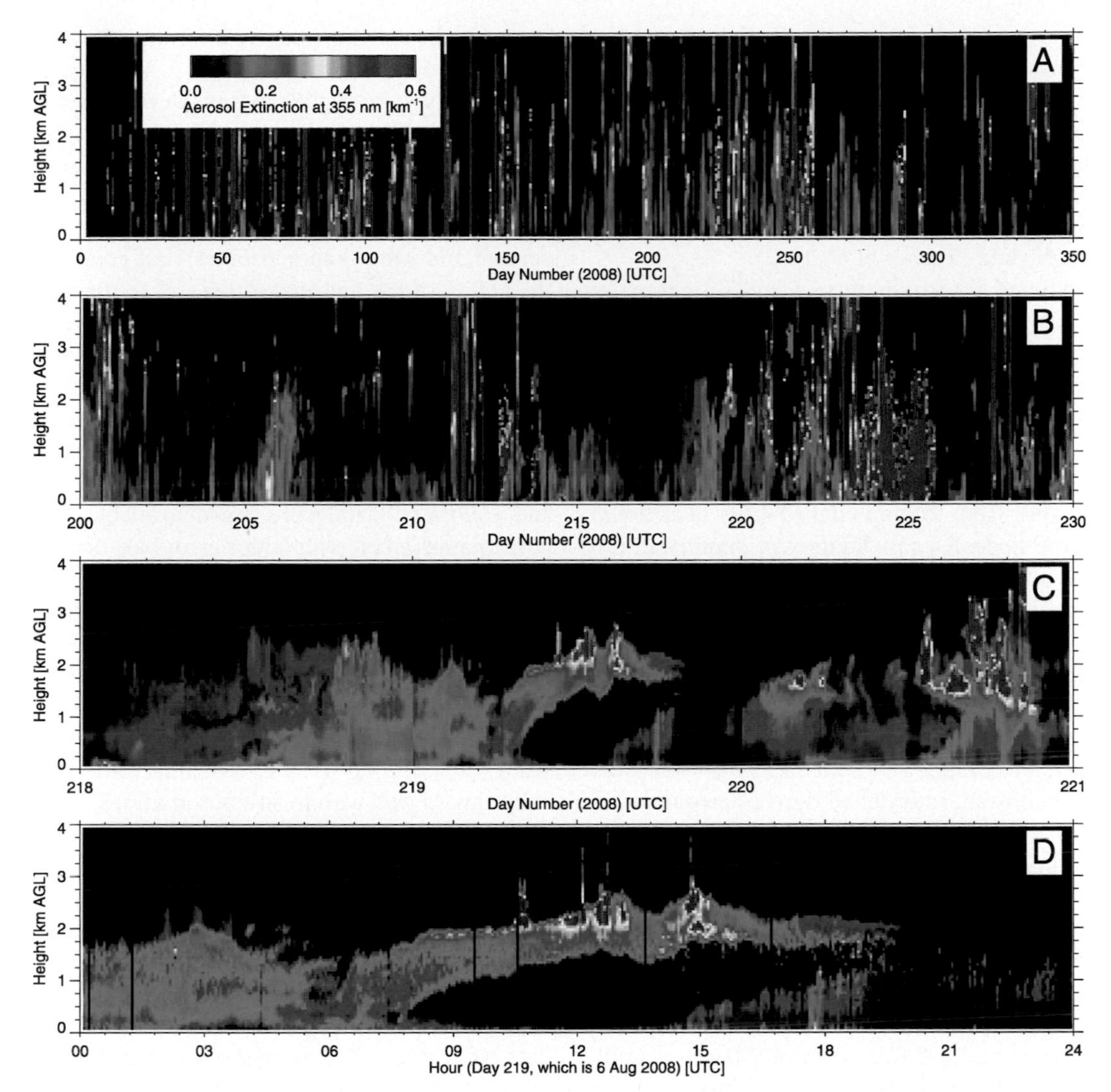

FIG. 18-4. (a)–(d) Raman lidar aerosol extinction time–height cross sections for the same time periods as shown in Fig. 18-3. The temporal resolution of (a)–(c) is 10 min, whereas the temporal resolution in (d) is 1 min. Areas of high extinction (purple) are primarily due to extinction by clouds.

some design choices in the construction of CARL that would turn out to hinder the derivation of aerosol scattering ratio (and hence backscatter coefficient) and extinction coefficient from the data.

The main "problem" was the orientation of the copolarized elastic return (i.e., the channel sensitive to the return at 355 nm) in the narrow FOV (henceforth called the high aerosol channel). The layout of the aft optics for the system in circa 1996 is shown in the top of Fig. 18-1. The separation of the co- versus cross-polarized elastic return was done with a polarizing cube, and thus the co- versus cross-polarized light would leave this optical element at positions 90° angles from each other. Because of space limitations on the optical breadboard, the high aerosol channel was oriented normal to the optical table (i.e., sticking up in the air). However, this made the channel susceptible to vibrations in the lidar enclosure (such as when the outer door was shut) and also thermal variations that were different than the other channels. Thus, the periodic alignment tweaks affected the alignment of this particular channel differently than the other channels that were all lying flat on the optical table.

The automated analysis software, which is run in a postprocessing mode in the ARM Data Management Facility, needed a solution. The approach used the wide FOV aerosol scattering ratio (ASR), which is defined as the ratio of the elastic return divided by the nitrogen return that has been calibrated to be unity in aerosol-free air, to determine the overlap correction for the narrow FOV (Turner et al. 2002). Each alignment tweak period was treated independently, and this approach (after numerous attempts to bullet-proof the routines) led to accurate ASR profiles throughout the troposphere. The same approach was used to correct for the overlap in the narrow FOV nitrogen channel, and thus allow the derivation of aerosol extinction profiles from the narrow FOV nitrogen data, which is less noisy than the wide FOV data (Turner et al. 2002).

The automated measurements of aerosol extinction and water vapor mixing ratio were new for the scientific community; up to this point, all Raman lidars were operated in episodic fashion, typically by a dedicated group of lidar scientists and engineers. The operational nature of CARL allowed a large dataset to be quickly collected, which led to the first climatologies of aerosol extinction and water vapor from Raman lidar (Turner et al. 2001). Furthermore, since Raman lidars can simultaneously measure aerosol extinction and backscatter coefficients directly (the latter is easily derived from the ASR profile), studies on the variability of the "lidar ratio," which is the ratio of the extinction to backscatter coefficients, could be performed. The lidar ratio is an intensive property of the aerosols and is sensitive to aerosol size, composition, and shape. The lidar ratio was usually assumed to be constant in order to derive aerosol extinction from single wavelength lidars, such as the ARM micropulse lidar (Campbell et al. 2002). Ferrare et al. (2001) showed for the first time that this was often a poor assumption over the central part of the United States, with approximately 30% of CARL observations showing significant changes in the lidar ratio as function of height. The aerosol data from the Raman lidar were very useful in characterizing different aerosol layers over the SGP site when smoke from fires in Central America advected over the SGP site (Peppler et al. 2000).

In the late 1990s, ARM began to place more emphasis on characterizing the aerosol properties above its sites (McComiskey and Ferrare 2016, chapter 21). Earlier work, supported both by the program (e.g., Kato et al. 2000) and by others (e.g., Masonis et al. 2002), demonstrated that Raman lidar measurements of aerosol extinction were typically 30% larger than airborne in situ measurements. To resolve this discrepancy and more completely describe the aerosol optical, microphysical, and chemical properties above the SGP site, the ARM Program conducted the aerosol IOP in May 2003 (McComiskey and Ferrare 2016, chapter 21). This experiment would bring a large array of aerosol instruments to the site to complement the routine ARM observations and would be one of the most expensive IOPs conducted by the ARM Program at the time.

Unfortunately, CARL had problems during this IOP. The system had been experiencing a slow degradation of its sensitivity starting in early 2002, and this degradation was not noticed until after this IOP. The impacts on the data quality were significant, as the random noise level had increased by a factor of 2–4 from its baseline values in the late 1990s by the time of the aerosol IOP, depending on the data product (Ferrare et al. 2006). The higher noise level reduced the maximum range of the aerosol and water vapor profiles, and greatly impacted the ability of the automated aerosol routines to correct for alignment issues in the narrow FOV with the wide FOV data. Because of the importance of the aerosol IOP, a massive analysis effort was made by Ferrare and Turner to advance the logic in the automated scripts in order to handle the adverse effects of the noisier raw data (Ferrare et al. 2006). This effort was largely successful, and calibrated data were made available to the IOP participants (albeit with larger error bars than normal).

The reprocessed CARL data were used in a large number of different studies, which testifies to the utility and value of the Raman lidar aerosol and water vapor products. Ferrare et al. (2006) showed that the daytime water vapor measurements were about 5%–10% moister than other observations, and retrievals of single scatter albedo that used CARL lidar ratio values typically agreed well with in situ observations. Schmid et al. (2006) demonstrated that the lidar-observed aerosol extinction and optical depth were still larger than in situ and sun photometer observations, although they expressed hope that if the lidar's sensitivity was its normal value would likely be good agreement between the lidar and sun photometer (this would lead to another field experiment in 2007). Pahlow et al. (2006) used CARL data to characterize the hygroscopic growth of aerosol particles as the relative humidity increases. They showed that the lidar observations are consistent with in situ nephelometer observations, but that the lidar observations can be made over a much higher relative humidity range (i.e., closer to saturation) and do not suffer from as many sampling artifacts experienced by in situ systems. A different study used the CARL observations together with surface in situ measurements to retrieve cloud condensation nuclei concentrations at cloud base (Ghan et al. 2006).

d. Refurbishment and upgrades in 2004

The unanticipated finding of the large degradation in CARL's sensitivity during the analysis of the aerosol IOP stimulated increased activity to identify the source of the problem and correct it. After testing a variety of optical components in both the transmitter and receiver chains, it was concluded that the problem was associated with a decrease in the efficiency of the telescope. The telescope was removed from the system in the spring of 2004 in order to be refurbished; this process took several months. The installation of the refurbished telescope restored the sensitivity of the lidar to its original levels.

During the early 2000s, a company in Berlin (Licel GbR) developed a new set of detection electronics that combined both analog-to-digital and photon-counting electronics into single package. The two detection systems have different strengths: AD is well suited for large signals (such as those typical for the lower troposphere), but PC is better suited for small signals (such as those in the upper troposphere or from weak scattering processes). The early studies by SNL and GSFC demonstrated that PC detection was superior over most altitude ranges, and hence PC detection was used in the initial development of CARL. The new Licel electronics were first tested in the GSFC SRL (Whiteman et al. 2006), and they worked well and resulted in improved data quality. The Raman lidar instrument mentors decided to switch to the new Licel detection system in CARL; these units were installed into the system while the telescope was being refurbished. Since the combined AD and PC electronics have a larger dynamic range, this allowed the removal of some of the neutral density filters in some channels of the lidar, which were originally in place to prevent the saturation of original PC electronics. The removal of the neutral density filters greatly increased the signal strengths in the aerosol and nitrogen channels by factors of 10–20 (Ferrare et al. 2006), which greatly improved the noise levels of the water vapor and aerosol data products. To properly use the Licel data, the AD and PC datastreams needed to be "glued" together, and an automated technique was developed (Newsom et al. 2009).

During that same time period, additional modifications were made. Scientists at GSFC had demonstrated two intriguing new measurements that could be made with a Raman lidar like CARL: (a) atmospheric temperature measurements using rotational Raman scattering by nitrogen and oxygen molecules (Di Girolamo et al. 2004) and (b) measurements of cloud liquid water using Raman scattering (Melfi et al. 1997; Whiteman and Melfi 1999). Thus, the decision was made to add three new channels (two for the rotational Raman scattering and one for liquid water) to CARL. The optical layout of the aft optics after the addition of these new channels is shown in Fig. 18-1 (bottom). Note that when these extra channels were added to the system, the high aerosol and depolarization channels were rearranged so that both of them were lying flat on the optical table just like the rest of the channels. This greatly improved the stability and calibration of the derived aerosol scattering ratio data.

The liquid water Raman measurements were deemed to have a lot of scientific potential because, if they worked, we would be able to measure both cloud and aerosol properties with the same system (same volume and time), which would be beneficial for studying aerosol–cloud interactions. The liquid water Raman scattering cross section is spectrally broad, and thus it was quickly realized that a wide-bandpass filter (6 nm instead of the typical 0.3 nm) would be needed in order to capture scattering across the liquid water Raman spectrum to improve the signal-to-noise ratio. However, the wide bandpass filter was inadequate for the daytime because the solar background over such a large spectral range swamped the signal, and thus the liquid water measurements were only possible during the night. The nighttime liquid water Raman scattering signal was still weak and required five minutes of averaging to get a good signal-to-noise ratio, but during this time the cloud properties changed markedly (even for stratus clouds), making the study of cloud–aerosol interactions problematic. However, Sakai et al. (2013) were able to use these observations to evaluate the strength of the liquid water Raman scattering cross section using AERI data.

Wang et al. (2004) used data from the liquid Raman scattering channel to look at the ice water Raman scattering signal; the liquid water and ice Raman scattering spectra largely overlap in the spectral domain. Wang et al. used these observations to derive ice water content in cirrus clouds, and the technique compared well with more traditional radar–lidar analysis techniques.

The addition of the rotational Raman scattering channels provides an excellent way to get ambient temperature profiles in the same volume as the water vapor mixing ratio, and thus improve measurements such as relative humidity from CARL. However, these channels were only added to the narrow FOV detection system, and thus there were two challenges. The first challenge was that the algorithm would need to use the narrow FOV data much lower to the surface than was done for any the other data products (since we desired temperature profiles over the same range as water vapor data). This made the derived temperature profile very sensitive to alignment of the outgoing laser beam in the detector's FOV. Very quickly, we realized that the 3-hourly alignment tweaks wreaked havoc on the derived temperature profiles. Fortunately, Licel also had developed a very nice and easy-to-use boresight alignment detector, which could be integrated into the Raman lidar system and mated into the LabView software that operated the lidar. This boresight sensor greatly improved the alignment stability of the Raman lidar, and we could use it to continuously maintain the alignment of the system. The accuracy of the aerosol products were also improved greatly, since the potential for large changes with the older 3-hourly alignment tweaks was eliminated.

The second challenge was how to routinely calibrate the ratio of the rotational Raman lidar signals to provide profiles of ambient temperature. The routine radiosonde launches were used for this purpose. However, the analysis also determined that there was a significant diurnal bias in the calibration that was related to the magnitude of the solar background. An approach was developed that accounts for the solar background in the calibration, thereby providing calibrated data across the diurnal cycle away from radiosonde launch times (Newsom et al. 2013).

The installation of the Licel detection electronics, and the subsequent reduction in the amount of neutral density attenuation in some of the detector channels, opened up a new area of research for the Raman lidar. The maximum temporal resolution for the water vapor mixing ratio and aerosol scattering ratio (and hence backscatter coefficient) was 1 min prior to the upgrade in 2004; the new temporal resolution (having the same approximate signal-to-noise levels) after the upgrade was 10 s. This resolution is fast enough to potentially resolve turbulent eddies in the convective boundary layer; the question was "Is the accuracy and resolution of state-of-the-art water vapor Raman lidar systems sufficient to derive higher-order moments of turbulence in the boundary layer?" Wulfmeyer et al. (2010) demonstrated that indeed CARL is able to resolve profiles of the second and third moments of water vapor in the convective boundary layer, and thus variance and skewness profiles can be derived with the accuracy needed to study turbulence. Furthermore, the Raman lidar measurements of water vapor variance and skewness were compared with in situ data collected during the Routine AAF Clouds with Low Optical Water Depths (CLOWD) Optical Radiative Observations (RACORO) field campaign that was conducted over the SGP site during the early part of 2009 (Vogelmann et al. 2012); these results showed good agreement between the two very different techniques (Turner et al. 2014a). The advantage of using an automated system like CARL to study turbulence is that multiple years of data can be included to build a climatology and investigate relationships between different variables (Turner et al. 2014b); this is true for many other processes beyond turbulence.

There are also some new exciting research areas that have resulted from the higher temporal resolution, especially when observations from other instruments are included in the analysis. These include deriving water vapor flux observations using coincident CARL and Doppler lidar measurements, and characterizing entrainment in cumulus clouds using Raman lidar, AERI, cloud radar, microwave radiometer, and surface measurements (Wagner et al. 2013).

e. Current status and future outlook

The primary goal of the Raman lidar within the ARM Program was to provide routine measurements of water vapor through the boundary layer across the diurnal cycle. The original programmatic dream was that water vapor and temperature profile be remotely sensed and that the ARM Program could move away from the routine launching of radiosondes. While the program will probably always need to launch radiosondes to help calibrate the lidar and provide profiles above clouds that attenuate the lidar's laser beam (and provide wind information aloft), the ARM Program has actually realized its dream of using an advanced remote sensor to provide measurements of these key thermodynamic variables at high time and spatial resolution.

The unique and powerful measurements from the SGP Raman lidar have been used in an extremely wide range of research—a much larger range of research than was originally anticipated. This success of the SGP system, both in terms of its operational uptime and its potential to open up new areas of study and contribute to others, led to the decision to build and deploy a new Raman lidar that was almost identical to CARL at the Tropical Western Pacific (TWP) site in Darwin using funds from the American Recovery and Reinvestment Act (ARRA) from 2009. This system became operational in December 2010. Data from this system are already being used to develop climatologies of aerosol extinction and water vapor mixing ratio as a function of synoptic classification, turbulence within the tropical boundary layer, and cirrus cloud macrophysical and optical properties in the tropics where the tropopause is very high (e.g., Thorsen et al. 2013).

The ARM Program took a chance, and invested heavily to advance the Raman lidar technology from a research-only tool into the world's first operational water vapor and aerosol Raman lidar. The success of the ARM Raman lidar enterprise is perhaps best captured in a quote from Reichardt et al. (2012, 8111–8112): "The Cloud and Radiation Testbed Raman Lidar CARL at the Atmospheric Radiation Measurement program's Southern Great Plains site in Oklahoma stood out because it added another layer of complexity (i.e., it was monitoring tropospheric humidity and clouds continuously and autonomously). Its success enticed meteorological services around the world to develop and operate similar instruments." Since that time, the community has seen the development of other automated Raman lidars (e.g., in Germany, Switzerland, and the Netherlands), but the multiyear record of near continuous observations made by CARL (Fig. 18-2) is totally unique and unprecedented and is clearly one of the shining achievements of the ARM Program.

The ARM Program recently constructed an additional Raman lidar system that was deployed at the new ARM site at Oliktok Point along the northern slope of Alaska in the fall of 2014. Furthermore, the ARM Program recently elected to close its TWP sites, and the Raman lidar at Darwin was relocated to the new ARM site in the Azores in 2015. There will be challenges running these advanced lidar systems in these harsh environments, but we have confidence that these challenges will be overcome and that the scientific benefit will be huge. This will result in three autonomous water vapor and aerosol Raman lidars operating in the ARM Program, which is quite an achievement given the uncertainties surrounding whether a Raman lidar could be made operational in the early 1990s.

Acknowledgments. There have been many people who have contributed to the success of the ARM Raman lidar. The Raman lidar has had a number of different "instrument mentors" over its history; these folks have been responsible for overseeing the lidar's operation from a scientific and engineering perspective. Tim Tooman, Rob Newsom, and the authors have all served (or are serving) as mentors for these lidars. We also would like to thank the superb staff at the SGP site, and in particular Chris Martin, as well as the TWP/Darwin staff for routine attention that any lidar system (operational or not) needs. We have benefitted greatly from our interactions with the Raman lidar group at NASA/GSFC over the years, and in particular Harvey Melfi and David Whiteman.

REFERENCES

Ansmann, A., M. Riebesell, and C. Weitkamp, 1990: Measurement of atmospheric aerosol extinction with a Raman lidar. *Opt. Lett.*, **15**, 746–748, doi:10.1364/OL.15.000746.

——, ——, U. Wandinger, C. Weitkamp, E. Voss, W. Lahmann, and W. Michaelis, 1992: Combined Raman elastic backscatter lidar for vertical profiling of moisture, aerosol extinction, backscatter, and lidar ratio. *Appl. Phys.*, **55B**, 18–28, doi:10.1007/BF00348608.

Bisson, S. E., and J. E. M. Goldsmith, 1993: Daytime tropospheric water vapor profile measurements with a Raman lidar. *Opt. Remote Sens. Atmos. Tech. Dig.*, Optical Society of America, 19–22.

——, ——, and M. G. Mitchell, 1999: Narrow-band, narrow-field-of-view Raman lidar with combined day and night capability for tropospheric water-vapor profile measurements. *Appl. Opt.*, **38**, 1841–1849, doi:10.1364/AO.38.001841.

Browell, E. V., T. D. Wilkerson, and T. J. McIlrath, 1979: Water vapor differential absorption lidar development and evaluation. *Appl. Opt.*, **18**, 3474–3483, doi:10.1364/AO.18.003474.

——, and Coauthors, 1997: LASE validation experiment. *Advances in Atmospheric Remote Sensing with Lidar*, A. Ansmann et al., Eds., Springer-Verlag, 289–295.

Campbell, J. R., D. L. Hlavka, E. J. Welton, C. J. Flynn, D. D. Turner, J. D. Spinhirne, V. S. Scott, and I. H. Hwang, 2002: Full-time, eye-safe cloud and aerosol lidar observations at Atmospheric Radiation Measurement Program sites: Instruments and data processing. *J. Atmos. Oceanic Technol.*, **19**, 431–442, doi:10.1175/1520-0426(2002)019<0431:FTESCA>2.0.CO;2.

Comstock, J. M., T. P. Ackerman, and D. D. Turner, 2004: Evidence of high ice supersaturation in cirrus clouds using ARM Raman lidar measurements. *Geophys. Res. Lett.*, **31**, L11106, doi:10.1029/2004GL019705.

Cooney, J. A., 1968: Measurements on the Raman component of laser atmospheric backscatter. *Appl. Phys. Lett.*, **12**, 40, doi:10.1063/1.1651884.

Di Girolamo, P., R. Marchese, D. N. Whiteman, and B. B. Demoz, 2004: Rotational Raman lidar measurements of atmospheric temperature in the UV. *Geophys. Res. Lett.*, **31**, L01106, doi:10.1029/2003GL018342.

DOE, 1990: Atmospheric Radiation Measurement Program Plan. U.S. Department of Energy Rep. DOE/ER-0441, 121 pp.

Ehret, G., C. Kiemle, W. Renger, and G. Simmet, 1993: Airborne remote sensing of tropospheric water vapor with a near-infrared differential absorption lidar system. *Appl. Opt.*, **32**, 4534–4551, doi:10.1364/AO.32.004534.

Ferrare, R. A., S. H. Melfi, D. N. Whiteman, K. D. Evans, F. J. Schmidlin, and D. O'C. Starr, 1995: A comparison of water vapor measurements made by Raman lidar and radiosondes. *J. Atmos. Oceanic Technol.*, **12**, 1177–1195, doi:10.1175/1520-0426(1995)012<1177:ACOWVM>2.0.CO;2.

——, D. D. Turner, L. A. Heilman Brasseur, W. F. Feltz, O. Dubovik, and T. P. Tooman, 2001: Raman lidar observations of the aerosol extinction-to-backscatter ratio over the southern Great Plains. *J. Geophys. Res.*, **106**, 20 333–20 348, doi:10.1029/2000JD000144.

——, and Coauthors, 2004: Characterization of upper-tropospheric water vapor measurements during AFWEX using LASE. *J. Atmos. Oceanic Technol.*, **21**, 1790–1808, doi:10.1175/JTECH-1652.1.

——, and Coauthors, 2006: Evaluation of daytime measurements of aerosols and water vapor made by an operational Raman lidar over the southern Great Plains. *J. Geophys. Res.*, **111**, D05S08, doi:10.1029/2005JD005836.

Ghan, S. J., and Coauthors, 2006: Use of in situ cloud condensation nuclei, extinction, and aerosol size distribution measurements to test a method for retrieving cloud condensation nuclei profiles from surface measurements. *J. Geophys. Res.*, **111**, D05S10, doi:10.1029/2004JD005752.

Goldsmith, J. E. M., and R. A. Ferrare, 1992: Performance modeling of daytime Raman lidar systems for profiling atmospheric water vapor. Sixteenth International Laser Radar Conference Proceedings, M. P. McCormick, Ed., NASA Publ. 3158, Part 2, 667–670.

——, S. E. Bisson, R. A. Ferrare, K. D. Evans, D. N. Whiteman, and S. H. Melfi, 1994: Raman lidar profiling of atmospheric water vapor: Simultaneous measurements with two collocated systems. *Bull. Amer. Meteor. Soc.*, **75**, 975–982, doi:10.1175/1520-0477(1994)075<0975:RLPOAW>2.0.CO;2.

——, F. H. Blair, S. E. Bisson, and D. D. Turner, 1998: Turn-key Raman lidar for profiling atmospheric water vapor, clouds, and aerosols. *Appl. Opt.*, **37**, 4979–4990, doi:10.1364/AO.37.004979.

Grant, W. B., 1991: Differential absorption and Raman lidar for water vapor profile measurements: A review. *Opt. Eng.*, **30**, 40–48, doi:10.1117/12.55772.

Ismail, S., and E. V. Browell, 1989: Airborne and spaceborne lidar measurements of water vapor profiles: A sensitivity analysis. *Appl. Opt.*, **28**, 3603–3615, doi:10.1364/AO.28.003603.

Kato, S., and Coauthors, 2000: A comparison of the aerosol optical thickness derived from ground-based and airborne measurements. *J. Geophys. Res.*, **105**, 14 701–14 717, doi:10.1029/2000JD900013.

Leonard, D. A., 1967: Observation of Raman scattering from the atmosphere using a pulsed nitrogen ultraviolet laser. *Nature*, **216**, 142–143, doi:10.1038/216142a0.

Linne, H., D. D. Turner, J. E. M. Goldsmith, T. P. Tooman, J. Bösenberg, K. Ertel, and S. Lehmann, 2000: Intercomparison of DIAL and Raman lidar measurements of humidity profiles. *Advances in Laser Remote Sensing: Selected Papers Presented at the 20th International Laser Radar Conference,* D. Dabas, C. Loth, and J. Pelon, Eds., Ecole Polytechnique, 293–298.

Masonis, S. J., K. Franke, A. Ansmann, D. Müller, D. Althausen, J. A. Ogren, A. Jefferson, and P. J. Sheridan, 2002: An intercomparison of aerosol light extinction and 180° backscatter as derived using in situ instruments and Raman lidar during the INDOEX field campaign. *J. Geophys. Res.*, **107**, 8014, doi:10.1029/2000JD000035.

McComiskey, A., and R. A. Ferrare, 2016: Aerosol physical and optical properties and processes in the ARM Program. *The Atmospheric Radiation Measurement (ARM) Program: The First 20 Years, Meteor. Monogr.*, No. 57, Amer. Meteor. Soc., doi:10.1175/AMSMONOGRAPHS-D-15-0028.1.

Mecikalski, J. R., K. M. Bedka, D. D. Turner, W. F. Feltz, and S. J. Peach, 2006: Ability to quantify coherent turbulent structures in the convective boundary layer using thermodynamic profiling instruments. *J. Geophys. Res.*, **111**, D12203, doi:10.1029/2005JD006456.

Melfi, S. H., 1972: Remote measurement of the atmosphere using Raman scattering. *Appl. Opt.*, **11**, 1605–1610, doi:10.1364/AO.11.001605.

——, D. N. Whiteman, and R. A. Ferrare, 1989: Observation of atmospheric fronts using Raman lidar moisture measurements. *J. Appl. Meteor.*, **28**, 789–806, doi:10.1175/1520-0450(1989)028<0789:OOAFUR>2.0.CO;2.

——, K. D. Evans, J. Li, D. N. Whiteman, R. A. Ferrare, and G. Schwemmer, 1997: Observations of Raman scattering by cloud droplets in the atmosphere. *Appl. Opt.*, **36**, 3551–3559, doi:10.1364/AO.36.003551.

Newsom, R. K., D. D. Turner, B. Mielke, M. Clayton, R. A. Ferrare, and C. Sivaraman, 2009: The use of simultaneous analog and photon counting detection for Raman lidar. *Appl. Opt.*, **48**, 3903–3914, doi:10.1364/AO.48.003903.

——, ——, and J. E. M. Goldsmith, 2013: Long-term evaluation of temperature profiles measured by an operational Raman lidar. *J. Atmos. Oceanic Technol.*, **30**, 1616–1634, doi:10.1175/JTECH-D-12-00138.1.

Pahlow, M., and Coauthors, 2006: Comparison between lidar and nephelometer measurements of aerosol hygroscopicity at the Southern Great Plains Atmospheric Radiation Measurement site. *J. Geophys. Res.*, **111**, D05S15, doi:10.1029/2004JD005646.

Peppler, R. A., and Coauthors, 2000: ARM Southern Great Plains observations of the smoke pall associated with the 1998 Central American fires. *Bull. Amer. Meteor. Soc.*, **81**, 2563–2591, doi:10.1175/1520-0477(2000)081<2563:ASGPSO>2.3.CO;2.

Reichardt, J., U. Wandinger, M. Serwazi, and C. Weitkamp, 1996: Combined Raman lidar for aerosol, ozone, and moisture measurements. *Opt. Eng.*, **35**, 1457–1465, doi:10.1117/1.600681.

——, ——, V. Klein, I. Mattis, B. Hilber, and R. Begbie, 2012: RAMSES: German meteorological service autonomous Raman lidar for water vapor, temperature, aerosol, and cloud measurements. *Appl. Opt.*, **51**, 8111–8131, doi:10.1364/AO.51.008111.

Renaut, D., J. C. Pourney, and R. Capitani, 1980: Daytime Raman-lidar measurements of water vapor. *Opt. Lett.*, **5**, 233–235, doi:10.1364/OL.5.000233.

Revercomb, H. E., and Coauthors, 2003: The ARM Program's water vapor intensive observations periods: Overview, initial accomplishments, and future challenges. *Bull. Amer. Meteor. Soc.*, **84**, 217–236, doi:10.1175/BAMS-84-2-217.

Sakai, T., D. N. Whiteman, F. Russo, D. D. Turner, I. Veselovskii, S. H. Melfi, and T. Nagai, 2013: Liquid water cloud measurements using the Raman lidar technique: Current understanding and future research needs. *J. Atmos. Oceanic Technol.*, **30**, 1337–1353, doi:10.1175/JTECH-D-12-00099.1.

Schmid, B., and Coauthors, 2006: How well do state-of-the-art techniques measuring the vertical profile of tropospheric aerosol extinction compare? *J. Geophys. Res.*, **111**, D05S07, doi:10.1029/2005JD005837.

Smullin, L. D., and G. Fiocco, 1962: Optical echoes from the moon. *Nature*, **194**, 1267, doi:10.1038/1941267a0.

Soden, B. J., S. A. Ackerman, D. O'C. Starr, S. H. Melfi, and R. A. Ferrare, 1994: Comparison of upper tropospheric water vapor from GOES, Raman lidar, and cross-chain loran atmospheric sounding system measurements. *J. Geophys. Res.*, **99**, 21 005–21 016, doi:10.1029/94JD01721.

——, D. D. Turner, B. M. Lesht, and L. M. Miloshevich, 2004: An analysis of satellite, radiosonde, and lidar observations of upper tropospheric water vapor from the Atmospheric Radiation Measurement Program. *J. Geophys. Res.*, **109**, D04105, doi:10.1029/2003JD003828.

Stokes, G. M., 2016: Original ARM concept and launch. *The Atmospheric Radiation Measurement (ARM) Program: The First 20 Years, Meteor. Monogr.*, No. 57, Amer. Meteor. Soc., doi:10.1175/AMSMONOGRAPHS-D-15-0021.1.

Thorsen, T. J., Q. Fu, J. M. Comstock, C. Sivaraman, M. A. Vaughn, D. M. Winker, and D. D. Turner, 2013: Macrophysical properties of tropical cirrus clouds from the CALIPSO satellite and from ground-based micropulse and Raman lidars. *J. Geophys. Res. Atmos.*, **118**, 9209–9220, doi:10.1002/jgrd.50691.

Turner, D. D., and J. E. M. Goldsmith, 1999: Twenty-four-hour Raman lidar water vapor measurements during the Atmospheric Radiation Measurement Program's 1996 and 1997 water vapor intensive observations periods. *J. Atmos. Oceanic Technol.*, **16**, 1062–1076, doi:10.1175/1520-0426(1999)016<1062:TFHRLW>2.0.CO;2.

——, and D. N. Whiteman, 2002: Remote Raman spectroscopy. Profiling water vapor and aerosols in the troposphere using Raman lidars. *Handbook of Vibrational Spectroscopy*, Vol. 4, J. M. Chalmers and P. R. Griffiths, Eds., Wiley and Sons, 2857–2878.

——, W. F. Feltz, and R. A. Ferrare, 2000: Continuous water vapor profiles from operational ground-based active and passive remote sensors. *Bull. Amer. Meteor. Soc.*, **81**, 1301–1317, doi:10.1175/1520-0477(2000)081<1301:CWBPFO>2.3.CO;2.

——, R. A. Ferrare, and L. A. Brasseur, 2001: Average aerosol extinction and water vapor profiles over the southern Great Plains. *Geophys. Res. Lett.*, **28**, 4441–4444, doi:10.1029/2001GL013691.

——, ——, L. A. Heilman Brasseur, W. F. Feltz, and T. P. Tooman, 2002: Automated retrieval of water vapor and aerosol profiles from an operational Raman lidar. *J. Atmos. Oceanic Technol.*, **19**, 37–50, doi:10.1175/1520-0426(2002)019<0037:AROWVA>2.0.CO;2.

——, B. M. Lesht, S. A. Clough, J. C. Liljegren, H. E. Revercomb, and D. C. Tobin, 2003: Dry bias and variability in Vaisala radiosondes: The ARM experience. *J. Atmos. Oceanic Technol.*, **20**, 117–132, doi:10.1175/1520-0426(2003)020<0117:DBAVIV>2.0.CO;2.

——, and Coauthors, 2004: The QME AERI LBLRTM: A closure experiment for downwelling high spectral resolution infrared radiance. *J. Atmos. Sci.*, **61**, 2657–2675, doi:10.1175/JAS3300.1.

——, R. A. Ferrare, V. Wulfmeyer, and A. J. Scarino, 2014a: Aircraft evaluation of ground-based Raman lidar water vapor turbulence profiles in convective mixed layers. *J. Atmos. Oceanic Technol.*, **31**, 1078–1088, doi:10.1175/JTECH-D-13-00075.1.

——, V. Wulfmeyer, L. K. Berg, and J. H. Schween, 2014b: Water vapor turbulence profiles in stationary continental convective mixed layers. *J. Geophys. Res. Atmos.*, **119**, 11 151–11 165, doi:10.1002/2014JD022202.

——, E. J. Mlawer, and H. E. Revercomb, 2016: Water vapor observations in the ARM Program. *The Atmospheric Radiation Measurement (ARM) Program: The First 20 Years, Meteor. Monogr.*, No. 57, Amer. Meteor. Soc., doi:10.1175/AMSMONOGRAPHS-D-15-0025.1.

Vogelmann, A. M., and Coauthors, 2012: RACORO extended-term aircraft observations of boundary layer clouds. *Bull. Amer. Meteor. Soc.*, **93**, 861–878, doi:10.1175/BAMS-D-11-00189.1.

Wagner, T. J., D. D. Turner, L. K. Berg, and S. K. Krueger, 2013: Ground-based remote retrievals of cumulus entrainment rates. *J. Atmos. Oceanic Technol.*, **30**, 1460–1471, doi:10.1175/JTECH-D-12-00187.1.

Wang, Z., D. N. Whiteman, B. B. Demoz, and I. Veselovskii, 2004: A new way to measure cirrus cloud ice water content by using ice Raman scatter with Raman lidar. *Geophys. Res. Lett.*, **31**, L15101, doi:10.1029/2004GL020004.

Weckwerth, T. M., V. Wulfmeyer, R. M. Wakimoto, R. M. Hardesty, J. W. Wilson, and R. M. Banta, 1999: NCAR–NOAA lower-tropospheric water vapor workshop. *Bull. Amer. Meteor. Soc.*, **80**, 2339–2357, doi:10.1175/1520-0477(1999)080<2339:NNLTWV>2.0.CO;2.

Whiteman, D. N., and S. H. Melfi, 1999: Cloud liquid water, mean droplet radius, and number density measurements using a Raman lidar. *J. Geophys. Res.*, **104**, 31 411–31 419, doi:10.1029/1999JD901004.

——, ——, and R. A. Ferrare, 1992: Raman lidar system for the measurement of water vapor and aerosols in the Earth's atmosphere. *Appl. Opt.*, **31**, 3068–3082, doi:10.1364/AO.31.003068.

——, ——, R. A. Campbell, and K. D. Evans, 1993: Solar blind Raman scattering measurements of water vapor using a KrF excimer laser. *Opt. Remote Sens. Atmos. Tech. Dig.*, Optical Society of America, 165–168.

——, and Coauthors, 2006: Raman lidar measurements during the international H_2O project. Part I: Instrumentation and analysis techniques. *J. Atmos. Oceanic Technol.*, **23**, 157–169, doi:10.1175/JTECH1838.1.

Wulfmeyer, V., and J. Bösenberg, 1998: Ground-based differential absorption lidar for water vapor profiling: Assessment of accuracy, resolution, and meteorological applications. *Appl. Opt.*, **37**, 3825–3844, doi:10.1364/AO.37.003825.

——, S. Pal, D. D. Turner, and E. Wagner, 2010: Can water vapour Raman lidar resolve profiles of turbulent variables in the convective boundary layer? *Bound.-Layer Meteor.*, **136**, 253–284, doi:10.1007/s10546-010-9494-z.

Chapter 19

Cloud Property Retrievals in the ARM Program

MATTHEW D. SHUPE

Cooperative Institute for Research in Environmental Science, University of Colorado, and NOAA/Earth System Research Laboratory, Boulder, Colorado

JENNIFER M. COMSTOCK

Pacific Northwest National Laboratory, Richland, Washington

DAVID D. TURNER

NOAA/National Severe Storms Laboratory, Norman, Oklahoma

GERALD G. MACE

Department of Atmospheric Science, University of Utah, Salt Lake City, Utah

1. Introduction

Cloud feedbacks and processes have been clearly highlighted as a leading source of uncertainty for understanding global climate sensitivity (IPCC 2007). Clouds play fundamental and complex roles in the climate system by redistributing heat and moisture through modulation of atmospheric radiation, latent heating processes, and serving as a critical link in the hydrological cycle. They are affected by aerosol properties, large-scale circulation patterns, interactions with the surface, and tropospheric thermodynamic structure. Importantly, cloud systems are entwined in many feedbacks acting on both large and small scales (Stephens 2005). At the crux of the significant uncertainty associated with cloud processes is the fact that all of these properties, processes, and interactions of clouds with the earth's climate system vary widely across the globe leading to a diversity, variability, and complexity of cloud systems that is difficult to represent using numerical models.

Reaching climate equilibrium after some perturbation to the system can take decades of model simulated time; however, Gregory and Webb (2008) and Andrews and Forster (2008) found that changes in clouds that are typically attributed to feedbacks (i.e., Soden et al. 2004; Dufresne and Bony 2008) are actually realized rapidly after a sudden CO_2 doubling. These findings are consistent with Williams and Tselioudis (2007) who show that much of the disparity between models and observations is because models fail to represent clouds accurately within present-day meteorological regimes. This distinction is important because feedbacks are largely beyond the reach of observations. However, a rapid response by clouds to altered meteorological regimes suggests that much of the uncertainty in what is normally termed cloud feedbacks arises because of differences in how the parameterized macro- and microphysical properties of clouds respond to changes locally. This further implies that improved understanding of the interactions between cloud processes and the broader climate system in the present climate can have a direct bearing on the fidelity of climate change predictions if that understanding can be encoded in models.

Thus, understanding cloud systems at a physical process level is critical to advancing numerical earth system modeling abilities. Such a need for process level understanding is becoming increasingly important as numerical models move to finer and finer spatial scales. It is not unreasonable to assume that global

Corresponding author address: Matthew D. Shupe, Cooperative Institute for Research in Environmental Science, University of Colorado, R/PSD3, 325 Broadway, Boulder, CO 80305.
E-mail: matthew.shupe@noaa.gov

DOI: 10.1175/AMSMONOGRAPHS-D-15-0030.1

cloud resolving models (Satoh et al. 2008) will be the preferred modeling tools in the future. As convective parameterizations are abandoned in favor of explicit microphysical parameterizations, it is the microphysical parameterizations that are increasingly being recognized as the weak link (Bryan and Morrison 2012; Han et al. 2013). Advancing numerical earth system modeling abilities requires tackling the complexities of clouds, including the following:

- formation processes and particle nucleation;
- transfer of water among three phases;
- evolution of particle size distributions related to growth, autoconversion, dynamics, and entrainment;
- growth regimes that lead to cloud particles of different shapes, sizes, and fall speeds; and
- scattering and absorption of atmospheric radiation.

To date many of these processes have been difficult to unravel and describe with predictive skill because they are difficult to observe. Many processes, such as entrainment and cloud particle formation, are largely hidden from direct observational capabilities. Further, the dimensionality of the problem (i.e., complexity of cloud systems) is much greater than the available observational constraints. These deficiencies together call for dedicated efforts to expand and improve cloud observational abilities in the form of new measurement technologies and sophisticated analytical tools that are better able to probe deeply into cloud processes and ultimately better constrain the cloud problem.

Over the past two decades, the U.S. Department of Energy (DOE) Atmospheric Radiation Measurement (ARM) Program (Stokes and Schwartz 1994; Ackerman and Stokes 2003) has taken a leading role in addressing the observational and modeling complexities of clouds. One foundational objective of the ARM Program has been to improve the representation of clouds in numerical models. To support this objective, the program has developed and operated diverse suites of ground-based cloud-sensing instrumentation at several geographically diverse locations. These efforts have produced publicly available cloud observational records that are historically unprecedented in length, continuity, and sophistication.

A major thrust of ARM's cloud activities has been in the development of cloud retrieval algorithms wherein instrument-level measurements are exploited to derive the geophysical properties of clouds that are needed to understand and represent cloud processes. These retrieval algorithms are a critical step in bridging the gap between basic measurements and model improvement. They provide input to observational and modeling process studies that are used to synthesize and generalize cloud knowledge in order to support model parameterization development.

Here we provide an historical overview of ARM's contributions toward cloud property retrievals. The overview covers advances in using specific passive and active instruments that are in operation at ARM sites and the expansion of multisensor retrieval approaches. It also addresses the general topic of uncertainty, how it is evaluated, and what it means for our ability to use cloud information derived from ground-based sensors. While there have been many individual contributions over the past two decades from investigators supported by ARM, the focus here is primarily on the larger movements and novel accomplishments that distinguish the ARM Program as a global leader in cloud research and retrieval development. We conclude with a look at the pathway forward, considering the possibilities presented by new enhanced measurements, the need to develop sophisticated forward models relating physical processes to observational parameters, and how these together can further advance cloud retrievals.

2. Cloud observing instruments

Since the early 1990s, the ARM Program has been a pioneer in designing and continuously operating ground-based atmospheric observatories. One strength of the ARM approach is the diverse, collocated, and complementary suites of well-characterized measurements that can be used individually and/or jointly to derive cloud properties in many atmospheric conditions. The value of this approach is that the diverse instruments provide measurements that constrain different aspects of cloud properties. In essence, the extreme dimensionality of the cloud problem can be made more tractable by using measurements that provide independent information about a cloudy volume or vertical column. Furthermore, by operating this instrument suite continuously, the ARM Program has been able to build datasets that extend over multiple years to characterize clouds in all seasons and conditions, and for accumulating the representative statistics needed for model development. In this regard, the ARM approach has been a model for other observational facilities around the globe (i.e., Haeffelin et al. 2005, 2016, chapter 29; Illingworth et al. 2007; Shupe et al. 2013).

Here we briefly summarize the basic set of instruments that compose the typical cloud-sensing component of each ARM site and serve as the data sources for cloud retrievals (see Table 19-1). More detailed information on all instruments is provided at www.arm.gov. These cloud-relevant sensors are characterized broadly into

TABLE 19-1. Core ARM cloud-sensing instrumentation available at most sites. Specific information on instrument specifications, operation parameters, deployment locations, periods of operation, data quality, and other relevant information are provided on the ARM web page at www.arm.gov.

Instrument	Measurements	Reference
Atmospheric Emitted Radiance Interferometer (AERI)	Spectral radiance, 3–25 μm	Knuteson et al. (2004a,b)
Ceilometer	Backscatter, 905 nm	—
Microwave Radiometer (MWR)	Brightness temperatures, 20–31, 90 GHz	Liljegren et al. (2001); Turner et al. (2007a)
Millimeter Cloud Radar (MMCR)	Reflectivity, Doppler spectra, Doppler moments, 35 GHz	Moran et al. (1998); Kollias et al. (2007)
W-band ARM Cloud Radar (WACR)	Reflectivity, Doppler spectra, Doppler moments, 95 GHz	Mead and Widener (2005)
Micropulse lidar (MPL)	Backscatter, depolarization ratio, 523 nm	Spinhirne (1993); Campbell et al. (2002)
Raman lidar	Backscatter and depolarization ratio, 355, 387, and 408 nm	Goldsmith et al. (1998); Turner et al. (2016, chapter 18)
Radiosondes	Temperature, humidity, winds	—
MFRSR family	Irradiance at multiple channels between 415 and 940 nm	Harrison et al. (1994)
Hemispheric Broadband pyranometer and pyrgeometer	Broadband irradiance, 0.3–3 and 4–50 μm	Michalsky and Long (2016, chapter 16)

two groups: passive sensors, where radiation emitted by the atmosphere at different wavelengths is simply measured by the instrument; and active sensors, where the instrument transmits a signal to the atmosphere and measures its return. Active sensors include millimeter-wavelength radars that are sensitive to cloud-sized hydrometeors (see Kollias et al. 2016, chapter 17), and measure the total radar reflectivity, as well as Doppler information on particle and air motions. Recent advances have allowed for measurement of the full radar Doppler spectrum as well. ARM radars include the Millimeter Cloud Radar (MMCR) and the W-band ARM Cloud Radar (WACR), in addition to recent upgrades and modifications to these systems to improve robustness and allow for scanning (Mather and Voyles 2013). Lidars that operate at visible or near-visible wavelengths (e.g., Campbell et al. 2002; Turner et al. 2016, chapter 18) are also critically important as they measure backscatter and depolarization ratio, which together contain information on cloud optical properties, particle shape, and hydrometeor phase, among others. ARM lidars include the Micropulse lidar (MPL), Raman lidar, and more recently the High Spectral Resolution lidar (HSRL). Each ARM site also includes a laser ceilometer, which operates on principles similar to the lidars.

Passive measurements provide critical radiative constraints and signatures of many cloud properties, thus ARM sites include passive instruments operating at a variety of targeted wavelengths. Microwave radiometers measure downwelling atmospheric radiances in the 23–31-GHz range that are sensitive to the total precipitable water vapor and condensed liquid water path. Higher-frequency channels near 90 and 150 GHz are sometimes used for enhanced sensitivity in particularly dry conditions. Moving to shorter wavelengths, the ARM Program operates the Atmospheric Emitted Radiance Interferometer (AERI) to obtain spectral infrared radiances. Among other properties, these spectral measurements contain information on the optical and microphysical properties of optically thin clouds. At visible and near-infrared wavelengths, the Multifilter Rotating Shadowband Radiometer (MFRSR), and related radiometers, measure solar irradiance at multiple narrowband channels, which provide information on cloud optical properties. Last, hemispheric broadband radiometers operating in both solar and thermal infrared spectral ranges offer constraints on bulk cloud properties and a linkage between clouds and net surface radiation.

The ARM approach has been to use instrument technologies that can operate continuously over long periods of time to provide robust, long-term datasets. As a result, most of the cloud-relevant instruments initially deployed at ARM facilities used a fixed viewing-orientation, typically pointing vertically. The assumption with this "column" viewing perspective is that long-term statistical analyses provide an ample representation of regional processes. However, this may not be the case if there are nearby geographic features that can influence the cloud field. Recent improvements in scanning technologies and the operational robustness of scanning, cloud-sensing instruments have allowed ARM to integrate these capabilities into its observational suite (e.g., Mather and Voyles 2013). These new observations, in coordination with the long-term vertically pointing measurements, offer the ability to evaluate the representativeness and relative capabilities of these two observational approaches toward characterizing cloud properties and processes.

3. Advances in cloud retrievals

The long-term, comprehensive datasets obtained from the ARM sites have laid the foundation for a wealth of cloud retrieval development. In some cases, the retrieval development has fed back to instrument development activities, leading to new measurements that offer additional information to constrain cloud properties. While many of the fundamental instruments and analytical techniques predate the ARM Program, the program has contributed to significant advances along a number of pathways in instrument development and data analysis.

a. Passive sensor systems

Passive microwave measurements have long been used to derive properties of the atmosphere. Through its emphasis on long-term operational measurements, the ARM Program has contributed to specific advances in ensuring robust, automated calibration for these microwave measurements (Liljegren 2000). Additionally, several advances have been made in retrieval algorithms. ARM investigators expanded on traditional microwave retrieval methods, which initially relied on statistical representations of atmospheric vertical structure to relate measured microwave radiances to the geophysical parameters of interest, such as the cloud liquid water path (LWP). These statistical retrievals were further constrained using local meteorological conditions and an estimate of the cloud temperature derived from collocated measurements (Liljegren et al. 2001). Additionally, a bias-offset technique was developed to account for clear-sky LWP biases related to variability of local conditions (Turner et al. 2007a). These enhancements resulted in a significant decrease in LWP retrieval bias and spread under clear-sky conditions, the latter of which is a nominal estimate of retrieval uncertainty (see Fig. 19-1). LWP retrievals were further constrained and enhanced with the addition of a physical, iterative approach that incorporates a priori information on atmospheric temperature and water vapor profiles from contemporaneous radiosonde measurements, showing improvement under certain conditions and confirming the quality of the enhanced statistical approach (Marchand et al. 2003; Turner et al. 2007a). This programmatic focus on deriving LWP in all conditions, including for very thin and cold clouds (e.g., Turner et al. 2007b; Wang 2007), promoted the development and integration of higher-frequency microwave channels into LWP retrievals. The utility of the MWRs have forced the program to evaluate the accuracy of the underlying liquid water absorption models, which led to the conclusion that the initial liquid water absorption model used by the program had a large bias (Westwater et al. 2001) and that the current absorption model is inadequate in supercooled liquid water clouds (Cadeddu and Turner 2011; Kneifel et al. 2014).

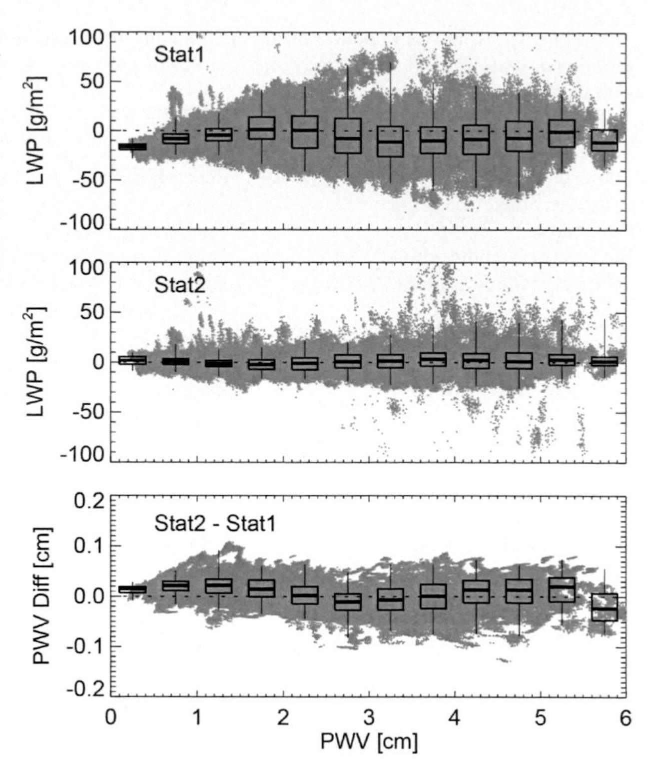

FIG. 19-1. Distribution of clear-sky liquid water path (LWP) from the Southern Great Plains site from September 1996 to December 2005, computed using the (top) original statistical method and the (middle) improved statistical method with brightness temperature offsets applied. (bottom) The difference in the retrieved precipitable water vapor (PWV) between the two methods is also shown. The box-and-whisker plots show the 25th and 75th percentiles (lower and upper boundaries of the box, respectively), the median value (thick line in the middle of the box), and the ends of the whiskers denote the first and ninety-ninth percentile points in the distribution. The size of each PWV bin is 0.5 cm. [Figure from Turner et al. (2007a), courtesy of *IEEE Transactions on Geoscience and Remote Sensing*.]

The ARM Program has contributed to important advances in interpreting spectral infrared measurements from the AERI toward characterizing clouds. While infrared spectra contain information on the presence, concentration, and vertical profile of various gaseous constituents that are active in the thermal infrared, the microwindows between gaseous absorption lines can offer insight into the properties of clouds. Radiances in these microwindows have been used to derive the

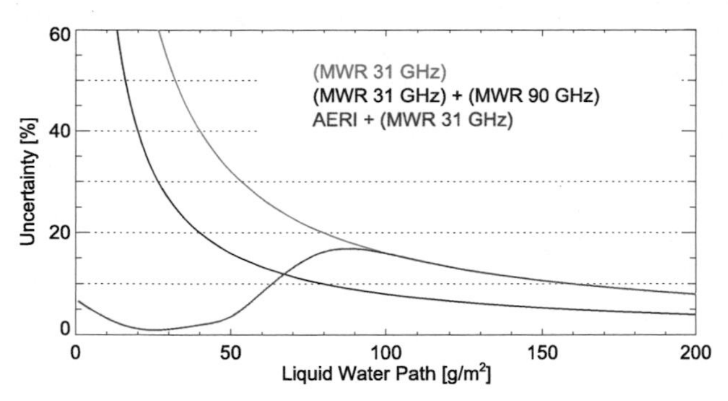

FIG. 19-2. Uncertainty in derived cloud liquid water path as a function of liquid water path for various combinations of measurements (microwave measurements at 31 and 90 GHz, and infrared measurements by the AERI) used in the retrieval process.

infrared cloud optical depth (DeSlover et al. 1999; Mitchell et al. 2006). Additionally, differential absorption of ice versus liquid water across the infrared spectrum has been exploited to determine cloud phase (Turner et al. 2003). Phase-dependent spectral microwindow signatures also form the basis for the retrieval of optical depth and hydrometeor effective radius of both liquid and ice components of mixed-phase clouds using a physically iterative optimal estimation approach (Turner 2005; Turner and Eloranta 2008). These retrievals can be performed only when the visible optical depth is greater than about 0.1 and less than about 6, such that the 8–13-μm band is semitransparent (i.e., not opaque) and, therefore, contains information on cloud properties. The high accuracy of AERI-based LWP retrievals in thin clouds has been combined with microwave radiometer-based LWP retrievals that perform better in thicker clouds to provide optimal retrievals over the full range of observed LWP (see Fig. 19-2; Turner 2007).

A variety of other passive retrievals for cloud optical depth using transmission of solar and near-IR radiation have been developed using ARM sensors, including: MFRSR (Min et al. 2004b; Min and Harrison 1996), broadband shortwave radiometers (Barnard and Long 2004), sun photometer (Marshak et al. 2004; Chiu et al. 2006), and oxygen A-band spectrometer (Min et al. 2004a). A technique for obtaining cloud optical depth in optically thick cirrus also was developed using the passive solar background signal from MPL systems (Chiu et al. 2007). This unique approach allows for the simultaneous retrieval of aerosol properties and cloud optical depth in broken low-level clouds.

b. Active sensor systems

Millimeter wavelength cloud radar has been an early centerpiece of the ARM approach, and ARM has contributed to significant advances in the development of operational, millimeter wavelength, cloud radars (e.g., Kollias et al. 2016, chapter 17). Prior operational radars were typically longer-wavelength precipitation-observing systems, while most prior cloud radar measurements were limited largely to a few, targeted campaigns using research-grade instruments that were not operated operationally. ARM's continuous radar operations in multiple locations have provided first-of-a-kind datasets that are fertile for radar-based cloud retrieval development.

These longer-term datasets have been used to better constrain the traditional radar power-law relationships that relate radar reflectivity Z to geophysical parameters such as the cloud ice water content (IWC) (e.g., Matrosov et al. 2003; Shupe et al. 2005) or liquid water content (LWC; e.g., Matrosov et al. 2004; Dunn et al. 2011). By capitalizing on cloud radar Doppler abilities and utilizing relationships between ice particle terminal fall speed, particle size, condensed mass, and backscatter cross section, information on ice crystal characteristic size and IWC can be derived simultaneously (e.g., Mace et al. 2002; Matrosov et al. 2002). Additionally, radar-based retrievals of cloud optical depth have even shown some potential in ice clouds (Matrosov et al. 2003).

Cloud radar-based retrievals of precipitation are another area in which the ARM Program has made important contributions. While typically derived using longer wavelength radars, the ability to retrieve precipitation properties from cloud radars allows for studies of the evolution and relation between clouds and precipitation properties above a single location. For snowfall rate S, the traditional reflectivity-based Z–S relationship formulation has been applied to millimeter-wavelength observations in dry snowfall, comparing to within a factor of 2 with independent snow gauge measurements (Matrosov et al. 2008). Such traditional reflectivity-based relationships are not possible for cloud radars operated in rain due to strong attenuation and non-Rayleigh scattering effects. However, these limitations have been capitalized upon by the development of attenuation-based retrievals for rainfall rate (Matrosov 2005; Matrosov et al. 2006) and raindrop size distribution retrievals using non-Rayleigh scattering signatures in radar Doppler spectra (Kollias et al. 2002, 2003; Giangrande et al. 2010).

The broader use of cloud radar Doppler spectra has been a game changer and is arguably the area in which ARM has made the largest and most distinctive contribution to cloud radar retrievals. In typical applications, only the first one to three moments of the Doppler spectrum are utilized in cloud retrievals. However, advances in computational power and data storage have allowed for the routine collection of the full Doppler

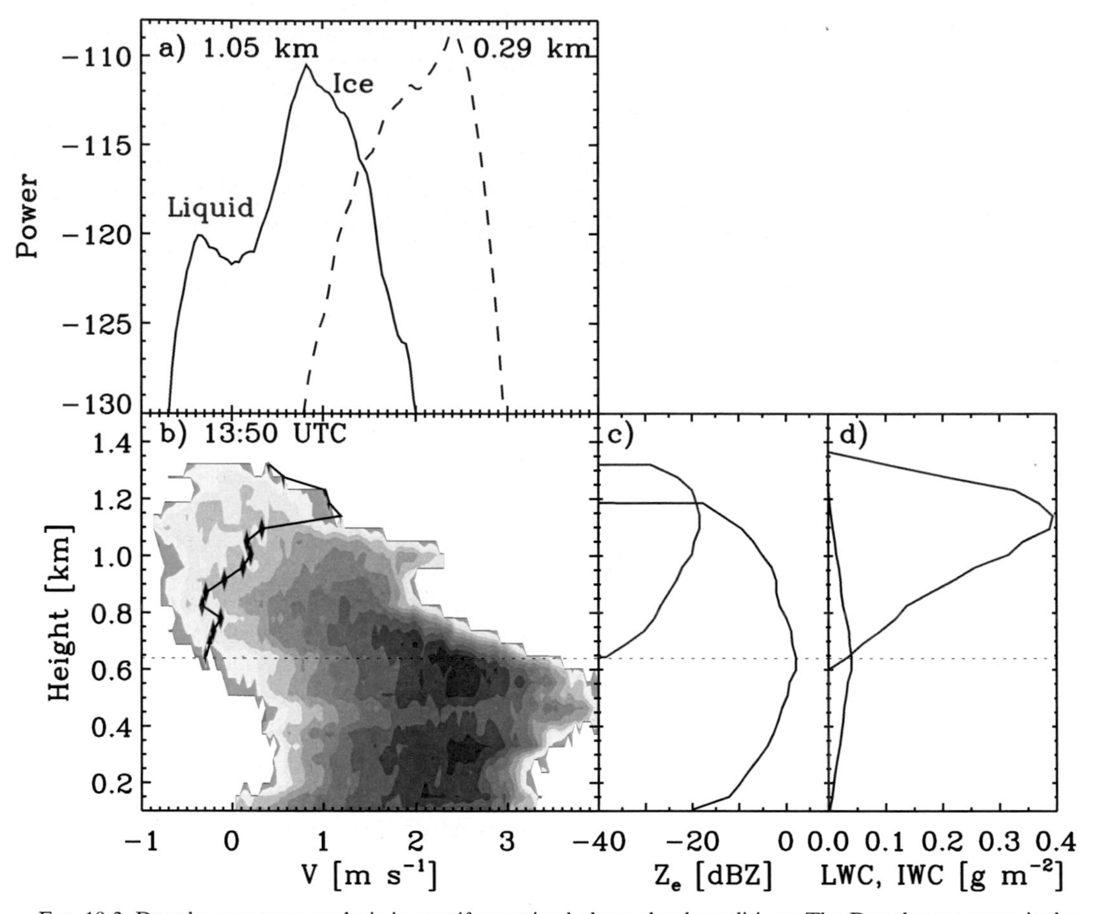

FIG. 19-3. Doppler spectrum analysis in stratiform mixed-phase cloud conditions. The Doppler spectrum is the distribution of returned radar power as a function of the radial velocity of the targets in the radar volume (positive velocity is downward). (a) A bimodal spectrum (solid) found in mixed-phase conditions near cloud top and a unimodal, ice-only spectrum (dashed) found below the liquid cloud base. These spectra are horizontal slices through the spectrograph in (b), which shows contours of returned power as a function of velocity and altitude (redder colors indicate higher power). A manually determined line distinguishes the liquid and ice phases in the spectrograph. (c) Based on the distinction of phase contributions to the spectrograph, individual profiles of liquid (red) and ice (blue) reflectivity, which are the total power in each mode, are computed. (d) Example profiles of liquid (red) and ice (blue) water contents derived from the distinct liquid and ice radar reflectivity profiles. [Figure from Shupe et al. (2008a), courtesy of the *Bulletin of the American Meteorological Society*.]

spectrum, unlocking a wealth of new information. Full spectra often contain multiple modes and different shapes that have been used to identify the presence of supercooled liquid water layers in mixed-phase clouds (Luke et al. 2010). Further, these complex spectra can facilitate independent characterization of liquid and ice components in mixed-phase clouds (Fig. 19-3; Shupe et al. 2004, 2008a; Rambukkange et al. 2011) and the distinction of cloud droplet populations from drizzle (Luke and Kollias 2013). Higher-order spectral moments, such as skewness and kurtosis, have been harnessed to identify processes such as drizzle formation (Kollias et al. 2011; Luke and Kollias 2013) and form the basis for a higher-order-derived data product called the Microphysical–Active Remote Sensing of Cloud Layers (MICRO-ARSCL) that reveals a great deal of information on microphysical processes and spatial structure within cloud layers.

c. Combined sensor approaches

Multisensor cloud retrievals increase the dimensionality of the input information, and thereby provide a stronger constraint on complex, multidimensional cloud properties. Moreover, such retrievals harness the strength of ARM sites with their extensive suites of coordinated measurements. Since clouds occur over a wide range of microphysical conditions related to phase, shape, number concentration, and size, no single instrument is optimally specified to detect all cloud conditions. For example, radar backscatter signals are proportional to the particle size to the sixth power, while lidar backscatter signals are proportional to the particle size squared, making these two instruments sensitive to different moments of the hydrometeor size distribution. Thus, to even simply observe the fractional occurrence of clouds and their heights requires a combined sensor approach (e.g., Comstock et al. 2002; Shupe et al. 2011; Borg et al. 2011).

One of the first approaches for building combined remote sensor cloud climatologies of cloud presence was developed early on within ARM. The ARSCL (Clothiaux et al. 2000) product optimally combined information from collocated radar and lidar to identify the vertical locations of all cloud layers. This first-order cloud product has been a widely used cornerstone of the ARM Program, and has enabled many long-term studies for understanding basic properties, such as cloud overlap (Mace and Benson-Troth 2002), and for evaluating models. Furthermore, data products like ARSCL have enabled the ARM Program to continue to support other multisensor retrieval development.

Cloud phase is a critical detail for studying and understanding cloud processes and the interactions of clouds with the climate system. Additionally, identification of cloud phase is often a prerequisite to the application of cloud retrieval techniques, which are typically designed for a specific cloud type usually characterized by phase. The diverse collection of ARM measurements contains complementary, phase-specific signatures that can be used together to constrain cloud phase. Multisensor, threshold-based techniques have been developed for classifying clouds according to meteorological type (Wang and Sassen 2001) and phase type (Shupe 2007), the latter of which is specifically designed to inform retrieval algorithms. However, cloud processes are not always best quantified by discrete thresholds (i.e., phase transitions do not necessarily occur at specific temperatures or radar reflectivities), such that more flexible classification criteria might be warranted. To meet this challenge, a neural network–based classification approach was developed to recognize phase-specific patterns within cloud radar Doppler spectra based on a training dataset of depolarization lidar measurements (Luke et al. 2010).

Characterizing the microphysical properties of clouds is critical for understanding internal cloud processes and cloud impacts on radiation. Multisensor techniques have been developed using various combinations of ARM measurements to target the microphysical properties of specific cloud classes, often pairing an active measurement with an additional constraint offered by passive radiative measurements. For example, in thin, nonattenuating cirrus clouds the lidar-infrared radiometer (LIRAD) method joins lidar integrated backscatter coefficient with the IR absorption derived from passive radiances at 10–12 μm to derive cloud IR emittance and visible optical depth. This technique pioneered in the early 1970s (Platt 1973) has been applied to extended ARM datasets in the tropics (Comstock and Sassen 2001; Comstock et al. 2002). A similar combination of IR radiances with radar reflectivity in cirrus clouds (Mace et al. 1998; Matrosov 1999) has been used to derive both layer-averaged and vertically resolve estimates of IWC and particle characteristic size (e.g., Fig. 19-4).

For nondrizzling liquid water clouds, microwave measurements have been combined with broadband shortwave radiation to iteratively retrieve layer-mean cloud properties with the aid of a radiative transfer algorithm (Dong et al. 1997). Profile information on LWC and droplet effective size in this same type of clouds has been derived by pairing the profile information from radar reflectivity with the column constraint of microwave brightness temperatures (Frisch et al. 1995; McFarlane et al. 2002). Mace and Sassen (2000) combined the constraints provided by radar, passive microwave, and solar flux to examine the vertical properties of nondrizzling stratocumulus. Last, retrievals based on two active sensors also have been developed, wherein the different responses of lidar and radar backscatter to hydrometeor size have been exploited to derive information on particle size and other properties (Wang and Sassen 2002). Zhao et al. (2011) developed a method that combined the first two radar Doppler moments with information from a Raman lidar in a Bayesian algorithm to explore the bimodality of cirrus particle size distributions.

Mixed-phase clouds are difficult to characterize due to the presence of both liquid and ice hydrometeors in the same cloud layer. Attempts to meet this challenge have been made using unique applications of multisensor ARM data that take further advantage of the differential response of different instruments to the properties of cloud liquid and ice. For stratiform mixed-phase clouds, Wang et al. (2004) applied a combined radar–lidar method to characterize precipitating ice crystals, then used infrared radiances and an iterative-minimization approach to derive the cloud liquid water properties (e.g., Fig. 19-5). Others have used radar or radar-plus-lidar measurements to characterize the ice component and microwave radiometer to characterize the liquid component, in some cases implementing an assumption of adiabatic liquid water distribution (Shupe et al. 2006; de Boer et al. 2009).

A final class of multisensor retrievals is the combined sensor, all-cloud, all-condition retrieval suite that is designed for operational application to all observations made over extended time periods. Such suites not only rely upon multiple sensors but also combine a number of different cloud retrieval techniques that may be designed for specific conditions. ARM's initial movement toward

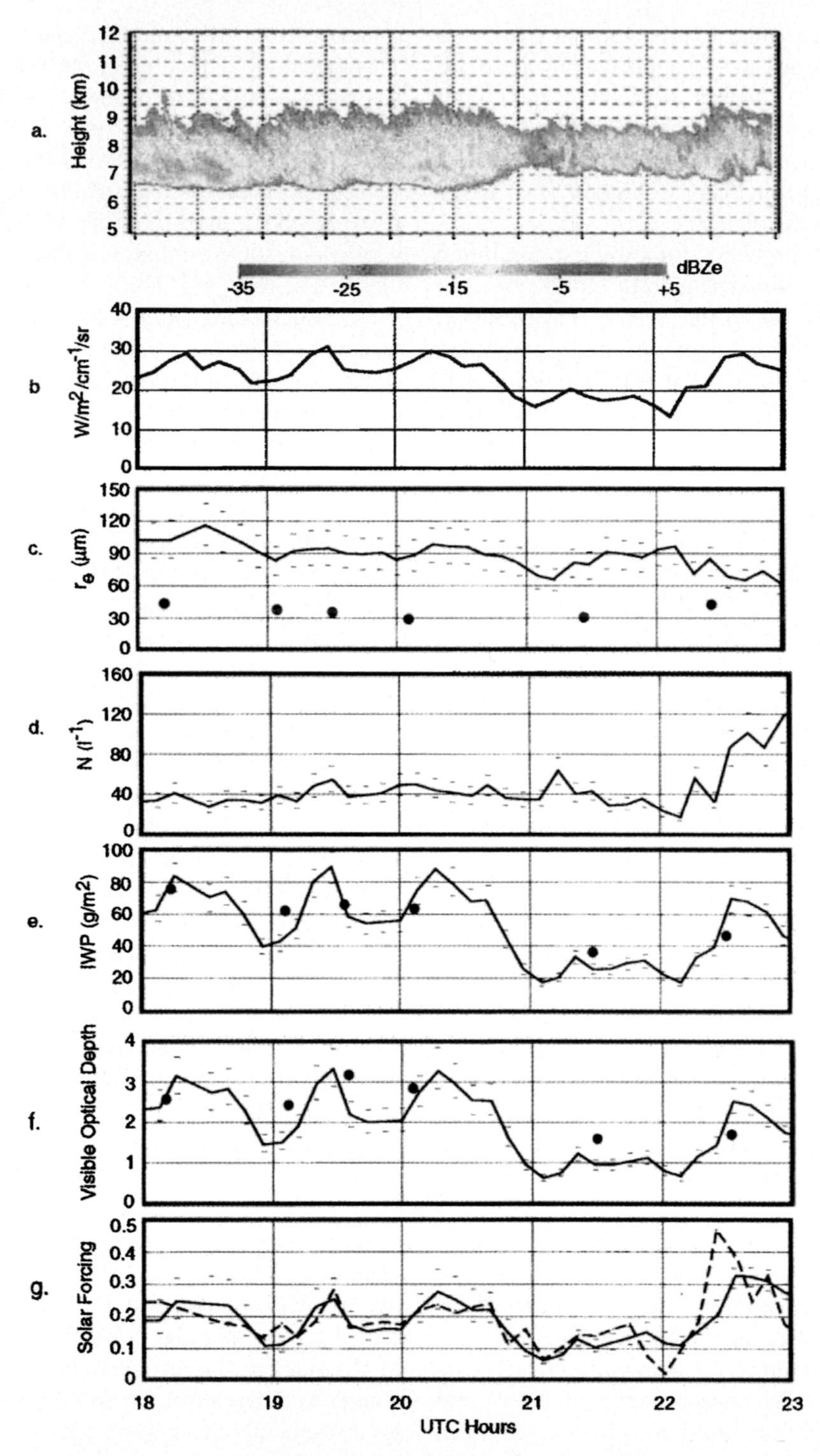

FIG. 19-4. Demonstration of radar reflectivity–infrared radiance retrieval of ice cloud properties for a case on 5 Apr 1996 at the Southern Great Plains site. (a) Time–height section of 94-GHz radar reflectivity, (b) 11-μm radiance from AERI, (c) retrieved effective radius, (d) retrieved particle concentration, (e) retrieved ice water path, (f) retrieved visible optical depth, and (g) comparison of the observed solar flux (dashed line) with the solar flux calculated using the retrieval results in (e) and (c). The flux is expressed as the fraction of the clear-sky flux removed by the cloud layer. (c)–(g) Solid lines show results from the radar-radiance algorithm, the horizontal tick marks show the uncertainty determined from the standard deviations of the radar reflectivity and AERI radiance, and the circles show results derived from geostationary satellite radiances. [Figure from Mace et al. (1998), courtesy of the *Journal of Geophysical Research*.]

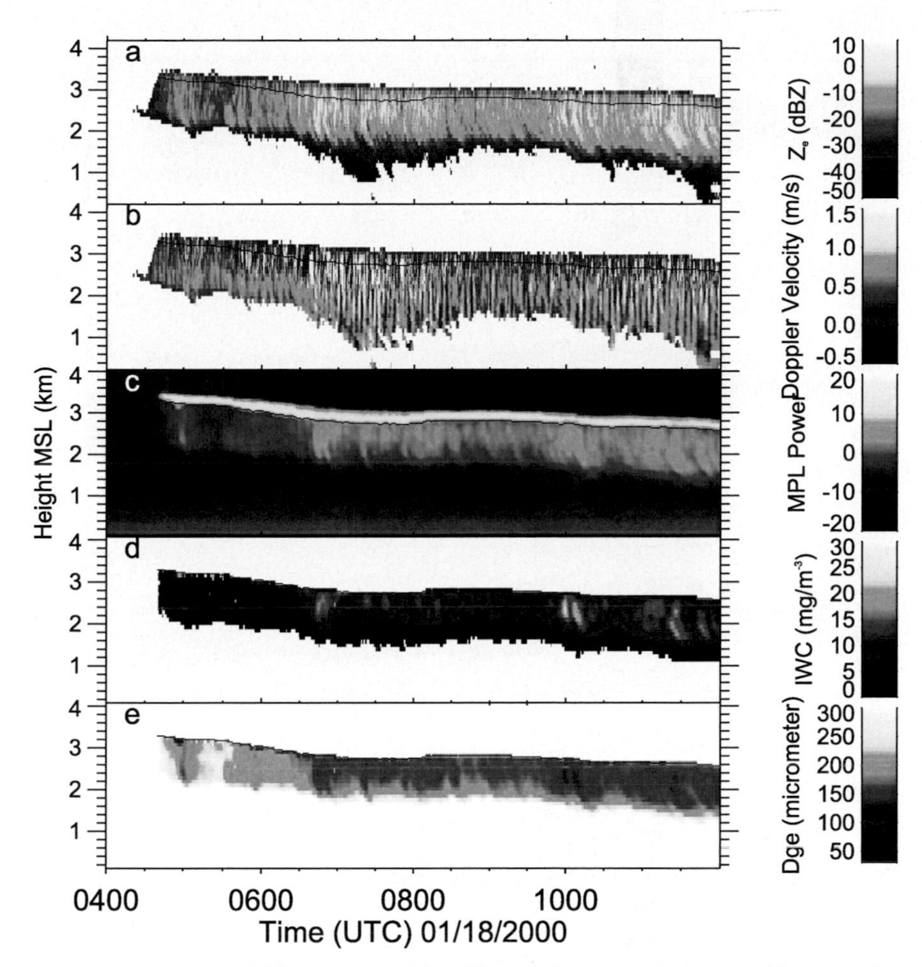

FIG. 19-5. Display of (a) radar reflectivity, (b) mean Doppler velocity, and (c) micropulse lidar return power, and the (d) retrieved ice water content and (e) effective diameter of ice virga on 18 Jan 2000 at the North Slope of Alaska site. Black lines in (a)–(c) denote the base of supercooled water cloud determined by lidar measurements. [Figure from Wang et al. (2004), courtesy of the *Journal of Applied Meteorology*.]

this type of operational retrieval was through the baseline cloud microphysics product called MICROBASE (Miller et al. 2003; Dunn et al. 2011), which applied a collection of simple yet widely applicable methods. Since that initial development, several other retrieval systems have been produced that implement more complex retrievals (e.g., Mace et al. 2006; Shupe et al. 2015; see Zhao et al. 2012 for a summary of others), often showing improved performance relative to certain metrics such as radiative closure at the surface.

d. Cloud-scale dynamics retrievals

In addition to deriving cloud macro- and microphysical properties, ARM has invested significant efforts toward observing and understanding the cloud-scale dynamics that impact cloud processes. Modeling groups both within and external to ARM have called repeatedly for enhanced information on vertical motions and turbulence within a variety of cloud types. To address this need in a concerted fashion, ARM developed the Vertical Velocity Focus group in the mid-2000s.

Since different cloud conditions have unique vertical motion characteristics and signatures, methods have been developed to specifically target distinct cloud types. Multiple retrievals have been developed for cirrus clouds, including an optimal estimation retrieval that uses the first three moments of the radar Doppler spectrum to jointly derive vertical velocity and cloud microphysical properties (Deng and Mace 2008) and a decomposition of radar Doppler velocities into reflectivity-weighted particle-fall velocities and vertical-air velocities using linear regressions between measured reflectivity and velocity at specific heights (see Fig. 19-6; Protat and Williams 2011; Kalesse and Kollias 2013). For low-level liquid clouds, the fact that cloud droplets trace air motions can be exploited to estimate vertical air velocity

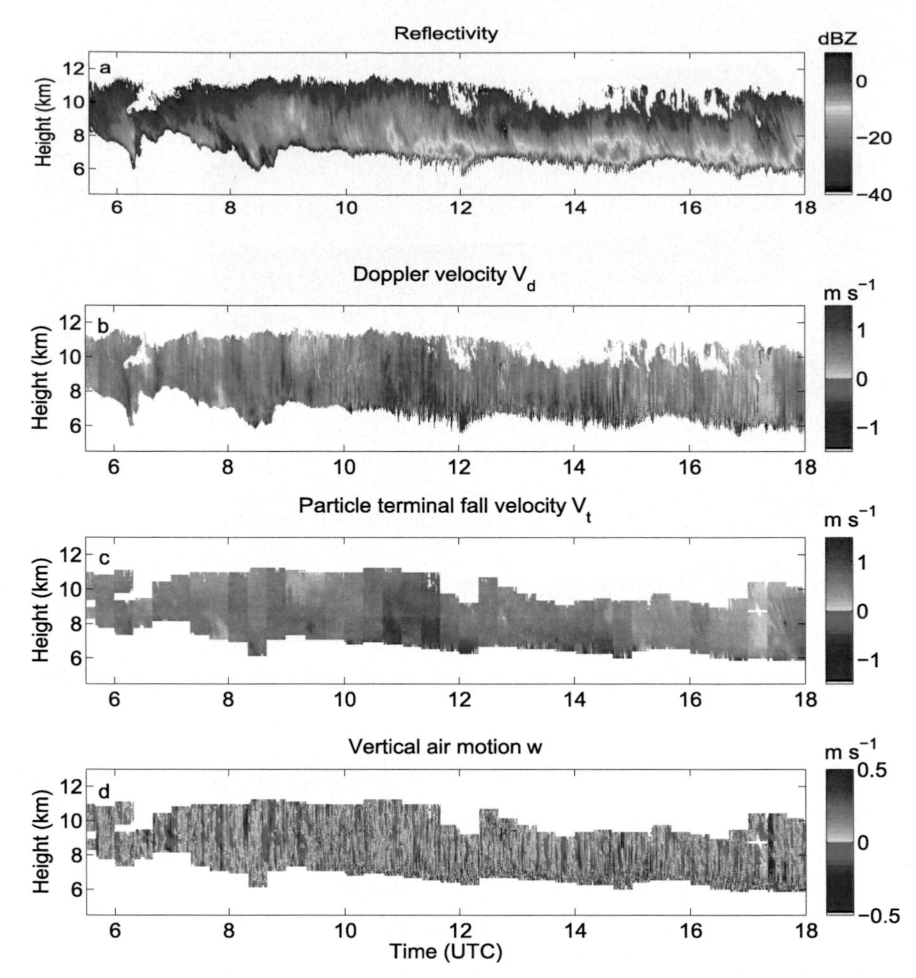

FIG. 19-6. Example of radar Doppler velocity decomposition into particle terminal fall velocity and vertical air motion on 8 Dec 2004 at the Southern Great Plains site. Positive velocity values indicate downward motion. (a) Radar reflectivity, (b) mean Doppler velocity, (c) particle fall speed, and (d) vertical air motion. [Figure from Kalesse and Kollias (2013), courtesy of the *Journal of Climate*.]

directly from radar Doppler velocities (Kollias et al. 2001; Ghate et al. 2011). This same basic principle also has been exploited in mixed-phase clouds, although full Doppler spectra are used to isolate the velocity signal from the liquid droplets (Shupe et al. 2008b). Vertical air motions in deep precipitating systems have been derived by using non-Rayleigh scattering signatures from large raindrops in cloud-radar Doppler spectra (Kollias et al. 2002, 2003; Giangrande et al. 2010). Finally, an estimation of the turbulent dissipation rate associated with cloud-scale dynamics has been adapted to high-temporal-resolution cloud radar velocity measurements (Shupe et al. 2008b, 2012).

4. Quantifying uncertainty and retrieval evaluation

Uncertainty quantification is an important component of any retrieval algorithm framework. Characterization of cloud retrieval accuracy is needed to understand the utility of retrieval results for conducting scientific process studies, developing model parameterizations, or addressing important climate questions. Several techniques for estimating and quantifying the uncertainty in retrieved cloud properties have been used extensively over the ARM Program's history, including algorithm intercomparisons, comparisons with aircraft in situ measurements, radiative closure studies, and more recently the use of optimal estimation techniques. All of these techniques face their own challenges and continue to be areas of active research and development toward improving retrieval uncertainty quantification.

a. Aircraft evaluation

In the early years of the ARM Program, several aircraft based field experiments were undertaken to sample cloud properties over the ARM Southern Great Plains

(SGP) site to help understand the radiative, dynamic, and microphysical properties of clouds. Modeled in part after the First International Satellite Cloud Climatology Project (ISCCP) Regional Experiment (FIRE) field programs of the late 1980s and early 1990s, which focused on satellite-based validation, these early ARM aircraft campaigns focused their efforts on supporting the growing ground-based Cloud and Radiation Testbed (CART) in the central plains of the United States. An important component of these early aircraft studies was to provide datasets for developing and evaluating ground-based retrievals of cloud properties and allowing a detailed look at cloud characteristics that are difficult to discern from remote sensors, such as droplet or ice crystal number concentration and size distribution, and bulk water content. The Subsonic Aircraft: Contrail and Cloud Effects Special Study (SUCCESS) campaign (Toon and Miake-Lye 1998), supported jointly by DOE ARM and NASA and the spring cloud intensive observing periods in 1998 and 2000, which focused on the properties of contrail cirrus, cirrus, and low-level liquid clouds (Dong et al. 2002), provided some of the first intercomparison datasets for ground-based remote sensors such as lidar and MMCR.

Subsequent field experiments tackled more complex cloud systems in different climatic regimes and clouds that are difficult to detect with single wavelength ground-based sensors. Arctic mixed-phase clouds present unique challenges for retrieval algorithms in that they require the simultaneous distinction between ice and liquid in the same volume. Aircraft measurements obtained during the Mixed-Phase Arctic Clouds Experiment (M-PACE; Verlinde et al. 2007) and the Indirect and Semi-Direct Aerosol Campaign (ISDAC; McFarquhar et al. 2011) were critical in helping interpret the radar Doppler spectra measurements and evaluating mixed-phase cloud retrievals. The Tropical Warm Pool–International Cloud Experiment (TWP-ICE) provided additional datasets in undersampled tropical cloud systems, and made progress in understanding the role of ice crystal shattering on aircraft-based probe inlets (McFarquhar et al. 2007).

An integral component of retrieval algorithm development is building a database of cloud properties measured in situ under various seasons and atmospheric conditions. ARM developed a unique approach to building these statistics using extended field campaigns. The Routine ARM Aerial Facility (AAF) Clouds with Low Liquid Water Depths (CLOWD) Optical Radiative Observations (RACORO; Vogelmann et al. 2012) aircraft campaign set out to document the statistical properties of tenuous liquid clouds in the boundary layer through routine measurements over a six-month period. Because of their thin optical depth, small droplet size, and often-small cloud fraction, low liquid water path clouds are difficult to detect with standard radiometric instruments, as well as cloud radar. Data obtained during RACORO helped develop the missing link in the characterization of these properties and provided the much needed evaluation dataset that helped develop the leading microwave and infrared radiometer retrievals. Likewise, the Small Particles In Cirrus (SPARTICUS) campaign accomplished a similar task as RACORO, but focused its efforts on cirrus clouds (Deng et al. 2013). The DOE ARM Program continues to be committed to building these important in situ datasets for retrieval algorithm development.

b. Algorithm intercomparisons and radiative closure techniques

As various datasets of cloud microphysical properties became available, there was a clear need to determine the sensitivity and accuracy of each algorithm and to begin assigning uncertainty to retrieved quantities. A pair of cloud retrieval algorithm intercomparisons examined retrievals of low-level, low LWP clouds (Turner et al. 2007b) and cirrus clouds (Comstock et al. 2007). Turner et al. (2007b) compared the LWP, effective radius (r_e), and optical depth of low LWP clouds derived from 18 different algorithms using different types of measurements (i.e., passive versus active sensors). Comparing retrieved cloud properties directly provides understanding of the spread (minimum/maximum) of the algorithms, but does not necessarily provide an independent measure of uncertainty. To help quantify the uncertainty, they used two forward model closure tests. First, the retrieved cloud properties were inserted into a radiative transfer model to compare computed surface shortwave diffuse fluxes with those observed using broadband radiometers. Second, the computed and observed cloud radar reflectivities were compared. These two different comparison methods independently tested the retrieved r_e and LWP. The main findings of Turner et al. (2007b) revealed that the large spread in the retrieved properties presents a continuing challenge for retrieval developers. This example, which focused on the simple case (single-layer, stratiform liquid clouds) also suggested that the measurements themselves require improved sensitivity or additional detection channels [i.e., the 90-GHz channel in the new 3-channel microwave radiometer; Cadeddu et al. (2013)] to improve the comparison with independent observations.

The cirrus community likewise examined a case study from the March 2000 cloud intensive observing period, where 14 different ice cloud retrieval algorithms were compared (Comstock et al. 2007). Independent satellite

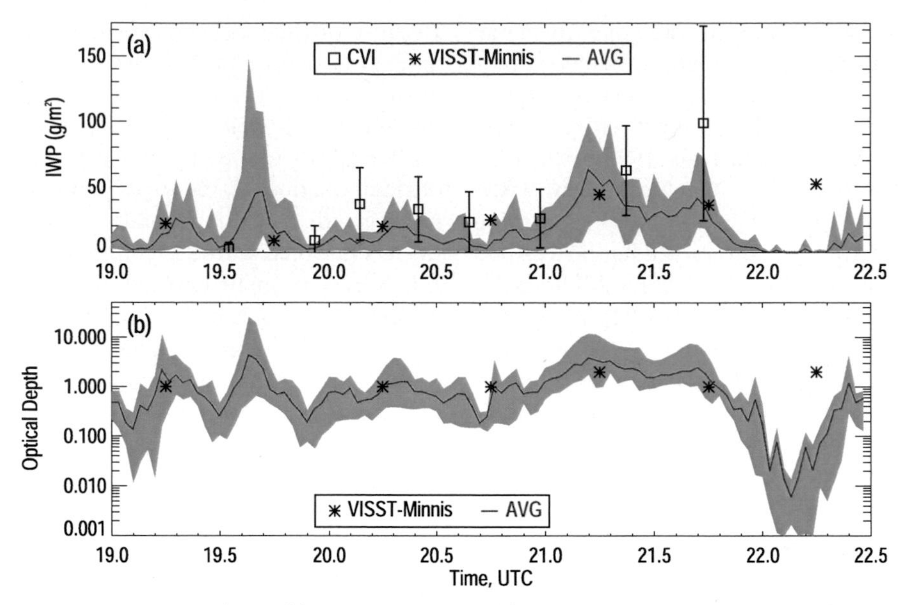

FIG. 19-7. Comparison of (a) ice water path and (b) optical depth derived from 14 different ice cloud retrieval algorithms. The blue shading represents the min and max and the solid red line indicates the mean for all retrievals. Results are compared to in situ (CVI; box) and satellite-derived (VISST-Minnis; asterisk) quantities. [Figure from Comstock et al. (2007), courtesy of the *Bulletin of the American Meteorological Society*.]

and aircraft-based measurements of ice water path (IWP) and visible optical depth τ were used to help evaluate the retrievals. The spread in the retrieved IWP was as large as $100\,\mathrm{g\,m^{-2}}$, and sometimes larger than an order of magnitude in τ (Fig. 19-7). The mean IWP and τ compared reasonably well to satellite retrievals, although most retrievals underestimated IWP compared to in situ observations.

Similar to the Turner et al. (2007b) study, Comstock et al. (2007) also used radiative closure to determine the uncertainty in retrieved cloud properties. They compared computed surface downwelling shortwave flux with observations from broadband radiometers located at the SGP site. Their findings show that for optically thin cirrus ($\tau < 1$) the retrieved flux is typically smaller than the observed flux, suggesting that τ in thin clouds is overestimated. In contrast for $\tau > 1$, the mean retrieved flux is in better agreement with the observed flux (Comstock et al. 2007).

Clearly radiative closure is an important tool for evaluating retrieval algorithms; however, there are several assumptions that are inherent to the radiative transfer models. Assumptions regarding ice crystal shape, particle size distribution, surface albedo, and aerosol loading will all contribute to the uncertainty in these comparisons, but can also help identify specific aspects of retrievals where improvement is needed. The above-mentioned comparisons focused their efforts on single case studies. More recently, longer-term datasets were examined to further quantify the uncertainty in retrieved cloud properties. Several ARM-sponsored focus groups have helped champion these efforts. In particular, the CLOWD focus group compared passive infrared and microwave retrieval techniques using observations at the ARM mobile facility deployment at Pt. Reyes, California. This location provided an extensive dataset of low liquid water path, stratiform clouds for algorithm evaluation (Turner 2007). By examining the mean shortwave flux difference (Fig. 19-8, top) as a function of LWP, Turner (2007) found that both small ($<30\,\mathrm{g\,m^{-2}}$) and large ($>60\,\mathrm{g\,m^{-2}}$) LWP clouds produce large deviations from the observed fluxes. However, the two retrieval methods evaluated demonstrated that the AERI-based method resulted in significantly less scatter (as shown by the variance in the middle panel of Fig. 19-8) than the MWR-based method, and thus was deemed to be a more accurate retrieval even though both retrievals had approximately the same bias.

An additional long-term study compared cirrus cloud properties derived from radar and combined radar–lidar algorithms using three years of observations from the ARM site located in Darwin, Australia (Comstock et al. 2013). These comparisons focused on retrievals from active remote sensors that provide vertical profiles of cloud properties. Demonstrating another variation of radiative closure by examining the transmittance

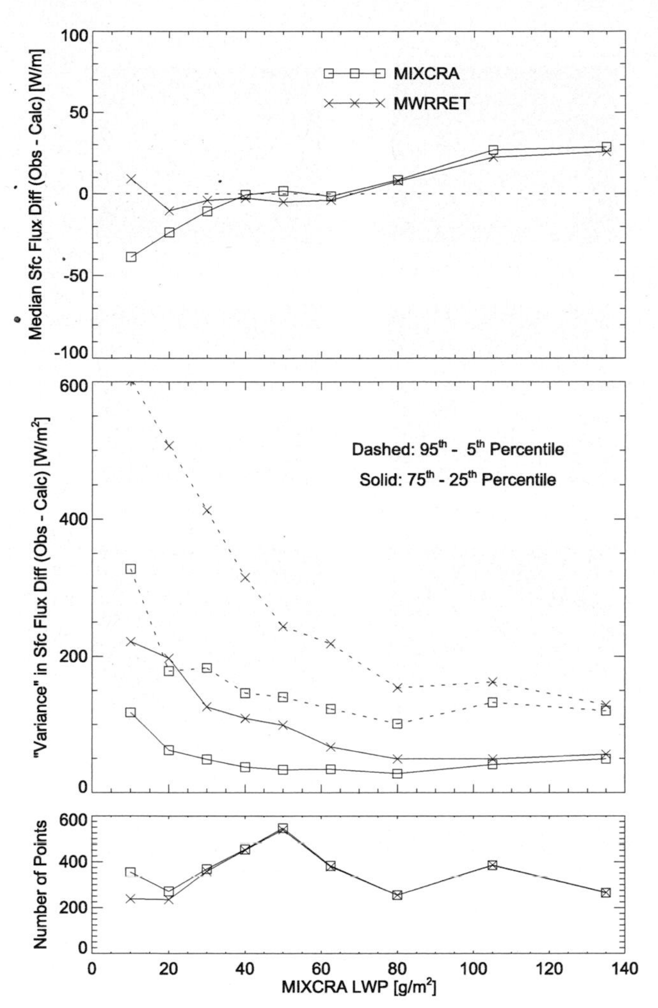

FIG. 19-8. Downwelling surface shortwave flux difference (observed − computed) as a function of cloud liquid water path. The median (top) and variance (middle) show the sources of uncertainty in liquid water path retrievals using the AERI-based MIXCRA method vs the microwave-based MWRRET method. (bottom) The number of points included in the analysis. [Figure from Turner (2007), courtesy of the *Journal of Geophysical Research*.]

difference (Fig. 19-9), Comstock et al. (2013) found that lidar–radar retrievals are generally less biased relative to the observed transmittance for clouds detected by both radar and lidar, as expected. The reflectivity–Doppler velocity (Z–V) algorithms also tend to be less biased but have larger variance. Primary differences between algorithms are related to the particle shape assumptions. Overall, the studies of Turner (2007) and Comstock et al. (2013) provide a basis for long-term evaluation of retrieval algorithms using the radiative closure approach.

However, radiative closure evaluation must also be used with care. For example, Protat et al. (2014) compared the long-term record of derived cloud properties and radiative forcing collected at the Darwin ARM site with similar estimates of cloud properties and radiative forcing derived from A-Train satellite measurements (Stephens et al. 2008). They found that, because of ubiquitous underlying cloud cover and sensitivity limitations of the ARM ground-based remote sensors, the majority of thin cirrus are not sensed by the ARM remote sensors. Conversely, they also document that the A-Train tends to miss a significant fraction of the low-level clouds. Each of these discrepancies results in unique biases with respect to cloud radiative effects. The ARM data at Darwin, because it misrepresents tropical cirrus, has a large heating rate and infrared radiative forcing biases in the upper troposphere, while the A-Train incorrectly characterizes the solar forcing due to boundary layer clouds. At the same time, another study by Thorsen et al. (2013) suggests that the bias in observing tropical cirrus might be much less when using ground-based Raman lidar. These studies highlight the potential difficulties of using radiative closure in assessing uncertainties in cloud property characterizations and emphasize the need for a critical assessment of any remote sensing dataset that is to be used for further scientific applications.

c. *Optimal estimation/Bayesian techniques*

Although radiative closure and aircraft comparisons provide independent measures of uncertainty, these techniques themselves have associated uncertainties. Evaluations of atmospheric model simulations require rigorous uncertainty quantification for retrieved cloud properties. To address this, Bayesian or optimal estimation approaches have become popular as a way to uniquely quantify the uncertainty in measurements, retrieval assumptions, and forward models. Rodgers (2000) outlines the capacity for inverse methods to simultaneously retrieve cloud properties and their uncertainty by assuming that the measurements, the assumptions used to implement forward models, and all prior information can be represented probabilistically. The resulting solution then is also derived as a probability distribution with the mean of the solution representing the best estimate of the atmospheric state and the breadth of the distribution representing the uncertainty in the retrieval.

While the popular optimal estimation technique is essentially limited to assumptions of Gaussian statistics, other probabilistic techniques such as Markov chain Monte Carlo (Posselt and Mace 2014) are not thereby limited—the trade-off being in computational expense. These straightforward, but computationally expensive, approaches have been applied in several studies using ARM data. The more flexible Bayesian approach, which allows for a non-Gaussian distribution, has been applied

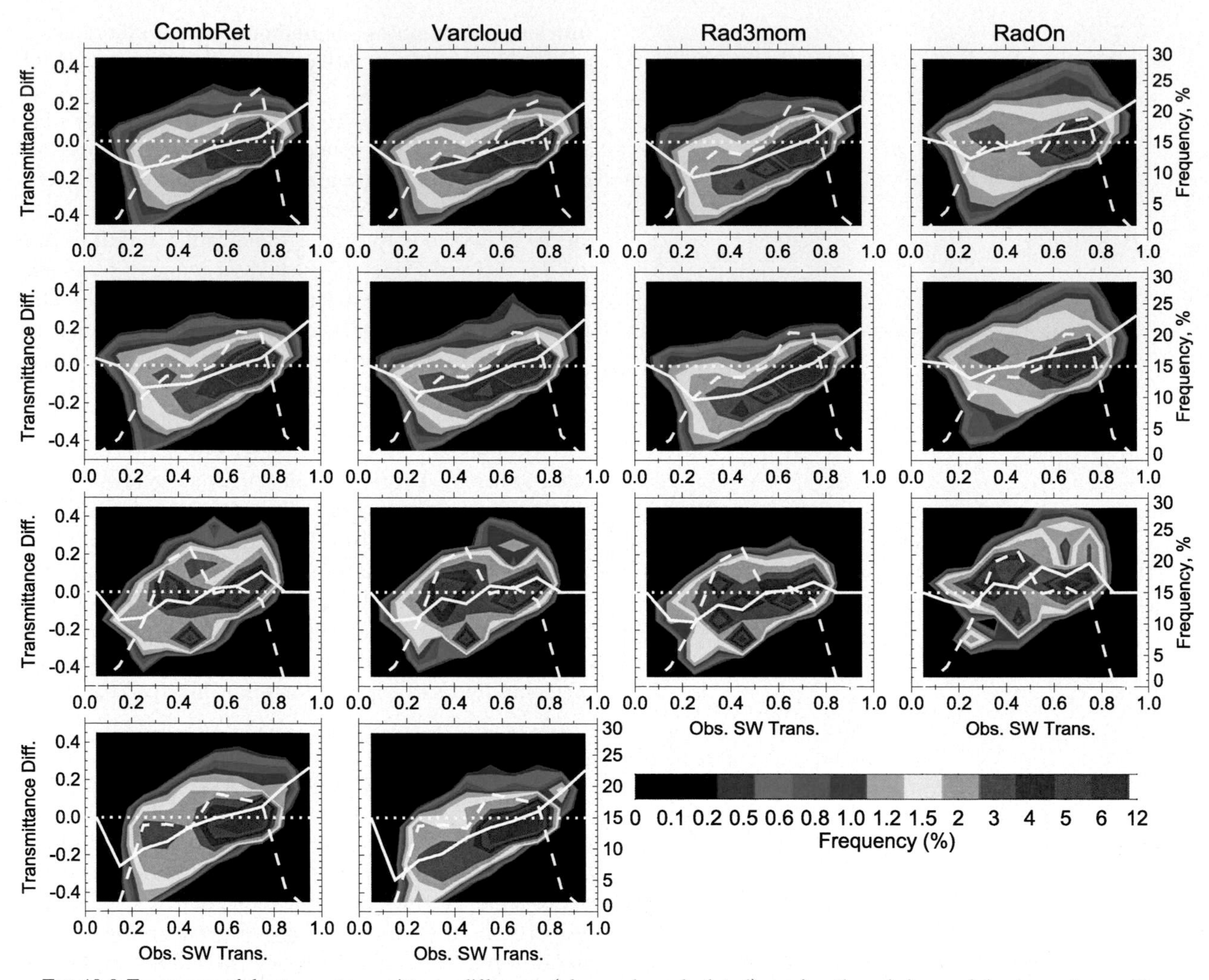

FIG. 19-9. Frequency of shortwave transmittance difference (observed − calculated) as a function of observed shortwave transmittance. Comparisons are for radiative transfer calculations using cloud microphysical results from two radar-lidar algorithms (CombRet and Varcloud) and two radar reflectivity-Doppler velocity algorithms (Rad3mom and RadOn). Each row (from top to bottom) represents: all retrievals; clouds only detected by radar; clouds detected by radar and lidar; and clouds only detected by lidar. The mean is given as a white line in each panel. [Figure from Comstock et al. (2013), courtesy of the *Journal of Geophysical Research.*]

to radar reflectivity and microwave radiometer measurements to retrieve LWC and r_e in low-level liquid clouds observed at the ARM Nauru site (McFarlane et al. 2002). The optimal estimation approach has been applied to cirrus (Mace et al. 2002; Delanoë and Hogan 2008; Deng and Mace 2008), low-level liquid clouds (Turner 2007; Cadeddu et al. 2013), and mixed-phase clouds (Turner 2005).

As our understanding of the prominent role of clouds in the climate system evolves, and numerical model capabilities push toward higher resolutions, information on cloud properties and processes is needed at increasingly finer detail. Cloud retrievals developed by ARM and other programs over the past couple of decades have offered a wealth of first-order information on basic cloud properties and their role in the climate system; this information has been used to evaluate and develop numerical models. The evolving demands for cloud products at higher spatial and temporal resolutions are placing stronger requirements on cloud retrievals and especially on quantifying retrieval uncertainties. As a result, more focus is needed in a number of key directions to both improve the overall cloud retrievals and better characterize their uncertainties. These directions include the following: identifying sources of uncertainty within both the measurements and retrieval framework, adding new measurement constraints and improving the accuracy of measurements, further developing forward models that more accurately represent the mapping of physical cloud-system properties to observational constraints, and building statistics concerning important physical processes upon which the retrievals are based (i.e., mass–dimension relationships and particle-size distributions in ice clouds). A path forward for improving retrieval algorithms and quantifying the associated uncertainty will be the focus of future efforts led, in part, by the recently formed Quantification of Uncertainty in Cloud Retrievals (QUICR) focus group.

5. ARM cloud retrievals into the future

Building on these significant contributions to cloud retrieval development over the past two decades, ARM is poised to make further advances. Recent innovation in observational technology and the acquisition of new instruments (Mather and Voyles 2013) will provide exciting measurements to stimulate a new surge of retrieval algorithm development in a number of directions.

These new ARM instruments present an array of additional constraints for basic retrievals of cloud microphysical properties. For example, each ARM site now contains multifrequency cloud and precipitation radars, such that each site has two to five different radar wavelengths that can be used simultaneously. Multiwavelength retrievals have been developed in the past (e.g., Gosset and Sauvageot 1992; Huang et al. 2009), but running these radar suites operationally year-round is sure to promote the development of new retrieval algorithms. Similarly, the addition of high spectral resolution (Eloranta 2005) and Raman lidars (Goldsmith et al. 1998; Turner et al. 2016, chapter 18) to many sites is a significant advancement that will bring much improved sensitivity at each site (e.g., Thorsen et al. 2013), and will be able to deliver true calibrated backscatter, which will facilitate the quantitative use of these measurements in advanced lidar-based and multisensor cloud retrievals. Other new instruments, such as the higher-frequency microwave radiometer observations (Cadeddu et al. 2013) and the shortwave spectrometer will play an important role in future cloud retrieval algorithms, providing critical new radiative constraints in bands that have not previously been exploited. However, the forward models used to simulate these observations must first be improved [e.g., improving the accuracy of the temperature dependence of liquid water absorption at microwave frequencies; Kneifel et al. (2014)].

Cloud dynamical properties also will be a focus of the near future. ARM has installed Doppler lidars at most sites that offer insight into low-level wind fields, vertical motions, and turbulence. These new measurements can be paired with radar-based information on vertical velocity and turbulence from within clouds to characterize the full cloud and subcloud dynamical environment. Importantly, this dynamical information can be combined with derived microphysical properties to study detailed cloud processes (e.g., Ghate et al. 2011; Luke and Kollias 2013). Additionally, enhanced radar Doppler spectra deconvolution techniques can be developed to distinguish the microphysical and turbulent contributions within identical sample volumes.

The introduction of scanning cloud radars to ARM sites will help the program to break free of the narrow zenith column above the sites and to better evaluate how representative column measurements are of regional processes. Scanning allows for innovative techniques to track clouds (e.g., Fielding et al. 2013) and to characterize spatial variability in new ways. Additionally, scanning measurements can be used to probe ice particle habit (e.g., Matrosov et al. 2012; Marchand et al. 2013), which can enhance our understanding of ice particle size distributions, particle growth regimes, fall speeds, and the assumptions used in bulk cloud retrievals. Scanning millimeter-wavelength cloud radars have never been operated in a continuous mode for extended periods of time, thus offering the unprecedented opportunity to examine multidimensional cloud structure in all seasons.

Precipitation is another exciting area of expansion and future development within ARM, as all sites have now been instrumented with scanning, polarimetric precipitation radars and enhanced surface-based disdrometers for quantifying precipitation properties. These additions offer the first operational polarimetric measurements at the ARM sites and the ability to evaluate and compare precipitation retrievals developed using cloud radars with those from traditional precipitation radars. Further, these enhancements present the possibility to simultaneously characterize both cloud and precipitation properties (e.g., Matrosov 2010), which will be useful for understanding precipitation formation processes, the mesoscale organization of storms, and other features.

One of the great challenges for cloud retrieval advancement will be in improving and developing the appropriate forward models through which the physical cloud–atmosphere system can be mapped onto these many new and existing measurements. For example, the temperature dependence of liquid water absorption at microwave frequencies must be improved over the full range of temperatures applicable for cloud processes of interest (Kneifel et al. 2014). Similar needs exist for representing new sophisticated lidar and radar measurements. For example, improved representations of lidar depolarization signals related to particle phase and habit are an important area of development. Robust polarimetric radar forward models, particularly for shorter wavelength radars that are sensitive to clouds, are also increasingly in demand. Additionally, simulators of radar Doppler spectra that can account for convolved cloud microphysical and dynamical conditions will help to unlock the vast information held in these spectra.

Together, these multiple avenues of advancement on cloud and precipitation retrievals will contribute toward improved operational, all-condition retrieval frameworks to obtain continuous, time–height cloud and precipitation products. Some of these methods will continue to combine disparate retrievals within a single framework that requires sophisticated logic for the correct application of techniques (i.e., the classification problem). However, new optimal estimation approaches also should be developed to combine larger suites of instruments within a single, broadly applicable retrieval algorithm to produce continuous estimates of both microphysical and dynamical properties. A further challenge moving forward will be to extend beyond bulk cloud properties and derive higher-order information on liquid droplet and ice particle size distributions.

Over the past two decades, ARM has established itself as a leader in developing and operating ground-based suites of instruments to characterize clouds. In that time, the program has grown and cloud-sensing instruments have become more robust and sophisticated. Paralleling this instrumental maturation, the program has supported similar progress in cloud retrieval development, leading to evermore advanced, complex, and comprehensive retrievals. The future trajectory of DOE-sponsored cloud retrieval development activities will build on ARM's extensive infrastructure and past accomplishments through coordination with the new DOE Atmospheric System Research Program and through enhanced collaboration with similar activities in Europe such as Cloudnet (Illingworth et al. 2007; Haeffelin et al. 2016, chapter 29). Together these programs will ensure that DOE continues to produce high-quality cloud properties datasets that are appropriate for studying cloud processes, evaluating models, constructing model parameterizations, and addressing key climate science questions.

Acknowledgments. This manuscript summarizes a great deal of research supported by numerous grants from the U.S. Department of Energy's Atmospheric Research Program. Support for writing the manuscript was provided by the DOE Atmospheric System Research Program Grant DE-SC0011918.

REFERENCES

Ackerman, T. P., and G. M. Stokes, 2003: The Atmospheric Radiation Measurement Program. *Phys. Today*, **56**, 38–44, doi:10.1063/1.1554135.

Andrews, T., and P. M. Forster, 2008: CO_2 forcing induces semi-direct effects with consequences for climate feedback interpretations. *Geophys. Res. Lett.*, **35**, L04802, doi:10.1029/2007GL032273.

Barnard, J. C., and C. N. Long, 2004: A simple empirical equation to calculate cloud optical thickness using shortwave broadband measurements. *J. Appl. Meteor.*, **43**, 1057–1066, doi:10.1175/1520-0450(2004)043<1057:ASEETC>2.0.CO;2.

Borg, L. A., R. E. Holz, and D. D. Turner, 2011: Investigating cloud radar sensitivity to optically thin cirrus using collocated Raman lidar observations. *Geophys. Res. Lett.*, **38**, L05807, doi:10.1029/2010GL046365.

Bryan, G. H., and H. Morrison, 2012: Sensitivity of a simulated squall line to horizontal resolution and parameterization of microphysics. *Mon. Wea. Rev.*, **140**, 202–226, doi:10.1175/MWR-D-11-00046.1.

Cadeddu, M. P., and D. D. Turner, 2011: Evaluation of water permittivity models from ground-based observations of cold clouds at frequencies between 23 and 170 GHz. *IEEE Trans. Geosci. Remote Sens.*, **49**, 2999–3008, doi:10.1109/TGRS.2011.2121074.

——, J. C. Liljegren, and D. D. Turner, 2013: The Atmospheric Radiation Measurement (ARM) Program network of microwave radiometers: Instrumentation, data, and retrievals. *Atmos. Meas. Tech.*, **6**, 2359–2372, doi:10.5194/amt-6-2359-2013.

Campbell, J. R., D. L. Hlavka, E. J. Welton, C. J. Flynn, D. D. Turner, J. D. Spinhirne, V. S. Scott, and I. H. Hwang, 2002: Full-time, eye-safe cloud and aerosol lidar observations at Atmospheric Radiation Measurement Program sites: Instruments and data processing. *J. Atmos. Oceanic Technol.*, **19**, 431–442, doi:10.1175/1520-0426(2002)019<0431:FTESCA>2.0.CO;2.

Chiu, J. C., A. Marshak, Y. Knyazikhin, W. Wiscombe, H. Barker, J. C. Barnard, and Y. Luo, 2006: Remote sensing of cloud properties using ground-based measurements of zenith radiance. *J. Geophys. Res.*, **111**, D16201, doi:10.1029/2005JD006843.

——, ——, W. Wiscombe, S. C. Valencia, and E. J. Welton, 2007: Cloud optical depth retrievals from solar background "signals" of micropulse lidars. *Geosci. Remote Sens. Lett.*, **4**, 456–460, doi:10.1109/LGRS.2007.896722.

Clothiaux, E. E., T. P. Ackerman, G. G. Mace, K. P. Moran, R. T. Marchand, M. Miller, and B. E. Martner, 2000: Objective determination of cloud heights and radar reflectivities using a combination of active sensors at the ARM CART sites. *J. Appl. Meteor.*, **39**, 645–665, doi:10.1175/1520-0450(2000)039<0645:ODOCHA>2.0.CO;2.

Comstock, J. M., and K. Sassen, 2001: Retrieval of cirrus cloud radiative and backscattering properties using combined lidar and infrared radiometer (LIRAD) measurements. *J. Atmos. Oceanic Technol.*, **18**, 1658–1673, doi:10.1175/1520-0426(2001)018<1658:ROCCRA>2.0.CO;2.

——, T. P. Ackerman, and G. G. Mace, 2002: Ground-based lidar and radar remote sensing of tropical cirrus clouds at Nauru Island: Cloud statistics and radiative impacts. *J. Geophys. Res.*, **107**, 4714, doi:10.1029/2002JD002203.

——, and Coauthors, 2007: An intercomparison of microphysical retrieval algorithms for upper-tropospheric ice clouds. *Bull. Amer. Meteor. Soc.*, **88**, 191–204, doi:10.1175/BAMS-88-2-191.

——, A. Protat, S. A. McFarlane, J. Delanoë, and M. Deng, 2013: Assessment of uncertainty in cloud radiative effects and heating rates through retrieval algorithm differences: Analysis using 3 years of ARM data at Darwin, Australia. *J. Geophys. Res. Atmos.*, **118**, 4549–4571, doi:10.1002/jgrd.50404.

de Boer, G., E. W. Eloranta, and M. D. Shupe, 2009: Arctic mixed-phase stratiform cloud properties from multiple years of surface-based measurements at two high-latitude locations. *J. Atmos. Sci.*, **66**, 2874–2887, doi:10.1175/2009JAS3029.1.

Delanoë, J., and R. J. Hogan, 2008: A variational scheme for retrieving ice cloud properties from combined radar, lidar, and infrared radiometer. *J. Geophys. Res.*, **113**, D07204, doi:10.1029/2007JD009000.

Deng, M., and G. G. Mace, 2008: Cirrus cloud microphysical properties and air motion statistics using cloud radar Doppler moments: Water content, particle size, and sedimentation relationships. *Geophys. Res. Lett.*, **35**, L17808, doi:10.1029/2008GL035054.

——, ——, Z. Wang, and R. P. Lawson, 2013: Evaluation of several A-Train ice cloud retrieval products with in situ measurements collected during the SPARTICUS campaign. *J. Appl. Meteor. Climatol.*, **52**, 1014–1028, doi:10.1175/JAMC-D-12-054.1.

DeSlover, D. H., W. L. Smith, P. K. Piironen, and E. W. Eloranta, 1999: A methodology for measurement cirrus cloud visible-to-infrared spectral optical depth ratios. *J. Atmos. Oceanic Technol.*, **16**, 251–262, doi:10.1175/1520-0426(1999)016<0251:AMFMCC>2.0.CO;2.

Dong, X., T. P. Ackerman, E. E. Clothiaux, P. Pilewskie, and Y. Han, 1997: Microphysical and radiative properties of boundary layer stratiform clouds deduced from ground-based measurements. *J. Geophys. Res.*, **102**, 23 829–23 843, doi:10.1029/97JD02119.

——, P. Minnis, G. G. Mace, W. L. Smith Jr., M. Poellot, R. Marchand, and A. Rapp, 2002: Comparison of stratus cloud properties deduced from surface, GOES, and aircraft data during the March 2000 ARM Cloud IOP. *J. Atmos. Sci.*, **59**, 3265–3284, doi:10.1175/1520-0469(2002)059<3265:COSCPD>2.0.CO;2.

Dufresne, J.-L., and S. Bony, 2008: An assessment of the primary sources of spread of global warming estimates from coupled atmosphere–ocean models. *J. Climate*, **21**, 5135–5144, doi:10.1175/2008JCLI2239.1.

Dunn, M., K. Johnson, and M. Jensen, 2011: The Microbase Value Added Product: A baseline retrieval of cloud microphysical properties. U.S. Department of Energy Tech. Rep. DOE/SC-ARM/TR-095, 34 pp.

Eloranta, E. W., 2005: High spectral resolution lidar. *Lidar: Range-Resolved Optical Remote Sensing of the Atmosphere*, K. Weitkamp, Ed., Springer-Verlag, 143–163.

Fielding, M. D., J. C. Chiu, R. J. Hogan, and G. Feingold, 2013: 3D cloud reconstructions: Evaluation of scanning radar scan strategy with a view to surface shortwave radiation closure. *J. Geophys. Res. Atmos.*, **118**, 9153–9167, doi:10.1002/jgrd.50614.

Frisch, A. S., C. W. Fairall, and J. B. Snider, 1995: Measurement of stratus cloud and drizzle parameters in ASTEX with a K_α-band Doppler radar and microwave radiometer. *J. Atmos. Sci.*, **52**, 2788–2799, doi:10.1175/1520-0469(1995)052<2788:MOSCAD>2.0.CO;2.

Ghate, V. P., M. A. Miller, and L. DiPretore, 2011: Vertical velocity structure of marine boundary layer trade wind cumulus clouds. *J. Geophys. Res.*, **116**, D16206, doi:10.1029/2010JD015344.

Giangrande, S. E., E. P. Luke, and P. Kollias, 2010: Automated retrievals of precipitation parameters using non-Rayleigh scattering at 95 GHz. *J. Atmos. Oceanic Technol.*, **27**, 1490–1503, doi:10.1175/2010JTECHA1343.1.

Goldsmith, J. E., F. H. Blair, S. E. Bisson, and D. D. Turner, 1998: Turn-key Raman lidar for profiling atmospheric water vapor, clouds, and aerosols. *Appl. Opt.*, **37**, 4979–4990, doi:10.1364/AO.37.004979.

Gosset, M., and H. Sauvageot, 1992: A dual-wavelength radar method for ice-water characterization in mixed-phase clouds. *J. Atmos. Oceanic Technol.*, **9**, 538–547, doi:10.1175/1520-0426(1992)009<0538:ADWRMF>2.0.CO;2.

Gregory, J., and M. Webb, 2008: Tropospheric adjustment induces a cloud component in CO_2 forcing. *J. Climate*, **21**, 58–71, doi:10.1175/2007JCLI1834.1.

Haeffelin, M., and Coauthors, 2005: SIRTA, a ground-based atmospheric observatory for cloud and aerosol research. *Ann. Geophys.*, **23**, 253–275, doi:10.5194/angeo-23-253-2005.

——, and Coauthors, 2016: Parallel developments and formal collaboration between European atmospheric profiling observatories and the U.S. ARM research program. *The Atmospheric Radiation Measurement (ARM) Program: The First 20 Years, Meteor. Monogr.*, No. 57, Amer. Meteor. Soc., doi:10.1175/AMSMONOGRAPHS-D-15-0045.1.

Han, M., S. A. Braun, T. Matsui, and C. R. Williams, 2013: Evaluation of cloud microphysics schemes in simulations of a winter storm using radar and radiometer measurements. *J. Geophys. Res. Atmos.*, **118**, 1401–1419, doi:10.1002/jgrd.50115.

Harrison, L., J. Michalsky, and J. Berndt, 1994: Automated multifilter rotating shadow-band radiometer: An instrument for optical depth and radiation measurements. *Appl. Opt.*, **33**, 5118–5125, doi:10.1364/AO.33.005118.

Huang, D., K. Johnson, Y. Liu, and W. Wiscombe, 2009: High resolution retrieval of liquid water vertical distributions using collocated Ka-band and W-band cloud radars. *Geophys. Res. Lett.*, **36**, L24807, doi:10.1029/2009GL041364.

Illingworth, A. J., and Coauthors, 2007: Cloudnet: Continuous evaluation of cloud profiles in seven operational models using ground-based observations. *Bull. Amer. Meteor. Soc.*, **88**, 883–898, doi:10.1175/BAMS-88-6-883.

IPCC, 2007: *Climate Change 2007: The Physical Science Basis*. Cambridge University Press, 996 pp.

Kalesse, H., and P. Kollias, 2013: Climatology of high cloud dynamics using profiling ARM Doppler radar observations. *J. Climate*, **26**, 6340–6359, doi:10.1175/JCLI-D-12-00695.1.

Kneifel, S., S. Redl, E. Orlandi, U. Löhnert, M. P. Cadeddu, D. D. Turner, and M.-T. Chen, 2014: Absorption properties of supercooled liquid water between 31 and 225 GHz: Evaluation of absorption models using ground-based observations. *J. Appl. Meteor. Climatol.*, **53**, 1028–1045, doi:10.1175/JAMC-D-13-0214.1.

Knuteson, R. O., and Coauthors, 2004a: Atmospheric Emitted Radiance Interferometer. Part I: Instrument design. *J. Atmos. Oceanic Technol.*, **21**, 1763–1777, doi:10.1175/JTECH-1662.1.

——, and Coauthors, 2004b: Atmospheric Emitted Radiance Interferometer. Part II: Instrument performance. *J. Atmos. Oceanic Technol.*, **21**, 1777–1789, doi:10.1175/JTECH-1663.1.

Kollias, P., B. A. Albrecht, R. Lhermitte, and A. Savtchenko, 2001: Radar observations of updrafts, downdrafts, and turbulence in fair weather cumuli. *J. Atmos. Sci.*, **58**, 1750–1766, doi:10.1175/1520-0469(2001)058<1750:ROOUDA>2.0.CO;2.

——, ——, and F. Marks Jr., 2002: Why Mie? *Bull. Amer. Meteor. Soc.*, **83**, 1471–1483, doi:10.1175/BAMS-83-10-1471.

——, ——, and ——, 2003: Cloud radar observations of vertical drafts and microphysics in convective rain. *J. Geophys. Res.*, **108**, 4053, doi:10.1029/2001JD002033.

——, E. E. Clothiaux, M. A. Miller, E. P. Luke, K. L. Johnson, K. P. Moran, K. B. Widener, and B. A. Albrecht, 2007: The Atmospheric Radiation Measurement Program cloud profiling radars:

Second-generation sampling strategies, processing, and cloud data products. *J. Atmos. Oceanic Technol.*, **24**, 1199–1214, doi:10.1175/JTECH2033.1.

——, W. Szyrmer, J. Rémillard, and E. Luke, 2011: Cloud radar Doppler spectra in drizzling stratiform clouds: 1. Forward modeling and remote sensing applications. *J. Geophys. Res.*, **116**, D13201, doi:10.1029/2010JD015237.

——, and Coauthors, 2016: Development and applications of ARM millimeter-wavelength cloud radars. *The Atmospheric Radiation Measurement (ARM) Program: The First 20 Years, Meteor. Monogr.*, No. 57, Amer. Meteor. Soc., doi:10.1175/AMSMONOGRAPHS-D-15-0037.1.

Liljegren, J. C., 2000: Automated self-calibration of ARM microwave radiometers. *Microwave Radiometry and Remote Sensing of the Earth's Surface and Atmosphere*, P. Pampaloni and S. Paloscia, Eds., VSP Press, 433–443.

——, E. E. Clothiaux, G. G. Mace, S. Kato, and X. Dong, 2001: A new retrieval for cloud liquid water path using a ground-based microwave radiometer and measurements of cloud temperature. *J. Geophys. Res.*, **106**, 14 485–14 500, doi:10.1029/2000JD900817.

Luke, E. P., and P. Kollias, 2013: Separating cloud and drizzle radar moments during precipitation onset using Doppler spectra. *J. Atmos. Oceanic Technol.*, **30**, 1656–1671, doi:10.1175/JTECH-D-11-00195.1.

——, ——, and M. D. Shupe, 2010: Detection of supercooled liquid in mixed-phase clouds using radar Doppler spectra. *J. Geophys. Res.*, **115**, D19201, doi:10.1029/2009JD012884.

Mace, G. G., and K. Sassen, 2000: A constrained algorithm for retrieval of stratocumulus cloud properties using solar radiation, microwave radiometer and millimeter cloud radar data. *J. Geophys. Res.*, **105**, 29 099–29 108, doi:10.1029/2000JD900403.

——, and S. Benson-Troth, 2002: Cloud overlap characteristics derived from long-term cloud radar data. *J. Climate*, **15**, 2505–2515, doi:10.1175/1520-0442(2002)015<2505:CLOCDF>2.0.CO;2.

——, T. P. Ackerman, P. Minnis, and D. F. Young, 1998: Cirrus layer microphysical properties derived from surface-based millimeter radar and infrared interferometer data. *J. Geophys. Res.*, **103**, 23 207–23 216, doi:10.1029/98JD02117.

——, A. J. Heymsfield, and M. R. Poellot, 2002: On retrieving the microphysical properties of cirrus clouds using the moments of the millimeter-wavelength Doppler spectrum. *J. Geophys. Res.*, **107**, 4815, doi:10.1029/2001JD001308.

——, and Coauthors, 2006: Cloud radiative forcing at the Atmospheric Radiation Measurement Program Climate Research Facility: 1. Technique, validation, and comparison to satellite-derived diagnostic quantities. *J. Geophys. Res.*, **111**, D11S90, doi:10.1029/2005JD005921.

Marchand, R., T. Ackerman, E. R. Westwater, S. A. Clough, K. Cady-Pereira, and J. C. Liljegren, 2003: An assessment of microwave absorption models and retrievals of cloud liquid water using clear-sky data. *J. Geophys. Res.*, **108**, 4773, doi:10.1029/2003JD003843.

——, G. G. Mace, A. G. Hallar, I. B. McCubbin, S. Y. Matrosov, and M. D. Shupe, 2013: Enhanced radar backscattering due to oriented ice particles at 95-GHz during StormVex. *J. Atmos. Oceanic Technol.*, **30**, 2336–2351, doi:10.1175/JTECH-D-13-00005.1.

Marshak, A., Y. Knyazikhin, K. D. Evans, and W. J. Wiscombe, 2004: The "red versus NIR" plane to retrieve broken-cloud optical depth from ground-based measurements. *J. Atmos. Sci.*, **61**, 1911–1925, doi:10.1175/1520-0469(2004)061<1911:TRVNPT>2.0.CO;2.

Mather, J. H., and J. W. Voyles, 2013: The ARM Climate Research Facility: A review of structure and capabilities. *Bull. Amer. Meteor. Soc.*, **94**, 377–392, doi:10.1175/BAMS-D-11-00218.1.

Matrosov, S. Y., 1999: Retrievals of vertical profiles of ice cloud microphysics from radar and IR measurements using tuned regressions between reflectivity and cloud parameters. *J. Geophys. Res.*, **104**, 16 741–16 753, doi:10.1029/1999JD900244.

——, 2005: Attenuation-based estimates of rainfall rates aloft with vertically pointing Ka-band radars. *J. Atmos. Oceanic Technol.*, **22**, 43–54, doi:10.1175/JTECH-1677.1.

——, 2010: Synergetic use of millimeter- and centimeter-wavelength radars for retrievals of cloud and rainfall parameters. *Atmos. Chem. Phys.*, **10**, 3321–3331, doi:10.5194/acp-10-3321-2010.

——, A. V. Korolev, and A. J. Heymsfield, 2002: Profiling cloud ice mass and particle characteristic size from Doppler radar measurements. *J. Atmos. Oceanic Technol.*, **19**, 1003–1018, doi:10.1175/1520-0426(2002)019<1003:PCIMAP>2.0.CO;2.

——, M. D. Shupe, A. J. Heymsfield, and P. Zuidema, 2003: Ice cloud optical thickness and extinction estimates from radar measurements. *J. Appl. Meteor.*, **42**, 1584–1597, doi:10.1175/1520-0450(2003)042<1584:ICOTAE>2.0.CO;2.

——, T. Uttal, and D. A. Hazen, 2004: Evaluation of radar reflectivity-based estimates of water content in stratiform marine clouds. *J. Appl. Meteor.*, **43**, 405–419, doi:10.1175/1520-0450(2004)043<0405:EORREO>2.0.CO;2.

——, P. T. May, and M. D. Shupe, 2006: Rainfall profiling using Atmospheric Radiation Measurement Program vertically-pointing 8-mm wavelength radars. *J. Atmos. Oceanic Technol.*, **23**, 1478–1491, doi:10.1175/JTECH1957.1.

——, M. D. Shupe, and I. V. Djalalova, 2008: Snowfall retrievals using millimeter-wavelength cloud radars. *J. Appl. Meteor. Climatol.*, **47**, 769–777, doi:10.1175/2007JAMC1768.1.

——, G. G. Mace, R. Marchand, M. D. Shupe, A. G. Hallar, and I. B. McCubbin, 2012: Influence of ice hydrometeor habits on scanning polarimetric cloud radar measurements. *J. Atmos. Oceanic Technol.*, **29**, 989–1008, doi:10.1175/JTECH-D-11-00131.1.

McFarlane, S. A., K. F. Evans, and A. S. Ackerman, 2002: A Bayesian algorithm for the retrieval of liquid water cloud properties from microwave radiometer and millimeter radar data. *J. Geophys. Res.*, **107**, 4317, doi:10.1029/2001JD001011.

McFarquhar, G. M., J. Um, M. Freer, D. Baumgardner, G. L. Kok, and G. Mace, 2007: Importance of small ice crystals to cirrus properties: Observations from the Tropical Warm Pool International Cloud Experiment (TWP-ICE). *Geophys. Res. Lett.*, **34**, L13803, doi:10.1029/2007GL029865.

——, and Coauthors, 2011: Indirect and Semi-Direct Aerosol Campaign: The impact of Arctic aerosols on clouds. *Bull. Amer. Meteor. Soc.*, **92**, 183–201, doi:10.1175/2010BAMS2935.1.

Mead, J. B., and K. Widener, 2005: W-band ARM cloud radar. *32nd Conf. on Radar Meteorology*, Albuquerque, NM, Amer. Meteor. Soc., P1R.3. [Available online at https://ams.confex.com/ams/32Rad11Meso/techprogram/paper_95978.htm.]

Michalsky, J. J., and C. N. Long, 2016: ARM solar and infrared broadband and filter radiometry. *The Atmospheric Radiation Measurement (ARM) Program: The First 20 Years, Meteor. Monogr.*, No. 57, Amer. Meteor. Soc., doi:10.1175/AMSMONOGRAPHS-D-15-0031.1.

Miller, M. A., K. L. Johnson, D. T. Troyan, E. E. Clothiaux, E. J. Mlawer, and G. G. Mace, 2003: ARM value-added cloud products: Description and status. *Proc. 13th Atmospheric Radiation Measurement (ARM) Science Team Meeting*, Broomfield, CO, ARM. [Available online at www.arm.gov/publications/proceedings/conf13/extended_abs/miller-ma.pdf.]

Min, Q., and L. C. Harrison, 1996: Cloud properties derived from surface MFRSR measurements and comparison with GOES results at the ARM SGP site. *Geophys. Res. Lett.*, **23**, 1641–1644, doi:10.1029/96GL01488.

——, ——, P. Kiedron, J. Berndt, and E. Joseph, 2004a: A high resolution oxygen A-band and water vapor band spectrometer. *J. Geophys. Res.*, **109**, D02202, doi:10.1029/2003JD003540.

——, E. Joseph, and M. Duan, 2004b: Retrievals of thin cloud optical depth from a multifilter rotating shadowband radiometer. *J. Geophys. Res.*, **109**, D02201, doi:10.1029/2003JD003964.

Mitchell, D. L., R. P. D'Entremont, and D. H. DeSlover, 2006: Passive thermal retrievals of ice and liquid water path, effective size and optical depth and their dependence on particle and size distribution shape. *12th Conf. on Atmospheric Radiation*, Madison, WI, Amer. Meteor. Soc., 12.5. [Available online at https://ams.confex.com/ams/Madison2006/techprogram/paper_113248.htm.]

Moran, K. P., B. E. Martner, M. J. Post, R. A. Kropfli, D. C. Welsh, and K. B. Widener, 1998: An unattended cloud-profiling radar for use in climate research. *Bull. Amer. Meteor. Soc.*, **79**, 443–455, doi:10.1175/1520-0477(1998)079<0443:AUCPRF>2.0.CO;2.

Platt, C. M. R., 1973: Lidar and radiometric observations of cirrus clouds. *J. Atmos. Sci.*, **30**, 1191–1204, doi:10.1175/1520-0469(1973)030<1191:LAROOC>2.0.CO;2.

Posselt, D., and G. G. Mace, 2014: MCMC-based assessment of the error characteristics of a surface-based combined radar-passive microwave cloud property retrieval. *J. Appl. Meteor. Climatol.*, **53**, 2034–2057, doi:10.1175/JAMC-D-13-0237.1.

Protat, A., and C. R. Williams, 2011: The accuracy of radar estimates of ice terminal fall speed from vertically pointing Doppler radar measurements. *J. Appl. Meteor. Climatol.*, **50**, 2120–2138, doi:10.1175/JAMC-D-10-05031.1.

——, and Coauthors, 2014: Reconciling ground-based and space-based estimates of the frequency of occurrence and radiative effect of clouds around Darwin, Australia. *J. Appl. Meteor. Climatol.*, **53**, 456–478, doi:10.1175/JAMC-D-13-072.1.

Rambukkange, M. P., J. Verlinde, E. W. Eloranta, C. J. Flynn, and E. E. Clothiaux, 2011: Using Doppler spectra to separate hydrometeor populations and analyze ice precipitation in multilayered mixed-phase clouds. *Geosci. Remote Sens. Lett.*, **8**, 108–112, doi:10.1109/LGRS.2010.2052781.

Rodgers, C. D., 2000: *Inverse Methods for Atmospheric Sounding: Theory and Practice*. Series on Atmospheric, Oceanic and Planetary Physics, Vol. 2, World Scientific, 238 pp.

Satoh, M., T. Matsuno, H. Tomita, H. Miura, T. Nasuno, and S. Iga, 2008: Nonhydrostatic Icosahedral Atmospheric Model (NICAM) for global cloud resolving simulations. *J. Comput. Phys.*, **227**, 3486–3514, doi:10.1016/j.jcp.2007.02.006.

Shupe, M. D., 2007: A ground-based multisensor cloud phase classifier. *Geophys. Res. Lett.*, **34**, L22809, doi:10.1029/2007GL031008.

——, P. Kollias, S. Y. Matrosov, and T. L. Schneider, 2004: Deriving mixed-phase cloud properties from Doppler radar spectra. *J. Atmos. Oceanic Technol.*, **21**, 660–670, doi:10.1175/1520-0426(2004)021<0660:DMCPFD>2.0.CO;2.

——, T. Uttal, and S. Y. Matrosov, 2005: Arctic cloud microphysics retrievals from surface-based remote sensors at SHEBA. *J. Appl. Meteor.*, **44**, 1544–1562, doi:10.1175/JAM2297.1.

——, S. Y. Matrosov, and T. Uttal, 2006: Arctic mixed-phase cloud properties derived from surface-based sensors at SHEBA. *J. Atmos. Sci.*, **63**, 697–711, doi:10.1175/JAS3659.1.

——, and Coauthors, 2008a: A focus on mixed-phase clouds: The status of ground-based observational methods. *Bull. Amer. Meteor. Soc.*, **89**, 1549–1562, doi:10.1175/2008BAMS2378.1.

——, P. Kollias, M. Poellot, and E. Eloranta, 2008b: On deriving vertical air motions from cloud radar Doppler spectra. *J. Atmos. Oceanic Technol.*, **25**, 547–557, doi:10.1175/2007JTECHA1007.1.

——, V. P. Walden, E. W. Eloranta, T. Uttal, J. R. Campbell, S. M. Starkweather, and M. Shiobara, 2011: Clouds at Arctic atmospheric observatories. Part I: Occurrence and macrophysical properties. *J. Appl. Meteor. Climatol.*, **50**, 626–644, doi:10.1175/2010JAMC2467.1.

——, I. Brooks, and G. Canut, 2012: Evaluation of turbulent dissipation rate retrievals from Doppler cloud radar. *Atmos. Meas. Tech.*, **5**, 1375–1385, doi:10.5194/amt-5-1375-2012.

——, and Coauthors, 2013: High and dry: New observations of tropospheric and cloud properties above the Greenland Ice Sheet. *Bull. Amer. Meteor. Soc.*, **94**, 169–186, doi:10.1175/BAMS-D-11-00249.1.

——, D. D. Turner, A. Zwink, M. M. Theimann, M. J. Mlawer, and T. R. Shippert, 2015: Deriving Arctic cloud microphysics at Barrow: Algorithms, results, and radiative closure. *J. Appl. Meteor. Climatol.*, **54**, 1675–1689, doi:10.1175/JAMC-D-15-0054.1.

Soden, B. J., A. J. Broccoli, and R. S. Hemler, 2004: On the use of cloud forcing to estimate cloud feedback. *J. Climate*, **17**, 3661–3665, doi:10.1175/1520-0442(2004)017<3661:OTUOCF>2.0.CO;2.

Spinhirne, J. D., 1993: Micro pulse lidar. *IEEE Trans. Geosci. Remote Sens.*, **31**, 48–55, doi:10.1109/36.210443.

Stephens, G. L., 2005: Cloud feedbacks in the climate system: A critical review. *J. Climate*, **18**, 237–273, doi:10.1175/JCLI-3243.1.

——, and Coauthors, 2008: CloudSat mission: Performance and early science after the first year of operation. *J. Geophys. Res.*, **113**, D00A18, doi:10.1029/2008JD009982.

Stokes, G. M., and S. E. Schwartz, 1994: The Atmospheric Radiation Measurement (ARM) Program: Programmatic background and design of the cloud and radiation testbed. *Bull. Amer. Meteor. Soc.*, **75**, 1201–1221, doi:10.1175/1520-0477(1994)075<1201:TARMPP>2.0.CO;2.

Thorsen, T. J., Q. Fu, J. M. Comstock, C. Sivaraman, M. A. Vaughn, D. M. Winker, and D. D. Turner, 2013: Macrophysical properties of tropical cirrus clouds from the CALIPSO satellite and from ground-based micropulse and Raman lidars. *J. Geophys. Res. Atmos.*, **118**, 9209–9220, doi:10.1002/jgrd.50691.

Toon, O. B., and R. C. Miake-Lye, 1998: Subsonic Aircraft: Contrail and Cloud Effects Special Study (SUCCESS). *Geophys. Res. Lett.*, **25**, 1109–1112, doi:10.1029/98GL00839.

Turner, D. D., 2005: Arctic mixed-phase cloud properties from AERI-lidar observations: Algorithm and results from SHEBA. *J. Appl. Meteor.*, **44**, 427–444, doi:10.1175/JAM2208.1.

——, 2007: Improved ground-based liquid water path retrievals using a combined infrared and microwave approach. *J. Geophys. Res.*, **112**, D15204, doi:10.1029/2007JD008530.

——, and E. Eloranta, 2008: Validating mixed-phase cloud optical depth retrieved from infrared observations with high spectral resolution lidar. *Geosci. Remote Sens. Lett.*, **5**, 285–288, doi:10.1109/LGRS.2008.915940.

——, S. A. Ackerman, B. A. Baum, H. E. Revercomb, and P. Yang, 2003: Cloud phase determination using ground-based AERI observations at SHEBA. *J. Appl. Meteor.*, **42**, 701–715, doi:10.1175/1520-0450(2003)042<0701:CPDUGA>2.0.CO;2.

——, S. A. Clough, J. C. Liljegren, E. E. Clothiaux, K. Cady-Pereira, and K. L. Gaustad, 2007a: Retrieving liquid water path and precipitable water vapor from the Atmospheric Radiation Measurement (ARM) microwave radiometers. *IEEE Trans. Geosci. Remote Sens.*, **45**, 3680–3690, doi:10.1109/TGRS.2007.903703.

——, and Coauthors, 2007b: Thin liquid water clouds: Their importance and our challenge. *Bull. Amer. Meteor. Soc.*, **88**, 177–190, doi:10.1175/BAMS-88-2-177.

——, J. E. M. Goldsmith, and R. A. Ferrare, 2016: Development and applications of the ARM Raman lidar. *The Atmospheric Radiation Measurement (ARM) Program: The First 20 Years, Meteor. Monogr.*, No. 57, Amer. Meteor. Soc., doi:10.1175/AMSMONOGRAPHS-D-15-0026.1.

Verlinde, J., and Coauthors, 2007: The Mixed-Phase Arctic Cloud Experiment. *Bull. Amer. Meteor. Soc.*, **88**, 205–221, doi:10.1175/BAMS-88-2-205.

Vogelmann, A. M., and Coauthors, 2012: RACORO extended-term aircraft observations of boundary layer clouds. *Bull. Amer. Meteor. Soc.*, **93**, 861–878, doi:10.1175/BAMS-D-11-00189.1.

Wang, Z., 2007: A refined two-channel microwave radiometer liquid water path retrieval for cold regions by using multiple-sensor measurements. *Geosci. Remote Sens. Lett.*, **4**, 591–595, doi:10.1109/LGRS.2007.900752.

——, and K. Sassen, 2001: Cloud type and macrophysical property retrieval using multiple remote sensors. *J. Appl. Meteor.*, **40**, 1665–1683, doi:10.1175/1520-0450(2001)040<1665:CTAMPR>2.0.CO;2.

——, and ——, 2002: Cirrus cloud microphysical property retrieval using lidar and radar measurements. Part I: Algorithm description and comparison with in situ data. *J. Appl. Meteor.*, **41**, 218–229, doi:10.1175/1520-0450(2002)041<0218:CCMPRU>2.0.CO;2.

——, ——, D. N. Whiteman, and B. B. Demoz, 2004: Studying altocumulus with ice virga using ground-based active and passive remote sensors. *J. Appl. Meteor.*, **43**, 449–460, doi:10.1175/1520-0450(2004)043<0449:SAWIVU>2.0.CO;2.

Westwater, E. R., Y. Han, M. D. Shupe, and S. Matrosov, 2001: Analysis of integrated cloud liquid and precipitable water vapor retrievals from microwave radiometers during the Surface Heat Budget of the Arctic Ocean project. *J. Geophys. Res.*, **106**, 32 019–32 030, doi:10.1029/2000JD000055.

Williams, K. D., and G. Tselioudis, 2007: GCM intercomparison of global cloud regimes: Present-day evaluation and climate change response. *Climate Dyn.*, **29**, 231–250, doi:10.1007/s00382-007-0232-2.

Zhao, C., and Coauthors, 2012: Toward understanding of differences in current cloud retrievals of ARM ground-based measurements. *J. Geophys. Res.*, **117**, D10206, doi:10.1029/2011JD016792.

Zhao, Y., G. G. Mace, and J. M. Comstock, 2011: The occurrence of particles size distribution bimodality in middle latitude cirrus as inferred from ground-based remote sensing data. *J. Atmos. Sci.*, **68**, 1162–1179, doi:10.1175/2010JAS3354.1.

Chapter 20

ARM's Progress on Improving Atmospheric Broadband Radiative Fluxes and Heating Rates

SALLY A. MCFARLANE

Office of Biological and Environmental Research, U.S. Department of Energy, Washington, D.C.

JAMES H. MATHER

Pacific Northwest National Laboratory, Richland, Washington

ELI J. MLAWER

Atmospheric and Environmental Research, Inc., Lexington, Massachusetts

1. Introduction and motivation

The geographical and vertical distribution of absorption of solar radiation and emission of longwave (LW) radiation in the atmosphere and at the surface are major drivers of the large-scale circulation and the hydrological cycle. Understanding and modeling the processes by which atmospheric gases, clouds, and aerosol particles affect the distribution of radiant energy in the atmosphere are essential to accurate simulations of Earth's climate. Figure 20-1 shows the key elements of Earth's energy budget (Trenberth et al. 2009). While satellites provide measurements of the global distribution of reflected and emitted broadband fluxes at the top of the atmosphere (TOA), less information is available on the radiative budget at the surface and the vertical distribution of absorption and emission in the atmosphere. Key goals of the Atmospheric Radiation Measurement (ARM) Program are to quantify the radiative energy balance profile in Earth's atmosphere, to understand the physical processes controlling this balance, and to improve the representation of these processes in global climate models (GCMs; DOE 1990).

To address these questions, the ARM site measurements were planned to provide a full characterization of the radiatively important properties of the atmospheric column, as well as measurements of the radiative fluxes themselves. The simultaneous measurement of surface broadband radiative fluxes (Michalsky and Long 2016, chapter 16) and cloud/aerosol properties (Shupe et al. 2016, chapter 19; McComisky and Ferrare 2016, chapter 21) enables both assessment of the overall impact of clouds and aerosol on the surface radiation budget as well as analysis of how those impacts vary with changes in cloud and aerosol properties. Additionally, comparison of measured broadband surface fluxes to those calculated using atmospheric properties measured or retrieved from other ARM instruments provides a test of how well the ARM measurements are able to characterize the full set of cloud, aerosol, atmospheric state, and surface properties in the vertical column. Such "radiative flux closure" exercises can identify weaknesses in measurements, retrievals, or radiative transfer models and indicate under what conditions ARM measurements are representative of the large-scale atmospheric conditions.

While spectral radiance and fluxes are important for understanding details of specific atmospheric processes that affect the radiative balance (Mlawer and Turner 2016, chapter 14), accurately measuring and modeling broadband radiative fluxes is critical since they represent the integrated impact of atmospheric processes on Earth's energy budget. In this chapter, we describe the contributions that the ARM Program has made toward understanding and characterizing the processes that control broadband radiative fluxes and radiative heating

Corresponding author address: Sally A. McFarlane, U.S. Department of Energy, Office of Biological and Environmental Research, 1000 Independence Ave. SW, Washington, DC 20585.
E-mail: sally.anne.mcfarlane@gmail.com

DOI: 10.1175/AMSMONOGRAPHS-D-15-0046.1

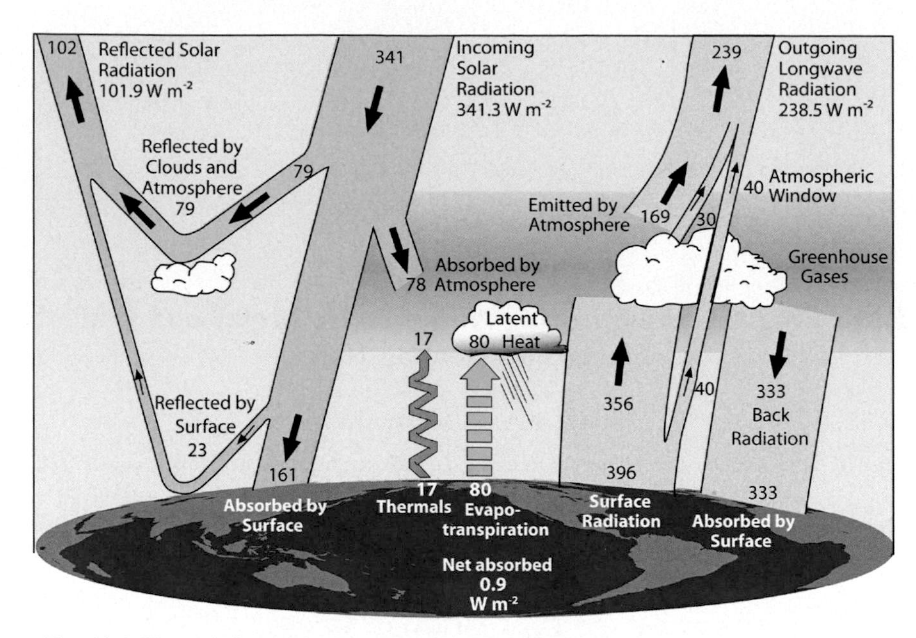

FIG. 20-1. The global annual mean Earth's energy budget for the March 2000 to May 2004 period (W m^{-2}; from Trenberth et al. 2009).

in the atmosphere. In particular, we focus on the impact of clouds on broadband surface fluxes and atmospheric heating rate profiles.

2. Early efforts to close the radiation budget

Around the time that the ARM Program was being established, several studies (Cess et al. 1995; Pilewskie and Valero 1995; Ramanathan et al. 1995; Zhang et al. 1997) reported large differences between estimates of broadband shortwave (SW) absorption in the column derived from radiative transfer calculations and those derived from combining satellite and ground-based observations of broadband radiative fluxes. In these studies, radiative transfer models typically underestimated the amount of absorption in the atmosphere relative to the observationally derived estimates, by up to 40% (Cess et al. 1995). These results led to the theory that there was an unknown source (or sources) of atmospheric absorption being neglected in radiative transfer models. The detailed observations of clouds and radiative fluxes at the ARM site, as well as several dedicated field experiments, were instrumental in helping resolve the "anomalous absorption" issue.

a. Clear sky

Using ARM data, several detailed clear-sky broadband flux closure studies were performed to explore whether clear-sky conditions were contributing to the uncertainty in modeled cloudy-sky absorption. Several studies (e.g., Kato et al. 1997; Halthore et al. 1997; Halthore and Schwartz 2000) used collocated measurements of atmospheric state, aerosol optical depth (AOD), and downwelling direct and diffuse SW irradiances from the ARM Southern Great Plains (SGP) site to compare modeled and observed fluxes under noncloudy conditions. These papers found that radiative transfer models overestimated total downward SW irradiance by up to 5% and that the overestimate was almost entirely in the diffuse flux. Discrepancies in clear-sky radiative transfer contributed to the cloudy-sky flux anomaly but were not the major factor. However, detailed investigations of potential uncertainties in the model calculations did not identify a clear cause for the clear-sky diffuse discrepancy. The results of these studies prompted further examination of pyranometer measurements and measurements of input parameters to the radiative transfer models (particularly aerosol and surface albedo properties).

Several ARM studies (e.g., Haeffelin et al. 2001; Dutton et al. 2001) identified and developed correction procedures for large offsets in pyranometer measurements due to thermal loss in the instruments. These instrument offsets could produce underestimates in measured diffuse SW of up to 10 W m^{-2}, which may have accounted for some of the model/measurement discrepancies in the earlier studies. While Halthore and Schwartz (2000) did account for this thermal offset, they noted that their conclusions showed some dependence on the method used to estimate the offset. Detailed studies of the offset issue by ARM scientists (see Michalsky and Long 2016, chapter 16) led to correction procedures for the pyranometer measurements (Younkin and Long 2003) as well identification of a new standard for diffuse measurements.

Later clear-sky radiative closure studies at the SGP site performed as part of ARM's Broadband Heating Rate Profile (BBHRP) project (Mlawer et al. 2002, 2003), discussed in depth below, and by Michalsky et al. (2006) were able to achieve agreement with measured diffuse fluxes (after corrections for nighttime offsets). The latter study examined several different radiative transfer models and found significant effects on the calculated diffuse flux from the assumptions used to treat the spectral dependence of AOD and surface albedo in the radiative transfer models. Using new observations of wavelength-dependent AOD and surface albedo, they were able to achieve agreement to within 1.9% of measured diffuse fluxes and concluded that overestimates of diffuse fluxes in previous studies were due primarily to details of the specification of the spectral dependence of aerosol and surface properties.

b. Cloudy sky

To improve understanding of SW absorption in cloudy skies, ARM sponsored many theoretical studies, as well as two observationally focused experiments. Marshak et al. (1997, 1999b) performed a series of theoretical radiative transfer studies to examine various issues (e.g., horizontal and vertical offsets between aircraft measurements, cloud inhomogeneity, and method of estimating absorption) associated with estimating cloud absorption from measurements above and below clouds and identified methods to minimize these biases in analyzing observational results. Li and Trishchenko (2001) found that surface estimates of cloud radiative effects (CREs) were much more sensitive to clear-sky scene identification than satellite estimates, potentially leading to overestimates in atmospheric absorption derived from combined satellite/surface measurements. Several other studies (e.g., O'Hirok and Gautier 1998; Marshak et al. 1998;, Titov 1998; Fu et al. 2000b) examined the potential impact of neglecting three-dimensional (3D) effects in the radiative transfer calculations of cloud absorption and found that including 3D radiative transfer effects in the calculations resulted in small increases in calculated column absorption, but not enough to account for the reported discrepancies.

The ARM Enhanced Shortwave Experiment (ARESE) consisted of collocated measurements of upwelling and downwelling broadband fluxes on two aircraft flying above and below the cloud (Valero et al. 1997a). Initial results (Valero et al. 1997b; Zender et al. 1997; O'Hirok et al. 2000) appeared to confirm that clouds absorbed significantly more SW radiation than theoretical estimates. However, further analysis of the ARESE results indicated potential calibration problems with the aircraft and satellite measurements that may have affected the estimates of column absorption from the observations (Li et al. 1999; Minnis et al. 2002). To address these issues, a second campaign (ARESE II) was sponsored by the ARM Program. Three sets of broadband solar radiometers were mounted on the Twin Otter aircraft and used in combination with the ground-based instruments at the SGP site to examine atmospheric absorption. Instrument redundancy and cross-calibration of measurements were key components of this campaign (Michalsky et al. 2002). Ackerman et al. (2003), O'Hirok and Gautier (2003), and Asano et al. (2004) analyzed measurements and radiative transfer calculations for several case studies observed during ARESE II (Fig. 20-2). Calculated and observed fluxes agreed within 10% for the clear-sky days and two of the three cloudy-sky days and within 14% for the third cloudy day. Including AOD in the cloudy-sky calculations reduced differences to approximately 5%. Around the same time, Li et al. (2002) was able to match observed surface and TOA fluxes for other cases at the SGP by including realistic representations of surface albedo, particularly in the near infrared, in the radiative transfer calculations. These studies indicated that although there were still some disagreements between the modeled and estimated absorption, they were much less than those identified in the early studies, and could generally be attributed to sampling issues, inhomogeneity of the cloud field, or were within the known uncertainties of the observations and model inputs.

3. Broadband Heating Rate Profile

Motivated by both the value and limitations of previous ARM radiative closure studies, ARM investigators initiated the BBHRP project in the early 2000s (Mlawer et al. 2002). In the decade before BBHRP began, several important achievements resulted from an ARM Quality Measurement Experiment (QME; Turner et al. 2004; Mlawer and Turner 2016, chapter 14) in which many years of radiance measurements from the Atmospheric Emitted Radiance Interferometer (AERI) at SGP were compared to corresponding calculations by a line-by-line radiation code. This AERI QME led to improvements to the instrument, radiation code, and the specification of the atmospheric profile above the site. Despite these successes, the focus on radiance rather than flux limited this effort's significance to the climate modeling community. BBHRP was designed to be a comprehensive flux-based expansion of ARM's

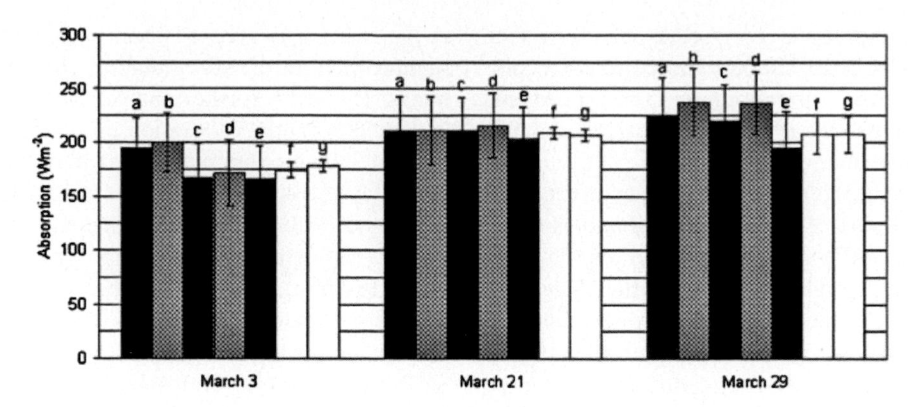

FIG. 20-2. Cloudy-sky absorption values for the atmospheric column from the surface to 7 km from 3 days during the ARESE II experiment (from Ackerman et al. 2003). The bars labeled a, c, and e are calculated using aircraft data from three different instruments at 7 km and the surface albedo measured by the 10-m tower at the SGP central facility. Bars b and d are calculated from two of the aircraft instruments at 7 km and their respective albedo values calculated from low-level aircraft passes. Bars f and g are model calculations using two different radiative transfer models.

successful radiative closure initiatives, taking advantage of the long time series of ARM observations to run extended measurement–calculation comparisons for a range of atmospheric, aerosol, and cloud conditions. By running a long-term closure experiment instead of focusing on case studies, more statistical evaluation of closure results could be performed, leading to identification of issues in input data streams and/or radiative transfer models, as well as providing better understanding of atmospheric conditions for which closure worked well and the ARM observations could be considered representative of the large-scale atmospheric state.

The goals of the BBHRP project were to

- extend long-term flux closure studies, which had been limited to clear-sky surface longwave flux at SGP, to cloudy conditions, TOA flux, the shortwave spectral region, and sites other than SGP;
- generate radiative heating rates directly from measurement-based specifications of the atmospheric state using a validated radiation code;
- generate a long-term dataset of calculated radiative fluxes and heating rates along with the corresponding atmospheric state; and
- establish a "test suite" that would allow researchers to use radiative closure to evaluate new approaches with respect to any of the components of the closure study.

This last goal was especially targeted at the evaluation of cloud retrieval algorithms.

The breadth of the BBHRP project led to the involvement of multiple ARM working groups, resulting in the development of several "best estimate" products to specify the needed input for the radiative transfer calculations. The Instantaneous Radiative Flux Working Group provided the overall leadership as well as the radiation code [the rapid radiative transfer model (RRTM); Mlawer et al. 1997], flux measurements at the surface and TOA, microwave retrievals of precipitable water vapor and liquid water path, spectral surface albedo, and trace gas profiles. The Aerosol Working Group provided a best estimate vertical profile of aerosol optical properties (Flynn et al. 2012). The Cloud Properties Working Group (CPWG) developed a baseline best estimate cloud retrieval product, called Microbase (Dunn et al. 2011; also see Shupe et al. 2016, chapter 19), and a product that provided a continuous specification of the temperature and water vapor profiles above the site (Mergesonde; Troyan 2010). The radiative closure studies performed within the BBHRP project were then used to further develop and refine all of these best estimate products, particularly Microbase. Although different methodologies were explored for the closure studies, the final methodology chosen was to use each individual cloud retrieval at 60-s resolution as input to a separate radiation calculation, average the calculated flux values over a 30-min period, and compare to the instantaneous flux measurement at the middle of that time period. The 30-min averaging was determined to give optimal results for this type of measurement–calculation comparison for the different cloud types at SGP (Mlawer et al. 2007).

In a series of conference presentations and papers, the results from the radiative closure studies were presented. Even though the first dataset analyzed included cases from only a single month, March 2000, notable conclusions were attained. The average measurement–calculation

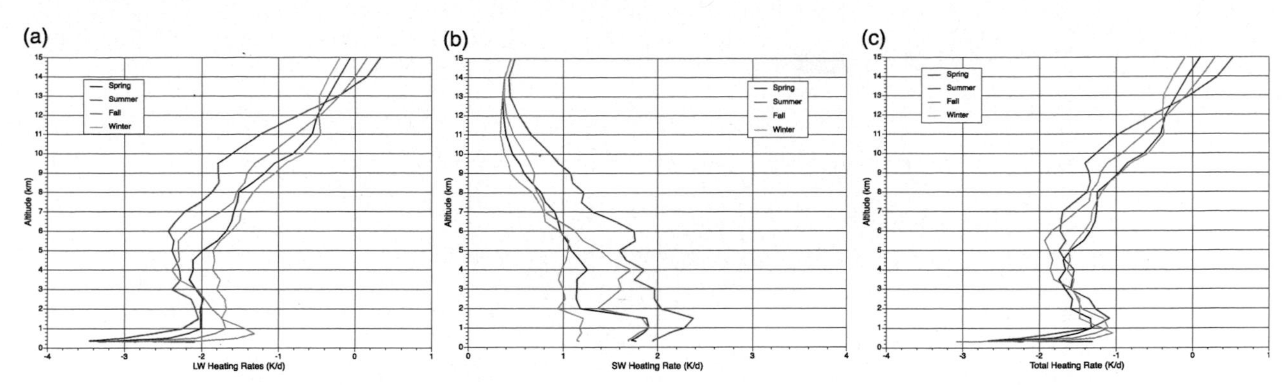

FIG. 20-3. Seasonal averages of calculated clear-sky radiative heating rate profiles from BBHRP dataset for 2000 at SGP for (a) longwave, (b) shortwave, and (c) total.

differences for clear-sky cases were 4.2 W m^{-2} [standard deviation (std dev) 5.9 W m^{-2}] for SW diffuse and 5.1 W m^{-2} (std dev 8.1) for direct normal, demonstrating agreement within the overall uncertainty and providing persuasive evidence against clear-sky anomalous absorption (Mlawer et al. 2003). Analysis of this same dataset showed that the radiative closure results for ice clouds were much improved if the effects of aerosols also were included in the calculation. The expansion of this dataset to a full year (March 2000 to February 2001) established that vertical profiles of clear-sky total radiative heating rates, especially in the lower troposphere, had relatively little seasonal dependence despite the extensive seasonal variability of the LW and SW heating rates (Fig. 20-3; Mlawer et al. 2006). In subsequent years the SGP dataset was extended to 5 years and ARM's North Slope of Alaska (NSA) site dataset was developed for 2 years. These longer datasets firmly established that radiative transfer model calculations using the ARM best estimate products as inputs were well able to reproduce clear-sky surface LW and SW diffuse fluxes, two fields that were key motivators for the genesis of the BBHRP project. For these fields, the measurement–model difference over the multiple years of data averaged to just 2.4 W m^{-2} (>22 600 cases, std dev 4.1 W m^{-2}) for clear-sky surface LW fluxes and −2.4 W m^{-2} (>7700 cases, std dev 4.0 W m^{-2}) for clear-sky surface SW diffuse fluxes.

The detailed and systematic analysis performed as part of the BBHRP project detected several problems with various ARM data streams that were not obvious from other analyses. Examples of these issues (which were all corrected) include calibration issues impacting measurements by the Normal Incidence Pyrheliometer and LW pyrgeometers, systematic errors in the AOD obtained from the multifilter rotating shadowband radiometer (MFRSR), issues with microwave retrievals of liquid water path, and systematic time-dependent inaccuracies in the TOA fluxes obtained from the Geostationary Operational Environmental Satellite (GOES). Although detection of these issues proved very helpful to the ARM Program and led to improvements in the best estimate products, their resolution led to significant delays in the BBHRP project.

Extensive analysis was performed of the closure results to assess different cloud retrieval methods. Analysis of the Microbase cloud retrieval method determined that the ice cloud optical depths provided by this retrieval approach were too large (Fig. 20-4). Here, the power of using the long-term statistics to evaluate cloud retrievals is evident as the large scatter indicates that a few case studies probably would not have been able to show the systematic error in the calculated fluxes as a function of ice water path. The BBHRP test bed was also used to examine the performance of five other cloud retrieval algorithms at SGP (provided by the CPWG) and one mixed-phase cloud property retrieval at NSA compared to Microbase (Shupe et al. 2015). Radiative closure results from these studies established that the Microbase liquid cloud retrieval was superior to the radar-based liquid cloud retrieval of Frisch et al. (1995), but other methods with simple, but different, assumptions than Microbase about the cloud's microphysical properties provided more or less the same level of agreement as Microbase (Mlawer et al. 2008). This result may reflect a limit in retrieval quality for approaches involving simple microphysical assumptions, or the inherent cloud inhomogeneity might impose a limit on the ability of a flux-based analysis like BBHRP to distinguish between more and less accurate retrieval approaches.

At the NSA, the comparison for mixed-phase clouds between the Microbase retrieval provided more definitive results, with the multisensor Shupe–Turner retrieval (Shupe et al. 2015) showing superior closure results, particularly in the downwelling LW (Fig. 20-5).

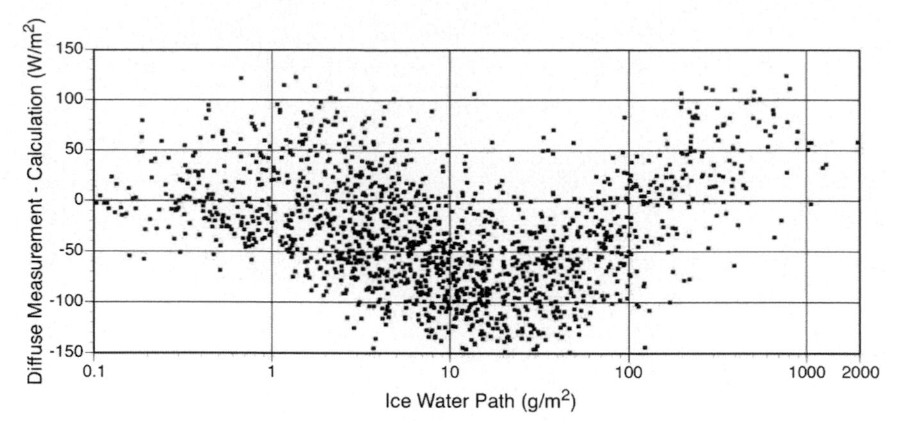

FIG. 20-4. For 1667 overcast ice cases from March 2000 to February 2004 in the BBHRP dataset, measurement-calculation residuals for shortwave diffuse flux plotted vs ice water path. The results are consistent with ice cloud optical depths that are too high. Also, the residuals for thin clouds suggest that the aerosol optical depths are too low.

Two primary reasons for the superior results of the Shupe–Turner algorithm were that it used a multisensor methodology, rather than a simple temperature threshold, to determine cloud phase and the location of liquid water and it used improved liquid water path retrievals (Shupe et al. 2015). Additional studies inspired by BBHRP, and using similar methodology, have examined ice cloud retrievals at the Darwin site (Comstock et al. 2013).

The BBHRP dataset was also used to evaluate advances in radiation parameterizations used in GCMs. Closure results (Mlawer et al. 2007) demonstrated that the GCM version of RRTM (RRTMG) using the Monte Carlo Independent Column Approximation (McICA; Pincus et al. 2003; Barker et al. 2008) stochastic approach to cloud overlap had similar residuals as the reference approach used with RRTM in BBHRP, but performed significantly better than the existing European Centre for Medium-Range Weather Forecasts (ECMWF) operational shortwave code. ECMWF subsequently adopted RRTMG/McICA as its new shortwave code (Ahlgrimm et al. 2016, chapter 28). In addition, the BBHRP dataset also provided most of the cases used in the Continuous Intercomparison of Radiation Codes (CIRC; Oreopoulos and Mlawer 2010), an initiative sponsored by the Global Energy and Water Exchanges (GEWEX) Radiation Panel and the International Radiation Commission to evaluate the quality of radiation codes used in global models. The measurement–calculation radiative closure for LW and SW flux at both the surface and TOA for the BBHRP cases used in CIRC was critical to providing confidence in the validity of the input profiles and reference line-by-line calculations used in the study.

BBHRP uses ARM observations both to derive the atmospheric state inputs to the radiative transfer calculation and to provide radiative fluxes for evaluating those radiative transfer calculations. The Clouds and the Earth's Radiant Energy System (CERES)/ARM/GEWEX Experiment (CAGEX; Charlock and Alberta 1996; Rutan et al. 2006) used a combination of ARM measurements and satellite data for the purpose of evaluating satellite-based retrievals of the surface and atmospheric radiative fluxes. The CAGEX investigators considered the model inputs, the evaluation datasets, and the model itself as an integrated framework—any component of which needed to be evaluated as a possible source of error. Issues investigated included 3D CREs; challenges with satellite measurements, including calibration and narrow band to broadband conversions; the representativeness of the surface validation measurements; and potential issues with the radiative transfer code, including the water vapor continuum. The CAGEX studies and evaluation methodology were important in developing improved algorithms and quantifying uncertainties in the CERES Surface and Atmospheric Radiation Budget (SARB) product.

4. Impact of clouds on broadband surface fluxes and radiative heating profiles

Clouds impact the climate system directly through their reflection of SW radiation and absorption and emission of LW radiation, but also indirectly through feedbacks associated with their effects on surface and atmospheric heating that in turn affect the large-scale circulation (Stephens 2005; Mace et al. 2006b). To accurately simulate cloud feedbacks, models must

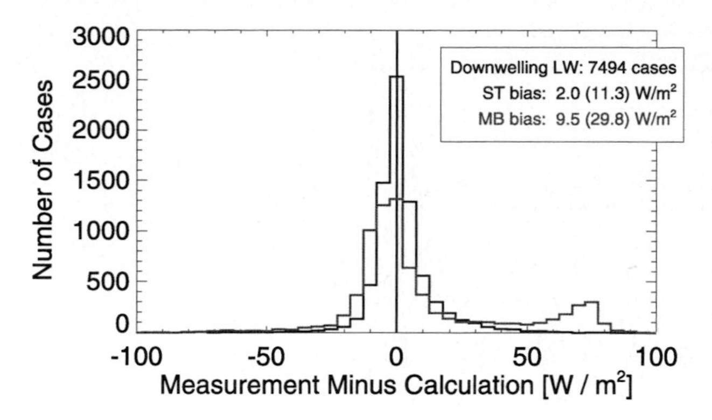

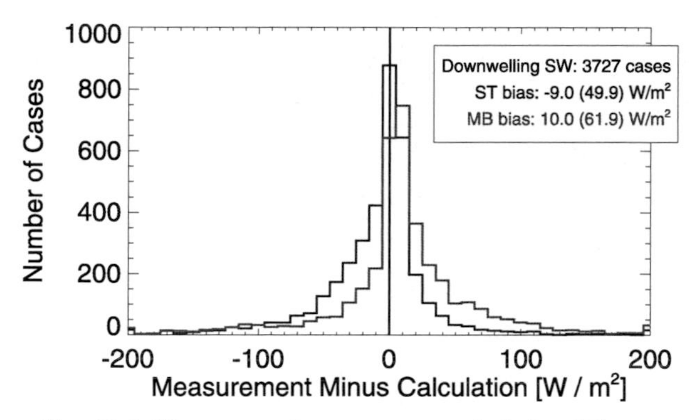

FIG. 20-5. Histograms of measurement-calculation differences for surface (top) LW and (bottom) SW diffuse fluxes for the Microbase (red) and Shupe–Turner (blue) retrievals for overcast mixed-phase clouds. This analysis uses two years of data from NSA (March 2004 to February 2006) and includes only cases where both algorithms indicated single-layer clouds. Number of cases and mean (RMS) biases are given in the figure legend. Figure courtesy of Dave Turner, NOAA.

accurately partition the cloud radiative impact between the surface and the atmosphere and how that partitioning might change as temperature and humidity profiles change in a warmer climate. A given cloud has different effects on the TOA, surface, and atmospheric radiation budgets, and the sign and magnitude of these radiative effects are complex functions of cloud temperature, phase, and optical properties and also depend on the geographic location, atmospheric profile, and surface properties. The effect of clouds on the TOA radiation budget has been documented both globally and regionally by satellite instruments such as the Earth Radiation Budget Experiment (ERBE) and CERES. However, the effect of clouds on the radiation budget at the surface and within the atmosphere is less well known because of the relative lack of surface observations and detailed observations of cloud vertical structure. Studies indicate that accurate simulation of the TOA radiation budget is not adequate to constrain cloud feedbacks because multiple vertical structures with different distributions of atmospheric and surface heating can produce the same TOA budget (Webb et al. 2001). Stephens (2005) presents a review of cloud–climate feedbacks and emphasizes that studies of cloud feedbacks must move beyond examination of surface and TOA budgets to consider perturbations to the atmospheric radiative heating, which "dictate the eventual response of the global-mean hydrological cycle of the climate model to climate forcing."

The long-term observations at the ARM sites have been used to make significant advances in quantifying our understanding of the impact of clouds on broadband surface fluxes and atmospheric heating rate profiles and have provided important datasets for evaluating model simulations. As discussed in Stokes (2016, chapter 2), each of the ARM fixed sites was located in a climatically important geographic location where CREs need to be better understood. The simultaneous observations of the vertical distribution of clouds and the broadband surface fluxes across the diurnal cycle at various geographic locations provide a unique dataset for studying how the macrophysical (cloud base and top height, cloud vertical structure) and microphysical (water content, phase, particle size) properties of clouds alter the clear-sky radiative fluxes. In the remainder of this section, we discuss results from studies using ARM data to quantify the impact of clouds on surface fluxes and heating rate profiles, which we refer to as the CRE. Throughout this field, investigators have used different terminology and methods to quantify the impacts of clouds on the energy budget, which in some cases makes it difficult to directly compare studies at different sites. In describing the results of each study, we try to clarify the methods and terminology used, although in some cases readers may need to refer to the original paper to obtain all of the details of the methodology.

a. Surface cloud radiative effects

Long and Ackerman (2000) pioneered a method of deriving SW surface clear-sky radiation from broadband radiometer measurements, enabling estimation of SW CREs at the surface from observations alone. Further work by Long et al. (2006) developed a method for retrieving cloud fraction from the SW radiometer measurements, and Long and Turner (2008) extended the methodology to broadband LW measurements, providing estimates of LW CREs. These radiative flux analysis methods have been applied to all ARM sites for multiple years, resulting in a comprehensive dataset of the magnitude and variability of cloud SW and LW CREs at the surface in a variety of different geographic regions.

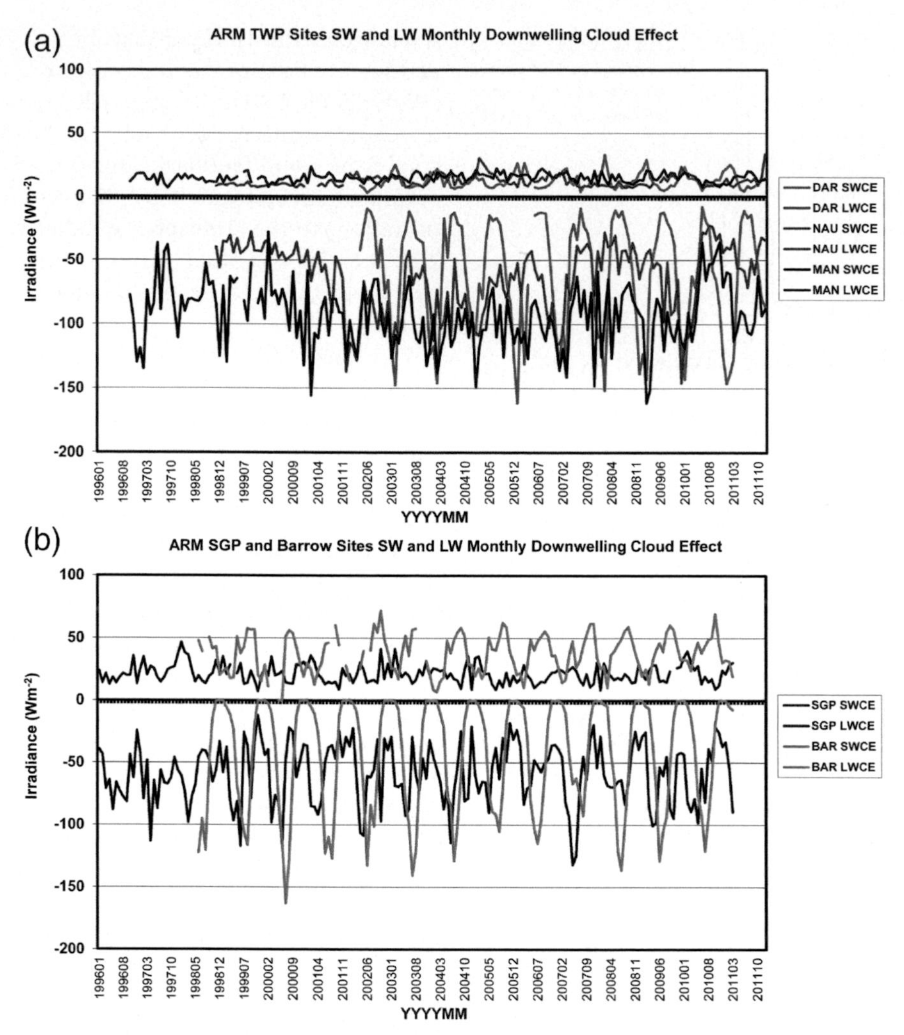

FIG. 20-6. Monthly LW (top panels) and SW (bottom panels) CRE at (a) the three TWP sites and (b) SGP and Barrow. Cloud radiative effect is calculated as all-sky downwelling flux minus clear-sky downwelling flux. Figure courtesy of Chuck Long, Pacific Northwest National Laboratory.

Figure 20-6 shows the monthly averages of LW and SW CREs at the surface from each of the ARM fixed sites. Here we define CRE as the difference between the measured all-sky and the estimated clear-sky fluxes at each time. The LW CRE is relatively small and fairly constant at the three tropical sites, because of the large amounts of water vapor in the tropical atmosphere. However, LW CRE is larger for SGP, with some seasonal and interannual variability. The LW CRE dominates the energy budget at Barrow, especially during the winter months when there is little solar radiation. The SW CRE also varies substantially across the sites. Manus has the largest-magnitude SW CRE, and the values are fairly steady across all the years of data. Nauru shows significant interannual variability in CREs—some years are similar in magnitude to Manus while other periods have much smaller magnitude of the SW CREs. Darwin shows strong interseasonal variability associated with the Australian monsoon—values of SW CREs are as large as those at Manus during the wet season but significantly smaller during the dry season. SGP and Barrow also show significant seasonal cycles, with the SW CREs going to zero during the winter at Barrow.

To further understand how cloud properties impact the magnitude and variability of surface fluxes, ARM scientists have combined the CRE estimates from the broadband radiometer measurements with measurements from other instruments (e.g., radars, lidars, microwave radiometers, and ceilometers) that provide information on cloud macrophysical and microphysical properties. Studies have been conducted at all of the ARM fixed sites to relate the observed broadband CREs

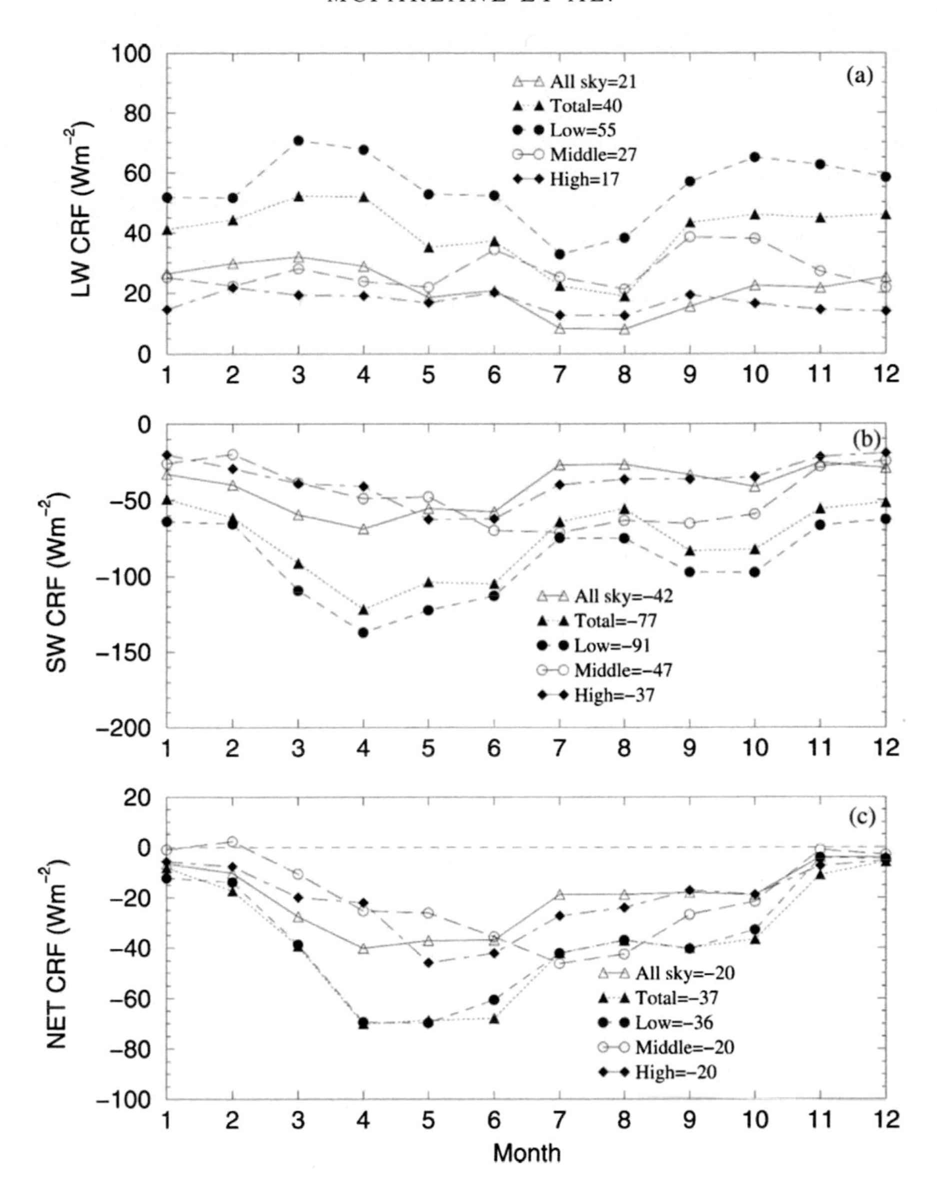

FIG. 20-7. Observed monthly mean (a) LW, (b) SW, and (c) net cloud radiative forcing (CRF) for overcast clouds as a function of cloud type at the ARM SGP site from 1997 to 2002. Monthly average cloudy- and clear-sky surface fluxes were calculated by the following procedure: 1) bin observed fluxes by sky conditions (clear or cloudy), 2) average binned fluxes over 1-h intervals, 3) calculate the monthly mean diurnal cycle by averaging all of the means corresponding to the same hour of the day within the month, and 4) average over the mean diurnal cycle to produce a monthly mean flux. CRF was then calculated as the difference between the monthly averaged net surface fluxes (downwelling minus upwelling) for observed cloudy and clear conditions (figure from Dong et al. 2006).

to specific cloud characteristics, such as cloud fraction, type, altitude, phase, and optical depth (e.g., Dong and Mace 2003; Dong et al. 2006; McFarlane and Evans 2004; McFarlane et al. 2013).

1) SOUTHERN GREAT PLAINS

Dong et al. (2006) performed a comprehensive study that examined the surface radiative impact of single-layer clouds at the SGP and found large seasonal and diurnal variability in cloud amount and radiative impact as a function of cloud type (Fig. 20-7). Here, they include upwelling surface fluxes when defining the CRE. Overall, they find that the net CRE at the surface was $-20\,\mathrm{W\,m^{-2}}$, and the magnitude of SW cooling at the surface was approximately twice as large as the LW heating. Low clouds dominated the total surface CRE, and corresponding to seasonal changes in cloud frequency, their impact was greatest during the spring and fall and least during summer. Dong et al. (2006) also quantified the impact of changes in the environment associated with clear and cloudy conditions on the CRE. Typical estimates of CREs from observations alone simply subtract average observed clear-sky fluxes from average observed cloudy-sky fluxes, neglecting the fact that atmospheric profiles are often different in cloudy and clear skies. Over the ARM SGP site, for the single-layer cloud cases examined in Dong et al. (2006), changes in humidity and surface albedo between clear and cloudy conditions offset 20% of the net radiative impact of the clouds alone. Their analysis indicated that for monthly averaged CREs, changes in cloud properties dominate the SW CREs, while the increase in water vapor between clear and cloudy conditions was a significant contributor to the LW flux changes.

Berg et al. (2011) focused specifically on the radiative impact of midlatitude shallow cumuli, a cloud type that is often not captured well in large-scale models or satellite measurements because of its small-scale variability. Although the net effect of the shallow cumuli was to decrease the SW flux at the surface, the spatial and temporal inhomogeneity of cumuli also resulted in periodic episodes of cloud-induced enhancement of the surface SW flux. These events occurred approximately 20% of the time that cumuli existed and produced occurrences of positive SW CREs with instantaneous values as large as $+75\,\mathrm{W\,m^{-2}}$.

2) North Slope of Alaska

The Arctic is one of the most rapidly changing regions on the planet, and clouds play an important role in climate feedbacks in this region (Curry et al. 1996; Stamnes et al. 1999). The melting of snow and sea ice in the Arctic is influenced strongly by the amount of downwelling LW radiation at the surface, which depends primarily on cloud amount and microphysical properties. Determining the characteristics of Arctic clouds from satellite observations is difficult because the highly reflective surface makes it difficult to distinguish clouds from the surface with visible wavelengths, and the nearly constant presence of a surface-based inversion makes thermal techniques challenging, as the cloud and the surface can be nearly the same temperature. Therefore, ground-based instruments provide a critical set of observations for determining properties of Arctic clouds and their concurrent impacts on the surface radiative budget. The Surface Heat Budget of the Arctic (SHEBA) campaign (in which ARM participated) provided an unprecedented dataset on cloud and radiative properties over the Arctic sea ice for a year (e.g., Shupe and Intrieri 2004) while the ARM NSA site near Barrow, Alaska, which has been operating since 1997, provides a unique source of long-term data for studying the seasonal and interannual variation of Arctic cloud properties and their impact on the surface radiative budget (Verlinde et al. 2016, chapter 8).

At NSA, single-layer overcast low-level stratus clouds show a significant annual cycle, with cloud fraction increasing significantly in the spring, staying high and relatively constant in the summer, and then decreasing from November to the following March (Dong and Mace 2003; Dong et al. 2010). The surface LW CRE, which is also influenced by changes in cloud liquid water path and temperature, shows a corresponding strong seasonal variation with a maximum in August and minimum in March. The SW CRE has maximum values during July and August. Although the annual average net surface CRE is only $3.5\,\mathrm{W\,m^{-2}}$, there is a seasonal dependence with low-level stratus clouds cooling the surface during the summer and warming the surface during spring and fall (Dong and Mace 2003; Dong et al. 2010). The Arctic stratus clouds appear to produce a positive feedback on Arctic change; relative to cloud-free skies, the presence of clouds increases the downwelling LW at the surface, which acts to enhance the melting of snow and ice in the spring and slows the freezing of the ice in the fall.

Mixed-phase clouds and aerosol indirect effects also play an important role in the Arctic surface radiation budget. Mixed-phase stratiform clouds are a prevalent cloud type in the Arctic and one of the most difficult for models to capture correctly. Lubin and Vogelmann (2011) used unique spectral SW irradiance measurements made during the ARM Indirect and Semi-Direct Aerosol Campaign (ISDAC) to examine the influence of mixed-phase stratiform clouds on the surface SW irradiance. Compared with liquid-water clouds, mixed-phase clouds during the Arctic spring cause a greater reduction of SW irradiance at the surface. Two studies in *Nature* (Garrett and Zhao 2006; Lubin and Vogelmann 2006) used aerosol and radiation measurements from the ARM NSA site to show that enhanced aerosol concentrations in the Arctic affect the properties of Arctic clouds and hence increase the downwelling LW surface fluxes under cloudy skies by $3.3\text{–}5.2\,\mathrm{W\,m^{-2}}$. Given the small average net surface CRE, these studies indicate that small changes in cloud or aerosol properties may have a large impact on the surface radiative budget in the Arctic.

3) Tropical Western Pacific

Understanding cloud radiative impacts in the Tropical Western Pacific (TWP) is extremely important as the

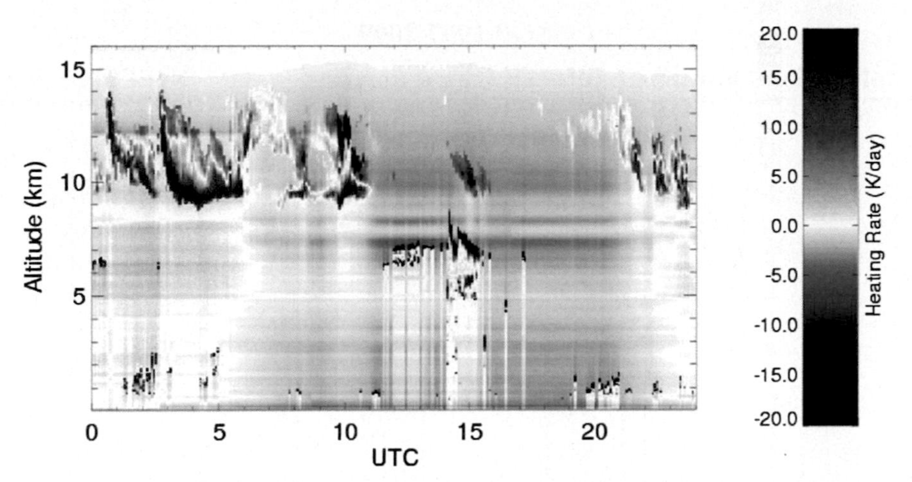

FIG. 20-8. Example of detailed structure in atmospheric radiative heating rate profile that can be calculated from ARM measurements. Figure shows calculated net heating rate for 10 March 2000 at Manus. Blues represent net cooling while reds represent net warming. Very dark colors represent heating/cooling rates greater than ± 20 K day^{-1}. To aid in interpreting the figure, note that local solar noon at Manus occurred at 0220 UTC.

absorption of solar energy in the tropics and transport of the excess energy to the poles is a primary driver of the meridional general circulation. However, because of the remote location and difficulty of finding observational sites in the tropics with suitable infrastructure, most understanding of tropical cloud radiative impacts was derived from short-term field experiments or satellite measurements. The establishment of the ARM tropical sites on Manus (1996) and Nauru (1998) provided the first long-term, detailed measurements of collocated cloud and radiative properties in the oceanic tropical western Pacific (Long et al. 2016, chapter 7). The extension of the ARM measurements to the Darwin site (2002) provided contrasting data from a tropical land site that exhibits strong seasonal variability associated with the Australian summer monsoon.

Because of their locations, the three ARM TWP sites exhibit very different time series of cloud properties and surface radiative fluxes. McFarlane et al. (2013) characterize the climatological mean and variability of the surface CRE at the three ARM tropical locations and examine differences in the CREs as a function of simple cloud type definitions that could be applied easily to climate models. The Nauru and Darwin sites show significant variability in sky cover, downwelling radiative fluxes, and surface CREs due to El Niño–Southern Oscillation (ENSO) and the Australian monsoon, respectively, while the Manus site (which is located in the tropical warm pool) shows little intraseasonal or interannual variability (Fig. 20-6). As at the other sites, clouds with low bases are the primary contributors (approximately 70%) to the surface SW and LW CREs, although clouds with midlevel and high bases also have important impacts on the surface radiative budget.

b. Impacts on atmospheric heating profiles

Clouds impact not only surface and TOA fluxes, but also redistribute energy within the atmosphere, which is important for climate feedbacks on the large-scale circulation and hydrological cycle. Before the ARM Program was established, understanding of the details of this redistribution of energy was limited by the lack of information on cloud vertical structure. Estimates of radiative heating profiles were limited to theoretical examination of idealized cases, a few case studies from field campaigns, and estimates from satellite measurements, which required large assumptions about cloud vertical structure. ARM investigators pioneered the use of combining retrievals of cloud microphysical properties from remote sensing instruments (Shupe et al. 2016, chapter 19) with detailed radiative transfer calculations to study the impacts of cloud microphysical properties and cloud vertical structure on atmospheric heating rate profiles. Figure 20-8 illustrates an example of the detailed structure of the atmospheric heating rate profile that could be calculated from the high-resolution ARM measurements.

Since there are very few observational constraints on radiative heating profiles, the flux closure concept described above is generally used to validate the calculated broadband surface and/or TOA fluxes in these studies, providing confidence in the derived cloud properties and the radiative transfer models used to calculate the heating profiles. However, the flux closure method

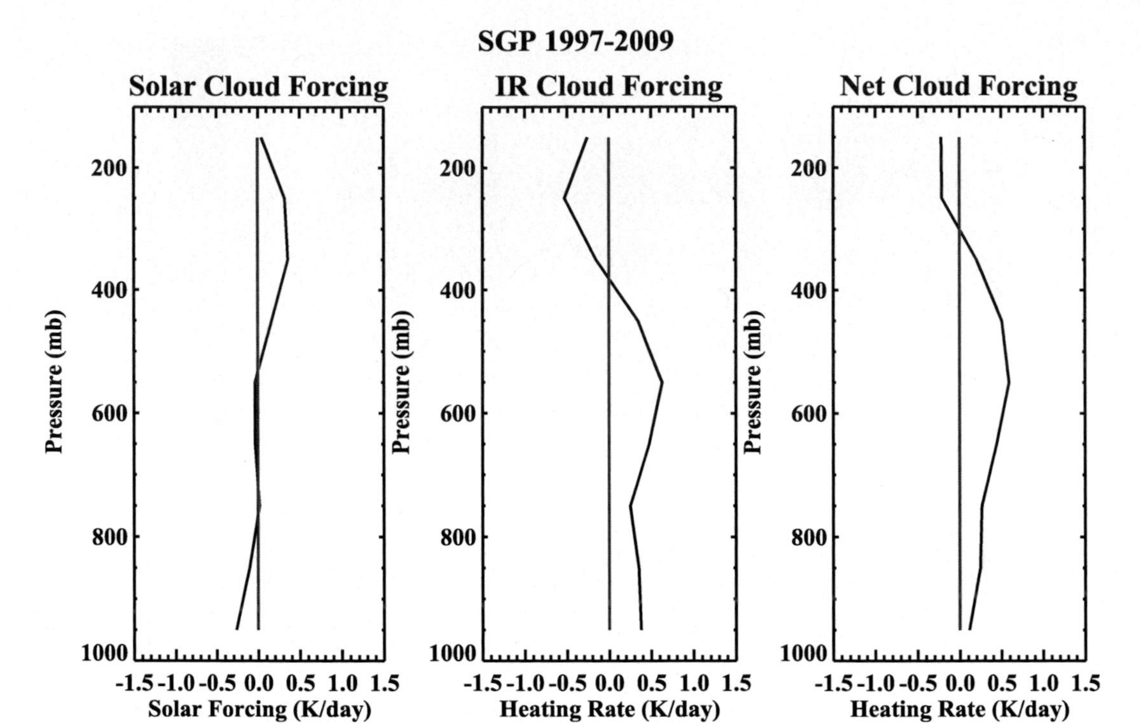

FIG. 20-9. Average CRE (black line) on atmospheric (left) SW, (center) LW, and (right) net heating profile calculated from over 20 years of data at SGP. Zero CRE is indicated in each panel by the red line. See text for more details. Figure courtesy of Jay Mace and Sally Benson, University of Utah.

cannot definitively evaluate the shape of the derived heating rate profiles, as there might be multiple profiles that produce the same fluxes. Early ARM studies examined the vertical structure of radiative heating profiles for case studies of common cloud types, including tropical cirrus (Comstock et al. 2002) and tropical precipitating convection (Jensen and Del Genio 2003). As the ARM sites gathered multiple years of data, later studies moved beyond the case study approach and used the multiple years of ARM measurements to characterize the vertical structure of SW and LW radiative heating profiles at the SGP (Fig. 20-9; Mace et al. 2006a,b; Mace and Benson 2008), the tropical sites (Fig. 20-10; Fueglistaler and Fu 2006; Mather et al. 2007; McFarlane et al. 2008), NSA site (Zwink 2013), and during ARM Mobile Facility deployments (Powell et al. 2012).

These ARM studies were the first to use observations to examine the detailed redistribution of energy between the surface and atmosphere and within the atmosphere by clouds on diurnal, monthly, and seasonal time scales. Applying similar methodologies to the datasets measured at the different ARM sites provides important information on similarities and differences in the impacts of clouds on atmospheric heating profiles at different geographic locations. Many of these radiative heating datasets have been released to the community as ARM Principal Investigator (PI) data products (http://www.arm.gov/data/pi).

Figure 20-9 shows the average CRE on the atmospheric heating profiles from 12 years of data at the ARM SGP site, while Fig. 20-10 illustrates the average CRE profiles at the TWP sites (Manus, Nauru, and Darwin). For both of these figures, CRE is defined as the heating profile calculated using a radiative transfer model with thermodynamic and cloud profiles derived from ARM observations minus the heating profile calculated using the same atmospheric thermodynamic profile, but with no clouds included in the calculations. At all of these ARM sites, clouds have a small net influence on the column atmospheric heating, but produce a significant vertical redistribution of radiant energy within the atmosphere. Different cloud types have quite different effects on the magnitude and location of heating and cooling within the atmosphere, with upper tropospheric clouds producing a net heating of the upper troposphere and boundary layer clouds producing a net cooling of thc lower troposphere (Mace and Benson 2008; Mather et al. 2007). At the tropical sites, midlevel clouds associated with detrainment near the freezing level also have a large impact on the radiative heating profiles. Detailed studies of the heating

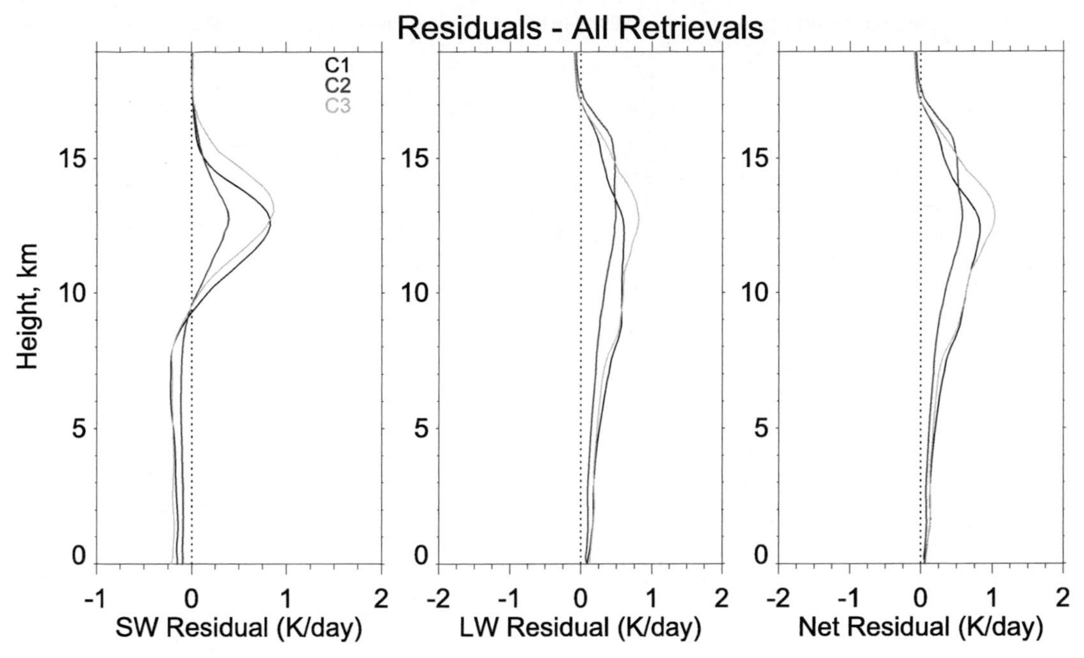

FIG. 20-10. Average CRE on (left) SW, (center) LW, and (right) net atmospheric heating profiles from Manus (C1, blue line), Nauru (C2, red line), and Darwin (C3, yellow line). Calculations include data from 2003 to 2012 at Manus, 2002 to 2009 at Nauru, and 2005 to 2011 at Darwin. See text for more details. Figure courtesy of Jennifer Comstock, Pacific Northwest National Laboratory.

rate profiles associated with different cloud types (Mather and McFarlane 2009) also illustrate that the existence of multiple cloud layers can have complex impacts on the resulting radiative heating profiles; reflection of SW by overlying cirrus can reduce the expected SW heating by water vapor at lower levels, and reduced upwelling LW emission from the surface due to low clouds can reduce the expected LW heating at the base of a cirrus layer. Several studies (Fueglistaler and Fu 2006; Mather et al. 2007) examined the potential impact of tropical cloud radiative heating on atmospheric dynamics. Fueglistaler and Fu (2006) found that the impact of clouds relative to clear-sky conditions was particularly large around 60 hPa, which is the base of the "tropical pipe." Their results suggested that gradients in upper tropospheric radiative heating rates may be partially responsible for stratospheric mixing. More recent studies (Thorsen et al. 2013; Protat et al. 2014) have combined the ARM ground-based data with satellite data to more fully characterize upper tropospheric tropical clouds and heating profiles.

All of the ARM sites exhibit substantial variability in radiative heating profiles at various time scales. At the SGP, although the TOA LW CRE is relatively constant over the annual cycle, the details of the radiative heating profile within the atmosphere change substantially over the year as the relative frequency of low and high clouds changes (Mace and Benson 2008). At the tropical sites, the radiative heating and atmospheric absorption profiles vary considerably depending on the amount of convection, associated with ENSO at Nauru and the Australian monsoon at Darwin (Mather et al. 2007; McFarlane et al. 2008). The sites also show distinct diurnal cycles in clouds and associated heating rates that vary with the large-scale dynamics and resulting cloud properties. At the NSA site, understanding the relative impact of liquid and mixed-phase clouds on radiative heating profiles (Fig. 20-11) is complicated by the concurrent changes in cloud properties (phase, depth, and liquid water content) and solar zenith angle with season. These results indicate that changes in the vertical distribution of clouds on seasonal and interannual time scales have important impacts on the redistribution of heating within the troposphere, and simply reproducing the average net surface and TOA effects of clouds in models will not capture essential cloud feedbacks on the general circulation.

Since the ARM sites have limited spatial sampling, an important question is how representative conclusions drawn at the sites are for the larger area. Jakob et al. (2005) used tropics-wide satellite data to define four tropical cloud regimes (two convective and two more suppressed regimes) and found that all four regimes exist at the Manus site and showed distinct differences in

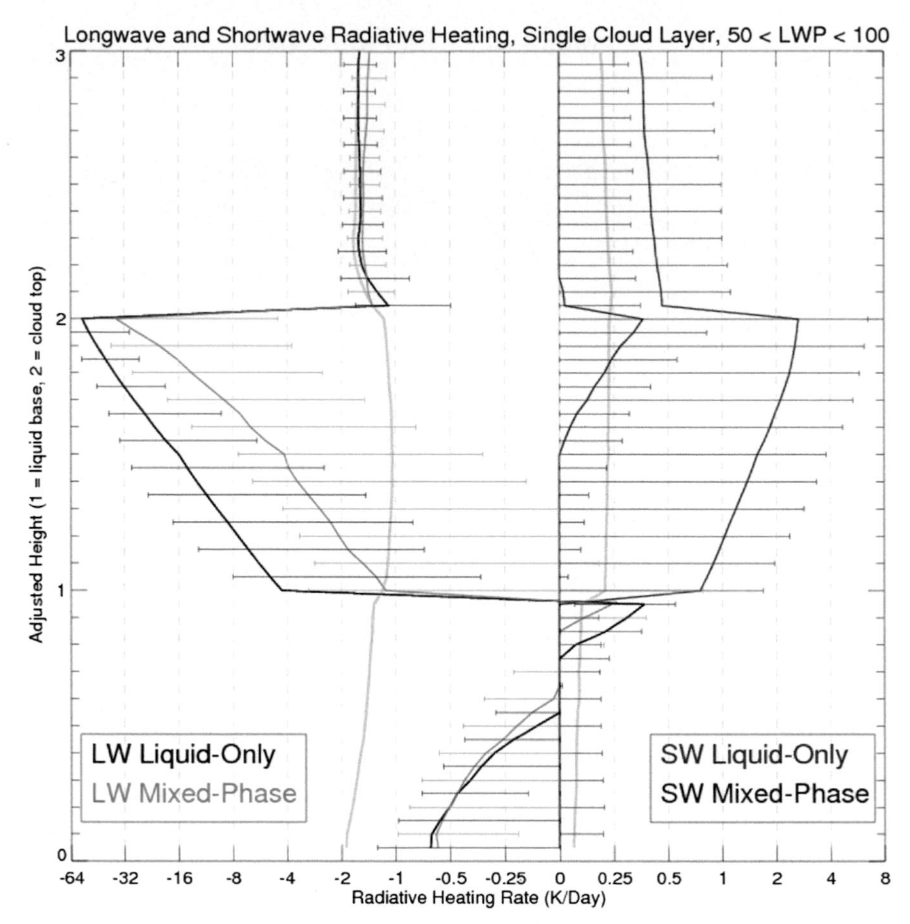

FIG. 20-11. Characteristic heating rate profiles for single-layer liquid-only and mixed-phase clouds, where the clouds have LWP between 50 and 100 g m^{-2}, calculated from 2 years of data at NSA. Cloud boundaries have been normalized. Note that the x axis is on a logarithmic scale. Figure courtesy of Dave Turner (NOAA) and Alex Zwink (University of Oklahoma).

cloud and radiative properties measured at the site. Mather and McFarlane (2009) examined differences in the structure of the radiative heating profiles at Manus and Nauru and found that the frequency of specific cloud types, and therefore the mean radiative heating profile, differed between the two sites. However, the characteristic heating profiles of individual cloud types at the two sites were remarkably similar. These studies indicated that if the frequency of different cloud types across a region were known from another source, such as satellite measurements, the cloud and heating rate profiles derived from the ARM data for each cloud type could be applied to estimate the vertical structure of cloud and heating rates over a larger area.

Another important question is the relative contribution of radiative heating to the total diabatic heating profile. Jensen and Del Genio (2003) estimated that the maximum radiative heating was 10%–30% of the maximum latent heating profiles for cases of mature deep convection near local solar noon. Li et al. (2013) combined the ARM-derived tropical radiative heating profiles with latent heating profiles derived from the Tropical Rainfall Measuring Mission (TRMM) satellite to examine the relative contributions of radiative and latent heating to the total diabatic heating profile of tropical clouds. They found that radiative heating of tropical upper-level clouds contributed 20% of the total column-integrated diabatic heating. A model simulation forced with the derived radiative and latent heating profiles suggested that the impact of radiative heating on large-scale tropical circulation was primarily from indirect impacts on convective feedbacks.

c. ARM Mobile Facility

The advent of the ARM Mobile Facility (AMF) in 2005 allowed ARM investigators to explore the impact of clouds on broadband fluxes and heating rates in additional geographical locations, including Pt. Reyes,

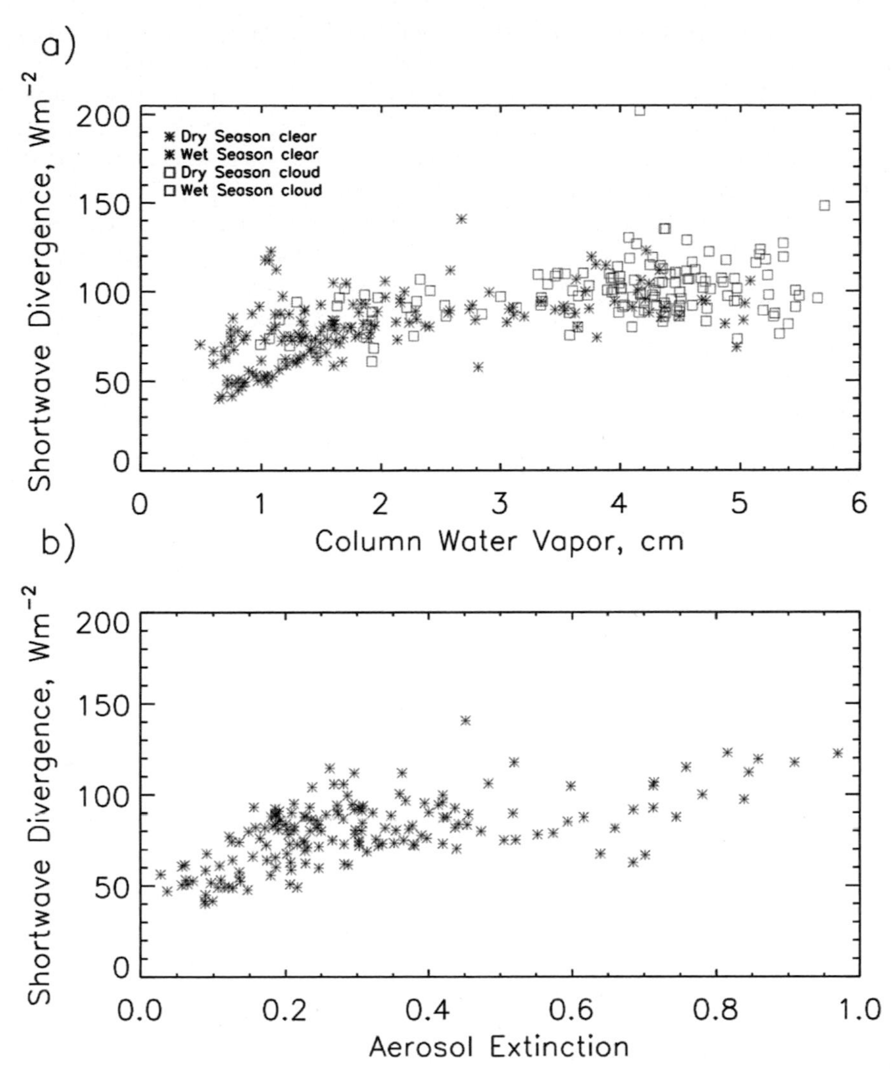

FIG. 20-12. Scatterplots of the atmospheric shortwave divergence at Niamey vs (a) column water vapor and (b) aerosol extinction at 870 nm for data measured during the wet season (blue) and dry season (red). In (a), cloudy points are indicated by open squares (figure from Slingo et al. 2009).

California; Niamey, Niger (Slingo et al. 2009); the Black Forest in Germany (Ebell et al. 2011); and the Maldives (Feng et al. 2014).

In particular, the AMF deployment to Niamey, Niger, was designed specifically to study the radiative flux divergence in the atmosphere in a region (the Sahel) with large variability in water vapor column amounts, aerosol loading, and cloud cover (Miller and Slingo 2007). An advantage of the Niamey deployment for examining atmospheric radiative heating was the location of the Niamey site under the Geostationary Earth Radiation Budget (GERB) instrument on the *Meteosat-8* European weather satellite. GERB provides broadband SW and LW fluxes every 15 min at about 50-km resolution. The combination of the TOA flux measurements from GERB and the surface flux measurements from ARM allowed the first direct estimation of broadband radiative flux divergence through the atmosphere at high temporal resolution for such an extended period of time. Slingo et al. (2009) examine the factors that control the surface and TOA fluxes, as well as the atmospheric radiative flux divergence at Niamey (Fig. 20-12). The high aerosol loadings throughout the year affect both the SW and LW fluxes while the LW fluxes also are impacted significantly by the large seasonal changes in column water vapor (CWV) and temperature between the wet and dry seasons. These effects on the LW fluxes somewhat counteract each other, as the highest temperatures occur at the end of the dry season when the CWV is lowest while the lowest temperatures in the wet season occur at the same time as the highest CWV values. The SW fluxes are affected strongly by clouds and by the seasonal changes in CWV. The LW total atmospheric divergence shows relatively small variations through the year, because of compensation between the seasonal variations in the outgoing longwave radiation (OLR) and surface net LW radiation due to the changes in temperature and CWV. The SW atmospheric divergence is mainly determined by the CWV and aerosol loadings (Fig. 20-12), while the effect of clouds is much smaller than on the surface and TOA fluxes.

Additional studies based on the AMF deployment examined other aspects of the radiative budget at Niamey, including comparisons of observed and modeled LW fluxes (Bharmal et al. 2009), the effect of heterogeneities on radiative divergence (Settle et al. 2008), detailed studies of aerosol impacts on SW and LW surface fluxes (Slingo et al. 2006; McFarlane et al. 2009; Turner 2008), diurnal and seasonal cloud radiative impacts (Bouniol et al. 2012), radiative heating of convective anvils (Powell et al. 2012), seasonal contrasts in the components of the surface energy balance (Miller et al. 2009), and the ability of GCMs to reproduce the observed cloud radiative effects and atmospheric divergence (Miller et al. 2012).

d. Evaluating model representations of surface and atmospheric radiative fluxes

The climatologies of surface radiation and cloud properties from the ARM sites, such as those described above, have been used for evaluating simulations of the surface radiation budget in a wide variety of models, including operational forecast models (e.g., Hinkelman et al. 1999; Morcrette 2002; Yang et al. 2006; Ahlgrimm and Forbes 2014), reanalysis models (Allan 2000; Kennedy et al. 2011), single-column models (Somerville and Iacobellis 1999; Lane et al. 2000; Iacobellis et al. 2003), regional models (Pinto et al. 1999), and GCMs (Xie et al. 2003; Qian et al. 2012; Miller et al. 2012). While multiple sites around the world have surface radiative flux measurements that can be used to evaluate model radiative fluxes (e.g., the Baseline Surface Radiation Network), the simultaneous measurements of cloud, aerosol, and atmospheric properties within the column along with the surface radiative fluxes at the ARM sites allow scientists to not only identify errors in the models' surface radiative budgets, but to diagnose which aspects of the model contribute to those errors.

These studies have generally indicated that model errors in simulating surface radiation arise from a combination of sources. For example, Hinkelman et al. (1999) found biases of $50\,\mathrm{W\,m^{-2}}$, on average, in the downwelling SW radiation at the surface in the Eta model and attributed half of the excess to errors in the treatment of water vapor and aerosol in the model and the other half to errors in the treatment of clouds. Similarly, Kennedy et al. (2011) showed that the North American Regional Reanalysis has significant positive biases in downwelling SW and negative biases in downwelling LW under both clear-sky and all-sky conditions. The biases were found to result from a combination of errors in aerosol, water vapor, and clouds in the model. Qian et al. (2012) used ARM observations at multiple sites to examine the relationships between cloud fraction and surface radiative flux in the GCMs used in the Intergovernmental Panel on Climate Change assessment reports. They find that errors in modeled total cloud fraction and normalized cloud radiative effect are similar in magnitude, but both are larger than model errors in downwelling SW radiation, indicating that the reasonable agreement in surface radiation is due to compensating errors in cloud vertical structure, overlap assumption, optical depth, and/or cloud fraction. These studies indicate the utility of the ARM observations in diagnosing reasons for errors in model-simulated fluxes.

ARM observations also have been used to evaluate how well models are able to represent broadband atmospheric radiative heating profiles. McFarlane et al. (2007) compared the ARM-derived vertical profiles of tropical cloud properties and associated BBHRP to results from two global atmospheric models and found large differences in the vertical profiles and diurnal cycle of cloud amount, water content, and radiative heating profiles. The differences in the heating rates between the models and the ARM-derived values were due to a combination of differences in the cloud properties (cloud frequency, vertical location of clouds, and cloud optical thickness) and the radiative transfer calculations (parameterization of LW absorption coefficient as a function of particle size). Powell et al. (2012) examined the ability of several microphysical schemes to reproduce the vertical structure of cloud properties and radiative heating observed in anvil clouds at Niamey. They found that all of the schemes underestimate the optical thickness of thin anvils and cirrus, resulting in a bias of excessive net anvil heating in all of the simulations.

An advantage of ARM datasets compared to shorter-term field experiments is that the multiple years of measurements provide enough samples to classify the data into different meteorological regimes. Regime analysis allows investigation of the effects of particular dynamical or meteorological conditions on cloud properties and radiative impacts, but it also provides a dataset for model evaluation that can be used to separate dynamical errors and parameterization errors (e.g., Marchand et al. 2009). Marchand et al. (2006), Evans et al. (2012), and Mülmenstädt et al. (2012) use objective methodologies to classify the multiple years of observations at the SGP, Darwin, and NSA sites, respectively, into distinct meteorological regimes. At each site, significant differences in cloud properties and the associated surface radiation fields are found between the different categories, indicating the potential utility of the classification for examining how dynamics influence cloud and radiative properties in observations and models.

5. Three-dimensional radiative transfer

The 3D radiative transfer effects cannot be included in climate models because of computational limitations that require radiative transfer calculations to be treated efficiently, and because GCMs do not contain information on the full 3D cloud scene within a grid box. Therefore, climate models have to make simplifying assumptions about cloud horizontal structure within a model grid box and how clouds in different vertical layers overlap, and they must use 1D radiative transfer models to calculate gridbox-average fluxes and heating rates. ARM has made significant advances in using sophisticated 3D radiative transfer models to understand the impacts of these assumptions on calculated broadband fluxes and heating rates and developing methodologies to reduce these biases in GCMs. More detail about development of new methods for radiative transfer in GCMs is given in Mlawer et al. (2016, chapter 15); here we focus primarily on ARM studies that used state-of-the-art radiative transfer models to quantify the impact of neglecting 3D effects on calculated fluxes and heating rates as well as the use of ARM data to understand inhomogeneity in real clouds.

a. Biases due to neglect of 3D effects

As discussed previously, ARM scientists conducted several important theoretical radiative transfer modeling studies that illustrated that errors due to neglecting 3D effects when deriving estimates of atmospheric absorption derived from surface and satellite (or aircraft) observations would contribute to the discrepancy between observations and 1D models, but could not explain all of it (e.g., O'Hirok and Gautier 1998; Marshak et al. 1998; Titov 1998; Fu et al. 2000b). ARM scientists also performed key studies characterizing the accuracy of the independent column approximation (ICA), a method to move beyond plane-parallel radiative transfer. In the ICA, a domain is divided into individual columns (or pixels), 1D radiative transfer is calculated on the cloud properties within each column, and then the contributions of each column are summed to obtain the domain-averaged fluxes of heating rates. ARM studies showed that GCM-scale domain-averaged errors in surface and TOA fluxes calculated using the ICA were generally less than 20 W m^{-2}, and those in atmospheric heating rates were typically less than 3% (Marshak et al. 1995, 1999a; Barker et al. 1999; Cole et al. 2005a).

The detailed radar and lidar observations from the ARM sites provided key information for studies on the impact of horizontal and vertical inhomogeneity of real clouds, rather than idealized or modeled clouds, on radiative fluxes (Fu et al. 2000a; Carlin et al. 2002). Várnai (2010) used multiple years of ARM radar data at the SGP, NSA, and TWP sites to perform a comprehensive study on the impact of horizontal photon transport effects on solar radiative heating calculations. The results show that average 2D effects are fairly small, but individual cases can have much larger impacts, especially for cases of high sun or convective clouds. Additionally, 2D effects at oblique sun angles often enhance surface heating. Assumptions about vertical inhomogeneity are also important for radiative transfer calculations, and cloud overlap statistics from the ARM radar data (e.g., Mace and Benson-Troth 2002) have been used to examine formulations of vertical overlap assumptions used in climate models (Stephens et al. 2004).

To examine multidimensional radiative transfer effects using the ARM vertically pointing radar and lidar data, researchers typically used the "frozen turbulence" assumption, in which a horizontal advection velocity is used to convert the temporal dimension of the data into a spatial dimension in order to create 2D cloud fields. Using model-simulated cloud fields, Pincus et al. (2005) show that because there are fewer cloud edges in 2D than 3D scenes, such studies systematically underestimate the magnitude of the 3D effect. Additionally, the limited sampling observed by the vertically pointing radar produces noise of up to 20% in estimates of the 3D effect. To move beyond these limitations of sampling from vertically pointing data, ARM scientists developed methods for stochastically generating multidimensional cloud fields that match observed statistics of optical depth and cloud structure (Evans and Wiscombe 2004; Prigarin and Marshak 2009).

While the studies of 3D radiative transfer described above focused primarily on the effects on SW fluxes and heating rates, a comprehensive set of studies by Ellingson and colleagues examined the impacts of 3D cloud structure on LW radiative transfer and found that they could be important for cloud fractions between 0.2 and 0.8. These studies included theoretical calculations to quantify the impact on LW fluxes and heating rates using idealized (Killen and Ellingson 1994) and observed (Han and Ellingson 1999) cloud geometries, identifying conditions under which LW scattering needs to be included in calculations (Takara and Ellingson 1996, 2000), and testing parameterizations for LW radiative effects against ARM observations (Han and Ellingson 2000; Taylor and Ellingson 2008).

b. Impact of 3D effects on cloud evolution

While models need to produce accurate domain-averaged fluxes and heating rates in order to correctly simulate interactions with the large-scale circulation, subgrid-scale interactions between clouds and radiation also can have important influences on cloud evolution. Several ARM studies using high-resolution cloud models have examined the impact of neglecting interactions between radiation and clouds on the evolution of different types of cloud fields. These studies indicate that the impact of neglecting 3D radiative transfer effects in models depends in part on the resolution of the model simulations.

Simulations of tropical cumulus convection indicated that direct radiative–convective interactions through LW cloud-top cooling and SW radiative heating are more important in determining the diurnal cycle of tropical precipitation than indirect radiative–dynamical–convective interactions through differential heating of clear and cloudy regions (Xu and Randall 1995). Vogelmann et al. (2001) found that cloud-top height variability and the structure of cloud edges in deep convective clouds could produce regions of intense local SW heating that occupied only a small area of the climate model grid cell but dominated the grid-mean value. Explicitly resolving these hot spots requires model grid sizes of approximately 20–30 km^2. Models that cannot resolve these hot spots and use typical cloud overlap assumptions to represent subgrid variability overestimate the solar heating above the cloud and underestimate it below the cloud, impacting the turbulence and vertical velocity within the cloud layers and having potential consequences for simulation of cloud evolution and feedback in weak dynamical regimes in GCMs. In high-resolution (horizontal resolutions of 50–100 m) model simulations, Mechem et al. (2008) found that neglecting multidimensional LW radiative effects had only small impacts on the cloud dynamics of boundary layer cloud systems. Although the spatial structure of the radiative forcing was changed significantly in model simulations using full 3D radiative transfer calculations, there were few systematic differences in the overall cloud fields, indicating the ICA approach was sufficiently accurate for these simulations.

Two studies examined the neglect of subgrid-scale cloud–radiation interactions on GCM scales. Gu and Liou (2006) implemented a cirrus inhomogeneity factor in a GCM simulation and found that it increased the global mean net solar flux at TOA by 5 $\mathrm{W\,m^{-2}}$. Inclusion of inhomogeneity also produced geographic differences in cloudiness, radiation, and precipitation in the model, indicating the indirect effects of cloud–radiation interactions in cirrus clouds on the climate simulation. Cole et al. (2005b) found unresolved cloud–radiation interactions on the scale of a GCM grid box significantly affected the statistics of cloud fraction and cloud radiative effects. When local cloud–radiation interactions were neglected, high cloud amount at most latitudes tended to be larger while marine stratocumulus cloud fractions were smaller. For high clouds, the small-scale structure in cloud-top radiative cooling entrains dry air into the cloud layer, reducing the cloud amount, while for marine stratocumulus the cloud-top radiative cooling tends to maintain the layer.

c. New radiative transfer methods

Along with applying state-of-the-art radiative transfer methods to understand cloud impacts on the radiative budget (as discussed above) and improving the representation of radiative transfer in GCMs (as discussed in Mlawer et al. 2016, chapter 15), ARM investigators have played a significant role in developing new state-of-the-art radiative transfer methods that calculate fluxes and heating rates more efficiently or more accurately. Improvements to 1D radiative transfer models supported by ARM include computationally efficient methods for solving the two-stream equation (Gabriel et al. 1993), improved correlated-k distribution methods (Mlawer et al. 1997; Kato et al. 1999), and new methods for treating horizontal variability (Wood et al. 2005). Improvements to 3D radiative transfer modeling techniques include the development of the spherical harmonic spatial grid method (Evans 1993), a new spectral radiative transfer method (Gabriel et al. 1993), and applications of photon diffusion theory and Green's function to remote sensing problems (e.g., Davis and Marshak 2002). ARM scientists have played key roles in developing stochastic radiative transfer methods for atmospheric applications (Lane et al. 2002; Kassianov 2003; Lane-Veron and Somerville 2004; Kassianov and Veron 2011) and have used ARM data both to develop statistics of cloud properties for stochastic models (Lane-Veron and Somerville 2004) and to evaluate their performance (Foster and Veron 2008; Kassianov et al. 2003). ARM also contributed to the development of easy-to-use, community radiative transfer codes, including the Santa Barbara Discrete

Ordinate Radiative Transfer (DISORT) Atmospheric Radiative Transfer (SBDART) 1D model (Ricchiazzi et al. 1998) and the community Monte Carlo 3D model (Pincus and Evans 2009).

d. Other 3D radiative transfer activities

ARM and NASA jointly support the Intercomparison of 3D Radiation Codes (I3RC) project (Cahalan et al. 2005), which aims to improve the treatment of 3D radiative transfer (RT) through "documentation of errors and limitations of 3D methods, sharing and development of 3D tools, and atmospheric science education in 3D RT." Many ARM scientists have participated in the I3RC intercomparison studies, and one of the intercomparison cases is based on ARM radar data. ARM also provided support for a monograph on 3D radiative transfer techniques in cloudy atmospheres (Marshak and Davis 2005), and over half of the contributing authors were members of the ARM Science Team.

6. Conclusions

As described in this chapter, ARM has made important advances in quantifying the role of clouds on broadband surface fluxes and atmospheric heating rate profiles and providing important information for model evaluation and improvement. We see two primary areas where ARM will continue to move forward in this area: the BBHRP test bed and use of 3D cloud fields from scanning radar data.

The recent automation of the BBHRP procedure (e.g., McFarlane et al. 2011) allows scientists to quickly and systematically evaluate the accuracy of cloud retrievals with respect to surface broadband flux and test the impact of changing the assumptions made in the retrievals on both the surface fluxes and broadband heating rates. Initial studies have started to use the BBHRP methodology to study an ensemble of cloud retrievals to provide information on retrieval uncertainty to cloud modelers (S. Xie 2013, personal communication), and we anticipate this type of activity will continue.

ARM scientists have already made important contributions to understanding the 3D radiative effects of clouds through the use of model-simulated 3D cloud fields and 2D cloud fields derived from vertically pointing radar instruments. However, the new ARM scanning cloud radars (Ackerman et al. 2016, chapter 3; Mather et al. 2016, chapter 4) will provide unprecedented information on the 3D structure of real cloud fields. An early study has already been conducted to quantify errors in calculated surface fluxes from radar-produced 3D cloud scenes associated with radar sensitivity, scan strategy, and method of scene reconstruction (Fielding et al. 2014).

REFERENCES

Ackerman, T. P., D. M. Flynn, and R. T. Marchand, 2003: Quantifying the magnitude of anomalous solar absorption. *J. Geophys. Res.*, **108**, 4273, doi:10.1029/2002JD002674.

——, T. S. Cress, W. R. Ferrell, J. H. Mather, and D. D. Turner, 2016: The programmatic maturation of the ARM Program. *The Atmospheric Radiation Measurement (ARM) Program: The First 20 Years, Meteor. Monogr.*, No. 57, Amer. Meteor. Soc., doi:10.1175/AMSMONOGRAPHS-D-15-0054.1.

Ahlgrimm, M., and R. Forbes, 2014: Improving the representation of low clouds and drizzle in the ECMWF model based on ARM observations from the Azores. *Mon. Wea. Rev.*, **142**, 668–685, doi:10.1175/MWR-D-13-00153.1.

——, ——, J.-J. Morcrette, and R. Neggers, 2016: ARM's impact on numerical weather prediction at ECMWF. *The Atmospheric Radiation Measurement (ARM) Program: The First 20 Years, Meteor. Monogr.*, No. 57, Amer. Meteor. Soc., doi:10.1175/AMSMONOGRAPHS-D-15-0032.1.

Allan, R. P., 2000: Evaluation of simulated clear-sky longwave radiation using ground-based observations. *J. Climate*, **13**, 1951–1964, doi:10.1175/1520-0442(2000)013<1951:EOSCSL>2.0.CO;2.

Asano, S., A. Uchiyama, A. Yamazaki, and K. Kuchiki, 2004: Solar radiation budget from the MRI radiometers for clear and cloudy air columns within ARESE II. *J. Atmos. Sci.*, **61**, 3082–3096, doi:10.1175/JAS-3288.1.

Barker, H. W., G. L. Stephens, and Q. Fu, 1999: The sensitivity of domain-averaged solar fluxes to assumptions about cloud geometry. *Quart. J. Roy. Meteor. Soc.*, **125**, 2127, doi:10.1002/qj.49712555810.

——, J. N. S. Cole, J.-J. Morcrette, R. Pincus, P. Räisänen, K. von Salzen, and P. A. Vaillancourt, 2008: The Monte Carlo Independent Column Approximation: An assessment using several global atmospheric models. *Quart. J. Roy. Meteor. Soc.*, **134**, 1463–1478, doi:10.1002/qj.303.

Berg, L. K., E. I. Kassianov, C. N. Long, and D. L. Mills Jr., 2011: Surface summertime radiative forcing by shallow cumuli at the Atmospheric Radiation Measurement Southern Great Plains site. *J. Geophys. Res.*, **116**, D01202, doi:10.1029/2010JD014593.

Bharmal, N. A., A. Slingo, G. J. Robinson, and J. J. Settle, 2009: Simulation of surface and top of atmosphere thermal fluxes and radiances from the radiative atmospheric divergence using the ARM mobile facility, GERB data, and AMMA stations experiment. *J. Geophys. Res.*, **114**, D00E07, doi:10.1029/2008JD010504.

Bouniol, D., F. Couvreux, P. H. Kamsu-Tamo, M. Leplay, F. Guichard, F. Favot, and E. J. O'Connor, 2012: Diurnal and seasonal cycles of cloud occurrences, types, and radiative impact over West Africa. *J. Appl. Meteor. Climatol.*, **51**, 534–553, doi:10.1175/JAMC-D-11-051.1.

Cahalan, R. F., and Coauthors, 2005: The I3RC: Bringing together the most advanced radiative transfer tools for cloudy atmospheres. *Bull. Amer. Meteor. Soc.*, **86**, 1275–1293, doi:10.1175/BAMS-86-9-1275.

Carlin, B., Q. Fu, U. Lohmann, G. Mace, K. Sassen, and J. M. Comstock, 2002: High-cloud horizontal inhomogeneity and solar albedo bias. *J. Climate*, **15**, 2321–2339, doi:10.1175/1520-0442(2002)015<2321:HCHIAS>2.0.CO;2.

Cess, R. D., and Coauthors, 1995: Absorption of solar radiation by clouds: Observations versus models. *Science*, **267**, 496–499, doi:10.1126/science.267.5197.496.

Charlock, T. P., and T. L. Alberta, 1996: The CERES/ARM/GEWEX Experiment (CAGEX) for retrieval of radiative fluxes with satellite data. *Bull. Amer. Meteor. Soc.*, **77**, 2673–2683, doi:10.1175/1520-0477(1996)077<2673:TCEFTR>2.0.CO;2.

Cole, J. N. S., H. W. Barker, W. O'Hirok, E. E. Clothiaux, M. F. Khairoutdinov, and D. A. Randall, 2005a: Atmospheric radiative transfer through global arrays of 2D clouds. *Geophys. Res. Lett.*, **32**, L19817, doi:10.1029/2005GL023329.

——, ——, D. A. Randall, M. F. Khairoutdinov, and E. E. Clothiaux, 2005b: Global consequences of interactions between clouds and radiation at scales unresolved by global climate models. *Geophys. Res. Lett.*, **32**, L06703, doi:10.1029/2004GL020945.

Comstock, J. M., T. P. Ackerman, and G. G. Mace, 2002: Ground-based lidar and radar remote sensing of tropical cirrus clouds at Nauru Island: Cloud statistics and radiative impacts. *J. Geophys. Res.*, **107**, D4714, doi:10.1029/2002JD002203.

——, A. Protat, S. A. McFarlane, J. Delanoe, and M. Deng, 2013: Assessment of uncertainty in cloud radiative effects and heating rates through retrieval algorithm differences: Analysis using 3-years of ARM data at Darwin, Australia. *J. Geophys. Res. Atmos.*, **118**, 4549–4571, doi:10.1002/jgrd.50404.

Curry, J. A., W. B. Rossow, D. Randall, and J. L. Schramm, 1996: Overview of Arctic cloud and radiation characteristics. *J. Climate*, **9**, 1731–1764, doi:10.1175/1520-0442(1996)009<1731:OOACAR>2.0.CO;2.

Davis, A., and A. Marshak, 2002: Space-time characteristics of light transmitted through dense clouds: A Green's function analysis. *J. Atmos. Sci.*, **59**, 2713–2727, doi:10.1175/1520-0469(2002)059<2713:STCOLT>2.0.CO;2.

Dong, X. Q., and G. G. Mace, 2003: Arctic stratus cloud properties and radiative forcing derived from ground-based data collected at Barrow, Alaska. *J. Climate*, **16**, 445–461, doi:10.1175/1520-0442(2003)016<0445:ASCPAR>2.0.CO;2.

——, B. K. Xi, and P. Minnis, 2006: A climatology of midlatitude continental clouds from the ARM SGP central facility. Part II: Cloud fraction and surface radiative forcing. *J. Climate*, **19**, 1765–1783, doi:10.1175/JCLI3710.1.

——, ——, K. Crosby, C. N. Long, R. S. Stone, and M. D. Shupe, 2010: A 10 year climatology of Arctic cloud fraction and radiative forcing at Barrow, Alaska. *J. Geophys. Res.*, **115**, D17212, doi:10.1029/2009JD013489.

DOE, 1990: Atmospheric Radiation Measurement Program plan. Rep. DOE/ER 0441, U.S. Dept. of Energy, 124 pp. [Available online at https://www.arm.gov/publications/doe-er-0441.pdf.]

Dunn, M., K. Johnson, and M. Jensen, 2011: The microbase value-added product: A baseline retrieval of cloud microphysical properties. Rep. DOE/SC-ARM/TR-095, 34 pp. [Available online at http://www.arm.gov/publications/tech_reports/doe-sc-arm-tr-095.pdf.]

Dutton, E. G., J. J. Michalsky, T. Stoffel, B. W. Forgan, J. Hickey, D. W. Nelson, T. L. Alberta, and I. Reda, 2001: Measurement of broadband diffuse solar irradiance using current commercial instrumentation with a correction for thermal offset errors. *J. Atmos. Oceanic Technol.*, **18**, 297–314, doi:10.1175/1520-0426(2001)018<0297:MOBDSI>2.0.CO;2.

Ebell, K., S. Crewell, U. Lohnert, D. D. Turner, and E. J. O'Connor, 2011: Cloud statistics and cloud radiative effect for a low-mountain site. *Quart. J. Roy. Meteor. Soc.*, **137**, 306–324, doi:10.1002/qj.748.

Evans, K. F., 1993: Two-dimensional radiative transfer in cloudy atmospheres: The spherical harmonic spatial grid method. *J. Atmos. Sci.*, **50**, 3111–3124, doi:10.1175/1520-0469(1993)050<3111:TDRTIC>2.0.CO;2.

——, and W. J. Wiscombe, 2004: An algorithm for generating stochastic cloud fields from radar profile statistics. *Atmos. Res.*, **72**, 263–289, doi:10.1016/j.atmosres.2004.03.016.

Evans, S. M., R. T. Marchand, T. P. Ackerman, and N. Beagley, 2012: Identification and analysis of atmospheric states and associated cloud properties for Darwin, Australia. *J. Geophys. Res.*, **117**, D06204, doi:10.1029/2011JD017010.

Feng, Z., S. A. McFarlane, C. Schumacher, S. Ellis, and N. Bharadwaj, 2014: Constructing a merged cloud-precipitation radar dataset for tropical convective clouds during the DYNAMO/AMIE Experiment at Addu Atoll. *J. Atmos. Oceanic Technol.*, **31**, 1021–1042, doi:10.1175/JTECH-D-13-00132.1.

Fielding, M. D., J. C. Chiu, R. J. Hogan, and G. Feingold, 2014: A novel ensemble method for retrieving properties of warm cloud in 3-D using ground-based scanning radar and zenith radiances. *J. Geophys. Res.*, **119**, 10 912–10 930, doi:10.1002/2014JD021742.

Flynn, C., D. D. Turner, A. Koontz, D. Chand, and C. Sivaraman, 2012: Aerosol Best Estimate (AEROSOLBE) value-added product. Rep. DOE/SC-ARM/TR-115, 29 pp. [Available online at http://www.arm.gov/publications/tech_reports/doe-sc-arm-tr-115.pdf.]

Foster, M. J., and D. E. Veron, 2008: Evaluating the stochastic approach to shortwave radiative transfer in the tropical western Pacific. *J. Geophys. Res.*, **113**, D22205, doi:10.1029/2007JD009581.

Frisch, A. S., C. W. Fairall, and J. B. Snider, 1995: Measurement of stratus cloud and drizzle parameters in ASTEX with a Ka-band Doppler radar and a microwave radiometer. *J. Atmos. Sci.*, **52**, 2788–2799, doi:10.1175/1520-0469(1995)052<2788:MOSCAD>2.0.CO;2.

Fu, Q., B. Carlin, and G. Mace, 2000a: Cirrus horizontal inhomogeneity and OLR bias. *Geophys. Res. Lett.*, **27**, 3341–3344, doi:10.1029/2000GL011944.

——, M. C. Cribb, H. W. Barker, S. K. Krueger, and A. Grossman, 2000b: Cloud geometry effects on atmospheric solar absorption. *J. Atmos. Sci.*, **57**, 1156–1168, doi:10.1175/1520-0469(2000)057<1156:CGEOAS>2.0.CO;2.

Fueglistaler, S., and Q. Fu, 2006: Impact of clouds on radiative heating rates in the tropical lower stratosphere. *J. Geophys. Res.*, **111**, D23202, doi:10.1029/2006JD007273.

Gabriel, P. M., S. C. Tsay, and G. L. Stephens, 1993: A Fourier–Riccati approach to radiative transfer. Part I: Foundation. *J. Atmos. Sci.*, **50**, 3125–3147, doi:10.1175/1520-0469(1993)050<3125:AFATRT>2.0.CO;2.

Garrett, T. J., and C. F. Zhao, 2006: Increased Arctic cloud long-wave emissivity associated with pollution from mid-latitudes. *Nature*, **440**, 787–789, doi:10.1038/nature04636.

Gu, Y., and K. N. Liou, 2006: Cirrus cloud horizontal and vertical inhomogeneity effects in a GCM. *Meteor. Atmos. Phys.*, **91**, 223–235, doi:10.1007/s00703-004-0099-2.

Haeffelin, M., S. Kato, A. M. Smith, C. K. Rutledge, T. P. Charlock, and J. R. Mahan, 2001: Determination of the thermal offset of the Eppley precision spectral pyranometer. *Appl. Opt.*, **40**, 472–484, doi:10.1364/AO.40.000472.

Halthore, R. N., and S. E. Schwartz, 2000: Comparison of model estimated and measured diffuse downward surface irradiance in cloud-free skies. *J. Geophys. Res.*, **105**, 20 165–20 077, doi:10.1029/2000JD900224.

——, ——, J. J. Michalsky, G. P. Anderson, R. A. Ferrare, B. N. Holben, and H. M. ten Brink, 1997: Comparison of model estimated and measured direct-normal solar irradiance. *J. Geophys. Res.*, **102**, 29 991–30 002, doi:10.1029/97JD02628.

Han, D. J., and R. G. Ellingson, 1999: Cumulus cloud formulations for longwave radiation calculations. *J. Atmos. Sci.*, **56**, 837–851, doi:10.1175/1520-0469(1999)056<0837:CCFFLR>2.0.CO;2.

——, and ——, 2000: An experimental technique for testing the validity of cumulus cloud parameterizations for longwave radiation calculations. *J. Appl. Meteor.*, **39**, 1147–1159, doi:10.1175/1520-0450(2000)039<1147:AETFTT>2.0.CO;2.

Hinkelman, L. M., T. P. Ackerman, and R. T. Marchand, 1999: An evaluation of NCEP Eta Model predictions of surface energy budget and cloud properties by comparison to measured ARM data. *J. Geophys. Res.*, **104**, 19 535–19 549, doi:10.1029/1999JD900120.

Iacobellis, S. F., G. M. McFarquhar, D. L. Mitchell, and R. C. J. Somerville, 2003: The sensitivity of radiative fluxes to parameterized cloud microphysics. *J. Climate*, **16**, 2979–2996, doi:10.1175/1520-0442(2003)016<2979:TSORFT>2.0.CO;2.

Jakob, C., G. Tselioudis, and T. Hume, 2005: The radiative, cloud, and thermodynamic properties of the major Tropical Western Pacific cloud regimes. *J. Climate*, **18**, 1203–1215, doi:10.1175/JCLI3326.1.

Jensen, M. P., and A. D. Del Genio, 2003: Radiative and microphysical characteristics of deep convective systems in the tropical western Pacific. *J. Appl. Meteor.*, **42**, 1234–1254, doi:10.1175/1520-0450(2003)042<1234:RAMCOD>2.0.CO;2.

Kassianov, E., 2003: Stochastic radiative transfer in multilayer broken clouds. Part I: Markovian approach. *J. Quant. Spectrosc. Radiat. Transfer*, **77**, 373–393, doi:10.1016/S0022-4073(02)00170-X.

——, and D. Veron, 2011: Stochastic radiative transfer in Markovian mixtures: Past, present, and future. *J. Quant. Spectrosc. Radiat. Transfer*, **112**, 566–576, doi:10.1016/j.jqsrt.2010.06.011.

——, T. Ackerman, R. Marchand, and M. Ovtchinnikov, 2003: Stochastic radiative transfer in multilayer broken clouds. Part II: Validation tests. *J. Quant. Spectrosc. Radiat. Transfer*, **77**, 395–416, doi:10.1016/S0022-4073(02)00171-1.

Kato, S., T. P. Ackerman, and E. E. Clothiaux, 1997: Uncertainties in modeled and measured clear-sky surface shortwave irradiances. *J. Geophys. Res.*, **102**, 25 881–25 898, doi:10.1029/97JD01841.

——, ——, J. H. Mather, and E. E. Clothiaux, 1999: The k-distribution method and correlated-k approximation for a shortwave radiative transfer model. *J. Quant. Spectrosc. Radiat. Transfer*, **62**, 109–121, doi:10.1016/S0022-4073(98)00075-2.

Kennedy, A., X. Dong, B. Xi, S. Xie, Y. Zhang, and J. Chen, 2011: A comparison of MERRA and NARR reanalyses with the DOE ARM SGP data. *J. Climate*, **24**, 4541–4557, doi:10.1175/2011JCLI3978.1.

Killen, R. M., and R. G. Ellingson, 1994: The effect of shape and spatial distribution of cumulus clouds on longwave irradiance. *J. Atmos. Sci.*, **51**, 2123–2136, doi:10.1175/1520-0469(1994)051<2123:TEOSAS>2.0.CO;2.

Lane, D. E., R. C. J. Somerville, and S. F. Iacobellis, 2000: Sensitivity of cloud and radiation parameterizations to changes in vertical resolution. *J. Climate*, **13**, 915–922, doi:10.1175/1520-0442(2000)013<0915:SOCARP>2.0.CO;2.

——, K. Goris, and R. C. J. Somerville, 2002: Radiative transfer trough broken clouds: Observations and model validation. *J. Climate*, **15**, 2921–2933, doi:10.1175/1520-0442(2002)015<2921:RTTBCO>2.0.CO;2.

Lane-Veron, D. E., and R. C. J. Somerville, 2004: Stochastic theory of radiative transfer through generalized cloud fields. *J. Geophys. Res.*, **109**, D18113, doi:10.1029/2004JD004524.

Li, W., C. Shucmacher, and S. A. McFarlane, 2013: Radiative heating of the ISCCP upper level cloud regimes and its impact on the large-scale tropical circulation. *J. Geophys. Res. Atmos.*, **118**, 592–604, doi:10.1002/jgrd.50114.

Li, Z., and A. P. Trishchenko, 2001: Quantifying uncertainties in determining SW cloud radiative forcing and cloud absorption due to variability in atmospheric conditions. *J. Atmos. Sci.*, **58**, 376–389, doi:10.1175/1520-0469(2001)058<0376:QUIDSC>2.0.CO;2.

——, ——, H. W. Barker, G. L. Stephens, and P. Partain, 1999: Analyses of Atmospheric Radiation Measurement (ARM) program's Enhanced Shortwave Experiment (ARESE) multiple data sets for studying cloud absorption. *J. Geophys. Res.*, **104**, 19 127–19 134, doi:10.1029/1999JD900308.

——, M. C. Cribb, and A. P. Trishchenko, 2002: Impact of surface inhomogeneity on solar radiative transfer under overcast conditions. *J. Geophys. Res.*, **107**, D4294, doi:10.1029/2001JD000976.

Long, C. N., and T. P. Ackerman, 2000: Identification of clear skies from broadband pyranometer measurements and calculation of downwelling shortwave cloud effects. *J. Geophys. Res.*, **105**, 15 609–15 626, doi:10.1029/2000JD900077.

——, and D. D. Turner, 2008: A method for continuous estimation of clear-sky downwelling longwave radiative flux developed using ARM surface measurements. *J. Geophys. Res.*, **113**, D18206, doi:10.1029/2008JD009936.

——, T. P. Ackerman, K. L. Gaustad, and J. N. S. Cole, 2006: Estimation of fractional sky cover from broadband shortwave radiometer measurements. *J. Geophys. Res.*, **111**, D11204, doi:10.1029/2005JD006475.

——, J. H. Mather, and T. P. Ackerman, 2016: The ARM Tropical Western Pacific (TWP) sites. *The Atmospheric Radiation Measurement (ARM) Program: The First 20 Years, Meteor. Monogr.*, No. 57, Amer. Meteor. Soc., doi:10.1175/AMSMONOGRAPHS-D-15-0024.1.

Lubin, D., and A. M. Vogelmann, 2006: A climatologically significant aerosol longwave indirect effect in the Arctic. *Nature*, **439**, 453–456, doi:10.1038/nature04449.

——, and ——, 2011: The influence of mixed-phase clouds on surface shortwave irradiance during the Arctic spring. *J. Geophys. Res.*, **116**, D00T05, doi:10.1029/2011JD015761.

Mace, G. G., and S. Benson-Troth, 2002: Cloud-layer overlap characteristics derived from long-term cloud radar data. *J. Climate*, **15**, 2505–2515, doi:10.1175/1520-0442(2002)015<2505:CLOCDF>2.0.CO;2.

——, and S. Benson, 2008: The vertical structure of cloud occurrence and radiative forcing at the SGP ARM site as revealed by 8 years of continuous data. *J. Climate*, **21**, 2591–2610, doi:10.1175/2007JCLI1987.1.

——, and Coauthors, 2006a: Cloud radiative forcing at the Atmospheric Radiation Measurement Program Climate Research Facility: 1. Technique, validation, and comparison to satellite-derived diagnostic quantities. *J. Geophys. Res.*, **111**, D11S90, doi:10.1029/2005JD005921.

——, and Coauthors, 2006b: Cloud radiative forcing at the Atmospheric Radiation Measurement Program Climate Research Facility: 2. Vertical redistribution of radiant energy by clouds. *J. Geophys. Res.*, **111**, D11S91, doi:10.1029/2005JD005922.

Marchand, R., N. Beagley, S. E. Thompson, T. P. Ackerman, and D. E. Schultz, 2006: A bootstrap technique for testing the relationship between local-scale radar observations of cloud occurrence and large-scale atmospheric fields. *J. Atmos. Sci.*, **63**, 2813–2830, doi:10.1175/JAS3772.1.

——, ——, and T. P. Ackerman, 2009: Evaluation of hydrometeor occurrence profiles in the multiscale modeling framework climate model using atmospheric classification. *J. Climate*, **22**, 4557–4573, doi:10.1175/2009JCLI2638.1.

Marshak, A., and A. Davis, Eds., 2005: *3D Radiative Transfer in Cloudy Atmospheres*. Springer, 686 pp.

——, ——, W. Wiscombe, and G. Titov, 1995: The verisimilitude of the independent pixel approximation used in cloud remote sensing. *Remote Sens. Environ.*, **52**, 71–78, doi:10.1016/0034-4257(95)00016-T.

——, ——, ——, and R. Cahalan, 1997: Inhomogeneity effects on cloud shortwave absorption: Two-aircraft simulations. *J. Geophys. Res.*, **102**, 16 619–16 637, doi:10.1029/97JD01153.

——, ——, ——, W. Ridgway, and R. Cahalan, 1998: Biases in shortwave column absorption in the presence of fractal clouds. *J. Climate*, **11**, 431–446, doi:10.1175/1520-0442(1998)011<0431:BISCAI>2.0.CO;2.

——, L. Oreopoulos, and A. B. Davis, 1999a: Horizontal radiative fluxes in clouds and accuracy of the Independent Pixel Approximation at absorbing wavelengths. *Geophys. Res. Lett.*, **26**, 1585–1588, doi:10.1029/1999GL900306.

——, W. J. Wiscombe, A. B. Davis, L. Oreopoulos, and R. F. Cahalan, 1999b: On the removal of the effect of horizontal fluxes in two-aircraft measurements of cloud absorption. *Quart. J. Roy. Meteor. Soc.*, **125**, 2153–2170, doi:10.1002/qj.49712555811.

Mather, J. H., and S. A. McFarlane, 2009: Cloud classes and radiative heating profiles at the Manus and Nauru Atmospheric Radiation Measurement (ARM) sites. *J. Geophys. Res.*, **114**, D19204, doi:10.1029/2009JD011703.

——, ——, M. A. Miller, and K. L. Johnson, 2007: Cloud properties and associated radiative heating rates in the tropical western Pacific. *J. Geophys. Res.*, **112**, D05201, doi:10.1029/2006JD007555.

——, D. D. Turner, and T. P. Ackerman, 2016: Scientific maturation of the ARM Program. *The Atmospheric Radiation Measurement (ARM) Program: The First 20 Years, Meteor. Monogr.*, No. 57, Amer. Meteor. Soc., doi:10.1175/AMSMONOGRAPHS-D-15-0053.1.

McComisky, A., and R. A. Ferrare, 2016: Aerosol physical and optical properties and processes in the ARM Program. *The Atmospheric Radiation Measurement (ARM) Program: The First 20 Years, Meteor. Monogr.*, No. 57, Amer. Meteor. Soc., doi:10.1175/AMSMONOGRAPHS-D-15-0028.1.

McFarlane, S. A., and K. F. Evans, 2004: Clouds and shortwave fluxes at Nauru. Part II: Shortwave flux closure. *J. Atmos. Sci.*, **61**, 2602–2615, doi:10.1175/JAS3299.1.

——, J. H. Mather, and T. P. Ackerman, 2007: Analysis of tropical radiative heating profiles: A comparison of models and observations. *J. Geophys. Res.*, **112**, D14218, doi:10.1029/2006JD008290.

——, ——, ——, and Z. Liu, 2008: Effect of clouds on calculated shortwave column absorption in the Tropics. *J. Geophys. Res.*, **113**, D18203, doi:10.1029/2008JD009791.

——, E. I. Kassianov, J. Barnard, C. Flynn, and T. Ackerman, 2009: Surface shortwave aerosol radiative forcing during the ARM Mobile Facility deployment in Niamey, Niger. *J. Geophys. Res.*, **114**, D00E06, doi:10.1029/2008JD010491.

——, T. Shippert, and J. Mather, 2011: Radiatively Important Parameters Best Estimate (RIPBE): An ARM value-added product. Rep. DOE/SC-ARM/TR-097, 27 pp. [Available online at http://www.arm.gov/publications/tech_reports/doe-sc-arm-tr-097.pdf.]

——, C. N. Long, and J. Flaherty, 2013: A climatology of cloud radiative effects at the ARM Tropical Western Pacific sites. *J. Appl. Meteor. Climatol.*, **52**, 996–1013, doi:10.1175/JAMC-D-12-0189.1.

Mechem, D. B., Y. L. Kogan, M. Ovtchinnikov, A. B. Davis, K. F. Evans, and R. G. Ellingson, 2008: Multidimensional longwave forcing of boundary layer cloud systems. *J. Atmos. Sci.*, **65**, 3963–3977, doi:10.1175/2008JAS2733.1.

Michalsky, J. J., and C. N. Long, 2016: ARM solar and infrared broadband and filter radiometry. *The Atmospheric Radiation Measurement (ARM) Program: The First 20 Years, Meteor. Monogr.*, No. 57, Amer. Meteor. Soc., doi:10.1175/AMSMONOGRAPHS-D-15-0031.1.

——, and Coauthors, 2002: Broadband shortwave calibration results from Atmospheric Radiation Measurements Enhanced Shortwave Experiment II. *J. Geophys. Res.*, **107**, 4307, doi:10.1029/2001JD001231.

——, and Coauthors, 2006: Shortwave radiative closure studies for clear skies during the Atmospheric Radiation Measurement 2003 aerosol intensive observation period. *J. Geophys. Res.*, **111**, D14S90, doi:10.1029/2005JD006341.

Miller, M. A., and A. Slingo, 2007: The Atmospheric Radiation Measurement (ARM) Mobile Facility (AMF) and its first international deployment: Measuring radiative flux divergence in West Africa. *Bull. Amer. Meteor. Soc.*, **88**, 1229–1244, doi:10.1175/BAMS-88-8-1229.

——, V. P. Ghate, and R. K. Zahn, 2012: The radiation budget of the West African Sahel and its controls: A perspective from observations and global climate models. *J. Climate*, **25**, 5976–5996, doi:10.1175/JCLI-D-11-00072.1.

Miller, R. L., A. Slingo, J. C. Barnard, and E. Kassianov, 2009: Seasonal contrast in the surface energy balance of the Sahel. *J. Geophys. Res.*, **114**, D00E05, doi:10.1029/2008JD010521.

Minnis, P., L. Nguyen, D. R. Doelling, D. F. Young, W. F. Miller, and D. P. Kratz, 2002: Rapid calibration of operational and research meteorological satellite imagers. Part I: Evaluation of research satellite visible channels as references. *J. Atmos. Oceanic Technol.*, **19**, 1233–1249, doi:10.1175/1520-0426(2002)019<1233:RCOOAR>2.0.CO;2.

Mlawer, E. J., and D. D. Turner, 2016: Spectral radiation measurements and analysis in the ARM Program. *The Atmospheric Radiation Measurement (ARM) Program: The First 20 Years, Meteor. Monogr.*, No. 57, Amer. Meteor. Soc., doi:10.1175/AMSMONOGRAPHS-D-15-0027.1.

——, S. J. Taubman, P. D. Brown, M. J. Iacono, and S. A. Clough, 1997: Radiative transfer for inhomogeneous atmospheres: RRTM, a validated correlated-k model for the longwave. *J. Geophys. Res.*, **102**, 16 663–16 682, doi:10.1029/97JD00237.

——, and Coauthors, 2002: The broadband heating rate profile (BBHRP) VAP. *Proc. 12th Atmospheric Radiation Measurement (ARM) Science Team Meeting*, St. Petersburg, FL, U.S. Department of Energy, 12 pp. [Available online at https://www.arm.gov/publications/proceedings/conf12/extended_abs/mlawer-ej.pdf.]

——, and Coauthors, 2003: Recent developments on the broadband heating rate profile value-added product. *Proc. 13th Atmospheric Radiation Measurement (ARM) Science Team Meeting*, Broomfield, CO, U.S. Department of Energy, 16 pp. [Available online at https://www.arm.gov/publications/proceedings/conf13/extended_abs/mlawer-ej.pdf.]

——, and Coauthors, 2006: Status of the Broadband Heating Rate Profile (BBHRP) VAP. *Proc. 16th Atmospheric Radiation Measurement (ARM) Science Team Meeting*, Albuquerque, NM, U.S. Department of Energy. [Available online at https://www.arm.gov/publications/proceedings/conf14/poster_abs/P00181.]

——, and Coauthors, 2007: Advances in the Broadband Heating Rate Profile (BBHRP) VAP. *Proc. 17th Atmospheric Radiation Measurement (ARM) Science Team Meeting*, Monterey, CA, U.S. Department of Energy. [Available online at https://www.arm.gov/publications/proceedings/conf17/poster/P00075.pdf.]

——, and Coauthors, 2008: Evaluating cloud retrieval algorithms within the ARM BBHRP framework. *Proc. 18th Atmospheric Radiation Measurement (ARM) Science Team Meeting*, Norfolk, VA, U.S. Department of Energy. [Available online at https://www.arm.gov/publications/proceedings/conf18/display?id=NTE5.]

——, M. Iacono, R. Pincus, H. Barker, L. Oreopoulos, and D. Mitchell, 2016: Contributions of the ARM Program to radiative transfer modeling for climate and weather applications. *The Atmospheric Radiation Measurement (ARM) Program: The First 20 Years, Meteor. Monogr.*, No. 57, Amer. Meteor. Soc., doi:10.1175/AMSMONOGRAPHS-D-15-0041.1.

Morcrette, J.-J., 2002: Assessment of the ECMWF model cloudiness and surface radiation fields at the ARM SGP site. *Mon. Wea. Rev.*, **130**, 257–277, doi:10.1175/1520-0493(2002)130<0257:AOTEMC>2.0.CO;2.

Mülmenstädt, J., D. Lubin, L. M. Russell, and A. M. Vogelmann, 2012: Cloud properties over the North Slope of Alaska: Identifying the prevailing meteorological regimes. *J. Climate*, **25**, 8238–8258, doi:10.1175/JCLI-D-11-00636.1.

O'Hirok, W., and C. Gautier, 1998: A three-dimensional radiative transfer model to investigate the solar radiation within a cloudy atmosphere. Part I: Spatial effects. *J. Atmos. Sci.*, **55**, 2162–2179, doi:10.1175/1520-0469(1998)055<2162:ATDRTM>2.0.CO;2.

——, and ——, 2003: Absorption of shortwave radiation in a cloudy atmosphere: Observed and theoretical estimates during the second Atmospheric Radiation Measurement Enhanced Shortwave Experiment (ARESE). *J. Geophys. Res.*, **108**, 4412, doi:10.1029/2002JD002818.

——, ——, and P. Ricchiazzi, 2000: Spectral signature of column solar radiation absorption during the Atmospheric Radiation Measurement Enhanced Shortwave Experiment (ARESE). *J. Geophys. Res.*, **105**, 17 471–17 480, doi:10.1029/2000JD900190.

Oreopoulos, L., and E. Mlawer, 2010: The Continual Intercomparison of Radiation Codes (CIRC). *Bull. Amer. Meteor. Soc.*, **91**, 305–310, doi:10.1175/2009BAMS2732.1.

Pilewskie, P., and F. P. J. Valero, 1995: Direct observations of excess solar absorption by clouds. *Science*, **267**, 1626–1629, doi:10.1126/science.267.5204.1626.

Pincus, R., and K. F. Evans, 2009: Computational cost and accuracy in calculating three-dimensional radiative transfer: Results for new implementations of Monte Carlo and SHDOM. *J. Atmos. Sci.*, **66**, 3131–3146, doi:10.1175/2009JAS3137.1.

——, H. W. Barker, and J.-J. Morcrette, 2003: A fast, flexible, approximate technique for computing radiative transfer in inhomogeneous cloud fields. *J. Geophys. Res.*, **108**, 4376, doi:10.1029/2002JD003322.

——, C. Hannay, and K. F. Evans, 2005: The accuracy of determining three-dimensional radiative transfer effects in cumulus clouds using ground-based profiling instruments. *J. Atmos. Sci.*, **62**, 2284–2293, doi:10.1175/JAS3464.1.

Pinto, J. O., J. A. Curry, A. H. Lynch, and P. O. G. Persson, 1999: Modeling clouds and radiation for the November 1997 period of SHEBA using a column climate model. *J. Geophys. Res.*, **104**, 6661–6678, doi:10.1029/98JD02517.

Powell, S. W., R. A. Houze Jr., A. Kumar, and S. A. McFarlane, 2012: Comparison of simulated and observed continental tropical anvil clouds and their radiative heating profiles. *J. Atmos. Sci.*, **69**, 2662–2681, doi:10.1175/JAS-D-11-0251.1.

Prigarin, S. M., and A. Marshak, 2009: A simple stochastic model for generating broken cloud optical depth and cloud-top height fields. *J. Atmos. Sci.*, **66**, 92–104, doi:10.1175/2008JAS2699.1.

Protat, A., and Coauthors, 2014: Reconciling ground-based and space-based estimates of the frequency of occurrence and radiative effect of clouds around Darwin, Australia. *J. Appl. Meteor. Climatol.*, **53**, 456–478, doi:10.1175/JAMC-D-13-072.1.

Qian, Y., C. Long, H. Wang, J. Comstock, S. McFarlane, and S. Xie, 2012: Evaluation of cloud fraction and its radiative effect simulated by IPCC AR4 global models against ARM surface observations. *Atmos. Chem. Phys.*, **12**, 1785–1810, doi:10.5194/acp-12-1785-2012.

Ramanathan, V., B. Subasilar, G. J. Zhang, W. Conant, R. D. Cess, J. T. Kiehl, H. Grassi, and L. Shi, 1995: Warm pool heat budget and shortwave cloud forcing: A missing physics? *Science*, **267**, 499–503, doi:10.1126/science.267.5197.499.

Ricchiazzi, P., S. R. Yang, C. Gautier, and D. Sowle, 1998: SBDART: A research and teaching software tool for plane-parallel radiative transfer in the Earth's atmosphere. *Bull. Amer. Meteor. Soc.*, **79**, 2101–2114, doi:10.1175/1520-0477(1998)079<2101:SARATS>2.0.CO;2.

Rutan, D., F. Rose, T. Charlock, E. Mlawer, T. Shippert, and S. Kato, 2006: Broadband heating rate product flux profiles compared to Clouds and the Earth's Radiant Energy System radiation transfer product. *Proc. 16th Atmospheric Radiation Measurement (ARM) Science Team Meeting*, Albuquerque, NM, U.S. Department of Energy, 11 pp. [Available online at https://www.arm.gov/publications/proceedings/conf16/extended_abs/rutan_d.pdf.]

Settle, J. J., N. A. Bharmal, G. J. Robinson, and A. Slingo, 2008: Sampling uncertainties in surface radiation budget calculations in RADAGAST. *J. Geophys. Res.*, **113**, D00E02, doi:10.1029/2008JD010509.

Shupe, M. D., and J. M. Intrieri, 2004: Cloud radiative forcing of the Arctic surface: The influence of cloud properties, surface albedo, and solar zenith angle. *J. Climate*, **17**, 616–628, doi:10.1175/1520-0442(2004)017<0616:CRFOTA>2.0.CO;2.

——, D. D. Turner, A. B. Zwink, M. M. Thieman, E. J. Mlawer, and T. R. Shippert, 2015: Deriving Arctic cloud microphysics at Barrow, Alaska: Algorithms, results, and radiative closure. *J. Appl. Meteor. Climatol.*, **54**, 1675–1689, doi:10.1175/JAMC-D-15-0054.1.

——, J. M. Comstock, D. D. Turner, and G. G. Mace, 2016: Cloud property retrievals in the ARM Program. *The Atmospheric Radiation Measurement (ARM) Program: The First 20 Years, Meteor. Monogr.*, No. 57, Amer. Meteor. Soc., doi:10.1175/AMSMONOGRAPHS-D-15-0030.1.

Slingo, A., and Coauthors, 2006: Observations of the impact of a major Saharan dust storm on the Earth's radiation balance. *Geophys. Res. Lett.*, **33**, L24817, doi:10.1029/2006GL027869.

——, H. E. White, N. A. Bharmal, and G. J. Robinson, 2009: Overview of observations from the RADAGAST experiment in Niamey, Niger: 2. Radiative fluxes and divergences. *J. Geophys. Res.*, **114**, D00E04, doi:10.1029/2008JD010497.

Somerville, R. C. J., and S. F. Iacobellis, 1999: Single-column models, ARM observations, and GCM cloud-radiation schemes. *Phys. Chem. Earth*, **24B**, 733–740, doi:10.1016/S1464-1909(99)00074-X.

Stamnes, K., R. G. Ellingson, J. A. Curry, J. E. Walsh, and B. D. Zak, 1999: Review of science issues, deployment strategy, and status for the ARM North Slope of Alaska–Adjacent Arctic Ocean Climate Research Site. *J. Climate*, **12**, 46–63, doi:10.1175/1520-0442-12.1.46.

Stephens, G. L., 2005: Cloud feedbacks in the climate system: A critical review. *J. Climate*, **18**, 237–273, doi:10.1175/JCLI-3243.1.

——, N. B. Wood, and P. M. Gabriel, 2004: An assessment of the parameterization of subgrid-scale cloud effects on radiative transfer. Part I: Vertical overlap. *J. Atmos. Sci.*, **61**, 715–732, doi:10.1175/1520-0469(2004)061<0715:AAOTPO>2.0.CO;2.

Stokes, G. M., 2016: Original ARM concept and launch. *The Atmospheric Radiation Measurement (ARM) Program: The First 20 Years, Meteor. Monogr.*, No. 57, Amer. Meteor. Soc., doi:10.1175/AMSMONOGRAPHS-D-15-0021.1.

Takara, E. E., and R. G. Ellingson, 1996: Scattering effects on longwave fluxes in broken cloud fields. *J. Atmos. Sci.*, **53**, 1464–1476, doi:10.1175/1520-0469(1996)053<1464:SEOLFI>2.0.CO;2.

——, and ——, 2000: Broken cloud field longwave-scattering effects. *J. Atmos. Sci.*, **57**, 1298–1310, doi:10.1175/1520-0469(2000)057<1298:BCFLSE>2.0.CO;2.

Taylor, P. C., and R. G. Ellingson, 2008: A study of the probability of clear line of sight through single-layer cumulus cloud fields in the Tropical Western Pacific. *J. Atmos. Sci.*, **65**, 3497–3512, doi:10.1175/2008JAS2620.1.

Thorsen, T. J., Q. Fu, J. M. Comstock, C. Sivaraman, M. A. Vaughan, D. Winker, and D. D. Turner, 2013: Macrophysical properties of tropical cirrus clouds from the CALIPSO satellite and from ground-based micropulse and Raman lidars. *J. Geophys. Res. Atmos.*, **118**, 9209–9220, doi:10.1002/jgrd.50691.

Titov, G. A., 1998: Radiative horizontal transport and absorption in stratocumulus clouds. *J. Atmos. Sci.*, **55**, 2549–2560, doi:10.1175/1520-0469(1998)055<2549:RHTAAI>2.0.CO;2.

Trenberth, K. E., J. T. Fasullo, and J. Kiehl, 2009: Earth's global energy budget. *Bull. Amer. Meteor. Soc.*, **90**, 311–323, doi:10.1175/2008BAMS2634.1.

Troyan, D., 2010: Merged sounding value-added product. Rep. DOE/SC-ARM/TR-087, 19 pp. [Available online at http://www.arm.gov/publications/tech_reports/doe-sc-arm-tr-087.pdf.]

Turner, D. D., 2008: Ground-based infrared retrievals of optical depth, effective radius, and composition of airborne mineral dust above the Sahel. *J. Geophys. Res.*, **113**, D00E03, doi:10.1029/2008JD010054.

——, and Coauthors, 2004: The QME AERI LBLRTM: A closure experiment for downwelling high spectral resolution infrared radiance. *J. Atmos. Sci.*, **61**, 2657–2675, doi:10.1175/JAS3300.1.

Valero, F. P. J., A. Bucholtz, B. Bush, S. Pope, W. Collins, P. Flatau, A. Strawa, and W. Gore, 1997a: The Atmospheric Radiation Measurements Enhanced Shortwave Experiment (ARESE): Experimental and data details. *J. Geophys. Res.*, **102**, 29 929–29 327, doi:10.1029/97JD02434.

——, R. D. Cess, M. H. Zhang, S. K. Pope, A. Bucholtz, B. Bush, and J. Vitko, 1997b: Absorption of solar radiation by the cloudy atmosphere: Interpretations of collocated aircraft measurements. *J. Geophys. Res.*, **102**, 29 917–29 927, doi:10.1029/97JD01782.

Várnai, T., 2010: Multiyear statistics of 2D shortwave radiative effects at three ARM sites. *J. Atmos. Sci.*, **67**, 3757–3762, doi:10.1175/2010JAS3506.1.

Verlinde, J., B. Zak, M. D. Shupe, M. Ivey, and K. Stamnes, 2016: The ARM North Slope of Alaska (NSA) sites. *The Atmospheric Radiation Measurement (ARM) Program: The First 20 Years, Meteor. Monogr.*, No. 57, Amer. Meteor. Soc., doi:10.1175/AMSMONOGRAPHS-D-15-0023.1.

Vogelmann, A. M., V. Ramanathan, and I. A. Podgorny, 2001: Scale dependence of solar heating rates in convective cloud systems with implications to general circulation models. *J. Climate*, **14**, 1738–1752, doi:10.1175/1520-0442(2001)014<1738:SDOSHR>2.0.CO;2.

Webb, M. J., C. Senior, S. Bony, and J.-J. Morcrette, 2001: Combining ERBE and ISCCP data to assess clouds in the Hadley Centre, ECMWF, and LMD atmospheric climate models. *Climate Dyn.*, **17**, 905–922, doi:10.1007/s003820100157.

Wood, N. B., P. M. Gabriel, and G. L. Stephens, 2005: An assessment of the parameterization of subgrid-scale cloud effects on radiative transfer. Part II: Horizontal inhomogeneity. *J. Atmos. Sci.*, **62**, 2895–2909, doi:10.1175/JAS3498.1.

Xie, S., R. T. Cederwall, M. Zhang, and J. J. Yio, 2003: Comparison of SCM and CSRM forcing data derived from the ECMWF model and from objective analysis at the ARM SGP site. *J. Geophys. Res.*, **108**, 4499, doi:10.1029/2003JD003541.

Xu, K.-M., and D. A. Randall, 1995: Impact of interactive radiative transfer on the macroscopic behavior of cumulus ensembles. Part II: Mechanisms for cloud-radiation interactions. *J. Atmos. Sci.*, **52**, 800–817, doi:10.1175/1520-0469(1995)052<0800:IOIRTO>2.0.CO;2.

Yang, F., H.-L. Pan, S. K. Krueger, S. Moorthi, and S. J. Lord, 2006: Evaluation of the NCEP Global Forecast System at the ARM SGP Site. *Mon. Wea. Rev.*, **134**, 3668–3690, doi:10.1175/MWR3264.1.

Younkin, K., and C. N. Long, 2003: Improved correction of IR loss in diffuse shortwave measurements: An ARM value-added product. Rep. DOE/SC-ARM/TR-009, 50 pp. [Available online at http://www.arm.gov/publications/tech_reports/arm-tr-009.pdf.]

Zender, C. S., B. Bush, S. K. Pope, A. Bucholtz, W. D. Collins, J. T. Kiehl, F. P. J. Valero, and J. Vitko, 1997: Atmospheric absorption during the Atmospheric Radiation Measurement (ARM) Enhanced Shortwave Experiment (ARESE). *J. Geophys. Res.*, **102**, 29 901–29 915, doi:10.1029/97JD01781.

Zhang, M., R. D. Cess, and X. Jing, 1997: Concerning the interpretation of enhanced cloud shortwave absorption using monthly-mean Earth Radiation Budget Experiment/Global Energy Balance Archive Measurements. *J. Geophys. Res.*, **102**, 25 899–25 906, doi:10.1029/97JD02196.

Zwink, A. B., 2013: The radiative flux divergence of different Arctic cloud types. Master's Thesis, School of Meteorology, University of Oklahoma, 111 pp.

Chapter 21

Aerosol Physical and Optical Properties and Processes in the ARM Program

ALLISON MCCOMISKEY

Cooperative Institute for Research in Environmental Sciences, University of Colorado Boulder, and NOAA Earth System Research Laboratory, Boulder, Colorado

RICHARD A. FERRARE

NASA Langley Research Center, Hampton, Virginia

1. Introduction

At the inception of the Atmospheric Radiation Measurement (ARM) Program, anthropogenic aerosols were becoming widely recognized for their influence on radiation transfer in the atmosphere and their role in climate change. The seminal article by Charlson et al. (1992) estimated the global mean aerosol radiative forcing by anthropogenic sulfate to be $-2\,\mathrm{W\,m^{-2}}$. At the same time, Penner et al. (1992) estimated a similar forcing by biomass burning aerosol under the assumption that it is a primarily light-scattering aerosol. Despite the recognition that there would be an attendant effect from absorbing aerosol, no estimate could be made for lack of knowledge of their mass, optical properties, and distribution. While containing substantial uncertainty, these estimates of cooling by aerosol are of a magnitude that could offset the warming caused by increases in greenhouse gases over the industrial period.

The tenet of the ARM Program at the outset was to improve the representation of properties and processes impacting atmospheric radiation in climate models, with an emphasis on clouds and water vapor. The prevailing approach was a hierarchical method of examination of processes that rested at its base on experimental verification in the form of radiative closure at the surface. The structure of ARM measurements was designed to implement this process evaluation and radiative closure. Historically, the inclusion of a uniform background tropospheric aerosol was considered adequate for closure studies. However, improved knowledge of source strengths of anthropogenic aerosols came with the realization that aerosol loading and optical properties are highly variable both seasonally and geographically. It became evident that inclusion of the local aerosol was required for closure studies and for the characterization of regional and global impacts necessary for improved understanding of climate change. Thus, it became necessary for ARM to address more realistic aerosol contributions to atmospheric radiation transfer.

With a complexity rivaling that of clouds, the characterization of local- to regional-scale aerosol properties would be a challenge to ARM and the larger community, requiring a sound scientific plan and experimental resources. Calling on those at the forefront of aerosol research, ARM formed the Aerosol Working Group (AWG) to develop research questions and recommend a suite of measurements that would allow for the characterization of relevant aerosol properties. The scope of this effort was defined in terms of the pertinent processes, properties, and measurable quantities for aerosol radiative forcing outlined in Charlson et al. (1992) and with regard to the major objectives of ARM (Fig. 21-1).

Several research objectives were developed to guide measurement needs, which encompassed aerosol–radiation interactions in both clear and cloudy skies. These included understanding relationships among cloud condensation nuclei concentrations, cloud droplet concentrations, and droplet size; defining relationships between aerosol chemical and physical properties (size

Corresponding author address: Allison McComiskey, NOAA Earth System Research Laboratory, Global Monitoring Division, 325 Broadway, Boulder, CO 80305.
E-mail: allison.mccomiskey@noaa.gov

DOI: 10.1175/AMSMONOGRAPHS-D-15-0028.1

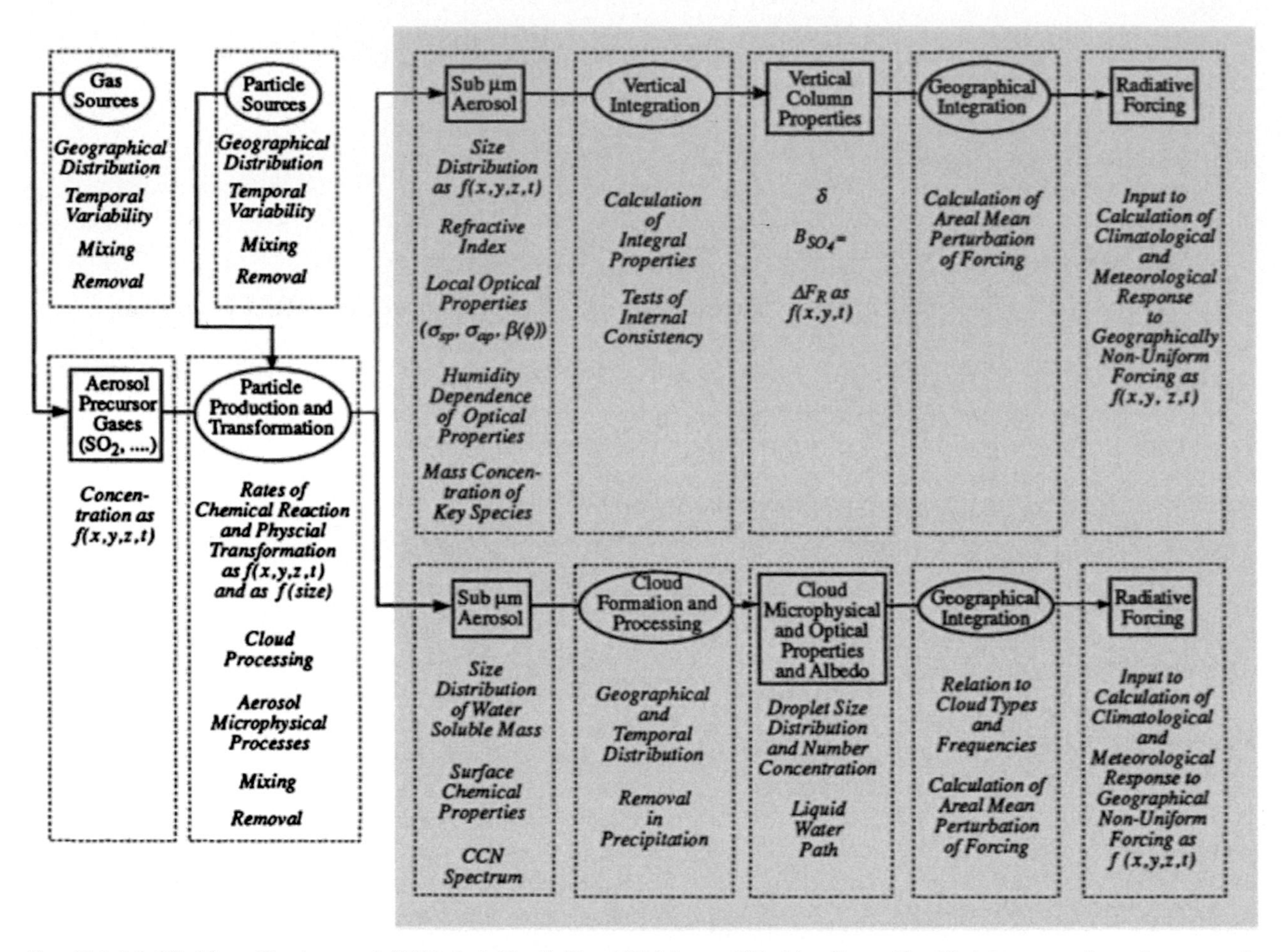

FIG. 21-1. Modified from Charlson et al. (1992, their Fig. 4). The ARM Aerosol Working Group identified the properties and processes in the gray box as related to the larger scope of the ARM Program, which then suggested the priority measurements to be made.

distribution, composition, morphology) and optical properties; defining relationships of aerosol optical and microphysical properties to radiative properties; providing parameterizations for climate models; and defining uncertainties and drivers of variability in these properties and processes and their importance for atmospheric radiation. These objectives were reflected in the suite of recommended measurements that would allow for characterization of instantaneous and temporally averaged aerosol properties within the full column for determination of radiative impacts on a range of scales.

The first ARM measurements were made at the Southern Great Plains (SGP) site in 1992 but, despite the best laid schemes of the AWG, no systematic measurements of aerosols were made until 1996. Given the clear need for inclusion of aerosols in closing the radiation budget and the obvious importance of aerosols on the climate it may seem surprising that ARM did not invest in these measurements at the outset. At this time, focus was placed on the measurement of clouds and clear-sky radiative processes. Understanding cloud–radiation processes was such a challenging effort that, with only so many resources available to the program, tension developed between the clouds and water vapor researchers and those interested in aerosol impacts. The latter—who soon became known as "the dirt guys"—was a group who could potentially absorb valuable program resources. However, in time, the plans from the 1992 AWG became reality and the first Aerosol Observing System (AOS) was built for deployment at the SGP in 1996.

Over the past two decades ARM has provided prodigious support for aerosol measurements and theoretical research contributing to understanding the role of aerosols in climate change in various regimes. While building climatologies was not the original intent of the program, these data are invaluable for characterizing trends and variability in aerosol properties and their radiative forcing. The ARM investment in this discipline continues to grow in step with the desire to address the increasing complexity as climate models become more

TABLE 21-1. Fundamental aerosol properties, measurements, and recommended instrumentation determined by the ARM Aerosol Working Group 1992 for improved understanding of aerosol radiative forcing in clear and cloudy skies. Some measurements were recommended for continuous long-term sampling at the ARM fixed sites while other more complex measurements were recommended for intensive periods only.

	Property	Measurement	Instrument
Local microscale	Size distributed chemical composition	Particle composition in two size classes (0.1–1 μm; 1–10 μm)	Filter samplers; impactor
		Total mass; major ions; elemental composition, organic and elemental carbon	Electrobalance; ion chromatography, particle-induced X-ray emission spectrometry (PIXE), volatilization/oxidation, CO_2 or CH_4 analysis
	Size distributed number concentration	Size distribution 0.01–5 μm (coarse resolution); size distribution 0.001–10 μm (high resolution)	Optical particle counter (OPC); CN counter (CPC) Differential mobility analyzer (DMA), aerodynamic particle sizer (APS)
	Size distributed number concentration as a function of relative humidity	Relative humidity profile Hygroscopic growth	Tandem differential mobility analyzer (TDMA); humidity-controlled nephelometer
	Size distributed cloud drop number concentration		
Local integrated	Scattering coefficient	3 wavelengths 450–850 nm; low reference humidity; as a function of scattering angle	Integrating nephelometer; polar nephelometer
	Backscattering coefficient	3 wavelengths 450–850 nm; low reference humidity	Integrating nephelometer with backscatter shutter
	Absorption coefficient		Filter deposit light attenuation
	Scattering coefficient as a function of relative humidity		
	Size distributed cloud condensation nucleus as a function of supersaturation	CCN at single supersaturation CCN spectrum	Thermal diffusion cloud chamber
Column	Optical depth	Multispectral optical depth	Radiometer
		Aerosol vertical distribution	Lidar
	Absorption optical depth		
	Backscattering optical depth		
	Cloud optical depth	Optical depth at several wavelengths	Sun photometer, shadowband radiometer
		Cloud-base height	Ceilometer
	Cloud albedo		Narrow and broadband radiometers

detailed and we strive to represent climate processes with greater accuracy. This chapter outlines the history of these achievements.

2. Measurements and value-added products

The AWG presented to ARM a list of necessary properties to be measured, associated observational characteristics, and recommended instrumentation to meet the stated research objectives (Penner et al. 1992). These are summarized in Table 21-1. Properties were organized by local integrated, size distributed, and column properties and addressed both clear-sky and cloudy-sky conditions. They included measurements to be made continuously at the fixed ground sites as well as more complex measurements for airborne intensive observation periods (IOPs). The column properties were largely already being addressed by the program in relation to meeting existing objectives through ground-based remote sensing by lidar and shortwave radiometry (Michalsky and Long 2016, chapter 16) and the surface properties would provide further characterization of aerosol properties that cannot be easily gained from remote sensing.

FIG. 21-2. (left) The original Aerosol Observing System (AOS) at the SGP in 1996. (right) Updated AOS racks at the SGP. Courtesy of John A. Ogren.

While the program has invested in developing and improving retrievals of pertinent properties from remote sensing at the surface, airborne in situ measurements in campaign mode are still critical for understanding aerosol properties and processes. Primary properties that are measured routinely at the surface are aerosol light scattering, backscattering, absorption, number concentration, hygroscopicity, size distribution, and chemical composition. Some properties can only be understood in detail through in situ measurements; however, the need for improved remote sensing capabilities is recognized in order to extrapolate knowledge of the finescale nature of aerosol to their regional- to global-scale radiative impacts. Aerosol optical depth at several wavelengths and backscatter and depolarization profiles from active remote sensing are measured routinely at all sites; at the SGP and Tropical Western Pacific (TWP) Darwin sites aerosol extinction profiles also are measured using the Raman technique. Through the years, as discussed below, the aerosol program has focused increasingly on the synergy between in situ and remote sensing techniques for better characterization of aerosols throughout the atmospheric column and representations in models.

a. The Aerosol Observing System

To address a subset of the measurement needs defined by the AWG, an integrated system of instruments for measuring aerosol microphysical and optical properties in situ at the surface was developed by the U.S. Department of Energy (DOE) Environmental Measurement Program (Leifer et al. 1993). The instruments were housed in an exclusive, self-contained shelter with an intake stack that sampled at 10 m above the ground to avoid surface turbidity. A common manifold supplied all instruments with the same sample air. The system included five instruments: a condensation particle counter (CPC), optical particle counter (OPC), single wavelength nephelometer (550 nm), particle soot absorption photometer (PSAP; 565 nm), and three-wavelength nephelometer with backscatter shutter (450, 550, and 700 nm). The system design provided space for expansion in the future. A 10-μm impactor limited the aerosol size sampled by each of the instruments exclusive of the OPC. The first AOS was deployed at the SGP in 1996 and has been making measurements continuously to the present day (Fig. 21-2).

Operation and development of the SGP AOS was assumed by the ARM-funded instrument mentors at the National Oceanic and Atmospheric Administration (NOAA) Climate Monitoring and Diagnostics Laboratory (now the NOAA Earth System Research Laboratory Global Monitoring Division), and within a year several upgrades were initiated. A switched impactor system was installed in 1997 to sample at two size cuts

(<1 μm and <10 μm) (Sheridan et al. 2001). Sample air is gently heated when necessary to provide measurements at a "dry" (<40%) relative humidity (RH), avoiding evaporation of volatile species while maintaining a reference state that is comparable to similar measurements made in any conditions at other times or locations. A second three-wavelength nephelometer was added in 1998 with an associated humidograph system to measure the change in aerosol scattering with water uptake. The system incrementally increases the RH of the sample over a 30-min period from 40% to ~85% RH. In 2000, monitoring of aerosol chemical composition by filter sample and ion chromatography began and continued until 2008 (Quinn et al. 2002). In 2005 a humidified tandem differential mobility analyzer (HTDMA), designed and deployed by Texas A&M (Santarpia et al. 2004), was installed in the AOS shelter for size distributions and hygroscopic growth. The PSAP was upgraded from a one- to a three-wavelength instrument and in 2006 a cloud condensation nucleus counter (CCNC) was added (Roberts and Nenes 2005).

At the time the AOS was established at the SGP, NOAA CMDL had been operating an aerosol measurement system at the NOAA Barrow, Alaska, observatory since 1976. In 1997, joint contributions from NOAA and DOE ARM were made to upgrade, maintain, and operate this site as the NOAA Barrow/ARM North Slope of Alaska (NSA) AOS system. Recreating the design and instrumentation from the SGP has allowed quantitative comparison of aerosol properties in these two distinct aerosol regimes. The SGP and NSA data records, both longer than 15 years now, constitute the two long-term records of aerosol properties within DOE. These are valuable for their insight into seasonal variability and long-term trends in distinct regimes and for deeper probing into processes that dictate aerosol optical properties and radiative transfer.

The first new fixed-site AOS for long-term monitoring in 15 years has now been established in a remote marine environment on Graciosa Island in the Azores. This system, mentored by the DOE Brookhaven National Laboratory, will mimic previous AOS but with expanded capabilities for measuring aerosol-size distribution, chemical composition, and gases. The Aerosol Chemical Speciation Monitor (ACSM; Ng et al. 2011) is being deployed for understanding episodic influences of long-range transport of dust and urban/industrial aerosol over the northern Atlantic. The site has a scientific focus on aerosol–cloud interactions, and researchers will take advantage of the synergy of surface measurements with remote sensors to understand the relationships of aerosol at the surface, cloud base, and within cloud.

b. In situ aerosol profiles

Knowledge of the extent to which aerosol sampled at the surface is representative of properties in the full column above is critical if these long-term surface observations are to be used to determine aerosol radiative forcing. Long-term, continuous observations are generally relegated to the surface where monitoring is logistically feasible. As previously stated, this approach to monitoring is indispensable for characterizing geographical and seasonal or shorter-term variability in aerosol as well as long-term trends. However, full column properties that dictate radiative fluxes at the surface and top of the atmosphere are often not well correlated to properties at the surface (Andrews et al. 2004). Intensive observations are typically made over short durations providing highly comprehensive characterization of aerosol properties in the column but do not resolve geographical and temporal variability. In 1998 the AWG met in Oak Ridge, Tennessee, and originated a concept for continuous yet cost-effective monitoring of the vertical profile of aerosol properties in clear skies. Robust statistical characterization of the vertical profile of aerosol properties comparable to those measured at the surface would be performed over the SGP for several years.

The long-term campaign called the In Situ Aerosol Profiles (IAP; Andrews et al. 2004, 2011) experiment flew a Cessna 172 two to three times a week over the SGP ground site beginning in 2000. Observations were made from an airborne version of AOS, with <1-μm light scattering and backscattering at three wavelengths and light absorption at one wavelength at <40% RH. During each flight, level legs were flown at nine altitudes (465, 610, 915, 1220, 1525, 1830, 2440, 3050, and 3660 m) above mean sea level during daylight hours and in clear-sky conditions. In 2005, an upgrade to a Cessna 206 Turbo was made and the instrument package was updated to measure light absorption at three wavelengths (467, 530, and 660 nm) plus dry scattering at <5 μm. The number of flight levels increased to 12 and the maximum altitude increased to ~4575 m. Figure 21-3 presents the measured light scattering contoured over the full IAP deployment showing the distinct annual cycle but with variability within this cycle among years. The IAP was a first, showing that long-term, routine aircraft sampling could be accomplished in a cost-effective and efficient manner and characterized the vertical structure of aerosol optical properties with greater than seven years of regular data at SGP.

c. ARM Mobile Facility

Aerosol chemical and microphysical properties, and thus radiative forcing, exhibit a geographical dependence

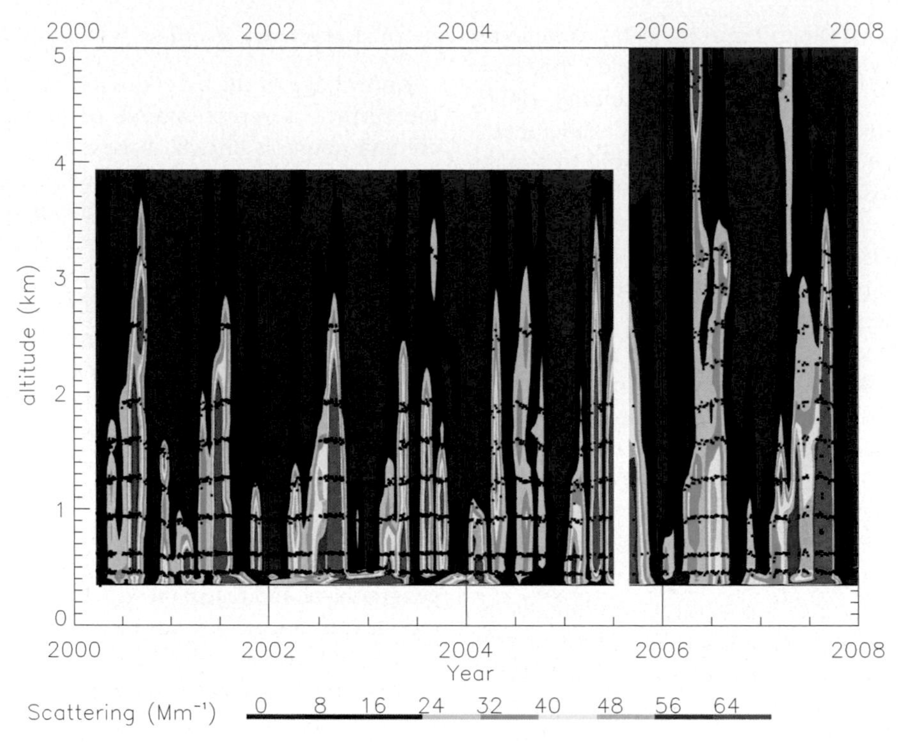

FIG. 21-3. Contoured light scattering from the nephelometer during the IAP campaign. The black dots represent individual flight segments at their altitude above mean sea level. From Andrews et al. (2011, their Fig. 1).

as a function of meteorology and proximity to different sources. Characterization of the regional variability in chemical, physical, and radiative properties as well as process understanding related to different aerosol types requires measurements in locales outside of the ARM fixed sites. To explore scientific questions outside those represented at ARM's long-term, fixed sites, an ARM Mobile Facility (AMF) was developed in 2005 for deployment around the world for 6- to 18-month periods (Miller and Slingo 2007). The AMF was equipped with an AOS system that contained the baseline instruments: CPC, nephelometer, humidified nephelometer, PSAP, and CCNC. This system was encompassed in a self-contained, portable container with a collapsible inlet stack and shock-mounted instrument racks for safe transport to any location chosen by ARM for deployment. Since 2005, the AMF has been deployed in distinct aerosol and meteorological regimes including Point Reyes, California; the Black Forest, Germany; Niamey, Niger; Shouxian, China; the Ganges valley, India; Graciosa Island, Azores; and Cape Cod, Massachusetts. As a component of the larger AMF, the AOS is always accompanied by active and passive remote sensors that provide information on the column aerosol, cloud properties, and thermodynamic state of the atmosphere, allowing for comprehensive investigations of aerosol radiative impacts in clear and cloudy conditions.

While the first AMF (AMF-1) has contributed greatly to characterizing aerosol variability over land surfaces, there has long been recognition that marine aerosol–cloud–radiation interactions are critical because of the vast area they represent, but also that these regions are highly undersampled. A second marine-capable AMF (AMF-2) was built mimicking the design of the AMF-1 but with expanded capabilities including an HTDMA. Its first deployment in 2010 was land-based in Steamboat Springs, Colorado, but since then the AMF-2 has spent nearly a year traveling the Pacific Ocean between Los Angeles, California, and Honolulu, Hawaii, sampling a gradient of aerosol between these urban/industrial centers and the remote marine environment as part of the Marine ARM GPCI Investigation of Clouds (MAGIC) Campaign [Lewis and Wiscombe 2012; GPCI is the GEWEX Cloud Systems Study (GCSS) Pacific Cross-section Intercomparison].

In 2009 the American Recovery and Reinvestment Act (ARRA) provided for additional infrastructure that has allowed ARM to delve deeper into processes controlling aerosol radiative properties. Aerosol optical properties are essentially a function of aerosol composition, size, and morphology, which are related to both sources and sinks and the intervening chemical and physical processing. Knowledge of how composition (source) is related to the processes that determine optical properties is critical for representing radiative forcing with fidelity in climate models. The fixed sites and two mobile facilities provide a baseline set of measurements from which aerosol optical properties have been well characterized. Combined with a greater understanding of how these properties are related to composition, size, and morphology will contribute to improved representation of aerosol processes in climate models. The Mobile Aerosol Observing System (MAOS) was developed by Brookhaven National Laboratory to serve as an accompaniment to either ARM fixed sites during intensive campaigns or mobile facilities with an aerosol focus. The MAOS comprises two separate containers, one containing the baseline aerosol instrumentation (MAOS-A) of the standard AOS with extended measurements of aerosol size distributions, chemical composition, and aerosol morphology while the second container, MAOS-C, contains enhanced chemistry measurements.

d. ARM Aerial Facility

Despite the heart of the ARM Program being located on the ground, the need for airborne measurements was fully recognized from the beginning. For aerosol research, critical applications include characterization of the vertical structure of aerosol properties, support for development and evaluation of remote sensing retrievals from the ground and space, and measurements of properties in situ that can advance understanding of aerosol life cycle processes not always accessible from the ground. In the latter case, understanding aerosol properties at cloud base and interstitial aerosol within cloud are essential for characterizing aerosol–cloud interactions, aerosol convective transport, and aerosol sinks. The ARM Aerial Facility (AAF) supports both routine observations such as the IAP program as well as airborne components of intensive campaigns. As emphasis on studying aerosol–cloud interactions increased in the program, the Routine AAF CLOWD Optical Radiative Observations (RACORO) Campaign (CLOWD = Clouds with Low Optical Water Depths), a second routine airborne program, was conducted over six months in 2009 on the Center for Interdisciplinary Remotely-Piloted Aircraft Studies (CIRPAS) Twin Otter to provide statistically robust datasets relevant to the properties of low liquid water boundary layer clouds and their relationship to aerosol (Vogelmann et al. 2012). Primary aerosol measurements made were cloud condensation nuclei (CCN) concentrations and aerosol size distributions above and below cloud.

Campaigns with significant airborne aerosol emphasis include the Indirect and Semi-Direct Aerosol Campaign (ISDAC) from the NRC Convair to better understand cloud and aerosol processes in Arctic mixed-phase clouds (McFarquhar et al. 2011) and the Carbonaceous Aerosols and Radiative Effects Study (CARES), a coordinated multiagency campaign in California to examine the aerosol processes and properties resulting from interactions of urban and biogenic emissions (Zaveri et al. 2012; Shilling et al. 2013). During CARES the DOE G-1 was heavily instrumented with an ARM suite of instrumentation that was expanded through ARRA and also with many guest instruments. The benefits of this comprehensive suite of instruments are evident in many findings from the campaign related to the evolution of secondary organic and black carbon aerosol and their climate impacts near a major urban center (e.g., Cahill et al. 2012; Moffet et al. 2013).

e. Remote sensing of aerosol properties

1) PASSIVE REMOTE SENSING

Broadband and spectral radiometry supported by ARM is described by Michalsky and Long (2016, chapter 16) along with a review of the utility in determining aerosol optical depth. The multifilter rotating shadowband radiometer (MFRSR; Harrison et al. 1994) has long played a key role in providing aerosol properties for applications concerning aerosol radiative impacts. This narrowband radiometer measures the total and diffuse horizontal components of the downwelling solar irradiance in seven wavelength bands and produces the direct normal component by difference. Langley calibrated measurements of the different irradiance components made by the same sensor allow for retrievals of optical depths with reduced uncertainties relative to using several different instruments calibrated by standard lamps. Together with the normal incidence multifilter radiometer (NIMFR), accurate (± 0.01) climatologies of aerosol optical depth have been produced at ARM sites continuously since 1992 (Michalsky et al. 2010). ARM's extensive use of MFRSRs is advantageous for linking with the large global networks of radiometry used to measure aerosol radiative properties such as the National Aeronautics and Space Administration (NASA) Aerosol Robotic Network (AERONET).

Harrison and Michalsky (1994) presented an automated algorithm for determining aerosol optical depth time series from the MFRSR with potential accuracy to 0.003 for averaging periods of 1 to 5 min. Subsequent work has extended and refined calibration and retrieval methods for the MFRSR providing for accurate retrievals under a larger range of conditions such as partly cloudy skies, which would result in enhanced climatologies, and estimates of additional properties such as effective particle size (Alexandrov et al. 2002a,b; Kassianov et al. 2005; Alexandrov et al. 2007). Kassianov et al. (2007) used an iterative approach with the direct irradiance and direct-to-diffuse ratios to simultaneously retrieve single-scatter albedo and asymmetry parameter. While this technique requires assumptions about the aerosol-size distribution and inputs for surface albedo, it produces the suite of first order optical properties required for calculating aerosol radiative fluxes or forcing integrated over the column. These data are useful for producing climatologies as well as evaluating independent measures of these properties from in situ, and other active and passive space- and ground-based remote sensing approaches. Validation of this procedure at the SGP resulted in radiative closure with direct and diffuse irradiances to within $5\,\mathrm{W\,m^{-2}}$. The algorithms have been brought together in the Column Intensive Properties (CIP) Value-Added Product (VAP), which is now operational and available on the ARM Data Archive for the SGP and is planned for other sites.

2) Active remote sensing

While passive remote sensing from radiometers provides column-integrated aerosol and surface property inputs adequate for many applications, it cannot provide information on the vertical structure of these properties known to have a significant impact on aerosol radiative transfer and atmospheric heating. Short of in situ measurements from aircraft, active remote sensing provides the only method for obtaining this information and is the only method for providing continuous, long-term measurements of vertical profiles. One of ARM's greatest accomplishments has been to produce continuous operational measurements of profiles of aerosol properties and water vapor from the Raman lidar (Turner et al. 2016, chapter 18). Beginning in 1998, the Raman lidar at the SGP site has collected continuous profiles of aerosol backscattering, extinction, and depolarization during daytime and nighttime operations excluding only expected downtimes for instrumental issues and maintenance. These measurements were used to develop and study mean and seasonal water vapor and aerosol extinction profiles over this site (Turner et al. 2001) as well as to study the vertical variability of aerosol above the SGP site (Ferrare et al. 2001). Peppler et al. (2000) demonstrated the utility of the Raman lidar in tracking the vertical distribution of forest and brush fire plumes from Central America over SGP with its ability to provide both aerosol extinction profiles and discrimination of biomass burning from background aerosol through extinction/backscatter ratios.

Several efforts outlined in the following sections have revolved around these measurements to better understand the relationship between lidar retrievals of aerosol properties and those measured by different approaches. Another Raman lidar with similar aerosol and water vapor measurement capabilities was installed at the TWP Darwin site and became operational in 2010. In 2011, a High Spectral Resolution Lidar (HSRL) was deployed at NSA for long-term operation in the Arctic. As with the Raman lidar, the HSRL provides vertical profiles of optical depth, scattering cross sections, and depolarization and is absolutely calibrated, which reduces the need for assumption about aerosol scattering properties for retrievals. Currently, HSRLs reside at the NSA and AMF-2 sites. A very recent addition includes three Doppler lidars enabling the 3D mapping of aerosol properties.

f. Modeling, quality measurement experiments, and value-added products

Early modeling activities associated with ARM provided information where measurements were lacking. Empirically based estimates of large-scale aerosol radiative forcing were impossible due to the dearth of observations of various anthropogenic species. Many studies extrapolated from known properties of sulfate (Charlson et al. 1992; Nemesure et al. 1995; Nemesure et al. 1997), biomass burning (Penner et al. 1992), and other absorbing aerosol (Nemesure and Schwartz 1998) to provide global or hemispheric estimates. The primary modeling objective stated in the original AWG report in 1992 was to examine consistency between measured aerosol size distribution and chemical composition and the associated scattering and absorption coefficients and CCN concentrations—the relationships on which all of these estimates depend. Today, much of the DOE Atmospheric System Research Program (ASR) Aerosol Life Cycle Working Group effort is dedicated to refining our understanding of these relationships. Later, as datasets became more developed and extensive, they could be used for direct evaluation of model performance; for example, the evaluation of the AeroCom models' ability to reproduce aerosol profiles as measured by the Raman lidar at SGP (Ferrare et al. 2005).

Despite the comprehensive instrumentation deployed by ARM, many quantities required in understanding the atmospheric system cannot be measured directly. Rather, these geophysical quantities of interest must be derived from direct measurements of different quantities, often with the benefit of models. Processing and compilation of these data into functional, coherent, and easily accessible datasets is a routine activity in ARM resulting in numerous VAPs. A special class of VAPs, called Quality Measurement Experiments (QMEs), was developed to identify issues with and improve the quality of existing measurements. QMEs are long time series (multiyear) radiative closure experiments; the first was for spectral longwave radiance using the Atmospheric Emitted Radiance Interferometer (AERI) and led to improvements in specification of model inputs for atmospheric state variables (Mlawer and Turner 2016, chapter 14). This success led to the Broadband Heating Rate Profile (BBHRP) effort that facilitates closure in the longwave and shortwave for all broadband components of radiative fluxes (McFarlane et al. 2016, chapter 20).

Expansion of the QME concept to address shortwave clear-sky fluxes required aerosol and surface albedo inputs, and thus the Aerosol Best Estimate (ABE) VAP was developed. The objective of ABE was to provide vertical profiles of aerosol extinction, single-scatter albedo, and asymmetry parameters (Turner et al. 2005). Climatologies of vertical profiles of extinction are taken from Raman lidar and used to extrapolate properties measured at the surface through the vertical column (Sivaraman et al. 2004; Flynn et al. 2012). Efforts to improve spectral measurements of surface albedo, critical for accuracy in modeling aerosol effects, are discussed by Michalsky and Long (2016, chapter 16). Inclusion of ABE in BBHRP did show improvement in simulating shortwave radiation at the surface; however, the exercise raised questions as to whether aerosol properties were extrapolated adequately in the vertical and across the spectrum to the ultraviolet and near-infrared where measurements were not made (Delamere et al. 2008).

In some cases, measurement–model comparisons pointed to instrumental problems with the radiometers themselves and contributed to improvements and higher-quality measurements at ARM sites over the long term. The bulk of the error, however, derives from the vertically resolved profiles of the aerosol single scatter albedo and asymmetry parameters, which are not currently determined from remote sensing or surface-based in situ measurement approaches. The Raman lidar provides profiles of extinction and backscattering very well, but quantifying the vertical profiles of aerosol absorption, size, and humidification effects required for understanding the effects of aerosol on atmospheric fluxes and heating requires additional measurements. Currently, ABE simply carries values for single-scatter albedo and asymmetry parameters measured at the surface up through the column. Efforts to derive profiles of each of the relevant properties require a compilation of various data sources and significant assumptions, an area of active effort as more sophisticated measurements become available and the sophistication of retrieval algorithms improves concordantly. Further, efforts to retrieve profiles of aerosol properties directly using multiwavelength Raman and HSRL lidars show great promise (e.g., Müller et al. 2001, 2014) and are being pursued actively in the larger community.

3. Long-term and intensive aerosol characterization at ARM sites and beyond

a. 1994–97 aerosol IOP at SGP

Despite the AOS not being fielded at the SGP until 1996, interest in aerosol measurements was evident from earlier campaigns such as the 1994 Remote Cloud Study IOP. Aerosol backscattering and extinction profiles as well as aerosol intensive properties (real refractive index, single-scattering albedo, and humidification factor) were derived from measurements acquired by the NASA Goddard Space Flight Center scanning Raman lidar and aircraft in situ measurements of size distribution; these represented nighttime aerosol at heights between 0.1 and 5 km (Ferrare et al. 1998). Characterization of the aerosol extinction and backscattering profiles as well as optical depths were examined in comparison to tower-mounted nephelometer measurements, daytime sunphotometer measurements, and airborne size distributions. Similarities and differences seen among these various approaches provided information on the strengths and weaknesses of each that served to drive measurement and retrieval science of aerosol properties into the future.

By 1997, investments had been made in the AOS and a suite of IOPs was conducted at the SGP to address many topics related to aerosol radiative forcing. Much of the work focused on intercomparison of measurement approaches for various aerosol properties to provide a better understanding of these approaches, their utility, and their limitations, in some cases spurring improvements in measurements and retrieval algorithms. Kato et al. (2000) performed a comprehensive comparison of airborne in situ scattering and absorption measurements in the column to ground-based radiometry including the MFRSR, Cimel sunphotometer, and Raman lidar. It was found that, under dry conditions the differences among these approaches were within instrumental uncertainties but became significant for humid conditions, precluding our ability to use measurements to quantify aerosol direct radiative forcing at the top-of-the-atmosphere with sufficient accuracy at this time.

b. Reno aerosol optics study

Recognizing the importance of accurate measures of aerosol absorption for quantifying aerosol forcing mechanisms, the Reno Aerosol Optics Study (RAOS) was conducted in June 2002 to compare the performance of existing and new instrumentation for measuring aerosol light absorption at the Desert Research Institute in Reno, Nevada (Sheridan et al. 2005). Instruments of interest included cavity ring-down extinction instruments (Moosmüller et al. 2005; Sheridan et al. 2005; Strawa et al. 2003), a folded-path optical extinction cell (Virkkula et al. 2005), integrating nephelometers, photoacoustic spectrometers, and filter-based instruments (Arnott et al. 2005) including the PSAP, which is deployed at all ARM sites. Because of their wide use, the project focused on determining how well these filter-based measurements represent absorption and found that correction schemes were required to improve their accuracy. Otherwise, good agreement was found among the various methods. While the RAOS experiments were highly valuable in moving our understanding of absorption measurements forward, the conditions under which aerosol were sampled was limited, consisting of mixtures of kerosene soot and ammonium sulfate at low humidity (~15%–25%) only. More recent laboratory and field analyses have revealed that filter-based absorption measurement may be biased under conditions of high organics loadings (Arnott et al. 2003; Cappa et al. 2008; Lack et al. 2008) and the ARM and ASR Programs continue to support improved characterization of these measurements.

c. 2003 ARM aerosol IOP

The laboratory intercomparisons of aerosol optical properties in RAOS were a preamble to the comprehensive Aerosol Intensive Observation Period of 2003 (AIOP 2003) field-based experiment undertaken in May at the SGP site. At this time is was clear that the largest contributor to uncertainty in forcing of climate change came from the incomplete knowledge of the relationships among aerosol composition, microphysical, and optical properties, leading to uncertainty in their clear-sky radiative effects and influence on the radiative properties of clouds. Comprehensive, redundant measurements of aerosol optical properties as well as broadband and spectral radiation were made from the surface and airborne platforms by in situ and active and passive remote sensing systems and were examined in relation to models to provide a better understanding of the source of these uncertainties.

Measurement verification and validation experiments included flying similar instrument systems in tandem for scattering, absorption, and extinction including nephelometers, PSAPs (Hallar et al. 2006), the Cadenza cavity ring-down extinction plus reciprocal nephelometer measurements (Strawa et al. 2006), and the first airborne photoacoustic measurements of absorption (Arnott et al. 2006). Airborne in situ derived optical properties were also compared to retrievals from ground-based radiance measurements from the Cimel sunphotometer and direct to diffuse ratios of irradiance calculated from these properties were compared to ground-based measurements from the MFRSR (Ricchiazzi et al. 2006). In general, all approaches agreed favorably for determining optical properties during the IOP. When larger differences were found, they were usually attributed to instrument response time, varying size ranges sampled due to differences in aircraft inlets, or conditions of low aerosol loading where sensitivity of some instruments was not adequate.

Other comparisons included those for the asymmetry parameter derived from a variety of methods (Andrews et al. 2006). In this study, high correlations were found between surface and airborne in situ values, but when these were compared to retrievals from both surface and airborne remote sensing measurements the correlations decreased. Pahlow et al. (2006) developed an approach for deriving aerosol hygroscopicity from lidar measurements and compared them to values derived from ground-based humidified nephelometry. While the lidar and nephelometer approaches showed reasonably good correlation, the lidar growth curves were much steeper with better sensitivity at high RH. This provided a possible synergy between these two measurement approaches to improve hygroscopicity estimates at and above the surface.

A large portion of the AIOP 2003 was devoted to better understanding of cloud condensation nuclei measurements. Gasparini et al. (2006) used a (tandem) differential mobility analyzer (DMA/TDMA) to measure aerosol size distribution and size-resolved hygroscopicity and modeled a multicomponent aerosol with these data as constraints. Composition dependent growth factors were determined and closure experiments were performed with measured optical properties and CCN. While reasonable agreement was found with optical properties, CCN were consistently slightly overpredicted. Rissman et al. (2006) also conducted an aerosol/CCN closure study, but in reverse to predict aerosol composition and mixing state. While these studies contributed to our understanding of aerosol properties that promote CCN, Ghan et al. (2006) developed a method for long-term, continuous estimation of CCN at cloud base, essential for understanding aerosol–cloud interactions from surface in situ measurements. This technique is now an operational VAP available in the ARM Data Archive. New methods for quantifying aerosol–cloud interactions from the ARM suite of surface in situ and remote sensing measurements was presented in Feingold et al. (2006), and the topic is further addressed by Feingold and McComiskey (2016, chapter 22).

Two studies served as compendia of sorts of the AIOP 2003. Schmid et al. (2006) provided a complete comparison of aerosol extinction profiles from all available instrumentation at the SGP site. Airborne instrumentation included the NASA Ames Airborne Tracking 14-channel sunphotometer (AATS-14), the IAP nephelometer and PSAP, and the Cadenza cavity ring-down instrument; the Raman lidar and two micropulse lidars were deployed on the ground. Relative to the AATS-14, which was used as the benchmark measurement, airborne in situ measurements were biased low (11%–17%) and ground-based lidar measurements were biased high (13%–54%). The high 54% bias occurring for the Raman lidar identified a previously undetected slow loss in sensitivity of the instrument leading up to the AIOP which subsequently underwent a full refurbishment and upgrade accompanied by an improved processing algorithm.

Prior to the AIOP 2003, efforts to obtain closure for downwelling shortwave diffuse irradiance were met with difficulty (Mlawer et al. 2000) and indicated large uncertainties in aerosol optical properties, specifically single-scattering albedo. Michalsky et al. (2006) performed a shortwave radiative closure using six different radiative transfer models and the well-validated, redundant measurements of aerosol properties from the campaign. In this study, closure was achieved across all models and at a range of solar zenith angles to 1% for direct irradiances and to within 1.9% for diffuse irradiances, a large improvement over former efforts, attributed to more accurate inputs for aerosol optical properties and better irradiance measurements.

These results raised the question: with what accuracy must we be able to measure aerosol optical properties to calculate radiative fluxes or forcing to a desired accuracy? The ARM AWG at the time discussed this issue and developed an approach to provide guidelines for measurement accuracy at the three ARM fixed sites. The sensitivity of direct radiative forcing to inputs of aerosol optical properties (optical depth, single scattering albedo, asymmetry parameter, and Ångström exponents) and environmental variables (solar geometry and surface albedo) were calculated for diurnally averaged and instantaneous quantities (McComiskey et al. 2008). Given typical measurement uncertainties for these inputs, relative uncertainties in radiative forcing were 20%–80%, with higher values at high latitudes where fluxes are lowest and single-scattering albedo as the largest contributor. This work has served as a guide to where efforts should be placed to make the greatest reductions in uncertainty of aerosol radiative forcing.

d. *The Aerosol Lidar Validation Experiment*

Given the importance and capability of lidar measurements for representing continuous profiles of aerosol properties and the refurbishment and upgrade of the Raman lidar after the AIOP in 2003, a collaborative effort between NASA and ARM in September 2005 at the SGP served as a follow-on validation experiment. The Aerosol Lidar Validation Experiment (ALIVE) focused on validation of the Raman and micropulse lidars, again using the AATS-14 as a benchmark. Flights were made by the IAP Cessna equipped with the suite of in situ aerosol measurement to extend these validation efforts as well as the Research Scanning Polarimeter (Knobelspiesse et al. 2008; Waquet et al. 2009) flown with the AATS-14. During ALIVE, agreement between aerosol extinction profiles derived from AATS-14 and Raman lidar measurements improved to 6% (Schmid et al. 2009).

e. *AMF Niamey*

A year-long deployment of the AMF in Niamey, Niger, in 2006 provided an excellent opportunity to examine the radiative impacts of dust and biomass burning aerosol. The Sahelian region experiences monsoon conditions throughout the winter, but the dry periods promote large loadings (0.08–2.5 optical depths over the deployment) of both aerosol types (McFarlane et al. 2009). Aerosol optical properties were retrieved from passive remote sensing of the column (MFRSR; Kassianov et al. 2007) and profiles of extinction were determined by the micropulse lidar and MFRSR. Clear-sky, diurnally averaged surface aerosol radiative forcing calculated for the year was $21.1 \pm 14.3\,\mathrm{W\,m^{-2}}$ with surface closure revealing a mean difference between observed and calculated surface fluxes to be $5\,\mathrm{W\,m^{-2}}$. Thus, it was determined that aerosol optical properties retrieved by this method were reasonable and that a 10% variation in these properties would produce closure.

Turner (2008) used data from the AERI, an infrared interferometer measuring downwelling radiation from $530\text{–}3050\,\mathrm{cm^{-1}}$ at $1\,\mathrm{cm^{-1}}$ resolution, at the Niamey site to develop an original algorithm for retrieving airborne mineral dust composition. The algorithm is based on differential absorption bands for the different minerologies. This information can be used to infer aerosol optical properties contributing to improvements in radiative forcing calculations. Turner found that, during the AMF deployment, kaolinite and gypsum fit most of the data and that the varying amounts of gypsum correlated with air mass origin and trajectory. While this method cannot determine aerosol hematite concentrations due to the lack of an absorption band in the AERI wavelength region, the shortwave radiation signatures examined in McFarlane et al. (2009) suggest a hematite component. Bringing these approaches together will provide more comprehensive information on mineral aerosol and their radiative forcing.

f. Long-term characterizations of aerosol physical and optical properties

Many of the intensive studies described above were dedicated to improved understanding of various approaches to measuring aerosol properties. Observations have improved through better measurement protocols, calibrations, and retrieval methods, and by understanding how these different approaches may complement each other. All of these efforts have contributed to our ability to construct longer-term characterization from continuous measurements at the fixed sites.

1) Southern Great Plains

At the SGP, understanding of the aerosol climatology has come from the combined ground-based in situ, airborne, and remote sensing measurements at the site. A 4-yr (1996–2000) statistical analysis of the early surface AOS data (Sheridan et al. 2001) showed the aerosol at SGP to be a complex mix of aerosol types influenced on a range of scales from local-to-regional scale agricultural activities to synoptic flows. These data show an average annual peak in aerosol scattering in August with a secondary peak in February while absorption was greatest in the summer and fall. Together these patterns resulted in a decrease in single scattering albedo in the fall months. Delene and Ogren (2002) examined the impact of this variability in aerosol optical properties on surface radiative forcing and found less than 10% variability in forcing annually and no significant variation over the diurnal cycle. The profiles from the IAP airborne profile observations reflected these patterns for the column (Andrews et al. 2004, 2011) while showing that the single scattering albedo and Ångström exponent were fairly invariant with altitude. Over the long term, it was found that variability in aerosol properties in the column as measured by IAP was represented by the surface data, but that shorter-term (e.g., daily) variations might not be as well represented. To understand the conditions under which the surface data might be more representative of the column, Delle Monache et al. (2004) examined the IAP data in relation to boundary layer height and mixing and found that a well-mixed boundary layer did not improve correlations between the airborne observations and those at the surface, but that the two were relatively well correlated under a range of conditions.

A full representation of the column aerosol properties can only be made using remote sensing. Michalsky et al. (2010) developed a 12-yr climatology (1992–2008) of aerosol optical depth and Ångström exponent from the MFRSR at SGP (Fig. 21-4) that reflected the summertime peak in scattering from the AOS data. The optical depth time series (top, Fig. 21-4), consisting of over 4000 daily average values, indicated a high degree of variability in the magnitude of the summertime peak whereas the winter minimum was relatively constant and showed a general lack of diurnal variability. The time series of Ångström exponent also showed a robust annual cycle with higher values (smaller particles) in the summer and minima (larger particles) in both December and April, the latter likely due to long-term transport of Asian dust. An outstanding feature in the Ångström exponent is the low values representative of the Pinatubo eruption at the beginning of the time series, and recovery to more typical continental aerosol values within a few years.

2) North Slope of Alaska

The phenomenon of Arctic haze, where aerosol concentration from anthropogenic sources at lower latitudes becomes highly concentrated over the Arctic in the winter and early spring, has been observed and documented for decades. However, quantifying the radiative impacts of these aerosols has been a challenge due to a lack of knowledge of their composition, optical properties, and interaction with radiation in the unique environmental conditions of the Arctic, characterized by high surface albedo and low solar angles. Nowhere have the long-term filter-based aerosol chemistry measurements been as important as at the NSA site in Barrow. Measurements have been made alongside optical properties at the surface since 1998 and have yielded a wealth of information on Arctic haze. Quinn et al. (2002) presented seasonal cycles of aerosol components, suggesting their sources, and their relationship to light scattering and absorption, and indicating the radiative effects of these different aerosol types. Both anthropogenic and natural (fine mode sea salt) aerosol were found to peak in winter and spring, all from long-term transport. In the summer, coarse mode sea salt and marine biogenic aerosols were found in greater concentration. Sea salt contributed most to light scattering in the winter and non–sea salt sulfate in the spring, while both were important over the summer. An analysis of trends at Barrow from 1976 to 2008 (Quinn et al. 2009) revealed decreases in anthropogenic aerosol of ~60% in the Arctic, although it was determined that

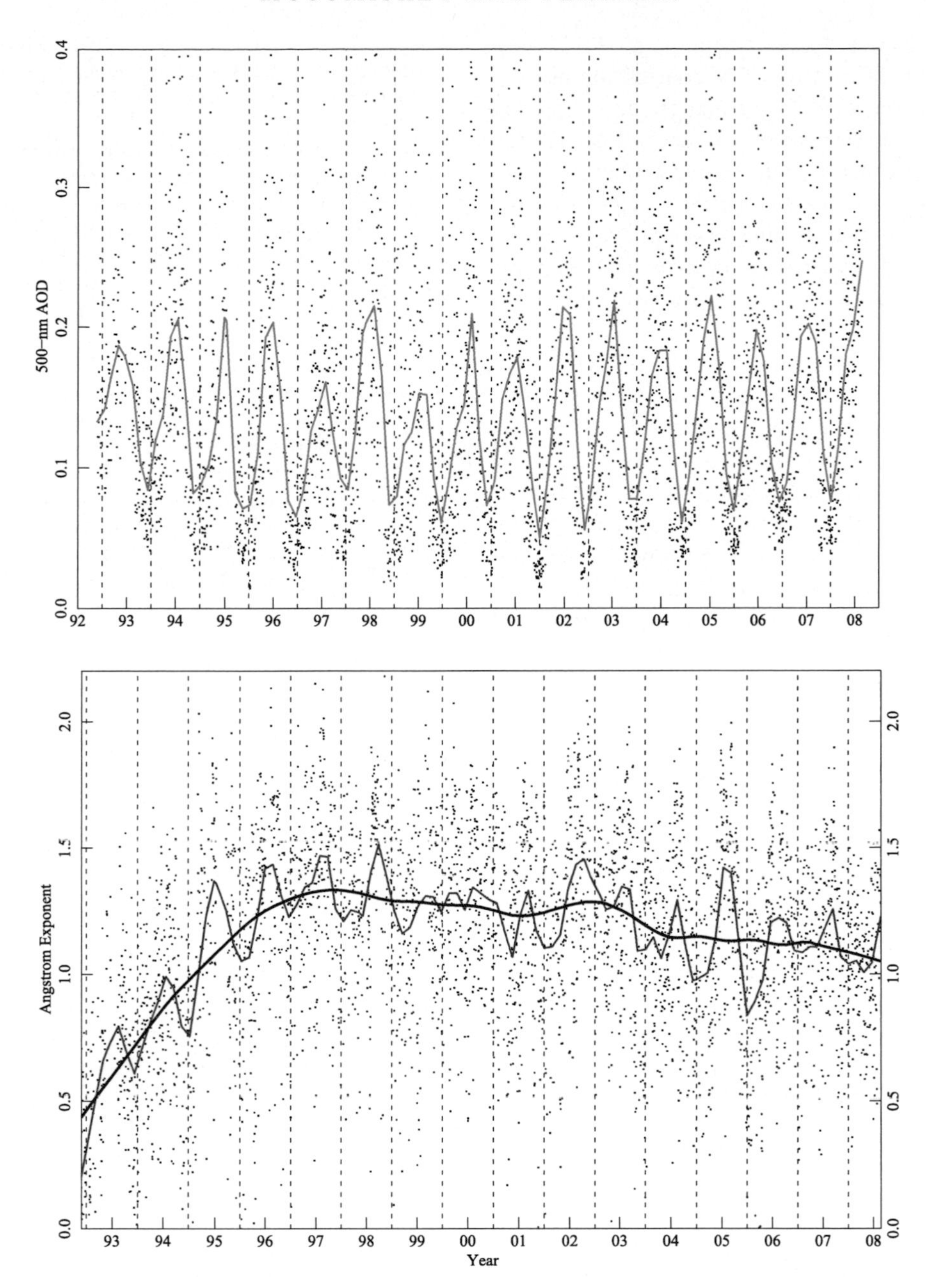

FIG. 21-4. (top) Aerosol optical depth (500 nm) and (bottom) Ångström exponent climatology (1992–2008) from the MFRSR at SGP. Black points represent daily averages and the green and red lines the locally weighted smoothed estimate. From Michalsky et al. (2010, their Figs. 8a and 10, respectively).

the source regions themselves remained similar. At the same time, marine biogenic aerosol has increased in the summertime, which has been found in other studies (O'Dwyer et al. 2000) to be correlated with loss of sea ice extent and increase in sea surface temperature.

4. Looking forward

In the past five years, the ARM Program has achieved a tremendous expansion of its aerosol observational capabilities with the recognition of the role and uncertainty of aerosol in the climate system. The integration of measurements of aerosol chemical, physical, and optical properties for improved model representation and parameterization development has been a focus. Further, ABE is seeing renewed development with the promise of direct retrievals of aerosol optical properties from active remote sensing (e.g., Müller et al. 2014). New foci in the program include understanding aerosol mixing state and its impact on radiative fluxes (Cappa et al. 2012), new particle formation related to CCN concentrations, secondary organic aerosol (SOA) formation, and improved understanding of measurements of aerosol absorption from various platforms. Obtaining comprehensive geographical coverage of aerosol measurements with the AMFs, with an eye to capturing seasonal cycles by deploying for at least a year, is critical to understanding aerosol processes and remains a priority within the ARM Program.

The seminal article by Charlson et al. (1992) that motivated much of the structure of the aerosol program within the ARM Program enjoys more than 2500 citations 20 years later. The science questions that the ARM AWG outlined that year are relevant today, but the details have changed and the knowledge required to understand processes at the scales represented in models has exploded. Capabilities for detailed, in situ observations at increasingly finer scales and for specific aspects of the aerosol system are continuously being added by the program, for example measurement of black carbon by the Single-Particle Soot Photometer (SP2) in the United States and India (Sedlacek et al. 2012) and new particle formation using the newly acquired nano-SMPS and SO_2 analyzer at SGP. ASR now has an Aerosol Life Cycle Working Group to address some aspect of the problems outlined here and a Cloud–Aerosol–Precipitation Interactions Working Group to address others. Contributions of the ARM Program in the latter category are addressed by Feingold and McComiskey (2016, chapter 22).

REFERENCES

Alexandrov, M., A. A. Lacis, B. E. Carlson, and B. Cairns, 2002a: Remote sensing of atmospheric aerosols and trace gases by means of multifilter rotating shadowband radiometer. Part I: Retrieval algorithm. *J. Atmos. Sci.*, **59**, 524–543, doi:10.1175/1520-0469(2002)059<0524:RSOAAA>2.0.CO;2.

——, ——, ——, and ——, 2002b: Remote sensing of atmospheric aerosols and trace gases by means of multifilter rotating shadowband radiometer. Part II: Climatological applications. *J. Atmos. Sci.*, **59**, 544–566, doi:10.1175/1520-0469(2002)059<0544:RSOAAA>2.0.CO;2.

——, P. Kiedron, J. J. Michalsky, G. Hodges, C. J. Flynn, and A. A. Lacis, 2007: Optical depth measurements by shadow-band radiometers and their uncertainties. *Appl. Opt.*, **46**, 8027–8038, doi:10.1364/AO.46.008027.

Andrews, E., P. J. Sheridan, J. A. Ogren, and R. Ferrare, 2004: In situ aerosol profiles over the Southern Great Plains cloud and radiation test bed site: 1. Aerosol optical properties. *J. Geophys. Res.*, **109**, D06208, doi:10.1029/2003JD004025.

——, and Coauthors, 2006: Comparison of methods for deriving aerosol asymmetry parameter. *J. Geophys. Res.*, **111**, D05504, doi:10.1029/2004JD005734.

——, P. J. Sheridan, and J. A. Ogren, 2011: Seasonal differences in the vertical profiles of aerosol optical properties over rural Oklahoma. *Atmos. Chem. Phys.*, **11**, 10 661–10 676, doi:10.5194/acp-11-10661-2011.

Arnott, W. P., and Coauthors, 2003: Photoacoustic and filter-based ambient aerosol light absorption measurements: Instrument comparisons and the role of relative humidity. *J. Geophys. Res.*, **108**, 4034, doi:10.1029/2002JD002165.

——, K. Hamasha, H. Moosmüller, P. Sheridan, and J. Ogren, 2005: Towards aerosol light-absorption measurements with a 7-wavelength aethalometer: Evaluation with a photoacoustic instrument and 3-wavelength nephelometer. *Aerosol Sci. Technol.*, **39**, 17–29, doi:10.1080/027868290901972.

——, and Coauthors, 2006: Photoacoustic insight for aerosol light absorption aloft from meteorological aircraft and comparison with particle soot absorption photometer measurements: DOE Southern Great Plains climate research facility and the coastal stratocumulus imposed perturbation experiments. *J. Geophys. Res.*, **111**, D05S02, doi:10.1029/2005JD005964.

Cahill, J. F., K. Suski, J. H. Seinfeld, R. A. Zaveri, and K. A. Prather, 2012: 2012: The mixing state of carbonaceous aerosol particles in northern and southern California measured during CARES and CalNex 2010. *Atmos. Chem. Phys.*, **12**, 10 989–11 002, doi:10.5194/acp-12-10989-2012.

Cappa C. D., D. A. Lack, J. B. Burkholder, and A. R. Ravishankara, 2008: Bias in filter-based aerosol light absorption measurements due to organic aerosol loading: Evidence from laboratory measurements. *Aerosol Sci. Technol.*, **42**, 1022–1032, doi:10.1080/02786820802389285.

——, and Coauthors, 2012: Radiative absorption enhancements due to the mixing state of atmospheric black carbon. *Science*, **337**, 1078–1081, doi:10.1126/science.1223447.

Charlson, R. J., S. E. Schwartz, J. M. Hales, R. D. Cess, J. A. Coakley Jr., J. E. Hansen, and D. J. Hofmann, 1992: Climate forcing by anthropogenic aerosols. *Science*, **255**, 423–430, doi:10.1126/science.255.5043.423.

Delamere, J., and Coauthors, 2008: Shortwave spectral radiative closure studies at the ARM Climate Research Facility Southern Great Plains site. *Proc. 18th ARM Science Team Meeting, ARM-CONF-2008*, Norfolk, VA, U.S. Department of Energy. [Available online at http://www.arm.gov/publications/proceedings/conf18/poster/P00131.pdf.]

Delene, D. J., and J. A. Ogren, 2002: Variability of aerosol optical properties at four North American surface monitoring sites. *J. Atmos. Sci.*, **59**, 1135–1150, doi:10.1175/1520-0469(2002)059<1135:VOAOPA>2.0.CO;2.

Delle Monache, L., K. D. Perry, R. T. Cederwall, and J. A. Ogren, 2004: In situ aerosol profiles over the Southern Great Plains cloud and radiation test bed site: 2. Effects of mixing height on aerosol properties. *J. Geophys. Res.*, **109**, D06209, doi:10.1029/2003JD004024.

Feingold, G., and A. McComiskey, 2016: ARM's aerosol–cloud–precipitation research (aerosol indirect effects). *The Atmospheric Radiation Measurement (ARM) Program: The First 20 Years, Meteor. Monogr.*, No. 57, Amer. Meteor. Soc., doi:10.1175/AMSMONOGRAPHS-D-15-0022.1.

——, R. Furrer, P. Pilewskie, L. A. Remer, Q. Min, and H. Jonsson, 2006: Aerosol indirect effect studies at Southern Great Plains during the May 2003 intensive operations period. *J. Geophys. Res.*, **111**, D05S14, doi:10.1029/2004JD005648.

Ferrare, R. A., S. H. Melfi, D. N. Whiteman, K. D. Evans, M. Poellot, and Y. J. Kaufman, 1998: Raman lidar measurements of aerosol extinction and backscattering: 2. Derivation of aerosol real refractive index, single-scattering albedo, and humidification factor using Raman lidar and aircraft size distribution. *J. Geophys. Res.*, **103**, 19 673–19 689, doi:10.1029/98JD01647.

——, D. D. Turner, L. A. Heilman Brasseur, W. F. Feltz, O. Dubovik, and T. P. Tooman, 2001: Raman lidar observations of the aerosol extinction-to-backscatter ratio over the Southern Great Plains. *J. Geophys. Res.*, **106**, 20 333–20 348, doi:10.1029/2000JD000144.

——, ——, M. Clayton, S. Guibert, M. Schulz, and M. Chin, 2005: The vertical distribution on aerosols over the Atmospheric Radiation Measurement Southern Great Plains site: Measured versus modeled. *Proc. 15th ARM Science Team Meeting, ARM-CONF-2005*, Daytona Beach, FL, U.S. Department of Energy. [Available online at http://www.arm.gov/publications/proceedings/conf15/extended_abs/ferrare_r.pdf?id=71.]

Flynn, C., D. Turner, A. Koontz, D. Chand, and C. Sivaraman, 2012: Aerosol Best Estimate (AEROSOLBE) Value-Added Product. DOE/SC-ARM/TR-115, 28 pp. [Available online at http://www.arm.gov/publications/tech_reports/doe-sc-arm-tr-115.pdf.]

Gasparini, R., R. Li, D. R. Collins, R. A. Ferrare, and V. G. Brackett, 2006: Application of aerosol hygroscopicity measured at the Atmospheric Radiation Measurement Program's Southern Great Plains site to examine composition and evolution. *J. Geophys. Res.*, **111**, D05S12, doi:10.1029/2004JD005448.

Ghan, S. J., and Coauthors, 2006: Use of in situ cloud condensation nuclei, extinction, and aerosol size distribution measurements to test a method for retrieving cloud condensation nuclei profiles from surface measurements. *J. Geophys. Res.*, **111**, D05S10, doi:10.1029/2004JD005752.

Hallar, A. G., and Coauthors, 2006: Atmospheric Radiation Measurements Aerosol Intensive Operating Period: Comparison of aerosol scattering during coordinated flights. *J. Geophys. Res.*, **111**, D05S09, doi:10.1029/2005JD006250.

Harrison, L., and J. Michalsky, 1994: Objective algorithms for the retrieval of optical depths from ground-based measurements. *Appl. Opt.*, **33**, 5126–5132, doi:10.1364/AO.33.005126.

——, ——, and J. Berndt, 1994: Automated multifilter rotating shadow-band radiometer: An instrument for optical depth and radiation measurements. *Appl. Opt.*, **33**, 5118–5125, doi:10.1364/AO.33.005118.

Kassianov, E. I., J. C. Barnard, and T. P. Ackerman, 2005: Retrieval of aerosol microphysical properties using surface Multifilter Rotating Shadowband Radiometer (MFRSR) data: Modeling and observations. *J. Geophys. Res.*, **110**, D09201, doi:10.1029/2004JD005337.

——, C. J. Flynn, T. P. Ackerman, and J. C. Barnard, 2007: Aerosol single-scattering albedo and asymmetry parameter from MFRSR observations during the ARM Aerosol IOP 2003. *Atmos. Chem. Phys.*, **7**, 3341–3351, doi:10.5194/acp-7-3341-2007.

Kato, S., and Coauthors, 2000: A comparison of the aerosol thickness derived from ground-based and airborne measurements. *J. Geophys. Res.*, **105**, 14 701–14 717, doi:10.1029/2000JD900013.

Knobelspiesse, K. D., B. Cairns, B. Schmid, M. O. Román, and C. B. Schaaf, 2008: Surface BRDF estimation from an aircraft compared to MODIS and ground estimates at the Southern Great Plains site. *J. Geophys. Res.*, **113**, D20105, doi:10.1029/2008JD010062.

Lack, D. A., and Coauthors, 2008: Bias in filter-based aerosol light absorption measurements due to organic aerosol loading: Evidence from ambient measurements. *Aerosol Sci. Technol.*, **42**, 1033–1041, doi:10.1080/02786820802389277.

Leifer, R., R. Knuth, and H.-N. Lee, 1993: Surface aerosol measurements at Lamont, Oklahoma. *Proc. Third ARM Science Team Meeting, ARM-CONF-1993*, Norman, OK, U.S. Department of Energy. [Available online at http://www.arm.gov/publications/proceedings/conf03/extended_abs/leifer_r.pdf?id=16.]

Lewis, E., and W. Wiscombe, 2012: MAGIC: Marine ARM GPCI Investigation of Clouds. DOE-SC-ARM-12-020, 12 pp. [Available online at http://www.arm.gov/publications/programdocs/doe-sc-arm-12-020.pdf.]

McComiskey, A., S. E. Schwartz, B. Schmid, H. Guan, E. R. Lewis, P. Ricchiazzi, and J. A. Ogren, 2008: Direct aerosol forcing: Calculation from observables and sensitivities to inputs. *J. Geophys. Res.*, **113**, D09202, doi:10.1029/2007JD009170.

McFarlane, S. A., J. H. Mather, and E. J. Mlawer, 2016: ARM's progress on improving atmospheric broadband radiative fluxes and heating rates. *The Atmospheric Radiation Measurement (ARM) Program: The First 20 Years, Meteor. Monogr.*, No. 57, Amer. Meteor. Soc., doi:10.1175/AMSMONOGRAPHS-D-15-0046.1.

——, E. I. Kassianov, J. Barnard, C. Flynn, and T. P. Ackerman, 2009: Surface shortwave aerosol radiative forcing during the Atmospheric Radiation Measurement Mobile Facility deployment in Niamey, Niger. *J. Geophys. Res.*, **114**, D00E06, doi:10.1029/2008JD010491.

McFarquhar, G. M., and Coauthors, 2011: Indirect and Semi-Direct Aerosol Campaign: The impact of Arctic aerosols on clouds. *Bull. Amer. Meteor. Soc.*, **92**, 183–201, doi:10.1175/2010BAMS2935.1.

Michalsky, J. J., and C. N. Long, 2016: ARM solar and infrared broadband and filter radiometry. *The Atmospheric Radiation Measurement (ARM) Program: The First 20 Years, Meteor. Monogr.*, No. 57, Amer. Meteor. Soc., doi:10.1175/AMSMONOGRAPHS-D-15-0031.1.

——, and Coauthors, 2006: Shortwave radiative closure studies for clear skies during the Atmospheric Radiation Measurement 2003 Aerosol Intensive Observation Period. *J. Geophys. Res.*, **111**, D14S90, doi:10.1029/2005JD006341.

——, F. Denn, C. Flynn, G. Hodges, P. Kiedron, A. Koontz, J. Schlemmer, and S. E. Schwartz, 2010: Climatology of aerosol optical depth in north central Oklahoma: 1992–2008. *J. Geophys. Res.*, **115**, D07203, doi:10.1029/2009JD012197.

Miller, M., and A. Slingo, 2007: The ARM Mobile Facility and its first international deployment: Measuring radiative flux divergence in West Africa. *Bull. Amer. Meteor. Soc.*, **88**, 1229–1244, doi:10.1175/BAMS-88-8-1229.

Mlawer, E. J., and D. D. Turner, 2016: Spectral radiation measurements and analysis in the ARM Program. *The Atmospheric Radiation Measurement (ARM) Program: The First 20 Years, Meteor. Monogr.*, No. 57, Amer. Meteor. Soc., doi:10.1175/AMSMONOGRAPHS-D-15-0027.1.

——, P. D. Brown, S. A. Clough, L. C. Harrison, J. J. Michalsky, P. W. Kiedron, and T. Shippert, 2000: Comparison of spectral direct and diffuse solar irradiance measurements and calculations for cloud-free conditions. *Geophys. Res. Lett.*, **27**, 2653–2656, doi:10.1029/2000GL011498.

Moffet, R. C., and Coauthors, 2013: Spectro-microscopic studies of carbonaceous aerosol aging in central California. *Atmos. Chem. Phys.*, **13**, 10 445–10 459, doi:10.5194/acp-13-10445-2013.

Moosmüller, H., R. Varma, and W. Arnott, 2005: Cavity ring-down and cavity-enhanced detection techniques for the measurement of aerosol extinction. *Aerosol Sci. Technol.*, **39**, 30–39, doi:10.1080/027868290903880.

Müller, D., U. Wandinger, D. Althausen, and M. Fiebig, 2001: Comprehensive particle characterization from three-wavelength Raman-lidar observations: Case study. *Appl. Opt.*, **40**, 4863–4869, doi:10.1364/AO.40.004863.

——, and Coauthors, 2014: Airborne multiwavelength High Spectral Resolution Lidar (HSRL-2) observations during TCAP 2012: Vertical profiles of optical and microphysical properties of a smoke/urban haze plume over the northeastern coast of the US. *Atmos. Meas. Tech.*, **7**, 3487–3496, doi:10.5194/amt-7-3487-2014.

Nemesure, S., and S. E. Schwartz, 1998: Effect of absorbing aerosol on shortwave radiative forcing of climate. *Proc. Eighth ARM Science Team Meeting, ARM-CONF-1998*, Tucson, AZ, U.S. Department of Energy, 531–535. [Available online at http://www.arm.gov/publications/proceedings/conf08/extended_abs/nemesure_sn.pdf?id=35.]

——, R. Wagner, and S. E. Schwartz, 1995: Direct shortwave forcing of climate by the anthropogenic sulfate aerosol: Sensitivity to particle size, composition, and relative humidity. *J. Geophys. Res.*, **100**, 26 105–26 116, doi:10.1029/95JD02897.

——, C. M. Benkovitz, and S. E. Schwartz, 1997: Aerosol sulfate loading and shortave direct radiative forcing over the North Atlantic. *Proc. Seventh ARM Science Team Meeting, ARM-CONF-1997*, San Antonio, TX, U.S. Department of Energy. [Available online at http://www.arm.gov/publications/proceedings/conf07/extended_abs/nemesure_s.pdf?id=68.]

Ng, N. L., and Coauthors, 2011: An aerosol chemical speciation monitor (ACSM) for routine monitoring of the composition and mass concentrations of ambient aerosol. *Aerosol Sci. Technol.*, **45**, 780–794, doi:10.1080/02786826.2011.560211.

O'Dwyer, J., E. Isaksson, T. Vinje, T. Jauhiainen, J. Moore, V. Pohjola, R. Vaikmae, and R. S. W. van de Wal, 2000: Methanesulfonic acid in a Svalbard ice core as an indicator of ocean climate. *Geophys. Res. Lett.*, **27**, 1159–1162, doi:10.1029/1999GL011106.

Pahlow, M., and Coauthors, 2006: Comparison between lidar and nephelometer measurements of aerosol hygroscopicity at the Southern Great Plains Atmospheric Radiation Measurement site. *J. Geophys. Res.*, **111**, D05S15, doi:10.1029/2004JD005646.

Penner, J. E., R. E. Dickinson, and C. A. O'Neill, 1992: Effects of aerosol from biomass burning on the global radiation budget. *Science*, **256**, 1432–1434, doi:10.1126/science.256.5062.1432.

Peppler, R. A., and Coauthors, 2000: ARM Southern Great Plains observations of the smoke pall associated with the 1998 Central American fires. *Bull. Amer. Meteor. Soc.*, **81**, 2563–2591, doi:10.1175/1520-0477(2000)081<2563:ASGPSO>2.3.CO;2.

Quinn, P. K., T. L. Miller, T. S. Bates, J. A. Ogren, E. Andrews, and G. E. Shaw, 2002: 3-year record of simultaneously measured aerosol chemical and optical properties at Barrow, Alaska. *J. Geophys. Res.*, **107**, 4103, doi:10.1029/2001JD001248.

——, T. S. Bates, K. Schulz, and G. E. Shaw, 2009: Decadal trends in aerosol chemical composition at Barrow, Alaska: 1976–2008. *Atmos. Chem. Phys.*, **9**, 8883–8888, doi:10.5194/acp-9-8883-2009.

Ricchiazzi, P., C. Gautier, J. A. Ogren, and B. Schmid, 2006: A comparison of aerosol optical properties obtained from in situ measurements and retrieved from sun and sky radiance observations during the May 2003 ARM Aerosol Intensive Observation Period. *J. Geophys. Res.*, **111**, D05S06, doi:10.1029/2005JD005863.

Rissman, T. A., and Coauthors, 2006: Characterization of ambient aerosol from measurements of cloud condensation nuclei during the 2003 Atmospheric Radiation Measurement Aerosol Intensive Observational Period at the Southern Great Plains site in Oklahoma. *J. Geophys. Res.*, **111**, D05S11, doi:10.1029/2004JD005695.

Roberts, G. C., and A. Nenes, 2005: A continuous-flow streamwise thermal-gradient CCN chamber for atmospheric measurements. *Aerosol Sci. Technol.*, **39**, 206–221, doi:10.1080/027868290913988.

Santarpia, J. L., R. Li, and D. R. Collins, 2004: Direct measurement of the hydration state of ambient aerosol populations. *J. Geophys. Res.*, **109**, D18209, doi:10.1029/2004JD004653.

Schmid, B., and Coauthors, 2006: How well do state-of-the-art techniques measuring the vertical profile of tropospheric aerosol extinction compare? *J. Geophys. Res.*, **111**, D05S07, doi:10.1029/2005JD005837.

——, and Coauthors, 2009: Validation of aerosol extinction and water vapor profiles from routine Atmospheric Radiation Measurement Program Climate Research Facility measurements. *J. Geophys. Res.*, **114**, D22207, doi:10.1029/2009JD012682.

Sedlacek, A. J., III, E. R. Lewis, L. Kleinman, J. Xu, and Q. Zhang, 2012: Determination of and evidence for non-core-shell structure of particles containing black carbon using the Single-Particle Soot Photometer (SP2). *Geophys. Res. Lett.*, **39**, L06802, doi:10.1029/2012GL050905.

Sheridan, P. J., D. J. Delene, and J. A. Ogren, 2001: Four years of continuous surface aerosol measurements from the Department of Energy's Atmospheric Radiation Measurement Program Southern Great Plains Cloud and Radiation Testbed site. *J. Geophys. Res.*, **106**, 20 735–20 747, doi:10.1029/2001JD000785.

——, and Coauthors, 2005: The Reno Aerosol Optics Study: An evaluation of aerosol absorption measurement methods. *Aerosol Sci. Technol.*, **39**, 1–16, doi:10.1080/027868290901891.

Shilling, J. E., and Coauthors, 2013: Enhanced SOA formation from mixed anthropogenic and biogenic emissions during the CARES campaign. *Atmos. Chem. Phys.*, **13**, 2091–2113, doi:10.5194/acp-13-2091-2013.

Sivaraman, C., D. D. Turner, and C. J. Flynn, 2004: Techniques and methods used to determine the aerosol best estimate value added product at SGP central facility. *Proc. 14th ARM Science Team Meeting, ARM-CONF-2004*, Albuquerque, NM, U.S. Department of Energy. [Available online at http://www.arm.gov/publications/proceedings/conf14/extended_abs/sivaraman-c.pdf.]

Strawa, A. W., R. Castaneda, T. Owano, D. S. Baer, and B. A. Paldus, 2003: The measurement of aerosol optical properties using continuous wave cavity ring-down techniques. *J. Atmos. Oceanic Technol.*, **20**, 454–465, doi:10.1175/1520-0426(2003)20<454:TMOAOP>2.0.CO;2.

——, and Coauthors, 2006: Comparison of in situ aerosol extinction and scattering coefficient measurements during the Aerosol Intensive Operations Period. *J. Geophys. Res.*, **111**, D05S03, doi:10.1029/2005JD006056.

Turner, D. D., 2008: Ground-based infrared retrievals of optical depth, effective radius, and composition of airborne mineral dust above the Sahel. *J. Geophys. Res.*, **113**, D00E03, doi:10.1029/2008JD010054.

——, R. A. Ferrare, and L. A. Brasseur, 2001: Average aerosol extinction and water vapor profiles over the southern Great Plains. *Geophys. Res. Lett.*, **28**, 4441–4444, doi:10.1029/2001GL013691.

——, C. Sivaraman, C. Flynn, and E. Mlawer, 2005: The SGP Aerosol Best-Estimate Value-Added Procedure and its Impact on the BHRP Project. *Proc. 15th ARM Science Team Meeting, ARM-CONF-2005*, Daytona Beach, FL, U.S. Department of Energy. [Available online at http://www.arm.gov/publications/proceedings/conf15/poster_abs/P00022.]

——, J. E. M. Goldsmith, and R. A. Ferrare, 2016: Development and applications of the ARM Raman lidar. *The Atmospheric Radiation Measurement (ARM) Program: The First 20 Years, Meteor. Monogr.*, No. 57, Amer. Meteor. Soc., doi:10.1175/AMSMONOGRAPHS-D-15-0026.1.

Virkkula, A., N. Ahlquist, D. Covert, P. Sheridan, W. Arnott, and J. Ogren, 2005: A three-wavelength optical extinction call for measuring aerosol light extinction and its application to determining light absorption coefficient. *Aerosol Sci. Technol.*, **39**, 52–67, doi:10.1080/027868290901918.

Vogelmann, A. M., and Coauthors, 2012: RACORO extended-term aircraft observations of boundary layer clouds. *Bull. Amer. Meteor. Soc.*, **93**, 861–878, doi:10.1175/BAMS-D-11-00189.1.

Waquet, F., B. Cairns, K. Knobelspiesse, J. Chowdhary, L. D. Travis, B. Schmid, and M. I. Mishchenko, 2009: Polarimetric remote sensing of aerosols over land. *J. Geophys. Res.*, **114**, D01206, doi:10.1029/2008JD010619.

Zaveri, R. A., and Coauthors, 2012: Overview of the 2010 Carbonaceous Aerosols and Radiative Effects Study (CARES). *Atmos. Chem. Phys.*, **12**, 7647–7687, doi:10.5194/acp-12-7647-2012.

Chapter 22

ARM's Aerosol–Cloud–Precipitation Research (Aerosol Indirect Effects)

GRAHAM FEINGOLD

Chemical Sciences Division, NOAA/Earth System Research Laboratory, Boulder, Colorado

ALLISON MCCOMISKEY

Global Monitoring Division, NOAA/Earth System Research Laboratory, Boulder, Colorado

1. Introduction/historical perspective

The conception of the Department of Energy's (DOE's) Atmospheric Radiation Measurement (ARM) Program over 20 years ago demonstrated prescience on the part of a number of astute scientists, many of whose words fill the pages of this monograph. The early years focused on a handful of cloud and radiation measurements and activities of relatively limited scope. The intervening decades have seen these efforts expanded to some of the finest instrumentation in the world to measure aerosol, clouds, radiation, and precipitation, accompanied by a substantial modeling effort. Together these have allowed the United States and international communities to tackle one of the thorniest problems associated with climate change, namely the influence of aerosol particles on cloud microphysics, precipitation, and cloud radiative properties (aerosol indirect effects). ARM research was at the forefront of aerosol indirect efforts from the outset, but because instrumentation was not readily in place and retrieval methodologies were still in their infancy, these were necessarily modeling efforts (e.g., Ghan et al. 1990; Feingold and Heymsfield 1992; Kim and Cess 1993) that addressed subsets of the problem. These early endeavors joined other key studies highlighting the climate forcing potential of tropospheric aerosol (e.g., Charlson et al. 1992) in setting the stage for a research effort that is, to this day, one of the cornerstones of ARM and the Atmospheric System Research Program (ASR).

Corresponding author address: Graham Feingold, Chemical Sciences Division, NOAA/Earth System Research Laboratory, 325 Broadway, Boulder, CO 80305.
E-mail: graham.feingold@noaa.gov

DOI: 10.1175/AMSMONOGRAPHS-D-15-0022.1

The goal of this chapter is to summarize ARM and ASR efforts in this realm. Because this chapter deals with measurement and modeling capabilities pertaining to each of the components of the aerosol–cloud–precipitation–radiation system, it rests heavily on other chapters that deal more specifically with each individual component. At the outset we note that the term "aerosol indirect effects" is often used loosely to include all aspects of aerosol–cloud interactions, whereas, by definition, the indirect effect is the radiative effect or forcing associated with these interactions. We will therefore introduce the term aerosol–cloud interactions (ACI) when we refer primarily to the microphysical/dynamical aspects of the problem and reserve "indirect effects" for the radiative forcing. ARM's early focus on atmospheric radiation measurements, followed by years of refinement of microphysical retrievals, has placed it in an excellent position to address both ACI and the associated indirect effects.

ACI or aerosol indirect effects are often used to convey a few underlying microphysical processes. The first is the "albedo effect" (Twomey 1977), which states that an increase in the number of aerosol particles results in more cloud condensation nuclei (CCN), a higher droplet concentration, and, all else being equal (particularly liquid water content), smaller drops and a more reflective cloud.[1] It is a fundamental expression of the ability of aerosol particles to generate a larger drop surface area to volume ratio. The second, the "lifetime effect" (Albrecht 1989), proposes that aerosol suppression

[1] An exception to the rule is the addition of giant CCN, the presence of which might lead to more collision/coalescence, fewer larger drops, and a less reflective cloud.

of collision-coalescence will suppress precipitation and allow clouds to sustain higher liquid water path (LWP) for a longer duration and therefore enhance cloud albedo. In mixed-phase clouds and ice clouds, the physics becomes more complex. For example, mixed-phase clouds could be expected to be more reflective because the reduction in droplet size resulting from an increase in aerosol reduces snow riming rates (Borys et al. 2003) and ice loss through precipitation (the riming effect). On the other hand, an increase in the aerosol might result in an increase in ice nuclei (IN), more efficient precipitation, lower cloud cover, and more solar absorption (the glaciation effect; Lohmann 2002). Finally, in deep convective clouds, there is evidence of an association between the aerosol and cloud-top height, updraft velocity, and lightning activity (the invigoration effect; Koren et al. 2005). Of all of these effects, the albedo effect has the strongest theoretical underpinnings; when the same amount of liquid water is divided among more drops, the cloud albedo must increase. The problem is that many other cloud processes act to obscure detection and quantification, rendering the magnitude uncertain. The others are less certain, mostly because they allow for realism in the form of the existence of multiple, simultaneous microphysical–dynamical interactions that occur while the system adjusts to the aerosol perturbation.

In addition to documenting ARM's history of ACI and indirect effect studies, a secondary goal of this chapter is to shift the community's thinking away from a linear superposition of these physical processes, and to encourage the reader to think more broadly about how all of these microphysical processes interact and adjust within the cloud system. This then places more emphasis on understanding how cloud systems work, and the role of meteorology in shaping cloud system evolution. It is in this spirit that the Intergovernmental Panel on Climate Change (IPCC 2013, p. 578) adopted the terminology ERFaci (effective radiative forcing due to aerosol–cloud interactions) to replace albedo and lifetime effects. This is a more inclusive term for the radiative forcing that occurs when the aerosol–cloud system is allowed to adjust to its environment, thereby allowing for changes in cloud fraction (f_c) and LWP or ice water path (IWP).

2. Development of crucial components that enabled ACI studies

a. Measurements

Shupe et al. (2016, chapter 19) and McComiskey and Ferrare (2016, chapter 21) describe more specifically the relevant measurement and retrieval capabilities; therefore we focus on the geophysical measurements themselves, with infrequent reference to instrument and instrument/measurement issues. This chapter is biased toward warm clouds (liquid water only), primarily because of the importance of shallow, warm convective clouds for the shortwave forcing of the climate system (Bony and Dufresne 2005),[2] but also because of the inherent difficulties in quantifying cloud microphysical properties of the mixed phase (ice + liquid). It is also biased toward surface remote sensing of clouds, arguably one of the major strengths of ARM, whose early architects recognized the value of long-term datasets as an anchor to infrequent in situ aircraft sampling, and less flexible satellite remote sensing. While airborne instruments were developed earlier for direct measurement of microphysical properties, many surface remote sensing measurements still required refinement for quantitative ACI work. The emphasis on surface remote sensing does not detract from ARM's significant collaborative efforts with other agencies in the field of satellite remote sensing (e.g., Minnis et al. 1992; Greenwald et al. 1999), and efforts in airborne campaigns in various parts of the world that have contributed greatly to the field. The 2009 RACORO[3] campaign (Vogelmann et al. 2012), conducted over the course of nearly six months at the Southern Great Plains (SGP) site, deserves special mention as an example of long-term in situ measurements addressing, among other topics, ACI and indirect effects.

Central to the study of ACI was capacity building of three primary measurement components:

1) *Cloud microphysical properties* such as drop effective radius r_e, drop concentration N_d, and cloud optical depth τ_c (Frisch et al. 1995; Dong et al. 1997; Chiu et al. 2007; 2010). Of these, τ_c is related closely to cloud reflectance, or cloud albedo, and is thus particularly relevant for ACI studies. Drop size retrievals typically separate into those that are averaged over height (e.g., derived from optical depth and liquid water path; Min and Harrison 1996; Kim et al. 2003; Chiu et al. 2012) or those that provide r_e profiles (e.g., from radar reflectivity and liquid water path; Frisch et al. 1995). Tradeoffs include the fact that optical measurements are more relevant to radiation than radar reflectivity (sixth moment of the drop size distribution) but can only be measured during daylight, whereas radar reflectivity is measurable at any

[2] Shallow clouds radiate at approximately the same temperature as the surface so that there is little longwave (LW) compensation for shortwave (SW) forcing. An exception is high-latitude wintertime clouds, where LW effects can be important, as discussed later.

[3] RACORO: Routine AAF CLOWD Optical Radiative Observations; CLOWD: Clouds with Low Optical Water Depths.

time of the day but is highly sensitive to the largest drops, the presence of which biases r_e retrievals.

2) *Aerosol microphysical properties* such as aerosol number concentration (typically at diameters >60 nm), CCN concentrations at specific supersaturation set points, aerosol extinction, total scatter, aerosol optical depth τ_a, backscatter, and aerosol index (Ångström exponent $\times$ τ_a) have all been used as measures of aerosol influence on clouds. Surface aerosol measurements may not always be representative of the aerosol entering the cloud; corrections can be made (e.g., Ghan and Collins 2004) or lidar measurements can be used to retrieve the aerosol properties below cloud base at a level where it is more likely that the aerosol is entering the cloud (Feingold et al. 2003). Aerosol chemical composition is being measured increasingly at ARM sites (McComiskey and Ferrare 2016, chapter 21). Although less important than aerosol number and/or size vis-à-vis influence on cloud microphysics (e.g., Feingold 2003; McFiggans et al. 2006), composition provides important insight into aerosol formation, growth, and removal processes, as well as the total aerosol budget. Aerosol composition is related closely to hygroscopicity, which is particularly relevant to the ability of an aerosol particle to act as a CCN, and to scatter light. The broader perspective of aerosol radiative forcing therefore requires consideration of aerosol number concentration, size, and composition. Moreover, aerosol indirect effects can be hard to separate from direct effects (McComiskey and Ferrare 2016, chapter 21) in the vicinity of clouds because the delineation between cloudy air and cloud-free air is often difficult to draw, and because the aerosol–cloud mix presents a different radiative forcing from the sum of the independent aerosol and cloud components (e.g., Schmidt et al. 2009). This means that key aerosol optical properties such as extinction (α), single scattering albedo, and asymmetry parameter are also important to the total radiative forcing (e.g., McComiskey et al. 2008).

3) *LWP measurements* are derived from microwave radiometers (Liljegren et al. 2001; Turner et al. 2007; Cadeddu et al. 2009) and for low LWP conditions, the Atmospheric Emitted Radiance Interferometer (AERI; Turner et al. 2007; 2016, chapter 13). Satellite sensors such as the Moderate Resolution Imaging Spectroradiometer (MODIS) also provide LWP measurements through combination of visible and near-IR measurements. Essential to all ACI and indirect effect evaluations, LWP is a bulk property that provides some reference against which to evaluate the influence of the aerosol on the cloud microphysical and/or radiative properties. Twomey (1977) likely had this in mind when formulating the principle of the albedo effect, which postulates that an increase in the number concentration of aerosol particles results in higher N_d, and all else being equal (i.e., LWP), smaller r_e. It is worth noting that this is not an assumption of constant LWP but rather a logical and necessary way to stratify the data within the context of Twomey's albedo effect. Also central to this discussion is the concept of cloud albedo susceptibility $S_o = \partial A/\partial N_d|_{\mathrm{LWP}}$, where A = cloud albedo (Platnick and Twomey 1994; Platnick and Oreopoulos 2008; Oreopoulos and Platnick, 2008), which assesses the magnitude of a cloud albedo response to an increase in N_d, and, by inference, aerosol concentration (N_a) at constant LWP. Below we will also discuss the idea of precipitation susceptibility $S_{o,R} = -\partial \ln R/\partial \ln N_d|_{\mathrm{LWP}}$ (Feingold and Siebert 2009; Sorooshian et al. 2009), which attempts to quantify the extent to which aerosol, via its influence on N_d, can modify rain rate R. Again, LWP is used as a reference. Thus both S_o and $S_{o,R}$ represent the potential rather than the actual influence of the aerosol, and in both cases LWP is an essential determinant of the possible effect.

While these three components form the backbone of ACI efforts, cloud-base vertical velocity is another key measurement relevant to ACI. Drop concentration is a function not only of aerosol properties, but also of the ability of a cloud to generate supersaturation, which is in turn a function of updraft velocity w. Doppler radars (in cloud) and lidars (below cloud) have made great strides in providing these measurements (Miller and Albrecht 1995; Clothiaux et al. 2000; Kollias and Albrecht 2000; Ghate et al. 2010).

b. Modeling of ACI

In parallel with deployment, testing, and refinement of aerosol and cloud measurement capabilities, development of cloud models appropriate to the study of ACI have been a significant ARM/ASR effort. Large-eddy simulation (LES) as a tool for studying ACI experienced significant development in the early stages of the ARM Program. LES is attractive in that it resolves the spatiotemporal scales relevant to aerosol activation and cloud turbulence. By coupling LES to detailed cloud microphysical models, several groups began to assess the ability of their models to simulate observed cloud structure and to provide a framework in which to test hypotheses, as well as being a simulation world to test retrieval methods (Kogan et al. 1994; Feingold et al. 1994; Ackerman et al. 2004; Stevens et al. 1996; Duda et al. 1996). The number of groups performing LES with microphysical models of varying complexity has grown significantly over the ARM/ASR lifetime and has facilitated the rigorous intercomparison of models to assess robustness of responses (e.g., Ackerman et al. 2009). Significant efforts also have been expended on developing cloud-resolving models (CRMs; Krueger et al. 2016, chapter 25) that simulate aerosol–precipitation interaction in deep clouds (Morrison and Grabowski 2011; van den Heever et al. 2011; Grabowski 2006; Fan et al. 2009).

3. Warm clouds

a. *Demonstrations of surface remote sensing of aerosol–cloud interactions*

Early demonstrations of ACI were performed at SGP using two different approaches, each with its own merits and drawbacks. The first approach (Feingold et al. 2003) considered a cloudy column of air and combined subcloud aerosol extinction (Raman lidar), a profile of r_e (cloud radar and microwave radiometer), w (Doppler cloud radar), and LWP (microwave radiometer) to produce plots of a cloud average r_e versus subcloud aerosol extinction α. A small sample of seven cases was analyzed as a proof of concept. Theoretical considerations indicate that if, as is observed, $N_d \propto N_a^b$ $(0 < b < 1)$[4] then, for constant LWP and drop-size dispersion, $r_e \propto N_a^{b/3}$. It then follows, assuming that α is a good proxy for the aerosol influencing the cloud, that $r_e \propto \alpha^b$. This power-law dependence was confirmed by the linear dependence of $\log r_e$ on $\log \alpha$. The data were sorted into different LWP bins in accord with the precept of the albedo effect. Values of b were calculated at different w and shown to be dependent on w, as expected from theory.

Simultaneously a second group (Kim et al. 2003) took a different approach using observations from SGP and derived a cloud-mean r_e from cloud optical depth [multifilter rotating shadowband radiometer (MFRSR)] and LWP (microwave radiometer) and used surface aerosol measurements of light scattering as a proxy for the aerosol concentration affecting the cloud.[5] They too quantified b and showed the strong correlation between LWP and τ_c for a much larger sample size; on average more than 60% of the variance in τ_c was due to variance in LWP. Garrett et al. (2004) used similar instrumentation to quantify b at the North Slope of Alaska (NSA), arguing that the aged aerosol particles reaching the Arctic are likely to be better CCN and therefore have a stronger ACI than at midlatitude locations such as SGP. Figure 22-1 shows some early examples of ACI measurements at SGP and NSA from the aforementioned studies.

The parameter b represents the magnitude of the microphysical response to an aerosol perturbation that is the basis of the albedo effect. Values tend to be ~0.2 to 0.6 (i.e., somewhat lower than one might expect due to aerosol activation alone). Both the magnitude and the variability suggest the presence of multiple cloud processes including collision/coalescence, entrainment, and wet scavenging (McComiskey et al. 2009). These values may be biased low when surface aerosol measurements are not representative of the aerosol entering cloud base.

b. *Surface, satellite, and aircraft in situ approaches*

Both approaches described above are attempts to achieve, from a single fixed location on the surface, what stacked aircraft can achieve by flying below, within, and above cloud (e.g., Brenguier et al. 2000; Wilcox et al. 2006; Roberts et al. 2008) or what is routinely attempted with space-based remote sensing. Logistically, surface remote sensing is a much simpler approach than stacked aircraft; however, it cannot measure the cloud radiance from above. Surface-based remote sensing provides a more controlled environment for measurement of ACI than does space-based remote sensing; sample volumes are much smaller, the temporal resolution is much higher, and meteorology can be better characterized. It is also amenable to collocated aerosol and cloud measurements; however, it lacks the huge global sampling advantage of satellite remote sensing. When using measurements from space, the aerosol in cloud-free pixels adjacent to cloudy pixels is assumed to be representative of the aerosol feeding into the base of the target clouds. There are several concerns with this approach, particularly because it is not trivial to identify cloud-free air (e.g., Koren et al. 2007; Charlson et al. 2007) and because humidification near clouds enhances τ_a, the typical proxy for subcloud CCN, in "cloud-free" pixels. The problem is exacerbated when addressing layer clouds of large spatial extent, since the cloud-free pixel may be a significant distance from the target clouds and at some point not particularly relevant to ACI. Even if one could identify purely cloud-free pixels, horizontal

[4] Note: various symbols and acronyms have been used to quantify the power or slope of cloud versus aerosol parameter power-law fits. Earlier work (e.g., Feingold et al. 2003) used "IE" (indirect effect), which is misleading because the authors addressed aerosol–cloud interactions, not the radiative response. Later, "ACI" was adopted (McComiskey et al. 2009) but, since we currently use this acronym to represent aerosol–cloud interactions, we use a more neutral b parameter.

[5] Note that the use of surface aerosol measurements as a proxy for the aerosol entering cloud base is only appropriate when the boundary layer is well mixed. Experience in the 19-month Azores AMF deployment in 2009/10 Clouds, Aerosol, and Precipitation in the Marine Boundary Layer (CAP-MBL) cautions against broad usage.

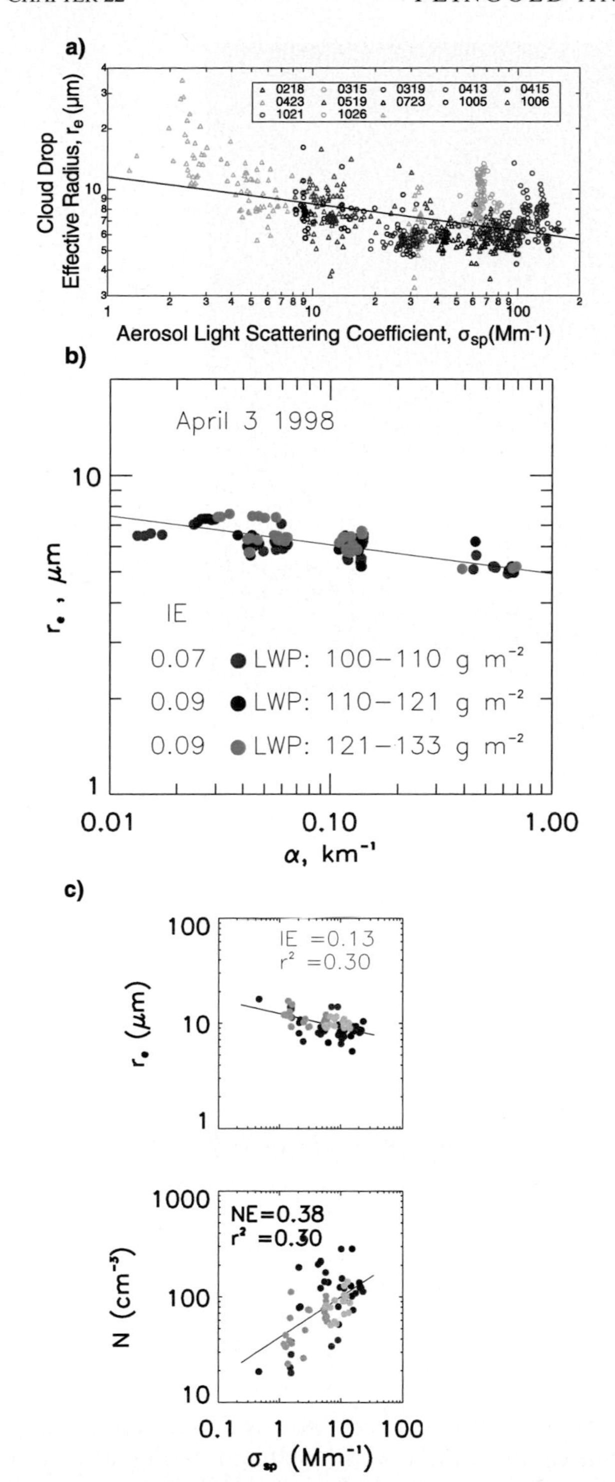

FIG. 22-1. First surface remote sensing demonstrations of the influence of the aerosol on cloud microphysical properties at ARM sites: (a) drop effective radius r_e as a function of surface aerosol light scattering σ_{sp} at SGP (after Kim et al. 2003), (b) r_e as a function of subcloud aerosol extinction α derived from the Raman lidar at SGP (after Feingold et al. 2003), and (c) r_e and drop concentration N as a function of surface σ_{sp} at NSA (after Garrett et al. 2004). See text and original articles for further details on the r_e and N retrievals.

light scattering from adjacent clouds has also been shown to enhance τ_a (Wen et al. 2007; Kassianov et al. 2009), which would exacerbate quantification of ACI.

Figure 22-2 provides a schematic showing some of the key differences in the approaches of satellite versus surface remote sensing to measurement of ACI.

c. *Quantification of ACI in warm clouds*

The various approaches to assessing ACI have differing strengths and weaknesses that have been explored in follow-up studies (e.g., McComiskey et al. 2009; Kim et al. 2008; Berg et al. 2011). Primary lessons are the sensitivity of b to binning data by LWP; the positive correlation between b and adiabaticity (i.e., how close the LWP is to the adiabatic value; Kim et al. 2008); use of subcloud aerosol as a more relevant CCN proxy to surface aerosol in decoupled boundary layers; the sensitivity of b to updraft w (Feingold et al. 2003); and the large influence of spatiotemporal scale on quantification of b (McComiskey and Feingold 2012). In many studies, b represents the full diversity of cloud microphysical processes—activation, collision/coalescence, entrainment/mixing, scavenging, and sedimentation—rather than simply the activation. This is particularly relevant in the context of climate models, which sometimes use b (or equivalent) as a representation of drop activation alone, thus biasing radiative forcing estimates. For example, b values are sometimes derived from remote sensing without sorting the data by LWP and used to represent the albedo effect in climate models. While this is clearly not a measure of the ACI associated with the albedo effect, it is more representative of effective ACI and associated forcing (ERFaci).

d. *ARM aircraft campaigns*

While various early field campaigns to some extent addressed ACI, we discuss several field campaigns that had a particularly strong focus on ACI. In chronological order these are MASE (2005), CLASIC/CHAPS (2007), VOCALS (2008), and RACORO (2009).[6]

MASE took place in July (2005) to overlap with the first ARM Mobile Facility (AMF) deployment at Point Reyes, California. The DOE G-1 flew aerosol and cloud microphysical payloads for addressing ACI in stratocumulus. DOE also partnered with the California Institute of Technology (CalTech), who operated the Center for

[6] MASE: Marine Stratus/Stratocumulus Experiment; CLASIC: Cloud Land Surface Interaction Campaign; CHAPS: Cumulus Humilis Aerosol Processing Study; VOCALS: Variability of American Monsoon Systems (VAMOS) Ocean–Cloud–Atmosphere–Land Study.

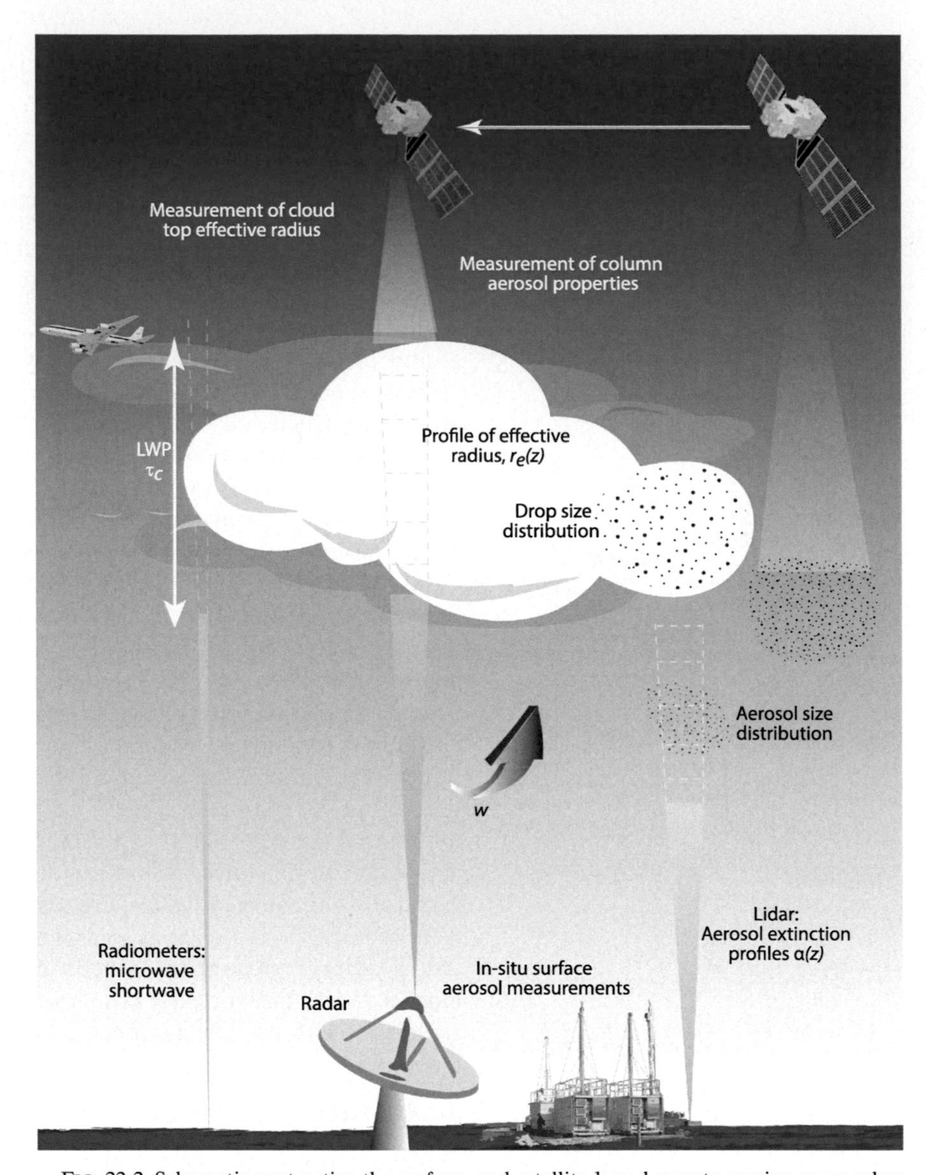

FIG. 22-2. Schematic contrasting the surface- and satellite-based remote sensing approaches to measurement of aerosol–cloud interactions. Some of the key surface instrumentation is also identified. Surface remote sensing has the advantage of being able to identify the aerosol entering the clouds from below while satellite remote sensing must contend with measuring column-integrated aerosol in cloud-free columns and cloud microphysics in adjacent cloudy columns. Satellite data have the advantage of global coverage while surface remote sensing is limited to a few sites, although at high temporal resolution.

Interdisciplinary Remotely-Piloted Aircraft Studies (CIRPAS) Twin Otter with a similar aerosol/cloud payload. The G-1 data acquired proved useful for testing theoretical formulations of the self-collection of cloud droplets to drizzle (autoconversion; Liu et al. 2005) and for identification of drizzle initiation at cloud top where LWC is highest. A notable aspect was the analysis of the droplet dispersion and its dependence on aerosol and drop concentrations (Lu et al. 2007), which influences cloud albedo susceptibility S_o. Some earlier studies showed dispersion decreasing with increasing drop concentration (Miles et al. 2000), which translates to an increase in S_o (Feingold et al. 1997). Others showed dispersion increasing with increasing aerosol concentration (Liu and Daum 2002) and therefore a decrease in S_o. The enhancement in S_o is expected in the coalescence-dominated growth regime, whereas the reduction should exist for condensation-dominated growth. Using Twin Otter MASE observations, Lu et al. (2007) showed decreasing dispersion with increasing

N_d but increasing dispersion with increasing N_a and therefore did not completely resolve this issue. Both coalescence- and condensation-dominated regimes exist, so conclusions are hard to draw about the influence of aerosol concentration on S_o.

CHAPS (Berg et al. 2009) is interesting because it used the DOE G-1 to address cloud processing of aerosol and gases, a closely related topic to ACI since the mutual effects of these two elements cannot be separated from one another. It was also one of several new studies to adopt use of (insoluble) CO as a proxy for CCN (Berg et al. 2011) and thus avoid the difficulty of characterizing the aerosol close to cloud, or in situations where wet removal may influence aerosol measurements. By performing flight legs downwind of source regions, CHAPS verified that even relatively small cities like Oklahoma City have a detectable ACI signal.

The DOE G-1 deployed to the VOCALS Regional Experiment (VOCALS-Rex) field campaign in Chile in the austral spring of 2008 for a large international deployment that included five aircraft, ground stations, and a number of ships targeting the stratocumulus regime in the southeast Pacific. ACI was an important goal of VOCALS-Rex; however, a large component of the campaign was to understand the meteorological environment of this stratocumulus regime. Because of the enormous scope of VOCALS, we touch briefly on the DOE G-1 activities and refer the reader to Wood et al. (2011) for an overview of the experiment. The G-1flew repeated legs along 20°S to collect statistics of aerosol and cloud properties over the course of a month; these will undoubtedly contribute to future documentation of the cloud field properties and boundary layer structure (Kleinman et al. 2012). G-1 data also have been used to evaluate satellite remote sensing (Min et al. 2012).

ARM undertook an ambitious task through deployment of the CIRPAS Twin Otter for RACORO—a long-duration (>5 month) field program in 2009 addressing shallow convective clouds (Vogelmann et al. 2012) and their radiative influences. The long duration and range of conditions encountered produced a dataset that is more statistically representative than the typical one-month intensive deployment. The deployment also provided a valuable test of surface remote sensing retrievals of aerosol and cloud retrievals, and important information on the relevance of surface CCN measurements to the clouds sampled. Lu et al. (2012) studied a large sample (568) of shallow cumulus and found a positive relationship between updraft velocity and N_d. They also found a negative correlation between N_d and relative droplet dispersion, supporting similar results from stratocumulus (Miles et al. 2000; Lu et al. 2007). Continuing analyses of ACI and indirect effects are currently in progress through a combination of LES and measurements of RACORO cases, as part of the Fast-physics System Testbed and Research (FASTER) project.

e. Modeling of ACI in warm clouds

Modeling of ACI in warm clouds has seen significant effort, and at a range of scales. Given our interest in microphysical processes and the capability of ARM to represent these properties at these scales, we focus on results from LES and CRMs coupled to microphysical models with either two-moment or bin microphysics schemes (i.e., schemes that carry information on drop size and therefore the potential for the aerosol to influence cloud system evolution). Early modeling efforts (Feingold et al. 1994; Kogan et al. 1994; Stevens et al. 1996) attempted to demonstrate the original constructs of "more aerosol particles lead to more/smaller droplets (all else equal)" (Twomey 1977) and "more aerosol particles lead to suppression of collision/coalescence and reduced rainfall" (Albrecht 1989). Perhaps the most important lesson learned from these studies is that, although the underlying "Twomey" and "Albrecht" constructs are physically sound, the cloud system often behaves in a much more nuanced manner because these aerosol perturbations occur within a dynamical framework that responds to, and sometimes buffers, the aerosol perturbation (e.g., Wang et al. 2003; Jiang et al. 2002; Ackerman et al. 2004; Xue et al. 2008; Stevens and Feingold 2009). Cancellation of aerosol effects has also been demonstrated in mixed-layer models (Wood 2007) lending more confidence to these ideas. Nevertheless, the cloudy boundary layer also may exhibit strong sensitivity to the aerosol (e.g., Ackerman et al. 2003; Christensen and Stephens 2011); in severely CCN-limited situations, boundary layers can even collapse because cloud dissipates as fast as it is generated. It therefore becomes important to identify conditions in which the cloudy boundary layer is either sensitive or insensitive to the aerosol. The general picture that has emerged is that, under clean conditions that are apt to generate precipitation, increases in the aerosol do indeed increase cloud fraction and LWP, but in non-precipitating conditions clouds tend to thin in response to increasing aerosol through a combination of droplet sedimentation (Bretherton et al. 2007) and evaporation/entrainment adjustments (e.g., Ackerman et al. 2004; Hill et al. 2009). These results are supported by satellite studies of ship tracks (Christensen and Stephens 2011).

f. Organization in shallow convection

Since the dawn of the satellite era in the 1960s, images of cloud field organization have become increasingly accessible. In addition to organization associated with precipitating deep convective cloud systems, shallow cloud fields also have been shown to organize into characteristic patterns (e.g., Agee 1984). Interagency field experiments such as the Eastern Pacific Investigation of Climate (EPIC; Bretherton et al. 2004), the Dynamics and Chemistry of Marine Stratocumulus Phase II (DYCOMS-II; Stevens et al. 2005), and VOCALS (Wood et al. 2011) have shown that marine stratocumulus cloud fields tend to prefer one of two states: the closed cellular state and the open cellular state. Closed cells have high cloud fraction and high albedo and tend to be nonprecipitating. Open cells are characterized by low cloud fraction, low albedo, and precipitation. Given the role of the aerosol in regulating precipitation, there is a clear pathway for the aerosol to select which of these is the preferred state, and therefore the albedo of the system. ARM has been a strong supporter of both observational (Sharon et al. 2006) and modeling studies (Xue et al. 2008; Wang and Feingold 2009a,b; Wang et al. 2010) that have clearly identified drizzle, whether a result of low CCN concentrations or of thick clouds, as playing a key role in the formation of open cells. Kazil et al. (2011) also identified aerosol replenishment mechanisms (nucleation of new particles, surface production of aerosol in cold pools, and entrainment of free tropospheric aerosol) that help make the open-cell system resilient. Transitions between closed- and open-cell states continue to be a topic of great interest because of the strong influence on cloud and planetary albedo. The importance of marine boundary layer clouds led ARM to support the first marine-based deployment of the AMF to the Pacific Ocean in 2012 [Marine ARM GCSS Pacific Cross-Section Intercomparison (GPCI) Investigation of Clouds (MAGIC)] to make regular observations of the stratocumulus to trade cumulus transition.

g. Precipitation susceptibility ($S_{o,R}$)

The precipitation susceptibility (Feingold and Siebert 2009) was proposed as a means to quantify the sensitivity of rain rate to aerosol perturbations. It has primarily been applied to warm clouds, but also to mixed-phase convection (Seifert et al. 2012). In warm clouds, raindrops are generated by two processes: 1) the self-collection of cloud droplets (autoconversion), which generates the first raindrop embryos, and 2) the collection of cloud droplets by rain drops (accretion). Autoconversion is a function of droplet (and cloud liquid water) concentration, whereas accretion is only a function of cloud and rain liquid water concentration. Therefore the potential for the aerosol to influence rain formation is via autoconversion, and the magnitude of $S_{o,R}$ is an indication of the relative balance of autoconversion and accretion.

Several ARM scientists have analyzed $S_{o,R}$ with parcel models (Feingold and Siebert 2009), heuristic one-dimensional models (Wood et al. 2009), LES (Jiang et al. 2010), and climate models (Wang et al. 2012). Observations have targeted warm clouds using satellite remote sensing (Sorooshian et al. 2009) and in situ aircraft data from VOCALS (Terai et al. 2012). LWP has been identified as having a strong control over $S_{o,R}$.

Two primary responses have been identified:

1) Parcel models, LES, and satellite remote sensing suggest that at low LWP $S_{o,R}$ increases with increasing LWP, reaches a maximum, and then decreases with further increases in LWP. The location and magnitude of the maximum depend on spatiotemporal averaging but also appears to depend on the background aerosol conditions, among other factors.
2) Heuristic models and in situ aircraft observations suggest a monotonic decrease in $S_{o,R}$ with increasing LWP. Ongoing efforts are attempting to explain these differences.

In addition to $S_{o,R}$ being useful for understanding rain-forming processes, it also provides an important way to evaluate the balance of these processes in global climate models (GCMs); the larger the value of $S_{o,R}$, the longer the cloud lifetime and the stronger the effective radiative forcing ERFaci. Quaas et al. (2009) evaluated a large number of GCMs and highlighted the sensitivity of indirect effects to parameters such as b and an equivalent parameter to $S_{o,R}$. As expected, models with larger b or $S_{o,R}$ have stronger indirect forcing. It is noteworthy that AMF Point Reyes data played an important role in this GCM evaluation study.

4. Arctic mixed-phase clouds

Arctic mixed-phase clouds are another particularly important component of the climate system. They also have been the focal point of several large ARM field campaigns such as M-PACE (2004) and ISDAC (2008).[7] They will be discussed relatively briefly, in part, because aerosol influences on these clouds are relatively poorly known.

The nucleation of ice is a key unknown in the maintenance of mixed-phase Arctic stratus. Numerical models are being confronted increasingly by aircraft measurements such as those from M-PACE and ISDAC

[7] M-PACE: Mixed-Phase Arctic Clouds Experiment; ISDAC: Indirect and Semi-Direct Aerosol Campaign.

(McFarquhar et al. 2011) and ground-based observations such as those at NSA (van Diedenhoven et al. 2009) or Eureka (de Boer et al. 2009). Key issues with simulating mixed-phase clouds are 1) understanding ice-nucleating mechanisms (e.g., Fridlind et al. 2007); 2) resolving disparities between observed (low) ice nuclei and (higher) ice crystal concentrations (Fridlind et al. 2007); and 3) simulating the observed relative amounts of liquid and ice in clouds of different thickness (Klein et al. 2009; Morrison et al. 2009).

To this point, the discussion has centered on microphysical processes that are likely to change the shortwave forcing of the planet. However, in the wintertime, investigations at NSA show that the aerosol has an influence on the longwave (LW) emissivity of Arctic clouds. In thin clouds with low LWP, an increase in the aerosol results in an increase in LW emissivity and a warming of the surface (Lubin and Vogelmann 2006; Garrett and Zhao 2006). An analog to the shortwave albedo susceptibility developed by Garrett and Zhao (2006) [$S_{\mathrm{LW}} = -(1 - \varepsilon)\ln(1 - \varepsilon)/3N_d$] shows that, in the longwave, the maximum susceptibility is at $\varepsilon = 0.6$ and low N_d. Thicker clouds act as black bodies ($\varepsilon = 1$), whereas thinner clouds are simply too thin to have much of an effect.

The Arctic boundary layer often presents one of two states: an opaque cloudy state and a cloud-free state. The mixed-phase Arctic stratus cloud appears to be a particularly robust state, often persisting for many days at a time, in spite of the inherently unstable mixture of ice and water. Lidar and radar imagery at NSA show ice precipitating from the base of these clouds and yet the clouds are not consumed. Why is this so?

Work based on many years of observations and modeling studies (Morrison et al. 2012; Fridlind et al. 2012a) suggests that the following factors, singly or in combination, might all play a role in maintaining the cloudy state:

- relatively low ambient IN concentrations (on the order of 1 per liter),
- a self-regulating ice crystal concentration,
- longwave cooling of the water layer driving sufficiently large turbulence to sustain the liquid cloud layer (but not strong enough to remix ice and water),
- moist inversions above cloud top, and
- long-range transport of water vapor.

The disappearance of the cloudy state seems to be caused by changes in the large-scale meteorology (Stramler et al. 2011)—particularly when accompanied by higher surface pressure. The Arctic boundary layer therefore appears to follow a slow manifold (a slowly evolving surface in phase space) with faster microphysical processes slaving the system to the slow manifold. Just how much microphysical versus dynamical processes contribute to the switch between these slow manifolds is unclear. Regardless, the Arctic cloudy boundary layer tends to be resilient, and when transitions between states do occur they seem to be controlled to some extent by changes in meteorology. Addressing questions like these requires significant effort. Continuous observations in the Arctic are sparse, and ARM continues to commit resources to extensive, continuous monitoring in this region from the ground and the air.

5. Deep convective clouds

a. Background

As with Arctic mixed-phase clouds, the influence of aerosol on deep convective clouds is a far more complex topic than it is for warm clouds, and although much work has been done over the past two decades, there is still a significant amount of disagreement on the extent to which the aerosol influences various cloud field properties. The complexity emanates from the multitude of liquid and ice microphysical pathways that become possible in the presence of ice. Indeed, a variety of possible responses probably exist depending, among other things, on environmental humidity (Khain et al. 2008) and shear (Fan et al. 2009). It has been hypothesized that, by delaying the onset of freezing to colder temperatures, the latent heat of freezing will be released at higher altitudes, thereby generating clouds with higher vertical velocities and higher cloud tops. This chain of events is often termed "aerosol invigoration of convection" (Koren et al. 2005). Invigoration tends to be used as a "catch-all" to describe various aspects of the aerosol influence, such as stronger convection, more rain, heavier rain, stronger cold pool outflow (Khain et al. 2005; Lee et al. 2008), and higher cloud tops. Given the complexity of the system it is possible that some of these responses occur in different cloud systems, and at different stages of cloud system evolution. Satellite studies seem to confirm this picture by showing a correlation between cloud-top height and aerosol loading (e.g., Koren et al. 2005). Concerns about potential "false correlation" have been raised because higher cloud tops often occur in the presence of higher instability and higher moisture, and therefore higher τ_a (e.g., Quaas et al. 2010). Efforts to address this issue (e.g., Koren et al. 2010) have used both chemical transport models and reanalysis data to place the observations in the context of meteorology. The reader is referred to Tao et al. (2012) for a broad review of this topic, which also covers work done at ARM sites (e.g., Li et al. 2011; see below).

b. Process studies

The physical processes behind "invigoration" describe a plausible response of an individual convective entity to an aerosol perturbation. However, given the myriad microphysical and dynamical internal feedbacks, what is more important is how a larger cloud system will respond to the aerosol perturbation and at time scales much longer than the lifetime of individual convective cells. For example, modeling by van den Heever et al. (2006) illustrated how the response of surface precipitation to aerosol perturbations may change during the course of a 12-h simulation. More recent work extending simulations to two weeks (Morrison and Grabowski 2011) and to multimonth radiative convective equilibrium (Grabowski 2006; van den Heever et al. 2011) underscores the fact that conclusions regarding aerosol influences on radiation at the top of the atmosphere cannot be drawn from short simulations. Whether the interest is climate-related or associated with an individual high aerosol event will dictate the required length of the simulation. This requires consideration of the duration and magnitude of the aerosol perturbation. For example, radiative convective equilibrium simulations with exceptionally high aerosol loadings may not be relevant for climate studies since aerosol perturbations typically subside as aerosol is removed by wet and dry deposition (e.g., Lee and Feingold 2010) or simply replaced by cleaner air through advection.

Sometimes responses ascribed to invigoration may not necessarily be part of the originally hypothesized chain of events. For example, Morrison and Grabowski (2011) showed that, in spite of a small aerosol-induced weakening in convection, cloud tops are higher because the resulting smaller ice particles have smaller fall velocities and can be lofted higher. Parsing out the various potential responses attributed to the aerosol, and their etiologies, would appear to be an important direction for research. The GCSS Tropical Warm Pool International Cloud Experiment (TWP-ICE) modeling study (Fridlind et al. 2012b), based on ARM measurements collected at the tropical western Pacific Darwin site in early 2006, has taken this essential step toward understanding sensitivity of results to microphysical complexity. Extensive environmental, cloud, and precipitation measurements to address convective processes, along with aerosol measurements, provide the necessary linkages for aerosol–deep convective cloud studies.

c. Surface-based observations

In addition to studies mentioned briefly above, surface-based measurements over a 10-yr period at SGP have also suggested elements of invigoration (Li et al. 2011). Using a combination of surface aerosol measurements (condensation nuclei, which include the very numerous nanometer sized particles) and ground-based remote sensors the authors concluded that, for mixed-phase clouds, aerosol perturbations cause increases in cloud depth and cloud-top height if the cloud base is warm. In contrast they found no such effect in clouds with cold bases and no ice. The authors also found evidence of aerosol influences on rain frequency and amount. To bolster conclusions, future studies will benefit from a more extensive set of precipitation measurements from radars at SGP, a closer focus on meteorological controls on convection, and CCN as opposed to condensation nuclei (which often do not participate as CCN) to represent the aerosol influencing the cloud.

6. Summary

As ARM moves toward the next 20 years, it is worth taking stock of some key successes and challenges that have arisen over the past 20 years. First, by developing and deploying a first-class array of aerosol and cloud instrumentation, the program has been able to demonstrate the ability of surface-based remote sensing to detect some expected correlations between aerosol and cloud microphysical properties. It has, together with other agencies, supported model development that has been crucial for understanding how clouds of various types respond to aerosol perturbations. Along with surface-based remote sensing data, ARM's airborne experiments have produced datasets that will for years to come provide data against which to evaluate models, at a range of scales.

One of the most difficult aspects of ACI studies is separating an observed correlation between aerosol and cloud parameters from a causal relationship between the two. That is, has the aerosol caused a change in the cloud properties, or are we merely observing a correlation between the two that might be related to each being correlated with the meteorological conditions? While models have played an important role in helping resolve this ambiguity, it is difficult to contend with the complexity of the models and the large number of internal feedbacks. As noted at the outset, the emphasis has continued and should continue to shift from identifying correlations to understanding the meteorological controls on cloud systems and the ways that the aerosol might perturb the system. From an observational perspective, characterizing the meteorological context of ACI is essential and will help determine whether the cloud system is resilient to aerosol perturbation or not. Refinement in variational analysis methodologies for model forcing and development of best estimate products for various ARM sites (e.g., Xie et al. 2010) are proving particularly useful. Methodologies for identifying drivers of ACI also are increasingly being applied.

In assessing the achievements, one does notice that more effort has been placed on detecting ACI compared to the radiative manifestation of ACI (i.e., indirect effects). Some important progress has been made. An understanding of the relationship between the parameter b ($N_d \propto N_a^b$) and radiative forcing by a cloud of known LWP has been established for plane-parallel clouds (McComiskey and Feingold 2008). ARM/ASR has advanced the study of 3D radiative transfer in a field of clouds (e.g., Barker and Marshak 2001); however, the forcing of the "aerosol-cloud soup" has received less attention. One approach is that taken by Schmidt et al. (2009), who compared spectral measurements of downwelling irradiance in a polluted cumulus cloud field with calculations of the spectral irradiance based on LES modeled cloud fields. In doing so the authors identified the importance of adequate characterization of aerosol properties such as hygroscopicity and absorption, alongside cloud optical depth, cloud size distribution, and other cloud-field properties. Improved understanding of how a cloud field influences aerosol radiative properties (Wen et al. 2007), together with the simultaneous influences of the aerosol on both microphysics and macrophysics, would be worthy of more attention. This is, after all, the essence of the "indirect effect." To this end ARM continues to develop spectral radiation measurement capabilities. New retrievals of cloud fraction and cloud albedo from surface broadband radiation measurements (Liu et al. 2011) will also be of great value for linking ACI to radiative effects.

Finally, with the recognition that the decades-long record of observations of clouds, aerosol, and radiation at megasites such as SGP or NSA could be used more fruitfully for model evaluation, ARM is creating a framework to further facilitate comparison between models and observations. Modelers tend to focus on a few choice case studies because of the difficulties of evaluating their models for a broad range of conditions. ARM and ASR are together engaging in plans for regular high-resolution modeling to complement regular, high-resolution observations. The benefits would include rigorous testing of models under different cloud conditions, comparison with observations at the appropriate scales, and testing, improvement, and development of new parameterizations for climate models (e.g., Neggers et al. 2012). The observationally constrained model output from regular high-resolution modeling will generate datasets that can be used for a variety of applications, including aerosol–cloud interactions. Efforts such as these will support efforts to quantify aerosol indirect effects and enhance ARM's legacy in the field.

Acknowledgments. The authors gratefully acknowledge funding from DOE ARM/ASR and the opportunity to participate in this exciting program over the past years. The reviewers and editors of this monograph are also thanked for their thoughtful comments.

REFERENCES

Ackerman, A. S., O. B. Toon, D. E. Stevens, and J. A. Coakley Jr., 2003: Enhancement of cloud cover and suppression of nocturnal drizzle in stratocumulus polluted by haze. *Geophys. Res. Lett.*, **30**, 1381, doi:10.1029/2002GL016634.

——, M. P. Kirkpatrick, D. E. Stevens, and O. B. Toon, 2004: The impact of humidity above stratiform clouds on indirect aerosol climate forcing. *Nature*, **432**, 1014–1017, doi:10.1038/nature03174.

——, and Coauthors, 2009: Large-eddy simulations of a drizzling, stratocumulus-topped marine boundary layer. *Mon. Wea. Rev.*, **137**, 1083–1110, doi:10.1175/2008MWR2582.1.

Agee, E. M., 1984: Observations from space and thermal convection: A historical perspective. *Bull. Amer. Meteor. Soc.*, **65**, 938–949, doi:10.1175/1520-0477(1984)065<0938:OFSATC>2.0.CO;2.

Albrecht, B. A., 1989: Aerosols, cloud microphysics, and fractional cloudiness. *Science*, **245**, 1227–1230, doi:10.1126/science.245.4923.1227.

Barker, H. W., and A. Marshak, 2001: Inferring optical depth of broken clouds above green vegetation using surface solar radiometric measurements. *J. Atmos. Sci.*, **58**, 2989–3006, doi:10.1175/1520-0469(2001)058<2989:IODOBC>2.0.CO;2.

Berg, L. K., and Coauthors, 2009: Overview of the cumulus humilis aerosol processing study. *Bull. Amer. Meteor. Soc.*, **90**, 1653–1667, doi:10.1175/2009BAMS2760.1.

——, C. M. Berkowitz, J. C. Barnard, G. Senum, and S. R. Springston, 2011: Observations of the first aerosol indirect effect in shallow cumuli. *Geophys. Res. Lett.*, **38**, L03809, doi:10.1029/2010GL046047.

Bony, S., and J.-L. Dufresne, 2005: Marine boundary layer clouds at the heart of tropical cloud feedback uncertainties in climate models. *Geophys. Res. Lett.*, **32**, L20806, doi:10.1029/2005GL023851.

Borys, R. D., D. H. Lowenthal, S. A. Cohn, and W. O. J. Brown, 2003: Mountaintop and radar measurements of anthropogenic aerosol effects on snow growth and snowfall rate. *Geophys. Res. Lett.*, **30**, 1538, doi:10.1029/2002GL016855.

Brenguier, J. L., and Coauthors, 2000: An overview of the ACE-2 CLOUDYCOLUMN closure experiment. *Tellus*, **52B**, 815–827, doi:10.1034/j.1600-0889.2000.00047.x.

Bretherton, C. S., and Coauthors, 2004: The EPIC 2001 stratocumulus study. *Bull. Amer. Meteor. Soc.*, **85**, 967–977, doi:10.1175/BAMS-85-7-967.

——, P. N. Blossey, and J. Uchida, 2007: Cloud droplet sedimentation, entrainment efficiency, and subtropical stratocumulus albedo. *Geophys. Res. Lett.*, **34**, L03813, doi:10.1029/2006GL027648.

Cadeddu, M. P., D. D. Turner, and J. C. Liljegren, 2009: A neural network for real-time retrievals of PWV and LWP from Arctic millimeter-wave ground-based observations. *IEEE Trans. Geosci. Remote Sens.*, **47**, 1887–1900, doi:10.1109/TGRS.2009.2013205.

Charlson, R. J., S. E. Schwartz, J. M. Hales, R. D. Cess, J. A. Coakley Jr., J. E. Hansen, and D. J. Hofmann, 1992: Climate forcing by anthropogenic aerosols. *Science*, **255**, 423–430, doi:10.1126/science.255.5043.423.

——, A. S. Ackerman, F. A. M. Bender, T. L. Anderson, and Z. Liu, 2007: On the climate forcing consequences of the albedo continuum between cloudy and clear air. *Tellus*, **59B**, 715–727, doi:10.1111/j.1600-0889.2007.00297.x.

Chiu, J. C., A. Marshak, W. Wiscombe, S. C. Valencia, and E. J. Welton, 2007: Cloud optical depth retrievals from solar background signal of micropulse lidars. *Geosci.Remote Sens. Lett.*, **4**, 456–460, doi:10.1109/LGRS.2007.896722.

——, C. H. Huang, A. Marshak, I. Slutsker, D. M. Giles, B. N. Holben, Y. Knyazikhin, and W. J. Wiscombe, 2010: Cloud optical depth retrievals from the Aerosol Robotic Network (AERONET) cloud mode observations. *J. Geophys. Res.*, **115**, D14202, doi:10.1029/2009JD013121.

——, and Coauthors, 2012: Cloud droplet size and liquid water path retrievals from zenith radiance measurements: Examples from the Atmospheric Radiation Measurement Program and the Aerosol Robotic Network. *Atmos. Chem. Phys.*, **12**, 10 313–10 329, doi:10.5194/acp-12-10313-2012.

Christensen, M. W., and G. L. Stephens, 2011: Microphysical and macrophysical responses of marine stratocumulus polluted by underlying ships: Evidence of cloud deepening. *J. Geophys. Res.*, **116**, D03201, doi:10.1029/2010JD014638.

Clothiaux, E. E., T. P. Ackerman, G. G. Mace, K. P. Moran, R. T. Marchand, M. Miller, and B. E. Martner, 2000: Objective determination of cloud heights and radar reflectivities using a combination of active remote sensors at the ARM CART sites. *J. Appl. Meteor.*, **39**, 645–665, doi:10.1175/1520-0450(2000)039<0645:ODOCHA>2.0.CO;2.

de Boer, G., E. W. Eloranta, and M. D. Shupe, 2009: Arctic mixed-phase stratiform cloud properties from multiple years of surface-based measurements at two high-latitude locations. *J. Atmos. Sci.*, **66**, 2874–2887, doi:10.1175/2009JAS3029.1.

Dong, X. Q., T. P. Ackerman, E. E. Clothiaux, P. Pilewskie, and Y. Han, 1997: Microphysical and radiative properties of boundary layer stratiform clouds deduced from ground-based measurements. *J. Geophys. Res.*, **102**, 23 829–23 843, doi:10.1029/97JD02119.

Duda, D. P., G. L. Stephens, B. Stevens, and W. R. Cotton, 1996: Effects of aerosol and horizontal inhomogeneity on the broadband albedo of marine stratus: Numerical simulations. *J. Atmos. Sci.*, **53**, 3757–3769, doi:10.1175/1520-0469(1996)053<3757:EOAAHI>2.0.CO;2.

Fan, J., and Coauthors, 2009: Dominant role by vertical wind shear in regulating aerosol effects on deep convective clouds. *J. Geophys. Res.*, **114**, D22206, doi:10.1029/2009JD012352.

Feingold, G., 2003: Modeling of the first indirect effect: Analysis of measurement requirements. *Geophys. Res. Lett.*, **30**, 1997, doi:10.1029/2003GL017967.

——, and A. J. Heymsfield, 1992: Parameterizations of condensational growth of droplets for use in general circulation models. *J. Atmos. Sci.*, **49**, 2325–2342, doi:10.1175/1520-0469(1992)049<2325:POCGOD>2.0.CO;2.

——, and H. Siebert, 2009: Cloud–aerosol interactions from the micro to the cloud scale. *Clouds in the Perturbed Climate System: Their Relationship to Energy Balance, Atmospheric Dynamics, and Precipitation*, J. Heintzenberg and R. J. Charlson, Eds., MIT Press, 319–338.

——, B. Stevens, W. R. Cotton, and R. L. Walko, 1994: An explicit microphysics/LES model designed to simulate the Twomey effect. *Atmos. Res.*, **33**, 207–233, doi:10.1016/0169-8095(94)90021-3.

——, R. Boers, B. Stevens, and W. Cotton, 1997: A modeling study of the effect of drizzle on cloud optical depth and susceptibility. *J. Geophys. Res.*, **102**, 13 527–13 534, doi:10.1029/97JD00963.

——, W. L. Eberhard, D. E. Veron, and M. Previdi, 2003: First measurements of the Twomey indirect effect using ground-based remote sensors. *Geophys. Res. Lett.*, **30**, 1287, doi:10.1029/2002GL016633.

Fridlind, A. M., A. S. Ackerman, G. McFarquhar, G. Zhang, M. R. Poellot, P. J. DeMott, A. J. Prenni, and A. J. Heymsfield, 2007: Ice properties of single-layer stratocumulus during the Mixed-Phase Arctic Cloud Experiment: 2. Model results. *J. Geophys. Res.*, **112**, D24202, doi:10.1029/2007JD008646.

——, B. van Diedenhoven, A. S. Ackerman, A. Avramov, A. Mrowiec, H. Morrison, P. Zuidema, and M. D. Shupe, 2012a: A FIRE-ACE/SHEBA case study of mixed-phase Arctic boundary layer clouds: Entrainment rate limitations on rapid primary ice nucleation processes. *J. Atmos. Sci.*, **69**, 365–389, doi:10.1175/JAS-D-11-052.1.

——, and Coauthors, 2012b: A comparison of TWP-ICE observational data with cloud-resolving model results. *J. Geophys. Res.*, **117**, D05204, doi:10.1029/2011JD016595.

Frisch, A. S., C. W. Fairall, and J. B. Snider, 1995: Measurement of stratus cloud and drizzle parameters in ASTEX with a Kα-band Doppler radar and a microwave radiometer. *J. Atmos. Sci.*, **52**, 2788–2799, doi:10.1175/1520-0469(1995)052<2788:MOSCAD>2.0.CO;2.

Garrett, T. J., and C. F. Zhao, 2006: Increased Arctic cloud longwave emissivity associated with pollution from mid-latitudes. *Nature*, **440**, 787–789, doi:10.1038/nature04636.

——, C. Zhao, X. Dong, G. G. Mace, and P. V. Hobbs, 2004: Effects of varying aerosol regimes on low-level Arctic stratus. *Geophys. Res. Lett.*, **31**, L17105, doi:10.1029/2004GL019928.

Ghan, S. J., and D. R. Collins, 2004: Use of in situ data to test a Raman lidar–based cloud condensation nuclei remote sensing method. *J. Atmos. Oceanic Technol.*, **21**, 387–394, doi:10.1175/1520-0426(2004)021<0387:UOISDT>2.0.CO;2.

——, K. E. Taylor, J. E. Penner, and D. J. Erickson, 1990: Model test of CCN-cloud albedo climate forcing. *Geophys. Res. Lett.*, **17**, 607–610, doi:10.1029/GL017i005p00607.

Ghate, V. P., B. A. Albrecht, and P. Kollias, 2010: Vertical velocity structure of nonprecipitating continental boundary layer stratocumulus clouds. *J. Geophys. Res.*, **115**, D13204, doi:10.1029/2009JD013091.

Grabowski, W. W., 2006: Indirect impact of atmospheric aerosols in idealized simulations of convective–radiative quasi equilibrium. *J. Climate*, **19**, 4664–4682, doi:10.1175/JCLI3857.1.

Greenwald, T. J., S. A. Christopher, J. Chou, and J. Liljegren, 1999: Intercomparison of cloud liquid water path derived from the GOES 9 imager and ground-based microwave radiometers for continental stratocumulus. *J. Geophys. Res.*, **104**, 9251–9260, doi:10.1029/1999JD900037.

Hill, A., G. Feingold, and H. Jiang, 2009: The influence of entrainment and mixing assumption on aerosol–cloud interactions in marine stratocumulus. *J. Atmos. Sci.*, **66**, 1450–1464, doi:10.1175/2008JAS2909.1.

IPCC, 2013: *Climate Change 2013: The Physical Science Basis.* Cambridge University Press, 1535 pp.

Jiang, H., G. Feingold, and W. Cotton, 2002: Simulations of aerosol–cloud–dynamical feedbacks resulting from entrainment of aerosol into the marine boundary layer during the Atlantic Stratocumulus Transition Experiment. *J. Geophys. Res.*, **107**, 4813, doi:10.1029/2001JD001502.

——, ——, and A. Sorooshian, 2010: Effect of aerosol on the susceptibility and efficiency of precipitation in warm trade cumulus clouds. *J. Atmos. Sci.*, **67**, 3525–3540, doi:10.1175/2010JAS3484.1.

Kassianov, E. I., M. Ovtchinnikov, L. K. Berg, and C. J. Flynn, 2009: Retrieval of aerosol optical depth in the vicinity of broken clouds from reflectance ratios: Sensitivity study. *J. Quant. Spectrosc. Radiat. Transfer*, **110**, 1677–1689, doi:10.1016/j.jqsrt.2009.01.014.

Kazil, J., H. Wang, G. Feingold, A. D. Clarke, J. R. Snider, and A. R. Bandy, 2011: Modeling chemical and aerosol processes in the transition from closed to open cells during VOCALS-REx. *Atmos. Chem. Phys.*, **11**, 7491–7514, doi:10.5194/acp-11-7491-2011.

Khain, A. P., D. Rosenfeld, and A. Pokrovsky, 2005: Aerosol impact on the dynamics and microphysics of deep convective clouds. *Quart. J. Roy. Meteor. Soc.*, **131**, 2639–2663, doi:10.1256/qj.04.62.

——, N. BenMoshe, and A. Pokrovsky, 2008: Factors determining the impact of aerosols on surface precipitation from clouds: An attempt at classification. *J. Atmos. Sci.*, **65**, 1721–1748, doi:10.1175/2007JAS2515.1.

Kim, B.-G., S. E. Schwartz, M. A. Miller, and Q. Min, 2003: Effective radius of cloud droplets by ground-based remote sensing: Relationship to aerosol. *J. Geophys. Res.*, **108**, 4740, doi:10.1029/2003JD003721.

——, M. A. Miller, S. E. Schwartz, Y. Liu, and Q. Min, 2008: The role of adiabaticity in the aerosol first indirect effect. *J. Geophys. Res.*, **113**, D05210, doi:10.1029/2007JD008961.

Kim, Y., and R. D. Cess, 1993: Effect of anthropogenic sulfate aerosols on low-level cloud albedo over oceans. *J. Geophys. Res.*, **98**, 14 883–14 885, doi:10.1029/93JD01211.

Klein, S. A., and Coauthors, 2009: Intercomparison of model simulations of mixed-phase clouds observed during the ARM Mixed Phase Arctic Cloud Experiment. I: Single-layer cloud. *Quart. J. Roy. Meteor. Soc.*, **135**, 979–1002, doi:10.1002/qj.416.

Kleinman, L. I., and Coauthors, 2012: Aerosol concentration and size distribution measured below, in, and above cloud from the DOE G-1 during VOCALS-REx. *Atmos. Chem. Phys.*, **12**, 207–223, doi:10.5194/acp-12-207-2012.

Kogan, Y. L., D. K. Lilly, Z. N. Kogan, and V. V. Filyushkin, 1994: The effect of CCN regeneration on the evolution of stratocumulus cloud layers. *Atmos. Res.*, **33**, 137–150, doi:10.1016/0169-8095(94)90017-5.

Kollias, P., and B. Albrecht, 2000: The turbulence structure in a continental stratocumulus cloud from millimeter-wavelength radar observations. *J. Atmos. Sci.*, **57**, 2417–2434, doi:10.1175/1520-0469(2000)057<2417:TTSIAC>2.0.CO;2.

Koren, I., Y. J. Kaufman, D. Rosenfeld, L. A. Remer, and Y. Rudich, 2005: Aerosol invigoration and restructuring of Atlantic convective clouds. *Geophys. Res. Lett.*, **32**, L14828, doi:10.1029/2005GL023187.

——, L. A. Remer, Y. J. Kaufman, Y. Rudich, and J. V. Martins, 2007: On the twilight zone between clouds and aerosols. *Geophys. Res. Lett.*, **34**, L08805, doi:10.1029/2007GL029253.

——, G. Feingold, and L. A. Remer, 2010: The invigoration of deep convective clouds over the Atlantic: Aerosol effect, meteorology or retrieval artifact? *Atmos. Chem. Phys.*, **10**, 8855–8872, doi:10.5194/acp-10-8855-2010.

Krueger, S. K., H. Morrison, and A. M. Fridlind, 2016: Cloud-resolving modeling: ARM and the GCSS story. *The Atmospheric Radiation Measurement (ARM) Program: The First 20 Years, Meteor. Monogr.*, No. 57, Amer. Meteor. Soc., doi:10.1175/AMSMONOGRAPHS-D-15-0047.1.

Lee, S.-S., and G. Feingold, 2010: Precipitating cloud-system response to aerosol perturbations. *Geophys. Res. Lett.*, **37**, L23806, doi:10.1029/2010GL045596.

——, L. J. Donner, V. T. J. Phillips, and Y. Ming, 2008: The dependence of aerosol effects on clouds and precipitation on cloud-system organization, shear and stability. *J. Geophys. Res.*, **113**, D16202, doi:10.1029/2007JD009224.

Li, Z., F. Niu, J. Fan, Y. Liu, D. Rosenfeld, and Y. Ding, 2011: Long-term impacts of aerosols on the vertical development of clouds and precipitation. *Nat. Geosci.*, **4**, 888–894, doi:10.1038/ngeo1313.

Liljegren, J. C., E. E. Clothiaux, G. G. Mace, S. Kato, and X. Dong, 2001: A new retrieval for cloud liquid water path using a ground-based microwave radiometer and measurements of cloud temperature. *J. Geophys. Res.*, **106**, 14 485–14 500, doi:10.1029/2000JD900817.

Liu, Y., and P. H. Daum, 2002: Anthropogenic aerosols: Indirect warming effect from dispersion forcing. *Nature*, **419**, 580–581, doi:10.1038/419580a.

——, ——, and R. L. McGraw, 2005: Size truncation effect, threshold behavior, and a new type of autoconversion parameterization. *Geophys. Res. Lett.*, **32**, L11811, doi:10.1029/2005GL022636.

——, W. Wu, M. P. Jensen, and T. Toto, 2011: Relationship between cloud radiative forcing, cloud fraction and cloud albedo, and new surface-based approach for determining cloud albedo. *Atmos. Chem. Phys.*, **11**, 7155–7170, doi:10.5194/acp-11-7155-2011.

Lohmann, U., 2002: A glaciation indirect aerosol effect caused by soot aerosols. *Geophys. Res. Lett.*, **29**, 1052, doi:10.1029/2001GL014357.

Lu, C., Y. Liu, S. Niu, and A. M. Vogelmann, 2012: Observed impacts of vertical velocity on cloud microphysics and implications for aerosol indirect effects. *Geophys. Res. Lett.*, **39**, L21808, doi:10.1029/2012GL053599.

Lu, M.-L., W. C. Conant, H. H. Jonsson, V. Varutbangkul, R. C. Flagan, and J. H. Seinfeld, 2007: The Marine Stratus/Stratocumulus Experiment (MASE): Aerosol–cloud relationships in marine stratocumulus. *J. Geophys. Res.*, **112**, D10209, doi:10.1029/2006JD007985.

Lubin, D., and A. M. Vogelmann, 2006: A climatologically significant aerosol longwave indirect effect in the Arctic. *Nature*, **439**, 453–456, doi:10.1038/nature04449.

McComiskey, A., and G. Feingold, 2008: Quantifying error in the radiative forcing of the first aerosol indirect effect. *Geophys. Res. Lett.*, **35**, L02810, doi:10.1029/2007GL032667.

——, and ——, 2012: The scale problem in quantifying aerosol indirect effects. *Atmos. Chem. Phys.*, **12**, 1031–1049, doi:10.5194/acp-12-1031-2012.

——, and R. A. Ferrare, 2016: Aerosol physical and optical properties and processes in the ARM Program. *The Atmospheric Radiation Measurement (ARM) Program: The First 20 Years, Meteor. Monogr.*, No. 57, Amer. Meteor. Soc., doi:10.1175/AMSMONOGRAPHS-D-15-0028.1.

——, S. E. Schwartz, B. Schmid, H. Guan, E. R. Lewis, P. Ricchiazzi, and J. A. Ogren, 2008: Direct aerosol forcing: Calculation from observables and sensitivities to inputs. *J. Geophys. Res.*, **113**, D09202, doi:10.1029/2007JD009170.

——, G. Feingold, A. S. Frisch, D. D. Turner, M. A. Miller, J. C. Chiu, Q. Min, and J. A. Ogren, 2009: An assessment of aerosol-cloud interactions in marine stratus clouds based on surface remote sensing. *J. Geophys. Res.*, **114**, D09203, doi:10.1029/2008JD011006.

McFarquhar, G. M., and Coauthors, 2011: Indirect and Semi-Direct Aerosol Campaign: The impact of Arctic aerosols on clouds. *Bull. Amer. Meteor. Soc.*, **92**, 183–201, doi:10.1175/2010BAMS2935.1.

McFiggans, G., and Coauthors, 2006: The effect of physical and chemical aerosol properties on warm cloud droplet activation. *Atmos. Chem. Phys.*, **6**, 2593–2649, doi:10.5194/acp-6-2593-2006.

Miles, N. L., J. Verlinde, and E. E. Clothiaux, 2000: Cloud droplet size distributions in low-level stratiform clouds. *J. Atmos. Sci.*, **57**, 295–311, doi:10.1175/1520-0469(2000)057<0295:CDSDIL>2.0.CO;2.

Miller, M. A., and B. A. Albrecht, 1995: Surface-based remote sensing of mesoscale cumulus–stratocumulus interaction during ASTEX. *J. Atmos. Sci.*, **52**, 2809–2826, doi:10.1175/1520-0469(1995)052<2809:SBOOMC>2.0.CO;2.

Min, Q., and L. C. Harrison, 1996: Cloud properties derived from surface MFRSR measurements and comparison with GOES results at the ARM SGP site. *Geophys. Res. Lett.*, **23**, 1641–1644, doi:10.1029/96GL01488.

——, and Coauthors, 2012: Comparison of MODIS cloud microphysical properties with in-situ measurements over the southeast Pacific. *Atmos. Chem. Phys.*, **12**, 11 261–11 273, doi:10.5194/acp-12-11261-2012.

Minnis, P., P. W. Heck, D. F. Young, C. W. Fairall, and J. B. Snider, 1992: Stratocumulus cloud properties derived from simultaneous satellite and island-based instrumentation during FIRE. *J. Appl. Meteor.*, **31**, 317–339, doi:10.1175/1520-0450(1992)031<0317:SCPDFS>2.0.CO;2.

Morrison, H., and W. W. Grabowski, 2011: Cloud-system resolving model simulations of aerosol indirect effects on tropical deep convection and its thermodynamic environment. *Atmos. Chem. Phys.*, **11**, 10 503–10 523, doi:10.5194/acp-11-10503-2011.

——, and Coauthors, 2009: Intercomparison of model simulations of mixed-phase clouds observed during the ARM Mixed-Phase Arctic Cloud Experiment. II: Multilayer cloud. *Quart. J. Roy. Meteor. Soc.*, **135**, 1003–1019, doi:10.1002/qj.415.

——, G. de Boer, G. Feingold, J. Harrington, M. D. Shupe, and K. Sulia, 2012: Resilience of persistent Arctic mixed-phase clouds. *Nat. Geosci.*, **5**, 11–17, doi:10.1038/ngeo1332.

Neggers, R. A. J., A. P. Siebesma, and T. Heus, 2012: Continuous single-column model evaluation at a permanent meteorological supersite. *Bull. Amer. Meteor. Soc.*, **93**, 1389–1400, doi:10.1175/BAMS-D-11-00162.1.

Oreopoulos, L., and S. Platnick, 2008: Radiative susceptibility of cloudy atmospheres to droplet number perturbations: 2. Global analysis from MODIS. *J. Geophys. Res.*, **113**, D14S21, doi:10.1029/2007JD009655.

Platnick, S., and S. Twomey, 1994: Determining the susceptibility of cloud albedo to changes in droplet concentration with the Advanced Very High Resolution Radiometer. *J. Appl. Meteor.*, **33**, 334–347, doi:10.1175/1520-0450(1994)033<0334:DTSOCA>2.0.CO;2.

——, and L. Oreopoulos, 2008: Radiative susceptibility of cloudy atmospheres to droplet number perturbations: 1. Theoretical analysis and examples from MODIS. *J. Geophys. Res.*, **113**, D14S20, doi:10.1029/2007JD009654.

Quaas J., and Coauthors, 2009: Aerosol indirect effects—General circulation model intercomparison and evaluation with satellite data. *Atmos. Chem. Phys.*, **9**, 8697–8717, doi:10.5194/acp-9-8697-2009.

——, B. Stevens, P. Stier, and U. Lohmann, 2010: Interpreting the cloud cover – aerosol optical depth relationship found in satellite data using a general circulation model. *Atmos. Chem. Phys.*, **10**, 6129–6135, doi:10.5194/acp-10-6129-2010.

Roberts, G. C., M. V. Ramana, C. Corrigan, D. Kim, and V. Ramanathan, 2008: Simultaneous observations of aerosol–cloud–albedo interactions with three stacked unmanned aerial vehicles. *Proc. Natl. Acad. Sci. USA*, **105**, 7370–7375, doi:10.1073/pnas.0710308105.

Schmidt, K., G. Feingold, P. Pilewskie, H. Jiang, O. Coddington, and M. Wendisch, 2009: Irradiance in polluted cumulus fields: Measured and modeled cloud-aerosol effects. *Geophys. Res. Lett.*, **36**, L07804, doi:10.1029/2008GL036848.

Seifert, A., C. Köhler, and K. D. Beheng, 2012: Aerosol-cloud-precipitation effects over Germany as simulated by a convective-scale numerical weather prediction model. *Atmos. Chem. Phys.*, **12**, 709–725, doi:10.5194/acp-12-709-2012.

Sharon, T. M., and Coauthors, 2006: Aerosol and cloud microphysical characteristics of rifts and gradients in maritime stratocumulus clouds. *J. Atmos. Sci.*, **63**, 983–997, doi:10.1175/JAS3667.1.

Shupe, M. D., J. M. Comstock, D. D. Turner, and G. G. Mace, 2016: Cloud property retrievals in the ARM Program. *The Atmospheric Radiation Measurement (ARM) Program: The First 20 Years, Meteor. Monogr.*, No. 57, Amer. Meteor. Soc., doi:10.1175/AMSMONOGRAPHS-D-15-0030.1.

Sorooshian, A., G. Feingold, M. Lebsock, H. Jiang, and G. L. Stephens, 2009: On the precipitation susceptibility of clouds to aerosol perturbations. *Geophys. Res. Lett.*, **36**, L13803, doi:10.1029/2009GL038993.

Stevens, B., and G. Feingold, 2009: Untangling aerosol effects on clouds and precipitation in a buffered system. *Nature*, **461**, 607–613, doi:10.1038/nature08281.

——, ——, W. Cotton, and R. Walko, 1996: Elements of the microphysical structure of numerically simulated nonprecipitating stratocumulus. *J. Atmos. Sci.*, **53**, 980–1006, doi:10.1175/1520-0469(1996)053<0980:EOTMSO>2.0.CO;2.

——, and Coauthors, 2005: Pockets of open cells and drizzle in marine stratocumulus. *Bull. Amer. Meteor. Soc.*, **86**, 51–57, doi:10.1175/BAMS-86-1-51.

Stramler, K., A. D. Del Genio, and W. B. Rossow, 2011: Synoptically driven Arctic winter states. *J. Climate*, **24**, 1747–1762, doi:10.1175/2010JCLI3817.1.

Tao, W.-K., J.-P. Chen, Z. Li, C. Wang, and C. Zhang, 2012: Impact of aerosols on convective clouds and precipitation. *Rev. Geophys.*, **50**, RG2001, doi:10.1029/2011RG000369.

Terai, C. R., R. Wood, D. C. Leon, and P. Zuidema, 2012: Does precipitation susceptibility vary with increasing cloud thickness in marine stratocumulus? *Atmos. Chem. Phys.*, **12**, 4567–4583, doi:10.5194/acp-12-4567-2012.

Turner, D. D., S. A. Clough, J. C. Liljegren, E. E. Clothiaux, K. E. Cady-Pereira, and K. L. Gaustad, 2007: Retrieving liquid water path and precipitable water vapor from the Atmospheric Radiation Measurement (ARM) microwave radiometers. *IEEE Trans. Geosci. Remote Sens.*, **45**, 3680–3690, doi:10.1109/TGRS.2007.903703.

——, E. J. Mlawer, and H. E. Revercomb, 2016: Water vapor observations in the ARM Program. *The Atmospheric Radiation Measurement (ARM) Program: The First 20 Years, Meteor. Monogr.*, No. 57, Amer. Meteor. Soc., doi:10.1175/AMSMONOGRAPHS-D-15-0025.1.

Twomey, S., 1977: Influence of pollution on shortwave albedo of clouds. *J. Atmos. Sci.*, **34**, 1149–1152, doi:10.1175/1520-0469(1977)034<1149:TIOPOT>2.0.CO;2.

van den Heever, S. C., G. G. Carrió, W. R. Cotton, P. J. DeMott, and A. J. Prenni, 2006: Impacts of nucleating aerosol on Florida storms. Part I: Mesoscale simulations. *J. Atmos. Sci.*, **63**, 1752–1775, doi:10.1175/JAS3713.1.

——, G. L. Stephens, and N. B. Wood, 2011: Aerosol indirect effects on tropical convection characteristics under conditions of radiative–convective equilibrium. *J. Atmos. Sci.*, **68**, 699–718, doi:10.1175/2010JAS3603.1.

van Diedenhoven, B., A. M. Fridlind, A. S. Ackerman, E. W. Eloranta, and G. M. McFarquhar, 2009: An evaluation of ice formation in large-eddy simulations of supercooled Arctic stratocumulus using ground-based lidar and cloud radar. *J. Geophys. Res.*, **114**, D10203, doi:10.1029/2008JD011198.

Vogelmann, A. M., and Coauthors, 2012: RACORO extended-term aircraft observations of boundary layer clouds. *Bull. Amer. Meteor. Soc.*, **93**, 861–878, doi:10.1175/BAMS-D-11-00189.1.

Wang, H., and G. Feingold, 2009a: Modeling mesoscale cellular structures and drizzle in marine stratocumulus. Part I: Impact of drizzle on the formation and evolution of open cells. *J. Atmos. Sci.*, **66**, 3237–3256, doi:10.1175/2009JAS3022.1.

——, and ——, 2009b: Modeling mesoscale cellular structures and drizzle in marine stratocumulus. Part II: The microphysics and dynamics of the boundary region between open and closed cells. *J. Atmos. Sci.*, **66**, 3257–3275, doi:10.1175/2009JAS3120.1.

——, ——, R. Wood, and J. Kazil, 2010: Modelling microphysical and meteorological controls on precipitation and cloud cellular structures in southeast Pacific stratocumulus. *Atmos. Chem. Phys.*, **10**, 6347–6362, doi:10.5194/acp-10-6347-2010.

Wang, M., and Coauthors, 2012: Constraining cloud lifetime effects of aerosols using A-Train satellite observations. *Geophys. Res. Lett.*, **39**, L15709, doi:10.1029/2012GL052204.

Wang, S., Q. Wang, and G. Feingold, 2003: Turbulence, condensation, and liquid water transport in numerically simulated nonprecipitating stratocumulus clouds. *J. Atmos. Sci.*, **60**, 262–278, doi:10.1175/1520-0469(2003)060<0262:TCALWT>2.0.CO;2.

Wen, G., A. Marshak, R. F. Cahalan, L. A. Remer, and R. G. Kleidman, 2007: 3-D aerosol-cloud radiative interaction observed in collocated MODIS and ASTER images of cumulus cloud fields. *J. Geophys. Res.*, **112**, D13204, doi:10.1029/2006JD008267.

Wilcox, E. M., G. Roberts, and V. Ramanathan, 2006: Influence of aerosols on the shortwave cloud radiative forcing from North Pacific oceanic clouds: Results from the Cloud Indirect Forcing Experiment (CIFEX). *Geophys. Res. Lett.*, **33**, L21804, doi:10.1029/2006GL027150.

Wood, R., 2007: Cancellation of aerosol indirect effects in marine stratocumulus through cloud thinning. *J. Atmos. Sci.*, **64**, 2657–2669, doi:10.1175/JAS3942.1.

——, T. L. Kubar, and D. L. Hartmann, 2009: Understanding the importance of microphysics and macrophysics for warm rain in marine low clouds. Part II: Heuristic models of rain formation. *J. Atmos. Sci.*, **66**, 2973–2990, doi:10.1175/2009JAS3072.1.

——, and Coauthors, 2011: The VAMOS Ocean–Cloud–Atmosphere–Land Study Regional Experiment (VOCALS-REx): Goals, platforms, and field operations. *Atmos. Chem. Phys.*, **11**, 627–654, doi:10.5194/acp-11-627-2011.

Xie, S., and Coauthors, 2010: Clouds and more: ARM climate modeling best estimate data. *Bull. Amer. Meteor. Soc.*, **91**, 13–20, doi:10.1175/2009BAMS2891.1.

Xue, H., G. Feingold, and B. Stevens, 2008: Aerosol effects on clouds, precipitation, and the organization of shallow cumulus convection. *J. Atmos. Sci.*, **65**, 392–406, doi:10.1175/2007JAS2428.1.

Chapter 23

Surface Properties and Interactions: Coupling the Land and Atmosphere within the ARM Program

LARRY K. BERG

Pacific Northwest National Laboratory, Richland, Washington

PETER J. LAMB*

Cooperative Institute for Mesoscale Meteorological Studies, University of Oklahoma, Norman, Oklahoma

1. Setting the research agenda

It is well known that the exchange of heat and moisture between the surface and atmosphere plays a key role in the earth's climate system (e.g., Randall et al. 2007). Science questions related to land–atmosphere interactions have remained an active topic of research, both inside and outside of the ARM Program, for a considerable period of time (e.g., Betts et al. 1996; Betts 2003, 2004; Dirmeyer et al. 2006; Betts 2009; Santanello et al. 2009; Betts and Silva Dias 2010). Given the sustained interest by the scientific community in regards to the impacts of land–atmosphere interactions on both the climate and the hydrologic cycle, this focus is likely to remain for the foreseeable future.

The ARM Program has played an important role in the research community by collecting data relevant for understanding land–atmosphere interactions and by supporting research efforts designed to address questions related to the coupling of land and atmosphere. At the start of the ARM Program, Stokes and Schwartz (1994) listed three different projects that were related to land surface processes: point–area relationships for global climate modeling led by Chris Doran; area-representative estimates of surface heat flux led by Richard Coulter; and remote sensing of surface fluxes important for cloud development led by Fairley Barnes. Two additional projects related to boundary layer clouds were led by Steve Ghan and the team of Ronald Stull (of which the lead author of this chapter was a member). The ARM Program's research efforts have since grown from these original projects to include diverse teams of scientists working at National Laboratories, academia, and private industry. Most of this chapter will focus on two research areas in which the ARM Program has made significant progress:

- Understanding the impact of small-scale variations in the surface fluxes, how these variations should be represented in large-scale atmospheric models, and land-use-induced circulation patterns driven by differential land use.
- Determining how to properly represent subgrid variability in the planetary boundary layer and how that variability is related to the onset and maintenance of boundary layer clouds.

While the central focus of this chapter is on these two areas, significant progress has been made in understanding several other important atmospheric processes including understanding the amount of downwelling radiation at the surface and the surface albedo, understanding the carbon cycle, and evaluating a wide range of atmospheric and coupled land–atmosphere models. These latter topics will be covered in a more abbreviated fashion.

2. Relevant ARM instrument systems

The ARM Program deployed several critical instrument systems to enable studies related to land–atmosphere

* Deceased.

Corresponding author address: Larry K. Berg, Pacific Northwest National Laboratory, 902 Battelle Blvd., K9-30, Richland, WA 99352.
E-mail: larry.berg@pnnl.gov

DOI: 10.1175/AMSMONOGRAPHS-D-15-0044.1

interactions. Often, these instrument systems are not as high profile as other systems, but they form the basis of many of the results presented in the rest of this chapter. Measurements of the most relevant properties (e.g., surface sensible and latent heat fluxes, soil surface heat flux, soil moisture, near-surface air temperature, and humidity) are made using a suite of instrument systems deployed at extended facilities around the primary central facility of each site [e.g., the Southern Great Plains (SGP); Sisterson et al. (2016, chapter 6)]. The Surface Meteorological Observations Systems (SMOS), later called Surface Meteorological stations (SMET) systems, provide measurements of the near-surface temperature, humidity, atmospheric pressure, and winds (Richardson et al. 2000). In the case of the SGP site, these measurements are augmented by measurements made by the Oklahoma Mesonet (Brock et al. 1995).

One challenge at the start of the ARM Program was to determine the locations for the extended facilities. Initially, a regular grid of extended facilities was envisioned at the SGP, but this view changed as the nature of both meteorological parameters and land use became better understood. Under the leadership of Marv Wesley, a team of scientists examined the variation of temperature and moisture across the SGP site. They documented an east–west gradient in precipitation (with irrigation often being required in the western half of the site). The team found that the wintertime precipitation in the northern half of the domain was frequently snow, while rain dominated in the southern half of the domain. These findings ultimately led to dividing the SGP ARM Climate Research Facility (ACRF) into four quadrants in order to capture these variations. The team also documented that the dominant land-use categories across the area were pasture and cropland, as well as a relatively small amount of wooded land. The extended facilities were distributed across the various land-use types in each quadrant to ensure that the measurements captured the heterogeneity across the site.

The ARM Program measures the surface sensible and latent heat fluxes using two different instrument systems: the energy balance Bowen ratio (EBBR) and the eddy correlation flux measurement (ECOR). The EBBR systems estimate the fluxes based on measurements of near-surface temperature and humidity gradients and the surface energy balance [net radiation plus soil heat flux; e.g., Stull (1988)]. One advantage of the EBBR system is that the surface energy budget is, by definition, closed. EBBR systems were installed first at the SGP site, followed by the addition of ECOR systems. The ECOR (Brotzge and Crawford 2003) systems use ultrasonic anemometers and fast response water vapor/carbon dioxide analyzers to infer carbon dioxide, sensible and latent heat fluxes, in addition to momentum flux. The EBBR systems were deployed at undisturbed sites, and thus they tended to be operated in either pasture or range land, while the ECOR systems were sited in locations that included disturbed cropland. The need for long-term deployment precluded the setup of the ECOR within the fields, thus they were placed along the predominantly downwind edge of the field. The neighboring land could have different use, so extra care is needed to properly interpret the surface flux data.

Soil moisture measurements are made over the relatively large spatial scales of the SGP using Soil Water and Temperature Systems (SWATS). While networks of surface measurements are relatively common, measurements of soil moisture profiles through the root zone are not common. The SWATS are collocated with the other ARM surface observations to enable measurement of the impact of soil moisture on the surface energy budget (e.g., Schneider et al. 2003). These measurements are used in a number of the studies identified in the next section and can be used to evaluate land surface and fully coupled models. Soil moisture measurements can be used to nudge model simulations, enabling researchers to more easily evaluate the representation of other processes in models. One example is the work of Berg and Zhong (2005), who used ARM-observed soil moisture measurements to adjust modeled soil moisture, so that the parameterization of the surface fluxes used in the model could be evaluated.

In addition to the routine measurements at the ARM sites, intensive field studies have been a hallmark of the program from its initiation. During the summer of 2007, the ARM community performed a major field study with the goal of better understanding land surface interactions and boundary layer clouds. This study was called the Cloud and Land Surface Interaction Campaign (CLASIC; Miller et al. 2007). The study used a suite of seven research aircraft as well as enhanced surface measurements to investigate the links between surface properties, the thermodynamic state of the atmosphere, and summertime, fair-weather clouds. Additional measurements made during CLASIC included passive and active remote sensing retrievals of soil moisture that were made from one of the research aircraft (Bindlish et al. 2009). Additional measurements were made during the DOE Atmospheric Science Program (ASP) Cumulus Humilis Aerosol Processing Study (CHAPS; Berg et al. 2009).

3. Improving our understanding of the land surface and its impact on weather and climate

a. *Local- and regional-scale fluxes*

Scaling point measurements of the surface sensible and latent heat fluxes up to larger spatial/temporal

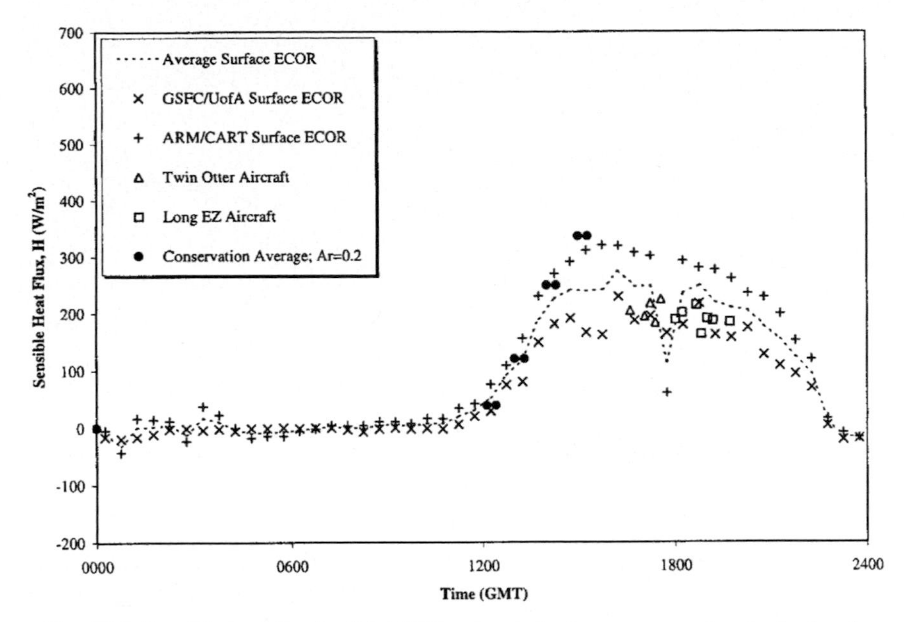

FIG. 23-1. Sensible heat flux for 5 Jul 1997, estimated using a conservation method and an assumed value of heat flux near the top of the boundary layer. Solid circles indicate the average values using different methods to estimate advection. Other symbols indicate measurements of the heat flux from different surface and airborne platforms. Note that the conservation estimates can only be defined when the tethersonde passed through the top of the boundary layer, leading to missing values in the afternoon when the boundary layer depth was greater than 1000 m. [Courtesy of Peters-Lidard and Davis (2000).]

scales has remained an important challenge in climate science (e.g., Sellers et al. 1995). The issue can be divided into two aspects: estimating the grid-scale flux, and the relative importance of subgrid variability of the flux and small-scale circulations that could be introduced by surface heterogeneity.

In practice, two different methods can be used to determine average regional-scale fluxes, either by scaling up based on the application of boundary layer scales (e.g., Brutsaert and Kustas 1987; Sugita and Brutsaert 1990) or based on the conservation of scalars within the boundary layer (e.g., McNaughton and Spriggs 1986). ARM SGP measurements have been used for a large number of studies, because of the extensive distribution of surface stations and surface flux measurements. Stull (1994) proposed a new method to represent the surface fluxes by using a "buoyancy velocity scale" that could be applied in cases with very small mean wind speeds, a case when traditional bulk parameterizations fail. This method also could be used to derive surface fluxes from radiometric measurements that are made over large spatial domains. A series of studies, conducted by Qin Xu and his team at the University of Oklahoma, focused on ways to improve estimates of the regional-scale surface fluxes. Xu and Qiu (1997) developed a variational method by combining similarity theory (e.g., Brutsaert and Kustas 1987) with surface flux measurements. Their method improved the simulation of the surface fluxes relative to the measurements. This work was extended by Xu et al. (1999) with the addition of two new constraints: the inclusion of 1) standard deviation of the wind and temperature and 2) soil moisture and a simple soil–vegetation model. The model was further refined by Zhou and Xu (1999) to retrieve soil moisture and surface skin temperature. The SGP 1997 Hydrology Experiment was an ARM-supported field study that included a component related to the computation of regional-scale fluxes. In their work, Peters-Lidard and Davis (2000) used the conservation approach and compared their results to surface flux measurements made across the SGP site. They found good agreement between the estimated and measured sensible heat flux (Fig. 23-1) for periods in which mixed-layer depth was sufficiently shallow that it could be sampled with a tethersonde system.

Simplified models can also be used to gain insight into the effect of soil moisture on regional-scale fluxes. Santanello et al. (2007) used ARM radiosonde and surface flux measurements in conjunction with a one-dimensional atmospheric model to investigate the effect of soil moisture on a number of different boundary layer variables. This led to a methodology for describing the regional-scale fluxes. Based on their analysis they described two limiting cases: instances where conditions

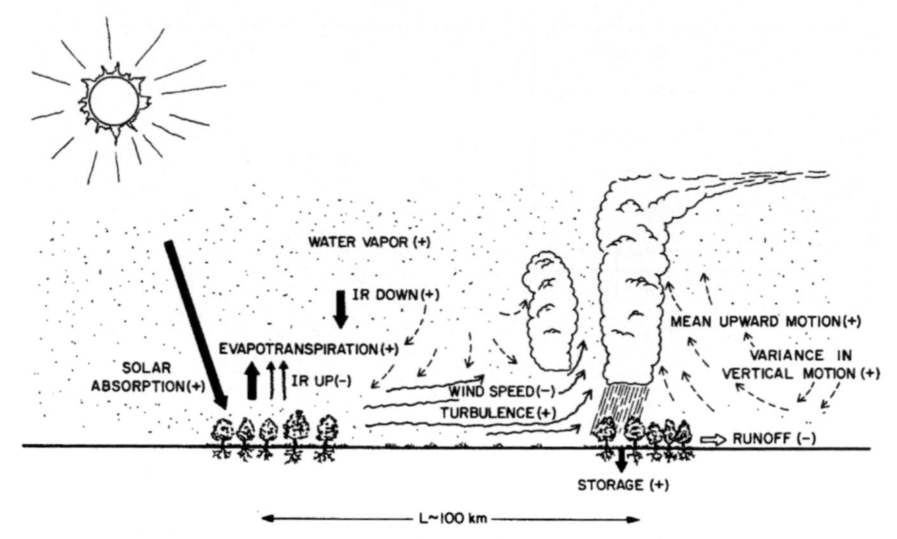

FIG. 23-2. Anticipated effect of bands of vegetation and dry soil in a semiarid region. Changes in the effect, either positive or negative, are indicated by the pluses or minuses. [Courtesy of Anthes (1984).]

are limited by the thermodynamic properties of the atmosphere and instances where conditions are limited by soil properties. In the case defined as atmosphere limited, they found a strong relationship between soil water content and sensible and latent heat fluxes only when the soil water content was less than 10%. For greater soil water content they found that the correlation was much weaker. They postulated that this weaker correlation was the result of an essentially unlimited supply of water at the surface, plus the balance of moistening of the boundary layer by surface evaporation and drying by entrainment of dry air from aloft.

b. Land-use-induced circulation patterns

Prior to 1995, it was speculated that a heterogeneous landscape with a pattern of alternating land use and soil moisture, such as broad bands of alternating crops, could give rise to organized mesoscale circulations. It was thought that accurately accounting for the subgrid spatial variability of surface fluxes could be important for climate simulations (e.g., Anthes 1984; Ookouchi et al. 1984; Yan and Anthes 1988; Pinty et al. 1989; Pielke et al. 1991; Avissar and Chen 1993). In this conceptual model, alternating bands of convection are expected to form over relatively moist areas (Fig. 23-2), similar to the circulation patterns associated with sea-breeze circulations common around oceans or large lakes. In their work, Avissar and Chen (1993) pointed out that atmospheric models in use at that time did not account for subgrid mesoscale circulations that could develop as a result of differences in land use within the model grid box. However, the majority of these studies were based on model simulations of an idealized atmosphere.

With early support from the ARM Program, Doran et al. (1995) conducted a field study over eastern Oregon, in an area with an alternating patchwork of native semiarid steppe and irrigated farmland, to investigate the nature of induced circulations in the real atmosphere. While not one of the official ARM sites, this area appeared ideal for investigating mesoscale flows because of the small amount of annual rainfall in the Columbia basin and the large extent of irrigation. During the course of the study, they measured sensible heat flux over nonirrigated steppe that was a factor of 4 greater than was measured over irrigated fields. Their work focused on two case studies with different ambient wind speeds. For the case with relatively weak ambient wind speeds (4–7 $m\,s^{-1}$) evidence of a local circulation was observed, while for the case of moderate wind speed (7–10 $m\,s^{-1}$) no local circulations were observed. These results led them to conclude that advection and large-scale forcing play a key role in the formation of local circulations. To investigate further, Zhong and Doran (1995) conducted additional analyses, including running a suite of simulations using the Regional Atmospheric Modeling System (RAMS) model. The simulations included several different cases: simulations with real topography and land use, simulations with real topography and uniform land use, uniform topography and real land use, and uniform topography and land use. An example simulation is shown in Fig. 23-3, which highlights regions of confluence and difluence within the model domain. Based on the observations and simulations, they concluded that while local circulation

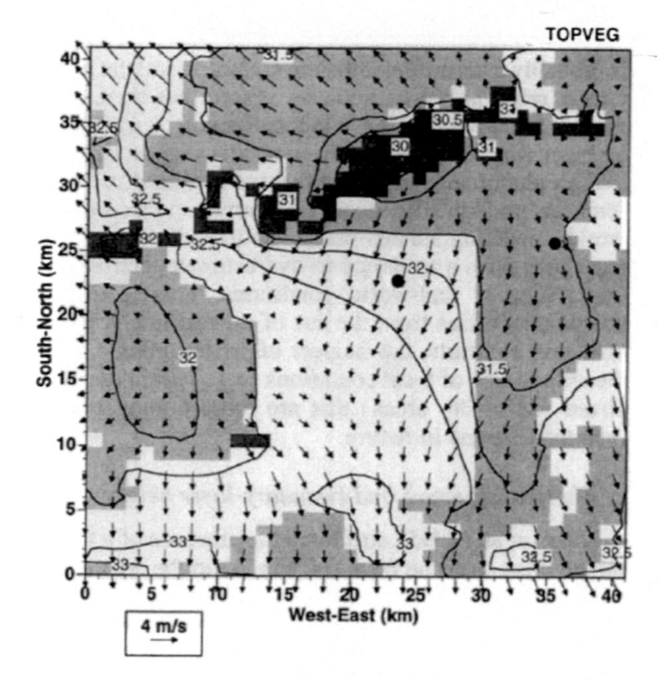

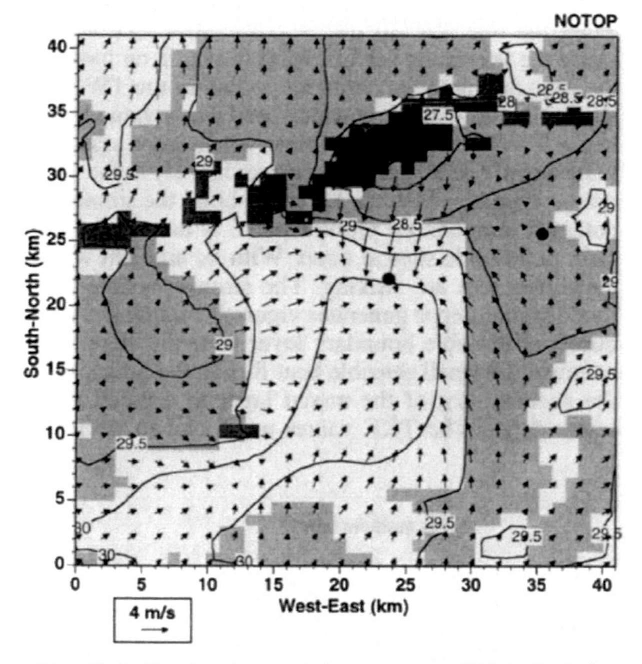

FIG. 23-3. Simulated potential temperature (°C) and winds at 1400 LST for RAMS simulations using (top) both realistic topography and vegetation and (bottom) vegetation only. Gray indicates irrigated farmland, white indicates native shrub steppe, and black indicates the Columbia River. [Courtesy of Zhong and Doran (1995).]

patterns are possible, they can be masked by the complex interplay of advection, large-scale subsidence, and shear-generated turbulence.

The deployment of ARM instruments at the SGP site, as well as advances in regional-scale land surface models, provided the opportunity to investigate the role of land-use differences in the development of local-scale circulations over the SGP. During the summer of 1996, a field study was conducted (Hubbe et al. 1997) to investigate local-scale circulations over the SGP region. This study included additional radiosonde launches at three additional sites, each approximately 150 km from the central facility: Elk Falls, Kansas; Meeker, Oklahoma; and Plenva, Kansas (Fig. 23-4). These locations were selected because of the differences in land use between them. In addition to the surface flux measurements and the calculation of the fluxes from changes in the thermodynamics profile, the Simple Biosphere Model (SiB2; Sellers et al. 1996) was used to estimate surface fluxes over the entire SGP site. Two days (7 and 12 July 1995) were selected because of large contrasts in the surface flux measurements between locations (Fig. 23-5), which ranged from approximately 600 $\mathrm{W\,m^{-2}}$ near the central facility, to near 0 $\mathrm{W\,m^{-2}}$ in the northeast corner of the domain. Based on their findings, Hubbe et al. (1997) concluded that local-scale circulations were nonexistent during their study period. In additional simulations of the same field study using RAMS driven with surface fluxes generated by SiB2, Zhong and Doran (1998) showed that the perturbations to the velocity field associated with land-use-induced circulations were quite small, on the order of 1 $\mathrm{m\,s^{-1}}$ in the northeast corner of the domain (cf. an ambient flow of approximately 4 $\mathrm{m\,s^{-1}}$; Fig. 23-6). They also found that the spatial variability of surface fluxes did not have an impact on the average temperature and humidity profiles calculated over spatial scales representative of a global climate model (GCM) grid box. As part of their study, they repeated the analysis using an idealized checkerboard pattern for representing the surface fluxes and different ambient wind speeds. They obtained similar results; the subgrid land-use pattern did not have a significant impact on the average fluxes.

c. *Variability in the planetary boundary layer and boundary layer clouds*

By their nature, atmospheric models predict the gridbox mean values of temperature and moisture. Within the convective planetary boundary layer (PBL), however, there can be significant horizontal variability, even at small spatial scales. The subgrid variability is thought to be very important for determining the onset of boundary layer clouds (e.g., Wilde et al. 1985). Using data from the Hydrologic Atmospheric Pilot Experiment (HAPEX), Schrieber et al. (1996) presented joint frequency distributions (JFDs) of temperature and humidity and found that their size and shape could be related to the underlying surface properties. For long flight

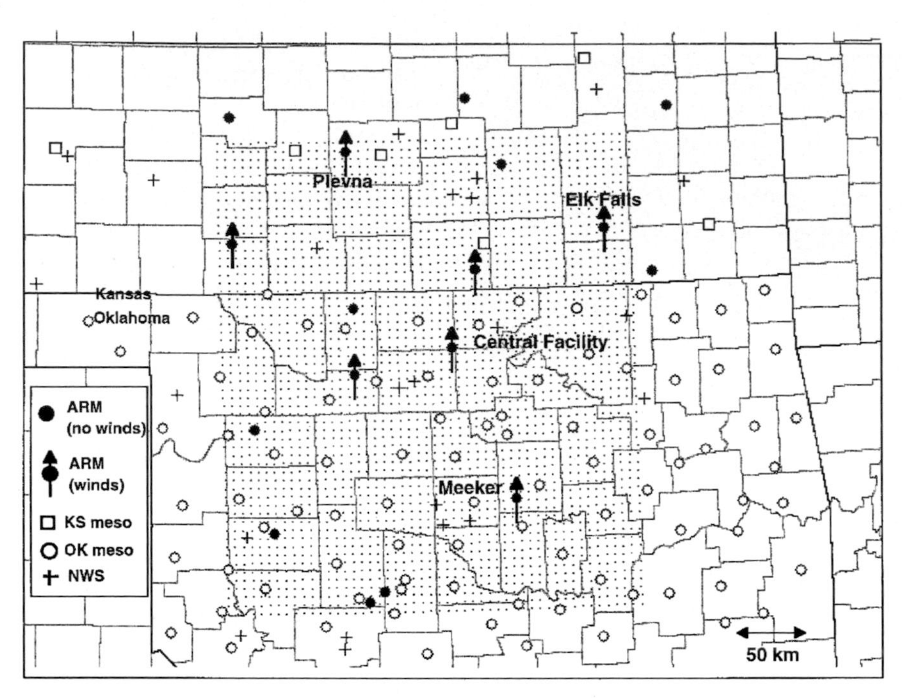

FIG. 23-4. Map of the SGP ACRF as it was configured during the study of Hubbe et al. (1997), showing the location of ACRF extended facilities, Oklahoma and Kansas State Mesonets, and NWS sites. Winds were only measured at the indicated sites and extra radiosondes were launched from Plevna and Elk Falls, KS; and Meeker, OK. [Courtesy of Hubbe et al. (1997).]

legs, the JFDs were quite complicated, but when the flight pattern was broken up by land-use type, the observed JFDs were monomodal and better behaved (Fig. 23-7). Their worked stopped short of developing a parameterization for the size and shape of the JFDs that could be derived from atmospheric models. Stull's research team addressed this issue by relating the size and shape of the distributions to the jump in temperature

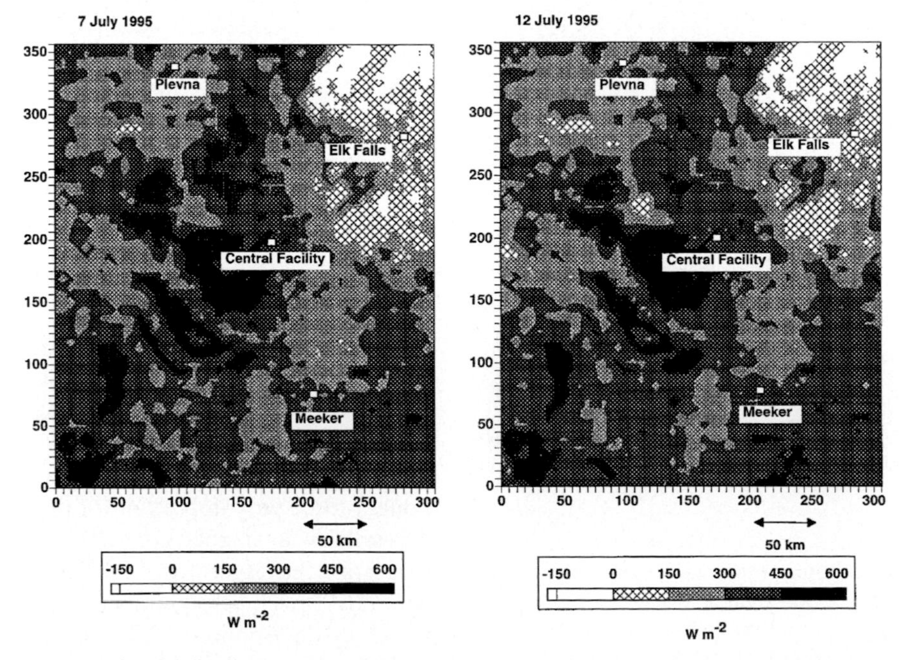

FIG. 23-5. Spatial distribution of sensible heat fluxes over the SGP ACRF, calculated using the SiB2 model at 1300 LST 7 and 12 Jul 1995. [Courtesy of Hubbe et al. (1997).]

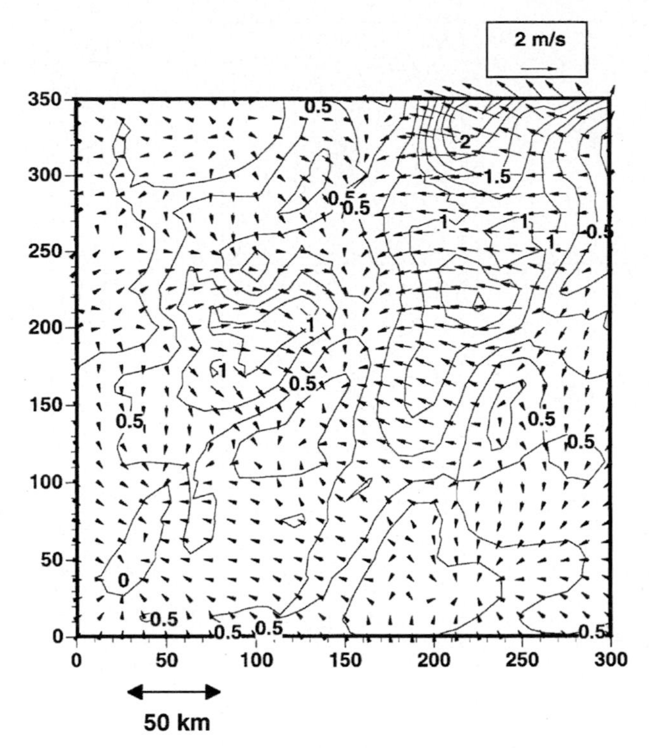

FIG. 23-6. Simulated perturbation surface wind field over the SGP ACRF at 1400 LST 12 Jul 1995. Wind speed contours are $0.25\ \mathrm{m\,s^{-1}}$. [Courtesy of Zhong and Doran (1998).]

and moisture at the surface and at the top of the PBL (Berg and Stull 2004). Their conceptual model is shown in Fig. 23-8 for a JFD of potential temperature θ and water vapor mixing ratio r. A number of different processes act to change the mixed-layer mean values of θ and r, including the surface fluxes, entrainment flux at the boundary layer top, and advection. Equipped with data collected during Boundary Layer Experiment 1996 (BLX96) using a research aircraft over the SGP site (Stull et al. 1997), they used their new method to represent the size and shape of the JFDs as a function of height within the PBL (Fig. 23-9).

Distributions such as these were related ultimately to the formation of boundary layer clouds by Berg and Stull (2005) with their development of the Cumulus Potential (CuP) scheme for shallow clouds. They used the JFDs of θ and r to define the distribution of thermodynamic properties in the mixed layer. In their scheme, the virtual potential temperature θ_v is compared to the mixed-layer mean value of θ_v to determine which parcels rise. Of the parcels that rise, those that reach their lifting condensation level (z_{LCL}) will form clouds. The cloud fraction is computed as a function of the fraction of parcels represented by the JFD that form clouds, the cloud-base height is the height of the z_{LCL}, and the cloud-top height is determined by the altitude at which the clouds reach their level of neutral buoyancy. Their parameterization was consistent with the results of Zhu and Albrecht (2002), who reported that the formation of the shallow clouds was a complicated function of processes within the boundary layer, including the fluxes at the top of the convective boundary layer.

Other ARM-supported studies included efforts to evaluate the performance of large-eddy simulation (LES) models and their simulation of boundary layer clouds. In intermodel comparisons of an idealized case of shallow boundary layer clouds over the SGP, Brown et al. (2002) found that subcloud conditions simulated by various LES models agreed well with standard boundary layer scales. They also pointed out the important impact of cloud-layer stability in the formation and maintenance of cloud fields. Using data from the same case study, Neggers et al. (2004) evaluated the mass flux closure scheme using both an LES and a single-column model to help understand the inner workings of the respective models.

Other researchers have focused on the formation of nocturnal stratus clouds over the SGP site. Zhu et al. (2001) used LES to investigate the role of turbulence in the formation of nocturnal clouds. They found that the height of the lifting condensation level and the critical level, which they defined to be a function of the Monin–Obukhov length, could be used to help define the onset of clouds.

d. Carbon cycle

Given its location in the central United States, and long-term record of a wide range of meteorological parameters, the SGP site has attracted the attention of scientists interested in the carbon cycle, including the exchange of carbon at the surface and at the top of the planetary boundary layer. Since 2001, the ARM Carbon Project has been conducted by scientists from Lawrence Berkeley National Laboratory (LBNL). This effort has included both surface measurements (Billesbach et al. 2004) and a unique long-term airborne component operated in collaboration with aerosol measurements made by scientists from the National Oceanic and Atmospheric Administration (NOAA) Earth System Laboratory (Andrews et al. 2004; Biraud et al. 2013). Rather than the short-duration aircraft studies commonly used in atmospheric research, the LBNL and NOAA groups utilized a small single-engine aircraft that could be operated on a regular schedule (biweekly weather conditions allowing) from the Ponca City, Oklahoma, airport. This type of deployment is unique, and at the time the program started, it was the only such effort over the central United States. Two different measurement systems have been deployed during the course of the study: a flask-based system to measure bulk CO_2 properties along horizontal flight legs; and starting in June 2007, a continuous CO_2 analyzer (Biraud

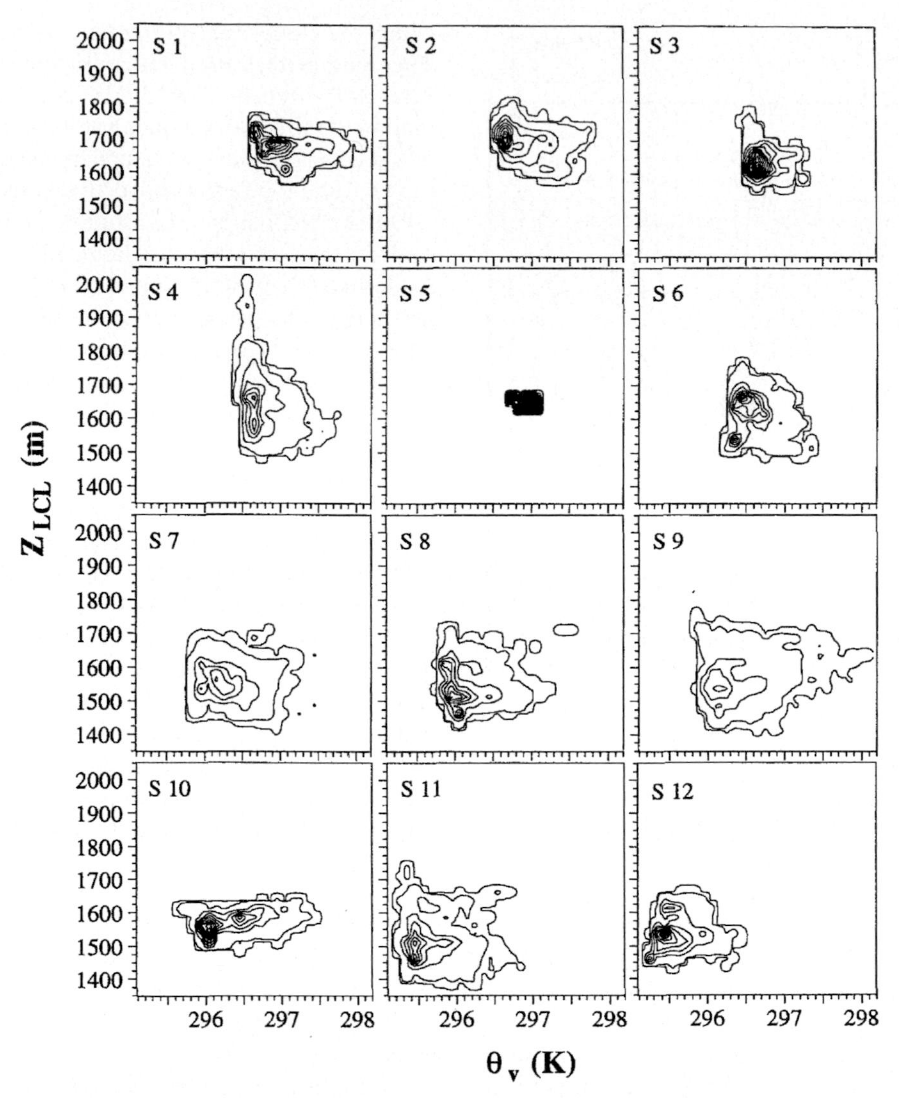

FIG. 23-7. Joint frequency distributions for subsets of a HAPEX flight leg on 9 Jun 1986. Values are sorted into bins of sizes 0.1 K (θ_v) and 25 m (z_{LCL}). The contours indicate frequencies of 0.000 01, 0.0001, 0.001, 0.01, and every 0.01 thereafter. Segment numbers are indicated in the top-left corner of each plot. [Courtesy of Schrieber et al. (1996).]

et al. 2013). The long-term measurements show a systematic increase in CO_2 concentration over the SGP ACRF, as well as a strong seasonal cycle (Fig. 23-10). Figure 23-10 indicates more variability in CO_2 concentration within the PBL than above it, which is consistent with the turbulent nature of transport within the PBL. Fischer et al. (2007) used surface measurements from near the SGP Central Facility to document the effects of crop type, land use, and soil moisture on the carbon exchange for the land types that were included in their study.

e. Model evaluation

Long-term, high-quality measurements of surface sensible and latent heat fluxes over a wide geographic area are relatively rare. The data collected by the ARM Program helps to fill this void and has been used by several researchers to evaluate the performance of mesoscale atmospheric models. Some of these (e.g., Oncley and Dudhia 1995; Berg and Zhong 2005) focused directly on the ability of regional-scale models to predict the surface fluxes. Oncley and Dudhia (1995) showed that the fifth-generation Pennsylvania State University–National Center for Atmospheric Research Mesoscale Model (MM5), when using the Blackadar boundary layer scheme, could successfully predict the surface fluxes if the correct values of surface roughness and soil moisture were used. Berg and Zhong (2005) also used MM5 and evaluated the surface fluxes predicted by

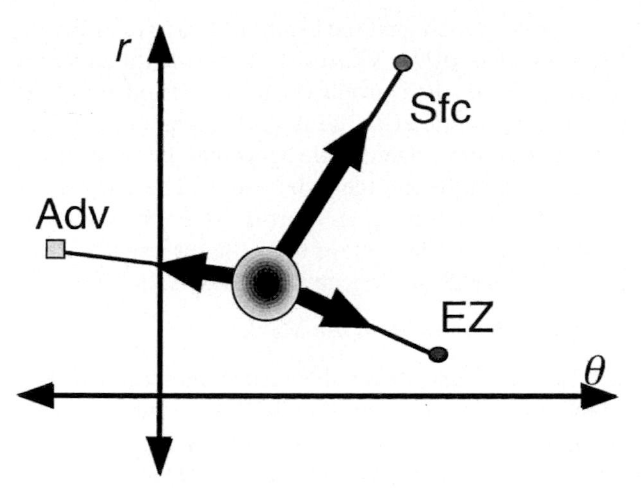

FIG. 23-8. Mixing diagram of potential temperature θ vs water vapor mixing ratio r. The large circle represents the mixed-layer mean, small circles represent the surface (Sfc) and entrainment-zone properties (EZ), and the square represents advection (Adv). The large arrows indicate the "direction" that the different processes pull on the mixed-layer mean values. [Courtesy of Berg and Stull (2004).]

three different turbulence parameterizations, including both local turbulence kinetic energy (TKE)-based schemes and nonlocal parameterizations designed for convective conditions. In addition to the ARM surface flux measurements, they also used airborne observations to evaluate the fluxes aloft and the PBL depth. They found that while the modeled latent heat fluxes agreed reasonably well with measurements, sensible heat flux was overestimated because of an overestimate of net radiation (Fig. 23-11). Cooley et al. (2005) used MM5 coupled with the Land Surface Model 1 to investigate the impact of the winter wheat harvest in the region around the SGP site. They found that an early wheat harvest, indicative of warmer and dryer conditions, produced a positive feedback on the climate system and even more drying.

Other studies have focused on the representation of the linkages between the land or ocean and the atmosphere in GCMs. Sud et al. (2001) used measurements from the SGP site to evaluate the Goddard Earth Observing System (GEOS-2) GCM single-column model. They investigated linkages between surface properties and rainfall, and precipitation recycling. Dirmeyer et al. (2006) evaluated the performance of 12 different climate models in regards to the simulation of the surface fluxes using measurements from the SGP, a flux site at Bondville, Illinois, and a number of surface stations located in Europe. They found that the individual models often did a poor job of simulating the surface fluxes and soil moisture.

More recently, SGP measurements have been used to evaluate land–atmosphere interactions in cloud-resolving models (CRMs). Zeng et al. (2007) compared results from two-dimensional and three-dimensional CRMs. Their study was unique because of the relatively long 20-day simulation periods and the detailed attention that was paid to the land surface model. They found that the treatment of the damping of the vertical velocity in the two-dimensional models was suspect, leading to rapid changes in the surface precipitation and unrealistic drying aloft compared to the three-dimensional simulations.

f. Surface radiation

The ARM Program has made important contributions to the understanding of radiative transfer within the atmospheric column, as described in detail in McFarlane et al. (2016, chapter 20). For land–atmosphere interactions,

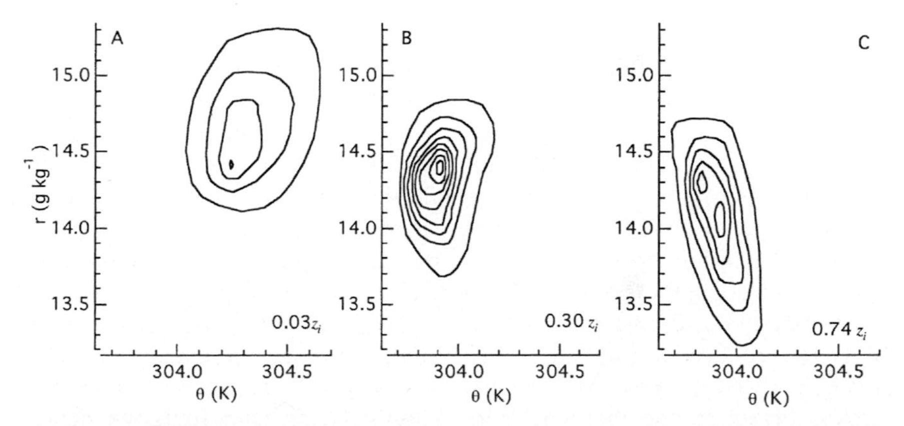

FIG. 23-9. JFDs of potential temperature θ and water vapor mixing ratio r observed at (a) $0.03z_i$, (b) $0.30z_i$, and (c) $0.74z_i$ on 16 Jul 1996, where z_i is the height of the boundary layer. The first contour indicates a normalized frequency of 0.005 and the contour interval is 0.01. [Courtesy of Berg and Stull (2004).]

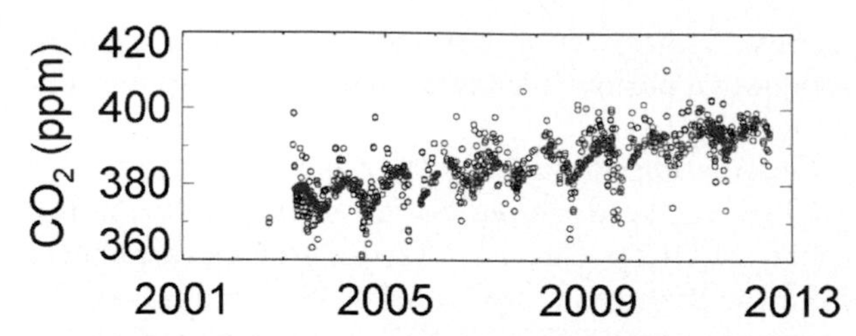

FIG. 23-10. Time series of CO_2 collected from September 2002 to July 2012 using the flask system on the research aircraft at altitudes of 3000 m (red) and 1000 m (black). These heights were selected to be above and within the PBL. [Courtesy of Biraud et al. (2013).]

the focus is primarily on issues related to the surface energy budget and the surface albedo. In most cases, the majority of solar radiation is absorbed at the surface, making our understanding of the amount of downwelling radiation and the surface albedo critical to our interpretation of the surface energy budget and the partitioning of sensible, latent, and soil surface heat flux. The normalized difference vegetation index (NDVI), which is computed using radiances measured in the near-IR and visible parts of the spectrum, is a commonly used metric that provides information about the leaf-area index, biomass, and other land-use parameters. Gao et al. (1998) used NDVI to estimate the surface roughness in their model Parameterization of Subgrid-Scale Processes (PASS). Regional-scale sensible and latent heat fluxes were derived and were compared to the SGP EBBR surface flux measurements.

SGP measurements have been used to evaluate the impact of solar zenith angle on the surface albedo and to examine the assumption that the albedo is only a function of the solar zenith angle, and not other surface properties. Using surface, as well as satellite measurements, Minnis et al. (1997) found that surface albedo was consistently greater in the morning than was observed during periods in the afternoon with the same solar zenith angle. This result was attributed to dew on the surface in the morning (Fig. 23-12). In subsequent work, Yang et al. (2008) used measurements collected at the SGP and Tropical Western Pacific (TWP) sites, in addition to NOAA SURFRAD sites, to derive the surface albedo from the Moderate Resolution Imaging Spectroradiometer (MODIS) using parameters commonly applied in models, including those from the National Centers for Environmental Prediction (NCEP) Global Forecast System (GFS). In contrast to the results presented by Minnis et al. (1997), they found that the albedo was not always higher in the morning (Fig. 23-13). They offered several possible explanations for the difference in results, in particular that the data record they used was significantly longer (ranging from 2 to 8 years depending on the location) than was available to Minnis et al. (1997). Gao et al. (1998) identified biases in the MODIS data products and suggested that it is possible to parameterize the albedo only as a function of the solar zenith angle, while ignoring the land cover (assuming that the surface is snow free). The contrasting results found when using short-duration versus long-duration data records prove the value of long-term deployments like those associated with the ARM Program.

g. *Results from beyond the SGP*

While the majority of the results presented in this chapter are focused on the SGP site, other ARM-supported research efforts related to land–atmosphere interactions have been conducted at the TWP and North Slope of Alaska (NSA) sites as well as at deployments of the ARM Mobile Facility (AMF), which was first deployed at Pt. Reyes, California, during 2005.

The TWP sites included the deployment of measurement systems at a number of remote island locations. When using data collected at these sites, it is important to understand the impact of the island on both the in situ and remote sensing measurements as well as on the populations of clouds themselves (e.g., McFarlane et al. 2005). These efforts are described in detail in Long et al. (2016, chapter 7), and the interested reader should review that chapter to understand the important progress that has been made for these sites.

The development of the AMF marked an opportunity to examine the coupling of land and atmosphere at sites well removed from the fixed ARM sites. The Convective and Orographically Induced Precipitation Study (COPS; Wulfmeyer et al. 2011) focused on improving our understanding of precipitation in regions of complex terrain and included the deployment of the AMF to the village of Heselback in the Murg Valley of Germany. Using soil moisture data collected at 47 stations within the COPS domain, Hauck et al. (2011) showed that the model they applied generally had a dry bias. They documented that the bias has a large impact on the simulated precipitation, although no systematic relationships were found. Similar results were reported by Barthlott and Kalthoff (2011), who found consistent changes in the near-surface moisture and latent heat fluxes, but only saw a systematic increase in precipitation for simulations that were relatively dry.

ARM-supported research has demonstrated that land–atmosphere coupling is important in the Artic, and it can have a measurable impact on boundary layer clouds. Using data from the microwave radiometers (MWRs) deployed at Barrow and Atqasuk, Alaska, Doran et al. (2002) showed that the cloud liquid water path (LWP) associated with low clouds was generally

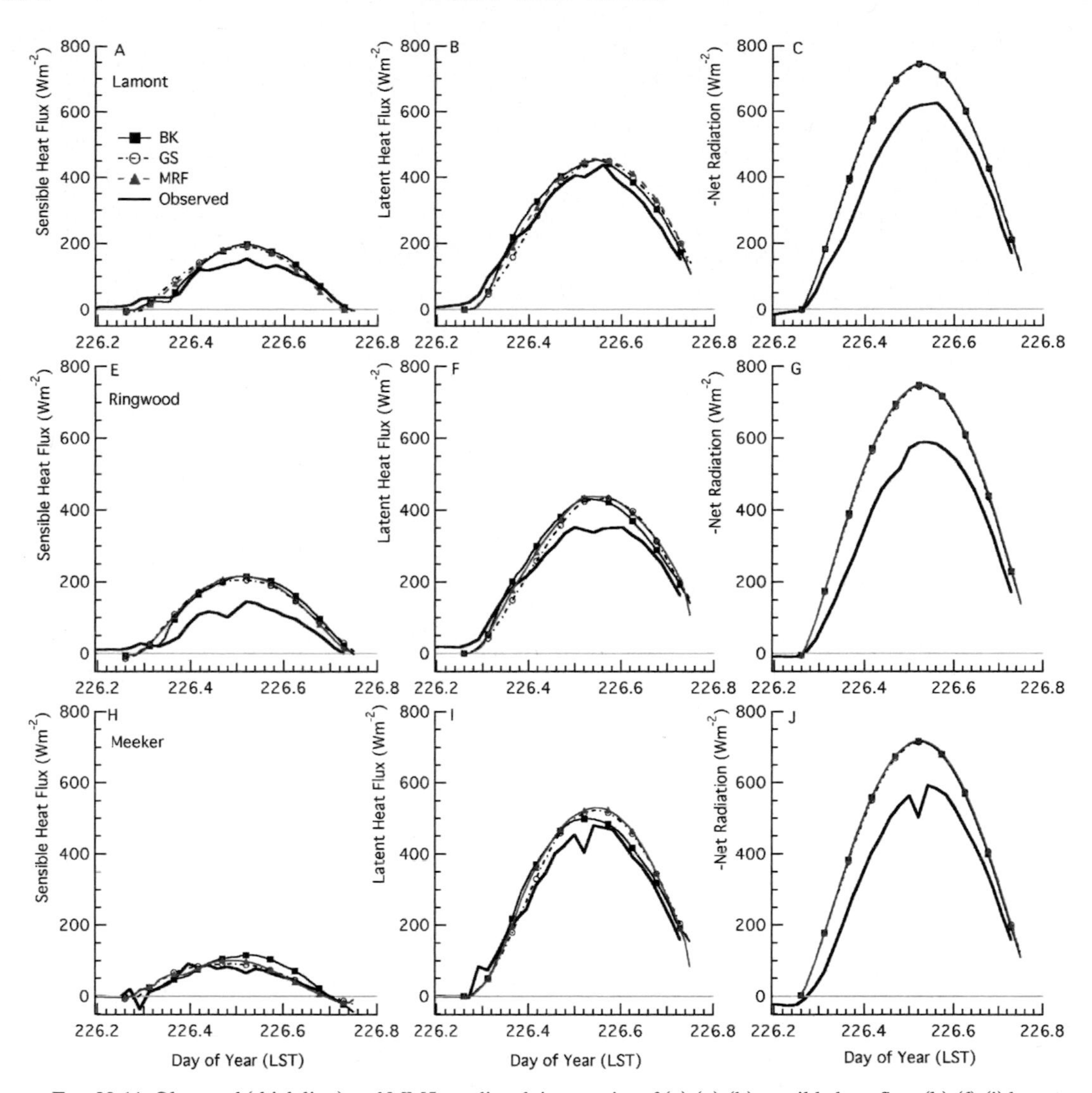

FIG. 23-11. Observed (thick line) and MM5-predicted times series of (a),(e),(h) sensible heat flux; (b),(f),(i) latent heat flux; and (c),(g),(j) net radiation using the Blackadar (BK; squares), Gayno–Seaman (GS; circles), and Medium Range Forecast (MRF; triangles) parameterizations on 13 Aug 1996 near (top) Lamont, OK; (middle) Ringwood KS; and (bottom) Meeker, OK. Note that net radiation is defined as positive away from the surface. [Courtesy of Berg and Zhong (2005).]

larger near the ocean at Barrow than 100 km inland at the Atqasuk site, but this relation was a function of the month. They also compared the ARM observations to simulations using the European Centre for Medium-Range Weather Forecasts (ECMWF) model and found that the model had a dry bias in regards to relative humidity and tended to underpredict the cloud LWP (Doran et al. 2002). This work was extended to include a multiyear study period, and an analysis of the cloud optical depth by Doran et al. (2006). In contrast to their earlier study, they found that the clouds at Atqasuk had a larger optical depth and cloud LWP for cases with onshore flow. They attributed this behavior to an increase in the sensible and latent heat fluxes associated with air parcels moving inland over the relatively warm tundra.

4. Looking to the future

The focus of this chapter has been an overview of research conducted prior to the merger of the ASP and ARM Science Programs into the Atmospheric System Research (ASR) Program (Mather et al. 2016, chapter 4). While the ASP and ARM Science Programs led to significant advances and increases in our scientific understanding, they do not mark the end of studies of land–atmosphere interactions supported by the ARM Program. Research related to land–atmosphere interactions

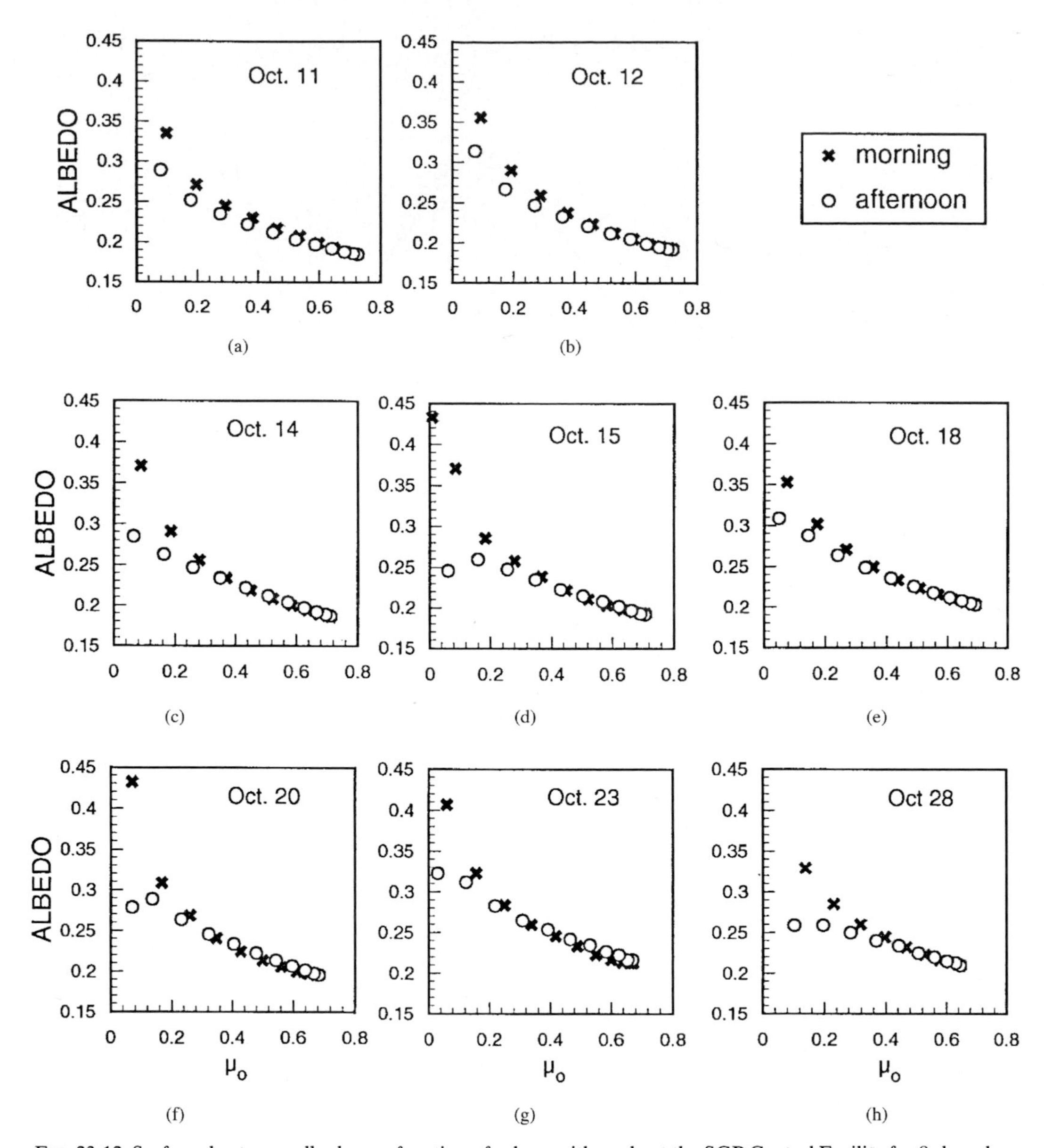

FIG. 23-12. Surface shortwave albedo as a function of solar zenith angle at the SGP Central Facility for 8 clear days in October 1995. [Courtesy of Minnis et al. (1997).]

has continued and more recent research has made use of a range of old and new ARM data streams to address relevant science questions.

One major focus of recent ASR research has been to improve our understanding of the life cycle and impact of boundary layer clouds. Using radiometric measurements from the SGP site, Berg et al. (2011) documented the impact of shallow clouds on the surface energy budget and found that the radiative forcing at the surface associated with shallow cumuli was $-45.5\,\mathrm{W\,m^{-2}}$ (out of $612\,\mathrm{W\,m^{-2}}$ estimated for clear-sky conditions). Measurements from the ARM cloud radar and micropulse lidar has been combined into the active remotely sensed clouds locations (ARSCL) data product developed by Clothiaux et al. (2000). This product has been used to investigate the diurnal macrophysical properties (cloud cover, cloud-base, and cloud-top heights) of fair-weather cumuli (Berg and Kassianov 2008). Using measurements from ARM instrument systems, Zhang and Klein (2010) documented systematic differences in the thermodynamic structure of the atmosphere associated with the transition from shallow to deep convection. Measurements from the SGP have been used to document improvements in the prediction of shallow cumuli. Berg et al. (2013) showed significant improvement in the forecast of downwelling shortwave

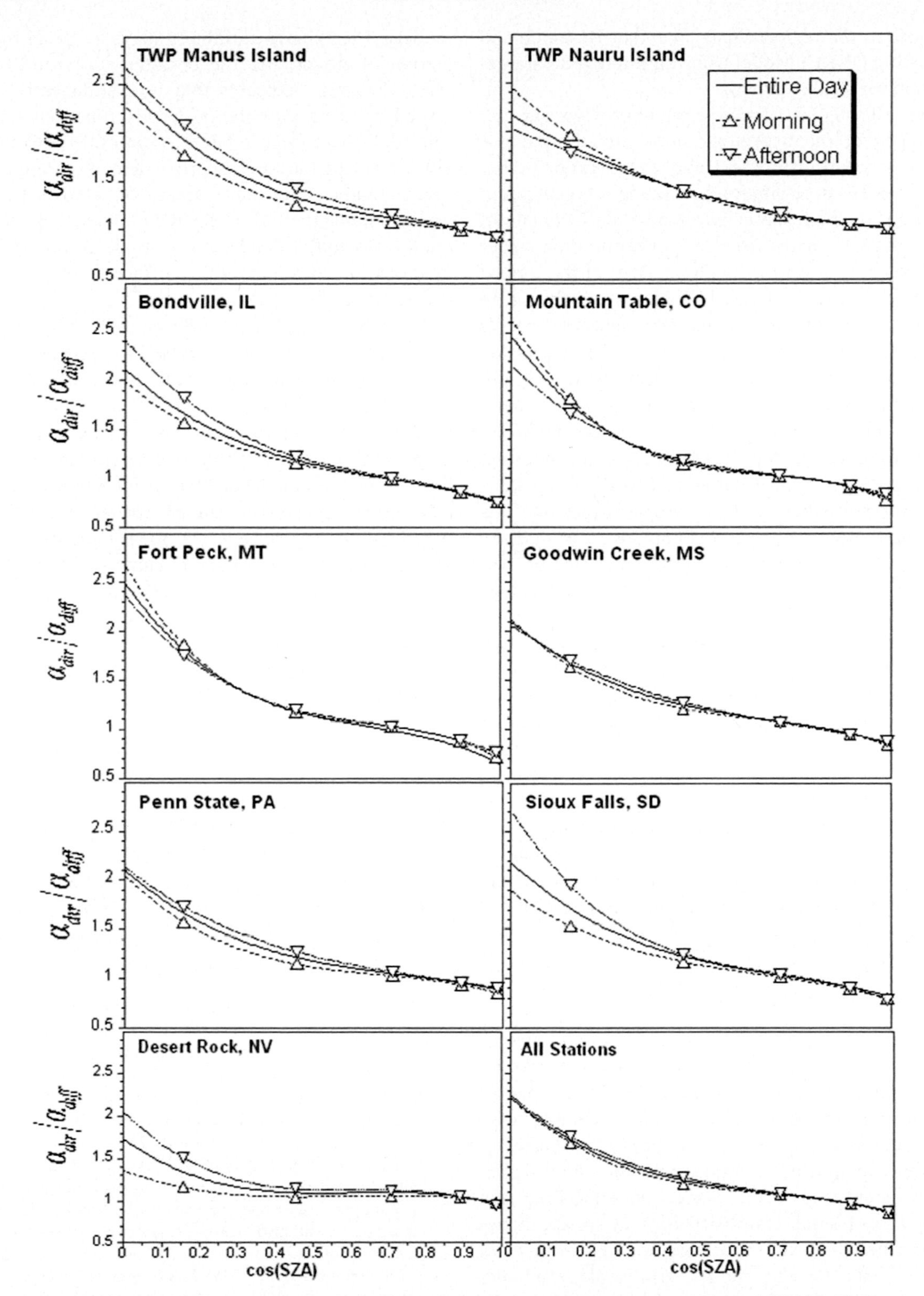

FIG. 23-13. Surface shortwave albedo as a function of solar zenith angle measured at ARM and SURFRAD stations. [Courtesy of Yang et al. (2008).]

irradiance in the regional-scale Weather Research and Forecasting (WRF) Model using an improved parameterization for shallow clouds.

Other efforts have focused on improving our understanding of the hydrologic cycle and the balance between evapotranspiration and precipitation, commonly called precipitation recycling. Precipitation recycling is defined as precipitation within a domain of interest that originates from evapotranspiration rather than through transport through the sides of the control volume (e.g., Trenberth and Guillemot 1998). Lamb et al. (2012) used measurements and reanalysis products to document the relative importance of precipitation recycling for years that were both very wet (as was observed during the CLASIC field campaign in 2007) and very dry. They found that, with the exception of the very wet conditions during the CLASIC period, precipitation is smaller than evapotranspiration and that the difference in them was balanced by horizontal moisture flux divergence within the control volume over the SGP. In contrast, during the CLASIC year, precipitation was greater than evapotranspiration. One important change in land use in the central United States has been an increase in the use of irrigation on agricultural land. Irrigation adds moisture to the surface, and leads to a decrease in the Bowen ratio, but many modeling studies currently ignore this potentially important moisture source. To address this shortcoming, Qian et al. (2013) implemented a simple representation of irrigation in WRF and simulated the water budget and water recycling for the same periods analyzed by Lamb et al. (2012). They documented changes in the simulated surface sensible and latent heat fluxes, PBL height, and z_{LCL}, but found no systematic change in the simulated precipitation.

5. Conclusions and summary

Throughout its history, the ARM Program has supported, via the collection of detailed datasets and the direct support of research scientists, a wide range of projects aimed at improving our understanding of land–atmosphere interactions. In this chapter, we have focused on two specific scientific areas for which significant progress has been made: understanding the importance of small-scale variations of surface fluxes and how best to scale those variations up from point measurements to the model grid scale, and understanding the role of subgrid variability on the initiation and maintenance of shallow clouds. Results showed that much of the details of the small-scale variability of the surface fluxes, such as spatial variability in the grid box, can be ignored within the context of large-scale models. The subgrid variability plays a role in the formation of clouds, but the research supported by the ARM Program indicates that it is sufficiently represented by using probability density functions (PDFs). The PDFs have been related to the grid-resolved variables through the development of new parameterizations. Other important areas of study, such as understanding the CO_2 cycle, the detailed evaluation of models (including land surface models), and the best methods for representing the surface albedo also were discussed briefly.

After the merger of the ASP and ARM Science Programs into ASR, research related to land–atmosphere interactions has continued, but with a renewed focus on the various aspects of boundary layer clouds, including their radiative impact, their macrophysical properties derived from a suite of remote sensing instruments, their formation, and the transition from shallow to deep convection. Long-term studies also have focused on precipitation recycling using measurements and reanalysis products, and have documented the effect of irrigation on surface sensible and latent heat fluxes, PBL height, and z_{LCL}. While important progress has been made in these areas of study, significant hurdles remain, and research related to land–atmosphere interactions will likely continue well into the future.

Acknowledgments. This work was supported by the DOE Office of Biological and Environmental Research (OBER) of the U.S. Department of Energy as part of the Atmospheric Radiation Measurement (ARM). The Pacific Northwest National Laboratory (PNNL) is operated by Battelle for the DOE under Contract DE-A06-76RLO 1830.

REFERENCES

Andrews, E., P. J. Sheridan, J. A. Ogren, and R. Ferrare, 2004: In situ aerosol profiles over the Southern Great Plains cloud and radiation test bed site: 1. Aerosol optical properties. *J. Geophys. Res.*, **109**, D06208, doi:10.1029/2003JD004025.

Anthes, R. A., 1984: Enhancement of convective precipitation by mesoscale variations in vegetative covering in semiarid regions. *J. Climate Appl. Meteor.*, **23**, 541–554, doi:10.1175/1520-0450(1984)023<0541:EOCPBM>2.0.CO;2.

Avissar, R., and F. Chen, 1993: Development and analysis of prognostic equations for mesoscale kinetic energy and mesoscale (subgrid scale) fluxes for large-scale atmospheric models. *J. Atmos. Sci.*, **50**, 3751–3774, doi:10.1175/1520-0469(1993)050<3751:DAAOPE>2.0.CO;2.

Barthlott, C., and N. Kalthoff, 2011: A numerical sensitivity study on the impact of soil moisture on convection-related parameters and convective precipitation over complex terrain. *J. Atmos. Sci.*, **68**, 2971–2987, doi:10.1175/JAS-D-11-027.1.

Berg, L. K., and R. B. Stull, 2004: Parameterization of joint frequency distributions of potential temperature and water vapor mixing ratio in the daytime convective boundary layer. *J. Atmos. Sci.*, **61**, 813–828, doi:10.1175/1520-0469(2004)061<0813:POJFDO>2.0.CO;2.

——, and ——, 2005: A simple parameterization coupling the convective daytime boundary layer and fair-weather cumuli. *J. Atmos. Sci.*, **62**, 1976–1988, doi:10.1175/JAS3437.1.

——, and S. Zhong, 2005: Sensitivity of MM5-simulated boundary layer characteristics to turbulence parameterizations. *J. Appl. Meteor.*, **44**, 1467–1483, doi:10.1175/JAM2292.1.

——, and E. I. Kassianov, 2008: Temporal variability of fair-weather cumulus statistics at the ACRF SGP site. *J. Climate*, **21**, 3344–3358, doi:10.1175/2007JCLI2266.1.

——, and Coauthors, 2009: Overview of the cumulus humilis aerosol processing study. *Bull. Amer. Meteor. Soc.*, **90**, 1653–1667, doi:10.1175/2009BAMS2760.1.

——, E. I. Kassianov, C. N. Long, and D. L. Mills Jr., 2011: Surface summertime radiative forcing by shallow cumuli at the Atmospheric Radiation Measurement Southern Great Plains site. *J. Geophys. Res.*, **116**, D01202, doi:10.1029/2010JD014593.

——, W. I. Gustafson, E. I. Kassianov, and L. Deng, 2013: Evaluation of a modified scheme for shallow convection: Implementation of CuP and case studies. *Mon. Wea. Rev.*, **141**, 134–147, doi:10.1175/MWR-D-12-00136.1.

Betts, A. K., 2003: The diurnal cycle over land. *Forests at the Land-Atmosphere Interface*, M. Mencuccini et al., Eds., CABI Publishing, 79–93.

——, 2004: Understanding hydrometeorology using global models. *Bull. Amer. Meteor. Soc.*, **85**, 1673–1688, doi:10.1175/BAMS-85-11-1673.

——, 2009: Land-surface-atmosphere coupling in observations and models. *J. Adv. Model. Earth Syst.*, **1** (4), doi:10.3894/JAMES.2009.1.4.

——, and M. A. F. Silva Dias, 2010: Progress in understanding land-surface-atmosphere coupling from LBA research. *J. Adv. Model. Earth Syst.*, **2** (6), doi:10.3894/JAMES.2010.2.6.

——, J. H. Ball, A. C. M. Beljaars, M. J. Miller, and P. A. Viterbo, 1996: The land surface-atmosphere interaction: A review based on observational and global modeling perspectives. *J. Geophys. Res.*, **101**, 7209–7225, doi:10.1029/95JD02135.

Billesbach, D. P., M. L. Fischer, M. S. Torn, and J. A. Berry, 2004: A portable eddy covariance system for the measurement of ecosystem–atmosphere exchange of CO_2, water vapor, and energy. *J. Atmos. Oceanic Technol.*, **21**, 639–650, doi:10.1175/1520-0426(2004)021<0639:APECSF>2.0.CO;2.

Bindlish, R., T. Jackson, S. Ruijing, M. Cosh, S. Yueh, and S. Dinardo, 2009: Combined passive and active microwave observations of soil moisture during CLASIC. *IEEE Geosci. Remote Sens. Lett.*, **6**, 644–648, doi:10.1109/LGRS.2009.2028441.

Biraud, S. C., M. S. Torn, J. R. Smith, C. Sweeney, W. J. Riley, and P. P. Tans, 2013: A multi-year record of airborne CO_2 observations in the US Southern Great Plains. *Atmos. Meas. Tech.*, **6**, 751–763, doi:10.5194/amt-6-751-2013.

Brock, F. V., K. C. Crawford, R. L. Elliott, G. W. Cuperus, S. J. Stadler, H. L. Johnson, and M. D. Eilts, 1995: The Oklahoma MESONET—A technical overview. *J. Atmos. Oceanic Technol.*, **12**, 5–19, doi:10.1175/1520-0426(1995)012<0005:TOMATO>2.0.CO;2.

Brotzge, J. A., and K. C. Crawford, 2003: Examination of the surface energy budget: A comparison of eddy correlation and Bowen ratio measurement systems. *J. Hydrometeor.*, **4**, 160–178, doi:10.1175/1525-7541(2003)4<160:EOTSEB>2.0.CO;2.

Brown, A. R., and Coauthors, 2002: Large-eddy simulation of the diurnal cycle of shallow cumulus convection over land. *Quart. J. Roy. Meteor. Soc.*, **128**, 1075–1093, doi:10.1256/003590002320373210.

Brutsaert, W., and W. P. Kustas, 1987: Surface water vapor and momentum fluxes under unstable conditions from a rugged-complex area. *J. Atmos. Sci.*, **44**, 421–431, doi:10.1175/1520-0469(1987)044<0421:SWVAMF>2.0.CO;2.

Clothiaux, E. E., T. P. Ackerman, G. G. Mace, K. P. Moran, R. T. Marchand, M. A. Miller, and B. E. Martner, 2000: Objective determination of cloud heights and radar reflectivities using a combination of active remote sensors at the ARM CART sites. *J. Appl. Meteor.*, **39**, 645–665, doi:10.1175/1520-0450(2000)039<0645:ODOCHA>2.0.CO;2.

Cooley, H. S., W. J. Riley, M. S. Torn, and Y. He, 2005: Impact of agricultural practice on regional climate in a coupled land surface mesoscale model. *J. Geophys. Res.*, **110**, D03113, doi:10.1029/2004JD005160.

Dirmeyer, P. A., R. D. Koster, and Z. Guo, 2006: Do global models properly represent the feedback between land and atmosphere? *J. Hydrometeor.*, **7**, 1177–1198, doi:10.1175/JHM532.1.

Doran, J. C., W. J. Shaw, and J. M. Hubbe, 1995: Boundary layer characteristics over areas of inhomogeneous surface fluxes. *J. Appl. Meteor.*, **34**, 559–571, doi:10.1175/1520-0450-34.2.559.

——, S. Zhong, J. C. Liljegren, and C. Jakob, 2002: A comparison of cloud properties at a coastal and inland site at the North Slope of Alaska. *J. Geophys. Res.*, **107**, 4120, doi:10.1029/2001JD000819.

——, J. C. Barnard, and W. J. Shaw, 2006: Modification of summertime Arctic cloud characteristics between a coastal and inland site. *J. Climate*, **19**, 3207–3219, doi:10.1175/JCLI3782.1.

Fischer, M. L., D. P. Billesbach, J. A. Berry, W. J. Riley, and M. S. Torn, 2007: Spatiotemporal variations in growing season exchanges of CO_2, H_2O, and sensible heat in agricultural fields of the Southern Great Plains. *Earth Interact.*, **11**, doi:10.1175/EI231.1.

Gao, W., R. L. Coulter, B. M. Lesht, J. Qiu, and M. L. Wesely, 1998: Estimating clear-sky regional surface fluxes in the Southern Great Plains Atmospheric Radiation Measurement site with ground measurements and satellite observations. *J. Appl. Meteor.*, **37**, 5–22, doi:10.1175/1520-0450(1998)037<0005:ECSRSF>2.0.CO;2.

Hauck, C., C. Barthlott, L. Krauss, and N. Kalthoff, 2011: Soil moisture variability and its influence on convective precipitation over complex terrain. *Quart. J. Roy. Meteor. Soc.*, **137**, 42–56, doi:10.1002/qj.766.

Hubbe, J. M., J. C. Doran, J. C. Liljegren, and W. J. Shaw, 1997: Observations of spatial variations of boundary layer structure over the Southern Great Plains Cloud and Radiation Testbed. *J. Appl. Meteor.*, **36**, 1221–1231, doi:10.1175/1520-0450(1997)036<1221:OOSVOB>2.0.CO;2.

Lamb, P. J., D. H. Portis, and A. Zangvil, 2012: Investigation of large-scale atmospheric moisture budget and land surface interactions over U.S. Southern Great Plains including for CLASIC (June 2007). *J. Hydrometeor.*, **13**, 1719–1738, doi:10.1175/JHM-D-12-01.1.

Long, C. N., J. H. Mather, and T. P. Ackerman, 2016: The ARM Tropical Western Pacific (TWP) sites. *The Atmospheric Radiation Measurement (ARM) Program: The First 20 Years*, *Meteor. Monogr.*, No. 57, Amer. Meteor. Soc., doi:10.1175/AMSMONOGRAPHS-D-15-0024.1.

Mather, J. H., D. D. Turner, and T. P. Ackerman, 2016: Scientific maturation of the ARM Program. *The Atmospheric Radiation*

Measurement (ARM) Program: The First 20 Years, Meteor. Monogr., No. 57, Amer. Meteor. Soc., doi:10.1175/AMSMONOGRAPHS-D-15-0024.1.

McFarlane, S. A., C. N. Long, and D. M. Flynn, 2005: Impact of island-induced clouds on surface measurements: Analysis of the ARM Nauru Island Effect Study data. *J. Appl. Meteor.*, **44**, 1045–1065, doi:10.1175/JAM2241.1.

——, J. H. Mather, and E. J. Mlawer, 2016: ARM's progress on improving atmospheric broadband radiative fluxes and heating rates. *The Atmospheric Radiation Measurement (ARM) Program: The First 20 Years, Meteor. Monogr.*, No. 57, Amer. Meteor. Soc., doi:10.1175/AMSMONOGRAPHS-D-15-0046.1.

McNaughton, K. G., and T. W. Spriggs, 1986: A mixed-layer model for regional evaporation. *Bound.-Layer Meteor.*, **34**, 243–262, doi:10.1007/BF00122381.

Miller, M. A., and Coauthors, 2007: SGP Cloud and Land Surface Interaction Campaign (CLASIC): Science and implementation plan. DOE/SC-ARM-0703, 14 pp. [Available online at https://www.arm.gov/publications/programdocs/doe-sc-arm-0703.pdf.]

Minnis, P., S. Mayor, W. L. Smith, and D. F. Young, 1997: Asymmetry in the diurnal variation of surface albedo. *IEEE Trans. Geosci. Remote Sens.*, **35**, 879–890, doi:10.1109/36.602530.

Neggers, R. A. J., A. P. Siebesma, G. Lenderink, and A. A. M. Holtslag, 2004: An evaluation of mass flux closures for diurnal cycles of shallow cumulus. *Mon. Wea. Rev.*, **132**, 2525–2538, doi:10.1175/MWR2776.1.

Oncley, S. P., and J. Dudhia, 1995: Evaluation of surface fluxes from MM5 using observations. *Mon. Wea. Rev.*, **123**, 3344–3357, doi:10.1175/1520-0493(1995)123<3344:EOSFFM>2.0.CO;2.

Ookouchi, Y., M. Segal, R. C. Kessler, and R. A. Pielke, 1984: Evaluation of soil moisture effects on the generation and modification of mesoscale circulations. *Mon. Wea. Rev.*, **112**, 2281–2292, doi:10.1175/1520-0493(1984)112<2281:EOSMEO>2.0.CO;2.

Peters-Lidard, C. D., and L. H. Davis, 2000: Regional flux estimation in a convective boundary layer using a conservation approach. *J. Hydrometeor.*, **1**, 170–182, doi:10.1175/1525-7541(2000)001<0170:RFEIAC>2.0.CO;2.

Pielke, R. A., G. A. Dalu, J. S. Snook, T. J. Lee, and T. G. F. Kittel, 1991: Nonlinear influence of mesoscale land use on weather and climate. *J. Climate*, **4**, 1053–1069, doi:10.1175/1520-0442(1991)004<1053:NIOMLU>2.0.CO;2.

Pinty, J.-P., P. Mascart, E. Richard, and R. Rosset, 1989: An investigation of mesoscale flows induced by vegetation inhomogeneities using an evapotranspiration model calibrated against HAPEX-MOBILHY data. *J. Appl. Meteor.*, **28**, 976–992, doi:10.1175/1520-0450(1989)028<0976:AIOMFI>2.0.CO;2.

Qian, Y., M. Huang, B. Yang, and L. K. Berg, 2013: A modeling study of irrigation effects on surface fluxes and land–air–cloud interactions in the Southern Great Plains. *J. Hydrometeor.*, **14**, 700–721, doi:10.1175/JHM-D-12-0134.1.

Randall, D. A., and Coauthors, 2007: Climate models and their evaluation. *Climate Change 2007: The Physical Science Basis*, S. Solomon et al., Eds., Cambridge University Press, 591–662.

Richardson, S. J., M. E. Splitt, and B. M. Lesht, 2000: Enhancement of ARM surface meteorological observations during the Fall 1996 Water Vapor Intensive Observation Period. *J. Atmos. Oceanic Technol.*, **17**, 312–322, doi:10.1175/1520-0426(2000)017<0312:EOASMO>2.0.CO;2.

Santanello, J. A., M. A. Friedl, and M. B. Ek, 2007: Convective planetary boundary layer interactions with the land surface at diurnal time scales: Diagnostics and feedbacks. *J. Hydrometeor.*, **8**, 1082–1097, doi:10.1175/JHM614.1.

——, C. D. Peters-Lidard, S. V. Kumar, C. Alonge, and W.-K. Tao, 2009: A modeling and observational framework for diagnosing local land–atmosphere coupling on diurnal time scales. *J. Hydrometeor.*, **10**, 577–599, doi:10.1175/2009JHM1066.1.

Schneider, J. M., D. K. Fisher, R. L. Elliott, G. O. Brown, and C. P. Bahrmann, 2003: Spatiotemporal variations in soil water: First results from the ARM SGP CART network. *J. Hydrometeor.*, **4**, 106–120, doi:10.1175/1525-7541(2003)004<0106:SVISWF>2.0.CO;2.

Schrieber, K., R. Stull, and Q. Zhang, 1996: Distributions of surface-layer buoyancy versus lifting condensation level over a heterogeneous land surface. *J. Atmos. Sci.*, **53**, 1086–1107, doi:10.1175/1520-0469(1996)053<1086:DOSLBV>2.0.CO;2.

Sellers, P. J., and Coauthors, 1995: Effects of spatial variability in topography, vegetation cover and soil moisture on area-averaged surface fluxes: A case study using the FIFE 1989 data. *J. Geophys. Res.*, **100**, 25 607–25 629, doi:10.1029/95JD02205.

——, and Coauthors, 1996: A revised land surface parameterization (SiB2) for atmospheric GCMS. Part I: Model formulation. *J. Climate*, **9**, 676–705, doi:10.1175/1520-0442(1996)009<0676:ARLSPF>2.0.CO;2.

Sisterson, D., R. Peppler, T. S. Cress, P. Lamb, and D. D. Turner, 2016: The ARM Southern Great Plains (SGP) site. *The Atmospheric Radiation Measurement (ARM) Program: The First 20 Years, Meteor. Monogr.*, No. 57, Amer. Meteor. Soc., doi:10.1175/AMSMONOGRAPHS-D-16-0004.1.

Stokes, G. M., and S. E. Schwartz, 1994: The Atmospheric Radiation Measurement (ARM) Program: Programmatic background and design of the cloud and radiation test bed. *Bull. Amer. Meteor. Soc.*, **75**, 1201–1221, doi:10.1175/1520-0477(1994)075<1201:TARMPP>2.0.CO;2.

Stull, R. B., 1988: *An Introduction to Boundary Layer Meteorology.* Kluwer Academic Publishers, 666 pp.

——, 1994: A convective transport theory for surface fluxes. *J. Atmos. Sci.*, **51**, 3–22, doi:10.1175/1520-0469(1994)051<0003:ACTTFS>2.0.CO;2.

——, E. Santoso, L. Berg, and J. Hacker, 1997: Boundary Layer Experiment 1996 (BLX96). *Bull. Amer. Meteor. Soc.*, **78**, 1149–1158, doi:10.1175/1520-0477(1997)078<1149:BLEB>2.0.CO;2.

Sud, Y. C., D. M. Mocko, G. K. Walker, and R. D. Koster, 2001: Influence of land surface fluxes on precipitation: Inferences from simulations forced with four ARM–CART SCM datasets. *J. Climate*, **14**, 3666–3691, doi:10.1175/1520-0442(2001)014<3666:IOLSFO>2.0.CO;2.

Sugita, M., and W. Brutsaert, 1990: Regional surface fluxes from remotely sensed skin temperature and lower boundary layer measurements. *Water Resour. Res.*, **26**, 2937–2944, doi:10.1029/WR026i012p02937.

Trenberth, K. E., and C. J. Guillemot, 1998: Evaluation of the atmospheric moisture and hydrological cycle in the NCEP/NCAR reanalyses. *Climate Dyn.*, **14**, 213–231, doi:10.1007/s003820050219.

Wilde, N. P., R. B. Stull, and E. W. Eloranta, 1985: The LCL zone and cumulus onset. *J. Climate Appl. Meteor.*, **24**, 640–657, doi:10.1175/1520-0450(1985)024<0640:TLZACO>2.0.CO;2.

Wulfmeyer, V., and Coauthors, 2011: The Convective and Orographically-induced Precipitation Study (COPS): The scientific strategy, the field phase, and research highlights. *Quart. J. Roy. Meteor. Soc.*, **137**, 3–30, doi:10.1002/qj.752.

Xu, Q., and C.-J. Qiu, 1997: A variational method for computing surface heat fluxes from ARM surface energy and radiation balance systems. *J. Appl. Meteor.*, **36**, 3–11, doi:10.1175/1520-0450(1997)036<0003:AVMFCS>2.0.CO;2.

——, B. Zhou, S. D. Burk, and E. H. Barker, 1999: An air–soil layer coupled scheme for computing surface heat fluxes. *J. Appl. Meteor.*, **38**, 211–223, doi:10.1175/1520-0450(1999)038<0211:AASLCS>2.0.CO;2.

Yan, H., and R. A. Anthes, 1988: The effect of variations in surface moisture on mesoscale circulation. *Mon. Wea. Rev.*, **116**, 192–208, doi:10.1175/1520-0493(1988)116<0192:TEOVIS>2.0.CO;2.

Yang, F., K. Mitchell, Y.-T. Hou, Y. Dai, X. Zeng, Z. Wang, and X.-Z. Liang, 2008: Dependence of land surface albedo on solar zenith angle: Observations and model parameterization. *J. Appl. Meteor. Climatol.*, **47**, 2963–2982, doi:10.1175/2008JAMC1843.1.

Zeng, X., and Coauthors, 2007: Evaluating clouds in long-term cloud-resolving model simulations with observational data. *J. Atmos. Sci.*, **64**, 4153–4177, doi:10.1175/2007JAS2170.1.

Zhang, Y., and S. A. Klein, 2010: Mechanisms affecting the transition from shallow to deep convection over land: Inferences from observations of the diurnal cycle collected at the ARM Southern Great Plains site. *J. Atmos. Sci.*, **67**, 2943–2959, doi:10.1175/2010JAS3366.1.

Zhong, S., and J. C. Doran, 1995: A modeling study of the effects of inhomogeneous surface fluxes on boundary-layer properties. *J. Atmos. Sci.*, **52**, 3129–3142, doi:10.1175/1520-0469(1995)052<3129:AMSOTE>2.0.CO;2.

——, and ——, 1998: An evaluation of the importance of surface flux variability on GCM-scale boundary-layer characteristics using realistic meteorological and surface forcing. *J. Climate*, **11**, 2774–2788, doi:10.1175/1520-0442(1998)011<2774:AEOTIO>2.0.CO;2.

Zhou, B., and Q. Xu, 1999: Computing surface fluxes from Mesonet data. *J. Appl. Meteor.*, **38**, 1370–1383, doi:10.1175/1520-0450(1999)038<1370:CSFFMD>2.0.CO;2.

Zhu, P., and B. Albrecht, 2002: A theoretical and observational analysis on the formation of fair-weather cumuli. *J. Atmos. Sci.*, **59**, 1983–2005, doi:10.1175/1520-0469(2002)059<1983:ATAOAO>2.0.CO;2.

——, ——, and J. Gottschalck, 2001: Formation and development of nocturnal boundary layer clouds over the Southern Great Plains. *J. Atmos. Sci.*, **58**, 1409–1426, doi:10.1175/1520-0469(2001)058<1409:FADONB>2.0.CO;2.

Chapter 24

The SCM Concept and Creation of ARM Forcing Datasets

MINGHUA ZHANG

State University of New York at Stony Brook, Stony Brook, New York

RICHARD C. J. SOMERVILLE

Scripps Institution of Oceanography, University of California, San Diego, La Jolla, California

SHAOCHENG XIE

Lawrence Livermore National Laboratory, Livermore, California

1. The concept of SCM for ARM

Two papers published in the early 1990s significantly influenced the subsequent design of ARM and its adoption of the single-column model (SCM) approach. The first paper, by Cess et al. (1990), showed a threefold difference in the sensitivity of climate models in a surrogate climate change that is attributed largely to cloud–climate feedbacks. The second paper, by Ellingson et al. (1991), reported 10%–20% difference in the calculated broadband radiation budget and 30%–40% difference in the radiative forcing of greenhouse gases in the radiation codes of climate models. At that time, the U.S. Department of Energy (DOE) had a program to study the climate impact of the increasing amount of carbon dioxide in the atmosphere. Results from these two papers pointed to the major uncertainties in climate forcing and feedbacks of climate models.

Radiation and clouds, therefore, emerged as a focus in the DOE ARM Program to improve models. To simulate clouds, information is needed about the atmospheric dynamics beyond the vertical profiles of atmospheric thermodynamics and winds at a single station. By the early 1990s, analyses of observed atmospheric dynamics from the 1974 Global Atmosphere Research Experiment (GARP) Atlantic Tropical Experiment (GATE) had been completed (e.g., Ooyama 1987). GATE used a well-coordinated sounding array to derive the atmospheric dynamics. It was shown to be feasible to simulate atmospheric moist processes by using observed atmospheric dynamics from a well-configured measurement array (e.g., Gregory and Rowntree 1990). The concept of using SCMs to represent a single grid box of a climate model was well suited to address the uncertainties in radiation and clouds in the models. To represent different cloud regimes in a global model, ARM decided to establish three sites: the Southern Great Plains (SGP), the Northern Slope of Alaska (NSA), and the Tropical Western Pacific (TWP) (Stokes and Schwartz 1994).

An SCM is a one-dimensional (vertical) computational model of a specific columnar region of the atmosphere. It may be thought of as being extracted from the array of such columns, which make up the atmospheric portion of a global climate model or general circulation model (GCM). In the GCM, this atmospheric model column interacts at each vertical level and at every time step with neighboring columns, providing horizontal fluxes of heat, water, and momentum to and from these neighbors. By contrast, an SCM requires these fluxes to be specified, either from model data, observations, or some combination of the two. If the fluxes are set to zero, the SCM becomes one type of a radiative–convective model (RCM). One way to think of an RCM (Ramanathan and Coakley 1978) is as a horizontally averaged GCM, with the horizontal averaging over a global domain resulting in zero horizontal flux convergence. The horizontal fluxes also can be applied to cloud-resolving models (CRMs) and large-eddy simulation (LES)

Corresponding author address: Minghua Zhang, School of Marine and Atmospheric Sciences, State University of New York, Stony Brook, 145 Endeavor Building, Stony Brook, NY 11794.
E-mail: minghua.zhang@stonybrook.edu

DOI: 10.1175/AMSMONOGRAPHS-D-15-0040.1

models, which allows CRM and LES results to be used to evaluate the SCMs, especially when the physical quantities cannot be measured observationally.

Many climate modeling research groups have developed and used SCMs as tools for parameterization development. ARM has historically been a significant source of support to the small community of SCM modelers by providing research grants, observational data, and computational resources. In particular, observational data from the ARM SGP site have been a major stimulus to the development and use of SCMs because of the wealth of available measurements. The ARM's success with research involving SCMs has encouraged several modeling groups to make their own internally developed SCMs publicly available, sometimes with professionally programmed and well-documented codes. The National Center for Atmospheric Research (NCAR) climate modeling group, for example, has made single-column versions of its global climate models available for many years.

A great deal of experience has been gained in using SCMs with ARM data, and, over time, the role of SCMs in climate research has been expanded and clarified. The use of SCMs clearly has a valuable place in the hierarchy of modeling approaches, which is needed to improve the realism and trustworthiness of climate models. Of course, a wide variety of techniques has long been employed to test and validate physical process parameterizations in both weather and climate models. One straightforward method is to compare the results of full three-dimensional GCM simulations, using different parameterizations, against global observations. Another is to carry out numerical weather prediction (NWP) experiments initialized with observations and to compare the effects of different parameterizations on short- and medium-range forecast skill. Both of these approaches are important. However, carrying out a carefully coordinated model parameterization intercomparison program with three-dimensional models, even when the same basic model is used as the vehicle, is time consuming and computationally intensive, especially when the parameter space is large. Because the SCM has only one space dimension (vertical), it is very fast, and it is practical to explore large segments of parameter space by making hundreds or even thousands of integrations, which is impossible with a full GCM.

2. Early studies using SCMs

The semiprognostic model of Lord (1982) and the convective adjustment tests of Betts and Miller (1986) are early examples of the idea of using a model of a single atmospheric column. In ARM, SCMs have been widely used to investigate parameterizations of cloud-radiation processes. This approach typically involves evaluating parameterizations directly against measurements from ARM field programs and using this validation to tune existing parameterizations and to guide the development of new ones. The single-column model is thus used to make the link between observations and parameterizations. Surface and satellite measurements are both used to provide an initial evaluation of the performance of the different parameterizations. The results of this evaluation are then used to develop improved cloud-precipitation schemes, and finally these schemes are tested in GCM experiments (e.g., Lee et al. 1997). An early example of using a single-column model in this way was described by Iacobellis and Somerville (1991a,b).

A major result of ARM is that SCMs have proven themselves capable of validating parameterization results directly against ARM measurements. Climatically critical observable quantities, such as cloud liquid water and downwelling surface shortwave and longwave radiation, can be both derived from SCM results and inferred from observations at the SGP site. It is safe to say that, with extensive examples of this type of research in ARM, a major step has been achieved in fulfilling the original promise of the SCM approach.

The SCM is a convenient test bed for examining many aspects of the ways in which GCMs treat subgrid physical processes. For example, Lane et al. (2000) found strong sensitivity to vertical resolution in several test integrations in which they increased the number of layers substantially. Several possibilities are raised by this result. One possibility is that the parameterizations are constructed around implicit assumptions as to how many layers are involved so that they do not generalize to arbitrary vertical resolution and converge at sufficiently small vertical grid sizes. Another possibility is that typical GCM and NWP vertical resolutions are simply inadequate for some aspects of parameterized subgrid physics, such as marine stratocumulus clouds, although they may generally be satisfactory from the viewpoint of large-scale dynamics.

The SCM approach also has been used by the Global Energy and Water Cycle Experiment (GEWEX) Cloud System Study (GCSS)/Global Atmospheric System Study (GASS) to study cloud processes in which CRMs and LES models are also used. Most of the GCSS/GASS studies employed idealized horizontal forcing data that represent certain aspects of the observations. Early GCSS SCM studies include the modeling of stratocumulus-topped boundary layer (Bechtold et al. 1996; Zhu et al. 2005), the smoke cloud case (Bretherton et al. 1999b), the Atlantic Stratocumulus Transition Experiment (ASTEX; Bretherton et al. 1999a), and the diurnal cycle of shallow cumulus over land (Lenderink et al. 2004; Guichard et al. 2004). The GCSS/GASS case studies have enriched the

SCM studies in ARM (e.g., Guichard et al. 2004; Fridlind et al. 2012; Petch et al. 2014).

3. Requirement on large-scale forcing data

In the simplest setting, the SCMs calculate the time evolution of the vertical distributions of temperature and water vapor, schematically written as follows:

$$\frac{\partial \theta_m}{\partial t} = \left(\frac{\partial \theta_m}{\partial t}\right)_{\text{phy}} - (\mathbf{V} \cdot \nabla\theta)_{\text{LS}} - \omega_{\text{LS}}\frac{\partial \theta_m}{\partial p} \quad \text{and} \tag{24-1}$$

$$\frac{\partial q_m}{\partial t} = \left(\frac{\partial q_m}{\partial t}\right)_{\text{phy}} - (\mathbf{V} \cdot \nabla q)_{\text{LS}} - \omega_{\text{LS}}\frac{\partial q_m}{\partial p}, \tag{24-2}$$

where θ and q are potential temperature and water vapor mixing ratio; subscript m denotes model values; LS stands for prescribed large-scale fields; phy represents physical parameterizations; and other symbols are as commonly used. In the vertical advection terms of Eqs. (24-1) and (24-2) (i.e., the last term on the right-hand side), the simulated profiles of θ and q are used, so the vertical advection terms retain some feedback of the simulated fields to the forcing fields. The horizontal advective tendencies $-(\mathbf{V} \cdot \nabla\boldsymbol{\theta})_{\text{LS}}$ and the vertical velocity ω_{LS} are the large-scale forcing (Randall and Cripe 1999). It is sometimes referred to as 2D forcing. In another formulation, the observed profiles of θ and q are used in the vertical advection term. Therefore $-(\mathbf{V} \cdot \nabla\boldsymbol{\theta})_{\text{LS}} - [\omega(\partial\theta/\partial p)]_{\text{LS}}$ is prescribed as the large-scale forcing, which is often referred as 3D forcing. In calculating the physical tendencies of θ_m and q_m [i.e., the first terms on the right-hand sides of Eqs. (24-1) and (24-2)], SCMs compute the clouds, convection, precipitation, radiation, and turbulent mixing that can be compared with observations.

The same forcing fields also can be applied to CRM or LES models. The CRMs and LESs simulate θ and q or sometimes their corresponding conservative variables of liquid water potential temperature and total liquid water. The forcing terms in Eqs. (24-1) and (24-2) are applied to all model grids in CRMs and LESs, but with θ_m and q_m in the vertical advection terms replaced by domain-averaged values.

The derivations of the large-scale forcing data from field measurements are subject to uncertainties that can directly impact the simulated cloud and radiation fields by the SCMs. These uncertainties originate from two sources. One is the instrument and measurement errors. The second is errors from scale aliasing, or sampling biases. Both error types depend on scales because horizontal derivatives are involved in the calculation of the horizontal fluxes. Generally speaking, the smaller the scale is, the larger the errors in the derivative fields.

If the accuracy requirements of the physical parameterization terms in Eqs. (24-1)–(24-2) are 1 K day^{-1} and 1 g kg^{-1} day^{-1}, the comparable accuracy requirements on the errors of the horizontal differences of temperature ($\Delta\theta$) and humidity (Δq) over a distance Δx can be estimated as the following:

$$\mathbf{V} \cdot \nabla\theta \sim \left|u\frac{\Delta\theta}{\Delta x}\right| \le 1\,\text{K day}^{-1},$$

$$|\Delta\theta|_{\min} \le 1(\text{K day}^{-1})\frac{\Delta x}{|u|_{\max}} \quad \text{and}$$

$$\mathbf{V} \cdot \nabla q \sim \left|u\frac{\Delta q}{\Delta x}\right| \le 1\,\text{g kg}^{-1}\,\text{day}^{-1},$$

$$|\Delta q|_{\min} \le 1(\text{g kg}^{-1}\,\text{day}^{-1})\frac{\Delta x}{|u|_{\max}}.$$

Given $|u|_{\max} \sim 10\,\text{m s}^{-1}$ and over a distance of 200 km, the above inequalities require that

$$|\Delta\theta|_{\min} \le 0.23\,\text{K} \quad \text{and} \tag{24-3}$$

$$|\Delta q|_{\min} \le 0.23\,\text{g kg}^{-1}. \tag{24-4}$$

It should be emphasized that these are the relative errors across the distance of Δx. The requirement on the pressure vertical velocity error is

$$\left|\omega\frac{\partial\theta}{\partial p}\right| \le 1\,\text{K day}^{-1},$$

$$|\omega|_{\min} \le 1(\text{K day}^{-1})\frac{pg}{R_d T}\Big/\left(\frac{\partial\theta}{\partial z}\right) \approx 10\,\text{mb day}^{-1} \quad \text{or}$$

$$\left|\omega\frac{\partial q}{\partial p}\right| \le 1\,\text{g kg}^{-1}\,\text{day}^{-1},$$

$$|\omega|_{\min} \le 1(\text{g kg}^{-1}\,\text{day}^{-1})\Big/\left(\frac{\partial q}{\partial p}\right)_{\max} \approx 10\,\text{mb day}^{-1}.$$

The requirement on the difference of horizontal winds across the domain that corresponds to the above error in vertical velocity can be estimated as follows:

$$\frac{\Delta u}{\Delta x} \sim \frac{\partial\omega}{\partial p}.$$

Assuming a vertical layer of 100 hPa, we get

$$|\Delta u|_{\min} \le 0.12\,\text{m s}^{-1} \tag{24-5}$$

The error bounds of the spatial differences in Eqs. (24-3)–(24-5), corresponding to an accuracy requirement of 1 K day^{-1} and 1 g kg^{-1} day^{-1} in the forcing data, need to be scaled proportionally if the horizontal scale is different from 200 km. These magnitudes are comparable to instrument errors (Zhang and Lin 1997), but since they are relative errors across the space, the systematic instrument errors are reduced if the same equipment is used over the domain. The random errors may be suppressed by averaging over vertical levels. The more problematic errors are those caused by scale aliasing or sampling bias. These errors are often handled by using statistical approaches. In ARM, they are dealt with additionally by using known physical constraints.

Because of the errors in the forcing data, when integrated for multiple days, the temperature and moisture errors in the SCMs can build up, leading to the drift of SCMs into a very different climate regime that is no longer useful as a diagnostic. A common practice to remedy the drift is to reinitialize the SCM to conduct short periods of integrations and then concatenate them into a multiday long period. Another practice is to weakly apply relaxation terms to the model to observed temperature and humidity, in which case the temperature and water vapor fields are no longer good measures of model performance.

4. Forcing data from field experiments prior to ARM

In the objective analysis of field experimental data of a sounding array, in theory both the horizontal advective tendencies and the large-scale vertical velocity can be obtained by using finite difference approximation of the horizontal derivatives when the input data are regularly spaced. Since balloon sounding stations are never regularly distributed, interpolations and extrapolations are needed to preprocess the atmospheric temperature, water vapor, and winds into a regular set of grids. This method is referred to as the "regular grid method" in Zhang et al. (2001). In this method, the forcing data are calculated at each grid, and area averages are performed to obtain forcing for the study domain. An alternate method is to write the advective tendencies in flux form. The horizontal flux divergence terms, when averaged over a domain, are calculated by line integrals at the lateral boundaries of the study domain. This approach is referred to as the "line-integral method."

A key element in the regular grid method is the fitting of atmospheric state variables to the desired grids. The fitting results depend on the choice of the assumed functional form, which can be quite subjective. Commonly used methods are linear fitting and the quadratic and spline fittings (Davies-Jones 1993; Thompson et al. 1979). The more convenient algorithms are the Barnes (1964) and the Cressman (1959) schemes (Lin and Johnson 1996). In these schemes, a background field (or initial guess) is used at the observational locations; the difference between the observation and the initial guess field is then interpolated to the regular set of grids to adjust the background fields at these grids. The calculation can be performed iteratively to reach the desired corrections. Both the interpolation method and the number of integrations can affect the final analysis. A more sophisticated method uses a statistical interpolation scheme, such as that of Ooyama (1987).

The line-integral method depends on the number of atmospheric measurements at the boundary of the study domain. It therefore contains fewer subjective assumptions than the regular grid method. An important requirement is that there need to be a sufficient number of measurement stations at the domain boundary. The line-integral method typically does not use measurements inside the study domain in calculating the lateral boundary fluxes.

The regular grid method is more suited to analyze data with many scattered measurement stations, while the line-integral method is more suited for a well-positioned sounding array with few measurement stations. Zhang et al. (2001) presented a hybrid approach in which the regular grid method is used to improve the lateral boundary fluxes in the line-integral method.

Both methods have been used in the past to derive SCM large-scale forcing data in field experiments. One of the most widely used legacy datasets was from the GATE in 1974. Ooyama (1987) derived the GATE objective analysis by designing a statistical regular grid method onto which a penalty function is imposed to ensure smoothness of the fields. While the GATE data by Ooyama (1987) have been used widely, a standard analysis algorithm is not available because many subjective procedures and judgments were made through trial and error tests for each data point.

Another widely used SCM forcing dataset was from the Tropical Ocean and Global Atmosphere Coupled Ocean–Atmospheric Response Experiment (TOGA COARE) from November 1992 to February 1993. Lin and Johnson (1996) used the Barnes analysis and the regular grid method to derive the forcing data. Frank et al. (1996) analyzed the TOGA COARE data using the line-integral method. The difference of the moisture budgets from these two analyses over the Intensive Flux Array (IFA) was large. The time-averaged diagnosed precipitation over the experiment period is 5.7–6.1 mm day^{-1} in Lin and Johnson (1996) and 10.5–11.8 mm day^{-1} in Frank et al. (1996). Therefore, although the analyzed data can be used to study the qualitative temporal variation of large-scale atmospheric phenomena, such as the Madden–Julian oscillation (MJO), their use to simulate the observed cloud fields for direct comparison with transient measurements of clouds can have large errors from the forcing data.

SCM forcing data have been calculated for other shorter field experiments. Many of these are in regions of Asian and Australian monsoons. They were summarized in Zhang et al. (2001). Uncertainties of the analyzed data are likely similar to those in TOGA COARE. These uncertainties represent fundamental limits of data from the balloon sounding arrays caused by scale aliasing, as stated succinctly by Ooyama (1987, p. 2501): "To make gold, one must start with gold."

Atmospheric reanalysis or operational analysis also can be used to obtain the large-scale forcing. However, because the operational models suffer from biases of cloud and precipitation parameterizations that ARM aims to improve, these products are not always suited for SCM results to be compared with observations. For example, operational models typically cannot simulate the timing and magnitude of observed precipitation, which results in a bias in the large-scale vertical velocity in these products. Large-scale forcing in the European Center for Medium-Range Weather Forecasts (ECMWF) operational analysis and the North American Regional Reanalysis (NARR) has been evaluated in Xie et al. (2003, 2006) and Kennedy et al. (2011), and it was shown that cloud fields and vertical velocity in these products contain large errors during precipitation events.

5. The ARM variational analysis method

Recognizing the accuracy limit in large-scale forcing data and the need for transient forcing data in ARM, Zhang and Lin (1997) developed a constrained variational algorithm to incorporate more measurements to improve the SCM forcing data. Physical constraints are enforced. These constraints include column-integrated conservations of atmospheric masses of moist air and water vapor as well as heat and momentum. They are written as follows:

$$\frac{1}{g}\frac{\partial p_s}{\partial t} = -\langle \nabla \cdot \mathbf{V} \rangle; \tag{24-6}$$

$$\frac{\partial \langle q \rangle}{\partial t} = -\langle \nabla \cdot \mathbf{V}q \rangle + E_s - \mathrm{Prec}; \tag{24-7}$$

$$\frac{\partial \langle s \rangle}{\partial t} + \langle \nabla \cdot \mathbf{V}s \rangle = R_{\mathrm{TOA}} - R_{\mathrm{SRF}} + L_v\mathrm{Prec} + \mathrm{SH} + L_v\frac{\partial \langle q_l \rangle}{\partial t}; \text{ and} \tag{24-8}$$

$$\frac{\partial \langle \mathbf{V} \rangle}{\partial t} + \langle \nabla \cdot \mathbf{VV} \rangle + f\mathbf{k} \times \langle \mathbf{V} \rangle + \nabla \langle \phi \rangle = \boldsymbol{\tau}_s. \tag{24-9}$$

In the above equations, $\mathbf{V}$, s, and q are the atmospheric state variables of the wind vector, dry static energy, and water vapor; p_s is the surface pressure; q_l is the cloud liquid water content; and ϕ denotes the geopotential height. The brackets represent vertical integration. The surface evaporation is represented by E_s. The surface precipitation is represented by Prec. The net downward radiative flux is represented by R; the subscripts TOA and SRF represent the top of the atmosphere and the surface, respectively. The latent heat is represented by L_v; SH is the surface sensible heat flux; and $\boldsymbol{\tau}_s$ denotes the wind stress at the surface. Other variables are as commonly used.

The final analysis is obtained by minimizing the cost function of

$$I(t) = (u - u_o)^{\mathrm{T}}\mathbf{B}_u^{-1}(u - u_o) + (v - v_o)^{\mathrm{T}}\mathbf{B}_v^{-1}(v - v_o) + (s - s_o)^{\mathrm{T}}\mathbf{B}_s^{-1}(s - s_o) + (q - q_o)^{\mathrm{T}}\mathbf{B}_q^{-1}(q - q_o), \tag{24-10}$$

where variables with the subscript o represent the first guess from preprocessed balloon sounding and wind profiler measurements or operational analysis; $\mathbf{B}$ is the error covariance matrix of the state variable.

Terms on the right-hand sides of Eqs. (24-6)–(24-9) are obtained from ARM and satellite measurements at the surface and TOA. Area-averaged precipitation is from radar measurements. Other surface variables are from the suite of stations deployed within a sounding array. These are described in section 6. In some cases, fluxes are derived from statistical interpolation between the limited number of stations and the background fields from the reanalysis products. Since each field experiment has different instrumentation and measurement configurations, the preprocessing of surface and atmospheric measurements is often specific to different experiments, and visual inspections of all input data are necessary.

The minimization of the cost function in Eq. (24-10) requires the specification of the error covariance matrices. These errors are taken as the sum of instruments and measurement biases and sampling biases in Zhang et al. (2001). The instrument and measurement biases are assumed to be $0.5\,\mathrm{m\,s^{-1}}$ for winds, 0.2 K for temperature, and 3% of the specific humidity for water vapor. These were estimated by instrument mentors. The sampling biases are estimated to be 20% of the temporal variances of the fields. In past analyses, the errors are assumed to be independent among different locations and variables. This assumption is being revised to allow for error covariance. Zhang and Lin (1997) described the minimization algorithm of Eq. (24-10).

The final analysis, therefore, is the closest to balloon sounding and wind profiler data or operational analysis that satisfies the required constraints of Eqs. (24-6)–(24-9). The divergence terms in these equation terms are calculated by using the line-integral method. The atmospheric state variables at the boundary stations are preprocessed by using the regular grid method so that data from all profiling stations are used.

The constraining requirements ensure that what enters into the atmospheric column is equal to what exits from the column and at the TOA as well as at the surface after adjusting for column-integrated temporal change. The forcing data can be considered as a better fitting of the atmospheric analysis to more observational measurements.

The terms on the right-hand side of Eqs. (24-6)–(24-9) are currently treated as known fields. Sensitivities of the analyzed fields to their uncertainties are used to characterize the errors in the forcing data (Zhang et al. 2001). In theory, these constraining variables also can be subject to variational adjustments based on their uncertainties. The imposed constraints can be expanded to include other known physical relationships and measurements, such as clear-sky water vapor and thermodynamic equations and radiance measurements at various wavelengths. The atmospheric state variables should ideally also include cloud hydrometeors, in which case radar reflectivity and cloudy-sky radiance measurements can be used as constraints. Additionally, the error covariance matrices in the cost function should be calculated better. Research is ongoing to make improvements in all these aspects.

6. Input data

The input data for the ARM variational analysis include measurements of both adjustment variables and constraint fields. The adjusted variables are the large-scale state variables: namely, winds, temperature, and humidity. The constraints include surface pressure, surface latent and sensible heat fluxes, wind stress, precipitation, net radiation at the surface and TOA, and column total cloud liquid water. The momentum constraint of Eq. (24-9) was not imposed in the existing analysis because of large sensitivity of the pressure gradient force to errors in temperature, the treatment of which is still under investigation.

The large-scale state variables are obtained primarily from balloonborne sounding measurements. Ideally, high-frequency soundings at least once every 3 h over a well-positioned array are needed to specify the SCM forcing. However, because of logistical difficulties and large expenses of operating coordinated balloon soundings, only during special ARM Intensive Operational Periods (IOPs) were radiosondes launched at 3-hourly or 6-hourly intervals to measure the vertical profiles of winds, temperature, and water vapor mixing ratio.

The majority of these IOPs were conducted at SGP. At the ARM TWP and NSA sites, coordinated balloon sounding measurements are more difficult to make; hence, only one and two IOPs have been conducted at these sites, respectively. ARM also deployed many surface stations at its sites, including various radiometers and surface flux stations. The stations are intended to characterize the water and energy budgets within the domain represented by a GCM grid box. The domain-averaged fluxes also are used in the variational analysis as constraints. The SGP has many more of these surface stations than do the other ARM sites.

At the SGP, hourly profiler measurements of winds are also available at the National Oceanic and Atmospheric Administration (NOAA) wind profiler stations, which are merged with the soundings in the analysis. The SGP site has five ARM sounding stations: the Central Facility (C1) and four boundary facilities (B1, B4, B5, and B6), as well as several NOAA wind profiler sites to provide the needed upper-air measurements (Fig. 24-1a). These boundary sites and wind profilers have not been and are not always available. For the SGP, the variational analysis scheme processes the original upper-air measurements from radiosondes and wind profilers over the analysis grid points (Fig. 24-1a) using the Cressman interpolation scheme (Cressman 1959), which requires a background field from an NWP model's operational analyses. Current variational analysis uses the operational analyses from the NOAA mesoscale model Rapid Update Cycle (RUC) for SGP (Fig. 24-1b) and the ECMWF for other ARM sites. The required constraint variables are derived from measurements of surface observational networks and satellites. Around the ARM SGP site, there is a dense surface network (Fig. 24-1c). The observation platforms include the following:

- Surface meteorological observation system (SMOS) measuring surface precipitation, surface pressure, surface winds, temperature, and relative humidity.
- Energy budget Bowen ratio (EBBR) stations measuring surface latent and sensible heat fluxes and surface broadband net radiative flux.
- Eddy correlation flux measurement system (ECOR) providing in situ averages of the surface vertical fluxes of momentum, sensible heat flux, and latent heat flux.
- Oklahoma and Kansas Mesonet stations (OKM and KAM) measuring surface precipitation, pressure, winds, and temperature.
- Microwave radiometer (MWR) stations measuring the column precipitable water and total cloud liquid water. The stations have experienced changes over the past 20 years, including decommissioning at the ARM boundary facilities.
- Solar and infrared radiation station (SIRS) providing continuous measurements of broadband shortwave (solar) and longwave (atmospheric or infrared) irradiances for downwelling and upwelling components.
- National Weather Service WSR-88D NEXRAD radar and rain gauge providing hourly surface precipitation data to the Arkansas-Red basin River Forecast Center (ABRFC).

The Geostationary Operational Environmental Satellite (GOES) provides satellite measurements, clouds,

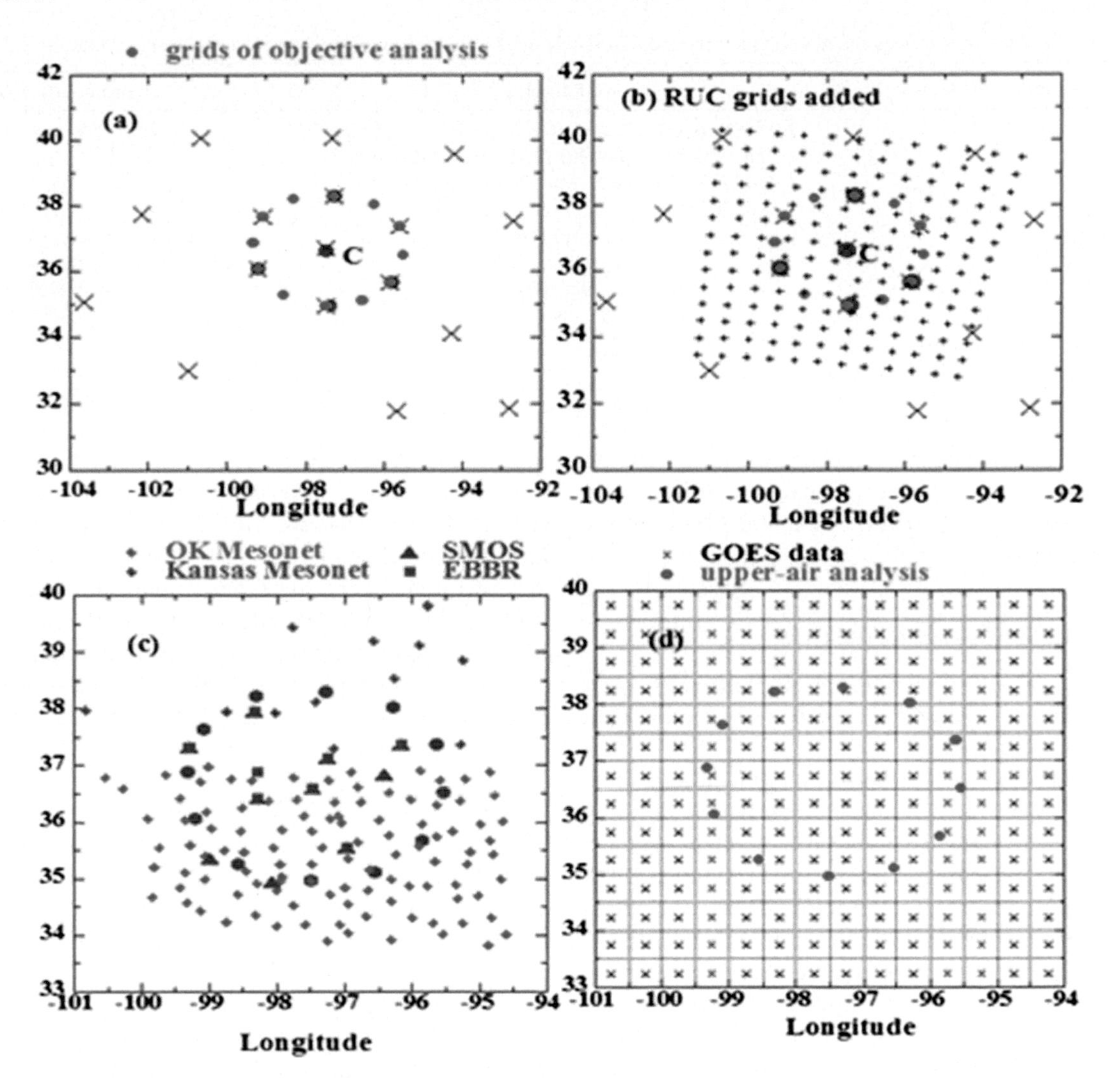

FIG. 24-1. Locations of the ARM upper-air data streams and the analysis grid points at the SGP site in the July 1997 IOP: (a) sounding stations (red circle), wind profilers (blue crosses), and final analysis grids (green circles); (b) RUC grids overlaid on other grids; (c) ARM surface data streams (see text for complete instrument names); and (d) GOES grids over the analysis domain. Adapted from Zhang et al. (2001).

and broadband radiative fluxes at TOA over the 0.5° × 0.5° grids (Fig. 24-1d) (Minnis et al. 1995). All the constraint variables are area-averaged quantities over the analysis domain. To avoid biases of using overcrowding measurement stations in some areas, the algorithm first lays the 0.5° × 0.5° GOES grids over the analysis domain and then derives the required quantities in each small grid box. If there are actual measurements within the subgrid box, simple arithmetic averaging is used to obtain the subgrid box means. Some variables are available from several instruments, as indicated above. They are merged in the arithmetic averaging process. If there is no actual measurement in the small box, the Barnes scheme (Barnes 1964) is used to fill in the missing data. Domain averages of these constraint quantities are obtained by using values from the 0.5° × 0.5° grid boxes within the analysis domain.

To enable SCM research beyond IOPs with coordinated balloon soundings, ARM also used hourly operational analysis of atmospheric fields as input to the constrained variational analysis to derive the so-called continuous forcing dataset (Xie et al. 2004). The operational analyses were constrained by the observed surface observations from ARM and TOA fields from GOES. The advantage of the continuous forcing is that it can be derived for long periods and over most regions of the globe as long as surface and TOA measurements are available. The use of the observed constraints in the analysis can significantly improve the accuracy of the forcing data derived from NWP analyses (Xie et al.

TABLE 24-1. Available ARM variational analysis forcing datasets. The given months and years are when the forcing data are available.

Location	IOP forcing	Continuous forcing
SGP	Jul 1995, Apr 1997, Jun 1997, Sep 1997, Apr 1998, Jan 1999, Mar 1999, Jul 1999, Mar 2000, Sep 2000, Nov 2000, Nov 2002, May 2003, Jun 2007, Apr 2012	Jan 1999–Jun 2011
TWP Darwin	Jan 2006	Three wet seasons between 2004 and 2007
NSA	Oct 2004	Apr 2008
AMF China	—	Nov 2008
AMF ARM MJO Experiment (AMIE) Gan	—	Nov 2011

2004). This method enabled the derivation of forcing data at ARM sites that lack coordinated high-frequency soundings, such as at the TWP, NSA, and ARM Mobile Facilities (AMFs).

7. Available ARM forcing datasets

The constrained variational analysis method has been applied to routinely derive the large-scale forcing data from ARM measurements for SCM and CRM studies. There are two types of variational analysis forcing data products available for the ARM permanent research sites and AMFs. The first is the IOP forcing, which is derived using sounding data collected during ARM major IOPs. The second is the continuous forcing, which is derived using NWP operational analyses for multiyear continuous periods where sounding measurements are not available. For both types of the forcing datasets, the large-scale state variables are constrained with surface and satellite observations.

Table 24-1 lists the available ARM variational analysis forcing datasets. These forcing datasets can be obtained from the ARM data archive (http://iop.archive.arm.gov/arm-iop/0eval-data/xie/scm-forcing). Over the past two decades, ARM has conducted numerous field campaigns in diverse climate regimes around the world to collect detailed observations of clouds and radiation, as well as related atmospheric variables for climate model evaluation and improvement. The majority of these field campaigns were conducted at the ARM SGP site, probably the largest and most extensive climate research site in the world. Major field campaigns at SGP include the June–July 1997 SCM IOP for midlatitude land convection, the March 2000 cloud IOP for frontal systems, the June 2007 Cloud Land Surface Interaction Campaign (CLASIC), and the April–May 2011 Midlatitude Continental Convective Clouds Experiment (MC3E). At the ARM NSA site, ARM conducted the Mixed-Phase Arctic Cloud Experiment (M-PACE) in October 2004 to study mixed-phase clouds and the Indirect and Semi-Direct Aerosol Campaign (ISDAC) in April 2008 to study the aerosol–cloud interaction in the Arctic region. In the tropics, the Tropical Warm Pool–International Cloud Experiment (TWP-ICE) took place in January and February 2006 around the ARM TWP Darwin site to improve understanding of the interaction of tropical convection with its environment. In addition, ARM regularly deploys its AMF in various climate regimes not previously explored. More details about these field campaigns can be found via the ARM website (http://www.arm.gov/campaigns).

The variational analysis forcing data products have been developed for all the major field campaigns conducted at the ARM permanent research sites and some of the AMF deployments. For the SGP and TWP Darwin sites, ARM has also created continuous forcing data over multiple years. These large-scale forcing datasets provide the needed initial and boundary conditions for SCMs and CRMs in studying various observed cloud systems and testing physical parameterizations in climate models.

8. Applications of SCMs and ARM data to understanding and improving models

It is now well recognized in the GCM and NWP communities that SCMs are tools that have a valuable role to play in testing and improving parameterizations by evaluating them empirically against field observations. Early surveys of SCM research have been published by Randall et al. (1996) and Somerville (2000). Since then, many papers have been published by using SCMs and ARM forcing data.

Most of these studies contribute to model improvements in one of the following three ways. The first is the evaluation of the performances of physical parameterizations in operational and global climate models (e.g., Yang et al. 2006; Kennedy et al. 2011; Song et al. 2013). The second is the validation, improvement, and development of parameterizations, including the triggering and closure assumptions of convection parameterizations (e.g., Xie and Zhang 2000; Zhang 2003; Guichard et al. 2004; Petch et al. 2007); cloud macrophysical schemes (e.g., Zhang et al. 2003); the mass flux parameterization of deep convection (Wu et al. 2009); the parameterization of shallow convection and boundary layer turbulence (Sušelj et al. 2012); and the parameterization of vertical velocity in shallow convection (Wang and Zhang 2014), among others. The third is to use SCMs with ARM forcing data to improve understanding of processes, including growth of ice particles (Comstock et al. 2008), cloud feedbacks (Del Genio et al. 2005), the interaction of deep and shallow convections (Wang and Zhang 2013), and land–atmosphere interactions (Sud et al. 2001). It should be noted that model development and improvement using SCMs is often done in conjunction with CRM or LES simulations under the same large-scale forcing.

Several ARM SCM case studies have been organized, with multiauthored publications. The first case used data from the ARM June 1995 SGP IOP (Ghan et al. 2000). SCMs and CRMs were used to simulate summertime continental convection. This case study settled the subsequent methods of how SCMs are run using observationally derived forcing data. Main results from the paper include the relative superior performance of the CRMs to SCMs, thus justifying the use of CRM results to improve SCMs.

The second ARM SCM case study by Xie et al. (2002) used the summer 1997 SGP IOP as a follow-up of Ghan et al. (2000). It was shown that deficiencies in convective triggering mechanisms were one of the major reasons for model biases. Using a triggering mechanism based solely on the vertical integral of parcel buoyant energy results in overactive convection, which in turn leads to large systematic warm/dry biases in the troposphere. It is also shown that a nonpenetrative convection scheme can underestimate the depth of instability for midlatitude convection, which leads to large systematic cold/moist biases in the troposphere. All models significantly underestimate the surface stratiform precipitation.

The third ARM SCM case study was published by Guichard et al. (2004) using an idealization of ARM June 1997 measurements to evaluate the simulation of deep convection. They found that the SCMs tend to simulate the onset of convection too early, while CRMs tend to simulate the onset too late.

The fourth ARM SCM case study was led by Xie et al. (2005) who used the ARM March 2000 SGP cloud IOP to investigate the parameterizations of frontal clouds. Most SCMs were found to underestimate cloud water and to contain huge biases in cloud ice of both signs. The SCMs underestimated the amount of midlevel clouds, which also appeared in CRMs. They attributed some of these biases to the lack of subgrid-scale dynamical forcing.

A case study for M-PACE was reported by Morrison et al. (2009). They showed that, for single-layer clouds, the simulated ice water path is generally consistent with observed values, and the median SCM and CRM liquid water path is a factor of 3 smaller than observed. For multilevel clouds, however, the models generally overestimate liquid water path and strongly underestimate ice water path. Models with more sophisticated, two-moment treatment of cloud microphysics were found to produce a somewhat smaller liquid water path closer to observations.

The case study for TWP-ICE has been reported in the work of Fridlind et al. (2012), who used CRMs with ARM forcing data. SCM intercomparison results have been reported in Davies et al. (2013) and Petch et al. (2014). All these studies showed how the ice microphysical parameterizations impact the simulated ice contents.

These cases serve as test beds for new parameterizations and as benchmarks for future studies. They are typically accompanied with comprehensive observational datasets that are processed and documented for models. Many other SCM studies by individual investigators have used ARM data. The ones listed in this chapter are a sample of them.

As the ARM Program evolves to have new measurement capabilities and new datasets (Mather and Voyles 2013), more future cases are expected that will target specific processes or scientific problems. An example is the recent case for the Routine ARM Aerial Facility Clouds with Low Optical Water Depths (CLOWD) Optical Radiative Observations (RACORO) campaign (e.g., Vogelmann et al. 2015; Lin et al. 2015).

9. Discussion

It is important to point out some of the limitations in using SCMs with prescribed forcing. First, because the large-scale vertical velocity is prescribed, the simulated precipitation is not the most appropriate field to use to evaluate model physics, as the SCM results are better suited to evaluate fields such as temperature, water vapor, clouds, and radiation. Second, in the case of propagating deep convections or middle-latitude cyclones with differential advection, SCMs should be viewed only as a way to describe a constrained balance of the model physics with the prescribed large-scale condition with possibly little insights on how the model physics are initiated. Third, SCMs cannot capture the interaction of model physics with large-scale dynamics. Some recent studies have attempted to overcome this limitation for tropical convective regimes where weak temperature gradient can be used to approximately parameterize the vertical velocity with dynamic heating in the SCMs (e.g., Sobel et al. 2001; Kuang 2012). Finally, results from SCMs are not necessarily transferable to global models because of the lack of feedbacks between physics and dynamics.

SCM (and CRM/LES) output is typically very sensitive to the imposed large-scale forcing. In some cases, given the nonlinear behavior of some parameterizations, a small change in the initial condition or forcing can cause the model to drift to a completely different state (Hack and Pedretti 2000). Given the inevitable uncertainties in the forcing data and initial conditions, one strategy is to use ensemble forcing data (Hume and Jakob 2005). Research is ongoing to develop ensemble forcing data that incorporate various uncertainties in the ARM variational algorithm.

Regardless of these limitations, however, many researchers have used SCMs to gain insights on physical parameterizations because they provide a convenient test of models and allow researchers to compare model results with observations and CRM/LES simulations. Results from the SCMs have motivated and led to many updates and modifications to the parameterizations in the current generation of global climate models. With the insights from SCMs, hypotheses and improvements can first be tested in SCMs, then evaluated against CRMs/LESs/observations and tested/implemented in global models.

Acknowledgments. The authors wish to thank the two reviewers and the editors, who provided detailed constructive comments that led to great improvements to an earlier version of this chapter.

REFERENCES

Barnes, S. L., 1964: A technique for maximizing details in numerical weather map analysis. *J. Appl. Meteor.*, **3**, 396–409, doi:10.1175/1520-0450(1964)003<0396:ATFMDI>2.0.CO;2.

Bechtold, P., S. K. Krueger, W. S. Lewellen, E. van Meijgaard, C. H. Moeng, D. A. Randall, A. van Ulden, and S. Wang, 1996: Modeling a stratocumulus-topped PBL: Intercomparison among different one-dimensional codes and with large-eddy simulation. *Bull. Amer. Meteor. Soc.*, **77**, 2033–2042, doi:10.1175/1520-0477(1996)077<2033:MASTPI>2.0.CO;2.

Betts, A. K., and M. J. Miller, 1986: A new convective adjustment scheme. Part II: Single column tests using GATE wave, BOMEX, ATEX, and arctic air mass data sets. *Quart. J. Roy. Meteor. Soc.*, **112**, 693–709, doi:10.1002/qj.49711247308.

Bretherton, C. S., S. K. Krueger, M. C. Wyant, P. Bechtold, E. van Meijgaard, B. Stevens, and J. Teixeira, 1999a: A GCSS boundary-layer cloud model intercomparison study of the first ASTEX Lagrangian experiment. *Bound.-Layer Meteor.*, **93**, 341–380, doi:10.1023/A:1002005429969.

——, and Coauthors, 1999b: An intercomparison of radiatively driven entrainment and turbulence in a smoke cloud, as simulated by different numerical models. *Quart. J. Roy. Meteor. Soc.*, **125**, 391–423, doi:10.1002/qj.49712555402.

Cess, R. D., and Coauthors, 1990: Intercomparison and interpretation of climate feedback processes in 19 atmospheric general circulation models. *J. Geophys. Res.*, **95**, 16 601–16 615, doi:10.1029/JD095iD10p16601.

Comstock, J. M., R.-F. Lin, D. O'C. Starr, and P. Yang, 2008: Understanding ice supersaturation, particle growth, and number concentration in cirrus clouds. *J. Geophys. Res.*, **113**, D23211, doi:10.1029/2008JD010332.

Cressman, G. P., 1959: An operational objective analysis scheme. *Mon. Wea. Rev.*, **87**, 367–374, doi:10.1175/1520-0493(1959)087<0367:AOOAS>2.0.CO;2.

Davies, L., and Coauthors, 2013: A single-column model ensemble approach applied to the TWP-ICE experiment. *J. Geophys. Res.*, **118**, 6544–6563, doi:10.1002/jgrd.50450.

Davies-Jones, R. P., 1993: Useful formulas for computing divergence, vorticity, and their errors from three or more stations. *Mon. Wea. Rev.*, **121**, 713–725, doi:10.1175/1520-0493(1993)121<0713:UFFCDV>2.0.CO;2.

Del Genio, A. D., A. Wolf, and M.-S. Yao, 2005: Evaluation of regional cloud feedbacks using single-column models. *J. Geophys. Res.*, **110**, D15S13, doi:10.1029/2004JD005011.

Ellingson, R. G., J. Ellis, and S. Fels, 1991: The intercomparison of radiation codes used in climate models: Long wave results. *J. Geophys. Res.*, **96**, 8929–8953, doi:10.1029/90JD01450.

Frank, W. M., H. Wang, and J. L. McBridge, 1996: Rawinsonde budget analyses during TOGA COARE IOP. *J. Atmos. Sci.*, **53**, 1761–1780, doi:10.1175/1520-0469(1996)053<1761:RBADTT>2.0.CO;2.

Fridlind, A. M., and Coauthors, 2012: A comparison of TWP-ICE observational data with cloud-resolving model results. *J. Geophys. Res.*, **117**, D05204, doi:10.1029/2011JD016595.

Ghan, S. J., and Coauthors, 2000: A comparison of single column model simulations of summertime midlatitude continental convection. *J. Geophys. Res.*, **105**, 2091–2124, doi:10.1029/1999JD900971.

Gregory, D., and P. R. Rowntree, 1990: A mass flux convection scheme with representation of cloud ensemble characteristics and stability-dependent closure. *Mon. Wea. Rev.*, **118**, 1483–1506, doi:10.1175/1520-0493(1990)118<1483:AMFCSW>2.0.CO;2.

Guichard, F., and Coauthors, 2004: Modelling the diurnal cycle of deep precipitating convection over land with cloud-resolving models and single-column models. *Quart. J. Roy. Meteor. Soc.*, **130**, 3139–3172, doi:10.1256/qj.03.145.

Hack, J. J., and J. A. Pedretti, 2000: Assessment of solution uncertainties in single-column modeling frameworks. *J. Climate*, **13**, 352–365, doi:10.1175/1520-0442(2000)013<0352:AOSUIS>2.0.CO;2.

Hume, T., and C. Jakob, 2005: Ensemble single column modeling (ESCM) in the tropical western Pacific: Forcing data sets and uncertainty analysis. *J. Geophys. Res.*, **110**, D13109, doi:10.1029/2004JD005704.

Iacobellis, S., and R. C. J. Somerville, 1991a: Diagnostic modeling of the Indian monsoon onset. Part I: Model description and validation. *J. Atmos. Sci.*, **48**, 1948–1959, doi:10.1175/1520-0469(1991)048<1948:DMOTIM>2.0.CO;2.

——, and ——, 1991b: Diagnostic modeling of the Indian monsoon onset. Part II: Budget and sensitivity studies. *J. Atmos. Sci.*, **48**, 1960–1971, doi:10.1175/1520-0469(1991)048<1960:DMOTIM>2.0.CO;2.

Kennedy, A. D., D. Xiquan, X. Baike, X. Shaocheng, Z. Yunyan, and C. Junye, 2011: A comparison of MERRA and NARR reanalyses with the DOE ARM SGP data. *J. Climate*, **24**, 4541–4557, doi:10.1175/2011JCLI3978.1.

Kuang, Z., 2012: Weakly forced mock-Walker cells. *J. Atmos. Sci.*, **69**, 2759–2786, doi:10.1175/JAS-D-11-0307.1.

Lane, D. E., R. C. J. Somerville, and S. F. Iacobellis, 2000: Sensitivity of cloud and radiation parameterizations to changes in vertical resolution. *J. Climate*, **13**, 915–922, doi:10.1175/1520-0442(2000)013<0915:SOCARP>2.0.CO;2.

Lee, W.-H., S. F. Iacobellis, and R. C. J. Somerville, 1997: Cloud radiation forcings and feedbacks: General circulation model tests and observational validation. *J. Climate*, **10**, 2479–2496, doi:10.1175/1520-0442(1997)010<2479:CRFAFG>2.0.CO;2.

Lenderink, G., and Coauthors, 2004: The diurnal cycle of shallow cumulus clouds over land: A single-column model intercomparison study. *Quart. J. Roy. Meteor. Soc.*, **130**, 3339–3364, doi:10.1256/qj.03.122.

Lin, W., and Coauthors, 2015: RACORO continental boundary layer cloud investigations. Part III: Separation of parameterization biases in single-column model CAM5 simulations of shallow cumulus. *J. Geophys. Res.*, **120**, 6015–6033, doi:10.1002/2014JD022524.

Lin, X., and R. H. Johnson, 1996: Kinematic and thermodynamic characteristics of the flow over the western Pacific warm pool during TOGA COARE. *J. Atmos. Sci.*, **53**, 695–715, doi:10.1175/1520-0469(1996)053<0695:KATCOT>2.0.CO;2.

Lord, S. J., 1982: Interaction of a cumulus cloud ensemble with the large-scale environment. Part III: Semi-prognostic test of the Arakawa–Schubert cumulus parameterization. *J. Atmos. Sci.*, **39**, 88–103, doi:10.1175/1520-0469(1982)039<0088:IOACCE>2.0.CO;2.

Mather, J. H., and J. W. Voyles, 2013: The ARM Climate Research Facility: A review of structure and capabilities. *Bull. Amer. Meteor. Soc.*, **94**, 377–392, doi:10.1175/BAMS-D-11-00218.1.

Minnis, P., W. L. Smith Jr., D. P. Garber, J. K. Ayers, and D. R. Doelling, 1995: Cloud properties derived from GOES-7 for spring 1994 ARM Intensive Observing Period using version 1.0.0 of the ARM Satellite Data Analysis Program. NASA Reference Publication 1366, 59 pp. [Available online at http://ntrs.nasa.gov/archive/nasa/casi.ntrs.nasa.gov/19960021096.pdf.]

Morrison, H., and Coauthors, 2009: Intercomparison of model simulations of mixed-phase clouds observed during the ARM Mixed-Phase Arctic Cloud Experiment. II: Multilayer cloud. *Quart. J. Roy. Meteor. Soc.*, **135**, 1003–1019, doi:10.1002/qj.415.

Ooyama, K., 1987: Scale-controlled objective analysis. *Mon. Wea. Rev.*, **115**, 2479–2506, doi:10.1175/1520-0493(1987)115<2479:SCOA>2.0.CO;2.

Petch, J. C., M. Willett, R. Y. Wong, and S. J. Woolnough, 2007: Modelling suppressed and active convection. Comparing a numerical weather prediction, cloud-resolving and single-column model. *Quart. J. Roy. Meteor. Soc.*, **133**, 1087–1100, doi:10.1002/qj.109.

——, A. Hill, L. Davies, A. Fridlind, C. Jakob, Y. Lin, S. Xie, and P. Zhu, 2014: Evaluation of intercomparisons of four different types of model simulating TWP-ICE. *Quart. J. Roy. Meteor. Soc.*, **140**, 826–837, doi:10.1002/qj.2192.

Ramanathan, V., and J. A. Coakley Jr., 1978: Climate modeling through radiative–convective models. *Rev. Geophys.*, **16**, 465–489, doi:10.1029/RG016i004p00465.

Randall, D. A., and D. G. Cripe, 1999: Alternative methods for specification of observed forcing in single-column models and cloud system models. *J. Geophys. Res.*, **104**, 24 527–24 545, doi:10.1029/1999JD900765.

——, K.-M. Xu, R. C. J. Somerville, and S. Iacobellis, 1996: Single-column models and cloud ensemble models as links between observations and climate models. *J. Climate*, **9**, 1683–1697, doi:10.1175/1520-0442(1996)009<1683:SCMACE>2.0.CO;2.

Sobel, A. H., J. Nilsson, and L. M. Polvani, 2001: The weak temperature gradient approximation and balanced tropical moisture waves. *J. Atmos. Sci.*, **58**, 3650–3665, doi:10.1175/1520-0469(2001)058<3650:TWTGAA>2.0.CO;2.

Somerville, R. C. J., 2000: Using single-column models to improve cloud-radiation parameterizations. *General Circulation Model Development: Past, Present, and Future*. International Geophysics Series, Vol. 40, Academic Press, 641–657.

Song, H., W. Lin, Y. Lin, A. B. Wolf, R. Neggers, L. J. Donner, A. D. Del Genio, and Y. Liu, 2013: Evaluation of precipitation by seven SCMs against ARM observations at the SGP site. *J. Climate*, **26**, doi:10.1175/JCLI-D-12-00263.1.

Stokes, G. M., and S. E. Schwartz, 1994: The Atmospheric Radiation Measurement (ARM) Program: Programmatic background and design of the cloud and radiation test bed. *Bull. Amer. Meteor. Soc.*, **75**, 1201–1221, doi:10.1175/1520-0477(1994)075<1201:TARMPP>2.0.CO;2.

Sud, Y. C., D. M. Mocko, G. K. Walker, and R. D. Koster, 2001: Influence of land surface fluxes on precipitation: Inferences from simulations forced with four ARM-CART SCM datasets. *J. Climate*, **14**, 3666–3691, doi:10.1175/1520-0442(2001)014<3666:IOLSFO>2.0.CO;2.

Šuselj, K., J. Teixeira, and G. Matheou, 2012: Eddy diffusivity/mass flux and shallow cumulus boundary layer: An updraft PDF multiple mass flux scheme. *J. Atmos. Sci.*, **69**, 1513–1533, doi:10.1175/JAS-D-11-090.1.

Thompson, R. M., S. W. Payne, E. E. Recker, and R. J. Reed, 1979: Structure and properties of synoptic-scale wave disturbances in the intertropical convergence zone of the eastern Atlantic. *J. Atmos. Sci.*, **36**, 53–72, doi:10.1175/1520-0469(1979)036<0053:SAPOSS>2.0.CO;2.

Vogelmann, A., and Coauthors, 2015: RACORO continental boundary layer cloud processes: 1. Case study generation and ensemble large-scale forcings. *J. Geophys. Res. Atmos.*, **120**, 5962–5992, doi:10.1002/2014JD022713.

Wang, X., and M. Zhang, 2013: An analysis of parameterization interactions and sensitivity of single-column model simulations to convection schemes in CAM4 and CAM5. *J. Geophys. Res. Atmos.*, **118**, 8869–8880, doi:10.1002/jgrd.50690.

——, and ——, 2014: Vertical velocity in shallow convection for different plume types. *J. Adv. Model. Earth Syst.*, **6**, 478–489, doi:10.1002/2014MS000318.

Wu, J., A. D. Del Genio, M.-S. Yao, and A. B. Wolf, 2009: WRF and GISS SCM simulations of convective updraft properties during TWP-ICE. *J. Geophys. Res.*, **114**, D04206, doi:10.1029/2008JD010851.

Xie, S., and M. H. Zhang, 2000: Impact of the convective triggering function on single-column model simulations. *J. Geophys. Res.*, **105**, 14 983–14 996, doi:10.1029/2000JD900170.

——, and Coauthors, 2002: Intercomparison and evaluation of cumulus parameterizations under summertime midlatitude continental conditions. *Quart. J. Roy. Meteor. Soc.*, **128**, 1095–1136, doi:10.1256/003590002320373229.

——, R. T. Cederwall, M. Zhang, and J. J. Yio, 2003: Comparison of SCM and CSRM forcing data derived from the ECMWF model and from objective analysis at the ARM

SGP site. *J. Geophys. Res.*, **108**, 4499, doi:10.1029/2003JD003541.

——, ——, and ——, 2004: Developing long-term single-column model/cloud system–resolving model forcing data using numerical weather prediction products constrained by surface and top of the atmosphere observations. *J. Geophys. Res.*, **109**, D01104, doi:10.1029/2003JD004045.

——, and Coauthors, 2005: Simulations of midlatitude frontal clouds by single-column and cloud-resolving models during the Atmospheric Radiation Measurement March 2000 cloud intensive operational period. *J. Geophys. Res.*, **110**, D15S03, doi:10.1029/2004JD005119.

——, S. A. Klein, J. J. Yio, A. C. M. Beljaars, C. N. Long, and M. Zhang, 2006: An assessment of ECMWF analyses and model forecasts over the North Slope of Alaska using observations from the ARM Mixed-Phase Arctic Cloud Experiment. *J. Geophys. Res.*, **111**, D05107, doi:10.1029/2005JD006509.

Yang, F. A., H. B. Pan, S. C. Krueger, S. D. Moorthi, and S. E. Lord, 2006: Evaluation of the NCAP Global Forecast System at the ARM SGP site. *Mon. Wea. Rev.*, **134**, 3668–3690, doi:10.1175/MWR3264.1.

Zhang, G. J., 2003: Convective quasi-equilibrium in the tropical western Pacific: Comparison with midlatitude continental environment. *J. Geophys. Res.*, **108**, 4592, doi:10.1029/2003JD003520.

Zhang, M. H., and J. L. Lin, 1997: Constrained variational analysis of sounding data based on column-integrated budgets of mass, heat, moisture and momentum: Approach and application to ARM measurements. *J. Atmos. Sci.*, **54**, 1503–1524, doi:10.1175/1520-0469(1997)054<1503:CVAOSD>2.0.CO;2.

——, ——, R. Cederwall, J. Yio, and S. C. Xie, 2001: Objective analysis of ARM IOP data: Method and sensitivity. *Mon. Wea. Rev.*, **129**, 295–311, doi:10.1175/1520-0493(2001)129<0295:OAOAID>2.0.CO;2.

——, W. Lin, C. Bretherton, J. J. Hack, and P. J. Rasch, 2003: A modified formulation of fractional stratiform condensation rate in the NCAR Community Atmospheric Model (CAM2). *J. Geophys. Res.*, **108**, 4035, doi:10.1029/2002JD002523.

Zhu, P., and Coauthors, 2005: Intercomparison and interpretation of single-column model simulations of a nocturnal stratocumulus-topped marine boundary layer. *Mon. Wea. Rev.*, **133**, 2741–2758, doi:10.1175/MWR2997.1.

Chapter 25

Cloud-Resolving Modeling: ARM and the GCSS Story

STEVEN K. KRUEGER

University of Utah, Salt Lake City, Utah

HUGH MORRISON

National Center for Atmospheric Research, Boulder, Colorado

ANN M. FRIDLIND

NASA Goddard Institute for Space Sciences, New York, New York

1. The GEWEX Cloud System Study

The Global Energy and Water Cycle Experiment (GEWEX) Cloud System Study (GCSS) was created in 1992. As described by Browning et al. (1993, p. 387), "The focus of GCSS is on cloud systems spanning the mesoscale rather than on individual clouds. Observations from field programs will be used to develop and validate the cloud-resolving models, which in turn will be used as test-beds to develop the parameterizations for the large-scale models." The most important activities that GCSS promoted were the following:

- Identify key questions about cloud systems relating to parameterization issues and suggest approaches to address them, and
- Organize model intercomparison studies relevant to cloud parameterization.

Four different cloud system types were chosen for GCSS to study: boundary layer, cirrus, frontal, and deep precipitating convective. A working group (WG) was formed for each of the cloud system types. The WGs organized model intercomparison studies and meetings to present results of the intercomparisons. The first such intercomparison study took place in 1994 (Moeng et al. 1996; Bechtold et al. 1996).

The GCSS approach uses cloud-resolving model (CRM) simulations to estimate cloud system structure, vertical fluxes, and other characteristics from large-scale atmospheric conditions. A CRM is a 2D or 3D nonhydrostatic numerical model that resolves cloud-scale motions while simulating a cloud system. For example, to simulate a convective cloud system that contains both cumulus-scale and mesoscale circulations, a CRM would typically have a horizontal grid size of about 2 km and a horizontal domain size of about 400 km.[1] The GCSS approach is an example of dynamical downscaling, which uses larger-scale conditions to drive a smaller-scale numerical model at higher spatial resolution, which in turn is able to simulate atmospheric phenomena in greater detail.

It is interesting to compare the time and space scales of CRMs used to simulate convective cloud systems to those of global climate models (GCMs; see Table 25-1). The scales are set by the characteristics of the dominant resolved eddies: cumulus clouds in CRMs and baroclinic eddies in GCMs. The scales of cumulus clouds are about a hundredth of those of baroclinic eddies. Thus, a 30-day CRM simulation is equivalent to a 10-yr GCM simulation. CRMs have appropriately been called "cloud GCMs."

Corresponding author address: Professor Steven K. Krueger, Department of Atmospheric Sciences, University of Utah, 135 South 1460 East, Room 819, Salt Lake City, UT 84112.
E-mail: steven.krueger@utah.edu

DOI: 10.1175/AMSMONOGRAPHS-D-15-0047.1

[1] Krueger (2000) described CRMs in greater detail and provided examples of using them to study several types of cloud systems.

TABLE 25-1. CRMs and GCMs: A scale comparison. (From Krueger 2000.)

Aspect	CRM	GCM
Eddies	Cumulus clouds	Baroclinic eddies
Eddy time scale	3×10^3 s	3×10^5 s
Forcing time scale	3–4 days	365 days
Domain size	400 km	40 000 km
Horizontal grid size	2 km	200 km
Time step	10 s	10^3 s

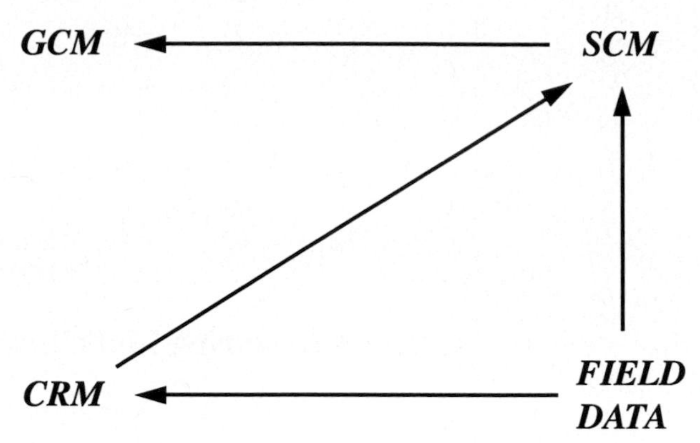

FIG. 25-1. Diagram illustrating how an SCM and a CRM can be used together to relate field data to a GCM. (From Randall et al. 1996.)

The results of the CRM simulations are synthetic or virtual datasets of cloud system realizations that can be used to test, improve, and develop parameterizations for large-scale models. However, because CRMs also contain parameterizations of unresolved processes (including turbulence, microphysics, and radiation), the results of such CRM simulations should be evaluated before being used for large-scale-model parameterization development.

The GCSS approach thus requires observations for both the downscaling and for the evaluation (Randall et al. 1996, 2003). Downscaling requires observations in order to provide a CRM with the same large-scale "forcing" that an atmospheric column in a GCM would have, while evaluation requires observations of macrophysical and microphysical cloud system properties. Large-scale forcing (Zhang et al. 2016, chapter 24) includes the profiles of the large-scale 1) horizontal advective tendencies of temperature and water vapor, 2) vertical velocity, 3) advective tendencies of the horizontal wind components, and 4) horizontal pressure gradient, as well as the surface and top-of-atmosphere boundary conditions, all varying in time. Because of the difficulties of determining the large-scale horizontal wind's advective tendency and the large-scale horizontal pressure gradient, the large-scale forcing for the horizontal wind is usually specified as a relaxation of the CRM's horizontally averaged wind towards the observed wind, with a relaxation time scale of about 2 h. This relaxation term also represents the Coriolis acceleration.

Obviously, accurate large-scale forcing will minimize simulation errors due to the forcing and thereby facilitate identification of errors due to CRM physics. Forcing errors can be detected through model intercomparison studies because such errors typically produce similar impacts in all of the simulations. Once they are detected, the only recourse is to improve the accuracy of the forcing. ARM contributed greatly to producing accurate large-scale forcing datasets by collecting high-frequency vertical profiles of temperature, water vapor, and wind with an array of balloonborne sounding systems (rawinsondes) during intensive observation periods (IOPs) and by using the constrained variational analysis method to process the sounding array profiles.

2. ARM data and single-column modeling

The ARM Program was established to improve the understanding of atmospheric radiation and its interaction with clouds and cloud processes. The goals were to make measurements of atmospheric radiation and the atmospheric properties that affect it at five (later reduced to three) fixed sites for up to 10 years and to use the measurements to test cloud and radiation parameterizations of varying complexity. Randall et al. (1996) proposed to use field data such as those collected by ARM together with single-column models (SCMs) and CRMs to test physical parameterizations used or to be used in GCMs, as shown in Fig. 25-1.

a. *Single-column model IOPs*

The first ARM single-column model IOP took place in winter 1994 and was led by David Randall. Starting with this IOP, seasonal SCM IOPs were conducted at the Southern Great Plains (SGP) site (Zhang et al. 2016, chapter 24) to enhance the frequency of observations for SCM uses, particularly vertical soundings of temperature, water vapor, and winds. These SCM IOPs (listed in Table 25-2) were conducted for a period of 21 days each. During each IOP, radiosondes were launched at the Central Facility and the four boundary facilities eight times per day, seven days per week. The data were required for quantifying boundary forcing and column response.

b. *Variational analysis*

The constrained variational analysis method was developed by Zhang and Lin (1997) for deriving large-scale vertical velocity and advective tendencies from

TABLE 25-2. ARM single-column model IOPs at the Southern Great Plains site from 1994 to 1998.

Year	Winter	Spring	Summer	Fall
1994	X	X	X	X
1995		X	X	X
1996		X	X	
1997		X	X	X
1998	X	X		

sounding arrays. A history of the development of this technique is provided in Zhang et al. (2016, chapter 24). It is used to process atmospheric soundings of winds, temperature, and water vapor mixing ratio over a network of a small number of stations. Given the inevitable uncertainties in the original data, the basic idea in this objective analysis approach is to adjust these atmospheric state variables by the smallest possible amounts to conserve column-integrated mass, moisture, dry static energy, and momentum. The analysis products include both the large-scale forcing terms and the evaluation fields, which can be used for driving SCMs and CRMs and for evaluating model simulations. The first variational analysis dataset produced for ARM was based on the Summer 1995 SCM IOP. This was followed by ones for the Spring, Summer, and Fall 1997 SCM IOPs. The Summer 1995 dataset was used for the first ARM SCM intercomparison study (Cederwall et al. 1998), which was used to evaluate the adequacy of the forcing dataset and to investigate various prescriptions of advective forcing.[2]

c. *Active Remote Sensing of Clouds*

Once ARM began producing accurate large-scale forcing data for the SGP site, the next task was to evaluate the results of CRM (and SCM) simulations that were based on these forcing data. This was accomplished with the extensive array of ARM instruments at the SGP site, particularly those designed to remotely sense cloud properties from the ground. The Active Remote Sensing of Clouds (ARSCL; Clothiaux et al. 2000; Kollias et al. 2016, chapter 17) value-added product combines data from millimeter cloud radars, laser ceilometers, microwave radiometers, and micropulse lidars to produce a time series of vertical distributions of cloud hydrometeors over the ARM sites. A preliminary version of ARSCL first became available for the SGP site in November 1996, and a stable, general version became available on 1 April 1997. Other measurements used to directly evaluate CRM and SCM results included surface broadband radiative fluxes, surface turbulent fluxes, surface precipitation rate, and profiles of temperature and water vapor.

[2] ARM SCM intercomparison cases: http://science.arm.gov/wg/cpm/scm/intercomparison.html.

3. Continental deep convection: Bringing GCSS and ARM together

The inaugural model intercomparison meeting of GCSS Working Group 4 (precipitating convective cloud systems) was held 21–23 October 1996 in Annapolis, Maryland. Moncrieff et al. (1997) described the first two model intercomparison projects organized by Working Group 4, which were designed in accord with Randall et al.'s (1996) proposed method of using SCMs and CRMs to test and develop physically based parameterizations. The main actions and recommendations from the GCSS Science Panel Sixth Session (GCSS Science Team 1998) included this action item for Working Group 4:

> Investigate the possibility of developing a continental case study built on data from the USA, Department of Energy, Atmospheric Radiation Measurement (ARM) Program. Include the possible options for such a case in the Working Group 4 report at the 1998 GCSS meeting.

The GCSS Science Panel Seventh Session (GCSS Science Team 1999) recommended that:

> WG 4 should proceed with a continental deep convection case drawn from data taken at the ARM Southern Great Plains experimental site during July 1997. Specialized instrumentation including a millimeter cloud radar and an extensive array of other meteorological instruments were operational at that time. This case will be done in collaboration with the ARM Single-Column Modeling Group as a means of involving more of the SCM community to participate in the process and to gain support of the ARM Data and Science Integration Team in the provision of forcing data and the compilation of results submitted by the modeling groups.

The first joint GCSS and ARM model intercomparison case was WG 4 Case 3: Summer 1997 ARM SCM IOP. GCSS and ARM collaborated on several major studies of cloud systems during the next 10 years.

4. GCSS and ARM: Confronting models with data

a. *Continental deep convection*

Ric Cederwall (Lawrence Livermore National Laboratory), Steve Krueger (University of Utah), and Dave Randall (Colorado State University) initiated a collaborative intercomparison of SCMs and CRMs that focused on the Summer 1997 SCM IOP at the SGP site (Cederwall et al. 1999). The study involved the ARM Single-Column Model Working Group, the ARM Cloud Working Group, GCSS WG 4 (precipitating convective cloud systems), and the National Centers for Environmental Prediction (NCEP) Environmental Modeling Center. The Summer 1997 IOP provided a 30-day dataset that includes a range of meteorological conditions but was dominated by deep convection.

Large-scale modelers at NCEP and the European Centre for Medium-Range Weather Forecasts (ECMWF) identified the following processes as two of those most in need of improved representation in their parameterizations of precipitating convective cloud systems:

- The occurrence (frequency and intensity) of deep convection, including the diurnal cycle over land and other interactions with the boundary layer
- The production of upper-tropospheric stratiform clouds by deep convection and a related issue: How much microphysical complexity is required in GCMs?

Continental deep convection is strongly coupled to the surface and boundary layer processes. The methodology applied by WG 4 was to start with idealized approaches (for instance, by prescribing surface fluxes) and then to proceed toward more realistic approaches after building physical understanding and developing/improving parameterizations of relevant physical processes. For instance, the diurnal cycle of continental convection, with shallow convection early in the day giving way to the deep convection in the afternoon, could be studied initially without considering the mesoscale variability of surface fluxes, without considering the impact of surface topography on convection development, and without coupling to a land surface model. Moreover, the impact on convection development of larger-scale features (e.g., synoptic-scale fronts and dry lines) that are abundant over midlatitude continents and challenging for the single-column modeling framework underlying the GCSS approach had to be addressed as well.

This model intercomparison project evaluated 8 2D CRMs, 2 3D CRMs, and 11 SCMs by testing their abilities to determine the large-scale statistics of precipitating convective cloud systems for three multiday periods during the Summer 1997 SCM IOP at the SGP site (Xu et al. 2002; Xie et al. 2002). The time-averaged CRM results for these periods of deep convection showed consistently smaller biases of time-averaged temperature and water vapor than did the SCM results. The time-averaged CRM cloud fraction profiles were in reasonable agreement with the observations from the cloud radar, while many of the SCM profiles were not. The CRM and SCM convective mass flux profiles differed significantly in the lower troposphere, and the CRM surface precipitation rates were better correlated with the observations than those from the SCMs. The results confirmed that, as expected, the CRMs are better able to reproduce the observed cloud system properties than are the SCMs. The CRM results for cloud system properties such as convective mass flux that are difficult to observe can therefore be used to evaluate and improve the parameterizations (of convective mass flux, for example) used in the SCMs.

Khairoutdinov and Randall (2003) used a new 3D CRM to simulate the 28-day evolution of clouds over the ARM SGP site during the Summer 1997 IOP. The sensitivity of the results to the domain dimensionality and size, horizontal grid resolution, and parameterization of microphysics was tested. In addition, the sensitivity to perturbed initial conditions was also tested using a 20-member ensemble of runs. The ensemble runs revealed that the uncertainty of the simulated precipitable water due to the fundamental uncertainty of the initial conditions can be as large as 25% of the mean value. Even though the precipitation rates averaged over the whole simulation period were virtually identical among the ensemble members, the timing uncertainty of the onset and the precipitation maximum were as long as one full day. Despite these predictability limitations, the simulation statistics were found to be almost insensitive to the uncertainty of the initial conditions.

The overall effects of the third spatial (second horizontal) dimension were found to be minor for the simulated mean fields and scalar fluxes but were considerable for the velocity and scalar variances. Neither changes over a rather wide range of domain sizes nor in the horizontal resolution had any significant impacts on the simulations. Although a rather strong sensitivity of the mean hydrometeor profiles and, consequently, cloud fraction to the microphysics parameters was found, the effects on the predicted mean temperature and humidity profiles were modest. The spreads among the time series of the simulated cloud fraction, precipitable water, and surface precipitation rate due to changes in the microphysics parameters were within the uncertainty of the ensemble runs.

b. Continental shallow convection

The second joint GCSS and ARM model intercomparison case was referred to as WG 1 ARM, which was an intercomparison study for CRMs and SCMs of the diurnal cycle of shallow cumulus over land based on an idealization of measurements at the ARM SGP site on 21 June 1997. The case coordinator was

Andy Brown from the Met Office. Brown et al. (2002) described the large-eddy simulation (LES)[3] results from this case, while Lenderink et al. (2004) described SCM results from the same case.

Many characteristics of the cumulus layer previously found in simulations of quasi-steady convection over the sea were reproduced in this more strongly surface-forced, unsteady case. Furthermore, the results were encouragingly robust, with similar results obtained with eight independent models and also across a range of numerical resolutions. The LES results for the subcloud layer were consistent with well-established convective boundary layer scalings. Direct validation of the cloud-layer results was difficult, although the tendency of the models to have cloud cover decreasing with time is at least broadly consistent with the observations. In the cloud layer, many of the results previously found in oversea cases are still applicable. For example, cloud cover and convective mass flux decrease with height. The similarity of the cloud-layer structures in the oversea and ARM cases was encouraging, because it suggested that there is no fundamental reason why a parameterization that performs well for one case should not also do so for the other. However, an intercomparison of single-column model results for the present ARM case (Lenderink et al. 2004) revealed the following general deficiencies: values of cloud cover and cloud liquid water that are too large, unrealistic thermodynamic profiles, and significant numerical noise. The results were also strongly dependent on vertical resolution. In part motivated by these results, several of the SCMs have been updated successfully with new parameterization schemes, and/or their present schemes have been modified successfully.

c. *Midlatitude frontal clouds*

Xie et al. (2005) and Xu et al. (2005) described the CRM and SCM intercomparison study of midlatitude frontal clouds based on the ARM March 2000 Cloud IOP. Xie et al. evaluated the overall performance of nine SCMs and four 2D CRMs in simulating a strong midlatitude frontal cloud system during the IOP, while Xu et al. focused on a 27-h period when only shallow frontal clouds were observed.

Xie et al. (2005) found that all models captured the bulk characteristics of the frontal system and the frontal precipitation. However, there were significant differences in the detailed structures of the frontal clouds. All models overestimated high thin cirrus cloud amounts before the main frontal passage. During the frontal passage with strong upward motion, the CRMs underestimated middle and low cloud amounts, while the SCMs overestimated cloud amounts at levels above 765 hPa. All CRMs and some SCMs also underestimated the middle cloud amounts after the frontal passage. In general, the CRM-simulated cloud water and ice contents were comparable to the observations, while most SCMs underestimated cloud water content. SCMs showed very large biases of cloud ice content. Many of the model biases could be traced to the lack of subgrid-scale dynamical structure in the applied forcing fields and to the lack of organized mesoscale hydrometeor advection.

Xu et al. found that all CRMs and SCMs correctly simulated clouds in the observed shallow frontal cloud layer. Most SCMs also produced clouds in the middle and upper troposphere while only shallow clouds were observed. Intermodel differences in the SCM-simulated cloud properties and their profiles were as large as those for summertime continental convection (Xie et al. 2002; Xu et al. 2002), but the intermodel differences among the thermodynamic profiles were comparable between the CRMs and SCMs for this case study. Overall, some CRMs did not perform better than the SCMs did.

d. *Midlatitude cirrus clouds*

The first two WG 2 (cirrus cloud) projects were the Idealized Cirrus Model Comparison (ICMC) Project, developed and led by David Starr (NASA GSFC), and the Cirrus Parcel Model Comparison (CPMC) Project, developed and led by Ruei-Fong Lin (NASA GSFC).

The ICMC Project involved the comparison of simulations of cirrus development and dissipation in two idealized baseline environments: a "warm" cirrus case and a "cold" cirrus case (Starr et al. 2000). The cloud was generated in an ice supersaturated layer about 1 km in depth. Continuing cloud formation was forced via an imposed diabatic cooling representing a 3 cm s^{-1} uplift over a 4-h time span. The simulations then proceeded for an additional 2 h to enable assessment of the cloud dissipation phase of the cloud life cycle among the models. This is a critical issue in that cirrus clouds are commonly observed to be long lasting.

The disagreements between the results from the 16 models submitted for the ICMC Project were substantial. The spread in ice water path (IWP) among the models revealed that at least some of the models had previously unexposed major deficiencies in their representations of ice water precipitation. The differences among the CRMs exceeded what had been expected based on previous studies. The SCMs exhibited a similar

[3] An LES is a 3D simulation that resolves the large, energy-containing turbulent eddies and thus does not rely on a turbulence parameterization to represent the effects of turbulence, unlike a CRM.

range. It was concluded that the models, which spanned a considerable range in terms of their design and heritage, should be able to be tested under much better circumstances as new information becomes available. Advances in observational capabilities, including those in measurement of small ice crystal populations, should be able to adequately resolve the shape of the ice water content profile and the overall ice water path. The result that internal cloud circulation intensity is highly correlated with the ice crystal size distribution should allow for an additional confirming test. At the time of this study, observations of bulk ice water fall speed were just starting to be derived from millimeter-wavelength Doppler radar.

The purpose of the CPMC Project was to compare cirrus simulations by parcel models as well as by parcel-model versions of multidimensional models (Lin et al. 2002). The CPMC Project directly addressed the primary source of dispersion found in the results of the ICMC Project: the development of the ice crystal size distribution. The models used microphysical schemes in which the size distribution of each class of particles (aerosols and ice crystals) is represented by bins or the evolution of each individual particle is calculated. The concept was to look at relatively simple calculations and then to investigate more complex issues, such as effects of gross changes in the aerosol particle size distribution, effects of aerosol particle composition, and direct radiation effects on ice crystal growth rate. The initial focus was on any and all ice nucleation mechanisms included in cirrus parcel models. A second focus was on the homogeneous nucleation process operating in isolation.

There was qualitative agreement for the homogeneous-nucleation-only simulations. For example, the number density of nucleated ice crystals increased with the strength of the prescribed updraft. However, significant quantitative differences were found. The differences in cloud microphysical properties between state-of-the-art models of ice crystal initiation were significant. Intermodel differences in the case of all-nucleation-mode simulations were even greater than in the case of homogeneous nucleation acting alone. It was concluded that definitive laboratory and atmospheric benchmark data are needed to improve the treatment of heterogeneous nucleation processes.

GCSS did not organize any modeling studies of cirrus clouds after 2002 because the cirrus community became heavily involved with convectively generated cirrus in the tropics and the ice microphysics community with Arctic mixed-phase clouds (see section 3). However, two modeling studies of midlatitude cirrus clouds observed at the ARM SGP site were made independently of GCSS under ARM support and are described next.

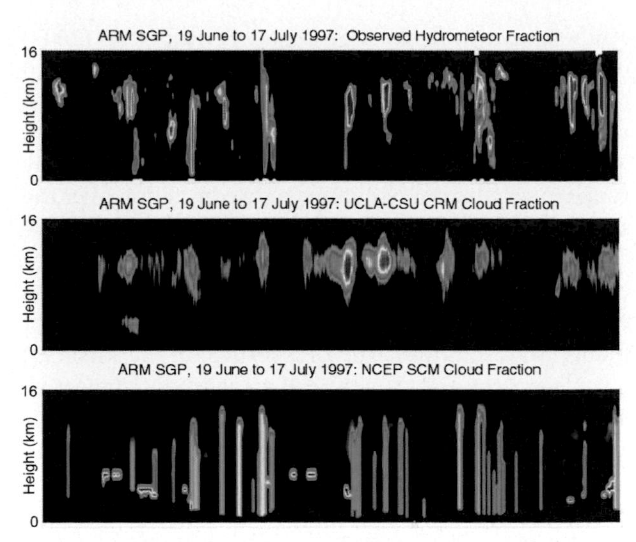

FIG. 25-2. Time–height cloud fraction at ARM SGP, 19 Jun–17 Jul 1997, surface to 16 km: (top) observed by millimeter cloud radar (3-h averages), (middle) simulated by the UCLA–CSU CRM (1-h averages), and (bottom) simulated by the NCEP SCM (3-h averages). Color indicates cloud fraction, which ranges from 0 (violet) to 1 (red).

ARM scientists Yali Luo and Steve Krueger (both at the University of Utah) noticed that the observed and CRM-simulated cloudiness during WG 4 Case 3 was dominated by high clouds (Fig. 25-2). Many of these clouds were generated initially by deep convection, while others formed in situ because of large-scale ascent in the upper troposphere. Luo et al. (2003) compared cirrus macrophysical and microphysical properties obtained from the University of California, Los Angeles–Colorado State University (UCLA–CSU) CRM simulation for the Summer 1997 SCM IOP at the SGP site to those from Mace et al.'s (2001) retrievals based on millimeter cloud radar (MMCR) measurements. Examples of such comparisons are shown in Fig. 25-3. From these comparisons, Luo et al. (2003) concluded that the CRM was able to reproduce the important aspects of the observed properties of cirrus clouds but that the fall speed of large ice crystals was probably too large in the CRM. The study demonstrated that CRM results can be sampled in a way that allows for direct comparison to ARM cirrus cloud property retrievals. This type of comparison allowed an unprecedented evaluation of a CRM's representation of cirrus cloud physics.

Using cloud radar observations of cirrus cloud properties obtained at the ARM SGP site and results from a CRM simulation, Luo et al. (2005) evaluated the cirrus properties simulated by a SCM. The SCM was based on the NCEP Medium-Range Forecast Model (MRF). Luo et al. used SCM and CRM simulations

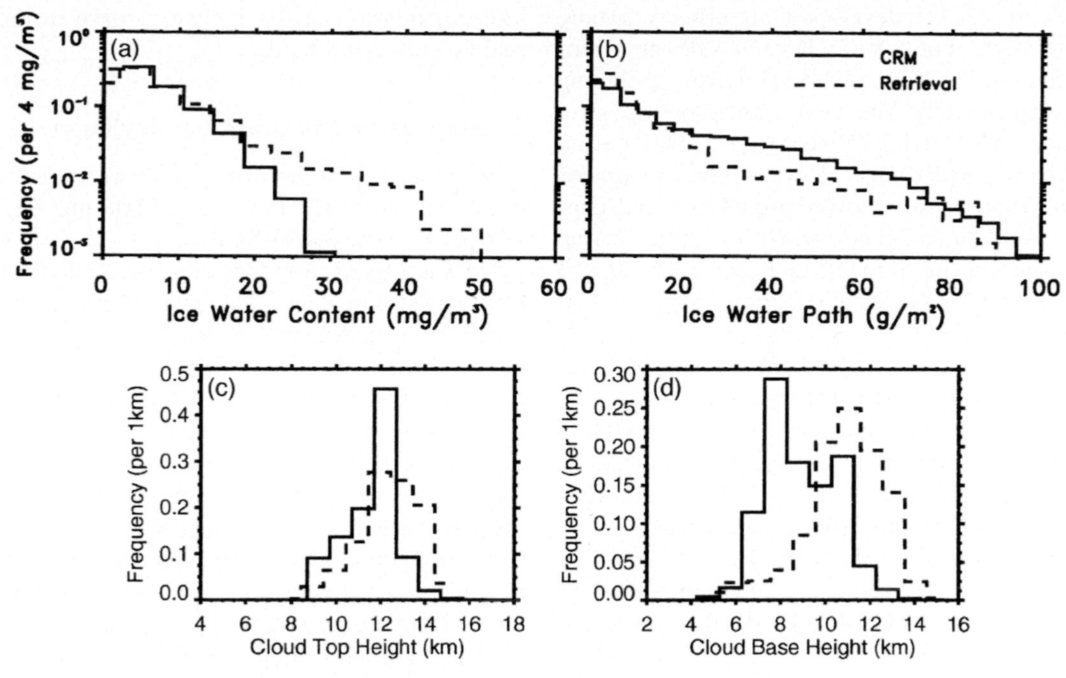

FIG. 25-3. Thin cirrus cloud microphysical and macrophysical properties from (solid line) a UCLA–CSU CRM simulation of a 29-day period (19 Jun–17 Jul 1997) and from (dashed line) retrievals based on summer 1997 (June, July, and August) millimeter cloud radar measurements. (a) Frequency distributions of layer-mean ice water content. (b) Frequency distributions of ice water path. (c) Frequency distributions of cloud-top height. (d) Frequency distributions of cloud-base height. (From Luo et al. 2003.)

based on intensive observations made at the ARM SGP site from 19 June to 17 July 1997. During this period, cirrus clouds, many generated by deep convection, were observed about 30% of the time by the cloud radar. There are significant differences between the NCEP SCM and observed cloud fraction profiles, most notably in the SCM's underestimate of cloud fraction at high levels (Fig. 25-2).

To produce cirrus statistics from the SCM results that are comparable to the cloud radar observations, Luo et al. (2005) employed a method described by Klein and Jakob (1999) that uses the SCM cloud fraction profile and the SCM's overlap assumption (random or maximum/random) to create a synthetic cloud field. Luo et al. sampled the synthetic cloud fields as would a cloud radar to determine the statistical properties of cirrus. They then compared the SCM's cirrus cloud properties to those obtained by Mace et al. (2001) using the MMCR and to a corresponding CRM simulation. Luo et al. found that cirrus properties simulated by the SCM significantly differed from cloud radar observations, while the CRM simulation reproduced most of the cirrus properties as revealed by the observations.

Luo et al. (2006) used the CRM results to evaluate the SCM's representation of the physical processes determining the simulated cirrus: specifically, cumulus detrainment and ice microphysics. They found that in the SCM (i) detrainment occurs too infrequently at a single level at a time, though the detrainment rate profile averaged over the entire IOP is comparable to that in the CRM simulation; (ii) too much detrained ice is sublimated when first detrained; and (iii) snow sublimates over too deep of a layer.

Solch and Karcher (2011) applied a cloud-resolving model with explicit aerosol and ice microphysics and Lagrangian ice particle tracking to simulate the evolution of a cirrus cloud field observed during the ARM IOP in March 2000. This dataset includes remote sensing, radiosonde, and aircraft measurements of a midlatitude cirrus cloud system, supported by estimates of the dynamical cloud forcing. The dataset allowed Solch and Karcher to evaluate and study in great detail the process-oriented representation of the microphysical processes relevant to the formation and evolution of deep, stratiform cirrus (in particular, ice crystal sedimentation and aggregation). The suite of explicitly resolved physical processes in their model enabled them to better understand the sensitivity of the simulated cirrus properties to a large number of microphysical and environmental parameters.

Yang et al. (2012) developed an observationally based case study that is suitable for a Global Atmospheric System Studies (GASS)[4] cirrus modeling intercomparison study. The case is based on measurements made on 9 March 2000 at the ARM SGP during an IOP. The retrievals of ice water content, ice number concentration, and fall velocity provide several constraints to evaluate model performances. Initial testing of the case using the Met Office Large Eddy Model suggests that the case is appropriate for a model intercomparison study.

A midlatitude cirrus case based on ARM measurements was developed in 2012 and included as part of the joint 8th International Cloud Modeling Workshop (Muhlbauer et al. 2013) and GASS Cirrus Model Intercomparison Project led by Andreas Muhlbauer and Thomas Ackerman (both at the University of Washington). The case describes a synoptically driven thick cirrus developing over the ARM SGP site on 1 April 2010 during the Small Particles in Cirrus (SPARTICUS) field campaign. The goal of the project was to investigate the microphysical and macrophysical evolution and life cycle of a synoptically driven cirrus and to compare simulated cirrus cloud properties and radiative effects among models. Special emphasis was placed on the contribution of small ice crystals in cirrus and the role of homogeneous and heterogeneous ice nucleation. Simulations were compared and evaluated with in situ aircraft observations and with various ground-based and spaceborne remote sensors. This project specifically targeted CRM, LES models, and SCMs with advanced cloud microphysics schemes, such as multimoment bulk microphysics parameterizations or bin microphysics schemes.

Preliminary analysis of the modeling results (Muhlbauer et al. 2013) suggests that models have difficulty in predicting the observed ice number concentrations and representing the vertical structure of the cirrus. Ice number concentrations are overestimated in the homogeneous freezing regime and at cold temperatures but are underestimated in the warmer temperature regime in which ice is initiated by heterogeneous ice nucleation mechanisms. The modeling results also indicate deficiencies in correctly representing the observed vertical profiles of ice water content and radar reflectivity and show an underestimation of the ice water path in the mesoscale cirrus cloud field.

[4] In 2011, GCSS became a part of the GEWEX GASS Panel. It also abandoned its working group structure and began operating instead through projects that could be initiated by any member of the community.

5. Long-term cloud-resolving simulations

Wu et al. (2007) performed a CRM simulation for a period of 26 days (22 June–17 July) during the Summer 1997 IOP at the ARM SGP site. A 2D version of the CRM was aligned east to west and used a domain 600 km long and 40 km deep. The model employed a 3-km horizontal grid size and a stretched grid in the vertical (100 m at the surface, increasing to 1500 m at the model top). The CRM differences from the ARM measurements, when averaged over the entire period, were less than 5 W m^2 in both longwave and shortwave radiative fluxes at the top of the atmosphere and surface. Using the CRM simulation as a benchmark, it was demonstrated that the conventional GCM radiation calculation greatly underestimates the shortwave and longwave fluxes at the top of atmosphere and surface because of the use of homogeneous cloud condensates and unrealistic random overlap assumptions.

Xie et al. (2004) developed long-term single-column model/cloud-system-resolving model forcing datasets using numerical weather prediction products constrained by surface and top-of-atmosphere observations. Wu et al. (2008) performed a year-long 2D CRM simulation using the variationally constrained large-scale forcing produced by Xie et al. for the year 2000 at the ARM SGP site. The CRM-simulated year-long cloud liquid water path and cloud (liquid and ice) optical depth are in good agreement with the ARM retrievals over the SGP site. The cloudy-sky total (shortwave and longwave) radiative heating profile shows a dipole pattern (cooling above and warming below) during spring and summer, while a second peak of cloud radiative cooling appears near the surface during winter and fall.

Zeng et al. (2007) simulated two 20-day continental midlatitude cases with a 3D CRM and compared the results to ARM data. The first case was the ARM Spring 2000 IOP (Xie et al. 2005; Xu et al. 2005). The second case covered the period from 25 May to 14 June 2002 for the same ARM domain. The two cases represent springtime and summertime midlatitude clouds, respectively. Large-scale forcing was based on the variational analysis approach. The surface fluxes were specified using sitewide averages of observed fluxes from the ARM Energy Balance Bowen Ratio (EBBR) stations or estimated from a land data assimilation system. Evaluation of the simulations shows that the model overpredicts cloud amounts, especially in the upper troposphere. Comparisons of 2D and 3D simulations showed that the 2D CRM not only had unrealistic rapid fluctuations in surface precipitation but also a spurious dehumidification (or a decrease in cloud amount). ARM surface fluxes were obtained from the EBBR stations, which use the Bowen ratio to partition the fluxes. When Land Information System (LIS) surface fluxes replaced ARM data in the CRM simulations, similar results were obtained, but LIS fluxes produced a better simulation of diurnal cloud variation in the lower troposphere.

6. Convective and cloud processes during TWP-ICE: A multimodel evaluation project

Accurate representation of the characteristics and effects of tropical deep convection remains a leading challenge in global modeling. Cirrus are also ubiquitous in the tropics and have major radiative impacts on their environment, but the properties of these clouds and their connections to deep convection processes are poorly understood. The Tropical Warm Pool–International Cloud Experiment (TWP-ICE) was aimed at describing the properties of tropical cirrus and the convection that leads to their formation (May et al. 2008). TWP-ICE took place 21 January–13 February 2006 in the region near the ARM fixed site at Darwin, northern Australia. The experiment was a collaborative effort between ARM, the Australian Bureau of Meteorology, NASA, the European Commission Directorate-General for Research and Innovation, and several U.S., Australian, Canadian, and European universities. Measurements included a five-station, 3-hourly sounding array, with one station aboard ship offshore. The sounding array enclosed a coastal area roughly the size of a coarse-resolution GCM grid box. Multiple surface energy budget sites characterized maritime and continental surface sensible, latent, and radiative heat fluxes. Scanning precipitation radars provided domainwide polarimetric retrievals of near-surface rain rate and dual-Doppler retrievals of vertical winds over a smaller interior region. Xie et al. (2010) applied the variational analysis approach of Zhang and Lin (1997) to this extensive set of measurements to perform an objective analysis from which the large-scale vertical velocity and the advective tendencies of temperature and water vapor were derived. These were used as the large-scale forcing for the CRM and SCM simulations. TWP-ICE overlapped with the Aerosol and Chemical Transport in Tropical Convection (ACTIVE) program at Darwin, which provided extensive collocated airborne aerosol measurements during TWP-ICE (Vaughan et al. 2008). A chief outcome from TWP-ICE was a dataset suitable to provide the forcing and testing required by modern cloud-resolving models and parameterizations in GCMs. More than 50 observational and modeling papers have been based primarily on TWP-ICE observations. Here, we briefly discuss papers that included cloud-resolving simulations and associated model intercomparison studies.

Because the outstanding questions regarding tropical deep convection (e.g., how do individual cirrus cloud particles form and evolve, and what is the structure and evolution of ascent and descent regions in mesoscale convective systems?) range from the microscale to the mesoscale and larger, the TWP-ICE experimental data motivated a series of interlocking model intercomparison studies that were coordinated by the GCSS Precipitating Cloud System (PCS) group led by Jon Petch (Met Office), following an approach previously tested at the Met Office (Petch et al. 2007). The TWP-ICE model intercomparison studies included one for CRMs with periodic boundary conditions led by Ann Fridlind (NASA Goddard Institute for Space Studies), one for SCMs led by Laura Davies (Monash University), one for limited-area models (LAMs) with open boundary conditions and nested grids led by Ping Zhu (Florida International University), and one for global atmospheric models led by Yanluan Lin (NOAA/Geophysical Fluid Dynamics Laboratory). By coordinating the days simulated to insure spatiotemporal overlap, the four studies allowed direct comparison of results. Each study optimized its approach by requesting diagnostics that focused on its individual primary research questions, which were selected from the wide range of possible relevant issues. Thus, four separate studies emerged, each with some modeling aspects that could be considered cloud resolving (albeit coarse in the GCM case).

The SCM study utilized a 100-member ensemble of large-scale forcings that was produced by considering experimental uncertainty of the retrieved surface rainfall rate that was an input to the variational analysis code (Davies et al. 2013). Since computational expense was not an obstacle, the SCM study spanned the full experiment period, including sequential active monsoon, suppressed monsoon, and break periods. Based on results submitted from 11 SCMs plus a subset of ensemble members from two CRM models (one 2D and one 3D), the largest range of model sensitivity occurred under weak forcing conditions. Overall, SCMs and CRMs differed substantially in predicted surface evaporation. In addition, the vertical structure of cloud variables was relatively insensitive across ensemble members for any given model but demonstrated pronounced differences from one model to the next.

The CRM study emphasized rigorous comparison of simulation results with observations, including forward simulation of Rayleigh radar reflectivity from the 3D model output fields for evaluation of convective and stratiform rain areas (Varble et al. 2011; Fridlind et al. 2012a). Owing to computational expense, simulations were limited to the first 16 days of active and suppressed monsoon. A single optional sensitivity test included domain-mean relaxation to observed conditions, which was intended to facilitate comparison with observations by limiting drift of the simulated moist static energy from that observed. In 10 submitted 3D simulations, predicted convective rain area fractions were highly correlated with observations. However, convective areas were systematically overestimated, and stratiform rain areas varied widely around observed values. Considering the handful of domainwide observables during active and suppressed periods (obtained via satellite and scanning radar), the strongest rank correlation was found between large stratiform-area and small top-of-atmosphere outgoing longwave radiation during the active period. Fridlind et al. (2012a) concluded that CRMs require closer observational evaluation of the often poorly predicted and radiatively important stratiform properties and the factors controlling them. Focusing on the source of differences between simulations, Varble et al. (2011) concluded that varying degrees of overestimation among the models of ice radar reflectivity could be attributed more to different assumptions about hydrometeor size distributions among the models than to differences in predicted mass mixing ratios among the models.

The LAM study focused on six simulations from LAMs that used three different large-scale input fields (i.e., different boundary conditions) and compared them with one another and with an ensemble from the CRM intercomparison results (Zhu et al. 2012; Varble et al. 2014a,b). All of the LAM simulations reproduced the observed large-scale wind, temperature, and water vapor fields quite successfully, but the predicted rain rates and cloud properties varied widely, especially in the stratiform rain regions. In a detailed comparison of simulated stratiform rain properties to disdrometer and profiling radar datasets, Varble et al. (2014b) found that biases in rain properties, precipitation radar reflectivities, and mean Doppler fall speeds relative to the observations were similar in LAM and CRM simulations that used similar microphysics schemes. Comparison with observations revealed errors associated with specific one- or two-moment microphysics scheme components, but also indicated the presence of a systematic underestimation of stratiform ice mixing ratio at the melting level in the LAM and CRM simulations. In a companion paper that described a detailed comparison of convective updraft properties with dual-Doppler retrievals, Varble et al. (2014a) indicated that there was a systematic overestimation of updraft speeds in the LAM and CRM simulations. Excessive updraft speeds were most pronounced in the upper troposphere but likely also significant near the surface, where excessive radar reflectivity was collocated with excessive lofting of rain above the freezing level. Together, the results supported a previously reported connection between decreased stratiform detrainment and increased detrainment height.

Petch et al. (2014) compared the LAM, CRM, and SCM results with an intercomparison of nine global atmospheric model simulations of the TWP-ICE monsoon and break periods (Lin et al. 2012). The global model study, which included two simulations with roughly 20-km horizontal resolution, found that predicted cloud properties varied widely across models but were not strongly sensitive to increased horizontal resolution. Petch et al. (2014) speculated that the cloud-resolving nature of the CRM and LAM simulations served to constrain predicted liquid water contents, in contrast to the highly parameterized schemes in SCM and global models, but that ice water contents remained strongly dependent on the representation of the microphysics across all model types.

Although the four independent TWP-ICE intercomparison studies resulted in a wide range of progress and conclusions, a disadvantage of this approach was that substantial differences in large-scale forcing methods and diagnostic definitions complicated direct comparisons across the various simulations and observations. By weighing the benefits of loose coordination against the benefits and costs of tighter coordination, in future studies it may be possible to better optimize the intercomparison approach by identifying a subset of well-constructed diagnostics and a limited set of goals that span the component studies without compromising structural differences that serve independent goals.

Several follow-on studies used the CRM intercomparison simulations: Rio et al. (2013) refined a cold pool parameterization; Mrowiec et al. (2012) reported roughly linear relationships between simulated updraft and downdraft mass fluxes and convective area fractions; and Mrowiec et al. (2015) applied an isentropic analysis approach to remove substantial gravity wave contributions to calculated convective mass fluxes for parameterization development.

Other studies that used independent 2D or 3D CRM simulations with TWP-ICE observations reported a variety of results, including the following:

- Substantial effects of cloud condensation nuclei concentrations, ice nuclei concentrations, freezing parameterizations, and wind shear on simulated anvil area and water budget terms (Fan et al. 2009b, 2010a,b; Zeng et al. 2009a,b, 2013);

- Increased cloud condensation nuclei concentrations leading to weaker convection because of changes in anvil ice characteristics and subsequent upper-tropospheric radiative heating and weaker tropospheric destabilization in one study (Morrison and Grabowski 2011);
- Weak enhancement of surface precipitation but substantial effects on its spatiotemporal distribution (Lee and Feingold 2010, 2013);
- Relaxation of water vapor toward saturation in overshooting convection (Hassim and Lane 2010);
- Forward simulation of multidirectional polarized reflectance to evaluate anvil-top ice properties (van Diedenhoven et al. 2012).

Other studies using independent LAM simulations reported a strong sensitivity to the choice of bulk parameterization of ice microphysics (Wang et al. 2009), identified the triggering mechanisms of Hector convection[5] (Ferretti and Gentile 2009; Zhu et al. 2013), and identified a relatively shallow entrainment-dominated layer in simulated deep updrafts in contrast to convective parameterization assumptions (Wang and Liu 2009). The studies also tested entrainment parameterization schemes against Weather Research and Forecast (WRF) Model simulations (Wu et al. 2009; Del Genio and Wu 2010), demonstrated the skill of WRF simulations to reproduce observed rainfall statistics (Wapler et al. 2010), and reported improvement of WRF simulations using a 3D variational data assimilation (3DVar) system or another observation incorporation technique (Yeh and Fu 2011; Zhu et al. 2012). Del Genio et al. (2012) used cloud-resolving simulations of TWP-ICE to characterize mesoscale organization processes and to provide parameterization guidance.

Taken as a group, these TWP-ICE cloud-resolving simulation studies emphasized understanding the mechanistic behavior of tropical convection, its sensitivity to environmental conditions, and the degree to which various modeling choices give divergent results or reproduce observations. The need to improve the simulation of ice microphysical processes—including the role of dynamics in determining ice properties—is evident.

7. Arctic clouds

Accelerated warming and rapid environmental change highlight the Arctic as a region particularly sensitive to climate change. Studies have linked this sensitivity to various feedbacks operating in the region, with changes in cloud properties central to these feedbacks. Mixed-phase clouds comprise a large fraction of clouds occurring in the Arctic and are a critical component of the regional climate, but the relevant physical processes and their parameterization in models remain uncertain. Several field experiments (led by ARM or with significant ARM contributions) conducted over the last 15 years have provided measurements to test and improve models. These have included the 1997–98 Surface Heat Budget of the Arctic Ocean (SHEBA) experiment (Uttal et al. 2002), the 2004 Mixed-Phase Arctic Cloud Experiment (MPACE; Verlinde et al. 2007), and the 2008 Indirect and Semi-Direct Aerosol Campaign (ISDAC; McFarquhar et al. 2011). Observations from these three field experiments have been used to develop case studies for ARM/GCSS/GASS model intercomparison projects.

The first intercomparison project was led by Stephen Klein (Lawrence Livermore National Laboratory) and Hugh Morrison (National Center for Atmospheric Research). It comprised two cases derived from MPACE observations gathered over northern Alaska and the adjacent Beaufort Sea: 1) a single-layer, boundary layer mixed-phase cloud system associated with cold air outflow from the Arctic pack ice southward across open ocean (Klein et al. 2009) and 2) a deeper multilayered mixed-phase cloud system driven by midlevel mesoscale and synoptic-scale forcing as well as cold air outflow over open ocean (Morrison et al. 2009). Xie et al. (2006) developed SCM and CRM large-scale forcing data for MPACE from sounding array data collected by ARM using an objective variational analysis approach. Multiple LES, cloud-resolving, and single-column models participated in the project. The single-layer case included 28 submissions from 17 SCMs and 9 CRMs, the most of any GCSS/GASS intercomparison to date. There was a wide spread of the simulated liquid and ice water paths among the models, leading to large differences in the cloud radiative forcings at the surface. A majority of the models underpredicted liquid water path for the single-layer case, with the opposite for the multilayer case. Models with more sophisticated microphysics schemes tended to produce more realistic results, although there was considerable scatter. There was also a range of several orders of magnitude in the modeled ice crystal concentrations that likely contributed to the large spread in the results. These cases also served as the basis for several additional publications focusing on LES, cloud-resolving, or mesoscale modeling, including Fridlind et al. (2007), Prenni et al. (2007), Luo et al. (2008a,b), Morrison et al. (2008), Fan et al.

[5] Thunderstorms that form regularly over the Tiwi Islands just to the north of Darwin, Australia.

(2009a, 2011), Sednev et al. (2009), and Solomon et al. (2009).

The next GCSS intercomparison project was based on observations of a single-layer mixed-phase case from SHEBA over the central Arctic Ocean and led by Hugh Morrison (Morrison et al. 2011). This case differed from the single-layer MPACE case in several ways: colder cloud temperatures, a sea-ice-covered surface instead of open ocean, and relatively polluted aerosol characteristics. This work built upon the MPACE intercomparison project with the goal of further exploring differences in model results and relationships between the dynamics, radiation, and microphysics driving this spread. To further simplify the model setup and analysis, ice nucleation was constrained in the simulations in a way that held the ice crystal concentrations approximately fixed, with two sets of sensitivity runs in addition to the baseline simulations utilizing different specified ice nucleus concentrations. Simulations clustered into two distinct quasi-steady states consisting of persistent mixed-phase clouds or all-ice clouds. Transitions from the mixed-phase to the all-ice state were accelerated by feedbacks between the dynamics, microphysics, and radiation when the bulk deposition rate of water vapor onto ice exceeded a threshold value. Additional publications based on modeling results from this case include Morrison and Pinto (2005, 2006), de Boer et al. (2010, 2013), van Diedenhoven et al. (2009, 2011), and Fridlind et al. (2012b).

The third and most recent intercomparison project was developed from observations of a single-layer, mixed-phase cloud system over northern Alaska and the adjacent Beaufort Sea observed during ISDAC (McFarquhar et al. 2011) and led by Mikhail Ovchinnikov (Pacific Northwest National Laboratory). This case was similar to the SHEBA single-layer case, but cloud temperatures were warmer, and the cloud-topped mixed layer was decoupled instead of coupled with the surface. Measurements of this cloud system were also much more comprehensive compared to the SHEBA case. The goal of the ISDAC intercomparison was to build upon the MPACE and SHEBA intercomparisons and further examine causes of differences in large-eddy simulations of mixed-phase clouds. To this end, simulations were further constrained compared to the previous intercomparisons. This was done by using the same ice microphysical characteristics (bulk density, capacitance, and fall speed relationships), simplified radiation scheme, and horizontal and vertical grid spacings in all models. Despite these constraints, there was still a fairly large spread of simulated cloud characteristics, including liquid water path and surface precipitation. This was attributed to differences in model dynamics, numerics, and subgrid-scale mixing schemes as well as in representations of the ice particle size distribution shape. Additional LES and cloud-resolving modeling publications based on ISDAC observations include Solomon et al. (2011, 2014), Ovchinnikov et al. (2011), Avramov et al. (2011), and Fan et al. (2011).

8. Summary

In 1993, GCSS promoted a two-stage methodology by which field observations can be combined with simulations using CRMs to develop parameterizations of cloud and precipitation processes for use in global models (Browning et al. 1993). The first stage is to evaluate and improve the CRMs with the help of observational field experiments. The second stage is to use CRMs to develop parameterizations for large-scale models. In 1994, while GCSS was organizing its first model intercomparison, ARM was carrying out its first SCM IOPs. The first variational analysis dataset produced by ARM was based on the Summer 1995 SCM IOP. The first stable version of ARSCL, based primarily on the MMCR, became available in 1997. The first joint GCSS and ARM model intercomparison case was WG 4 Case 3 based on the Summer 1997 ARM SCM IOP. As described in this chapter, GCSS and ARM collaborated on several major studies of cloud systems during the next 10 years. These included studies of the diurnal cycle of shallow cumulus based on an idealization of measurements at the ARM SGP site, midlatitude frontal clouds based on the ARM March 2000 Cloud IOP, and idealized cirrus clouds via the Idealized Cirrus Model Comparison (ICMC) and Cirrus Parcel Model Comparison (CPMC) Projects. They also included synoptically driven thick cirrus observed over the ARM SGP site on 1 April 2010 during the Small Particles in Cirrus (SPARTICUS) field campaign and deep tropical convection during the Tropical Warm Pool–International Cloud Experiment (TWP-ICE) that took place during 2006 at the ARM site in Darwin, Australia. Arctic cloud studies were based on cases from three field experiments: the 1997–98 Surface Heat Budget of the Arctic Ocean (SHEBA) experiment, the 2004 Mixed-Phase Arctic Cloud Experiment (MPACE), and the 2008 Indirect and Semi-Direct Aerosol Campaign (ISDAC).

Other GCSS-type modeling projects described in this chapter were carried out independently by ARM scientists and included studies of midlatitude cirrus generated by deep convection during the Summer 1997 ARM SCM IOP, the evolution of a cirrus cloud field observed during the ARM IOP in March 2000, the seasonal variation of cloud systems over ARM SGP for the year 2000, and midlatitude clouds during the ARM Spring 2000 IOP and the May–June 2002 IOP.

When GCSS and ARM began collaborating, CRMs were 2D and used single-moment microphysics. Now, 20 years later, 3D CRMs have become standard, and two-moment microphysical schemes are commonly used. Another change has been a shift from the single-column modeling approach to a nested modeling approach, in which the highest-resolution domain is embedded within one or more larger and lower-resolution domains that provide lateral boundary conditions. ARM's observational capabilities also have increased with the deployment of improved cloud and precipitation radars.

There will always be a need for parameterizations in atmospheric models. The need for cumulus parameterization has been removed from some global models by embedding CRMs within each grid column or by increasing the horizontal resolution to that of a CRM. However, these approaches are computationally expensive, so improving conventional cumulus parameterizations remains a priority. Furthermore, the need for turbulence parameterization and especially for microphysics parameterization in global models and cloud-resolving models will not go away any time soon. Improving these parameterizations will remain challenging tasks for GASS and Atmospheric System Research (ASR; Mather et al. 2016, chapter 4) in the years ahead.

The impacts of GCSS-style modeling projects are wide but diffuse and difficult to summarize strictly in terms of scientific results in papers. GCSS promoted the use of CRMs (and LES models) to better understand the cloud-scale processes that must be parameterized in GCMs. This methodology (the SCM approach) requires high-quality datasets for forcing and evaluating the SCMs and the CRMs, and it requires CRMs that reproduce the observations to an adequate degree, in order to convince parameterization developers to use the results of the CRMs for parameterization development and improvement. What may not be so obvious is that observationally based model intercomparison projects do not necessarily lead directly to parameterization improvements. More often, they establish the capabilities or lack thereof of the CRMs, and are followed by idealized and simplified model intercomparison cases, which have two important advantages over strictly observationally based ones: by simplifying a case (by limiting the physical processes involved), the CRMs could reach a consensus; and by idealizing a case, a particular aspect of a cloud system could be studied without extraneous complications. There were many examples of first trying a realistic case, then simplifying and idealizing it. Insights that led most rapidly to parameterization improvements were almost always obtained from the idealized cases.

An indirect but ultimately perhaps very important impact of GCSS on parameterization development was and remains that of building a community of cloud-resolving modelers who are interested in parameterization development and know how to use CRMs and observations for that purpose.

Acknowledgments. The authors acknowledge the support and efforts of the ARM Climate Research Facility site and field campaign teams whose measurements are the basis of the modeling studies described in this chapter. In addition, the first author in particular acknowledges David A. Randall, Minghua Zhang, Richard T. Cederwall, and Shaocheng Xie for instituting the first SCM IOPs and for developing the methods used to create high-quality large-scale forcing data sets from SCM IOP measurements.

REFERENCES

Avramov, A., and Coauthors, 2011: Towards ice formation closure in Arctic mixed-phase boundary layer clouds during ISDAC. *J. Geophys. Res.*, **116**, D00T08, doi:10.1029/2011JD015910.

Bechtold, P., S. K. Krueger, W. S. Lewellen, E. van Meijgaard, C.-H. Moeng, D. A. Randall, A. van Ulden, and S. Wang, 1996: Modeling a stratocumulus-topped PBL: Intercomparison among different one-dimensional codes and with large eddy simulation. *Bull. Amer. Meteor. Soc.*, **77**, 2033–2042, doi:10.1175/1520-0477(1996)077<2033:MASTPI>2.0.CO;2.

Brown, A. R., and Coauthors, 2002: Large-eddy simulation of the diurnal cycle of shallow cumulus convection over land. *Quart. J. Roy. Meteor. Soc.*, **128**, 1075–1093, doi:10.1256/003590002320373210.

Browning, K. A., and Coauthors, 1993: The GEWEX Cloud System Study (GCSS). *Bull. Amer. Meteor. Soc.*, **74**, 387–399, doi:10.1175/1520-0477(1993)074<0387:TGCSS>2.0.CO;2.

Cederwall, R. T., J. J. Yio, and S. K. Krueger, 1998: The ARM SCM intercomparison study—Overview and preliminary results for case 1. *Proc. Eighth Atmospheric Radiation Measurement (ARM) Science Team Meeting*, Tucson, AZ, U.S. Department of Energy, 123–126. [Available online at http://www.arm.gov/publications/proceedings/conf08/extended_abs/cederwall_rt.pdf.]

——, S. K. Krueger, D. J. Rodriguez, and D. A. Randall, 1999: The ARM-GCSS Intercomparison Study of Single-Column Models and Cloud System Models. *Proc. Ninth Atmospheric Radiation Measurement (ARM) Science Team Meeting*, San Antonio, TX, U.S. Department of Energy, 1–10. [Available online at http://www.arm.gov/publications/proceedings/conf09/extended_abs/cederwall_rt.pdf.]

Clothiaux, E. E., T. P. Ackerman, G. G. Mace, K. P. Moran, R. T. Marchand, M. A. Miller, and B. E. Martner, 2000: Objective determination of cloud heights and radar reflectivities using a combination of active remote sensors at the ARM CART sites. *J. Appl. Meteor.*, **39**, 645–665, doi:10.1175/1520-0450(2000)039<0645:ODOCHA>2.0.CO;2.

Davies, L., and Coauthors, 2013: A single-column model ensemble approach applied to the TWP-ICE experiment. *J. Geophys. Res.*, **118**, 6544–6563, doi:10.1002/jgrd.50450.

de Boer, G., T. Hashino, and G. J. Tripoli, 2010: Ice nucleation through immersion freezing in mixed-phase stratiform clouds: Theory and numerical simulations. *Atmos. Res.*, **96**, 315–324, doi:10.1016/j.atmosres.2009.09.012.

——, and Coauthors, 2013: Near-surface meteorology during the Arctic Summer Cloud Ocean Study (ASCOS): Evaluation of reanalyses and global climate models. *Atmos. Chem. Phys. Discuss.*, **13**, 19 421–19 470, doi:10.5194/acpd-13-19421-2013.

Del Genio, A. D., and J. Wu, 2010: The role of entrainment in the diurnal cycle of continental convection. *J. Climate*, **23**, 2722–2738, doi:10.1175/2009JCLI3340.1.

——, ——, and Y. Chen, 2012: Characteristics of mesoscale organization in WRF simulations of convection during TWP-ICE. *J. Climate*, **25**, 5666–5688, doi:10.1175/JCLI-D-11-00422.1.

Fan, J., M. Ovtchinnikov, J. M. Comstock, S. A. McFarlane, and A. Khain, 2009a: Ice formation in Arctic mixed-phase clouds: Insights from a 3-D cloud-resolving model with size-resolved aerosol and cloud microphysics. *J. Geophys. Res.*, **114**, D04205, doi:10.1029/2008JD010782.

——, and Coauthors, 2009b: Dominant role by vertical wind shear in regulating aerosol effects on deep convective clouds. *J. Geophys. Res.*, **114**, D22206, doi:10.1029/2009JD012352.

——, J. M. Comstock, and M. Ovchinnikov, 2010a: Tropical anvil characteristics and water vapor of the tropical tropopause layer: Impact of heterogeneous and homogeneous freezing parameterizations. *J. Geophys. Res.*, **115**, D12201, doi:10.1029/2009JD012696.

——, ——, and ——, 2010b: The cloud condensation nuclei and ice nuclei effects on tropical anvil characteristics and water vapor of the tropical tropopause layer. *Environ. Res. Lett.*, **5**, 044005, doi:10.1088/1748-9326/5/4/044005.

——, S. Ghan, M. Ovchinnikov, X. Liu, P. Rasch, and A. Korolev, 2011: Representation of Arctic mixed-phase clouds and the Wegener–Bergeron–Findeisen process in climate models: Perspectives from a cloud-resolving study. *J. Geophys. Res.*, **116**, D00T07, doi:10.1029/2010JD015375.

Ferretti, R., and S. Gentile, 2009: A study of the triggering mechanisms for deep convection in the tropics using a mesoscale model: Hector events during SCOUT-O_3 and TWP-ICE campaigns. *Atmos. Res.*, **93**, 247–269, doi:10.1016/j.atmosres.2008.11.004.

Fridlind, A. M., A. S. Ackerman, G. McFarquhar, G. Zhang, M. R. Poellot, P. J. DeMott, A. J. Prenni, and A. J. Heymsfield, 2007: Ice properties of single-layer stratocumulus during the Mixed-Phase Arctic Cloud Experiment: 2. Model results. *J. Geophys. Res.*, **112**, D24202, doi:10.1029/2007JD008646.

——, and Coauthors, 2012a: A comparison of TWP-ICE observational data with cloud-resolving model results. *J. Geophys. Res.*, **117**, D05204, doi:10.1029/2011JD016595.

——, B. van Diedenhoven, A. S. Ackerman, A. Avramov, A. Mrowiec, H. Morrison, P. Zuidema, and M. D. Shupe, 2012b: A FIRE-ACE/SHEBA case study of mixed-phase Arctic boundary-layer clouds: Entrainment rate limitations on rapid primary ice nucleation processes. *J. Atmos. Sci.*, **69**, 365–389, doi:10.1175/JAS-D-11-052.1.

GCSS Science Team, 1998: Summary of progress and main developments from the Sixth Meeting of the GEWEX Cloud System Study (GCSS) Science Team, WCRP Rep., WMO.

——, 1999: Summary of progress and main developments from the Seventh Meeting of the GEWEX Cloud System Study (GCSS) Science Team. WCRP Rep., WMO.

Hassim, M. E. E., and T. P. Lane, 2010: A model study on the influence of overshooting convection on TTL water vapour. *Atmos. Chem. Phys.*, **10**, 9833–9849, doi:10.5194/acp-10-9833-2010.

Khairoutdinov, M. F., and D. A. Randall, 2003: Cloud-resolving modeling of the ARM summer 1997 IOP: Model formulation, results, uncertainties, and sensitivities. *J. Atmos. Sci.*, **60**, 607–625, doi:10.1175/1520-0469(2003)060<0607:CRMOTA>2.0.CO;2.

Klein, S. A., and C. Jakob, 1999: Validation and sensitivities of frontal clouds simulated by the ECMWF model. *Mon. Wea. Rev.*, **127**, 2514–2531, doi:10.1175/1520-0493(1999)127<2514:VASOFC>2.0.CO;2.

——, and Coauthors, 2009: Intercomparison of model simulations of mixed-phase clouds observed during the ARM Mixed-Phase Arctic Cloud Experiment. I: Single-layer cloud. *Quart. J. Roy. Meteor. Soc.*, **135**, 979–1002, doi:10.1002/qj.416.

Kollias, P., and Coauthors, 2016: Development and applications of ARM millimeter-wavelength cloud radars. *The Atmospheric Radiation Measurement (ARM) Program: The First 20 Years, Meteor. Monogr.*, No. 57, Amer. Meteor. Soc., doi:10.1175/AMSMONOGRAPHS-D-15-0037.1.

Krueger, S. K., 2000: Cloud system modeling. *General Circulation Model Development*. D. A. Randall, Ed., International Geophysics Series, Vol. 70, Academic Press, 605–640.

Lee, S.-S., and G. Feingold, 2010: Precipitating cloud-system response to aerosol perturbations. *Geophys. Res. Lett.*, **37**, L23806, doi:10.1029/2010GL045596.

——, and ——, 2013: Aerosol effects on the cloud-field properties of tropical convective clouds. *Atmos. Chem. Phys.*, **13**, 6713–6726, doi:10.5194/acp-13-6713-2013.

Lenderink, G., and Coauthors, 2004: The diurnal cycle of shallow cumulus clouds over land: A single-column model intercomparison study. *Quart. J. Roy. Meteor. Soc.*, **130**, 3339–3364, doi:10.1256/qj.03.122.

Lin, R.-F., D. Starr, P. J. DeMott, R. Cotton, K. Sassen, E. Jensen, B. Kärcher, and X. Liu, 2002: Cirrus parcel model comparison project. Phase 1: The critical components to simulate cirrus initiation explicitly. *J. Atmos. Sci.*, **59**, 2305–2319, doi:10.1175/1520-0469(2002)059<2305:CPMCPP>2.0.CO;2.

Lin, Y., and Coauthors, 2012: TWP-ICE global atmospheric model intercomparison: Convection responsiveness and resolution impact. *J. Geophys. Res.*, **117**, D09111, doi:10.1029/2011JD017018.

Luo, Y., S. K. Krueger, G. G. Mace, and K.-M. Xu, 2003: Cirrus cloud properties from a cloud-resolving model simulation compared to cloud radar observations. *J. Atmos. Sci.*, **60**, 510–525, doi:10.1175/1520-0469(2003)060<0510:CCPFAC>2.0.CO;2.

——, ——, and S. Moorthi, 2005: Cloud properties simulated by a single-column model. Part I: Comparison to cloud radar observations of cirrus clouds. *J. Atmos. Sci.*, **62**, 1428–1445, doi:10.1175/JAS3425.1.

——, ——, and K.-M. Xu, 2006: Cloud properties simulated by a single-column model. Part II: Evaluation of cumulus detrainment and ice-phase microphysics using a cloud-resolving model. *J. Atmos. Sci.*, **63**, 2831–2847, doi:10.1175/JAS3785.1.

——, K.-M. Xu, H. Morrison, and G. M. McFarquhar, 2008a: Arctic mixed-phase clouds simulated by a cloud-resolving model: Comparison with ARM observations and sensitivity to microphysics parameterizations. *J. Atmos. Sci.*, **65**, 1285–1303, doi:10.1175/2007JAS2467.1.

——, ——, ——, ——, Z. Wang, and G. Zhang, 2008b: Multi-layer Arctic mixed-phase clouds simulated by a cloud-resolving model: Comparison with ARM observations and sensitivity experiments. *J. Geophys. Res.*, **113**, D12208, doi:10.1029/2007JD009563.

Mace, G. G., E. E. Clothiaux, and T. P. Ackerman, 2001: The composite characteristics of cirrus clouds: Bulk properties revealed by one year of continuous cloud radar data. *J. Climate*, **14**, 2185–2203, doi:10.1175/1520-0442(2001)014<2185:TCCOCC>2.0.CO;2.

Mather, J. H., D. D. Turner, and T. P. Ackerman, 2016: Scientific maturation of the ARM Program. *The Atmospheric Radiation Measurement (ARM) Program: The First 20 Years, Meteor. Monogr.*, No. 57, Amer. Meteor. Soc., doi:10.1175/AMSMONOGRAPHS-D-15-0053.1.

May, P. T., J. H. Mather, G. Vaughan, and C. Jakob, 2008: Characterizing oceanic convective cloud systems: The Tropical Warm Pool International Cloud Experiment. *Bull. Amer. Meteor. Soc.*, **154**, 153–155, doi:10.1175/BAMS-89-2-153.

McFarquhar, G. M., and Coauthors, 2011: Indirect and Semi-Direct Aerosol Campaign: The impact of Arctic aerosols on clouds. *Bull. Amer. Meteor. Soc.*, **92**, 183–201, doi:10.1175/2010BAMS2935.1.

Moeng, C.-H., and Coauthors, 1996: Simulation of a stratocumulus-topped PBL: Intercomparison among different numerical codes. *Bull. Amer. Meteor. Soc.*, **77**, 261–278, doi:10.1175/1520-0477(1996)077<0261:SOASTP>2.0.CO;2.

Moncrieff, M. W., S. K. Krueger, D. Gregory, J.-L. Redelsperger, and W.-K. Tao, 1997: GEWEX Cloud System Study (GCSS) Working Group 4: Precipitating convective cloud systems. *Bull. Amer. Meteor. Soc.*, **78**, 831–845, doi:10.1175/1520-0477(1997)078<0831:GCSSGW>2.0.CO;2.

Morrison, H., and J. O. Pinto, 2005: Mesoscale modeling of springtime Arctic mixed-phase stratiform clouds using a new two-moment bulk microphysics scheme. *J. Atmos. Sci.*, **62**, 3683–3704, doi:10.1175/JAS3564.1.

——, and ——, 2006: Intercomparison of bulk cloud microphysics schemes in mesoscale simulations of springtime Arctic mixed-phase stratiform clouds. *Mon. Wea. Rev.*, **134**, 1880–1900, doi:10.1175/MWR3154.1.

——, and W. W. Grabowski, 2011: Cloud-system resolving model simulations of aerosol indirect effects on tropical deep convection and its thermodynamic environment. *Atmos. Chem. Phys.*, **11**, 10 503–10 523, doi:10.5194/acp-11-10503-2011.

——, J. O. Pinto, J. A. Curry, and G. M. McFarquhar, 2008: Sensitivity of modeled Arctic mixed-phase stratocumulus to cloud condensation and ice nuclei over regionally varying surface conditions. *J. Geophys. Res.*, **113**, D05203, doi:10.1029/2007JD008729.

——, and Coauthors, 2009: Intercomparison of model simulations of mixed-phase clouds observed during the ARM Mixed-Phase Arctic Cloud Experiment. II: Multilayer cloud. *Quart. J. Roy. Meteor. Soc.*, **135**, 1003–1019, doi:10.1002/qj.415.

——, and Coauthors, 2011: Intercomparison of cloud model simulations of Arctic mixed-phase boundary layer clouds observed during SHEBA/FIRE-ACE. *J. Adv. Model. Earth Syst.*, **3**, M06003, doi:10.1029/2011MS000066.

Mrowiec, A. A., C. Rio, A. M. Fridlind, A. S. Ackerman, A. D. Del Genio, O. M. Pauluis, A. C. Varble, and J. Fan, 2012: Analysis of cloud-resolving simulations of a tropical mesoscale convective system observed during TWP-ICE: Vertical fluxes and draft properties in convective and stratiform regions. *J. Geophys. Res.*, **117**, D19201, doi:10.1029/2012JD017759.

——, O. M. Pauluis, A. M. Fridlind, and A. S. Ackerman, 2015: Properties of a mesoscale convective system in the context of an isentropic analysis. *J. Atmos. Sci.*, **72**, 1945–1962, doi:10.1175/JAS-D-14-0139.1.

Muhlbauer, A., and Coauthors, 2013: Reexamination of the state of the art cloud modeling shows real improvements. *Bull. Amer. Meteor. Soc.*, **94**, ES45–ES48, doi:10.1175/BAMS-D-12-00188.1.

Ovchinnikov, M., A. Korolev, and J. Fan, 2011: Effects of ice number concentration on dynamics of a shallow mixed-phase stratiform cloud. *J. Geophys. Res.*, **116**, D00T06, doi:10.1029/2011JD015888.

Petch, J., M. Willett, R. Y. Wong, and S. J. Woolnough, 2007: Modelling suppressed and active convection. Comparing a numerical weather prediction, cloud-resolving and single-column model. *Quart. J. Roy. Meteor. Soc.*, **133**, 1087–1100, doi:10.1002/qj.109.

——, A. Hill, L. Davies, A. Fridlind, C. Jakob, Y. Lin, S. Xie, and P. Zhu, 2014: Evaluation of intercomparisons of four different types of model simulating TWP-ICE. *Quart. J. Roy. Meteor. Soc.*, **140**, 826–837, doi:10.1002/qj.2192.

Prenni, A. J., and Coauthors, 2007: Can ice-nucleating aerosols affect Arctic seasonal climate? *Bull. Amer. Meteor. Soc.*, **88**, 541–550, doi:10.1175/BAMS-88-4-541.

Randall, D. A., K.-M. Xu, R. J. C. Somerville, and S. Iacobellis, 1996: Single-column models and cloud ensemble models as links between observations and climate models. *J. Climate*, **9**, 1683–1697, doi:10.1175/1520-0442(1996)009<1683:SCMACE>2.0.CO;2.

——, and Coauthors, 2003: Confronting models with data: The GEWEX Cloud Systems Study. *Bull. Amer. Meteor. Soc.*, **84**, 455–469, doi:10.1175/BAMS-84-4-455.

Rio, C., and Coauthors, 2013: Control of deep convection by sub-cloud lifting processes: The ALP closure in the LMDZ5B general circulation model. *Climate Dyn.*, **40**, 2271–2292, doi:10.1007/s00382-012-1506-x.

Sednev, I., S. Menon, and G. M. McFarquhar, 2009: Simulating mixed-phase Arctic stratus clouds: Sensitivity to ice initiation mechanisms. *Atmos. Chem. Phys.*, **9**, 4747–4773, doi:10.5194/acp-9-4747-2009.

Solch, I., and B. Karcher, 2011: Process-oriented large-eddy simulations of a midlatitude cirrus cloud system based on observations. *Quart. J. Roy. Meteor. Soc.*, **137**, 374–393, doi:10.1002/qj.764.

Solomon, A., H. Morrison, P. O. G. Persson, M. D. Shupe, and J.-W. Bao, 2009: Investigation of microphysical parameterizations of snow and ice in Arctic clouds during M-PACE through model–observation comparisons. *Mon. Wea. Rev.*, **137**, 3110–3128, doi:10.1175/2009MWR2688.1.

——, M. D. Shupe, P. O. G. Persson, and H. Morrison, 2011: Moisture and dynamical interactions maintaining decoupled Arctic mixed-phase stratocumulus in the presence of a humidity inversion. *Atmos. Chem. Phys.*, **11**, 10 127–10 148, doi:10.5194/acp-11-10127-2011.

——, ——, ——, ——, T. Yamaguchi, P. M. Caldwell, and G. de Boer, 2014: The sensitivity of springtime Arctic mixed-phase stratocumulus clouds to surface layer and cloud-top inversion layer moisture sources. *J. Atmos. Sci.*, **71**, 574–595, doi:10.1175/JAS-D-13-0179.1.

Starr, D. O. C., and Coauthors, 2000: Comparison of cirrus cloud models: A project of the GEWEX Cloud System Study (GCSS) working group on cirrus cloud systems. *Proc. 13th Int. Conf. on Clouds and Precipitation*, Reno, NV, 1–4.

Uttal, T., and Coauthors, 2002: Surface heat budget of the Arctic Ocean. *Bull. Amer. Meteor. Soc.*, **83**, 255–275, doi:10.1175/1520-0477(2002)083<0255:SHBOTA>2.3.CO;2.

van Diedenhoven, B., A. M. Fridlind, A. S. Ackerman, E. W. Eloranta, and G. M. McFarquhar, 2009: An evaluation of ice formation in large-eddy simulations of supercooled Arctic stratocumulus using ground-based lidar and cloud radar. *J. Geophys. Res.*, **114**, D10203, doi:10.1029/2008JD011198.

——, ——, and ——, 2011: Influence of humidified aerosol on lidar depolarization measurements below ice-precipitating Arctic stratus. *J. Appl. Meteor. Climatol.*, **50**, 2184–2192, doi:10.1175/JAMC-D-11-037.1.

——, ——, ——, and B. Cairns, 2012: Evaluation of hydrometeor phase and ice properties in cloud-resolving model simulations of tropical deep convection using radiance and polarization measurements. *J. Atmos. Sci.*, **69**, 3290–3314, doi:10.1175/JAS-D-11-0314.1.

Varble, A., and Coauthors, 2011: Evaluation of cloud-resolving model intercomparison simulations using TWP-ICE observations: Precipitation and cloud structure. *J. Geophys. Res.*, **116**, D12206, doi:10.1029/2010JD015180.

——, and Coauthors, 2014a: Evaluation of cloud-resolving and limited area model intercomparison simulations using TWP-ICE observations: 1. Deep convective updraft properties. *J. Geophys. Res.*, **119**, 13 891–13 918, doi:10.1002/2013JD021371.

——, and Coauthors, 2014b: Evaluation of cloud-resolving and limited area model intercomparison simulations using TWP-ICE observations: 2. Precipitation microphysics. *J. Geophys. Res.*, **119**, 13 919–13 945, doi:10.1002/2013JD021372.

Vaughan, G., K. Bower, C. Schiller, A. R. MacKenzie, T. Peter, H. Schlager, N. R. P. Harris, and P. T. May, 2008: SCOUT-O3/ACTIVE: High-altitude aircraft measurements around deep tropical convection. *Bull. Amer. Meteor. Soc.*, **89**, 647–662, doi:10.1175/BAMS-89-5-647.

Verlinde, J., and Coauthors, 2007: The Mixed-Phase Arctic Cloud Experiment (M-PACE). *Bull. Amer. Meteor. Soc.*, **88**, 205–221, doi:10.1175/BAMS-88-2-205.

Wang, W., and X. Liu, 2009: Evaluating deep updraft formulation in NCAR CAM3 with high-resolution WRF simulations during ARM TWP-ICE. *Geophys. Res. Lett.*, **36**, L04701, doi:10.1029/2008GL036692.

Wang, Y., C. N. Long, L. R. Leung, J. Dudhia, S. A. McFarlane, J. H. Mather, S. J. Ghan, and X. Liu, 2009: Evaluating regional cloud-permitting simulations of the WRF model for the Tropical Warm Pool International Cloud Experiment (TWP-ICE), Darwin, 2006. *J. Geophys. Res.*, **114**, D21203, doi:10.1029/2009JD012729.

Wapler, K., T. P. Lane, P. T. May, C. Jakob, M. J. Manton, and S. T. Siems, 2010: Cloud-system-resolving model simulations of tropical cloud systems observed during the Tropical Warm Pool–International Cloud Experiment. *Mon. Wea. Rev.*, **138**, 55–73, doi:10.1175/2009MWR2993.1.

Wu, J., A. D. Del Genio, M.-S. Yao, and A. B. Wolf, 2009: WRF and GISS SCM simulations of convective updraft properties during TWP-ICE. *J. Geophys. Res.*, **114**, D04206, doi:10.1029/2008JD010851.

Wu, X., X.-Z. Liang, and S. Park, 2007: Cloud-resolving model simulations over the ARM SGP. *Mon. Wea. Rev.*, **135**, 2841–2853, doi:10.1175/MWR3438.1.

——, S. Park, and Q. Min, 2008: Seasonal variation of cloud systems over ARM SGP. *J. Atmos. Sci.*, **65**, 2107–2129, doi:10.1175/2007JAS2394.1.

Xie, S., and Coauthors, 2002: Intercomparison and evaluation of cumulus parametrizations under summertime midlatitude continental conditions. *Quart. J. Roy. Meteor. Soc.*, **128**, 1095–1136, doi:10.1256/003590002320373229.

——, R. T. Cederwall, and M. H. Zhang, 2004: Developing long-term single-column model/cloud system–resolving model forcing data using numerical weather prediction products constrained by surface and top of the atmosphere observations. *J. Geophys. Res.*, **109**, D01104, doi:10.1029/2003JD004045.

——, and Coauthors, 2005: Simulations of midlatitude frontal clouds by single-column and cloud-resolving models during the Atmospheric Radiation Measurement March 2000 cloud intensive operational period. *J. Geophys. Res.*, **110**, D15S03, doi:10.1029/2004JD005119.

——, S. A. Klein, M. Zhang, J. J. Yio, R. T. Cederwall, and R. McCoy, 2006: Developing large-scale forcing data for single-column and cloud-resolving models from the Mixed-Phase Arctic Cloud Experiment. *J. Geophys. Res.*, **111**, D19104, doi:10.1029/2005JD006950.

——, T. Hume, C. Jakob, S. A. Klein, R. B. McCoy, and M. Zhang, 2010: Observed large-scale structures and diabatic heating and drying profiles during TWP-ICE. *J. Climate*, **23**, 57–79, doi:10.1175/2009JCLI3071.1.

Xu, K.-M., and Coauthors, 2002: An intercomparison of cloud-resolving models with the Atmospheric Radiation Measurement summer 1997 Intensive Observation Period data. *Quart. J. Roy. Meteor. Soc.*, **128**, 593–624, doi:10.1256/003590002321042117.

——, and Coauthors, 2005: Modeling springtime shallow frontal clouds with cloud-resolving and single-column models. *J. Geophys. Res.*, **110**, D15S04, doi:10.1029/2004JD005153.

Yang, H., S. Dobbie, G. G. Mace, A. Ross, and M. Quante, 2012: GEWEX Cloud System Study (GCSS) cirrus cloud working group: Development of an observation-based case study for model evaluation. *Geosci. Model Dev.*, **5**, 829–843, doi:10.5194/gmd-5-829-2012.

Yeh, H.-C., and X. Fu, 2011: Incorporating additional sounding observations in weather analysis and rainfall prediction during the Intensive Observing Period of 2006 TWP-ICE. *Terr. Atmos. Oceanic Sci.*, **22**, 421–434, doi:10.3319/TAO.2011.04.11.01(A).

Zeng, X., and Coauthors, 2007: Evaluating clouds in long-term cloud-resolving model simulations with observational data. *J. Atmos. Sci.*, **64**, 4153–4177, doi:10.1175/2007JAS2170.1.

——, and Coauthors, 2009a: A contribution by ice nuclei to global warming. *Quart. J. Roy. Meteor. Soc.*, **135**, 1614–1629, doi:10.1002/qj.449.

——, and Coauthors, 2009b: An indirect effect of ice nuclei on atmospheric radiation. *J. Atmos. Sci.*, **66**, 41–61, doi:10.1175/2008JAS2778.1.

——, W.-K. Tao, S. W. Powell, R. A. Houze Jr., P. Ciesielski, N. Guy, H. Pierce, and T. Matsui, 2013: A comparison of the water budgets between clouds from AMMA and TWP-ICE. *J. Atmos. Sci.*, **70**, 487–503, doi:10.1175/JAS-D-12-050.1.

Zhang, M., and J. L. Lin, 1997: Constrained variational analysis of sounding data based on column-integrated budgets of mass, heat, moisture, and momentum: Approach and application to ARM measurements. *J. Atmos. Sci.*, **54**, 1503–1524, doi:10.1175/1520-0469(1997)054<1503:CVAOSD>2.0.CO;2.

——, R. C. J. Somerville, and S. Xie, 2016: The SCM concept and creation of ARM forcing datasets. *The Atmospheric Radiation Measurement (ARM) Program: The First 20 Years, Meteor. Monogr.*, No. 57, Amer. Meteor. Soc., doi:10.1175/AMSMONOGRAPHS-D-15-0040.1.

Zhu, M., P. Connolly, G. Vaughan, T. Choularton, and P. T. May, 2013: Numerical simulation of tropical island thunderstorms (Hectors) during the ACTIVE campaign. *Meteor. Appl.*, **20**, 357–370, doi:10.1002/met.1295.

Zhu, P., and Coauthors, 2012: A limited area model (LAM) intercomparison study of a TWP-ICE active monsoon mesoscale convective event. *J. Geophys. Res.*, **117**, D11208, doi:10.1029/2011JD016447.

Chapter 26

The Impact of ARM on Climate Modeling

DAVID A. RANDALL
Colorado State University, Fort Collins, Colorado

ANTHONY D. DEL GENIO
National Aeronautics and Space Administration, New York, New York

LEO J. DONNER
Geophysical Fluid Dynamics Laboratory, Princeton, New Jersey

WILLIAM D. COLLINS
Lawrence Berkeley National Laboratory, Berkeley, California

STEPHEN A. KLEIN
Lawrence Livermore National Laboratory, Livermore, California

1. What is a climate model?

Climate models are among humanity's most ambitious and elaborate creations. They are designed to simulate the interactions of the atmosphere, ocean, land surface, and cryosphere on time scales far beyond the limits of deterministic predictability and including the effects of time-dependent external forcings. The processes involved include radiative transfer, fluid dynamics, microphysics, and some aspects of geochemistry, biology, and ecology. The models explicitly simulate processes on spatial scales ranging from the circumference of Earth down to 100 km or smaller and implicitly include the effects of processes on even smaller scales down to a micron or so. The atmospheric component of a climate model can be called an atmospheric global circulation model (AGCM).

In an AGCM, calculations are done on a three-dimensional grid, which in some of today's climate models consists of several million grid cells.[1] For each grid cell, about a dozen variables are "time stepped" as the model integrates forward from its initial conditions. These so-called prognostic variables have special importance because they are the only things that a model remembers from one time step to the next; everything else is recreated on each time step by starting from the prognostic variables and the boundary conditions. The prognostic variables typically include information about the mass of dry air, the temperature, the wind components, water vapor, various condensed-water species, and at least a few chemical species, such as ozone.

A good way to understand how climate models work is to consider the lengthy and complex process used to develop one. Let us imagine that a new AGCM is to be created, starting from a blank piece of paper. The model may be intended for a particular class of applications (e.g., high-resolution simulations on time scales of a few decades). Before a single line of code is written, the conceptual foundation of the model must be designed

Corresponding author address: David A. Randall, Colorado State University, Department of Atmospheric Science, 1371 Campus Delivery, Fort Collins, CO 80523.
E-mail: randall@atmos.colostate.edu

DOI: 10.1175/AMSMONOGRAPHS-D-15-0050.1

[1] The AGCMs used for weather prediction employ much finer grids.

through a creative envisioning that starts from the intended application and is based on current understanding of how the atmosphere works and the inventory of mathematical methods available. The design process can be viewed as an ordered sequence of choices:

- Where (how high) should the top of the model be placed?
- What range of processes should be included? For example, what chemical and biological processes are needed?
- What horizontal and vertical resolutions are needed? The answers will strongly influence the choice of the continuous equation system and the nature of the physical parameterizations. The available computer power determines what range of resolutions can be considered.
- What set of continuous equations should be used to describe the fluid dynamics? For example, should vertically propagating sound waves be included or filtered out? What conservation properties should the continuous system have?
- What approach should be used to discretize the model's domain? For example, should we use a latitude–longitude grid or a cubed sphere grid or a geodesic grid? Should the vertical coordinate be terrain following or not? Should the vertical coordinate move up and down, following the air, as in the case of isentropic coordinates?
- What approach should be used to discretize the equations? Possibilities include spectral, finite-volume, semi-Lagrangian, and spectral-element methods.
- Which variables should be prognosed (time stepped)? For example, should the model prognose temperature or potential temperature? With the continuous system of equations the choice does not matter, but with the discrete system it does.
- How should the variables of the model be arranged on the horizontal and/or vertical grids? There can be good reasons to place different variables in different locations.
- What vertical resolution is needed, and how should it vary with height?
- What conservation properties of the continuous system should be carried over to the discrete system? Possibilities include the mass of dry air, the mass of total water, the total energy content of the air, and the potential vorticity. Conservation is known to be particularly important in long simulations, such as those needed to explore climate change scenarios.
- What order of accuracy should be built into the discrete system? High accuracy can be beneficial but is computationally expensive.
- How many prognostic variables are needed for the representation of clouds? For example, is information about particle size needed, in addition to information about condensed-water mass?
- How should the many important subgrid-scale processes be parameterized, and how should those parameterizations be coupled to each other and to the resolved-scale fluid dynamics? A key goal of the Atmospheric Radiation Measurement (ARM) Program has been to enhance and facilitate the process of parameterization development, especially for parameterizations of clouds and radiation. Parameterizations are important because they are needed to enable simulations with models and also because they are based on simplified models that encapsulate our understanding of how those processes interact with larger-scale weather systems. In some distant future, it may be possible to explicitly simulate many processes that are parameterized today; even then, parameterizations will be needed to understand why the simulations turned out as they did.

The answers to the questions listed above define the scientific architecture of the model, which should be documented in journal articles, technical reports, and web pages that explain not only what choices have been made, but why.

Next, the scientific architecture described above must be combined with and complemented by a computational architecture. The form of the computational architecture will be dictated in part by the scientific architecture and in part by the characteristics of the machines that will be used to run the model. This is where software engineering comes in.

2. Improving the models

Although AGCMs are sometimes created "from scratch," as outlined above, existing AGCMs are updated routinely to incorporate new understanding and to address inadequacies of their formulations (Jakob 2003). Key steps are to identify model deficiencies through comparison with observations, attribute these deficiencies to particular defects of the model's formulation, and test new modeling concepts at the component level, in the same way that the engines, airframe, and other components of a new type of aircraft are tested individually before an actual flight is attempted. The data collected by ARM are used primarily to test individual model components, especially parameterizations.

The ARM Program has a particular interest in the parameterization of atmospheric radiative heating and cooling. As discussed in the next section, ARM has supported the development of greatly improved radiation parameterizations, which are now in use at modeling centers throughout the world. Even a perfect radiation parameterization needs realistic inputs to produce realistic heating and cooling rates. The required inputs include information about cloudiness, water vapor, and aerosols. ARM has therefore devoted a lot of attention to the parameterization of clouds and aerosols and to the effects of cloud processes on the distribution of water vapor.

During the first five years of ARM, the program's emphasis was on its first priority (i.e., precisely measuring the surface radiation field), the effect of clouds on radiation and the atmospheric state, and less on its second priority of measuring cloud properties (Stokes 2016, chapter 2). During that time, ARM provided crucial support for cloud parameterization development, but cloud observations and value-added products did not yet exist for most fields of interest to modelers. Within a few years, it became clear that, although global climate model (GCM) radiative transfer schemes were not as accurate as they needed to be (Ellingson et al. 1991; Mlawer et al. 2016, chapter 15), this source of simulated radiation errors was dwarfed by the effect of uncertainties in GCM predictions of the occurrence of clouds and their macrophysical and microphysical properties.

In the early years, only the ARM Southern Great Plains (SGP) site was operational (Cress and Sisterson 2016, chapter 5), which limited the ability of ARM data to address questions about clouds in the tropics, the global oceans, and the polar sea ice regions that are now known to account for most of the spread in GCM estimates of climate sensitivity (Bony and Dufresne 2005; Zelinka et al. 2012). The single-column modeling (SCM) concept (Randall et al. 1996; Zhang et al. 2016, chapter 24) was being implemented at the SGP, based on intensive observing periods (IOPs), during which frequent soundings over a GCM gridbox-sized area provided estimates of large-scale advective tendencies of temperature and humidity to force the parameterizations in a GCM column. It took several years for the limitations in the advective products to be understood (Ghan et al. 2000) and to develop a strategy to improve them (Zhang and Lin 1997). Early SCM case studies were used to understand how different ways of specifying the observed forcing (Randall and Cripe 1999) determined what could be learned and how nondeterministic behavior could develop as SCM solutions drifted from reality (Hack and Pedretti 2000).

Despite this, the early datasets shed light on several outstanding cloud-climate issues. As the SCM framework matured at the end of ARM's first decade, cloud-resolving models (CRMs) began to be used as intermediaries between ARM data and SCMs to identify parameterization deficiencies that were not obvious from the observations alone.

At the same time, the internationally based GEWEX Cloud System Study (GCSS) was also being planned (Randall et al. 2003). The idea of GCSS was that CRMs, which simulated spatial scales on which clouds form, could be compared more directly to observations than SCMs, while providing information on the small-scale motions that underlie stratiform cloud and cumulus parameterization assumptions. In this way, CRMs would serve as a bridge to identify and remedy parameterization errors. ARM and GCSS worked very well together; GCSS undertook model intercomparisons based on ARM data, and ARM benefitted from the expanded use of its data products by the international community at no direct cost to the program.

The use of ARM data by GCM developers also has been facilitated through ARM's sponsorship, along with the U.S. Department of Energy (DOE)'s Climate Change Prediction Program (CCPP), of the CCPP-ARM Parameterization Testbed (CAPT; Phillips et al. 2004). CAPT was designed to bridge the gap between ARM data used to develop, test, and improve parameterizations of physical processes and the GCMs where improved parameterizations are used. In CAPT, full atmospheric GCMs are integrated in weather forecast mode, like numerical weather prediction models, by initializing them with analyses produced by weather prediction centers. Short forecast simulations are performed, and simulation output is compared directly to ARM data to diagnose errors related to the parameterizations. By using results after short integration times, parameterization deficiencies can be identified before they are masked by compensations due to multiple error sources. Examining physical parameterizations in this way is a good complement to the use of SCMs because it allows for interactions between the large-scale dynamics and physics in ways that an SCM cannot do. Integrating GCMs in forecast mode has been applied widely to the Community Earth System Model (CESM; Williamson et al. 2005; Xie et al. 2008; Boyle and Klein 2010; to name just a few) and the Geophysical Fluid Dynamics Laboratory (GFDL) model (examples given below).

As active remote sensing cloud products emerged and IOPs began to be conducted at the other ARM sites during ARM's second decade, more direct evaluations of GCM fields were enabled. Some of these efforts have only borne fruit (as published papers and/or model improvements) since the ARM era ended in 2009 and the joint Atmospheric System Research (ASR)–ARM era began (Mather et al. 2016, chapter 4). However, papers continue to be published based on datasets acquired during the first 20 years of ARM. This attests to the continuing impact of the innovative ARM observational strategy.

3. The role of ARM in improving the Community Earth System Model

a. Improved parameterization of radiative transfer

One of the significant contributions from the ARM Program to climate studies is the development and introduction of highly accurate parameterizations of radiative processes into the CESM (Hurrell et al. 2013). These parameterizations, known as the Rapid Radiative Transfer Model for GCMs (RRTMG; Mlawer et al. 1997), are tested continuously against observations from the ARM Program (Oreopoulos et al. 2012). They also are updated routinely relative to benchmark line-by-line models of radiative transfer using the latest spectroscopic databases and empirical formulations of continuum absorption by water vapor and carbon dioxide (Clough et al. 2005; Mlawer et al. 2012, 2016, chapter 15). The RRTMG parameterizations are demonstrably more accurate than the traditional band models they replaced (Oreopoulos et al. 2012) and have thereby improved the simulation of both present-day climate and its response to future anthropogenic forcing (Iacono et al. 2008).

These changes are important for climate studies because the CESM, a model jointly developed by the DOE and the National Science Foundation (NSF), is used by over 3000 scientists and groups worldwide. The source code, input data, simulation output, and model documentation are freely available to the global community. The CESM supports a large community of researchers studying the dynamics and consequences of climate change and reporting these findings in major national and international reports. The CESM is one of several U.S. models used to produce the large suite of simulations analyzed by the Intergovernmental Panel on Climate Change (IPCC) in their assessment reports [e.g., the Fourth and current Fifth Assessment Reports (AR4 and AR5; IPCC 2007, 2013)]. Projections from the CESM also represent a key source for the most recent U.S. National Assessment (Melillo et al. 2014). The introduction of RRTMG represents an important enhancement to the physical fidelity of the Community Atmosphere Model, version 5 (CAM5), the component of CESM that simulates atmospheric processes (Gettelman et al. 2012).

Before the introduction of RRTMG, radiative processes at CESM had been treated using parameterizations based upon traditional band formulations developed by the National Center for Atmospheric Research (NCAR; Kiehl and Briegleb 1991, 1993). While periodic updates (Collins 2001; Collins et al. 2002a) to these parameterizations maintained reasonable absolute accuracy relative to benchmark radiative codes (Feldman et al. 2011), these parameterizations suffered from several shortcomings inherent in their band formulation. First, it proved difficult to maintain and continually update the accuracy of most of the radiatively active species in the longwave parameterization because of the complex formulation of the absorptivity and emissivity terms in that scheme. The computation of these same terms scaled quadratically with the number of vertical levels, thereby imposing a major barrier to increasing the vertical resolution of CAM. In addition, the band formulations in both the shortwave and longwave proved quite difficult to extend to incorporate additional radiatively active compounds [e.g., volcanic and speciated anthropogenic aerosols (Meehl et al. 2012; Collins et al. 2002b, 2006b)]. Finally, several groups had demonstrated the appreciable technical, computational, and scientific advantages readily available from an alternate formulation of radiative transfer (Lacis and Oinas 1991; Fu and Liou 1992). This alternative is based upon the correlated-*k* formalism for the spectral integrations required to compute broadband fluxes. Correlated-*k* treatments can be readily derived and updated from line-by-line codes applied to periodically updated spectroscopic databases, such as the HITRAN compilations of line properties (Rothman et al. 2013). It is also much easier to extend correlated-*k* parameterizations to include new radiatively active species (e.g., NF_3) as their potential climatic significance is demonstrated (Prather and Hsu 2008).

In response to these considerations, members of the ARM community introduced the RRTMG family of parameterizations into the Community Climate Model (CCM; the predecessor to CAM) on an experimental basis. Simulations run with the band codes and with RRTMG were compared to quantify the impact of RRTMG on the radiative fluxes and climatological state simulated by CCM (Iacono et al. 2000). These comparisons demonstrated that introduction of RRTMG would appreciably improve the longwave fluxes by reducing the outgoing longwave radiation by 6–9 $W\,m^2$, enhance longwave atmospheric cooling rates by 0.2–0.4 $K\,d^{-1}$, and thereby reduce a number of systematic temperature biases in the model. These changes were attributed to the updated treatment of spectral and continuum absorption by water vapor in RRTMG relative to the band models. Changes of comparable magnitude were obtained when the effects of near-infrared absorption by water vapor were updated in accordance with modern spectroscopic databases and continuum formulations (Collins et al. 2006a). The sensitivity of CCM, CAM, and other GCMs to the radiative properties of water vapor follows from its roles as both the most important greenhouse gas, accounting for roughly 60% of the clear-sky greenhouse effect and the most important absorber of near-infrared radiation, contributing almost 75% of the clear-sky shortwave atmospheric heating rate (Kiehl and Trenberth 1997).

ARM also has improved radiation transfer under cloudy-sky conditions. At the beginning of the ARM era, the optical properties of liquid water clouds were well described by Mie theory with suitable parameterizations thereof (e.g., Slingo 1989), but radiation transfer through ice clouds [which cover about 19% of the planet (Chen et al. 2000; Hartmann et al. 1992)] posed a serious challenge, since no theory of radiation–particle interactions addressed the complex geometry of atmospheric ice particles. Not only are the optical properties of single ice crystals a challenge; so are the optical properties of the ice particle size distribution (PSD) that are not parameterized easily even with perfect knowledge of the former (Mitchell et al. 2011).

Over a decade of ARM research yielded an accurate means of treating ice cloud optical properties in terms of the physical attributes of both the PSD and the ice particles (for any given shape) within an analytical framework. This produced a considerable improvement over the previous ice optics scheme in CCSM in regards to LW radiation, where the mass absorption coefficient in the atmospheric window region was reduced by ~50% for cirrus clouds (Mitchell et al. 2006). This is primarily a consequence of optically describing a particle in terms of its volume-to-projected-area ratio instead of describing it as an equivalent-area sphere (appropriate only for extinction), as was done before in CCSM.

These results, along with the advantages of a modern correlated-*k* formulation, led the CESM Atmospheric Model Working Group (AMWG) to adopt the RRTMG parameterizations for CAM5 in the first version of the new CESM (CESM1). While several other teams developing weather and climate codes had already adopted the longwave component of RRTMG [e.g., the European Centre for Medium-Range Weather Forecasts (ECMWF), as described in Ahlgrimm et al. (2016, chapter 28)], CESM1 was the first climate model to adopt the shortwave component as well. The first comprehensive suite of historical and future climate simulations produced with CESM1 for the fifth phase of the Coupled Model Intercomparison Project (CMIP5) was assessed as part of the IPCC AR5.

The RRTMG family of parameterizations is now one of the core physical parameterizations in the CAM and CESM. Its adoption by the CESM science team, as well as by leading international groups such as the ECMWF, represents a significant advance in the treatment of radiative processes in numerical forecasts. This advance represents a major contribution from the ARM Program to the operational weather and Earth system modeling communities.

b. *The early development of superparameterizations*

Grabowski and Smolarkiewicz (1999) described a simplified GCM in which the physical processes associated with clouds were represented by running a simplified cloud-resolving model within each grid column of a low-resolution AGCM. Parameterizations of radiation, cloud microphysics, and turbulence (including small clouds) are included in the CRM, which explicitly simulates the larger clouds and some mesoscale processes. The model successfully simulated some aspects of organized tropical convection, which many other models had failed to capture. In particular, the model produced a signal resembling the Madden–Julian oscillation (MJO; Madden and Julian 1971, 1972), which is an eastward-propagating tropical disturbance characterized by a large zonal extent and a period of about 40–50 days. The MJO has proven very difficult to simulate with AGCMs (e.g., Lin et al. 2006; Kim et al. 2009).

Inspired by the results of Grabowski and Smolarkiewicz (1999), and with the support of the ARM Program, Khairoutdinov and Randall (2001) created a superparameterized version of the CAM (SP-CAM), in which the CAM's parameterizations were replaced, in each CAM grid column, by a simplified version of Khairoutdinov's CRM (Khairoutdinov and Randall 2003). One copy of the CRM runs in each grid column of the CAM. The CRM is two-dimensional (one horizontal dimension, plus the vertical) and uses periodic lateral boundary conditions.

The ARM Program's early support of the SP-CAM made it possible to explore the behavior of the model in more detail. In 2006, the National Science Foundation created a Science and Technology Center (STC) focused on continuing development and applications of the SP-CAM. In effect, the STC was incubated by ARM. Over the past decade, the SP-CAM has been coupled with an ocean model (Stan et al. 2010) and used in studies of the MJO (Benedict and Randall 2009, 2011), monsoons (DeMott et al. 2011, 2013), the diurnal cycle of precipitation (Pritchard and Somerville 2009 a,b; Pritchard et al. 2011; Kooperman et al. 2013), African easterly waves (McCrary 2012), and climate change (Wyant et al.

2006, 2012; Arnold et al. 2013). Further discussion is given by Randall (2013).

4. The role of ARM in improving the GISS model

ARM data have strongly influenced parameterization evaluation and development in the Goddard Institute for Space Studies (GISS) AGCM. Here, we discuss how the data have been used for model components relating to low-cloud feedbacks, cloud phase, and convective entrainment and downdrafts.

a. Low-cloud feedbacks

At the dawn of the ARM era, cloud optical property feedbacks were just being recognized as a serious climate issue. Early GCMs had fixed cloud optical thicknesses or albedos. However, Somerville and Remer (1984) and Betts and Harshvardhan (1987) argued that liquid water content (LWC) and thus cloud albedo should increase with temperature, providing a negative cloud feedback. In the first Atmospheric Model Intercomparison Project (Cess et al. 1989), a number of GCMs assumed such behavior as a parameterization. Meanwhile, several GCMs were implementing prognostic cloud water budgets, producing different cloud feedbacks depending on specific process representations (e.g., Mitchell et al. 1989; Roeckner et al. 1987). Satellite datasets were showing that, except at cold temperatures, liquid water path (LWP) and low-cloud optical thickness were correlated negatively with temperature (Tselioudis et al. 1992; Greenwald et al. 1995), although there were concerns that this might be an artifact of the satellite sensors' resolution. The GISS GCM reproduced the satellite behavior because of liquid water sinks (cloud-top entrainment and precipitation) and varying cloud physical thickness (Tselioudis et al. 1998), but it was not known whether these were responsible for the observed behavior. The resulting positive optical thickness feedback increased the climate sensitivity by 0.35°C (Yao and Del Genio 1999).

Although cloud radars had not yet been deployed at the SGP, early ARM data permitted a preliminary study of continental midlatitude low-cloud optical properties (Del Genio and Wolf 2000). The ARM microwave radiometer (MWR) was used to obtain LWP, the ceilometer for cloud-base height, satellite brightness temperatures and soundings for cloud-top height, surface meteorology observations for relative humidity, and surface weather reports of cloud type. From these, cloud physical thickness and LWC were derived, along with indices of boundary layer structure.

The results documented the midlatitudes as a transition region between the satellite-observed low- and high-latitude behaviors. Low-cloud LWP was invariant with temperature during winter but decreased with temperature in summer. LWC showed no temperature dependence, but clouds physically thinned with temperature, especially during summer and in the warm sector of baroclinic waves. This was due primarily to a rising cloud base with warming as relative humidity decreased and the lifting condensation level increased. The temperature dependence of cloud thickness only occurred in well-mixed or decoupled boundary layers and was, in part, the result of a shift in the relative frequency of convective and stable boundary layers. Dong et al. (2005) revisited this analysis with accurate radar-derived cloud-top heights and a more recent MWR processing and found that LWC decreased with increasing temperature instead, but overall they agreed with the conclusions of Del Genio and Wolf (2000).

b. Cloud phase

Changes in the relative occurrence of cloud ice and liquid as climate warms exert a negative feedback on climate change, because of their different particle sizes and scattering phase functions and thus in the condensate retained rather than precipitated out (Mitchell et al. 1989). The feedback depends on the temperature range over which the transition (in a statistical sense) from liquid to ice occurs. In principle, both phases can exist from temperatures ~0°C down to the homogeneous ice nucleation threshold of ~−38°C. Which phase exists at a given temperature within this range depends on the cloud-scale dynamics, the resulting degree of supersaturation, the availability of ice nuclei, and the age of the cloud. Some GCMs use single-moment cloud microphysics parameterizations that diagnose cloud phase from grid-scale properties. Others use two-moment schemes that determine phase from parameterized microphysical processes that estimate nucleation rates of liquid and ice and conversions between them. Model comparisons to ARM observations during the Mixed-Phase Arctic Cloud Experiment (M-PACE) IOP at the ARM North Slope of Alaska (NSA) site in 2004 showed significant scatter in the amounts of ice and liquid and a tendency for the liquid phase to be underpredicted in boundary layer stratocumulus (Klein et al. 2009) but overpredicted in a frontal multilayer cloud (Morrison et al. 2009).

Parameterizations of cloud phase during the ARM era had been influenced by midlatitude aircraft observations in the frontal regions of baroclinic storms (Bower et al. 1996). These data suggested that liquid water was rare at temperatures <−15°C, whereas earlier aircraft data (Feigelson 1978) had liquid present down to −40°C. Naud et al. (2010) used the ARM SGP Raman lidar and

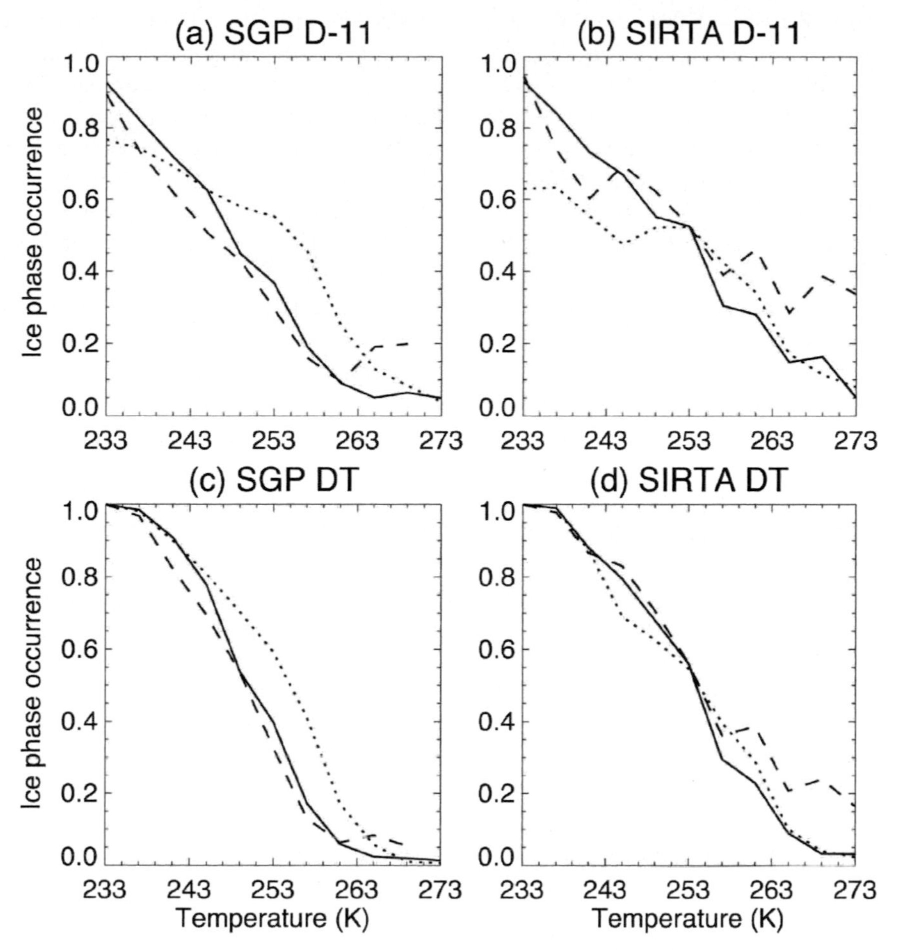

FIG. 26-1. Lidar-based temperature dependence of the fractional occurrence of the ice phase in optically thin clouds at the (a),(c) SGP and (b),(d) SIRTA sites (Naud et al. 2010). (top), (bottom) Two different approaches to specifying the depolarization ratio threshold that separates ice from liquid are represented. The solid curves show the temperature dependence at the median level of the cloud, while the dashed and dotted curves represent the phase at cloud top and cloud base, respectively.

the SIRTA lidar in France (Haeffelin et al. 2016, chapter 29) to compile statistics of cloud phase based on the lidar depolarization ratio. Lidar phase profiles are restricted to optically thinner clouds, such as altocumulus that often occur behind fronts, a different sampling than that of Bower et al. (1996).

Naud et al. (2010) found that liquid persists in these clouds down to ~−40°C, depending on the lidar and depolarization threshold used (Fig. 26-1), much colder than in the Bower et al. data. Likewise, the temperature at which ice and liquid occur equally is much colder in the lidar data (~−20°C) than in the Bower et al. data (−6.5°C). The GISS GCM at that time used a hybrid diagnostic scheme (Del Genio et al. 1996) in which cloud phase at nucleation varies probabilistically with temperature down to −38°C, but with Bergeron–Findeisen glaciation of supercooled cloud liquid by falling snow possible as the cloud ages. The overall resulting dependence of cloud phase on temperature in the GCM appears realistic, but the GCM analysis was not performed separately for thick frontal and thinner postfrontal clouds.

c. *Convective downdrafts*

The GATE field experiment showed that convective downdrafts are important to the energy and water budgets of convective systems (Houze and Betts 1981). Downdrafts were neglected in early cumulus parameterizations, though. By the time ARM began, some GCMs had included simple representations of downdrafts, including GISS (Del Genio and Yao 1988).

The first GCSS case study to examine midlatitude continental convection was based on the ARM summer 1997 SCM IOP. CRMs diagnosed updraft and downdraft mass fluxes (Xu et al. 2002), and these were compared to those parameterized in 15 SCMs (Xie et al. 2002). The SCM and CRM updraft mass fluxes were in reasonable agreement. Downdraft mass fluxes were much weaker in the SCMs than in the CRMs, however. Several possible reasons for this were suggested by Xie et al. (2002). First, the cumulus parameterizations only accounted for convective downdrafts, while the CRMs included both convective and mesoscale downdrafts. Second, some parameterizations (including that used by GISS) prescribed a single downdraft with a prescribed fraction of the updraft mass flux and/or did not allow downdrafts below cloud base.

Third, and perhaps most important, is that in most GCMs a stronger downdraft erroneously suppresses future convection. This occurs because in most GCMs, low moist static energy downdraft air immediately mixes with the ambient high moist static energy boundary layer air that gave rise to the convection, prematurely stabilizing the boundary layer. Downdrafts actually form boundary layer cold pools that remain distinct from the ambient air for hours (Houze and Betts 1981; Tompkins 2001). As the cold pools spread, high moist static energy air at the cold pool leading edge is lifted, triggering the next generation of convection rather than shutting it down. Indeed, several years earlier Mapes (2000) had made the point that GCM downdraft parameterizations were perhaps doing more harm than good because of this behavior.

The Xie et al. (2002) result led to several attempts to strengthen the GISS downdraft. For CMIP3 (Schmidt et al. 2006), the downdraft mass flux was increased by adding entrainment and extending the downdraft below cloud base. For CMIP5 (Schmidt et al. 2014), multiple downdrafts were added whenever an equal mixture of cloud and environment air was negatively buoyant. Buoyancy was based only on temperature, rather than on virtual temperature with precipitation loading, because the latter created an excessive downdraft mass flux. Post-CMIP5, as part of an effort to create realistic GCM intraseasonal variability, convective rain reevaporation was strengthened. This sufficiently moistened the environment that downdraft negative buoyancies were reduced, and it finally became possible to include the precipitation loading effect (Del Genio et al. 2012). Recently, a downdraft cold pool parameterization has been developed (Del Genio et al. 2013), with some effect on convective occurrence frequency.

d. Convective entrainment and vertical velocities

By 2006, cloud radars were standard at all ARM sites, and the Active Remotely Sensed Cloud Locations (ARSCL) value-added product (Clothiaux et al. 2000; Kollias et al. 2016, chapter 17) had become ARM's signature contribution to the evaluation of GCM cloud parameterizations. That year ARM conducted its first full-scale tropical IOP in Darwin, Australia, the Tropical Warm Pool–International Cloud Experiment (TWP-ICE; May et al. 2008). During TWP-ICE, Darwin experienced changes in weather regime that are characteristic of the Australian winter monsoon season: an active monsoon period of onshore flow and extensive rain; a suppressed monsoon period with drier midlevel conditions and isolated, moderate depth convection; an even drier fully suppressed period of mostly clear skies; and a monsoon break period of building instability and occasional but vigorous deep convection. These regime shifts provided an ideal opportunity to test model convection behavior, and intercomparisons of SCMs (Davies et al. 2013), CRMs (Varble et al. 2011; Fridlind et al. 2012), and GCMs (Lin et al. 2012) followed.

Before TWP-ICE, convective entrainment had been identified as a glaring shortcoming of cumulus parameterizations. This was based on a GCSS case study of the ARM summer 1997 IOP (Guichard et al. 2004) that showed that SCMs triggered continental deep convection too early in the day and a tropical ocean case study (Derbyshire et al. 2004) that showed that CRM convection depth was much more sensitive to environmental humidity in CRMs than SCMs. This behavior was traced to weak entrainment, a remnant of early cumulus parameterization history in which simulating convection that reached the tropopause was one of the few observational constraints. ARM ARSCL data at the Nauru Island site had verified that the depth of cumulus congestus was indeed sensitive to midtropospheric humidity (Jensen and Del Genio 2006).

By the time of TWP-ICE, the GISS GCM was using the Gregory (2001) entrainment parameterization, which is based on convective turbulence scalings. The Gregory scheme diagnoses updraft speed w and parameterizes entrainment ε as a function of parcel buoyancy B and updraft speed: $\varepsilon = CB/w^2$. The proportionality constant C indicates the fraction of buoyant turbulent kinetic energy available for use by entrainment. TWP-ICE data documented the more maritime character of active period convection (lower radar reflectivities and less graupel above the melting level, less lightning) relative to the stronger, more continental convection during the break period. Wu et al. (2009) showed that the Weather Research and Forecasting (WRF) Model, run at convection-resolving resolution, simulated stronger updraft speeds during the break period than during the active period, consistent with the indirect observational inferences.

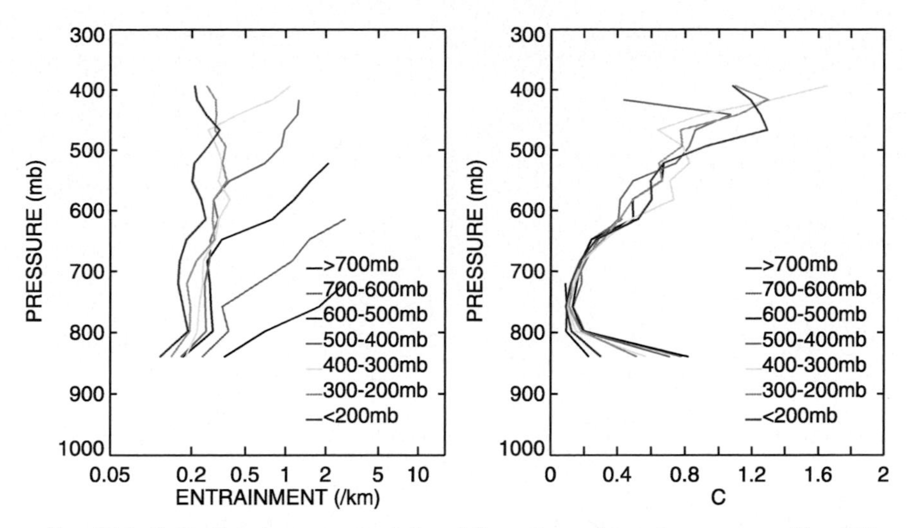

FIG. 26-2. (left) Entrainment rates inferred from the moist static energy profile within convective columns penetrating to different pressure levels, as simulated by the WRF Model for the TWP-ICE break period (Del Genio and Wu 2010). (right) Parameterization test from the same simulation showing that a single vertical profile of the proportionality constant in the Gregory (2001) entrainment parameterization works generally for all types of convection.

Del Genio et al. (2007) implemented the Gregory parameterization in the GISS GCM, with different values of the proportionality constant to represent more- and less-entraining parts of the cumulus spectrum. The parameterization was evaluated by Wu et al. (2009) in SCM tests against the WRF-derived TWP-ICE updraft speeds. The SCM reproduced the difference in convection strength between the active and break periods but overestimated updraft speeds in the upper troposphere. A WRF study of the TWP-ICE break period diurnal cycle tested various proposed parameterizations of entrainment (Del Genio and Wu 2010). The entrainment rate inferred from the thermodynamic structure in convecting grid boxes decreased over the afternoon as shallow convection gradually gave way to congestus and then predominantly deep convection (Fig. 26-2, left panel). To see whether these variations were consistent with the Gregory scheme, w, B, and ε were derived from the WRF fields and the implied values of C for different convection depths calculated from these. The results (Fig. 26-2, right panel) suggest that a single profile of C applies to convection of varying depths, except near cloud base, where the deeper events have smaller C than the shallow events. This suggests that the Gregory scheme is, in general, a good predictor of entrainment but that the SCM shortcomings seen by Wu et al. (2009) may be due to changes in convective parcel properties that the Gregory scheme by itself cannot anticipate (e.g., larger parcel sizes or nonturbulent sources of lifting as convection deepens). If so, then the operational GISS GCM approach of allowing weakly and strongly entraining plumes (smaller and larger C) to coexist at all times needs to be reconsidered. Tests with the cold pool parameterization (Del Genio et al. 2013), in which the less-entraining plume exists only after cold pools form, is more in keeping with the WRF inferences and produces some improvement, but entrainment remains an ongoing focus of research.

5. The role of ARM in improving the GFDL model

During the ARM era, the AGCM of the GFDL has undergone extensive development. The ARM data were particularly important for the research that led from the earlier version of the model, called the Atmospheric Model version 2 (AM2), to the newer version, called AM3.

As described above, observations of temperature and moisture advection, and their refinement to provide forcing for SCMs, were among the key achievements of ARM during the late 1990s. In addition to forcing observations, ARM has provided increasingly comprehensive characterization of other aspects of the atmospheric state, including important details of the microphysical and dynamical structure of clouds. In-cloud vertical velocities for both shallow and deep convective systems have recently become available, based on profiling and multiple Doppler radars (Collis et al. 2013). These observations have been used to evaluate and develop parameterizations for clouds and convection in GFDL models, with the goal of driving cloud microphysics and aerosol–cloud interactions with physically realistic vertical velocities.

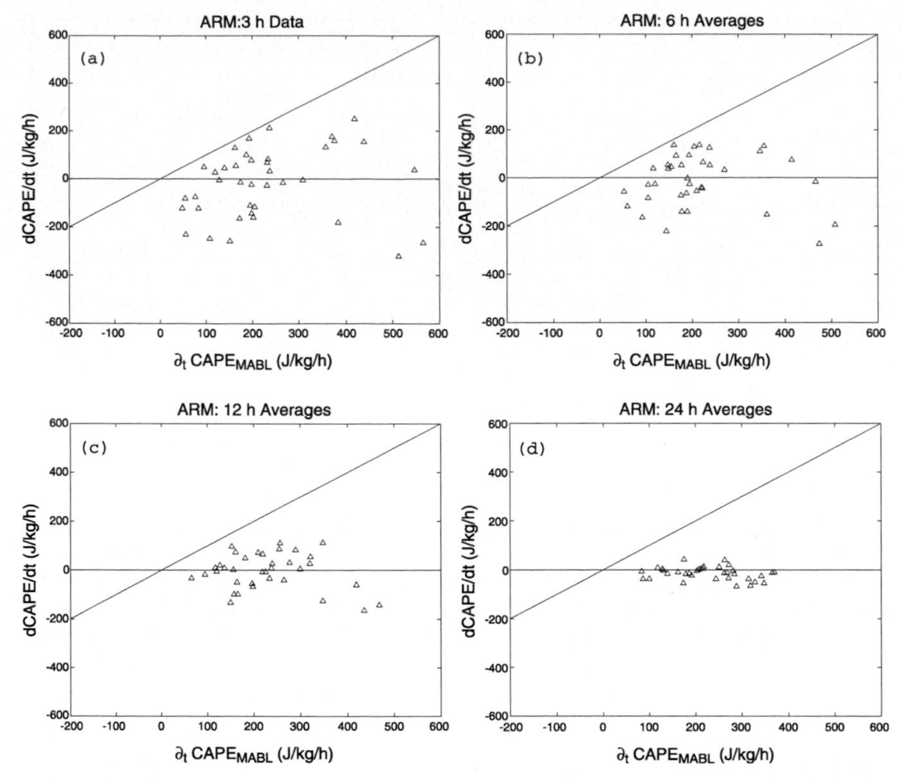

FIG. 26-3. Observations from the ARM SGP site have been used to examine quasi equilibrium for deep convection (Donner and Phillips 2003). Under quasi equilibrium, the rate at which CAPE changes, plotted on the vertical axis, should be small relative to the rate at which CAPE changes by advection averaged over large spatial scales and by boundary layer processes, plotted on the horizontal axis. At the ARM SGP, quasi equilibrium holds over daily time averages, but less so over shorter, subdiurnal periods. A consequence is that cumulus parameterizations using quasi-equilibrium closures often do not simulate the diurnal cycle of convection well. Results like these have motivated the development of closures that recognize the importance of nonequilibrium convection (Bechtold et al. 2014).

An ongoing challenge for the evolving GFDL climate models has been triggering and closure for cumulus parameterizations. Donner and Phillips (2003) used observations of changes in convective available potential energy (CAPE) due to boundary layer processes and large-scale-average advective tendencies from the ARM SGP site and other field programs focused on deep convection to provide empirical guidance for choosing closures for cumulus parameterization (Fig. 26-3). Fast changes in CAPE tied to boundary layer processes were found to violate quasi equilibrium, and a closure excluding boundary layer contributions to CAPE change was more consistent with ARM SGP observations, though not with observations from some of the other field programs. Benedict et al. (2013) incorporated this closure into GFDL AM3 (Donner et al. 2011) and found that it substantially improved AM3's simulation of tropical variability, including the MJO.

Both AM2 and AM3 use stochastically generated subcolumns to represent cloud structure, especially vertical overlap for clouds and radiation (Pincus et al. 2006). Its implementation was supported by ARM, as was much of the research that underpins the approach, including studies of the nonlinear effects of cloud heterogeneity on cloud microphysics (Pincus and Klein 2000), analysis of cloud overlap in CRMs (Pincus et al. 2005), analysis of total water variance and skewness in CRMs (Klein et al. 2005), and analysis of observed cloud heterogeneity at the SGP site (Kim et al. 2005).

AM3's parameterizations for shallow and deep convection provide multivariate probability density functions (PDFs) for in-cloud dynamics, thermodynamics, and microphysics. Vertical velocities play an especially important role in activating aerosols and the subsequent microphysical evolution of clouds, with important implications for cloud–aerosol and cloud–radiative interactions. The vertical-velocity PDFs in the Donner (1993) deep cumulus parameterization used in AM3 have been subject to only limited observational constraints, but the emergence of ARM vertical-velocity observations (Collis et al. 2013) will permit considerably more robust evaluation and enable further development of this parameterization approach.

AM3 includes PDFs of stratiform vertical velocity for aerosol activation, taken as normally distributed with a standard deviation related to turbulence. Experiments in AM3 with multivariate PDFs for stratiform clouds and boundary layers using higher-order closure with an assumed distribution have shown promising prospects for improving simulation of marine stratocumulus clouds (Guo et al. 2010) and aerosol–cloud interactions involving turbulence and cloud dynamics (Guo et al. 2011). PDFs of vertical velocity from ARM cloud radars at the ARM SGP site have been compared with those from AM3 and a modified version of AM3 that uses the boundary layer and cloud parameterizations of Guo et al. (2010). The latter agree with ARM radar observations in producing a binormal vertical-velocity PDF, in contrast to the normal distribution in AM3.

Relative to AM2 and AM3, GFDL's highest-resolution models adopt simplified parameterizations for cumulus convection and stratiform clouds. The simplified PDF cloud parameterization was developed with ARM support (Zhao et al. 2009).

AM2 (Anderson et al. 2004) has been evaluated extensively by comparing SCM integrations against ARM observations for deep convection at the SGP site during June 1995 and June–July 1997 (Xie et al. 2002), midlatitude stratiform frontal clouds at the SGP site (Xie et al. 2005), and mixed-phase clouds during M-PACE (Klein et al. 2009). AM2 also has been evaluated against ARM observations by integrating it in forecast (CAPT) mode for M-PACE (Xie et al. 2008); the Tropical Ocean Global Atmosphere Coupled Ocean–Atmosphere Response Experiment (Boyle et al. 2008); the east Pacific Investigation of Climate Cruise (Hannay et al. 2009); and, along with AM3, TWP-ICE (Lin et al. 2012). Simulations of cloud fraction and precipitation in AM2 and AM3 SCMs also have been compared against observations at the ARM SGP site (Song et al. 2013). These evaluations have yielded many insights on the behaviors of AM2 and AM3. For example, the summer forecast experiments using SGP observations revealed AM2's warm bias in that region to be due to its inability to simulate enough precipitation and not due to deficient soil moisture or unrealistic radiation (Klein et al. 2006). In another example, the forecast experiments during the dry period of TWP-ICE suggested that sensitivity of the cumulus parameterization to free-tropospheric humidity was important for successful simulation. Experiments with increased dependence of convective entrainment on humidity are currently underway at GFDL.

6. Concluding discussion

The main subject of this chapter is ARM's influence on climate model development. ARM has directly funded model development activities, and it has collected data that make it possible to test the models in ways that could only be dreamed of before ARM started.

We have presented a few examples to show how ARM data have led to improvements in climate models. Additional examples are provided in other chapters of this monograph (e.g., Ghan and Penner 2016, chapter 27).

It is important to recognize that the process that leads from data collected in the field to improvements in climate simulations is not at all straightforward or even predictable. The record shows that ARM data have suggested ideas, supported ideas, and ruled out ideas. Modelers have been challenged to devise ways of using the data to test the models, for example through the SCM strategy, which took years to develop and is still evolving. Climate modelers continue to use data collected in the early stages of the ARM Program, as well as the more detailed data collected later in the program. ARM will continue to influence climate model development and evaluation for many decades to come.

Acknowledgments. The efforts of Stephen A. Klein were funded by the Atmospheric System Research and Regional and Global Climate Modeling programs of DOE's Office of Science and were performed under the auspices of the U. S. Department of Energy by Lawrence Livermore National Laboratory under Contract DE-AC52-07NA27344.

REFERENCES

Ahlgrimm, M., R. Forbes, J.-J. Morcrette, and R. Neggers, 2016: ARM's impact on numerical weather prediction at ECMWF. *The Atmospheric Radiation Measurement (ARM) Program: The First 20 Years, Meteor. Monogr.*, No. 57, Amer. Meteor. Soc., doi:10.1175/AMSMONOGRAPHS-D-15-0032.1.

Anderson, J. L., and Coauthors, 2004: The new GFDL global atmosphere and land model AM2–LM2: Evaluation with prescribed SST simulations. *J. Climate*, **17**, 4641–4673, doi:10.1175/JCLI-3223.1.

Arnold, N. P., Z. Kuang, and E. Tziperman, 2013: Enhanced MJO-like variability at high SST. *J. Climate*, **26**, 988–1001, doi:10.1175/JCLI-D-12-00272.1.

Bechtold, P., N. Semane, P. Lopez, J.-P. Chaboureau, A. Beljaars, and N. Bormann, 2014: Representing equilibrium and non-equilibrium convection in large-scale models. *J. Atmos. Sci.*, **71**, 734–753, doi:10.1175/JAS-D-13-0163.1.

Benedict, J. J., and D. A. Randall, 2009: Structure of the Madden–Julian oscillation in the Superparameterized CAM. *J. Atmos. Sci.*, **66**, 3277–3296, doi:10.1175/2009JAS3030.1.

——, and ——, 2011: Impacts of idealized air–sea coupling on Madden–Julian oscillation structure in the superparameterized CAM. *J. Atmos. Sci.*, **68**, 1990–2008, doi:10.1175/JAS-D-11-04.1.

——, E. D. Maloney, A. H. Sobel, D. M. Frierson, and L. J. Donner, 2013: Tropical intraseasonal variability in version 3 of the GFDL atmosphere model. *J. Climate*, **26**, 426–449, doi:10.1175/JCLI-D-12-00103.1.

Betts, A. K., and Harshvardhan, 1987: Thermodynamic constraint on the cloud liquid water feedback in climate models. *J. Geophys. Res.*, **92**, 8483–8485, doi:10.1029/JD092iD07p08483.

Bony, S., and J.-L. Dufresne, 2005: Marine boundary layer clouds at the heart of tropical cloud feedback uncertainties in climate models. *Geophys. Res. Lett.*, **32**, L20806, doi:10.1029/2005GL023851.

Bower, K. N., S. J. Moss, D. W. Johnson, T. W. Choularton, J. Latham, P. R. A. Brown, A. M. Blyth, and J. Cardwell, 1996: A parameterization of the ice water content observed in frontal and convective clouds. *Quart. J. Roy. Meteor. Soc.*, **122**, 1815–1844, doi:10.1002/qj.49712253605.

Boyle, J., and S. Klein, 2010: Impact of horizontal resolution on climate model forecasts of tropical precipitation and diabatic heating for the TWP-ICE period. *J. Geophys. Res.*, **115**, D23113, doi:10.1029/2010JD014262.

——, ——, G. Zhang, S. Xie, and X. Wei, 2008: Climate model forecast experiments for TOGA COARE. *Mon. Wea. Rev.*, **136**, 808–832, doi:10.1175/2007MWR2145.1.

Cess, R. D., and Coauthors, 1989: Interpretation of cloud–climate feedback as produced by 14 atmospheric general circulation models. *Science*, **245**, 513–516, doi:10.1126/science.245.4917.513.

Chen, T., W. Rossow, and Y. Zhang, 2000: Radiative effects of cloud-type variations. *J. Climate*, **13**, 264–286, doi:10.1175/1520-0442(2000)013<0264:REOCTV>2.0.CO;2.

Clothiaux, E. E., T. P. Ackerman, G. G. Mace, K. P. Moran, R. T. Marchand, M. Miller, and B. E. Martner, 2000: Objective determination of cloud heights and radar reflectivities using a combination of active remote sensors at the ARM CART sites. *J. Appl. Meteor.*, **39**, 645–665, doi:10.1175/1520-0450(2000)039<0645:ODOCHA>2.0.CO;2.

Clough, S. A., M. W. Shephard, E. Mlawer, J. S. Delamere, M. Iacono, K. Cady-Pereira, S. Boukabara, and P. D. Brown, 2005: Atmospheric radiative transfer modeling: A summary of the AER codes. *J. Quant. Spectrosc. Radiat. Transfer*, **91**, 233–244, doi:10.1016/j.jqsrt.2004.05.058.

Collins, W. D., 2001: Parameterization of generalized cloud overlap for radiative calculations in general circulation models. *J. Atmos. Sci.*, **58**, 3224–3242, doi:10.1175/1520-0469(2001)058<3224:POGCOF>2.0.CO;2.

——, J. K. Hackney, and D. P. Edwards, 2002a: An updated parameterization for infrared emission and absorption by water vapor in the National Center for Atmospheric Research Community Atmosphere Model. *J. Geophys. Res.*, **107**, 4664, doi:10.1029/2001JD001365.

——, P. J. Rasch, B. E. Eaton, D. W. Fillmore, J. T. Kiehl, C. T. Beck, and C. S. Zender, 2002b: Simulation of aerosol distributions and radiative forcing for INDOEX: Regional climate impacts. *J. Geophys. Res.*, **107**, 8028, doi:10.1029/2000JD000032.

——, J. M. Lee-Taylor, D. P. Edwards, and G. L. Francis, 2006a: Effects of increased near-infrared absorption by water vapor on the climate system. *J. Geophys. Res.*, **111**, D18109, doi:10.1029/2005JD006796.

——, and Coauthors, 2006b: The formulation and atmospheric simulation of the Community Atmosphere Model version 3 (CAM3). *J. Climate*, **19**, 2144–2161, doi:10.1175/JCLI3760.1.

Collis, S., A. Protat, P. T. May, and C. Williams, 2013: Statistics of storm updraft velocities from TWP-ICE including verification with profiling measurements. *J. Appl. Meteor. Climatol.*, **52**, 1909–1922, doi:10.1175/JAMC-D-12-0230.1.

Cress, T. S., and D. L. Sisterson, 2016: Deploying ARM sites and supporting infrastructure. *The Atmospheric Radiation Measurement (ARM) Program: The First 20 Years, Meteor. Monogr.*, No. 57, Amer. Meteor. Soc., doi:10.1175/AMSMONOGRAPHS-D-15-0049.1.

Davies, L., and Coauthors, 2013: A single-column model ensemble approach applied to the TWP-ICE experiment. *J. Geophys. Res.*, **118**, 6544–6563, doi:10.1002/jgrd.50450.

Del Genio, A. D., and M.-S. Yao, 1988: Sensitivity of a global climate model to the specification of convective updraft and downdraft mass fluxes. *J. Climate*, **45**, 2641–2668, doi:10.1175/1520-0469(1988)045<2641:SOAGCM>2.0.CO;2.

——, and A. B. Wolf, 2000: The temperature dependence of the liquid water path of low clouds in the Southern Great Plains. *J. Climate*, **13**, 3465–3486, doi:10.1175/1520-0442(2000)013<3465:TTDOTL>2.0.CO;2.

——, and J. Wu, 2010: The role of entrainment in the diurnal cycle of continental convection. *J. Climate*, **23**, 2722–2738, doi:10.1175/2009JCLI3340.1.

——, M.-S. Yao, W. Kovari, and K. K.-W. Lo, 1996: A prognostic cloud water parameterization for global climate models. *J. Climate*, **9**, 270–304, doi:10.1175/1520-0442(1996)009<0270:APCWPF>2.0.CO;2.

——, ——, and J. Jonas, 2007: Will moist convection be stronger in a warmer climate? *Geophys. Res. Lett.*, **34**, L16703, doi:10.1029/2007GL030525.

——, Y. Chen, D. Kim, and M.-S. Yao, 2012: The MJO transition from shallow to deep convection in CloudSat/CALIPSO data and GISS GCM simulations. *J. Climate*, **25**, 3755–3770, doi:10.1175/JCLI-D-11-00384.1.

——, M.-S. Yao, J. Wu, and A. Wolf, 2013: Cold pools—A first step in representing convective organization in GCMs. [Available online at http://asr.science.energy.gov/meetings/stm/posters/view?id=784.]

DeMott, C. A., C. Stan, D. A. Randall, J. L. Kinter III, and M. Khairoutdinov, 2011: The Asian monsoon in the superparameterized CCSM and its relation to tropical wave activity. *J. Climate*, **24**, 5134–5156, doi:10.1175/2011JCLI4202.1.

——, ——, and ——, 2013: Northward propagation mechanisms of the boreal summer intraseasonal oscillation in the ERA-Interim and SP-CCSM. *J. Climate*, **26**, 1973–1992, doi:10.1175/JCLI-D-12-00191.1.

Derbyshire, S., I. Beau, P. Bechtold, J.-Y. Grandpeix, J.-M. Piriou, J.-L. Redelsperger, and P. M. M. Soares, 2004: Sensitivity of moist convection to environmental humidity. *Quart. J. Roy. Meteor. Soc.*, **130**, 3055–3079, doi:10.1256/qj.03.130.

Dong, X., P. Minnis, and B. Xi, 2005: A climatology of midlatitude continental clouds from the ARM SGP central facility: Part I: Low-level cloud macrophysical, microphysical, and radiative properties. *J. Climate*, **18**, 1391–1410, doi:10.1175/JCLI3342.1.

Donner, L. J., 1993: A cumulus parameterization including mass fluxes, vertical momentum dynamics, and mesoscale effects. *J. Atmos. Sci.*, **50**, 889–906, doi:10.1175/1520-0469(1993)050<0889:ACPIMF>2.0.CO;2.

——, and V. T. Phillips, 2003: Boundary layer control on convective available potential energy: Implications for cumulus parameterization. *J. Geophys. Res.*, **108**, 4701, doi:10.1029/2003JD003773.

——, and Coauthors, 2011: The dynamical core, physical parameterizations, and basic simulation characteristics of the atmospheric component of the GFDL global coupled model CM3. *J. Climate*, **24**, 3484–3519, doi:10.1175/2011JCLI3955.1.

Ellingson, R. G., J. Ellis, and S. Fels, 1991: The intercomparison of radiation codes used in climate models: Longwave results. *J. Geophys. Res.*, **96**, 8929–8953, doi:10.1029/90JD01450.

Feigelson, E. M., 1978: Preliminary radiation model of a cloudy atmosphere. Part I—Structure of clouds and solar radiation. *Contrib. Atmos. Phys.*, **51**, 203–229.

Feldman, D. R., C. A. Algieri, J. R. Ong, and W. D. Collins, 2011: CLARREO shortwave observing system simulation experiments of the twenty-first century: Simulator design and implementation. *J. Geophys. Res.*, **116**, D10107, doi:10.1029/2010JD015350.

Fridlind, A. M., and Coauthors, 2012: A comparison of TWP-ICE observational data with cloud-resolving model results. *J. Geophys. Res.*, **117**, D05204, doi:10.1029/2011JD016595.

Fu, Q., and K. N. Liou, 1992: On the correlated *k*-distribution method for radiative transfer in nonhomogeneous atmospheres. *J. Atmos. Sci.*, **49**, 2139–2156, doi:10.1175/1520-0469(1992)049<2139:OTCDMF>2.0.CO;2.

Gettelman, A., J. E. Kay, and K. M. Shell, 2012: The evolution of climate sensitivity and climate feedbacks in the Community Atmosphere Model. *J. Climate*, **25**, 1453–1469, doi:10.1175/JCLI-D-11-00197.1.

Ghan, S., and J. Penner, 2016: ARM-led improvements in aerosols in climate and climate models. *The Atmospheric Radiation Measurement (ARM) Program: The First 20 Years, Meteor. Monogr.*, No. 57, Amer. Meteor. Soc., doi:10.1175/AMSMONOGRAPHS-D-15-0033.1.

——, and Coauthors, 2000: A comparison of single-column model simulations of summertime midlatitude continental convection. *J. Geophys. Res.*, **105**, 2091–2124, doi:10.1029/1999JD900971.

Grabowski, W. W., and P. K. Smolarkiewicz, 1999: CRCP: A Cloud Resolving Convection Parameterization for modeling the tropical convective atmosphere. *Physica D*, **133**, 171–178, doi:10.1016/S0167-2789(99)00104-9.

Greenwald, T. J., G. L. Stephens, S. A. Christopher, and T. H. Vonder Haar, 1995: Observations of the global characteristics and regional radiative effects of marine cloud liquid water. *J. Climate*, **8**, 2928–2946, doi:10.1175/1520-0442(1995)008<2928:OOTGCA>2.0.CO;2.

Gregory, D., 2001: Estimation of entrainment rate in simple models of convective clouds. *Quart. J. Roy. Meteor. Soc.*, **127**, 53–72, doi:10.1002/qj.49712757104.

Guichard, F., and Coauthors, 2004: Modelling the diurnal cycle of deep precipitating convection over land with cloud-resolving models and single-column models. *Quart. J. Roy. Meteor. Soc.*, **130**, 3139–3172, doi:10.1256/qj.03.145.

Guo, H., J.-C. Golaz, L. J. Donner, V. E. Larson, D. P. Schanen, and B. M. Griffin, 2010: A dynamic probability density function treatment of cloud mass and number concentrations for low level clouds in GFDL SCM/GCM. *Geosci. Model Dev. Discuss.*, **3**, 541–588, doi:10.5194/gmdd-3-541-2010.

——, ——, and ——, 2011: Aerosol effects on stratocumulus water paths in a PDF-based parameterization. *Geophys. Res. Lett.*, **38**, L17808, doi:10.1029/2011GL048611.

Hack, J. J., and J. A. Pedretti, 2000: Assessment of solution uncertainties in single-column modeling frameworks. *J. Climate*, **13**, 352–365, doi:10.1175/1520-0442(2000)013<0352:AOSUIS>2.0.CO;2.

Haeffelin, M., and Coauthors, 2016: Parallel developments and formal collaboration between European atmospheric profiling observatories and the U.S. ARM research program. *The Atmospheric Radiation Measurement (ARM) Program: The First 20 Years, Meteor. Monogr.*, No. 57, Amer. Meteor. Soc., doi:10.1175/AMSMONOGRAPHS-D-15-0045.1.

Hannay, C., D. L. Williamson, J. J. Hack, J. T. Kiehl, J. G. Olson, S. A. Klein, C. S. Bretherton, and M. Köhler, 2009: Evaluation of forecasted Southeast Pacific stratocumulus in the NCAR, GFDL, and ECMWF models. *J. Climate*, **22**, 2871–2889, doi:10.1175/2008JCLI2479.1.

Hartmann, D., M. Ockert-Bell, and M. Michelsen, 1992: The effect of cloud type on Earth's energy balance: Global analysis. *J. Climate*, **5**, 1281–1304, doi:10.1175/1520-0442(1992)005<1281:TEOCTO>2.0.CO;2.

Houze, R. A., Jr., and A. K. Betts, 1981: Convection in GATE. *Rev. Geophys.*, **19**, 541–576, doi:10.1029/RG019i004p00541.

Hurrell, J. W., and Coauthors, 2013: The Community Earth System Model: A framework for collaborative research. *Bull. Amer. Meteor. Soc.*, **94**, 1339–1360, doi:10.1175/BAMS-D-12-00121.1.

Iacono, M. J., E. J. Mlawer, S. A. Clough, and J.-J. Morcrette, 2000: Impact of an improved longwave radiation model, RRTM, on the energy budget and thermodynamic properties of the NCAR community climate model, CCM3. *J. Geophys. Res.*, **105**, 14 873–14 890, doi:10.1029/2000JD900091.

——, J. S. Delamere, E. J. Mlawer, M. W. Shephard, S. A. Clough, and W. D. Collins, 2008: Radiative forcing by long-lived greenhouse gases: Calculations with the AER radiative transfer models. *J. Geophys. Res.*, **113**, D13103, doi:10.1029/2008JD009944.

IPCC, 2007: *Climate Change 2007: The Physical Science Basis*. Cambridge University Press, 996 pp.

——, 2013: *Climate Change 2013: The Physical Science Basis*. Cambridge University Press, 1535 pp., doi:10.1017/CBO9781107415324.

Jakob, C., 2003: An improved strategy for the evaluation of cloud parameterizations in GCMS. *Bull. Amer. Meteor. Soc.*, **84**, 1387–1401, doi:10.1175/BAMS-84-10-1387.

Jensen, M. P., and A. D. Del Genio, 2006: Factors limiting convective cloud-top height at the ARM Nauru Island climate research facility. *J. Climate*, **19**, 2105–2117, doi:10.1175/JCLI3722.1.

Khairoutdinov, M. F., and D. A. Randall, 2001: A cloud resolving model as a cloud parameterization in the NCAR Community Climate System Model: Preliminary results. *Geophys. Res. Lett.*, **28**, 3617–3620, doi:10.1029/2001GL013552.

——, and ——, 2003: Cloud resolving modeling of ARM Summer 1997 IOP: Model formulation, results, uncertainties, and sensitivities. *J. Atmos. Sci.*, **60**, 607–625, doi:10.1175/1520-0469(2003)060<0607:CRMOTA>2.0.CO;2.

Kiehl, J. T., and B. P. Briegleb, 1991: A new parameterization of the absorptance due to the 15-μm band system of carbon dioxide. *J. Geophys. Res.*, **96**, 9013–9019, doi:10.1029/89JD00993.

——, and ——, 1993: The relative roles of sulfate aerosols and greenhouse gases in climate forcing. *Science*, **260**, 311–314, doi:10.1126/science.260.5106.311.

——, and K. E. Trenberth, 1997: Earth's annual global mean energy budget. *Bull. Amer. Meteor. Soc.*, **78**, 197–208, doi:10.1175/1520-0477(1997)078<0197:EAGMEB>2.0.CO;2.

Kim, B.-G., S. A. Klein, and J. R. Norris, 2005: Continental liquid water cloud variability and its parameterization using Atmospheric Radiation Measurement data. *J. Geophys. Res.*, **110**, D15S08, doi:10.1029/2004JD005122.

Kim, D., and Coauthors, 2009: Application of MJO simulation diagnostics to climate models. *J. Climate*, **22**, 6413–6436, doi:10.1175/2009JCLI3063.1.

Klein, S. A., R. Pincus, C. Hannay, and K.-M. Xu, 2005: How might a statistical cloud scheme be coupled to a mass-flux convection scheme? *J. Geophys. Res.*, **110**, D15S06, doi:10.1029/2004JD005017.

——, X. Jiang, J. Boyle, S. Malyshev, and S. Xie, 2006: Diagnosis of the summertime warm and dry bias over the U.S. Southern Great Plains in the GFDL climate model using a weather forecasting approach. *Geophys. Res. Lett.*, **33**, L18805, doi:10.1029/2006GL027567.

——, and Coauthors, 2009: Intercomparison of model simulations of mixed-phase clouds observed during the ARM Mixed-Phase Arctic Cloud Experiment. I: Single-layer cloud. *Quart. J. Roy. Meteor. Soc.*, **135**, 979–1002, doi:10.1002/qj.416.

Kollias, P., and Coauthors, 2016: Development and applications of the ARM millimeter-wavelength cloud radars. *The Atmospheric Radiation Measurement (ARM) Program: The First 20 Years, Meteor. Monogr.*, No. 57, Amer. Meteor. Soc., doi:10.1175/AMSMONOGRAPHS-D-15-0037.1.

Kooperman, G. J., M. S. Pritchard, and R. C. J. Somerville, 2013: Robustness and sensitivities of central U.S. summer convection in the super-parameterized CAM: Multi-model intercomparison with a new regional EOF index. *Geophys. Res. Lett.*, **40**, 3287–3291, doi:10.1002/grl.50597.

Lacis, A. A., and V. Oinas, 1991: A description of the correlated *k* distribution method for modeling nongray gaseous absorption, thermal emission, and multiple scattering in vertically inhomogeneous atmospheres. *J. Geophys. Res.*, **96**, 9027–9063, doi:10.1029/90JD01945.

Lin, J., and Coauthors, 2006: Tropical intraseasonal variability in 14 IPCC AR4 climate models. Part I: Convective signals. *J. Climate*, **19**, 2665–2690, doi:10.1175/JCLI3735.1.

Lin, Y., and Coauthors, 2012: TWP-ICE global atmospheric model intercomparison: Convection responsiveness and resolution impact. *J. Geophys. Res.*, **117**, D09111, doi:10.1029/2011JD017018.

Madden, R. A., and P. R. Julian, 1971: Detection of a 40–50 day oscillation in the zonal wind in the tropical Pacific. *J. Atmos. Sci.*, **28**, 702–708, doi:10.1175/1520-0469(1971)028<0702:DOADOI>2.0.CO;2.

——, and ——, 1972: Description of global scale circulation cells in the tropics with a 40–50 day period. *J. Atmos. Sci.*, **29**, 1109–1123, doi:10.1175/1520-0469(1972)029<1109:DOGSCC>2.0.CO;2.

Mapes, B. E., 2000: Convective inhibition, subgrid-scale triggering energy, and stratiform instability in a toy tropical wave model. *J. Atmos. Sci.*, **57**, 1515–1535, doi:10.1175/1520-0469(2000)057<1515:CISSTE>2.0.CO;2.

Mather, J. H., D. D. Turner, and T. P. Ackerman, 2016: Scientific maturation of the ARM Program. *The Atmospheric Radiation Measurement (ARM) Program: The First 20 Years, Meteor. Monogr.*, No. 57, Amer. Meteor. Soc., doi:10.1175/AMSMONOGRAPHS-D-15-0053.1.

May, P. T., J. H. Mather, G. Vaughan, C. Jakob, G. M. McFarquhar, K. N. Bower, and G. G. Mace, 2008: The Tropical Warm Pool International Cloud Experiment. *Bull. Amer. Meteor. Soc.*, **89**, 629–645, doi:10.1175/BAMS-89-5-629.

McCrary, R. R., 2012: Seasonal, synoptic and intraseasonal variability of the West African monsoon. Ph.D. dissertation, Colorado State University, 160 pp. [Available online at http://digitool.library.colostate.edu/R/?func=dbin-jump-full&object_id=212297.]

Meehl, G. A., and Coauthors, 2012: Climate system response to external forcings and climate change projections in CCSM4. *J. Climate*, **25**, 3661–3683, doi:10.1175/JCLI-D-11-00240.1.

Melillo, J. M., T. C. Richmond, and G. W. Yohe, Eds., 2014: *Climate Change Impacts in the United States: The Third National Climate Assessment.* U.S. Global Change Research Program, 841 pp., doi:10.7930/J0Z31WJ2.

Mitchell, D. L., A. J. Baran, W. P. Arnott, and C. Schmitt, 2006: Testing and comparing the modified anomalous diffraction approximation. *J. Atmos. Sci.*, **63**, 2948–2962, doi:10.1175/JAS3775.1.

——, R. P. Lawson, and B. Baker, 2011: Understanding effective diameter and its application to terrestrial radiation in ice clouds. *Atmos. Chem. Phys.*, **11**, 3417–3429, doi:10.5194/acp-11-3417-2011.

Mitchell, J. F. B., C. A. Senior, and W. J. Ingram, 1989: CO_2 and climate—A missing feedback. *Nature*, **341**, 132–134, doi:10.1038/341132a0.

Mlawer, E. J., S. J. Taubman, P. D. Brown, M. J. Iacono, and S. A. Clough, 1997: Radiative transfer for inhomogeneous atmospheres: RRTM, a validated correlated-k model for the longwave. *J. Geophys. Res.*, **102**, 16 663–16 682, doi:10.1029/97JD00237.

——, V. H. Payne, J.-L. Moncet, J. S. Delamere, M. J. Alvarado, and D. C. Tobin, 2012: Development and recent evaluation of the MT_CKD model of continuum absorption. *Philos. Trans. Roy. Soc. London*, **A370**, 2520–2556, doi:10.1098/rsta.2011.0295.

——, M. Iacono, R. Pincus, H. Barker, L. Oreopoulos, and D. Mitchell, 2016: Contributions of the ARM Program to radiative transfer modeling for climate and weather applications. *The Atmospheric Radiation Measurement (ARM) Program: The First 20 Years, Meteor. Monogr.*, No. 57, Amer. Meteor. Soc., doi:10.1175/AMSMONOGRAPHS-D-15-0041.1.

Morrison, H., and Coauthors, 2009: Intercomparison of model simulations of mixed-phase clouds observed during the ARM Mixed-Phase Arctic Cloud Experiment. II: Multilayer cloud. *Quart. J. Roy. Meteor. Soc.*, **135**, 1003–1019, doi:10.1002/qj.415.

Naud, C. M., and Coauthors, 2010: Thermodynamic phase profiles of optically thin midlatitude clouds and their relation to temperature. *J. Geophys. Res.*, **115**, D11202, doi:10.1029/2009JD012889.

Oreopoulos, L., and Coauthors, 2012: The continual intercomparison of radiation codes: Results from Phase I. *J. Geophys. Res.*, **117**, D06118, doi:10.1029/2011JD016821.

Phillips, T. J., and Coauthors, 2004: Evaluating parameterizations in general circulation models: Climate simulation meets weather prediction. *Bull. Amer. Meteor. Soc.*, **85**, 1903–1915, doi:10.1175/BAMS-85-12-1903.

Pincus, R., and S. A. Klein, 2000: Unresolved spatial variability and microphysical process rates in large-scale models. *J. Geophys. Res.*, **105**, 27 059–27 065, doi:10.1029/2000JD900504.

——, C. Hannay, S. A. Klein, K.-M. Xu, and R. Hemler, 2005: Overlap assumptions for assumed probability distribution function cloud schemes in large-scale models. *J. Geophys. Res.*, **110**, D15S09, doi:10.1029/2004JD005100.

——, R. Hemler, and S. A. Klein, 2006: Using stochastically generated subcolumns to represent cloud structure in a large-scale model. *Mon. Wea. Rev.*, **134**, 3644–3656, doi:10.1175/MWR3257.1.

Prather, M. J., and J. Hsu, 2008: NF_3, the greenhouse gas missing from Kyoto. *Geophys. Res. Lett.*, **35**, L12810, doi:10.1029/2008GL034542.

Pritchard, M. S., and R. C. J. Somerville, 2009a: Empirical orthogonal function analysis of the diurnal cycle of precipitation in a multi-scale climate model. *Geophys. Res. Lett.*, **36**, L05812, doi:10.1029/2008GL036964.

——, and ——, 2009b: Assessing the diurnal cycle of precipitation in a multi-scale climate model. *J. Adv. Model. Earth Syst.*, **1**, 12, doi:10.3894/JAMES.2009.1.12.

——, M. W. Moncrieff, and R. C. J. Somerville, 2011: Orogenic propagating precipitation systems over the United States in a global climate model with embedded explicit convection. *J. Atmos. Sci.*, **68**, 1821–1840, doi:10.1175/2011JAS3699.1.

Randall, D. A., 2013: Beyond deadlock. *Geophys. Res. Lett.*, **40**, 5970–5976, doi:10.1002/2013GL057998.

——, and D. G. Cripe, 1999: Alternative methods for specification of observed forcing in single-column models and cloud system models. *J. Geophys. Res.*, **104**, 24 527–24 545, doi:10.1029/1999JD900765.

——, K.-M. Xu, R. J. C. Somerville, and S. Iacobellis, 1996: Single-column models and cloud ensemble models as links between observations and climate models. *J. Climate*, **9**, 1683–1697, doi:10.1175/1520-0442(1996)009<1683:SCMACE>2.0.CO;2.

——, and Coauthors, 2003: Confronting models with data: The GEWEX Cloud Systems Study. *Bull. Amer. Meteor. Soc.*, **84**, 455–469, doi:10.1175/BAMS-84-4-455.

——, C. DeMott, C. Stan, M. Khairoutdinov, J. Benedict, R. McCrary, K. Thayer-Calder, and M. Branson, 2016: Simulations of the tropical general circulation with a multiscale global model. *Multiscale Convection-Coupled Systems in the Tropics: A Tribute to the Late Professor Yanai, Meteor. Monogr.*, No. 56, Amer. Meteor. Soc., doi:10.1175/AMSMONOGRAPHS-D-15-0016.1.

Roeckner, E., U. Schlese, J. Biercamp, and P. Loewe, 1987: Cloud optical depth feedbacks and climate modeling. *Nature*, **329**, 138–140, doi:10.1038/329138a0.

Rothman, L. S., and Coauthors, 2013: The HITRAN2012 molecular spectroscopic database. *J. Quant. Spectrosc. Radiat. Transfer*, **130**, 4–50, doi:10.1016/j.jqsrt.2013.07.002.

Schmidt, G. A., and Coauthors, 2006: Present-day atmospheric simulations using GISS ModelE: Comparison to in situ, satellite, and reanalysis data. *J. Climate*, **19**, 153–192, doi:10.1175/JCLI3612.1.

——, and Coauthors, 2014: Configuration and assessment of the GISS ModelE2 contributions to the CMIP5 archive. *J. Adv. Model Earth Syst.*, **6**, 141–184, doi:10.1002/2013MS000265.

Slingo, A., 1989: A GCM parameterization for the shortwave radiative properties of water clouds. *J. Atmos. Sci.*, **46**, 1419–1427, doi:10.1175/1520-0469(1989)046<1419:AGPFTS>2.0.CO;2.

Somerville, R. C. J., and L. A. Remer, 1984: Cloud optical thickness feedbacks in the CO_2 climate problem. *J. Geophys. Res.*, **89**, 9668–9672, doi:10.1029/JD089iD06p09668.

Song, H., W. Lin, Y. Lin, A. B. Wolf, R. Neggers, L. J. Donner, A. D. DelGenio, and Y. Liu, 2013: Evaluation of precipitation simulated by seven SCMs against the ARM observations at the SGP site. *J. Climate*, **26**, 5467–5492, doi:10.1175/JCLI-D-12-00263.1.

Stan, C., M. Khairoutdinov, C. A. DeMott, V. Krishnamurthy, D. M. Straus, D. A. Randall, J. L. Kinter III, and J. Shukla, 2010: An ocean–atmosphere climate simulation with an embedded cloud resolving model. *Geophys. Res. Lett.*, **37**, L01702, doi:10.1029/2009GL040822.

Stokes, G. M., 2016: Original ARM concept and launch. *The Atmospheric Radiation Measurement (ARM) Program: The First 20 Years, Meteor. Monogr.*, No. 57, Amer. Meteor. Soc., doi:10.1175/AMSMONOGRAPHS-D-15-0021.1.

Tompkins, A. M., 2001: Organization of convection in low vertical wind shears: The role of cold pools. *J. Atmos. Sci.*, **58**, 1650–1672, doi:10.1175/1520-0469(2001)058<1650:OOTCIL>2.0.CO;2.

Tselioudis, G., W. B. Rossow, and D. Rind, 1992: Global patterns of cloud optical thickness variation with temperature. *J. Climate*, **5**, 1484–1495, doi:10.1175/1520-0442(1992)005<1484:GPOCOT>2.0.CO;2.

——, A. D. Del Genio, W. Kovari Jr., and M.-S. Yao, 1998: Temperature dependence of low cloud optical thickness in the GISS GCM: Contributing mechanisms and climate implications. *J. Climate*, **11**, 3268–3281, doi:10.1175/1520-0442(1998)011<3268:TDOLCO>2.0.CO;2.

Varble, A., and Coauthors, 2011: Evaluation of cloud-resolving model intercomparison simulations using TWP-ICE observations: Precipitation and cloud structure. *J. Geophys. Res.*, **116**, D12206, doi:10.1029/2010JD015180.

Williamson, D. L., and Coauthors, 2005: Moisture and temperature balances at the Atmospheric Radiation Measurement Southern Great Plains Site in forecasts with the Community Atmosphere Model (CAM2). *J. Geophys. Res.*, **110**, D15S16, doi:10.1029/2004JD005109.

Wu, J., A. D. Del Genio, M.-S. Yao, and A. B. Wolf, 2009: WRF and GISS SCM simulations of convective updraft properties during TWP-ICE. *J. Geophys. Res.*, **114**, D04206, doi:10.1029/2008JD010851.

Wyant, M. C., M. Khairoutdinov, and C. S. Bretherton, 2006: Climate sensitivity and cloud response of a GCM with a superparameterization. *Geophys. Res. Lett.*, **33**, L06714, doi:10.1029/2005GL025464.

——, C. S. Bretherton, P. N. Blossey, and M. Khairoutdinov, 2012: Fast cloud adjustment to increasing CO_2 in a superparameterized climate model. *J. Adv. Model Earth Syst.*, **4**, M05001, doi:10.1029/2011MS000092.

Xie, S., and Coauthors, 2002: Intercomparison and evaluation of cumulus parametrizations under summertime midlatitude continental conditions. *Quart. J. Roy. Meteor. Soc.*, **128**, 1095–1135, doi:10.1256/003590002320373229.

——, and Coauthors, 2005: Simulations of midlatitude frontal clouds by single-column and cloud-resolving models during the Atmospheric Radiation Measurement March 2000 cloud intensive operational period. *J. Geophys. Res.*, **110**, D15S03, doi:10.1029/2004JD005119.

——, J. Boyle, S. A. Klein, X. Liu, and S. Ghan, 2008: Simulations of Arctic mixed-phase clouds in forecasts with CAM3 and AM2 for M-PACE. *J. Geophys. Res.*, **113**, D04211, doi:10.1029/2007JD009225.

Xu, K.-M., and Coauthors, 2002: An intercomparison of cloud-resolving models with the Atmospheric Radiation Measurement summer 1997 intensive observation period data. *Quart. J. Roy. Meteor. Soc.*, **128**, 593–624, doi:10.1256/003590002321042117.

Yao, M.-S., and A. D. Del Genio, 1999: Effects of cloud parameterization on the simulation of climate changes in the GISS GCM. *J. Climate*, **12**, 761–779, doi:10.1175/1520-0442(1999)012<0761:EOCPOT>2.0.CO;2.

Zelinka, M. D., S. A. Klein, and D. L. Hartmann, 2012: Computing and partitioning cloud feedbacks using cloud property histograms. Part II: Attribution to changes in cloud amount, altitude, and optical depth. *J. Climate*, **25**, 3736–3754, doi:10.1175/JCLI-D-11-00249.1.

Zhang, M. H., and J. L. Lin, 1997: Constrained variational analysis of sounding data based on column-integrated budgets of mass, heat, moisture, and momentum: Approach and application to ARM measurements. *J. Atmos. Sci.*, **54**, 1503–1524, doi:10.1175/1520-0469(1997)054<1503:CVAOSD>2.0.CO;2.

——, R. C. J. Somerville, and S. Xie, 2016: The SCM concept and creation of ARM forcing datasets. *The Atmospheric Radiation Measurement (ARM) Program: The First 20 Years, Meteor. Monogr.*, No. 57, Amer. Meteor. Soc., doi:10.1175/AMSMONOGRAPHS-D-15-0040.1.

Zhao, M., I. M. Held, S.-J. Lin, and G. A. Vecchi, 2009: Simulations of global hurricane climatology, interannual variability, and response to global warming using a 50-km resolution GCM. *J. Climate*, **22**, 6653–6678, doi:10.1175/2009JCLI3049.1.

Chapter 27

ARM-Led Improvements in Aerosols in Climate and Climate Models

STEVEN GHAN

Pacific Northwest National Laboratory, Richland, Washington

JOYCE E. PENNER

University of Michigan, Ann Arbor, Michigan

1. Introduction

The climate science community has been aware for several decades of the role of aerosols in the climate system and of the potential influence of anthropogenic aerosols in driving climate change (Penner et al. 1994). In the Fifth Assessment Report of climate change by the Intergovernmental Panel on Climate Change (IPCC), Boucher et al. (2013) introduced new terminology that more clearly distinguishes the key mechanisms by which anthropogenic aerosols alter the energy balance of the earth. The term radiative forcing (RF) refers to the impact of anthropogenic aerosols on the shortwave and longwave radiative fluxes without considering the adjustment of clouds to the aerosol. RFari is the component of RF due to aerosol–radiation interactions, specifically scattering and absorption of radiation, while RFaci is the component of RF due to aerosol–cloud interactions, specifically aerosol effects on droplet and ice crystal number but not liquid water or ice mass concentration. Radiative forcing is the sum of RFari and RFaci. ERFari refers to the effective radiative forcing due to aerosol–radiative interactions, including the adjustment of clouds to the aerosol scattering and absorption as well as RFari. ERFaci refers to the effective radiative forcing due to aerosol–climate interactions, including the adjustment of cloud microphysical processes and properties to the aerosol. The sum of ERFari and ERFaci is the effective radiative forcing (ERF). RFari was formerly called aerosol direct radiative forcing and RFaci was called the cloud albedo effect. ERFari includes both the direct effect and what was formerly called the semidirect effect, in which clouds adjusted to the scattering and absorption by the aerosol. ERFaci includes both the cloud albedo effect (RFaci) and what was formerly called the cloud lifetime effect. Ghan (2013) showed how the partitioning of ERF into (i) RFari and (ii) the sum of ERFaci and the semidirect effect (total effective forcing due to aerosol effects on clouds) can be diagnosed from the radiation balance calculated with and without scattering and absorption by aerosols in the model calculation.

The DOE ARM Program has played a foundational role in efforts to quantify aerosol effects on climate, beginning with the early back-of-the-envelope estimates of RFari by anthropogenic sulfate and biomass burning aerosol (Penner et al. 1994). This led to one of the first aerosol model intercomparison studies (Penner et al. 2001). In this chapter we review the role that ARM has played in subsequent detailed estimates of RFari, RFaci, and ERFaci based on physically based representations of aerosols in climate models. Only recently have other DOE programs applied the aerosol modeling capability to simulate the climate response to the radiative forcing.

2. Aerosol challenges for climate models

All estimates of aerosol radiative forcing begin with emissions of aerosol and aerosol precursor gases and with a representation of the aerosol life cycle. In the simplest back-of-the-envelope estimates the aerosol life cycle is expressed in terms of a prescribed value for the aerosol lifetime in the atmosphere, which converts global emissions into a global mean atmospheric burden.

Corresponding author address: Steven Ghan, Pacific Northwest National Laboratory, 902 Battelle Blvd., P.O. Box 999, MSIN K9-24, Richland, WA 99352.
E-mail: steve.ghan@pnnl.gov

DOI: 10.1175/AMSMONOGRAPHS-D-15-0033.1

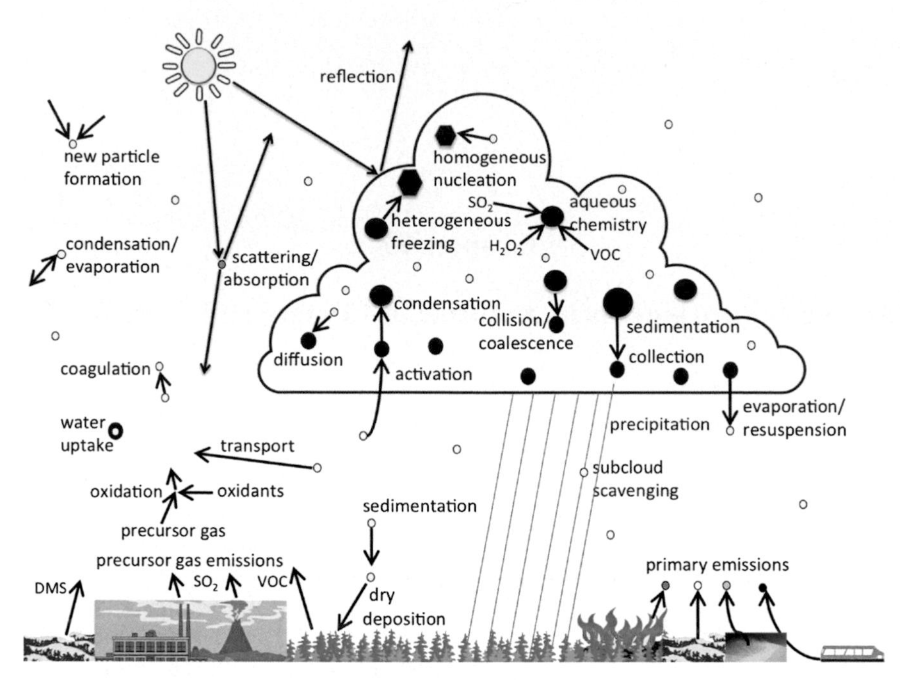

FIG. 27-1. Atmospheric processes important for producing aerosol effects on climate (after Ghan and Schwartz 2007).

In climate models, the aerosol life cycle is explicitly simulated, accounting for transport, transformation, sedimentation, and removal by wet and dry deposition to produce simulations of atmospheric concentrations of the aerosol of interest. Figure 27-1 illustrates the processes that must be represented in physically based models of aerosol effects on climate.

Since anthropogenic aerosols affect the earth's energy balance through scattering and absorption of solar radiation, and through effects on droplet number and cloud albedo, the ARM Program first focused on those aspects of aerosols that determine these effects. Most anthropogenic aerosols are composed of sulfate and carbonaceous material, so the focus has been on radiative forcing by those aerosol components.

Early in the program it was recognized that effects of one aerosol component on cloud droplet formation depends upon competition with all other components that contribute significantly to cloud condensation nuclei (CCN) concentration, natural as well as anthropogenic (Ghan et al. 1998), and that the influence depends on the size distribution of particles as well as their mass concentration (Ghan et al. 1993, 1995). This led to efforts to simulate the size distribution of all important components of CCN (Chuang et al. 2002; Easter et al. 2004; X. Liu et al. 2005). Since a significant fraction of the CCN is produced by gas-to-particle conversion, the global aerosol model must simulate emissions and oxidation of the secondary aerosol precursor gases such as dimethyl sulfide (DMS), SO_2, and volatile organic compounds (VOC). In the earliest simulations, the chemistry included in the model was simplified considerably. During the calculation of sulfur oxidation, the ozone concentration was prescribed because the aerosol does not affect it significantly, while hydrogen peroxide concentrations were allowed to respond to the significant loss during aqueous phase oxidation of SO_2 dissolved in cloud droplets. However, it is now recognized that the contribution of VOC oxidation to secondary organic aerosol formation requires that a full treatment of chemistry be included (Lin et al. 2012, 2014).

The importance of realistically representing clouds, particularly cloud microphysics, was also recognized early in the ARM Program. Droplet number concentration is determined by a balance of processes that include droplet nucleation, droplet collision/coalescence, collection by drizzle and rain droplets, and evaporation. To account for all of these processes Ghan et al. (1997a,b) introduced prognostic droplet number into a climate model, and later Morrison et al. (2005) generalized the approach to represent all processes that control number and mass concentrations of rain, snow, and ice as well as cloud droplets. The initial focus of aerosol–cloud interactions has been on the microphysics of stratiform clouds, but more recently attention has been directed to aerosol impacts on the microphysics of convective clouds (Song and Zhang 2011).

Just as estimates of cloud feedback by climate models depend on the cloud parameterizations in the models, so do estimates of ERFaci. The effort of the ARM Program to improve cloud parameterizations is therefore relevant for ERFaci as well as cloud feedbacks. The representation of droplet collision/coalescence, which is sensitive to droplet size and hence droplet number, is particularly critical, as it is the gateway to cloud microphysical adjustments to the aerosol, which then influence cloud fraction and cloud liquid and ice water contents. The treatment of turbulence and entrainment are also important, as those processes play important roles in both aerosol activation (Ghan et al. 1997b; Barahona and Nenes 2007) and in the response of cloud liquid water path to aerosol-induced changes in sedimentation (Ackerman et al. 2004).

Climate and aerosol models developed and applied by the ARM Program have been characterized by physically based rather than empirical representations of aerosol and cloud properties and processes. Rather than relate droplet number concentration to sulfate mass or aerosol number concentration as in some models, ARM researchers developed physically based representations of droplet formation (Ghan et al. 1993, 1995; Chuang et al. 1997; Abdul-Razzak et al. 1998; Abdul-Razzak and Ghan 2000), ice nucleation (Liu and Penner 2005), as well as the freezing of liquid to ice in mixed-phase clouds (DeMott et al. 2010; Yun and Penner 2012). Such physically based schemes offered the generality necessary for application to models with complete representations of the aerosol composition and size distribution and also offer insights into uncertainties und biases introduced by approximations.

For global climate simulations, it is not practical to represent the full complexity of aerosol and cloud properties and processes in climate models, so approximations were introduced to make the representation affordable. Ghan and Easter (1992), for example, introduced the diagnostic precipitation approximation, which made it possible to apply complex cloud microphysics schemes to climate models without excessive computational cost, although recent studies suggest the approximation adversely affects simulated aerosol effects on clouds (Lee and Penner 2010; Wang et al. 2012).

Similarly, the complexity of the distribution of size and composition of the aerosol population has been approximated by the use of a small set of aerosol modes, each mode composed of internal mixtures of aerosol components with the same proportionality for all particles within a mode (Easter et al. 2004). The distribution of number with size for each mode is usually represented by a lognormal distribution, with the width of the distribution prescribed and the mode radius either prescribed or diagnosed from the total volume concentration of the mode and a separate prediction of the total number concentration of the mode. Here again, though, simple internal mixture approximations may cause large differences in aerosol forcing compared to models that make fewer approximations (Zhou and Penner 2014).

The use of field and laboratory data to improve and test the representation of aerosol, cloud, and cloud–aerosol interaction processes is essential to increase confidence and reduce uncertainty in aerosol radiative forcing estimates. The strategy of founding and testing process models with field experiment data, applying the process models to regional and global models of the aerosol and cloud life cycles, using field experiment data to test the life cycle models, and using the models to estimate ERFari and ERFaci was laid out by Ghan and Schwartz (2007) and was followed during the development of all of the global models described in the next section. Retrievals of aerosol optical depth (AOD), spectral dependence of AOD, and single-scatter albedo from the surface at ARM and other surface sites and from satellite, as well as measurements of aerosol mass and CCN concentrations from aircraft have all been used to constrain and evaluate simulations of the present-day aerosol life cycle by global and regional aerosol models (Ghan et al. 2001a,b; Easter et al. 2004; X. Liu et al. 2005, 2007a; Wang et al. 2009). In parallel, simulated cloud–aerosol relationships have been evaluated using satellite retrievals of aerosol and clouds (Quaas et al. 2009; Wang et al. 2012).

Estimating anthropogenic aerosol impacts on climate also requires quantifying aerosol and aerosol precursor gas emissions and their life cycle and radiative and cloud impacts at preindustrial times. Evaluating aerosol simulations for preindustrial times is most directly done by comparing historical simulations with deposition rates determined from ice core measurements (which has not been addressed by the ARM Program), or by comparing simulations with measurements in remote regions (Penner et al. 2012), which the ARM Program can provide.

3. Process model development

To address the challenge of representing aerosol effects on climate in models adequately, it has been necessary to develop models of a variety processes involved in the link between emissions and climate impacts. As an atmospheric process science program, ARM has played a vital role developing models of several such processes.

It is straightforward to determine the optical properties from physical property measurements for externally mixed aerosol composed of separate particle types. However, since most aerosol particles are known to be composed of internal mixtures, more difficulties arise. To determine the optical properties of internal mixtures of multiple hydrated aerosol components in aerosol modes with variable size distribution, Ghan et al. (2001a) and Ghan and Zaveri (2007) assumed particles are spheres and used the volume mean hygroscopicity and the kappa Kohler theory of Petters and Kreidenweis (2007) to diagnose the impact of water uptake on the wet mode radius of each mode. Parameterized Mie calculations then provide aerosol extinction, absorption, and the asymmetry parameter in terms of the wet mode radius and the volume mean wet refractive index of the aerosol components.

Optical properties and radiative impacts also are modified when mixing aerosol and cloud particles. To account for enhanced solar absorption by black carbon in cloud droplets, Chuang et al. (2002) used the effective medium approximation with the Maxwell Garnett mixing rule to diagnose the refractive index of black carbon-containing cloud droplets, and then used geometric optics to approximate the droplet single-scatter albedo in terms of droplet radius and the complex refractive index.

After the optical properties of the cloud and aerosol layers are determined, these have to be used in radiative transfer models. The radiative transfer in simulated atmospheres containing aerosol, clouds, and gases including interactions with the surface was made substantially more robust with the development of the Rapid Radiative Transfer Model for GCMs (RRTMG) (Mlawer et al. 1997; Iacono et al. 2003, 2008); the history of this model is described in Mlawer et al. (2016, chapter 15).

To treat aerosol indirect effects through the aerosol influence on droplet nucleation it was necessary to develop models of the aerosol activation process in terms of aerosol properties simulated by climate models. Twomey's (1959) expression for the number activated in terms of a power of supersaturation, while very popular for analysis of measurements, is not easily related to the aerosol quantities that can be predicted in climate models. Ghan et al. (1993, 1995) addressed this need by deriving an expression for the number activated in terms of the parameters of multiple lognormal aerosol size distributions. The parameters of the lognormal distribution can be related to aerosol mass and number concentrations for each mode, which can be simulated by climate models. Chuang et al. (1997, 2002) modified the Ghan et al. (1993) expression to account for different assumptions about the sulfate size distribution and aerosol mixing state over land and ocean. Abdul-Razzak et al. (1998) extended the generality of the Ghan et al. (1993) scheme by accounting for the influence of droplet growth after activation, and Abdul-Razzak and Ghan (2000) generalized the single-mode scheme to the multimode case. The latter scheme is used widely in climate models today. Abdul-Razzak and Ghan (2004) extended their scheme to treat the influence of surfactants on activation, and Abdul-Razzak and Ghan (2005) explored the role of slightly soluble components. Ghan et al. (2011) summarized the aerosol activation schemes developed since the Ghan et al. (1993) publication.

The droplet collision/coalescence process, also known as autoconversion in bulk cloud microphysics parameterizations, also has received considerable attention from the ARM Program because of its dependence on droplet size (and hence droplet number concentration) and because of its influence on cloud liquid water content (and hence cloud albedo). The ARM Program has developed two very different treatments of this process. Khairoutdinov and Kogan (2000) developed an empirical scheme using results from large-eddy simulations with size-resolved microphysics. The result, a simple expression proportional to droplet number concentration to the power -1.79, is used widely in cloud-resolving as well as general circulation models (GCMs).

A second treatment of droplet collision/coalescence was developed and described in a series of manuscripts by Y. Liu. Liu et al. (2004) began with the widely used Kessler (1969) autoconversion scheme, which is a discontinuous function of mean droplet radius, and used kinetic potential theory to derive an analytic expression for the critical droplet radius at the threshold between autoconversion and no autoconversion. Y. Liu et al. (2005, 2006a) related the scheme to earlier continuous ad hoc schemes, and Liu et al. (2006c) related the degree of continuity in the expression to the dispersion in the droplet size distribution.

In parallel, Y. Liu also developed representations of the droplet dispersion. Liu and Daum (2000) used in situ measurements to show that droplet dispersion is positively correlated with droplet number concentration, and Liu and Daum (2002) suggested that the correlation could diminish the aerosol cloud albedo effect; this was later quantified by Chen and Penner (2005). Liu et al. (2006b) provided theoretical support for the relationship in terms of activation of an aerosol size distribution, but employed Twomey's (1959) power law rather than lognormal size distributions.

As noted above, droplet nucleation and droplet collision/coalescence are two of numerous processes controlling cloud microphysics properties. The ARM Program has supported the development of packages of cloud microphysics that are well suited to quantifying aerosol indirect effects. Ghan et al. (1997a,b) introduced the droplet number balance to account for droplet collection, turbulent transport, and evaporation as well as activation and collision/coalescence. This double-moment treatment of cloud droplets provided a physical basis for determining droplet number and the aerosol impact on the albedo of stratiform clouds. Morrison et al. (2005) applied the double-moment approach to cloud ice, rain, and snow as well as droplets, thus permitting treatment of aerosol effects on ice clouds and the potential for a more realistic treatment of aerosol resuspension as raindrops evaporate below clouds. Liu et al. (2007b) introduced parameterizations of ice nucleation mechanisms suitable for global models.

4. Development of a new generation of climate models

When the ARM Program started in 1990, the only climate models that simulated the aerosol life cycle were those designed first to study the climate impacts of aerosols produced by urban fires from a thermonuclear war. The GRANTOUR model (Walton et al. 1988) was a Lagrangian model consisting of 50 000 aerosol-containing air parcels coupled to meteorology simulated first by the Oregon State University two-level general circulation model (Ghan et al. 1988) and subsequently by the Community Climate Model (CCM1; Penner et al. 1991). Coupling was only through the radiative heating by the aerosol and through transport and removal of the aerosol by winds and precipitation, respectively. Chuang et al. (1997) first applied GRANTOUR to tropospheric sulfate aerosol for present-day and preindustrial emissions to estimate direct and indirect effects of anthropogenic sulfate. The preindustrial sulfate aerosol size distribution was prescribed, and two different treatments of anthropogenic sulfate were considered: (i) an internally mixed treatment, in which anthropogenic sulfate mass condenses on the preindustrial aerosol; and (ii) an externally mixed treatment, in which the anthropogenic sulfate aerosol has a different prescribed size distribution. The influence of water uptake on scattering by sulfate particles was introduced following the Köhler and Mie theory. Droplet number was diagnosed from an elaboration of the Ghan et al. (1993) scheme. Penner et al. (1998) then applied GRANTOUR to carbonaceous as well as sulfate aerosol, and Chuang et al. (2002) extended it to simulate all climatically important tropospheric aerosol (sulfate, black carbon, organic carbon, mineral dust, and sea salt), with a treatment of solar absorption by black carbon in cloud droplets. Chuang et al. (2002) used surface measurements to evaluate the aerosol concentrations simulated by GRANTOUR.

A second global aerosol model developed with ARM funding was the Model for Integrated Research on Atmospheric Global Exchanges (MIRAGE) developed at the Pacific Northwest National Laboratory (PNNL). MIRAGE (Easter et al. 2004) represented the aerosol-size distribution as the sum of four lognormal distributions, with the width of each distribution prescribed but the number and mass concentrations predicted from expressions for emissions, oxidation, new particle formation, condensation, coagulation, sedimentation, transport by resolved winds, turbulence and convection, aqueous-phase oxidation, and wet and dry removal. Predicting both number and mass for each mode is important for distinguishing between the effects of processes that primarily affect aerosol number (nucleation, coagulation) from those that primarily affect aerosol mass (condensation, aqueous production). All aerosol components (sulfate, organic carbon, black carbon, mineral dust, and sea salt) are assumed to be internally mixed within each mode, and concentrations of the cloud-borne as well as interstitial aerosol were predicted separately to ensure consistency of nucleation scavenging and cloud processing of the aerosol. The aerosol in MIRAGE interacted in each time step with the meteorology in CCM1 through absorption and scattering of solar radiation, droplet nucleation (which influenced cloud liquid water content as well as cloud albedo), aqueous chemistry, and removal by nucleation scavenging and impaction scavenging. Droplet nucleation was parameterized by the Abdul-Razzak and Ghan (2000) physically based scheme, and the optical properties of the internally mixed hydrated aerosol in each mode were parameterized in terms of the wet mode radius and volume mean wet refractive index (Ghan and Zaveri 2007). Droplet number was determined from the full droplet number balance (Ghan et al. 1997a,b). Easter et al. (2004) used surface and airborne measurements to evaluate the aerosol simulated by MIRAGE. Ghan et al. (2001a) used ARM and other surface-based and satellite retrievals to evaluate the simulated AOD and RFari, while Ghan et al. (2001b) evaluated the CCN concentration and cloud properties using aircraft measurements and satellite retrievals.

A third ARM-sponsored model is the Integrated Massively Parallel Atmospheric Chemical Transport (IMPACT) model (Liu and Penner 2002; X. Liu et al. 2005; Wang et al. 2009) developed jointly by Lawrence Livermore National Laboratory (LLNL) and the University of Michigan. IMPACT can be run as both an offline modal aerosol model driven by meteorology from the Goddard Earth Observing System (GEOS1.3; X. Liu et al. 2005) or inline together with the Community Atmosphere Model, versions 3 and 5 (CAM3 or

CAM5; Wang et al. 2009; Zhou and Penner 2014). Like MIRAGE, it predicts total number and mass concentrations of all important aerosol species in multiple aerosol modes (Herzog et al. 2004), but also includes a pure sulfate aerosol component, causing large differences in predicted sulfate number. It also does not explicitly predict the cloud-borne aerosol. The offline mode of calculation neglects the feedback on clouds and therefore does not include the cloud lifetime effect of the aerosol, but it simplifies estimates of RFaci because natural variability is not introduced (Wang and Penner 2009). Feng and Penner (2007) added nitrate and ammonium aerosols to the basic aerosol representation used by Chuang et al. (2002) in GRANTOUR. X. Liu et al. (2005) used surface measurements to evaluate the aerosol mass concentrations simulated by IMPACT, and Wang et al. (2009) used surface and aircraft observations to evaluate the simulated aerosol number concentration and size distribution in the coupled CAM3/IMPACT model.

Much of the aerosol and cloud physics developed for GRANTOUR, MIRAGE, and IMPACT was incorporated in the CAM5 (Liu et al. 2012). Two different modal configurations were introduced: a "benchmark" version with seven modes, and a simple version with only three modes. This model adopts an internally mixed representation of the aerosol. Total number concentration and mass concentrations of all important aerosol components are predicted for each mode, and interstitial as well as cloud-borne aerosol are predicted separately (though cloud-borne aerosol is not transported by the large-scale winds). Water uptake is based on Köhler theory using the volume-mean hygroscopicity, optical properties are expressed in terms of the wet mode radius and volume mean refractive index (Ghan and Zaveri 2007), and droplet nucleation is parameterized in terms of subgrid updraft velocity and the size distribution and volume-mean hygroscopicity of all aerosol modes (Abdul-Razzak and Ghan 2000). In addition, treatments of droplet collision/coalescence (Khairoutdinov and Kogan 2000) and homogeneous and heterogeneous ice nucleation (Liu and Penner 2005) were introduced as part of a double-moment cloud microphysics parameterization (Morrison and Gettelman 2008; Gettelman et al. 2008) based on the Morrison et al. (2005) double-moment cloud microphysics scheme developed with ARM funding. Liu et al. (2007b) introduced a treatment of ice supersaturation and vapor deposition on ice in CAM, and compared simulations with ARM retrievals of ice water content and in situ measurements of ice supersaturation. All of these capabilities are needed to quantify aerosol indirect effects through their role as droplet and crystal nuclei. Other ARM contributions to CAM5 include the RRTMG solar and infrared radiative transfer scheme (Iacono et al. 2003, 2008) and a shallow cumulus scheme (McCaa and Bretherton 2004; Bretherton et al. 2004; Park and Bretherton 2009; Bretherton and Park 2009). Liu et al. (2012) used ARM and other surface observations and satellite retrievals to evaluate the aerosol mass concentrations and AOD simulated by CAM5. Estimates of ERFari and ERFaci by CAM5 are described by Ghan et al. (2012). CAM5 was used as part of the Community Earth System Model (CESM) in simulations of anthropogenic climate change for the IPCC Fifth Assessment Report, and as a community model is widely used for studies of aerosol effects on climate.

5. Estimates of aerosol radiative forcing

Aerosol radiative forcing estimates have come a long way since the first back-of-the-envelope estimates by Penner et al. (1994), both in terms of aerosol speciation and processes represented. Estimates of the global mean ERF have converged slowly from a wide range between -0.4 and $-3\,\mathrm{W\,m^{-2}}$ in the early estimates to between -0.5 and $-2.4\,\mathrm{W\,m^{-2}}$ (Lohmann et al. 2010) as additional forcing mechanisms are introduced and estimates are constrained more effectively with data. In this section, we summarize what has been learned from aerosol radiative forcing studies funded by the ARM Program.

Penner et al. (1998) estimated RFari of carbonaceous and sulfate aerosol using GRANTOUR driven by meteorology from CCM1. Carbonaceous aerosol was assumed to be an internal mixture of black carbon and organic carbon, with refractive indices determined from volume mixing. Internal mixing of sulfate and carbonaceous aerosol was neglected. The radiative forcing by carbonaceous aerosol was found to be smaller than $0.1\,\mathrm{W\,m^{-2}}$, with compensation between $0.2\,\mathrm{W\,m^{-2}}$ radiative warming and $-0.2\,\mathrm{W\,m^{-2}}$ cooling by carbonaceous aerosol from fossil fuel and biomass burning aerosol, respectively. Consequently, total RFari was dominated by radiative cooling from anthropogenic sulfate aerosol, estimated to be -0.5 to $-0.8\,\mathrm{W\,m^{-2}}$.

Boucher et al. (1998) compared sulfate aerosol RFari estimates by different radiative transfer models and found that the estimates of forcing normalized by sulfate column burden agreed to within about 10% for most conditions. However, Boucher et al. did not consider the influence of clouds on the radiative forcing. More recent work (Stier et al. 2013) using prescribed aerosols with clouds simulated by climate models found much larger diversity for absorbing aerosol arising from differences in the simulated cloud distributions.

Chuang et al. (1997) produced the first estimates of RFaci of anthropogenic sulfate with a physically based aerosol activation scheme, using the GRANTOUR model. Estimates ranged from -0.4 to $-1.6\,\mathrm{W\,m^{-2}}$, depending on assumptions about how sulfate is distributed on preindustrial aerosol. Chuang et al. (2002) extended the model to include indirect forcing by carbonaceous aerosol, and simulated the life cycles of natural aerosol (sulfate, organic matter, sea salt, mineral dust) as well as anthropogenic aerosol. The total RFaci was estimated to be $-1.85\,\mathrm{W\,m^{-2}}$, with most of that from biomass burning. Significant dependence of anthropogenic forcing on emissions of natural aerosol was found because of the nonlinear dependence of droplet nucleation on aerosol concentration. Much smaller RFaci are calculated for nitrate and ammonium aerosols (i.e., $-0.09\,\mathrm{W\,m^{-2}}$) (Xu and Penner 2012).

The radiative influence of the correlation between droplet dispersion and mean droplet number was explored by Rotstayn and Liu (2003, 2009). When the Liu and Daum (2000) empirical relationship between dispersion and droplet number was applied to the Commonwealth Scientific and Industrial Research Organisation (CSIRO) GCM, the estimated RFaci was reduced by 12%–35%. When a more general empirical relationship between dispersion and the ratio of liquid water content/droplet number (Liu et al. 2008) was used, the cloud albedo effect was reduced by 42%.

Uncertainty in RFaci was systematically explored by Chen and Penner (2005) using prescribed aerosol, analyzed relative humidity, parameterizations of cloud fraction and cloud liquid water path, and an offline radiative transfer model.

From a reference estimate of $-1.3\,\mathrm{W\,m^{-2}}$, they found the following:

- 20% uncertainty in RFaci due to anthropogenic emissions, preindustrial aerosol, and aerosol activation parameterization.
- 30%–60% uncertainty due to aerosol mass concentration.
- 50% uncertainty due to cloud fraction.
- 15%–40% uncertainty due to droplet dispersion.
- 5% uncertainty due to cloud liquid water path, updraft velocity, and aerosol mode radius and standard deviation.

The dependence of RFaci on the treatment of new particle formation was addressed with the IMPACT model by Wang and Penner (2009). Interestingly, when an empirical treatment of new particle formation replaces the binary homogeneous nucleation treatment in the boundary layer, RFaci changes little, from -1.55 to $-1.49\,\mathrm{W\,m^{-2}}$, when primary emissions of sulfur are neglected, but changes much more, from -2.03 to $-1.65\,\mathrm{W\,m^{-2}}$, when primary emissions of sulfate are included. As might be expected, both the total RFaci from sulfur and the enhancement when primary sulfur emissions are included, are weaker when the empirical nucleation scheme is added, because the empirical scheme produces more natural CCN in the boundary layer.

One of the first estimates of indirect forcing that included the cloud lifetime effect was by MIRAGE (Ghan et al. 2001c). The ERFaci by anthropogenic sulfate was estimated to be $-1.7\,\mathrm{W\,m^{-2}}$, but a much larger estimate, $-3.2\,\mathrm{W\,m^{-2}}$, was found when a treatment of autoconversion that is a discontinuous function of droplet number was used. ERFaci also was about 30% larger when the aerosol number was diagnosed from the simulated mass and a prescribed radius for each mode, rather than simulated independently. This sensitivity is to be expected as much of the sulfate condensation should go into making particles larger rather than making more particles. Such large negative estimates exceed the radiative warming by increasing greenhouse gases and hence are inconsistent with the observed warming of the earth over the last 150 years, but indicate the magnitude of uncertainty in bottom-up estimates of the radiative forcing.

The strong dependence on the parameterization of autoconversion was further explored in an ARM-funded AeroCom intercomparison of indirect effects estimates by different climate models (Penner et al. 2006). Even when the distribution of aerosol concentration was prescribed, the addition of a common autoconversion scheme that depends on droplet number yielded a wide range of responses of cloud lifetime effect by different climate models, with ERFaci increasing by as little as 22% to as much as a factor of 3 compared with estimates without the cloud lifetime effect. The particular choice of autoconversion scheme also yielded large differences, with the total indirect effect differing by just 10% when the same autoconversion scheme is used in each model but differing by a factor of 3 when different autoconversion schemes are used in each model. As might be expected, the one model that used an autoconversion scheme that is a continuous function of droplet number produced the smallest estimates of ERFaci.

Most global studies of aerosol effects on clouds have focused on stratiform clouds and have neglected aerosol effects on the microphysics of cumulus clouds. Menon and Rotstayn (2006) applied an empirical relationship between aerosol number and droplet number to cumulus clouds in two different global models, and added a simple treatment of the dependence of precipitation formation on droplet number. They found very different responses of cloud liquid water path and radiative forcing to similar treatments of aerosol effects on cumulus clouds, with the CSIRO climate model producing a large liquid water path and radiative forcing response and the Goddard Institute for Space Studies (GISS) model producing a negligible response. This suggests considerable dependence of aerosol–cumulus interactions on details of the cumulus cloud parameterizations.

Semidirect aerosol effects, in which solar absorption by anthropogenic aerosol influences the energy balance through changes in clouds, were estimated by Chuang et al. (2002) and Penner et al. (2003) using GRANTOUR. Chuang et al. (2002) estimated the semidirect effect of cloud-borne black carbon on the planetary energy balance to be $0.07\,W\,m^{-2}$. Penner et al. (2003) found similar values for the solar semidirect effect, but found larger radiative cooling from the impact of the cloud changes on the longwave energy balance as cloud dissipation enhanced longwave emission to space.

In the last decade the ARM Program has supported efforts to quantify anthropogenic effects on ice clouds. Liu et al. (2009) coupled the IMPACT aerosol model to CAM3 using the Liu and Penner (2005) ice nucleation scheme and quantified radiative impacts of anthropogenic aerosol via homogeneous and heterogeneous ice nucleation. Anthropogenic sulfate was estimated to produce $0.5\,W\,m^{-2}$ of radiative warming through homogeneous nucleation, with positive contributions from both shortwave and longwave radiation. Anthropogenic black carbon, which Liu and Penner assumed to be capable of serving as ice nuclei through the immersion nucleation mechanism, increased crystal number and cloud ice, making shortwave cloud forcing $1.5\,W\,m^{-2}$ more negative and longwave cloud forcing $1.5\,W\,m^{-2}$ more positive, for a small net forcing. The anthropogenic aerosol effect on longwave cloud forcing was estimated to be $0.5\,W\,m^{-2}$ in CAM5 (Ghan et al. 2012) with $\sim 0.3\,W\,m^{-2}$ resulting from homogenous ice nucleation of anthropogenic sulfate and the remainder from immersion freezing of droplets detrained from deep convection. This estimate agrees well with a more recent estimate of $0.27\,W\,m^{-2}$ for ERFaci through ice clouds (Gettelman et al. 2012). Yun and Penner (2012) and Yun et al. (2013) examined forcing in mixed-phase clouds, finding values between 0.16 and $0.93\,W\,m^{-2}$. Total forcing using the IMPACT aerosol model was $-2.36\,W\,m^{-2}$.

Climate models are designed for future projections of climate change, and those projections depend on future emission and radiative forcing. Menon et al. (2008) explored changes in aerosol radiative forcing between years 1980, 1995, and 2030 using the GISS ModelE and projected emissions for the IPCC A1B scenario. The aerosol radiative cooling was projected to be greatest for the year 2030 with the projected larger sulfur emissions from China.

6. Cooperating with other programs

While the ARM Program played an instrumental role in collecting the data, developing and evaluating the process models and cloud and aerosol life cycle models, and using the models to estimate aerosol radiative forcing, other federal programs also played essential roles.

The DOE Atmospheric Science Program supported the development of empirical parameterizations of aerosol nucleation (Kuang et al. 2008), of the Y. Liu autoconversion parameterization (Liu and Daum 2004; Liu et al. 2006a), and of the MOSAIC aerosol thermodynamics model (Zaveri et al. 2005a,b; 2008) that have been or are being applied to global aerosol models. The DOE Science Discovery through Advanced Chemistry (SciDAC) program funded the development of CAM5, which incorporated many of the advances described in this chapter.

The National Aeronautics and Space Administration (NASA) Interdisciplinary Science Program supported the development of MIRAGE. The NASA Global Aerosol Climatology Project and Radiation Science Program partly funded development and application of GRANTOUR. The NASA Atmospheric Chemistry Modeling and Analysis Program (ACMAP) partly funded the AeroCom intercomparison and the development and application of the IMPACT model.

The National Science Foundation provided partial funding to couple the IMPACT model and the CAM model as well as to incorporate new treatments of ice and mixed-phase aerosol/cloud microphysics.

7. Summary and future progress

The ARM Program has played a vital role in the emergence of global modeling of the aerosol life cycle and aerosol effects on the earth's energy balance. Key insights into aerosol radiative forcing mechanisms and ways to represent them efficiently in climate models have been supported by ARM. The importance of representing all aerosol components, natural as well as anthropogenic, was established by ARM-funded scientists. This understanding led to the development of the treatments of the aerosol life cycle and cloud–aerosol interactions in the CESM1 used for simulations that contributed to the IPCC Fifth Assessment Report (Boucher et al. 2013).

However, the uncertainty in estimates of aerosol radiative forcing still drives an almost threefold uncertainty in total radiative forcing of climate change. Although the relative uncertainty in future forcing will decline as carbon dioxide accumulates in the atmosphere and anthropogenic aerosol emissions are reduced, a threefold uncertainty in radiative forcing between present-day and preindustrial conditions renders the observed warming of little use in constraining estimates of climate sensitivity.

How then, to reduce that uncertainty? Ghan and Schwartz (2007) outline a strategy that involves greater complexity in process models constrained with more complete, accurate, and sensitive measurements of the aerosol. Missing aerosol components, such as nitrate, need to be added, and crudely represented components, such as secondary organic aerosol, need to be improved. A more general representation of thermodynamic equilibrium that is suitable for all important aerosol components is needed. The influence of aging by condensation on the aerosol mixing state needs to be represented more realistically, as well as emissions of primary black carbon and organic carbon and secondary organic carbon. Better understanding of new particle formation is needed as well. The treatment of wet removal of aerosol needs to be constrained better with measurements. The microphysical effects on cumulus clouds need to be represented, and better understanding and representation of heterogeneous ice nucleation on dust and black carbon are also needed. ARM can play a central role in each of these process studies.

Finally, while most of the emphasis of aerosol research during the last 20 years has focused on the impact of anthropogenic aerosol on the planetary energy balance, studies of the influence on precipitation and snow albedo are just beginning.

Acknowledgments. JEP is grateful for funding from the DOE ARM Program and through Grant DOE DE-SC0008486. SG was supported by the DOE ARM Program for many years and more recently by the DOE Atmospheric Systems Research program. The Pacific Northwest National Laboratory is operated for DOE by Battelle Memorial Institute under Contract DE-AC06-76RLO 1830.

REFERENCES

Abdul-Razzak, H., and S. J. Ghan, 2000: A parameterization of aerosol activation. Part 2: Multiple aerosol types. *J. Geophys. Res.*, **105**, 6837–6844, doi:10.1029/1999JD901161.

——, and ——, 2004: Parameterization of the influence of organic surfactants on aerosol activation. *J. Geophys. Res.*, **109**, D03205, doi:10.1029/2003JD004043.

——, and ——, 2005: Influence of slightly soluble organics on aerosol activation. *J. Geophys. Res.*, **110**, D06206, doi:10.1029/2004JD005324.

——, ——, and C. Rivera-Carpio, 1998: A parameterization of aerosol activation. Part I: Single aerosol type. *J. Geophys. Res.*, **103**, 6123–6132, doi:10.1029/97JD03735.

Ackerman, A. S., M. P. Kirkpatrick, D. E. Stevens, and O. B. Toon, 2004: The impact of humidity above stratiform clouds on indirect aerosol climate forcing. *Nature*, **432**, 1014–1017, doi:10.1038/nature03174.

Barahona, D., and A. Nenes, 2007: Parameterization of cloud droplet formation in large-scale models: Including effects of entrainment. *J. Geophys. Res.*, **112**, D16206, doi:10.1029/2007JD008473.

Boucher, O., and Coauthors, 1998: Intercomparison of models representing direct shortwave radiative forcing by sulfate aerosols. *J. Geophys. Res.*, **103**, 16 979–16 998, doi:10.1029/98JD00997.

——, and Coauthors, 2013: Clouds and aerosols. *Climate Change 2013: The Physical Science Basis*, T. F. Stocker et al., Eds., Cambridge University Press, 571–658.

Bretherton, C. S., and S. Park, 2009: A new moist turbulence parameterization in the Community Atmosphere Model. *J. Climate*, **22**, 3422–3448, doi:10.1175/2008JCLI2556.1.

——, J. R. McCaa, and H. Grenier, 2004: A new parameterization for shallow cumulus convection and its application to marine subtropical cloud-topped boundary layers. Part I: Description and 1D results. *Mon. Wea. Rev.*, **132**, 864–882, doi:10.1175/1520-0493(2004)132<0864:ANPFSC>2.0.CO;2.

Chen, Y., and J. E. Penner, 2005: Uncertainty analysis of the first indirect aerosol effect. *Atmos. Chem. Phys.*, **5**, 2935–2948, doi:10.5194/acp-5-2935-2005.

Chuang, C. C., J. E. Penner, K. E. Taylor, A. S. Grossman, and J. J. Walton, 1997: An assessment of the radiative effects of anthropogenic sulfate. *J. Geophys. Res.*, **102**, 3761–3778, doi:10.1029/96JD03087.

——, ——, J. M. Prospero, K. E. Grant, G. H. Rau, and K. Kawamoto, 2002: Cloud susceptibility and the first aerosol indirect forcing: Sensitivity to black carbon and aerosol concentrations. *J. Geophys. Res.*, **107**, 4564, doi:10.1029/2000JD000215.

DeMott, P. J., and Coauthors, 2010: Predicting global atmospheric ice nuclei distributions and their impacts on climate. *Proc. Natl. Acad. Sci. USA*, **107**, 11 217–11 222, doi:10.1073/pnas.0910818107.

Easter, R. C., and Coauthors, 2004: MIRAGE: Model description and evaluation of aerosols and trace gases. *J. Geophys. Res.*, **109**, D20210, doi:10.1029/2004JD004571.

Feng, Y., and J. E. Penner, 2007: Global modeling of nitrate and ammonium: Interaction of aerosols and tropospheric chemistry. *J. Geophys. Res.*, **112**, D01304, doi:10.1029/2005JD006404.

Gettelman, A., H. Morrison, and S. J. Ghan, 2008: A new two-moment bulk stratiform cloud microphysics scheme in the NCAR Community Atmosphere Model (CAM3). Part II: Single-column and global results. *J. Climate*, **21**, 3660–3679, doi:10.1175/2008JCLI2116.1.

——, X. Liu, D. Barahona, U. Lohmann, and C. Chen, 2012: Climate impacts of ice nucleation. *J. Geophys. Res.*, **117**, D20201, doi:10.1029/2012JD017950.

Ghan, S. J., 2013: Technical note: Estimating aerosol effects on cloud radiative forcing. *Atmos. Chem. Phys.*, **13**, 9971–9974, doi:10.5194/acp-13-9971-2013.

——, and R. C. Easter, 1992: Computationally efficient approximations to stratiform cloud microphysics parameterization. *Mon. Wea. Rev.*, **120**, 1572–1582, doi:10.1175/1520-0493(1992)120<1572:CEATSC>2.0.CO;2.

——, and S. E. Schwartz, 2007: Aerosol properties and processes: A path from field and laboratory measurements to global climate models. *Bull. Amer. Meteor. Soc.*, **88**, 1059–1083, doi:10.1175/BAMS-88-7-1059.

——, and R. A. Zaveri, 2007: Parameterization of optical properties for hydrated internally mixed aerosol. *J. Geophys. Res.*, **112**, D10201, doi:10.1029/2006JD007927.

——, M. C. MacCracken, and J. J. Walton, 1988: The climatic response to large atmospheric smoke injections: Sensitivity studies with a tropospheric general circulation model. *J. Geophys. Res.*, **93**, 8315–8337, doi:10.1029/JD093iD07p08315.

——, C. C. Chuang, and J. E. Penner, 1993: A parameterization of cloud droplet nucleation. Part I: Single aerosol type. *Atmos. Res.*, **30**, 198–221, doi:10.1016/0169-8095(93)90024-I.

——, ——, R. C. Easter, and J. E. Penner, 1995: A parameterization of cloud droplet nucleation. Part II: Multiple aerosol types. *Atmos. Res.*, **36**, 39–54, doi:10.1016/0169-8095(94)00005-X.

——, L. R. Leung, and Q. Hu, 1997a: Application of cloud microphysics to NCAR CCM2. *J. Geophys. Res.*, **102**, 16 507–16 528, doi:10.1029/97JD00703.

——, ——, R. C. Easter, and H. Abdul-Razzak, 1997b: Prediction of droplet number in a general circulation model. *J. Geophys. Res.*, **102**, 21 777–21 794, doi:10.1029/97JD01810.

——, G. Guzman, and H. Abdul-Razzak, 1998: Competition between sea salt and sulfate particles as cloud condensation nuclei. *J. Atmos. Sci.*, **55**, 3340–3347, doi:10.1175/1520-0469(1998)055<3340:CBSSAS>2.0.CO;2.

——, N. Laulainen, R. Easter, R. Wagener, S. Nemesure, E. Chapman, Y. Zhang, and R. Leung, 2001a: Evaluation of aerosol direct radiative forcing in MIRAGE. *J. Geophys. Res.*, **106**, 5295–5316, doi:10.1029/2000JD900502.

——, R. C. Easter, J. Hudson, and F.-M. Breon, 2001b: Evaluation of aerosol indirect radiative forcing in MIRAGE. *J. Geophys. Res.*, **106**, 5317–5334, doi:10.1029/2000JD900501.

——, and Coauthors, 2001c: A physically-based estimate of radiative forcing by anthropogenic sulfate aerosol. *J. Geophys. Res.*, **106**, 5279–5294, doi:10.1029/2000JD900503.

——, and Coauthors, 2011: Droplet nucleation: Physically-based parameterizations and comparative evaluation. *J. Adv. Model. Earth Syst.*, **3**, M10001, doi:10.1029/2011MS000074.

——, X. Liu, R. C. Easter, R. Zaveri, P. J. Rasch, J.-H. Yoon, and B. Eaton, 2012: Toward a minimal representation of aerosols in climate models: Comparative decomposition of aerosol direct, semidirect, and indirect radiative forcing. *J. Climate*, **25**, 6461–6476, doi:10.1175/JCLI-D-11-00650.1.

Herzog, M., D. K. Weisenstein, and J. E. Penner, 2004: A dynamic aerosol module for global chemical transport models: Model description and impact of nonsulfate aerosol particles. *J. Geophys. Res.*, **109**, D18202, doi:10.1029/2003JD004405.

Iacono, M. J., J. S. Delamere, E. J. Mlawer, and S. A. Clough, 2003: Evaluation of upper tropospheric water vapor in the NCAR Community Climate Model (CCM3) using modeled and observed HIRS radiances. *J. Geophys. Res.*, **108**, 4037, doi:10.1029/2002JD002539.

——, ——, ——, M. W. Shephard, S. A. Clough, and W. D. Collins, 2008: Radiative forcing by long-lived greenhouse gases: Calculations with the AER radiative transfer models. *J. Geophys. Res.*, **113**, D13103, doi:10.1029/2008JD009944.

Kessler, E., 1969: *On the Distribution and Continuity of Water Substance in Atmospheric Circulation. Meteor. Monogr.*, No. 10, Amer. Meteor. Soc., 84 pp.

Khairoutdinov, M. F., and Y. Kogan, 2000: A new cloud physics parameterization in a large-eddy simulation model of marine stratocumulus. *Mon. Wea. Rev.*, **128**, 229–243, doi:10.1175/1520-0493(2000)128<0229:ANCPPI>2.0.CO;2.

Kuang, C., P. H. McMurry, A. V. McCormick, and F. L. Eisele, 2008: Dependence of nucleation rates on sulfuric acid vapor concentration in diverse atmospheric locations. *J. Geophys. Res.*, **113**, D10209, doi:10.1029/2007JD009253.

Lee, S. S., and J. E. Penner, 2010: Comparison of a global-climate model to a cloud-system resolving model for the long-term response of thin stratocumulus clouds to preindustrial and present-day aerosol conditions. *Atmos. Chem. Phys.*, **10**, 6371–6389, doi:10.5194/acp-10-6371-2010.

Lin, G., J. E. Penner, S. Sillman, D. Taraborrelli, and J. Lelieveld, 2012: Global modeling of SOA formation from dicarbonyls, epoxides, organic nitrates and peroxides. *Atmos. Chem. Phys.*, **12**, 4743–4774, doi:10.5194/acp-12-4743-2012.

——, S. Silman, J. E. Penner, and A. Ito, 2014: Global modeling of SOA: The use of different mechanisms for aqueous-phase formation. *Atmos. Chem. Phys.*, **14**, 5451–5475, doi:10.5194/acp-14-5451-2014.

Liu, X., and J. E. Penner, 2002: Effect of Mount Pinatubo H_2SO_4/H_2O aerosol on ice nucleation in the upper troposphere using a global chemistry and transport model (IMPACT). *J. Geophys. Res.*, **107**, 4141, doi:10.1029/2001JD000455.

——, and ——, 2005: Ice nucleation parameterization for global models. *Meteor. Z.*, **14**, 499–514, doi:10.1127/0941-2948/2005/0059.

——, ——, and M. Herzog, 2005: Global modeling of aerosol dynamics: Model description, evaluation, and interactions between sulfate and nonsulfate aerosols. *J. Geophys. Res.*, **110**, D18206, doi:10.1029/2004JD005674.

——, ——, B. Das, D. Bergmann, J. M. Rodriguez, S. Strahan, M. Wang, and Y. Feng, 2007a: Uncertainties in global aerosol simulations: Assessment using three meteorological data sets. *J. Geophys. Res.*, **112**, D11212, doi:10.1029/2006JD008216.

——, ——, S. J. Ghan, and M. Wang, 2007b: Inclusion of ice microphysics in the NCAR Community Atmospheric Model, version 3 (CAM3). *J. Climate*, **20**, 4526–4547, doi:10.1175/JCLI4264.1.

——, ——, and M. Wang, 2009: Influence of anthropogenic sulfate and black carbon on upper tropospheric clouds in the NCAR CAM3 model coupled to the IMPACT global aerosol model. *J. Geophys. Res.*, **114**, D03204, doi:10.1029/2008JD010492.

——, and Coauthors, 2012: Toward a minimal representation of aerosols in climate models: Description and evaluation in the Community Atmosphere Model CAM5. *Geosci. Model Dev.*, **5**, 709–739, doi:10.5194/gmd-5-709-2012.

Liu, Y., and P. H. Daum, 2000: Spectral dispersion of cloud droplet size distributions and the parameterization of cloud droplet effective radius. *Geophys. Res. Lett.*, **27**, 1903–1906, doi:10.1029/1999GL011011.

——, and ——, 2002: Indirect warming effect from dispersion forcing. *Nature*, **419**, 580–581, doi:10.1038/419580a.

——, and ——, 2004: Parameterization of the autoconversion process. Part I: Analytical formulation of the Kessler-type parameterizations. *J. Atmos. Sci.*, **61**, 1539–1548, doi:10.1175/1520-0469(2004)061<1539:POTAPI>2.0.CO;2.

——, ——, and R. McGraw, 2004: An analytical expression for predicting the critical radius in the autoconversion parameterization. *Geophys. Res. Lett.*, **31**, L06121, doi:10.1029/2003GL019117.

——, ——, and R. L. McGraw, 2005: Size truncation effect, threshold behavior, and a new type of autoconversion parameterization. *Geophys. Res. Lett.*, **32**, L11811, doi:10.1029/2005GL022636.

——, ——, R. McGraw, and R. Wood, 2006a: Parameterization of the autoconversion process. Part II: Generalization of Sundqvist-type parameterizations. *J. Atmos. Sci.*, **63**, 1103–1109, doi:10.1175/JAS3675.1.

——, ——, and S. Yum, 2006b: Analytical expression for the relative dispersion of the cloud droplet size distribution. *Geophys. Res. Lett.*, **33**, L02810, doi: 10.1029/2005GL024052.

——, ——, R. McGraw, and M. Miller, 2006c: Generalized threshold function accounting for effect of relative dispersion on threshold behavior of autoconversion process. *Geophys. Res. Lett.*, **33**, L11804, doi:10.1029/2005GL025500.

——, ——, H. Guo, and Y. Peng, 2008: Dispersion bias, dispersion effect and aerosol-cloud conundrum. *Environ. Res. Lett.*, **3**, 045021, doi:10.1088/1748-9326/3/4/045021.

Lohmann, U., and Coauthors, 2010: Total aerosol effect: Radiative forcing or radiative flux perturbation? *Atmos. Chem. Phys.*, **10**, 3235–3246, doi:10.5194/acp-10-3235-2010.

McCaa, J. R., and C. S. Bretherton, 2004: A new parameterization for shallow cumulus convection and its application to marine subtropical cloud-topped boundary layers. Part II: Regional simulations of marine boundary layer clouds. *Mon. Wea. Rev.*, **132**, 883–896, doi:10.1175/1520-0493(2004)132<0883:ANPFSC>2.0.CO;2.

Menon, S., and L. Rotstayn, 2006: The radiative influence of aerosol effects on liquid-phase cumulus and stratiform clouds based on sensitivity studies with two climate models. *Climate Dyn.*, **27**, 345–356, doi:10.1007/s00382-006-0139-3.

——, N. Unger, D. Koch, J. Francis, T. Garrett, I. Sednev, D. Shindell, and D. Streets, 2008: Aerosol climate effects and air quality impacts from 1980 to 2030. *Environ. Res. Lett.*, **3**, 024004, doi:10.1088/1748-9326/3/2/024004.

Mlawer, E. J., S. J. Taubman, P. D. Brown, M. J. Iacono, and S. A. Clough, 1997: Radiative transfer for inhomogeneous atmospheres: RRTM, a validated correlated-k model for the longwave. *J. Geophys. Res.*, **102**, 16 663–16 682, doi:10.1029/97JD00237.

——, M. Iacono, R. Pincus, H. Barker, L. Oreopoulos, and D. Mitchell, 2016: Contributions of the ARM Program to radiative transfer modeling for climate and weather applications. *The Atmospheric Radiation Measurement (ARM) Program: The First 20 Years, Meteor. Monogr.*, No. 57, Amer. Meteor. Soc., doi:10.1175/AMSMONOGRAPHS-D-15-0041.1.

Morrison, H., and A. Gettelman, 2008: A new two-moment bulk stratiform cloud microphysics scheme in the Community Atmosphere Model, version 3 (CAM3). Part I: Description and tests. *J. Climate*, **21**, 3642–3659, doi:10.1175/2008JCLI2105.1.

——, J. A. Curry, and V. I. Khvorostyanov, 2005: A new double-moment microphysics scheme for application in cloud and climate models. Part I: Description. *J. Atmos. Sci.*, **62**, 1665–1677, doi:10.1175/JAS3446.1.

Park, S., and C. S. Bretherton, 2009: The University of Washington shallow convection scheme and moist turbulence schemes and their impact on climate simulations with the Community Atmosphere Model. *J. Climate*, **22**, 3449–3469, doi:10.1175/2008JCLI2557.1.

Penner, J. E., C. S. Atherton, J. Dignon, S. J. Ghan, J. J. Walton, and S. Hameed, 1991: Tropospheric nitrogen: A three-dimensional study of sources, distributions, and deposition. *J. Geophys. Res.*, **96**, 959–990, doi:10.1029/90JD02228.

——, and Coauthors, 1994: Quantifying and minimizing uncertainty in climate forcing by anthropogenic aerosols. *Bull. Amer. Meteor. Soc.*, **75**, 375–400, doi:10.1175/1520-0477(1994)075<0375:QAMUOC>2.0.CO;2.

——, C. C. Chuang, and K. E. Grant, 1998: Climate forcing by carbonaceous and sulfate aerosols. *Climate Dyn.*, **14**, 839–851, doi:10.1007/s003820050259.

——, and Coauthors, 2001: Aerosols: Their direct and indirect effects. *Climatic Change 2001: The Scientific Basis*, J. T. Houghton et al., Eds., Cambridge University Press, 289–348.

——, S. Y. Zhang, and C. C. Chuang, 2003: Soot and smoke aerosol may not warm climate. *J. Geophys. Res.*, **108**, 4657, doi:10.1029/2003JD003409.

——, and Coauthors, 2006: Model intercomparison of indirect aerosol effects. *Atmos. Chem. Phys.*, **6**, 3391–3405, doi:10.5194/acp-6-3391-2006.

——, C. Zhou, and L. Xu, 2012: Consistent estimates from satellites and models for the first aerosol indirect forcing. *Geophys. Res. Lett.*, **39**, L13810, doi:10.1029/2012GL051870.

Petters, M. D., and S. M. Kreidenweis, 2007: A single parameter representation of hygroscopic growth and cloud condensation nucleus activity. *Atmos. Chem. Phys.*, **7**, 1961–1971, doi:10.5194/acp-7-1961-2007.

Quaas, J., and Coauthors, 2009: Aerosol indirect effects—General circulation model intercomparison and evaluation with satellite data. *Atmos. Chem. Phys.*, **9**, 8697–8717, doi:10.5194/acp-9-8697-2009.

Rotstayn, L. D., and Y. Liu, 2003: Sensitivity of the first indirect aerosol effect to an increase of cloud droplet spectral dispersion with droplet number concentration. *J. Climate*, **16**, 3476–3481, doi:10.1175/1520-0442(2003)016<3476:SOTFIA>2.0.CO;2.

——, and ——, 2009: Cloud droplet spectral dispersion and the indirect aerosol effect: Comparison of two treatments in a GCM. *Geophys. Res. Lett.*, **36**, L10801, doi:10.1029/2009GL038216.

Song, X., and G. J. Zhang, 2011: Microphysics parameterization for convective clouds in a global climate model: Description and single-column model tests. *J. Geophys. Res.*, **116**, D02201, doi:10.1029/2010JD014833.

Stier, P., and Coauthors, 2013: Host model uncertainties in aerosol radiative forcing estimates: Results from the AeroCom prescribed intercomparison study. *Atmos. Chem. Phys.*, **13**, 3245–3270, doi:10.5194/acp-13-3245-2013.

Twomey, S., 1959: The nuclei of natural cloud formation. Part II: The supersaturation in natural clouds and the variation of droplet concentration. *Pure Appl. Geophys.*, **43**, 243–249, doi:10.1007/BF01993560.

Walton, J. J., M. C. MacCracken, and S. J. Ghan, 1988: A global-scale Lagrangian trace species model of transport, transformation, and removal processes. *J. Geophys. Res.*, **93**, 8339–8354, doi:10.1029/JD093iD07p08339.

Wang, M., and J. E. Penner, 2009: Aerosol indirect forcing in a global model with particle nucleation. *Atmos. Chem. Phys.*, **9**, 239–260, doi:10.5194/acp-9-239-2009.

——, ——, and X. Liu, 2009: Coupled IMPACT aerosol and NCAR CAM3 model: Evaluation of predicted aerosol number and size distribution. *J. Geophys. Res.*, **114**, D06302, doi:10.1029/2008JD010459.

——, and Coauthors, 2012: Constraining cloud lifetime effects of aerosols using A-Train satellite observations. *Geophys. Res. Lett.*, **39**, L15709, doi:10.1029/2012GL052204.

Xu, L., and J. E. Penner, 2012: Global simulations of nitrate and ammonium aerosols and their radiative effects. *Atmos. Chem. Phys.*, **12**, 9479–9504, doi:10.5194/acp-12-9479-2012.

Yun, Y., and J. E. Penner, 2012: Global model comparison of heterogeneous ice nucleation parameterizations in mixed phase clouds. *J. Geophys. Res.*, **117**, D07203, doi:10.1029/2011JD016506.

——, ——, and O. Popovicheva, 2013: The effects of hygroscopicity on ice nucleation of fossil fuel combustion aerosols in mixed-phase clouds. *Atmos. Chem. Phys.*, **13**, 4339–4348, doi:10.5194/acp-13-4339-2013.

Zaveri, R. A., R. C. Easter, and A. S. Wexler, 2005a: A new method for multicomponent activity coefficients of electrolytes in aqueous atmospheric aerosols. *J. Geophys. Res.*, **110**, D02201, doi:10.1029/2004JD004681.

——, ——, and L. K. Peters, 2005b: A computationally efficient Multicomponent Equilibrium Solver for Aerosols (MESA). *J. Geophys. Res.*, **110**, D24203, doi:10.1029/2004JD005618.

——, ——, J. D. Fast, and L. K. Peters, 2008: Model for Simulating Aerosol Interactions and Chemistry (MOSAIC). *J. Geophys. Res.*, **113**, D13204, doi:10.1029/2007JD008782.

Zhou, C., and J. E. Penner, 2014: Aircraft soot indirect effect on large-scale cirrus clouds: Is the indirect forcing by aircraft soot positive or negative? *J. Geophys. Res. Atmos.*, **119**, 11 303–11 320, doi:10.1002/2014JD021914.

Chapter 28

ARM's Impact on Numerical Weather Prediction at ECMWF

MAIKE AHLGRIMM, RICHARD M. FORBES, AND JEAN-JACQUES MORCRETTE

European Centre for Medium-Range Weather Forecasts, Reading, United Kingdom

ROEL A. J. NEGGERS

Institute for Geophysics and Meteorology, University of Cologne, Cologne, Germany

1. Introduction

The European Centre for Medium-Range Weather Forecasts (ECMWF) is one of the leading centers in operational global numerical weather prediction (NWP) and provides forecasts for days to monthly and seasonal time scales across a range of resolutions. Parameterization of subgrid physical processes are a key part of the model [the Integrated Forecast System (IFS)] and, for global application, must be appropriate for all meteorological regimes and regions across a wide range of spatial and temporal scales. The parameterizations of cloud, precipitation, and radiation and their interactions are of particular importance for the effects on atmospheric heating and cooling, the hydrological cycle, and the dynamics of the atmosphere. To make progress and continue to increase the accuracy of model forecasts for NWP and longer time scales (model climate), a wide range of observations is required for evaluation of model systematic errors and for inspiring innovative developments to model parameterizations.

The ARM Program historically has played a key role in all of these areas. There are many examples where the ECMWF model has benefitted from comparison with ARM observations from the Arctic, midlatitudes, and tropics over the past two decades (e.g., Mace et al. 1998; Morcrette 2002b; Comstock and Jakob 2004; Tselioudis and Kollias 2007; Klein et al. 2009; Agustí-Panareda et al. 2010), as well as for exploration of new data assimilation techniques (Lopez et al. 2006). Cheinet et al. (2005) performed an evaluation of the ECMWF model at the ARM Southern Great Plains (SGP) site, which provides an example of how the comprehensive instrumentation, high-quality data, and extensive temporal coverage of ARM site data can characterize deficiencies in the model and give insight into potential model parameterization improvement. Not every aspect of ARM's impact on NWP at ECMWF can be covered in detail here. The focus of this chapter is therefore to relate three primary examples of model innovation, parameterization development, and evaluation at ECMWF that were directly motivated and facilitated by the ARM Program:

1) the development of an innovative boundary layer parameterization,
2) the operational implementation of new accurate and efficient radiation parameterizations, and
3) a process-oriented model evaluation with ARM observations to guide development of cloud and radiation parameterizations.

All three areas have contributed to the ongoing drive for improvement of the ECMWF model in medium-range weather forecasting. Readers interested in ECMWF operational model changes beyond the scope of this paper are referred to the ECMWF web page (http://www.ecmwf.int/en/forecasts/documentation-and-support/changes-ecmwf-model) for a brief summary of model cycles and links to further information.

2. Boundary layer parameterization innovation inspired by ARM observations

a. Background

The first generation of boundary layer schemes in global climate models (GCMs) were designed to

Corresponding author address: Maike Ahlgrimm, European Centre for Medium-Range Weather Forecasts, Shinfield Park, Reading RG2 9AX, United Kingdom.
E-mail: maike.ahlgrimm@ecmwf.int

DOI: 10.1175/AMSMONOGRAPHS-D-15-0032.1

reproduce the first-order impact of unresolved turbulence and convection on the vertical transport of heat, moisture, and momentum and their contribution to fractional cloudiness and condensate. This approach has led to demonstrable improvement in weather and climate predictions (e.g., Tiedtke 1989). The initial success inspired the further sophistication and development of these boundary layer schemes. Typically, schemes were developed and tested for certain cases based on relevant observational data obtained during meteorological field campaigns (e.g., Holland and Rasmusson 1973; Yanai et al. 1973). An often-applied technique in model development has been the time integration of the suite of subgrid parameterizations in a single-column model (SCM) in an offline, "isolated" mode, using prescribed large-scale forcings and boundary conditions (Randall et al. 1996). The absence of interaction with the larger scale simplifies the model analysis, giving insight into the behavior of parameterizations at process level.

The advent of large-eddy simulation (LES; e.g., Sommeria 1976) has enhanced the opportunities for the evaluation and development of boundary layer parameterizations significantly. In pre-LES days, progress in model development was often hampered by lack of key observations of small-scale variability related to turbulence and clouds. The capability of LES to provide four-dimensional fields (in space and time) at high turbulence- and cloud-resolving resolutions has now been demonstrated and implies that it can act as a virtual laboratory, giving access to relevant data for parameterization development at a level that is still unprecedented and unmatched by present-day instrumentation. LES simulations can thus "fill the gap" in the instrument coverage of key parameters.

Together, the SCM and LES can form a powerful set of tools in parameterization development. A continuing effort to combine both techniques has been applied by the Boundary Layer Working Group (BLWG) of the GEWEX Cloud System Studies (GCSS). Their strategy has been to bring together the international boundary layer modeling community by organizing model intercomparison studies for SCM and LES. Observational datasets have played a key role in this process 1) for building "prototype" cases for boundary layer regimes that still provide challenges for boundary layer schemes in GCMs and 2) for confronting SCM and LES results on turbulent transport and clouds. To this purpose, "golden days" were identified at some locations of interest where relevant measurements were available.

Many of the first prototype intercomparison cases focused on marine quasi-steady-state boundary layers. The assumption of a balanced prognostic budget allows application of equilibrium assumptions, which could be powerful tools for the modeler as they might enhance numerical model stability. Also, because the boundary layer does not change fundamentally in time, the tendency has been to design schemes in such a way that different modes or regimes are represented by unique combinations of hard-coded settings. In practice, this has led to the representation of separate convective boundary layer regimes by means of separate schemes; examples are the dry convective boundary layer, the stratocumulus-capped boundary layer, and shallow-cumulus-capped boundary layer. Examples of past GCSS intercomparison studies on steady-state cases are those based on the Barbados Oceanographic and Meteorological Experiment (BOMEX) (Siebesma et al. 2003) and the Atlantic Tradewind Experiment (ATEX; Stevens et al. 2001) field campaigns.

b. The ARM case

Compared to steady-state situations, the case of a boundary layer over land represents a much more difficult situation for parameterization schemes. The reason is that the convective boundary layer is highly transient, experiencing a diurnal cycle of initiation at sunrise, growth or deepening during the day, and decay at sunset when the surface-driven turbulence dies out. In this daily life cycle, the boundary layer can experience a series of transitions between separately defined regimes as mentioned above. The GCSS BLWG case based on ARM observational data for 21 June 1997 at the SGP site was developed to address the question of how both LES and SCM perform for this situation (Brown et al. 2002; Lenderink et al. 2004). Figure 28-1 presents some measurements and LES data to highlight the key aspects of the boundary layer on this day. Since its formulation, this case has become used so frequently as a testing ground for model development that within the community it is often simply referred to as "the ARM case."

One of the main results of the GCSS intercomparison study on the ARM case showed that while LES codes agree reasonably well on the vertical structure and time development of the thermodynamic and cloudy state of the cumulus-capped boundary layer (Brown et al. 2002), the participating SCMs have problems reproducing these LES results (Lenderink et al. 2004). For example, Fig. 28-2 illustrates the significant scatter that exists among SCM codes for the total cloud cover. For some SCM realizations, the oscillations were linked to the occurrence of discrete transitions between regimes, as a result of the use of separate schemes for different boundary layer regimes in the GCM. These artificial transitions can then lead to unrealistic, often numerical, instability.

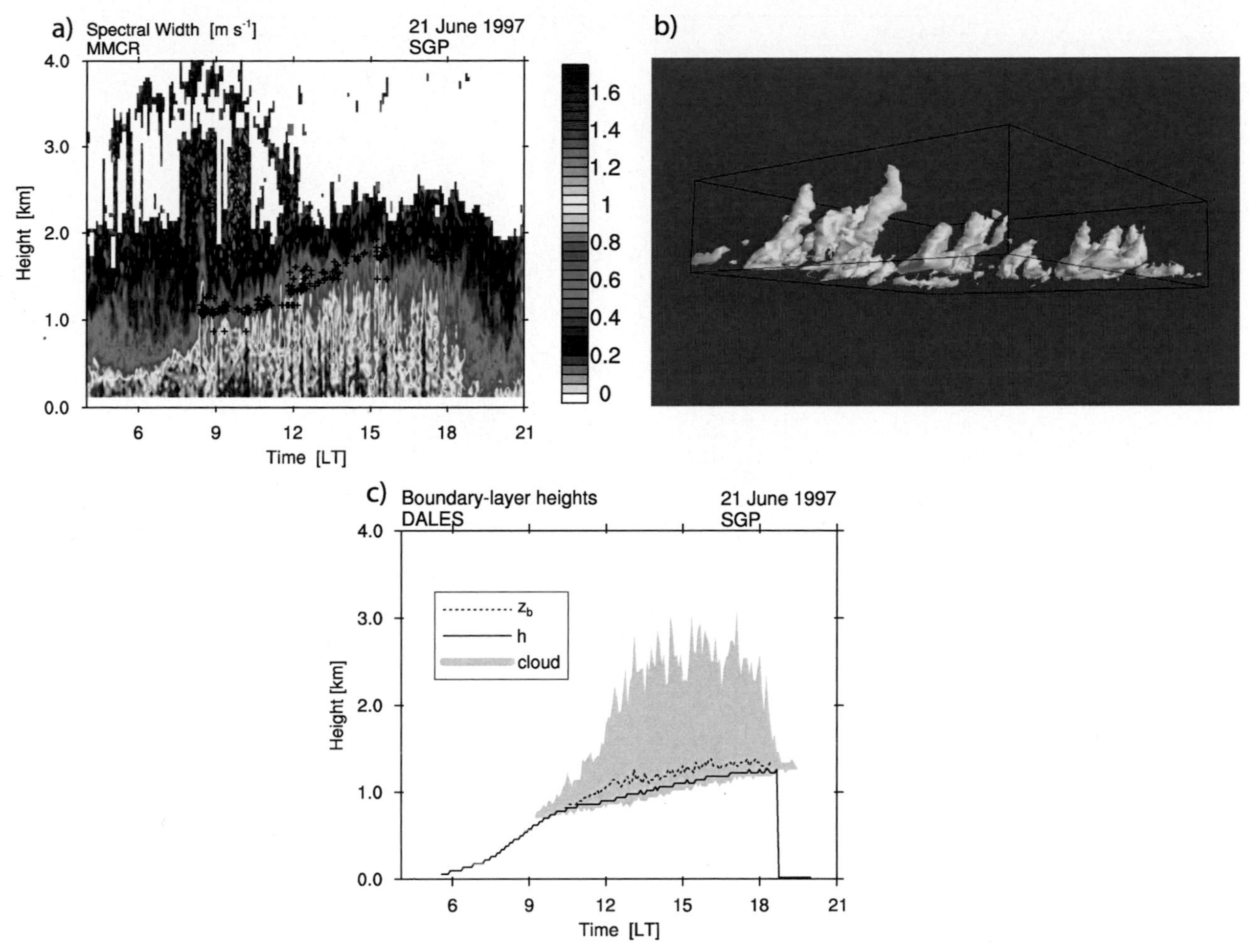

FIG. 28-1. Observations and LES simulation for 21 June 1997 at the ARM SGP site. (a) Measurements by the millimeter cloud radar (MMCR) of the spectral width, overplotted with the best estimate of the ceilometer-observed cloud base height [Active Remote Sensing of Clouds (ARSCL)]. (b) Snapshot at 1330 LT of the cumulus cloud field as generated by the Dutch Atmospheric Large-Eddy Simulation (DALES; Heus et al. 2010) code. (c) Time series of heights in the atmospheric boundary layer as diagnosed in the LES, including the level of minimum buoyancy flux (solid black), the level of maximum cloud core fraction (dotted black), and the layer enveloped between the lowest cloud base and the highest cloud top (shaded gray).

The poor SCM performance for the transient ARM case has motivated a structural rethink of the design of boundary layer schemes in the last decade. In particular, it has driven a move toward model unification; this is the idea that all regimes are represented by a single, internally consistent conceptual model, which is theoretically and practically capable of representing smooth transitions in response to smooth variations in the applied forcing. Operational forecast centers such as the ECMWF can provide a rigorous testing ground for new parameterizations as the model is confronted continuously with observations during data analysis, thus maintaining a model state close to reality. This allows a direct, day-to-day comparison between modeled clouds and observations such as those from the ARM sites. Forecast scores also routinely provide an objective measure of overall model performance to complement more process-oriented model evaluation.

c. *An innovation for boundary layer parameterization*

The boundary layer schemes in most operational GCMs are at the core formulated in terms of only two basic models for vertical transport. These are the eddy diffusivity (ED) model and the mass flux (MF) model. While the ED model represents small-scale motions that transport properties exclusively in the downgradient direction, the MF model has an advective nature and can transport properties against gradients. In the IFS these two models are combined into a single framework, named EDMF (Siebesma and Teixeira 2000), in which

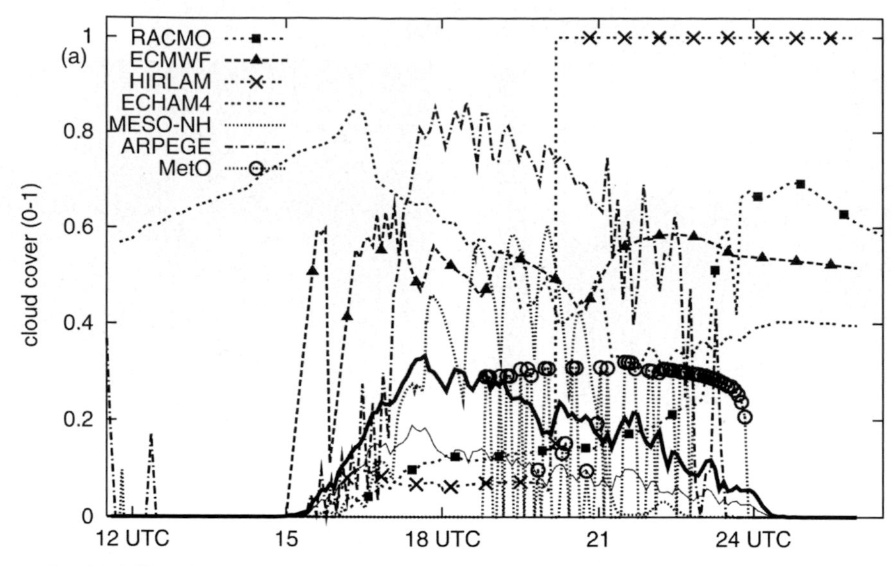

FIG. 28-2. The cloud cover as produced by various SCM simulations for 21 June 1997 at the ARM SGP site. The LES results are included for reference (solid black line). Note that local time at the SGP site lags UTC time by 6 h. Figure reproduced from Lenderink et al. (2004), copyright Royal Meteorological Society (RMETS). Reproduced with permission of John Wiley and Sons Ltd. on behalf of the RMETS.

the volumetric turbulent flux of a property ϕ that is conserved for moist adiabatic ascent is formulated as

$$\overline{w'\phi'} = -K\frac{\partial\overline{\phi}}{\partial z} + \sum_{i=1}^{I} M_i(\phi_i - \overline{\phi}), \qquad (28\text{-}1)$$

where K is the eddy diffusivity coefficient, M_i is the volumetric mass flux of an updraft i, the overbar indicates the gridbox mean, and I is the number of updrafts that are explicitly modeled by the mass flux approach (Neggers et al. 2009).

In most mass flux schemes to date, only one advective updraft is modeled ($I = 1$), argued to represent the total transport done by all unresolved updrafts within the gridbox. This effective updraft is commonly referred to as the "bulk" updraft. For (quasi) steady-state situations, in which the ensemble of (cloudy) updrafts does not change that much in time, this choice could be defended; the constants of proportionality applied in the parameterization of the amplitude and vertical structure of the updraft-related terms in Eq. (28-1) might be representative within a single regime. However, the GCSS intercomparison study on the ARM case emphasized that in a transient situation, like the diurnal cycle over land, the ensemble of updrafts can change significantly in time. Starting out as a completely dry collection of thermals the ensemble experiences partial condensation later in the day, the degree of which in itself is highly time dependent (Neggers et al. 2009). One wonders if the number of degrees of freedom provided by the single bulk updraft limit of Eq. (28-1) is actually too small to allow a parameterization to capture this gradual variation.

The goal of this ARM-funded project was to address this problem by revisiting the basic design of the EDMF boundary layer scheme, in particular the number of degrees of freedom included. The approach was guided by the following basic question: What is the minimum level of complexity required in a mass flux scheme to enable the representation of the major convective boundary layer regimes, including potentially smooth transitions between them?

When external forcings vary smoothly in time, any transitions between regimes also should evolve smoothly, and ideally a scheme should be able to reproduce this for the right reasons. As argued by Neggers et al. (2009), a unified representation of all boundary layer regimes by a mass flux scheme can be achieved by expanding the number of updrafts in Eq. (28-1). On the other hand, the limited computational efficiency of supercomputers still constrains this number. For these reasons, a dual mass flux (DualM) configuration is proposed, as schematically illustrated in Fig. 28-3a. This configuration contains two transporting bulk updrafts: one representing all dry thermals and the other all moist (i.e., cloudy) thermals. Flexibility is introduced by assigning flexible weights to each updraft category (see Fig. 28-3b). This in principle enables a gradual onset and decay of the two main updraft modes, as well as their simultaneous coexistence. It can be argued that this configuration allows the representation of transport in

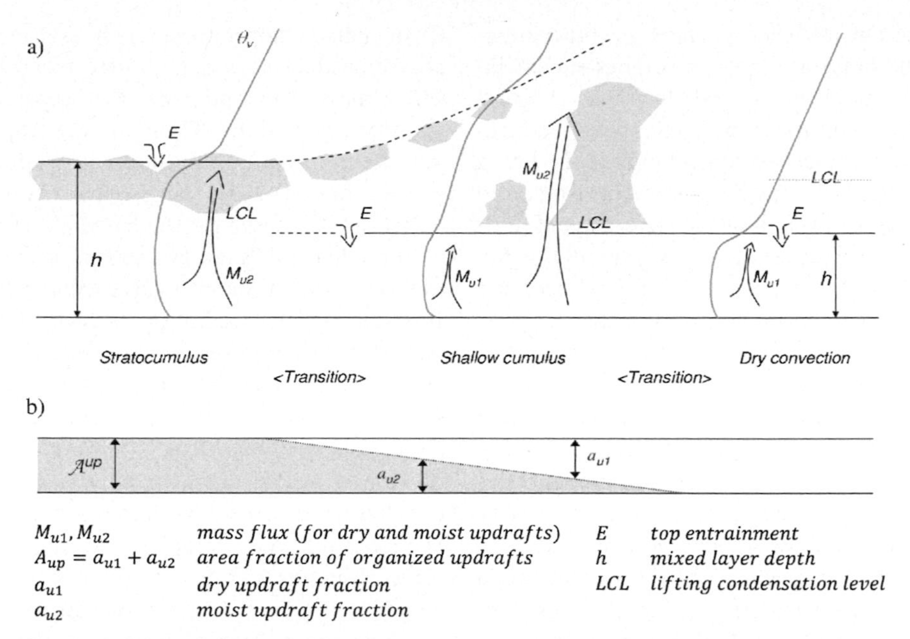

FIG. 28-3. Schematic illustration of the configuration of the DualM scheme for various boundary layer regimes and transitions between them. Figure reproduced from Neggers et al. (2009).

well-mixed boundary layers as well as internally decoupled boundary layers and transitions between the two.

Figure 28-4 shows SCM results with the DualM setup of the EDMF for the ARM case. The figure illustrates that key aspects in the time development of the transient cumulus-capped boundary layer are reproduced, such as 1) the development of the moist updraft lifting condensation level and termination height, as well as 2) the total cloud cover. Comparison against the LES shows that the deepening of the cloud layer is well captured by the DualM scheme. Comparison against various available measurements of the cloud cover shows that both the LES and the DualM scheme in the SCM reproduce the gradual decay of cloud cover in the second part of the day; note that the observations of cloud presence between 1400 and 1600 UTC also include some clouds situated at much higher altitudes that were not part of the convective boundary layer.

It is instructive to interpret these results from a broader perspective. The SCM simulations for the ARM case show that a dual bulk-updraft setup of EDMF, including flexible weights, already contains enough degrees of freedom to conceptually enable the simulation of a gradual onset, amplification, and decay of the cloudy (cumulus) part of the updraft ensemble during highly transient situations. This capability is arguably the key element of progress

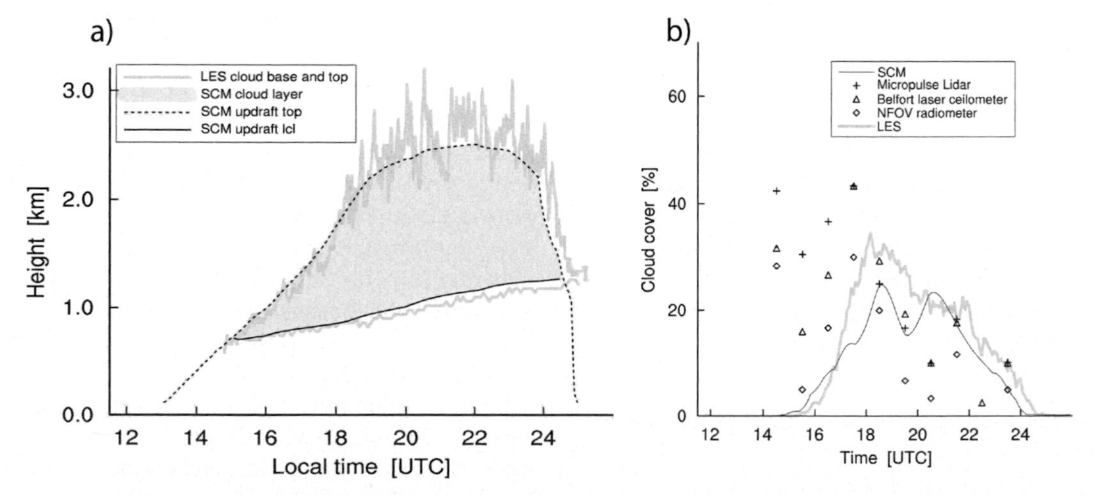

FIG. 28-4. SCM results for the DualM scheme for (a) boundary layer heights and (b) cloud cover. The label "SCM" refers to the DualM scheme as implemented. Panel (a) reproduced from Neggers et al. (2009).

introduced by the DualM scheme. There are other situations in which the boundary layer undergoes similar internal transitions between updraft regimes. A good example highly relevant for climate science is the transition within the marine subtropical boundary layer, changing from well mixed and stratocumulus capped to internally decoupled and shallow-cumulus capped (e.g., Albrecht et al. 1995; Bretherton and Wyant 1997). Accordingly, the DualM limit significantly enhances the general applicability of the EDMF approach.

d. Outlook

The ARM Program has played a key role in model development by providing observational datasets that can be used to build prototype cases as well as long time series to test and evaluate new parameterizations for a wide range of situations. Its observational database has been used by international projects such as GCSS to 1) build cases that can act as a testing ground during parameterization development and 2) evaluate schemes and constrain parametric functions. What this ARM-funded project at ECMWF has illustrated is that this strategy can be effective in driving the modeling effort forward, both in a theoretical and applied engineering sense.

The focus on continental transient cumulus cases has forced the modeling community to study more complex situations than considered previously. Transient cases, including the ARM case, have by now become key testing grounds for existing and new parameterization schemes (e.g., Bretherton et al. 1999). The realization that nature seldom behaves like an idealized "golden day" but exhibits great day-to-day variability has motivated the modeling community to move from single case studies to continuous, long-term SCM and LES evaluation (e.g., Jakob 2010; Neggers et al. 2012; Neggers and Siebesma 2013). The benefit of this strategy is that the modeling result maintains its statistical significance, allowing the SCM result to be representative of the full model climate. On the other hand, the strategy still preserves the high model transparency typical of single-case model studies at process level. The continuous and long-term measurements required for this approach are typically provided by the various ARM permanent and Mobile Facility sites established at locations from the Arctic to the tropics considered to represent key regimes in the earth's climate and make ARM data invaluable for such long-term evaluation studies.

3. Advances in the parameterization of radiation transfer

Radiative transfer calculations are an expensive part of the forecast model in terms of computational cost. Cost-saving compromises such as longer time steps, coarser grid resolution, or limited spectral resolution are often made to cope with the expense of radiative transfer calculations. Thus, the development of faster, yet more accurate methods, to calculate radiative transfer has been and remains an ongoing development effort in GCMs and for the ECMWF IFS global model.

In the late 1990s, an evaluation of the IFS downward longwave radiation with observations from ARM, the Baseline Surface Radiation Network (BSRN), and the National Oceanic and Atmospheric Administration (NOAA) Surface Radiation Network (SURFRAD) sites highlighted systematic model errors present under cloudy as well as clear conditions (Chevallier and Morcrette 2000). Surface shortwave irradiance was overestimated, usually with the error linked primarily to cloudy conditions, while the surface downward longwave radiation was underestimated in clear as well as cloudy conditions. In addition, an increase in vertical resolution from 37 to 60 levels was planned for the IFS. The cost to run the model's emissivity scheme operational at the time (Morcrette 1991) scaled quadratically with the number of model levels; thus leading to an unacceptable increase in computational cost for the envisaged increase in vertical resolution.

a. The Rapid Radiative Transfer Model

The Rapid Radiative Transfer Model (RRTM) developed by Atmospheric and Environmental Research (AER), Inc. (Mlawer et al. 1997, 2016, chapter 15) promised better performance at lower cost. This model is conceptually much more closely related to a line-by-line radiation scheme, has 16 spectral bands (versus 6 in the previous scheme), and the two-stream calculation of the RRTM scales linearly with the number of model levels—an attractive property with the prospect of an increase in vertical resolution. The RRTM's performance compared favorably with line-by-line radiation models and ARM spectral surface observations (Mlawer et al. 1996). It was tested successfully as a replacement for the longwave radiation scheme in the IFS (Morcrette et al. 1998) and other models (Iacono et al. 2000) and eventually implemented operationally on 27 June 2000. The comparative performance of the old and the new longwave radiation scheme in the IFS was documented in Morcrette (2002a) and showed that the RRTM longwave scheme greatly reduced the surface downward longwave radiation bias under clear-sky conditions.

Further evaluation of the model's surface radiation was performed at the ARM SGP site (Morcrette 2002b). This study confirmed the good performance of the RRTM longwave scheme for clear-sky and overcast conditions (when cloud base height is predicted correctly), but highlighted problems in surface shortwave irradiance, which was overestimated under all conditions. The clear-sky bias in surface shortwave irradiance was attributed primarily to a somewhat outdated shortwave parameterization, which slightly underestimated water vapor absorption and did not adequately separate between the UV and visible spectral bands. Inadequate representation of aerosol also was implicated as a secondary factor. Under cloudy conditions, the model's assumed cloud properties, such as a fixed effective radius of 10 μm, likely contributed to the model error. To address these issues, the spectral intervals in the shortwave scheme were increased (from 4 to 6), and a slightly more sophisticated diagnostic scheme for liquid effective radius was introduced (Martin et al. 1994).

b. Representing subgrid inhomogeneity: The Monte Carlo independent column approximation

Another challenge in radiative transfer is the representation of cloud inhomogeneity within a model grid box. Initially, radiative transfer was calculated on plane-parallel clouds, ignoring inhomogeneity within a model grid box and layer. Tiedtke's cloud scheme (Tiedtke 1993) used in the IFS accounts for one aspect of cloud inhomogeneity—namely, the partitioning of the model grid box into a cloudy and a clear fraction—with a prognostic cloud fraction variable. However, cloud liquid was assumed initially to be horizontally and vertically uniform within the cloudy part of the grid box, ignoring the natural variability of liquid water content observed in clouds. This plane-parallel homogenous assumption leads to an albedo bias in the radiative transfer calculations (Cahalan et al. 1994, 1995). To account for the albedo bias, a correction factor was introduced in the longwave and shortwave radiative transfer calculations in the operational IFS on 16 December 1997 (Tiedtke 1996) based on Cahalan et al.'s (1994) study of cloud inhomogeneity in stratocumulus clouds. While improving results, this fixed parameter was merely a stop-gap measure, failing to relate to the actually modeled cloudy conditions.

If the variability of the cloud liquid water is known, the albedo bias can be avoided largely by dividing the model grid box into subcolumns, with each subcolumn representing part of the liquid water distribution, and performing radiative transfer calculations for each of the subcolumns [independent column approximation (ICA); Cahalan et al. 1994; Barker et al. 1999], then averaging the result. While accurate, this approach is expensive, increasing the number of radiative transfer calculations by a factor equal to the number of subcolumns chosen.

The Monte Carlo Independent Column Approximation (McICA; Barker et al. 2003; Pincus et al. 2003; Barker et al. 2008), developed as part of ARM-funded projects, offers a less expensive yet elegant solution to this problem (Mlawer et al. 2016, chapter 15). As for the ICA, the grid box is subdivided into independent columns. For each model level, some columns are cloudy and some clear, reflecting the predicted cloud fraction at this level and allowing further sampling of the cloud's water distribution. However, in the McICA approach, spectral intervals also are assigned randomly one per column, and radiative transfer is calculated in a plane-parallel fashion for each column. Thus, the subcolumns collectively sample both the radiation spectrum and the cloud's water distribution at the same time. This elegant approach allows a more physical, situation-dependent representation of cloud inhomogeneity at limited additional expense. The McICA scheme was implemented in the IFS on 5 June 2007 together with the shortwave version of the RRTM (Iacono et al. 2004). This combined radiation upgrade (McRad; Morcrette et al. 2008) improved model performance on shorter and longer time scales (for seasonal/climate applications). The new scheme modified the relative vertical distribution of the shortwave and longwave heating as well as the amount of surface shortwave irradiance, with consequences for the boundary layer structure and the strength of convection in the tropics. The altered tropical convection had an impact on the global circulation and resulted in improved low-level winds in the tropics and overall better model performance (Morcrette et al. 2008). Figure 28-5 illustrates one aspect of the improvements: a better agreement with the National Aeronautics and Space Administration (NASA) Clouds and the Earth's Radiant Energy System (CERES) observations for top-of-the-atmosphere longwave cloud forcing. The model results in this figure are from a small three-member ensemble of year-long forecasts, showing that the improvements to the model's energy budget extend into the long range and are not confined to the short-term forecast range alone.

While the McICA scheme provides a framework that allows a better representation of cloud inhomogeneity in the IFS, the current model does not yet take full advantage of this capability. A single inhomogeneity function is applied currently to the cloud water content within each grid box, irrespective of the cloud type or regime, and while this function is used in the radiative transfer calculations, the cloud water content is still assumed to be uniform for microphysical processes. Further improvements to the treatment of cloud inhomogeneity are part of the IFS development plans, and the ARM Program provides a rich set of observations to guide future changes.

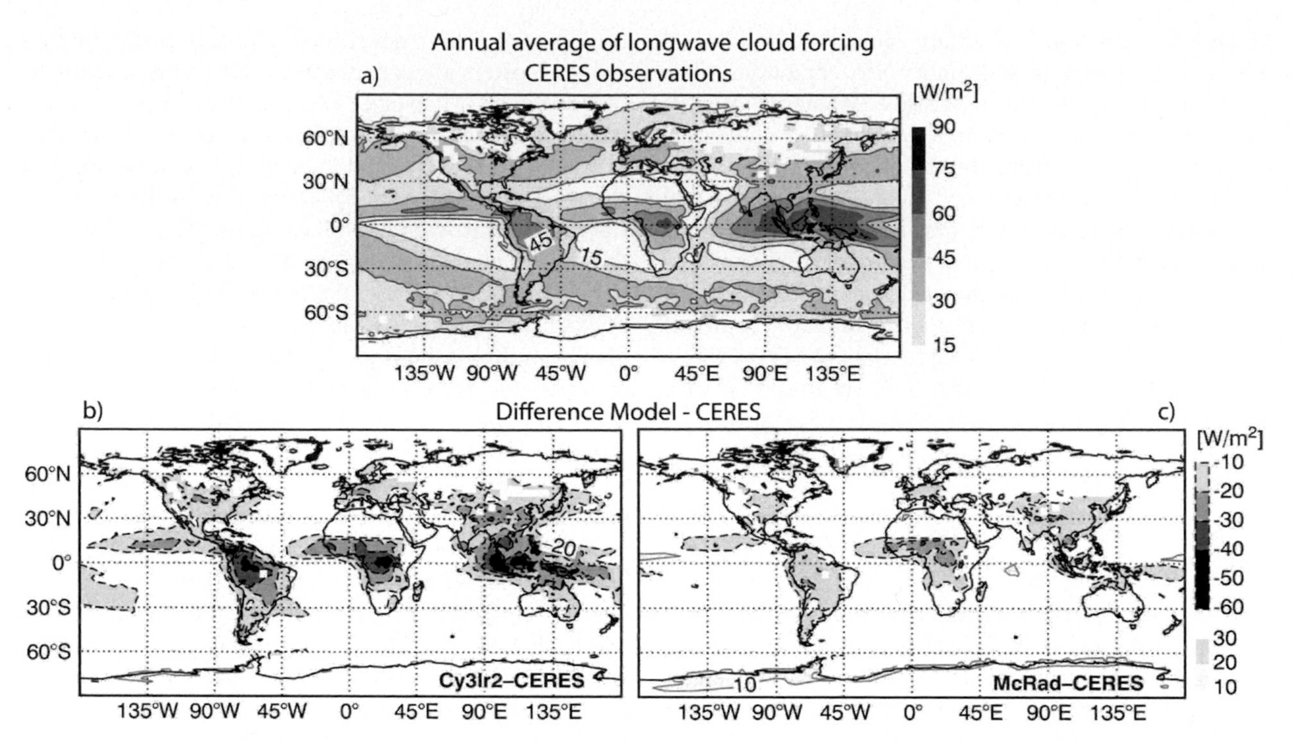

FIG. 28-5. Annual average of longwave cloud forcing (W m^{-2}) from (a) the CERES observations, (b) the difference between the operational model and CERES observations, and (c) the difference between the model with McRad radiation package and CERES observations (i.e., before and after the change to the McRad radiation package). For the model, results are for averages over three simulations starting 24 h apart, with output parameters averaged over the period September 2000 to August 2001. Reproduced from Morcrette et al. (2008, their Fig. 4).

4. Process-oriented model evaluation with ARM observations in the IFS: Reducing systematic errors

A key area for parameterization improvement is to reduce the systematic errors associated with the representation of cloud and interactions with radiation. Since 1995, the cloud and precipitation parameterization scheme in the IFS was based on Tiedtke (1993), parameterizing the sources and sinks of cloud and precipitation due to the main generation and destruction processes from convection and microphysics. Although modified in several ways over the years, the basic structure remained with three prognostic variables: water vapor, a combined liquid–ice cloud condensate, and a cloud fraction. Although performing well in the context of an NWP model, further improvements could be made to address some of the shortcomings of the cloud scheme. Two examples are highlighted here for mixed-phase and warm-phase cloud processes.

a. An ARM case study for improving the representation of supercooled liquid water

A model intercomparison study (Klein et al. 2009) based on the Mixed-Phase Arctic Cloud Experiment (MPACE) at the ARM North Slope of Alaska site highlighted an issue with the representation of mixed-phase cloud in many GCMs, including the IFS. The single prognostic cloud condensate variable required a diagnostic partitioning into cloud liquid and ice, and a temperature-dependent function was used to determine the liquid fraction for temperatures between 0°C and −23°C, assuming all condensate to be frozen below −23°C. Considering that temperature generally decreases with height, this type of function will diagnose a larger liquid fraction at cloud base than at cloud top. Despite the ability of the IFS to predict the large-scale state of the atmosphere well during MPACE (Xie et al. 2006), the diagnostic phase partitioning left the model unable to represent the observed mixed-phase structure of the Arctic clouds with liquid layers near the top of the cloud and led to an underestimation of the liquid fraction. The limitations of the representation of the mixed phase by the original Tiedtke scheme, together with several other issues (precipitation advection on an increasingly finer horizontal grid, numerical issues, physical realism) prompted an extension of the cloud scheme to include separate prognostic variables for cloud liquid and ice and additional prognostic variables for snow and rain (Forbes

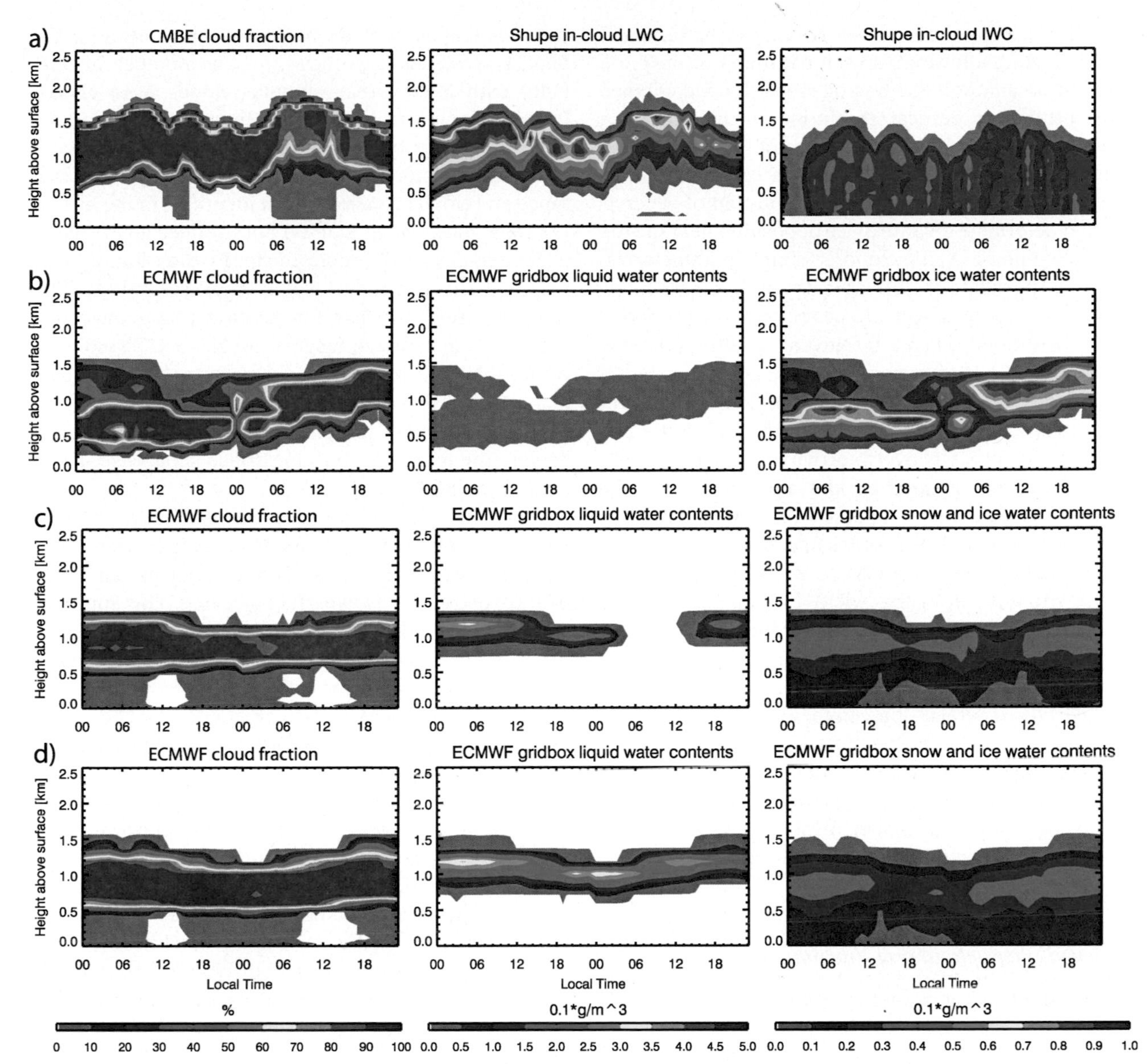

FIG. 28-6. (left) Cloud fraction, (center) cloud liquid water content, and (right) cloud ice water content for a single-layer mixed-phase cloud observed during 8 and 9 Oct 2004 at the NSA site during MPACE. (a) Observations, (b) IFS with previous cloud scheme with diagnostic mixed phase, (c) IFS with new cloud scheme with separate prognostic liquid and ice variables, and (d) the new cloud scheme with revised ice deposition rate. From Forbes and Ahlgrimm (2014).

and Tompkins 2011; Forbes et al. 2011). Separate variables for liquid and ice provide additional degrees of freedom to represent the variability of supercooled liquid water observed at a given temperature and also provide new opportunities to revisit processes such as ice deposition, sedimentation, etc., that act as sources and sinks for the prognostic cloud variables.

Shortly after the operational implementation of the new cloud scheme in November 2010, a cold bias in 2-m temperature was identified in wintertime northern Europe, and an investigation found a link with a lack of supercooled liquid in clouds over wintertime Europe. A revisit of the MPACE case study illustrates the problem (Fig. 28-6). For these cold clouds between 0° and −23°C, the diagnostic scheme with a temperature-dependent mixed phase by definition has small amounts of supercooled liquid water present, although much too little and with the wrong vertical structure. The new cloud scheme with separate prognostic liquid and ice variables, however, produces a qualitatively better cloud structure with supercooled liquid in the upper half but is unable to sustain the liquid layer throughout the test case period

(Fig. 28-6c). Thus, the longwave cooling at the surface is overestimated, allowing 2-m temperatures to drop too low. The assumption of a homogenous cloud, combined with insufficient vertical resolution, means that the model is unable to represent the small-scale processes adequately, such as cloud top-driven turbulence, variability of ice nucleation, and sedimentation of ice crystals that determine the rate of ice deposition near the top of a mixed-phase Arctic cloud. A simple parameterization to encapsulate the effect of these processes in reducing the ice deposition rate near cloud top was added to the new cloud scheme in November 2010, enabling the model to better sustain supercooled liquid layers (Fig. 28-6d). The model's improved ability to represent mixed-phase clouds was reflected in more accurate 2-m temperature forecasts over Northern Hemisphere continents, but also improved the radiation over the Southern Ocean where supercooled topped boundary layer cloud is abundant (Forbes and Ahlgrimm 2014).

Case studies such as MPACE, using remote sensing and in situ data from the ARM Program, proved invaluable to evaluate the model and inform the improved representation of specific processes in the model cloud parameterization in the IFS. However, the ARM Data Archive also contains long time series of continuous radiation and cloud observations. Conditional sampling of these time series and subsequent compositing can help to establish a link between model errors and particular aspects of the model formulation. These long time series are explored in the following section.

b. Long time series of ARM observations to improve warm-phase cloud and precipitation

It is common for model errors to compensate to a degree, and while this may be immediately beneficial for a better performance, it poses a problem for future development. An improvement of one model aspect can easily result in a deterioration of the forecast when another existing error is no longer compensated. The following example illustrates how the long-term observations from the ARM SGP site, after thoughtful compositing, helped to identify specific aspects of the boundary layer and shallow convection parameterizations in need of improvement.

The observational record at the ARM SGP site shows that the ECMWF model has a long-standing bias in surface shortwave irradiance, allowing too much shortwave radiation to reach the surface. Previous work comparing the model to observations from SGP for a month-long period (Cheinet et al. 2005) suggested that a lack of shallow convective clouds in the model, such as those commonly observed in the summertime fair-weather cumulus regime at SGP, might be contributing to the bias. To test this hypothesis, a large number of days (146) with fair-weather cumulus clouds were chosen from the long-term record and composited, and cloud occurrence, properties, and surface radiation were examined. Compensating errors between cloud occurrence and cloud properties were identified in the model. On some days, the model did not produce any cloud at all, but on days with predicted clouds, their liquid water path and effective radius would be overestimated. This suggested several areas for revision (e.g., convection triggering, moisture transport, assumed CCN concentration), but since the errors in surface shortwave irradiance from clear and cloudy days largely compensated for the fair-weather cumulus regime, this could not fully explain the long-term bias in surface shortwave irradiance at the SGP site.

To identify the cause of the bias without an a priori guess, the observed and modeled clouds were then classified by type and ranked by their contribution to the overall surface shortwave irradiance bias. This approach highlighted cloud regimes that are particularly relevant to the surface shortwave irradiance bias and revealed that low-cloud conditions do contribute significantly, though it is during overcast conditions that the model overestimates the surface shortwave irradiance. Cloud cover is often underpredicted when overcast low clouds are observed, and liquid water path is too low (Ahlgrimm and Forbes 2012). For broken low-cloud conditions, on the other hand, liquid water path tends to be overestimated, as was evident during the fair-weather cumulus days. The assumption of a fixed cloud water inhomogeneity distribution for all conditions proved to be ill suited to represent both broken and overcast low-cloud regimes. Thus, surface shortwave irradiance errors from broken and overcast low-cloud regimes partially compensate, with the bias from the overcast regime dominating. These contrasting model errors for broken and overcast low-cloud conditions may explain why the surface shortwave irradiance bias has persisted in the model for many years and through many model changes. A simple increase or decrease of cloud cover or liquid water path for low clouds would not address the underlying problem fully. Instead, broken low clouds, generally produced by the shallow convection parameterization in the IFS, and overcast low clouds, produced by the stratocumulus parameterization embedded in the boundary layer scheme, or the cloud scheme, need to be addressed individually.

The IFS, in common with other global models, also overestimates the occurrence of light precipitation and can underestimate the occurrence of heavier precipitation. The 19-month-long observational dataset

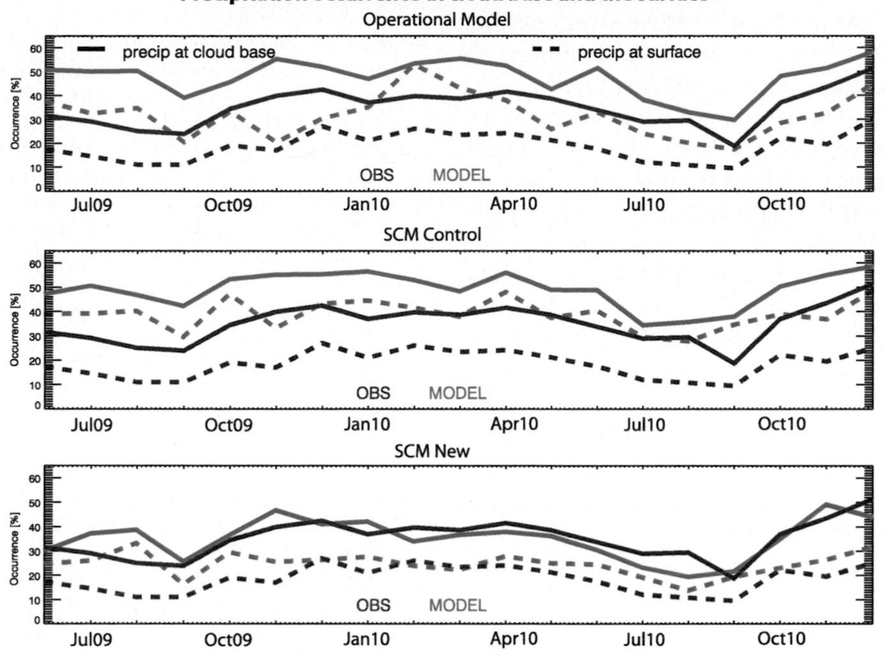

FIG. 28-7. Monthly precipitation occurrence at the cloud base (solid lines) and at the surface (dashed lines) from observations gathered over 19 months on Graciosa Island (Azores), from (top) the operational model forecasts and (middle) a control version of the SCM and (bottom) for a modified version of the SCM including changes to the boundary layer scheme, autoconversion, accretion, and evaporation parameterizations showing improved precipitation occurrence at cloud base and at the surface. Figure adapted from Ahlgrimm and Forbes (2014).

obtained from the ARM Mobile Facility located on Graciosa Island in the Azores (Rémillard et al. 2012) provided an opportunity to evaluate the representation of precipitating marine boundary layer clouds in the IFS. In contrast to satellite-based observations, a comparison with the detailed profiles of precipitation retrieved from ground-based radar allows an attribution of the model's overestimate in surface precipitation occurrence to in-cloud generation and subcloud evaporation. As anticipated, the model overestimated monthly mean precipitation occurrence both at cloud base and at the surface compared to the observations (Fig. 28-7). The overestimate was similar in both the full IFS and a single-column version of the model (SCM). Thus, the SCM could be used to test parameterization changes for the entire 19-month period, an undertaking that would be too expensive to perform with the full model. Several parameterization changes were tested to improve the precipitation distribution with an initial focus on warm-phase rain formation from boundary layer clouds. An alternative parameterization of the autoconversion and accretion processes (Khairoutdinov and Kogan 2000) led to reduced generation of light precipitation. A parameterization of rain evaporation that represents more realistically the higher evaporation rates of small droplets (Abel and Boutle 2012) also reduced the occurrence of light rain at the surface.

The impact of combined parameterization changes to the parcel entrainment in the boundary layer (based on the insights gained from the ARM SGP low cloud investigation), as well as to the warm rain autoconversion–accretion and evaporation parameterizations significantly improved the precipitation occurrence (Fig. 28-7), low-cloud cover, and agreement with observed surface radiation in the SCM and in the full model (Ahlgrimm and Forbes 2014).

Only two examples are discussed here in detail, but observational data from other sites, such as the Tropical Western Pacific (Lin et al. 2012) and the ARM Mobile Facility in Niamey (Agustí-Panareda et al. 2010), have also contributed to our understanding of model biases in the IFS.

5. Conclusions

The long-standing interaction between the ARM Program and ECMWF has shaped the development of the IFS global model over many years and contributed directly to the improvement of NWP at ECMWF and indirectly to developments in the wider weather and climate modeling community. This is arguably most evident in the RRTM and McICA schemes for radiative transfer, but ARM observations have continued to provide inspiration and valuable insights into parameterization formulations, from novel boundary layer parameterization developments such as the DualM scheme to warm-phase and mixed-phase cloud improvements in the IFS. By providing the necessary observational datasets to develop and test new model parameterizations, whether at ECMWF or in the wider research community, ARM has enabled process-oriented evaluation efforts to identify and target systematic model errors. In turn, the NWP environment at ECMWF, with stringent requirements for forecast performance and synoptic weather systems always close to reality in the short-term forecasts, allows model evaluation in a deterministic way to make the most of observations from the ARM Program.

REFERENCES

Abel, S. J., and I. A. Boutle, 2012: An improved representation of the raindrop size distribution for single-moment microphysics schemes. *Quart. J. Roy. Meteor. Soc.*, **138**, 2151–2162, doi:10.1002/qj.1949.

Agustí-Panareda, A., and Coauthors, 2010: The ECMWF re-analysis for the AMMA observational campaign. *Quart. J. Roy. Meteor. Soc.*, **136**, 1457–1472, doi:10.1002/qj.662.

Ahlgrimm, M., and R. Forbes, 2012: The impact of low clouds on surface shortwave radiation in the ECMWF model. *Mon. Wea. Rev.*, **140**, 3783–3794, doi:10.1175/MWR-D-11-00316.1.

——, and ——, 2014: Improving the representation of low clouds and drizzle in the ECMWF model based on ARM observations from the Azores. *Mon. Wea. Rev.*, **142**, 668–685, doi:10.1175/MWR-D-13-00153.1.

Albrecht, B. A., C. S. Bretherton, D. Johnson, W. H. Scubert, and A. S. Frisch, 1995: The Atlantic Stratocumulus Transition Experiment—ASTEX. *Bull. Amer. Meteor. Soc.*, **76**, 889–904, doi:10.1175/1520-0477(1995)076<0889:TASTE>2.0.CO;2.

Barker, H. W., G. L. Stephens, and Q. Fu, 1999: The sensitivity of domain-averaged solar fluxes to assumptions about cloud geometry. *Quart. J. Roy. Meteor. Soc.*, **125**, 2127–2152, doi:10.1002/qj.49712555810.

——, R. Pincus, and J.-J. Morcrette, 2003: The Monte Carlo independent column approximation: Application within large-scale models. *Proc. GCSS/ARM Workshop on the Representation of Cloud Systems in Large-Scale Models*, Kananaskis, AB, Canada, GEWEX, 10 pp.

——, J. N. S. Cole, J.-J. Morcrette, R. Pincus, P. Räisänen, K. von Salzen, and P. Vaillancourt, 2008: The Monte-Carlo independent column approximation: An assessment using several global atmospheric models. *Quart. J. Roy. Meteor. Soc.*, **134**, 1463–1478, doi:10.1002/qj.303.

Bretherton, C. S., and M. C. Wyant, 1997: Moisture transport, lower-tropospheric stability, and decoupling of cloud-topped boundary layers. *J. Atmos. Sci.*, **54**, 148–167, doi:10.1175/1520-0469(1997)054<0148:MTLTSA>2.0.CO;2.

——, and Coauthors, 1999: A GCSS boundary-layer cloud model intercomparison study of the first ASTEX Lagrangian experiment. *Bound.-Layer Meteor.*, **93**, 341–380, doi:10.1023/A:1002005429969.

Brown, A. R., and Coauthors, 2002: Large-eddy simulation of the diurnal cycle of shallow cumulus convection over land. *Quart. J. Roy. Meteor. Soc.*, **128**, 1075–1094, doi:10.1256/003590002320373210.

Cahalan, R. F., W. Ridgway, W. J. Wiscombe, T. L. Bell, and J. B. Snider, 1994: The albedo of fractal stratocumulus clouds. *J. Atmos. Sci.*, **51**, 2434–2455, doi:10.1175/1520-0469(1994)051<2434:TAOFSC>2.0.CO;2.

——, D. Silberstein, and J. B. Snider, 1995: Liquid water path and plane-parallel albedo bias during ASTEX. *J. Atmos. Sci.*, **52**, 3002–3012, doi:10.1175/1520-0469(1995)052<3002:LWPAPP>2.0.CO;2.

Cheinet, S., A. Beljaars, M. Köhler, J.-J. Morcrette, and P. Viterbo, 2005: Assessing physical processes in the ECMWF model forecasts using the ARM SGP observations. ECMWF ARM Rep. 1, 25 pp.

Chevallier, F., and J.-J. Morcrette, 2000: Comparison of model fluxes with surface and top-of-the-atmosphere observations. *Mon. Wea. Rev.*, **128**, 3839–3852, doi:10.1175/1520-0493(2001)129<3839:COMFWS>2.0.CO;2.

Comstock, J. M., and C. Jakob, 2004: Evaluation of tropical cirrus cloud properties derived from ECMWF model output and ground based measurements over Nauru Island. *Geophys. Res. Lett.*, **31**, L10106, doi:10.1029/2004GL019539.

Forbes, R., and A. Tompkins, 2011: An improved representation of cloud and precipitation. *ECMWF Newsletter*, No. 129, ECMWF, Reading, United Kingdom, 13–18.

——, and M. Ahlgrimm, 2014: On the representation of high-latitude boundary layer mixed-phase cloud in the ECMWF global model. *Mon. Wea. Rev.*, **142**, 3425–3445, doi:10.1175/MWR-D-13-00325.1.

——, A. M. Tompkins, and A. Untch, 2011: A new prognostic bulk microphysics scheme for the IFS. ECMWF Tech. Memo. 649, 30 pp. [Available online at http://www.ecmwf.int/sites/default/files/elibrary/2011/9441-new-prognostic-bulk-microphysics-scheme-ifs.pdf.]

Heus, T., and Coauthors, 2010: Formulation of the Dutch Atmospheric Large-Eddy Simulation (DALES) and overview of its applications. *Geosci. Model Dev.*, **3**, 415–444, doi:10.5194/gmd-3-415-2010.

Holland, J. Z., and E. M. Rasmusson, 1973: Measurement of atmospheric mass, energy, and momentum budgets over a 500-kilometer square of tropical ocean. *Mon. Wea. Rev.*, **101**, 44–55, doi:10.1175/1520-0493(1973)101<0044:MOTAME>2.3.CO;2.

Iacono, M. J., E. J. Mlawer, S. A. Clough, and J.-J. Morcrette, 2000: Impact of an improved longwave radiation model, RRTM, on the energy budget and thermodynamic properties of NCAR Community Climate Model, CCM3. *J. Geophys. Res.*, **105**, 14 873–14 890, doi:10.1029/2000JD900091.

——, J. S. Delamere, E. J. Mlawer, S. A. Clough, J.-J. Morcrette, and Y.-T. Hou, 2004: Development and evaluation of RRTMG_SW, a shortwave radiative transfer model for general circulation model applications. *Proc. 14th Atmospheric Radiation Measurement (ARM) Science Team Meeting*, Albuquerque, NM, DOE, 10 pp. [Available online at https://www.arm.gov/publications/proceedings/conf14/extended_abs/iacono-mj.pdf.]

Jakob, C., 2010: Accelerating progress in global atmospheric model development through improved parameterizations: Challenges, opportunities, and strategies. *Bull. Amer. Meteor. Soc.*, **91**, 869–875, doi:10.1175/2009BAMS2898.1.

Khairoutdinov, M., and Y. Kogan, 2000: A new cloud physics parameterization in a large-eddy simulation model of marine stratocumulus. *Mon. Wea. Rev.*, **128**, 229–243, doi:10.1175/1520-0493(2000)128<0229:ANCPPI>2.0.CO;2.

Klein, S. A., and Coauthors, 2009: Intercomparison of model simulations of mixed-phase clouds observed during the ARM Mixed-Phase Arctic Cloud Experiment. I: Single-layer cloud. *Quart. J. Roy. Meteor. Soc.*, **135**, 979–1002, doi:10.1002/qj.416.

Lenderink, G., and Coauthors, 2004: The diurnal cycle of shallow cumulus clouds over land: A single-column model intercomparison study. *Quart. J. Roy. Meteor. Soc.*, **130**, 3339–3364, doi:10.1256/qj.03.122.

Lin, Y., and Coauthors, 2012: TWP-ICE global atmospheric model intercomparison: Convection responsiveness and resolution impact. *J. Geophys. Res.*, **117**, D09111, doi:10.1029/2011JD017018.

Lopez, P., A. Benedetti, P. Bauer, M. Janisková, and M. Köhler, 2006: Experimental 2D-Var assimilation of ARM cloud and precipitation observations. *Quart. J. Roy. Meteor. Soc.*, **132**, 1325–1347, doi:10.1256/qj.05.24.

Mace, G. G., C. Jakob, and K. P. Moran, 1998: Validation of hydrometeor occurrence predicted by the ECMWF model using millimeter wave radar data. *Geophys. Res. Lett.*, **25**, 1645–1648, doi:10.1029/98GL00845.

Martin, G. M., D. W. Johnson, and A. Spice, 1994: The measurement and parameterization of effective radius of droplets in warm stratocumulus clouds. *J. Atmos. Sci.*, **51**, 1823–1842, doi:10.1175/1520-0469(1994)051<1823:TMAPOE>2.0.CO;2.

Mlawer, E. J., S. J. Traubman, and S. A. Clough, 1996: RRTM: A Rapid Radiative Transfer Model. *Proc. Fifth Atmospheric Radiation Measurement (ARM) Science Team Meeting*, San Diego, CA, DOE, 219–222.

——, S. J. Taubman, P. D. Brown, M. J. Iacono, and S. A. Clough, 1997: Radiative transfer for inhomogeneous atmospheres: RRTM, a validated correlated-k model for the longwave. *J. Geophys. Res.*, **102**, 16 663–16 682, doi:10.1029/97JD00237.

——, M. Iacono, R. Pincus, H. Barker, L. Oreopoulos, and D. Mitchell, 2016: Contributions of the ARM Program to radiative transfer modeling for climate and weather applications. *The Atmospheric Radiation Measurement (ARM) Program: The First 20 Years, Meteor. Monogr.*, No. 57, Amer. Meteor. Soc., doi:10.1175/AMSMONOGRAPHS-D-15-0041.1.

Morcrette, J.-J., 1991: Radiation and cloud radiative properties in the European Centre for Medium Range Weather Forecasts forecasting system. *J. Geophys. Res.*, **96**, 9121–9132, doi:10.1029/89JD01597.

——, 2002a: The surface downward longwave radiation in the ECMWF forecast system. *J. Climate*, **15**, 1875–1892, doi:10.1175/1520-0442(2002)015<1875:TSDLRI>2.0.CO;2.

——, 2002b: Assessment of the ECMWF model cloudiness and surface radiation fields at the ARM-SGP site. *Mon. Wea. Rev.*, **130**, 257–277, doi:10.1175/1520-0493(2002)130<0257:AOTEMC>2.0.CO;2.

——, S. A. Clough, E. J. Mlawer, and M. J. Iacono, 1998: Impact of a validated radiative transfer scheme, RRTM, on the ECMWF model climate and 10-day forecasts. ECMWF Tech. Memo. 252, 47 pp.

——, H. W. Barker, J. N. S. Cole, M. J. Iacono, and R. Pincus, 2008: Impact of a new radiation package, McRad, in the ECMWF Integrated Forecasting System. *Mon. Wea. Rev.*, **136**, 4773–4798, doi:10.1175/2008MWR2363.1.

Neggers, R. A. J., and A. P. Siebesma, 2013: Constraining a system of interacting parameterizations through multiple-parameter evaluation: Tracing a compensating error between cloud vertical structure and cloud overlap. *J. Climate*, **26**, 6698–6715, doi:10.1175/JCLI-D-12-00779.1.

——, M. Köhler, and A. C. M. Beljaars, 2009: A dual mass flux framework for boundary layer convection. Part I: Transport. *J. Atmos. Sci.*, **66**, 1465–1487, doi:10.1175/2008JAS2635.1.

——, A. P. Siebesma, and T. Heus, 2012: Continuous single-column model evaluation at a permanent meteorological supersite. *Bull. Amer. Meteor. Soc.*, **93**, 1389–1400, doi:10.1175/BAMS-D-11-00162.1.

Pincus, R., H. W. Barker, and J.-J. Morcrette, 2003: A fast, flexible, approximate technique for computing radiative transfer in inhomogeneous clouds. *J. Geophys. Res.*, **108**, 4376, doi:10.1029/2002JD003322.

Randall, D. A., K. Xu, R. J. C. Somerville, and S. Iacobellis, 1996: Single-column models and cloud ensemble models as links between observations and climate models. *J. Climate*, **9**, 1683–1697, doi:10.1175/1520-0442(1996)009<1683:SCMACE>2.0.CO;2.

Rémillard, J., P. Kollias, E. Luke, and R. Wood, 2012: Marine boundary layer cloud observations in the Azores. *J. Climate*, **25**, 7381–7398, doi:10.1175/JCLI-D-11-00610.1.

Siebesma, A. P., and J. Teixeira, 2000: An advection–diffusion scheme for the convective boundary layer: Description and 1D results. Preprints, *14th Symp. on Boundary Layer and Turbulence*, Aspen, CO, Amer. Meteor. Soc., 133–140.

——, and Coauthors, 2003: A large eddy simulation intercomparison study of shallow cumulus convection. *J. Atmos. Sci.*, **60**, 1201–1219, doi:10.1175/1520-0469(2003)60<1201:ALESIS>2.0.CO;2.

Sommeria, G., 1976: Three-dimensional simulation of turbulent processes in an undisturbed trade wind boundary layer. *J. Atmos. Sci.*, **33**, 216–241, doi:10.1175/1520-0469(1976)033<0216:TDSOTP>2.0.CO;2.

Stevens, B., and Coauthors, 2001: Simulations of trade wind cumuli under a strong inversion. *J. Atmos. Sci.*, **58**, 1870–1891, doi:10.1175/1520-0469(2001)058<1870:SOTWCU>2.0.CO;2.

Tiedtke, M., 1989: A comprehensive mass flux scheme for cumulus parameterization in large-scale models. *Mon. Wea. Rev.*, **117**, 1779–1800, doi:10.1175/1520-0493(1989)117<1779:ACMFSF>2.0.CO;2.

——, 1993: Representation of clouds in large-scale models. *Mon. Wea. Rev.*, **121**, 3040–3061, doi:10.1175/1520-0493(1993)121<3040:ROCILS>2.0.CO;2.

——, 1996: An extension of cloud-radiation parameterization in the ECMWF Model: The representation of subgrid-scale variations of optical depth. *Mon. Wea. Rev.*, **124**, 745–750, doi:10.1175/1520-0493(1996)124<0745:AEOCRP>2.0.CO;2.

Tselioudis, G., and P. Kollias, 2007: Evaluation of ECMWF cloud type simulations at the ARM Southern Great Plains site using a new cloud type climatology. *Geophys. Res. Lett.*, **34**, L03803, doi:10.1029/2006GL027314.

Xie, S., S. A. Klein, J. J. Yio, A. C. M. Beljaars, C. N. Long, and M. Zhang, 2006: An assessment of ECMWF analyses and model forecasts over the North Slope of Alaska using observations from the ARM Mixed-Phase Arctic Cloud Experiment. *J. Geophys. Res.*, **111**, D05107, doi:10.1029/2005JD006509.

Yanai, M., S. Esbensen, and J.-H. Chu, 1973: Determination of bulk properties of tropical cloud clusters from large-scale heat and moisture budgets. *J. Atmos. Sci.*, **30**, 611–627, doi:10.1175/1520-0469(1973)030<0611:DOBPOT>2.0.CO;2.

Chapter 29

Parallel Developments and Formal Collaboration between European Atmospheric Profiling Observatories and the U.S. ARM Research Program

M. HAEFFELIN,* S. CREWELL,[+] A. J. ILLINGWORTH,[#] G. PAPPALARDO,[@] H. RUSSCHENBERG,[&] M. CHIRIACO,** K. EBELL,[+] R. J. HOGAN,[#] AND F. MADONNA[@]

** L'Institut Pierre-Simon Laplace, Centre National de la Recherche Scientifique, Palaiseau, France*
[+] Institute for Geophysics and Meteorology, University of Cologne, Cologne, Germany
[#] Department of Meteorology, University of Reading, Reading, United Kingdom
[@] Consiglio Nazionale delle Ricerche–Istituto di Metodologie per l'Analisi Ambientale, Potenza, Italy
[&] Department of Geoscience and Remote Sensing, Delft University of Technology, Delft, Netherlands
*** Laboratoire Atmosphère, Milieu, Observations Spatiales, Université Versailles Saint Quentin en Yvelines, Guyancourt, France*

1. Introduction

The climate research community aims to better characterize climate forcings such as aerosols, reactive gases, and greenhouse gases, and to better understand the responses of the climate system to these forcings. Such investigations rely in part on monitoring, studying, and understanding essential climate variables such as temperature, water vapor, clouds, radiation, and perturbations of aerosols and reactive gases. According to Dufresne and Bony (2008), the parameters that play a predominant role in radiative feedbacks of the climate system are atmospheric humidity, adiabatic thermal gradients, clouds, and surface albedo. Interactions between humidity, clouds, aerosols, and radiation make climate predictions more complex.

The climate research community has long recognized the link between climate prediction uncertainty and atmospheric process complexity. For more than 20 years, it has demonstrated the necessity to perform collocated long-term observations of thermodynamic parameters (temperature, humidity, wind) and atmospheric constituents (gases, aerosols, clouds) distributed along the entire atmospheric column (surface to stratosphere) and associated radiative components.

As a result, the U.S. Department of Energy (DOE) launched the Atmospheric Radiation Measurement (ARM) Program in the 1990s (Ackerman and Stokes 2003; Stokes 2016, chapter 2). Four atmospheric profiling observation facilities were developed to gather in situ and remote sensing instruments to monitor physical processes in the atmospheric column. A large research community of observation experts and climate modelers was funded to exploit the observation data. Similar atmospheric profiling observation facilities associated with large scientific communities emerged in Europe at the end of the 1990s. Several European initiatives were triggered or encouraged through bilateral collaborations between U.S. and European Union (EU) scientists or through participation of EU scientists in ARM projects (e.g., Cabauw observatory in the Netherlands; Palaiseau observatory in France; Jülich observatory in Germany).

Atmospheric profiling observatories provide scientists with the most resolved description of the atmospheric column. In Europe, as in the United States, these observatories have been collecting data every minute daily for more than a decade, allowing links to be established between processes occurring at diurnal or finer temporal scales and phenomenon occurring at climate scales. The limitation of an atmospheric profiling observatory is that it can only document one location of the globe with its specific atmospheric properties. The aerosol distributions, meteorological anomalies, and cloud properties observed at that location are representative of a limited spatial domain. Hence, atmospheric profiling observatories are needed at many locations around the globe to cover climatically diverse areas: near coasts, in continental

Corresponding author address: Martial Haeffelin, L'Institut Pierre-Simon Laplace, Palaiseau, 91128, France.
E-mail: martial.haeffelin@ipsl.polytechnique.fr

DOI: 10.1175/AMSMONOGRAPHS-D-15-0045.1

plains, mountains, and urban environments. The U.S. ARM Program was designed initially to cover three distinct climatic regions (Cress and Sisterson 2016, chapter 5): the Arctic (Alaska), midlatitudes [U.S. southern Great Plains (SGP)], and the tropics [tropical western Pacific (TWP) Ocean]. Atmospheric profiling observatories in Europe were developed primarily over the European continent, extending from locations around the Mediterranean Basin to the Arctic, and including coastal, continental, urban, and mountain sites.

The European Commission established several funding mechanisms to develop collaborations between researchers in Europe, to promote development of harmonized research infrastructures, and to reduce fragmentation in European research investments. As a result, in the past 10 years Europe was able to build an infrastructure essential to a large community of users by harmonizing aerosol, cloud, and trace gas observations across Europe.

As infrastructures, measurement techniques, data interpretation algorithms, and scientific expertise developed on both sides of the Atlantic, scientists became interested in the added benefits of collaboration and cross-fertilization between the U.S. ARM Program and EU atmospheric profiling research observatories. To expand investigations beyond existing atmospheric observatories, U.S. ARM scientists and ARM Mobile Facility (AMF) infrastructures participated in field experiments initiated by EU programs. EU and U.S. ARM scientists developed collaborations to harmonize data interpretation algorithms and to exploit jointly U.S. and EU observation datasets. Further development of formal collaboration between U.S. ARM and EU programs would enhance the ability of scientists worldwide to take on science challenges about climate change.

This chapter presents several European atmospheric profiling research observatories, development of European networking, and the current European research infrastructure (section 2). Section 3 presents EU program initiatives of interest for future collaboration with the ARM Program. Section 4 highlights collaborations that were developed subsequently between the U.S. ARM Program and its European counterparts. In section 5, we present an outlook toward future U.S.–EU collaborations around climate change challenges and observations.

2. European atmospheric profiling research observatories

Atmospheric profiling capabilities using active and passive remote sensing were developed as independent national initiatives in several European countries in the 1990s. Meteorological services and research institutes gathered several remote sensing systems, collocated them, and started to develop capacities to perform continuous measurements of atmospheric profiles and to store data for scientific research (section 2a). Through different initiatives of the European commission, several projects emerged in the early 2000s to coordinate atmospheric remote sensing activities across multiple European countries (section 2b). At the end of the 2000s, these coordination efforts were taken one step further to create a European research infrastructure initiative dedicated to a Europewide coordination of atmospheric profiling of aerosols, clouds, and trace gases for scientific research (section 2c).

a. National atmospheric profiling research observatories

Atmospheric profiling research observatories (APRO) with remote sensing capabilities were developed in Europe toward the end of the 1990s, a few years after the start of the U.S. ARM Program. Some APROs were developed by National Hydrological and Meteorological Services and their partners around existing meteorological facilities. Weather observations started in 1905 at the Meteorologisches Observatorium Lindenberg, now called the Richard Assmann Observatory, which became an atmospheric profiling observatory with remote sensing capabilities operated by the German Weather Service (DWD) in the mid-1990s. Similarly, the Royal Netherlands Meteorological Office (KNMI) founded a meteorological observatory in the early 1970s, which was upgraded in the early 2000s with many remote sensing instruments to become one of the more prominent European facilities for atmospheric research. Another example is the Payerne aerological station of the Swiss Meteorological Institute located in the western part of the Swiss midland.

Other observatories were developed by national research communities by bringing together atmospheric and climate scientists, who were experts in different remote sensing techniques. Some national research communities were connected to the ARM research community through participations in ARM projects or through bilateral collaborations with ARM scientists. This was the case of the SIRTA Observatory near Paris, France, which started from the initiative of a scientist in the 1990s. The development of the site was boosted in the early 2000s through collaboration with ARM scientists and participation in EU networks. Fifteen years later it has become a prominent European facility operating more than 100 sensors from 10 different institutes. In 1975, the National University of Ireland (Galway) established the Atmospheric Research station at Mace Head on the west coast of Ireland. The major observatory has been

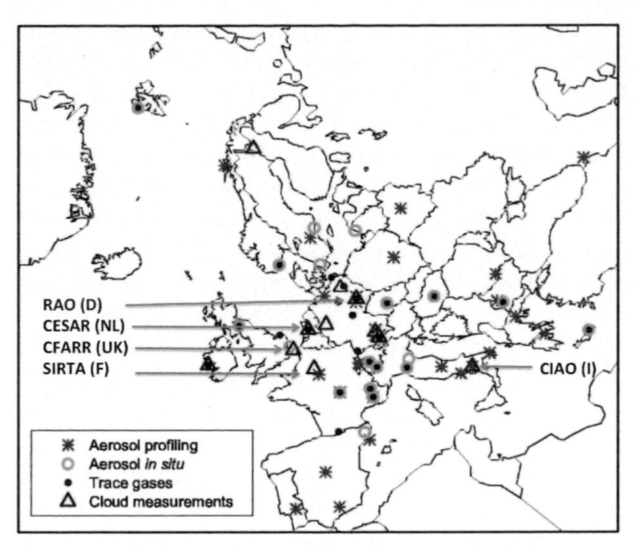

FIG. 29-1. Map of aerosol, cloud, and trace gas profiling and in situ measurement infrastructures in Europe, extending from the Mediterranean Basin to the polar regions (in 2011). Blue arrows indicate the geographical locations of the five European atmospheric observatories presented in section 2a.

used as a background baseline research station for over 50 years. (Aerosol measurements started in 1958 at a location nearby.)

Figure 29-1 shows the geographical distribution of atmospheric observatories in Europe dedicated to aerosol, cloud, and trace gas monitoring. Figure 29-1 highlights five prominent European atmospheric research observatories that contribute to many international networks, like the Baseline Surface Radiation Network (BSRN); the European Aerosol Research Lidar Network (EARLINET); Cloudnet; Aerosols, Clouds, and Trace Gases Research Infrastructure (ACTRIS) network; and Global Climate Observing System Upper-Air Reference Network (GRUAN). Their facilities, instruments, developments, and activities are presented in the following five subsections.

Atmospheric profiling observation activities in Europe were given a major boost in 1998 when the European Space Agency financed the 1998 Cloud Lidar and Radar Experiment (CLARE'98) field campaign. This campaign involved flying three instrumented aircraft from Germany, France, and the United Kingdom equipped with in situ sampling instruments, cloud radar, and lidars over the ground-based 94-GHz cloud radar at the Chilbolton observatory in the United Kingdom. This campaign demonstrated the ability of cloud radars and lidars to infer cloud properties leading to the selection of the joint European–Japanese Earth Clouds, Aerosol and Radiation Explorer (EarthCARE) satellite mission, which is scheduled to be launched in 2017. More recently, national meteorological and atmospheric research communities realized that activities around atmospheric profiling measurement and scientific research exploiting these measurements could be coordinated at regional or national levels, which led to construction of national networks of atmospheric profiling observatories. One example is a German network whose goal is to harmonize activities of several observatories around the High Definition Clouds and Precipitation for Climate Prediction project [HD(CP)^2]. Another example is the French Réseau d'Observatoires pour la Surveillance de l'Eau Atmosphérique (ROSEA), a network of five observatories dedicated to atmospheric water profiling. The geographical distributions of these two national networks are shown in Fig. 29-2.

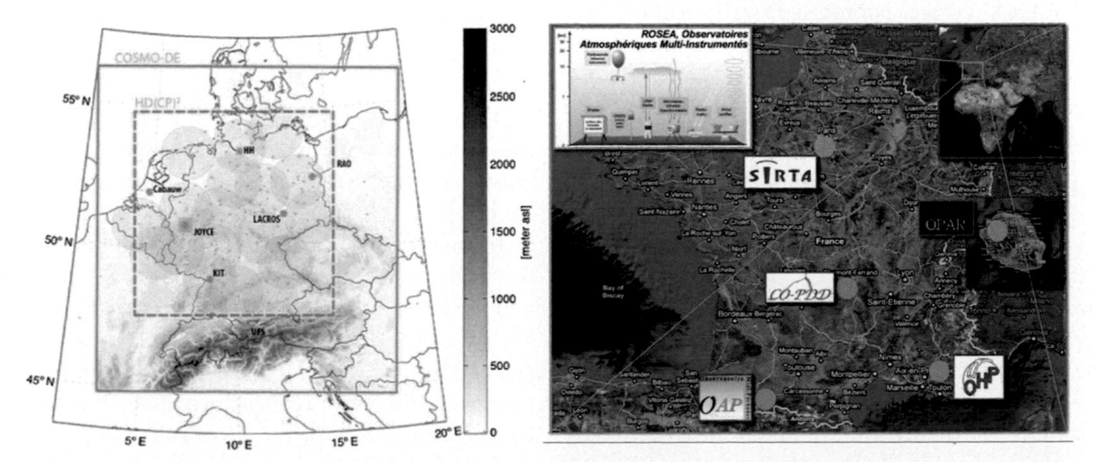

FIG. 29-2. Geographical locations of (a) German atmospheric profiling research observatories part of HD(CP)^2 and (b) French network of observatories for atmospheric water and aerosol profiling, including four observatories in continental France and one on Réunion Island (Indian Ocean).

1) The Cabauw Experimental Site for Atmospheric Research

The Cabauw Experimental Site for Atmospheric Research (CESAR) observatory is located in the western part of the Netherlands (NL; 51.97°N, 4.92°E). The site is located close to the sea and to some of the major European industrial and populated areas. The site is exposed to a large variety of airmass types. In 1973, a 213-m-high meteorological mast was built at the Cabauw site for the study of the atmospheric boundary layer (ABL), land surface conditions, and the general weather situation. Also, well-kept observation fields are onsite for micrometeorological observations, including soil heat flux, soil temperatures, and various radiation measurements (including a BSRN station). Within a 40-km radius, there are four major synoptic weather stations, ensuring a permanent supporting mesoscale network. Since 2000, remote sensing observations have been performed on clouds, rain, aerosols, and radiation (see Table 29-1 and Fig. 29-3b). Since 2002 the CESAR Observatory has been a national facility with commitments from eight research institutes and universities.

The CESAR site is used for

- monitoring long-term tendencies in atmospheric changes;
- studying atmospheric and land surface processes for climate and weather modeling;
- validating spaceborne observations;
- developing and implementing new measurement techniques;
- training young scientists at postdoctoral, Ph.D., and Masters levels.

Selected research highlights are presented in Table 29-2. The observatory is also used by the industry to test new technologies, either for comparison with similar instruments or for long-term endurance tests. All data are freely available through the CESAR data portal (www.cesar-observatory.nl), which also lists all publications that report on the use of CESAR data.

2) The Richard Assmann Observatory and German observatory network

The Meteorological Observatory Lindenberg–Richard Assmann Observatory (MOL-RAO) at Lindenberg operated by the DWD was originally founded in 1905. Since 1991, the MOL-RAO has been part of the DWD with extensive facilities. MOL-RAO serves as a regional reference station for many international programs and projects (Neisser et al. 2002). MOL-RAO (52.17°N, 14.12°E) is located in a rural environment dominated by farmland about 60 km to the southeast of Berlin (see Fig. 29-3a and Table 29-1). The midlatitude site is characterized by moderate climate in the transition zone between maritime and continental climate. In addition to the MOL-RAO, several advanced atmospheric profiling sites have become operational in Germany (see Fig. 29-2a). The Jülich Observatory for Cloud Evolution (JOYCE; Löhnert et al. 2014), located in the westernmost part of Germany (50.91°N, 6.41°E, 111 m MSL), was established in 2011 to characterize boundary layer clouds in the environment in which they form and decay. The Environmental Research Station Schneefernerhaus (UFS) is a unique research station located at an elevation of 2650 m in the Bavarian Alps just 300 m below the peak of the Zugspitze mountain (Germany's highest mountain). Originally set up for atmospheric trace gas measurements, it has now turned into a multipurpose station managed as a virtual institute for altitude, environment, and climate research by the Bavarian State Ministry of the Environment. Two mobile atmospheric profiling facilities [i.e., Leipzig Aerosol and Cloud Remote Observations System (LACROS) by the Leibniz Institute for Tropospheric Research, and the Karlsruhe Institute of Technology's (KIT) KITCube] were also developed. In total, seven K-band cloud radars operate continuously, giving Germany the world's densest cloud radar network. Selected MOL-RAO and JOYCE research highlights are presented in Table 29-3.

3) The Chilbolton Facility for Atmospheric and Radio Research

The Chilbolton observatory, located in Hampshire, United Kingdom (51.14°N, 1.44°W), was opened in 1967 when the construction of the 25-m dish was completed, and it now hosts the Chilbolton Facility for Atmospheric and Radio Research (CFARR). The S-band 3-GHz Advanced Meteorological Radar (CAMRa) installed on the big dish is the largest fully steerable meteorological radar in the world and is able to probe clouds and storms with unparalleled sensitivity and resolution. In 1980, it provided the first demonstration of improved radar estimates of rainfall by transmitting and receiving pulses alternately polarized in the horizontal and vertical (Hall et al. 1984). CFARR now comprises 20 major instruments (Fig. 29-3e), 10 of which are new since 2005, for studying clouds, rainfall, boundary layer processes, and aerosols (see Table 29-1). Many instruments operate 24–7 including the 35-GHz cloud radar, ceilometer, and microwave radiometer to provide continuous monitoring of the vertical structure of clouds and aerosol backscatter as part of the Cloudnet activity described in section 2b. Meteorological instruments include high-resolution rain gauges and disdrometers to measure raindrop spectra. All data are archived at the British

TABLE 29-1. List of main instruments operating continuously at European observatories for atmospheric profiling of temperature, humidity, clouds, aerosols, dynamics, and radiation. All instruments are typically found at U.S. ARM sites, except aerosol multiwavelength Raman lidars.

Instrument	MOL-RAO Lindenberg	CESAR Cabauw	SIRTA Palaiseau	CFARR Chilbolton	CIAO Potenza
Aerosol multiwavelength Raman lidar	—	—	355, 532, 1064 nm 387-, 607-nm Raman 355-db polarization	—	355, 532, 1064 nm 387-, 607-nm Raman 532-db polarization
Water vapor Raman or differential absorption lidar (DIAL)	408-nm Raman	408-nm Raman	408-nm Raman	Raman	408-nm Raman
Doppler cloud radar (35 or 95 GHz)	35-GHz pulsed	35 GHz pulsed	95-GHz frequency-modulated continuous-wave (FMCW) Laboratory developed	35- and 95-GHz pulsed	35-GHz pulsed
Microwave T and H profiler	Radiometrics 20 and 50 GHz	HATPRO 20 and 50 GHz	Radiometer Physics Gmbh HATPRO 20 and 50 GHz	Radiometrics MP 1516A	Radiometrics MP3014 22 and 60 GHz
Automatic backscatter lidar or profiling ceilometer	Jenoptik CHM 15k + CHM15kx	CT75 LD40 (905 nm) ALS450 (355 nm)	CL31 (905 nm) CHM15k (1064 nm)	CT75K (905 nm) ALS450 (355 nm)	CHM15k (1064 nm) CT25k (905 nm)
Radar/sodar Wind profiler	482-MHz wind profiler/ RASS sodar/RASS	1.29-GHz UHF radar	Degreanne L-band radar Remtech sodar	—	—
Lidar wind profiler	1.5-μm HALO-Photonics Streamline	—	Leosphere WLS70	Halo photonics Doppler lidar	Foreseen for 2014
Atmospheric radiosouding systems	Vaisala RS92 4 times daily	Vaisala RS92 0000 UTC daily	Modem M10 Twice daily	—	Vaisala autosonde; Weekly
Ozonesounding	Vaisala (Science Pump) ECC	At De Bilt	—	—	Vaisala MW21 Monthly
Multiwavelength sun photometer	Cimel CE-318, PFR	Yes	Cimel CE-318 340–1640 nm	CE-318	CE-318 340–1640 nm
Surface radiation station SW, LW, up, down	Kipp and Zonen (K&Z) CH1, CM22, CG4 (BSRN)	K&Z CH1, CM22, CG4 (BSRN)	K&Z CH1, CM22, CG4 (BSRN)	K&Z CH1, CM21, CMP21, CG4	K&Z CH1, CM22, CG4 (BSRN)
Surface sensible and latent heat fluxes	Metek USA-1 LiCor LI7500	Yes	Metek sonic Licor 7500	Metek sonic Licor 7500	Foreseen for 2014
Surface meteorological station	1–99-m mast	Yes	In ground, 1–30 m	Yes; many +distrometers	Vaisala MILOS520
GPS receiver	JPS Legacy, Ashtech Z-18	Yes	Trimble RET9	—	Trimble L1/L2 with chokerings
All-sky imager	WSI	Yes	Yes TSI-440	JVC KY55-BE	ORION StarShoot AllSky Camera II

FIG. 29-3. National atmospheric profiling research observatories in Europe: (a) RAO, Lindenberg, Germany; (b) CESAR, the Netherlands; (c) SIRTA atmospheric research observatory, Palaiseau, France; (d) CNR-IMAA atmospheric observatory, Potenza, Italy; (e) CFARR, United Kingdom.

Atmospheric Data Centre and are publicly available. Papers published from 1996 to 2011 have accrued 2307 citations. CFARR plays an important role in student training and was used by 9 Ph.D. students in 2013. Selected scientific highlights and campaigns over the past decade in areas such as cloud overlap, ice cloud physics, mixed-phase clouds, boundary layer dynamics, and volcanic ash are summarized in Table 29-4.

4) The Instrumental Site for Atmospheric Remote Sensing Research in Palaiseau

SIRTA is a French national atmospheric research observatory developed by L'Institut Pierre-Simon Laplace (IPSL; a research institute in environmental and climate sciences in the Paris metropolitan area) and its partners since the late 1990s (Haeffelin et al. 2005). The observatory is operated by staff from Centre National de la Recherche Scientifique, Ecole Polytechnique, Université Versailles Saint Quentin, Electricité de France, and Météo-France, and supported by the French Space Agency. SIRTA is located in a semiurban environment, 25 km south of the Paris city center (48.72°N, 2.21°E; see Fig. 29-3c). It operates over 100 sensors, monitoring ground conditions, surface fluxes, and profiles of atmospheric constituents and physical processes (see Table 29-1). Research objectives of SIRTA are to develop comprehensive

TABLE 29-2. Selected research highlights at the CESAR Observatory.

Theme	Highlight description	References
Liquid water clouds	Microphysical properties of water clouds are retrieved and validated with ground-based shortwave flux measurements.	Brandau et al. (2010), Wang et al. (2011)
KNMI Test Bed	Single-column models and LES are confronted with long-term and continuous observations for a statistical evaluation of model performance.	Neggers et al. (2012)
Volcanic ash	Optical properties of the Eyjafjallajökull ash cloud were characterized by detailed lidar measurements.	Donovan and Apituley (2013)
Climate model evaluation	Aerosol properties in the aerosol–climate model ECHAM5-HAM were evaluated with CESAR data.	Roelofs et al. (2010)

TABLE 29-3. Selected research highlights at RAO and the German observatory network.

Theme	Highlight description	References
Reference networks	MOL-RAO is the lead center for GRUAN and a WMO–Commission on Instruments and Methods of Observation (CIMO) Lead Centre on process-oriented observations. Furthermore, it contributes to EUMETNET/E-PROFILE, BSRN, Instruments and Methods of Observation Programme (IMOP)/CIMO, GEWEX–Coordinated Enhanced Observation Period (CEOP), and GEWEX Atmospheric Boundary Layer Studies (GABLS).	Engelbart and Steinhagen (2001), Neisser et al. (2002)
Boundary layer structure	At MOL-RAO, intensive campaigns to investigate boundary layer structure with additional in situ (including unmanned aerial vehicles) and remote sensing have been carried out, e.g., for entrainment studies. At JOYCE, the typical cumulus cloud-topped boundary layer is analyzed with respect to stability, turbulence, and cloud properties.	Martin et al. (2014), Löhnert et al. (2014)
Snowfall	Ground-based remote sensing and in situ measurements used in synergy at the Environmental Station Schneefernerhouse help to characterize the vertical distribution of snowfall necessary for satellite retrieval applications as well as for numerical model evaluation.	Löhnert et al. (2011)
Mobile stations	The KITcube consists of in situ and remote sensing systems including a scanningan X-band rain radar. It was deployed fully for the first time on the French island of Corsica during the Hydrological Cycle in the Mediterranean Experiment (HyMeX). Together with LACROS, it was deployed within the HOPE campaign a triangle of APROs around JOYCE.	Kalthoff et al. (2013), Bühl et al. (2013)

long-term atmospheric observations based on remote sensing and in situ sensors to study atmospheric processes and to analyze regional climate variability. The location of the observatory is designed to study both local/regional-scale processes typical to the urban–rural transition such as the formation mechanisms of gaseous and particulate pollution (Freutel et al. 2013) or the effects of aerosols on fog and shallow cumulus (Haeffelin et al. 2010) under high-pressure situations and larger-scale cloud–aerosol processes associated with baroclinic fronts. The SIRTA database is also geared toward global circulation model and numerical weather prediction model evaluations (e.g., Cheruy et al. 2013). Atmospheric process studies frequently take advantage of possible ground and satellite remote sensing synergies (e.g., Protat et al. 2009; Dupont et al. 2010). Selected SIRTA research highlights are presented in Table 29-5.

5) THE CNR-IMAA ATMOSPHERIC OBSERVATORY IN POTENZA

The Institute of Methodologies for Environmental Analysis (IMAA) of the National Research Council of Italy (CNR) runs the CNR-IMAA Atmospheric Observatory (CIAO). CIAO is located in Tito Scalo, 6 km from Potenza, in southern Italy, on the Apennine Mountains (40.60°N, 15.72°E, 760 m MSL) and less than 150 km from the west, south, and east coasts. The site is in a plain surrounded by low mountains (<1100 m MSL; see Fig. 29-3d). The observatory operates in a typical mountain weather environment strongly influenced by Mediterranean atmospheric circulation, resulting in generally dry, hot summers and cold winters, and is affected by a large number of Saharan dust intrusions each year (Mona et al. 2006).

CIAO represents the most equipped ground-based remote sensing station in the Mediterranean Basin for atmospheric profiling (see Table 29-1; Madonna et al. 2011; Boselli et al. 2012). Since 2000, CIAO is collecting systematic observations of aerosol, water vapor, and clouds. The main scientific objective is the long-term measurement for the climatology of aerosol and cloud properties to provide quality-assured measurements for satellite validation (Mona et al. 2009; Wetzel et al. 2013) and model evaluation (Pappalardo et al. 2004; Villani et al. 2006; Meier et al. 2012) and to fully exploit the synergy and integration of the active and passive sensors for the improvement of the atmospheric profiling (Madonna et al. 2010; Mona et al. 2012). CIAO provides access to data, services, and the research facility for conducting measurements campaigns, and instrument testing, with hundreds of users each year. Selected CIAO research highlights are presented in Table 29-6.

TABLE 29-4. Selected research highlights at the CFARR Observatory.

Theme	Highlight description	References
EarthCARE algorithm validation	CLARE'98 Campaign. Aircraft flights to validate radar/lidar retrievals of clouds for future EarthCARE mission.	Hogan et al. (2003b)
Rain rates from polarization radar	First demonstration of improved rainfall estimates and hydrometeor identification using polarization diversity radar; these techniques now implemented on operational radars worldwide.	Hall et al. (1984)
Cloud overlap	Measurements and parameterization of the degree of overlap of clouds and IWC; results implemented in many climate and weather forecast models worldwide.	Hogan and Illingworth (2000, 2003)
Ice cloud physics	Doppler radar demonstration that dominant growth mechanism in ice clouds is aggregation.	Westbrook et al. (2010)
Ice particle shattering	Demonstration using Doppler lidar that high concentrations of ice particles reported from aircraft are an artifact due to shattering.	Westbrook and Illingworth (2009)
Mixed-phase clouds	95% of ice in clouds warmer than $-20°C$ originates via the freezing of liquid supercooled droplets.	Westbrook and Illingworth (2011)
Ice nucleation	Observations demonstrating that ice nucleation in supercooled layer clouds is stochastic with a seemingly inexhaustible supply of ice nuclei.	Westbrook and Illingworth (2013)
Boundary layer dynamics	Use of Doppler lidar observations of vertical velocity skewness and variance to infer the upward and downward convective forcing in cloud-topped boundary layer.	Hogan et al. (2009b)
Turbulence measurement.	Using Doppler lidar observations in the boundary layer to infer turbulent kinetic energy dissipation rates.	O'Connor et al. (2010)
Volcanic ash	Monitoring of ash with lidar and photometers during Eyjafjallajokull eruption and validation of Met Office dispersion model.	Dacre et al. (2011)
Convective clouds	Cloud Storm Initiation Project (CSIP) 2004–05. Large international experiment based at Chilbolton to study convective cloud initiation.	Browning et al. (2007)

b. European networks of atmospheric profiling observations

1) TOOLS TO STRUCTURE EUROPEAN ATMOSPHERIC SCIENCE RESEARCH

The EU Framework Programme for Research and Technological Development is the main instrument for funding research in Europe (Defazio et al. 2009). By funding collaborative projects across Europe, the EU Framework Programme contributed significantly to develop collaboration between atmospheric research communities specializing in profiling atmospheric aerosols, clouds, and radiation in the early 2000s. Three initiatives that allowed construction of durable collaboration on aerosol and cloud profiling across Europe are presented in the three subsections below.

European Cooperation in Science and Technology (COST) is an intergovernmental framework whose goal is to reduce fragmentation in European research investments. COST helps develop cooperation between scientists and researchers across Europe by increasing their mobility through travel funds for meeting and short-term missions. COST Action 720 (2000–06), entitled "Integrated Ground-Based Remote Sensing Stations for Atmospheric Profiling," supported researchers from 12 countries (Engelbart et al. 2009). The main objective of the action was the development and assessment of cost-effective integrated ground-based remote sensing stations for atmospheric profiling of wind, humidity, and clouds. It made important contributions to the development of techniques for integrated profiling systems. COST Action ES0702 (2008–12), entitled "European Ground-Based Observations of Essential Variables for Climate and Operational Meteorology" (EG-CLIMET), supported researchers from 18 countries. The main objective of the EG-CLIMET action was the specification, development, and demonstration of cost-effective ground-based integrated profiling systems suitable for future networks providing essential atmospheric observations for both climate and weather. Following conclusions from the EG-CLIMET action, the European network of national hydrological and meteorological services (EUMETNET) launched a new program called E-PROFILE that will aim at coordinating the provision of calibrated aerosol and cloud profiling data from profiling ceilometers across Europe. This EUMETNET initiative will be accompanied by a new COST action (ES1303, 2013–17) entitled "Towards Operational Ground Based Profiling with Ceilometers, Doppler Lidars and Microwave Radiometers for Improving Weather Forecasts" (TOPROF). The TOPROF

TABLE 29-5. Selected research highlights at the SIRTA Observatory.

Theme	Highlight description	References
Cloud and fog processes	Subsidence and lifting of low stratus clouds can be driven by four different processes: coupling with the surface, changes in cloud-top radiative cooling, drizzle and precipitation rate, or large-scale subsidence.	Dupont and Haeffelin (2008), Dupont et al. (2012), Haeffelin et al. 2013
Origin of pollution	1/3 of regional particulate matter concentrations are due to local emissions, while 2/3 originate from continental transport. The proportion of the transported contribution increases in situations of high particulate matter concentrations.	Zhang et al. (2013)
GCM parameterization evaluation	Biases in temperature and humidity can be explained by biases in the partition between surface sensible and latent heat, underestimation of boundary layer clouds, and insufficient turbulent transport in the surface layer.	Cheruy et al. (2013)
Boundary layer structure	Synergy between lidar backscatter profiles and a Monin–Obukov length classification derived from sonic anemometer measurements to reduce uncertainties in daytime and nighttime mixing-height retrievals by more than a factor 2 compared to lidar retrievals alone.	Haeffelin et al. (2012) Pal et al. (2013) Cimini et al. (2013)
Access to the observatory	The SIRTA Observatory provides nearly 1000 accesses per year, where an access is defined as 1 user (researcher, student, visitor) for 1 day. Users access the observatory mainly (50%) in the framework of continuous long-term observation programs but also (25%) for shorter deployments such as field campaigns [e.g., Megacities: Emissions, Urban, Regional and Global Atmospheric Pollution and Climate Effects, and Integrated Tools for Assessment and Mitigation (MEGAPOLI), ParisFog], and 25% for experimental teaching sessions and outreach. Each year more than 2500 student hours of teaching are performed on the observatory.	Freutel et al. (2013) Haeffelin et al. (2010)

action aims at developing the procedures to harmonize the provision of data from profiling ceilometers, microwave radiometers, and Doppler lidars.

The last decade has shown rapid advancement in ground-based remote sensing instrumentation being first implemented at reference sites with high potential for larger networks. Because the principles and applications of these instruments are not reflected in past and current university curricula, training activities on various educational levels are required. In addition to training future users, this training also is interesting for small and medium enterprises with growing demand for well-trained personnel. The European Marie Curie Initial Training Network on Atmospheric Remote Sensing (ITARS) aims to bridge the gap between the specialized development of single instruments and atmospheric applications by providing individual training, courses, and summer schools with focus on sensor synergy for early stage and experienced researchers.[1]

[1] For the first time ever, ARM hosted a summer workshop in 2015 to train graduate students to use data from ground-based remote sensors. The ARM summer workshop was a follow-on activity from the 2014 ITaRS summer school, which included several ARM principal investigators as instructors.

2) THE FP5 CLIWA-NET PROJECT

The Cloud Liquid Water Network (CLIWA-NET) project (2000–03) was initiated in the context of the EU Baltic Sea Experiment (BALTEX). The objectives of CLIWA-NET were to improve parameterizations of cloud processes in atmospheric models with a focus on vertically integrated cloud liquid water path (LWP) and vertical structure of clouds. To achieve this goal, a prototype of a European Cloud Observation Network was set up, which consisted of 12 ground-based stations and satellite measurements. Because microwave radiometry is the most accurate way to measure liquid water path, more than 10 different microwave radiometers from European universities and research organizations operated successfully during three enhanced observation phases—all part of BRIDGE, the major field experiment of BALTEX. Most importantly, the BALTEX BRIDGE Campaign (BBC; Crewell et al. 2004) included multiple aircraft observations and a microwave intercomparison campaign that served as a baseline to develop an operational microwave radiometer for LWP and thermodynamic profiles (HATPRO; Rose et al. 2005). Methodologies focusing on the evaluation of model-predicted cloud parameters with CLIWA-NET inferred observations were developed and examined in various applications, for example, a statistical evaluation

TABLE 29-6. Selected research highlights at CIAO.

Theme	Highlight description	References
Aerosol	Characterization of aerosol optical and microphysical properties using lidar sun photometer and radar measurements. Climatological studies, long-range transport events, Saharan dust outbreaks, plumes from volcanic eruptions and for model evaluation and satellite data validation and integration.	Madonna et al. (2013), Mona et al. (2012)
Aerosol–cloud interactions	Study the variability of aerosol optical properties, relative humidity, updrafts, and downdrafts in broken thin liquid water clouds with the aim to gain a better insight in droplet activation process using Raman lidar, Doppler radar, and microwave radiometer observations.	Rosoldi et al. (2013)
Aerosol transport	Analysis of the physical and dynamical processes related to aerosol transport as well the validation of the main transport modeled [Dust Regional Atmospheric Model (DREAM), Navy Aerosol Analysis and Prediction System (NAAPS), HYSPLIT] using advanced lidar observations for different aerosol types (e.g., Saharan, volcanic, biomass burning).	Villani et al. (2006), Sawamura et al. (2012), Pappalardo et al. (2013)
Satellite calibration/validation	A strategy for EARLINET correlative measurements for *CALIPSO* has been developed at CIAO, allowing a reliable statistical analysis and validation of *CALIPSO* data.	Mona et al. (2009), Pappalardo et al. (2010)
Advanced statistical analysis of atmosphere thermodynamics	General and versatile statistical modeling approach to understand to what extent measurement uncertainty and redundancy are related to environmental factors, height, and distance has been elaborated using data from the main highly instrumented station available worldwide.	Fassò et al. (2014), Madonna et al. (2014)
Upper-air measurements	In situ and ground-based remote sensing measurements in the upper troposphere are routinely performed to assess long-term trends and variability, providing traceable measurements with their uncertainty budget.	Mona et al. (2007)
Access to the observatory	CIAO provides nearly 500 physical accesses per year, where an access is defined as 1 user (researcher, student, visitor) for 1 day accessing the infrastructure. CIAO provides also open access to its data archive and to specific services on request. 60% of the access is provided to European and international users through calibration services, data processing services, access to data, and physical access for specific experiments and training activities. International large field campaigns are organized with international partners in the framework of EU and international projects/programs. Access to new users is promoted through dissemination activities (per review articles, presentations at European and international conferences). Access to young scientists is promoted through Marie Curie Actions and European and international schools.	Madonna et al. (2011) www.ciao.imaa.cnr.it

of LWP (van Meijgaard and Crewell 2005), the representation of vertically distributed liquid water content (Willen et al. 2005), and comparisons of model-predicted LWP fields with satellite retrieved spatial distributions. Activities to improve temperature and humidity profile retrievals from microwave radiometers were initiated during CLIWA-NET and are described in section 3a.

3) THE FP5 CLOUDNET PROJECT

Originally an EU-funded project running from 2001 to 2005 [Fifth Framework Programme (FP5)], the aim of Cloudnet is to provide a systematic evaluation of clouds in forecast models (Illingworth et al. 2007). This evaluation has been achieved by establishing several ground-based remote sensing sites within Europe, which, like those of the U.S. ARM Program, are equipped with an array of instrumentation using active sensors such as lidar and Doppler millimeter-wave radar. These ground-based remote sensing sites provide vertical profiles at high spatial and temporal resolution of the main cloud variables used in forecast models, namely cloud cover and cloud ice and liquid water contents. Previously, the efforts to improve clouds in forecast models had been hampered by the difficulty of making accurate and continuous observations of clouds. Aircraft studies by their nature provide incomplete spatial and temporal studies, and published papers concentrating on case studies may be atypical.

Following the ethos of the ARM Program, these sites have operated continuously for many years in order to gain statistics and sample the full range of weather phenomena. An important aspect of Cloudnet was the involvement of a number of European operational forecast centers in a cooperative effort to evaluate and improve their skill in cloud predictions. These centers provided profiles of cloud properties hourly for the model grid box over the three original Cloudnet observing stations (see map in Fig. 29-1): CESAR (the Netherlands), CFARR (United Kingdom), and SIRTA (France), but more recently extended to MOL-RAO (Germany) and many other sites as discussed in section 3c. The procedure for deriving cloud properties from ground-based observations for evaluating models is not trivial (e.g., see Shupe et al. 2016, chapter 19). Each of the sites has a different mix of instruments, so a crucial part of Cloudnet has been to devise a uniform set of procedures and data formats to enable the algorithms to be applied at all sites and used to test all models. Cloudnet algorithm developments are presented in section 3a. The core instruments for use in cloud retrievals at each site are a Doppler cloud radar, a lidar ceilometer, a dual- or multiwavelength microware radiometer, and a rain gauge, all operating 24 hours each day. A crucial aspect is to have a common calibration standard for the instruments, so techniques were developed for automatically calibrating cloud lidars (O'Connor et al. 2004) and cloud radar (Hogan et al. 2003a) using the properties of the meteorological targets themselves.

The evaluation of the representation of clouds in seven European operational forecast models as reported by Illingworth et al. (2007) and Bouniol et al. (2010) were quite revealing. In 2003, several gross errors in cloud fraction were identified in some models, but analysis of updated models for the year 2004 showed a considerable improvement. However, a common shortcoming of all models was the lack of midlevel cloud and the inability of many models to produce sufficient occasions when there was 100% cloud cover. Results are provided in section 3c.

4) THE FP5 AND FP6 EARLINET PROJECTS

EARLINET was established in 2000 as a research project funded by the European Commission, within the Fifth Framework Programme, with the main goal to provide a comprehensive, quantitative, and statistically significant database for the aerosol distribution on a continental scale. After the end of this project, the network activity continued based on a voluntary association. The 5-yr (2006–11) project EARLINET-Advanced Sustainable Observation System (ASOS) in the Sixth Framework Programme (FP6), starting on the EARLINET infrastructure, has contributed strongly to optimize the operation of the network.

The network started to perform measurements on 1 May 2000 with 22 lidar stations distributed over 14 European countries. Since then, the network has grown both in number of stations and observational capability. Currently, EARLINET consists of 27 lidar stations: 10 single backscatter lidar stations, 8 Raman lidar stations with the UV Raman channel for independent measurements of aerosol extinction and backscatter, and 9 multiwavelength Raman lidar stations (elastic channel at 1064, 532, and 355 nm; Raman channels at 532 and 355 nm; plus a depolarization channel at 532 nm). (A complete list of stations can be found at www.earlinet.org. The locations of these stations are shown as red stars in Fig. 29-1.)

Lidar observations within the network are performed on a regular schedule of one daytime measurement per week around noon, when the boundary layer is usually well developed, and two nighttime measurements per week, with low background light, in order to perform Raman extinction measurements (Matthias et al. 2004a). In addition to the routine measurements, further observations are devoted to monitor special events such as Saharan dust outbreaks (Ansmann et al. 2003; Papayannis et al. 2008), forest fires (Balis et al. 2003) and volcano eruptions (Pappalardo et al. 2013). Since June 2006, additional measurements have been performed at EARLINET stations in coincidence with *CALIPSO* overpasses according to a strategy for correlative measurements developed within EARLINET (Pappalardo et al. 2010).

Data quality has been assured by instrument intercomparisons using the reference transportable systems (Matthias et al. 2004b). The quality assurance also included the intercomparison of the retrieval algorithms for both backscatter and Raman lidar data (Böckmann et al. 2004; Pappalardo et al. 2004). Moreover, ad hoc tools for the continuous quality check of the instruments and algorithms are used regularly.

The EARLINET database is an important source of data that contributes to the quantification of anthropogenic and biogenic emissions and concentrations of aerosols, quantification of their budgets, radiative properties, and prediction of future trends. It contributes therefore to the improvement of the understanding of physical and chemical processes related to aerosols, their long-range transport and deposition, and their interaction with clouds (e.g., Guibert et al. 2005; Meier et al. 2012).

c. *Network of networks*

Since 2000, significant efforts have been made in Europe to establish research infrastructures and networks for atmospheric research. However, only in the EU Seventh Framework Programme was a coordinated research infrastructure for these observations established.

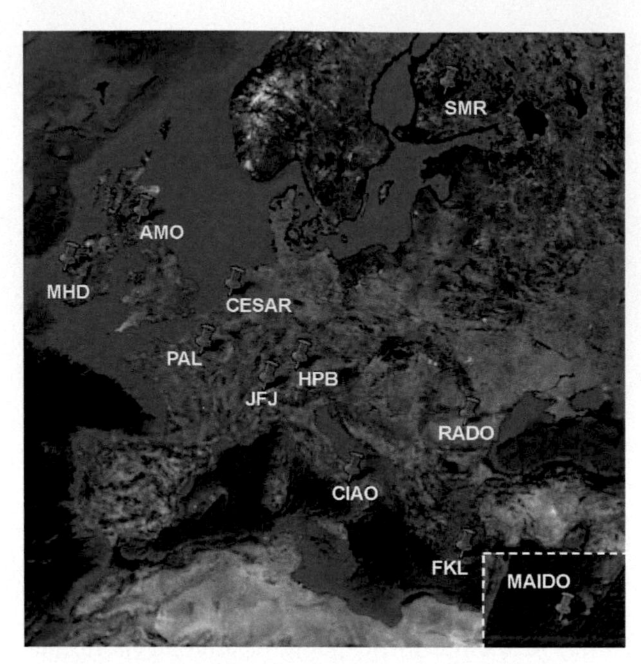

FIG. 29-4. ACTRIS sites offering transnational access.

The ACTRIS network is an outstanding research infrastructure launched in 2011 that aims to coordinate the European ground-based network of stations equipped with advanced atmospheric probing instrumentation for aerosols, clouds, and short-lived trace gases. The main objectives of ACTRIS are the following:

- To provide long-term observational data relevant to climate and air quality research on the regional scale produced with standardized or comparable procedures throughout the network (Fig. 29-1).
- To provide a coordinated framework to support transnational access to large infrastructures (Fig. 29-4) strengthening high-quality collaboration in and outside the European Union and access to high-quality information and services for the user communities (research, environmental protection agencies, etc.).
- To develop new integration tools to fully exploit the use of multiple atmospheric techniques at ground-based stations, in particular for the calibration/validation/integration of satellite sensors and for the improvement of the parameterizations used in global and regional-scale climate and air quality models.
- To enhance training of new scientists and new users in particular students, young scientists, and scientists from eastern European and non-EU developing countries in the field of atmospheric observation.
- To promote development of new technologies for atmospheric observation of aerosols, clouds, and trace gases through close partnership with EU small and medium enterprises (SMEs).

A key for ACTRIS success is to build a new research infrastructure on the basis of a consortium joining existing networks/observatories that are already providing consistent datasets of observations and that are performed using state-of-the-art measurement technology and data processing.

In particular, the ACTRIS consortium merges two existing research infrastructures funded by the European Commission under FP6: European Supersites for Atmospheric Aerosol Research (EUSAAR) and EARLINET (section 2b). ACTRIS also includes the distributed infrastructure on aerosol interaction existing from the Cloudnet EU research project (section 2b) and by grouping the existing EU ground-based monitoring capacity for short-lived trace gases, which currently is not coordinated at any level—except for the European Monitoring and Evaluation Programme (EMEP) and the Global Atmosphere Watch (GAW) caring for a few specific compounds. Therefore, ACTRIS represents an unprecedented effort toward integration of a distributed network of ground-based stations, covering most climatic regions of Europe, and responding to a strong demand from the atmospheric research community. ACTRIS is a step toward better integration of aerosol, cloud, and trace gases communities in Europe necessary to match the integration of high-quality long-term observations of aerosol, clouds, and short-lived gas-phase species and for assessing their impact on climate and environment. ACTRIS outcomes will be used for supporting decisions in a wide range of policy areas, including air quality, health, international protocols, and research requirements.

3. EU program initiatives opening to areas of collaboration with the U.S. ARM Program

Projects funded by the European Commission, presented in sections 2b and 2c, allowed European countries to develop and harmonize observation infrastructures. These projects also allowed important scientific developments by supporting the improvement of retrieval methods and algorithms to derive essential climate variables (section 3a), the reanalysis of long-term atmospheric profiling observations to produce quality controlled and harmonized datasets to study climate variability and related atmospheric processes (section 3b), and the development of frameworks including better tools and methods to evaluate weather forecast and climate prediction models (section 3c).

a. Retrieval algorithm developments

EU research programs associated with atmospheric profiling observatories have focused on the development

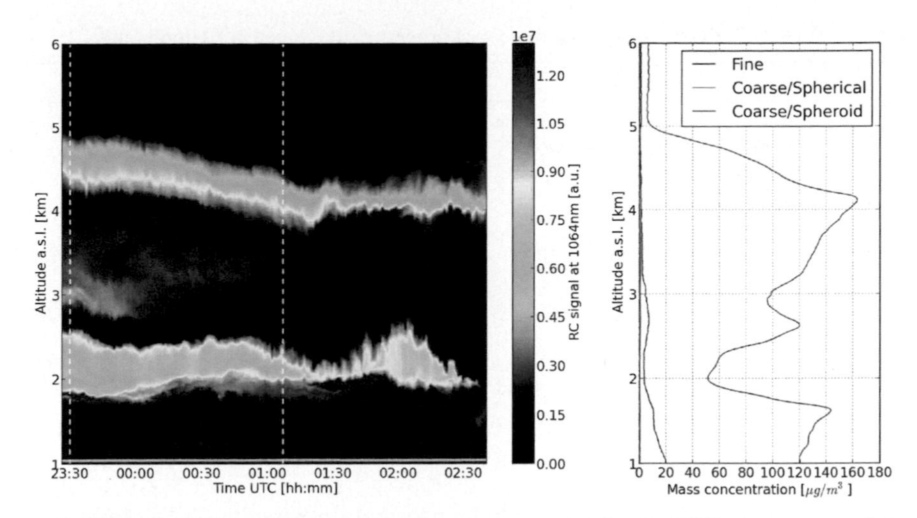

FIG. 29-5. (left) Time evolution of the lidar range-corrected signal at 1064 nm as measured at CIAO on 4 Sep 2011 during a Saharan dust outbreak. (right) Corresponding mass concentration profiles for fine (blue) and coarse particles, both spherical (green) and spheroid (red) as retrieved using collocated multiwavelength backscatter (355, 532, and 1064 nm) and depolarization (532 nm) lidar and sky-scanning radiometer observations. Dashed white lines in (left) indicate the time window for the retrieval reported in (right) (2330 UTC 4 Sep–0107 UTC 5 Sep).

of algorithms to retrieve aerosol properties; temperature and humidity profiles; boundary layer height; and cloud properties from radars, lidars, and microwave radiometers. Developments focused on retrievals from sophisticated systems such as multiwavelength Raman lidars and polarized Doppler cloud radars. Recently, low-cost low-power elastic backscatter lidars (profiling ceilometers), profiling microwave radiometers, and continuous-emission cloud radars became available. In Europe alone, several hundred profiling ceilometers are gathering aerosol and cloud backscattering data continuously as national weather services started to build up networks of ceilometers (e.g., Flentje et al. 2010). About 30 microwave profilers are also available, and the potential for low-cost continuous-emission cloud radar networks to develop is high. Hence research developments now also focus on assessing the performance of the low-cost instruments and developing specific retrieval algorithms.

1) AEROSOL PROFILE RETRIEVALS

Detailed knowledge of optical, microphysical, and radiative properties of aerosol particles is required to understand their role in atmospheric processes as well as their impact on human health and the environment (Forster et al. 2007). The properties must be monitored as a function of time and space, where the vertical dimension is of particular importance because of high variability. Lidar techniques are ideal for collecting range-resolved data for the characterization of aerosol particles.

EU programs such as EARLINET and ACTRIS provided collaboration frameworks within Europe and strongly supported developments of multiwavelength Raman lidar. These programs also motivated algorithm developments to retrieve aerosol optical properties (backscatter and extinction profiles) as well as microphysical properties (size, shape) and types from Raman lidars. Recent developments now take advantage of the synergy between multiwavelength measurements of lidars and sunphotometers, as illustrated in Fig. 29-5. Examples of developments are presented in Table 29-7.

EU COST actions such as EG-CLIMET and TOPROF also provided useful collaboration frameworks to exploit existing, yet underexploited, low-power automatic backscatter lidars and profiling ceilometers (ALCs). Following spring 2010 when air traffic was disrupted in Europe because of the presence of volcanic ash plumes (e.g., Pappalardo et al. 2013), a renewed interest was gained in the potential of ALCs to retrieve aerosol properties. Techniques for calibrating ALCs and for retrieving backscatter profiles from ALCs developed in the framework of EU programs are presented in Table 29-7.

2) TEMPERATURE AND HUMIDITY PROFILE RETRIEVALS

Tropospheric temperature and humidity are basic meteorological quantities that determine atmospheric stability. Therefore thermodynamic profiling with high

TABLE 29-7. Retrieval methods of aerosol properties developed in EU programs.

Retrieved variables	Input data	References
Backscatter coefficient (from ceilometers)	Ceilometer-attenuated backscatter profile and optical depth from sunphotometer	O' Connor et al. (2004), Markowicz et al. (2008), Flentje et al. (2010), Heese et al. (2010), Morille et al. (2007), Wiegner and Geiß (2012), Wiegner et al. (2014)
Aerosol backscatter and extinction	Raman lidar	Ansmann et al. (1990, 1992)
Microphysical properties	Multiwavelength Raman lidars	Müller et al. (1999), Veselovskii et al. (2002), Böckmann et al. (2005), Ansmann et al. (2012)
Microphysical properties and aerosol typing	Multiwavelength Raman lidars and sun photometers	Müller et al. (2004, 2007), Wiegner et al. (2008), Gasteiger et al. (2011), Mona et al. (2012), Chaikovsky et al. (2012), Pappalardo et al. (2013), Wagner et al. (2013), Lopatin et al. (2013)
Giant aerosol	Multiwavelength Raman lidars and millimiter-wavelength radars	Madonna et al. (2010, 2013)

temporal and spatial resolution is of high importance for many applications in atmospheric sciences, such as initialization of weather forecasting, model evaluation, and process studies. Radiosonde soundings can provide high vertical resolution profiles along the balloon trajectory but are limited to time intervals of typically 12 h. Therefore, continuous profile observations by unattended remote sensing instruments are of high interest (Carbone et al. 2012) but suffer some drawbacks in vertical resolution and accuracy.

Microwave radiometry is commonly used to derive temperature and humidity profiles from brightness temperature (BT) measurements by applying regression-based retrieval algorithms relying on a comprehensive prior dataset. BT measurements typically in zenith direction are made at several frequencies along absorption complexes, that is, water vapor and oxygen, requiring a good knowledge on atmospheric absorption characteristics. Kadygrov and Pick (1998) introduced a single frequency technique for boundary layer temperature profiling where different opacities are realized via different elevation angles. To improve accuracy and vertical resolution multifrequency and multiangle measurements can be combined (Crewell et al. 2009). A major advantage of microwave radiometer retrievals is that they are mostly independent on the occurrence of clouds, except for cases of heavy precipitation where saturation effects may occur or when the measurement is influenced by rainwater on the microwave radiometer radome. Infrared spectrometers also can provide thermodynamic profiles but are limited to clear-sky conditions where they are more accurate than the microwave retrievals (Löhnert et al. 2009; Fig. 29-6). However, in the lowest 500-m microwave-derived temperature profiles, derived from elevation scans are as accurate as the infrared retrievals. To optimally exploit the information content of microwave radiometers, variational techniques that combine BT measurements with a priori knowledge and/or auxiliary information have been developed for physically consistent temperature and humidity profiling (Hewison 2007; Cimini et al. 2006). Table 29-8 presents temperature and humidity profile retrieval methods based on microwave radiances developed in the framework of EU programs. Further developments through collaboration with the ARM Program are presented in section 4b.

3) MIXING-HEIGHT RETRIEVALS

The atmospheric mixing height is the height of the layer adjacent to the ground over which constituents emitted within this layer or entrained into it become vertically dispersed by convection or mechanical turbulence within a time scale of about one hour (Seibert et al. 2000). During daytime the mixing layer tends to be unstable as a result of convection and is capped by an entrainment zone. At night a shallow stable layer forms near the surface in which mixing occurs through intermittent turbulence, leaving a residual layer above. Mixing height is a necessary parameter to relate boundary layer concentrations of gases to upstream fluxes and to scale dispersion of trace gases and aerosols for air quality applications.

As pointed out in Seibert et al. (2000), there is no "mixing-height meter" able to determine the mixing height without uncertainties and assumptions. Furthermore, the definitions of mixing layer depend on the geophysical quantity employed in the definition. Because of the importance of this parameter, in the past 20 years, no less than five EU COST actions were at least partially dedicated to better understanding and improving mixing-height retrieval techniques. Table 29-9 provides references to retrieval methods based on radio sounding, lidar, sodar, radar, and microwave radiometers derived in the framework of EU COST actions. The use of instrument synergy allows objective retrievals to be developed as illustrated in Fig. 29-7 (Pal et al. 2013).

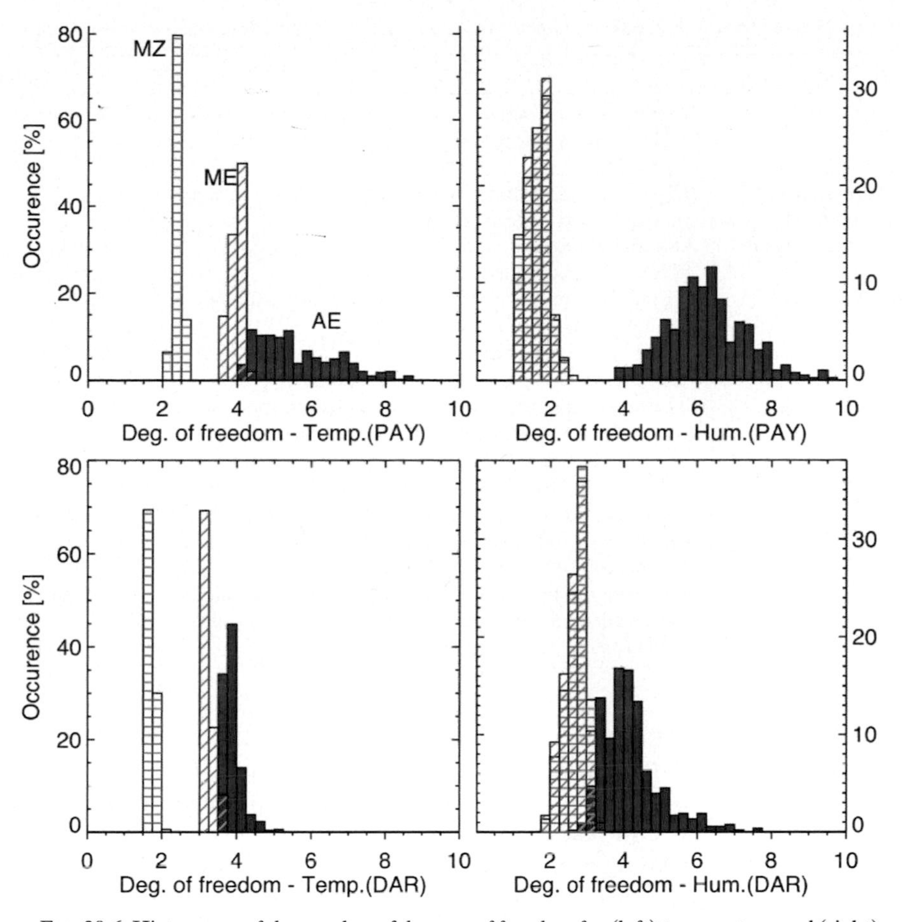

FIG. 29-6. Histograms of the number of degrees of freedom for (left) temperature and (right) humidity retrievals at the (top) Payerne and (bottom) Darwin sites. The different shading indicates the retrieval methods: microwave radiometer zenith only (MZ) in green horizontal lines, microwave radiometer with variable elevation angles (ME) in green slanted lines, and Atmospheric Emitted Radiance Interferometer (AE) in red.

The multi-instrument retrieval techniques could be of interest to derive mixing heights over ARM sites.

4) Cloud profile retrievals

Cloud property retrievals derived from cloud radars and lidars were developed in the framework of the EU Cloudnet project (presented in section 2b). These retrieval algorithms use continuous observations by millimeter-cloud radars, lidar ceilometers, microwave radiometers, and rain gauges to derive values of cloud fraction, ice, and liquid cloud water content (Illingworth et al. 2007). The overall retrieval framework consists of two steps, with the target classification being performed first followed by the microphysical retrievals. The retrieval algorithms were chosen based on their ability to be applied robustly to long periods of data with well-characterized errors. The first step in processing is to perform 30-s averaging from each site with the instrument vertical resolution of 30 or 60 m, followed by classifying the target in terms of liquid cloud, ice cloud, rain, aerosol, insects, and combinations thereof. The target classification then guides the retrieval of ice and liquid water content at the instrument resolution. Values of cloud fraction, liquid water content, and ice water content (see Table 29-10 for details) are derived and then averaged onto the vertical grid of each forecast model, and also averaged in time by an amount equivalent to the horizontal resolution of the model given the profile of wind speed. Application of this retrieval scheme to ARM Program measurements is presented in section 4c.

As well as being used for model evaluation, the target classification and microphysical retrievals have been used to study cloud processes. For example, the identification of supercooled water clouds has been used in an analysis (Fig. 29-8) of four years of data at Chilbolton to reveal that 95% of ice forming at temperatures warmer than −20°C originates via the freezing of liquid drops in supercooled clouds.

TABLE 29-8. Retrieval methods of temperature and humidity profiles developed in EU programs.

Retrieved variables	Input data	References
Temperature profile (by microwave radiometry)	Brightness temperatures at several frequencies along 60-GHz oxygen absorption complex taken in zenith direction	Westwater et al. (2005)
Temperature profile with improved vertical resolution in boundary layer (by microwave radiometry)	Brightness temperatures taken at several elevation angles along 60-GHz oxygen absorption bands	
	At a single frequency	Kadygrov and Pick (1998)
	At several frequencies	Löhnert and Maier (2012)
Humidity profile (by microwave radiometry)	Brightness temperatures at several frequencies along water vapor absorption bands	
	22-GHz water vapor line	Güldner and Spänkuch (2001)
	183-GHz water vapor line (dry conditions)	Ricaud et al. (2010)
Temperature and humidity profiling (by infrared interferometry)	Spectral infrared radiance in different bands, spectral observations: 612–713 and 2223–2260 cm^{-1} (i.e., 15- and 4.3-mm CO_2 bands, respectively) for temperature profiling; 538–588 and 1250–1350 cm cm^{-1} for water vapor	Spänkuch et al. (1996)
Temperature and humidity profiles (1D VAR method)	Brightness temperatures along 22.235-GHz water vapor absorption and 60 GHz, ambient temperature and humidity, infrared temperature	Hewison (2007) Cimini et al. (2006)
Temperature and humidity profiles, LWC (IPT method)	Brightness temperatures along 22.235-GHz water vapor absorption and 60-GHz oxygen absorption complex, cloud radar reflectivity profile	Löhnert et al. (2008)

5) SYNERGETIC LIQUID CLOUD PROFILE RETRIEVALS AND BLIND TEST INITIATIVE

State-of-the-art liquid cloud profile retrievals typically use information from cloud radar, microwave radiometer (MWR) and lidar to retrieve liquid cloud parameters like liquid water content, cloud droplet number concentration (N), effective radius (R_{eff}), and cloud optical depth (COD). Various methods to retrieve these properties exist and may differ in the measurements used and assumptions made. Some methods combine cloud radar and MWR information, for example, the Technical University Delft Remotely-Sensed Cloud Property Profiles (TUD-RSCCP) algorithm (Brandau et al. 2010) or the integrated profiling technique (IPT; Löhnert et al. 2004; Löhnert et al. 2008). In contrast to TUD-RSCCP

TABLE 29-9. Mixing-height retrieval methods developed in EU programs.

Retrieved variables	Input data	References
Mixing height from numerical model output	Parameterizations in meteorological preprocessors	COST Action 710 Fisher et al. (1998), Seibert et al. (2000)
Mixing height from measurements	Radiosonde profiles, sodar, and wind profiler measurements	COST Action 710 Seibert et al. (2000)
Urban mixing height from numerical model output and measurements	Mesoscale numerical simulations	COST Action 715
	Radiosonde profiles and sodar measurements	Fisher et al. (2001) Piringer et al. (2007)
Mixing height from surface-based remote sensing	Sodar, radar, and lidar profiling measurements	Emeis et al. (2008)
Mixing height traced by aerosols	Attenuated backscatter profiles measured by automatic lidars and ceilometers	COST Action ES0702 [Structure of the Atmosphere (STRAT) methods]
	Backscatter profiles alone	Haeffelin et al. (2012)
	Backscatter and surface stability conditions derived from sonic anemometers	Pal et al. (2013)
Mixing height traced by temperature profilers	Microwave radiometer temperature profiles	COST Action ES0702
	Microwave brightness temperatures	Cimini et al. (2013)
Mixing-height dynamics using Doppler lidar	Vertical velocity profiles and velocity variance profiles	Barlow et al. (2011)
Boundary layer types	Vertical velocity profiles and velocity variance profiles	Harvey et al. (2013)

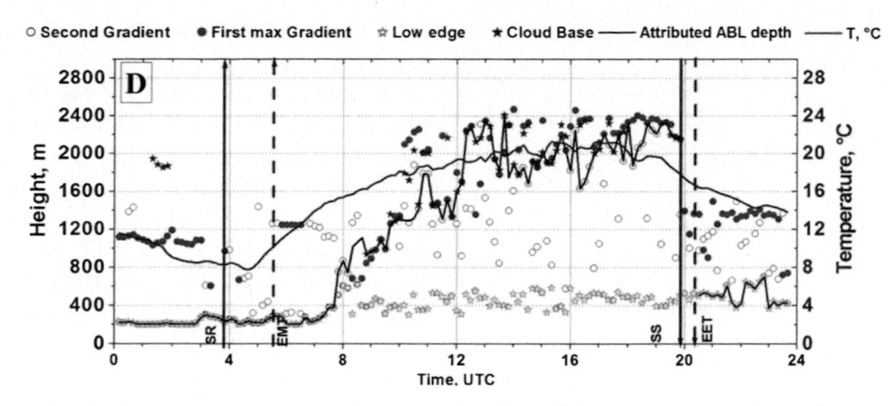

FIG. 29-7. Ceilometer-attenuated backscatter gradients (black circles, red circles, and green stars); cloud-base height (blue stars); attributed mixing height (black line); air temperature (blue line); sunrise (SR) and sunset (SS)—vertical blue solid lines; early morning transition (EMT) and early evening transition (EET)—vertical blue dashed lines. Parameters are derived from lidar-attenuated backscatter and sonic anemometer measurements at the SIRTA Observatory.

and IPT, the synergistic remote sensing of cloud (SYRSOC) algorithm (Martucci and O'Dowd 2011; Martucci et al. 2012) also makes use of lidar observations (see Table 29-11 for more information on the algorithms). Within the EG-CLIMET COST action (http://www.cost.eu/COST_Actions/essem/ES0702), the above-listed algorithms were assessed thoroughly via an observing system simulation experiment (OSSE). Using synthetic observations based on scenes from cloud-resolving model output, an independent evaluation of the different retrieval algorithms was conducted. All methods are very sensitive to the correct description of the cloud boundaries and the correct discrimination between cloud droplets and precipitation. The accuracy of the SYRSOC liquid water content depends on the accuracy of the retrieved lidar extinction. For nonprecipitating cases, the TUD-RSCCP method provides the best results with accuracy in liquid water content of ~15% (Fig. 29-9). In precipitating cases, drizzle drops dominate the radar reflectivity factor signal resulting in an overestimation (underestimation) of the effective radius (droplet number concentration). However, both IPT and TUD-RSCCP still provide robust results for the liquid water content with errors in the range of 20%–50%.

During the EU–DOE Ground-Based Cloud and Precipitation Retrieval Workshop, which took place on 13–14 May 2013 in Cologne, it was decided that an extended experiment within the same framework that would also include DOE ARM retrieval algorithms would be conducted in the future.

b. Long-term climate datasets

Atmospheric profiling observatories are useful for modeling applications and climate studies, in particular because local processes can be used to explain the seasonal and interannual variability of climate (e.g., Chiriaco et al. 2014). Nevertheless, climate trends or variability cannot be detected in a dataset if the climate signal is less than the measurement biases. These biases must be reduced using specific procedures. The data from each APRO must be reprocessed carefully to include better quality control and better retrieval algorithms, to make use of instrument synergy, to reduce biases, and to evaluate uncertainties and spatial representativeness. Further, APRO data must be harmonized in temporal and vertical grids and must follow naming conventions and commonly adopted user-friendly formats. This work consists in reanalyzing the original data to reach a high level of harmonization and standardization.

Ad hoc activities within the U.S. and European atmospheric observation communities have been initiated to produce comprehensive datasets of clouds, radiation,

TABLE 29-10. Retrievals of cloud profiles developed in the EU Cloudnet program.

Retrieval variables	Input data	References
Cloud fraction	Cloud radar, ceilometer; fraction of pixels in model grid box classified as cloud	Illingworth et al. (2007)
LWC	Cloud radar and ceilometer for cloud top and base: assume linear LWC with height scaled to agree with water path from radiometer	Illingworth et al. (2007)
IWC	Cloud radar reflectivity corrected for attenuation by LWC and humidity	Hogan et al. (2006)
Drizzle rate	Radar reflectivity and lidar backscatter	O'Connor et al. (2005)

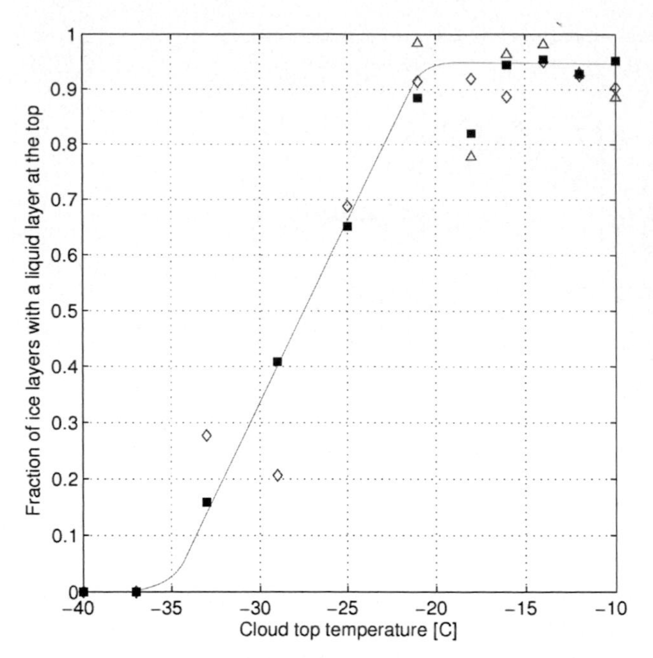

FIG. 29-8. The fraction of ice clouds containing liquidlike layers as a function of cloud-top temperature derived from four years of continuous observations at CFARR. To test the sensitivity of the identification of supercooled water and ice from radar and lidar observations, the diamonds are for data when the dBZ radar threshold was increased from −20 to −10 dBZ to ensure that no liquid droplet clouds are being diagnosed as containing ice. The triangles are for cases when the presence of ice was confirmed by specular reflection from oriented ice particles.

and atmospheric profiles for driving and evaluating large-eddy simulation (LES) models, for assessing climate model simulations and NWP forecasts, and for performing climate studies and climate feedback analyses. The U.S. ARM Program developed a dataset specifically designed for modelers to evaluate climate models. This dataset, called ARM Climate Modeling Best Estimate (CMBE; Xie et al. 2010), consists of a dozen cloud and radiation quantities provided as hourly averages, standard deviations within each hour, and quality control flags to qualify data quality and temporal variability. CMBE products were derived from data at each ARM observation facility.

In Europe, a similar initiative was started in the frame of the European Union Cloud Intercomparison, Process Study and Evaluation (EUCLIPSE) Project (see section 3c). A dataset was developed from data gathered at the observatories of CESAR (only EUCLIPSE period for now), CFARR (only EUCLIPSE period for now), and SIRTA (all the available period, starting in 2002 for the earliest observations). The European Climate Testbed Dataset (ECTD) includes meteorological parameters, cloud and surface fluxes parameters, and instrument observables. For each parameter, a retrieval algorithm was identified to harmonize data interpretation across the three observatories. A quality control procedure was developed for each parameter. Spatial representativeness was evaluated over a 50-km domain around the observatory using observations from standard meteorological stations. Similarly to the CMBE dataset, ECTD provides data as hourly averages, standard deviations within each hour, and quality control flags to qualify data quality and temporal and spatial variability. Data files are in netCDF format, which includes all necessary metadata associated with each parameter. An important feature is that the ECTD data nomenclature (names of geophysical variables) is made consistent with the ARM CMBE nomenclature and the nomenclature used by CMIP5 climate models. A description of the ECTD of the SIRTA Observatory is provided in Cheruy et al. (2013) and Chiriaco et al. (2014), including a description of the quality control procedure. Table 29-12 provides the content of the SIRTA file (available online at http://sirta.ipsl.polytechnique.fr/sirta.old/reobs.html).

EU and U.S. scientists have recognized that there is a strong need to have these activities coordinated in a better way so that U.S. and EU datasets have common retrieval methods, data formats, naming conventions, common grids, etc. This coordination would help increase the number of studies that make combined use of EU and U.S. APRO datasets. As suggested during the

TABLE 29-11. Liquid cloud retrievals that participated in the OSSE within the EG-CLIMET COST Action.

Retrieved variables	Input data	References
LWC, N, R_{eff}, COD, aerosol indirect effect index	Lidar extinction profile, T and q profiles from MWR, LWP from MWR, Z and linear depolarization ratio (LDR) from cloud radar	SYRSOC (Martucci and O'Dowd 2011; Martucci et al. 2012)
LWC	T and p profiles, LWP from MWR, Cloudnet classification product	Cloudnet scaled-adiabatic LWC (Illingworth et al. 2007)
LWC, N, R_{eff}, COD	Z from cloud radar, cloud base from lidar, LWP from MWR	TUD-RSCCP (Brandau et al. 2010)
LWC, LWP, T, and q profiles	Z from cloud radar, brightness temperatures from MWR, prior information on LWC, T, and q profiles	IPT (Löhnert et al. 2004, 2008)

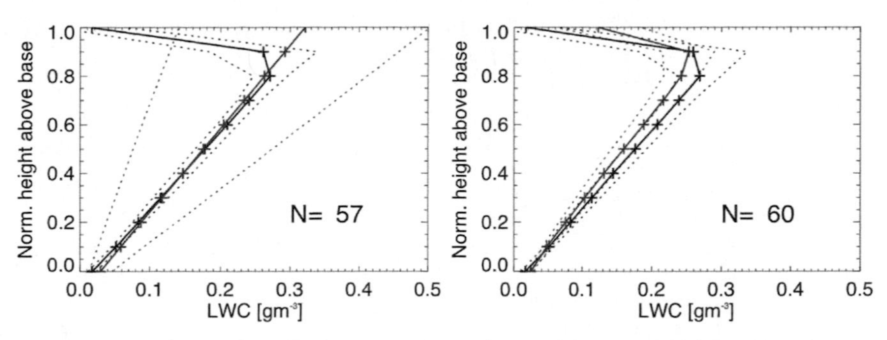

FIG. 29-9. Mean LWC profile from cloud-resolving model output (thick black) and corresponding 1σ range (dotted black) for a simulation initialized over the CESAR Observatory (the Netherlands). The red lines show results and corresponding 1σ ranges from (left) the Cloudnet scaled-adiabatic method and (right) the TUD-RSCCP method.

U.S.–EU workshop (DOE-Climate and Environmental Sciences Division 2013), a data-harmonization working group should be created to address these data issues, specifically for applications that require use of long-term multiparameter datasets.

c. Climate and weather model evaluation initiatives

Existing observations from routine measurements or long field campaigns carried out at atmospheric profiling observatories can be used to evaluate models on synoptic, seasonal, interannual, and now even climatic time scales at a relatively low cost. Such continuous evaluation is complementary to detailed case studies performed with 1D versions of climate models or with cloud-resolving models and large-eddy simulations carried out on highly documented cases obtained during focused field experiments. Long-term or continuous evaluations offer more representative evaluations to identify limitations in physical parameterizations of models, to evaluate the impact of modified parameterizations, and to confront the behavior of different models on different observatories. We present three examples of model evaluation frameworks developed in Europe that use long-term observations from atmospheric profiling observatories: the Cloudnet framework to evaluate NWP models, the KNMI Parameterization Test Bed (KPT) framework for single-column and climate model evaluations, and the EUCLIPSE framework to develop datasets for the International Climate Model Intercomparison Project.

1) CLOUDNET NWP MODEL EVALUATIONS

A framework for continuous evaluation of NWP models was developed in the EU Cloudnet project described in sections 2a and 2b. As an example of Cloudnet model evaluation, Fig. 29-10b shows that in 2004 the profiles of mean ice water content (IWC) and the probability distribution of IWC in European operational forecast models were generally in fair agreement with the observations. It can be seen that the Met Office mesoscale and the ECMWF model reproduce the mean IWC within the uncertainty of the IWC retrieval. Below 0.1 g m^{-3}, the DWD model has the best representation of the PDF, but because it treats falling snow as a separate noncloud variable, it predicts virtually no IWC above this, thus the mean IWC below 7 km is substantially underestimated. Both the Météo-France and Met Office global models have too low a mean value of IWC mainly because they are simulating too narrow a distribution of the IWC. As part of the ACTRIS FP7 project, the Cloudnet analysis system is being extended to cover more sites within Europe and to implement new model evaluation metrics. Many European forecast models are now carrying aerosol loading as prognostic variables, and the first steps are now being made to compare the forward-modeled lidar backscatter profiles of the aerosols with those observed by Cloudnet. This also raises the possibility of assimilating the observations in real time.

New techniques have been developed for evaluating models. For example, Barrett et al. (2009) compared diurnal composites of observed and modeled stratocumulus clouds and found that models with a nonlocal mixing scheme and an explicit formulation for cloud-top entrainment had the best diurnal cycle of cloud occurrence. New approaches have been developed for evaluating not just the climatological occurrence of clouds in models but their ability to forecast them at the right time and location. The equitable threat score (ETS) is used widely in forecast verification but Hogan et al. (2009a, 2010) pointed to several inherent problems with ETS. Most important is that the ETS value depends upon the frequency of occurrence and tends to zero for increasingly rare events. Cloud occurrence decreases rapidly toward the troposphere leading to a misleading drop in the value of ETS. They proposed a new metric, the symmetric extreme dependency score, which avoids these problems and is being implemented within Cloudnet.

TABLE 29-12. List of parameters currently included in the ECTD. ECTD variable names, equivalent ARM CMBE nomenclature, units, and description.

Variable	ARM CMBE name	Description	Period of obs.	Reference
tas	T_sfc	2-m air temperature, K	2003–16	—
hurs	rh_sfc	2-m relative humidity, %	2003–16	—
huss	—	2-m specific humidity, kg kg^{-1}	2003–16	—
psl	—	Sea-level pressure, Pa	2003–16	—
sfcWind	wspd_sfc	2-m wind speed, m s^{-1}	2003–16	—
vas	v_sfc	2-m northward wind, m s^{-1}	2003–16	—
uas	u_sfc	2-m eastward wind, m s^{-1}	2003–16	—
pr	prec_sfc	precipitation at surface, kg m^{-2} s^{-1}	2003–16	—
visi	—	visibility, m	2010–16	—
rlds	lwdn	Surface downwelling longwave radiation, W m^{-2}	2003–16	—
rlus	lwup	Surface upwelling longwave radiation, W m^{-2}	2007–16	—
rsds	swdn	Surface downwelling shortwave radiation, W m^{-2}	2003–16	—
rsus	swup	Surface upwelling shortwave radiation, W m^{-2}	2007–16	—
hfss	SH	Surface upward sensible heat flux, W W m^{-2}	2006–16	—
hfls	LH	Surface upward latent heat flux, W m^{-2}	2006–16	—
saa	—	solar azimuthal angle, °	2003–16	—
sza	—	solar zenithal angle, °	2003–16	—
Stx[a]	—	Soil temperature x cm below ground level, K	2007–16	—
Smx[a]	—	Soil moisture x cm below ground level, g cm^{-3}	2007–16	—
channel_x_mean[b]	—	Mean brightness temperature from MSG at x μm, K	2005–10	—
cf_nfov	—	Lidar cloud fraction	2008–13	Morille et al. (2007)
rsdscs	—	Surface downwelling shortwave radiation for clear sky, W m^{-2}	2003–16	Long et al. (2006)
rldscs	—	Surface downwelling longwave radiation for clear sky, W m^{-2}	2003–16	Long et al. (2006)
tot_cld_tsi	tot_cld_tsi	Cloud fraction from sky imager	2009–16	—
cflw	—	Cloud fraction from longwave radiation	2003–16	Long et al. (2006)
cfsw	—	Cloud fraction from shortwave radiation	2003–16	Long et al. (2006)
Cbhx[c]	—	Lidar cloud base heigh, m	2008–13	Morille et al. (2007)
aot_x[d]	—	Aerosol optical thickness at x nm	2003–16	Holben et al. (1998)
lwp	—	liquid water path, g m^{-2}	2010–16	—
mld	—	mixing layer depth, m	2008–13	Pal et al. (2015)
water	—	Clear sky integrated water vapor, kg m^{-2}	2003–16	Holben et al. (1998)
x_yangstrom[e]	—	Angstrom exponent between x and y nm, nm	2003–16	Holben et al. (1998)
cld_frac	—	Percentage cloudy pixels over 15 × 15 pixels	2005–10	Roebeling et al. (2006)
clwp_mean	—	Mean cloud liquid water path over 15 × 15 pixels, g m^{-2}	2005–10	Roebeling et al. (2006)
ctt_mean	—	Mean cloud top temperature over 15 × 15 pixels, K	2005–10	Roebeling et al. (2006)
reff_mean	—	Mean cloud effective radius over 15 × 15 pixels, mm	2005–10	Roebeling et al. (2006)
tau_mean	—	Mean cloud optical thickness over 15 × 15 pixels, g m^{-2}	2005–10	Roebeling et al. (2006)
SR	—	Lidar scattering ratio vertical histograms	2003–16	—
Strat	—	Lidar STRAT classification vertical histograms	2003–16	Morille et al. (2007)
Molecular	—	Lidar molecular profile	2003–16	—
Alt norm	—	Altitude of normalisation of lidar profiles, m	2003–16	—

[a] x is 5, 10, 20, 30, 50 cm
[b] x is 12, 0.6, 0.8, 1.6, 3.8, 10.8 μm
[c] x is first layer (1), second layer (2), third layer (3)
[d] x is 1020, 870, 675, 500, 440, 380, 340 nm
[e] x and y are the interval between [d] values.

Hogan et al. (2009a) used this score to show that the "half life" of a cloud forecast (the time into the forecast at which, on average, the score fell to half of its initial value) was 2.5–4.5 days rather than around 9 days for a pressure forecast. Operational forecast models within Europe are introducing more advanced cloud and aerosol parameterizations with additional variables, but there is a risk that if the new variables are not constrained by observations, they can actually degrade the forecast. Comparison of skill scores of forecasts with and without the new variables should reveal if they are leading to a more realistic representation of cloud/aerosol interactions.

2) KNMI PARAMETERIZATION TEST BED

Diabatic processes like turbulence, convection, clouds, and radiation still are represented insufficiently in

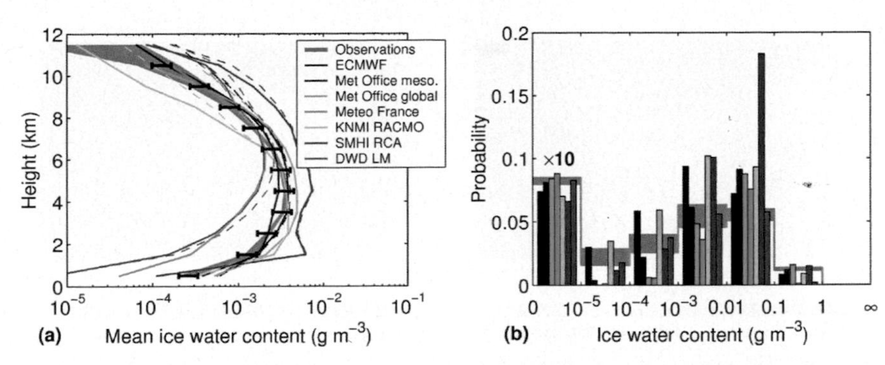

FIG. 29-10. (a) Mean IWC at the three Cloudnet observatories (CESAR, SIRTA, and CFARR) for the year 2004 from the observations and seven models. Two lines are shown for each model: the thick solid lines show the model after filtering to remove ice clouds too tenuous for the radar to detect, while the thin dashed lines are for all model clouds. The error bars indicate the uncertainty resulting from possible radar calibration errors and uncertainties in the mass–size relationship. (b) Corresponding histograms of observed and model IWC for clouds between 3- and 7-km altitude. Note that the bars in the lowest bin are shown at a tenth of their true height.

weather and climate models making the development and improvement of scale-adaptive parameterizations a necessity. Measurements obtained from permanent profiling sites can help to constrain these insufficiencies but require a framework that brings together simulations and observations in an appropriate manner. Neggers et al. (2012) developed such a platform, the KPT, where models and measurements can be evaluated and compared interactively. Here data streams from the CESAR site are used for evaluation of continuous single-column model (SCM) and LES runs at multiple time scales. In this way, both typical long-term model behavior and process-level case studies can be investigated. KPT proved its value by successfully identifying a compensating error between cloud vertical structure and cloud overlap (Neggers and Siebesma 2013). The test bed approach being at the interface of the observational and the modeling community helps to efficiently exploit observations for atmospheric model improvement. Currently the KPT is extended to the Integrated Scale-Adaptive Parameterization and Evaluation (InScAPE) project centered at the JOYCE observatory with the potential for transfer to further profiling sites (http://gop.meteo.uni-koeln.de/~neggers/InScAPE/).

3) EUCLIPSE CMIP MODEL EVALUATIONS

The EUCLIPSE project is a European collaborative effort, funded by the Seventh Framework Program of the European Commission, dedicated to improve the evaluation, understanding, and description of the role of clouds in Earth's climate. The central focus of the project is to reduce the uncertainty in the representation of cloud processes and feedbacks in the new generation of earth system models.

Cheruy et al. (2013) used the harmonized ECTD (presented in section 3b) to evaluate the standard and new parameterizations of boundary layer, convection, and clouds in the Earth System Model of L'Institut Pierre-Simon Laplace. Realistic coupling with the surface is an essential element of 3D simulations over a continental site. Hence two different land surface hydrology parameterizations were considered to analyze different land–atmosphere interactions. For this evaluation, the multiparameter characteristic of atmospheric profiling observatories is essential. It allows separate components of the system to be constrained simultaneously, such as radiative fluxes, latent and sensible heat fluxes, the height of the mixing layer, temperature and humidity in the boundary layer and in the soil, and properties of boundary layer clouds. Ten-year simulations of the coupled land surface–atmospheric modules were compared to observations collected at the SIRTA Observatory. Simulations were conducted with a stretched grid in the vicinity of the SIRTA Observatory, in a nudged mode to enable comparisons with observed parameters following a methodology developed by Coindreau et al. (2007). The study highlights how identified biases in temperature and humidity can be explained by biases in the partition between surface sensible and latent heat, by underestimation of boundary layer clouds, and insufficient turbulent transport in the surface layer. In addition, the approach allowed the authors to test how new parameterizations can reduce biases in the different components. Stegehuis et al. (2013) suggest that the partition between surface sensible and latent heat is of particular importance if climate prediction models are to correctly predict summertime heat waves over Europe. Campoy et al. (2013) suggest

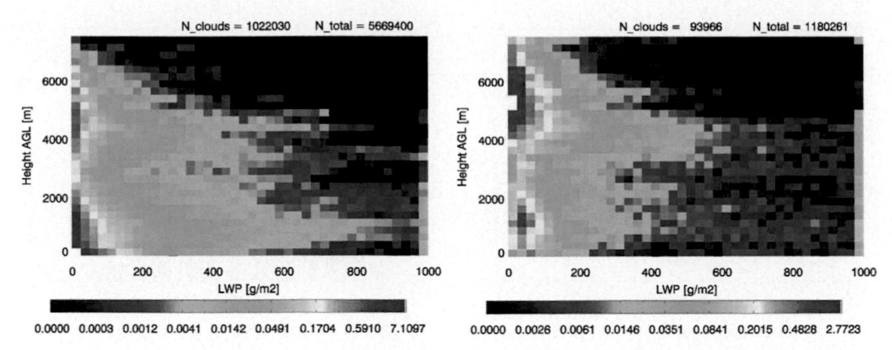

FIG. 29-11. Joint histogram of LWP and cloud-base height for 2006 for (left) Djougou and (right) Niamey.

that the description of groundwater in land surface models should be improved to obtain better predictions of summertime heat waves.

4. Collaborations between U.S. ARM and EU programs

The scientific programs, improvements in retrieval algorithm, extensive datasets, and efficient frameworks for model evaluations developed by European scientists, have triggered significant interest in the U.S. ARM scientific community. The conduct of observational field campaigns/experiments is an area where collaboration between the ARM and EU programs can be found. The participation of U.S. investigators in EU field campaigns was reinforced after the AMF was developed (section 4a). Collaborations between ARM and EU atmospheric profiling research observatories strengthened through the development of harmonized data interpretation algorithms (section 4b) and of model evaluation frameworks (section 4c). Common use of ARM–EU APRO datasets in scientific investigations is also an identified avenue of collaboration (section 4d). Collaboration was developed mostly outside any formal framework through bilateral collaboration between U.S. and EU scientists. These initiatives resulted in significant cross-fertilization between the ARM and EU programs.

a. Common field campaigns

To complement its permanent sites, the ARM Program developed the AMF to collect data in additional regions of interest to the general atmospheric science community (Miller et al. 2016, chapter 9). An open call for proposals for deployment periods of 6–12 months is issued each year. The European Community successfully applied twice to complement major field experiments with AMF proposals. The first AMF deployment occurred in Niamey, Niger, in 2006. In 2007, the AMF was deployed in the Black Forest, Germany.

The AMF deployment in Niamey (13.5°N, 2.1°E) was associated with two large international campaigns: the African Monsoon Multidisciplinary Analysis (AMMA; Lebel et al. 2010) and the Geostationary Earth Radiation Budget (GERB; Harries et al. 2005) experiment. The proposal to the ARM Program leading to this deployment was titled Radiative Atmospheric Divergence Using the AMF, GERB Data, and AMMA Stations (RADAGAST). The proposal represented an international effort to measure continuously the radiative fluxes at the surface and top-of-the-atmosphere through the seasonal progression of the West African monsoon (Miller and Slingo 2007). Because precipitation in Niamey is limited to the monsoon period from June to September, a strong seasonality in the surface energy balance is obvious (Miller et al. 2009). The site is also well-suited to study the impact of Saharan dust, biomass burning, and deep convection.

The AMF deployment in Niamey was an integral part of the AMMA north–south transect that allowed the monsoon progression to be studied in detail. The most southern station Djougou, Benin (9.6°N, 1.7°E), is under monsoon influence already in April, while the most northern station Gourma, Mali (16.0°N, 1.5°W), becomes affected by moist air masses usually after June. Therefore annual precipitation in Djougou is much stronger (1124 mm in 2006) than in Niamey (384 mm). As shown in Fig. 29-11, the difference in low-level clouds is also quite pronounced with only few clouds bearing more than 200 $g\,m^{-2}$ liquid water path in Niamey, while above Djougou such values typical for daytime boundary layer development are found much more frequently (Pospichal 2009). Both sites show the frequent occurrence of midlevel clouds located at the top of the Saharan air layer (Bouniol et al. 2012).

The Convective and Orographically Induced Precipitation Study (COPS; Wulfmeyer et al. 2011) in summer 2007 was motivated by the need to advance the quality of forecasts of orographically induced convective precipitation. To identify the physical and chemical

FIG. 29-12. ARM Mobile Facility deployment in the Murg Valley, Germany.

processes responsible for the forecast deficiencies, COPS combined 4D observations and high-resolution modeling in a strong international collaboration. In situ and remote sensing networks were installed in southwestern Germany, 10 research aircraft were operated, and a synergy of multiwavelength passive and active remote sensing instruments such as lidars, radars and radiometers were operated at five supersites. The AMF located in the Murg Valley (Fig. 29-12) was an integral part of the supersite transect across the Black Forest, the Rhine Valley, and the Vosges Mountains. The continuous measurements over the 9-month deployment period allowed long-term cloud statistics to be derived (Ebell et al. 2011) and supplemented the 37 COPS intensive operation days. The broad frequency coverage of microwave radiometers from ARM and European partners at the AMF also were exploited to improve the description of water vapor continuum absorption (Turner et al. 2009). The COPS special issue in the *Quarterly Journal of the Royal Meteorological Society* in January 2011 with 21 contributions nicely represents the breadth of activities including the effects of soil moisture, surface energy budget, convective initiation and enhancement, multiscale interactions, long-range dust transport, aerosol and cloud microphysics, data assimilation, and forecast studies.

In addition to the complete AMF deployments, ARM–EU collaborations also took place on smaller scales via less formal participation in field campaigns. One example is the second field experiment of the Radiative Heating in Underexplored Bands Campaigns (RHUBC) project that took place in the Atacama Desert at 5300 m above sea level (Turner and Mlawer 2010). Here, the University of Cologne participated with a microwave radiometer to complement measurements across the full spectral range. In addition, the campaign demonstrated the superiority of more recent water vapor absorption models for climate simulations (Turner et al. 2012) and improved our knowledge in microwave calibration techniques (Maschwitz et al. 2013).

b. ARM–EU collaboration on retrieval algorithm development

Partly triggered by the activities of CLIWA-NET (section 2b), strong collaboration on microwave radiometry between ARM and EU scientists has been developed over more than a decade. In this period microwave radiometers developed from research instruments to operational tools for profiling atmospheric temperature and humidity and observing the columnar amount of liquid water. Scientists from ARM and EU have written reviews jointly on microwave radiometry (Westwater et al. 2005), worked on various processing challenges that affect the accuracy of the derived products, and participated in joint field experiments (see section 4a).

Maschwitz et al. (2013) assessed the different sources of uncertainty involved in the calibration of microwave radiometers. This includes the effects of antenna beamwidth, which is especially important for elevation scans used in the tipping curve calibration, as well as the impact of channel bandpass characteristics, which were investigated

TABLE 29-13. Collaborative work on retrieval development.

Topic	Details	References
Microwave radiometer measurement uncertainty	Calibration assessment	Maschwitz et al. (2013)
	Liquid nitrogen calibration	Paine et al. (2014)
	Instrument cross validation	Cimini et al. (2009)
Microwave absorption models	Supercooled liquid water continuum water vapor absorption	Kneifel et al. (2014) Turner et al. (2012, 2009)
Microwave retrieval uncertainty	Effect of instrument parameters	Meunier et al. (2013)
	Uncertainty in ground truthing	Mattioli et al. (2008)
Microwave retrieval of integrated quantities	Integrated water vapor and liquid water for Arctic observations	Cimini et al. (2007)
Microwave thermodynamic profiling	1D VAR for continuous profiling of temperature and humidity for 2010 Winter Olympics	Cimini et al. (2011)
Infrared retrievals	Uncertainty of thermodynamic profiles and cloud properties	Turner and Löhnert (2014)
Sensor synergy	Uncertainty in the retrieval of cloud liquid water from active and passive microwave observations	Ebell et al. (2010)
Sensor synergy	Thermodynamic profile retrieval from combined spectral microwave and infrared	Löhnert et al. (2009)
Cloud profile retrieval	Feasibility of liquid water profile retrieval from passive microwave radiometer measurements	Crewell et al. (2009)

in detail by Meunier et al. (2013). In terms of converting measured brightness temperatures into geophysical products, the absorption characteristics of atmospheric gases and hydrometeors are important parameters for modeling the radiative transfer. Observations across the globe have been used in a collaborative effort (Turner et al. 2009; Kneifel et al. 2014) to test and further improve these models. In addition, the microwave observations were evaluated using different ground-truth data at profiling sites (Mattioli et al. 2008; Cimini. et al. 2009). Selected collaborative developments are presented in Table 29-13. To facilitate the exchange of information in the microwave radiometer user community the Microwave Radiometer Network (MWRNET) was established in 2009 by the EG-CLIMET COST action. The MWRNET connects people worldwide working with ground-based microwave radiometers to ultimately establish an operational network sharing knowledge, software, procedures, formats, and quality control. Collaboration with radar processing is starting first with work on Doppler spectra processing (Maahn and Kollias 2012) and formalizing information exchange.

c. *ARM–EU collaboration on model evaluation*

Following presentations of the Cloudnet radar–lidar analysis scheme and NWP model evaluation framework at ARM science team meetings, the Cloudnet scheme was included in the Fast Physics Testbed and Research (FASTER) project of the DOE Earth System Modeling (ESM) program that aims to evaluate and improve the fast-physics processes, particularly those associated with clouds, in various atmospheric models. The Cloudnet analysis scheme (see section 3a for a full description) was implemented on the observations from the various ARM sites worldwide, including the AMF at its numerous deployments. The Cloudnet model evaluation framework (described in section 3c) was implemented as an integral part of FASTER's Single Column Model Testbed (SCM-Testbed) and Numerical Weather Prediction Testbed (NWP-Testbed). In the SCM-Testbed, various SCMs are run over the ARM sites and compared to the observations. Since these models are very fast to run, it is straightforward to carry out reruns to test the impact of different physical parameterizations and to test how they affect the performance in terms of cloud properties.

In the NWP-Testbed, the performance of NWP models has been assessed in a much wider range of climate regimes and over longer periods compared to the original Cloudnet project. Figure 29-13 shows the time series of the symmetric extreme dependency score (SEDS) that gauges the skill of the various forecast models to predict cloud fraction above 5% in the right place at the right time. As discussed in section 3c, SEDS has the advantage over the traditional ETS that the value does not depend upon the frequency of the event. The skill scores for cloud fraction are plotted in Fig. 29-13a for the ARM SGP site from 2001 to 2010 and in Fig. 29-13b for the ARM Darwin site. Over the SGP site, the models show considerably higher skill in the winter than the summer, presumably because the location of convective clouds is more difficult to predict than clouds associated with wintertime synoptic disturbances. Also, all models show considerably higher skill than achieved by a 24-h persistence forecast. The picture is different over Darwin in Fig. 29-13b. While

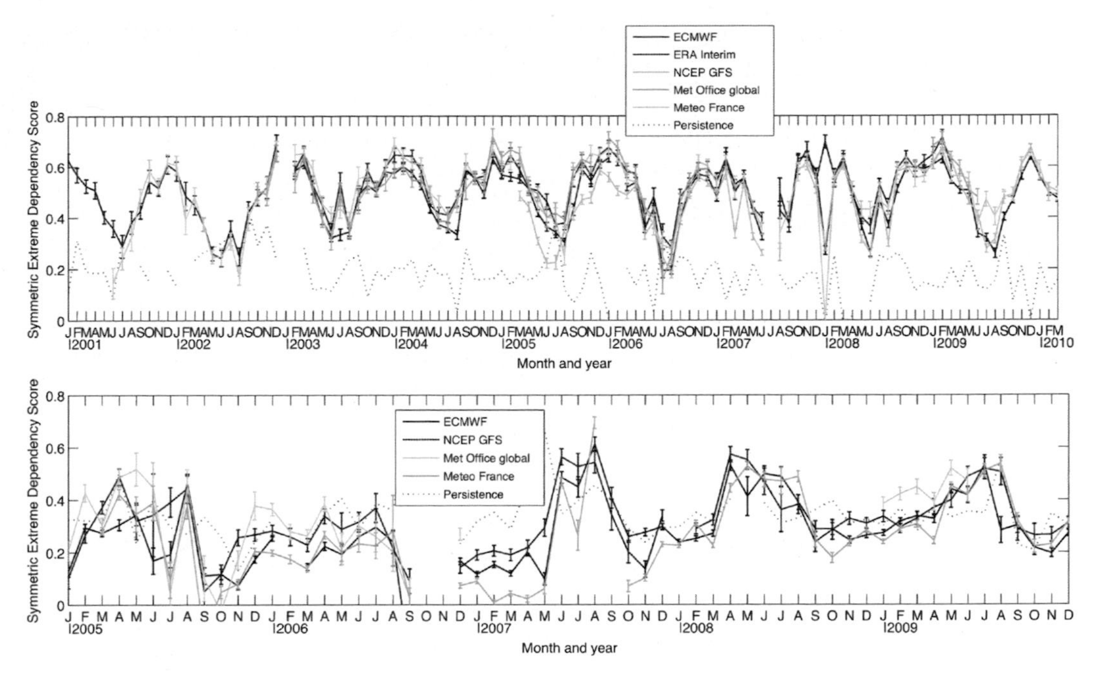

FIG. 29-13. The skill of various numerical weather prediction models in predicting cloud fraction greater than 0.05, as measured by the SEDS, for the (top) ARM SGP site 2001–10 and (bottom) Darwin site 2005–09. "Persistence" refers to using the observations from the previous 24 h as the prediction.

the models show generally more skill during the May–August peak of the dry season, there is considerably more year-to-year variability. Moreover, the challenge of tropical forecasting is highlighted by the fact that all models struggle to perform better than a persistence forecast at this location.

Having a decade of data makes it possible to determine whether cloud forecasts have improved in this period, but in order to account for natural variability in the predictability of weather systems from year to year, it is necessary to compare the skill to that from a reanalysis, in which the forecast system was kept constant. The reference in Fig. 29-13a is the ERA-Interim. Over this period, the NCEP, ECMWF, and Met Office forecasts appear to show no significant improvement relative to the reanalysis, in spite of the concerted research effort over recent years.

d. Common use of ARM–EU data in scientific investigations

A dozen publications are identified where EU–ARM collaboration was established to carry out algorithm developments, data validation, process studies, and other analyses using observations from both European APRO and ARM programs. A dozen, compared to several hundred publications using data from European APRO programs, and a similar number using ARM Program data, is a limited number. A review of the publications allows us to shed some light on the issue. The list of investigations and related keywords are presented in Table 29-14.

Nearly all first authors of these publications are principal investigators of European APRO programs. Hence they are all familiar with the European APRO data. All publications include coauthors who are principal investigators of the ARM Program or have been involved in a formal EU–ARM collaboration cited in this chapter. Hence these publications result from collaborations between authors who are familiar with the content and the benefits of both EU and ARM ground-based atmospheric profiling observations.

Half the publications rely on combined analyses of ground-based and satellite observations both for validation studies and for comprehensive process studies. Chepfer et al. (1999; 2000) and Naud et al. (2006) both evaluated retrievals of cloud properties (e.g., cloud altitude, cloud thermodynamic phase) from spaceborne passive radiometers using ground-based active remote sensors (e.g., cloud lidars and radars). Both used data from ARM (SGP and TWP) and Europe (CFARR, United Kingdom; SIRTA, France) to show if the uncertainties in satellite retrievals are site dependent. The use of multiple validation sites is particularly important to assess retrievals that are available globally. Pougatchev et al. (2007) developed a mathematical model to evaluate the contribution to bias and noise due to spatial mismatch between satellite and ground-based observations in intercomparison studies. They illustrated their model using ARM (SGP and TWP) and EU (MOL-RAO, Germany) radiosonde profiles. The A-Train *CloudSat* and *CALIPSO*

TABLE 29-14. Scientific publications highlighting EU/U.S. collaborations using observations from both European APROs and U.S. ARM sites.

Project framework (if relevant)	Reference	Key words	Observatories used
Polarization and Directionality of Earth Reflectances (POLDER)	Chepfer et al. (1999) Chepfer et al. (2000)	Satellite validation, cloud altitude, cloud phase, cirrus, ground-based lidars	ARM: SGP, TWP EU: SIRTA
Along-Track Scanning Radiometer (ATSR)-2	Naud et al. (2006)	Satellite validation, cloud-top altitude, ground-based radar	ARM: SGP EU: CFARR
	Pougatchev et al. (2007)	Satellite validation, bias and noise in satellite retrieval, radiosonde measurements	ARM: SGP, TWP EU: RAO
CloudSat	Protat et al. (2009) Protat et al. (2010)	Satellite validation, cloud base, cloud top, cloud thickness, cloud reflectivity, cloud microphysics, ground-based cloud radars	ARM: Darwin, AMF (AMMA, COPS) EU: SIRTA, RAO
CloudSat	Protat et al. (2011)	Calibration of ground-based radars	ARM: NSA EU: CESAR
CALIPSO	Dupont et al. (2010)	Satellite validation, cloud base, cloud top, cloud thickness, optical depth, ground-based lidars	ARM: SGP EU: SIRTA Other: Observatoire de Haute Provence (OHP), CERES Ocean Validation Experiment (COVE)
Cloudnet	van Zadelhoff et al. (2004)	Retrieval of ice cloud properties, ground-based radars	ARM: SGP EU: CFARR, CESAR
Cloudnet	Hogan et al. (2005)	Retrieval of cloud liquid water content, ground-based radars	ARM: SGP EU: CFARR
	Dupont et al. (2008, 2009)	Cirrus cloud radiative effects, broadband radiometers, GPS, sunphotometers, lidars	ARM: SGP, TWP, NSA EU: SIRTA
COPS field experiment	Ebell et al. (2011)	Retrieval of cloud fraction, cloud heights, cloud LWP, cloud phase	ARM: AMF (COPS) EU: CFARR, CESAR
	Naud et al. (2010)	Vertical profiles of temperature and cloud phase, GCM, ground-based lidars	ARM: SGP EU: SIRTA
	Tonttila et al. (2011)	Cloud vertical velocity, AROME mesoscale model, ground-based radars	ARM: SGP EU: RAO

programs triggered numerous validations studies, among which a few relied on combined analyses of ARM and EU APRO data. Protat et al. (2009; 2011) assessed cloud-base height, top height, geometrical thickness, and reflectivity of ice clouds derived from *CloudSat* using airborne and five ground-based cloud radars (at ARM Darwin, EU MOL-RAO, and SIRTA observatories). They extended their investigation to evaluate retrievals of cloud microphysical properties (Protat et al. 2010). Dupont et al. (2010) assessed cloud-base height, top height, geometrical thickness, and optical depth of cirrus clouds from four midlatitude ground-based lidar datasets (at ARM SGP and EU SIRTA observatories). Later, Protat et al. (2011) showed that since the *CloudSat* cloud radar reflectivity had been calibrated using multiple references (Protat et al. 2009), the satellite cloud radar could be used in turn to calibrate cloud radars of the ground-based network that had not yet been intercalibrated [e.g., ARM North Slope of Alaska (NSA) and EU CESAR, the Netherlands].

Another topic of collaboration is the development of retrieval algorithms. Early work by van Zadelhoff et al. (2004) compared retrieval of ice cloud properties from radar measurements in the ARM Program and the EU Cloudnet program (see section 2b). They found that the relationships between radar reflectivity and ice water content were consistent between the European Union (CFARR and CESAR) and ARM SGP. However the relationship between radar reflectivity and droplet radius did not show such trans-Atlantic consistency. Hogan et al. (2005) showed that the liquid water content in stratocumulus can be retrieved by using the differential absorption between a 35- and a 94-GHz radar. To prove the efficiency of the technique, the authors apply their method to dual radar datasets collected in both Europe (CFARR) and ARM SGP.

Several authors also use the multiprogram datasets to explore processes over different climate regions to study potential regional differences or to make their findings more universal if they are consistent at different locations. The added value of using multiprogram datasets is that authors can develop a complex analysis method that relies typically on multiple collocated observations and then apply this method on measurements from several observatories. This requires that the different observing programs offer consistent observing datasets. Several studies concern radiative effects of clouds. Dupont et al. (2009) investigated shortwave and longwave radiative effects of cirrus clouds using broadband radiometers, sun photometers, GPS, and lidars from EU SIRTA, SGP, TWP, and NSA. They showed that cloud radiative effects on surface shortwave and longwave irradiance varied greatly from the tropics to the midlatitudes and the Arctic. Ebell et al. (2011) investigated cloud properties and cloud radiative effects in a European mountain site using the AMF. They found that cloud liquid water path and radiative effects in the continental mountain site are significantly less than at EU CFARR and CESAR maritime site. Other authors used multiprogram datasets to study cloud processes in the observations and in atmospheric models, either climate models or numerical weather prediction models. Naud et al. (2010) used lidar and radiosonde measurements at ARM SGP and EU SIRTA observatories to study vertical profiles of temperature and their relationship to thermodynamic phase of optically thin cirrus. Tonttila et al. (2011) found significantly higher variability in observed cloud-base vertical velocity in ARM SGP and EU MOL-RAO data than in the Applications of Research to Operations at Mesoscale (AROME) mesoscale model.

We can conclude that there is a real motivation for carrying out investigations that rely on datasets developed by completely independent programs to expand the geographic coverage, to explore the validity of results across several locations (satellite and model evaluation), to explore process in different climate zones (process studies), to consolidate results (algorithm developments), and to prove the usefulness of the study. However, until now this has required significant skill, knowledge, and effort on the part of coauthors because EU and ARM APRO data are fully not harmonized. As datasets become more harmonized a larger number of publications can be expected to rely on multiple datasets.

5. Outlook toward future collaborations

Clouds, aerosol, and precipitation still pose key challenges for the prediction of future climate. Detailed ground-based profiling observations by APROs have unique potential to advance our understanding, but the full amount of information available across the globe is not fully exploited yet, as pointed out in section 4c. In November 2012, the Department of Energy Climate and Environmental Sciences Division hosted a joint workshop bringing together participants from the various European Union programs and the DOE Atmospheric Radiation Measurement Program to explore "Climate Change Challenges and Observations" (DOE-Climate and Environmental Sciences Division 2013). The workshop identified six outstanding science questions and discussed observation strategies to tackle them.

1) What is the distribution of aerosol properties for the Atmospheric Model Intercomparison Project period (i.e., since 1979)?
2) What is the coupling among microphysics, aerosols, and cloud dynamics as a function of scale and regime (e.g., vertical velocity or stability)?
3) How are precipitation, water vapor, and cloudiness coupled, and what roles does organization play in this coupling?
4) How do clouds and precipitation couple with surface properties?
5) What is the response of clouds to warming?
6) What is the response of the probability density function of precipitation to warming?

Clearly answering all questions would benefit from an enhanced collaboration between the EU and ARM communities. Within the discussions four collaboration topics emerged that are promising opportunities for joint activities.

a. Collaboration topic 1: Retrieval algorithms and uncertainty

Most importantly, the EU and ARM observing stations should develop integrated datasets with similar standards that are made available in a common location. These datasets should include both measured and retrieved atmospheric properties. For high-quality measurements, common methods for calibrating instruments must be developed—a good example is the already ongoing work on microwave radiometry within MWRNET (see section 4b). In response to the potential collaboration, a second workshop was organized to focus on retrieval algorithms and uncertainty. This workshop was held in May 2013 at the University of Cologne, Germany. There were 20 participants from both the ARM and EU partners. They discussed common algorithm frameworks and paths forward for improving and/or implementing and evaluating retrieval algorithms across EU and ARM observing stations. As a first step, a joint paper is being written to provide a general overview on retrieval algorithms and identifying important sources of uncertainty that need to be quantified in all retrieval algorithms (D. D. Turner et al. 2015, unpublished manuscipt).

b. Collaboration topic 2: Field experiments and cruises

Field campaigns like the HD(CP)^2 Observational Prototype Experiment (HOPE) in April/May 2013 in Germany, the Biogenic Aerosols–Effects on Clouds and Climate (BAECC) in Hyytiälä (Finland) in 2014, and the Green Ocean Amazon experiment (GOAMAZON) in Brazil 2015 provide other opportunities for collaboration. The HOPE campaign that combined three profiling sites within less than a 10-km range to investigate clouds at high resolution could serve as a test bed for LES models (see below) while the combination of airborne and ground-based observations seems promising for GOAMAZON. Bridging the Atlantic can be achieved by linking the atmospheric profiling site in Barbados (MPI Hamburg), the ARM site in the Azores, and transects of the *Meteor* and *Polarstern* research vessels. Future field campaigns, for example, Arctic sea ice study or clouds in the Southern Ocean, could benefit strongly from an early stage joint planning phase.

c. Collaboration topic 3: Improving the link between models and observations

The operational use of LES at profiling sites as done in the KPT (section 3c) is highly promising to match the scales of observations and models and should be made transferable to various sites. Model evaluation approaches developed in the United States and the European Union (section 3c) could be extended to include instrument simulators, for example, cloud radar simulators. For the larger-scale (see section 3b) a common observational dataset to be used for CMIP5 modeling evaluation should be developed.

d. Collaboration topic 4: Standardization and organization

On the more technical side, the architecture, standards, and framework for an integrated portal for metadata, products, and related information have been discussed. First steps have been taken already in terms of data integration between ACTRIS and ARM as a network of networks (section 2c). Aerosol profiles, water vapor, and liquid water will be the first geophysical parameters to test the full cycle from data harmonization via retrieval algorithms and uncertainty, value-added, and synthesis products.

Currently collaboration between the ARM Program and EU atmospheric observation programs rely on voluntary initiatives of motivated researchers in the United States and Europe. Coordination between U.S. and European funding agencies would be greatly beneficial to strengthen collaboration between the ARM Program and EU atmospheric observation programs. Such coordination would encourage the organization of common field campaigns and raise the level of scientific achievements. In addition, a bottom-up process building on mutual exchange visits by early career scientists, participation in summer schools, and sabbaticals has already proven to be efficient in enhancing scientific collaboration. European and U.S. researchers are ready for intensified collaborations in the future, which should be encouraged by both EU and ARM Programs.

Acknowledgments. The authors acknowledge the following people for their contributions and helpful remarks: Frank Beyrich, Sandrine Bony, Nico Cimini, Frédérique Cheruy, Volker Lehmann, Ulrich Löhnert, Sandip Pal, and Bernhard Pospichal. The financial support for EARLINET by the European Union under Grant RICA 025991 in the Sixth Framework Programme is gratefully acknowledged. Since 2011 EARLINET is integrated in the ACTRIS Research Infrastructure Project supported by the European Union Seventh Framework Programme (FP7/2007-2013) under Grant Agreement 262254. We also acknowledge the funding support within the research initiative High Definition Clouds and Precipitation for Climate Prediction HD(CP)^2 (first phase 2012–16) by the German Ministry for Education and Research (BMBF).

REFERENCES

Ackerman, T. P., and G. M. Stokes, 2003: The Atmospheric Radiation Measurement Program. *Phys. Today*, **56**, 38–46, doi:10.1063/1.1554135.

Ansmann, A., M. Riebesell, and C. Weitkamp, 1990: Measurements of atmospheric aerosol extinction profiles with Raman lidar. *Opt. Lett.*, **15**, 746–748, doi:10.1364/OL.15.000746.

——, U. Wandinger, M. Riebesell, C. Weitkamp, and W. Michaelis, 1992: Independent measurement of extinction and backscatter profiles in cirrus clouds by using a combined Raman elastic-backscatter lidar. *Appl. Opt.*, **31**, 7113–7131, doi:10.1364/AO.31.007113.

——, and Coauthors, 2003: Long-range transport of Saharan dust to northern Europe: The 11–16 October 2001 outbreak with EARLINET. *J. Geophys. Res.*, **108**, 4783, doi:10.1029/2003JD003757.

——, P. Seifert, M. Tesche, and U. Wandinger, 2012: Profiling of fine and coarse particle mass: Case studies of Saharan dust and Eyjafjallajökull/Grimsvötn volcanic plumes. *Atmos. Chem. Phys.*, **12**, 9399–9415, doi:10.5194/acp-12-9399-2012.

Balis, D., and Coauthors, 2003: Raman lidar and sunphotometric measurements of aerosol optical properties over Thessaloniki, Greece during a biomass burning episode. *Atmos. Environ.*, **37**, 4529–4538, doi:10.1016/S1352-2310(03)00581-8.

Barlow, J. F., T. M. Dunbar, E. G. Nemitz, C. R. Wood, M. W. Gallagher, F. Davies, E. O'Connor, and R. M. Harrison, 2011:

Boundary layer dynamics over London, UK, as observed using Doppler lidar during REPARTEE-II. *Atmos. Chem. Phys.*, **11**, 2111–2125, doi:10.5194/acp-11-2111-2011.

Barrett, A. I., R. J. Hogan, and E. J. O'Connor, 2009: Evaluating forecasts of the evolution of the cloudy boundary layer using diurnal composites of radar and lidar observations. *Geophys. Res. Lett.*, **36**, L17811, doi:10.1029/2009GL038919.

Böckmann, C., and Coauthors, 2004: Aerosol lidar intercomparison in the framework of the EARLINET project. 2. Aerosol backscatter algorithms. *Appl. Opt.*, **43**, 977–989, doi:10.1364/AO.43.000977.

——, D. Miranova, D. Müller, L. Scheidenbach, and R. Nessler, 2005: Microphysical aerosol parameters from multiwavelength lidar. *J. Opt. Soc. Amer.*, **22**, 518–528, doi:10.1364/JOSAA.22.000518.

Boselli, A., R. Caggiano, C. Cornacchia, F. Madonna, M. Macchiato, L. Mona, G. Pappalardo, and S. Trippetta, 2012: Multi year sunphotometer measurements for aerosol characterization in a central Mediterranean site. *Atmos. Res.*, **104**, 98–110, doi:10.1016/j.atmosres.2011.08.002.

Bouniol, D., and Coauthors, 2010: Using continuous ground-based radar and lidar measurements for evaluating the representation of clouds in four operational models. *J. Appl. Meteor. Climatol.*, **49**, 1971–1991, doi:10.1175/2010JAMC2333.1.

——, F. Couvreux, P. H. Kamsu-Tamo, M. Leplay, F. Guichard, F. Favot, and E. J. O'Connor, 2012: Diurnal and seasonal cycles of cloud occurrences, types, and radiative impact over West Africa. *J. Appl. Meteor. Climatol.*, **51**, 534–553, doi:10.1175/JAMC-D-11-051.1.

Brandau, C. L., H. W. J. Russchenberg, and H. W. Knap, 2010: Evaluation of ground-based remotely sensed liquid water cloud properties using shortwave radiation measurements. *Atmos. Res.*, **96**, 366–377, doi:10.1016/j.atmosres.2010.01.009.

Browning, K. A., and Coauthors, 2007: The Convective Storm Initiation Project. *Bull. Amer. Meteor. Soc.*, **88**, 1939–1955, doi:10.1175/BAMS-88-12-1939.

Bühl, J., and Coauthors, 2013: LACROS: The Leipzig Aerosol and Cloud Remote Observations System. *Remote Sensing of Clouds and the Atmosphere XVIII; and Optics in Atmospheric Propagation and Adaptive Systems XVI*, A. Comeron et al., Eds., International Society for Optical Engineering (SPIE Proceedings, Vol. 8890), doi:10.1117/12.2030911.

Campoy, A., A. Ducharne, F. Chéruy, F. Hourdin, J. Polcher, and J. C. Dupont, 2013: Response of land surface fluxes and precipitation to different soil bottom hydrological conditions in a general circulation model. *J. Geophys. Res. Atmos.*, **118**, 10 725–10 739, doi:10.1002/jgrd.50627.

Carbone, R. E., and Coauthors, 2012: Thermodynamic Profiling Technologies Workshop report to the National Science Foundation and the National Weather Service. NCAR Tech. Note NCAR/TN-488+STR, 80 pp., doi:10.5065/D6SQ8XCF.

Chaikovsky, A., and Coauthors, 2012: Algorithm and software for the retrieval of vertical aerosol properties using combined lidar/radiometer data: Dissemination in EARLINET. *Proc. 26th Int. Laser and Radar Conf.*, Vol. 1, Porto Heli, Greece, ICLAS/IRC, 399–402.

Chepfer, H., P. Goloub, L. Sauvage, P. H. Flamant, G. Brogniez, J. Spinhirne, M. Lavorato, and N. Sugimoto, 1999: Validation of POLDER/ADEOS data using a ground-based lidar network: Preliminary results for semi-transparent clouds. *Phys. Chem. Earth*, **24**, 203–206, doi:10.1016/S1464-1909(98)00038-0.

——, ——, J. Spinhirne, P. H. Flamant, M. Lavorato, L. Sauvage, G. Brogniez, and J. Pelon, 2000: Cirrus cloud properties derived from POLDER-1/ADEOS polarized radiances: First validation using a ground-based lidar network. *J. Appl. Meteor.*, **39**, 154–168, doi:10.1175/1520-0450(2000)039<0154:CCPDFP>2.0.CO;2.

Cheruy, F., A. Campoy, J. C. Dupont, A. Ducharne, F. Hourdin, M. Haeffelin, M. Chiriaco, and A. Idelkadi, 2013: Combined influence of atmospheric physics and soil hydrology on the simulated meteorology at the SIRTA atmospheric observatory. *Climate Dyn.*, **40**, 2251–2269, doi:10.1007/s00382-012-1469-y.

Chiriaco, M., S. Bastin, P. Yiou, M. Haeffelin, J. C. Dupont, and M. Stéfanon, 2014: European heatwave in July 2006: Observations and modeling showing how local processes amplify conducive large-scale conditions. *Geophys. Res. Lett.*, **41**, 5644–5652, doi:10.1002/2014GL060205.

Cimini, D., T. J. Hewison, L. Martin, J. Güldner, C. Gaffard, and F. S. Marzano, 2006: Temperature and humidity profile retrievals from ground-based microwave radiometers during TUC. *Meteor. Z.*, **15**, 45–56, doi:10.1127/0941-2948/2006/0099.

——, E. R. Westwater, A. J. Gasiewski, M. Klein, V. Leusky, and J. Liljegren, 2007: Ground-based millimeter- and submillimiter-wave observations of low vapor and liquid water contents. *IEEE Trans. Geosci. Remote Sens.*, **45**, 2169–2180, doi:10.1109/TGRS.2007.897450.

——, F. Nasir, E. R. Westwater, V. H. Payne, D. D. Turner, E. J. Milawer, M. L. Exner, and M. P. Cadeddu, 2009: Comparison of ground-based millimeter-wave observations and simulations in the Arctic winter. *IEEE Trans. Geosci. Remote Sens.*, **47**, 3098–3106, doi:10.1109/TGRS.2009.2020743.

——, and Coauthors, 2011: Thermodynamic atmospheric profiling during the 2010 Winter Olympics using ground-based microwave radiometry. *IEEE Trans. Geosci. Remote Sens.*, **49**, 4959–4969, doi:10.1109/TGRS.2011.2154337.

——, F. De Angelis, J. C. Dupont, S. Pal, and M. Haeffelin, 2013: Mixing layer height retrievals by multichannel microwave radiometer observations. *Atmos. Meas. Tech.*, **6**, 2941–2951, doi:10.5194/amt-6-2941-2013.

Coindreau, O., F. Hourdin, M. Haeffelin, A. Mathieu, and C. Rio, 2007: Assessment of physical parameterizations using a global climate model with stretchable grid and nudging. *Mon. Wea. Rev.*, **135**, 1474–1489, doi:10.1175/MWR3338.1.

Cress, T. S., and D. L. Sisterson, 2016: Deploying the ARM sites and supporting infrastructure. *The Atmospheric Radiation Measurement (ARM) Program: The First 20 Years, Meteor. Monogr.*, No. 57, Amer. Meteor. Soc., doi:10.1175/AMSMONOGRAPHS-D-15-0049.1.

Crewell, S., and Coauthors, 2004: The BALTEX Bridge Campaign: An integrated approach for a better understanding of clouds. *Bull. Amer. Meteor. Soc.*, **85**, 1565–1584, doi:10.1175/BAMS-85-10-1565.

——, K. Ebell, U. Löhnert, and D. D. Turner, 2009: Can liquid water profiles be retrieved from passive microwave zenith observations? *Geophys. Res. Lett.*, **36**, L06803, doi:10.1029/2008GL036934.

Dacre, H. F., and Coauthors, 2011: Evaluating the structure and magnitude of the ash plume during the initial phase of the 2010 Eyjafjallajokull eruption using lidar observations and NAME simulations. *J. Geophys. Res.*, **116**, D00U03, doi:10.1029/2011JD015608.

Defazio, D., A. Lockett, and M. Wright, 2009: Funding incentives, collaborative dynamics and scientific productivity: Evidence from the EU framework program. *Res. Policy*, **38**, 293–305, doi:10.1016/j.respol.2008.11.008.

DOE-Climate and Environmental Sciences Division, 2013: Biological and environmental research. U.S./European Workshop on Climate Change Challenges and Observations, U.S. Department of Energy Office of Science Rep. DOE/SC-0154, 40 pp. [Available online at http://science.energy.gov/~/media/ber/pdf/CESD_EUworkshop_report.pdf.]

Donovan, D. P., and A. Apituley, 2013: Practical depolarization-ratio-based inversion procedure: Lidar measurements of the Eyjafjallajokull ash cloud over the Netherlands. *Appl. Opt.*, **52**, 2394–2415, doi:10.1364/AO.52.002394.

Dufresne, J. L., and S. Bony, 2008: An assessment of the primary sources of spread of global warming estimates from coupled atmosphere–ocean models. *J. Climate*, **21**, 5135–5144, doi:10.1175/2008JCLI2239.1.

Dupont, J.-C., and M. Haeffelin, 2008: Observed instantaneous cirrus radiative effect on surface-level shortwave and longwave irradiances. *J. Geophys. Res.*, **113**, D21202, doi:10.1029/2008JD009838.

——, ——, and C. N. Long, 2008: Evaluation of cloudless-sky periods detected by shortwave and longwave algorithms using lidar measurements. *Geophys. Res. Lett.*, **35**, L10815, doi:10.1029/2008GL033658.

——, ——, and ——, 2009: Cirrus cloud radiative effect on surface-level shortwave and longwave irradiances at regional and global scale. *Atmos. Chem. Phys. Discuss.*, **9**, 26 777–26 832, doi:10.5194/acpd-9-26777-2009.

——, and Coauthors, 2010: Macrophysical and optical properties of midlatitude high-altitude clouds from four ground-based lidars and collocated CALIOP observations. *J. Geophys. Res.*, **115**, D00H24, doi:10.1029/2009JD011943.

——, M. Haeffelin, A. Protat, D. Bouniol, N. Boyouk, and Y. Morille, 2012: Stratus fog formation and dissipation: A 6-day case study. *Bound.-Layer Meteor.*, **143**, 207–225, doi:10.1007/s10546-012-9699-4.

Ebell, K., U. Löhnert, S. Crewell, and D. Turner, 2010: On characterizing the error in a remotely sensed liquid water content profile. *Atmos. Res.*, **98**, 57–68, doi:10.1016/j.atmosres.2010.06.002.

——, S. Crewell, U. Löhnert, D. Turner, and E. O'Connor, 2011: Cloud statistics and cloud radiative effect for a low mountain site. *Quart. J. Roy. Meteor. Soc.*, **137**, 306–324, doi:10.1002/qj.748.

Emeis, S., K. Schäfer, and C. Münkel, 2008: Surface-based remote sensing of the mixing-layer height—A review. *Meteor. Z.*, **17**, 621–630, doi:10.1127/0941-2948/2008/0312.

Engelbart, D. A. M., and H. Steinhagen, 2001: Ground-based remote sensing of atmospheric parameters using integrated profiling stations. *Phys. Chem. Earth*, **26B**, 219–223.

——, W. A. Monna, J. Nash, and C. Matzler, 2009: Integrated ground-based remote sensing stations for atmospheric profiling. Earth System Science and Environmental Management (ESSEM), COST Action 720. [Available online at www.cost.eu/domains_actions/essem/Actions/720.]

Fassò, A., R. Ignaccolo, F. Madonna, B. B. Demoz, and M. Franco-Villoria, 2014: Statistical modelling of collocation uncertainty in atmospheric thermodynamic profiles. *Atmos. Meas. Tech.*, **7**, 1803–1816, doi:10.5194/amt-7-1803-2014.

Fisher, B. E. A., et al., Eds., 1998: Harmonisation of the preprocessing of meteorological data for atmospheric dispersion models. Office for Official Publications of the European Communities, Earth System Science and Environmental Management (ESSEM), COST Action 710, 431 pp.

——, J. Kukkonen, and M. Schatzmann, 2001: Meteorology applied to urban air pollution problems: COST 715. *Int. J. Environ. Pollut.*, **16**, 560–570, doi:10.1504/IJEP.2001.000650.

Flentje, H., B. Heese, J. Reichardt, and W. Thomas, 2010: Aerosol profiling using the ceilometer network of the German Meteorological Service. *Atmos. Meas. Tech. Discuss.*, **3**, 3643–3673, doi:10.5194/amtd-3-3643-2010.

Forster, P., and Coauthors, 2007: Changes in atmospheric constituents and in radiative forcing. *Climate Change 2007: The Physical Science Basis*, S. Solomon et al., Eds., Cambridge University Press, 129–234.

Freutel, F., and Coauthors, 2013: Aerosol particle measurements at three stationary sites in the megacity of Paris during summer 2009: Meteorology and air mass origin dominate aerosol particle composition and size distribution. *Atmos. Chem. Phys.*, **13**, 933–959, doi:10.5194/acp-13-933-2013.

Gasteiger, J., S. Groß, V. Freudenthaler, and M. Wiegner, 2011: Volcanic ash from Iceland over Munich: Mass concentration retrieved from ground-based remote sensing measurements. *Atmos. Chem. Phys.*, **11**, 2209–2223, doi:10.5194/acp-11-2209-2011.

Guibert, S., and Coauthors, 2005: The vertical distribution of aerosol over Europe: Synthesis of one year of EARLINET aerosol lidar measurements and aerosol transport modeling with LMDzT-INCA. *Atmos. Environ.*, **39**, 2933–2943, doi:10.1016/j.atmosenv.2004.12.046.

Güldner, J., and D. Spänkuch, 2001: Remote sensing of the thermodynamic state of the atmospheric boundary layer by ground-based microwave radiometry. *J. Atmos. Oceanic Technol.*, **18**, 925–933, doi:10.1175/1520-0426(2001)018<0925:RSOTTS>2.0.CO;2.

Haeffelin, M., and Coauthors, 2005: SIRTA, a ground-based atmospheric observatory for cloud and aerosol research. *Ann. Geophys.*, **23**, 253–275, doi:10.5194/angeo-23-253-2005.

——, and Coauthors, 2010: PARISFOG: Shedding new light on fog physical processes. *Bull. Amer. Meteor. Soc.*, **91**, 767–783, doi:10.1175/2009BAMS2671.1.

——, and Coauthors, 2012: Evaluation of mixing height retrievals from automatic profiling lidars and ceilometers in view of future integrated networks in Europe. *Bound.-Layer Meteor.*, **143**, 49–75, doi:10.1007/s10546-011-9643-z.

——, J. C. Dupont, N. Boyouk, D. Baumgardner, L. Gomes, G. Roberts, and T. Elias, 2013: A comparative study of radiation fog and quasi-fog formation processes during the ParisFog Field Experiment 2007. *Pure Appl. Geophys.*, **170**, 2283–2303, doi:10.1007/s00024-013-0672-z.

Hall, M. P. M., J. W. F. Goddard, and S. M. Cherry, 1984: Identification of hydrometeors and other targets by dual-polarization radar. *Radio Sci.*, **19**, 132–140, doi:10.1029/RS019i001p00132.

Harries, J. E., and Coauthors, 2005: The Geostationary Earth Radiation Budget (GERB) experiment. *Bull. Amer. Meteor. Soc.*, **86**, 945–960, doi:10.1175/BAMS-86-7-945.

Harvey, N. J., R. J. Hogan, and H. F. Dacre, 2013: A method to diagnose boundary-layer type using Doppler lidar. *Quart. J. Roy. Meteor. Soc.*, **139**, 1681–1693, doi:10.1002/qj.2068.

Heese, B., H. Flentje, D. Althausen, A. Ansmann, and S. Frey, 2010: Ceilometer lidar comparison: Backscatter coefficient retrieval and signal-to-noise ratio determination. *Atmos. Meas. Tech.*, **3**, 1763–1770, doi:10.5194/amt-3-1763-2010.

Hewison, T., 2007: 1D-VAR retrievals of temperature and humidity profiles from a ground-based microwave radiometer. *IEEE Trans. Geosci. Remote Sens.*, **45**, 2163–2168, doi:10.1109/TGRS.2007.898091.

Hogan, R. J., and A. J. Illingworth, 2000: Deriving cloud overlap statistics from radar. *Quart. J. Roy. Meteor. Soc.*, **126**, 2903–2909, doi:10.1002/qj.49712656914.

——, and ——, 2003: Parameterizing ice cloud inhomogeneity and the overlap of inhomogeneities using cloud radar data. *J. Atmos. Sci.*, **60**, 756–767, doi:10.1175/1520-0469(2003)060<0756:PICIAT>2.0.CO;2.

——, D. Bouniol, D. N. Ladd, E. J. O'Connor, and A. J. Illingworth, 2003a: Absolute calibration of 94/95-GHz radars using rain. *J. Atmos. Oceanic Technol.*, **20**, 572–580, doi:10.1175/1520-0426(2003)20<572:ACOGRU>2.0.CO;2.

——, P. N. Francis, H. Flentje, A. J. Illingworth, M. Quante, and J. Pelon, 2003b: Characteristics of mixed-phase clouds. I: Lidar, radar and aircraft observations from CLARE'98. *Quart. J. Roy. Meteor. Soc.*, **129**, 2089–2116, doi:10.1256/rj.01.208.

——, N. Gaussiat, and A. J. Illingworth, 2005: Stratocumulus liquid water content from dual-wavelength radar. *J. Atmos. Oceanic Technol.*, **22**, 1207–1218, doi:10.1175/JTECH1768.1.

——, M. P. Mittermaier, and A. J. Illingworth, 2006: The retrieval of ice water content from radar reflectivity factor and temperature and its use in the evaluation of a mesoscale model. *J. Appl. Meteor. Climatol.*, **45**, 301–317, doi:10.1175/JAM2340.1.

——, E. J. O'Connor, and A. J. Illingworth, 2009a: Verification of cloud-fraction forecasts. *Quart. J. Roy. Meteor. Soc.*, **135**, 1494–1511, doi:10.1002/qj.481.

——, A. L. M. Grant, A. J. Illingworth, G. N. Pearson, and E. J. O'Connor, 2009b: Vertical velocity variance and skewness in clear and cloud-topped boundary layers as revealed by Doppler lidar. *Quart. J. Roy. Meteor. Soc.*, **135**, 635–643, doi:10.1002/qj.413.

——, C. A. T. Ferro, I. T. Jolliffe, and D. B. Stephenson, 2010: Equitability revisited: Why the "equitable threat score" is not equitable. *Wea. Forecasting*, **25**, 710–726, doi:10.1175/2009WAF2222350.1.

Holben, B. N., and Coauthors, 1998: AERONET—A federated instrument network and data archive for aerosol characterization. *Remote Sens. Environ.*, **66**, 1–16, doi:10.1016/S0034-4257(98)00031-5.

Illingworth, A. J., and Coauthors, 2007: Cloudnet: Continuous evaluation of cloud profiles in seven operational models using ground-based observations. *Bull. Amer. Meteor. Soc.*, **88**, 883–898, doi:10.1175/BAMS-88-6-883.

Kadygrov, E. N., and D. R. Pick, 1998: The potential for temperature retrieval from an angular-scanning single-channel microwave radiometer and some comparison with in situ observation. *Appl. Meteor.*, **5**, 393–404, doi:10.1017/S1350482798001054.

Kalthoff, N., and Coauthors, 2013: KITcube—A mobile observation platform for convection studies deployed during HyMeX. *Meteor. Z.*, **22**, 633–647, doi:10.1127/0941-2948/2013/0542.

Kneifel, S., S. Redl, E. Orlandi, U. Löhnert, M. P. Cadeddu, D. D. Turner, and M.-T. Chen, 2014: Absorption properties of supercooled liquid water between 31 and 225 GHz: Evaluation of absorption models using ground-based observations. *J. Appl. Meteor. Climatol.*, **53**, 1028–1045, doi:10.1175/JAMC-D-13-0214.1.

Lebel, T., and Coauthors, 2010: The AMMA field campaigns: Multiscale and multidisciplinary observations in the West African region. *Quart. J. Roy. Meteor. Soc.*, **136**, doi:10.1002/qj.486.

Löhnert, U., and O. Maier, 2012: Operational profiling of temperature using ground-based microwave radiometry at Payerne: Prospects and challenges. *Atmos. Meas. Tech.*, **5**, 1121–1134, doi:10.5194/amt-5-1121-2012.

——, S. Crewell, and C. Simmer, 2004: An integrated approach toward retrieving physically consistent profiles of temperature, humidity, and cloud liquid water. *J. Appl. Meteor.*, **43**, 1295–1307, doi:10.1175/1520-0450(2004)043<1295:AIATRP>2.0.CO;2.

——, ——, O. Krasnov, E. J. O'Connor, and H. W. J. Russchenberg, 2008: Advances in continuously profiling the thermodynamic state of the boundary layer: Integration of measurements and methods. *J. Atmos. Oceanic Technol.*, **25**, 1251–1266, doi:10.1175/2007JTECHA961.1.

——, D. Turner, and S. Crewell, 2009: Ground-based temperature and humidity profiling using spectral infrared and microwave observations. Part I: Simulated retrieval performance in clear sky conditions. *J. Appl. Meteor. Climatol.*, **48**, 1017–1032, doi:10.1175/2008JAMC2060.1.

——, S. Kneifel, A. Battaglia, M. Hagen, L. Hirsch, and S. Crewell, 2011: A multisensor approach toward a better understanding of snowfall microphysics: The TOSCA project. *Bull. Amer. Meteor. Soc.*, **92**, 613–628, doi:10.1175/2010BAMS2909.1.

——, and Coauthors, 2014: JOYCE: Jülich Observatory for Cloud Evolution. *Bull. Amer. Meteor. Soc.*, **96**, 1157–1174, doi:10.1175/BAMS-D-14-00105.1.

Long, C. N., T. P. Ackerman, K. L. Gaustad, and J. N. S. Cole, 2006: Estimation of fractional sky cover from broadband shortwave radiometer measurements. *J. Geophys. Res.*, **111**, D11204, doi:10.1029/2005JD006475.

Lopatin, A., O. Dubovik, A. Chaikovsky, P. Goloub, T. Lapyonok, D. Tanré, and P. Litvinov, 2013: Enhancement of aerosol characterization using synergy of lidar and sun-photometer coincident observations: The GARRLiC algorithm. *Atmos. Meas. Tech.*, **6**, 2065–2088, doi:10.5194/amt-6-2065-2013.

Maahn, M., and P. Kollias, 2012: Improved micro rain radar snow measurements using Doppler spectra post-processing. *Atmos. Meas. Tech.*, **5**, 2661–2673, doi:10.5194/amt-5-2661-2012.

Madonna, F., A. Amodeo, G. D'Amico, L. Mona, and G. Pappalardo, 2010: Observation of non-spherical ultragiant aerosol using a microwave radar. *Geophys. Res. Lett.*, **37**, L21814, doi:10.1029/2010GL044999.

——, and Coauthors, 2011: CIAO: The CNR-IMAA advanced observatory for atmospheric research. *Atmos. Meas. Tech.*, **4**, 1191–1208, doi:10.5194/amt-4-1191-2011.

——, A. Amodeo, G. D'Amico, and G. Pappalardo, 2013: A study on the use of radar and lidar for characterizing ultragiant aerosol. *J. Geophys. Res. Atmos.*, **118**, 10 056–10 071, doi:10.1002/jgrd.50789.

——, M. Rosoldi, J. Güldner, A. Haefele, R. Kivi, M. P. Cadeddu, D. Sisterson, and G. Pappalardo, 2014: Quantifying the value of redundant measurements at GCOS Reference Upper-Air Network sites. *Atmos. Meas. Tech. Discuss.*, **7**, 6327–6357, doi:10.5194/amtd-7-6327-2014.

Markowicz, K. M., P. J. Flatau, A. E. Kardas, J. Remiszewska, K. Stelmaszczyk, and L. Woeste, 2008: Ceilometer retrieval of the boundary layer vertical aerosol extinction structure. *J. Atmos. Oceanic Technol.*, **25**, 928–944, doi:10.1175/2007JTECHA1016.1.

Martin, S., F. Beyrich, and J. Bange, 2014: Observing entrainment processes using a small unmanned aerial vehicle: A feasibility study. *Bound.-Layer Meteor.*, **150**, 449–467, doi:10.1007/s10546-013-9880-4.

Martucci, G., and C. D. O'Dowd, 2011: Ground-based retrieval of continental and marine warm cloud microphysics. *Atmos. Meas. Tech.*, **4**, 2749–2765, doi:10.5194/amt-4-2749-2011.

——, J. Ovadnevaite, D. Ceburnis, H. Berresheim, S. Varghese, D. Martin, R. Flanagan, and C. D. O'Dowd, 2012: Impact of volcanic ash plume aerosol on cloud microphysics. *Atmos. Environ.*, **48**, 205–218, doi:10.1016/j.atmosenv.2011.12.033.

Maschwitz, G., U. Löhnert, S. Crewell, T. Rose, and D. D. Turner, 2013: Investigation of ground-based microwave radiometer calibration techniques at 530 hPa. *Atmos. Meas. Tech.*, **6**, 2641–2658, doi:10.5194/amt-6-2641-2013.

Matthias, V., and Coauthors, 2004a: Vertical aerosol distribution over Europe: Statistical analysis of Raman lidar data from 10 European Aerosol Research Lidar Network (EARLINET) stations. *J. Geophys. Res.*, **109**, D18201, doi:10.1029/2004JD004638.

——, and Coauthors, 2004b: Aerosol lidar intercomparison in the framework of the EARLINET Project. 1. Instruments. *Appl. Opt.*, **43**, 961–976, doi:10.1364/AO.43.000961.

Mattioli, V., E. R. Westwater, D. Cimini, A. J. Gasiewski, M. Klein, and V. Y. Leuski, 2008: Microwave and millimeter-wave radiometric and radiosonde observations in an Arctic environment. *J. Atmos. Oceanic Technol.*, **25**, 1768–1777, doi:10.1175/2008JTECHA1078.1.

Meier, J., and Coauthors, 2012: A regional model of European aerosol transport: Evaluation with sun photometer, lidar and air quality data. *Atmos. Environ.*, **47**, 519–532, doi:10.1016/j.atmosenv.2011.09.029.

Meunier, V., U. Löhnert, P. Kollias, and S. Crewell, 2013: Biases caused by the instrument bandwidth and beam width on simulated brightness temperature measurements from scanning microwave radiometers. *Atmos. Meas. Tech.*, **6**, 1171–1187, doi:10.5194/amt-6-1171-2013.

Miller, M., and A. Slingo, 2007: The ARM Mobile Facility and its first international deployment: Measuring radiative flux divergence in West Africa. *Bull. Amer. Meteor. Soc.*, **88**, 1229–1244, doi:10.1175/BAMS-88-8-1229.

Miller, M. A., K. Nitschke, T. P. Ackerman, W. R. Ferrell, N. Hickmon, and M. Ivey, 2016: The ARM Mobile Facilities. *The Atmospheric Radiation Measurement (ARM) Program: The First 20 Years, Meteor. Monogr.*, No. 57, Amer. Meteor. Soc., doi:10.1175/AMSMONOGRAPHS-D-15-0051.1.

Miller, R. L., A. Slingo, J. C. Barnard, and E. Kassianov, 2009: Seasonal contrast in the surface energy balance of the Sahel. *J. Geophys. Res.*, **114**, D00E05, doi:10.1029/2008JD010521.

Mona, L., A. Amodeo, M. Pandolfi, and G. Pappalardo, 2006: Saharan dust intrusions in the Mediterranean area: Three years of Raman lidar measurements. *J. Geophys. Res.*, **111**, D16203, doi:10.1029/2005JD006569.

——, and Coauthors, 2007: Characterization of the variability of the humidity and cloud fields as observed from a cluster of ground-based lidar systems. *Quart. J. Roy. Meteor. Soc.*, **133**, 257–271, doi:10.1002/qj.160.

——, and Coauthors, 2009: One year of CNR-IMAA multi-wavelength Raman lidar measurements in correspondence of CALIPSO overpass: Level 1 products comparison. *Atmos. Chem. Phys.*, **9**, 7213–7228, doi:10.5194/acp-9-7213-2009.

——, A. Amodeo, G. D'Amico, A. Giunta, F. Madonna, and G. Pappalardo, 2012: Multi-wavelength Raman lidar observations of the Eyjafjallajökull volcanic cloud over Potenza, southern Italy. *Atmos. Chem. Phys.*, **12**, 2229–2244, doi:10.5194/acp-12-2229-2012.

Morille, Y., M. Haeffelin, P. Drobinski, and J. Pelon, 2007: STRAT: An automated algorithm to retrieve the vertical structure of the atmosphere from single-channel lidar data. *J. Atmos. Ocean. Technol.*, **24**, 761–775, doi: 10.1175/JTECH2008.1.

Müller, D., U. Wandinger, and A. Ansmann, 1999: Microphysical particle parameters from extinction and backscatter lidar data by inversion with regularization: Theory. *Appl. Opt.*, **38**, 2346–2357, doi:10.1364/AO.38.002346.

——, I. Mattis, A. Ansmann, B. Wehner, D. Althausen, U. Wandinger, and O. Dubovik, 2004: Closure study on optical and microphysical properties of a mixed urban and Arctic haze air mass observed with Raman lidar and Sun photometer. *J. Geophys. Res.*, **109**, D13206, doi:10.1029/2003JD004200.

——, ——, ——, U. Wandinger, C. Ritter, and D. Kaiser, 2007: Multiwavelength Raman lidar observations of particle growth during long-range transport of forest-fire smoke in the free troposphere. *Geophys. Res. Lett.*, **34**, L05803, doi:10.1029/2006GL027936.

Naud, C., J. P. Muller, and E. E. Clothiaux, 2006: Assessment of multispectral ATSR2 stereo cloud-top height retrievals. *Remote Sens. Environ.*, **104**, 337–345, doi:10.1016/j.rse.2006.05.008.

——, and Coauthors, 2010: Thermodynamic phase profiles of optically thin midlatitude clouds and their relation to temperature. *J. Geophys. Res.*, **115**, D11202, doi:10.1029/2009JD012889.

Neggers, R. A. J., and A. P. Siebesma, 2013: Constraining a system of interacting parameterizations through multiple-parameter evaluation: Tracing a compensating error between cloud vertical structure and cloud overlap. *J. Climate*, **26**, 6698–6715, doi:10.1175/JCLI-D-12-00779.1.

——, ——, and T. Heus, 2012: Continuous single-column model evaluation at a permanent meteorological supersite. *Bull. Amer. Meteor. Soc.*, **93**, 1389–1400, doi:10.1175/BAMS-D-11-00162.1.

Neisser, J., W. Adam, F. Beyrich, U. Leiterer, and H. Steinhagen, 2002: Atmospheric boundary layer monitoring at the Meteorological Observatory Lindenberg as a part of the "Lindenberg Column": Facilities and selected results. *Meteor. Z.*, **11**, 241–253, doi:10.1127/0941-2948/2002/0011-0241.

O'Connor, E. J., A. J. Illingworth, and R. J. Hogan, 2004: A technique for autocalibration of cloud lidar. *J. Atmos. Oceanic Technol.*, **21**, 777–786, doi:10.1175/1520-0426(2004)021<0777:ATFAOC>2.0.CO;2.

——, R. J. Hogan, and A. J. Illingworth, 2005: Retrieving stratocumulus drizzle parameters using Doppler radar and lidar. *J. Appl. Meteor.*, **44**, 14–27, doi:10.1175/JAM-2181.1.

——, A. J. Illingworth, I. M. Brooks, C. D. Westbrook, R. J. Hogan, F. Davies, and B. J. Brooks, 2010: A method for estimating the turbulent kinetic energy dissipation rate from a vertically pointing Doppler lidar, and independent evaluation from balloon-borne in situ measurements. *J. Atmos. Oceanic Technol.*, **27**, 1652–1664, doi:10.1175/2010JTECHA1455.1.

Paine, S., D. D. Turner, and N. Küchler, 2014: Understanding thermal drift in liquid nitrogen loads used for radiometric calibration in the field. *J. Atmos. Oceanic Technol.*, **31**, 647–655, doi:10.1175/JTECH-D-13-00171.1.

Pal, S., M. Haeffelin, and E. Batchvarova, 2013: Exploring a geophysical process-based attribution technique for the determination of the atmospheric boundary layer depth using aerosol lidar and near-surface meteorological measurements. *J. Geophys. Res. Atmos.*, **118**, 9277–9295, doi:10.1002/jgrd.50710.

Papayannis, A., and Coauthors, 2008: Systematic lidar observations of Saharan dust over Europe in the frame of EARLINET (2000–2002). *J. Geophys. Res.*, **113**, D10204, doi:10.1029/2007JD009028.

Pappalardo, G., and Coauthors, 2004: Aerosol lidar intercomparison in the framework of the EARLINET project. 3. Raman lidar algorithm for aerosol extinction, backscatter and lidar ratio. *Appl. Opt.*, **43**, 5370–5385, doi:10.1364/AO.43.005370.

——, and Coauthors, 2010: EARLINET correlative measurements for CALIPSO: First intercomparison results. *J. Geophys. Res.*, **115**, D00H19, doi:10.1029/2009JD012147.

——, and Coauthors, 2013: Four-dimensional distribution of the 2010 Eyjafjallajökull volcanic cloud over Europe observed by EARLINET. *Atmos. Chem. Phys.*, **13**, 4429–4450, doi:10.5194/acp-13-4429-2013.

Piringer, M., and Coauthors, 2007: The surface energy balance and the mixing height in urban areas—Activities and recommendations of COST-Action 715. *Bound.-Layer Meteor.*, **124**, 3–24, doi:10.1007/s10546-007-9170-0.

Pospichal, B., 2009: Diurnal to annual variability of the atmospheric boundary layer over West Africa: A comprehensive view by remote sensing observations. Doctoral thesis, Faculty of Mathematics and Natural Sciences, University of Cologne, 120 pp. [Available online at http://kups.ub.uni-koeln.de/2985/1/dissertation_pospichal.pdf.]

Pougatchev, N., G. Bingham, D. Seidel, and F. Berger, 2007: Statistical approach to validation of satellite atmospheric retrievals. *Remote Sensing of Clouds and the Atmosphere XII*, A. Comerón et al., Eds., International Society for Optical Engineering (SPIE Proceedings, Vol. 6745), doi:10.1117/12.737943.

Protat, A., and Coauthors, 2009: Assessment of Cloudsat reflectivity measurements and ice cloud properties using ground-based and airborne cloud radar observations. *J. Atmos. Oceanic Technol.*, **26**, 1717–1741, doi:10.1175/2009JTECHA1246.1.

——, J. Delanoë, E. J. O'Connor, and T. S. L'Ecuyer, 2010: The evaluation of CloudSat and CALIPSO ice microphysical products using ground-based cloud radar and lidar observations. *J. Atmos. Oceanic Technol.*, **27**, 793–810, doi:10.1175/2009JTECHA1397.1.

——, D. Bouniol, E. J. O'Connor, H. K. Baltink, J. Verlinde, and K. Widener, 2011: Cloudsat as a global radar calibrator. *J. Atmos. Oceanic Technol.*, **28**, 445–452, doi:10.1175/2010JTECHA1443.1.

Ricaud, P., B. Gabard, S. Derrien, J. L. Attié, T. Rose, and H. Czekala, 2010: Validation of tropospheric water vapor as measured by the 183-GHz HAMSTRAD radiometer over the Pyrenees Mountains, France. *IEEE Trans. Geosci. Remote Sens.*, **48**, 2189–2203, doi:10.1109/TGRS.2009.2037920.

Roebeling, R. A., A. J. Feijt, and P. Stammes, 2006: Cloud property retrievals for climate monitoring: Implications of differences between Spinning Enhanced Visible and Infrared Imager (SEVIRI) on METEOSAT-8 and Advanced Very High Resolution Radiometer (AVHRR) on NOAA-17. *J. Geophys. Res.*, **111**, D20210, doi:10.1029/2005JD006990.

Roelofs, G.-J., H. ten Brink, A. Kiendler-Scharr, G. de Leeuw, A. Mensah, A. Minikin, and R. Otjes, 2010: Evaluation of simulated aerosol properties with the aerosol-climate model ECHAM5-HAM using observations from the IMPACT field campaign. *Atmos. Chem. Phys.*, **10**, 7709–7722, doi:10.5194/acp-10-7709-2010.

Rose, T., S. Crewell, U. Löhnert, and C. Simmer, 2005: A network suitable microwave radiometer for operational monitoring of the cloudy atmosphere. *Atmos. Res.*, **75**, 183–200, doi:10.1016/j.atmosres.2004.12.005.

Rosoldi, M., and Coauthors, 2013: Study of thin clouds at CNR-IMAA Atmospheric Observatory (CIAO). *Ann. Geophys.*, **56**, doi:10.4401/ag-6337.

Sawamura, P., and Coauthors, 2012: Stratospheric AOD after the 2011 eruption of Nabro volcano measured by lidars over the Northern Hemisphere. *Environ. Res. Lett.*, **7**, 034013, doi:10.1088/1748-9326/7/3/034013.

Seibert, P., F. Beyrich, S. Gryning, S. Joffre, A. Rasmussen, and P. Tercier, 2000: Review and intercomparison of operational methods for the determination of the mixing height. *Atmos. Environ.*, **34**, 1001–1027, doi:10.1016/S1352-2310(99)00349-0.

Shupe, M. D., J. M. Comstock, D. D. Turner, and G. G. Mace, 2016: Cloud property retrievals in the ARM Program. *The Atmospheric Radiation Measurement (ARM) Program: The First 20 Years, Meteor. Monogr.*, No. 57, Amer. Meteor. Soc., doi:10.1175/AMSMONOGRAPHS-D-15-0030.1.

Spänkuch, D., W. Döhler, J. Güldner, and A. Keens, 1996: Ground-based passive atmospheric remote sounding by FTIR emission spectroscopy: First results with EISAR. *Atmos. Phys*, **69**, 97–111.

Stegehuis, A. I., R. Vautard, P. Ciais, A. J. Teuling, M. Jung, and P. Yiou, 2013: Summer temperatures in Europe and land heat fluxes in observation-based data and regional climate model simulations. *Climate Dyn.*, **41**, 455–477, doi:10.1007/s00382-012-1559-x.

Stokes, G. M., 2016: Original ARM concept and launch. *The Atmospheric Radiation Measurement (ARM) Program: The First 20 Years, Meteor. Monogr.*, No. 57, Amer. Meteor. Soc., doi:10.1175/AMSMONOGRAPHS-D-15-0021.1.

Tonttila, J., E. J. O'Connor, S. Niemela, P. Raisanen, and H. Jarvinen, 2011: Cloud base vertical velocity statistics: A comparison between an atmospheric mesoscale model and remote sensing observations. *Atmos. Chem. Phys.*, **11**, 9207–9218, doi:10.5194/acp-11-9207-2011.

Turner, D. D., and E. J. Mlawer, 2010: The Radiative Heating in Underexplored Bands Campaigns. *Bull. Amer. Meteor. Soc.*, **91**, 911–923, doi:10.1175/2010BAMS2904.1.

——, and U. Löhnert, 2014: Information content and uncertainties in thermodynamic profiles and liquid cloud properties retrieved from the ground-based Atmospheric Emitted Radiance Interferometer (AERI). *J. Appl. Meteor. Climatol.*, **53**, 752–771, doi:10.1175/JAMC-D-13-0126.1.

——, M. P. Caddedu, U. Löhnert, S. Crewell, and A. M. Vogelmann, 2009: Modifications to the water vapor continuum in the microwave suggested by ground-based 150 GHz Observations. *IEEE Trans. Geosci. Remote Sens.*, **47**, 3326–3337, doi:10.1109/TGRS.2009.2022262.

——, and Coauthors, 2012: Ground-based high spectral resolution observations of the entire terrestrial spectrum under extremely dry conditions. *Geophys. Res. Lett.*, **39**, L10801, doi:10.1029/2012GL051542.

van Meijgaard, E., and S. Crewell, 2005: Comparison of model predicted liquid water path with ground-based measurements during CLIWA-NET. *Atmos. Res.*, **75**, 201–226, doi:10.1016/j.atmosres.2004.12.006.

van Zadelhoff, G. J., D. P. Donovan, H. K. Baltink, and R. Boers, 2004: Comparing ice cloud microphysical properties using Cloudnet and Atmospheric Radiation Measurement Program data. *J. Geophys. Res.*, **109**, D24214, doi:10.1029/2004JD004967.

Veselovskii, I., A. Kolgotin, V. Griaznov, D. Müller, U. Wandinger, and D. Whiteman, 2002: Inversion with regularization for the retrieval of tropospheric aerosol parameters from multiwavelength lidar sounding. *Appl. Opt.*, **41**, 3685–3699, doi:10.1364/AO.41.003685.

Villani, M. G., and Coauthors, 2006: Transport of volcanic aerosol in the troposphere: the case study of the 2002 Etna plume. *J. Geophys. Res.*, **111**, D21102, doi:10.1029/2006JD007126.

Wagner, J., A. Ansmann, U. Wandinger, P. Seifert, A. Schwarz, M. Tesche, A. Chaikovsky, and O. Dubovik, 2013: Evaluation of the Lidar/Radiometer Inversion Code (LIRIC) to determine microphysical properties of volcanic and desert dust. *Atmos. Meas. Tech.*, **6**, 1707–1724, doi:10.5194/amt-6-1707-2013.

Wang, P., and Coauthors, 2011: Cloudy sky shortwave radiative closure for a Baseline Surface Radiation Network site. *J. Geophys. Res.*, **116**, D08202, doi:10.1029/2010JD015141.

Westbrook, C. D., and A. J. Illingworth, 2009: Testing the influence of small crystals on ice size spectra using Doppler lidar observations. *Geophys. Res. Lett.*, **36**, L12810, doi:10.1029/2009GL038186.

——, and ——, 2011: Evidence that ice forms primarily in supercooled liquid clouds at temperatures $>-27°C$. *Geophys. Res. Lett.*, **38**, L14808, doi:10.1029/2011GL048021.

——, and ——, 2013: The formation of ice in a long-lived supercooled layer cloud. *Quart. J. Roy. Meteor. Soc.*, **139**, 2209–2221, doi:10.1002/qj.2096.

——, ——, E. J. O'Connor, and R. J. Hogan, 2010: Doppler lidar measurements of oriented planar ice crystals falling from supercooled and glaciated layer clouds. *Quart. J. Roy. Meteor. Soc.*, **136**, 260–276, doi:10.1002/qj.528.

Westwater, E. R., S. Crewell, C. Mätzler, and D. Cimini, 2005: Principles of surface-based microwave and millimeter wave radiometric remote sensing of the troposphere. *Quad. Soc. Ital. Elettromagn.*, **1** (3), 50–90.

Wetzel, G., and Coauthors, 2013: Validation of MIPAS-ENVISAT H2O operational data collected between July 2002 and March 2004. *Atmos. Chem. Phys.*, **13**, 5791–5811, doi:10.5194/acp-13-5791-2013.

Wiegner, M., and A. Geiß, 2012: Aerosol profiling with the Jenoptik ceilometer CHM15kx. *Atmos. Meas. Tech.*, **5**, 1953–1964, doi:10.5194/amt-5-1953-2012.

Wiegner, M. J., and Coauthors, 2008: Numerical simulations of optical properties of Saharan dust aerosols with special emphasis on the linear depolarization ratio. *Tellus*, **61B**, 180–194, doi:10.1111/j.1600-0889.2008.00381.x.

Wiegner, M., and Coauthors, 2014: What is the benefit of ceilometers for aerosol remote sensing? An answer from EARLINET. *Atmos. Meas. Tech.*, **7**, 1979–1997, doi:10.5194/amt-7-1979-2014.

Willen, U., S. Crewell, H. K. Baltink, and O. Sievers, 2005: Assessing model predicted vertical cloud structure and cloud overlap with radar and lidar ceilometer observations for the Baltex Bridge Campaign of CLIWA-NET. *Atmos. Res.*, **75**, 227–255, doi:10.1016/j.atmosres.2004.12.008.

Wulfmeyer, V., and Coauthors, 2011: The Convective and Orographically Induced Precipitation Study (COPS): The scientific strategy, the field phase, and research highlights. *Quart. J. Roy. Meteor. Soc.*, **137**, 3–30, doi:10.1002/qj.752.

Xie, S., and Coauthors, 2010: Clouds and more: ARM climate modeling best estimate data: A new data product for climate studies. *Bull. Amer. Meteor. Soc.*, **91**, 13–20, doi:10.1175/2009BAMS2891.1.

Zhang, Q. J., and Coauthors, 2013: Formation of organic aerosol in the Paris region during the MEGAPOLI summer campaign: Evaluation of the volatility-basis-set approach within the CHIMERE model. *Atmos. Chem. Phys.*, **13**, 5767–5790, doi:10.5194/acp-13-5767-2013.

Chapter 30

ARM and Satellite Cloud Validation

ROGER MARCHAND

Department of Atmospheric Sciences, University of Washington, Seattle, Washington

1. Introduction

Many of the same factors that helped create the ARM Program, including the need to better constrain climate models, also helped lead to an expansion of Earth observing satellites in recent decades. In particular, starting in the late 1990s the National Aeronautics and Space Administration (NASA) Earth Observing System (EOS) program (Kaufman et al. 1998) began launching a series of polar-orbiting and low-inclination satellites for long-term global observations of the land surface, biosphere, atmosphere, and oceans.[1] Several of these missions had, and continue to have, a strong emphasis on clouds, aerosols, and radiation.

Among others, these missions included several large multi-instrument platforms, namely the Tropical Rainfall Measuring Mission (TRMM, launched November 1997), *Terra* (also known as *EOS-AM-1*, launched December 1999), and *Aqua* (*EOS-PM-1*, launched May 2002). All three of these large multi-instrument missions include measurements of broadband shortwave and longwave radiances from the Clouds and the Earth's Radiant Energy System (CERES) instruments, as well as multiple-wavelength narrow-band imagers—namely, the Visible and Infrared Scanner (VIRS) on board TRMM, the Multiangle Imaging SpectroRadiometer (MISR) on board *Terra*, and the Moderate Resolution Imaging Spectroradiometer (MODIS) on board both *Terra* and *Aqua*. A major advance of these new imagers is the improved calibration compared to previous generation observations due to the use of solar diffusers, a strong emphasis on vicarious calibration activities, and other factors. In combination with geostationary and polar-orbiting weather satellite observations, the CERES and NASA imager measurements have substantially improved knowledge of top-of-the-atmosphere radiative fluxes, including the variability of these fluxes and the large impact of clouds.

The MISR and MODIS imagers also make measurements at wavelengths not collected by weather satellites, at higher spatial resolutions (up to 250 m at some wavelengths), and in the case of MISR, at nine view angles along the satellite flight direction. These additional capabilities enable application of algorithms for the detection of clouds, as well as determination of their radiative or microphysical properties (such as the amount of water or ice) that were not previously possible. New does not necessarily mean better, and ARM played a pivotal role in characterizing or validating (as it is often called in the satellite community) retrievals of cloud properties from NASA EOS, NOAA weather (including geostationary), and other satellites.

Much of the ARM Program's role in satellite cloud retrieval validation stems from the program's dedication to developing and improving ground-based instruments, such as millimeter-wavelength cloud radar and Raman lidar, as well as the development of retrievals based on combinations of ground-based instruments including passive microwave radiometers and infrared interferometers (e.g., Shupe et al. 2008). Equally important has been the program's commitment to making measurements continuously at several sites (including tropical and high-latitude locations) over many years. These measurements have provided a large number of samples across a wide variety of meteorological conditions and enabled evaluation of satellite datasets with many differing orbits, as well as helping

[1] NASA's Earth Observing System Project Science Office: http://eospso.gsfc.nasa.gov/.

Corresponding author address: Roger Marchand, Dept. of Atmospheric Sciences, University of Washington, 3737 Brooklyn Ave. NE, Seattle, WA 98105.
E-mail: rojmarch@u.washington.edu

DOI: 10.1175/AMSMONOGRAPHS-D-15-0038.1

to understand limitations in the diurnal sampling of nongeostationary sensors.

While not the focus of this chapter, the retrieval algorithms that ARM helped developed have not only provided validation data, but also helped in the development of radar, lidar, and combined instrument retrievals from space (e.g., Wang and Sassen 2002). In April 2006, NASA launched the *CloudSat* and *CALIPSO* missions, placing into orbit a millimeter-wavelength (W band, 95 GHz) cloud radar and dual-wavelength (532 and 1064 nm) lidar. In a process known as formation flying, these two platforms fly along nearly the same orbit as the *Aqua* platform, making observations separated by less than 2 min, and are controlled carefully so that the narrow radar and lidar beams gather data from overlapping ground points about 90% of the time. These three satellites, along with the NASA *Aura* and *Orbiting Carbon Observatory 2* (*OCO-2*) satellites, the French *Polarization and Anisotropy of Reflectances for Atmospheric Sciences Coupled with Observations from a Lidar* (*PARASOL*) satellite, and the Japanese *Global Change Observation Mission–Water* (*GCOM-W1*), form the satellite afternoon constellation or A-Train.[2]

In this chapter we summarize results from a variety of satellite cloud validation studies. Observations at the ARM sites have also made important contributions in validating surface and atmospheric properties such as surface radiation (e.g., Charlock and Alberta 1996), surface albedo (e.g., Jin et al. 2003; Trishchenko et al. 2008), microwave emissivity and soil moisture (e.g., Lin and Minnis 2000), precipitation (e.g., Sapiano and Arkin 2009), and water vapor (e.g., Tobin et al. 2006). However, here we focus on clouds since ARM has played a central role in validation of these quantities with few other programs providing as comprehensive a range of cloud observations and none with a longer cloud (millimeter wavelength) radar record. Our intent here is to provide representative examples of ARM contributions to satellite validation rather than trying to provide an exhaustive list. The examples focus on long-term measurements, but case study analyses have also provided important insights, especially for algorithm development.

2. Passive sensors and clouds

a. Cloud macrophysics: Cloud detection

In most satellite retrievals, the first step is to determine if a cloud is present; dividing the observed image into a set of image pixels that contain cloud (on which one might apply a retrieval for cloud microphysical or radiative properties, as discussed later) and a set that are cloud free (on which one might apply surface and/or aerosol property retrievals). The cloud fraction (i.e., the number of cloudy pixels relative to total pixels) is perhaps the cloud property used most frequently in the evaluation of climate models. Thus, understanding how well and under what conditions clouds can be identified is critical.

Typically, the most difficult clouds to identify are optically thin high-altitude cirrus clouds and small-scale (often subimager resolution) cumulus clouds, because both of these cloud types may have only a small effect on observed visible and infrared images. That is, these clouds produce little contrast in the observed imagery between cloudy and cloud-free conditions, which means that the accuracy of satellite cloud detection schemes is dependent upon the underlying surface, making it important to have multiple validation sites.

At high latitudes, where the surface is often snow covered much of the year and the sun is low (near the horizon), cloud detection has traditionally been particularly challenging. This is because visibly bright snow-covered surfaces and shadowing by clouds greatly reduce the contrast between clouds and the surface at visible wavelengths, while cold surface temperatures with strong and frequent temperature inversions (colder air near the surface than above) complicate the use of infrared channels. In response to this challenge, cloud detection techniques have been developed combining observations at several infrared and shortwave-infrared wavelengths (the latter meaning wavelengths where both thermal emission and solar scattering contribute significantly). In particular, several wavelengths (or bands) were included in the MODIS instrument to improve surface retrievals and improve discrimination of cloud over snow and sea ice (Schueler and Barnes 1998).

Observations collected by the ARM Program at its North Slope of Alaska (NSA) sites at Barrow and Atqasuk, Alaska, have been used by several researchers to evaluate satellite cloud detection. For example, Berendes et al. (2004) compared MODIS (MOD35 product, collection 4) cloud detection with several ARM datasets including the ARM Active Remote Sensing of Cloud Layers (ARSCL) product, which combines data from the ARM vertically pointing cloud radar and lidar systems (Clothiaux et al. 2000). Berendes et al. (2004) generated a scatterplot of MODIS cloud fraction calculated over a 15-km region surrounding the ARM Barrow site against cloud fraction calculated from ARSCL data collected within 15 min of the MODIS overpass (Fig. 30-1). The ARM radar and lidar systems used are vertically pointing instruments with a very narrow field of view. Thus, these instruments produce a vertically resolved time series (sometimes called a time–height plot) of clouds as they pass over the ARM site.

[2] Stephens et al. (2002); http://atrain.nasa.gov/intro.php.

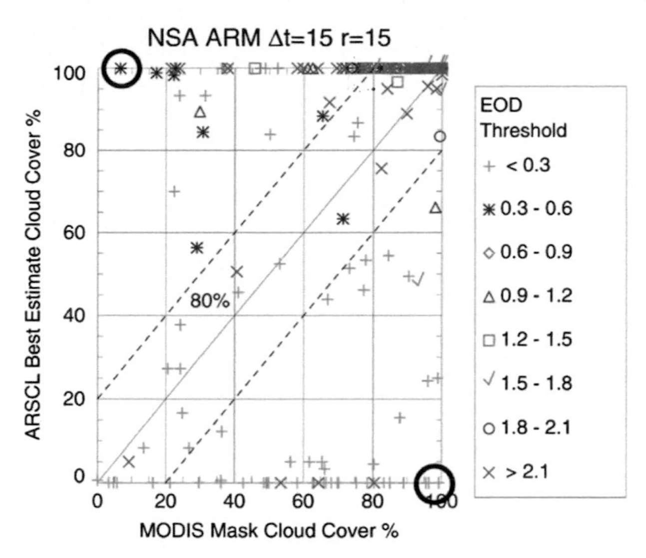

FIG. 30-1. Comparison of MODIS (MOD35 collection 4) cloud detection with ARM ARSCL (radar–lidar) product for the Barrow site. Symbols represent different optical depth (EOD) categories. Daytime data only. Taken from Berendes et al. (2004). Black circles denote cases examined in greater detail by Berendes et al. that are not discussed here.

In this figure, the symbols show an estimate for the cloud optical depth (a measure of how much direct sunlight is reflected or absorbed by the cloud before it reaches the surface). Small values of optical depth (less than 2) imply an optically thin cloud through which you could easily see the solar disk. These authors found agreement between the individual MODIS and ARSCL cloud fractions to better than 20%, more than 80% of the time. When the two datasets disagree, the cloud was typically either a broken low cloud over snow (asterisks in Fig. 30-1, which had a 15-min mean optical depth typically less than about 0.6) that was not detected by MODIS or optically thin high-altitude cirrus (with an optical depth less than about 0.3) that was in some cases not detected by the low-power micropulse lidar used in this study and in other cases not detected by MODIS.

This impressive level of cloud detection accuracy is similar to that found at other sites using MODIS, including at the ARM Southern Great Plains (SGP) site, where Ackerman et al. (2008) found agreement was approximately 85% between MODIS and the ARSCL in identifying cloud and clear scenes. In their approach, Ackerman et al. limited themselves to those cases that were identified by ARSCL over a 5-min window to be either cloudy at least 95% of the time (which were taken to be cloudy), or cloudy less than 5% of the time (which were taken to be clear). This analysis likewise identified thin cirrus as a primary driver of differences in cloud detection between MODIS and ground-based lidar systems. Ackerman et al. (2008) found that when the Arctic High-Spectral Resolution Lidar detected a cloud and MODIS indicated clear, more than 60% of the time the optical depth (estimated from the lidar data) was less than 0.2 and 90% of the time the optical depth was less than 0.4. This is similar to what Marchand et al. (2007) found with MISR cloud detection using lidar observations from the ARM Barrow, SGP, and Nauru sites (see Fig. 30-2).

It is worth noting that the analysis of Berendes et al. (2004) was restricted to daytime conditions. Liu et al. (2004) found similarly good agreement in daytime cloud detection by MODIS (collection 4) but also that MODIS failed to detect more than 40% of the nighttime cloud cover over the ARM sites. Based on ARM and other ground-based datasets, Liu et al. were able to develop additional cloud tests (primarily utilizing the MODIS 7.2-μm water vapor and 14.2-μm carbon dioxide bands) reducing the misidentification of cloud from about 40% to 16% at the ARM sites at night. These tests were included subsequently in the next version of MODIS processing (collection 5).

b. *Cloud macrophysics: Cloud-top height (CTH)*

A climatically critical cloud property that satellites are in a particularly advantageous position to monitor is cloud-top height. As with cloud detection, the ARM ARSCL datasets have been extremely helpful in characterizing the accuracy of satellite imager and sounder retrievals of cloud-top height (or equivalently, cloud-top pressure) owing to the unambiguous range-resolved nature of the cloud radar and lidar observations. The ARM radars and lidars have typically operated with a range resolution of 90 m or better. This includes analysis of retrievals from many satellite imagers (e.g., Hollars et al. 2004; Smith et al. 2008; Mace et al. 2006, 2011), including the NASA MODIS and MISR imagers (e.g., Naud et al. 2002, 2005; Mace et al. 2005; Marchand et al. 2007), as well as infrared sounders (e.g., Hawkinson et al. 2005; Kahn et al. 2007). A detailed discussion of the various satellite retrieval techniques is beyond the scope of what can be included here and so we refer interested readers to the recent review article by Marchand et al. (2010) for a detailed description of the retrieval approaches and their errors for imagers.

Broadly, these analyses have found that, for *high-altitude clouds*, imager or sounder-based retrievals of cloud-top height tend to be located below "true" cloud top as detected by sensitive lidar systems. That is, many high-altitude clouds do not have a sharp upper boundary but rather have low amounts of condensate near cloud top (with low or modest values of optical extinction near cloud top). In this situation, the cloud top identified by imagers represents a "radiative-effective cloud top." The exact position of this radiative-effective level varies

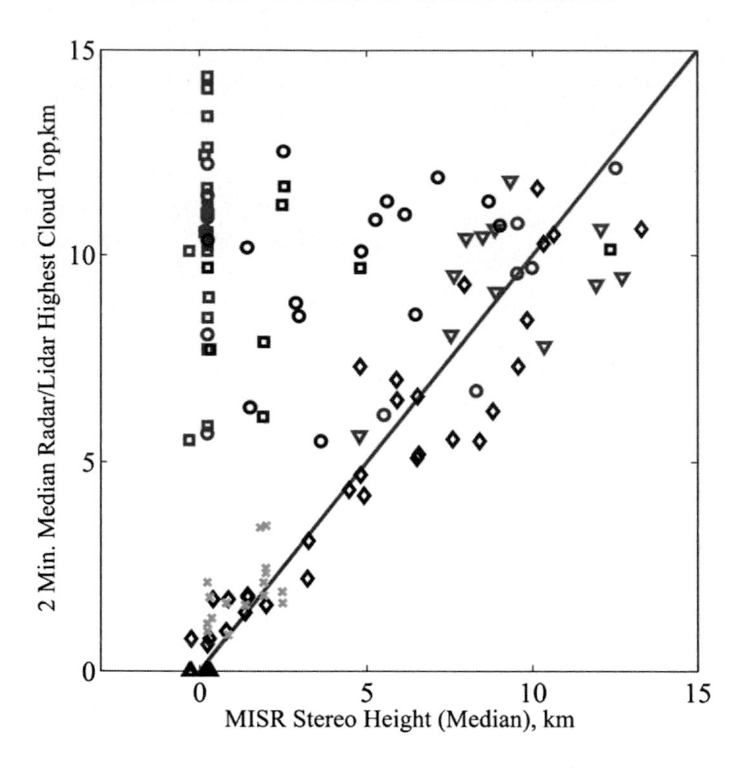

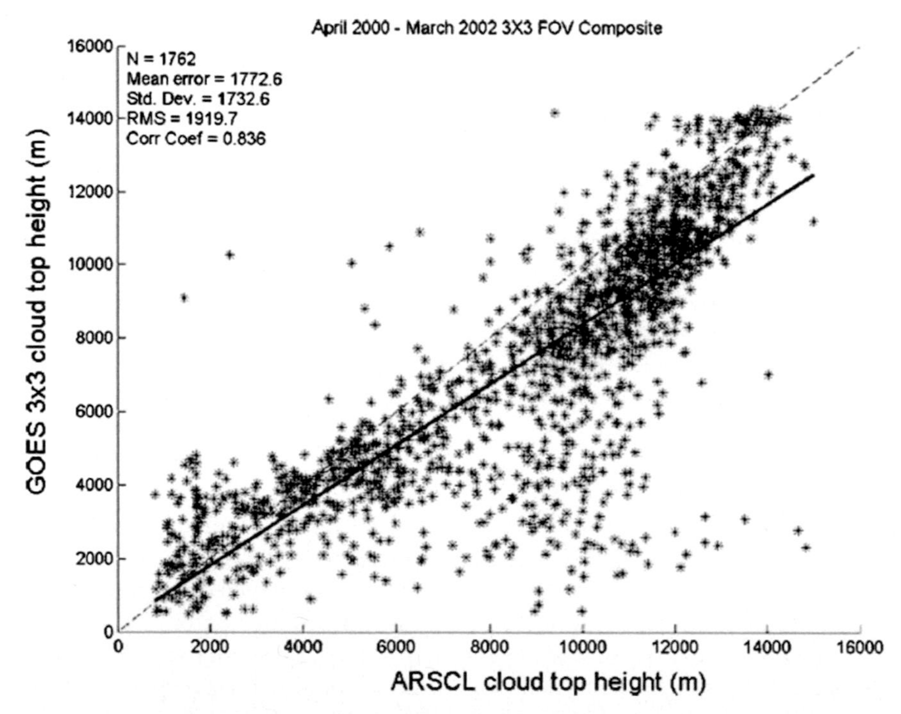

FIG. 30-2. Two comparisons of ARM radar–lidar and satellite-retrieved cloud-top height at SGP site. (top) MISR stereo-imaging retrieval [taken from Marchand et al. (2007)]. (bottom) GOES sounder CO_2 slicing retrieval [taken from Hawkinson et al. (2005)]. Symbols in (top) represent different cloud types. Red symbols are for very optically thin clouds (ARM retrieved optical depth $< \sim 0.3$), which MISR usually fails to detect. Black symbols are multilayer clouds, where the upper-level cloud is optically thin (ARM-retrieved optical depth <1–2) and MISR has returned the height of a lower cloud layer. Other symbols represent various types of optically thick clouds. Blue is stratiform cloud (fills 11-km patch), green is broken boundary layer cloud, and magenta is cloud with diffuse (low condensate) cloud top. Uncertainty is higher at high altitudes (RMS $\sim$ 1 km) than at lower altitudes ($\sim$500 m).

with the details of the retrievals and the vertical profile of cloud extinction. For vertically extensive clouds, it tends to be located within the cloud, often a kilometer below lidar cloud top.

For clouds with high values of optical extinction near cloud top, which includes most clouds with liquid water at the top, retrievals of cloud-top height are in reasonably good agreement with ARM radar and lidar datasets, with little bias and uncertainties (RMS errors) typically between 500 m and 1 km. One notable exception is that infrared-temperature techniques applied to boundary layer clouds under strong temperature inversions (typical in stratocumulus clouds) are often biased, placing cloud much too high in altitude. Figure 30-2 shows results taken from Marchand et al. (2007) and Hawkinson et al. (2005) comparing the MISR stereo-height algorithm and GOES sounder (CO_2 slicing technique) with ARM ARSCL. For the MISR example, red symbols are clouds that are too thin to be detected, while black symbols represent multilayer clouds with a thin upper-level cloud that MISR does not detect, and instead finds the altitude of a lower-altitude cloud layer.

Multilayer clouds present a major challenge to retrievals from imagers and are associated frequently with larger errors (Marchand et al. 2010). Algorithms specifically designed to detect multilayer clouds remain a topic of active research, and ARM datasets have been and will continue to be used in the evaluation of these algorithms (e.g., Baum et al. 2003; Chang and Li 2005; Minnis et al. 2007).

c. *Cloud microphysics and optical properties*

As discussed in Shupe et al. (2016, chapter 19), the development of retrievals based on radar and lidar data for cloud microphysical properties (principally cloud water content and effective particle size) and optical properties (principally optical depth) has been a topic of intensive research within the ARM community. Like satellite retrievals, ground-based retrievals make many simplifying assumptions and have their own characteristic errors, which must be considered carefully in any comparisons. Nonetheless, the ground-based observations often have finer vertical resolution and better sensitivity, because they have the advantage of being much closer to the object of study. ARM retrievals are also usually based on a combination of measurements from different sensors, and in particular, often include measurements of downward shortwave or infrared radiation that provides information on the transmissivity of the clouds (i.e., an integral constraint on the total cloud optical depth). In situ data are often used directly to validate satellite and ground-based retrievals as part of case studies (e.g., Dong et al. 2002) or in the specification of some retrieval parameters such as mixtures of cloud particle habits or shapes [see Nasiri et al. (2002)]. However, these datasets are fundamentally sparse and more difficult to match up with satellites than ground-based retrievals, which provide for much larger datasets and better overall statistical measures.

Two examples of satellite microphysical and optical property validation studies are provided by Mace et al. (2005) for high-altitude ice clouds and Dong et al. (2008) for low clouds.

Mace et al. (2005) use a combination of matchup points (direct comparison of results from two dozen cirrus with coincident measurements) and broader regional statistics (comparing data gathered over a longer period and a larger area but not coincident in time or space) for the ARM SGP region. In the direct comparison, wind-profiler data were used to identify coincident areas for matching the time–height data gathered by ARM with MODIS imagery, and clouds were required to cover this full area. Two separate satellite retrievals (one derived from MODIS data by the MODIS Atmospheres Team and one derived from MODIS data by the CERES Science Team) were compared with results from two ground-based properties retrievals. Overall, Mace et al. show that there is a positive correlation in the effective particle size, the optical thickness, and the ice water path (vertical sum of ice water content/density) between the various methods, although sometimes there are significant biases.

Figure 30-3 shows several comparisons made by Mace et al. between the MODIS team retrieval (known as MOD06) and results from one of the ARM retrievals (known as the Z-radiance technique). The top three panels show results from the individual matchup points and the bottom two panels show distributions from the statistical analysis. The top three panels of Fig. 30-3 demonstrate a reasonable correspondence between the MOD06 and Z-radiance results for effective radius and ice water path. Summary statistics for these comparisons are given by Mace et al. but are not discussed here. The uncertainty bars show spatial and temporal variability over the matchup box. The MOD06 optical depths are biased slightly high with respect to the Z-radiance results. Not shown, the statistical regional analysis further bears out this bias and shows that MOD06 underrepresents the amount of cirrus with optical depths less than 1.0 (especially less than 0.5). However, for clouds with optical depths between 1 and 3, the distribution of cloud particle size and ice water path is quite good, as shown in the bottom two panels.

Dong et al. (2008) compared MODIS (CERES Team) and ARM-based retrieval of stratus cloud microphysical properties for the SGP site. The ARM retrievals were

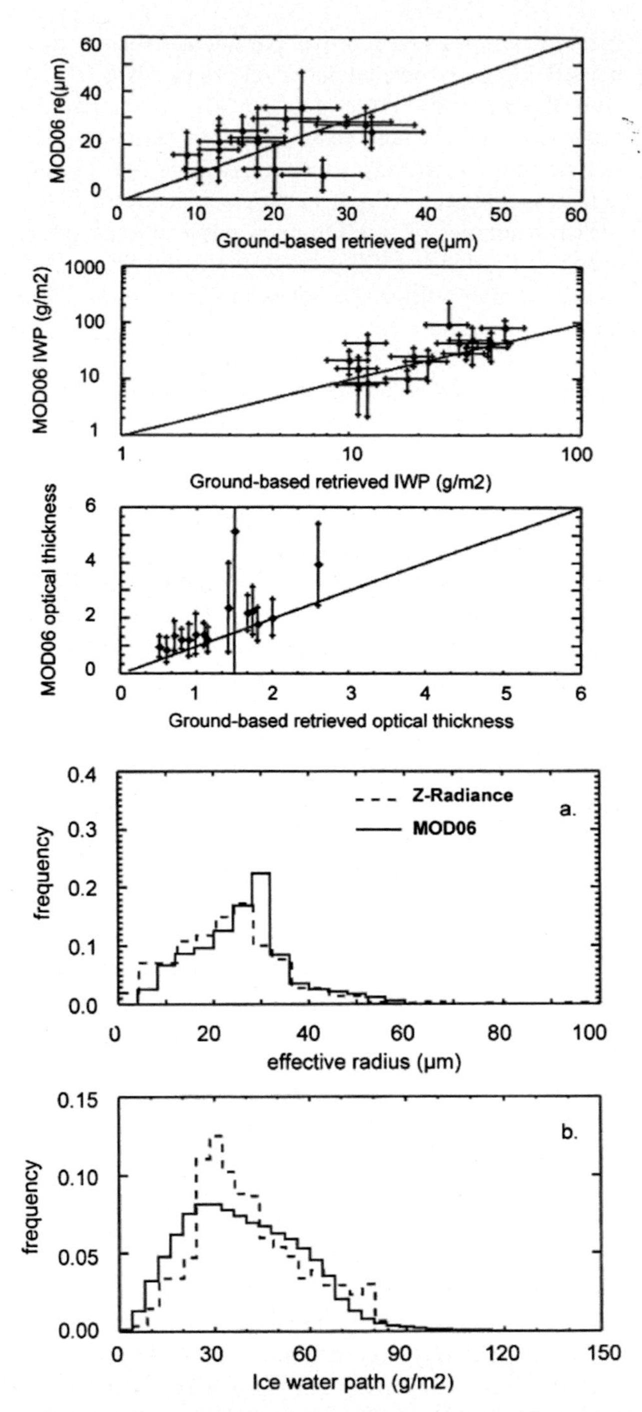

FIG. 30-3. (top three panels) Scatterplots of MOD06 (collection 4) vs *Z*-radiances microphysical retrievals for matchup points. (bottom two panels) The distributions of effective radius and ice water path for clouds with an optical depth between 1.0 and 3.0. MOD06 underrepresents clouds with and optical depth less than 1.0 (especially those with optical depth less than 0.5). Taken from Mace et al. (2005, see their Figs. 10 and 15).

based on a combination of ARM measurements from a pyranometer (downward hemispheric shortwave broadband surface flux), microwave radiometer, cloud radar, and micropulse lidar. For the period March 2000–December 2004, these authors identified 64 daytime cases with stratus clouds during the *Terra* overpass (~1030 local time) and 45 cases during the *Aqua* overpass (~1330 local time). These overpasses were further restricted to ensure clouds that were single layered with relatively large optical depths (ODs; OD > 10 for ARM retrieval, OD > 5 for MODIS), and unbroken (no clear satellite pixel within 30 km × 30 km box and no gaps in ARM radar data over a 1-h period), reducing the number of matchup points to 33 and 21, respectively. Figure 30-4 shows scatterplots of microphysical properties for the *Terra* overpasses for this restricted set.

The differences (bias ± standard deviation) between the *Terra* and ARM retrievals of effective radius, optical depth, and liquid water path for single-layer stratus are modest: $0.1 \pm 1.9\,\mu$m ($1.2\% \pm 23.5\%$), 1.3 ± 9.5 ($-3.6\% \pm 26.2\%$), and $0.6 \pm 49.9\,\mathrm{g\,m^{-2}}$ ($0.3\% \pm 27\%$), respectively, with percentages relative to the mean. The corresponding correlation coefficients are high for optical depth and liquid water path, 0.87 and 0.89, respectively, but relatively poor for effective radius at 0.44. Dong et al. suggest this poorer agreement in effective radius may be due the variability in the vertical profile of effective radius, which is not determined by the satellite retrievals. On the other hand, imager-based retrievals of effective radius are known to be sensitive to horizontal variability of cloud properties and this may be an issue even for stratocumulus clouds, which are relatively homogenous compared to cumulus and most other clouds (e.g., Marshak et al. 2006). The overall statistics for the *Aqua* overpasses (not shown), as well as when taken across all stratus cases (without stringent homogeneity conditions) are similar [see Table 4 in Dong et al. (2008)].

3. *CloudSat*, *CALIPSO*, and multi-instrument retrievals

The advent of space-borne cloud radar (*CloudSat*; Stephens et al. 2002) coupled with a two-wavelength lidar (*CALIPSO*; Winker et al. 2009) is providing an unprecedented look into the vertical structure of hydrometeor systems around the globe. As with the ARM radar and lidars that preceded them, these instruments have become a focus for the development of many macro and microphysical retrievals. The *CloudSat* and *CALIPSO* instruments do not scan, but collect data only along a very narrow swath or curtain as the satellites move along their orbits. This small sampling volume

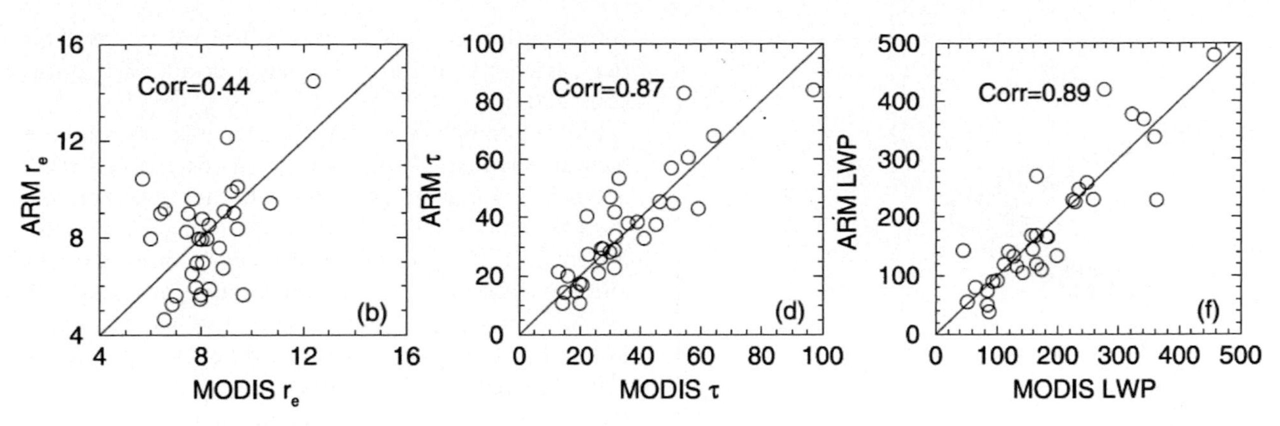

FIG. 30-4. Comparison of MODIS-retrieved (CERES Team) stratus properties and ARM retrieval over the SGP site. Taken from Dong et al. (2008, see their Fig. 5).

provides many fewer samples than satellite imagers, making comparison of these data with ground-based instruments more difficult. Nonetheless, statistical comparisons using several years of observations have been published recently.

Direct comparison of *CloudSat* cloud fractions and reflectivity distributions for nonprecipitating clouds show good agreement with both ARM radars and airborne radars (Liu et al. 2010; Protat et al. 2009), while comparison of ARM and *CALIPSO* lidar have produced valuable insights into the detection capabilities of these systems for very optically thin tropical cirrus (Thorsen et al. 2013). With regard to cloud radar, precipitating systems cause considerable attenuation of the ARM cloud radars due to water on the radome and for both systems within the atmosphere, making comparisons in precipitating conditions problematic.

Figure 30-5 shows a comparison of ARM and *CloudSat*-derived cloud occurrence with height (top panel) and measured reflectivity distribution collected at all altitudes above 1 km (bottom panel) for non- or lightly precipitating clouds. ARM data are shown in black and *CloudSat* data are shown as a white line with uncertainty bars (due to sampling) depicted with gray shading. Uncertainties due to sampling by the ARM system are very small (not shown) due to the abundant observations a fixed site provides. This figure shows a high degree of correspondence, with *CloudSat* detecting only slightly more low-reflectivity clouds, primarily at high altitudes. This additional cloud is due in part to lower resolution of the *CloudSat* radar (cf. vertical ~480 m and horizontal ~2 km for *CloudSat* and ~90 m × 10 s for ARM), which tends to make clouds look a bit larger than they are, and in part to attenuation suffered by the ARM system.

Evaluation of microphysical retrievals based on *CloudSat* and *CALIPSO* measurements is an area of ongoing research. Protat et al. (2010, 2011) published a comparison of radar–lidar retrievals using ARM observations of tropical ice clouds at Darwin, Australia, with the two most basic *CloudSat* operational retrievals, known as the radar-only product (2B-CWC-RO) and the radar optical depth product (2B-CWC-RVOD).

Figure 30-6 compares distributions of retrieved ice water content, extinction, effective radius, and cloud particle number concentration between the *CloudSat* radar-only (RO) retrieval, ground-based radar–lidar (gray line), and simple ground-based radar temperature approach (dashed line) for ice-only clouds at the ARM Darwin site. The *CloudSat* radar-only retrieval uses only the measured cloud reflectivity, and so might not be expected to produce very accurate results compared with multi-instrument retrievals. Nonetheless, the distribution of *CloudSat* radar-only retrieved ice water content (top-left panel) is in remarkably good agreement. However, it is worth noting that this distribution includes clouds at altitudes ranging from 5 to 16 km above the surface. While the distribution aggregated from data at all altitudes compares favorably, there is a clear overestimation of ice water contents above 12 km and an underestimation of ice water contents below 10 km, which is discussed by Protat et al. (2010) and not shown here. Perhaps not surprisingly, the *CloudSat* radar-only effective radius and number concentration retrievals appear to have significant biases. Protat et al. (2010) indicate that errors in cloud microphysics using the radar optical depth (RVOD) retrieval are similar to those shown here for the radar-only retrieval. Of course, multiple-instrument retrievals have been developed for *CloudSat*, and ARM will no doubt be involved in evaluating these products in the future, as well as to examine the impact of diurnal variability, which is not observed by *CloudSat* and other sun-synchronous satellites.

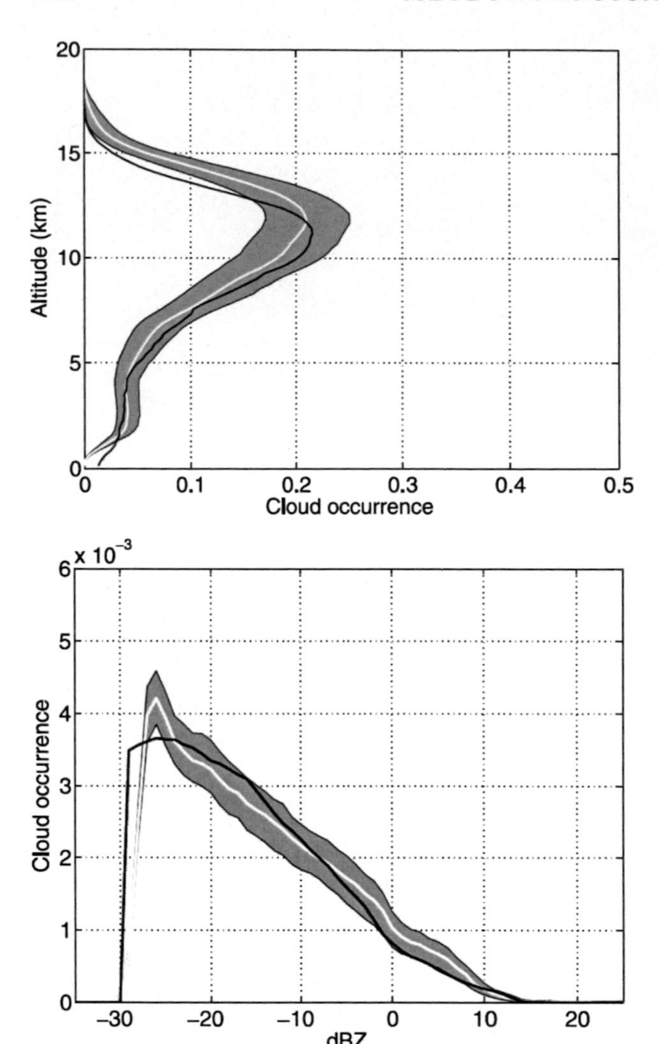

FIG. 30-5. Comparison of ARM 35-GHz cloud radar with *CloudSat* 95-GHz systems for nonprecipitating clouds using *CloudSat* data collected within 4° of the ARM site in Darwin, Australia. (top) The occurrence of clouds with reflectivity greater than −30 dBZ (cloud fraction with height) and (bottom) the distribution of reflectivity at all altitudes. Taken from Liu et al. (2010).

4. Discussion and outlook

In summary, ARM has played a key role in the validation of satellite cloud detection schemes, as well as satellite retrievals of cloud-top height, microphysical, and optical properties, and examples of each of these activities were provided. While the examples shown here focused largely on NASA Earth observing satellites, specifically MODIS, MISR, and *CloudSat*, these examples represent only a small subset of the cloud validation research that has been undertaken. ARM data have been used in the evaluation of retrievals from many other satellites including those on geostationary (e.g., Dong et al. 2002; Hollars et al. 2004; Smith et al. 2008), as well as European platforms (e.g., Riedi et al. 2001; Vanbauce et al. 2003; Larar et al. 2009).

It is not coincidental that the examples shown here have focused largely on horizontally extensive stratus and cirrus clouds. In part, this is because horizontally uniform clouds reduce difficulties (variability) associated with trying to match the nearly instantaneous and typically lower-resolution satellite datasets with the time–height data produced by the nadir-pointing radar and lidar systems that ARM began deploying in 1996. Comparisons of satellite datasets with broken or otherwise less homogenous clouds generally show larger differences including significant biases, especially for microphysical and optical quantities such as cloud optical depth. These differences may well be due to effects of "subpixel" or unresolved variability in cloud properties. Retrievals nearly universally assume constant cloud properties over whatever resolution the observations are taken (e.g., Mace et al. 2011). However, both satellite and ground-based retrievals are also expected to be of lower quality and in some cases are known to have significant biases for clouds that are spatially inhomogeneous even at resolved scales (especially near cloud edges) owing to the widespread use of one-dimensional radiative transfer in modeling cloud reflectivity or transmissivity at all wavelengths. Variability of cloud properties also complicates validation of retrievals against in situ data, making direct assessment of errors due to spatial variability difficult.

The ARM Program is currently in the process of deploying scanning dual-polarization multiple-wavelength cloud radars (K- and W- or K- and X-band systems) at all of its fixed and mobile sites (Mather and Voyles 2013). These scanning systems will lead to the development of new retrievals that will capture spatial variability at scales resolved by the radars. In addition to providing improved retrievals and enabling more effective evaluations against in situ measurements, the scanning radars are expected to provide insights into the effects of spatial variability on established (vertically pointing) algorithms.

It is also not coincidental that the examples discussed in this chapter avoided precipitating systems. While there has been some research into retrieving rain rates and other microphysical quantities from ARM vertically pointing radars (e.g., Matrosov et al. 2006), surface precipitation corrupts observations from many ARM instruments and the program has naturally focused on nonprecipitating conditions. The ARM Program is also in the process of deploying a variety of scanning polarimetric precipitation radars (C- and

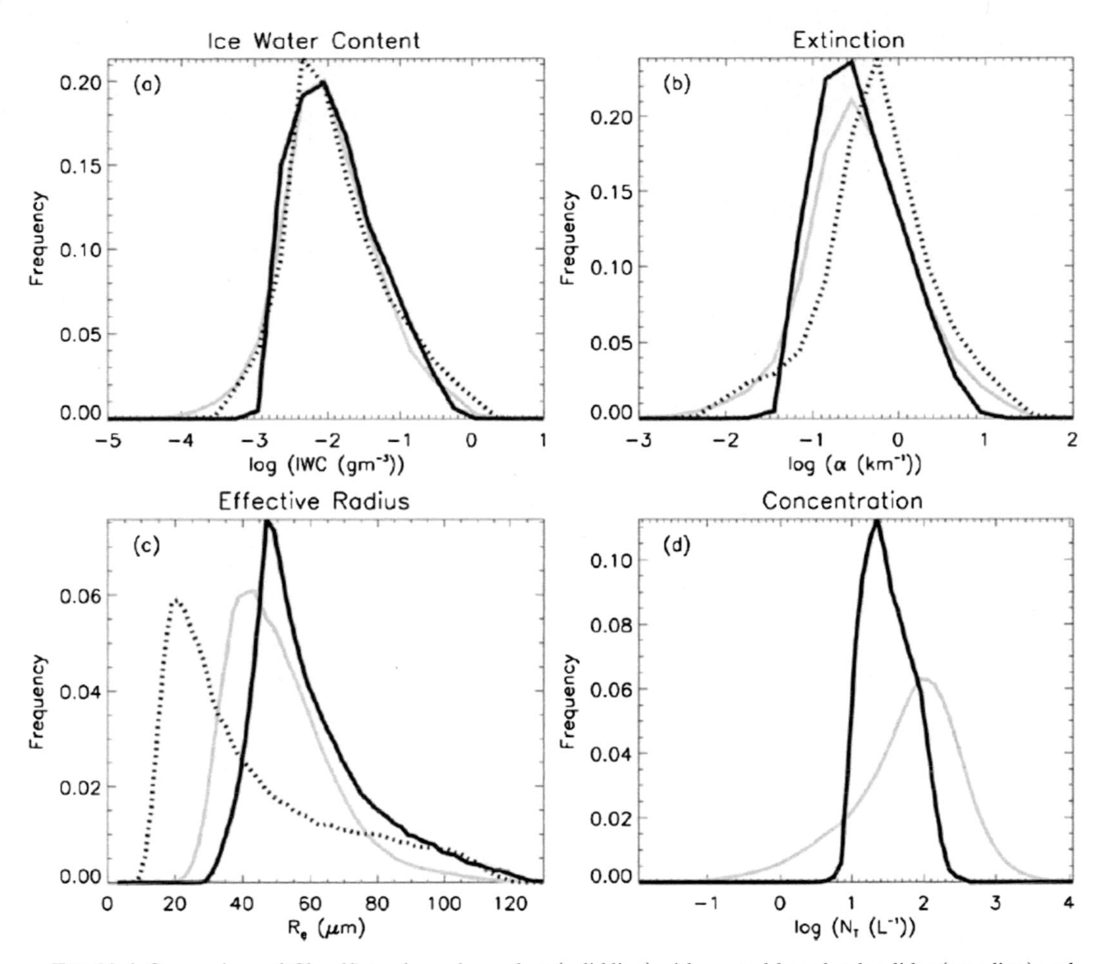

FIG. 30-6. Comparison of *CloudSat* radar-only product (solid line) with ground-based radar–lidar (gray line) and simple ground-based radar temperature approach (dashed line) for ice-only clouds at the ARM Darwin site. Taken from Protat et al. (2010).

X-band systems). The combination of scanning precipitation and cloud radar systems will no doubt lead to further advancements in combined cloud and precipitation microphysical retrievals and a future role for the ARM Program in the validation of satellite-based precipitation retrievals, such as those from the NASA Global Precipitation Measurement (GPM) mission. ARM data gathered during the recent ARM Midlatitude Convective Continental Clouds Experiment (M3CE) in combination with NASA aircraft data were used in the development of GPM algorithms before launch of this satellite in 2014 (e.g., Petersen and Jensen 2011). With regard to cloud properties, it is hoped that the combination of cloud radar with polarimetric C- and X-band data will provide constraints on ice crystal habit, which is a source of great uncertainty in both ground-based and satellite retrievals (McFarlane and Marchand 2008) and help constrain retrievals in mixed-phase clouds, which remain rudimentary except perhaps for thin Artic stratus.

While the presence of *CALIPSO* and *CloudSat* (cloud radar and lidar in space) and the global perspective these data bring have to some degree reduced the importance of ARM radar and lidar data for cloud validation, these sensors are approaching the end of their mission lives and ARM still has much to contribute to future satellite validation. In addition to advances that will likely come from scanning radar discussed above, the complete diurnal sampling provided by ARM is critical to evaluating the representativeness of retrievals from polar-orbiting sensors (including future radar and lidar missions such as EarthCARE)[3], and will be of great value in the direct evaluation of the next generation of geostationary weather satellites, GOES-R.[4]

[3] Bezy et al. (2005); http://www.esa.int/Our_Activities/Observing_the_Earth/The_Living_Planet_Programme/Earth_Explorers/EarthCARE.

[4] Schmit et al. (2005); http://www.goes-r.gov/.

REFERENCES

Ackerman, S. A., R. E. Holz, R. Frey, E. W. Eloranta, B. C. Maddux, and M. McGill, 2008: Cloud detection with MODIS. Part II: Validation. *J. Atmos. Oceanic Technol.*, **25**, 1073–1086, doi:10.1175/2007JTECHA1053.1.

Baum, B. A., R. A. Frey, G. G. Mace, M. K. Harkey, and P. Yang, 2003: Nighttime multilayered cloud detection using MODIS and ARM data. *J. Appl. Meteor.*, **42**, 905–919, doi:10.1175/1520-0450(2003)042<0905:NMCDUM>2.0.CO;2.

Berendes, T. A., D. A. Berendes, R. M. Welch, E. G. Dutton, T. Uttal, and E. E. Clothiaux, 2004: Cloud cover comparisons of the MODIS daytime cloud mask with surface instruments at the north slope of Alaska ARM site. *IEEE Trans. Geosci. Remote Sens.*, **42**, 2584–2593, doi:10.1109/TGRS.2004.835226.

Bezy, J.-L., W. Leibrandt, A. Heliere, P. Silvestrin, C.-C. Lin, P. Ingmann, T. Kimura, and H. Kumagai, 2005: The ESA Earth Explorer EarthCARE mission. *Earth Observing Systems X*, J. J. Butler, Ed., International Society for Optical Engineering (SPIE Proceedings, Vol. 5882), doi:10.1117/12.619438.

Chang, F., and Z. Li, 2005: A new method for detection of cirrus overlapping water clouds and determination of their optical properties. *J. Atmos. Sci.*, **62**, 3993–4009, doi:10.1175/JAS3578.1.

Charlock, T. P., and T. L. Alberta, 1996: The CERES/ARM/GEWEX Experiment (CAGEX) for the retrieval of radiative fluxes with satellite data. *Bull. Amer. Meteor. Soc.*, **77**, 2673–2683, doi:10.1175/1520-0477(1996)077<2673:TCEFTR>2.0.CO;2.

Clothiaux, E. E., T. P. Ackerman, G. G. Mace, K. P. Moran, R. T. Marchand, M. A. Miller, and B. E. Martner, 2000: Objective determination of cloud heights and radar reflectivities using a combination of active remote sensors at the ARM CART sites. *J. Appl. Meteor.*, **39**, 645–665, doi:10.1175/1520-0450(2000)039<0645:ODOCHA>2.0.CO;2.

Dong, X., G. Mace, P. Minnis, W. L. Smith, M. Poellot, R. T. Marchand, and A. D. Rapp, 2002: Comparison of stratus cloud properties deduced from surface, GOES, and aircraft data during the March 2000 ARM Cloud IOP. *J. Atmos. Sci.*, **59**, 3265–3284, doi:10.1175/1520-0469(2002)059<3265:COSCPD>2.0.CO;2.

——, P. Minnis, B. Xi, S. Sun-Mack, and Y. Chen, 2008: Comparison of CERES-MODIS stratus cloud properties with ground-based measurements at the DOE ARM Southern Great Plains site. *J. Geophys. Res.*, **113**, D03204, doi:10.1029/2007JD008438.

Hawkinson, J. A., W. Feltz, and S. A. Ackerman, 2005: A comparison of GOES sounder- and cloud lidar- and radar-retrieved cloud-top heights. *J. Appl. Meteor.*, **44**, 1234–1242, doi:10.1175/JAM2269.1.

Hollars, S., Q. Fu, J. Comstock, and T. Ackerman, 2004: Comparison of cloud-top height retrievals from ground-based 35 GHz MMCR and GMS-5 satellite observations at ARM TWP Manus site. *Atmos. Res.*, **72**, 169–186, doi:10.1016/j.atmosres.2004.03.015.

Jin, Y., C. B. Schaaf, C. E. Woodstock, F. Gao, X. Li, A. H. Strahler, W. Lucht, and S. Liang, 2003: Consistency of MODIS surface bidirectional reflectance distribution function and albedo retrievals: 2. Validation. *J. Geophys. Res.*, **108**, 4159, doi:10.1029/2002JD002804.

Kahn, B. H., A. Eldering, A. J. Braverman, E. J. Fetzer, J. H. Jiang, E. Fishbein, and D. L. Wu, 2007: Toward the characterization of upper tropospheric clouds using Atmospheric Infrared Sounder and Microwave Limb Sounder observations. *J. Geophys. Res.*, **112**, D05202, doi:10.1029/2006JD007336.

Kaufman, Y. J., D. D. Herrring, K. J. Ranson, and G. J. Collatz, 1998: Earth observing system AM1 mission to Earth. *IEEE Trans. Geosci. Remote Sens.*, **36**, 1045–1055, doi:10.1109/36.700989.

Larar, A. M., W. L. Smith, D. K. Zhou, X. Liu, H. Revercomb, J. P. Taylor, S. M. Newman, and P. Schlüssel, 2009: IASI spectral radiance performance validation: Case study assessment from the JAIVEx field campaign. *Atmos. Chem. Phys. Discuss.*, **9**, 10 193–10 234, doi:10.5194/acpd-9-10193-2009.

Lin, B., and P. Minnis, 2000: Temporal variations of land surface microwave emissivities over the Atmospheric Radiation Measurement Program Southern Great Plains site. *J. Appl. Meteor.*, **39**, 1103–1116, doi:10.1175/1520-0450(2000)039<1103:TVOLSM>2.0.CO;2.

Liu, Y., J. R. Key, R. A. Frey, S. A. Ackerman, and W. P. Menzel, 2004: Nighttime polar cloud detection with MODIS. *Remote Sens. Environ.*, **92**, 181–194, doi:10.1016/j.rse.2004.06.004.

Liu, Z., R. Marchand, and T. Ackerman, 2010: A comparison of observations in the tropical western Pacific from ground-based and satellite millimeter-wavelength cloud radars. *J. Geophys. Res.*, **115**, D24206, doi:10.1029/2009JD013575.

Mace, G. G., Y. Zhang, S. Platnick, M. D. King, P. Minnis, and P. Yang, 2005: Evaluation of cirrus cloud properties derived from MODIS data using cloud properties derived from ground-based observations collected at the ARM SGP site. *J. Appl. Meteor.*, **44**, 221–240, doi:10.1175/JAM2193.1.

——, and Coauthors, 2006: Cloud radiative forcing at the Atmospheric Radiation Measurement Program Climate Research Facility: 1. Technique, validation, and comparison to satellite-derived diagnostic quantities. *J. Geophys. Res.*, **111**, D11S90, doi:10.1029/2005JD005921.

——, S. Houser, S. Benson, S. A. Klein, and Q. Min, 2011: Critical evaluation of the ISCCP simulator using ground-based remote sensing data. *J. Climate*, **24**, 1598–1612, doi:10.1175/2010JCLI3517.1.

Marchand, R. T., T. P. Ackerman, and C. Moroney, 2007: An assessment of Multiangle Imaging Spectroradiometer (MISR) stereo-derived cloud top heights and cloud top winds using ground-based radar, lidar, and microwave radiometers. *J. Geophys. Res.*, **112**, D06204, doi:10.1029/2006JD007091.

——, ——, M. Smyth, and W. B. Rossow, 2010: A review of cloud top height and optical depth histograms from MISR, ISCCP, and MODIS. *J. Geophys. Res.*, **115**, D16206, doi:10.1029/2009JD013422.

Marshak, A., S. Platnick, T. Varnai, G. Wen, and R. F. Cahalan, 2006: Impact of three-dimensional radiative effects on satellite retrievals of cloud droplet sizes. *J. Geophys. Res.*, **111**, D09207, doi:10.1029/2005JD006686.

Mather, J. H., and J. W. Voyles, 2013: The ARM Climate Research Facility: A review of structure and capabilities. *Bull. Amer. Meteor. Soc.*, **94**, 377–392, doi:10.1175/BAMS-D-11-00218.1.

Matrosov, S. Y., P. T. May, and M. D. Shupe, 2006: Rainfall profiling using Atmospheric Radiation Measurement Program vertically pointing 8-mm wavelength radars. *J. Atmos. Oceanic Technol.*, **23**, 1478–1491, doi:10.1175/JTECH1957.1.

McFarlane, S. A., and R. T. Marchand, 2008: Analysis of ice crystal habits derived from MISR and MODIS observations over the ARM Southern Great Plains site. *J. Geophys. Res.*, **113**, D07209, doi:10.1029/2007JD009191.

Minnis, P., J. Huang, B. Lin, Y. Yi, R. F. Arduini, T.-F. Fan, J. K. Ayers, and G. G. Mace, 2007: Ice cloud properties in ice-over-water cloud systems using Tropical Rainfall Measuring Mission (TRMM) visible and infrared scanner and TRMM Microwave Imager data. *J. Geophys. Res.*, **112**, D06206, doi:10.1029/2006JD007626.

Nasiri, S. L., B. A. Baum, A. J. Heymsfield, P. Yang, M. Poellot, D. P. Kratz, and Y. X. Hu, 2002: Development of midlatitude cirrus models for MODIS using FIRE-I, FIRE-II, and ARM in situ data. *J. Appl. Meteor.*, **41**, 197–217, doi:10.1175/1520-0450(2002)041<0197:TDOMCM>2.0.CO;2.

Naud, C. M., J.-P. Muller, and E. E. Clothiaux, 2002: Comparison of cloud top heights derived from MISR stereo and MODIS CO_2-slicing. *Geophys. Res. Lett.*, **29**, doi:10.1029/2002GL015460.

——, ——, ——, B. A. Baum, and W. P. Menzel, 2005: Intercomparison of multiple years of MODIS, MISR and radar cloud-top heights. *Ann. Geophys.*, **23**, 2415–2424, doi:10.5194/angeo-23-2415-2005.

Petersen, W. A., and M. P. Jensen, 2011: Physical validation of GPM retrieval algorithms over land: An overview of the Mid-Latitude Continental Convective Clouds Experiment (MC3E). *2011 Fall Meeting*, San Francisco, CA, Amer. Geophys. Union, Abstract H42E-01.

Protat, A., and Coauthors, 2009: Assessment of Cloudsat reflectivity measurements and ice cloud properties using ground-based and airborne cloud radar observations. *J. Atmos. Oceanic Technol.*, **26**, 1717–1741, doi:10.1175/2009JTECHA1246.1.

——, J. Delanoë, E. J. O'Connor, and T. S. L'Ecuyer, 2010: The evaluation of *CloudSat* and CALIPSO ice microphysical products using ground-based cloud radar and lidar observations. *J. Atmos. Oceanic Technol.*, **27**, 793–810, doi:10.1175/2009JTECHA1397.1; Corrigendum, **28**, 734–735, doi:10.1175/2010JTECHA1504.1.

Riedi, J., P. Goloub, and R. T. Marchand, 2001: Comparison of POLDER cloud phase retrievals to active remote sensors measurements at the ARM SGP site. *Geophys. Res. Lett.*, **28**, 2185–2188, doi:10.1029/2000GL012758.

Sapiano, M. R. P., and P. A. Arkin, 2009: An intercomparison and validation of high-resolution satellite precipitation estimates with 3-hourly gauge data. *J. Hydrometeor.*, **10**, 149–166, doi:10.1175/2008JHM1052.1.

Schmit, T. J., M. M. Gunshor, W. P. Menzel, J. J. Gurka, J. Li, and A. S. Bachmeier, 2005: Introducing the next-generation advanced baseline imager on GOES-R. *Bull. Amer. Meteor. Soc.*, **86**, 1079–1096, doi:10.1175/BAMS-86-8-1079.

Schueler, C. F., and W. L. Barnes, 1998: Next-generation MODIS for Polar Operational Environmental Satellites. *J. Atmos. Oceanic Technol.*, **15**, 430–439, doi:10.1175/1520-0426(1998)015<0430:NGMFPO>2.0.CO;2.

Shupe, M. D., and Coauthors, 2008: A focus on mixed-phase clouds: The status of ground-based observational methods. *Bull. Amer. Meteor. Soc.*, **89**, 1549–1562, doi:10.1175/2008BAMS2378.1.

——, J. M. Comstock, D. D. Turner, and G. G. Mace, 2016: Cloud property retrievals in the ARM Program. *The Atmospheric Radiation Measurement (ARM) Program: The First 20 Years, Meteor. Monogr.*, No. 57, Amer. Meteor. Soc., doi:10.1175/AMSMONOGRAPHS-D-15-0030.1.

Smith, W. L., Jr., P. Minnis, H. Finney, R. Palikonda, and M. M. Khaiyer, 2008: An evaluation of operational GOES-derived single-layer cloud top heights with ARSCL data over the ARM Southern Great Plains site. *Geophys. Res. Lett.*, **35**, L13820, doi:10.1029/2008GL034275.

Stephens, G. L., and Coauthors, 2002: The CloudSat mission and the A-Train: A new dimension of space-based observations of clouds and precipitation. *Bull. Amer. Meteor. Soc.*, **83**, 1771–1790, doi:10.1175/BAMS-83-12-1771.

Thorsen, T. J., Q. Fu, J. M. Comstock, C. Sivaraman, M. A. Vaughan, D. M. Winker, and D. D. Turner, 2013: Macrophysical properties of tropical cirrus clouds from the *CALIPSO* satellite and from ground-based micropulse and Raman lidars. *J. Geophys. Res. Atmos.*, **118**, 9209–9220, doi:10.1002/jgrd.50691.

Tobin, D. C., and Coauthors, 2006: Atmospheric Radiation Measurement site atmospheric state best estimates for Atmospheric Infrared Sounder temperature and water vapor retrieval validation. *J. Geophys. Res.*, **111**, D09S14, doi:10.1029/2005JD006103.

Trishchenko, A. P., Y. Luo, K. V. Khlopenkov, and S. Wang, 2008: A method to derive the multispectral surface albedo consistent with MODIS from historical AVHRR and VGT satellite data. *J. Appl. Meteor. Climatol.*, **47**, 1199–1221, doi:10.1175/2007JAMC1724.1.

Vanbauce, C., B. Cadet, and R. T. Marchand, 2003: Comparison of POLDER apparent and corrected oxygen pressure to ARM/MMCR cloud boundary pressures. *Geophys. Res. Lett.*, **30**, 1212, doi:10.1029/2002GL016449.

Wang, Z., and K. Sassen, 2002: Cirrus cloud microphysical property retrieval using lidar and radar measurements. Part I: Algorithm description and comparison with in situ data. *J. Appl. Meteor.*, **41**, 218–229, doi:10.1175/1520-0450(2002)041<0218:CCMPRU>2.0.CO;2.

Winker, D. M., M. A. Vaughan, A. Omar, Y. Hu, K. A. Powell, Z. Liu, W. H. Hunt, and S. A. Young, 2009: Overview of the CALIPSO mission and CALIOP data processing algorithms. *J. Atmos. Oceanic Technol.*, **26**, 2310–2323, doi:10.1175/2009JTECHA1281.1.

APPENDIX A

This appendix is the executive summary of the 1990 Atmospheric Radiation Measurement Program Plan (DOE/ER-0442; available online at https://www.arm.gov/publications/programdocs/doe-er-0442.pdf) sponsored by the U.S. Department of Energy, Office of Energy Research. The text has been edited to conform to the style of the American Meteorological Society, but the content is otherwise unchanged from the original document.

Executive Summary: Atmospheric Radiation Measurement Program Plan

Foreword

In 1978 the Department of Energy initiated the Carbon Dioxide Research Program to address climate change from the increasing concentration of carbon dioxide in the atmosphere. Over the years the program has studied the many facets of the issue, from the carbon cycle, the climate diagnostics, the vegetative effects, to the societal impacts. The program is presently the department's principal entry in the U.S. Global Change Research Program coordinated by the Committee on Earth Sciences (CES) of the Office of Science and Technology Policy (OSTP).

The recent heightened concern about global warming from an enhanced greenhouse effect has prompted the department to accelerate the research to improve predictions of climate change. The emphasis is on the timing and magnitude of climate change as well as on the regional characteristics of this change. The Atmospheric Radiation Measurement (ARM) Program was developed to supply an improved predictive capability, particularly as it relates to the cloud–climate feedback.

Scientists from the DOE National Laboratory community contributed to the preparation of the ARM Program Plan with input from members of the academic community, the private sector, and from scientists of other CES agencies. The plan was subjected to an extensive peer review and the many helpful comments we have received have been incorporated into this document. We believe that ARM will serve the CES objectives in global change research and support the DOE mission of formulating a national energy strategy that takes into account the potential for global climate change.

Objective

To understand energy's role in anthropogenic global climate change, significant reliance is being placed on general circulation models. A major goal of the department is to foster the development of general circulation models capable of predicting the timing and magnitude of greenhouse gas-induced global warming and the regional effects of such warming. DOE research has revealed that cloud radiative feedback is the single most important effect determining the magnitude of possible climate responses to human activity. However, cloud radiative forcing and feedbacks are not understood at the levels needed for reliable climate prediction.

The ARM Program will contribute to the DOE goal by improving the treatment of cloud radiative forcing and feedbacks in general circulation models. Two issues will be addressed: the radiation budget and its spectral dependence and the radiative and other properties of clouds. Understanding cloud properties and how to predict them is critical because cloud properties may very well change as climate changes.

The experimental objective of the ARM Program is to characterize empirically the radiative processes in Earth's atmosphere with improved resolution and accuracy. A key to this characterization is the effective treatment of cloud

DOI: 10.1175/AMSMONOGRAPHS-D-15-0036.1

formation and cloud properties in general circulation models. Through this characterization of radiative properties, it will be possible to understand both the forcing and feedback effects. General circulation model modelers will then be able to better identify the best approaches to improved parameterizations of radiative transfer effects. This is expected to greatly improve the accuracy of long-term general circulation model predictions and the efficacy of those predictions at the important regional scale, as the research community and DOE attempt to understand the effects of greenhouse gas emissions on Earth's climate.

The ARM initiative and field experiment

The DOE's ARM Program is a key component of the department's research strategy to address global climate change. The program is a direct continuation of DOE's decade-long effort to improve general circulation models (GCMs) and provide reliable simulations of regional and long-term climate change in response to increasing greenhouse gases.

The ARM Program is a highly focused observational and analytical research effort that will compare observations with model calculations in the interest of accelerating improvements in both observational methodology and GCMs. During the ARM Program, DOE will continue to collaborate extensively with existing global change programs at other agencies, including the National Oceanic and Atmospheric Administration (NOAA), the National Science Foundation (NSF), and the National Aeronautics and Space Administration (NASA).

The objective of the ARM Program is to provide an experimental test bed for the study of important atmospheric effects, particularly cloud and radiative processes, and testing parameterizations of these processes for use in atmospheric models. This effort will support the continued and rapid improvement of GCM predictive capability.

The state of the art

Over the past 10 years, the research programs of DOE and other agencies have made significant progress toward understanding the potential for global climate change and the resulting consequences. Rising concentrations of greenhouse gases, primarily carbon dioxide, have been well documented. Research programs are determining the relative roles of human activities and natural processes on the land, biosphere, and oceans. Models of the global climate system have advanced to include realistic geography, the annual cycle of the seasons, and varying cloud cover. Very recently, models have begun to include coupling of the ocean–atmosphere system. Results of climate models suggest that projected greenhouse gas emission patterns may lead to a global climate warming of 1.5° to 4.5°C and to significant changes in water availability during the next century.

However, this decade of research has also revealed that considerable uncertainties in model estimates remain. For example, although the 1980s have been especially warm, the extent of global warming over the past century may have been 2 to 3 times less than that estimated by current models. Further, when the results of different models are compared, there are substantial differences among their estimates of temperature and precipitation changes in response to doubled carbon dioxide. Significant climate change due to anthropogenic effects may be a plausible conclusion based on current GCMs. However, we do not know with sufficient accuracy how large the climatic changes will be, how rapidly the changes will occur, or how the changes will be distributed over the globe. We also know virtually nothing about the potential changes in the frequency of extreme climatic events.

Department of energy context for ARM

Greenhouse gases directly affect the radiation balance of the atmosphere. Theoretical models predict a net surface warming of the globe from the direct radiative forcing of these gases and, more importantly, the resulting series of feedbacks. These feedbacks directly affect many processes important to climate such as snow cover and sea ice melting, cloud formation, air–ocean interaction, and global circulation patterns. Consequently, a lack of understanding of the complex response of the atmosphere–ocean system to anthropogenic inputs allows much room for uncertainty about the future consequences of continued increases in the atmospheric concentration of greenhouse gases.

Decisions made in the next decade will determine the international response to projected anthropogenic global climate change. GCMs are the best scientific tool to estimate global climate change and its regional distribution. The results from such models are being used as a basis for formulating national and international policies, which could greatly influence the economies of the United States and the world.

Despite their weaknesses, the GCMs are the only tools available to provide the basis for policy formulation. It seems certain that GCMs will remain a part of the scientific basis for policy decisions during the 1990s and beyond.

Therefore, it is urgent that the scientific community promote the rapid improvement of the accuracy and predictive capability of GCMs.

The DOE has responsibility for preparing a National Energy Strategy (NES) that fully considers the environmental effects of energy-related activities. The potential climatic and ecological changes that may result from the continuing emissions of carbon dioxide, methane, and other greenhouse gases will be important considerations in forming the most environmentally compatible energy policy.

To address these considerations, the DOE has proposed a threefold initiative. One element will support the development of specialized GCM computing machines and another will promote the training of a new generation of climate scientists. The third, the ARM Program, will contribute to improved GCM predictions by improving the parameterization of model physics. All three will provide an improved scientific basis for the development of a responsible and appropriate national energy policy.

Science context for ARM

The interagency CES has identified cloud–climate interactions as the highest research priority within global change research to produce the needed improvements of GCMs. The ARM Program seeks to supplement ongoing cloud climatology and satellite cloud-radiation projects by contributing critical data and analyses from an intensive measurement and modeling program.

Changes in cloud cover and cloud characteristics, because of their intimate relationship with infrared and solar radiation, are a major factor in determining the magnitude of potential warming resulting from increased concentrations of greenhouse gases. Also, the accuracy of radiative calculations, including the treatment of clouds, affects the accuracy of estimates of climate sensitivity. Together they control the radiative forcing that drives some of the key feedbacks of the global climate system.

Recent satellite measurements have revealed the magnitude of the effects of clouds on solar and infrared radiation (Ramanathan et al. 1989). The measurements indicate that the global effects of clouds are large. The size of these effects is important in the following sense. Clouds affect both the incoming (solar) and outgoing (infrared) radiation in the atmosphere. Clouds affect the solar radiation by changing the amount of solar radiation that is reflected back to space, an effect which is currently thought to lead to a net cooling. On the other hand, clouds can trap infrared radiation (the greenhouse effect) and an increase in cloudiness could cause a heating of the troposphere. Current models suggest that the absolute magnitude of these two feedback effects is individually about 10 times the size of the direct radiative forcing due to a doubling of the atmospheric carbon dioxide concentration. The net effect, their difference, is about three times the magnitude of the direct radiative forcing. Small uncertainties in the modeling of cloudiness or cloud properties could produce predicted effects comparable to or larger than the relatively better understood anthropogenic radiative perturbation. Therefore, inferred changes in cloud distribution or properties are critical to understanding the temperature response of the entire system due to increased greenhouse gas concentrations.

Predictions of climatic response to changing greenhouse concentrations are also ambiguous because of uncertainties in estimating radiative forcing. There is a range of about 20% in the estimates of the radiative flux change at the tropopause from a doubling of carbon dioxide concentration among the different radiation models used in GCMs (Luther et al. 1988). There are other significant inaccuracies and disagreements due to inadequate modeling of specific effects within GCMs. Estimates of radiative perturbations due to changing water vapor concentrations and distribution are particularly uncertain.

These uncertainties along with uncertainties in our understanding of clouds contribute directly to the differences among GCM estimates of climate sensitivity (Cess et al. 1989; Gates 1987; Wang et al. 1988) and the consequent lack of confidence in GCM predictions at all levels, but particularly on the regional scale.

Limits to the current understanding of radiation and cloud interactions also contribute to many other uncertainties in estimating climate change. Radiative processes create the temperature differences that drive convective cloud-forming processes. These processes generate warm season precipitation, important for agriculture, and much of the cirrus cloud cover that can trap additional infrared radiation. Gates (1987) points out the necessity of properly characterizing major energy fluxes in climate models. This becomes even more critical as model grid resolution is increased to levels needed for regional prediction (~50 km) and when such features as coupled atmospheric and ocean processes are added. As model parameters change, through the addition of new effects or changes in scale, the model physics needs to be modified as well. This is particularly true for radiation and clouds, because of their intimate relationship to the overall energy budget.

Model and data intercomparisons suggest a definite focus for future GCM research. Grotch (1988) has compared GCMs with historical regional climatology and demonstrated that future GCM research needs to improve regional prediction. The failure of the current generation of GCMs to converge on accurate regional predictions is not surprising. Other studies point out that the treatment of the surface energy balance and its relationship to the hydrologic cycle (Wang et al. 1988) and radiative transfer (Luther et al. 1988) are still not adequate. Both of these studies show discrepancies among the models several times larger than the projected anthropogenic radiative forcing functions. In short, the models do not agree among themselves at climatologically significant levels in their treatment of the energy balance. Most importantly, Cess et al. (1989) show that there are significant disagreements among models in their estimates of cloud radiative forcing under closely controlled experimental conditions for the model intercomparisons.

The state of the lowest few kilometers of the atmosphere is the most crucial to determining the surface climate. It is this part of the atmosphere that contains most of the air, water, vapor, clouds, and other critical constituents, and into which man-made pollutants are directly injected. The direction of climate change, cooling or warming, and the degree of change caused by anthropogenic gases in the atmosphere, depends upon the detailed absorption and emission characteristics of the atmosphere. However, the radiative characteristics of the lower atmosphere have never been measured with any great detail; certainly not with the resolution and precision required to assist the development of accurate climate predictions on the regional scale needed from GCMs. The ARM Program results will be combined with results of other DOE programs; NOAA, NFS, and NASA programs; and interagency programs, such as the First ISCCP Regional Experiment (FIRE); and others, to specifically meet this important scientific need.

ARM Program requirements

A decade of research on the performance of GCMs, including several model intercomparison programs, has highlighted important areas of scientific need associated with the understanding and prediction of global climate change. Some of the most important needs fall in the general area of the treatment of physical processes that are not resolved in GCMs, particularly radiative transfer and cloud formation. In these two areas, the following scientific requirements emerge as the most critical for a program designed to remedy key weaknesses of current models:

1) A quantitative description of the spectral radiative energy balance profile under a wide range of meteorological conditions must be developed. Such descriptions must come from field measurements and must be quantified at a level consistent with climatologically significant energy flows of 1 to 2 $W m^{-2}$.
2) The processes controlling the radiative balance must be identified and investigated. Validation of our understanding of these processes must come from a direct and comprehensive comparison of field observations with detailed calculations of the radiation field and associated cloud and aerosol interactions.
3) The knowledge necessary to improve parameterizations of radiative properties of the atmosphere for use in GCMs must be developed. This requires intensive measurements at a variety of temporal and physical scales. A major emphasis must be placed on the role of clouds, including their distribution and microphysical properties.

The above requirements are direct consequences of the sensitivity of atmospheric equilibrium to changes in the radiation field. Current models indicate that if carbon dioxide were to instantaneously double, the outgoing long-wave radiation leaving the atmosphere (more precisely, the troposphere) would be temporarily reduced by about 4 $W m^{-2}$, until the climate system adjusted to restore the balance. Most GCMs suggest that, under these conditions, the globally averaged surface temperature would warm by about 1.5° to 4.5°C before a new climatic equilibrium would be reached.

In addition to the basic sensitivity of the climate system to radiative forcing, the intercomparison studies identify two other important needs for effective modeling of the terrestrial radiation field. First, clouds play a critical role in regulating the flow of both longwave and shortwave radiation within the troposphere. Changes in the distribution and physical characteristics of clouds can have major effects on climate sensitivity. Therefore, it is essential to account for the interaction of clouds and radiation for reliable prediction of climate change.

Second, the radiative transfer problem is not simply an energy balance problem. The greenhouse effect is a spectral redistribution process, in which the radiation absorbed by carbon dioxide and other radiatively important trace species is absorbed in particular parts of the spectrum. Carbon dioxide is particularly important in the greenhouse warming process because it absorbs near the peak of the blackbody radiation curve for the atmosphere. The energy absorbed heats the atmosphere, which redistributes the radiation to other wavelengths.

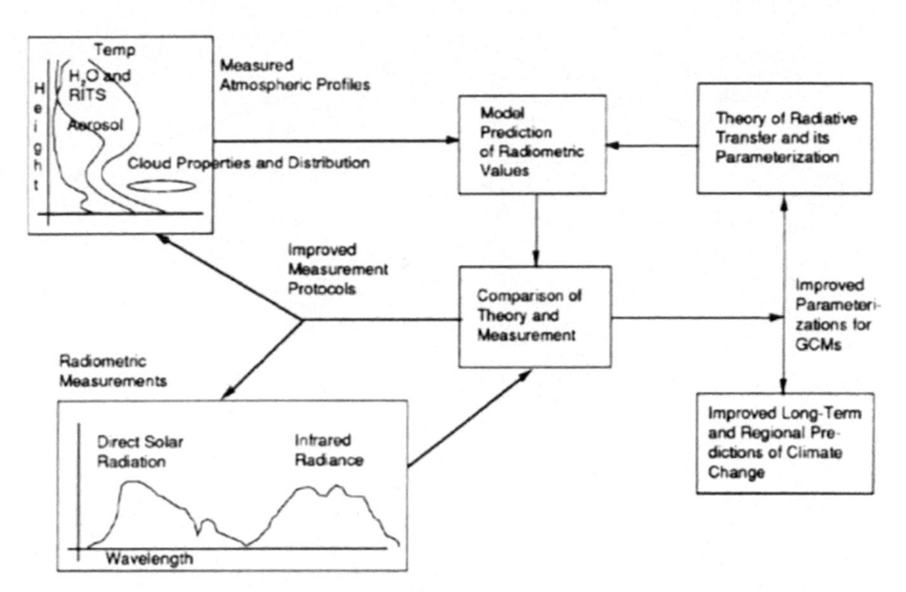

FIG. 1. The ARM approach to systematic integration of theory, measurement, and parameterization is shown for the special case of testing models of the radiation field. An analogous approach will be employed by ARM for the study of clouds and cloud models.

These considerations suggest that a comparison between the radiation field calculated in a model and actual observations of the spectral dependence radiation would constitute a sensitive test of the efficacy of the modeling process. As illustrated in Fig. 1, ARM is best viewed as a hypothesis testing approach.

This approach has three elements: a set of measurements of meteorological and other physical conditions that can be used as inputs to a radiative model or a cloud parameterization model; the models being tested, which predict atmospheric features, such as the direction and spectral-dependent radiation field or the cloud type and distribution; and a set of measurements designed to confirm the model predictions.

The goals of ARM are twofold. First, it will attempt to improve the treatment of radiative transport in GCMs for the clear-sky, general overcast, and broken cloud cases. Second, it will provide a test bed for cloud parameterization models used in GCMs. The measures of the quality of the models will include their ability to reproduce observed wavelength and direction-dependent fluxes of longwave and shortwave radiation and the time-varying distribution of cloud type and amount. Figure 1 illustrates the ARM experimental approach to the study of the radiation field. That approach, based on meteorological measurements made both to drive models and to confirm their predictions, will use those results to guide improvements in both the measurements and the models.

Experimental approach

ARM is an observational program driven by the theoretical and modeling requirements. The ARM Program must provide data that can improve and test the GCM parameterizations of clouds and their microphysical composition. The smallest domain explicitly represented in a GCM is the single grid cell. A GCM cell is orders of magnitude larger than the scale associated with important cloud characteristics. It is possible that over the next decade model resolution will increase substantially so that single grid cells will have dimensions of a few tens of kilometers (comparable to an ERBE pixel). Even so, since clouds can have dimensions less than a kilometer, subgrid parameterization will remain necessary.

Of all the subgrid-scale characteristics that may affect radiation, cloud inhomogeneities and surface albedo variations are most important. Uncertainties in climate models will be reduced substantially when a reliable cloud parameterization is developed that will consistently apply under important mean climate conditions. Data that characterizes the statistics of clouds on a subgrid-scale is necessary for the development of improved cloud models.

In response to the nature of the problem of studying subgrid phenomena, the experimental equipment will be deployed at a series of field settings. These settings will be chosen on the basis of their climatological significance and ability to support a systematic exploration of the performance of radiation cloud parameterization and cloud formation models under a wide range of climatologically significant conditions.

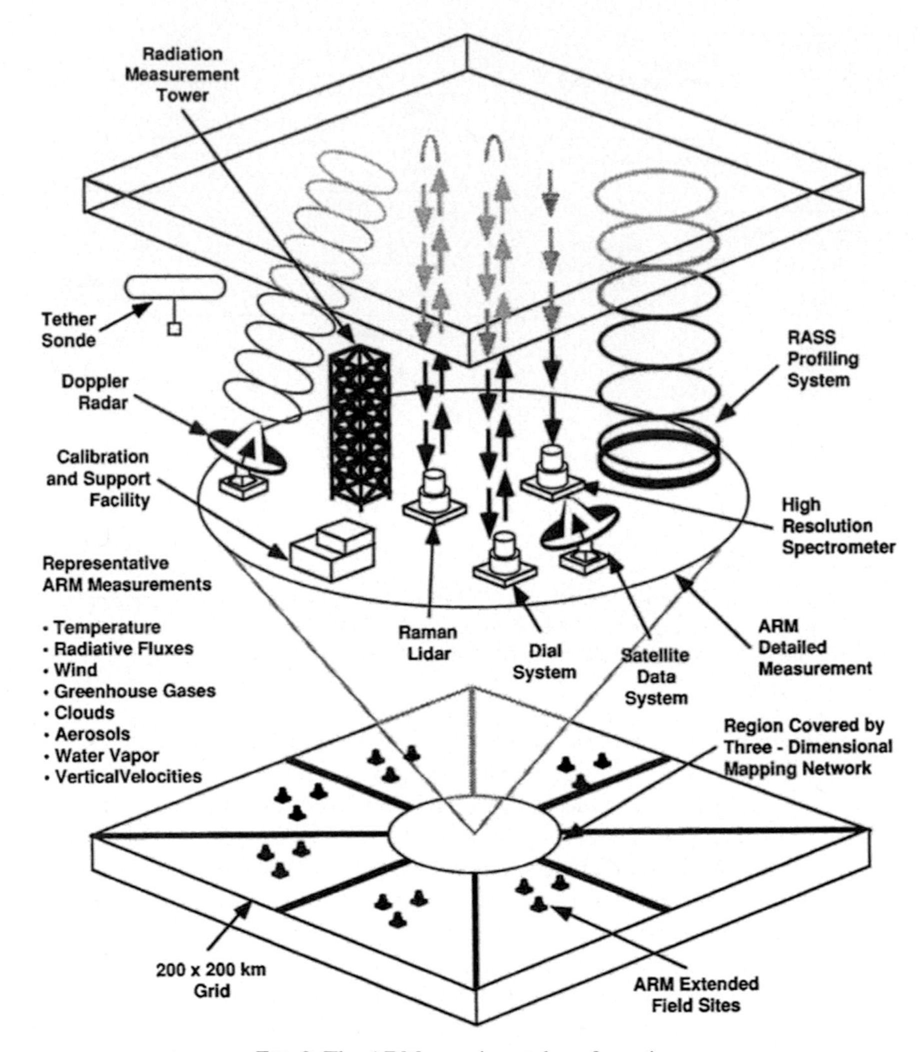

FIG. 2. The ARM experimental configuration.

The ARM experiment will consist of coordinated sets of instruments at each of four to six permanent base sites. These sites are the primary experimental resource of CART. Each ARM site will have three closely associated components. Figure 2 shows an artist's conception of an ARM site. Each component is briefly described here:

- The central facility. A critical experimental task of ARM is to make intensive measurements of the radiation field and the physical conditions that control the radiative transfer. Therefore, ARM will field two classes of equipment at the central facility: those for measuring the radiation field directly and those intended to characterize the local radiative circumstances, such as surface and cloud properties. In general, the base site complement of instruments will include more expensive pieces of equipment, some of which will be experimental in nature. The focus of the observations at the central facility will be the detailed characterization of the atmospheric column above the facility and high-spectral-resolution radiometric instruments.
- The three-dimensional mapping network. A series of auxiliary stations will surround the central facility within a 20-km radius (this radius was derived from consideration of the scale height of the atmosphere). These stations will contain instrumentation designed to measure the three-dimensional structure of the atmosphere near the base site and will make use of fundamental profiling equipment, as well as basic radiometric and meteorological equipment. A focus of the specialized stations will be the reconstruction of the cloud geometry surrounding the base site using state-of-the-art photogrammetric methods. This cloud "visualization system" will be supplemented with a system of wind profilers capable of measuring large-scale vertical velocities. These observations are critical to the study of cloud parameterization and cloud formation.

An ARM site will have three components as described in the text. The central facility, for which representative equipment is shown, will be supported by a system for mapping the three-dimensional distribution of meteorological variables. In addition, 16 to 25 sets of instrumentation will provide critical data for understanding how to generalize the results to the 200 km × 200 km GCM grid size.

- Extended observing network. Surrounding the central facility and the mapping network will be 16 to 25 extended observing stations. These stations will support the development and study of methods used to generalize detailed atmospheric models for use in GCMs and related models through the process of parameterization.

The extended observing area of a base site will include a region of the order of magnitude expected for GCM grid cells in the near future, approximately 200 km × 200 km. The instrumentation at these stations will be less extensive, less specialized, and capable of more autonomous operation than that at the base sites. The instruments at the extended stations will be designed to collect the basic radiometric information and conventional meteorological data needed to characterize the radiative transfer throughout the extended area. Only limited vertical information will be collected, with the more extensive and demanding profiling equipment reserved for the base sites and mapping stations. Wind profilers will, however, be employed on this scale as well to observe the general vertical motions associated with mesoscale phenomena.

The mobile observing system and campaign studies

In addition to the permanently placed equipment at the base and extended sites, ARM will maintain a mobile version of the basic experimental equipment found at the central facility and additional instrumentation for use in directed campaign studies. The ARM Program will take three approaches to planning campaign experiments that will involve the use of the mobile observing system. The first approach is to conduct short-term operations aimed at the exploration of specific physical mechanisms and processes. An example might be the deployment of the mobile system to support an intensive field experiment that is part of FIRE. The second approach will be through longer-term operation and data acquisition designed to reveal experimental anomalies at the base sites. The third approach will be to verify models for conditions intermediate to those of the base sites.

A campaign may also involve the addition of instruments to the basic instrument suite, for a finite time, in order to achieve a specific scientific objective. Throughout its duration, ARM will encounter a variety of circumstances in which it will be desirable to operate other instruments at the site to specifically supplement the routine data. This might be desirable, for instance, to perform a comprehensive calibration experiment on an experimental instrument or to take advantage of an extraordinary transient climate condition.

Measurement strategy and instrument selection

To meet the goals of ARM, the instrument selection must support

- the measurement of key aspects of the radiation field under a range of climatologically significant meteorological conditions sufficient to constrain detailed radiative calculations;
- detailed studies of atmospheric trace gas, aerosol, and water-vapor distributions;
- detailed studies of meteorological variables, including cloud type and distribution, wind field, temperature, etc.;
- measurement of large-scale vertical velocities;
- measurement of critical microphysical properties of clouds.

To support these measurements, it will be necessary to have a support infrastructure with

- near-real-time processing of data and execution of models;
- state-of-the-art calibration methods, including onsite calibration at facilities explicitly designed to support the measurement systems and redundant measurement suites providing near-real-time evaluation of instrument performance.

The intent of the measurements in ARM is to test the predictive power of the models. The instrumentation will be improved continuously. Specialized research instruments, either developed through this program or by others, may be brought to an operational state and then added to the complement of instruments. Observing protocols may also be changed to increase the quality of the tests. All critical measurements will be systematically replicated. The Science Team will have a critical role in the selection of instruments and their evolution at particular sites. The

TABLE 1. The spectrophotometric recommendations for ARM.

Instrument	Spectral range	Resolution
Interferometer #1	5–15 mm	0.02 cm^{-1}
Interferometer #2	4–16 mm	0.3 cm^{-1}
Solar interferometer	2–20 mm	0.002 cm^{-1}
Grating spectrometer	8–25 mm	0.5 cm^{-1}

instrument complement for a specific site may be tailored to individual site characteristics. In spite of these caveats, it is expected that the complement of instruments will look something like the following.

Central facility instrumentation

The primary mission of the central facility is the simultaneous measurement of the radiation field and the physical conditions that might control the radiative transfer. The instrument selection emphasizes redundant measurements and varied observing strategies.

In the longwave radiometric regime, four spectrometers have been tentatively identified as likely instruments. These include two interferometers, a grating spectrometer for measuring atmospheric emission and a much higher-resolution interferometer for measuring the solar infrared spectrum. The specific list of instruments is shown in Table 1. These types of spectrometers have been extensively field tested and thoroughly presented in the literature and at numerous meetings (e.g., Kunde et al. 1987; Brasunas et al. 1988; Murcray et al. 1984; Murcray 1984; Revercomb et al. 1988).

In the visible region, a spectrophotometer will be used for the spectrally resolved observations. If a shadowband spectrometer can be field proven, it will be included in the instrument complement. An automated filter photometer will also be employed to provide a moderate-resolution measurement comparable to that obtained using handheld sunphotometers. It would also be particularly useful if radiometers were included with spectral sensitivity similar to those chosen for use on the Earth Observing System (EOS).

The strategy for the broadband radiometric instrumentation is to duplicate exactly at the base site the complement of instrumentation selected for the extended sites. This instrumentation will support calibration and facilitate quality control. The radiometric instrumentation at the extended stations is discussed below.

Measurement of the meteorological conditions associated with radiative transfer is one of the principal tasks of ARM. Previous radiation studies have had to rely upon radiosonde or aircraft measurements of temperature, humidity, cloud, and aerosol profiles. However, the atmosphere is sufficiently dynamic that such profiles are rarely compatible with the requirements for modeling radiative processes. Radiation properties change with the instantaneous state of the atmosphere.

Fortunately, recent advances in surface-based profiling technology during the past decade have produced instruments capable of near-instantaneous measurement of vertical profiles. These are generally possible for important variables to altitudes of at least 5 km. In some cases, profiles to 10 km or more may be measured. ARM needs to employ those technologies that have been field proven and that give the best possible vertical resolution and accuracy.

The proposed complement of profiling systems is as follows:

- Raman Lidar and Differential Absorption Lidar (DIAL): These technologies are chosen to provide humidity distribution data required by the ARM Program to parameterize cloud formation and radiation balance (Grant 1990, manuscript submitted to *Opt. Eng.*; Wilkerson and Schwemmer 1988). They have undergone significant field tests, including ground-based measurements using Raman lidar (Melfi et al. 1989; Melfi and Whiteman 1985), and both ground-based (Browell et al. 1979; Cahen et al. 1982; Grant et al. 1987) and airborne (Browell 1983) studies using DIAL. Both techniques have comparable measurement accuracies for water vapor, that is, 5% to 10%, for acquiring data from the ground during nighttime, with ranges extending to roughly 7 km. DIAL technology presently is able to produce a profile in about 10 s at nighttime, and in about 15 s in daytime with roughly the same accuracy and resolution. At night, Raman systems have demonstrated the ability to acquire concentration profile data in several minutes. Practical implementation of daytime Raman lidar, expected in the near term, awaits planned experiments utilizing solar-blind detection combined with powerful XeCl excimer lasers. This system also will speed up nighttime Raman data acquisition. The expected values are roughly several minutes in the daytime and about 5 s at night. Both Raman and DIAL systems can achieve desirable range resolutions of about 100 to 200 m.
- Radio Acoustic Sounding System (RASS): The RASS provides good measurement of virtual temperature. It is also possible to get actual temperature by combining RASS data with humidity data from Raman lidar. ARM

plans to field a 400-MHz system with a 300-m to 3-km altitude range and a 50-MHz system with a 2- to 7-km range. The vertical resolution of these systems is 150 m, with an accuracy better than 0.55°C when the vertical wind component is below $0.25 \, m \, s^{-1}$. Otherwise the system accuracy is 15°C. The RASS will also be used in a wind profiling mode to obtain measurements of wind field and turbulence information with the same vertical resolution.

- Lidar: The Wave Propagation Laboratory of NOAA and NASA Langley have developed a wide variety of lidar systems for aerosol and cloud measurements. At present it is clear that a 10-mm carbon dioxide lidar would be highly desirable. Measurements from this instrument will eliminate the need to extrapolate aerosol properties from the visible wavelength spectra collected by most lidars.
- Tethersonde and tower system: These will provide for in situ pressure, temperature, humidity, and ozone measurements up to 2 km. Remote sensing systems are "blind" to the region just above the surface. Most of the radiation in the more opaque spectral bands will be coming from this near-field region. Tower- and sonde-based measurements will be invaluable for filling this data gap and for providing calibration points for the Raman lidar and RASS.
- Satellite data: Since surface-based and radiosonde profiling accuracy declines with altitude, satellite retrievals of temperature, humidity, and ozone will be relied on for information above the mid troposphere.
- In addition to the radiation and related meteorological measurements of ARM, a variety of other measurements will be taken at the central facility. Some of the additional equipment provisions necessary for these are described here.
- Trace gas concentrations: Trace-gas concentrations will be determined from a combination of flask samples and direct real-time sampling using commercial nondispersive infrared analyzers. The solar spectrometer data can be used to infer trace-gas column amounts.
- Surface aerosol concentration: Knollenberg counters, or the equivalent, can provide the aerosol data needed to impose an important boundary condition on the aerosol profile. Aerosol lidars, like other profiling systems, have a blind region near the surface.
- Aerosol optical depth and water vapor column amount: A variety of methods will be used to infer these important column densities. One risk associated with these methods, which include sunphotometers and radiometers, is that they rely on knowledge of radiative transfer for calibration and interpretation. Nevertheless, despite the question as to whether these are quantities that should be inputs to the radiative models or predicted by them, the measurements will have very useful corroborative value.
- Routine surface weather observations: It is particularly crucial to have routine data of surface pressure to calibrate the satellite data, which are expressed in pressure coordinates. The central site will duplicate the basic meteorological information available at the extended observing sites, adding appropriate other measurements as required.

Three-dimensional mapping instruments

There are no well-established systems for mapping the three-dimensional structure of the atmosphere in a reasonably automated fashion. An important part of ARM will be an equipment development activity, a major portion of which will focus in this area. The Cloud Lidar and Radar Exploratory Test (CLARET) experiment at the NOAA Wave Propagation Laboratory may provide some guidance for the development of this system. The most desirable solution would be a system based on imaging arrays of devices like charge-coupled devices (CCDs), scanning DIAL systems, dual-Doppler radar, and wind profilers. A system of this type offers far more automatic data processing options and should be able to take advantage of the many years of development of advanced photogrammetric techniques that have been applied to aircraft and satellite imagery. The quality of instrumentation in this area will have a direct effect on the ARM Program's ability to contribute to the understanding of parameterized cloud formation models.

Extended observing station instrumentation

The extended station instruments will be less extensive than the central facility equipment and must be capable of more autonomous operation. The primary mission of these instruments will be to collect basic radiometric information and conventional meteorological data. There will be only limited vertical information collected.

The ARM selection of extended station instrumentation is motivated by the desire to make the instrument complement as compatible as possible with that of the Global Baseline Surface Radiation Network (GBSRN), a program being designed by John DeLuisi of NOAA for the World Climate Research Program (WCRP).

TABLE 2. Aircraft-based measurement systems.

Quantity measured	Candidate techniques
Part I: Primary measurements	
Liquid water content	Heated wire, integrated size spectrum (see below), virtual impactor (see Part II, below)
Solid water content	Integrated size spectrum (see below), virtual impactor (see Part II, below)
Cloud-droplet size distribution	Optical probe
Raindrop size distribution	Optical probe
Ice morphology and size distribution	Optical array probe, Formvar replicator, foil impactor
Part II: Secondary measurements*	
Thermodynamic properties: temperature, aircraft pressure, humidity (1)	Standard research package: resistance thermometer, piezoelectric transducer, mirror hygrometer
Aerosol loading and size distribution (2)	Optical probe, optical particle counter, electrostaticmobility analyzer
Cloud condensation nucleus count (3)	Controlled humidity chamber-optical counting device
Ice nucleus count (4)	Controlled supercooled chamber device
Aerosol chemical content (3)	Low-pressure impactor
Cloud water chemical content (3)	Counterflow virtual impactor

* Flagging convention for secondary measurements is as follows: (1) important and easy to perform; (2) important but moderately difficult or expensive to perform well; (3) important but very difficult to perform well; (4) relatively unimportant. Categories (1) and (2) are recommended for routine application; category (3) is recommended for intensive campaigns, as deemed advisable to specific campaign objectives.

Instrumentation for the ARM extended observing network will include the basic instrumentation listed for a GBSRN station. The ARM Program will attempt to coordinate its final instrument selection with GBSRN, matching their choice of specific instruments to the greatest extent possible. The only exception is that a rotating shadowband radiometer will be substituted for the sunphotometer, pending comparison operation and calibration studies. The basic measurements and instrumentation for the extended sites are listed below.

Radiometric measurements and instrumentation will be

- pyranometers and tracking pryheliometers (several of each, some unfiltered and some filtered);
- pyrgeometer and low-resolution thermal infrared radiometer to cover both sides of the 9.6-mm ozone band (latter provides direct monitoring in the atmospheric "window" regions);
- upward and downward components of solar and longwave infrared radiation (includes longwave net radiometer);
- rotating shadowband radiometer for flux ratios (rotating shadowband spectrometer would be preferred and will be substituted for some of the radiometers if development is successful);
- spectral ultraviolet measurements.

Other instrumentation at the extended sites will be

- normal complement of weather station measurements such as surface temperature, relative humidity, winds, etc.;
- micrometeorological instrumentation for measuring the ratio of latent to sensible heat fluxes;
- whole-sky cameras for automatic measurement of cloud amount in coordination with satellite observations;
- lidar for measuring cloud ceiling at the site.

Other measurements to be conducted in conjunction with the operation of the network will be

- routine measurement of surface reflectivity surrounding the sites;
- regular soil moisture sampling.

Aircraft-borne operational and campaign measurements

In addition to the complement of fixed instruments that will be placed at the permanent sites, the ARM research program will require additional instruments that will be used on both an operational and a campaign basis. An

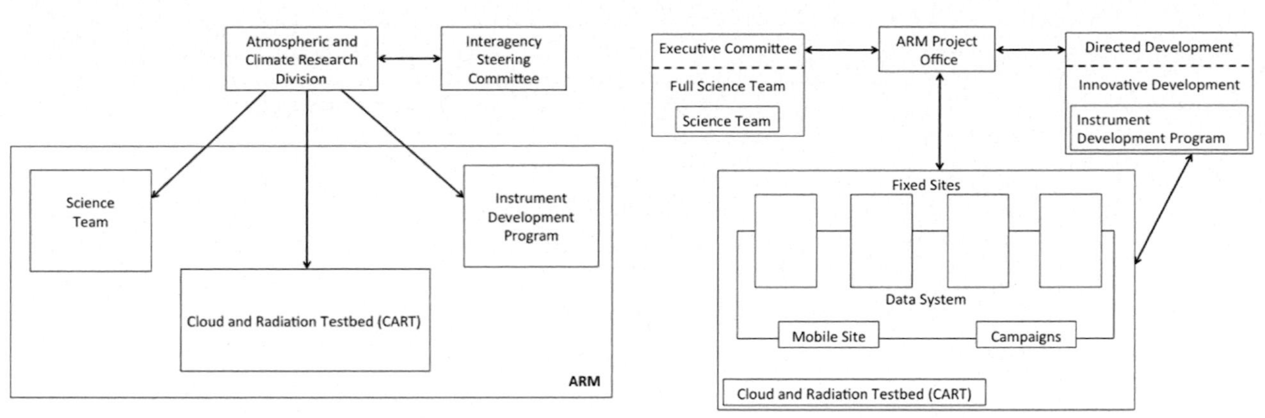

FIG. 3. DOE management oversight of ARM.

FIG. 4. Internal management of ARM.

important activity at the permanent sites will be the routine overflight of airborne sensors for measuring cloud microphysical properties. As has been described previously, this data will be central to the ARM mission.

The aircraft cloud-microphysics measurements of ARM can be subdivided into two types: primary and secondary. Primary measurements are those that pertain to cloud-physics features that directly influence radiative transfer. Secondary variables are those quantities that directly influence the primary features, but influence radiative transfer only indirectly. ARM will concentrate on the primary cloud-microphysics measurements and will perform selected secondary measurements as necessary. Key primary and secondary measurements are summarized in Table 2.

Site selection

Finally, the site-selection process for ARM will be complicated. The choices must incorporate the optimal combination of characteristics in several areas. The general groupings of the criteria are climatic significance, appropriate climatic sampling, synergistic potential with other programs, scientific viability, and logistical viability.

The focus of the ARM measurements is the basic physics of GCMs. However, the physics of the atmosphere are not immutable, as in the sense of a physical law. GCMs integrate elements from theory, basic physics, and observation. They are computational tools and, as such, only approximate reality.

This approximate treatment is very much at issue in the discussion of the parameterizations used in the models. Therefore, the use of ARM data is not only to confirm the details of the basic physical processes, but to understand what physical processes and effects must be preserved as the problem is solved in the highly unresolved GCM case.

The application of the first two criteria for site selection, climatic significance, and climatic sampling clearly show that multiple sites will be required. The parameterization of clouds in GCMs is so important that it is absolutely necessary to confirm observationally the correctness of those parameterizations in those regions of the globe that are important to climate modeling. More than one region is important.

Further, there is sufficient diversity among the climatically important parameters at different sites that no single site can be thought to adequately explore the meteorological envelope and ensure proper parameterization for GCMs.

The critically important choice of sites will be carried out by the Science Team under direction of the Atmospheric and Climate Research Division (ACRD).

Management of ARM

The planned management and organizational structure for the program appears in Figs. 3 and 4. The major features of the program's organization are fourfold:

1) Direct management of the program by the ACRD of DOE's Office of Health and Environmental Research supported by an interagency working group to ensure close coordination with other agency-led programs such as FIRE, GEWEX, and TOGA.
2) A strong Science Team will set the scientific and intellectual direction of the program. It is made up of two groups. The first, the project scientists, will be selected based on peer-reviewed proposals to conduct specific

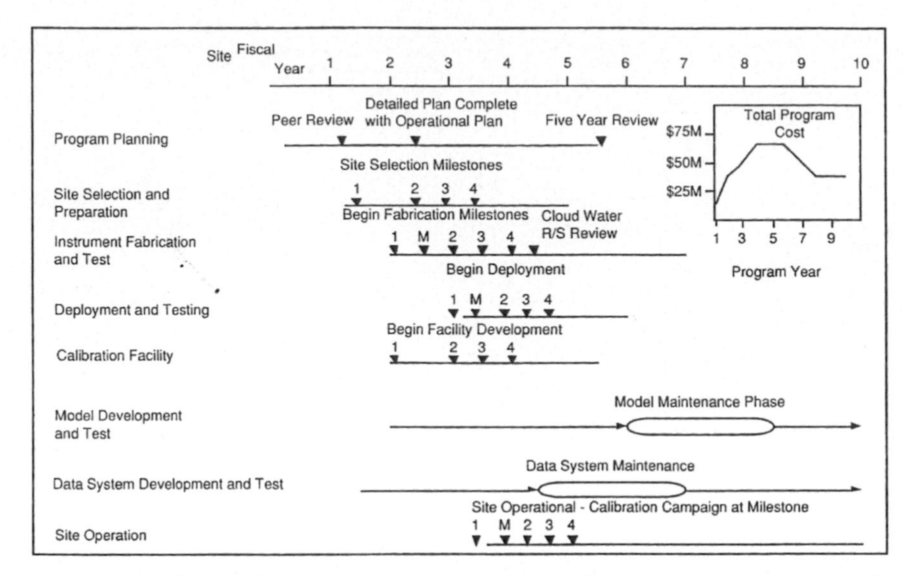

FIG. 5. Atmospheric radiation measurement program schedule and budget for four (1 to 4) fixed sites and one mobile (M) site.

scientific programs with the ARM facilities and data. The second group will be selected by DOE to provide an interface with existing programs both within DOE and other agencies.

3) The CART will serve as the experimental framework and infrastructure within ARM. CART will include fixed experimental sites, a mobile complement of instrumentation, and a series of focused campaigns aimed at particular scientific issues. All elements will be drawn together by a shared data system that will provide ready access to major experimental results for the Science Team and other investigators.
4) An Instrument Development Program will support ARM and the CART in two significant ways, as a place for new and innovative instrumentation to be developed in response to the needs of ARM and as a pathway for instruments developed outside of ARM and DOE to be introduced into the operational ARM environment.

The three internal elements of ARM, the Science Team, CART, and the Instrument Development Program, will by managed on a day-by-day basis through a project office that will be responsible for the general coordination and scheduling of major ARM activities. Final approval, oversight, and funding authority will be retained by ACRD.

The three elements of ARM will be funded independently by ACRD using a combination of competitive proposals, interagency transfers, and funding to the DOE laboratories. All Science Team research will proceed through a competitive peer-review process regardless of the status of the institutional affiliation of the principal investigator, be it university, private industry, DOE laboratory, or non-DOE laboratory. The Instrument Development Program will be funded through several processes, including the review of unsolicited proposals, directed development, and interagency transfer of funds to obtain the unique capabilities of other government agencies. The funding of the CART will follow a similar plan with overall management provided through the DOE laboratory system. However, individual sites or campaigns may well be operated by universities, other laboratories, or private contractors. The budget for ARM and the associated schedule is shown in Fig. 5.

The project office will employ several basic functions to meet its responsibilities.

Specifically these functions will be organized into a series of teams with specific tasks and charters.

- The modeling team will be responsible for the development and maintenance of a set of models to be used for data quality assurance and to serve as a set of "community models" for the Science Team. The selection and design of these models will be conducted under the guidance of the Science Team.
- The instrument teams will be formed by the project office around particular parts of the experimental program. These teams will be formed to ensure integration of particular parts of the experimental program within the program objectives and with the Instrument Development Program. There will be teams associated with the meteorological remote sensing, the radiometric instrumentation, the extended site instrumentation, and data management. The coordination of these teams will be managed by the project office. The goal of these teams is to

develop, deploy, and research sites and provide a smooth transition to the groups responsible for operation of the equipment and the data system. The final instrument complement will be approved by ACRD following recommendations from the Science Team and appropriate reviews.

- The data management team will be responsible for the design, development, and deployment of the data management and analysis system for the program. Unlike the operations team, which will be organized around the operation of a particular site, the data management team will have program-wide responsibility.
- The operation teams will be formed by the project office around the management and operation of each individual site and the mobile system. The goal of these teams is to provide for the operation of the individual sites. Responsibility for the operation of individual sites will be determined on the basis of logistical considerations and could be contracted, assigned to a DOE laboratory, or operated by another federal agency.
- Campaign teams will be formed on an ad hoc basis around the conduct of a particular campaign or coordinated activity with another program. The campaign team will be responsible for the development and maintenance of liaison with the operational teams as required to support campaign activities.

REFERENCES

Brasunas, J. C., V. G. Kunde, and L. W. Herath, 1988: Cryogenic Fourier spectrometer for measuring trace species in the lower stratosphere. *Appl. Opt.*, **27**, 4964–4976, doi:10.1364/AO.27.004964.

Browell, E. V., 1983: Remote sensing of tropospheric gases and aerosols with an airborne DIAL system. *Optical and Laser Remote Sensing*, D. K. Killinger and A. Mooradian, Eds., Springer Series in Optical Sciences, Vol. 39, Springer-Verlag, 138–147.

——, T. D. Wilkerson, and T. J. McIlrath, 1979: Water vapor differential absorption lidar development and evaluation. *Appl. Opt.*, **18**, 3474–1383, doi:10.1364/AO.18.003474.

Cahen, C., G. Megie, and P. Flamant, 1982: Lidar monitoring of water vapor cycle in the troposphere. *J. Appl. Meteor.*, **21**, 1506–1515, doi:10.1175/1520-0450(1982)021<1506:LMOTWV>2.0.CO;2.

Cess, R. D., and Coauthors, 1989: Interpretation of cloud-climate feedback as produced by 14 atmospheric general circulation models. *Science*, **245**, 513–516, doi:10.1126/science.245.4917.513.

Gates, L. W., 1987: Problems and prospects in climate modeling. *Toward Understanding Climate Change*, U. Radok, Ed., Westview Press, 5–34.

Grant, W. B., 1990: Optimizing lidar for water vapor measurements. *Opt. Eng.*, submitted.

——, J. S. Margolis, A. M. Brothers, and D. M. Tratt, 1987: CO_2 DIAL measurements of water vapor. *Appl. Opt.*, **26**, 3033–3042, doi:10.1364/AO.26.003033.

Grotch, S. L., 1988: Regional intercomparisons of general circulation model predictions and historic climate data. U.S. Department of Energy Tech. Rep. DOE-NBB-0084, 291 pp.

Kunde, V. G., and Coauthors, 1987: Infrared spectroscopy of the lower stratosphere with a balloon-borne cryogenic Fourier spectrometer. *Appl. Opt.*, **26**, 545–553, doi:10.1364/AO.26.000545.

Luther, F. A., R. G. Ellingson, Y. Fouquart, S. Fels, N. A. Scott, and W. J. Wiscombe, 1988: Intercomparison of Radiation Codes in Climate Models (ICRCCM): Longwave clear-sky results—A workshop summary. *Bull. Amer. Meteor. Soc.*, **69**, 40–48.

Melfi, S. H., and D. Whiteman, 1985: Observation of lower-atmospheric moisture structure and its evolution using a Raman lidar. *Bull. Amer. Meteor. Soc.*, **66**, 1288–1292, doi:10.1175/1520-0477(1985)066<1288:OOLAMS>2.0.CO;2.

——, ——, and R. Ferrarre. 1989: Observation of atmospheric fronts using Raman lidar moisture measurements. *J. Appl. Meteor.*, **28**, 789–806, doi:10.1175/1520-0450(1989)028<0789:OOAFUR>2.0.CO;2.

Murcray, D. G., 1984: Atmospheric transmission in the 750–200 cm^{-1} region. *J. Quant. Spectrosc. Radiat. Transfer*, **32**, 381–396, doi:10.1016/0022-4073(84)90035-9.

Murcray, F. H., F. J. Murcray, D. G. Murcray, J. Pritchard, G. Vanasse, and H. Sakai, 1984: Liquid nitrogen-cooled Fourier spectrometer system for measuring atmospheric emission at high altitudes. *J. Atmos. Oceanic Technol.*, **1**, 351–357, doi:10.1175/1520-0426(1984)001<0351:LNCFTS>2.0.CO;2.

Ramanathan, V., R. D. Cess, E. F. Harrison, P. Minnis, B. R. Barkstrom, E. Ahmad, and D. Hartmann, 1989: Cloud-radiative forcing and climate: Results from the Earth Radiation Budget Experiment. *Science*, **243**, 57–63, doi:10.1126/science.243.4887.57.

Revercomb, H. E., H. Buijs, H. B. Howell, D. D. LaPorte, W. L. Smith, and L. A. Sromovsky, 1988: Radiometric calibration of IR Fourier transform spectrometers: Solution to a problem with the High-Resolution Interferometer Sounder. *Appl. Opt.*, **27**, 3210–3218, doi:10.1364/AO.27.003210.

Wang, W. C., and Coauthors, 1988: Surface energy balance of three general circulation models: Current climate and response to increasing atmospheric carbon dioxide. U.S. Department of Energy Tech. Rep. DOE/ER/60422-H1, 119 pp.

Wilkerson, T. D., and G. K. Schwemmer, 1988: Lidar probing of tropospheric density, temperature, pressure, and humidity for ballistics corrections. *Proc. Lower Tropospheric Profiling: Needs and Technologies*, Boulder, CO, NCAR, NOAA/WPL, and Amer. Meteor. Soc.

APPENDIX B

This appendix is the executive summary of the 1996 Science Plan for the Atmospheric Radiation Measurement Program (ARM) (DOE/ER-0670T, UC-402; available online at https://www.arm.gov/publications/programdocs/doe-er-0670t.pdf) sponsored by the U.S. Department of Energy, Office of Energy Research, Office of Health and Environmental Research, Environmental Sciences Division. The text has been edited to conform to the style of the American Meteorological Society, but the content is otherwise unchanged from the original document.

Executive Summary: Science Plan for the Atmospheric Radiation Measurement Program (ARM)

The purpose of this Atmospheric Radiation Measurement (ARM) Science Plan is to articulate the scientific issues driving the ARM Program and to relate them to DOE's programmatic objectives for ARM, based on the experience and scientific progress gained over the past five years.

ARM programmatic objectives are to

1) relate observed radiative fluxes and radiances in the atmosphere, spectrally resolved and as a function of position and time, to the temperature and composition of the atmosphere, specifically including water vapor and clouds, and to surface properties, and sample sufficient variety of situations so as to span a wide range of climatologically relevant possibilities;
2) develop and test parameterizations that can be used to accurately predict the radiative properties and to model the radiative interactions involving water vapor and clouds within the atmosphere, with the objective of incorporating these parameterizations into general circulation models.

The achievement of these programmatic objectives should lead to the improvement of the treatment of atmospheric radiation in climate models, explicitly recognizing the crucial role of clouds in influencing this radiation and the consequent need for accurate description of the presence and properties of clouds in climate models. There are key scientific issues that must be resolved in order to achieve these objectives. The primary scientific questions are as follows:

1) What are the direct effects of temperature and atmospheric constituents, particularly clouds, water vapor, and aerosols, on the radiative flow of energy through the atmosphere and across Earth's surface?
2) What is the nature of the variability of radiation and the radiative properties of the atmosphere on climatically relevant space and time scales?
3) What are the primary interactions among the various dynamic, thermodynamic, and radiative processes that determine the radiative properties of an atmospheric column, including clouds and the underlying surface?
4) How do radiative processes interact with dynamical and hydrologic processes to produce cloud feedbacks that regulate climate change?

The programmatic objectives of ARM call for measurements suitable for testing parameterizations over a sufficiently wide variety of situations so as to span the range of climatologically relevant possibilities. To accomplish this, highly detailed measurements of radiation and optical properties are needed both at Earth's surface and inside the atmospheric column, and also at the top of the atmosphere. Among the most critical factors determining the optical properties of the atmosphere is the distribution of liquid water and ice (i.e., clouds) within the atmospheric column. It follows that ARM must obtain sufficiently detailed measurements of the clouds and their optical properties.

DOI: 10.1175/AMSMONOGRAPHS-D-15-0035.1

The primary observational method is remote sensing and other observations at the surface, particularly remote sensing of clouds, water vapor, and aerosols. It is impossible to meet ARM's objectives, however, without obtaining a large volume of detailed in situ measurements, some of which will have to be acquired from manned or unmanned aircraft; in addition, high-quality satellite observations are needed to measure the top-of-the-atmosphere radiation.

To obtain the requisite in situ and surface-based remote sensing data, ARM is making measurements, over a period of years, at three sites:

- the Southern Great Plains (SGP) site;
- the Tropical Western Pacific (TWP) site;
- the North Slope of Alaska/Adjacent Arctic Ocean (NSA/AAO) site.

These sites were selected from a longer list through a process of prioritization and resource allocation, in order to provide opportunities to observe a wide range of climatologically important meteorological conditions, as summarized in DOE (1990).

There are two primary strategies through which ARM plans to achieve its programmatic objectives and address its programmatic objectives and address its scientific issues. The first is called the "Instantaneous Radiative Flux" (IRF) strategy, which consists of collecting data on the distribution of radiation and the radiatively active constituents of the atmosphere and the radiative properties of the lower boundary. The second involves the use of single-column models (SCMs) to develop and test cloud formation parameterizations.

ARM programmatic objectives will be achieved as the testing of hypotheses leads to improved parameterizations for use in climate models. The activities of the IRF are central to achieving the ARM Program objective of relating observed radiative fluxes in the atmosphere to the temperature and composition of the atmosphere and to surface properties. Since the parameterizations of radiative processes play major roles in climate model forcing and feedback mechanisms, the radiative parameterizations developed by IRF studies will play essential functions in the development and testing of other parameterizations for predicting the distributions of properties that strongly affect atmospheric radiation.

Among the several methods for testing general circulation models (GCM) parameterizations by comparison with observations, the SCM approach has some unique advantages. SCM-based tests are inexpensive, and the results are not affected by errors arising from the other components of the model. Cloud ensemble models are a useful supplement to SCMs, and can be used in much the same way with essentially the same data requirements. SCMs in conjunction with large-eddy simulation (LES) and cloud ensemble models (CEMs) can be used to investigate basic physical questions, develop cloud amount parameterizations, and evaluate the sensitivity of model results to parameter changes. In support of ARM, SCMs and CEMs will be particularly valuable for testing parameterizations of cloud formation, maintenance, and dissipation.

The data required to drive the SCMs, LES models, and CEMs, and to evaluate their performance, are not easy to obtain. ARM has the potential to provide uniquely valuable data for SCM-based parameterization testing. Efforts are under way to "package" data collected at ARM's Southern Great Plains site in a form particularly convenient for use with SCMs and CEMs.

A key to the ultimate success of ARM is continued evaluation of the needs for new observing capabilities as progress is made in understanding the important scientific issues.

The goal of the Instrument Development Program (IDP) of ARM is to bring existing research instrumentation to the advanced state of development required to allow routine, highly accurate operation in remote areas of the world, and to develop new instrumentation as requirements are identified.

The evolving ARM IDP has combined components of basic research into improved remote sensor system and techniques (i.e., cloud retrieval) development, and an engineering effort intended to provide within the Cloud and Radiation Testbed (CART) setting a kernel of instruments to adequately characterize the local atmosphere. Currently, effort is directed toward placing these sensors at CART sites and validating the analysis approaches. The next step is to develop the means to convert the CART remote sensor data stream into the types of derived data quantities that are necessary to comprehend the effects the clouds and clear atmosphere on the IRF and GCM-class radiative transfer models.

The IRF scientific questions being addressed with data from the SGP are not really site specific; the same questions could be addressed with data from the other two sites. The site-specific scientific questions that the SCM strategy is addressing with SGP data include

- What processes control the formation, evolution, and dissipation of cloud systems in the Southern Great Plains?
- In the Southern Great Plains, what relative roles do the advection of air mass properties and variation in surface characteristics play in cloud development? How do these roles vary with season and short-term climatic regime?
- What aspects of cloud development are controlled by the low level jet, the return flow of moisture from the Gulf of Mexico during the winter and early spring months, the development of mesoscale convective complexes, frontal passages?
- What are the implications of the regional east–west gradients in altitude, soil type, vegetation, temperature, and precipitation on radiative fluxes?
- How important are seasonally varying distributions of aerosols and particulates (e.g., from regional oil refineries, or from burning of wheat fields) in the energy transfer processes?

The choice of ARM's Tropical Western Pacific (TWP) site was dictated, in large part, by the ocean warm pool and the deep convection associated with it. Satellite observations show that the ocean surface temperatures in the vicinity of the Maritime Continent are consistently the warmest and cloud top temperatures are the coldest found anywhere.

ARM has an important role to play in the TWP and that role has three distinct and critical elements:

1) Provide a long time series of basic observations at several locations that aid in understanding intra annual and interannual variability of surface radiation fluxes and cloud properties. These observations would also serve as truth points for satellite retrievals of surface and atmospheric quantities.
2) Augment radiation and cloud observations made in the context of intensive field campaigns to elucidate the role of deep convection in the tropics as it affects radiative processes.
3) Devise and implement a strategy for long-term measurements of ocean–atmosphere properties and fluxes.

The specific scientific objectives to be addressed at the ARM site in the North Slope of Alaska/Adjacent Arctic Ocean (NSA/AAO) focus on improving the performance of climate models at high latitudes by improving our understanding of specific physical processes.

The observational strategy proposed will allow us to improve our understanding of the cloud and radiation environment of the Arctic, over land and ocean. These observations will be used to initialize and validate cloud-resolving models, and as a basis for comparing parameterizations. These improved parameterizations will be incorporated into a regional climate model of the Arctic and global climate models. Collaboration with other programs, such as the Surface Heat Budget of the Arctic Ocean (SHEBA), the First ISCCP (International Satellite Cloud Climatology Experiment) Regional Experiment (FIRE), and Land-Atmosphere-Ice-Interactions (LAII) allows ARM to address its secondary science objectives; together, these programs will have a substantial impact on our ability to model the arctic climate, specifically the cloud-radiation feedback.

Interactions of the Data and Science Integration Team (DSIT) with Science Team members, individually and collectively, are the primary information exchange mechanism that drives the collection and management of data within the ARM Program. The interaction leads to translating the science needs into data needs. A critical part of this translation is the management and documentation of data quality. The stated goal of ARM is to produce data of "known and reasonable quality." This goal is translated into both actions to ensure that the instruments produce data of sufficient precision and accuracy to meet scientific needs and the obligation to produce a record of the calibration and operational history of instruments and their associated data streams sufficient to ensure an enduring record of data quality.

APPENDIX C

This appendix is the executive summary of the 2004 Atmospheric Radiation Measurement Program Science Plan: Current Status and Future Directions of the ARM Science Program (DOE/ER-ARM-0402; available online at https://www.arm.gov/publications/programdocs/doe-er-arm-0402.pdf) sponsored by the U.S. Department of Energy, Office of Science, Office of Biological and Environmental Research. The text has been edited to conform to the style of the American Meteorological Society, but the content is otherwise unchanged from the original document.

Executive Summary: Atmospheric Radiation Measurement Program Science Plan: Current Status and Future Directions of the ARM Science Program

The Atmospheric Radiation Measurement (ARM) Program has matured into one of the key programs in the U.S. Climate Change Science Program. The ARM Program has achieved considerable scientific success in a broad range of activities, including site and instrument development, atmospheric radiative transfer, aerosol science, determination of cloud properties, cloud modeling, and cloud parameterization testing and development. The focus of ARM science has naturally shifted during the last few years to an increasing emphasis on modeling and parameterization studies to take advantage of the long time series of data now available.

During the next 5 years, the principal focus of the ARM science program will be to

- maintain the data record at the fixed ARM sites for at least the next five years;
- improve significantly our understanding of and ability to parameterize the 3D cloud radiation problem at scales from the local atmospheric column to the global climate model (GCM) grid square;
- continue developing techniques to retrieve the properties of all clouds, with a special focus on ice clouds and mixed-phase clouds;
- develop a focused research effort on the indirect aerosol problem that spans observations, physical models, and climate model parameterizations;
- implement and evaluate an operational methodology to calculate broadband heating rates in the atmospheric columns at the ARM sites;
- develop and implement methodologies to use ARM data more effectively to test atmospheric models, both at the cloud-resolving model scale and the GCM scale;
- use these methodologies to diagnose cloud parameterization performance and then refine these parameterizations to improve the accuracy of climate model simulations.

In addition, the ARM Program is actively developing a new ARM Mobile Facility (AMF) that will be available for short deployments (several months to a year or more) in climatically important regions. The AMF will have much of the same instrumentation as the remote facilities at ARM's Tropical Western Pacific and the North Slope of Alaska sites. Over time, this new facility will extend ARM science to a much broader range of conditions for model testing.

DOI: 10.1175/AMSMONOGRAPHS-D-15-0034.1